Springer Monographs in Mathematics

Dietlinde Lau

Function Algebras on Finite Sets

A Basic Course on Many-Valued Logic and Clone Theory

With 42 Figures and 46 Tables

Springer

Dietlinde Lau
Institute for Mathematics
University of Rostock
Universitätsplatz 1
18055 Rostock, Germany
e-mail: dietlinde.lau@uni-rostock.de

Library of Congress Control Number: 2006929534

Mathematics Subject Classification (2000): 03B50, 08Axx, 08A40, 08A30, 08A05, 06A15

ISSN 1439-7382

ISBN-10 3-540-36022-0 Springer-Verlag Berlin Heidelberg New York
ISBN-13 978-3-540-36022-3 Springer-Verlag Berlin Heidelberg New York

Springer is a part of Springer Science+Business Media
springer.com

Printed in Germany

Typesetting by the author using a Springer TEX macro package
Production: LE-TEX Jelonek, Schmidt & Vöckler GbR, Leipzig
Cover design: Erich Kirchner, Heidelberg

Printed on acid-free paper 44/3100YL - 5 4 3 2 1 0

To my mother, Brigitte Lau

Preface

Functions (or operations), which are defined on finite sets, occur in almost all fields of mathematics. For more than 80 years, algebras (so-called function algebras), whose universes are such functions, have been studied. Particularly in Mathematical Logic, in Universal Algebra (more precise in the Clone Theory), and in parts of Computer Science, certain knowledge about these algebras are subject of the fundamental knowledge.
Currently only one book has been published about function algebras, apart from certain monographs or dissertations of specific themes, survey articles and books that contain sections about function algebras or clones. This book has been written by R. Pöschel und L. A. Kalužnin in the German language and gives a very good overview about the results achieved up to 1979.
During the last 26 years, many new results have been obtained; however, a new book about function algebras is overdue.

The aim of the present book is to introduce the reader to the theory of function algebras and to give the latest state of research for some selected fields. The author would like to acquaint the reader with proof of the fundamental theorems and the different proof methods, to enable research in the field of the function algebras.

This book is self-contained. All necessary fundamental concepts and facts are introduced, but some background knowledge about linear and abstract algebra would be helpful for readers.

In the following *Introduction*, the reader finds short summaries of the 26 chapters of this book.
The adjoined section *Preliminaries* explains abbreviations and some general symbols, which are more or less standard, and gives some facts from basic mathematics.

Part I of this book introduces the reader to Universal Algebra to provide almost every knowledge concerning other fields of mathematics. Moreover, this part of Universal Algebra informs the reader that many of the following results of function algebras reply to questions that arise upon studying an algebra. The structure of this book enables the reader to skip the first part and immediately start reading *Part II* Function Algebras.

The author provides the reader with new proofs concerning classic results of the theory of function algebras. The remaining proofs are adapted to the style of the book.
The theorems from Sections 14.10, 15.4, 18.2, 18.3, and from Chapter 17 have not yet been published.
Small mistakes from the original papers (including the papers of the author) have been corrected in this book without referring to the original mistakes.

During the writing process I have tried to solve open mathematics questions and problems, which I recognized during my study of the corresponding literature. In some cases colleagues helped me. In such a case their names are mentioned in the relevant places in the book.
I would like to thank my colleagues as well as the authors of the articles I referred to in my book.
Especially, I would like to thank my doctoral thesis supervisors, Prof. G. Burosch (Rövershagen), and Prof. V. B. Kudrjavcev (Moscow). I received first information on the complexity of themes in this book from their lectures.
I owe Prof. I. G. Rosenberg (Montreal) much gratitude, as I learned many proof methods while studying his papers.
I thank my colleagues Prof. Dr. K. Denecke (Potsdam), Prof. Dr. L. Haddad (Kingston), and Prof. Dr. R. Pöschel (Dresden) for their fruitful cooperation during many years.
In particular, Prof. Haddad was a very great aid during the finishing of the book. I also owe him the organization and the financing (by the Natural Sciences and Engineering Council of Canada) of a linguistic correction of my text by Ms Eryn Kirkwood (from RedInk Editors, Ottawa ON, Canada). Ms Kirkwood deserves my special thanks.

Rostock, June 2006 *Dietlinde Lau*

Contents

Introduction

The present book deals with a subarea of the Discrete Mathematics. We study functions, which are defined on finite sets, and we study the composition of these functions. Such functions are used, for example, in Computer Science (in particular, in the Switching Theory and in the Theory of Automata), in Mathematical Logic, and in Universal Algebra (in particular, the Clone Theory).[1]

In other words, we choose an arbitrary finite set A and study an algebra, whose universe is the set

$$P_A \quad (\text{or } P_k := P_{\{0,1,2,..,k-1\}})$$

of all n-ary mappings ($n \in \mathbb{N}$), which maps the Cartesian power A^n (of all ordered n-tuples of elements from A) into A, and whose operations are the so-called **superposition operations** that are described as follows:[2]

- permutation of variables
- identification of variables
- adding of fictitious variables and
- substitution of variables of a function by functions

Denote $F^{(n)}$ the set of all n-ary functions of $F \subseteq P_A$. Moreover, let Ω be the set of all superposition operations described above.
Then,

$$\mathbf{P_A} := (P_A; \Omega)$$

is called (full) **function algebra** (on A). The universe of a subalgebra of $\mathbf{P_A}$ is called **closed set** or a **subclass** (or briefly a **class**) of P_A. If F is a subclass of P_k then F is also called a **subclass of the k-valued logic**.
The set of all closed sets of P_k together with the set inclusion forms a lattice $\mathbb{L}_k$.

---

[1] See, for this purpose, also Section 1.5 of Part II.

[2] One finds an exact definition of these operations in Section 1.2 of Part II.

The closed sets of P_2 were already determined in the papers [Pos 20] and [Pos 41] by E. L. Post. For over 50 years, many papers have dealt with function algebras or closed sets of P_k for arbitrary k. One finds a survey of the essential articles, which were published up to the year 1978, on the topic in [Pös-K 79] with 730 references and in [Ros 77] with 464 references. To give the reader a first impression of the problems handled in the theory of the function algebras, some explanations are subsequently given to the **completeness problem**[3], which stood in the center of a line of investigations in many articles:

One finds a criterion, to decide if a set of functions that belong to P_k is sufficient for the construction of any arbitrary other function of P_k (by means of the superposition operations).

A general answer to that is given by the following criterion, which was formulated by E. L. Post in 1921, first, for Boolean functions, i.e., for functions of P_2:

A set $F \subseteq P_k$ is complete in P_k if and only if F is a subset of no maximal class of P_k.

A closed set $M \subset P_k$ is said to be **maximal in** P_k, if M can not be properly extended to a closed proper subset of P_k.
With regard to such and similar question formulations, one naturally deals with the structure of the subclasses of P_k or with the lattice of the subclasses of P_k. The first and most important result in this direction is Post's pioneering description of $\mathbb{L}_2$ ([Pos 41], see Figure 1), now known as **Post lattice**. One can describe the many Post's classes with the aid of the classes P_2, M, S, L, T_a $T_{a,\mu}$, $D \cup C$, $K \cup C$, $[P_2^{(1)}]$, C, $(a \in \{0,1\}, \mu \in \mathbb{N} \setminus \{1\})$ and through formation of certain intersections of these classes, where T_0, T_1, M, S, and L are exactly the maximal classes of P_2.[4] For some time, it was thought that $\mathbb{L}_3$ is equally simple. Then, Ju. I. Janov and A. A. Mučnik showed in the year 1959 the existence of subclasses of P_k for $k \geq 3$ with infinite and without basis, respectively, and this implies the existence of a non-countable infinite set of subclasses in $\mathbb{L}_k$ (see also [Ehr 55]). Thus, there are as many closed sets in P_k as there are subsets of P_k for $k \geq 3$. In contrast to $k = 2$, therefore, it seems hard to give an effective description of $\mathbb{L}_k$ for $k \geq 3$. Nevertheless, one could determine the maximal classes of P_k for arbitrary k. S. V. Jablonskij determined all 18 maximal classes of P_3 in [Jab 54] and [Jab 58]. Moreover, he extended his results by describing several types of maximal classes of P_k. This work was continued by V. V. Martynyuk [Mar 60], E. Ju. Zacharova [Zac 67], R. A. Bairamov [Bai 67] (some results incorrect) and V. L. Rvačev/ L. I. Šljarov [Rva-Š 67]. In [Zac-K-J 71] it has been reported that late A. I. Mal'tsev had all 82 maximal sets of P_4. I. G. Rosenberg published the missing maximal classes for P_k in [Ros 65] and

[3] In Section 1.5 of Part II, it is shown that this problem is a mathematical way to describe problems that occur during the construction of electronic circuits.

[4] See Chapter 3 of Part II for details.

he proved in the book [Ros 70a] that there can not be any further maximal classes for arbitrary k, whereby the general completeness problem for P_k was solved. Rosenberg's description of the maximal classes is based on the idea of a function preserving a relation; i.e., he could prove that every maximal class has the form $Pol_k\varrho$, where ϱ is a certain h-ary relation on E_k, $1 \leq h \leq k$, $k \geq 3$ and $Pol_k\varrho$ is the set of all functions of P_k, which preserve the relation ϱ. To find the maximal ones among classes of the form $Pol_k\varrho$, he designed a sieve method, which eliminates at each step every relation ϱ for which a relation σ is found in the list such that $Pol_k\varrho \subseteq Pol_k\sigma$. The process terminates when the candidate list contains only relations ϱ such that $Pol_k\varrho$ can be proved to be maximal. This approach was twofold in scope. To discover the maximal classes and, at the same time, to prove that one has a full list of them.
The idea to describe classes through relations kept on being developed to a Galois-theory for function algebras and relation algebra by V. G. Bodnarčuk, L. A. Kalužnin, V.N. Kotov, B. A. Romov (in [Bod-K-K-R 69]).
The interesting subclasses of P_k are not only the maximal ones, and so the Galois theory was and is an important aid for the study of subclasses of P_k. Namely, one can classify finite universal algebras and also combinatorial problems with the aid of the elements of $\mathbb{L}_k$ (see Section 1.5 of Part II). These and many other applications presuppose, however, precise knowledge about the subclasses of P_k. Therefore, the object of this book is to explain some methods to the finding of subclasses of P_k and to give a survey of subclasses and their properties, which were determined in the last years.

To familiarize the reader from the beginning also with the algebraic side of the function algebras and to show that the concepts introduced for function algebras are only special cases of more general concepts mostly from the Universal Algebra, we begin with an introduction to the Universal Algebra in the first part of this book. One finds supplements to this short introduction to the Universal Algebra in the books [Bur-S 81], [Coh 65], [Grä 68], [Ihr 93] (or [Ihr 2003]), [McK-M-T 87], [Wer 78] and [Lau 2004], volume 2.

In Part II, we deal only with the function algebras. The construction of the book is chosen in a way that one can immediately begin reading Part II and, if one needs concepts and facts from Part I, one can reference these. Unlike Part I, which is strongly linearly structured in that each chapter is based on the preceding chapter, after studying the first two chapters of Part II, all successive studies can also be studied.

We concentrate in Part II on the following topics:

- Basic concepts and notations
- Galois-connection between function algebras and relation algebras
- Post's results on the subclasses of P_2
- the maximal classes of P_k
- completeness criterions for P_k (in particular, the Rosenberg's completeness criterion)

- congruences on subclasses of P_k
- complexity measures for generating systems (for example as order or relation degree and dimension) of certain subclasses of P_k
- subclasses of linear functions of P_p (p prime number)
- a survey on subclasses of P_k, whose functions have at most two different values
- submaximal classes of P_3
- the description of all finite and countably infinite sublattices of the depth 1 or 2 of the lattice of all subclasses of P_3
- submaximal classes of P_k, which are subsets of maximal classes described by unary relations
- a survey of results to maximal classes of P_k for arbitrary k
- minimal clones
- partial function algebras.

One finds completions to this book and a survey on further topics of the function-algebras-theory in [Pös-K 79] and [Ros 84].

Subsequently the content of this text is described in brief without the necessary concepts and notations.

Part I

Chapter 1 begins with the definition of a **universal algebra** (briefly: **algebra**) as a pair (A;F) consists of a nonempty set A and a set F of certain operations on A and gives numerous examples of algebras (among that also the function algebras). In addition, one finds the very important concept of the **subalgebra** in this chapter.

Chapter 2 compiles needed order concepts within the framework of an introduction to the lattice theory. The usual definitions of a **lattice** are indicated, and the calculation in the lattice theory illustrates by means of the proof of a few classical theorems of the lattice theory.

Chapter 3 generalizes observations, which one can make when one examines more closely such concepts as *subalgebra* or *linear hull* of subsets of a vector space (or another closure operators of the classical algebra), to the concepts **hull system** and **closure operator**. Then, it is shown that these concepts deliver, roughly, the same. In addition, certain combinations to the lattice theory are given, and the lattices, formed from the subalgebras of an algebra, are uniquely characterized through certain properties.

Chapter 4 combines properties of such important concepts as the **homomorphic and isomorphic mappings** between universal algebras, which the readers surely know from the classical algebra here, and which are defined obviously for universal algebras. It is shown, how homomorphic mappings are ultimately determined by **congruence relations** (i.e., by equivalence relations, which are compatible with the operations of algebra). The (general) homomorphism theorem keeps on being proved and is shown by examples, as this theorem can be improved for concrete algebras. A section of Chapter 4 deals with **Galois connections** between set systems that are continued later for function algebras in Chapter 2 of the second part.

Chapter 5 shows how one can form new algebras from given algebras through **direct or subdirect products** and how one can recognize whether a given algebra is isomorphic to an algebra formed in this way. More precisely: we deal with the following two questions: Which algebras are smallest constituents of given algebras? How can one reduce a given algebra to its smallest constituents? Answers to these questions are given by theorems found by G. Birkhoff. We prove these theorems in Chapter 5. In particular, we prove the *representation theorems for algebras*: Each *finite* algebra is isomorphic to a direct product of directly irreducible algebras. *Each* algebra is isomorphic to a subdirect product of subdirectly irreducible algebras.

Chapter 6 deals with classes of algebras of the same type and some theorems on such classes, which were found also by G. Birkhoff.
At first we introduce so-called **varieties** as classes of algebras, which are closed in respect to the formation of subalgebras, homomorphic images and direct products. We then come to a method for the construction of algebra classes that strongly differs from the first method at first sight: Based upon certain equations from variables and operation symbols of a certain type τ, we form the class of all algebras of type τ that these equations fulfill. The result is so-called **equational classes**. We will see, however, that there is a close connection between the two methods of the algebra class construction: class of algebras is equational defined if and only if it is a variety.
In the section on equational classes, we will also treat such concepts as the **conclusion of an equational set**. In addition, we treat ways to receive such conclusions.
In Chapter 9 of Part II, we show that the results on congruences of a subclass $F \subseteq P_A$ imply results on subvarieties of the variety, which is generated by the algebra $(A; F)$ and vice versa.

Part II

Chapter 1 begins with the precise definition of function algebras $(P_A; \Omega)$, where Ω is an infinite set of operations, which describe the above superposition operations exactly, on P_A first. It is shown then how one can receive the operations of the set Ω through five elementary operations (so-called "Mal'tsev-Operationen") $\zeta, \tau, \Delta, \nabla$ and $\star$ by means of composition. The basis of further investigations is, then, the (full iterative) function algebra

$$\mathbf{P_A} := (P_A; \zeta, \tau, \Delta, \nabla, \star) \text{ bzw. } \mathbf{P_k} := \mathbf{P_{E_k}} \text{ mit } E_k := \{0, 1, ..., k-1\}.$$

A universe of a subalgebra of $\mathbf{P_k}$ is called **subclass** (or more briefly, **class**) in the following. If $[T]$ denotes the set of all functions of P_A, which one can form from the functions of $T \subseteq P_A$ by means of superposition operations, then we have $[F] = F$ for arbitrary subclass F of P_A.
A mapping defined by $e_i^{(n)}$: $E_A^n \longrightarrow A, (x_1, x_2, ..., x_n) \mapsto x_i$, where $i \in \{1, 2, ..., n\}$, is called a **projection**. A **clone** is a subclass of P_A, which contains all projections of P_A.
In Chapter 1 there are notations, concepts ... that are used in the following chapters repeatedly. In addition, it is shown that the set P_A and some subsets of P_A can be formed from binary functions of P_A by means of superposition operations.
At the end of Chapter 1, we briefly show how one can use the results of this book and how certain investigations are motivated in the theory of function algebras.

Chapter 2 provides tools (like the the concepts of "a function preserves a relation" and "a relation is an invariant for a function") by which classes, dealt in later chapters, can be effectively described. The set R_k of all n-ary relations on E_k ($n = 1, 2, 3, ...$) is defined and operations over the set R_k so that R_k with these operations forms a **relation algebra $\mathbf{R_k}$**. Each universe of a subalgebra of $\mathbf{R_k}$ is called a **co-clone** of R_k. The main result of Chapter 2 is the proof that the lattice of the clones of P_k is antiisomorphic to the lattice of all co-clones of R_k. The basis of this result is the Galois-connection (Pol, Inv), where $Pol : \mathfrak{P}(R_k) \longrightarrow \mathfrak{P}(P_k)$ and $Inv : \mathfrak{P}(P_k) \longrightarrow \mathfrak{P}(R_k)$ are mappings defined by

$$Pol_k Q := \{f \in P_k \mid f \text{ preserves each relation of } Q\},$$
$$Inv_k F := \{\varrho \in R_k \mid \varrho \text{ is an invariant for each function } f \text{ of } F\}$$

for arbitrary $Q \subseteq R_k$ and arbitrary $F \subseteq P_k$. The Galois-connection (Pol, Inv) makes it possible to consider the function algebras instead of relation algebras and vice versa. Moreover, one can handle function algebras and relation algebras with equal significance, as it happens in [Pös-K 79], for example. We deal, however, with the clones in this book and proceed only with the co-clones if the proof methods developed for function algebras are not sufficient.

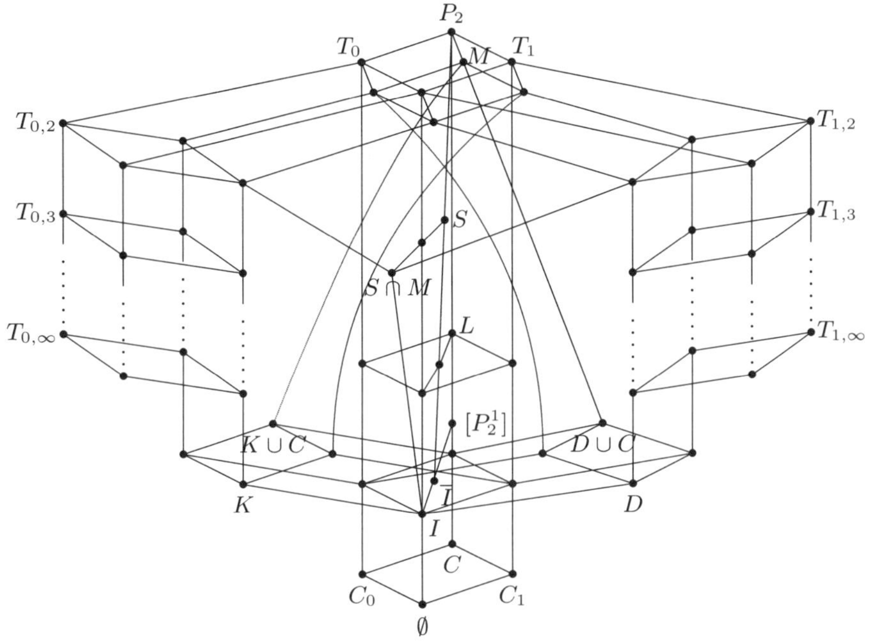

Fig. 1. The Post's lattice

The connection between algebras and relations was introduced by M. Krasner ([Kra 46] and [Kra 68/69]). He has completely developed the special Galois-connection between subsets of the permutation groups $\mathbf{S_n} := \{f \in P_k^{(1)} \mid f \text{ is bijective}\}$ and subsets of R_k. Later, Krasner also developed a similar theory for subsemigroups of $P_k^{(1)}$. Krasner's theory was generalized by V. G. Bodnarčuk, L. A. Kalužnin, V. N. Kotov, and B. A. Romov in [Bod-K-K-R 69].

Chapter 3 gives all subclasses from P_2 (the so-called **Post's classes**) and proves the completeness of the list. In addition, smallest generating systems ("bases") and the orders of the Post's classes are determined. The results of Chapter 3 were found by E. L. Post (see [Pos 20] and [Pos 41]). Figure 1 gives the Hasse-diagram of the lattice of the subclasses of P_2. A knot of the graph without denotation represents a class that can be described as an intersection of classes that lie above it.

Chapter 4 describes all classes that contain all unary functions of P_k. There exist $k+1$ of such classes, which form a chain in the lattice of all subclasses of P_k. The results of Chapter 4 were found by G. A. Burle (see [Bur 67]).

Chapter 5 begins with a coarse description of the maximal classes of P_k. The idea to describe classes this way was indicated already in [Kuz 59] by

A. V. Kuznecov. Then classes, from which we show later (in Section 5.3) that they are maximal classes of P_k, are defined. (In Chapter 6, the proof is all maximal classes are determined.) As in [Ros 70] the classes are divided in Section 5.2 into six types and characterized by their invariants (relations). We denote the relation sets needed in this case with with $\mathfrak{M}_k$, $\mathfrak{S}_k$, $\mathfrak{U}_k$, $\mathfrak{L}_k$, $\mathfrak{C}_k$ and $\mathfrak{B}_k$. Subsequently a maximal class is called a maximal class of the type $\mathfrak{X}$, when it can be described with the aid of a relation from set $\mathfrak{X}_k$, where $\mathfrak{X} \in \{\mathfrak{M}, \mathfrak{S}, \mathfrak{U}, \mathfrak{L}, \mathfrak{C}, \mathfrak{B}\}$. Next to the definitions of the maximal classes of P_k one finds the proofs of some properties of the maximal classes needed in later chapters, and the derivation of a recursion formula for the number determination of the maximal classes of P_k.

Chapter 6 gives the proof found by I. G. Rosenberg and published in [Ros 70a] that the classes defined in Section 5.2 are the only maximal classes of P_k for arbitrary k. In some parts, the original proof can be abbreviated by using proof ideas from some papers (for example from [Qua 82] and [Pös-K 79]). As already explained, there is also a solution of the completeness problem for P_k through the description of maximal classes of P_k.

Chapter 7 shows how one can derive some further completeness criteria for specific question formulations from the Rosenberg's completeness criterion 6.1. First of all we will handle a criterion for Sheffer functions which was found by G. Rousseau. [5] Then we will show how one can reduce the conditions from Theorem 6.1 if one considers only surjective functions. Finally, we deal with criteria indicate under which conditions a set ($\subseteq P_k$) consisting of certain unary functions and a Słupecki-function [6] is complete in P_k.

Chapter 8 explains the qualitative differences between the lattice $\mathbb{L}_2$ of the subclasses of P_2 and the lattice $\mathbb{L}_k$ of the subclasses of P_k for $k \geq 3$. While $\mathbb{L}_2$ is countable-infinite, $\mathbb{L}_k$ has the cardinality of the continuum (denoted by $\mathfrak{c}$) for $k \geq 3$. This results from the fact that, for every $k \geq 3$, the set P_k has a subclass with an infinite basis. One finds examples of such classes not only in Chapter 8, but also in Sections 12.3 and 14.10.
Moreover, we deal with cardinality statements about chains and antichains (in $\mathbb{L}_k$) and with the embedding of $\mathbb{L}_k$ into $\mathbb{L}_{k'}$ in Chapter 8. Then the question is clarified, which cardinalities the lattices $\mathbb{L}_k^{\downarrow}(A)$ [7] for maximal classes A of P_k have. In the last section of Chapter 8, the reader finds "strategies" for the determination of "manageable" sublattices of $\mathbb{L}_k$. In addition, two examples of theorems prove how one can determine certain sections of $\mathbb{L}_k$.

Chapter 9 deals with homomorphic mappings from a subclass of P_k onto a

[5] A Sheffer function is a function that every function of P_k can be formed from by means of the superposition operations.
[6] This is a function $f \in P_k \setminus [P_k^1]$ with $Im(f) = E_k$.
[7] $\mathbb{L}_k^{\downarrow}(A)$ denotes the set of all subclasses of A.

subclass of $P_{k'}$, As generally accepted,[8] one can characterize every homomorphism from a class A through a certain congruence relation[9] on A.
After some basic concepts are defined, all congruences on the subclasses of P_2 are determined in Chapter 9. It is the aim of the sections following then, to specify the general homomorphism theorem for function algebras and to find statements on the number of congruences on a subclass of P_k through determination of some general properties of the congruences on subclasses of P_k. Then, all congruences are determined for selected classes (among other things, these are the maximal classes of P_k and certain classes from linear functions). Criteria with which one can find out whether on a partial class of P_k only trivial congruences exist are in addition derived.
A further section deals with the connection between clone congruences and the fully invariant congruences on free algebras. The theorem found in this case has interesting inferences and is a bridge between certain investigations in the Universal Algebra and certain investigations in the theory of the function algebras.
At the end of this chapter, one can find some results on automorphisms. It is proven that P_k, the subclasses of P_2, and the maximal classes of P_k have only inner automorphisms.
The starting point and the basis of subsequent results are derived from an article by A. I. Mal'tsev (see [Mal 66]). Important contributions to the topic of Chapter 9 are also performed by V. V. Gorlov and I. A. Mal'tsev.

Chapter 10 deals with the relation degree and the dimension of subclasses A of P_k. The **relation degree** of A is the smallest number h so that the class A is unambiguously described by a relation set whose elements have the arity h at most. The **dimension** of A is the smallest arity of a relation that characterizes the class A unambiguously. Chapter 10 begins by investigating the connections between these two complexity measures of the relational description of classes. After that, the relation degrees and dimensions of the Post's classes, which were found by G. N. Blochina in [Blo 70], are proven. In Section 10.3, one can find further classes for which one knows the relation degree or the dimension.

Chapter 11 deals with a further measure (the so-called order of a class), with whose aid the complexity of a description of a subclass of P_k can be characterized. If the class $A \subseteq P_k$ is finitely generated, we call the smallest number r with $[A^{(n)}] = A$ the **order** of A. In the case that the subclass $A \subseteq P_k$ is not finitely generated we write $ord\ A = \infty$. Section 11.1 shows that one can determine the order easily for a class, if one knows the subclasses of the treated class. Therefore, order statements are often the "by-products" of considerations, which were actually used for determining certain classes or

[8] See also Chapter 4 of Part I.

[9] Those ones are the equivalence relations, which are compatible with the superposition operations on the considered class.

sublattices of P_k. So, for example, the orders of the Post's classes result from the considerations of Chapter 3 for verification of the Post's lattice. Further order statements are found in Chapters 13 and 14. Therefore, in Chapter 11, only the statements on the maximal classes that are described in Chapter 5 are determined. It is proven that, for $k \geq 3$, each maximal class of the type $\mathfrak{S}, \mathfrak{U}, \mathfrak{L}, \mathfrak{B}$ or $\mathfrak{C}^1$ has the order 2 and that, for the order of a class of the type $\mathfrak{C}^h$, where $h \geq 2$, the arity h of the descriptive relation of this class is an upper bound. If $M \subset P_k$ is a maximal class of the type $\mathfrak{M}$, we can only prove that M has the order 2 for the cases $2 \leq k \leq 7$ and for case that the descriptive binary relation ϱ of M has certain properties. For $k \geq 8$ there are maximal classes of the type $\mathfrak{M}$ that do not have a finite order. To prove this fact, we give an example, published by G. Tardos in [Tar 86].
If one has a finite generating class A, it is an interesting problem to clarify which cardinalities are possible for the bases of this class A. In Section 11.2 we show that the proof for $ord\ Pol_k \varrho \leq 3$, where $\varrho \in \mathfrak{C}_k^1 \cup \mathfrak{U}_k \cup \mathfrak{S}_k$, implies the existence of functions f_ϱ with $[f_\varrho] = Pol_k \varrho$. Following corresponding notation for functions from P_k, such function f_ϱ is called a **Sheffer function for** $Pol_k \varrho$.
The last section explains shortly, as one can determine the cardinalities of the possible bases of the class A, if $ord\ A < \infty$. Furthermore, some basic ideas of [Miy 71], [Miy-S-L-R 86], [Miy 88], and [Sto 87] are explained on basis classifications. For more information on the topic, we direct the reader to [Miy 88] and [Sto 87] of M. Miyakawa and I. Stojmenovic.

Chapter 12 summarizes the subclasses of the class

$$P_{k,2} := \{f \in P_k \mid f \text{ has only values of the set } \{0,1\}\},$$

which were found by G. Burosch, J. Dassow, N. Grünwald, W. Harnau, and the author. When one restricts the domain of a function $f^{(n)} \in P_{k,2}$ to the set E_2^n, a homomorphic mapping pr ("projection") from $P_{k,2}$ onto P_2 can be defined.
Since the image $pr A$ of a subclass A of $P_{k,2}$ is a subclass of P_2 and the subclasses of P_2 are known, one can hope to find of certain properties of the inverse images ($\subseteq P_{k,2}$) through the known properties of the images ($\subseteq P_2$). This hope confirmed itself in a certain sense (see, e.g., Theorem 12.2.5 and Theorem 9.7.6).
Conversely, $P_{k,2}$, $k \geq 3$, also reflects the negative properties of P_k because the examples of classes from Section 8.1.1 with infinite and without bases are subclasses of $P_{k,2}$.
Further, the functions of $P_{k,2}$ are important since they can be interpreted as predicates.
Chapter 12 is organized as follows: Section 12.1 contains the basic concepts and notations. Section 12.2 contains results on inverse images of subclasses of P_2 (with respect to the above projection). The remaining sections deal with the determination of cardinality and with the determination of the elements

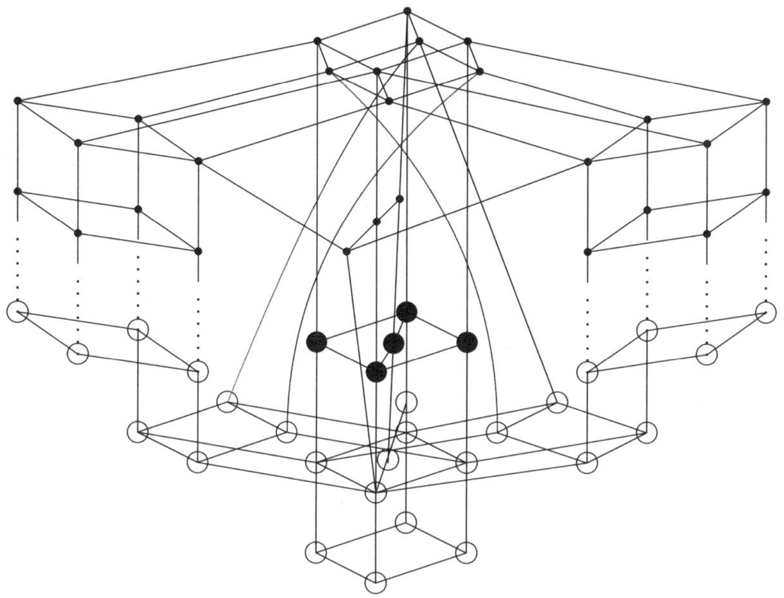

Fig. 2. A survey on $|\mathfrak{N}_k(B)|$ for $B \in \mathbb{L}_2$

of the set

$$\mathfrak{N}_k(B) := \{A \mid A \text{ is a subclass of } P_{k,2} \text{ with } pr\ A = B\}$$

for arbitrary subclass B of P_2. In Section 12.3, one can find some structure statements about the lattice of all subclass of $P_{k,2}$. It is also clarified in this section whether the set $\mathfrak{N}_k(B)$ is finite or infinite or has the cardinality of continuum. Section 12.4 generalizes some statements from Section 12.3 for $P_{k,l}$ with $2 \le l < k$. In Section 12.5, the maximal and the submaximal classes of $P_{k,2}$ are determined. Then, the investigations are continued from Section 12.3 for $k = 3$, i.e., for many classes $B \subseteq P_2$ with $|\mathfrak{N}_3(B)| \le \aleph_0$ the elements of the set $\mathfrak{N}_3(B)$ are determined. Figure 2 shows a survey of the obtained results. This figure shows the Post's graph, where the knots of this graph are differently labeled. If a class B of P_2 is marked in this graph through ○, this means that the set $\mathfrak{N}_k(B)$ has the cardinality of the continuum. The second marker ● means that $|\mathfrak{N}_k(B)| \ge \aleph_0$ and $|\mathfrak{N}_3(B)| = \aleph_0$ are valid. For the remaining classes B whose knots do not have any marker the set $\mathfrak{N}_k(B)$ is finite for arbitrary $k \in \mathbb{N}$.

Chapter 13 deals with classes of linear functions of P_k, i.e., with closed subsets of the set

$$L_k := \bigcup_{n\geq 1}\{f^{(n)} \in P_k \mid \exists a_0, ..., a_n \in E_k : \ f(\mathbf{x}) = a_0 + \sum_{i=1}^{n} a_i \cdot x_i \ (mod\ k)\}.$$

The lattice of the subclasses of L_k belongs to the earliest and best investigated sublattices of $\mathbb{L}_k$. For the case that k is a prime number, all subclasses, which are no subsets of $[L_k^1]$, were determined by A. A. Salomaa in [Sal 64]. The results of [Sal 64] were proven by J. Bagyinszki and J. Demetrovics in [Bag-D 82] and complemented with the remaining subclasses of $[L_k^1]$, $p \in \mathbb{P}$.
Many results about linear functions were obtained by Á. Szendrei. For example, she proved in [Sze 78] that L_k has only finitely many subclasses, if k is square-free. In addition, she showed that an arbitrary class has, at most, the order 2 (or 3), if k is quare-free and k is an odd number (or an even number), respectively.
For the case that k is not square-free, one easily finds a class that does not have any finite basis, whereby the set L_k has infinitely-many subclasses in this case.
Section 13.1 starts with properties of certain subclasses of the set

$$\begin{aligned} U_d \quad &:= \quad \textstyle\bigcup_{n\geq 1}\{f^{(n)} \in L_k \mid \exists a_0, ..., a_n \in E_k \ \exists j \in \{1, ..., n\} : \\ &\qquad f(\mathbf{x}) = a_0 + a_j \cdot x_j + d \cdot \textstyle\sum_{i=1, i\neq j} a_i \cdot x_i\}. \end{aligned}$$

This set is closed, if d is a divisor of k (notation: $d \mid k$). Then, with the aid of the results from the first section new proofs are indicated for the theorems of [Sal 64] and [Bag-D 82] in Section 13.2. Chapter 13 ends with a survey on further results about linear and quasi-linear functions.

Chapter 14 is the first chapter in this book of this book that deals only with submaximal classes. A subclass (or a subclone) of P_k is called **submaximal** if it is covered by a maximal class (clone). The concept *submaximal class* was introduced by I. G. Rosenberg in [Ros 74]. In [Ros 74] one finds also the first results about submaximal classes of P_k (see Chapter 17 for details).
The submaximal classes are interesting not only because of their position in the second layer below P_k, but also because further completeness criteria for P_k result from a list of all submaximal classes. For arbitrary k, the full list of the maximal classes of a maximal class A of P_k is only known if A has the type $\mathfrak{S}$ ([Ros-S 84], see Section 18.1), $\mathfrak{C}^1$ (see Chapters 16 and 17) or $A = Pol_k\varrho$, where $\varrho := E_{k-1}^2 \cup \{(k-1, k-1)\}$ (see Section 18.3). There are, however, some papers, in which one finds submaximal classes for specific k or only such submaximal classes that contain certain functions.
In Section 14.1, one finds a complete description of all submaximal classes of P_3 and some remarks about generalizations of the indicated theorems. The following papers provide a basis for this description: [Mac 79], [Mar-D-H 80], [Sal 64], [Bag-D 82] and [Lau 82a]. The theorems from Section 14.1 are proven then in Sections 14.2 – 14.9. In Section 14.10, we will prove that there are 5 submaximal classes with finitely many subclasses; 7 have countably many

subclasses, and the remaining 146 submaximal classes have uncountably many subclasses. All elements of the lattices $\mathbb{L}_3^{\downarrow}(A)$, where A is a submaximal class with$|\mathbb{L}_3^{\downarrow}(A)| \leq \aleph_0$ are determined in Chapter 15.

Chapter 15 gives all finite or countably infinite sublattices of depth 1 or 2 of the lattice of the subclasses of P_3, where the finite cases are easy conclusions from Chapter 13. We say that the lattice $\mathbb{L}_k^{\downarrow}(A)$ of subclasses of $A = [A] \subseteq P_k$ has the **depth** t, if t is the least integer for which there are some classes A_1, ..., $A_{t-1} \in \mathbb{L}_k$ with

$$A \subset A_1 \subset A_2 \subset ... \subset A_{t-1} \subset P_k.$$

For $k = 3$ by the Theorems 13.2.3 and 8.1.6 there exist finite and countably infinite sublattices of the depth 1 or 2. In Chapter 15 these sublattices shall be determined exactly.
The finite lattices $\mathbb{L}_3^{\downarrow}(L_3)$ of depth 1 are found in Section 15.1. This lattice is a conclusion from Chapter 13. In addition, by Section 14.10, this lattice is the only finite lattice of depth 1 and, further, this lattice contains all finite sublattices of $\mathbb{L}_3$ of depth 2.
Because of Theorem 14.10.1, there are exactly 7 submaximal classes with $|\mathbb{L}_3^{\downarrow}(A)| = \aleph_0$. One obtains these classes through formation of isomorphic pictures of the following two classes:

$$\begin{aligned}\mathcal{L}_3 = [P_3^{(1)}] \cup \textstyle\bigcup_{n\geq 1}\{f^{(n)} \in P_3 \mid \exists f_0, f_1, ..., f_n \in P_3^1 : f(x_1, ..., x_n) = \\ f_0(f_1(x_1) + f_2(x_2) + ... + f_n(x_n) \ (mod\ 2))\}\end{aligned}$$

and

$$\mathcal{M} := \bigcup_{n\geq 1}\{f^{(n)} \in P_3 \mid \exists f_1, ..., f_n \in O^1 :\ f(x_1, ..., x_n) = f_1(x_1) \vee ... \vee f_n(x)\},$$

where O^1 is the set of all unary monotonous functions (in respect to the total order $0 < 1 < 2$) of P_3 and $x \vee y := max\{x, y\}$.
Since $\mathcal{L}_3$ contains also all subclasses, which are generated from unary functions of P_3. First of all, the 1299 subsemigroups of the semigroup $(P_3^1; \star)$ are determined in Section 15.2. The list of these semigroups is then an important aid in Section 15.3 during the determination of the remaining subclasses of $\mathcal{L}_3$.
In Section 15.4 one finds all subclasses of $\mathcal{M}$.

Chapter 16 supplies the description (coming from [Sze 91] or [Lau 82b, 95a]) of all maximal classes of the subclass

$$T_Q := \bigcap_{a\in Q} Pol_k\{a\}$$

of P_k for arbitrary Q with $\emptyset \neq Q \subseteq E_k$, $k \geq 2$. With the aid of these classes, a completeness criterion for T_Q can be formulated easily. This criterion implies

necessary and sufficient conditions for whether a finite algebra is semi-primal and has only trivial subalgebras.
If $|Q| = 1$, then T_Q is a maximal class of P_k and the maximal classes of T_Q, given in Chapter 15, are submaximal classes of P_k.
Moreover, if $|Q| \geq 2$, we prove that every maximal class of T_Q is an intersection of T_Q with certain maximal classes of P_k or $Pol_k\{a\}$ $(a \in Q)$.

Chapter 17 continues the investigations of Chapter 16 and generalizes Theorem 14.1.3. For arbitrary $k, l \in \mathbb{N}$ with $2 \leq l \leq k-1$ all maximal classes of $Pol_k E_l$ are determined, where 9 relation sets are needed. It is important to note that, for the description of the maximal classes of $Pol_k\{a\}$, $a \in E_k$, one needs only 6 relation sets.
With the help of the maximal classes of $Pol_k E_l$, one can easily give a completeness criterion for $Pol_k E_l$.
The proofs given in Chapter 17 resemble those ones from Chapter 6, i.e., the results of this chapter were achieved with the means developed by I. G. Rosenberg in [Ros 70a].

Chapter 18 gives a survey (partial without proof) over further submaximal classes found until now. In supplement to Chapters 16 and 17, all submaximal classes of a maximal class of the type $\mathfrak{S}$ are described in this chapter. It is shown, then, how one can prove the special case $k \in \mathbb{P}$ of this general description easily. The rest of this chapter deals with submaximal classes of P_k that lie below a maximal class of the type $\mathfrak{U}$. In Section 18.2, one can find some maximal classes of $Pol_k \varrho$, where $\varrho \in \mathfrak{U}_k$ is arbitrary. Then, in Section 18.3, the list is completed from Section 18.2 to the list of all maximal classes of $Pol_k \varrho$ for $\varrho = E_{k-1}^2 \cup \{(k-1, k-1)\}$.

Chapter 19 Chapter 19 deals with classes of $\mathbb{L}_k$, which are either direct predecessors of the empty set (so-called *minimal classes*) or which are direct predecessors of the set of all projections (so-called *minimal clones*).
Consequently, it is not difficult to determine the minimal classes. However, for the minimal clones, only partial results can be given.
In Chapter 19, one finds a description of all minimal classes. Moreover, Rosenberg's classification of minimal clones is proven. At the end of Chapter 19, one finds a survey of further results about minimal clones and the description of all partial minimal clones.

Chapter 20 deals with partial function algebras. A partial n-ary function on E_k is a mapping from a subset of E_k^n into E_k, $n \in \mathbb{N}$. Let $\widetilde{P_k}$ the set of all such functions with $n \in \mathbb{N}$. Then one can introduce certain modified Mal'tsev-operations over the set $\widetilde{P_k}$ of all partial functions on E_k. Then, the set $\widetilde{P_k}$, together with these operations, forms a so-called (full) partial function algebra $(\widetilde{P_k}; \tau, \zeta, \Delta, \nabla, \star)$, which can similarly be examined like the function algebra $(P_k; \tau, \zeta, \Delta, \nabla, \star)$.
The choice of results on partial function algebras in this chapter focuses on questions already treated for P_k in previous chapters.

After a composition of some basic concepts in Section 20.1, Section 20.2 shows that the lattice of all partial clones of $\widetilde{P_k}$ is isomorphic to a certain sublattice of the lattice of all clones of P_{k+1}. Thus, one gets many properties of the partial clones from the properties of the clones, which were already found. However many results on clones that may be helpful to find certain partial clones with the aid of the above-mentioned isomorphism are missing. One could, for example, not solve the completeness problem for $\widetilde{P_k}$ with the help of an isomorphism.

In Section 20.3, we show how to describe partial clones by relations.

Sections 20.4 and 20.5 deal with the maximal partial clones, with whose help, as for P_k, the completeness problem of the partial logic is soluble. We will prove that $\widetilde{P_2}$ has exactly 8 and $\widetilde{P_3}$ has exactly 58 maximal partial clones. In Section 20.5, there is a complete list of all maximal clones of $\widetilde{P_k}$ for arbitrary $k \in \mathbb{N}$, found by L. Haddad and I. G. Rosenberg. The list is given without proof. In Section 20.6, we determine those descriptive relations of the maximal clones of P_k, which are also descriptive relations of the maximal partial clones of $\widetilde{P_k}$. In addition, we survey those papers that deal with the determination of the orders of maximal partial clones. Section 20.7 deals with the determination of the cardinality of the set $\mathcal{I}(A) := \{C \subseteq \widetilde{P_k} \mid C = [C] \wedge C \cap P_k = A\}$, where A is an arbitrary maximal clone of P_k. We will prove that, if A has the type $\mathfrak{U}$, $\mathfrak{S}$ or $\mathfrak{C}$, $\mathcal{I}(A)$ is a finite set. On the other hand, the set $\mathcal{I}(A)$ has the cardinality of continuum if A has the type $\mathfrak{L}$ or $\mathfrak{B}$. For the type $\mathfrak{M}$ we can give only partial results.

Section 20.8 surveys of the cardinalities of the sets $\mathcal{I}(A)$, where A is an arbitrary subclass of P_2.

In the last section we determine the congruences on the maximal partial clones. It is proven, particularly, that $\widetilde{P_k}$ has exactly 4 congruences, whereas a maximal partial clone has exactly 4, 8, or 10 pairwise distinct congruences.

Finally, one can find some **technical references**:

As already noted, the chapters of this book are not continuously numbered, except for the chapters of the first and second part. References to parts, chapters, sections, lemmas, theorems, ... are given through their numbers. If a reference of the first or second part is missing, then the part discussed is referred to.

About the basics of Universal Algebra there are many publications and books. Therefore, only some references for the theorems are given in Part I of this text.

In Part II of the book, references for a theorem are left out only if one can prove the theorem easily and if there are several references for this theorem. Normally, one finds references at the beginning of the chapters and with the theorems, whereby it is also clear that the lemmas appertaining to these theorems are based on considerations from the quoted references. For some few theorems, the proofs are taken over from the quoted references directly.

However new and shorter proofs are presented particularly for theorems that come from older books.

The end of a proof (or of a statement with easy proof) is marked by ■.

Some book parts that can be ignored during the first reading differ from the remaining text through a smaller writing.

Preliminaries

We assume that the reader has some knowledge about the basic concepts from the set theory[1], linear algebra, classical algebraic structures and mathematical logic. The following is a list of terms to which we assume the reader is familiar.

1) Logical Symbols

So that we can write down mathematical statements quickly and correctly, we use the following symbols from mathematical logic:

symbol	denotation for
$\wedge$	and
$\vee$	or
$\neg$	not
$\Longrightarrow$	implies; if - then
$\Longleftrightarrow$	if and only if; iff
$:=$	equal by definition
$:\Longleftrightarrow$	equivalent by definition
$\exists$	there exists; it exists at least
$\exists!$	it exists exactly
$\forall$	for all

In order to save parentheses, we write instead of $\exists x\ (\ E\ )$ (read: "There exists an x with the property E") shortly $\exists x\ :\ E$. We arrange analogous one for

[1] A naive theory of sets is sufficient for our purposes.

formulas that contain the sign $\forall$. In addition, $\forall x\ \exists y$ ... stands for $\forall x\ (\exists y\ (...))$, and so forth.

2) Symbols and Concepts of the Set Theory

In dealing with sets, we use the following standard notations: membership ($\in$), nonmembership ($\notin$), set-builder notations ($\{....\mid ....\}$ [2]), the empty set ($\emptyset$), inclusion ($\subseteq$), noninclusion ($\nsubseteq$), proper inclusion ($\subset$), intersection ($\cap$ and $\bigcap$), union ($\cup$ and $\bigcup$) and difference ($\setminus$).
The **power set** of a set A is the set $\{B \mid B \subseteq A\}$ of all subsets of A. It is denoted by $\mathfrak{P}(A)$.
$\mathbb{N}$, $\mathbb{N}_0$, $\mathbb{Z}$, $\mathbb{Q}$, $\mathbb{R}$, $\mathbb{C}$ denote respectively the set $\{1, 2, 3, ...\}$ of all natural numbers, the set $\mathbb{N} \cup \{0\}$, the set of all integers, the set of all rational numbers, the set of all real numbers, and the set of all complex numbers.

For $n \in \mathbb{N}$, we write the order **n-tuples** (briefly n-tuples) in the form $(x_1, x_2, ..., x_n)$. For two n-tuples $\mathbf{x} := (x_1, x_2, ..., x_n)$ and $\mathbf{y} := (y_1, y_2, ..., y_n)$ is $\mathbf{x} = \mathbf{y}$ iff $x_i = y_i$ for all $i \in \{1, 2, ..., n\}$.
$A \times B$ is the set of all 2-tuples (a, b) with $a \in A$ and $b \in B$ and is called the (**Cartesian** or **direct**) **product** of the sets A and B. For $n \in \mathbb{N}$, $\Pi_1^n A_i$ (or $\Pi_{i \in I} A_i$, where $I := \{1, 2, ..., n\}$) denotes the set of all $(a_1, a_2, ..., a_n)$ with $a_i \in A_i$ for all $i \in I$. [3] Further, let $A^n := \{(a_1, ..., a_n) \mid \forall i \in \{1, 2, ..., n\} : a_i \in A\}$ be the **direct n-powers** of the set A, $n \in \mathbb{N}$.

A **correspondence** F **of** A **into** B is a subset of $A \times B$, where $D(f) := \{a \in A \mid \exists b \in B :\ (a, b) \in F\}$ is the **domain** of F and $Im(f) := \{b \in b \mid \exists a \in A : (a, b) \in F\}$ is the **image** (or **range**) of F. If $D(F) = A$, then F is a correspondence **from** F. If $Im(F) = B$, then we say that F is a correspondence **onto** B.
The **inverse** (or **converse**) F^{-1} of a correspondence $F \subseteq A \times B$ is given by $F^{-1} := \{(b, a) \in B \times A \mid (a, b) \in F\}$.
If $F \subseteq A \times B$ and $G \subseteq B \times C$, then the **correspondence product** $F \Box G$ is defined by $F \Box G := \{(a, c) \in A \times C \mid \exists b \in B :\ (a, b) \in F\ \wedge\ (b, c) \in G\}$.
We remark that $(F \Box G)^{-1} = G^{-1} \Box F^{-1}$ and $\Box$ is associative.

A **mapping** (or **map**) f from A into B is a subset of $A \times B$ such that for each $a \in A$ there is exactly one $b \in B$ with $(a, b) \in f$. If f is a mapping from A into B, then we write $f :\ A \longrightarrow B$ and, if $(a, b) \in f$, $f(a) = b$, or $a \mapsto b$.
A mapping $f : A \longrightarrow B$ is called **injective** (or is an **injection**) iff $f(a) = f(a')$ implies $a = a'$ for all $a, a' \in A$. The mapping $f : A \longrightarrow B$ is **surjective** (or is a **surjection**) iff $Im(f) = B$. The mapping $f : A \longrightarrow B$ is **bijective** (or is a **bijection**) iff is both injective and surjective.

Let A be a set and $n \in \mathbb{N}$. Then, an **n-ary relation on** A (or an **n-ary**

[2] For example, we write $S := \{x \mid x \in A\ \wedge\ P(x)\}$ or briefly $S := \{x \in A \mid P(x)\}$ instead of "S is the set of all $x \in A$ with the property $P(x)$".

[3] For arbitrary I, we will define $\Pi_{i \in I} A_i$ in Part I, Section 5.

relation over A) is a subset of A^n. A 2-ary , 3-ary or 4-ary relation on A is called respectively a **binary**, **ternary** or **quaternary** relation on A.

Two sets A and B have the same **cardinality** (in symbol $|A| = |B|$) iff there exists a bijection from A onto B.
A set A is **infinite**, iff there is a subset $B \subset A$ with $|A| = |B|$. A set A is called **finite** iff A is not infinite.
If A is a finite set, then $|A|$ denotes the number of elements of A, i.e., $|\emptyset| = 0$ and $|x \cup \{x\}| = |x| + 1$ (or $n = |\{0, 1, 2, ..., n-1\}|$ for all $n \in \mathbb{N}_0$, where n is called a (finite) ordinal).
We put $\aleph_0 := |\mathbb{N}|$ and $\mathfrak{c} := |\mathbb{R}|$. The set A is called **countable** if $|A| = \aleph_0$. If $|A| = \mathfrak{c}$, we say that A has the **cardinality of the continuum**.
It is well-known that $\aleph_0$ is the least infinite ordinal and that $\mathbb{N}$ and $\mathbb{R}$ do not have the same cardinality. Further, it holds $|\mathbb{Q}| = \aleph_0$, $|\mathbb{C}| = \mathfrak{c}$, $|\mathfrak{P}(\mathbb{N})| = \mathfrak{c}$ and $|\bigcup_{n=1}^{\infty} A_n| = \aleph_0$ if $|A_n| = \aleph_0$ for all $n \in \mathbb{N}$.

Part I

Universal Algebra

The *Universal Algebra* is a mathematical discipline, which examines so-called universal algebras, where the classic algebraic structures – such as groups, rings, and lattices – are special cases of such algebras. Roughly said, a universal algebra is nothing other than a nonempty set together with a family of finite-ary operations over this set.
Although the concept *universal algebra* was used already at the end of the 19th century, the development of the theory *Universal Algebra* started only in the 1930's. This development has two historical roots. The first root is the classical algebra and the studies of properties that have all algebraic structures (such as semigroups, groups, rings, etc.). The second root is the *Mathematical Logic.*
In the last decades, the Universal Algebra has continued the classic examinations and has also developed new areas. An introduction to the bases of the Universal Algebra follows.
The topic of the following six chapters is based on what is needed later, during the investigation of the function algebras.
The first five chapters show that the Universal Algebra is a general theory of the algebraic partial disciplines. Chapter 6 treats problems with an impression of the mode of operation of the algebra and that have partial also contact points to the mathematical logic. [1]

Additions to the topics can be found in [Lau 2004] [2].
In detail one can study the Universal Algebra with the aid of the following books [Bur-S 81], [Coh 65], [Den-T 96], [Den 2003], [Grä 68], [Ihr 2003], [McK-M-T 87], [Wer 78]. The book [McK-M-T 87] can particularly be recommended.

In the preface, we have already noted that to understand the second part of the text need not be read absolutely. Thus, readers can select for themselves the required terms and theorems from Part I, during the reading of Part II. One night notice, after reading Part II, that some examinations are special cases of general problems.

[1] To realize this, reads [Bur-S 81] and [Ric 78].

[2] There, one finds a continuation to the chapters of Universal Algebra of this book and additional chapters.

1

Basic Concepts of Universal Algebra

1.1 Universal Algebras

First, we define the concept of an n-ary partial operation:
Let A be a nonempty set, $n \in \mathbb{N}$ and $\emptyset \subset \varrho \subseteq A^n$. An **$n$-ary partial operation** is a mapping f from ϱ into A. The number n is called **arity** of f and the arity of f is also denoted by af. To denote the arity of f we also write $f^{(n)}$ or briefly f^n, since for content-related reasons, the interpretation is impossible of f^n as Cartesian product in the following.
If $(a_1, ..., a_n) \in \varrho$ then let $f(a_1, ..., a_n)$ be the image of $(a_1, ..., a_n)$ under an n-ary operation f.
The set ϱ we call **domain** of f, and we denote the domain of f by $D(f, A)$ or briefly by $D(f)$.
Let $Im(f)$ be the set $\{f(a_1, ..., a_n) \mid (a_1, ..., a_n) \in D(f)\}$, which is called **image** or **range** of f.
For $n \in \mathbb{N}$ let $c_\infty^{(n)}$ be the n-ary partial operation with $D(c_\infty^{(n)}) = \emptyset$.[1]
If $\varrho = A^n$ and $n \geq 1$, then f is called an **n-ary operation** on A. If $D(f^{(n)}) \subset A^n$ we call $f^{(n)}$ a **proper partial operation**.
A **nullary operation** f on A is an element f of A with $af := 0$, $D(f) := \emptyset$ and $Im(f) := \{f\}$.

Example The operation $\circ$ of a group (see also Section 1.2) is a binary operation. One can understand the formation of inverse elements in a group as a unary operation $^{-1}$ and the identity e (unit element) of a group as a nullary operation.

As is generally accepted, one can form **compositions** of mappings in the following manner: If $f : M_1 \longrightarrow M_2$ and $g : M_2 \longrightarrow M_3$ are mappings, then we denote the mapping

[1] The necessity of such a definition is seen, if one forms superpositions of partial functions (see Part II, Chapter 1). For example, it holds $g \Box f = c_\infty^{(1)}$ for the unary operations f, g with $D(f) := \{a\}$ and $a \notin Im(g)$.

$$f\Box g : M_1 \longrightarrow M_3,\ x \mapsto g(f(x))$$

by $f\Box g$. It is well-known (or it is easy to see) that $\Box$ is associative. Thus we can renounce the putting of brackets in the case of more than two compositions. Furthermore, we write

$$g_1 g_2 ... g_r$$

instead of

$$g_1\Box g_2\Box ...\Box g_r\ (g_i : M_i \longrightarrow M_{i+1},\ i = 1, ..., r),$$

if the operation $\Box$ follows from the context.
We notice that the above definition of $\Box$ is a special case of the following definition, if one writes the mappings in the form of subsets of Cartesian products: Let $R \subseteq A \times B$ and $Q \subseteq B \times C$. Then let $R\Box Q := \{(a,c) \mid \exists b \in B : (a,b) \in R \wedge (b,c) \in Q\}$.
With the help of the partial operations, we can define the following concepts:

A **partial algebra** *is an ordered pair*

$$\mathbf{A} := (A; F),$$

where A is an arbitrary nonempty set and F is a set of partial operations on A. The set A is called **universe** *(or* **underlying set***) of* **A**. *The elements of F are called* **(partial) fundamental operations** *of* **A**.
A **universal algebra** *or briefly an* **algebra** *is a partial algebra $(A; F)$, where every element of F is an operation, i.e., F does not have any proper partial operations.*

Here we usually only deal with algebras. Therefore, we often introduce the following concepts only for algebras.

If $F = \{f_1, ..., f_r\}$ holds we write $(A; F)$ in the form

$$(A; f_1, ..., f_r),$$

where we often choose $af_1 \geq af_2 \geq ... \geq af_r$.
We also use the notation

$$(A; (f_i)_{i\in I}),$$

if $F = \{f_i \mid i \in I\}$ and I is a certain index set.

A (partial) algebra $(A; F)$ is called **finite**, if A is a finite set, otherwise **infinite**.

In order to receive a first classification of the algebras, it offers itself to carry out a classification according to the arities of the (partial) operations of the algebras. For this purpose, we define the concept **type of an algebra** $(A; (f_i)_{i\in I})$ as a sequence

$$\tau := (af_i \mid i \in I)$$

of the arities of their fundamental operations, if I is a finite or countable set. In the case that I is uncountable, we define as type of $\mathbf{A}$ the mapping $\tau : I \longrightarrow \mathbb{N},\ i \mapsto af_i$. If I is a finite set, then $(A; (f_i)_{i \in I})$ is called of **finite type**. If $\mathbf{A} = (A; f_1, ..., f_r)$, $af_1 \geq af_2 \geq ... \geq af_r$, then let $\tau = (af_1, af_2, ..., af_r)$. Two (partial) algebras $(A; F)$, $(B; G)$ are **of same type**, iff there is a bijection $\varphi : F \longrightarrow G,\ f \mapsto g$ with $af = ag$.

Examples[2]

1) A semigroup $(H, \circ)$ is an algebra of type (2) ($F := \{\circ\}$).

2) By the above remark, a group is an algebra of the type $(2, 1, 0)$.

3) A lattice (see 1.2.11) is an algebra of type $(2, 2)$.

Because of existence of a bijection between the operations of two algebras $\mathbf{A} := (A; F)$ *and* $\mathbf{B} := (B; G)$ *of same type, we often denote the operation sets of* $\mathbf{A}$ *and of* $\mathbf{B}$ *with the same symbol (*F *or* G*). In analog mode we deal with the denotation of the elements.*

If it is necessary to denote the operations of two different algebras $\mathbf{A}$ and $\mathbf{B}$ of the same type differently, then we write $f^{\mathbf{A}}$ or $f^{\mathbf{B}}$ (or also $f_{\mathbf{A}}$ or $f_{\mathbf{B}}$), respectively.

Before we deal with further basic terms from the theory of the algebras, let's define some specific algebras. We will study some of these specific algebras in detail later.

1.2 Examples of Universal Algebras

We start with some "classical algebras". In defining these algebras, we try to use no existence quantor. [3] We call the equations, which are given for the definition of an algebra, axioms. As usual, we will try to use only a few (independent) axioms. We differ from this principle in the following only in few places (see e.q. the definition of a lattice).

1.2.1 Gruppoids

An algebra $(A; \circ)$ of type (2) is called **gruppoid**. Thus, a gruppoid is not a different one as a nonempty set together with a binary operation.

---

[2] Section 1.2 contains the explanations to the concepts used.

[3] In Chapter 6, we see that this manner of description has interesting conclusions. In the following, we use the descriptions of the algebraic structures given in this section.

1.2.2 Semigroups

A gruppoid $(H; \circ)$ is called **semigroup** if the algebra $(H; \circ)$ fulfills the following axiom:

$$(A) \qquad \forall x, y, z \in H : (x \circ y) \circ z = x \circ (y \circ z) \qquad \text{(associativity)}.$$

A semigroup is called **commutative**, if it satisfies the following condition in addition:

$$(C) \qquad \forall x, y \in H : x \circ y = y \circ x.$$

1.2.3 Monoids

An algebra $(M; \circ, e)$ of type $(2, 0)$ is called **monoid**, if $(M; \circ)$ is a semigroup and

$$(E) \qquad \forall x \in M : \ e \circ x = x \circ e = x$$

holds. The (nullary) operation e is the the **neutral element** of M.

1.2.4 Groups

A **group** is an algebra $(G; \circ, ^{-1}, e)$ of type $(2, 1, 0)$, which fulfills the above axioms (A), (E) for $x, y, z, e \in G$ and

$$(I) \qquad \forall x \in G : \ x \circ x^{-1} = x^{-1} \circ x = e\,.$$

x^{-1} is called the **inverse element** of x. A group, which (C) fulfills in addition, is called **Abelian** (or **commutative**).
It is usual in Abelean groups to use $+, -x, 0$ instead of $\circ, x^{-1}, e$ (the **additive notation**).

1.2.5 Semirings

An algebra $(R; +, \cdot)$ of type $(2, 2)$ is called a **semiring** if $(R; +)$ is a commutative semigroup, $(R; \cdot)$ is a semigroup, and if the following distributive laws hold:

$$\begin{array}{ll} (D_1) & \forall x, y, z \in R : \ x \cdot (y + z) = (x \cdot y) + (x \cdot z), \\ (D_2) & \forall x, y, z \in R : (x + y) \cdot z = (x \cdot z) + (y \cdot z). \end{array}$$

1.2.6 Rings

An algebra $(R; +, \cdot, -, 0)$ of type $(2, 2, 1, 0)$ is called **ring** if $(R; +, -, 0)$ is an Abelean group and $(R; +, \cdot)$ is a semiring.
In order to save parentheses, we use the known rule subsequently "The dot bill is carried out before the stroke bill".
A **unitary ring** (or a **ring with a unit element**) is an algebra $(R; +, \cdot, -, 0, 1)$ of type $(2, 2, 1, 0, 0)$, where $(R; +, \cdot, -, 0)$ is a ring and (E) holds with $e = 1$ and $\circ = \cdot$.

1.2.7 Fields

A partial algebra $(K;+,\cdot,-,^{-1},0,1)$ is called **field** if the algebra $(K;+,\cdot,-,0,1)$ is an unitary ring and $(K\backslash\{0\};\cdot,^{-1},1)$ is an Abelean group. We remark that the operation $^{-1}$ on K is only a partial operation, since 0^{-1} is not defined.

1.2.8 Modules

Let $\mathbf{R} := (R;+,\cdot,-,0)$ be a ring. An algebra $(M;F)$ where $F := \{+,-,0\}\cup R$, $+$ is binary, 0 is a nullary and $-$ and all $r \in R$ are unary operations is called a R**-module** (or **module over the ring R**), if $(M;+,-,0)$ is an Abelean group and if for all $r, s \in R$ the following equations hold:

$$\begin{array}{ll} (M_1) & \forall x,y \in M : r(x+y) = r(x)+r(y), \\ (M_2) & \forall x,y \in M : (r+s)(x) = r(x)+s(x), \\ (M_3) & \forall x \in M : (r\cdot s)(x) = r(s(x)). \end{array}$$

Let $\mathbf{M} := (M;+,-,(r)_{r\in R},0)$ be the short notation for the $\mathbf{R}$-module defined above. Further, we say that $\mathbf{M}$ has the type $(2,1,(1)_{r\in R},0)$.
A **module over a unitary ring** $(R;+,\cdot,-,0,1)$ is an above module that also satisfies the following condition:

$$(M_4) \qquad \forall x \in M : 1(x) = x.$$

Three remarks to the above definitions:
– Instead of the unary operations $r \in R$ one can also define a mapping $\odot$: $R \times M \longrightarrow M$, $(r,x) \mapsto r \odot x$, which fulfills the axioms. However, a module would then not be a universal algebra.
– A module has infinite many operations if the ring R is infinite.
– The operation symbols $+,-,0$ have two different meanings: on the one hand, they describe the operations of the Abelean group $(R;+,-,0)$; on the other hand, they describe the operations of the Abelean group $(M;+,-,0)$.

1.2.9 Vector Spaces

Let $\mathbf{K} := (K;+,\cdot,-,0,1)$ be a field. Then every $\mathbf{K}$-module

$$(V;+,-,(k)_{k\in K},0)$$

is called **K-vector space** (or **vector space over the field K**).

1.2.10 Semilattices

A commutative semigroup $(S;\circ)$, in which

$$\forall x \in S : \; x \circ x = x$$

holds, is called **semilattice**.

1.2.11 Lattices

A **lattice** is an algebra $(L; \vee, \wedge)$ of type $(2, 2)$, in which for arbitrary $x, y, z \in L$ it holds:

$$\begin{array}{lll} (L_1) & x \vee y = y \vee x, x \wedge y = y \wedge x & \text{(commutativity)}, \\ (L_2) & x \vee (y \vee z) = (x \vee y) \vee z, & \\ & x \wedge (y \wedge z) = (x \wedge y) \wedge z & \text{(associativity)}, \\ (L_3) & x \vee x = x,\ x \wedge x = x & \text{(idempotency)}, \\ (L_4) & x \vee (x \wedge y) = x, x \wedge (x \vee y) = x & \text{(absorption)}. \end{array}$$

A **bounded lattice** (or a **lattice with 0 and 1**) is an algebra $(L; \vee, \wedge, 0, 1)$ of type $(2, 2, 0, 0)$ such that $(L; \vee, \wedge)$ is a lattice and furthermore the following equations hold for all $x \in L$:

$$(L_5) \qquad x \wedge 0 = 0,\ x \vee 1 = 1.$$

A lattice is called **distributive**, if the following distributive laws hold:

$$\begin{array}{ll} (DL_1) & \forall x, y, z \in L : x \wedge (y \vee z) = (x \wedge y) \vee (x \wedge z), \\ (DL_2) & \forall x, y, z \in L : x \vee (y \wedge z) = (x \vee y) \wedge (x \vee z). \end{array}$$

Remark: One can show (with the help of (L_1)–(L_4)) that (DL_1) and (DL_2) are equivalent; that is, it suffices to require either (D_1) or (D_2) in the definition of a distributive lattice.

1.2.12 Boolean Algebras

An algebra $(B; \vee, \wedge, ^-, 0, 1)$ of type $(2, 2, 1, 0, 0)$ is called **Boolean algebra** if $(B; \vee, \wedge, 0, 1)$ is a bounded distributive lattice and the following equalities hold for all $x \in B$:

$$\begin{array}{lll} (B_1) & x \wedge \overline{x} = 0, & x \vee \overline{x} = 1 \\ (B_2) & x \wedge 0 = 0, & x \vee 1 = 1, \end{array}$$

where $\overline{x} := {}^-(x)$.

1.2.13 Function Algebras

One can choose the set of all operations defined on A as a universe of an algebra and can then define operations on the operations on A. For the purpose of distinction we will subsequently replace the concept "operation (on the set A)" by the concept "*function (on the set A)*". In Part II, the set A is always a finite set. Therefore, the following concepts become explained only for a specific k-element set E_k. Put

$$E_k := \{0, 1, ..., k-1\},$$

$k \in \mathbb{N}\backslash\{1\}$. Let P_k^n be the set of all n-ary functions f^n, which map the n-fold Cartesian product E_k^n into E_k. Put $P_k := \bigcup_{n\geq 1} P_k^n$.
Elementary operations (called **Mal'tsev-operations**) over P_k are

$$\zeta, \tau, \Delta, \nabla \text{ (unary operations) and}$$

$$\star \text{ (a binary operation)}$$

defined by

$$(\zeta f)(x_1, ..., x_n) := f(x_2, x_3, ..., x_n, x_1),$$
$$(\tau f)(x_1, ..., x_n) := f(x_2, x_1, x_3, ..., x_n),$$
$$(\Delta f)(x_1, ..., x_{n-1}) := f(x_1, x_1, x_2, ..., x_{n-1}) \text{ if } n \geq 2,$$
$$\zeta f = \tau f = \Delta f := f \text{ if } n = 1,$$
$$(\nabla f)(x_1, ..., x_{n+1}) := f(x_2, x_3, ..., x_{n+1}),$$
$$(f \star g)(x_1, ..., x_{m+n-1}) := f(g(x_1, ..., x_m), x_{m+1}, ..., x_{m+n-1})$$

$(f^n, g^m \in P_k)$.
It holds the following (see Part II, Chapter 1):
With the help of the Mal'tsev-operations, one can form the following operations (for arbitrary functions and arbitrary variables of the functions):
– permutation of variables
– identification of variables
– adding of fictitious variables and
– substitution of variables of a function by functions
The set of all functions that can be obtained by a finite number of applications of the above operations from the functions of $F \subseteq P_k$ is called the **closure** of F, and is denoted by $[F]$. If $F = [F]$, then we say that F **is closed** or F is a **(sub)class** of P_k.
The algebra $(P_k; \zeta, \tau, \Delta, \nabla, \star)$ of type $(1, 1, 1, 1, 2)$ is called **(full) iterative function algebra**.
If A is a subclass of P_k then $(A; \zeta, \tau, \Delta, \nabla, \star)$ is called **function algebra**.

1.3 Subalgebras

Let $\mathbf{A} = (A; F)$ be a (partial) algebra and let B be a nonempty subset of A. The (partial) algebra $\mathbf{B} = (B; F)$ of the same type as $\mathbf{A}$ is called (partial) **subalgebra** of $\mathbf{A}$ (or $\mathbf{A}$ is an **extension** of $\mathbf{B}$) iff it holds $D(f_{\mathbf{B}}, B) \subseteq D(f_{\mathbf{A}}, A)$ for arbitrary $f \in F$ and

$$f_{\mathbf{A}}(x_1, ..., x_{af}) = f_{\mathbf{B}}(x_1, ..., x_{af})$$

for all $(x_1, ..., x_{af}) \in B^{af} \cap D(f, A)$. Then we write

$$\mathbf{B} \leq \mathbf{A}.$$

If **A** has a nullary operation $f \in A$ and **B** is a subalgebra of **A**, then f also belongs to B.
The following lemma summarizes easy consequences from the above definition of a subalgebra.

Lemma 1.3.1

(a) A subset B of the universe A of an algebra $(A; F)$ together with the restrictions $f_{|B}$ of the operations f of **A** *to B forms an algebra if and only if*

$$\begin{array}{l} f(b_1, ..., b_{af}) \in B \textit{ for all } b_1, ..., b_{af}, \textit{ if } af \geq 1, \\ \textit{and } f \in B \textit{ if } af = 0 \end{array}$$

holds for arbitrary $f \in F$.

(b) Let I be an arbitrary index set, $\mathbf{B_i} \leq \mathbf{A}$ for every $i \in I$ and $\bigcap_{i \in I} B_i \neq \emptyset$. Then $(\bigcap_{i \in I} B_i; F)$ is a subalgebra of **A**.

(c) For every nonempty subset T of the universe A of an algebra $\mathbf{A} = (A; F)$ there exists exactly one smallest subalgebra $\mathbf{T}'$ of **A**, *which contains T and which one can describe as follows:*

$$\mathbf{T}' = (T'; F) \textit{ with } T' := \bigcap \{B \mid \mathbf{B} \leq \mathbf{A} \;\; \textit{and} \;\; T \subseteq B\}.$$

■

The set T from Lemma 1.3.1, (c) is called **generating system** of the algebra $\mathbf{T}'$ and the universe T' of $\mathbf{T}'$ is denoted by $[T]_{\mathbf{A}}$ or $[T]_F$ or by $[T]_{f_1,...,f_r}$, if $F = \{f_1, ..., f_r\}$, or simply by $[T]$.
Example For the algebra $\mathbf{A} = (\{0, 1, 2, ..., k-1\}; f)$ of the type (3) with

$$f(x, y, z) = x + y - z \; (mod \; k)$$

for arbitrary $x, y, z \in A$, it holds e.g. $[\{a\}] = \{a\}$ for every $a \in A$ and $[\{0, 1\}] = A$.

If $[T] = T$ for a subset $T \subseteq A$ of an algebra $\mathbf{A} = (A; F)$, then we say that T **is closed**.
The following lemma gives another possibility of the description of the set $[T]$, defined above.

Lemma 1.3.2 *Let $\mathbf{A} = (A; F)$ be an algebra, T be a subset of A and F^0 be the set of all nullary operations of F. Then, for T we can define recursively the following subsets T_n of A as follows:*

$$T_0 := T \cup F^0$$

$$T_{n+1} := T_n \cup \{f(g_1, ..., g_{af}) \mid f \in F \backslash F^0 \;\; \textit{and} \;\; \{g_1, ..., g_{af}\} \subseteq T_n\}$$

$$(n \in \mathbb{N}_0).$$

Then

$$[T]_{\mathbf{A}} = \bigcup_{n\geq 0} T_n. \tag{1.1}$$

Proof. To prove "$\subseteq$" in (1.1), we have to show that $\bigcup_{n\geq 0} T_n$ is the universe of a subalgebra of $\mathbf{A}$. This would be shown if we have shown that $\bigcup_{n\geq 0} T_n$ is closed regarding the application of all operations of $F \setminus F^0$. Now let $f \in F \setminus F^0$ and $\{g_1, ..., g_{af}\} \subseteq \bigcup_{n\geq 0} T_n$ be arbitrary. Since $af \in \mathbb{N}$, there is an $m \in \mathbb{N}$ with $\{g_1, ..., g_{af}\} \subseteq T_m$. Consequently, $f(g_1, ..., g_{af})$ belongs to T_{m+1}, i.e., the set $\bigcup_{n\geq 0} T_n$ is closed.
"$\supseteq$" follows from $T_n \subseteq [T]_{\mathbf{A}}$ for all $n \geq 0$. This inclusion is easy to prove by induction on n, since $[T]_{\mathbf{A}}$ is the universe of a subalgebra of $\mathbf{A}$. ■

The set

$$\mathbf{S}(\mathbf{A}) := \{\mathbf{B} \mid \mathbf{B} \text{ is subalgebra of } \mathbf{A}\} \cup \{\emptyset\}$$

we call briefly **set of all subalgebras** of $\mathbf{A}$. Per definitionem (essentially for technical reasons) is also the empty set a subalgebra of each algebra.

Then (as the reader can easily verify) $\mathbf{S}(\mathbf{A})$ together with the operations

$$\wedge: \ \mathbf{S}(\mathbf{A}) \times \mathbf{S}(\mathbf{A}) \longrightarrow \mathbf{S}(\mathbf{A}), \ \mathbf{B_1} \wedge \mathbf{B_2} = (B_1 \cap B_2; F)$$

$$\vee: \ \mathbf{S}(\mathbf{A}) \times \mathbf{S}(\mathbf{A}) \longrightarrow \mathbf{S}(\mathbf{A}), \ \mathbf{B_1} \vee \mathbf{B_2} = ([B_1 \cup B_2]_F; F)$$

forms a lattice.[4]

[4] This property would not hold for all algebras, if the empty set was not an algebra.

2

Lattices

Lattices arise often in algebraic investigations. In the following we see that the concept "lattice" is always needed, if the elements of sets are ordered in a certain meaning. In addition, the lattice theory is an interesting branch of the Universal Algebra.
For space reasons, only the most important basis concepts and some proof ideas of the lattice theory can be indicated here. For a secondary study of the lattice theory, refer to the books on the Universal Algebra and to [Bir 48], [Ern 82], [Sko 73] and [Dav-P 90].

2.1 Two Definitions of a Lattice

There are two standard ways of defining lattices. One of these ways was already given in Section 1.2.11:

Definition (First Definition of a Lattice)

Let L be a nonempty set on which the binary operations $\vee$ (called "join") and $\wedge$ (called "meet") are defined.
$(L;\vee,\wedge)$ is called **lattice** if for arbitrary $x, y, z \in L$ the following identities hold:

(L_1a) $x \vee y = y \vee x$,
(L_1b) $x \wedge y = y \wedge x$ (commutativity),

(L_2a) $x \vee (y \vee z) = (x \vee y) \vee z$,
(L_2b) $x \wedge (y \wedge z) = (x \wedge y) \wedge z$ (associativity),

(L_3a) $x \vee x = x$,
(L_3b) $x \wedge x = x$ (idempotency),

(L_4a) $x \vee (x \wedge y) = x$,
(L_4b) $x \wedge (x \vee y) = x$ (absorption).

Since one receives an axiom again by exchanging of $\vee$ *and* $\wedge$ *in above axioms, results from every equation, which the axioms imply, by exchanging of* $\vee$ *and* $\wedge$ *a further valid equation. One names this procedure of deriving equations in lattices, using the* **duality principle of the lattice theory**.

In the following proofs we often use the equivalence

$$x \vee y = y \iff x \wedge y = x, \tag{2.1}$$

which is a conclusion from the absorption laws and from the commutative laws.

For the second definition of a lattice we need some concepts and notations:

Definitions

- A binary relation $\leq$ ($\subseteq A \times A$) is called a **partial order** on the set A, if it fulfills the following conditions:
  (O_1) $\forall a \in A : a \leq a$ (reflexivity),
  (O_2) $\forall a, b \in A : (a \leq b$ and $b \leq a) \Longrightarrow a = b$ (antisymmetry),
  (O_3) $\forall a, b, c \in A : (a \leq b$ and $b \leq c) \Longrightarrow a \leq c$ (transitivity).
- The pair $(A; \leq)$, where $\leq$ satisfies the above conditions, we call **partially ordered set** or, briefly, **poset**. In examples, we often declare the posets through **order diagrams**[1] (or **Hasse diagrams**).
- A poset $\mathbf{P} := (P; \leq)$ with $P \subseteq A \times A$, which besides fulfills the condition

  (O_4) $\forall a, b \in A : a \leq b$ or $b \leq a$,

  is called **totally ordered set** or **linearly ordered set** or, briefly, **chain**.

We also use the notation $a < b$ if $a \leq b$ and $a \neq b$ is valid, where $(L; \leq)$ is a poset and $a, b \in L$. We also write $b \geq a$ instead of $a \leq b$.

Definition Let Q be a subset of P, where $\mathbf{P} = (P; \leq)$ is a poset. The element $s \in P$ is said to be a **supremum** of Q (denoted by $\sup Q$) iff s has the following properties:

(S_1) $\forall q \in Q : q \leq s$;

(S_2) $\forall p \in P \; ((\forall q \in Q : q \leq p) \implies s \leq p)$.

Remark The supremum of Q does not exist in general for every subset Q of a poset P. Let e.g. $P = \{0, 1, 2, 3, 4\}$ and let $\leq$ defined by the following order diagram:

---

[1] Here, the elements of the poset are represented as points in the plane and, if $x < y$ and no z exists with $x < z < y$, we draw y higher up than x and connect x and y with a line segment.

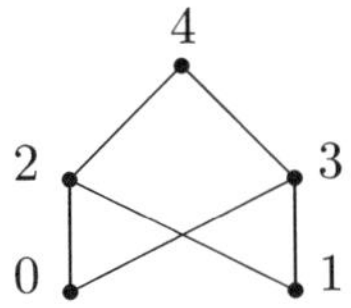

Fig. 2.1

Then, $\sup\{0,1\}$ does not exist, since for $p \in \{2,3\}$ $0 \le p$ and $1 \le p$ are valid, the elements 2 and 3 are not comparable in respect to $\le$.

Definition Let Q be a subset of P, where $\mathbf{P} = (P; \le)$ is a poset. The element $i \in P$ is said to be an **infimum** of Q (denoted by $\inf Q$) iff i has the following properties:

(I_1) $\forall q \in Q : i \le q$;

(I_2) $\forall p \in P \; ((\forall q \in Q : p \le q) \implies p \le i)$.

Definition (**Second Definition of a Lattice**)

A poset $\mathbf{L} := (L; \le)$ is called **lattice** iff for arbitrary $a, b \in L$ both $\sup\{a,b\}$ and $\inf\{a,b\}$ in L exist.

Some elementary properties of sup and inf in a lattice are summarized in the following lemma:

Lemma 2.1.1 *Let $(P; \le)$ be a lattice by the second definition. Then, for arbitrary $x, y, z, u, v \in P$ it holds:*
(a) $x \le y \implies \sup\{x,z\} \le \sup\{y,z\}$,
(b) $x \le y \implies \inf\{x,z\} \le \inf\{y,z\}$,
(c) $(x \le y \text{ and } u \le v) \implies \sup\{x,u\} \le \sup\{y,v\}$,
(d) $(x \le y \text{ and } u \le v) \implies \inf\{x,u\} \le \inf\{y,v\}$.

Proof. (a) and (b) are easy to check.
(c): Let $x \le y$ and $u \le v$. Then by (a) it holds $\sup\{x,u\} \le \sup\{y,u\}$ and $\sup\{y,u\} \le \sup\{y,v\}$. Since $\le$ is transitive, by this (c) follows.
One can prove (d) analogously to (c). ■

The next theorem shows how the two lattice definitions are associated.

Theorem 2.1.2 *It holds:*

(a) If $(L; \vee, \wedge)$ is a lattice by the first definition, then one can define by

$$a \le b \;:\Longleftrightarrow\; a = a \wedge b \tag{2.2}$$

a partial order $\le$, which together with L forms a lattice by the second definition.[2]

(b) Conversely, if $(L;\leq)$ is a lattice by the second definition, then one can define two binary operations $\vee$, $\wedge$ by

$$a \vee b := \sup\{a,b\} \qquad \textit{and}$$
$$a \wedge b := \inf\{a,b\}$$

and $(L;\vee,\wedge)$ is a lattice by the first definition.

Proof. (a): Let $(L;\vee,\wedge)$ be a lattice and $\leq$ is defined by (2.2). Then $a \wedge a = a$ holds and thus $a \leq a$ for every $a \in A$. Hence $\leq$ is reflexive.
If $a \leq b$ and $b \leq a$, we have $a = a \wedge b$ and $b = b \wedge a$. This and $(L_1 b)$ imply $a = b$. Thus $\leq$ is antisymmetric.
Let now $a \leq b$ and $b \leq c$. Then it holds $a = a \wedge b$ and $b = b \wedge c$ by definition of $\leq$. By this and by $(L_2 b)$ we have $a = a \wedge (b \wedge c) = (a \wedge b) \wedge c = a \wedge c$. Hence $a \leq c$ and thus $\leq$ is transitive.
It remains to show that there exist $\sup\{a,b\}$ and $\inf\{a,b\}$ for all $a,b \in L$.
For these let $a,b \in L$ be arbitrary. By $(L_4 a)$ and $(L_4 b)$ then we have $a = a \wedge (a \vee b)$ and $b = b \wedge (a \vee b)$. Thus $a \leq a \vee b$ and $b \leq a \vee b$. Let now $a \leq u$ and $b \leq u$ for a certain $u \in L$. Then the following equations hold:

$$a = a \wedge u, \;\; b = b \wedge u,$$
$$a \vee u = (a \wedge u) \vee u = u, \;\; b \vee u = (b \wedge u) \vee u = u, \text{ and}$$
$$(a \vee b) \vee u = (a \vee u) \vee (b \vee u) = u \vee u = u.$$

If one uses the equation $(a \vee b) \vee u = u$, which just was proven, then one gets:

$$(a \vee b) \wedge u = (a \vee b) \wedge ((a \vee b) \vee u) = a \vee b.$$

Therefore, we have $a \vee b \leq u$ and $\sup\{a,b\} = a \vee b$ holds by definition of sup.
Analogously, one can show $a \wedge b = \inf\{a,b\}$. For this, one mixed up $\wedge$ in above considerations by $\vee$ and one replaced $\leq$ by $\geq$.
Thus $(L;\leq)$ is a lattice.

(b): Let $(L;\leq)$ be a lattice, $a \vee b := \sup\{a,b\}$ and $a \wedge b := \inf\{a,b\}$. Obviously, the so defined operations $\vee$ and $\wedge$ are commutatively and idempotent. To show the validity of the associative law $\sup\{x, \sup\{y,z\}\} = \sup\{\sup\{x,y\},z\}$ one can prove (under use of Lemma 2.1.1) $\sup\{x,\sup\{y,z\}\} \leq \sup\{\sup\{x,y\},z\}$ and $\sup\{x,\sup\{y,z\}\} \geq \sup\{\sup\{x,y\},z\}$. Analogously, one can show the associativity of inf.
Because of $\sup\{x,\inf\{x,y\}\} \geq x$ and $\sup\{x,\inf\{x,y\}\} \leq x$ it holds obvious $\sup\{x,\inf\{x,y\}\} = x$. Analogously, one can show another absorption law. Consequently, $(L;\vee,\wedge)$ is a lattice by the first definition. ■

[2] Because of (2.1) one also could have defined the following:

$$a \leq b :\Longleftrightarrow a \vee b = b.$$

Because of Theorem 2.1.2 we can use the first or the second definition of an lattice subsequently according to requirement, where the statements of the Theorem 2.1.2 are used where appropriate.
If one wants the above-mentioned duality principle on inequalities of the form ... $\leq$ ... enlarge, one has to define the exchanging of $\leq$ through $\geq$ as an additional replacement rule.
In addition the implication

$$(x \leq y \;\; and \;\; u \leq v) \implies (x \vee u \leq y \vee v \;\; and \;\; x \wedge u \leq y \wedge v), \qquad (2.3)$$

resulting from Lemma 2.1.1 and from Theorem 2.1.2, will be an important aid in the below-given proofs.

2.2 Examples for Lattices

2.2.1 Let $L := \{0, 1\}$. Further, let $\vee$ be the conjunction on L and let $\wedge$ be the disjunction on L. Obviously, $(L; \vee, \wedge)$ is a lattice.

2.2.2 Let $L := \mathbb{N}_0$, let $a \vee b$ be the least common multiple and let $a \wedge b$ be greatest common divisor of the integers $a, b \in \mathbb{N}_0$. In this case, it is also easy to check that $(L; \vee, \wedge)$ is a lattice.

2.2.3 Examples for lattices by the second definition are:
$(\mathfrak{P}(A); \subseteq)$, where A is an arbitrary nonempty set;
$(\mathbb{R}; \leq)$, where $\leq$ is the usual order on $\mathbb{R}$.

One finds further examples in Section 2.4 and in the following chapters.

2.3 Isomorphic Lattices and Sublattices

Definition Two lattices $\mathbf{L_1}$, $\mathbf{L_2}$ are called **isomorphic**, iff there exists a bijective mapping α from L_1 onto L_2 such that for arbitrary $a, b \in L_1$ the following two equations hold:

$$\alpha(a \vee b) = \alpha(a) \vee \alpha(b) \qquad \text{and}$$
$$\alpha(a \wedge b) = \alpha(a) \wedge \alpha(b).$$

The mapping α is also called an **isomorphism**.

Definition Let $(P_1, \leq)$ and $(P_2; \leq)$ be posets. A mapping α from P_1 onto P_2 is called **order-preserving**, if the following holds:

$$\forall a, b \in P_1 : a \leq b \implies \alpha(a) \leq \alpha(b).$$

Figure 2.2 gives an **example** of an order-preserving mapping between the lattices $\mathbf{L_1}$ and $\mathbf{L_2}$:

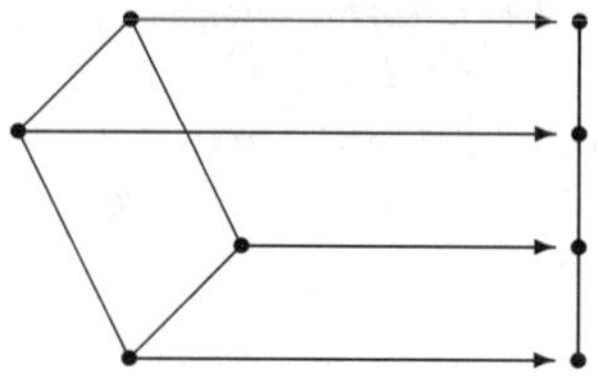

Fig. 2.2

Theorem 2.3.1 *Two lattices* $\mathbf{L_1}$ *and* $\mathbf{L_2}$ *are isomorphic iff there is a bijective mapping* α *from* L_1 *onto* L_2 *such that both* α *and* α^{-1} *are order-preserving.*

Proof. "$\Longrightarrow$": Let α be an isomorphism from $\mathbf{L_1}$ onto $\mathbf{L_2}$. Then for all $a, b \in L_1$ it holds:

$$a \leq b \Longrightarrow a = a \wedge b \Longrightarrow \alpha(a) = \alpha(a \wedge b) = \alpha(a) \wedge \alpha(b).$$

Consequently, we have $\alpha(a) \leq \alpha(b)$, i.e., α is order-preserving.
Let now $c, d \in L_2$ arbitrary with $c \leq d$. Then there exist $a, b \in L_1$ with $\alpha(a) = c$ and $\alpha(b) = d$. That α^{-1} is also order-preserving follows then from

$$\begin{aligned} c \leq d &\Longrightarrow \alpha(a) \leq \alpha(b) \Longrightarrow \alpha(a) \wedge \alpha(b) = \alpha(a) \Longrightarrow \alpha(a \wedge b) = \alpha(a) \\ &\Longrightarrow a \wedge b = a \Longrightarrow a \leq b. \end{aligned}$$

"$\Longleftarrow$": Let now α be a bijective mapping from L_1 onto L_2 with the property that both α and α^{-1} are order-preserving. Then for arbitrary $a, b \in L_1$ we have $a \leq a \vee b$ and $b \leq a \vee b$. Thus, $\alpha(a) \leq \alpha(a \vee b)$ and $\alpha(b) \leq \alpha(a \vee b)$. From this, it follows $\alpha(a) \vee \alpha(b) \leq \alpha(a \vee b)$. Further, for arbitrary $u \in L_2$ it holds

$$\begin{aligned} \alpha(a) \vee \alpha(b) \leq u &\Longrightarrow \alpha(a) \leq u \text{ and } \alpha(b) \leq u \\ &\Longrightarrow a \leq \alpha^{-1}(u) \text{ and } b \leq \alpha^{-1}(u) \\ &\Longrightarrow a \vee b \leq \alpha^{-1}(u) \\ &\Longrightarrow \alpha(a \vee b) \leq u. \end{aligned}$$

Consequently, $\alpha(a \vee b) = \sup\{\alpha(a), \alpha(b)\}$ and therefore $\alpha(a) \vee \alpha(b) = \alpha(a \vee b)$. Analogously, one can prove $\alpha(a) \wedge \alpha(b) = \alpha(a \wedge b)$. ■

Remark Figure 2.2 shows that one cannot leave the condition "α^{-1} order-preserving" from Theorem 2.3.1.

Theorem 2.3.2 *For every poset* $(P; \leq)$ *there is a set system* M_P *such that* $(P; \leq)$ *and* $(M_P; \subseteq)$ *are isomorphic; i.e., there exists a bijective mapping* α *from* P *onto* M_P *with the property that* α *and* α^{-1} *are order-preserving.*

Proof. For every $p \in P$ let $M(p) := \{x \in P \mid x \leq p\}$. Then the set $M_P := \{M(p) \mid p \in P\}$ is a set system with the properties claimed in the theorem. To show this, we study the mapping $\alpha : P \longrightarrow M_P$.
α is surjective by definition.
From $M(a) = M(b)$ it follows $a \in M(b)$ and $b \in M(a)$. Hence $a \leq b$ and $b \leq a$, i.e., $a = b$. Thus α is injective.
α is order-preserving, since $a \leq b$ and $x \in M(a)$ imply $x \leq a \leq b$. Therefore, $a \leq b$ implies $M(a) \subseteq M(b)$. Furthermore, $M(a) \subseteq M(b)$ implies $a \leq b$, i.e., α^{-1} is also order-preserving and thus α is an isomorphism. ■

The following definition is a special case of a definition from Section 1.3:

Definition If $\mathbf{L}$ is a lattice and L' is a subset of L with the property

$$\forall x, y \in L' : x \vee y \in L' \text{ and } x \wedge y \in L',$$

then $(L'; \vee, \wedge)$ is called a **sublattice** of $(L; \vee, \wedge)$.

Remark Let $(P; \leq)$ and $(Q; \leq)$ be posets with $Q \subseteq P$. Then, it is not valid in general that $(Q; \vee, \wedge)$ is a sublattice of the lattice $(P; \vee, \wedge)$, where $\vee, \wedge$ are defined in Theorem 2.1.2. An example is the poset $(\{0, 1, 2, 3, 4\}; \leq)$ defined by the following order diagram:

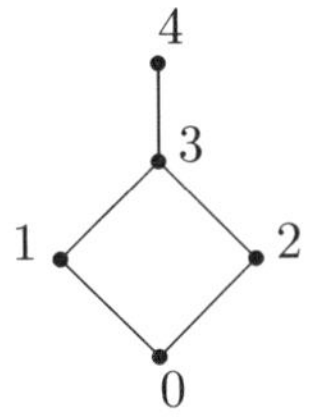

Fig. 2.3

Obviously, $(\{0, 1, 2, 4\}; \leq)$ is a poset, but $(\{0, 1, 2, 4\}; \vee, \wedge)$ is not a sublattice of $(\{0, 1, 2, 3, 4\}; \vee, \wedge)$, since $1 \vee 2 = 3 \notin \{0, 1, 2, 4\}$.

Definition A lattice $\mathbf{L_1}$ can be **embedded** into a lattice $\mathbf{L_2}$, if there is a sublattice of $\mathbf{L_2}$ isomorphic to $\mathbf{L_1}$.

2.4 Complete Lattices and Equivalence Relations

Definition A poset is called **complete** iff for every subset A of P both $\sup A$ (denoted by $\bigvee A$) and $\inf A$ (denoted by $\bigwedge A$) exist.

A lattice which is complete as poset is a **complete lattice**.

Examples
1) Obviously, each complete poset is also a lattice.
2) Every finite lattice is complete.
3) $(\mathbb{R}; \leq)$ is not complete.

One can find further examples in Theorems 2.4.3 and 2.4.5.

Theorem 2.4.1 *For an arbitrary poset $(L; \leq)$ with a greatest and a least element it holds:*

$$\forall A \subseteq L\ \exists \bigwedge A \iff \forall A \subseteq L\ \exists \bigvee A.$$

Proof. "$\Longrightarrow$": Let $A_o := \{x \in L \mid \forall a \in A :\ a \leq x\}$. Since L has a greatest element, we have $A_o \neq \emptyset$. It is easy to check that then $\bigvee A = \bigwedge A_o$ holds. Analogously, one can prove "$\Longleftarrow$". ∎

Theorem 2.4.2 *Let* **A** *be an algebra. Then the lattice $(S(\mathbf{A}); \subseteq)$ of all subalgebras of* **A** *is a complete lattice.*

Proof. The proof follows from Theorem 2.4.1 and the fact that every intersection of subalgebras of **A** is also a subalgebra of **A** (see Lemma 1.3.1, (b)). ∎

The following is a short summary of the equivalence relations:
Definitions A binary relation ϱ is called an **equivalence relation** on a (nonempty) set A, if ϱ fulfills the following conditions:

(E_1) $\{(a,a) \mid a \in A\} \subseteq \varrho$ (reflexivity);

(E_2) $\forall a, b \in A : (a,b) \in \varrho \implies (b,a) \in \varrho$ (symmetry);

(E_3) $\forall a, b, c \in A : ((a,b) \in \varrho$ and $(b,c) \in \varrho) \implies (a,c) \in \varrho$ (transitivity).

Let $Eq(A)$ be the set of all equivalence relations on A.

By defining of the operation $^{-1}$: $\mathfrak{P}(A \times A) \longrightarrow \mathfrak{P}(A \times A), \varrho \rightarrow \varrho^{-1}$ by

$$\varrho^{-1} := \{(b,a) \mid (a,b) \in \varrho\}$$

and of the binary operation

$$\Box : \mathfrak{P}(A \times A) \times \mathfrak{P}(A \times A) \longrightarrow \mathfrak{P}(A \times A),\ (\varrho, \varrho') \mapsto \varrho \circ \varrho'$$

by

$$\varrho \Box \varrho' := \{(a,c) \mid \exists b \in A : (a,b) \in \varrho \text{ and } (b,c) \in \varrho'\},$$

one can also write down the above conditions (E_2) and (E_3) as follows:

(E_2) $\varrho^{-1} = \varrho$,

(E_3) $\varrho \Box \varrho \subseteq \varrho$.

Instead of $(a, b) \in \varrho$ we often write $a\varrho b$ or $\varrho(a, b)$ or $a = b\ (mod\ \varrho)$ or $a \sim b\ (mod\ \varrho)$ (one say: a is equals (or equivalent) to b modulo ϱ).

For every set A there are two **trivial equivalence relations**:

$$\nabla_A(:= \kappa_1) := A^2 \text{ (\textbf{all relation})} \qquad \text{and}$$
$$\Delta_A(:= \kappa_0) := \{(a, a) \mid a \in A\} \text{ (\textbf{identity} or \textbf{diagonale})}.$$

Theorem 2.4.3 *$(Eq(A); \subseteq)$ is a complete lattice for every nonempty set A. One can determine the infimum and the supremum of an arbitrary subset $T := \{\varrho_i \mid i \in I\}$ of $Eq(A)$ as follows:*

$$\bigwedge T = \bigcap_{i\in I} \varrho_i,$$
$$\bigvee T = \bigcup_{i_0,\dots,i_t\in I; t<\aleph_0} \varrho_{i_0}\Box\varrho_{i_1}\Box\dots\Box\varrho_{i_t}.$$

In particular, it holds for the equivalence relations κ and μ:

$$\kappa \vee \mu = \kappa \cup (\kappa\Box\mu) \cup (\kappa\Box\mu\Box\kappa) \cup (\kappa\Box\mu\Box\kappa\Box\mu) \cup \dots$$

(i.e.,

$$(a, b) \in \kappa\vee\mu \iff \exists c_1, \dots, c_n \in A : (\forall i \in \{1, 2, \dots, n-1\} :$$
$$(c_i, c_{i+1}) \in \kappa \text{ or } (c_i, c_{i+1}) \in \mu) \text{ and } (a, b) = (c_1, c_n)).$$

Proof. The statement "$(Eq(A); \subseteq)$ is a complete lattice" follows from Theorem 2.4.1 and from the easily verifiable fact that the intersection of arbitrarily many equivalence relations on A is an equivalence relation on A again. The remaining statements one can easily check. ■

Definitions Let $\varrho \in Eq(A)$ and $a \in A$. The set

$$[a]_\varrho(:= a/\varrho) := \{x \in A \mid (a, x) \in \varrho\}$$

is called **equivalence class of** ϱ.
The set

$$A/\varrho := \{a/\varrho \mid a \in A\}$$

of all equivalence classes of ϱ is called **factor set of** A **by** ϱ.

The proof of the following theorem can be found in many books with sections on basic mathematics concepts.

Theorem 2.4.4 *Let A be a nonempty set. Then, for arbitrary $\varrho \in Eq(A)$ and arbitrary $a, b \in A$:*

(a) $A = \bigcup_{a\in A} a/\varrho$;

(b) $a/\varrho \neq b/\varrho \iff a/\varrho \cap b/\varrho = \emptyset$. ■

Definitions A **partition** π of the set A is a set of nonempty subsets A_i ($i \in I$) of A (called **blocks** of π) with the properties:

(Z_1) $\bigcup_{i \in I} A_i = A$

and

(Z_2) $\forall i, j \in I : A_i \cap A_j = \emptyset$ or $A_i = A_j$.

Let $\Pi(A)$ be the set of all partitions of A.

Regarding the contents, the following theorem, written with the help of concepts of the lattice theory, is identical to the well-known *Main Theorem over equivalence relations.*

Theorem 2.4.5 *If the partial order* $\leq$ *on* $\Pi(A)$ *is defined by*

$$\pi \leq \pi' :\Longleftrightarrow \forall b \in \pi \; \exists b' \in \pi' : b \subseteq b',$$

then the lattices $(Eq(A); \subseteq)$ *and* $(\Pi(A); \leq)$ *are isomorphic.*
An order-preserving mapping α *of* $Eq(A)$ *auf* $\Pi(A)$ *can be defined as follows: Let* $\kappa \in Eq(A)$ *and* $\pi := \{b_i \mid i \in I\} \in \Pi(A)$. *Then*

$$\alpha(\kappa) := \{x/\kappa \mid x \in A\}$$

and

$$\alpha^{-1}(\pi) := \{(x, y) \mid \exists i \in I : \{x, y\} \subseteq b_i\}.$$

■

3

Hull Systems and Closure Operators

This chapter generalized observations that can be made when examining such similar concepts as *subalgebra* or *linear hull* of subsets of a vector space (or another closure operators of the classical algebra), to the concepts *hull system* and *closure operator*. Then, it is shown that these concepts deliver roughly the same.

3.1 Basic Concepts

Definitions Let A be a nonempty set and let $\mathfrak{P}(A)$ be the power set of A. A subset H of $\mathfrak{P}(A)$ is a **hull system** (or a **closed set system**) on the set A if and only if

(i) $A \in H$, and

(ii) $\bigcap B := \bigcap\{b \mid b \in B\} \in H$ for every nonempty subset B of H.[1]

The elements of H are called **hulls**.

Examples

1) The set of all closed subsets of an algebra A is a hull system on A.

2) The set systems $\mathfrak{P}(A)$, $\{A\}$ and $\{A\} \cup \{E \in \mathfrak{P}(A) \mid E \text{ finite}\}$ are hull systems on A for every nonempty set A.

3) The set of all equivalence relations $Eq(A)$ is a hull system on A^2, $A \neq \emptyset$.

Definitions A mapping

$$C : \mathfrak{P}(A) \longrightarrow \mathfrak{P}(A)$$

is called **closure operator** on A if and only if the following three conditions hold for all $X, Y \subseteq A$:

(i) $X \subseteq C(X)$ (extensivity),

(ii) $X \subseteq Y \implies C(X) \subseteq C(Y)$ (monotony),

(iii) $C(C(X)) = C(X)$ (idempotency).

[1] Let $\bigcap\{b \mid b \in B\} := \bigcap_{b\in B} b := \{x \mid \forall b \in B : x \in b\}$.

Sets of the form $C(X)$ are called **closed**, and we say that $C(X)$ **is generated** of X.

Example By Section 1.3, for every algebra, the operator $[...]_{\mathbf{A}}$ is a closure operator.

Definition If a closure operator C fulfills the following condition in addition

(iv) $\forall X \subseteq A : C(X) = \bigcup\{C(Y) \mid Y \subseteq X$ and Y is finite$\}$,

then it is called **algebraic closure operator**.

3.2 Some Properties of Hull Systems and Closure Operators

Hull systems and closure operators are generally the same:

Theorem 3.2.1 (Main Theorem on Hull Systems and Closure Operators)

(a) Let H be a hull system on A. For all $X \subseteq A$ let

$$C_H(X) := \bigcap\{h \in H \mid X \subseteq h\}. \tag{3.1}$$

Then

$$C_H : \mathfrak{P}(A) \longrightarrow \mathfrak{P}(A),\ X \mapsto C_H(X)$$

is a closure operator on A, and the closed sets of C_H are exactly the hulls of H.

(b) Conversely, let C be a closure operator on A. Then

$$H_C := \{C(X) \mid X \subseteq A\} \tag{3.2}$$

is a hull system on A, and the hulls of H_C are exactly the closed sets of A.

(c) For every hull system H on A it holds

$$H_{(C_H)} = H, \tag{3.3}$$

and for every closure operator C on A it holds

$$C_{(H_C)} = C. \tag{3.4}$$

Proof. (a): The extensivity of the operator defined by (3.1) follows directly from the definition of $C_H(X)$. If $X \subseteq Y$, we have $U := \{h \in H \mid X \subseteq h\} \supseteq V := \{h \in$

$H \mid Y \subseteq h\}$. Thus, $C_H(X) = \bigcap\{h \in H \mid X \subseteq h\} \subseteq C_H(Y) = \bigcap\{h \in H \mid Y \subseteq h\}$. Therefore C_H is monotone. To prove the idempotency of C_H, first we show

$$Y \subseteq C_H(Z) \implies C_H(Y) \subseteq C_H(Z). \tag{3.5}$$

$Y \subseteq C_H(Z)$ implies $\{h \in H \mid Z \subseteq h\} \subseteq \{h \in H \mid Y \subseteq h\}$ and from this $C_H(Y) \subseteq C_H(Z)$, i.e., (3.5) holds. Choosing in (3.5) $Y = C_H(Z)$, then $C_H(C_H(Z)) \subseteq C_H(Z)$ follows from (3.5). By the extensivity of C_H we have $C_H(Z) \subseteq C_H(C_H(Z))$. Thus, C_H is also idempotent and therefore C_H is a closure operator.
Next we prove

$$X \in H \iff C_H(X) = X. \tag{3.6}$$

The statement "$\Longrightarrow$" in (3.6) follows directly from the definition of C_H. "$\Longleftarrow$" follows from the definition of C_H and the assumption (ii) (see Section 3.1). Thus (a) is proven.

(b): Let $C : \mathfrak{P}(A) \longrightarrow \mathfrak{P}(A)$ be a closure operator and H_C is defined as in (3.2). The following equivalences are easy to check:

$$\forall X \subseteq A \ (X \in H_C \iff C(X) = X). \tag{3.7}$$

The definition of C implies $A \subseteq C(A) \subseteq A$, i.e., $A = C(A) \in H_C$. For arbitrary $\mathfrak{B} \subseteq H_C$ we have further $\bigcap \mathfrak{B}$ $(= \bigcap_{X\in\mathfrak{B}} X) \in H_C$ to show. Since for every $B \in \mathfrak{B}$ we have $B = C(B)$ and $\bigcap_{X\in\mathfrak{B}} X \subseteq B$, the extensivity of C implies $C(\bigcap_{X\in\mathfrak{B}} X) \subseteq C(B) = B$ for every $B \in \mathfrak{B}$. Then, by this we have $C(\bigcap_{X\in\mathfrak{B}} X) \subseteq \bigcap_{B\in\mathfrak{B}} B$. On the other hand $\bigcap_{X\in\mathfrak{B}} X \subseteq C(\bigcap_{X\in\mathfrak{B}} X)$ also holds. Thus, $C(\bigcap \mathfrak{B}) = \bigcap \mathfrak{B} \in H_C$ is valid.

(c): By using (3.7) and (3.6) it follows $H_{(C_H)} = H$ from

$$X \in H_{(C_H)} \iff C_H(X) = X \iff X \in H.$$

To prove $C_{(H_C)} = C$, let $Y \subseteq A$ be arbitrary. We have $C_{(H_C)}(Y) = C(Y)$ to show. By definition it holds that $C_{(H_C)}(Y) = \bigcap\{Z \mid Z \in H_C, Y \subseteq Z\}$. Consequently, by (3.7), $C_{(H_C)}(Y) = \bigcap\{Z \mid C(Z) = Z, Y \subseteq Z\}$. Because of $C(Y) = C(C(Y))$, we have $\bigcap\{Z \mid C(Y) = Z, Y \subseteq Z\} \subseteq C(Y)$. On the other hand it follows from $Y \subseteq Z = C(Z)$ that $C(Y) \subseteq C(Z) = Z$. This implies $\bigcap\{Z \mid C(Z) = Z, Y \subseteq Z\} \supseteq C(Y)$. By summarizing, we obtain $\bigcap\{Z \mid C(Y) = Z, Y \subseteq Z\} = C(Y)$. ■

Theorem 3.2.2

(a) For every algebra $\mathbf{A} = (A; F)$, $[...]_{\mathbf{A}}$ is an algebraic closure operator on A.

(b) Conversely, for every algebraic closure operator C on the set A there is a set F of operations on A such that $\{C(X) \mid X \subseteq A\}$ is the set of all subalgebras of $\mathbf{A} = (A; F)$.

Proof. (a) follows from the description of the closure $[T]_{\mathbf{A}}$ of an arbitrary set $T \subseteq A$ (see Lemma 1.3.1).

(b): Let C be an algebraic closure operator. We have to define a certain set of operations F on A such that $C(X) = [X]_F$ holds for arbitrary $X \subseteq A$.
For every finite n-element subset $E := \{e_1, ..., e_n\}$ of A and for every $e \in C(E)$ we define $f_{E;e}$ as follows:

$$f^n_{E;e}(x_1, ..., x_n) := \begin{cases} e, & \text{if} \quad \{x_1, ..., x_n\} = E, \\ x_1 & \text{otherwise.} \end{cases}$$

We summarize the functions of type $f_{E;e}$ to the set F:

$$F := \bigcup_{n\in\mathbb{N}} \{f^n_{E;e} \mid |E| = n, E \subseteq A, e \in C(E)\}.$$

Then, by construction, we have $C(E) = [E]_F$.
Since C is an algebraic closure operator, from this $C(X) = [X]_F$ follows for every subset $X \subseteq A$. ■

Theorem 3.2.3

(a) Let C be a closure operator on A and let

$$L_C := \{X \subseteq A \mid C(X) = X\}.$$

Then L_C is a complete lattice with

$$\bigwedge_{i\in I} C(A_i) = \bigcap_{i\in I} C(A_i)$$

and

$$\bigvee_{i\in I} C(A_i) = C(\bigcup_{i\in I} A_i).$$

(b) Every complete lattice is isomorphic to the lattice $\mathbf{L_C}$ of all closed subsets of a certain set A with a closure operator C.

Proof. (a) is easy to check.
(b) Let $\mathbf{L}$ be a complete lattice. For $X \subseteq L$ we define

$$C(X) := \{a \in L \mid a \leq \sup X\}.$$

Then C is a closure operator on L and the mapping

$$\alpha : L \longrightarrow L_C,\ a \rightarrow \{b \in L \mid b \leq a\}$$

is an isomorphic mapping from L into L_C. ■

Definitions Let $X, Y \subseteq A$, let H be a hull system on A and let $h \in H$.

- X is called a **generating system** (or a **generating set**) of h, if $h = C_H(X)$ holds.
- h is **finitely generated**, if there is a finite set $E \subseteq h$ with $h = C_H(E)$.

- X is C_H**-independent** (or briefly: **independent**), if for all $x \in X$ it holds $x \notin C_H(X\backslash\{x\})$.
- X is called a C_H**-basis** (or briefly: **basis**) of h, if X is a C_H-independent generating set of h.

From general theorems about hull systems and closure operators, one gets the following statements (about algebras), which we will prove, only for the concrete hull system $S(\mathbf{A})$ and the corresponding closure operator $[...]_{\mathbf{A}}$.

Theorem 3.2.4

(a) Let $\mathbf{A}$ *be a finitely generated algebra. Then, for every generating set* T *of* $\mathbf{A}$*, there is a finitely generating set* $T' \subseteq T$ *of* $\mathbf{A}$.
(b) Every finitely generated algebra has a finite basis.
(c) Every algebra, which has not a basis, is not finitely generated.
(d) Let $\mathbf{A}$ *be an algebra that has a subalgebra* $\mathbf{A_1}$ *with countable infinite basis. Then* $S(\mathbf{A})$ *has a cardinality that is greater than or equal to the cardinality of the continuum.*

Proof. (a): Let $\mathbf{A} = (A; F)$ be finitely generated. Then, there is a finite generating set $B_0 = \{b_1, b_2, ..., b_r\} \subseteq A$ of the universe A with $[B_0]_{\mathbf{A}} = A$. Now let $T \subseteq A$ be an arbitrary generating set of the algebra $\mathbf{A}$. Because of $[T] = A$ and because of Lemma 1.3.2, one can get an arbitrary element $a \in A$ from elements of T by successively applying the operations of F. In particular, this also holds for the elements $b_1, ..., b_r$ of B_0. Obviously, the set of all elements of T needed for this is a finitely generating set $T' \subseteq T$ of $\mathbf{A}$.

(b): Let $\mathbf{A} = (A; F)$ be a finitely generated algebra. Then by (a) there exists with every generating set T of $\mathbf{A}$ a finite subset $T' \subseteq T$ that is already a generating set for the given algebra. If $t \notin [T'\backslash\{t\}]$ for all $t \in T'$, then T' is a finite basis. Otherwise, if $t_1 \in [T'\backslash\{t_1\}]$, $T'' := T' \setminus \{t_1\}$ is a generating set for $\mathbf{A}$. If now for all $t \in T''$ it holds $t \notin [T''\backslash\{t\}]$, then T'' is basis for $\mathbf{A}$. In opposite case, we deal with T'' like with T' by going over to T''', etc. By the finiteness of T, this process breaks off with the construction of a basis for $\mathbf{A}$, whereby (b) is proven.

(c) follows from (b).

(d): Let $T := \{t_1, t_2, ...\}$ be a countable infinite basis of the subalgebra $\mathbf{A_1}$ of $\mathbf{A}$. By definition, for all subsets $T' \subset T$, we have, then, that $a \in T\backslash T'$ implies $a \notin [T']$. Therefore, different subsets T' and T'' of T define different subalgebras $[\mathbf{T}']$ and $[\mathbf{T}'']$. Consequently, the cardinality of the set $S(\mathbf{A})$ is not less than the cardinality of the set $\mathfrak{P}(T)$. Now, it is well-known that $\mathfrak{P}(T)$ has the cardinality of continuum. Thus (c) holds. ■

4

Homomorphisms, Congruences and Galois Connections

It is clear that abstract algebraic properties of an algebra do not change if the elements and the operations of the algebra get other names. One can describe this fact mathematically using the concept *isomorphism* or *isomorphic mapping*. A generalization of this concept is *homomorphism* (or *homomorphic mapping*), with which one can describe similarities of algebras. *Congruences* and *factor algebras* are aids to determine homomorphisms.
A *Galois connection* is a pair of mappings (σ, τ) between two power sets $\mathfrak{P}(A)$ and $\mathfrak{P}(B)$, where the mappings τ and σ are *antitone* and *extensive* (see Section 4.4). With the aid of these mappings, one can define hull operators on A and on B. Then, these operators define closed set systems ($\subseteq \mathfrak{P}(A)$ or $\subseteq \mathfrak{P}(B)$, respectively). Part of the usefulness of such Galois connections resides in the possibility of drawing conclusions about a closed set system on the basis of information about the other system.
The Galois connection between function algebras and relation algebras, which we study in Chapter 2 of Part II, is an important aid in the solution of the completeness problem of the many-valued logic (see Part II, Chapter 5 and 6).

4.1 Homomorphisms and Isomorphisms

Definitions Let $(A; F)$ and $(B; G)$ be two universal algebras of the same type and $F = \{f_i \mid i \in I\}$ and $G = \{g_i \mid i \in I\}$ with $af_i = ag_i$ for every $i \in I$. We call the mapping $\varphi : A \longrightarrow B$ a **homomorphism** or a **homomorphic mapping from A into B**, if φ is compatible with all $f \in F$, $g \in G$; i.e., it holds for all $i \in I$ and all $a_1, ..., a_{af_i} \in A$:

$$\varphi(f_i(a_1, ..., a_{af_i})) = g_i(\varphi(a_1), \varphi(a_2), ..., \varphi(a_{af_i})),$$

if $af_i > 0$, and $\varphi(f_i) = g_i$ for all $f_i \in F$ with $af_i = 0$.

If φ in addition is bijective from $\mathbf{A}$ onto $\mathbf{B}$, then φ is called an **isomorphism** or an **isomorphic mapping** from $\mathbf{A}$ onto $\mathbf{B}$.
A homomorphism from an algebra $\mathbf{A}$ into $\mathbf{A}$ is called an **endomorphism** of $\mathbf{A}$. An isomorphism from $\mathbf{A}$ onto $\mathbf{A}$ is an **automorphism** of $\mathbf{A}$.

Obviously, φ^{-1} is an isomorphism from $\mathbf{B}$ onto $\mathbf{A}$, if φ is an isomorphism from $\mathbf{A}$ onto $\mathbf{B}$. Therefore, we can say: "$\mathbf{A}$ and $\mathbf{B}$ are isomorphic", we write in this case

$$\mathbf{A} \cong \mathbf{B}$$

(read: "$\mathbf{A}$ isomorphic $\mathbf{B}$") for the identification of this fact.
It is easy to see that the composition of homomorphic mappings $\varphi_1 \Box \varphi_2$ ($\varphi_1 : A_1 \longrightarrow A_2, \varphi_2 : A_2 \longrightarrow A_3$) is a homomorphism from $\mathbf{A_1} = (A_1; F_1)$ into $\mathbf{A_3} = (A_3; F_3)$. With the help of this property, one can show that the relation "$\cong$" is an equivalence relation on sets of algebras.

The following lemma summarizes further elementary properties of homomorphic mappings.

Lemma 4.1.1 *Let φ be a homomorphic mapping from $\mathbf{A} = (A; F)$ into $\mathbf{B} = (B; G)$. Then it holds:*

(a) $\varphi(A)$ is closed in respect to $\mathbf{B}$; i.e., $(\varphi(A); G)$ is a subalgebra of $\mathbf{B}$, which is called **homomorphic image** *of $\mathbf{A}$ by φ.*
(b) If $\mathbf{A}'$ is a subalgebra of $\mathbf{A}$, then $(\varphi(A'); G)$ is a subalgebra of $(\varphi(A); G)$.
(c) If $T \subseteq A$ is a generating system of A, then $\varphi(T)$ is a generating system of $\varphi(A)$.
(d) If $(B'; G)$ is a subalgebra of $(\varphi(A); G)$, then the so-called **inverse image** *$(\varphi^{-1}(B'); F)$ with $\varphi^{-1}(B') := \{a \in A \mid \varphi(a) \in B'\}$ is a subalgebra of $\mathbf{A}$.*
(e) Let ψ be a homomorphism from $\mathbf{A}$ into $\mathbf{B}$, which is identical with φ on a generating system of A. Then, $\varphi(a) = \psi(a)$ for each $a \in A$. ■

4.2 Congruence Relations and Factor Algebras of Algebras

Definitions A **congruence relation** (briefly: **congruence**) or a **kernel** of a homomorphic mapping φ of an algebra $\mathbf{A} = (A; F)$ is an equivalence relation κ_φ on A that is induced by the homomorphism φ from $\mathbf{A}$ into $\mathbf{B}$, that is, for all $a, a' \in A$ it holds:

$$(a, a') \in \kappa_\varphi \ :\Longleftrightarrow\ \varphi(a) = \varphi(a').$$

In the following, we also denote the relation κ_φ with $Ker\ \varphi$. Let

$$Con(\mathbf{A})$$

be the set of all congruences of $\mathbf{A}$.
Examples The identity mapping id_A with

$$\mathrm{id}_A : A \longrightarrow A, a \mapsto a$$

and the mapping φ_C with

$$\varphi_C : A \longrightarrow \{c\}, a \mapsto c$$

from **A** onto the 1-element algebra of the same type $(\{c\}; G)$, where $g(c, c, ..., c) = c$ for all $g \in G$, if $ag > 0$, and $g = c$, if $ag = 0$ holds, induce two so-called **trivial congruences**, the **zero-congruence** κ_0 and the **all-congruence** (or **one-congruence**) κ_1:

$$\kappa_0 := \{(a, a) \mid a \in A\},$$
$$\kappa_1 := A \times A.$$

Definition An algebra, which has only both congruences κ_0 and κ_1, is called **simple**.
Examples for simple algebras are groups of prime order and the fields.

As the following lemma demonstrates, it is possible to describe the concept "congruence" without use of the concept "homomorphism".

Lemma 4.2.1 *An equivalence relation κ on* **A** *is a congruence on the algebra* $\mathbf{A} = (A; F)$ *if and only if it is compatible with all not nullary operations of F; i.e., if for all $f^n \in F$ with $n = af > 0$ and arbitrary $a_1, ..., a_n, a'_1, ..., a'_n \in A$ it holds:*

$$\{(a_1, a'_1), ..., (a_n, a'_n)\} \subseteq \kappa \Longrightarrow (f(a_1, ..., a_n), f(a'_1, ..., a'_n)) \in \kappa.$$

Proof. Let
$$F := \{f_i^{n_i} \mid i \in I\},$$
where I denotes a certain index set.
"$\Longrightarrow$": Let κ be a congruence on A. Then, by the definition of a congruence, there exists an algebra $\mathbf{B} := (B; G)$, where

$$G := \{g_i^{n_i} \mid i \in I\},$$

and a homomorphic mapping
$$\varphi : A \longrightarrow B$$
with the property
$$\kappa = \{(a, a') \mid \varphi(a) = \varphi(a')\}.$$
We have to show that κ is compatible with all $f \in F \setminus F^0$. To show this let $f := f_i^{n_i} \in F \setminus F^0$ be arbitrary. For the purpose of simplification, we put $n := n_i$ and $g := g_i^{n_i}$. Furthermore, let

$$(a_1, a'_1), ..., (a_n, a'_n) \in \kappa$$

be arbitrary. Since φ is a homomorphism, then we have

$$\begin{aligned}\varphi(f(a_1,...,a_n)) &= g(\varphi(a_1),...,\varphi(a_n))\\ &= g(\varphi(a'_1),...,\varphi(a'_n))\\ &= \varphi(f(a'_1,...,a'_n)).\end{aligned}$$

Consequently, $(f(a_1,...,a_n), f(a'_1,...,a'_n)) \in \kappa$.

"$\Longleftarrow$": Conversely, let the equivalence relation κ on A be compatible with all operations of the algebra $\mathbf{A} = (A; F)$. We have to show: There is an algebra $\mathbf{B} = (B; G)$ of the same type as $\mathbf{A}$ and a homomorphic mapping $\varphi : A \longrightarrow B$ with the property

$$\{(x, y) \mid \varphi(x) = \varphi(y)\} = \kappa.$$

To construct the algebra $\mathbf{B}$, we define:

$$a/\kappa := \{x \in A \mid (x, a) \in \kappa\} \qquad (a \in A)$$

(i.e., a/κ is the notation of the equivalence class in which lies a) and

$$B := A/\kappa := \{a/\kappa \mid a \in A\}$$

(i.e., B is the set of all equivalence classes of κ).
Now, we are able to define the operations $g_i^{n_i}$ on the set B as follows:

$$g_i^{n_i}(a_1/\kappa, ..., a_{n_i}/\kappa) := f_i^{n_i}(a_1, ..., a_{n_i})/\kappa \tag{4.1}$$

(i.e., $g_i^{n_i}(a_1/\kappa, ..., a_{n_i}/\kappa)$ is exactly the equivalence class in which $f_i^{n_i}(a_1, ..., a_{n_i})$ lies). The above definition is possible, since f_i is compatible with κ. Detailed: Choosing $(a_1, a'_1), ..., (a_{n_i}, a'_{n_i}) \in \kappa$, then it holds $(a_1/\kappa, ..., a_{n_i}/\kappa) = (a'_1/\kappa, ..., a'_{n_i}/\kappa)$ and $f_i^{n_i}(a_1, ..., a_{n_i})/\kappa = f_i^{n_i}(a'_1, ..., a'_{n_i})/\kappa$.
We put

$$G := \{g_i^{n_i} \mid i \in I\}.$$

Then the mapping

$$\varphi : A \longrightarrow B,\ a \mapsto a/\kappa$$

is a homomorphic mapping from $\mathbf{A}$ onto $\mathbf{B} := (B; G)$, since by definition of κ

$$\varphi(f_i^{n_i}(a_1, ..., a_{n_i})) = f_i^{n_i}(a_1, ..., a_{n_i})/\kappa$$

holds and (by the compatibility of κ with $f_i^{n_i}$ and by the definition (4.1) of $g_i \in G$ further

$$f_i^{n_i}(a_1, ..., a_{n_i})/\kappa = g_i^{n_i}(a_1/\kappa, ..., a_{n_i}/\kappa) = g_i^{n_i}(\varphi(a_1), ..., \varphi(a_{n_i}))$$

holds, which implies

$$\varphi(f_i^{n_i}(a_1, ..., a_{n_i})) = g_i^{n_i}(\varphi(a_1), ..., \varphi(a_{n_i})).$$

Furthermore, by definition of φ and by the properties of an equivalence relation, it holds:

$$\{(x, y) \in A \times A \mid \varphi(x) = \varphi(y)\} = \{(x, y) \in A \times A \mid x/\kappa = y/\kappa\} = \kappa.$$

■

Definitions The above algebra, which was designed in the proof of the Lemma 4.2.1,

$$(A/\kappa; G)$$

of the so-called **congruence classes of** A **modulo** κ is called **factor algebra of** $(A;F)$ (or **quotient algebra**).
The homomorphic mapping

$$\varphi:\ A \longrightarrow A/\kappa,\ a \mapsto a/\kappa,$$

defined in the proof of Lemma 4.2.1, is called **natural homomorphism** (or **quotient homomorphism**) from $\mathbf{A}$ onto $\mathbf{A}/\kappa$.

We make following use of our agreement from the first chapter according to which the operations of algebras of the same type are described equal. Only then, if distinctions are necessary at the denotation, do we indicate the operations.
After these preparations, we can prove the following theorem (in generalization of analogous theorems over groups, rings, ...):

Theorem 4.2.2 (General Homomorphism Theorem)
For each homomorphism φ from an algebra $\mathbf{A} := (A;F)$ into an algebra of the same type $\mathbf{B} := (B;F)$ the algebra $\varphi(\mathbf{A}) := (\varphi(A);F)$ is isomorphic to the factor algebra $\mathbf{A}/\kappa_\varphi := (A/\kappa_\varphi;F)$, where

$$\kappa_\varphi = \{(a,a') \in A \times A \mid \varphi(a) = \varphi(a')\}.$$

Proof. It suffices to check that

$$\alpha : \varphi(A) \longrightarrow A/\kappa_\varphi, \varphi(a) \mapsto a/\kappa_\varphi$$

is an isomorphism. Since

$$a/\kappa_\varphi = a'/\kappa_\varphi \implies (a,a') \in \kappa_\varphi \implies \varphi(a) = \varphi(a'),$$

α is a mapping. Obviously, by definition, α is a surjection. The injectivity of α follows from the following:

$$\begin{aligned} & \alpha(\varphi(a)) = \alpha(\varphi(a')) \\ \implies & a/\kappa_\varphi = a'/\kappa_\varphi \\ \implies & (a,a') \in \kappa_\varphi \\ \implies & \varphi(a) = \varphi(a') \end{aligned}$$

$(a, a' \in A)$. Thus, α is bijective.
To prove that α is a homomorphism, let $f^n \in F$ and let $\varphi(a_1), ..., \varphi(a_n) \in \varphi(A)$ be arbitrary. Then it holds:

$$\begin{aligned}
&\alpha(f_{\varphi(\mathbf{A})}(\varphi(a_1),...,\varphi(a_n)))\\
&= \alpha(\varphi(f_{\mathbf{A}}(a_1,...,a_n))) && \text{(since } \varphi \text{ is a homomorphism)}\\
&= f(a_1,...,a_n)/\kappa_\varphi && \text{(by definition of } \alpha)\\
&= f_{\mathbf{A}/\kappa_\varphi}(a_1/\kappa_\varphi,...,a_n/\kappa_\varphi) && \text{(since } \kappa_\varphi \text{ is compatible with } f)\\
&= f_{\mathbf{A}/\kappa_\varphi}(\alpha(\varphi(a_1)),...,\alpha(\varphi(a_n))) && \text{(by definition of } \alpha),
\end{aligned}$$

Thus our bijective mapping α is an isomorphism. ■

The next lemma summarizes some elementary properties of congruences. The proof for this lemma is left to the reader.

Lemma 4.2.3 *It holds:*

(a) The intersection of arbitrary many congruences of an algebra $\mathbf{A}$ *is also a congruence on* $\mathbf{A}$.
(b) If κ *is a congruence on* $(A;F)$ *and* φ *is a homomorphism defined over* $\mathbf{A}$ *with* $\kappa_\varphi \subseteq \kappa$, *then* $\varphi(\kappa) := \{(\varphi(a),\varphi(a')) \mid (a,a') \in \kappa\}$ *is a congruence on* $(\varphi(A);F)$.
(c) If π *is a congruence on* $(\varphi(A);F)$ *and* φ *is a homomorphism defined on* $(A;F)$, *then*

$$\varphi^{-1}(\pi) := \{(a,a') \in A \times A \mid (\varphi(a),\varphi(a')) \in \pi\}$$

is a congruence on $(A;F)$, *which* κ_φ *includes.*
(d) If κ *is a congruence on* $\mathbf{A} = (A;F)$ *and* $\mathbf{B} = (B;F)$ *is a subalgebra of* $\mathbf{A}$, *then*

$$\kappa_{|B} := \{(b,b') \in \kappa \mid b,b' \in B\}$$

is a congruence on B.
(e) The set $Con\mathbf{A}$ *of all congruence relations of an algebra* $\mathbf{A}$ *is a complete lattice in respect to the inclusion* $\subseteq$ *with*

$$\inf\{\kappa_j \mid j \in J\} = \bigcap_{j\in J} \kappa_j,$$
$$\sup\{\kappa_j \mid j \in J\} = <\bigcup_{j\in J} \kappa_j >_{Con\mathbf{A}},$$

where $\{\kappa_j \mid j \in J\} \subseteq Con\mathbf{A}$ *and* $<\bigcup_{j\in J} \kappa_j >_{Con\mathbf{A}}$ *is the intersection of all congruences which contain* $\bigcup_{j\in J} \kappa_j$. ■

4.3 Examples for Congruence Relations and Some Homomorphism Theorems

In the following, we characterize the congruence relations by groups and rings more closely. In addition, we give the special homomorphism theorems that follow from these characterizations.

4.3.1 Congruences on Groups

Let $\mathbf{G} = (G;\circ,^{-1},e)$ be a group. A subgroup $\mathbf{N}$ of $\mathbf{G}$ is called a **normal subgroup** iff

$$\forall x \in G : x \circ N = N \circ x$$

holds. Let **NG** be the set of all normal subgroups of **G**. If **N** is a normal subgroup of **G**, one also writes

$$\mathbf{N} \trianglelefteq \mathbf{G}.$$

The following connections between normal subgroups and congruences on **G** can easily be checked:

(a) For every $\kappa \in Con\mathbf{G}$, e/κ is a normal subgroup of **G**, and for arbitrary $a, b \in G$ it holds:

$$(a, b) \in \kappa \iff a \circ b^{-1} \in e/\kappa.$$

(b) If **N** is a normal subgroup then one can obtain by

$$\kappa_N := \{(a, b) \mid a \circ b^{-1} \in N\}$$

a congruence on **G** with $e/\kappa = N$.

Consequently, the mapping

$$\alpha : Con\mathbf{G} \longrightarrow \mathbf{NG}, \ \kappa \mapsto e/\kappa$$

is an order-preserving bijection. Since for every homomorphism φ from **G** into a group $\mathbf{G}'$ the neutral element e is mapped to the neutral element of $\mathbf{G}'$, the properties (a), (b), and Theorem 4.2.2 imply the following

Theorem 4.3.1.1 (Homomorphism Theorem for Groups)
*For every homomorphism φ from a group **G** into a group $\mathbf{G}'$ there exists as so-called kernel (notation: ker φ) a normal subgroup **K** of **G**, which consists of all the elements of G which are mapped to the neutral element of G'. The group $\mathbf{G}'$ is isomorphic to the factor group*

$$\mathbf{G}/\mathbf{K} = (\{x \circ K \mid x \in G\}; \circ, ^{-1}, K)),$$

where

$$(x \circ K) \circ (y \circ K) := (x \circ y) \circ K$$

for arbitrary $x, y \in G$.
*Conversely, if **K** is a normal subgroup of **G**, then **K** is the kernel of the natural homomorphism from **G** onto the factor group **G/K** ($\forall x \in G : x \mapsto x \circ K$).* ■

We notice that the concept *kernel of a group homomorphism* differs from the general term *kernel of a homomorphism.* The corresponding is valid for the concept *kernel* from Theorem 4.3.2.1.

4.3.2 Congruences on Rings

Let $\mathbf{R} = (R; +, -, 0, \cdot)$ be a ring. A subgroup $\mathbf{I}$ of $(R; +, -, 0)$ is called an **ideal** of $\mathbf{R}$ iff

$$\forall x \in R: \ x \cdot I \subseteq I \ \wedge \ I \cdot x \subseteq I$$

holds. Let $\mathbf{IR}$ be the set of all ideals of $\mathbf{R}$.
Connections between congruences on R and ideals of $\mathbf{R}$ are the following:
For every $\kappa \in Con\mathbf{R}$, $0/\kappa$ is an ideal of $\mathbf{R}$ and we have for arbitrary $a, b \in R$:

$$(a, b) \in \kappa \iff a - b \in 0/\kappa.$$

If $\mathbf{I}$ is an ideal of $\mathbf{R}$ then one can obtain by

$$\kappa_I := \{(a, b) \mid a - b \in I\}$$

a congruence on R with $0/\kappa_I = I$. Consequently, the mapping $Con\mathbf{R} \longrightarrow \mathbf{IR}$, $\kappa \rightarrow 0/\kappa$ is an order-preserving bijection. Further it holds:

Theorem 4.3.2.1 (Homomorphism Theorem for Rings)
For every homomorphism φ from a ring $\mathbf{R}$ into a ring $\mathbf{R}'$ belongs as a **kernel** *(notation: ker φ) an ideal $\mathbf{I}$ of $\mathbf{R}$, which consists of all the elements of R which are mapped to the zero element of R'. The ring $\mathbf{R}'$ is isomorphic to the* **residue class ring**

$$\mathbf{R/I} = (\{x + I \mid x \in R\}; +, -, I, \cdot),$$

where $(x + I) + (y + I) := (x + y) + I$ and $(x + I) \cdot (y + I) := (x \cdot y) + I$ for arbitrary $x, y \in R$.
Conversely, if $\mathbf{I}$ is an ideal of $\mathbf{R}$ then I is the kernel of the natural homomorphism from $\mathbf{R}$ onto $\mathbf{R/I}$ ($\forall x \in R : x \mapsto x + I$). ■

It is obvious idea, according to a homomorphism theorem for rings, to find also a homomorphism theorem for fields. The next theorem shows, however, that such a theorem is only a trivial and special case of the general homomorphism theorem.

Theorem 4.3.2.2 *A ring $\mathbf{R}$, which is also a field, has only the two trivial ideals $\{0\}$ and R. In other words: A field has only the trivial congruences κ_0 and κ_1.*

Proof. Let $\mathbf{R}$ be a field and let κ be a congruence of $\mathbf{R}$, which is different from κ_0. Denote I the ideal belonging to κ. Since $\kappa \neq \kappa_0$, there is certain $a, b \in R$ with $a \neq b$ and $(a, b) \in \kappa$. Consequently, we have $(c, 0) := (a - b, b - b) \in \kappa$. Hence, I contains the invertable element c. Since I also contains all elements $r \cdot c$ for every $r \in R$ (specially, $r = r' \cdot c^{-1}$ with arbitrary $r' \in R$) we have $I = R$. Thus, $\kappa = \kappa_1$. ■

4.4 Galois Connections

Definitions A **Galois connection** (or a **Galois correspondence**) between the sets A and B is a pair (σ, τ) of mappings

$$\sigma : \mathfrak{P}(A) \longrightarrow \mathfrak{P}(B)$$

and

$$\tau : \mathfrak{P}(B) \longrightarrow \mathfrak{P}(A),$$

such that for all $X, X' \subseteq A$ and all $Y, Y' \subseteq B$ the following conditions are fulfilled:

$$(GC1) \qquad \left.\begin{array}{l} X \subseteq X' \Longrightarrow \sigma(X) \supseteq \sigma(X') \\ Y \subseteq Y' \Longrightarrow \tau(Y) \supseteq \tau(Y') \end{array}\right\} \quad \text{(antitony)}$$

$$(GC2) \qquad \left.\begin{array}{l} X \subseteq \tau(\sigma((X))) \\ Y \subseteq \sigma(\tau((Y))) \end{array}\right\} \quad \text{(extensivity).}$$

Let $(P; \leq)$ be a poset. The **dual order** $\leq^\delta$ to the order $\leq$ is defined by

$$x \leq^\delta y \; :\Longleftrightarrow \; y \leq x.$$

$(P; \leq)$ is called **dual isomorphic** (or **antiisomorphic**) to $(Q; \leq)$, iff $(P; \leq)$ is isomorphic to $(Q; \leq^\delta)$. The bijective mapping appertaining to this fact is called **dual isomorphism** (or **antiisomorphism**) .

Theorem 4.4.1 *Let the pair* (σ, τ) *of mappings* $\sigma\colon \mathfrak{P}(A) \longrightarrow \mathfrak{P}(B)$ *and* $\tau\colon \mathfrak{P}(B) \longrightarrow \mathfrak{P}(A)$ *be a Galois connection between* A *and* B. *Then it holds:*

(a) The mappings

$$\sigma\tau := \sigma\Box\tau : \; \mathfrak{P}(A) \longrightarrow \mathfrak{P}(A)$$

and

$$\tau\sigma := \tau\Box\sigma : \; \mathfrak{P}(B) \longrightarrow \mathfrak{P}(B)$$

are hull operators on A *or* B, *respectively.*

(b) The $\sigma\tau$*-closed sets are exactly the sets of the form* $\tau(Y)$, $Y \subseteq B$. *The* $\tau\sigma$*-closed sets are exactly the sets of the form* $\sigma(X)$, $X \subseteq A$.

(c) Let $\mathfrak{H}_{\sigma\tau}$ *and* $\mathfrak{H}_{\tau\sigma}$ *be the hull systems that are assigned to* $\sigma\tau$ *and* $\tau\sigma$, *respectively. Then the lattices* $(\mathfrak{H}_{\sigma\tau}; \subseteq)$ *and* $(\mathfrak{H}_{\tau\sigma}; \subseteq)$ *are dual isomorphic, and* σ *and* τ *are inversely dual isomorphisms to each other of these lattices.*

Proof. (a): The extensivity and the monotony of $\sigma\tau$ and $\tau\sigma$ immediately follow from $(GC1)$ and $(GC2)$, respectively.
Thus, for all $X \subseteq A$ we have $X \subseteq \tau(\sigma(X))$ and therefore $\sigma(X) \supseteq \sigma(\tau(\sigma(X)))$. On the other hand, $\sigma(X) \subseteq (\tau\sigma)(\sigma(X)) = \sigma(\tau(\sigma(X)))$ follows from $(GC2)$. Consequently, we have

$$\sigma(X) = \sigma(\tau(\sigma(X))) \tag{4.2}$$

and analogously

$$\tau(Y) = \tau(\sigma(\tau(Y))). \tag{4.3}$$

Then the equations

$$\tau(\sigma(X)) = \tau(\sigma(\tau(\sigma(X))))$$

and

$$\sigma(\tau(Y)) = \sigma(\tau(\sigma(\tau(Y))))$$

follow from these. Thus, $\sigma\tau$ and $\tau\sigma$ are idempotent.

(b): For a $\sigma\tau$-closed set X we have $X = \tau(\sigma(X))$; that is, X has the form $X = \tau(Y)$ with $Y := \sigma(X) \subseteq B$. Conversely, a set of the form $X := \tau(Y)$, $Y \subseteq B$, is $\sigma\tau$-closed by (4.3). Analogously, one can prove the assertion for $\tau\sigma$-closed sets.

(c): Because of (b) and Theorem 3.2.1 it holds $\mathfrak{H}_{\sigma\tau} = \{\tau(Y) \mid Y \subseteq B\}$ and $\mathfrak{H}_{\tau\sigma} = \{\sigma(X) \mid X \subseteq A\}$. Thus, we have $\sigma(\mathfrak{H}_{\sigma\tau}) := \{\sigma(\tau(Y)) \mid Y \subseteq B\} = \mathfrak{H}_{\tau\sigma}$ and $\tau(\mathfrak{H}_{\tau\sigma}) := \{\tau(\sigma(X)) \mid X \subseteq A\} = \mathfrak{H}_{\sigma\tau}$. By $(GC1)$ σ and τ are antitone, and therefore the restrictions of these mappings to $\mathfrak{H}_{\sigma\tau}$ and $\mathfrak{H}_{\tau\sigma}$ have also this property. It follows from the idempotence of $\sigma\tau$ that $\sigma\tau$ on $\mathfrak{H}_{\sigma\tau}$ is the identical mapping. Analogously, one can see that $\tau\sigma$ is also the identical mapping on $\mathfrak{H}_{\tau\sigma}$. Therefore, the mappings $\sigma : \mathfrak{H}_{\sigma\tau} \longrightarrow \mathfrak{H}_{\tau\sigma}$ and $\tau : \mathfrak{H}_{\tau\sigma} \longrightarrow \mathfrak{H}_{\sigma\tau}$ are bijective mappings and it holds $\sigma^{-1} = \tau$. Consequently, σ, τ are isomorphisms of the lattices $(\mathfrak{H}_{\sigma\tau}; \subseteq)$ and $(\mathfrak{H}_{\tau\sigma}; \subseteq^{\delta})$ ∎

An example of a Galois connection concludes this section. Further examples can be found in [Den-E-W 2004].

Theorem 4.4.2 *Let A, B be nonempty sets and let $R \subseteq A \times B$ with $R \neq \emptyset$. The mappings $\sigma : \mathfrak{P}(A) \longrightarrow \mathfrak{P}(B)$, $\tau : \mathfrak{P}(B) \longrightarrow \mathfrak{P}(A)$ are defined by*

$$\sigma(X) := \{y \in B \mid \forall x \in X : (x, y) \in R\},$$
$$\tau(Y) := \{x \in A \mid \forall y \in Y : (x, y) \in R\}.$$

Then the pair (σ, τ) is a Galois connection between A and B.

Proof. Because of symmetry of the assumptions, it suffices to show that σ is an antitone and $\tau\sigma$ is an extensive mapping.
Let $X \subseteq X' \subseteq A$. Then for every $y \in \sigma(X')$ we have: $(x, y) \in R$ for all $x \in X'$. Thus (by $X \subseteq X'$) we have also $(x, y) \in R$ for all $x \in X$. Therefore $\sigma(X') \subseteq \sigma(X)$ holds. The inclusion $X \subseteq \tau(\sigma(X))$ follows directly from $\tau(\sigma(X)) = \{x \in A \mid \forall y \in \sigma(X) : (x, y) \in R\}$ and from the definition of $\sigma(X)$. ∎

5

Direct and Subdirect Products

With the help of *direct* and *subdirect products*, it is possible to form new algebras with larger universes from given algebras. Of these constructions one immediately asks the following questions: Which algebras are smallest "constituents" of given algebras? How can one reduce a given algebra to its smallest "constituents"? First, we will show that every *finite* algebra is isomorphic to a direct product of directly irreducible algebras. Then, we prove that *every* algebra is isomorphic to a subdirect product of subdirectly irreducible algebras.

5.1 Direct Products

First, we consider direct products of two algebras.

Definitions Let $\mathbf{B} = (B; F)$ and $\mathbf{C} = (C; F)$ be algebras of the same type τ.

The algebra $\mathbf{A} := \mathbf{B} \times \mathbf{C}$ of type τ is called **direct product** of the algebras $\mathbf{B}$ and $\mathbf{C}$, iff $B \times C$ is the universe of the algebra $\mathbf{A}$ and the operations of $\mathbf{A}$ are defined as follows: If $f \in F$ is nullary, then let $f_{\mathbf{A}} := (f_{\mathbf{B}}, f_{\mathbf{C}})$. If $af \geq 1$, let $f_{\mathbf{A}}$ defined as follows:

$$f_{\mathbf{B}\times\mathbf{C}}((b_1, c_1), (b_2, c_2), ..., (b_{af}, c_{af})) := (f_{\mathbf{B}}(b_1, ..., b_{af}), f_{\mathbf{C}}(c_1, ..., c_{af})).$$

Obviously, $\mathbf{B}$ and $\mathbf{C}$ are homomorphic images of the algebra $\mathbf{B} \times \mathbf{C}$, since the (so-called **projection-)mappings**

$$pr_1 : B \times C \longrightarrow B, \ (b, c) \mapsto pr_1(b, c) := b \ \text{ and}$$

$$pr_2 : B \times C \longrightarrow C, \ (b, c) \mapsto pr_2(b, c) := c$$

are homomorphisms from $\mathbf{B} \times \mathbf{C}$ onto $\mathbf{B}$ or $\mathbf{C}$, respectively. The kernels of these projection mappings (congruences on $B \times C$)

$$Ker\ pr_i := \{((b,c),(b',c')) \in (B \times C)^2 \mid pr_i(b,c) = pr_i(b',c')\}$$

($i = 1, 2$) are distinguished from other congruences by certain properties (see Theorem 5.1.2). We can describe these properties with the help of the following

Definition Two equivalence relations κ, $\mu \in Eq(A)$ are called **permutable**, if $\kappa \square \mu = \mu \square \kappa$ holds; i.e.,

$$\forall x, z \in A :$$

$$(\exists y \in A:\ (x,y) \in \kappa \wedge (y,z) \in \mu) \iff (\exists y' \in A:\ (x,y') \in \mu \wedge (y',z) \in \kappa).$$

The following lemma gives some other definition possibilities for the above concept. In this lemma and in the following theorems, $(Eq; \subseteq)$, by Theorem 2.4.3, is a complete lattice and operations $\wedge = \cap$ ("infimum") and $\vee$ ("supremum") are defined on Eq (see Chapter 2).

Lemma 5.1.1 *For $\kappa, \mu \in Eq(A)$ the following statements are equivalent:*

(a) κ and μ are permutable
(b) $\kappa \square \mu \subseteq \mu \square \kappa$
(c) $\mu \square \kappa \subseteq \kappa \square \mu$
(d) $\kappa \vee \mu = \kappa \square \mu$
(e) $\mu \vee \kappa = \kappa \square \mu$

Proof. Because of the commutativity of $\vee$, we have (d)$\iff$(e). The equivalence (b)$\iff$(c) is a conclusion from (a)$\iff$(b). Thus, we have to show that (a)$\iff$(b) and (a)$\iff$(e) holds.
(a)$\Longrightarrow$(b) is trivial.
(b)$\Longrightarrow$(a): Let $\kappa \square \mu \subseteq \mu \square \kappa$. Then (by $(\kappa \square \mu)^{-1} = \mu^{-1} \square \kappa^{-1}$ and by the symmetry of κ and μ):

$$(\kappa \square \mu)^{-1} \subseteq (\mu \square \kappa)^{-1}$$
$$\Longrightarrow \underbrace{\mu^{-1} \square \kappa^{-1}}_{=\mu \square \kappa} \subseteq \underbrace{\kappa^{-1} \square \mu^{-1}}_{=\kappa \square \mu}.$$

Hence, $\kappa \square \mu = \mu \square \kappa$.
(a)$\Longrightarrow$(e): Let $\kappa \square \mu = \mu \square \kappa$. (e) is shown, if it is proven that $\kappa \square \mu$ is the smallest equivalence relation on A, which contains $\kappa \cup \mu$.
Because of

$$\forall (x,y) \in \kappa\ ((\ (x,y) \in \kappa\ \wedge\ (y,y) \in \mu) \implies (x,y) \in \kappa \square \mu),$$
$$\forall (x,y) \in \mu\ ((\ (x,x) \in \kappa\ \wedge\ (x,y) \in \mu) \implies (x,y) \in \kappa \square \mu)$$

we have $\kappa \cup \mu \subseteq \kappa \square \mu$. $\kappa \square \mu$ is reflexive, since κ (or μ) is reflexive. The symmetry of $\kappa \square \mu$ follows from $(\kappa \square \mu)^{-1} = \mu^{-1} \square \kappa^{-1} = \mu \square \kappa = \kappa \square \mu$. Because of

$$(\kappa \square \mu) \square (\kappa \square \mu) = \kappa \square \underbrace{(\mu \square \kappa)}_{=\kappa \square \mu} \square \mu = \underbrace{(\kappa \square \kappa)}_{\subseteq \kappa} \square \underbrace{(\mu \square \mu)}_{\subseteq \mu} \subseteq \kappa \square \mu$$

$\kappa \square \mu$ is also transitive. Thus, $\kappa \square \mu$ is an equivalence relation on A.
$\kappa \square \mu$ is the smallest equivalence relation of $Eq(A)$, which contains $\kappa \cup \mu$, since every

other equivalence relation, which contains $\kappa \cup \mu$, also contains the transitive closure of $\kappa \cup \mu$ and therefore contains $\kappa \Box \mu$.

(e)$\Longrightarrow$(a): Let $\kappa \vee \mu = \kappa \Box \mu$. Since $\kappa \vee \mu$ is an equivalence relation, the compatibility of κ and μ follows from

$$\kappa \Box \mu = \kappa \vee \mu = (\kappa \Box \mu)^{-1} = \mu^{-1} \Box \kappa^{-1} = \mu \Box \kappa.$$

■

Theorem 5.1.2 *For algebras* **B** *and* **C** *of the same type and the projection mappings* $pr_1\colon \mathbf{B} \times \mathbf{C} \longrightarrow \mathbf{B}$ *and* $pr_2\colon \mathbf{B} \times \mathbf{C} \longrightarrow \mathbf{C}$ *it holds:*

(a) $Ker\ pr_1 \wedge Ker\ pr_2 = \kappa_0$
(b) $Ker\ pr_1 \vee Ker\ pr_2 = \kappa_1$
(c) $Ker\ pr_1$ *and* $Ker\ pr_2$ *are permutable*

Proof. Let $((b, c), (b', c')) \in Ker\ pr_1 \cap Ker\ pr_2$ be arbitrary. Then, $pr_i(b, c) = pr_i(b', c')$ for $i = 1, 2$. This means, however, that $b = b'$ and $c = c'$. Therefore, (a) is shown.
For arbitrary $b, b' \in B$, $c, c' \in C$ we have

$$(\ (b, c), (b, c')\ ) \in Ker\ pr_1\ \wedge\ (\ (b, c'), (b', c')\ ) \in Ker\ pr_1.$$

Hence, $(\ (b, c), (b', c')\ ) \in \kappa \Box \mu$ for all $b, b' \in B$ and $c, c' \in C$ and thus

$$(Ker\ pr_1) \Box (Ker\ pr_2) = \kappa_1.$$

With the help of Lemma 5.1.1, the assertions (b) and (c) follow from this. ■

Now we will examine the conditions under which an algebra is a direct product of two other smaller algebras. Theorem 5.1.2 provides instructions on how to proceed.

Theorem 5.1.3 *Let* $\mathbf{A} = (A; F)$ *be an algebra and let* $\kappa, \mu \in Con\mathbf{A}$ *be two congruence relations with the following three properties:*

(a) $\kappa \wedge \mu = \kappa_0$
(b) $\kappa \vee \mu = \kappa_1$
(c) κ *and* μ *are permutable*

Then **A** *is isomorphic to the direct product of* $\mathbf{A}/\kappa$ *and* $\mathbf{A}/\mu$*. An isomorphism* $\varphi\colon \mathbf{A} \longrightarrow \mathbf{A}/\kappa \times \mathbf{A}/\mu$ *is defined by*

$$\forall a \in A:\ \varphi(a) := (a/\kappa, a/\mu).$$

Proof. φ is injective: Let $\varphi(a) = \varphi(b)$. Then, $a/\kappa = b/\kappa$ and $a/\mu = b/\mu$. From these $(a, b) \in \kappa \wedge \mu$ follows. Thus $a = b$ by (a).
φ is surjective: By (b) and (c) for every pair a, b there is a $c \in A$ with $(a, c) \in \kappa$ and $(c, b) \in \mu$. Consequently, $(a/\kappa, b/\mu) = (c/\kappa, c/\mu) = \varphi(c)$.
φ is an isomorphism: For all $f_A \in F$ ($af_A =: n$) and arbitrary $a_1, ..., a_n \in A$ it holds

$$\begin{aligned} \varphi(f_{\mathbf{A}}(a_1, ..., a_n)) &= (f_{\mathbf{A}}(a_1, ..., a_n)/\kappa, f_{\mathbf{A}}(a_1, .., a_n)/\mu) \\ &= (f_{\mathbf{A}/\kappa}(a_1/\kappa, ..., a_n/\kappa), f_{\mathbf{A}/\mu}(a_1/\mu, ..., a_n/\mu)) \\ &= f_{\mathbf{A}/\kappa \times \mathbf{A}/\mu}(\varphi(a_1), ..., \varphi(a_n)). \end{aligned}$$

■

Definition An algebra $\mathbf{A}$ is called **directly irreducible**, iff $\mathbf{A} \cong \mathbf{B} \times \mathbf{C}$ implies $|B| = 1$ or $|C| = 1$ for all algebras $\mathbf{B}$ and $\mathbf{C}$.

Example Obviously, every finite algebra $\mathbf{A}$ with $|A| \in \mathbb{P}$ is directly irreducible.
One finds further examples after Theorem 5.1.4.

Theorem 5.1.4 *An algebra* $\mathbf{A}$ *is directly irreducible iff* κ_0 *and* κ_1 *are the only one pair of congruences of* $Con\mathbf{A}$*, which satisfy the conditions (a)–(c) of Theorem 5.1.3.*

Proof. Let $\mathbf{A}$ be directly irreducible. Further, $\kappa, \mu \in Con\mathbf{A}$ fulfill the conditions (a)–(c) of Theorem 5.1.3. Then, by Theorem 5.1.3, we have $\mathbf{A} \cong \mathbf{A}/\kappa \times \mathbf{A}/\mu$. Thus, w.l.o.g., $|A/\kappa| = 1$. From this, $\kappa = \kappa_1$ follows, and then $\mu = \kappa_0$ by (a).
Conversely, Let κ_0 and κ_1 be the only one pair of congruences of $\mathbf{A}$ with the properties (a)–(c) of Theorem 5.1.3, and let $\mathbf{A} \cong \mathbf{B} \times \mathbf{C}$. Obviously, then κ_0 and κ_1 are the only one pair of congruence relations on $\mathbf{B} \times \mathbf{C}$ with (a)–(c). By Theorem 5.1.2 the kernels of the projection mappings pr_1 and pr_2 fulfill (a)–(c). Thus $Ker\ pr_1 = \kappa_0$ or $Ker\ pr_2 = \kappa_0$ holds and we have $|C| = 1$ or $|B| = 1$, respectively. ■

With the help of Theorem 5.1.4, one can prove the following statements:[1]

(a) Every simple algebra $\mathbf{A}$; i.e., every algebra $\mathbf{A}$ with $Con\ \mathbf{A} = \{\kappa_0, \kappa_1\}$, is directly irreducible.
(b) If $\mathbf{A}$ is a Boolean algebra then

$$\mathbf{A} \text{ is directly irreducible} \iff |A| \leq 2.$$

(c) The residue class group $(\mathbb{Z}_n; +, -, 0)$ is directly irreducible iff n is a prime number power.
(d) A vector space $\mathbf{V} := (V; +, -, K, 0)$ over the field K is directly irreducible iff $|V| = 1$ or $\mathbf{V}$ is 1-dimensional.

One can generalize our definition of the direct product of two algebras in an obvious way for finite many algebras of the same type. As the next definitions demonstrate, it is also possible to form direct products of arbitrarily many algebras of the same type.

Definitions The **Cartesian product** $\Pi_{j \in J} A_j$ of the sets A_j $(j \in J)$ is the set of all mappings α from J into $\bigcup_{j \in J} A_j$ with $\alpha(j) \in A_j$ for all $j \in J$. We write down the elements of $\Pi_{j \in J} A_j$ in the form $(x_j \mid j \in J)$.
In analog mode too, one can define then the algebra

$$(\Pi_{j \in J} A_j; (f_i)_{i \in I})$$

as a **direct product** $\mathbf{\Pi_{j \in J} A_j}$ of the algebras $\mathbf{A_j}$ $(j \in J)$, where

[1] The proof for the statements can be found in [Lau 2004], volume 2.

$$f_i((a_{j1}|j \in J), (a_{j2}|j \in J), ..., (a_{j,af_i}|j \in J)) := (f_{ji}(a_{j1}, a_{j2}, ..., a_{j,af_i})|j \in J)$$

for arbitrary $((a_{j1}|j \in J), ..., (a_{j,af_i}|j \in J) \in \Pi_{j\in J}A_j$, if $af_i > 0$, and $f_i = (f_{ji}|j \in J)$, if $af_i = 0$.
If $J = \emptyset$, then let $\mathbf{\Pi_{j\in J}A_j}$ be the 1-element algebra of the corresponding type. If $A_j = A$ for all $j \in J$, we write $\mathbf{A^J}$ instead of $\mathbf{\Pi_{j\in J}A_j}$. If $J = \{1, 2, .., n\}$, one also uses the denotation $\mathbf{A_1} \times ... \times \mathbf{A_n}$ for the direct product of the algebras $\mathbf{A_j}$.

Theorem 5.1.5 *Every finite algebra is isomorphic to a direct product of direct irreducible algebras.*

Proof. Induction over the cardinality of the universes of the algebras:
Obviously, every algebra $\mathbf{A}$ with $|A| = 1$ is directly irreducible. Now, let $\mathbf{A}$ be a finite algebra with $|A| \geq 2$, and assume that the assertion is proven for all algebras $\mathbf{A'}$ with $|A'| < |A|$. If $\mathbf{A}$ is directly irreducible, we have to show nothing more. If, however, $\mathbf{A} \cong \mathbf{B} \times \mathbf{C}$ with $|B| > 1$, $|C| > 1$ is valid, then we have $|B| < |A|$ and $|C| < |A|$; i.e., one can find direct irreducible algebras $\mathbf{B_1}, ..., \mathbf{B_m}, C_1, ..., C_n$ with

$$\mathbf{B} \cong \mathbf{B_1} \times ... \times \mathbf{B_m},$$
$$\mathbf{C} \cong \mathbf{C_1} \times ... \times \mathbf{C_n}.$$

Thus $\mathbf{A} \cong \mathbf{B_1} \times ... \times \mathbf{B_m} \times \mathbf{C_1} \times ... \times \mathbf{C_n}$. ■

The direct products of more than two algebras have similar properties, like the direct products of only two algebras. For example, the j_0-th projection pr_{j_0} from $\Pi_{j\in J}A_j$ onto A_j with

$$(a_j \mid j \in J) \mapsto a_{j_0}$$

is a homomorphism for every $j_0 \in J$.

One can easily check the following property of direct products.

Theorem 5.1.6 *For every family*[2] φ_i*:* $\mathbf{B} \longrightarrow \mathbf{A_i}$*,* $i \in I$*, of homomorphisms one can form a homomorphism* φ*:* $\mathbf{B} \longrightarrow \mathbf{\Pi_{i\in I}A_i}$ *by*

$$(\varphi(b))_i := \varphi_i(b).$$

■

[2] A **family** $(a_i \mid i \in I)$ of elements of a set A is a mapping $\varphi : I \to A$, $i \mapsto a_i$. This denotation is used around a certain selection of (not necessarily different) elements from A to characterize. Then, I is called **index set** of the family $(a_i \mid i \in I)$.

5.2 Subdirect Products

Unlike finite algebras, infinite algebras cannot always be represented by direct products of directly irreducible algebras.

Example As we already noticed above, a Boolean algebra **A** is directly irreducible iff $|A| \leq 2$. Furthermore, one can easily prove that every 2-element Boolean algebra is isomorphic to the algebra

$$\mathbf{B} = (\{0,1\}; \vee, \wedge, ^-, 0, 1).$$

An infinite direct product of **B** is not countable. Consequently, the countable Boolean algebra $\mathbf{C} = (C; \vee', \wedge', \neg)$ with

$$C := \{(a_1, a_2, ...) \in \{0,1\}^{\mathbb{N}} \mid |\{i \in \mathbb{N} \mid a_i = 0\}| < \aleph_0 \ \text{ or } \ |\{i \in \mathbb{N} \mid a_i = 1\}| < \aleph_0\},$$

$(a_1, a_2, ...) \circ' (b_1, b_2, ...) := (a_1 \circ b_1, a_2 \circ b_2, ...)$ for $\circ \in \{\vee, \wedge\}$ and $\neg(a_1, a_2, ...) := (\overline{a}_1, \overline{a}_2, ...)$ is not isomorphic to a direct product of directly irreducible algebras, however, it is a subalgebra of the direct product $\mathbf{B}^{\mathbb{N}}$.

In generalizing this example, we get the following new product concept:

Definition Let the algebras $\mathbf{A_i}$, $i \in I$, be of the same type. A subalgebra **B** of $\mathbf{\Pi_{i \in I} A_i}$ is called a **subdirect product** of the $\mathbf{A_i}$ iff

$$pr_j(\mathbf{B}) = \mathbf{A_j}$$

holds for all $j \in I$.

Example Obviously, every direct product is also a subdirect product.

Theorem 5.2.1 *For a subdirect product* **B** *of the algebras* $\mathbf{A_i}$, $i \in I$, *and the projection mappings* $pr_j \colon \mathbf{\Pi_{i \in I} A_i} \longrightarrow \mathbf{A_j}$ *it holds*

$$\bigcap_{j \in I} Ker(pr_j)_{|B} = \kappa_0.$$

Proof. If $(a, b) \in \bigcap_{j \in I} Ker(pr_j)_{|B}$, then $a_j = b_j$ follows for all $j \in I$ and therefore $a = b$. ∎

By the above theorem and by the fact that all $pr_{j_{|B}}$ are surjective, the subdirect products are already characterized:

Theorem 5.2.2 *Let* **A** *be an algebra. For certain congruences* $\kappa_i \in Con\mathbf{A}$, $i \in I$, *let*

$$\bigcap_{i \in I} \kappa_i = \kappa_0.$$

Then **A** *is isomorphic to a subdirect product of the algebras* $\mathbf{A}/\kappa_i$, $i \in I$. *By*

$$\varphi(a) := (a/\kappa_i \mid i \in I)$$

an injective homomorphism $\varphi \colon \mathbf{A} \longrightarrow \mathbf{\Pi_{i \in I}(A/\kappa_i)}$ *is defined, and* $\varphi(\mathbf{A})$ *is a subdirect product of the algebras* $\mathbf{A}/\kappa_i$.

Proof. By Theorem 5.1.6 φ is a homomorphism. φ is also injective: Let $\varphi(a) = \varphi(b)$. This implies $a/\kappa_i = b/\kappa_i$ and thus $(a,b) \in \kappa_i$ for all $i \in I$. Therefore, $(a,b) \in \bigcap_{i\in I} \kappa_i = \kappa_0$; i.e., $a = b$. Consequently, $\mathbf{A}$ and $\varphi(\mathbf{A})$ are isomorphic. Further, by definition of φ we have $pr_j(\varphi(\mathbf{A})) = \mathbf{A}/\kappa_\mathbf{j}$ for all $j \in I$. Hence, $\varphi(\mathbf{A})$ is a subdirect product of the algebras $\mathbf{A}/\kappa_\mathbf{i}$. ■

Definition An injective homomorphism (a so-called **embedding**)

$$\varphi\colon \mathbf{A} \longrightarrow \mathbf{\Pi_{i\in I}A_i}$$

is called a **subdirect representation** of $\mathbf{A}$, if $\varphi(\mathbf{A})$ is a subdirect product of the algebras $\mathbf{A_i}$.

Example The mapping φ of Theorem 5.2.2 is a subdirect representation.

Definition An algebra $\mathbf{A}$ is called **subdirectly irreducible**, if for every subdirect representation

$$\varphi : \mathbf{A} \longrightarrow \mathbf{\Pi_{i\in I}A_i}$$

there exists a $j \in I$ such that the mapping

$$\varphi\Box pr_j : \mathbf{A} \longrightarrow \mathbf{A_j}$$

is an isomorphism.

Thus, an algebra is subdirectly irreducible if and only if one gets by with a single component in every subdirect representation.

Theorem 5.2.3 *An algebra* $\mathbf{A}$ *is subdirectly irreducible, if and only if the universe of* $\mathbf{A}$ *contains at most an element, or if*

$$\bigcap(Con\mathbf{A}\backslash\{\kappa_0\}) \neq \kappa_0$$

holds. The latter holds obviously iff κ_0 *exactly has an upper neighbor in* $Con\mathbf{A}$*:*

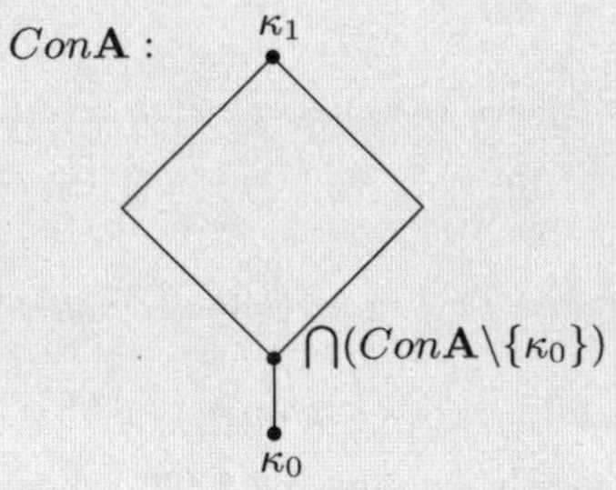

Proof. W.l.o.g. let $|A| \notin \{0,1\}$ in the following.

Suppose, $\bigcap(Con\mathbf{A}\backslash\{\kappa_0\}) = \kappa_0$. Put $I := Con\mathbf{A}\backslash\{\kappa_0\}$. Then, with the help of Theorem 5.2.2, one obtains a subdirect representation $\varphi\colon \mathbf{A} \longrightarrow \mathbf{\Pi_{\kappa\in I}(A/\kappa)}$. For every mapping pr_κ $(\kappa \in I)$ and all $a \in A$ it holds $(\varphi\Box pr_\kappa)(a) = a/\kappa$. By $\kappa_0 \notin I$, therefore $\varphi\Box pr_\kappa\colon \mathbf{A} \longrightarrow \mathbf{A}/\kappa$ is not injective (i.e., is not an isomorphism). Thus, $\mathbf{A}$ is not subdirectly irreducible.

Let now $\mu := \bigcap(Con\mathbf{A}\backslash\{\kappa_0\}) \neq \kappa_0$; i.e., there exists $(a,b) \in \mu \setminus \kappa_0$, and denote φ: $\mathbf{A} \longrightarrow \mathbf{\Pi_{i\in I}A_i}$ a subdirect representation of $\mathbf{A}$. We have to show that there exists an

$i \in I$ so that $\varphi \Box pr_i$ is an isomorphism from $\mathbf{A}$ onto $\mathbf{A_i}$. Since φ is injective and $a \neq b$, there exists a $j \in I$ with $pr_j(\varphi(a)) \neq pr_j(\varphi(b))$, whereby $(a,b) \notin Ker(\varphi \Box pr_j)$. Thus, by $(a,b) \in \mu$, we have $\mu \not\subseteq Ker(\varphi \Box pr_j)$. Then, by definition of μ, $Ker(\varphi \Box pr_j) = \kappa_0$ holds; i.e., $\varphi \Box pr_j$ is injective. Since $\varphi(A)$ is a subdirect product, $\varphi \Box pr_j$ is surjective. Therefore, $\varphi \Box pr_j$ is an isomorphism. Consequently, $\mathbf{A}$ is subdirectly irreducible. ■

With the help of Theorems 5.2.3 and 5.1.4, one can easily prove the following connection between the direct and the subdirectly irreducible algebras:

$$\mathbf{A} \text{ is subdirectly irreducible} \implies \mathbf{A} \text{ is directly irreducible.} \tag{5.1}$$

The reversal of the statement (5.1) is not valid, because one can prove with the help of a 3-element lattice which is directly irreducible but which is not subdirectly irreducible.

An essential aid for proving the following Theorem is Zorn's Lemma which is indicated without proof here. This lemma is equivalent to the *axiom of choice* (see e.g. [Her 55]).

Lemma 5.2.4 (Zorn's Lemma)
In every set system $\mathfrak{M}$ with the property

$$\forall T \subseteq \mathfrak{M} \; ((\forall X, Y \in T \; \exists Z \in T : \; X \cup Y \subseteq Z) \Longrightarrow \bigcup_{X \in T} X \in \mathfrak{M})$$

(i.e., $\mathfrak{M}$ is an **inductively set system***)*
there is a maximal element[3]*; i.e., an element $M \in \mathfrak{M}$ that is not contained in any proper subset of $\mathfrak{M}$.* ■

After these preparations we can prove the following theorem, published by G. Birkhoff in 1944.

Theorem 5.2.5 *Every algebra is isomorphic to a subdirect product of subdirectly irreducible algebras.*

Proof. Let $\mathbf{A}$ be an algebra. One can easily see that, for every pair $a, b \in A$ with $a \neq b$, the set

$$\mathfrak{M}_{a,b} := \{\kappa \in Con\mathbf{A} \mid (a,b) \notin \kappa\}$$

is an inductive set system. By Lemma 5.2.4 $\mathfrak{M}_{a,b}$ has a maximal element $\Phi(a,b)$. In the lattice $Con\mathbf{A}$ the element $\Phi(a,b)$ has exactly an upper neighbor, namely $\Phi(a,b) \vee \Omega(a,b)$, where $\Omega(a,b)$ is the congruence relation generated by (a,b). It is easy to check that the factor algebra $\mathbf{A}/\mathbf{\Phi}(\mathbf{a},\mathbf{b})$ is isomorphic to an interval

$$[\Phi(a,b), \kappa_1] := \{\kappa \in Con\mathbf{A} \mid \Phi(a,b) \subseteq \kappa \subseteq \kappa_1\}$$

[3] Let $(B; \leq)$ be a poset. A maximal element of the set $A \subseteq B$ is then an element $a \in A$ with

$$a < x \implies x \notin A$$

for all $x \in B$.

of $Con\mathbf{A}$. By Theorem 5.2.3 the algebra $\mathbf{A}/\boldsymbol{\Phi}(\mathbf{a},\mathbf{b})$ is subdirectly irreducible. From

$$\bigcap\{\Phi(a,b) \mid a,b \in A \wedge a \neq b\} = \kappa_0$$

and from Theorem 5.2.2, it follows that $\mathbf{A}$ is isomorphic to a subdirect product of subdirectly irreducible algebras (namely of the algebras $\mathbf{A}/\boldsymbol{\Phi}(\mathbf{a},\mathbf{b})$). ■

6

Varieties, Equational Classes, and Free Algebras

In this Chapter, only certain classes[1] of algebras of the same type shall be regarded.
First we introduce so-called *varieties* as classes of algebras, which are closed in respect to formation of subalgebras, homomorphic images, and direct products. We then come to a method for constructing algebra classes that strongly differs from the first method at first sight: Starting from certain equations from variables and operation symbols of a certain type τ, we form the class of all algebras of type τ that fulfill these equations. The result is an *equational class.* We will see, however, that there is a close connection between the two methods of algebra class construction: A class of algebras is equationally definable if and only if it is a variety.
Free algebras are "the most general" algebras within a variety or an equational class (or an equationally definable class). In the section on equational classes, we will also address such concepts such as *conclusion of an equational set.* In addition, we investigate methods to receive such conclusions.

6.1 Varieties

The following operators S, H, P, I map a class K of algebras of type τ to a class of algebras of the same type.

Let

$S(K)$ be the class of all subalgebras of algebras aus K,

$H(K)$ be the class of all homomorphic images of algebras of K,

$P(K)$ be the class of all direct products of families of algebras of K,

$I(K)$ be the class of all algebras which are isomorphic to algebras of K.

---

[1] The concept "class" is a generalization of the concept "set". Informally speaking, a class is a collection so large that subjecting it to the operations admissible for sets would lead to logical contradictions.

We denote the composition of the operators $Y, X \in \{H, S, P, I\}$ by XY; i.e., it holds $XY(K) := X(Y(K))$.

It is easy to check that S, H and IP are hull operators; i.e., for all classes K and L of algebras of the same type we have:

$$\forall X \in \{S, H, IP\} :$$
$$K \subseteq X(K) \ \wedge \ (K \subseteq L \Longrightarrow X(K) \subseteq X(L)) \ \wedge \ X(K) = X(X(K)).$$

The operator P is not idempotent:

For all $\mathbf{A}, \mathbf{B}, \mathbf{C} \in K$ it holds $(\mathbf{A} \times \mathbf{B}) \times \mathbf{C} \in P(P(K))$, but $(\mathbf{A} \times \mathbf{B}) \times \mathbf{C} \in P(K)$ generally does not hold. However, $(\mathbf{A} \times \mathbf{B}) \times \mathbf{C} \cong \mathbf{A} \times \mathbf{B} \times \mathbf{C} \in P(K)$ is right; i.e., $(\mathbf{A} \times \mathbf{B}) \times \mathbf{C} \in IP(K)$.

Definitions

- A class K of algebras of the same type is called under $X \in \{S, H, P\}$ **closed**, if $X(K) \subseteq K$ holds.
- If the class K of algebras of the same type is closed under the three operators S, H and P, then the class K is called a **variety**.

Examples

(1) It is easy to see that the class of all groups is a variety.
(2) Since the direct product of the field $\mathbb{Z}_2$ with the field $\mathbb{Z}_2$ because of

$$\forall x, y \in \mathbb{Z}_2 : \ (0,1) \cdot (x,y) = (0 \cdot x, 1 \cdot y) = (0, y) \neq (1,1)$$

is not a field, the class of all fields is not a variety.

Lemma 6.1.1 *For every class K of algebras of the same type it holds:*
(a) $SH(K) \subseteq HS(K)$,
(b) $PS(K) \subseteq SP(K)$,
(c) $PH(K) \subseteq HP(K)$.

Proof. (a): Let $\mathbf{A} \in SH(K)$; i.e., there is an algebra $\mathbf{B} \in H(K)$ with $\mathbf{A} \leq \mathbf{B}$ and $\mathbf{B}$ is homomorphic image of an algebra $\mathbf{C} \in K$. Let φ: $\mathbf{C} \longrightarrow \mathbf{B}$ be a surjective homomorphism. Then, for the subalgebra $\varphi^{-1}(\mathbf{A})$ of $\mathbf{C}$ it holds $\varphi(\varphi^{-1}(\mathbf{A})) = \mathbf{A}$. Consequently, we have $\mathbf{A} \in HS(K)$.

(b): Let $\mathbf{A} \in PS(K)$. Then it holds $\mathbf{A} = \Pi_{i \in I} \mathbf{B_i}$ with $\mathbf{B_i} \leq \mathbf{C_i} \in K$ for all $i \in I$. Since obviously $\Pi_{i \in I} \mathbf{B_i}$ is a subalgebra of $\Pi_{i \in I} \mathbf{C_i}$, we have $\mathbf{A} \in SP(K)$.

(c): Let $\mathbf{A} \in PH(K)$. Then, $\mathbf{A} = \Pi_{i \in I} \mathbf{B_i}$, where for every $i \in I$ there are an algebra $\mathbf{C_i}$ and a surjective homomorphism φ_i: $\mathbf{C_i} \longrightarrow \mathbf{B_i}$. If $pr_j : \Pi_{i \in I} \mathbf{C_i} \longrightarrow \mathbf{C_j}$ is the projection mapping, then $pr_j \Box \varphi_j$: $\Pi_{i \in I} \mathbf{C_i} \longrightarrow \mathbf{B_j}$ is a surjective homomorphism. By Theorem 5.1.6 we get by $\varphi(c)_j := (pr_j \Box \varphi_j)(c)$ a homomorphism φ: $\Pi_{i \in I} \mathbf{C_i} \longrightarrow \Pi_{i \in I} \mathbf{B_i}$, which is also surjective. Consequently, we have $\mathbf{A} \in HP(K)$. ∎

Theorem 6.1.2 *A class K of algebras of the same type is a variety if and only if $HSP(K) = K$ holds.*

Proof. If K is a variety, then obviously $HSP(K) = K$.
Let now $HSP(K) = K$. We have to show $H(K) \subseteq K$, $S(K) \subseteq K$ and $P(K) \subseteq K$. It holds:

$$H(K) = H(HSP(K)) = HSP(K),$$

since H is idempotent. Thus,

$$H(K) = K$$

by assumption. Further we have:

$$S(K) = S(HSP(K)) = SH(SP(K)) \subseteq HS(SP(K)) \subseteq HSP(K)$$

by Lemma 6.1.1, (a) and since S is idempotent. Therefore, $S(K) \subseteq K$ holds. Furthermore,

$$\begin{aligned} P(K) &= PHSP(K) \subseteq HPSP(K) \subseteq HSPP(K) \\ &\subseteq HSIPIP(K) = HSIP(K) \\ &\subseteq HSHP(K) \subseteq HHSP(K) = HSP(K) \end{aligned}$$

by Lemma 6.1.1 and since the operators IP and H are idempotent. Thus, $P(K) \subseteq K$ holds. Consequently, K is a variety. ■

Every variety is exactly determined by its elements, which are subdirect irreducible algebras:

Theorem 6.1.3 *Every algebra of a variety K is isomorphic to a subdirect product of subdirect irreducible algebras of K.*

Proof. By Theorem 5.2.5, every algebra $\mathbf{A}$ is isomorphic to a subdirect product of subdirect irreducible algebras $\mathbf{A_i}$, where every $\mathbf{A_i}$ is isomorphic to a factor algebra of $\mathbf{A}$; i.e., it holds $\mathbf{A_i} \in H(\mathbf{A})$. If $\mathbf{A}$ belongs to a variety K, then $\mathbf{A_i} \in H(K) \subseteq K$. ■

6.2 Terms, Term Algebras, and Term Functions

In Section 1.1, we had agreed to describe the operations of algebras of the same type by the same notations, if it is clearly from the context to which algebra a given operation belongs. Since we will often make use of this convention, we generalize the concept "type of an algebra" as follows:

Definitions A **type of algebras** is an ordered pair $(\mathfrak{F}, \tau)$, where $\mathfrak{F}$ is a set, whose elements are called **operation symbols**, and $\tau\colon \mathfrak{F} \longrightarrow \mathbb{N}_0$ is a mapping, which assigns the arity af to every operation $f \in \mathfrak{F}$.
Then, the algebra $\mathbf{A} = (A; F)$ with $F := \{f_{\mathbf{A}} \mid f \in \mathfrak{F}\}$ is called **algebra of type** $(\mathfrak{F}, \tau)$. Let $\mathfrak{F}^n$ be the set of the n-ary operation symbols of F.

Now let $(\mathfrak{F}, \tau)$ be a type of algebras and let

$$X$$

be a finite or countable infinite set, whose elements are called **variables**, where $X \cap \mathfrak{F}^0 = \emptyset$.

$$T(X)$$

denotes the smallest set with the following two properties:

(1) $X \cup \mathfrak{F}^0 \subseteq T(X)$,

(2) $(f \in \mathfrak{F}^n \ \wedge \ \{t_1, ..., t_n\} \subseteq T(X)) \implies f(t_1, ..., t_n) \in T(X)$.

One observes that $f(t_1, ..., t_n)$ is a syntactic expression (a symbol sequence) and not a function value.

The elements of $T(X)$ are called **terms of type** $(\mathfrak{F}, \tau)$ **over the alphabet** X.

Example Let $\mathfrak{F} := \{e, f\}$, $\tau(e) := 0$, $\tau(f) := 2$ and $X := \{x, y, z\}$. Then

$$\begin{aligned} T(X) &= \{e, x, y, z, f(e,e), f(e,x), ..., f(z,y), f(f(e,e),e), f(f(x,e),e), ..., \\ &\quad f(f(x,z), f(z,y)), ...\}. \end{aligned}$$

We agree that $(u \circ v) := f(u, v)$ for arbitrary $u, v \in T(X)$ and, furthermore, we do without outer brackets. Then the set $T(X)$ can be also written down as follows:

$$\{e, x, y, z, e \circ e, e \circ x, ..., z \circ y, (e \circ e) \circ e, (x \circ e) \circ e, ..., \\ (x \circ z) \circ (z \circ y), ...\}.$$

$T(X)$ is the universe of the so-called **term algebra**

$$\mathbf{T(X)} := (T(X); F)$$

of type $(\mathfrak{F}, \tau)$, where for every $f \in \mathfrak{F}^n$ $(n \in \mathbb{N} \cup \{0\})$ the operations of this algebra are defined as follows:

$$f_{\mathbf{T(X)}} := f,$$

if $n = 0$, and

$$\forall t_1, ..., t_n \in T(X) : f_{\mathbf{T(X)}}(t_1, ..., t_n) := f(t_1, ..., t_n)$$

for $n \geq 1$. A part of the operation table of the operation $f_{\mathbf{T(X)}}$ from the above example then looks as follows:

u	v	$f_{\mathbf{T(X)}}(u, v)$
e	e	$e \circ e$
e	$e \circ x$	$e \circ (e \circ x)$
$x \circ (y \circ e)$	$(x \circ x) \circ z$	$(x \circ (y \circ e)) \circ ((x \circ x) \circ z)$

The following lemma follows directly from the definition of $\mathbf{T(X)}$:

Lemma 6.2.1 *The term algebra* $\mathbf{T(X)}$ *is generated by* X*; i.e., it holds* $[X] = T(X)$. ∎

The following theorem gives an important property of the algebra $\mathbf{T(X)}$:

Theorem 6.2.2 *Let* $\mathbf{T(X)}$ *be the term algebra of type* $(\mathfrak{F}, \tau)$ *over* X*. Then, for every algebra* $\mathbf{A}$ *of type* $(\mathfrak{F}, \tau)$ *and for every mapping* $\varphi\colon X \longrightarrow A$ *there is exactly one homomorphism* $\widetilde{\varphi} : \mathbf{T(X)} \longrightarrow \mathbf{A}$*, which* φ *continues, i.e., for which* $\widetilde{\varphi}|_X = \varphi$ *holds.*

Proof. For a given algebra $\mathbf{A}$ of type $(\mathfrak{F}, \tau)$ and for a given mapping $\varphi\colon X \longrightarrow A$ let $\widetilde{\varphi}\colon T(X) \longrightarrow A$ be a mapping with the properties:

$$\forall x \in X : \widetilde{\varphi}(x) := \varphi(x)$$

and

$$\forall f(t_1, ..., t_n) \in T(X)\backslash X : \ \widetilde{\varphi}(f(t_1, ...t_n)) := f_{\mathbf{A}}(\widetilde{\varphi}(t_1), ..., \widetilde{\varphi}(t_n)).$$

Obviously, $\widetilde{\varphi}$ is defined over $T(X)$ by the above conditions, and it is the only possible continuation of φ over $T(X)$. ∎

Using the concepts from Part II, Chapter 1, we can briefly define the following

Definition The **term functions** of an algebra $\mathbf{A} = (A; F)$ of type $(\mathfrak{F}, \tau)$ are operations over A that can be formed by superposition from the fundamental operations of F and from the projections.

Without using concepts from Part II, Chapter 1, we can define the term functions an algebra $\mathbf{A}$ of type $(\mathfrak{F}, \tau)$ as follows:

Let t be a term of type $(\mathfrak{F}, \tau)$ over $X = \{x_1, ..., x_n\}$, and for $a_1, ..., a_n \in A$ let $\varphi_{a_1,...,a_n}\colon \mathbf{T(X)} \longrightarrow \mathbf{A}$ be the unique homomorphism with $x_i \rightarrow a_i$, $i = 1, 2, ..., n$. Then we can define an n-ary operation

$$t_{\mathbf{A}} : \ A^n \longrightarrow A$$

by

$$\forall a_1, ..., a_n \in A : \ t_{\mathbf{A}}(a_1, ..., a_n) := \varphi_{a_1,...,a_n}(t).$$

These operations are identical with the term functions already defined above. If t and $T_{\mathbf{A}}$ are defined as above, we say that the **term t induces the term function $t_{\mathbf{A}}$**.

Let

$$TF(\mathbf{A})$$

be the set of all term functions of $\mathbf{A}$.

One can prove the next two lemmas with properties of $TF(\mathbf{A})$ easily.

Lemma 6.2.3 *Let* $[..]$ *be the subalgebra-hull-operator (see Chapter 3). Then for every algebra* $\mathbf{A}$ *and every subset* B *of* A *it holds:*

$$[B] = \{t_{\mathbf{A}}(b_1, ..., b_n) \mid n \in \mathbb{N} \ \wedge \ t \in T(\{x_1, ..., x_n\}) \ \wedge \ \{b_1, ..., b_n\} \subseteq B\}.$$

∎

Lemma 6.2.4 *The algebras* $\mathbf{A}$, $\mathbf{B}$ *and the* n*-ary term* t *have the same type. Then, for every homomorphism* $\varphi\colon \mathbf{A} \longrightarrow \mathbf{B}$ *and all* $a_1, ..., a_n \in A$ *it holds:*

$$\varphi(t_{\mathbf{A}}(a_1, ..., a_n)) = t_{\mathbf{B}}(\varphi(a_1), ..., \varphi(a_n));$$

i.e., the term functions react just like the fundamental operations in respect to homomorphisms. ■

6.3 Equations and Equational Classes

In this section, let $T(X)$ be the set of all terms of type $(\mathfrak{F}, \tau)$.
To show that the variables of the term $t \in T(X)$ are of the set $\{x_1, ..., x_n\} \subseteq X$, we write

$$t < x_1, ..., x_n >$$

and set

$$t = t < x_1, ..., x_n > .$$

The notation

$$s := t < t_1, ..., t_n >$$

meant that the term s was formed from the term t by substituting every variable x_i $(1 \leq i \leq n)$ by t_i in every place the variable x_i in t appeared. We agree on an analogous notation for term functions.

Definitions

- The elements of $T(X) \times T(X)$ are called **equations** (or **identities**) over X and we write

$$s \approx t :\Longleftrightarrow (s,t) \in T(X) \times T(X).$$

- An algebra $\mathbf{A}$ of type $(\mathfrak{F}, \tau)$ **fulfills** the equation

$$s < x_1, ..., x_n > \approx t < x_1, ..., x_n >$$

(or the equation $s \approx t$ **holds in** A), if for all $a_1, ..., a_n \in A$

$$s_{\mathbf{A}} < a_1, ..., a_n > = t_{\mathbf{A}} < a_1, ..., a_n >$$

is right. In this case, we also write

$$\mathbf{A} \models s \approx t.$$

Further, we set

$$Id_X(\mathbf{A}) := \{(s,t) \in T(X) \times T(X) \mid \mathbf{A} \models s \approx t\}.$$

- Let be for $\Sigma \subseteq T(X) \times T(X)$ and classes K of algebras of the same type $(\mathfrak{F}, \tau)$:
$$\mathbf{A} \models \Sigma \;:\Longleftrightarrow\; (\forall s \approx t \in \Sigma : \mathbf{A} \models s \approx t).$$
- The class
$$Mod(\Sigma) := \{\mathbf{A} \mid \mathbf{A} \models \Sigma\}$$
is called the **set of all models of** Σ.

- Conversely, for every class K of algebras of type $(\mathfrak{F}, \tau)$ let
$$Id_X(K) := \{(s,t) \in T(X) \times T(X) \mid \forall \mathbf{A} \in K : \mathbf{A} \models s \approx t\}$$
be the **class of all equations over** X **that hold in all algebras of** K.
- A class K of algebras is **equationally definable**, if there exists a $\Sigma \subseteq T(X) \times T(X)$ with $Mod(\Sigma) = K$.
- A set $\Sigma \subseteq T(X) \times T(X)$ is called **equational theory** over X, if there is a class K of algebras with $\Sigma = Id_X(K)$.
- An equation $s \approx t$ is called a **conclusion of** $\Sigma \subseteq T(X) \times T(X)$, if $\mathbf{A} \models s \approx t$ holds for all $\mathbf{A} \in Mod(\Sigma)$.

  Let $Cons_X(\Sigma)$ be the set of all conclusions of Σ; i.e., it holds
$$Cons_X(\Sigma) := Id_X(Mod(\Sigma)).$$

Instead of $Y(Z(..))$, where $Y, Z \in \{Mod, Id_X, Cons_X\}$, we write briefly $YZ(..)$.

The following theorem summarizes elementary properties of the sets defined above and connections between the concepts just defined.

Theorem 6.3.1 *For arbitrary $\Sigma, \Sigma' \subseteq T(X) \times T(X)$ and arbitrary classes K, K' of algebras of type $\mathfrak{F}$ it holds:*

(1) $\Sigma \subseteq \Sigma' \Longrightarrow Mod(\Sigma') \subseteq Mod(\Sigma)$,
$K \subseteq K' \Longrightarrow Id_X(K') \subseteq Id_X(K)$;

(2) $\Sigma \subseteq Id_X Mod(\Sigma)$,
$K \subseteq ModId_X(K)$;

(3) $ModId_X Mod(\Sigma) = Mod(\Sigma)$,
$Id_X ModId_X(K) = Id_X(K)$;

(4) $\Sigma \subseteq Cons_X(\Sigma)$,
$\Sigma \subseteq \Sigma' \Longrightarrow Cons_X(\Sigma) \subseteq Cons_X(\Sigma')$,
$Cons_X Cons_X(\Sigma) = Cons_X(\Sigma)$;

(5) $K \subseteq ModId_X(K)$,
$K \subseteq K' \Longrightarrow ModId_X(K) \subseteq ModId_X(K')$,
$ModId_X ModId_X(K) = ModId_X(K)$;

(6) Σ *is equational theory* $\Longleftrightarrow \Sigma = Cons_X(\Sigma)$,
K *is equationally definable* $\Longleftrightarrow K = ModId_X(K)$.

Proof. (1) and (2) immediately follow from the definitions of Mod and Id_X.
(3): By (2) $\Sigma \subseteq Id_X Mod(\Sigma) =: \Sigma'$ holds. Thus, by means of (1), we have $ModId_X Mod(\Sigma) \subseteq Mod(\Sigma)$. Conversely, we have also by (2): $K := Mod(\Sigma) \subseteq ModId_X Mod(\Sigma)$. Therefore, $Mod(\Sigma) = ModId_X Mod(\Sigma)$.
Analogously, one can show $Id_X ModId_X(K) = Id_X(K)$.
(4) and (5) one can easily prove by means of (1)–(3).
(6): Let Σ be an equational theory; i.e., there is a class K of algebras of type $(\mathfrak{F}, \tau)$ with $\Sigma = Id_X(K)$. Then

$$Cons_X(\Sigma) = Id_X Mod(\Sigma) \overset{\text{assumption}}{=} Id_X ModId_X(K) \overset{(3)}{=} Id_X(K) \overset{\text{assumption}}{=} \Sigma.$$

Conversely, let $\Sigma = Cons_X(\Sigma)$. Then we have $\Sigma = Id_X Mod(\Sigma)$; i.e., Σ is an equational theory.
The statement over equational definable classes can be proven analogously. ■

If we neglect that the objects formed by classes need an exact definition, [2] then the following theorem is an immediate conclusion of the above theorem and of Theorem 4.4.1.

Theorem 6.3.2 *Let X be a countable infinite set, $Alg(\mathfrak{F}, \tau)$ the class of all algebras of type $(\mathcal{F}, \tau)$ over X and let $T(X)$ be the set of all terms of type $(\mathcal{F}, \tau)$. Then the pair (Id_X, Mod) forms a Galois connection between $\mathfrak{P}(T(X) \times T(X))$ and $\mathfrak{P}(Alg(\mathfrak{F}, \tau))$. Furthermore, the lattice of all equational classes of $Alg(\mathfrak{F}, \tau)$ is antiisomorphic to the lattice of all equational theories of type $(\mathfrak{F}, \tau)$.* ■

6.4 Free Algebras

To define a free algebra, we need the following properties of $\mathbf{T}(\mathbf{X})$:

Theorem 6.4.1 *Let K be a class of algebras of type $(\mathfrak{F}, \tau)$, and let $\mathbf{T}(\mathbf{X})$ be the term algebra of the same type over the alphabet X. Then*
(a) $Id_X(K) = \bigcap\{Ker\ \varphi \mid \exists \mathbf{A} \in K : \ \varphi : \mathbf{T}(\mathbf{X}) \longrightarrow \mathbf{A}$ is a homomorphic mapping$\}$,
(b) $Id_X(K) \in Con\mathbf{T}(\mathbf{X})$.

Proof. (a): Let $s, t \in T(X')$ with $X' := \{x_1, ..., x_n\} \subseteq X$. For every algebra $\mathbf{A} \in K$ and all $a_1, ..., a_n \in A$ there is by Theorem 6.2.2 a homomorphism φ: $\mathbf{T}(\mathbf{X}) \longrightarrow \mathbf{A}$ with $\varphi(x_i) = a_i$, $i = 1, ..., n$. For every φ it holds $\varphi(s) = s_{\mathbf{A}}(a_1, ..., a_n)$ and $\varphi(t) = t_{\mathbf{A}}(a_1, ..., a_n)$. Therefore we have $(s, t) \in Ker\ \varphi$ for all $\varphi : \mathbf{T}(\mathbf{X}) \longrightarrow \mathbf{A}$ with $\mathbf{A} \in K$ if and only if for all $\mathbf{A} \in K$ and all $a_1, .., a_n \in A$ the equation $s_{\mathbf{A}}(a_1, ..., a_n) = t_{\mathbf{A}}(a_1, ..., a_n)$ holds. But this is equivalently with $\mathbf{A} \models s \approx t$ for all $\mathbf{A} \in K$.
(b) follows directly from (a), since the intersection of congruences of an algebra is a congruence of the algebra, as is well-known. ■

[2] See, for example, [Sch 74], Chapter II.

By the above theorem, we can form the factor algebra:

$$\mathbf{T(X)/Id_X(K)} \qquad (6.1)$$

for an arbitrary class K of algebras of the same type and of a set X of variables.
Definitions If the factor algebra (6.1)belongs to K, then $\mathbf{T(X)/Id_X(K)}$ is called the **free algebra** of K with **free generating set** X. If (6.1) belongs to K, we describe (6.1) with

$$\mathbf{F_K(X)}.$$

In case $X = \{x_1, ..., x_n\}$ we also write $\mathbf{F_K(x_1, ..., x_n)}$ or briefly $\mathbf{F_K(n)}$, and, if $X = \{x_i \mid i \in \mathbb{N}\}$, we write $\mathbf{F_K(x_1, x_2, ...)}$ or $\mathbf{F_K}(\aleph_0)$ (or $\mathbf{F_K}(\omega)$).

We notice, that, strictly speaking, the free algebra $\mathbf{F_K(X)}$ is not generated by the set X but by the congruence classes $x/Id_X(K)$, $x \in X$. Nevertheless, one often writes x instead of $x/Id_X(K)$, since in a nontrivial class K of algebras (i.e., K contains not only 0- or 1-element algebras), the equation $x/Id_X(K) = y/Id_X(K)$ implies $x = y$.

The importance of the free algebras results from the following theorems with whose aid the main theorems to the equational theory are proven in Section 6.5.

The factor algebra $\mathbf{T(X)/Id_X(K)}$, in respect to the class K, has the same property as $\mathbf{T(X)}$ in respect to the class of all algebras of type $(\mathfrak{F}, \tau)$ (see Theorem 6.2.2):

Theorem 6.4.2 *Let K be a class of algebras of type $(\mathfrak{F}, \tau)$ and $\mathbf{T(X)}$ be the term algebra of the same type. Furthermore, let $\overline{x} := x/Id_X(K)$ and $\overline{X} := \{\overline{x} \mid x \in X\}$. Then there is for every algebra $\mathbf{A} \in K$ and every mapping $\varphi : \overline{X} \longrightarrow A$ exactly one homomorphism $\overline{\varphi} : \mathbf{T(X)/Id_X(K)} \longrightarrow \mathbf{A}$, which continues φ; i.e., for which $\overline{\varphi}|_X = \varphi$ holds.*

Proof. Let $\alpha : X \longrightarrow A$ be a mapping defined by $\alpha(x) := \varphi(\overline{x})$. Then, by Theorem 6.2.2 there is a homomorphism $\overline{\alpha} : \mathbf{T(X)} \longrightarrow \mathbf{A}$, which continues α. For the homomorphism $\pi : \mathbf{T(X)} \longrightarrow \mathbf{T(X)/Id_X(K)}$ with $Ker\ \pi = Id_X(K)$ we have by Theorem 6.4.1, (a) that $Ker\ \pi \subseteq Ker\ \overline{\alpha}$ holds; i.e., it holds: $\pi(s) = \pi(t) \implies \overline{\alpha}(s) = \overline{\alpha}(t)$. By $\overline{\varphi}(\pi(t)) := \overline{\alpha}(t)$ one receives a well-defined mapping $\overline{\varphi} : T(X)/Id_X(K) \longrightarrow A$. It is easy to see that $\overline{\varphi}$ is a homomorphism and that $\overline{\varphi}(\overline{x}) = \varphi(\overline{x})$ for all $x \in X$ holds. Because of $[\overline{X}] = [\pi(X)] = \pi[X] = \pi(T(X)) = T(X)/Id_X(K)$ the mapping $\overline{\varphi}$ is uniquely defined through the definition over $\overline{X}$. ■

Theorem 6.4.3 *For every class K of algebras of the same type and every variable set X it holds*

$$\mathbf{T(X)/Id_X(K)} \in ISP(K).$$

Proof. Let $T := T(X)$ and $\kappa := Id_X(K)$. By Theorem 6.4.1 (a) and with the help of Lemma 17.4.1 from [Lau 2004], volume 2 one can prove

$$\begin{aligned} &\bigcap\{(Ker\ \varphi)/\kappa \mid \exists \mathbf{A} \in K :\ \varphi : \mathbf{T} \longrightarrow \mathbf{A} \text{ is a homomorphic mapping}\} \\ = \ &\Delta_{T/\kappa}\ (= \kappa_0 \text{ on } T/\kappa). \end{aligned}$$

Because of Theorem 5.2.2, the algebra $\mathbf{T}/\kappa$ is isomorphic to a subdirect product of the algebras

$$(\mathbf{T}/\kappa)/((\mathbf{Ker}\ \varphi)/\kappa)$$

with $\varphi : \mathbf{T} \longrightarrow \mathbf{A}$, $\mathbf{A} \in K$. For every such φ it holds (by the First Isomorphism Theorem [3])

$$(\mathbf{T}/\kappa)/((\mathbf{Ker}\ \varphi)/\kappa) \cong \mathbf{T}/(\mathbf{Ker}\ \varphi) \cong \varphi(\mathbf{T}) \in S(K).$$

Therefore one gets altogether

$$\mathbf{T}/\kappa \in ISP(IS(K)) \subseteq ISP(S(K)) \subseteq ISP(K),$$

where, obviously, the first inclusion is valid and the second inclusion follows from Lemma 6.1.1, (b). ■

An immediate conclusion from Theorem 6.4.3 is as follows:

Theorem 6.4.4 *For every class K of algebras of the same type (in particular for a variety K) which is closed in respect to the operators I, S and P, it holds $\mathbf{T(X)/Id_X(K)} \in K$; i.e., K contains a free algebra $\mathbf{F_K(X)}$.* ■

Lemma 6.4.5 *Every free algebra $\mathbf{F_K(X)}$ of a variety is isomorphic to a subdirect product of certain algebras $\mathbf{F_K(E)}$, where $E \subseteq X$ is finite and $E \neq \emptyset$.*

Proof. For $x \in X$ let $\overline{x} := x/Id_X(K)$. Further, for every $E \subseteq X$ let $\overline{E} := \{\overline{e} \in F_K(X) \mid e \in E\}$. Denote $\mathbf{U}(\overline{\mathbf{E}})$ the subalgebra of $\mathbf{F_K(X)}$ which is generated by $\overline{E}$. It is easy to check that $\mathbf{U}(\overline{\mathbf{E}})$ and $\mathbf{F_K(E)}$ are isomorphic. Therefore, it is sufficient to show that $\mathbf{F_K}(\overline{\mathbf{X}})$ is isomorphic to a subdirect product of the algebras $\mathbf{U}(\overline{\mathbf{E}})$, where $E \subseteq X$ is nonempty and finite. For every such E, one chooses a surjective mapping $\varphi_E : \overline{X} \longrightarrow \overline{E}$ with $(\varphi_E)_{|\overline{E}} = \mathrm{id}_{\overline{E}}$. Then, the homomorphic continuation $\overline{\varphi}_E$ is surjective, and it holds $(\overline{\varphi}_E)_{|U(\overline{E})} = \mathrm{id}_{U(\overline{E})}$. Every term has only finite many variables. Thus for every pair $s, t \in \mathbf{F_K(X)}$, there is a finite subset $E \subseteq X$ with $s, t \in U(\overline{E})$. In the case $s \neq t$ we have even $(s,t) \notin Ker\ (\overline{\varphi}_E)$ because of $\overline{\varphi}_E(s) = s$ and $\overline{\varphi}_E(t) = t$. Therefore $\bigcap\{Kern(\overline{\varphi}_E) \mid \emptyset \subset E \subseteq X \ \wedge\ E \text{ is finite}\} = \kappa_0$. By Theorem 5.2.2 $\mathbf{F_K(X)}$ is also isomorphic to an subdirect product of $\mathbf{F_K(x)/Kern}(\overline{\varphi}_\mathbf{E})$. Because of $\mathbf{F_K(X)/Kern}(\overline{\varphi}_\mathbf{E}) \cong \mathbf{U}(\overline{\mathbf{E}})$, the assertion follows. ■

Theorem 6.4.6 *For every variety K it holds*

$$K = HSP(\{\mathbf{F_K(n)} \mid n \in \mathbb{N}\}) = HSP(\{\mathbf{F_K}(\omega)\}).$$

Proof. Every algebra $\mathbf{A} \in K$ is a homomorphic image of $\mathbf{F_K(X)}$, if $|X| \geq |A|$ (one chooses a mapping $\varphi : X \longrightarrow A$ and then one uses Theorem 6.4.2). Thus, the first equality sign in our theorem follows from Lemma 6.4.5, and the second equality sign follows from the fact that $\mathbf{F_K(n)}$ is isomorphic to a subalgebra of $\mathbf{F_K}(\omega)$ for all $n \in \mathbb{N}$. ■

[3] See for example [Wec 92], p. 140 or [Den-W 2002], Theorem 3.2.2 or Theorem 17.4.2 from [Lau 2004], volume 2.

6.5 Connections Between Varieties and Equational Defined Classes

We need the following statement.

Lemma 6.5.1 *Let K be a class of algebras of the same type. Then it holds for an arbitrary alphabet X that:*

(a) $\forall Op \in \{H, S, P\} : Op(K) \subseteq ModId_X(K)$;

(b) $ModId_X(K)$ is a variety.

Proof. (a): Let $Op = H$. First, we will show that $Id_X(K) \subseteq Id_X(H(K))$ is right. Let $s < x_1, ..., x_n > \approx t < x_1, ..., x_n >$ be an equation of $Id_X(K)$ with $\{x_1, ..., x_n\} \subseteq X$. Then, this equation also holds in an arbitrary algebra $\mathbf{B} \in H(K)$: If namely $\varphi(\mathbf{A}) = \mathbf{B}$ for a certain algebra $\mathbf{A} \in K$ and a surjective homomorphism φ, then for arbitrary $b_1, ..., b_n \in B$ there are $a_1, ..., a_n \in A$ with the property

$$\begin{aligned} s_{\mathbf{B}} < b_1, ..., b_n > &= s_{\mathbf{B}} < \varphi(a_1), ..., \varphi(a_n) > \\ &= \varphi(s_{\mathbf{A}} < a_1, ..., a_n >) \\ &= \varphi(t_{\mathbf{A}} < a_1, ..., a_n >) \\ &= t_{\mathbf{B}} < \varphi(a_1), ..., \varphi(a_n) > \\ &= t_{\mathbf{B}} < b_1, ..., b_n >, \end{aligned}$$

i.e., $s \approx t \in Id_X(\{\mathbf{B}\})$ holds, and thus we have $Id_X(K) \subseteq Id_X(H(K))$.
Now, if one uses Theorem 6.3.1, (1), then one gets $ModId_X(H(K)) \subseteq ModId_X(K)$. Furthermore, by Theorem 6.3.1, (2), we have $H(K) \subseteq ModId_X(H(K))$. Thus $H(K) \subseteq ModId_X(K)$.

Analogously one can prove (a) for $Op \in \{S, P\}$.

(b): Let $K^* = ModId_X(K)$. Then, by (a) and with the help of Theorem 6.3.1, (3) it holds for every $Op \in \{H, S, P\}$:

$$Op(K^*) \subseteq ModId_X(K^*) = Mod(Id_X ModId_X(K)) = ModId_X(K) = K^*.$$

Therefore, K^* is a variety. ■

Theorem 6.5.2 (First Main Theorem of the Equational Theory; *[Bir 35]***)**
A class K of algebras of the same type is a variety iff it is equationally definable; i.e., it holds (by Theorem 6.3.1, (6) and Theorem 6.1.2):

$$K = HSP(K) \iff \exists X : K = ModId_X(K).$$

Proof. "$\Longleftarrow$": By Lemma 6.5.1, (a) it holds $Op(K) \subseteq ModId_X(K)$ for every $Op \in \{H, S, P\}$. If now $K = ModId_X(K)$, then this implies $Op(K) \subseteq K$. Thus K is a variety.

"$\Longrightarrow$": Let K be a variety. By Lemma 6.5.1, (b) the class $K^* := ModId_X(K)$ is also a variety and we have for an arbitrary alphabet X:

$$\begin{aligned}\mathbf{F_{K^*}(X)} &= \mathbf{T(X)/Id_X(K^*)} \quad \text{(by definition)}\\ &= \mathbf{T(X)/Id_X(K)} \quad \text{(since by Theorem 6.3.1, (3) : } Id_X(K^*) = \\ &\qquad\qquad Id_X Mod Id_X(K) = Id_X(K))\\ &= \mathbf{F_K(X)} \quad \text{(by definition).}\end{aligned}$$

From that (with the aid of Theorem 6.4.6 and $X := \{x_1, x_2, ...\}$) we get the equations

$$K = HSP(\{\mathbf{F_K}(\omega)\}) = HSP(\{\mathbf{F_{K^*}}(\omega)\}) = K^*$$

Therefore, K is equationally definable. ■

6.6 Deductive Closure of Equation Sets and Equational Theory

With the following definition of the deductive closure, we generalize the usual procedure of deriving equations from already proven equations. At the end of this section, we will be able to prove that the deductive closure of an equation sets Σ is identical with the set of all conclusions of Σ.

Definitions Let $(\mathfrak{F}, \tau)$ be a type of algebras, $T(X)$ is defined as in Section 6.2 and $\Sigma \subseteq T(X) \times T(X)$. Then, the **deductive closure** $D(\Sigma)$ of Σ is the smallest subset of $T(X) \times T(X)$ containing Σ such that the following five conditions hold:

(R1) $\forall p \in T(X) : p \approx p \in D(\Sigma)$;

(R2) $\forall p, q \in T(X) : p \approx q \in D(\Sigma) \Longrightarrow q \approx p \in D(\Sigma)$;

(R3) $\forall p, q, r \in T(X) : (p \approx q \in D(\Sigma) \ \wedge \ q \approx r \in D(\Sigma) \Longrightarrow p \approx r \in D(\Sigma))$;

(Rep) $\forall f^n \in \mathfrak{F} \ \forall \{s_1 \approx t_1, ..., s_n \approx t_n\} \subseteq D(\Sigma) : f(s_1, ..., s_n) \approx f(t_1, ..., t_n) \in D(\Sigma)$;

("**replacement rule**");

(Sub) $\forall s < x_1, ..., x_n > \approx t < x_1, ..., x_n > \in D(\Sigma) \ \forall t_1, ..., t_n \in T(X) :$
$s < t_1, ..., t_n > \approx t < t_1, ..., t_n > \in D(\Sigma)$

("**substitution rule**").

$\Sigma \subseteq T(X) \times T(X)$ is called **deductively closed** if $D(\Sigma) = \Sigma$ holds.

Obviously, every set Σ of equations with $\Sigma = Id_X(K)$, where K denotes a class of algebras of type $(\mathfrak{F}, \tau)$, is deductively closed. In other words, if Σ is an equational theory of a class of algebras, then Σ is deductively closed. Furthermore, it holds that

$$D(\Sigma) \subseteq Cons_X(\Sigma).$$

The aim of the following considerations is the proof that the reversals of the above two statements are also right.
Exacter: It shall be shown that every deductively closed set of equations is

the equational theory of a certain class of algebras and that for every set Σ of equations, it holds $Cons_X(\Sigma) \subseteq D(\Sigma)$.

Obviously, a deductively closed set $\Sigma \subseteq T(X) \times T(X)$ can also be characterized as follows:

Because of (R1)–(R3) Σ is an equivalence relation,
because of (Rep) Σ is a congruence on $T(X)$, and
because of (Sub) Σ is compatible with every endomorphism of $\mathbf{T}(\mathbf{X})$ (this is a homomorphism from $\mathbf{T}(\mathbf{X})$ into $\mathbf{T}(\mathbf{X})$) (Proof: Let $t_1, ..., t_n \in T(X)$ be arbitrary. Then there exists an endomorphism φ of $\mathbf{T}(\mathbf{X})$ with $\varphi(x_1) = t_1$, $\varphi(x_2) = t_2$, ..., $\varphi(x_n) = t_n$, and for every such endomorphism, it holds that $\varphi(s) = s < t_1, ..., t_n >$ and $\varphi(t) = t < t_1, ..., t_n >$.).

Definition A congruence relation κ of an algebra $\mathbf{A}$ is called **fully invariant**, if it is compatible with all endomorphisms of $\mathbf{A}$; i.e., if for every endomorphism φ of $\mathbf{A}$, $(a, b) \in \kappa$ implies $(\varphi(a), \varphi(b)) \in \kappa$.

The following two lemmas follow immediately from this definition and the above considerations.

Lemma 6.6.1 *A set $\Sigma \subseteq T(X) \times T(X)$ is deductively closed iff Σ is a fully invariant congruence on $\mathbf{T}(\mathbf{X})$.* ■

Lemma 6.6.2 *For every class K of algebras of the same type and every variable set X, $Id_X(K)$ is a fully invariant congruence on $\mathbf{T}(\mathbf{X})$.* ■

The reversal of Lemma 6.6.2 is also valid:

Lemma 6.6.3 *For every fully invariant congruence κ over $\mathbf{T}(\mathbf{X})$ it holds that*

$$Id_X(\{\mathbf{T}(\mathbf{X})/\kappa\}) = \kappa,$$

i.e., for arbitrary $s, t \in T(X)$ we have:

$$(s, t) \in \kappa \Longleftrightarrow \mathbf{T}(\mathbf{X})/\kappa \models s \approx t.$$

In other words, an arbitrary fully invariant congruence κ on $\mathbf{T}(\mathbf{X})$ is an equational theory of the algebra $\mathbf{T}(\mathbf{X})/\kappa$.

Proof. "$\Longrightarrow$": Let $s = s < x_1, ..., x_n >$, $t = t < x_1, ..., x_n >$ and $(s, t) \in \kappa$. For arbitrary $t_1, ..., t_n \in T(X)$ it holds because of full invariance of κ: $(s < t_1, ..., t_n >, t < t_1, ..., t_n >) \in \kappa$. Consequently, we have:

$$s_{\mathbf{T}(\mathbf{X})/\kappa} < t_1/\kappa, ..., t_n/\kappa >= t_{\mathbf{T}(\mathbf{X})/\kappa} < t_1/\kappa, ..., t_n/\kappa >,$$

i.e., the equation $s \approx t$ holds in $\mathbf{T}(\mathbf{X})/\kappa$.
"$\Longleftarrow$": Let $s \approx t \in Id_X(\mathbf{T}(\mathbf{X})/\kappa)$. Then it holds

$$s_{\mathbf{T}(\mathbf{X})/\kappa} < x_1/\kappa, ..., x_n/\kappa >= t_{\mathbf{T}(\mathbf{X})/\kappa} < x_1/\kappa, ..., x_n/\kappa > .$$

Thus $(s_{\mathbf{T}(\mathbf{X})}, t_{\mathbf{T}(\mathbf{X})}) \in \kappa$ and $(s, t) \in \kappa$. ■

The following theorem is a conclusion from the Lemmas 6.6.2 and 6.6.3:

Theorem 6.6.4 (Second Main Theorem of the Equational Theory; *[Bir 35]***)**
A set $\Sigma \subseteq T(X) \times T(X)$ is an equational theory iff Σ is a fully invariant congruence on $T(X)$. ■

Because of Theorem 6.3.1, (6) and Lemma 6.6.1, one can also write Theorem 6.6.4 as follows:

Theorem 6.6.5 (Completeness Theorem for the Equational Logic; *[Bir 35]***)**
For an arbitrary alphabet X and an arbitrary $\Sigma \subseteq T(X) \times T(X)$ it holds:

(a) $\Sigma = Cons_X(\Sigma) \Longleftrightarrow D(\Sigma) = \Sigma$;

(b) $D(\Sigma) = Cons_X(\Sigma)$.

Proof. (a): "$\Longrightarrow$": Let $\Sigma = Cons_X(\Sigma)$. Then, we have $D(\Sigma) \subseteq Cons_X(\Sigma) = \Sigma$ and $\Sigma \subseteq D(\Sigma)$. Thus $D(\Sigma) = \Sigma$.
"$\Longleftarrow$": Let $D(\Sigma) = \Sigma$. Then, by Theorem 6.6.1, Σ is a fully invariant congruence on $T(X)$. With the help of Theorem 6.6.4 it follows from this that Σ is an equational theory. Therefore, by Theorem 6.3.1, (6) $\Sigma = Cons_X(\Sigma)$.
(b): Let $\Sigma_1 := D(\Sigma)$. Then we have $D(\Sigma_1) = \Sigma_1$, $\Sigma \subseteq \Sigma_1$ and

$$D(\Sigma) \subseteq Cons_X(\Sigma) \subseteq Cons_X(\Sigma_1). \tag{6.2}$$

By $D(\Sigma_1) = \Sigma_1$ it follows from (a): $Cons_X(\Sigma_1) = \Sigma_1$. Then, because of the idempotency of D, we have:

$$D(\Sigma) = D(D(\Sigma)) = D(\Sigma_1) = Cons_X(\Sigma_1).$$

This and (6.2) imply $D(\Sigma) = Cons_X(\Sigma)$. ■

6.7 Finite Axiomatizability of Algebras

The reader needs knowledge of the other sections of this chapter, as well as some knowledge of Part II for this section.
An old question in universal algebra is whether, for given algebra $\mathbf{A}$ of finite type, there is a finite set $\Sigma \subseteq Id_X(\mathbf{A})$ with $D(\Sigma) = Id_X(\mathbf{A})$, where $X := \{x_1, x_2, x_3, ...\}$. For the case that $D(\Sigma) = Id_X(\mathbf{A})$ holds for a finite set Σ, we say that $\mathbf{A}$ is **finitely axiomatizable** or **finitely based**.

The following theorem is easy to prove.

Theorem 6.7.1 *([Lyn 51]) Let* $\mathbf{A} := (A; F)$, $\mathbf{B} := (B; G)$ *be finite and equivalent algebras of finite types; i.e.,* $|A| < \aleph_0$, $A = B$, $|F| < \aleph_0$, $|G| < \aleph_0$ *and* $TF(\mathbf{A}) = TF(\mathbf{B})$ [4]. *Then* $\mathbf{A}$ *is finitely axiomatizable if and only if* $\mathbf{B}$ *is.*

The next theorem was founded by G. Birkhoff.

Theorem 6.7.2 *Let* $\mathbf{A}$ *be a finite algebra of finite type* $(\mathfrak{F}, \tau)$ *and let* $X_n := \{x_1, x_2, ..., x_n\}$ *be a finite set of variables. Then* $Id_{X_n}(\mathbf{A})$ *is finitely axiomatizable.*

Proof. Let $\kappa := \{(s,t) \in T(X_n) \times T(X_n) \mid \mathbf{A} \models s \approx t\}$. Then $\kappa \in Con\mathbf{T}(\mathbf{X_n})$ by Theorem 6.4.2, (b). The congruence κ has only a finitely many equivalence classes $\varepsilon_1, ..., \varepsilon_q$, since A, $\mathfrak{F}$ and X_n are finite, and since $(s,t) \in \kappa$ iff the induced term functions $s_{\mathbf{A}}$, $t_{\mathbf{A}}$ satisfy $s_{\mathbf{A}} = t_{\mathbf{A}}$.
For $t \in T(X)$ we denote by $\#t$ the number of operation symbols ($\in \mathfrak{F}$) occurring in t. In particular, if $t \in X$ then $\#t = 0$. Now, from each equivalence class ε_i ($i \in \{1, ..., q\}$) of κ, we choose one representative r_i with $\#r_i \leq \#t$ for all terms $t \in \varepsilon_i$. Set

$$M := \{r_1, ..., r_q\} \cup \{f(r_{i_1}, r_{i_2}, ..., r_{i_{af}}) \mid f \in \mathfrak{F} \backslash \mathfrak{F}^0,\ \{r_{i_1}, ..., r_{i_{af}}\} \subseteq \{r_1, ..., r_q\}\},$$

$m := \max\{\#\varphi \mid \varphi \in M\}$ and

$$\Sigma := \{s \approx t \in Id_{X_n}(\mathbf{A}) \mid \#s \leq m \ \wedge\ \#\, t \leq m\}.$$

By induction on α we prove that

$$\forall \alpha \in \mathbb{N}_0 : \underbrace{(s \approx t \in Id_{X_n}(\mathbf{A}) \ \wedge\ \#s \leq \alpha \ \wedge\ \#t \leq \alpha) \implies s \approx t \in D(\Sigma)}_{=:S(\alpha)}. \tag{6.3}$$

(I) If $\alpha \leq m$, then the statement $S(\alpha)$ is obviously valid.
(II) Assume, $S(\beta)$ holds for certain $\beta \geq m$. Let $s \approx t \in Id_{X_n}(\mathbf{A})$ be arbitrary with $\#s \leq \beta + 1$ and $\#t \leq \beta + 1$.
First, we consider the case

$$s := f(s_1, s_2, ..., s_{af}),\ t := g(t_1, t_2, ..., t_{ag}), \tag{6.4}$$

where $f, g \in \mathfrak{F}$ and $s_1, ..., s_{af}, t_1, ...t_{ag} \in T(X_n)$. Then $\#s_1 + ... + \#s_{af} \leq \beta$ and $\#t_1 + ... + \#t_{ag} \leq \beta$. By definition of $r_1, ..., r_q$ there exist $u_1, u_2, ..., u_{af}, v_1, v_2, ..., v_{ag}, w \in \{1, ..., q\}$ with $[s_1]_\kappa = [r_{u_1}]_\kappa$, ..., $[s_{af}]_\kappa = [r_{u_{af}}]_\kappa$, $[t_1]_\kappa =$

[4] In other words, the operations of algebra $\mathbf{A}$ can be represented as superpositions over the operations of algebra $\mathbf{B}$ and vice versa (see Part II, Section 1.5.1).

$[r_{v_1}]_\kappa$, ..., $[t_{ag}]_\kappa = [r_{v_{ag}}]_\kappa$, $[f(s_1,...,s_{af})]_\kappa = [r_w]_\kappa = [g(t_1,...,t_{ag})]_\kappa$. Then, by assumption, we have

$$s_i \approx r_{u_i},\ t_j \approx r_{v_j} \in D(\Sigma)$$

and (by definition of Σ)

$$f(r_{u_1},...,r_{u_{af}}) \approx r_w, g(r_{v_1},...,r_{v_{ag}}) \approx r_w \in \Sigma.$$

With the aid of these equations and the rules (R1)–(R3), (Rep) and (Sub) we get

$$\begin{aligned} & \Sigma \vdash f(s_1,...,s_{af}) \approx f(s_1,...,s_n) \\ \Longrightarrow\ & \Sigma \vdash f(s_1,...,s_{af}) \approx f(r_{u_1},...,r_{u_{af}}) \\ \Longrightarrow\ & \Sigma \vdash f(s_1,...,s_{af}) \approx r_w \\ \Longrightarrow\ & \Sigma \vdash f(s_1,...,s_{af}) \approx g(r_{v_1},...,r_{v_{ag}}) \\ \Longrightarrow\ & \Sigma \vdash f(s_1,...,s_{af}) \approx g(t_1,...,t_{ag}), \end{aligned}$$

i.e., $(s,t) \in D(\Sigma)$ in case (6.4).
In the remaining cases (i.e., $\{s,t\} \cap X_n \neq \emptyset$), one can prove the above in analog mode as well. Thus, (6.3) is right, whereby $Id_{X_n}(\mathbf{A})$ is finitely axiomatizable.
Remark: The above proof shows that one can choose, instead of Σ, the following finite set Σ':

$$\begin{aligned} \Sigma' := &\ \{x \approx y \mid x,y \in X_m \ \wedge\ (x,y) \in \kappa\} \cup \\ &\ \{x \approx r \mid x \in X_n \ \wedge\ r \in \{r_1,...,r_q\} \ \wedge\ (x,r) \in \kappa\} \cup \\ &\ \{f(g_1,...,g_{af}) \approx g \mid f \in \mathfrak{F} \ \wedge\ \{g_1,...,g_{af},g\} \subseteq \{r_1,...,r_q\} \ \wedge \\ &\ \qquad (f(g_1,...,g_{af}),g) \in \kappa\} \end{aligned}$$

■

As a conclusion of Theorems 6.7.1 and 6.7.2 we get:

Theorem 6.7.3 *Let* $\mathbf{A} := (A;F)$ *be a finite algebra of finite type with* $[A] \subseteq [P_A^{(1)}]$*, i.e, the operations of* F *have at most an essential variable (see Part II, Chapter 1). Then* $\mathbf{A}$ *is finitely axiomatizable.* ■

The equational theory and parts of the mathematical logic deal with similar problems. Therefore, one can use sometimes results of the one theory for the other and vice versa. For this purpose, the next theorem provides an example. Because of the better survey, we agree with the following notation: If $\mathbf{A} = (\{0,1\};F)$, $f \in F$ defined by

$$f(x) := \neg x \quad \text{or} \quad f(x,y) := x \circ y \tag{6.5}$$

($\circ \in \{\vee, \wedge, \Rightarrow\}$) and t is a term, whose operation symbols belong to F, then

$$\hat{t}$$

denotes a formula of *Prop* (see Part II, Section 1.5.2), which one obtains by (6.5).

Theorem 6.7.4 *([Lyn 51]) Let* $\mathbf{A} = (\{0,1\}; 0, 1, f_1, f_2, f_3, g)$ *be an algebra of the type* $(0,0,1,2,2,2)$ *with* $f_1(x) := \neg x$, $f_2(x,y) := x \wedge y$, $f_3(x,y) := x \vee y$ *and* $g(x,y) := x \Rightarrow y$ *(see Table 1.2 of Part II). Further, let* T *be a set of tautologies* ($\subseteq$ *Prop) with the property that every tautology has a derivation from* T *with the help of sub and modus ponens.* [5] *Then, the equations*

(1) $x \Rightarrow x \approx 1$ ($g(x,x) \approx 1$),

(2) $1 \Rightarrow x \approx x$ ($g(1,x) \approx x$),

(3) $(x \Rightarrow y) \Rightarrow y \approx (y \Rightarrow x) \Rightarrow x$ ($g(g(x,y),y) \approx g(g(y,x),x)$)

and the equations of the type

(4) $t \approx 1$ *für every* $t \in T$

form a system Σ *of equations with* $D(\Sigma) = Id_X(\mathbf{A})$ *(more precise:* $D(\Sigma) = \{\hat{\alpha} \mid \alpha \in Id_X(\mathbf{A})\}$*).*

Proof. First, we show that

$$(\varphi \in \mathit{Prop} \text{ is a tautology}) \text{ implies } (\ \varphi \approx 1 \in D(\Sigma)\ ). \tag{6.6}$$

By assumption, we can derive a tautology φ from T with the aid of *sub* and modus ponens. Consequently, we have to prove that it is possible to copy the modus ponens through the rules (R1)–(R3), (Rep) and (Sub).
Let $\sigma \approx 1$, $\sigma \Rightarrow \tau \approx 1 \in D(\Sigma)$. Then $\tau \approx 1 \in D(\Sigma)$ follows from

$$\sigma \approx 1,\ \sigma \Rightarrow \tau \approx 1 \overset{Sub}{\vdash} 1 \Rightarrow \tau \approx 1 \overset{(2)}{\vdash} \tau \approx 1.$$

Thus (6.6) holds.
Now, let $\alpha \approx \beta \in Id_X(\mathbf{A})$ be arbitrary; i.e., $\alpha \Rightarrow \beta$ and $\beta \Rightarrow \alpha$ are tautologies. We show $\alpha \approx \beta \in D(\Sigma)$. By (6.6) we have that

$$\begin{array}{l} \alpha \Rightarrow \beta \approx 1 \text{ and} \\ \beta \Rightarrow \alpha \approx 1 \end{array} \tag{6.7}$$

belong to $D(\Sigma)$. Furthermore, by (3) and (Rep):

$$(\beta \Rightarrow \alpha) \Rightarrow \alpha \approx (\alpha \Rightarrow \beta) \Rightarrow \beta \in D(\Sigma).$$

Thus

$$(\beta \Rightarrow \alpha) \Rightarrow \alpha \approx (\alpha \Rightarrow \beta) \Rightarrow \beta \overset{(Sub),(6.7)}{\vdash} 1 \Rightarrow \alpha \approx 1 \Rightarrow \beta \overset{(2),(Sub)}{\vdash} \alpha \approx \beta,$$

whereby $\alpha \approx \beta \in D(\Sigma)$ is shown. ■

R. L. Lyndon proven the following basic result with the aid of Theorem 6.7.1 and Post's theorem (see Part II, Theorem 3.1.1):

---

[5] see Part II, Section 1.5.2.

Theorem 6.7.5 *([Lyn 51], without proof)*
Every two-element algebra of finite type is finitely axiomatizable.

Notice that J. Berman gave a short proof for the above theorem in [Ber 80] with the aid of theorems by Baker ([Bak 77]) and McKenzie ([McK 78]).

R. C. Lyndon constructed a 7-element algebra of type $(0, 2)$ whose equations are not finitely based (see [Lyn 54]). The smallest such example was found by V. L. Murskij:

Theorem 6.7.6 *([Mur 65]; without proof) The algebra* $(\{0, 1, 2\}; \circ)$, *where* $\circ$ *is defined by*

$\circ$	0	1	2
0	0	0	0
1	0	0	1
2	0	2	2

is not finitely axiomatizable.

Notice that all finite groups, rings and lattices are finitely axiomatizable (see [Oat-P 65], [Kru 73], and [McK 70], respectively). The result for lattices was considerably generalized by K. Baker ([Bak 77]), who proven that every finite algebra whose generated variety is congruence distributive is finitely axiomatizable. In [Per 69] was proven that the multiplicative semigroup of all 2×2-matrices over a 2-element field has a 6-element subsemigroup with no finite basis.

One can find further important results on the topic in [McK 78] and [Wil 2001].

Part II

Function Algebras

1

Basic Concepts, Notations, and First Properties

In this chapter, we begin by investigating multi-digit operations, which are defined on a finite set A. We define some operations on the set of these operations. For the purpose of distinction, we subsequently replace the concept "operation (on the set A)" by the concept "*function (on the set A)*".

1.1 Functions on Finite Sets

Let A be a finite set with at least two elements. Often we choose the set

$$E_k := \{0, 1, 2, ..., k-1\}, \ k \geq 2$$

instead of A in the following.
We say that f is an **n-ary (n-digit) function on A** (or an **n-ary function of the $|A|$-valued logic**[1]), if f a mapping from the n-fold Cartesian product A^n into A, $n \geq 1$. For technical reasons (see Section 1.3), we renounce that, in this section, we consider also nullary functions. Otherwise, we use the concepts introduced in Chapter 1 of Part I for operations: af, f^n, $D(f)$, ...

Let P_A^n be the set of all n-ary functions on A [2], $n \geq 1$. Instead of $P_{E_k}^n$ we write also P_k^n. Further let

[1] One finds an explanation for this notation in Section 1.5.
[2] A confusion with the direct product is not possible for content-related reasons.

$$P_A := \bigcup_{n\geq 1} P_A^n,$$
$$F^n := F \cap P_A^n \text{ for every subset } F \text{ of } P_A,$$
$$P_k := \bigcup_{n\geq 1} P_k^n,$$
$$P_{A,B} := \{f \in P_A \mid Im(f) \subseteq B\},$$
$$P_{k,l} := P_{E_k,E_l},$$
$$P_A(l) := \{f \in P_A \mid |Im(f)| \leq l\},$$
$$P_k(l) := P_{E_k}(l),$$
$$P_A[l] := \{f \in P_A \mid |Im(f)| = l\} \text{ and}$$
$$P_k[l] := P_{E_k}[l] \ (2 \leq l \leq k).$$

With $(x_1, ..., x_n)$ (briefly $\mathbf{x^{(n)}}$) or $\mathbf{x}$ we denote an arbitrary n-tuple of A^n or E_k^n and usually we say that the x_i $(i = 1, 2, ..., n)$ are **variables**. If $n = 2$ or $n = 3$ we also write (x, y) or (x, y, z) instead of $(x_1..., x_n)$, respectively. We will define, subsequently, specific functions f^n from P_k either through a table of the form

Table 1.1

x_1	x_2	...	x_n	$f(x_1, x_2, ..., x_n)$
0	0	...	0	$f(0, 0, ..., 0)$
0	0	...	1	$f(0, 0, ..., 1)$
.	.	...	.	.
a_1	a_2	...	a_n	$f(a_1, a_2, ..., a_n)$
.	.	...	.	.
$k-1$	$k-1$	...	$k-1$	$f(k-1, k-1, ..., k-1)$

or through formulas, for example, of the form

$$\forall \mathbf{x} \in E_k^n : \ f(x_1, ..., x_n) := x_1 + ... + x_n \ (mod\ k), \tag{1.1}$$

on the (variable-) alphabet $\{x, y, z, x_1, x_2, ...\}$. We write often instead of (1.1) briefly

$$f(x_1, ..., x_n) := x_1 + ... + x_n (mod\ k) \tag{1.2}$$

or (if the arity of f is clear from the context or is without importance) we write still more briefly

$$\text{“}f \text{ is defined by } x_1 + ... + x_n (mod\ k)\text{”}; \tag{1.3}$$

i.e., we do not distinguish between a function and the formula (or term) defining it.[3]

[3] One finds the concept “term” explained in Part I, Section 6.2.

Two functions $f^n, g^m \in P_A$ are identical (we write $f^n = g^m$) iff $n = m$ and $f(\mathbf{x}) = g(\mathbf{x})$ for all $\mathbf{x} \in A^n$ hold.
Let $f^n \in P_A$ and $i \in \{1, 2, ..., n\}$. Then we say that the i-th variable (or the i-th place) of the function $f \in P_A$ is **essential**, iff there are n-tuples

$$\mathbf{a} = (a_1, ..., a_{i-1}, b, a_{i+1}, ..., a_n) \text{ and}$$
$$\mathbf{a}' = (a_1, ..., a_{i-1}, c, a_{i+1}, ..., a_n)$$

such that $b \neq c$ and $f(\mathbf{a}) \neq f(\mathbf{a}')$ hold. In the opposite case, one calls the i-th variable (or i-th place) of f **fictitious** (or **non-essential**).
If the i-th variable of f is not fictitious, we say that f **depends** on the i-th variable.

The function e_i^n defined by

$$e_i^n(x_1, ..., x_n) := x_i$$

($i \in \{1, ..., n\}$) is called **projection** or also **selector**. Let J_A (or J_k) be the set of all projections of P_A (or P_k), respectively.

A **constant function** (briefly, **constant**) is a function c_a^n defined by

$$c_a^n(x_1, ..., x_n) := a,$$

where $a \in A$.
Notations for certain functions of P_2, the **Boolean functions**, are given in the following table, where it is defined, as usual

$$\circ(x, y) := x \circ y \text{ if } \circ \in \{\wedge, \vee, +, \Rightarrow, \Longleftrightarrow\}$$

and

$$^{-}(x) := \overline{x}.$$

Table 1.2

x	$\overline{x}$	x	y	$x \wedge y$	$x \vee y$	$x + y$	$x \Rightarrow y$	$x \Longleftrightarrow y$
0	1	0	0	0	0	0	1	1
1	0	0	1	0	1	1	1	0
		1	0	0	1	1	0	0
		1	1	1	1	0	1	1

Instead of $x \wedge y$ we also write $x \cdot y$ or we write xy briefly. It can easily be shown that the above functions have the following properties:

Theorem 1.1.1 *It holds:*

(a) $\forall \circ \in \{\vee, \wedge, \Longleftrightarrow, +\} : x \circ (y \circ z) = (x \circ y) \circ z$;

(b) $x \vee x = x, x \wedge x = x, x \Longleftrightarrow x = 1, x \Rightarrow x = 1, x + x = 0,$
$x \vee 0 = x, x \wedge 1 = x, x \vee 1 = 1, x \wedge 0 = 0$;

(c) $\forall \circ \in \{\vee, \wedge, +, \Longleftrightarrow\} : x \circ y = y \circ x;$

(d) $x \wedge \overline{x} = 0,\ x \vee \overline{x} = 1, \overline{\overline{x}} = x,$
$\overline{x \vee y} = \overline{x} \wedge \overline{y},\ \overline{x \wedge y} = \overline{x} \vee \overline{y}$ *("de Morgan's laws");*

(e) $x \Rightarrow y = \overline{x} \vee y,\ \overline{x \Rightarrow y} = x \wedge \overline{y};$

(f) $x \wedge (y \vee z) = (x \wedge y) \vee (x \wedge z), x \vee (y \wedge z) = (x \vee y) \wedge (x \vee z);$

(g) $x \wedge (x \vee y) = x, x \vee (x \wedge y) = x.$ ■

1.2 Operations on P_A, Function Algebras

The "formula notation" of our functions from the first section motivates the determination of the following operations on P_A:

- permutation of variables,
- identification of variables,
- adding of fictitious variables and
- substitution of variables of a function by functions;

which are called **superposition operations** on P_A and which one can describe in different way exactly. We give only two possibilities here.
First we want describe the above operations through the following (infinite many) partial operations:

$$\pi_s : P_A^n \longrightarrow P_A^n$$
$$\Delta_t : P_A^n \longrightarrow P_A^r \ (r < n)$$
$$\nabla_q : P_A^n \longrightarrow P_A^u \ (u > n)$$
$$\star_i : P_A^n \times P_A^m \longrightarrow P_A^{n+m-1}$$

Let f^n, g^m be functions of P_A, let s be a permutation on the set $\{1, 2, ..., n\}$, let t be a mapping from $\{1, 2, ..., n\}$ onto $\{1, 2, ..., r\}$ $(r < n)$, let q be an injective mapping from $\{1, 2, ..., n\}$ into $\{1, 2, ..., u\}$ $(u > n)$ and let $i \in \{1, 2, ..., n\}$. Then, $\pi_s f \in P_A^n$, $\Delta_t f \in P_A^r$, $\nabla_q f \in P_A^u$, $f \star_i g \in P_A^{m+n-1}$ are defined by

$$(\pi_s f)(x_1, ..., x_n) := f(x_{s(1)}, x_{s(2)}, ..., x_{s(n)})$$
("permutation of variables of f"),
$$(\Delta_t f)(x_1, ..., x_r) := f(x_{t(1)}, x_{t(2)}, ..., x_{t(n)})$$
("identification of certain variables of f"),
$$(\nabla_q f)(x_1, x_2, ..., x_u) := f(x_{q(1)}, x_{q(2)}, ..., x_{q(n)})$$
("adding of certain fictitious variables")

and

$$(f \star_i g)(x_1, ..., x_{m+n-1}) :=$$

$$f(x_1, ..., x_{i-1}, g(x_i, ..., x_{i+m-1}), x_{i+m}, ..., x_{m+n-1})$$

("the replacement of the i-th variable of f through the function g and the changing of the denotation of variables of f").

For partial operations $\alpha \in \{\pi_s, \Delta_t, \nabla_q, \star_i\}$ defined above, one can continue to certain operations α' on P_A. For later investigations, however, it is better that we choose the minimal number of the operations on P_A. Therefore, next we consider that there are five **elementary operations** (or **Mal'tsev-operations**) $\zeta, \tau, \Delta, \nabla, \star$ on P_A, with which we can form certain continuations of the partial operations $\pi_s, \Delta_t, \nabla_q, \star_i$ for arbitrary s, t, q, i, n, m through composition. These operations were published by A. I. Mal'tsev in [Mal 66], and one can define these operations as follows for arbitrary $f^n, g^m \in P_A$:

$$\zeta f^n \in P_A^n, \tau f^n \in P_A^n, \Delta f^n \in P_A^{max\{1,n-1\}}, \nabla f^n \in P_A^{n+1}, f^n \star g^m \in P_A^{m+n-1}$$

and

$$(\zeta f)(x_1, ..., x_n) := f(x_2, x_3, ..., x_n, x_1),$$

$$(\tau f)(x_1, ..., x_n) := f(x_2, x_1, x_3, ..., x_n),$$

$$(\Delta f)(x_1, ..., x_{n-1}) := f(x_1, x_1, x_2, ..., x_{n-1}) \text{ if } n \geq 2,$$

$$\zeta f = \tau f = \Delta f = f \text{ if } n = 1,$$

$$(\nabla f)(x_1, ..., x_{n+1}) := f(x_2, x_3, ..., x_{n+1}),$$

$$(f \star g)(x_1, ..., x_{m+n-1}) := f(g(x_1, ..., x_m), x_{m+1}, ..., x_{m+n-1}).$$

To prove that one can describe the operation π_s (over P_A^n) by means of operations ζ, τ (over P_A), it suffices to show that the set S_n of all permutations on the set $\{1, 2, ..., n\}$ is generating by the permutations $(12...n)$ and (12) (given in cyclic description); that is, that $[\{(12...n), (12)\}]_\square = S_n$ holds, where $(s \square s')(x) := s'(s(x))$ for all $s, s' \in S_n$. But, this follows from the fact that every permutation $s \in S_n$ is a product of pairwise disjunct cycles:

$$s = (i_1...i_p)(j_1...j_q)...,$$

that every cycle is a product of transpositions

$$(i_1...i_p) = (i_1 i_2)(i_1 i_3)...(i_1 i_p),$$

and that

$$(ij) = (1i)(1j)(i1) \text{ for arbitrary } i > j, i > 1,$$

$$(1i) = (12)(23)...(i-1i)(i-1, i-2)...(21) \text{ and}$$

$$(i, i+1) = (12...n)^{n-i+1}(12)(12...n)^{i-1}$$

are valid. Evidently, then, the operations Δ_t or ∇_q or $\star_i$ can be created (in respect to $\square$) through the operations π_s $(s \in S_n), \Delta$ or π_s $(s \in S_n), \nabla$ or π_s $(s \in S_n), \star$ respectively. Thus we proven the following lemma.

Lemma 1.2.1 *It holds:*

(a) *For every permutation $s \in S_n$ there exists an operation $\widetilde{\pi} \in [\{\zeta, \tau\}]_\square$ with $\pi_s f = \widetilde{\pi}_s f$ for all $f \in P_A^n$.*

(b) *For every mapping t from $\{1, 2, ..., n\}$ onto $\{1, 2, ..., r\}$ $(r < n)$ there exists an operation $\widetilde{\Delta} \in [\{\zeta, \tau, \Delta\}]_\square$ with $\Delta_t f = \widetilde{\Delta}_t f$ for arbitrary $f \in P_A^n$.*

(c) *For every injective mapping q from $\{1, 2, ..., n\}$ into $\{1, 2, ..., u\}$ there exists an operation $\widetilde{\nabla}_q \in [\{\zeta, \tau, \nabla\}]_\square$ with $\nabla_q f = \widetilde{\nabla}_q f$ for arbitrary $f \in P_A^n$.*

(d) *For every $i \in \{1, 2, ..., n\}$, there exists an operation $\widetilde{\star}_i \in [\{\zeta, \tau, \star\}]_\square$ with $f \star_i g = f \widetilde{\star}_i g$ for arbitrary $f \in P_A^n$ and arbitrary $g \in P_A^m$.* ■

By means of the operations $\zeta, \tau, \Delta, \nabla, \star$ we can describe the subject of investigation of this chapter.
P_A together with the operations $e_1^2, \zeta, \tau, \Delta, \star$ forms an algebra

$$(P_A; e_1^2, \zeta, \tau, \Delta, \star)$$

of the type $(0, 1, 1, 1, 2)$, which is called **(full) function algebra** on A.

A little changed form of the full function algebra is the so-called **iterative (full) function algebra** $(P_A; \zeta, \tau, \Delta, \nabla, \star)$ of the type $(1, 1, 1, 1, 2)$. Since $\nabla f = f \star (\tau e_1^2)$ is valid, however, both algebras can be regarded as equivalent in a certain sense: If $(S; e_1^2, \zeta, \tau, \Delta, \star)$ is a subalgebra of $(P_A; e_1^2, \zeta, \tau, \Delta, \star)$, then $(S; \zeta, \tau, \Delta, \nabla, \star)$ is also a subalgebra of $(P_A; \zeta, \tau, \Delta, \nabla, \star)$. Conversely, if $(T; \zeta, \tau, \Delta, \nabla, \star)$ is a subalgebra of $(P_A; \zeta, \tau, \Delta, \nabla, \star)$, then $(T \cup J_A; e_1^2, \zeta, \tau, \Delta, \star)$ is a subalgebra of $(P_A; e_1^2, \zeta, \tau, \Delta, \star)$. Therefore, we often deal only with the algebra

$$\mathbf{P_A} = (P_A; \zeta, \tau, \Delta, \nabla, \star).$$

1.3 Superpositions, Subclasses, and Clones

A function $f \in P_A$ is called a **superposition** over F $(\subseteq P_A)$, if f can be obtained by a finite number of applications of the operations $\zeta, \tau, \Delta, \nabla, \star$ from the functions of F.
We describe a superposition f over F in the rarest cases f through a term over certain function symbols, $\zeta, \tau, \Delta, \nabla, \star$ and parentheses. We use the variable alphabet

$$\{x, y, z, x_1, x_2, ...\},$$

certain function symbols, commas, and parenthesis instead of this. In some cases, where an equation for the precise definition of the function would be necessary formally, we are satisfied with the the right side of the defining equation if the remaining information on the function results from the context. Further, if $f \in P_k^n$, $g_1, ...g_n \in P_k^m$ and the m-ary function $h \in P_k$ is defined by

$$h(x_1, ..., x_m) := f(g_1(x_1, ..., x_m), g_2(x_1, ..., x_m), ..., g_n(x_1, ..., x_m)),$$

then, we write briefly

$$h := f(g_1, ..., g_n).$$

The set of all superpositions over F ($\subseteq P_A$) is called **hull** or **closure** of F and it is denoted by $[F]$.
Obviously, $[..]$ is a hull operator on the set P_A. A set $F \subseteq P_A$ satisfying $[F] = F$ is called a **closed set** or a **subclass** or briefly **class** of P_A. We define that the empty set is also a closed set, i.e., $\emptyset = [\emptyset]$.
One can form many examples of closed sets with the aid of the following concept:
Let $T \subseteq P_k^m$ and $f \in P_k^n$. Then we say that f **preserves the set** T iff

$$\forall g_1, ..., g_n \in T : \ f(g_1, ..., g_n) \in T.$$

It is easy to see that a projection preserves every set $T \subseteq P_k^m$ and that the set of all functions, which preserve the set T, is closed.

The set $F \subseteq P_A$ is called a **clone** of P_A, if F is closed and $J_A \subseteq F$ holds.[4] Obviously, the subclasses of P_A are exactly the universes of subalgebras of $(P_A; \zeta, \tau, \Delta, \nabla, \star)$ and clones are exactly the universes of subalgebras of $(P_A; e_1^2, \zeta, \tau, \Delta, \star)$.
Let $\mathbb{L}_A$ be the set of all closed subsets of P_A. Put

$$\mathbb{L}_k := \mathbb{L}_{E_k}.$$

$(\mathbb{L}_A; \subseteq)$ is a lattice (see Part I, Chapter 2).
Further, let

$$\mathbb{L}_A^{\downarrow}(F) := \{F' \in \mathbb{L}_A \mid F' \subseteq F\}$$

and

$$\mathbb{L}_A^{\uparrow}(F) := \{F' \in \mathbb{L}_A \mid F \subseteq F'\}.$$

Analogously, one can define $\mathbb{L}_k^{\uparrow}(F)$ and $\mathbb{L}_k^{\downarrow}(F)$ for $A = E_k$. Further, let

$$\mathbb{L}_A(F; G) := \mathbb{L}_A^{\uparrow}(F) \cap \mathbb{L}_A^{\downarrow}(G),$$

where $F, G \in \mathbb{L}_A$ and $F \subset G$.

---

[4] If F is a class of P_A, then $J_A \subseteq F$ iff $e_1^1 \in F$.

If $[G] = F$ $(\subseteq P_A)$, then G is called **complete in** F. In particular, if $F = P_A$, we say, G is **complete** or G is a **complete set**.
A closed set F is called a **maximal subclass of the closed set** F', if $F \subset F'$ and $[F \cup \{f\}] = F'$ for every $f \in F' \backslash F$. If $F' = P_A$ then, we say briefly, F is a **maximal class**. The maximal classes of the maximal classes of P_A are called **submaximal classes**.
As usual, we call a subset F' of F a **generating system** of F, if $[F'] = F$. A generating system F' of F is called **basis** of the closed set F, if every proper subset of F' is not a generating system of F. If a subclass F of P_A has a finite generating system, then the **order** of F we denote with

$$ord\ F.$$

We understand from that, the smallest number with $[F^r] = F$. If F does not have any finite generating system, we write $ord\ F = \infty$.

1.4 Generating Systems for P_A

For the purpose of determining of generating systems for the set P_A, we consider some descriptions (so-called "**normal forms**") for an arbitrary function $f^n \in P_A$. These descriptions are superpositions over certain functions of P_A, which are to be described easily and which have small arities.
We use the following notations:

$$j_a(x) := \begin{cases} 1 \text{ if} & x = a, \\ 0 \text{ otherwise} & \end{cases}$$

$(a \in A)$ and

$$j_{\mathbf{a}}(x_1, ..., x_n) := \begin{cases} 1 \text{ if} & (x_1, ..., x_n) = \mathbf{a}, \\ 0 \text{ otherwise} & \end{cases}$$

$(\mathbf{a} \in A^n, n \in \mathbb{N})$.

Theorem 1.4.1 (Representation Theorem for Functions of P_A)
Let $0, 1 \in A$ and let $\wedge$, $\vee$ be two binary associative[5] operations on A with

$$a \wedge 1 = a,\ 0 \vee a = a \vee 0 = a \text{ and } a \wedge 0 = 0 \tag{1.4}$$

for each $a \in A$. Then, for every function $f^n \in P_A$ it holds:

$$f(\mathbf{x}) = (\bigvee_{i=1}^{m} f_{\mathbf{a}_i}(\mathbf{x}) :=) f_{\mathbf{a}_1}(\mathbf{x}) \vee f_{\mathbf{a}_2}(\mathbf{x}) \vee ... \vee f_{\mathbf{a}_m}(\mathbf{x}), \tag{1.5}$$

where $A^n := \{\mathbf{a}_1, ..., \mathbf{a}_m\}$, $m := |A|^n$ and

$$f_{\mathbf{a_i}}(\mathbf{x}) := c_{f(\mathbf{a_i})}(x_1) \wedge j_{\mathbf{a_i}}(\mathbf{x}) \ (i = 1, ..., m).$$

Furthermore:

$$j_{\mathbf{a_i}}(\mathbf{x}) = j_{a_{i1}}(x_1) \wedge j_{a_{i2}}(x_2) \wedge ... \wedge j_{a_{in}}(x_n),$$

where $\mathbf{a_i} := (a_{i1}, ..., a_{in})$.

Proof. The correctness of the equation (1.5) can be assured by checking that on both the left side and the right side of the formula, the same value stands for every $\mathbf{x}$. ■

If $A = \{0, 1\}$, the functions $\vee$, $\wedge$ $(= \cdot)$ are defined as in Table 1.2, $j_0(x) = \overline{x}$ and $j_1(x) = x$, then one receives the **disjunctive normal form** (or **DNF**) of an arbitrary Boolean function $f^n \in P_2$ as an conclusion from (1.5):

$$f(x_1, ..., x_n) = \bigvee_{\mathbf{a} \in E_2^n} f(a_1, ..., a_n) \cdot x_1^{a_1} \cdot x_2^{a_2} \cdot ... \cdot x_n^{a_n}, \qquad (1.6)$$

where

$$x^\alpha := \begin{cases} \overline{x} \text{ if } \alpha = 0, \\ x \text{ if } \alpha = 1 \end{cases}$$

$(\alpha \in E_2)$. If $f \neq c_0^n$, then we can write

$$f(x_1, ..., x_n) = \bigvee_{\mathbf{a} \in E_2^n, f(\mathbf{x})=1} x_1^{a_1} \cdot x_2^{a_2} \cdot ... \cdot x_n^{a_n} \qquad (1.7)$$

instead of (1.6). For example,

$$f(x, y, z) = \overline{x} \cdot y \cdot \overline{z} \vee x \cdot \overline{y} \cdot \overline{z} \vee x \cdot y \cdot z$$

is the DNF for the ternary function f defined by Table 1.3.

Table 1.3

x	y	z	$f(x,y,z)$
0	0	0	0
0	0	1	0
0	1	0	1
0	1	1	0
1	0	0	1
1	0	1	0
1	1	0	0
1	1	1	1

[5] The associativity can be renounced if the necessary parentheses are put in the following formulas.

If A is an arbitrary finite set, then one can choose $\wedge$ and $\vee$ as lattice operations with $0 = \bigwedge A$ and $1 = \bigvee A$. Then, by Theorem 2.1.2 of Part I, we have

$$\vee(x, y) = \sup_{\varrho}\{x, y\} \text{ and}$$
$$\wedge(x, y) = \inf_{\varrho}\{x, y\},$$

where ϱ is the partial order (appertaining to the lattice) over A with the greatest element 1 and the least element 0.
If $A = E_k$, then the functions

$$\vee := + \ (mod\ k) \text{ and } \wedge := \cdot \ (mod\ k)$$

also fulfill (1.4) and we get the following normal form for an arbitrary function $f^n \in P_k$:

$$f(\mathbf{x}) = \sum_{\mathbf{a} \in E_k^n} f(a_1, ..., a_n) \cdot j_{a_1}(x_1) \cdot ... \cdot j_{a_n}(x_n) \ (mod\ k) \tag{1.8}$$

The following theorem results from the above considerations immediately:

Theorem 1.4.2 *It holds:*

(a) Let $\vee$ and $\wedge$ be binary operations on A, which (1.4) fulfill. Then, $\{\vee, \wedge\} \cup \{c_a^1, j_a^1 \mid a \in A\}$ is a generating system for P_A.

In particular, if $A = E_2$, then $[\{\vee, \wedge, ^-\}] = P_2$ and (because of Theorem 1.1.1, (d)) $[\{\vee, ^-\}] = [\{\wedge, ^-\}] = P_2$.

(b) ord $P_A = 2$. ■

Theorem 1.4.3 *Let $A = E_k$ and let $k = p^m$ be a prime number power. Then one can define operations + and $\cdot$ on E_k so that $(E_k; +, \cdot)$ is a field with the neutral element o in respect to + and the neutral element e of the group $(E_k \backslash \{o\}; \cdot)$.*
Then, one can represent an arbitrary function $f^n \in P_k$ with the aid of these field operations as follows:

$$f(\mathbf{x}) = \sum_{(i_1,...,i_n) \in E_k^n} a_{i_1 i_2 ... i_n} \cdot x_1^{i_1} \cdot x_2^{i_2} \cdot ... \cdot x_n^{i_n} \tag{1.9}$$

($x^0 := e$; $a_{i_1 i_2 ... i_n} \in E_k$). This representation is unique except the order of addends, i.e., the equality of the corresponding coefficients results from the equality of two functions $\in P_k^n$.

Proof. The existence of a field $(E_k; +, \cdot)$ for $k = p^m$, $p \in \mathbb{P}$ and $m \in \mathbb{N}$ is well-known. (see for example [Lid-N 87] or [Lau 2004], volume 2).

Every polynomial of the form (1.9) is uniquely represented through the sequence of coefficients. Thus there are $k^{(k^n)}$ different formulas of the form (1.9). Since $|P_k^n| = k^{(k^n)}$, our theorem is proven if

$$f(\mathbf{x}) = \sum_{(i_1,...,i_n)\in E_k^n} a_{i_1 i_2 ... i_n} x_1^{i_1} x_2^{i_2} ... x_n^{i_n} \text{ and}$$
$$f(\mathbf{x}) = \sum_{(i_1,...,i_n)\in E_k^n} b_{i_1 i_2 ... i_n} x_1^{i_1} x_2^{i_2} ... x_n^{i_n}$$

implies $a_{i_1...i_n} = b_{i_1...i_n}$ for all $(i_1, ..., i_n) \in E_k^n$. This is clear for $(i_1, ..., i_n) = (o, o, ..., o)$ (one forms $f(o, ..., o)$!). Let $I := \{x_{i_j} \mid i_j \neq o \wedge j \in \{1, ..., n\}\}$ for the proof of $a_{i_1...i_n} = b_{i_1...i_n}$ in the case $(i_1, ..., i_n) \in E_k^n \backslash \{\mathbf{o}\}$. If one identifies now the variables in f from I with x and one replaces the remaining variables through $c_0(x)$, then one receives a unary function f' that can be represented as follows:

$$f'(x) = a_0 + a_1 \cdot x + a_2 \cdot x^2 + ... + a_{r-1} \cdot x^{r-1} \tag{1.10}$$

or

$$f'(x) = b_0 + b_1 \cdot x + b_2 \cdot x^2 + ... + b_{r-1} \cdot x^{r-1} \tag{1.11}$$

with $r-1 := |I|$, $a_{r-1} = a_{i_1...i_n}$ and $b_{r-1} = b_{i_1...i_n}$ for certain $a_0, ..., a_{r-2}, b_0, ..., b_{r-2}$. If one forms now in (1.10) and (1.11)

$$f'(\alpha_1),\ f'(\alpha_2),\ ...,\ f'(\alpha_r)$$

for pairwise distinct $\alpha_1, \alpha_2, ..., \alpha_r \in E_k$, then one sees that both $(a_0, ..., a_{r-1})^T$ as also $(b_0, ..., b_{r-1})^T$ is a solution of the matrix equation $\mathfrak{A} \cdot \mathfrak{x} = (f'(\alpha_1), ..., f'(\alpha_r))^T$ with

$$\mathfrak{A} := \begin{pmatrix} 1 & \alpha_1 & \alpha_1^2 & ... & \alpha_1^{r-1} \\ 1 & \alpha_2 & \alpha_2^2 & ... & \alpha_2^{r-1} \\ \multicolumn{5}{c}{\dots\dots\dots\dots\dots} \\ 1 & \alpha_r & \alpha_r^2 & ... & \alpha_r^{r-1} \end{pmatrix}.$$

But, because of $det\ \mathfrak{A} \neq o$,[6] this is only possible for $a_0 = b_0, ..., a_{r-1} = b_{r-1}$. ■

The next property of functions of P_k for $k \geq 3$ is not only useful while determining from generating systems for P_k; it also has interesting consequences, which we will deal with later. We need the following denotation:[7]

$$\iota_k^h := \{(a_1, ..., a_h) \in E_k^h \mid |\{a_1, ..., a_h\}| \leq h - 1\}\ (h \geq 2),$$
$$\delta^3_{\{\alpha,\beta\}} := \{(a_1, a_2, a_3) \in E_k^3 \mid a_\alpha = a_\beta\}\ (\alpha, \beta \in \{1, 2, 3\}) \text{ and}$$
$$\delta^3_{\{1,2,3\}} := \{(x, x, x) \mid x \in E_k\}.$$

[6] This follows from the fact that $\mathfrak{A}$ is a Vandermonde matrix (see for example [Lau 2004], volume 1).

[7] See also Chapter 2.

Furthermore, for arbitrary $\mathbf{r_i} := (r_{1i}, r_{2i}, ..., r_{hi}) \in E_k^h$, $i = 1, 2, .., n$, and $f \in P_k^n$ we put:

$$f(\mathbf{r_1}, ..., \mathbf{r_n}) := (f(r_{11}, r_{12}, ..., r_{1n}), f(r_{21}, r_{22}, ..., r_{2n}), .., f(r_{h1}, r_{h2}, ..., r_{hn})).$$

Theorem 1.4.4 *Let f be an n-ary function of P_k, which is essentially dependent of at least two variables (w.l.o.g. of x_1 and x_2) and which has q pairwise distinct values. Then:*

(a) $q \geq 3 \implies \exists \mathbf{r_1}, ..., \mathbf{r_n} \in \delta^3_{\{1,2\}} \cup \delta^3_{\{2,3\}} : f(\mathbf{r_1}, ..., \mathbf{r_n}) \in E_k^3 \setminus \iota_k^3$

("Fundamental Lemma of Jablonskij"); (Jab 58])

(b) $q \geq 3 \implies \exists \mathbf{r_1}, ..., \mathbf{r_n} \in \iota_k^q : f(\mathbf{r_1}, ..., \mathbf{r_n}) \in E_k^q \setminus \iota_k^q$;

(c) $q = 2 \implies \exists \mathbf{r_1}, ..., \mathbf{r_n} \in \delta^3_{\{1,2\}} \cup \delta^3_{\{2,3\}} : f(\mathbf{r_1}, ..., \mathbf{r_n}) \in \delta^3_{\{1,3\}} \setminus \delta^3_{\{1,2,3\}}$.

Proof. (a), (c): Since f depends on the variable x_1 essentially, there is an $\mathbf{a} := (a_2, ..., a_n) \in E_k^{n-1}$, so that

$$T_{\mathbf{a}} := \{f(x, a_2, ..., a_n) \mid x \in E_k\}$$

has at least two different elements. We distinguish two cases:
Case 1: $|T_{\mathbf{a}}| < q$.
In this case, one can find a tuple $\mathbf{c} = (c_1, ..., c_n)$ with $\gamma := f(\mathbf{c}) \notin T_{\mathbf{a}}$. Consequently, we have

$$f\begin{pmatrix} a_1 & a_2 & ... & a_n \\ c_1 & a_2 & ... & a_n \\ c_1 & c_2 & ... & c_n \end{pmatrix} := \begin{pmatrix} f(a_1, a_2, ..., a_n) \\ f(c_1, a_2, ..., a_n) \\ f(c_1, c_2, ..., c_n) \end{pmatrix} = \begin{pmatrix} \alpha \\ \beta \\ \gamma \end{pmatrix}$$

and $|\{\alpha, \beta, \gamma\}| = 3$ for certain $a_1 \in E_k$.
Case 2: $|T_{\mathbf{a}}| = q$.
Since f also depends on x_2 essentially, the function $f_1(x_1, ..., x_{n-1}) := f(d, x_1, ..., x_{n-1})$ is not a constant function for certain $d \in E_k$. Let now $\beta' := f(d, a_2, ..., a_n)$. Because of $f_1 \neq c_{\beta'}$ there are certain $c'_2, ..., c'_n$ and a γ' with $\gamma' := f(d, c'_2, ..., c'_n) \neq \beta'$. Since $|T_{\mathbf{a}}| = q$, then one can find an $a'_1 \in E_k$ with

$$\alpha' := f(a'_1, a_2, ..., a_n) = \begin{cases} \gamma' & \text{if } q = 2, \\ \alpha' \notin \{\beta', \gamma'\} & \text{if } q \geq 3. \end{cases}$$

Consequently, we have

$$f\begin{pmatrix} a'_1 & a_2 & ... & a_n \\ d & a_2 & ... & a_n \\ d & c'_2 & ... & c'_n \end{pmatrix} = \begin{pmatrix} \alpha' \\ \beta' \\ \gamma' \end{pmatrix}.$$

(b) follows easy from (a). ■

From the many conclusions from this theorem, we provide only the following, first.

Lemma 1.4.5 *Let f^n be a function of P_k, which is essentially dependent on two variables and which has $q \geq 3$ distinguish values. Then:*

$$P_{k,Im(f)} \subseteq [\{f\} \cup P_k(q-1)].$$

Proof. W.l.o.g. let $Im(f) = E_q$. By Theorem 1.4.4, (b) there exist $r_1, ..., r_n \in \iota_k^q$ with $f(r_1, ..., r_n) = (0, 1, ..., q-1)^T$ and

$$(r_1^T, ..., r_n^T) = \begin{pmatrix} a_{01} & a_{02} & \cdots & a_{0n} \\ a_{11} & a_{12} & \cdots & a_{1n} \\ \cdots & \cdots & \cdots & \cdots \\ a_{q-1,1} & a_{q-1,2} & \cdots & a_{q-1,n} \end{pmatrix}.$$

For an arbitrary function $g^m \in P_{k,Im(f)}$ let

$$g_j(x_1, ..., x_m) = a_{ij} \;:\Longleftrightarrow\; \exists i : g(x_1, ..., x_m) = i,$$

$(j = 1, 2, ..., n)$. Obviously, the functions $g_1, ..., g_n$ belong to $P_k(q-1)$ and it holds:

$$g(x_1, ..., x_m) = f(g_1(x_1, ..., x_m), ..., g_n(x_1, ..., x_m)).$$

Thus $g \in [\{f\} \cup P_k(q-1)]$. ■

Subsequently, we declare another possibility of the characterization of functions of P_k that we need later during the description of a certain type of maximal classes of P_k.

Lemma 1.4.6 *Let $\mathfrak{A}_i$ $(i \in E_{k'})$ be a partition of the set E_k and $a_i \in \mathfrak{A}_i$ $(i \in E_{k'})$. Furthermore, let*

$$x \diamond y := \begin{cases} y & \text{if } \exists i \in E_{k'} : x = a_i \text{ and } y \in \mathfrak{A}_i, \\ x & \text{otherwise.} \end{cases}$$

Then, an arbitrary function f^n of P_k is a superposition over the functions z, g_f, f_i $(i \in E_{k'})$ defined by

$$\begin{aligned} & z(x,y) := x \diamond y, \\ & g_f(x_1, ..., x_n) := a_i \iff f(x_1, ..., x_n) \in \mathfrak{A}_i, \\ & f_i(x_1, ..., x_n) := \begin{cases} f(x_1, ..., x_n) & \text{if } \quad f(x_1, ..., x_n) \in \mathfrak{A}_i, \\ a_i & \text{otherwise} \end{cases} \\ & (i \in E_{k'}), \end{aligned}$$

and

$$f(\mathbf{x}) = ((...((g_f(\mathbf{x}) \diamond f_0(\mathbf{x})) \diamond f_1(\mathbf{x})) \diamond ...) \diamond f_{k'-1}(\mathbf{x})) \tag{1.12}$$

is a representation of f^n. ■

1.5 Some Applications of the Function Algebras

It is the aim of this section to show by examples how results in function algebras can be used in other mathematical disciplines. In addition, some problems that motivate certain investigations with function algebras are explained.

1.5.1 Classification of Universal Algebras

In Part I, Chapter 1, we introduced the concept of the type of an algebra, and in Chapter 6 combinations (classes) of algebras of the same type, which fulfill certain equations. Such a decomposition of the algebras is, however, very coarse. The set $\mathcal{M}_A$ of all finite algebras on the same universal A can be more finely decomposed with the aid of the following equivalence relation $\mathcal{R}_A$: The algebras $(A; F)$ and $(A; G)$ with $F, G \subseteq P_A$ are called **equivalent** in respect to $\mathcal{R}_A$, iff $[F] = [G]$ holds, i.e., iff the operations of the one algebra can be represented as superpositions over the operations of the other algebra and vice versa.
The equivalence classes (blocks) of the relation $\mathcal{R}_A$ form a partition of the set $\mathcal{M}_A$, where the set of all algebras $(A; F)$ with $F \in \mathbb{L}_A$ is a representative system of these equivalence classes.

Reducts of $(A; F)$ are defined to be algebras of the form $(A; G)$ with $G \subseteq [F]$.

An algebra $(A; F)$ with $[F] = P_A$ is called **primal** ([Fos 59]).
For example, by Theorem 1.4.2, the two-element Boolean algebra is primal. The Rosenberg's completeness criterion from Chapter 6 can be used to obtain further examples of primal algebras (see also Chapter 7).

An algebra $(A; F)$ is called **preprimal** iff $[F]$ is a maximal clone of P_A ([Den 82], [Kno 85]).

For a description of further generalizations of "primal algebras", we need some concepts and results of Chapter 2 and the notations
$S(\mathbf{A})$ (the set of all subalgebras of the algebra $\mathbf{A}$),
$Aut(\mathbf{A})$ (the set of all automorphisms on $\mathbf{A}$),
$SubIso(\mathbf{A})$ (the set of all isomorphisms between subalgebras of $\mathbf{A}$) and
$Con(\mathbf{A})$ (the set of all congruences of $\mathbf{A}$).
Let $\mathbf{A} := (A; F)$ be an algebra, for which there is a relation set Q on A with $[F] = Pol_A Q$. Then $\mathbf{A}$ is called

- **semiprimal** iff $Q \subseteq R_k^1$, i.e., iff $Q = S(\mathbf{A})$ (or with other words: iff every operation on A which preserves all subalgebras of $\mathbf{A}$ belongs to $[F]$) [Fos-P 64];
- **demiprimal** iff $Q = Aut(\mathbf{A})$ ([Qua 71]);
- **infraprimal** (or **demisemiprimal**) iff $Q = S(\mathbf{A}) \cup Aut(\mathbf{A})$ [Qua 71];
- **quasiprimal** iff $Q = SubIso(\mathbf{A})$ ([Pix 71]);
- **hemiprimal** iff $Q = Con(\mathbf{A})$ ([Fos 70]);

For further concepts and properties of the above-defined algebras, refer to [Den 82], [Den-W 2002] and [Sze 86].
Since the present book describes many clones, one finds properties of the above-defined algebras, in many places in this book.

1.5.2 Propositional Logic and First Order Logic

A **proposition** is a "sentence" (of a natural or artificial language) for which it makes sense to ask whether it false (notation: 0) or true (notation: 1). At the basis of the concept "proposition" we have the two-value principle (also called **principle of the excluded middle**). This means that each proposition must be either false or true, there is no other possibility "in between".
The following are propositions:

- Rostock is a city in Germany.
- There are infinite many prime numbers.
- $2 \cdot 3 = 5$.
- There exists extra-terrestrial life.

The following are not propositions:

- Two chickens on the way after the day before yesterday.
- Everything that I say is false.
- It is a beautiful day.
- Be quiet!

Propositions will be denoted by capital letters $A, B, ...$. Instead of "A is an arbitrary proposition" we say "A is a **proportional variable**". Therefore, a propositional variable takes the values 0 and 1.
One can associate propositions ("sentences") in the informal language in multiple ways with each other (for example through such conjunctions as "and", "or", "if – then", ...). The result of this connection is normally, again, a proposition, whose value (0 or 1) is dependent from the values of associated single propositions. In the propositional logic, a part of the colloquial connections is modelled and defined exactly (unlike the informal language). Since we abstract from the content of a proposition during the consideration of a proposition (i.e., we have interest only in the so-called **truth value** 0 or 1 of the proposition), proposition combinations are multi-digit functions on $\{0, 1\}$, i.e., Boolean functions (or function of the 2-valued logic).
Interpretations of the Boolean functions that are defined in Table 1.2 are, for example,

- The **negation** of the proposition A: $\overline{A}$ ("not A").
  $\overline{A}$ is true if and only if A is false.
- The **conjunction** of the propositions A, B: $A \wedge B$ ("A and B").
  $A \wedge B$ is true if and only if A as well as B are true.
- The **disjunction** of the propositions A, B: $A \vee B$ ("A or B").
  $A \vee B$ is true if and only if either A or B or both are true.

- **contravalence**: $A + B$ ("either A or B").
  $A + B$ is true if and only if either A or B is true.
- **equivalence**: $A \Longleftrightarrow B$ ("A if and only if B").
  $A \Longleftrightarrow B$ is true if and only if A and B have the same truth value.
- **implication**: $A \Rightarrow B$ ("A implies B"; "If A, then B").
  $A \Rightarrow B$ is false if and only if A has the value 1 and B has the value 0.

To provide mathematics with a precise language, the mathematical logic creates an artificial, formal language. Next, we give a short introduction on proportional logic; that is, the logic that deals only with propositions. Later, we extend our treatment to the first order logic, which also takes properties of individuals into account.
The process of formalization of proportional logic consists of two stages: (i) present a formal language (ii) specify a procedure for obtaining *valid* or *true* propositions.
The language of propositional logic has an alphabet consisting of

- the set of proposition symbols

$$At := \{A, B, C, ..., A_1, B_1, C_1, ..., A_n, B_n, C_n, ...\}$$

  (the elements of At are called **atoms** or **atomic propositions**)
- the set of connectives $\mathfrak{J}_0 := \{\wedge, \vee, \neg, \Rightarrow, \Leftrightarrow\}$
- the set of auxilliary symbols $\{(,)\}$

The set $Prop$ is the smallest set X with the following three properties:

- $At \cup \{0, 1\} \subseteq X$
- if $\alpha, \beta \in X$, then $(\alpha \circ \beta) \in X$ for all $\circ \in \mathfrak{J}_0$
- if $\alpha \in X$, then $(\neg\alpha) \in X$

Notice that $Prop = T(At)$ (see Part I, Section 6.2).

A mapping $v : Prop \longrightarrow \{0, 1\}$ is called a **valuation** if $v(0) = 0$, $v(1) = 1$, $v(\neg\alpha) = \neg v(\alpha)$ and $v(\alpha \circ \beta) = v(\alpha) \circ v(\beta)$ for all $\alpha, \beta \in Prop$ and all $\circ \in \mathfrak{J}_0$. [8]

If $v(\alpha) = 1$ (or $v(\alpha) = 0$) then α is **true** (or **false**) under v, respectively.

If $\alpha = (\beta) \in Prop$, then we only write β instead of α in the following.

Obviously, we have: If $v_0 : At \longrightarrow \{0, 1\}$, then there exists a unique valuation v such that $v(\alpha) = v_0(\alpha)$ for all $\alpha \in At$.

$\alpha \in Prop$ is a **tautology** if $v(\alpha) = 1$ for all valuations v. We write

$$\models \alpha \quad (\text{or } \emptyset \models \alpha)$$

for "α is a tautology". If $\Sigma \subseteq Prop$ and $\alpha \in Prop$, then

[8] Notice that the right sides of the equations are determined by Table 1.2.

$$\Sigma \models \alpha$$

iff for all valuations v: $v(\sigma) = 1$ for all $\sigma \in \Sigma$ implies $v(\alpha) = 1$. If $\Sigma \models \alpha$ then α is called a **consequence** of Σ. Further, set

$$Cons(\Sigma) := \{\alpha \in Prop \mid \Sigma \models \alpha\}.$$

It is a classical problem of the propositional logic to find a system of axioms and rules with which one can determine all tautologies and all consequences from a set of propositions. In books about mathematical logic, one finds many solutions for this purpose. We declare a solution with proof here. The proof is chosen so that it can be used as proof for a corresponding theorem of the predicate logic as a basic idea.
It is normal to write down the rule "R: if $\alpha_1, ..., \alpha_r \in Prop$ are derivable, then α_{r+1} is derivable" in the following form:

$$\frac{\alpha_1, ..., \alpha_r}{\alpha_{r+1}} \; R$$

As an example, we give the **substitution rule** *sub*: Let $\alpha \in Prop$, which contains $x \in At$ (in symbol: $\alpha(..., x, ...)$). If α is a tautology, then one can form a tautology by replacing every occurrence of x by $\beta \in Prop$ in α:

$$\frac{\vdash \alpha(..., x, ...); \; \beta \in Form}{\vdash \alpha(..., \beta, ...)} \; sub \; .$$

The **Hilbert-type-calculus** for the **classical proportional logic** is defined by the following 13 axioms and 1 rule, where α, β, γ, σ, τ are arbitrary elements of $Prop$:
(I) Axioms are all formulas of the form

(A1) $\alpha \Rightarrow \alpha$
(A2) $\alpha \Rightarrow (\beta \Rightarrow \alpha)$
(A3) $(\alpha \Rightarrow \beta) \Rightarrow ((\beta \Rightarrow \gamma) \Rightarrow (\alpha \Rightarrow \gamma))$
(A4) $(\alpha \Rightarrow (\beta \Rightarrow \gamma)) \Rightarrow ((\alpha \Rightarrow \beta) \Rightarrow (\alpha \Rightarrow \gamma))$
(A5) $\alpha \Rightarrow (\alpha \vee \beta), \beta \Rightarrow (\alpha \vee \beta)$
(A6) $(\alpha \Rightarrow \gamma) \Rightarrow ((\beta \Rightarrow \gamma) \Rightarrow ((\alpha \vee \beta) \Rightarrow \gamma))$
(A7) $(\alpha \wedge \beta) \Rightarrow \alpha, (\alpha \wedge \beta) \Rightarrow \beta$
(A8) $(\gamma \Rightarrow \alpha) \Rightarrow ((\gamma \Rightarrow \beta) \Rightarrow (\gamma \Rightarrow (\alpha \wedge \beta)))$
(A9) $((\alpha \wedge \beta) \vee \gamma) \Rightarrow ((\alpha \vee \gamma) \wedge (\beta \vee \gamma)), ((\alpha \vee \gamma) \wedge (\beta \vee \gamma)) \Rightarrow ((\alpha \wedge \beta) \vee \gamma)$
(A10) $((\alpha \vee \beta) \wedge \gamma) \Rightarrow ((\alpha \wedge \gamma) \vee (\beta \wedge \gamma)), ((\alpha \wedge \gamma) \vee (\beta \wedge \gamma)) \Rightarrow ((\alpha \vee \beta) \wedge \gamma)$
(A11) $(\alpha \Rightarrow \beta) \Rightarrow (\neg\beta \Rightarrow \neg\alpha)$
(A12) $(\alpha \wedge \neg\alpha) \Rightarrow \beta$
(A13) $\beta \Rightarrow (\alpha \vee \neg\alpha)$. [9]

(II) Rules: There is only one rule: "from σ and $\sigma \Rightarrow \tau$ conclude τ", written as

[9] One can also choose α, β, γ as three different variables and the substitution rule *sub* as an additional rule.

$$\frac{\sigma, \sigma \Rightarrow \tau}{\tau} \quad (\text{“}\mathbf{modus\ ponens}\text{”})$$

A **derivation of** φ **from** Σ ($\subseteq Prop$) is a finite sequence $(\varphi_1, \varphi_2, ..., \varphi_n)$ of formulas with $\varphi_n = \varphi$, where each φ_i $(1 \leq i \leq n)$ is an axiom or is an element of Σ or is the result of the application of modus ponens on φ_u, φ_v $(u, v < i)$. Let $\vdash$ be the **derivation operator** defined in this way, i.e., we write

$$\Sigma \vdash \varphi$$

iff there is a derivation of φ from Σ. Put

$$cons(\Sigma) := \{\varphi \mid \Sigma \vdash \varphi\}$$

(the set of all **derivation consequences** from Σ). It is easy to check that $cons(\Sigma) \subseteq Cons(\Sigma)$, since the axioms are tautologies and since, for each valuation v, $v(\sigma) = 1$ and $v(\sigma \Rightarrow \tau) = 1$ imply $v(\tau) = 1$. For proof that $cons(\Sigma) \subset Cons(\Sigma)$ is false, we need the following notations and facts:
Since $Prop$ is a set of terms, one can form the term algebra

$$\mathbf{Prop} := (Prop; \vee, \wedge, \Rightarrow, \neg, 0, 1)$$

of the type $(2, 2, 2, 1, 0, 0)$.
For arbitrary $\alpha, \beta \in Prop$ let

$$\alpha \approx_\Sigma \beta \text{ iff } (\Sigma \vdash \alpha \Rightarrow \beta \text{ and } \Sigma \vdash \beta \Rightarrow \alpha).$$

Theorem 1.5.2.1 *The relation $\approx_\Sigma$ has the following properties for arbitrary Σ:*

(a) $\approx_\Sigma$ is a congruence of the algebra **Prop**.
(b) The factor algebra $(Prop_{/\approx_\Sigma}; \vee, \wedge, \neg, 0, 1)$ is a Boolean algebra.
(c) The set $cons(\Sigma)$ is the 1 of the Boolean algebra $(Prop_{/\approx_\Sigma}; \vee, \wedge, \neg, 0, 1)$.
(d) There exists a homomorphic mapping ν from **Prop** *onto* $\mathbf{Prop}_{/\approx_\Sigma}$ *with*

$$\nu(\varphi) = 1 \ \text{ iff } \ \Sigma \vdash \varphi.$$

Proof. (a): First we show that $\approx_\Sigma$ is an equivalence relation. By $\Sigma \vdash \alpha \Rightarrow \alpha$ (see (A1)) is $\approx_\Sigma$ reflexive. The symmetry of $\approx_\Sigma$ follows from the definition of $\approx_\Sigma$. One can prove the transitivity of $\approx_\Sigma$ with the aid of (A3) and the modus ponens as follows: Let $\alpha \approx_\Sigma \beta$ and $\beta \approx_\Sigma \gamma$ be arbitrary, i.e., it holds:

$$\Sigma \vdash \alpha \Rightarrow \beta, \ \Sigma \vdash \beta \Rightarrow \alpha, \ \Sigma \vdash \beta \Rightarrow \gamma, \ \Sigma \vdash \gamma \Rightarrow \beta.$$

By means of the modus pones ($\sigma := \alpha \Rightarrow \beta$, $\sigma \Rightarrow \tau :=$ (A3)) and (A3) one obtains

$$\Sigma \vdash (\beta \Rightarrow \gamma) \Rightarrow (\alpha \Rightarrow \gamma).$$

When one uses the modus pones ($\sigma := \beta \Rightarrow \gamma$) again, one receives $\Sigma \vdash \alpha \Rightarrow \gamma$. Analogously, one can show $\Sigma \vdash \gamma \Rightarrow \alpha$. Thus, $\approx_\Sigma$ is an equivalence relation. To prove the compatibility of $\approx_\Sigma$ with $\Rightarrow$ we consider

$$\varphi \approx_\Sigma \varphi', \psi \approx_\Sigma \psi',$$

$(\varphi, \varphi', \psi, \psi' \in \mathit{Prop})$, i.e., it holds

$$\Sigma \vdash \varphi \Rightarrow \varphi',\ \Sigma \vdash \varphi' \Rightarrow \varphi,\ \Sigma \vdash \psi \Rightarrow \psi',\ \Sigma \vdash \psi' \Rightarrow \psi.$$

$(\psi \Rightarrow \psi') \Rightarrow (\varphi \Rightarrow (\psi \Rightarrow \psi'))$ $(\alpha = \psi \Rightarrow \psi'$ and $\tau = \beta$ in (A2)) and the modus ponens imply

$$\Sigma \vdash \varphi \Rightarrow (\psi \Rightarrow \psi').$$

Furthermore, by (A4), we have $((\varphi \Rightarrow (\psi \Rightarrow \psi')) \Rightarrow ((\varphi \Rightarrow \psi) \Rightarrow (\varphi \Rightarrow \psi'))$. Hence we get

$$\Sigma \vdash ((\varphi \Rightarrow \psi) \Rightarrow (\varphi \Rightarrow \psi'))$$

with the aid of modus ponens. Analogously, one can show

$$\Sigma \vdash ((\varphi \Rightarrow \psi') \Rightarrow (\varphi \Rightarrow \psi)).$$

Consequently,

$$\varphi \Rightarrow \psi \approx_\Sigma \varphi \Rightarrow \psi'. \tag{1.13}$$

By (A3) we have

$$\Sigma \vdash (\varphi' \Rightarrow \varphi) \Rightarrow ((\varphi \Rightarrow \psi') \Rightarrow (\varphi' \Rightarrow \psi')),$$

whereby

$$\Sigma \vdash (\varphi \Rightarrow \psi') \Rightarrow (\varphi' \Rightarrow \psi').$$

Analogously,

$$\Sigma \vdash (\varphi' \Rightarrow \psi') \Rightarrow (\varphi \Rightarrow \psi'),$$

and therefore,

$$\varphi \Rightarrow \psi' \approx_\Sigma \varphi' \Rightarrow \psi'. \tag{1.14}$$

Then

$$\varphi \Rightarrow \psi \approx_\Sigma \varphi' \Rightarrow \psi',$$

since $\approx_\Sigma$ is transitive and (1.13) and (1.14) are valid. Hence, $\approx_\Sigma$ is compatible with $\Rightarrow$. To prove that $\approx_\Sigma$ is compatible with $\neg$ we assume $\alpha \approx_\Sigma \beta$; i.e., we have $\Sigma \vdash \alpha \Rightarrow \beta$ and $\Sigma \vdash \beta \Rightarrow \alpha$. Then, by (A11) and the modus ponens, we get $\Sigma \vdash \neg\beta \Rightarrow \neg\alpha$ and $\Sigma \vdash \neg\alpha \Rightarrow \neg\beta$, whereby $\neg\alpha \approx_\Sigma \neg\beta$. Because of this property of $\approx_\Sigma$, we can define a partial order on $\mathbf{Prop}_{/\approx_\Sigma}$ as follows:

$$[\varphi]_{\approx_\Sigma} \leq [\psi]_{\approx_\Sigma} \text{ iff } \Sigma \vdash \varphi \Rightarrow \psi,$$

where $[\varphi]_{\approx_\Sigma} = [\psi]_{\approx_\Sigma}$ means that $\varphi \approx_\Sigma \psi$ (i.e., $\Sigma \vdash \varphi \Rightarrow \psi$ and $\Sigma \vdash \psi \Rightarrow \varphi$) holds. (By (A1) is $\leq$ reflexive; (A3) implies the transitivity of $\leq$. The antisymmetry follows from the definition of $\approx_\Sigma$: I If $\Sigma \vdash \varphi \Rightarrow \psi$ and $\Sigma \vdash \psi \Rightarrow \varphi$, then we have $\varphi \approx_\Sigma \psi$.) Now (A5) shows that

$$[\alpha]_{\approx_\Sigma} \leq [\alpha \vee \beta]_{\approx_\Sigma} \text{ and } [\beta]_{\approx_\Sigma} \leq [\alpha \vee \beta]_{\approx_\Sigma}.$$

Assume there is a γ with $[\alpha]_{\approx_\Sigma} \leq [\gamma]_{\approx_\Sigma}$ and $[\beta]_{\approx_\Sigma} \leq [\gamma]_{\approx_\Sigma}$. By (A6) and the modus ponens, this implies

$$[\alpha \vee \beta]_{\approx_\Sigma} \leq [\gamma]_{\approx_\Sigma},$$

whereby

$$sup([\alpha]_{\approx_\Sigma}, [\beta]_{\approx_\Sigma}) = [\alpha]_{\approx_\Sigma} \vee [\beta]_{\approx_\Sigma} = [\alpha \vee \beta]_{\approx_\Sigma}$$

holds.
Analogously,

$$inf([\alpha]_{\approx_\Sigma}, [\beta]_{\approx_\Sigma}) = [\alpha]_{\approx_\Sigma} \wedge [\beta]_{\approx_\Sigma} = [\alpha \wedge \beta]_{\approx_\Sigma}$$

follows from (A7), (A8) and the modus ponens. Thus $\approx_\Sigma$ is also compatible with the operations $\vee$ and $\wedge$.
(b): (a) implies that $\mathbf{Prop}_{/\approx_\Sigma}$ is a lattice. Because of the axioms (A9) and (A10) this lattice is distributive.
Assume $\Sigma \vdash \varphi$ and $\Sigma \vdash \psi$. Then $[\varphi]_{\approx_\Sigma} = [\psi]_{\approx_\Sigma}$, since

$$\Sigma \vdash \psi \Rightarrow \varphi \text{ and } \Sigma \vdash \varphi \Rightarrow \psi$$

follows from this with the aid of (A2) and the modus ponens.
Because of (A2), we have $\Sigma \vdash (\beta \Rightarrow \alpha)$ for all $\beta \in Prop$ and all $\alpha \in Prop$ with $\Sigma \vdash \alpha$, i.e., $\beta \leq \alpha$. Therefore, $\{\alpha \mid \Sigma \vdash \alpha\}$ is the greatest element of the lattice $\mathbf{Prop}_{/\approx_\Sigma}$.
By (A12) the smallest element is the equivalence class $[\alpha \wedge \neg\alpha]_{\approx_\Sigma}$. Consequently, the algebra $\mathbf{Prop}_{/\approx_\Sigma}$ also fulfills the axioms (B_1) (see Section 1.2.12). (B_2) follows from the fact that $\approx_\Sigma$ is compatible with $\wedge$, $\vee$ and $\neg$ with the aid of (A13) as follows:

$$\begin{aligned}
&[\alpha]_{\approx_\Sigma} \wedge (\neg[\alpha]_{\approx_\Sigma}) = [\alpha \wedge (\neg\alpha)]_{\approx_\Sigma} = 0 \ \text{ (see above)},\\
&[\alpha]_{\approx_\Sigma} \vee (\neg[\alpha]_{\approx_\Sigma}) = [\alpha \vee (\neg\alpha)]_{\approx_\Sigma} = 1\\
&(\text{because of } \ [\beta]_{\approx_\Sigma} \leq [\alpha \vee \neg\alpha]_{\approx_\Sigma} \ \text{ for each } \ \beta \in Prop \ \text{ by } \ (A13)).
\end{aligned}$$

Hence, $\mathbf{Prop}_{/\approx_\Sigma}$ is a Boolean algebra.
(c): We have shown above that the 1 of $Prop_{/\approx_\Sigma}$ contains all formulas derivable from Σ. Further, if $\alpha \in Prop$ belongs to the equivalence class 1 of $Prop_{/\approx_\Sigma}$, then we get $\Sigma \vdash \varphi \Rightarrow \alpha$ for each φ with $\Sigma \vdash \varphi$ by the above shown. Thus (by modus ponens) we have $\Sigma \vdash \alpha$. Hence, (c) is proven.
(d) By Part I, Section 4.1, the natural homomorphism

$$\varphi : Prop \longrightarrow Prop_{/\approx_\Sigma}, \ \alpha \mapsto [\alpha]_{/\approx_\Sigma}$$

fulfills (d). ■

Theorem 1.5.2.2 (Completeness Theorem of Proportional Logic)
Let $\Sigma \subseteq Prop$. Then

$$cons(\Sigma) = Cons(\Sigma).$$

Proof. Obviously, $cons(\Sigma) \subseteq Cons(\Sigma)$. Assume there exists an $\alpha \in Cons(\Sigma) \setminus cons(\Sigma)$. Then $[\alpha]_{/\approx_\Sigma} \neq [1]_{/\approx_\Sigma}$ by Theorem 1.5.2.1, whereby α is an element of a certain prime ideal (maximal ideal) of the Boolean algebra $\mathbf{Prop}_{/\approx_\Sigma}$.[10] Consequently, there exists a homomorphism μ from $\mathbf{Prop}_{/\approx_\Sigma}$ onto the two-element Boolean algebra $\{\mathbf{0}, \mathbf{1}\}$ with $\mu([\alpha]_{/\approx_\Sigma}) = 0$. Then, with the aid of Theorem 1.5.2.1, (d), it is easy to see that $v := \nu\Box\mu$ is a valuation with $v(\alpha) = 0$ and $v(\sigma) = 1$ for all $\sigma \in \Sigma$. But, this is a contradiction to $\alpha \in Cons(\Sigma)$. ■

Next is a short introduction to predicate logic (or first order logic). Predicate logic is a language to describe statements about algebraic structures (to given signature). A **signature** is a pair

$$\delta := ((n_i)_{i\in I}, (m_j)_{j\in J})$$

with $n_i \in \mathbb{N}_0$, $m_j \in \mathbb{N}$ for all $i \in I$ and $j \in J$.
$\mathfrak{A} := (A; (f_i^A)_{i\in I}, (R_j^A)_{j\in J})$ is called a **structure** of signature δ, if $(A; (f_i^A)_{i\in I})$ an algebra of the type $(n_i)_{i\in I}$ and $R_j^A \subseteq A^{m_j}$ holds for all $j \in J$.

We say "$f_i^{\mathfrak{A}}$ is the interpretation of f_i" and "$R_j^{\mathfrak{A}}$ is the interpretation of R_j". With the aid of a mapping P from A^h into $\{0, 1\}$, one can describe an h-ary relation R on a set A (i.e., $R \subseteq A^h$) as follows:

$$P(a_1, .., a_h) = 1 \text{ iff } (a_1, ..., a_h) \in R.$$

Such a mapping P is called an h-ary **predicate**.
Let P, Q be predicates on A. Then one can understand the predicates as propositions and form $(\neg P)$ and $(P \circ Q)$ for each $\circ \in \mathfrak{J}_0$.
Therefore, let $P_j^{m_j}$ be the m_j-ary predicate with

$$P_j(a_1, .., a_{m_j}) = 1 \text{ iff } (a_1, ..., a_{m_j}) \in R_j^{m_j}$$

in the following.
The alphabet of the first order logic consists of the following symbols:

- the set $Var := \{x_0, x_1, x_2, ...\}$ of **variables**
- the set $\{P_j^{m_j} \mid j \in J\}$ of **predicate symbols**
- the set $\{f_i^{n_i} \mid i \in I\}$ of **operation symbols**
- the set of **connectives** $\mathfrak{J} := \{\wedge, \vee, \neg, \Rightarrow, \Leftrightarrow, \exists, \forall\}$
  ($\exists$ and $\forall$ are called the **existential** and **universal quantifier**)
- the set of **auxilliary symbols** $\{(,)\}$

As described in Part I, Section 6.2, we form the set of all terms $Term := T(Var)$ and the term algebra

$$\mathbf{Term} = (Term; (f_i)_{i\in I}).$$

Then, $FORM$ is the smallest set $\mathcal{X}$ with the following four properties:

[10] See books on Boolean algebras or universal algebras.

- $P_j(t_1, ..., t_{m_j}) \in \mathcal{X}$ for all $j \in J$ and all $t_1, ..., t_{m_j} \in Term$ ($P_j(t_1, ..., t_{m_j})$ is called an **atom**);
- if $\alpha, \beta \in \mathcal{X}$, then $(\alpha \circ \beta) \in \mathcal{X}$ for all $\circ \in \mathfrak{J}_0$
- if $\alpha \in \mathcal{X}$ then $(\neg\alpha) \in \mathcal{X}$
- if $\alpha \in \mathcal{X}$ then $(\exists x_k \alpha) \in \mathcal{X}$ and $(\forall x_k \alpha) \in \mathcal{X}$ for all $k \in \mathbb{N}_0$

If $t \in Term$, then $Var(t)$ denotes the set of all elements of X, which occur in T. For $\varphi \in FORM$ the set $Var(\varphi)$ of all **variables of** φ is defined by

- $Var(P_j(t_1, ..., t_{m_j})) := Var(t_1) \cup ... \cup Var(t_{m_j})$
- $Var(\neg\alpha) := Var(\alpha)$ and $Var(\alpha \circ \beta) := Var(\alpha) \circ Var(\beta)$ for all $\circ \in \mathfrak{J}_0$
- $Var(\exists x_k \alpha) = Var(\forall x_k \alpha) := Var(\alpha) \cup \{x_k\}$

($\alpha, \beta \in FORM$).
The set $fr(\varphi)$ of **free variables** of φ is defined by

- $fr(\varphi) := Var(\varphi)$ if φ is an atom
- $fr(\neg\alpha) := fr(\alpha)$ and $fr(\alpha \circ \beta) := fr(\alpha) \circ fr(\beta)$ for all $\circ \in \mathfrak{J}_0$
- $fr(\forall x_k \alpha) = fr(\exists x_k \alpha) := fr(\alpha) \backslash \{x_k\}$

($\alpha, \beta \in FORM$).
The set $bd(\varphi)$ of **bound variables of** φ is defined by

- $bd(\varphi) := \emptyset$ if φ is an atom
- $bd(\neg\alpha) := bd(\alpha)$ and $bd(\alpha \circ \beta) := bd(\alpha) \circ bd(\beta)$ for all $\circ \in \mathfrak{J}_0$
- $bd(\forall x_k \alpha) = bd(\exists x_k \alpha) := bd(\alpha) \cup \{x_k\}$

($\alpha, \beta \in FORM$).
– φ is called **open formula** if $bd(\varphi) = \emptyset$. φ is a **sentence** or **is closed**, if $fr(\varphi) = \emptyset$. Obviously, $Var(\varphi) = fr(\varphi) \cup bd(\varphi)$, but $fr(\alpha) \cap bd(\varphi)$ need not be empty.
Let u, u' be mappings from Var into A. Then we write $u =_{x_k} u'$ iff $u(x_j) = u'(x_j)$ for all $j \neq k$.
Let $\mathfrak{A} := (A; (f_i)_{i \in I}, (R_j)_{j \in J})$ be a structure of signature δ and let $u : Var \longrightarrow A$ be a mapping. For u, there is exactly one homomorphism $\widetilde{u} : \mathbf{T(Var)} \longrightarrow (A; (f_i)_{i \in I})$, which u continues (see Part I, Theorem 6.2.2). In the following manner, one can interpret the elements of $FORM$ with the aid of $\mathfrak{A}$, u and $\widetilde{u}$:
An **interpretation function** (or **valuation**) is a mapping $v_{\mathfrak{A},u}$ from $FORM$ into $\{0, 1\}$ defined by

- $v_{\mathfrak{A},u}(P_j(t_1, ..., t_{m_j})) = 1$ iff $(\widetilde{u}(t_1), ..., \widetilde{u}(t_{m_j})) \in R_j^{\mathfrak{A}}$
- $v_{\mathfrak{A},u}(\neg\alpha) := \neg(v_{\mathfrak{A},u}(\alpha))$ and $v_{\mathfrak{A},u}(\alpha \circ \beta) := v_{\mathfrak{A},u}(\alpha) \circ v_{\mathfrak{A},u}(\beta)$ for all $\circ \in \mathfrak{J}_0$
- $v_{\mathfrak{A},u}(\forall x_k \varphi) = 1$ iff $v_{\mathfrak{A},u'}(\varphi) = 1$ for all u' with $u =_{x_k} u'$
  (i.e., $v_{\mathfrak{A},u}(\forall x_k \varphi) = \bigwedge_{u',\ u =_{x_k} u'} v_{\mathfrak{A},u'}(\varphi)$)
- $v_{\mathfrak{A},u}(\exists x_k \varphi) = 1$ iff there is a u' with $u =_{x_k} u'$ and $v_{\mathfrak{A},u'}(\varphi) = 1$
  (i.e., $v_{\mathfrak{A},u}(\exists x_k \varphi) = \bigvee_{u',\ u =_{x_k} u'} v_{\mathfrak{A},u'}(\varphi)$)

$\varphi \in FORM$ is **satisfied by** $u : Var \longrightarrow A$ iff $v_{\mathfrak{A},u}(\varphi) = 1$.

φ is **true** (or **valid**) in $\mathfrak{A}$, iff $v_{\mathfrak{A},u}(\varphi) = 1$ for all $u : Var \longrightarrow A$.

$\mathfrak{A} \models \varphi$ stands for "φ is **true** in $\mathfrak{A}$".

$\mathfrak{A} \not\models \varphi$ stands for "φ is **not true** (or **false**) in $\mathfrak{A}$".

If $\mathfrak{A} \models \varphi$, we call $\mathfrak{A}$ a **model** of φ. In general, if $\mathfrak{A} \models \sigma$ for all $\sigma \in \Sigma \subseteq FORM$, we call $\mathfrak{A}$ a model of Σ.

Notice that a closed formula of $FORM$ is either true or false in $\mathfrak{A}$.

Formulas $\alpha, \beta \in FORM$ are **equivalent** (in symbol $\alpha \equiv \beta$) iff $v_{\mathfrak{A},u}(\alpha) = v_{\mathfrak{A},u}(\beta)$ for every structure $\mathfrak{A}$ and for every $u : Var \longrightarrow A$.
The following theorem is easy to prove:

Theorem 1.5.2.3 *Let* $\varphi, \psi \in FORM$. *Then:*
(1) $\neg\forall x_k\varphi \equiv \exists x_k\neg\varphi$, $\neg\exists x_k\varphi \equiv \forall x_k\neg\varphi$;
(2) if $x_k \notin fr(\psi)$ *then* $(Qx_k\varphi \circ \psi) \equiv Qx_k(\varphi \circ \psi)$ *for all* $Q \in \{\exists, \forall\}$ *and* $\circ \in \{\wedge, \vee\}$;
(3) $(\forall x_k\varphi \wedge \forall x_k\psi) \equiv \forall x_k(\varphi \wedge \psi)$,
$(\exists x_k\varphi \vee \exists x_k\psi) \equiv \exists x_k(\varphi \vee \psi)$;
(4) $\forall x_k\forall x_l\varphi \equiv \forall x_l\forall x_k\varphi$, $\exists x_k\exists x_l\varphi \equiv \exists x_l\exists x_k\varphi$. ■

Next, we define the mappings rep_τ, tut_τ and sub_τ from $FORM$ into $FORM$, where τ is a mapping from Var into $Term$. With rep_τ, one can describe the replacement of variables by some terms in a formula from $FORM$. The mapping tut_τ renames the bound variables of a formula so that there are no variables of the formula anymore, which is free and bound. sub_τ is a combination of rep_τ and tut_τ.
Let $\tau : Var \longrightarrow Term$ be an arbitrary mapping and $\widehat{\tau} : Term \longrightarrow Term$ the unique homomorphic continuation of τ. Then the mapping $rep_\tau : FORM \longrightarrow FORM$ is defined by

- $rep_\tau(P_j(t_1, ..., t_{m_j})) := P_j(\widehat{\tau}(t_1), ..., \widehat{\tau}(t_{m_j}))$
- $rep_\tau(\neg\alpha) := \neg rep_\tau(\alpha)$ and $rep_\tau(\alpha \circ \beta) := rep_\tau(\alpha) \circ rep_\tau(\beta)$ for all $\circ \in \mathfrak{J}_0$
- if $\varphi = Qx_k\alpha$ with $Q \in \{\forall, \exists\}$, then $rep_\tau(\varphi) := Qx_k rep_{\tau'}(\alpha)$, where

$$\tau'(x_i) := \begin{cases} x_k & \text{if } x_i = x_k, \\ \tau(x_i) & \text{if } x_i \neq x_k. \end{cases}$$

For the mapping $\tau : Var \longrightarrow Term$ and $\varphi \in FORM$ let $n(\tau, \varphi)$ be the smallest number $j_0 \in \mathbb{N}_0$ with $x_j \notin fr(\varphi) \cup Var(\tau(x))$ for all $x \in fr(\varphi)$.
Let $tut_\tau : FORM \longrightarrow FORM$ be defined by

- $tut_\tau(\varphi) = \varphi$ if φ is an atom
- $tut_\tau(\neg\varphi) = \neg(tut_\tau(\varphi))$ and $tut_\tau(\varphi \circ \psi) = tut_\tau(\varphi) \circ tut_\tau(\psi)$ for all $\circ \in \mathfrak{J}_0$;
- for $Q \in \{\forall, \exists\}$ let $tut_\tau(Qx_k\varphi) = Qx_j tut_\tau(rep_{\tau_{k,j}}(\varphi))$ with $j := n(\tau, \varphi)$ and

$$\tau_{k,j}(x_l) := \begin{cases} x_j & \text{if } \quad k = l, \\ x_l & \text{otherwise.} \end{cases}$$

For $\tau : Var \longrightarrow Term$ and arbitrary $\varphi \in FORM$ set

$$sub_\tau(\varphi) := rep_\tau(tut_\tau(\varphi)).$$

The following lemma gives some properties of the above mappings:

Lemma 1.5.2.4 *(without proof) Let $\mathfrak{A}$ be a structure of signature δ, $u : Var \longrightarrow A$ a mapping with the homomorphic continuation $\widehat{u} : Term \longrightarrow A$. Then:*

(a) $v_{\mathfrak{A},u}(\varphi) = v_{\mathfrak{A},u}(tut_\tau(\varphi))$;

(b) $fr(\varphi) = fr(tut_\tau(\varphi))$;

(c) $v_{\mathfrak{A},\tau\Box\widehat{u}}(\varphi) = v_{\mathfrak{A},u}(sub_\tau(\varphi))$.

With the aid of the above mapping, we can describe the **Hilbert-type-calculus for the classical predicative logic**, which consists of following axioms and rules:

(I) **axioms** are:

(a) the axioms (A1) - (A13) of the classical propositional logic;

(b) all formulas of the form

(A14) $(\forall x_k \varphi) \Rightarrow sub_\tau(\varphi)$, where $\tau =_{x_k} id$

(i.e, τ replaces something at most for the variable x_k);

(A15) $\exists x_k \varphi \Rightarrow \neg\forall x_k \neg\varphi$, $\neg\forall x_k \neg\varphi \Rightarrow \exists x_k \varphi$;

(A16) $sub_{id}(\varphi) \Rightarrow \varphi$.

(II) **rules are**:

(a) the modus ponens;

(b)

$$\frac{\varphi \Rightarrow sub_{\tau_{k,l}}(\psi)}{\varphi \Rightarrow \forall x_k \psi},$$

where

$$\tau_{k,l}(x_j) := \begin{cases} x_l & \text{if} \quad j = k, \\ x_j & \text{otherwise} \end{cases}$$

and $x_l \notin Var(\psi) \cup fr(\varphi)$ ("**$\forall$-introductory-rule**");

(c)

$$\frac{\varphi}{sub_\tau(\varphi)}$$

for all τ.

Analogously to the proportional logic, one can define $\vdash$ and $cons(\Sigma)$ for $\Sigma \in FORM$ in respect to the above calculus. The **completeness theorem of the first order logic** says that $\vdash = \models$ holds, i.e., $cons(\Sigma) = Cons(\Sigma)$ is valid for all $\Sigma \subseteq FORM$ and that there are axioms and rules, so that $\vdash = \models$ was proven for the first time by K. Gödel in his dissertation 1929. For the above calculus one can prove the completeness theorem for first order

logic similar to Theorem 1.5.2.2, where the Rasiowa-Sikorski-Tarski-Lemma is needed (see [Ric 78] for details).
The famous incompleteness theorems of Gödel say that second order logic [11] does not have a completeness theorem in general.

1.5.3 Many-Valued Logics

The above considerations can be generalized when one assigns the "sentences" certain values from the set E_k ($k \geq 3$). We receive then a so-called k-valued logic or many-valued logic. One finds a full introduction to the logic for example in [Got 89] or [Kre-G-S 88].
From a mathematical perspective it doesn't matter which interpretations have the elements of E_k. Nevertheless, it is shown which interpretations of the elements of E_k, for example, are possible. According to these interpretations, one can select certain functions of P_k, with which one can form many-valued logics. For reasons of space limitations, we have excluded information. These functions can be found in [Men 85], where literature on the topic is also provided.
1.5.2.1 The three values of a three-valued logic can be interpreted as follows:

"false", "indefinite", "true";
"false", "possible", "true";
"false", "undecidable", "true"

or

"invalid", "in part valid", "full-valid".

Legally relevant actions can be divided in

"punishable", "prohibited but not punishable", "allowed".

1.5.2.2 The 4-valued logic seems particularly suitable for analyzing problem areas logically, in which there are two different kinds of truth (or validity) and two different kinds of falsehood (or nullity). Examples:

"fact falsehood", "legal nullity", "legal validity", "fact truth" ;
"knowledge falsehood", "belief falsehood", "belief truth", "knowledge truth".

1.5.2.3 A legal interpretation of the 6-valued logic is, for example,

0: "logically false", 1: "legally invalid",
2: "false according to facts", 3: "true according to facts",
4: "legally valid", 5: "logically true".

[11] In this logic, quantifications over elements of a structure as well at quantifications over partial sets and relations of the structure are allowed.

1.5.2.4 Instead of E_k, one can also choose a finite subset of the real numbers x with $0 \le x \le 1$. Then, with the aid of such a set, a probability logic can be formed:
The value 1 corresponds to the certainty of the truth; the value 0.5 corresponds to the uncertainty, whether true or false; the value 0 corresponds to the certainty of the falsehood (or the impossibility); values between 1 and 0.5 correspond to degrees of higher probability; values between 0.5 and 0 correspond to degrees of low probability.

We notice that one could also have chosen the interpretation of the **Fuzzy Logic** instead of the above-mentioned probability logic (see for example [Ban-G 90] or [Til 92]).

The problems of all the above-mentioned logics correspond basically to those of the propositional logic.

1.5.4 Information Transformer

The functions $f^n \in P_A$ can be understood simply as mathematical models of objects that process information (see Figure 1.1). The "object" from Figure 1.1 receives the information $x_1, ..., x_n \in A$ at the entries, processed to the information $f(x_1..., x_n) \in A$. In this case, we neglect the time which is needed for the workmanship.

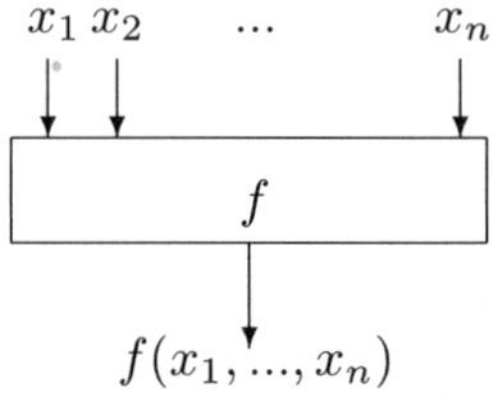

Fig. 1.1

Superpositions over functions of P_A correspond with this model to the "assembling" of such objects. For example, one can describe the diagram

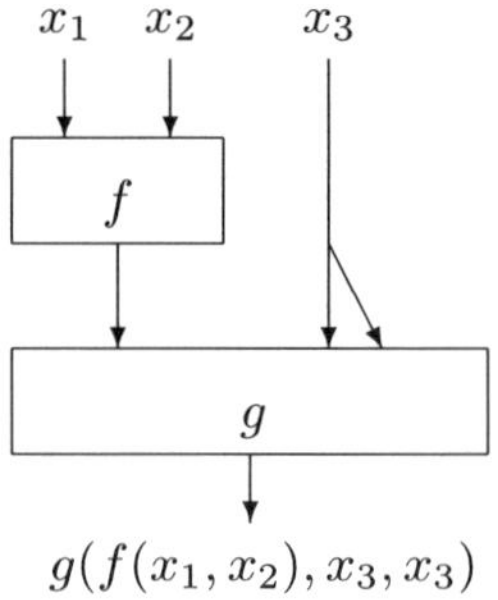

Fig. 1.2

through the formula $g(f(x_1, x_2), x_3, x_3)$. In particular, Boolean functions are used for the mathematical description of electrical circuits or components in computers. This mathematical description is independent of the concrete technical realization (as, for example, relay contact circuits or transistors). Naturally, one receives the so-called **completeness problem** from these interpretations of the functions from P_A:

> *A necessary and sufficient criterion is searched in order to be able to decide whether a system of certain selected functions (which correspond to certain elementary elements) produces all functions from P_A by means of superposition.*

One way of finding this criterion indicates the following theorem:

Theorem 1.5.4.1 *Let A be a subclass of P_k with the property that to every proper subclass A' of A there exists a certain maximal class M of A with $A' \subseteq M$. Furthermore, denote $\mathfrak{M}$ the set of all maximal classes of A. Then for an arbitrary subset T of A it holds:*

$$[T] = A \iff \forall M \in \mathfrak{M} : T \not\subseteq M. \tag{1.15}$$

Proof. "$\Longrightarrow$": Let $[T] = A$. Suppose there is an $M \in \mathfrak{M}$ with $T \subseteq M$. Then, a contradiction results; however, from that, immediately, $A = [T] \subseteq [M] = M \subset A$.
"$\Longleftarrow$": Let $T \subseteq P_k$ be no subset of a maximal class of A. Suppose, $[T] \subset A$. Then, by assumption, one can find a certain maximal class $M \in \mathfrak{M}$ with $[T] \subseteq M$, a contradiction to the supposition. ■

By Theorem 11.1.1, every finitely generated class fulfills the conditions of Lemma 1.5.3.1 and has only finitely many maximal classes, whereby (1.15) supplies a criterion of the wanted kind if one knows the maximal classes of A.

Further problems that result from interpreting the functions of P_A as information transformers are the following:

- Find minimal generating systems, where "minimal" is related to the number of the generating system's functions or to the arities of the generating system's functions.
- Find algorithms to construct a minimal realization of a given function by certain elementary functions.

One can find further problems and their solutions in [Rin 84] and [Pös-K 79].

1.5.5 Classification of Combinatorial Problems

In this section, we need some notations (R_k, Pol_k, Inv_k, ...) of Chapter 2.
Further, we need some concepts of the algorithm theory:[12]
Algorithms are techniques for solving problems, wherein the word "problem" is used in a very general sense: A **problem class** consists of infinitely many **instances** having a common structure.
A problem, whose solution is either "yes" or "no", is called a **decision problem**.
Often, finding a solution algorithm for the solution of a mathematical problem does not suffice. Statements about the complexity of the algorithms are important for the applicability of algorithms in particular. The complexity of an algorithm is a function $f(n)$, which gives the number of the arithmetic steps (e.g. value assignments, elementary arithmetic operations, comparison operations, ...) which are necessary to the estimating of a solution of a given problem with n master data. For $g:\ \mathbb{N} \longrightarrow \mathbb{R}_+$ let

$$O(g(n)) := \{f(n) \mid (f:\mathbb{N} \longrightarrow \mathbb{R}_+) \ \wedge \ (\exists c > 0 \ \forall n \in \mathbb{N}:\ f(n) \le c \cdot g(n))\}$$

We say that the **algorithm has the complexity** $O(g(n))$, if the complexity $f(n)$ of the algorithm belongs to $O(g(n))$. Generally one holds algorithms with a complexity $O(n^t)$ for "good" and these algorithms are called polynomial algorithms.
A decision problem with the complexity $O(n^t)$, where $n, t \in \mathbb{N}$, is called **tractable**.
Let **P** be the class of all tractable problems. **L** denotes the class of all problems with the complexity $O(\log n)$.
The class of decision problems for which a positive answer can be verified in polynomial time is denoted by **NP** (for "non-deterministic polynomial"). That is, we do not only require the answer "yes" or "no", but the explicit specification of a *certificate* which allows to verify the correctness of a positive answer.
LP denotes the class of decision problems for which a positive answer can be verified in logarithmic time.
Obviously, $\mathbf{P} \subseteq \mathbf{NP}$. Further, we know that

$$\mathbf{L} \subseteq \mathbf{NL} \subseteq \mathbf{P}.$$

It is the greatest problem of the algorithm theory to decipher whether $\mathbf{P} = \mathbf{NP}$ is valid. Most people believe, however, that $\mathbf{P} \neq \mathbf{NP}$.
A problem is called **NP-complete** if is in **NP** and if the polynomial solvability of this problem would imply that all problems in **NP** are solvable in polynomial time as well. In other words: Each problem in **NP** can be "transformed" (in polynomial time) to the given **NP**-complete problem, such that

[12] See e.g. [Pap 94] and [Jun 2005].

a solution of the **NP**-complete problem also gives a solution to that other problem in **NP**. Therefore, if one finds a polynomial algorithm for an **NP**-complete problem, this would imply that **P** = **NP**.

Next, is a short introduction to some papers that deal with classifications of combinatorial problems.

Definition A **CSP** (or a "**c**onstraint **s**atisfication **p**roblem") is a triple (V, D, C), where

- V is a set of variables
- D is a set of values which can take the variables
- $C := \{C_1, C_2, ..., C_q\}$ with $C_i := (s_i, \tau_i)$ for $i = 1, ..., q$, where s_i is an m_i-ary tuple of variables and τ_i is an m_i-ary relation on D (C_i is called **constraint**).

A **solution** of the CSP is a mapping

$$f : V \longrightarrow D$$

with the property

$$\forall\, C_i := ((x_{i_1}, x_{i_2}, ..., x_{i_{m_i}}), \tau_i) \in C : \ (f(x_{i_1}), f(x_{i_2}), ..., f(x_{i_{m_i}})) \in \tau_i.$$

Many combinatorial problems can be described as **CSP** (see [Jea 97] [13])

Example The problem SAT as **CSP**

An instance of the standard propositional satisfiability problem is specified by giving a formula on propositional logic, and asking whether there are values for the variables that make the formula true. For example, consider the formula

$$\begin{aligned} t(x_1, x_2, x_3, x_4) \ :=\ & (x_1 \vee x_2 \vee x_3 \vee x_4) \wedge (\overline{x_1} \vee \overline{x_2} \vee x_3) \wedge \\ & (\overline{x_3} \vee \overline{x_4} \vee x_1) \wedge (\overline{x_3} \vee \overline{x_2} \vee x_4) \wedge (\overline{x_1} \vee \overline{x_3}). \end{aligned}$$

The question is whether there are $a_1, a_2, a_3, a_4 \in \{0, 1\}$ with $t(a_1, a_2, a_3, a_4) = 1$. This problem can be expressed as the **CSP** as follows: Put

$$\begin{aligned} V &:= \{x_1, x_2, x_3, x_4\}, \\ D &:= \{0, 1\}, \\ C_1 &:= ((x_1, x_2, x_3, x_4), D^4 \setminus \{(0, 0, 0, 0)\}), \\ C_2 &:= ((x_1, x_2, x_3), D^3 \setminus \{(1, 1, 0)\}), \\ C_3 &:= ((x_3, x_4, x_1), D^3 \setminus \{(1, 1, 0)\}), \\ C_4 &:= ((x_3, x_2, x_4), D^3 \setminus \{(1, 1, 0)\}), \\ C_5 &:= ((x_1, x_3), D^2 \setminus \{(1, 1)\}), \\ C &:= \{C_1, c_2, C_3, C_4\}. \end{aligned}$$

The solutions of the above problem are

[13] In this paper are 20 combinatorial problems expressed as **CSP**.

	x_1	x_2	x_3	x_4
f_1 :	0	0	0	1
f_2 :	0	0	1	0
f_3 :	0	1	0	0
f_4 :	0	1	0	1
f_5 :	1	0	0	0
f_6 :	1	0	0	1

It is also possible to describe a **CSP** with the aid of the following notations:

Definitions Let V be a nonempty set and let $\tau_1, \tau_2, ..., \tau_t$ be relations on V. Then, $(V; \tau_1, \tau_2, ..., \tau_t)$ is called **relational structure.**
The mapping

$$\varrho : \{1, 2, ..., t\} \longrightarrow \mathbb{N}$$

is the **rank functions** of the relational structure $(V; \tau_1, \tau_2, ..., \tau_t)$ if $\varrho(i)$ is the arity of the relation τ_i $(i = 1, 2, ..., t)$.
A relational structure Σ is **similar** to a relational structure Σ' if they have the same rank function.
Let $\Sigma := (V; \tau_1, ..., \tau_t)$ and $\Sigma := (V'; \tau_1', ..., \tau_t')$ be two similar relational structure, and let ϱ be their common rank function. A **homomorphism** from Σ into Σ' is a mapping $h : V \longrightarrow V'$ such that

$$\forall i \in \{1, ..., t\} :$$
$$(a_1, .., a_{h_i}) \in \tau_i \implies (h(a_1), ..., h(a_{h_i})) \in \tau_i'.$$

Let $Hom(\Sigma, \Sigma')$ be the set of all homomorphisms from Σ into Σ'.
Obviously, it holds that

Theorem 1.5.5.1 *([Jea-C-P 98])*
(a) Let $\mathcal{P} := (V, D, C)$ be a **CSP** *with*

$$C := \{(s_1, \tau_1), (s_2, \tau_2), ..., (s_q, \tau_q)\}.$$

Then, the set of all solutions of $\mathcal{P}$ is the set $Hom(\Sigma, \Sigma')$, where $\Sigma := (V; \{s_1\}, \{s_2\} ..., \{s_q\})$ and $\Sigma' := (D; \tau_1, \tau_2, ..., \tau_q)$.
Conversely:
Let $\Sigma := (V; \tau_1, ..., \tau_t)$ and $\Sigma := (D; \tau_1', ..., \tau_t')$ be two similar relational structures. Then, $Hom(\Sigma, \Sigma')$ is the set of all solutions of the **CSP** *(V, D, S) with*

$$C := \bigcup_{i=1}^{t} \{(s, \tau_i') \mid s \in \tau_i\}.$$

The following theorem is our first example of an application of *Pol* and *Inv* (see Chapter 2) in algorithm theory.

Theorem 1.5.5.2 *Let $D := E_k$, $\varGamma := \{\varrho_1, ..., \varrho_q\} \subseteq R_k$, $V := \{x_1, ..., x_n\}$ and let M be a set of mappings from V into E_k. Then it holds:*
There exists a **CSP** *$(V^\star, D, C)$, where $V \subseteq V^\star$ and $C = \{(c_1, \varrho_1), ..., (c_q, \varrho_q)\}$, with the solution set S and $S_{|V} = M$ if and only if*

$$\varrho := \{(f(x_1), ..., f(x_n)) \mid f \in M\} \in Inv_k(Pol_k\varGamma).$$

The proof of the above theorem is explained by subsequent examples:
We choose: $D := E_2$, $\varrho_1 := \{(0,1,1),(1,0,0),(1,1,1)\}$, $\varrho_2 := \{(0,0),(1,0)\}$, $\varrho_3 := \{(1,1)\}$, $\varGamma := \{\varrho_1, \varrho_2, \varrho_3\}$, $V := \{x_1, x_2, x_3, x_4\}$ and $M := \{f_1, f_2\}$, where

x	$f_1(x)$	$f_2(x)$
x_1	0	1
x_2	1	0
x_3	1	1
x_4	1	1

"$\Longleftarrow$": Assume, $\varrho := \{(f(x_1), ..., f(x_n)) \mid f \in M\} \in Inv_k(Pol_k\varGamma)$. Then ϱ can be obtained by a finite number of applications of the elementary operations $\zeta, \tau, pr, \wedge$ and $\times$ from ϱ and $\delta^3_{k;\{1,2\}}$ (see Section 2.3). For example:

$$\begin{aligned}\varrho \;=\; & \{(x_1, x_2, x_3, x_4) \in E_2^4 \mid \exists x_5 \in E_2 : \\ & (x_1, x_5, x_3) \in \varrho_1 \;\wedge\; (x_5, x_2) \in \varrho_2 \;\wedge\; (x_5, x_4) \in \varrho_3\}.\end{aligned}$$

Therefore, $V^\star = \{x_1, x_2, x_3, x_4, x_5\}$ and $(V^\star, D, C)$, where

$$C := \{((x_1, x_5, x_3), \varrho_1), ((x_5, x_2), \varrho_2), ((x_5, x_4), \varrho_3)\},$$

is a **CSP** with the solution set $S = \{f_1', f_2'\}$, where

x	$f_1(x)$	$f_2(x)$
x_1	0	1
x_2	1	0
x_3	1	1
x_4	1	1
x_5	1	1

and $S_{|V} = M = \{f_1, f_2\}$.
"$\Longrightarrow$": Choose $V := \{x_1, x_2\}$ and $M := \{g_1, g_2\}$, where

x	$g_1(x)$	$g_2(x)$
x_1	0	1
x_2	0	0

$(\{x_1, x_2, x_3, x_4, x_5\}, E_2, C)$ with $C := \{((x_1, x_5, x_3), \varrho_1), ((x_5, x_2), \varrho_2), ((x_5, x_4), \varrho_3)\}$ is a **CSP** and it holds $S_{|V} = M = \{g_1, g_2\}$. Then

$$\begin{aligned}\varrho &= \{(0,0),(1,0)\}\\ &= \{(x_1,x_2)\in E_2^2 \mid \exists x_3\in E_2\ \exists x_4\in E_2\ \exists x_5\in E_2 :\\ &\qquad (x_1,x_5,x_3)\in\varrho_1\ \wedge\ (x_5,x_2)\in\varrho_2\ \wedge\ (x_5,x_4)\in\varrho_3\}\end{aligned}$$

Thus $\varrho \in Inv_k Pol_k \Gamma$ (see Chapter 2).

Next we show how one can use the above theorem for solving the following task:
Put

$$D := E_2,$$

$$\Gamma := \{\varrho \subseteq E_2^2 \mid \varrho \neq \emptyset\} =: \{\varrho_1, ..., \varrho_{15}\},$$

$M := \{f_1, f_2, f_3\}$, where

x	$f_1(x)$	$f_2(x)$	$f_3(x)$
x_1	0	0	0
x_2	0	0	1
x_3	0	1	0
x_4	1	0	0

Is there a **CSP** $(V^\star, E_2, C)$ with $V \subseteq V^\star$, $C = \{((c_1,\varrho_1)), ..., (c_{15\varrho_{15}})\}$ and the solution set S which ones $S_{|V} = M$ is valid for?
Because of the above theorem, this question is equivalent for the following question:

$$\begin{aligned}\varrho &:= \{(f(x_1,...,f(x_4))) \mid f\in M\}\\ &= \{(0,0,0,1),(0,0,1,0),(0,1,0,0)\}\\ &\in Inv_2 Pol_2 \Gamma\ ?\end{aligned}$$

With the help of the results of Chapter 3, it is easy to prove that

$$Pol_2\Gamma = \bigcap_{i=1}^{15} Pol_2\varrho_i = [h_2],$$

where $h_2(x,y,z) := (x\wedge y)\vee(x\wedge z)\wedge(y\vee z)$. Then, because of

$$h_2\begin{pmatrix}0&0&0\\0&0&1\\0&1&0\\1&0&0\end{pmatrix} = \begin{pmatrix}0\\0\\0\\0\end{pmatrix}$$

we have $\varrho \notin Inv_2 Pol_2 \varrho$; i.e., there is no **CSP** with the above properties.

Definitions The **general combinatorial problem** (**GCP**) is the decision problem with:
Instance: A pair (Σ_1, Σ_2) of similar finite relational structures Σ_1 and Σ_2.
Question: Is there a homomorphism from Σ_1 to Σ_2?
For any **GCP** instance $P := (\Sigma_1, \Sigma_2)$ a homomorphism from Σ_1 to Σ_2 will be called a **solution** of P.

The **GCP** (or **CSP**) is known to be **NP**-complete. However, certain restrictions may affect the complexity of **GCP** (or **CSP**). One of the natural possibilities for restricting **CSP** is by limiting the relations that can appear in constraints (or relational structures).

Definition Let Γ be a set of relations on the set D. Denote by **CSP**(Γ) the subclass of **CSP** defined by the following property: any constraint relation in any instance must belong to Γ.

The next theorem is the basis for the following theorems.

Theorem 1.5.5.3 *([Jea 98]; without proof)*
Let Γ and Ψ be finite sets of relation on D (w.l.o.g. we put $D := E_k$). If $\Psi \subseteq Inv_k(Pol_k\Gamma)$ then **GCP**(Ψ) *can be reduced in polynomial time to* **GCP**(Γ).
In other words:
For any finite set $\Gamma \subseteq R_k$, the complexity of **GCP**(Γ) *is determined, up to polynomial-time reductions, by $Pol_k\Gamma$.*

Theorem 1.5.5.4 *([Jea 98]; without proof)*
Let $D := E_k$, $k \geq 2$ and $\Gamma \subseteq R_k$. Then

(1) **GCP**(Γ) $\subseteq$ **P**, *if one of the following conditions is valid:*
   (a) $Pol_k\Gamma$ contains a constant function.
   (b) $Pol_k\Gamma$ contains a binary function which is associative, commutative and idempotent.
   (c) $r \in Pol_k\Gamma$, where $r(x,y,z) := x + y + z$ and $(E_k;+)$ is Abelian 2-group.

(2) If $Pol_k\Gamma$ contains a ternary function φ defined by

$$\varphi(x,y,z) := \begin{cases} y & \text{if} \quad y = z, \\ x & \text{otherwise,} \end{cases}$$

then **GCP**(Γ) $\subseteq$ **NL**.

(3) If each function of $Pol_k\Gamma$ is either a projection or a semiprojection, then **GCP**(Γ) *is* **NP**-*complete.*

With the aid of the above theorem and Post's description of all closed sets of Boolean functions, one can prove the following:

Theorem 1.5.5.5 *([Sch 78])*
For arbitrary $\Gamma \subseteq R_2$ it holds: **GCP**(Γ) *is* **NP**-*complete, if $Pol_2\Gamma \subseteq [P_2^1]$; in all other cases we have* **GCP**(Γ) $\subseteq$ **P**.

More can be found on the topic in [Jea 98], [Jea-C-P 98], [Bul-K-J 2000] and [Coh-J-G 2003].

2

The Galois-Connection Between Function- and Relation-Algebras

This section aims to develop a "suitable" means to describe function algebras or clones. I mean, suitable in the sense that "big" function algebras (or clones) can be described with an "expenditure" as small as possible.
Analogously to other fields of algebra we introduce invariants for our function algebras.
The two papers [Bod-K-K-R 68/69] of V. G. Bodnarčuk, L. A. Kalužnin, V. N. Kotov and B. A. Romov are basis of this so-called **Pol-Inv theory** (or **Galois theory for function- and relation-algebras**). These articles generalize results by M. Krasner on a Galois theory for groups (see [Kra 45], [Kra 68/69]).
A full representation of the Pol-Inv theory with many applications can be found in the monograph [Pös-K 79] by R. Pöschel and L. A. Kalužnin.

2.1 Relations

An h-**ary relation** ϱ on E_k is a subset of the h-fold Cartesian product E_k^h of the set E_k, $h \in \mathbb{N}$. The elements $(a_1, a_2, ..., a_h)$ of ϱ (we say "h-**tuple**") are written as columns

$$\begin{pmatrix} a_1 \\ a_2 \\ \cdots \\ a_h \end{pmatrix},$$

and then we also write

$$\begin{pmatrix} a_1 \\ a_2 \\ \cdots \\ a_h \end{pmatrix} \in \varrho.$$

The relation is written often as a matrix whose columns are the elements of the relation. For example,

$$\varrho := \begin{pmatrix} a_1 & a_1' & a_1'' \\ \ldots & \ldots & \ldots \\ a_h & a_h' & a_h'' \end{pmatrix}$$

instead of $\varrho := \{(a_1, ..., a_h), (a_1', ..., a_h'), (a_1'', ..., a_h'')\}$.
We think of this matrix representation of ϱ if we subsequently talk about the **length** h and the **width** $|\varrho|$ of the relation ϱ as well as about **rows** from ϱ.

Denote R_k^h the **set of all** h**-ary relations on** E_k and let

$$R_k := \bigcup_{h \geq 1} R_k^h.$$

We remark that the empty set is an element of R_k.
If $Q \subseteq R_k$ then let $Q^h := Q \cap R_k^h$.

2.2 Diagonal Relations

The simplest relations, in a certain sense[1], are the **diagonal relations** (or **diagonals**) defined as follows:
For an arbitrary equivalence relation ε on $\{1, 2, ..., h\}$ let

$$\delta_{k,\varepsilon}^h := \{(a_1, ..., a_h) \in E_k^h \mid (i,j) \in \varepsilon \implies a_i = a_j\}.$$

If h or k follows from the context, then we write only δ_ε or δ_ε^h or $\delta_{k,\varepsilon}$.
Every element of the set

$$D_k^h := \{\ \delta_{k,\varepsilon}^h \mid \varepsilon \text{ is an equivalence relation on } \{1, 2, ..., h\}\ \}.$$

is called a **diagonal** h**-ary relation**. Let

$$D_k := \{\emptyset\} \cup \bigcup_{h \geq 1} D_k^h$$

be the **set of all diagonal relations**.
For the purpose of a more simple description of $\delta_{k,\varepsilon}^h$, we often declare this relation in the form

$$\delta_{k;\varepsilon_1,...,\varepsilon_r}^h$$

or, briefly, by $\delta_{\varepsilon_1,...,\varepsilon_r}$, where $\varepsilon_1, ..., \varepsilon_r$ are exactly the equivalence classes of ε, which have at least two elements. In particular, we have

$$\delta_{k;}^h = E_k^h$$

and

$$\delta_{k;E_k}^h = \{(x, x, ..., x) \in E_k^h \mid x \in E_k\}.$$

---

[1] See Theorem 2.5.1.

2.3 Elementary Operations on R_k

We define in this section some operation $\zeta, \tau, pr, \wedge$ and $\times$ on R_k with whose aid we can form later more "complex" operations on R_k. We call, therefore, the operations $\zeta, \tau, pr, \wedge$ and $\times$ the **elementary operations on** R_k.

Let $\varrho \in R_k^h$ and $\varrho' \in R_k^{h'}$, where $h, h' \in \mathbb{N}$. Then, if $\varrho \neq \emptyset$ and $\varrho' \neq \emptyset$, let $\zeta\varrho \in R_k^h$, $\tau\varrho \in R_k^h$, $pr\,\varrho \in R_k^{h-1}$ for $h \geq 2$ and $pr\,\varrho = \emptyset$ for $h = 1$, $\varrho \times \varrho' \in R_k^{h+h'}$ and $\varrho \wedge \varrho' \in R_k^h$ (only for $h = h'$) defined by:

$$\begin{aligned}
\zeta\varrho &:= \{(a_2, a_3, ..., a_h, a_1) \mid (a_1, a_2, ..., a_h) \in \varrho\} \\
&\quad \textbf{(cyclical exchanging of the rows)}, \\
\tau\varrho &:= \{(a_2, a_1, a_3, ..., a_h) \mid (a_1, a_2, ..., a_h) \in \varrho\} \\
&\quad \textbf{(exchange of the first two rows)} \\
&\quad \text{for } h \geq 2 \text{ and } \zeta\varrho = \tau\varrho = \varrho \text{ for } h = 1 \text{ or } \varrho = \emptyset; \\
pr\ \varrho &:= \{(a_2, ..., a_h) \mid \exists a_1 \in E_k : (a_1, a_2, ..., a_h) \in \varrho\} \\
&\quad \textbf{(projection onto the 2th,..., } h\textbf{-th coordinate or} \\
&\quad \textbf{the strike of the first row)} \text{ for } h \geq 2, \\
\varrho \times \varrho' &:= \{(a_1, ..., a_h, b_1, ..., b_{h'}) \mid (a_1, a_2, ..., a_h) \in \varrho \\
&\qquad\qquad \wedge (b_1, b_2, ..., b_{h'}) \in \varrho'\} \\
&\quad \textbf{(Cartesian product of } \varrho \textbf{ and } \varrho'\textbf{)} \qquad \text{and} \\
\varrho \wedge \varrho' &:= \{(a_1, ..., a_h) \mid (a_1, ..., a_h) \in \varrho \cap \varrho'\} \\
&\quad \textbf{(intersection of the relations } \varrho \textbf{ and } \varrho'\textbf{)}.
\end{aligned}$$

2.4 Relation Algebras, Co-Clones, and Derivation of Relations

The algebra

$$\mathbf{R_k} := (R_k; \delta^3_{k;\{1,2\}}, \zeta, \tau, pr, \wedge, \times)$$

of type $(0, 1, 1, 1, 2, 2)$ is called **full relation algebra on** E_k. Every subalgebra Q of R_k (in symbol $Q \leq R_k$) is a **relation algebra on** E_k.
Let $Q \subseteq R_k$. Then, $[Q]$ denotes the set of all relations of R_k that can be obtained by a finite number of applications of the elementary operations $\zeta, \tau, pr, \wedge$ and $\times$ from the relations of Q and $\delta^3_{k;\{1,2\}}$,i.e., $[Q]$ is the universe of the smallest relation algebra, which contains Q.
If $[Q] = Q$ ($\subseteq R_k$) we say that Q **is closed** or Q is a **co-clone** of R_k.
Further, we say **a relation** ϱ' **can be derived from the relation** ϱ (or ϱ' **is** ϱ**-derivable**), if $\varrho' \in [\{\varrho\}]$. In this case we also write:

$$\varrho \vdash \varrho'.$$

2.5 Some Operations on R_k Derivable from the Elementary Operations

We say that an **operation** on R_k **is derivable** from the elementary operations $\zeta, \tau, pr, \wedge$ and $\times$ (or $\{\zeta, \tau, pr, \wedge, \times\}$**-derivable**), if their effect on an arbitrary relation $\varrho \in R_k$ can also be described by the effect of a finite composition of the elementary operations and $\delta^3_{k;\{1,2\}}$ on ϱ.
The following is a small list of such derivable operations. In this case, ϱ always describes an h-ary and ϱ' an h'-ary relation of R_k.

($\mathbf{O_1}$): **Permutation of coordinates (or permutations of rows).** As is known, the permutations

$$\begin{pmatrix} 1 & 2 & \dots & h-1 & h \\ 2 & 3 & \dots & h & 1 \end{pmatrix} \text{ and } \begin{pmatrix} 1 & 2 & 3 & \dots & h \\ 2 & 1 & 3 & \dots & h \end{pmatrix}$$

form a generating system for the symmetric group $\mathbf{S_h}$ (i.e., for the set of all permutations on $\{1, 2, ..., h\}$ with the operation $\Box$). Consequently one can realize all rearrangements of the rows of ϱ with the aid of ζ and τ. Thus, for every $s \in S_h$,

$$\sigma_s(\varrho) := \{(a_{s(1)}, ..., a_{s(h)}) \mid (a_1, ..., a_h) \in \varrho\}$$

is a $\{\zeta, \tau, pr, \wedge, \times\}$-derivable operation.

($\mathbf{O_2}$): **Projection onto the α_1-th, ..., α_t-th coordinates (or deleting of rows).** For $\{\alpha_1, ..., \alpha_t\} \subseteq \{1, 2, .., h\}$ it holds:

$$pr_{\alpha_1,...,\alpha_t}(\varrho) :=$$
$$\{(a_{\alpha_1}, ..., a_{\alpha_t}) \mid (a_1, .., a_h) \in \varrho\} = \underbrace{pr(pr(...(pr}_{h-t \text{ times}}(\sigma_s(\varrho)))...)),$$

where $s \in S_h$ and

$$s(\alpha_1) = h - t + 1,\ s(\alpha_2) = h - t + 2,\ ...,\ s(\alpha_t) = h.$$

In particular, we have

$$pr_{s(1),...,s(n)}(\varrho) = \sigma_s(\varrho).$$

We remark that, in the case where ϱ is given in the form $\{(a_0, a_1, ..., a_{h-1}) \mid ....\}$, we choose $\{\alpha_1, ..., \alpha_t\} \subseteq \{0, 1, 2, .., h-1\}$ and define $pr_{\alpha_1,...,\alpha_t}(\varrho)$ in analog manner, as above.

($\mathbf{O_3}$): **Identification of coordinates.** For $i, j \in \{1, 2, ..., h\}$ and $i \neq j$ let

$$\Delta_{i,j}(\varrho) := \{(a_1, ..., a_{j-1}, a_{j+1}, ..., a_h) \mid (a_1, ..., a_{j-1}, a_i, a_{j+1}, ..., a_h) \in \varrho\}$$

and $\Delta := \Delta_{1,2}$.
$\Delta_{i,j}$ can be formed as follows:
It is easy to prove that $pr_1(\delta^3_{k;\{1,2\}}) = E_k$ and
$\delta^h_{k;\{i,j\}} = pr_{1,...,i-1,h+1,i+1,...,j-1,h+2,j+1,...,k}(\varrho_1)$,
where $\varrho_1 := \underbrace{E_k \times ... \times E_k}_{h-2 \text{ times}} \times pr_{1,2}\delta^3_{k;\{1,2\}}$ and $i < j$. Consequently, $\Delta_{i,j}$ is derivable because of

$$\Delta_{i,j} = pr_{1,...,j-1,j+1,...,h}(\varrho \wedge \delta^h_{k,\{i,j\}}).$$

($\mathbf{O_4}$): **Doubling of coordinates (rows).** One receives a doubling of the i-th row of ϱ as follows:

$$\begin{aligned} \nu_i(\varrho) &:= \{(a_1, ..., a_{i-1}, a_i, a_i, a_{i+1}, ..., a_h) \mid (a_1, ..., a_h) \in \varrho\} \\ &= pr_{1,...,i-1,h,h+1,,i,...,h-1}(\Delta_{i,h+1}(\varrho \times pr(\delta^3_{k;\{2,3\}})). \end{aligned}$$

($\mathbf{O_5}$): **Adding of fictitious coordinates.** Let

$$\nabla\varrho := \{(a_1, ..., a_{h+1}) | a_1 \in E_k \wedge (a_2, ..., a_{h+1}) \in \varrho\}.$$

The first coordinate of $\nabla\varrho$ is a so-called **fictitious coordinate**. Then, with the aid of (O_1) one can derive the relation

$$\nabla_i\varrho :=$$
$$\{(a_1, ..., a_{i-1}, a_i, a_{i+1}, ..., a_{h+1}) \in E_k^{h+1} \mid (a_1, ..., a_{i-1}, a_{i+1}, ..., a_{h+1}) \in \varrho\}$$

for $i \in \{1, ..., h+1\}$.

($\mathbf{O_6}$): **General composition (relation product).** Let

$$\varrho \circ_t \varrho' := \{(a_1, ..., a_{h-t}, b_{t+1}, ..., b_{h'}) \mid \exists u_1, ..., u_t \in E_k :$$
$$(a_1, ..., a_{h-t}, u_1, ..., u_t) \in \varrho \wedge (u_1, ..., u_t, b_{t+1}, ..., b_{h'}) \in \varrho'\}$$

for $t \in \mathbb{N}$ with $t \leq h$ and $t \leq h'$.
In particular, $\varrho o \varrho' := \varrho o_1 \varrho'$.
For $h = h' = 2$, $\circ$ is the well-known relation product $\square$.

The following theorem results from the above considerations.

Theorem 2.5.1 *Every co-clone Q of R_k contains all diagonal relations and is closed in respect to*
- *permutation of coordinates (rows),*
- *projection onto coordinates (or deleting of rows),*
- *identification of coordinates (rows),*
- *doubling of coordinates (rows),*
- *adding of fictitious coordinates,*
- *(finite) intersection formation,*
- *Cartesian products* *and*
- *general composition.* ∎

2.6 The Preserving of Relations; Pol, Inv

We say that a function $f^n \in P_k$ **preserves the relation** $\varrho \in R_k^h$ (or ϱ **is invariant for** f or ϱ **is a invariant for** f), if

$$f\begin{pmatrix} a_{11} & a_{12} & \ldots & a_{1n} \\ a_{21} & a_{22} & \ldots & a_{2n} \\ \ldots & \ldots & \ldots & \ldots \\ a_{h1} & a_{h2} & \ldots & a_{hn} \end{pmatrix} := \begin{pmatrix} f(a_{11}, a_{12}, \ldots, a_{1n}) \\ f(a_{21}, a_{22}, \ldots, a_{2n}) \\ \ldots\ldots\ldots\ldots \\ f(a_{h1}, a_{h2}, \ldots, a_{hn}) \end{pmatrix} \in \varrho$$

for all

$$\begin{pmatrix} a_{11} \\ a_{21} \\ \ldots \\ a_{h1} \end{pmatrix}, \begin{pmatrix} a_{12} \\ a_{22} \\ \ldots \\ a_{h2} \end{pmatrix}, \ldots, \begin{pmatrix} a_{1n} \\ a_{2n} \\ \ldots \\ a_{hn} \end{pmatrix} \in \varrho.$$

The empty set $\emptyset$ is preserved by every function $f \in P_A$.
Note that f preserves ϱ iff ϱ is the universe of a subalgebra $(A; f)^h$.
By

$$Pol_k \varrho$$

or, briefly, $Pol\ \varrho$ we denote the set of all functions $f \in P_k$ that preserve the relation ϱ. For $Q \subseteq R_k$, we put

$$Pol_k Q := \bigcap_{\varrho \in Q} Pol_k \varrho.$$

$Pol_k \varrho$ or $Pol_k Q$ is a short-cut of **polymorphisms** of ϱ or Q, respectively.
The set of all relations $\varrho \in R_k$ that are preserved from the function $f \in P_k$ is

$$Inv_k f.$$

For $A \subseteq P_k$ let

$$Inv_k A := \bigcap_{f \in A} Inv_k f$$

be the **set of all invariants of A** and let

$$(Inv_k A)^n := (Inv_k A) \cap R_k^h$$

be the set of all n-ary invariants of A.
If k can be seen from the context, we write Inv instead of Inv_k.
Further notations used by us are

$$Pol^n Q := (Pol\ Q)^n\ (Q \subseteq R_k \text{ or } Q \in R_k)$$

and

$$Inv^n A := (Inv\ A)^n\ (A \subseteq P_k \text{ or } A \in P_k)$$

for $n \in \mathbb{N}$.
Elementary connections between Pol and Inv are summarized in the following theorem.

Theorem 2.6.1 *For arbitrary $A, B \subseteq P_k$ and arbitrary $S, T \subseteq R_k$, it holds:*

(a) $A \subseteq B \implies Inv\ B \subseteq Inv\ A$,

$S \subseteq T \implies Pol\ T \subseteq Pol\ S$;

(b) $A \subseteq Pol\ Inv\ A$,

$S \subseteq Inv\ Pol\ S$;

((a) and (b) mean that the pair (Pol, Inv) of mappings $Pol : \mathfrak{P}(P_k) \longrightarrow \mathfrak{P}(R_k)$, $A \mapsto Pol_k A$ and $Pol : \mathfrak{P}(R_k) \longrightarrow \mathfrak{P}(P_k)$, $Q \mapsto Inv_k Q$ is a Galois connection (see Part I, 4.4) between the sets P_k and R_k.)

(c) $Inv\ Pol\ Inv\ A = Inv\ A$,

$Pol\ Inv\ Pol\ S = Pol\ S$;

(d) $A \subseteq Pol\ S \iff S \subseteq Inv\ A$;

(e) $Pol\ (S \cup T) = Pol\ S \cap Pol\ T$,

$Inv\ A \cup B = Inv\ A \cap Inv\ B$.

Proof. (a), (b), (d), and (e) are direct conclusions from the definitions of Pol and Inv.
(c): Let $S := Inv\ A$. By (b) we have $S \subseteq Inv\ Pol\ S$. Conversely, it holds $A \subseteq Pol\ Inv\ A$ and thus by (a): $Inv\ Pol\ Inv\ A \subseteq Inv\ A$. Therefore, $Inv\ Pol\ Inv\ A = Inv\ A$. Analogously, one can show that $Pol\ Inv\ Pol\ S = Pol\ S$ holds. ■

Theorem 2.6.2 *For every $A \subseteq P_k$ and every $Q \subseteq R_k$ the sets $Inv\ A$ and $Pol\ Q$ are closed (in respect to the operations defined above, respectively); i.e., $Pol\ Q$ is a clone of P_k and $Inv\ A$ is a co-clone of R_k.*
Furthermore, it holds:

$$Inv\ [A] = Inv\ A \ \text{ and } \ Pol\ [Q] = Pol\ Q.$$

In particular we have:

$$Inv\ [\{e_1^2\}] = R_h \ \text{ and } \ Pol\ D_k = P_k.$$

Proof. Let $\varrho, \varrho' \in Inv\ A$ and $f \in A$ be arbitrary. Then, f preserves the ϱ-derivable relations $\zeta\varrho$, $\tau\varrho$, $pr\varrho$, $\varrho \wedge \varrho'$ and $\varrho \times \varrho'$ obviously. Thus $\{\zeta\varrho, \tau\varrho, pr\varrho, \varrho \wedge \varrho', \varrho \times \varrho'\} \subseteq Inv\ A$, whereby $Inv\ A$ is closed.
Now let $f^n, g^m \in Pol\ Q$ and $\varrho \in Q$ be arbitrary. Clear that, then the functions $\zeta f, \tau f$ and Δf also preserve ϱ. Further, the relation ϱ is also an invariant of $f \star g$, because $f(g(r_1, ..., r_m), r_{m+1}, ..., r_{m+n-1}) \in \varrho$ holds for all $r_1, ..., r_{m+n-1} \in \varrho$, since $g(r_1, .., r_m) \in \varrho$.
$Inv\ [A] = Inv\ A$ follows from

$$Inv\ [A] \subseteq Inv\ A$$

(because of $A \subseteq [A]$ and Theorem 2.6.1, (a)) and

$$Inv\ A \subseteq Inv\ [A].$$

(By Theorem 2.6.1, (a) we have $A \subseteq Pol\ Inv\ A$ indeed. Since $Pol\ Inv\ A$ is closed, this implies $[A] \subseteq Pol\ Inv\ A$. If one uses Theorem 2.6.1, (a) again and then 2.6.1, (c), one receives $Inv\ Pol\ Inv\ A = Inv\ A \subseteq Inv\ [A]$.)
Analogously, one can prove $Pol\ [Q] = Pol\ Q$. One can easily checks the remaining statements of the theorem. ■

A conclusion of Theorem 2.6.1, (a) and of the definition of $\vdash$ (see Section 2.4) is the following

Theorem 2.6.3 *For arbitrary relations $\varrho, \varrho' \in R_k$ it holds:*

$$\varrho \vdash \varrho' \implies Pol\ \varrho \subseteq Pol\ \varrho'.$$

■

2.7 The Relations χ_n and G_n

For arbitrary $n \in \mathbb{N}$ and $k \in \mathbb{N}\backslash\{1\}$ denote $\chi_{k;n}$ or – if k can be seen from the context – χ_n the k^n-ary relation, whose rows are just all $(x_1, ..., x_n) \in E_k^n$ that are arranged (we say "**lexicographical**") unambiguously according to the following regulation:

The tuple $(x_1, ..., x_n)$ is before the tuple $(y_1, ..., y_n)$, if the integer

$$x_1 \cdot k^{n-1} + x_2 \cdot k^{n-2} + ... + x_{n-1} \cdot k + x_n$$

is smaller than

$$y_1 \cdot k^{n-1} + y_2 \cdot k^{n-2} + ... + y_{n-1} \cdot k + y_n.$$

For example, the following is valid:

$$\chi_{2;3} := \begin{pmatrix} 0 & 0 & 0 \\ 0 & 0 & 1 \\ 0 & 1 & 0 \\ 0 & 1 & 1 \\ 1 & 0 & 0 \\ 1 & 0 & 1 \\ 1 & 1 & 0 \\ 1 & 1 & 1 \end{pmatrix}.$$

We denote the columns of χ_n with

$$\chi(1), ..., \chi(n).$$

Obviously, there is exactly a function $f_r \in P_k^n$ with $f(\chi_n) = r$ to every column $r \in E_k^{k^n}$.
The relation

$$G_n(A) := \{r \in E_k^{k^n} | f_r \in A^n\},$$

which one can form with the aid of the functions f_r, is called **n-th graphic of $A \subseteq P_k$**.
The following theorem summarizes elementary properties of the relation $G_n(A)$.

Theorem 2.7.1 *For an arbitrary clone $A \subseteq P_k$ it holds:*

(a) $\forall n \in \mathbb{N} : G_n(A) \in Inv\ A$;

(b) $f^n \in A^n \iff f^n \in Pol\ G_n(A)$;

(c) $A \subseteq ... \subseteq Pol\ G_n(A) \subseteq Pol\ G_{n-1}(A) \subseteq ... \subseteq Pol\ G_2(A) \subseteq Pol\ G_1(A)$;

(d) $A = \bigcap_{n \geq 1} Pol\ G_n(A)$;

(e) $\forall \varrho \in Inv\ A : \varrho \in [\bigcup_{n \geq 1}\{G_n(A)\}]$;

(f) $Inv\ A = [\bigcup_{n \geq 1}\{G_n(A)\}]$.

Proof. (a): Let $g^m \in A$ and $r_1, ..., r_m \in G_n(A)$ be arbitrary. Then $g(r_1, ..., r_m) = g(f_{r_1}, ..., f_{r_m}) = h(\chi_n)$, where

$$h(x_1, ..., x_n) := g(f_{r_1}(x_1, ..., x_n), ..., f_{r_m}(x_1, ..., x_n)).$$

Since h belongs to A^n, we obtain $h(\chi_n) \in G_n(A)$.
(b): If $f^n \in A$, we have $f^n \in Pol\ \varrho$ for every $\varrho \in Inv\ A$. Consequently, $f^n \in Pol\ G_n(A)$ because of (a). On the other hand, $f^n \in G_n(A)$ implies the existence of a certain $r \in G_n(A)$ with $f^n(\chi_n) = r$; thus $f = f_r \in A^n$ is valid.
(c) and (d) follow from (b) and from the clone properties.
(e): Let $\varrho \in Inv\ ^h A$ be arbitrary and $t := |\varrho|$. We show that ϱ can be derived from $G_t(A)$ using the operation $pr_{\alpha_1,...,\alpha_h}$. Since $J_k \subseteq A$, we have $\chi_t \subseteq G_t(A)$. For every $j \in \{1, ..., h\}$, one can find an α_j so that the j-th row of ϱ is identical with the α_j-th row of χ_t. Because of $\varrho \in Inv\ ^h A$, it follows $pr_{\alpha_1,...,\alpha_h} G_t(A) = \varrho \in [\{G_t(A)\}]$.
(f): By (e) we have $Inv\ A \subseteq [\bigcup_{n \geq 1}\{G_n(A)\}]$. Further, because of (a) and Theorem 2.6.2, we have $[\bigcup_{n \geq 1}\{G_n(A)\}] \subseteq Inv\ A$. Hence, (f) is valid. ∎

2.8 The Operator Γ_A

For arbitrary $A \subseteq P_k$ denote Γ_A a mapping from R_k into R_k, which is defined for $\sigma \in R_k^h$ as follows:

$$\Gamma_A(\sigma) := \bigcap\{\varrho \in R_k \mid \varrho \in Inv\ A \wedge \sigma \subseteq \varrho\}. \tag{2.1}$$

In the language of the Universal Algebra, $\Gamma_A(\sigma)$ is a subalgebra of $(E_k; A)^h$, which is generated by σ, or $\Gamma_A(\sigma)$ is the universe of the smallest subalgebra of the algebra $(E_k; A)^h$, which contains the set σ. Obviously, Γ_A is a hull operator.

Theorem 2.8.1 *For an arbitrary clone $A \subseteq P_k$ and every $n \in \mathbb{N}$ it holds:*

(a) $\Gamma_A(\chi_n) \in Inv\ A$;

(b) $\Gamma_A(\chi_n) = G_n(A)$;

(c) $A^n = \{f_r | r \in \Gamma_A(\chi_n)\}$.

Proof. (a) follows from $[Inv\ A] = Inv\ A$ (see Theorem 2.6.2) and the definition of $\Gamma_A(\chi_n)$.
(b): Since every projection e_i^n belongs to A^n, we have $\chi_n \subseteq G_n(A)$.
Denote now ϱ an arbitrary k^n-ary relation of R_k with $\chi_n \subseteq \varrho$. If $\varrho \in Inv\ A$, then $f(\chi_n) \in \varrho$ for every $f \in A^n$, i.e., $G_n(A) \subseteq \varrho$. Consequently, we have shown that $G_n(A) \subseteq \Gamma_A(\chi_n)$ holds. $\Gamma_A(\chi_n) \subseteq G_n(A)$ follows from $G_n(A) \in Inv\ A$ (see Theorem 2.7.1, (a)).
(c) is an easy conclusion from (b) and the definition of $G_n(A)$. ∎

2.9 The Galois Theory for Function- and Relation-Algebras

Theorem 2.9.1 *Let A be a clone of P_k. Then*

$$A = Pol\ Inv\ A.$$

Proof. By Theorem 2.6.1, (b) we have $A \subseteq Pol\ Inv\ A$. To prove that $Pol\ Inv\ A \subseteq A$ let $f^n \in Pol\ Inv\ A$ be arbitrary. Because of Theorem 2.7.1, (a) we have that $f \in Pol\ G_n(A)$ holds and (by Theorem 2.7.1, (b)) $f \in A^n$. Thus $A = Pol\ Inv\ A$. ■

Theorem 2.9.2 *Let Q be a co-clone of R_k. Then*

$$Q = Inv\ Pol\ Q.$$

Proof. Let $A := Pol\ Q$. Because of Theorem 2.6.1, (b) we have $Q \subseteq Inv\ A$. To prove that $Inv\ A \subseteq Q$ it is sufficient to show that $\Gamma(\chi_t) \in Q$ for arbitrary $t \in \mathbb{N}$, since $[\bigcup_{t \geq 1}\{\Gamma_A(\chi_t)\}] = Inv\ A$ (see Theorem 2.7.1, (f) and Theorem 2.8.1, (b)). Let now

$$\gamma := \bigcap\{\varrho \in Q | \chi_t \subseteq \varrho\}.$$

Because of $E_k^{k^t} \in Q$ and the fact that Q is closed in respect to $\cap$, we have $\chi_t \subseteq \gamma$ and γ is the smallest relation ($\in Q^{k^t}$), which contains χ_t, in respect to cardinality. Further, we have $\Gamma_A(\chi_t) \subseteq \gamma$, since $\gamma \in Q \subseteq Inv\ A$ (see (2.1)). Consequently, our theorem is proven, if we can show $\Gamma_A(\chi_t) = \gamma$. Suppose, $\Gamma_A(\chi_t) \subset \gamma$. Then, there is a column $r \in \gamma \backslash \Gamma_A(\chi_t)$. Because of $A^t = \{f_s | s \in \Gamma_A(\chi_t)\}$ (see Theorem 2.8.1, (c)) we have $f_r \notin A^t$. Consequently, there exists an m-ary relation $\beta \in Inv\ A$ and certain columns $r_1, ..., r_m \in \beta$ with $f(r_1, ..., r_m) \notin \beta$. Every row of the matrix $(r_1, ..., r_m)$ is also a row of the matrix χ_t. Denote i_j the number of a row of χ_t, which agrees with the j-th row of $(r_1, ..., r_m)$ $(j = 1, 2, ..., m)$. Let now

$$\gamma' := pr_{1,2,...,k^t}(\gamma \times \beta) \cap \delta^{k^t+m}_{\{i_1,k^t+1\},\{i_2,k^t+2\},...,\{i_m,k^t+m\}}.$$

Since Q is closed, γ' belongs to Q, and by construction of γ' we have $\chi_t \subseteq \gamma' \subseteq \gamma$. Furthermore, we have $r \in \gamma \backslash \gamma'$, since $r_1, ..., r_t \in \beta$, $f_r(r_1, ..., r_t) \notin \beta$ and $f_r(\chi_t) = r \in \gamma$. With γ' we received a contradiction to the choice of γ. Thus, $\gamma = \Gamma_A(\chi_t)$. ■

With the aid of Theorems 2.9.1 and 2.9.2, we can prove the important properties of the *Pol*-*Inv*-connection:

Theorem 2.9.3 (Theorem of V. G. Bodnarčuk, L. A. Kalužnin, V. N. Kotov and B. A. Romov; *[Bod-K-K-R 68/69]) indextheorem of Bodnarčuk, Kalužnin,. Kotov and Romov*
Let $\mathbb{L}(P_k)$ (or $\mathbb{L}(R_k)$) be the set of all clones (or co-clones) of P_k (or R_k) respectively. Then the mappings

$$Inv\ : \mathbb{L}(P_k) \longrightarrow \mathbb{L}(R_k), A \mapsto Inv\ A$$

and

$$Pol\ : \mathbb{L}(R_k) \longrightarrow \mathbb{L}(P_k), Q \mapsto Pol\ Q$$

are bijective mappings, which reverse the partial order $\subseteq$, i.e., it holds

$$\forall A, B \in \mathbb{L}(P_k) : A \subseteq B \Longrightarrow Inv\ B \subseteq Inv\ A$$

and

$$\forall S, T \in \mathbb{L}(R_k) : S \subseteq T \Longrightarrow Pol\ T \subseteq Pol\ S.$$

In other words:
The lattices $(\mathbb{L}(P_k), \subseteq)$ and $(\mathbb{L}(R_k), \subseteq)$ are antiisomorphic.

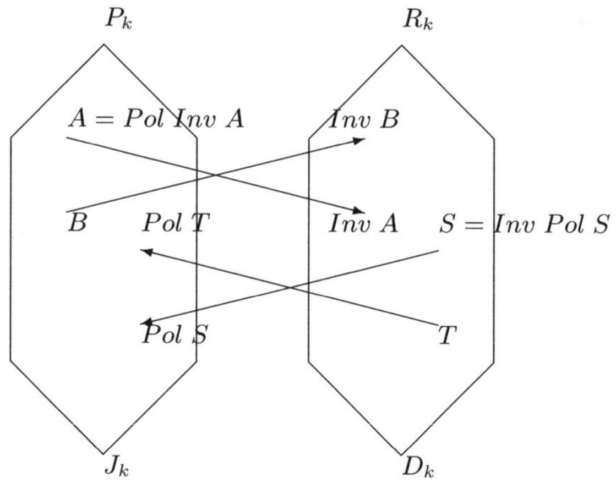

Fig. 2.1

Proof. By Theorem 2.6.2, the mappings Inv and Pol are mappings from $\mathbb{L}(P_k)$ (or $\mathbb{L}(R_k)$) into $\mathbb{L}(R_k)$ (or $\mathbb{L}(P_k)$), respectively. The surjectivity and the injectivity (and then the bijectivity) of these mappings are easy conclusions from Theorem 2.9.1 and Theorem 2.9.2 (with the help of Theorem 2.6.2). The "reversal property" of Pol and Inv (in respect to $\subseteq$) was already given in Theorem 2.6.1, (a). ■

2.10 Some Modifications of the *Pol*-*Inv*-Connection

2.10.1 Galois Theory for Finite Monoids and Finite Groups

It is well-known that every finite semigroup $\mathbf{H} = (H; \circ)$ is isomorphic to a certain subsemigroup of $(P^1_{|H|}; \star)$ and every finite group $\mathbf{G} = (G; \circ)$ is isomorphic to a certain subgroup of the group $(S_{|G|}; \star)$, where $S_k := P^1_k[k]$ for $k \in \mathbb{N}$. [2]

Furthermore we have

Lemma 2.10.1.1 *For an arbitrary subset A of $[P^1_k]$, the set A is a clone of $[P^1_k]$ if and only if $A = [A]_\nabla$ holds and $(A^1; \star)$ is a subsemigroup of $(P^1_k; \star)$ with the unit element e.* ■

In other words, a clone $A \subseteq [P^1_k]$ is completely determined by the monoid $(A^1; \star)$.
Because of the above properties, it is possible to derive a Galois theory for semigroups and groups from the Galois theory for function algebras (see Section 2.9). For this purpose, we define two new operations $\vee$ and $\neg$ on R_k: For arbitrary $\varrho, \varrho' \in R^h_k$ and $h \in \mathbb{N}$ we set

$$\varrho \vee \varrho' := \varrho \cup \varrho' \text{ (union)}$$

and

$$\neg\varrho := E^h_k \backslash \varrho \text{ (negation, complement)}.$$

The following lemma supplies a reason for these new relation operations.

Lemma 2.10.1.2 *It holds:*
(a) $\forall H \subseteq P^1_k : \; \varrho, \varrho' \in Inv^h\, H \implies \varrho \vee \varrho' \in Inv^h\, H;$
(b) $\forall G \subseteq S_k : \; \varrho \in Inv^h\, G \implies \neg\varrho \in Inv^h\, G.$

Proof. (a): If $f \in H \subseteq P^1_k$, $\{\varrho, \varrho'\} \subseteq Inv\, H$ and $r \in \varrho \vee \varrho'$, then we have obvious $r \in \varrho$ or $r \in \varrho'$, whereby $f(r) \in \varrho \vee \varrho'$, since f preserves the relations ϱ and ϱ'. Consequently, $\varrho \vee \varrho'$ is also an invariant of f.
(b): Let $f \in G \subseteq S_k$, $\varrho \in Inv\, G$ and $r \in \neg\varrho$ be arbitrary. Set $r' := f(r)$. Then r' belongs to $\neg\varrho$, since $f^{-1}(r') = r$, $f^{-1} \in [G]$ and $Inv\, G = Inv\, [G]$ (see Theorem 2.6.2). Consequently, f preserves the relation $\neg\varrho$. ■

An algebra of the form

$$(Q; \delta^3_{k;\{1,2\}}, \zeta, \tau, pr, \times, \wedge, \vee)$$

[2] For proof, one can assume w.l.o.g. $H = G = \{0, 1, ..., k-1\}$. Let now $f_a(x) := a \circ x \in P^1_k$ for arbitrary $a \in E_k$. Then, the mapping $\alpha : \; E_k \longrightarrow P^1_k,\; a \mapsto f_a$ is an isomorphic mapping from H (or G) onto the semigroup (or group) $(\{f_0, f_1, ..., f_{k-1}\}; \star)$, since $\alpha(a \circ b)(x) = f_{a\circ b}(x) = (a \circ b) \circ x \overset{(A)}{=} a \circ (b \circ x) = f_a(f_b(x)) = (f_a \star f_b)(x)$.

with $Q \subseteq R_k$ is called **Krasner-algebra of first kind** and an algebra of the form

$$(Q; \delta^3_{k;\{1,2\}}, \zeta, \tau, pr, \times, \wedge, \vee, \neg)$$

with $Q \subseteq R_k$ **Krasner-algebra of second kind**.
In Theorem 2.10.1.4, we will see that the algebras just defined are the right partners for semigroups (or groups) for a Galois-correspondence. Following our terminology from Section 2.4, we call this a clone, which is also closed, concerning the operation $\vee$, a **co-monoid**. Moreover, a co-monoid, which is closed concerning the operation $\neg$, is a **co-group**.
The following lemma summarizes all auxiliary statements that are necessary to prove Theorem 2.10.1.4.

Lemma 2.10.1.3 *Let $H, G \subseteq P^1_k$ and $Q, T \subseteq R_k$. Further, let $\mathbf{H} := (H; \star)$ be a monoid, $\mathbf{G} := (G; \star)$ be a group, Q be a co-monoid, and T be a co-group. Then*

(a)	$H = Pol^1\ Inv\ H$	*(a')*	$G = Pol^1\ Inv\ G$
(b)	$Pol\ Q \subseteq [P^1_k]$	*(b')*	$Pol\ T \subseteq [S_k]$
(c)	$Q = Inv\ Pol^1\ Q$	*(c')*	$T = Inv\ Pol^1\ T.$

Proof. (a), (a'): Let A be a monoid of P^1_k (or a group of S_k). Then, by Lemma 2.10.1.1, $[A]_\nabla$ is a clone of P_k and, because of Theorem 2.6.2, we have $[A]_\nabla = Pol\ Inv\ [A]_\nabla = Pol\ Inv\ A$. Consequently, $A = [A]^1 = Pol^1\ Inv\ A$.
(b): Since the diagonal relations belong to Q and since Q is closed in respect to $\vee$, the relation

$$\gamma := \delta^3_{k;\{1,2\}} \cup \delta^3_{k;\{2,3\}}$$

belongs to Q. With the help of Theorem 1.4.4, it can easily be shown that the relation is preserved of no function of P_k, which depends on at least two variables essentially. Thus, $Pol\ Q \subseteq [P^1_k]$.
(b'): Through (b) we already proven $Pol\ T \subseteq [P^1_k]$. Further, the diagonal relation $\iota^2_k := \{(x,x) | x \in E_k\}$ belongs to Q, whereby also $\neg\iota^2_k \in Q$. Obviously the relation $\neg\iota^2_k = \{(x,y) | x \neq y\}$ is preserved, however, only from the permutations of P^1_k. Consequently, we have $Pol\ T \subseteq [S_k]$.
(c), (c'): If Q is a co-monoid (or is a co-group) of R_k, then Q is also a co-clone of R_k and it is valid (because of $Pol^1\ Q \subseteq Pol\ Q$, Theorem 2.6.1, (a) and Theorem 2.9.2):

$$Q = Inv\ Pol\ Q \subseteq Inv\ Pol^1\ Q.$$

By (b) (or (b')) we have in addition $Pol\ Q \subseteq [Pol^1\ Q]$, whereby $Inv\ [Pol^1\ Q] \subseteq Inv\ Pol\ Q$ and (because of $Inv\ Pol^1\ Q = Inv\ [Pol^1\ Q]$ by Theorem 2.6.2 and $Inv\ Pol\ Q = Q$ by Theorem 2.9.2) $Inv\ Pol^1\ Q \subseteq Q$ follow. ∎

One obtains the following theorem as an easy conclusion in analogy to the above theorem from Lemmas 2.10.1.2 and 2.10.1.3.

Theorem 2.10.1.4 *([Kra 45], [Kra 68/69])*
Let $\mathfrak{M}$ be the set of all monoids of the form $(M;\star)$ with $M \subseteq P_k^1$ and $\mathfrak{G}$ be the set of all groups of the form $(G;\star)$ with $G \subseteq S_k$. Further, let $\mathfrak{K}_1$ be the set of all co-monoids of R_k and $\mathfrak{K}_2$ be the set of all co-groups of R_k. Then, the lattices $(\mathfrak{M};\subseteq)$ and $(\mathfrak{K}_1;\subseteq)$ and the lattices $(\mathfrak{G};\subseteq)$ and $(\mathfrak{K}_2;\subseteq)$ are antiisomorphic.

In particular, for $i \in \{1,2\}$, $\mathfrak{L}_1 := \mathfrak{M}$ and $\mathfrak{L}_2 := \mathfrak{G}$ are valid:
The mappings

$$Pol^1 : \mathfrak{K}_i \longrightarrow \mathfrak{L}_i,\ Q \mapsto Pol^1\ Q$$

and

$$Inv : \mathfrak{L}_i \longrightarrow \mathfrak{K}_i,\ H \mapsto Inv\ H$$

are bijective mappings, which reverse the partial order $\subseteq$, i.e., it holds

$$\forall H, G \subseteq \mathfrak{L}_i :\ H \subseteq G \Longrightarrow Inv\ G \subseteq Inv\ H$$

and

$$\forall S, T \subseteq \mathfrak{K}_i :\ S \subseteq T \Longrightarrow Pol^1\ T \subseteq Pol^1\ S.$$

■

2.10.2 Galois Theory for Iterative Function Algebras

Iterative function algebras are algebras of the form $(P_k;\zeta,\tau,\Delta,\nabla;\star)$. The universes of subalgebras of these algebras $\mathbf{P_k}$ ($k \in \mathbb{N}\backslash\{1\}$) are called subclasses or, briefly, classes of P_k.
Unlike clones of P_k, there are classes of P_k that do not contain the projections. Nevertheless, for the classes of P_k, one can also find certain relation algebras as partners for a Galois-correspondence. Subsequently, two possibilities for this purpose will be presented.
For the first possibility, we choose a certain class A of P_k that does not contain the function e_1^1 and which, therefore, is not a clone.
Let $\mathbb{L}_k^{\downarrow}(A)$ be the lattice of all subclasses of A and let $\mathbb{L}_k(J_k, J_k \cup A)$ be the lattice of all subclasses of $J_k \cup A$, which have J_k as a subset. Moreover, let $\mathbb{L}_k(Inv(J_k \cup A), R_k)$ be the set of all co-clones $\mathcal{C}$ of R_k with $Inv(J_k \cup A) \subseteq \mathcal{C}$. Obviously, the mapping

$$\alpha : \mathbb{L}_k^{\downarrow}(A) \longrightarrow \mathbb{L}_k(J_k, J_k \cup A),\ T \mapsto J_k \cup T$$

is an isomorphism from $\mathbb{L}_k^{\downarrow}$ onto $\mathbb{L}_k(J_k, J_k \cup A)$. Since the elements of $\mathbb{L}_k(J_k, J_k \cup A)$ are clones of P_k, the lattice $\mathbb{L}_k(J_k, J_k \cup A)$ is antiisomorphic to the lattice $\mathbb{L}_k(Inv(J_k \cup A), R_k)$ by Theorem 2.9.3. Consequently, we have

Theorem 2.10.2.1 *Let A be a subclass of P_k. Further, put*

$$Pol_A Q := A \cap Pol_k Q$$

for arbitrary $Q \subseteq R_k$. Then the mappings

$$Inv : \mathbb{L}_k^{\downarrow}(A) \longrightarrow L_k(Inv(J_k \cup A)),\ T \mapsto Inv\, T,$$

and

$$Pol_A : \mathbb{L}_k(Inv(J_k \cup A), R_k) \longrightarrow \mathbb{L}_k^{\downarrow}(A),\ Q \mapsto Pol_A Q$$

are bijective mappings, and the lattices $\mathbb{L}_k^{\downarrow}(A)$ and $\mathbb{L}_k(Inv(J_k \cup A), R_k)$ are antiisomorphic. ■

As W. Harnau in [Har 83] was pointing, another possibility to characterize subclasses of P_k consists that one establishes relation pairs and relation-pair-algebras. Let

$$Rp_k^h := \{(\varrho, \varrho') \in R_k^h \times R_k^h \mid \varrho' \subseteq \varrho\}$$

and

$$Rp_k := \bigcup_{h \geq 1} Rp_k^h.$$

The elements of Rp_k are called **relation pairs**.
We say that $f^n \in P_k$ **preserves the relation pair** $(\varrho, \varrho') \in Rp_k$, if $f(r_1, ..., r_n) \in \varrho'$ for arbitrary $r_1, ..., r_n \in \varrho$. By definition, each function of P_k preserves the pair $(\emptyset, \emptyset)$.
If $\varrho = \varrho'$, then the fact "f preserves the relation pair (ϱ, ϱ')" is identical with the fact "f preserves ϱ". Let $F \subseteq P_k$ and $Q \subseteq Rp_k$. Then, denote $Invp_k F$ the set of all relation pairs of Rp_k, which are preserved from every function of F. Further, let $Polp_k Q$ be the set of all functions of P_k, which preserve each relation pair of Q. It can easily be shown that the following lemma holds:

Lemma 2.10.2.2 *([Har 83])*
(a) $\forall Q \subseteq Rp_k : \ Polp_k Q = [Polp_k Q]_{\zeta,\tau,\Delta,\nabla,\star}$*;*
(b) $\forall (\varrho, \varrho') \in Rp_k : \ ([Pol(\varrho, \varrho')]_{e_1^2,\zeta,\tau,\Delta,\star} = Pol(\varrho, \varrho') \Longleftrightarrow \varrho = \varrho')$. ■

Analogous to Section 2.3, one can define certain operations on the set Rp_k: For $\alpha \in \{\zeta, \tau, \Delta, pr, \sim\}$ (see Section 2.3, $\sim \varrho^h := \nabla \varrho \wedge \delta_{\{1,2\}}^{h+1}$ ("the doubling of the first row of ϱ")) and $(\varrho, \varrho'), (\mu, \mu') \in Rp_k$ we set:

$$\alpha(\varrho, \varrho') := (\alpha\varrho, \alpha\varrho'),$$

$$\nabla(\varrho, \varrho') := \begin{cases} (\nabla\varrho, \nabla\varrho'), & \text{if } \varrho' \neq \emptyset, \\ (\nabla\varrho, \emptyset), & \text{if } \varrho' = \emptyset, \end{cases}$$

and

$$(\varrho, \varrho') \times (\mu, \mu') := (\varrho \times \mu, \varrho' \times \mu').$$

Further, put $E := \bigcup_{h \geq 1} E_k^h$ in the following. For each $\mathbf{a} \in E$ one can define two operations on Rp_k as follows:

$$\nu_{1,\mathbf{a}}(\varrho, \varrho') := \begin{cases} (\varrho \backslash \{\mathbf{a}\}, \varrho'), \text{ if } & \varrho' \subseteq \varrho \backslash \{\mathbf{a}\}, \\ (\varrho, \varrho') & \text{otherwise}, \end{cases}$$

and

$$\nu_{2,\mathbf{a}}(\varrho, \varrho') := \begin{cases} (\varrho, \varrho' \cup \{\mathbf{a}\}), \text{ if } & \mathbf{a} \in \varrho, \\ (\varrho, \varrho') & \text{otherwise} \end{cases}$$

$((\varrho, \varrho') \in Rp_k)$. [3]

The importance of the above-defined operations results from the following lemma, which one can easily check.

Lemma 2.10.2.3 *Let* (ϱ, ϱ') *be a relation pair which is derivable from* Q $(\subseteq Rp_k)$ *by means of the operations* $\zeta, \tau, \Delta, \nabla, \sim, pr, \times, \nu_{1,\mathbf{a}}, \nu_{2,\mathbf{a}}$ $(\mathbf{a} \in E)$. *Then*

$$Polp_k Q \subseteq Polp_k(\varrho, \varrho').$$

■

The algebra

$$\mathbf{Rp_k} := (Rp_k; (\emptyset, \emptyset), \zeta, \tau, \Delta, pr, \sim, \times, (\nu_{1,\mathbf{a}})_{\mathbf{a} \in E}, (\nu_{2,\mathbf{a}})_{\mathbf{a} \in E})$$

of the type (0,1,1,1,1,1,2,1,1,1,...) is called **full relation-pair algebra** on E_k. A **relation-pair algebras** is a subalgebra of this algebra. Further, a universe of a relation-pair algebra is called **co-class**.

Then, the following theorem can be proven with some expenditure (ultimately similar to the proof of Theorem 2.9.3):

Theorem 2.10.2.4 *([Har 83])*

Let $\mathbb{L}(P_k)$ *(or* $\mathbb{L}(Rp_k)$*) be the set of all classes (or co-classes) of* P_k *(or* Rp_k*), respectively. Then the mappings*

$$Inv_k : \mathbb{L}(P_k) \longrightarrow \mathbb{L}(Rp_k), \ A \mapsto Inv_k A$$

and

$$Polp_k : \mathbb{L}(Rp_k) \longrightarrow \mathbb{L}(P_k), \ Q \mapsto Polp_k Q$$

[3] In [Har 83] so-called multi-operators d_h and d_v are first defined instead of the operations $\nu_{1,\mathbf{a}}, \nu_{2,\mathbf{a}}$ $(\mathbf{a} \in E)$:
$d_h(\varrho, \varrho') := \{(\varrho, \mu) \in Rp_k | \varrho' \subseteq \mu \subseteq \varrho\}$, $d_v(\varrho, \varrho') := \{(\mu, \varrho') \in Rp_k | \varrho' \subseteq \mu \subseteq \varrho\}$.
Then, relation-matrix-algebras, in which these multi-operators are describable through two operations, are introduced. It is shown how one can repair the "lack" with the multi-operators by going over to the relation-matrix-algebras with finite many operations.

are bijective mappings with the properties

$$\forall A, B \in \mathbb{L}(P_k): \ A \subseteq B \Longrightarrow Inv_k B \subseteq Inv_k A,$$

$$\forall S, T \in \mathbb{L}(Rp_k): S \subseteq T \Longrightarrow Polp_k T \subseteq Polp_k S;$$

i.e., the lattices $(\mathbb{L}(P_k);\subseteq)$ *and* $(\mathbb{L}(Rp_k);\subseteq)$ *are antiisomorphic.* ■

2.11 Some Connections Between the Relation Operations

In this last section of Chapter 2, we want to clarify the order in which one can use the relation operations given in Section 2.3, to obtain all relations from the set $[Q]$ $(\subseteq R_k)$ as effectively as possible.
Let

$$\Delta'\varrho := \{(x_1, x_2, ..., x_h) \in \varrho \mid x_1 = x_2\} \quad (\varrho \in R_k^h).$$

This relation is obvious derivable from our elementary operations. For arbitrary $\alpha, \alpha_1, ..., \alpha_t \in \{\zeta, \tau, \Delta, \Delta', \nabla, pr\}$ we use the following notations:

$$\alpha^1\varrho := \alpha(\varrho), \ \alpha^i\varrho := \alpha(\alpha^{i-1}\varrho) \text{ for } i \in \mathbb{N},$$
$$\alpha_1\alpha_2...\alpha_t\varrho := \alpha_1(...(\alpha_{t-1}(\alpha_t(\varrho))...)$$
$$(\varrho \in R_k).$$

It can easily be shown that the Lemma 2.11.1 holds.

Lemma 2.11.1 *For arbitrary relations* $\varrho \in R_k^h$, $\varrho' \in R_k^{h'}$ *and* $\varrho'' \in R_k^{h''}$ *it is valid:*

(a) $\varrho \times \varrho' = (\zeta^{h'}\Delta^{h'}\varrho) \cap (\nabla^h\varrho');$

(b) $\emptyset = pr^h\varrho, \ \delta^3_{k;\{1,2\}} = \Delta'\nabla^3\emptyset;$

(c) $\zeta(pr\,\varrho) = pr(\zeta(\tau(\varrho))),$
$\tau(pr\,\varrho) = pr(\zeta^{h-1}(\tau(\zeta(\varrho)))),$
$\Delta'(pr\,\varrho) = pr(\zeta^{h-1}(\Delta'(\zeta(\varrho)))),$
$\nabla(pr\,\varrho) = pr(\tau((\nabla(\varrho))));$

(d) $\zeta(\varrho \wedge \varrho') = (\zeta\varrho) \wedge (\zeta\varrho'),$
$\tau(\varrho \wedge \varrho') = (\tau\varrho) \wedge (\tau\varrho'),$
$\Delta'(\varrho \wedge \varrho') = (\nabla\varrho) \wedge (\nabla\varrho')$, *if* $h = h'$;

(e) $(\varrho \vee \varrho') \wedge \varrho'' = (\varrho \wedge \varrho'') \vee (\varrho' \wedge \varrho'')$, *if* $h = h' = h''$. ■

Theorem 2.11.2 *Let* $\Omega := \{\delta^3_{k;\{1,2\}}, \zeta, \tau, pr, \times, \wedge\}$. *Then for all* $Q \subseteq R_k$ *it is valid:*

(a) $[Q]_\Omega = [Q]_{\zeta,\tau,\Delta',\nabla,pr,\wedge}$;

(b) $[Q]_{\zeta,\tau,\Delta',\nabla,pr,\wedge} = [\,[[Q]_{\zeta,\tau,\Delta',\nabla}]_\wedge\,]_{pr}$;

(c) $[Q]_{\Omega\cup\{\vee\}} = [Q]_{\zeta,\tau,\Delta',\nabla,pr,\wedge,\vee}$;

(d) $[Q]_{\zeta,\tau,\Delta',\nabla,pr,\wedge,\vee} = [\,[\,[\,[Q]_{\zeta,\tau,\Delta',\nabla}]_\wedge]_\vee]_{pr}$.

Proof. (a): Since Δ' and ∇ are Ω-derivable operations, we have $[Q]_{\zeta,\tau,\Delta',\nabla,pr,\wedge} \subseteq [Q]_\Omega$. The reversed inclusion $[Q]_\Omega \subseteq [Q]_{\zeta,\tau,\Delta',\nabla,pr,\wedge}$ follows from Lemma 2.11.1, (a), (b).
(b) follows from Lemma 2.11.1, (c), (d).
One proves the statements (c) and (d) analogously to (a) and (b) using Lemma 2.11.1, (e). ∎

A translation of these observations into the language of mathematical logic can be found in [Pös-K 79], p. 63–68.

3

The Subclasses of P_2

A basic result of many-valued logic is the description of all closed sets of Boolean functions given by E. L. Post in [Pos 20] and [Pos 41]. Since Post's proof is long and rather complicated, revisions (for instance [Jab-G-K 70] and [Ugo 88]) and new proofs (for instance [Ber 80], [McK-M-T 87] and [Res-D 89] or [Den-W 2002]) have been published. The new proof methods of the last years mainly result from the fact that parts of Post' results are special cases or conclusions of certain theorems of many-valued logic or universal algebra. In this chapter, we tried to verify Post's results in an elementary way by working out some essential basic ideas.

3.1 Definitions of the Subclasses of P_2 and Post's Theorem

We need some notations introduced in Chapter 1 to define the subclasses of P_2 and to describe certain generated sets of this subclasses.
We shall define functions of P_2 by formulae over the alphabet $\{x, y, z, x_1, x_2, ...\}$ and we use the usual symbols $\wedge$ ("conjunction" or "multiplication modulo 2"), $\vee$ ("disjunction"), $+$ ("addition modulo 2") and $^-$ ("negation"). By

$$\circ \in \{\wedge, \vee, +\},\ ^-,\ c_a^n\ (a \in E_2),\ e_i^n\ (1 \le i \le n),\ m^3,\ t^2,\ q^3,\ r^3,\ h_\mu^{\mu+1}\ (\mu \in \mathbb{N})$$

we denote functions of P_2 given by

$$\circ(x,y) := x \circ y,$$
$$^{-}(x) := \overline{x},$$
$$c_a^n(x_1,...,x_n) := a,$$
$$e_i^n(x_1,...,x_n) := x_i,$$
$$m(x,y,z) := x \wedge (y \vee z),$$
$$t(x,y) := x \wedge \overline{y},$$
$$q(x,y,z) := x \wedge (y \vee \overline{z}),$$
$$r(x,y,z) := x + y + z,$$

$$h_\mu(x_1,...,x_{\mu+1}) := \bigvee_{i=1}^{\mu+1}(x_1 \wedge x_2 \wedge ... \wedge x_{i-1} \wedge x_{i+1} \wedge ... \wedge x_{\mu+1}) =$$
$$\begin{cases} 1 \text{ if} & \exists i \in \{1,...,\mu+1\} : x_1 = ... = x_{i-1} = x_{i+1} = ... = x_{\mu+1} = 1, \\ 0 \text{ otherwise} & \end{cases}$$

($h^1 = \vee$), respectively.
We write $\mathbf{x}$ for $(x_1,...,x_n)$, α for $(\alpha,\alpha,...,\alpha)$ ($\alpha \in E_2$) and often xy instead of $x \wedge y$. Finally, let

$$x^\sigma := \begin{cases} \overline{x}, & \text{if } \sigma = 0, \\ x, & \text{if } \sigma = 1. \end{cases}$$

The mapping

$$\delta : P_2 \longrightarrow P_2, f^n \longrightarrow (f^\delta)^n$$

with

$$f^\delta(x_1,...,x_n) := \overline{f(\overline{x_1},\overline{x_2},...,\overline{x_n})}$$

is known to be an automorphism from $\mathbf{P_2}$, which we shall use to describe isomorphic closed subsets of P_2 (see Section 9.11), where $A^\delta := \{f^\delta \mid f \in A\}$. We remark that δ is the unique non-trivial isomorphism from a subalgebra of $\mathbf{P_2}$ to a subalgebra of $\mathbf{P_2}$ (see Theorem 9.12.5). In the following, we define some closed subsets of P_2 with the help of which we can describe all subclasses of P_2 with the applications of $\cap$ and $\cup$:

- $M := Pol\begin{pmatrix} 0 & 0 & 1 \\ 0 & 1 & 1 \end{pmatrix}$

  $= \bigcup_{n\geq 1}\{f^n \in P_2 \mid \forall \mathbf{a},\mathbf{b} \in E_2^n : \mathbf{a} \leq \mathbf{b} \Longrightarrow f(\mathbf{a}) \leq f(\mathbf{b})\}$

  (set of all non-decreasing **monotone functions**),

- $S := Pol\begin{pmatrix} 0 & 1 \\ 1 & 0 \end{pmatrix}$

  $= \bigcup_{n\geq 1}\{f^n \in P_2 \mid f(x_1,...,x_n) = \overline{f(\overline{x_1},\overline{x_2},...,\overline{x_n})}\}$

  (set of all **self-dual functions**),

- $L := \bigcup_{n\geq 1}\{f^n \in P_2 \mid \exists a_0, ..., a_n \in E_2 : f(\mathbf{x}) = a_0 + \Sigma_{n=1}^{n} a_i \cdot x_i\}$ (set of all **linear functions**),

- $T_{0,\mu} := Pol E_2^{\mu} \backslash \{\mathbf{1}\}$ if $\mu \in \mathbb{N}$

$$\begin{aligned}(f^n \in T_{0,\mu} \iff & (\forall \mathbf{a_1}, ..., \mathbf{a}_\mu \in E_2^{\mu} : \\ & (\forall i \in \{1, ..., \mu\} : \ f(\mathbf{a_i}) = 1 \text{ and } \mathbf{a_i} = (a_{i1}, ..., a_{in})) \Longrightarrow \\ & \exists\, j \in \{1, ..., n\} : a_{1j} = a_{2j} = ... = a_{\mu j} = 1) \\ \iff & (\forall \mathbf{a_1}, ..., \mathbf{a}_\mu \in E_2^{n}\ \exists\, j \in \{1, ..., n\} : \\ & \forall \mathbf{x} \in \{\mathbf{a_1}, ..., \mathbf{a}_\mu\} : f(\mathbf{x}) = x_j \wedge f(\mathbf{x}))),\end{aligned}$$

$T_{1,\mu} := T_{0,\mu}^{\delta} = Pol E_2^{\mu} \backslash \{\mathbf{0}\}$,

$T_a := T_{a,1}$, where $a \in E_2$,

$$\begin{aligned}T_{0,\infty} &:= \bigcap_{\mu\geq 1} T_{0,\mu} \\ &= \bigcup_{n\geq 1}\{f^n \in P_2 \mid \exists j \in \{1, ..., n\} \exists f' \in P_2 : f(\mathbf{x}) = x_j \wedge f'(\mathbf{x})\},\end{aligned}$$

$T_{1,\infty} := T_{0,\infty}^{\delta}$,

- $K := [\wedge]$ (set of all **conjunctions**),
- $D := K^{\delta} = [\vee]$ (set of all **disjunctions**),
- $C := [c_0, c_1]$ (set of all constant functions),
- $C_a := [c_a]$, $a \in E_2$,
- $I := [e_1^1]$ (set of all projections),
- $\overline{I} := [^-]$.

Theorem 3.1.1 (Post's Theorem; [Pos 41])

(1) The set of all subclasses of P_2 is countably infinite.

(2) The non-empty subclasses of P_2 are

$$P_2,\ S,\ M,\ L,$$
$$T_{a,\mu},\ T_{a,\mu}\cap T_{\overline{a}},\ T_{a,\mu}\cap M,\ T_{a,\mu}\cap M\cap T_{\overline{a}},$$
$$K\cup C, K\cup C_a,\ K,\ D\cup C,\ D\cup C_a,\ D,$$
$$S\cap T_0,\ S\cap M,\ S\cap L, S\cap L\cap T_0,\ L\cap T_a,$$
$$\overline{I}\cup C,\ I\cup C,\ \overline{I},\ I\cup C_a,\ I,\ C,\ C_a,$$

where $a \in E_2$ and $\mu \in \{1, 2, ..., \infty\}$.

(The Hasse-diagram of these classes is given in Figure 3.1.)

(3) In the set P_2, there exists exactly

(a) 9 closed subsets of order 1:

$$[P_2^1],\ I\cup C,\ \overline{I},\ I\cup C_0,\ I\cup C_1,\ I,\ C,\ C_0,\ C_1;$$

(b) 20 closed subsets of order 2:

$$P_2,\ T_0,\ T_1,\ M,\ L,\ M\cap T_0,\ M\cap T_1,\ L\cap T_0,\ L\cap T_1,$$
$$M\cap T_0\cap T_1,\ K\cup C,\ K\cup C_0,\ K\cup C_1,\ K,\ D\cup C,$$
$$D\cup C_0,\ D\cup C_1,\ D,\ T_{0,\infty},\ T_{1,\infty};$$

(c) 20 closed subsets of order 3:

$$S,\ S\cap T_0,\ S\cap M,\ S\cap L,\ S\cap L\cap T_0,$$
$$T_{0,2},\ T_{1,2},\ T_{0,2}\cap T_1,\ T_{1,2}\cap T_0,\ T_{0,2}\cap M,\ T_{1,2}\cap M,$$
$$T_{0,2}\cap M\cap T_1,\ T_{1,2}\cap T_0\cap M,$$
$$T_{0,\infty}\cap T_1,\ T_{1,\infty}\cap T_0,\ T_{0,\infty}\cap M,\ T_{1,\infty}\cap M,$$
$$T_{0,\infty}\cap M\cap T_1,\ T_{1,\infty}\cap M\cap T_0,\ T_0\cap T_1;$$

(d) 8 closed subsets of order $\mu+1$ ($\mu \geq 3$):

$$T_{a,\mu},\ T_{a,\mu}\cap T_{\overline{a}},\ T_{a,\mu}\cap M,\ T_{a,\mu}\cap M\cap T_{\overline{a}}\ (a\in E_2).$$

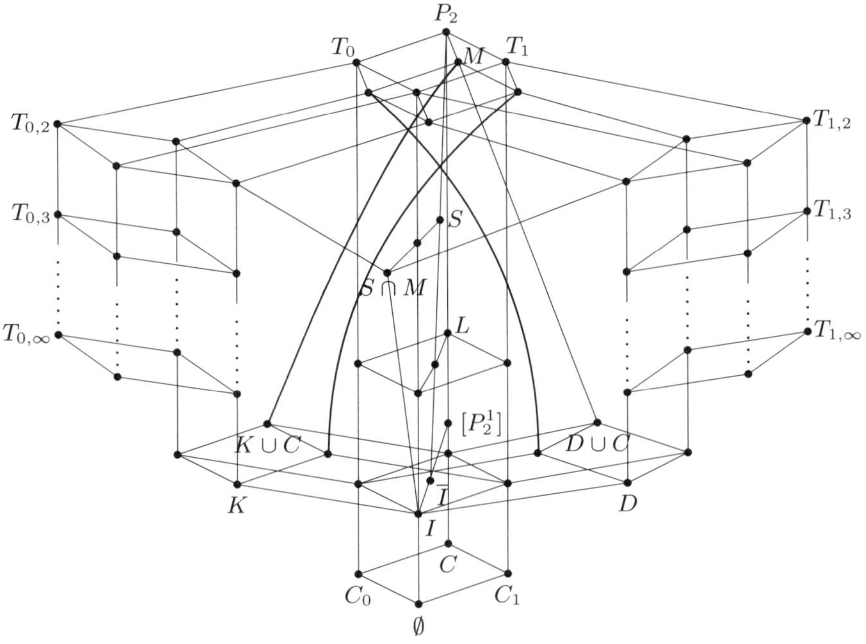

Fig. 3.1. The Post Lattice

3.2 A Proof for Post's Theorem

For any subclass A of P_2, the following three cases are possible:

Case 1: $A \nsubseteq L$ and $A \nsubseteq S$,
Case 2: $A \subseteq L$,
Case 3: $A \nsubseteq L$ and $A \subseteq S$.

By following this case distinction, all subclasses of P_2 and minimal generating subsets of these classes are determined as follows. The orders of the subclasses given are easy conclusions from the proven statements about generating sets of the subclasses. We start with:

3.2.1 The Subclasses A of P_2 with $A \nsubseteq L$ and $A \nsubseteq S$

Theorem 3.2.1.1 *The following holds:*
$P_2 = [\circ, ^-]$ *for* $\circ \in \{\wedge, \vee\}$, $M = [\vee, \wedge, c_0, c_1]$, $T_{0,\mu} = [h_\mu, t]$, $T_{0,\mu} \cap T_1 = [h_\mu, q]$, $T_{0,\mu} \cap M = [h_\mu, m, c_0]$ *(* $T_0 \cap M = [\vee, \wedge, c_0]$ *)*, $T_{0,\mu} \cap M \cap T_1 = [h_\mu, m]$, $T_{0,\infty} = [t]$, $T_{0,\infty} \cap T_1 = [q]$, $T_{0,\infty} \cap M = [m, c_0]$ *and* $T_{0,\infty} \cap M \cap T_1 = [m]$, *where* $\mu \in \mathbb{N}$.

Proof. Let be $f^n \in P_2$, for which $\mu+1$ distinct tuples $\mathbf{a}_1, ..., \mathbf{a}_{\mu+1}$ exist with $f(\mathbf{a}_1) = ... = f(\mathbf{a}_{\mu+1}) = 1$. Then, we have

$$f(\mathbf{x}) = h_\mu(f_{\mathbf{a}_1}(\mathbf{x}), f_{\mathbf{a}_2}(\mathbf{x}), ..., f_{\mathbf{a}_{\mu+1}}(\mathbf{x})), \tag{3.1}$$

where

$$f_{\mathbf{a}_i}(\mathbf{x}) := \begin{cases} 0 & \text{if} \quad \mathbf{x} = \mathbf{a}_i, \\ f(\mathbf{x}) & \text{otherwise} \end{cases}$$

$(i = 1, 2, ..., \mu + 1)$.
We call every function f^n of a subclass A of P_2 with

$$\{h_\mu, f_{\mathbf{a}_1}, ..., f_{\mathbf{a}_{\mu+1}}\} \subseteq A$$

and

$$f(\mathbf{a}_1) = ... = f(\mathbf{a}_{\mu+1}) = 1$$

for certain $\mathbf{a}_1, ..., \mathbf{a}_{\mu+1} \in E_2^n$ a **reducible** function. We denote by N_A the set of all not reducible functions f of a class A. Obviously, the set $N_A \cup \{h_\mu\}$ is a generating set for the class A, if $h_\mu \in A$ and $A \cap \{h_1, ..., h_{\mu-1}\} = \emptyset$ for certain $\mu \geq 1$. The Table 3.1 gives an easily verifiable description of the set $(N_A)^n$ for $A \in \{P_2, T_{0,m}, M, T_{0,m} \cap M, T_{0,m} \cap T_1, T_{0,m} \cap M \cap T_1\}$, $m \in \mathbb{N}$, and the minimal μ with $h_\mu \in A$. [1] The functions g_J and m_J for $J \subseteq E_2^n$ from Table 3.1 are defined by

$$g_J(\mathbf{x}) := \begin{cases} 1 & \text{if} \quad \mathbf{x} \in J, \\ 0 & \text{otherwise} \end{cases}$$

and

$$m_J(\mathbf{x}) := \begin{cases} 1 & \text{if} \quad \exists \mathbf{a} \in J : \mathbf{x} \geq \mathbf{a}, \\ 0 & \text{otherwise.} \end{cases}$$

Table 3.1

A	N_A^n	a minimal μ with $h_\mu \in A$
P_2	$\{g_J \mid \lvert J \rvert \leq 1\}$	1
M	$\{m_J \mid \lvert J \rvert \leq 1\}$	1
$T_{0,\mu}$	$\{g_J \mid \lvert J \rvert \leq \mu$ and $\exists t : g_J(\mathbf{x}) = x_t \wedge g_J(\mathbf{x})\}$	μ
$T_{0,\mu} \cap T_1$	$\{g_J \in N_{T_{0,\mu}}^n \mid \mathbf{1} \in J\}$	μ
$T_{0,\mu} \cap M$	$\{m_J \mid \lvert J \rvert \leq \mu$ and $\exists t : m_J(\mathbf{x}) = x_t \wedge m_J(\mathbf{x})\}$	μ
$T_{0,\mu} \cap M \cap T_1$	$\{m_J \in N_{T_{0,\mu} \cap M}^n \mid \mathbf{1} \in J\}$	μ

[1] One possible proof of Table 3.1 is as follows: We start with a "partition" of an arbitrary $f \in A$ in functions $f_{\mathbf{a_i}}$ by (3.1), then we repeat this construction for the functions $f_{\mathbf{a_i}}$ instead of f, if $f_{\mathbf{a_i}} \notin N_A$, etc. In case $A \subseteq M$ be let all tuples $\mathbf{a_i}$ minimal with respect to $\leq$ by this $f_{\mathbf{a_i}} \in M$.

Since

$$g^n_\emptyset = m^n_\emptyset = c^n_0,$$
$$g^n_{\{(a_1,\ldots,a_n)\}}(\mathbf{x}) = x_1^{\sigma_1} \wedge x_2^{\sigma_2} \wedge \ldots \wedge x_n^{\sigma_n},$$
$$m^n_{\{\mathbf{0}\}} = c^n_1,$$
$$m^n_{\{(a_1,\ldots,a_n)\}}(\mathbf{x}) = x_{i_1} \wedge x_{i_2} \wedge \ldots \wedge x_{i_\nu} \text{if } \{i_1,\ldots,i_\nu\} = \{i \mid a_i = 1\} \neq \emptyset,$$

we have

$$N_{P_2} \subseteq [\wedge, ^-], \qquad N_{T_0} \subseteq [c_0, \wedge, t], = [t],$$
$$N_M \subseteq [\wedge, c_0, c_1], \quad N_{M\cap T_0} \subseteq [\wedge, c_0] \text{ and}$$
$$N_{M\cap T_0\cap T_1} \subseteq [\wedge].$$

Consequently, $P_2 = [\vee, \wedge, ^-]$ (and $P_2 = [\circ, ^-]$ for $\circ \in \{\vee, \wedge\}$ by Morgan's laws), $T_0 = [\vee, t]$, $M = [\vee, \wedge, c_0, c_1]$, $M \cap T_0 = [\vee, \wedge, c_0]$ and $M \cap T_0 \cap T_1 = [\vee, \wedge]$.
Furthermore, Table 3.1 implies the following: if $A = T_{0,\mu} \cap B$ with $B \in \{P_2, T_1, M, T_1 \cap M\}$ and $\mu \geq 2$, then $A = [\{h_\mu\} \cup (T_{0,\infty} \cap B)]$ and $T_{0,\infty} \cap B = \bigcup_{n\geq 1}\{f^n \in P_2 \mid \exists i \in \{1,2,\ldots,n\} \; \exists\, f' \in B^n : \;\; f(\mathbf{x}) = x_i \wedge f'(\mathbf{x})\}$. By this fact, from a generating subset $\{f^n, g^m, \ldots\}$ of B, we get a generating subset of the form $\{\wedge \star f, \wedge \star g, \ldots\}$ for the class $T_{0,\infty} \cap B$. Thus, $T_{0,\infty} = [t]$, since $t(x, t(x,y)) = x \wedge y$ and $P_2 = [\wedge, ^-]$. $T_{0,\infty} \cap T_1 = [q]$ follows from $T_1 = T_0^\delta = [\wedge, t^\delta]$, $t^\delta(x,y) = x \vee \overline{y}$ and $\wedge \star \wedge, \wedge \star t^\delta \in [q]$. By $T_1 \cap M = [\wedge, \vee, c_1]$ and $\{\wedge \star \wedge, \wedge \star \vee, \wedge \star c_1^1 = e_2^2\} \subseteq [m]$ is $T_{0,\infty} \cap T_1 \cap M = [m]$. Finally, $T_{0,\infty} \cap M = [m, c_0]$, since $T_{0,\infty} \cap M = (T_{0,\infty} \cap M \cap T_1) \cup [c_0]$. ■

Lemma 3.2.1.2 *If $A = [A] \subseteq P_2$, $c_a \in A$ for certain $a \in E_2$ and $A \not\subseteq L$, then A contains a binary non-linear function.*

Proof. By $x \circ y + x + y = x \circ' y$ for $\{\circ, \circ'\} = \{\wedge, \vee\}$, $\Delta(+\star c_1) = ^-$ and Theorem 1.4.2 we have $[\circ, +, c_1] = P_2$ for $\circ \in \{\wedge, \vee\}$, i.e., every function $g^n \in P_2$ has a description of the form

$$g(\mathbf{x}) = a_0 + \sum_{\{i_1,i_2,\ldots,i_\nu\}\subseteq\{1,2,\ldots,n\}} a_{i_1 i_2 \ldots i_\nu} \circ x_{i_1} \circ x_{i_2} \circ \ldots \circ x_{i_\nu}$$

for some $a_0, a_{i_1 i_2 \ldots i_\nu} \in E_2$ (see Theorem 1.4.3).
Therefore, because $A \not\subseteq L$, there exists a function $f^n \in A$ with

$$f(\mathbf{x}) = a_0 + x_1 \circ x_2 \circ \ldots \circ x_r + \sum_{i=1}^{n} a_i \circ x_i + \sum_{\substack{i_1,\ldots,i_\nu \\ \in \{1,2,\ldots,n\},\; \nu \geq r, \\ \{i_1,\ldots,i_\nu\} \not\subseteq \{1,2,\ldots,r\}}} a_{i_1 \ldots i_\nu} \circ x_{i_1} \circ \ldots \circ x_{i_\nu},$$

where $r \geq 2$ and

$$\circ := \begin{cases} \vee & \text{if } a = 1, \\ \wedge & \text{if } a = 0. \end{cases}$$

Our statement follows from

$$f(x, \underbrace{y, y, ..., y}_{(r-1) \text{ times}}, c_a, ..., c_a) = b + x \circ y + c \circ x + d \circ y \in A \backslash L$$

for some $b, c, d \in E_2$. ■

Lemma 3.2.1.3 *Let A be a subclass of P_2, which is not a subset of L and not a subset of S. Then, the function $\wedge$ or the function $\vee$ belongs to A.*

Proof. It is easy to verify that $\vee$ or $\wedge$ is a superposition over a binary non-linear function of P_2. Thus, we have to show $A^2 \not\subseteq L$. Since $A \not\subseteq S$, there exists an $f \in A$ with

$$f\begin{pmatrix} 0 & 1 \\ 1 & 0 \end{pmatrix} = \begin{pmatrix} a \\ a \end{pmatrix},$$

$a \in E_2$, i.e., $f \in \{f_1, f_2, ..., f_8\}$ (see Table 3.2).

Table 3.2

x	y	f_1	f_2	f_3	f_4	f_5	f_6	f_7	f_8
0	0	0	1	1	0	0	1	0	1
0	1	0	1	0	1	0	0	1	1
1	0	0	1	0	1	0	0	1	1
1	1	0	1	1	0	1	0	1	0

The functions $f_5, ..., f_8$ are non-linear. If $f \in \{f_1, ..., f_4\}$ then Δf is a constant function and $A^2 \not\subseteq L$ follows from Lemma 3.2.1.2. ■

Lemma 3.2.1.4 *Let A be a subclass of P_2 with $\wedge \in A$. Then the following implications hold:*
(a) $(\exists a \in E_2 : A \not\subseteq T_a) \Longrightarrow c_{\overline{a}} \in A$,
(b) $(A \subseteq M$ *and* $A \not\subseteq K \cup C) \Longrightarrow m \in A$,
(c) $A \not\subseteq M \Longrightarrow q \in A$,
(d) $(A \not\subseteq K \cup C$ *and* $A \not\subseteq T_0) \Longrightarrow \vee \in A$,
(e) $(\exists \mu \in \mathbb{N} : A \subseteq T_{0,\mu}$ *and* $A \not\subseteq T_{0,\mu+1}) \Longrightarrow h_\mu \in A$.

Proof. (a): If $A \not\subseteq T_a$ there exists a $g^1 \in A$ with $g(a) = \overline{a}$, i.e., $g \in \{c_{\overline{a}}, ^-\}$. $c_{\overline{a}} \in A$ follows from $\wedge \in A$, $x \wedge \overline{x} = 0$ and $\overline{0} = 1$.
(b): f denotes an n-ary function of $A \backslash (K \cup C)$. Then there exist two tuple $\mathbf{a} = (a_1, ..., a_n)$ and $\mathbf{b} = (b_1, ..., b_n)$ with the following properties:

$$f(\mathbf{a}) = f(\mathbf{b}) = 1,$$
$$\mathbf{a} \not\leq \mathbf{b}, \mathbf{b} \not\leq \mathbf{a} \text{ and}$$
$$f(\mathbf{c}) = 0 \text{ for every } \mathbf{c} \text{ with } \mathbf{c} < \mathbf{a} \text{ or } \mathbf{c} < \mathbf{b}\ .$$

Defining functions $g_i^3 \in A$ by

$$g_i(x,y,z) := \begin{cases} x & \text{if } \begin{pmatrix} a_i \\ b_i \end{pmatrix} = \begin{pmatrix} 1 \\ 1 \end{pmatrix}, \\ y & \text{if } \begin{pmatrix} a_i \\ b_i \end{pmatrix} = \begin{pmatrix} 0 \\ 1 \end{pmatrix}, \\ z & \text{if } \begin{pmatrix} a_i \\ b_i \end{pmatrix} = \begin{pmatrix} 1 \\ 0 \end{pmatrix}, \\ yz & \text{if } \begin{pmatrix} a_i \\ b_i \end{pmatrix} = \begin{pmatrix} 0 \\ 0 \end{pmatrix} \end{cases}$$

$(i = 1, 2, ..., n)$ we obtain

$$x \wedge f(g_1(x,y,z), ..., g_n(x,y,z)) = m(x,y,z) \in A,$$

since $f \in M$.
(c): By $A \not\subseteq M$ we have a function h^3 in A with

$$h\begin{pmatrix} 1 & 0 & 0 \\ 1 & 0 & 1 \end{pmatrix} = \begin{pmatrix} 1 \\ 0 \end{pmatrix}.$$

Consequently, $h'(x,y,z) := x \wedge h(x,y,z) \in A$ and $h'(x,y,z) \in \{x \wedge \overline{y} \wedge \overline{z}, x(y{+}z{+}1), x{\wedge}\overline{z}, x(y{\vee}\overline{z})\}$. Since $x{\wedge}(\overline{y}{\wedge}z) = x(y{\vee}\overline{z})$ and $x(yz{+}z{+}1) = x(y{\vee}\overline{z})$, it holds $q \in A$.
(d): By (a), (b), (c) and $m(x,y,z) = q(x,y,q(x,y,z))$ it holds $\{c_1, m\} \subseteq A$ and hence $\Delta(m \star c_1^1) = \vee \in A$.
(e): If $A \not\subseteq T_{0,\mu+1}$, there is an $f \in A$ with $f(E_2^{\mu+1} \backslash \{\mathbf{1}\}) = \mathbf{1}$, i.e., w.l.o.g.:

$$f\begin{pmatrix} 0 & 1 & 1 & \dots & 1 & 0 & 0 & 0 & \dots & 0 & \dots\dots & 0 \\ 1 & 0 & 1 & \dots & 1 & 0 & 1 & 1 & \dots & 0 & \dots\dots & 0 \\ 1 & 1 & 0 & \dots & 1 & 1 & 0 & 1 & \dots & 0 & \dots\dots & 0 \\ \dots & & & & & & & & & & & \dots \\ \underbrace{1 \;\; 1 \;\; 1 \;\; \dots \;\; 0}_{(\mu+1) \text{ times}} & & & & & 1 & 1 & 1 & \dots & 1 & \dots\dots & 0 \end{pmatrix} = \begin{pmatrix} 1 \\ 1 \\ 1 \\ . \\ 1 \end{pmatrix}.$$

Since $\wedge \in A$, we have

$$f'(x_1, ..., x_{\mu+1}) := f(x_1, ..., x_{\mu+1}, x_1x_2, x_1x_3, ..., x_1x_2x_3, ..., x_1x_2...x_{\mu+1}) \in A.$$

By $f' \in T_{0,\mu}$ we have

$$f'(\mathbf{x}) = \begin{cases} h_\mu(\mathbf{x}) & \text{if } \mathbf{x} \neq \mathbf{1}, \\ a & \text{if } \mathbf{x} = \mathbf{1} \end{cases}$$

for a certain $a \in E_2$. If $a = 1$ then $f' = h_\mu \in A$. If $a = 0$ and $\mu \geq 2$ then

$$f'(x_1, ..., x_{\mu-1}, x_\mu x_{\mu+1}, f'(x_1, ..., x_{\mu+1})) = h_\mu(\mathbf{x}) \in A.$$

Finally, if $a = 0$ and $\mu = 1$, we have $f' = +$ and $h_1(x,y) = xy + x + z \in A$. ■

Theorem 3.2.1.5 *The subclasses A of P_2 with $A \not\subseteq S$ and $A \not\subseteq L$ are*

$$P_2,\ M,\ T_{a,\mu},\ T_{a,\mu} \cap B,\ K,\ K \cup C_a,\ K \cup C,\ D, D \cup C_a,\ D \cup C,$$

where $a \in E_2$, $\mu \in \{\infty, 1, 2, ...\}$ and $B \in \{T_{\overline{a}}, M, M \cap T_{\overline{a}}\}$.

Proof. Denote A a subclass of P_2 with $A \not\subseteq S$ and $A \not\subseteq L$. By Lemma 3.2.1.3, we have that A obtains K or D. W.l.o.g. $[\wedge] = K \subseteq A$. If $A \notin \{K, K \cup C_0, K \cup C_1, K \cup C\}$ then we can distinguish 8 cases, which are given in Table 3.3. Using Lemma 3.2.1.4 and Theorem 3.2.1.1, we see that A is a certain class, which is given in the last column of Table 3.3.

Table 3.3

$\exists \mu \in \{\infty, 1, 2, ...\}: A \subseteq T_{0,\mu} \wedge A \not\subseteq T_{0,\mu+1}$	$A \subseteq T_1$	$A \subseteq M$	conclusions from the assumptions	A
$-$	$-$	$-$	$\{c_1, c_0, q\} \subseteq A$	P_2
$-$	$-$	$+$	$\{c_1, c_0, m\} \subseteq A$	M
$-$	$+$	$-$	$\{c_1, q\} \subseteq A \subseteq T_1$	T_1
$-$	$+$	$+$	$\{c_1, m\} \subseteq A \subseteq T_1 \cap M$	$T_1 \cap M$
$+$	$-$	$-$	$\{h_\mu, c_0, q\} \subseteq A \subseteq T_{0,\mu}$	$T_{0,\mu}$
$+$	$-$	$+$	$\{h_\mu, c_0, m\} \subseteq A \subseteq T_{0,\mu} \cap M$	$T_{0,\mu} \cap M$
$+$	$+$	$-$	$\{h_\mu, q\} \subseteq A \subseteq T_{0,\mu} \cap T_1$	$T_{0,\mu} \cap T_1$
$+$	$+$	$+$	$\{h_\mu, m\} \subseteq A \subseteq T_{0,\mu} \cap T_1 \cap M$	$T_{0,\mu} \cap T_1 \cap M$

(+ stands for the truth of the assertion in the first row, – for the truth of the negated assertion.
Furthermore, let $T_{0,\infty+1} := \emptyset$ and $h_\infty := e_1^1$) ∎

3.2.2 The Subclasses of *L*

Obviously, all subclasses of $[P_2^1]$

$$[P_2^1], I \cup C, \overline{I}, I \cup C_0, I \cup C_1, C, C_0, C_1, I, \emptyset$$

are also subclasses of L. With the help of these sets, we can determine all subclasses of L with the following.

Lemma 3.2.2.1 *Let $\mathfrak{L}$ be subclass of L with $\mathfrak{L} \not\subseteq [P_2^1]$. Then*

$$\mathfrak{L} = [\mathfrak{L}^1 \cup \{r\}]$$

$(r(x, y, z) = x + y + z)$.

Proof. Let $\mathfrak{L} = [\mathfrak{L}] \subseteq L$ and $\mathfrak{L} \not\subseteq [P_2^1]$. Then, there is a function $g \in \mathfrak{L}$ with $g(x, y, z) = a + x + y + bz$ for some $a, b \in E_2$. Thus $g(g(x, y, z), z, z) = r(x, y, z) \in \mathfrak{L}$. By forming of the functions of the type $r \star r \star ... \star r$ and by identifying of variables in these functions, we get that every function $\sum_{i=1}^{n} b_i x_i$

with $b_1 + ... + b_n = 1$ belongs to $[r] \subseteq A$. Now, denote f an n-ary function of $\mathfrak{L}$ and let $f(\mathbf{x}) = a_0 + \sum_{i=1}^{n} a_i x_i$. Then, we have

$$f'(x, x_1, ..., x_n) := x + (a_2 + ... + a_n)x_1 + \sum_{i=2}^{n} a_i x_i \in [r]$$

and

$$f(\mathbf{x}) = f'(f(x_1, ..., x_1), x_1, ..., x_n).$$

Therefore $f \in [\{r\} \cup \mathfrak{L}^1]$ and $[\{r\} \cup \mathfrak{L}^1] = \mathfrak{L}$ is proven. ■

Looking for subsemigroups of $(P_2^1; \star)$, which have the property to be preserved by r, we get the following

Theorem 3.2.2.2 *All subclasses of L are*
L, $L \cap T_0 = [c_0, +]$, $L \cap T_1$, $L \cap S = [^-, r]$, $L \cap T_0 \cap S = [r]$,
$[P_2^1]$, $I \cup C$, $\overline{I}$, $I \cup C_0$, $I \cup C_1$, I, C, C_0, C_1, $\emptyset$. ■

3.2.3 The Subclasses of S, Which Are Not Subsets of L

Obviously, a function $f^n \in P_2$ $(n \geq 2)$ belongs to S iff there exists a function $F^{n-1} \in P_2$ with the property

$$f(x_1, ..., x_n) = x_1 F(x_2, ..., x_n) \vee x_1 \overline{F(\overline{x_2}, ..., \overline{x_n})}, \tag{3.2}$$

where

$$F(x_2, ..., x_n) := f(0, x_2, ..., x_n).$$

Consequently, we can define a bijective mapping α of $S' := S \backslash S^1$ onto P_2 as follows:

$$\alpha : f \longrightarrow F.$$

Lemma 3.2.3.1 *The mapping α has the following properties:*

(a) For the operations $\widehat{\zeta}, \widehat{\tau}, \widehat{\Delta}, \widehat{\nabla}$ and $\widehat{\star}$, defined by

$$\begin{aligned}
&(\widehat{\zeta} f)(x_1, ..., x_n) := f(x_1, x_3, x_4, ..., x_n, x_2),\\
&(\widehat{\tau} f)(x_1, ..., x_n) := f(x_1, x_3, x_2, x_4, ..., x_n),\\
&(\widehat{\Delta} f)(x_1, ..., x_{n-1}) := f(x_1, x_2, x_2, x_3, ..., x_{n-1}),\\
&(\widehat{\nabla} f)(x_1, ..., x_{n+1}) := f(x_1, x_3, x_4, ..., x_{n+1}) \text{ and}\\
&(f \widehat{\star} g)(x_1, ..., x_{m+n-2}) := f(x_1, g(x_1, ..., x_m), x_{m+1}, ..., x_{m+n-2})\\
&(n, m \geq 2),
\end{aligned}$$

it holds $\alpha(\widehat{\gamma} f) = \gamma(\alpha(f))$ for every $\gamma \in \{\zeta, \tau, \Delta, \nabla\}$ and $\alpha(f \widehat{\star} g) = \alpha(f) \star \alpha(g)$, i.e., the algebra $(S'; \widehat{\zeta}, \widehat{\tau}, \widehat{\Delta}, \widehat{\nabla}, \widehat{\star})$ is isomorphic to the algebra $(P_2; \zeta, \tau, \Delta, \nabla, \star)$.

(b) For every subclass A $(\neq \emptyset)$ of S, $\alpha(A)$ is a subclass of P_2, and it holds $\alpha(A) \not\subseteq S$, $A \subseteq \alpha(A)$ and $\alpha(A) \cap S = A$.

Proof. (a) is easy to check.
(b): Let A be a subclass of S. By (a) we have that $\alpha(A)$ is also a closed set. Assume $\alpha(A) \subseteq S$. Then $F(x_2, ..., x_n) = \overline{F(\overline{x_2}, ..., \overline{x_n})}$ for every $f^n \in A$. Thus by (3.2) we get that the variable x_1 is fictitious for every function $f^n \in A$. However, this is not possible. Hence $\alpha(A) \not\subseteq S$ holds.
Let $f^n \in A$. Then $\nabla f \in A$ and therefore $\alpha(\nabla f) = f \in \alpha(A)$, i.e., $A \subseteq \alpha(A)$. If $f^n \in S \cap \alpha(A)$, we have $\Delta(\alpha^{-1} f) = f \in A$ and thus $S \cap \alpha(A) \subseteq A$. From this, it follows that $A = S \cap \alpha(A)$, since $A \subseteq \alpha(A)$ and $A \subseteq S$. ∎

With the help of Lemma 3.2.3.1 and Theorem 3.2.1.5, it is not difficult to determine the missing subclasses of S. It holds

Theorem 3.2.3.2

(a) The sets $S \cap M$ and $S \cap T_0$ are the only proper subclasses of S, which are not subsets of L.
(b) It holds that

$$S = [h_2, ^{-}], \; S \cap T_0 = [h_2, r], \; S \cap M = [h_2].$$

Proof. (a): Let A be a subclass of S with $A \not\subseteq L$. Then it holds $\alpha(A) \not\subseteq L$. By Lemma 3.2.1.2 and Lemma 3.2.1.3, we have $\{\vee, \wedge\} \cap \alpha(A) \neq \emptyset$. Consequently, the function h_2 $(= \overline{x}yz \vee x(y \vee z))$ or the function $g(x, y, z) := \overline{x}(y \vee z) \vee xyz$ belong to A. Since it holds $g(g(x, y, z), y, z) = h_2(x, y, z)$, $h_2 \in A$. Further, the functions $\alpha(h_2) = \wedge$, $x \wedge h_2(x, y, z) = x(y \vee z)$ and $c_0^1 = \alpha(e_1^2)$ are elements of $\alpha(A)$. Thus, $T_{0,2} \cap M \subseteq \alpha(A)$ and by Lemma 3.2.3.1, (b) we have $T_{0,2} \cap M \cap S \subseteq A$. By Theorem 3.2.1.5, there exists only the following possibilities for A: $S \cap T_{0,2}$, $S \cap M = S \cap M \cap T_0$, $S \cap T_0$ and (a) follows from $S \cap T_{0,2} = S \cap M$. (This fact is easy to prove, for example, with the help of the relation product $\square$ and the property $Pol \varrho \cap Pol \varrho' \subseteq Pol \varrho \square \varrho'$:
Let $\varrho_1 := \begin{pmatrix} 1 & 0 \\ 0 & 1 \end{pmatrix}$, $\varrho_2 := \begin{pmatrix} 0 & 0 & 1 \\ 0 & 1 & 0 \end{pmatrix}$ and $\varrho_3 := \begin{pmatrix} 0 & 0 & 1 \\ 0 & 1 & 1 \end{pmatrix}$. Then $\varrho_2 \square \varrho_1 = \varrho_3$ and thus $S \cap T_{0,2} \subseteq S \cap M$. Conversely, $S \cap M \subseteq S \cap T_{0,2}$ holds, since $\varrho_3 \square \varrho_1 = \varrho_2$.)
(b) follows from Theorem 3.2.1.1 and Lemma 3.2.3.1. ∎

3.2.4 A Completeness Criterion for P_2

The following is a conclusion from Theorem 3.1.1:

Theorem 3.2.4.1 (Completeness Criterion for P_2)
Let $A \subseteq P_2$. Then

$$[A] = P_2 \iff \forall X \in \{T_0, T_1, M, S, L\} : A \not\subseteq X.$$

∎

Without using of Theorem 3.1.1, one can prove Theorem 3.2.4.1 with the help of Theorem 1.5.4.1 and the following:

Theorem 3.2.4.2 *P_2 has exactly five maximal classes: T_0, T_1, M, S, and L.*

Proof. It is easy to see that the sets T_0, T_1, M, S, L are pairwise distinct proper subclasses of P_2. Therefore, proof of the following suffices for the proof of our theorem:

$$\forall A \subseteq P_2 : ((\forall K \in \{T_0, T_1, M, S, L\} : A \not\subseteq K) \Longrightarrow [A] = P_2)$$

Let now $A \subseteq P_2$ with $\{f_0, f_1, f_M, f_S, f_L\} \subseteq A$, where

$$f_0 \notin T_0, f_1 \notin T_1, f_M \notin M, f_S \notin S \text{ and } f_L \notin L.$$

If one identifies all variables of f_0 with each other, then one gets an unary function $f_0' \in [A]$ with $f_0'(0) = 1$, i.e., $f_0' \in \{c_1, \overline{e_1^1}\}$.
Case 1: $f_0' = c_1$.
In this case, it holds that $f_1(c_1(x), ..., c_1(x)) = c_0(x) \in [A]$. Since $f_M \in A \backslash M$, there are some $(a_i, b_i) \in \{(0,0),(0,1),(1,1)\}$ $(i = 1, 2, ..., n)$ with $f(a_1, ..., a_n) > f(b_1, ..., b_n)$. Consequently, the function $\overline{e_1^1}$ is a superposition over $\{f_M, c_0, c_1\} \subseteq [A]$. Thus P_2^1 belongs to $[A]$. By the proof of Lemma 3.2.1.2 (see also Theorem 1.4.3), we can describe the function f_L^n with the help of a so-called Shegalkin polynom. Since f_L does not belong to L, we can assume w.l.o.g. that

$$f_L(\mathbf{x}) = a_0 + x_1 \cdot x_2 \cdot ... \cdot x_r + \sum_{i=1}^{n} a_i \cdot x_i + \sum_{\substack{i_1, ..., i_\nu \\ \in \{1,...,n\}, \nu \geq r, \\ \{i_1,...,i_\nu\} \not\subseteq \{1,...,r\}}} a_{i_1 ... i_\nu} \cdot x_{i_1} \cdot ... \cdot x_{i_\nu},$$

holds for $r \geq 2$.
Now, we consider the function

$$f_L'(x,y) := f_L(x, \underbrace{y, ..., y}_{(r-1) \text{ times}}, c_0(x), ..., c_0(x))$$

which has the form $f_L'(x,y) = a + b \cdot x + c \cdot y + x \cdot y$ for some $a, b, c \in \{0,1\}$. It is easy to check that $x \cdot y$ is a superposition over $\{f_L'\} \cup P_2^1$ $(\subseteq A)$. Consequently, by Theorem 1.4.2, we have $[A] = P_2$.
Case 2: $f_0' = \overline{e_1^1}$.
Since $f_S^n \notin S$, there exist some $a_1, ..., a_n \in \{0,1\}$ with $f_S(a_1, ..., a_n) = f_S(\overline{a_1}, ..., \overline{a_n})$. Thus, c_1 is a superposition over $\{f_S, \overline{e_1^1}\}$ and the Case 2 is put down to the Case 1. ■

4

The Subclasses of P_k Which Contain P_k^1

We start with the definitions of some subclasses of P_k. We show later that these classes are all classes A of P_k with $P_k^1 \subseteq A$.
Let

$$U_t := P_k(t) \cup [P_k^1]$$

for $t = 2, 3, ..., k$. In particular, $U_k = P_k$.
Further, let

$$L_k$$

be the set

$$[P_k^1] \cup \bigcup_{n\geq 1} \{f^n \in P_k \mid \exists a \in E_2 \; \exists f_0 \in P_k^1 \; \exists f_1, ..., f_n \in P_{k,2}^1 : \\ f(\mathbf{x}) = f_0(a + f_1(x_1) + f_2(x_2) + ... + f_n(x_n) \; (mod\; 2))\}.$$

For $k = 2$ L_k is the set L, already defined in Chapter 3.

The sets U_t and L_k can be described with the help of relations:

Lemma 4.1 *Let*

$$\iota_k^h := \{(a_0, ..., a_{h-1}) \in E_k^h \mid \exists i, j \in E_k : i \neq j \wedge a_i = a_j\}$$

and

$$\lambda_k := \{(a, a, b, b), (a, b, a, b), (a, b, b, a) \mid a, b \in E_k\}.$$

Then

(a) $U_t = Pol\; \iota_k^{t+1}$ *for each* $t \in \{2, 3, ..., k-1\}$

and

(b) $L_k = Pol\; \lambda_k$.

Proof. (a): Obviously, U_t is a subset of $Pol\; \iota_k^{t+1}$. The inclusion $U_t \subset Pol\; \iota_k^{t+1}$ (i.e., there is a function ($\in Pol\; \iota_k^{t+1}$) that essentially depends on at least two variables and which has at least $t+1$ different values) is false because of

Theorem 1.4.4, (b). Thus, $U_t = Pol\ \iota_k^{t+1}$.
(b): It is easy to check that $L_k \subseteq Pol\ \lambda_k$. For the proof of $Pol\ \lambda_k \subseteq L_k$, denote f^n an arbitrary function from $P_{k,2} \cap (Pol\ \lambda_k)$. First, we will show that then

$$f(x_1, 0, ..., 0) + f(0, x_2, ..., x_n) + f(0, 0, ..., 0) = f(x_1, ..., x_n)\ (mod\ 2) \quad (4.1)$$

holds. Suppose, for some $x_1 = a_1, ..., x_n = a_n$ is this false. Then, we have

$$f\begin{pmatrix} a_1 & 0 & \dots & 0 \\ 0 & a_2 & \dots & a_n \\ 0 & 0 & \dots & 0 \\ a_1 & a_2 & \dots & a_n \end{pmatrix} \notin \begin{pmatrix} a & a & a \\ a & b & b \\ b & a & b \\ b & b & a \end{pmatrix}$$

for arbitrary $a, b \in E_2$. However, this is a contradiction to $f \in Pol\ \lambda_k$. Therefore, (4.1) holds and this implies:

$$f(x_1, ..., x_n) = a + f_1(x_1) + ... + f_n(x_n)\ (mod\ 2),$$

where $a := (n-1) \cdot f(0, 0, ..., 0)\ (mod\ 2)$ and

$$f_i(x) := f(0, ..., 0, \underbrace{x}_{i\text{-th place}}, 0, ..., 0).$$

Because of $pr_{1,2,3}\lambda_k = \iota_k^3$, an arbitrary function g from $(Pol\ \lambda_k)\backslash[P_k^1]$ has exactly only two different values. Consequently, there exists a certain permutation $s \in S_k$ with $s(Im(g)) = \{0, 1\}$ such that g has the description $s^{-1} \star (s \star g)$ and $s \star g$ belongs to $P_{k,2} \cap Pol\ \lambda_k$. It is obvious then that $g \in L_k$. ■

Definition Let $E \subseteq E_k$ with $|E| \geq 2$. Then we can define a mapping pr_E from $P_{k,E}$ into P_E as follows:

$$pr_E f^n = g^m :\Longleftrightarrow (n = m\ \wedge\ \forall \mathbf{a} \in E^n :\ f(\mathbf{a}) = g(\mathbf{a}))$$

$(f^n \in P_{k,E},\ g^m \in P_E)$.

Lemma 4.2 *Let f^2 be a function from L_k with*

$$f(x, y) = x + y\ (mod\ 2)\ \textit{for all}\ x, y \in E_2,$$

let g^m be a function from $P_{k,2}$ with $pr_{E_2} g \notin L_2$ and let $h \in U_t \backslash U_{t-1}$. Then,

(a) $L_k = [P_k^1 \cup \{f\}]$,

(b) $U_2 = [P_k^1 \cup \{g\}]$ and

(c) $\forall t \in \{3, 4, ..., k\} : U_t = [U_{t-1} \cup \{h\}]$.

Proof. (a) follows directly from the definition of L_k.
(b): By Theorem 3.2.4.1, $[\{pr_{E_2} g\} \cup pr_{E_2} P_{k,2}^1] = P_2$. Therefore, some binary

functions $\wedge', +'$ with $pr_{E_2}\wedge' = \wedge$ and $pr_{E_2}+' = +$ belong to $[P_k^1 \cup \{g\}]$ and an arbitrary function u^t of $P_{k,2}$ is a superposition over $P_k^1 \cup \{\wedge', +'\}$:

$$u(x_1, ..., x_t) = \sum_{\substack{(a_1, ..., a_t) \\ \in E_k^t}} u(a_1, ..., a_t) \cdot j_{a_1}(x_1) \cdot ... \cdot j_{a_t}(x_t) \ (mod\ 2)$$

(see (1.8) from Section 1.4). This implies (b).
(c) is a consequence from Lemma 1.4.5. ∎

Theorem 4.3 (Burle's Theorem, *[Bur 67]***)**
The classes

$$[P_k^1], L_k, U_2, U_3, ..., U_{k-1}, P_k$$

with

$$[P_k^1] \subset L_k \subset U_2 \subset U_3 \subset ... \subset U_{k-1} \subset P_k$$

are the only subclasses of P_k *which contain* P_k^1.

Proof. Suppose there exists a class A of P_k with $P_k^1 \subset A$, which is different from the classes of the above theorem. Then A contains a certain function f^n, which has at least two essentially variables and at least $l \geq 2$ different values. Consequently, by Theorem 1.4.4, (a), (c), there exists some $a_1, ..., a_n, b_1, ..., b_n, \alpha, \beta, \gamma \in E_k$ with

$$f\begin{pmatrix} a_1 & a_2 & a_3 & ...a_n \\ a_1 & b_2 & b_3 & ...b_n \\ b_1 & b_2 & b_3 & ...b_n \end{pmatrix} = \begin{pmatrix} \alpha \\ \beta \\ \gamma \end{pmatrix},$$

where $|\{\alpha, \beta, \gamma\}| = 3$ for $l \geq 3$ and $\alpha = \gamma$ and $\alpha \neq \beta$ for $l = 2$.
Then, a binary function f' with $f'(x, y) := g_0(f(g_1(x), g_2(y), ..., g_n(y)))$ and

$$f'\begin{pmatrix} 0 & 0 \\ 0 & 1 \\ 1 & 1 \end{pmatrix} = \begin{pmatrix} 0 \\ 1 \\ 0 \end{pmatrix}$$

is a superposition over f and some $g_0, ..., g_n \in P_k^1$. We distinguish two cases:
Case 1: $f'(1, 0) = 0$.
In this case, the function $pr_{E_2}f'$ is nonlinear. Thus, by Lemma 4.2, (b) it holds $U_2 \subseteq A$. Since we have assumed, however, $A \neq U_t$ for each $t \in \{2, ..., k\}$, we obtain a contradiction with the aid of Lemma 4.2, (c).
Case 2: $f'(1, 0) = 1$.
In this case, by Lemma 4.2, (a) L_k is a subset of A and, because of $A \neq L_k$, there is a function $g \in A$, which does not preserve λ_k. Then, one can form a function $g' \in A \cap P_{k,2}$ with $pr_{E_2}g' \notin L_2$ as a superposition over g and some functions of L_k. Thus, by Case 1, we get a contradiction.

Consequently, the classes given in our theorem are the only subclasses of P_k that contain P_k^1.
The claimed chain property is an immediate conclusion from the definitions of these classes. The claimed chain property is a direct conclusion from the definitions of these classes. ■

5

The Maximal Classes of P_k

5.1 Introduction, a Rough Description of the Maximal Classes

A subclass A of P_k is **maximal in** P_k (or A is called **maximal class**)[1] if and only if no further classes of P_k exist between A and P_k. In other words:

$$A = [A] \subseteq P_k \text{ is maximal in } P_k \text{ if and only if } A \neq P_k \text{ and } [A \cup \{f\}] = P_k \text{ for each } f \in P_k \backslash A.$$

One is interested in these classes not only for structural reasons, but particularly because one can solve a central problem of the Many-Valued Logic (the so-called **Completeness Problem**) with the aid of these classes (see Theorem 1.5.4.1).
Inter alia, the following papers dealt with the determination and description of maximal classes: [Pos 41], [Jab 54], [Jab 58], [Ros 70;a], [Mart 60], [Lo 63;a–c], [Lo 64], [Zac 67], [Zac-K-J 69], [Bai 67] and [Šai 70;b].
One knows the maximal classes T_0, T_1, M, S, and L of P_2 (see Theorem 3.2.4.2) through the papers [Pos 20], [Pos 41] by E. L. Post.
Efforts to determine all maximal classes of P_k for $k \geq 3$ began more than 50 years ago. S. V. Jablonskij determined all 18 maximal classes of P_3 in [Jab 54]. A. I. Mal'tsev proved how in the paper [Zac-K-J 69] was mentioned that P_4 has exactly 82 maximal classes. I. G. Rosenberg was the first that succeeded in the description of all maximal classes of P_k for each $k \in \mathbb{N} \setminus \{1\}$ (see [Ros 65]), and he proved in [Ros 70a] that the list of the given classes is complete. Rosenberg defined six relation sets, which are subsequently denoted by us with $\mathfrak{U}_k, \mathfrak{M}_k, \mathfrak{S}_k, \mathfrak{L}_k, \mathfrak{C}_k$ and $\mathfrak{B}_k$, and he proved that the set

$$\{Pol_k \varrho \mid \varrho \in \mathfrak{U}_k \cup \mathfrak{M}_k \cup \mathfrak{S}_k \cup \mathfrak{L}_k \cup \mathfrak{C}_k \cup \mathfrak{B}_k\}$$

[1] One uses instead of "maximal class" the concept "precomplete class" in older papers.

is exactly the set of all maximal classes of P_k. He could use that some other authors had shown already the maximality of some classes $Pol\ \varrho$. For example, for classes of the form $Pol\ \varrho_1$ with $\varrho_1 \in \mathfrak{U}_k \cup \mathfrak{S}_k \cup \mathfrak{C}_k^1$ and $\varrho \in \mathfrak{L}_k$, where k is a prime number, it was proven in [Jab 58] that these classes are maximal classes. In [Mart 60], one finds maximal classes of the form $Pol\ \varrho_2$ with $\varrho_2 \in \mathfrak{M}_k$. For further details, we refer the reader to [Ros 70;a].

In Section 5.2, we will describe the maximal classes of P_k in the manner found from Rosenberg.[2] Further, we give first properties of the maximal classes, where these properties are consequences from the definitions more or less. Then, in Section 5.3, the maximality of the classes described in the theorem is proven. It turns out, in this case, that one manages with three basic ideas in the proof. Chapter 6 is dedicated to the proof of the completeness of the given set of maximal classes then. This most difficult part of the determination of maximal classes orientates itself strongly onto the proof given in [Ros 70;a]. One could abbreviate, however, the proof through transfer and modification of some ideas from the papers [Qua 82], [Pös- K 79] (p. 126-129) and [Lau 92a].

A first coarse description of the maximal classes supplies the following theorem basically already proven by A. V. Kuznezov in 1959.

Theorem 5.1.1 (*[Kuz 59], [Ros 70;a] (3.2.5), [But 60]*)

The class L_k for $k=2$ or U_{k-1} $(= Pol_k \iota_k^k)$ for $k \geq 3$ is the only maximal class of P_k, which contains P_k^1. For every maximal class A of P_k, which is different from L_2 and U_{k-1}, is valid:

$$A = Pol_k G_1(A).$$

Proof. Let A be an arbitrary maximal class of P_k. Then, the following two cases are possible:
Case 1: $A^1 = P_k^1$.
Then, by Theorem 4.3, A is the set L_2 for $k = 2$ or the set U_{k-1} for $k \geq 3$.
Case 2: $A^1 \subset P_k^1$.
In this case, $(A^1; \star)$ is a proper subsemigroup of $(P_k^1; \star)$ which $e := e_1^1$ contains, what one can prove as follows:
Suppose, $e \notin A^1$. Then $A^1 \cap [P_k^1[k]] = \emptyset$, since $s^k = e$ for all $s \in P_k^1[k]$. Consequently, we have $A \subset J_k \cup A = [J_k \cup A] \subset P_k$, which contradicts the presupposed maximality of A.
Thus A is a clone, for which, by Theorem 2.7.1, (c) $A \subseteq Pol_k G_1(A) \subseteq P_k$

[2] This is not the only means to describe maximal classes. In [Den-P 88] one can find descriptions of maximal classes of P_k through hyperidentities. One finds more about hyperidentities in the book [Den-W 2000] of K. Denecke and S. L. Wismath.

holds. Because of $A^1 \neq P_k^1$ and the maximality of A, this is possible only for $A = Pol_k G_1(A)$. ∎

5.2 Definitions of the Maximal Classes of P_k

The maximal classes are defined with the aid of the relation sets $\mathfrak{M}_k$, $\mathfrak{S}_k$, $\mathfrak{U}_k$, $\mathfrak{L}_k$, $\mathfrak{C}_k$ and $\mathfrak{B}_k$.
Indeed, by Theorem 5.1.1, one can describe every maximal class of P_k for $k \geq 3$ with the aid of a certain k-ary relation ϱ in the form $Pol\ \varrho$. If $Pol \varrho$ is a maximal class, then, $Pol\ \varrho = Pol\ \varrho'$ is valid for all ϱ-derivable non-diagonal relation ϱ'. Therefore, the elements of $\bigcup_{h=1}^{k} R_k^h \backslash D_k^h$ are possible descriptive relations for a maximal class $Pol_k \varrho$. The subsequently defined relations of $\mathfrak{M}_k, \mathfrak{S}_k, \mathfrak{U}_k, \mathfrak{L}_k, \mathfrak{C}_k$ and $\mathfrak{B}_k$ are (with few exceptions), with respect to the arity, minimally chosen relations, which one can use to describe the maximal classes (see Chapter 10).

We say that a maximal class A is a **class of type** $\mathfrak{X}$, if there exist an $\mathfrak{X} \in \{\mathfrak{M}, \mathfrak{S}, \mathfrak{U}, \mathfrak{L}, \mathfrak{C}, \mathfrak{B}\}$ and a $\varrho \in \mathfrak{X}_k$ with $A = Pol_k \varrho$.

5.2.1 Maximal Classes of Type $\mathfrak{M}$ (Maximal Classes of Monotone Functions)

Let $\mathfrak{M}_k$ be the set of all partial orders on E_k with a greatest and a least element. More exactly, a binary relation $\varrho \in R_k$ belongs to $\mathfrak{M}_k$ if and only if ϱ has the following four properties:
1) ϱ is reflexive (i.e., $\iota_k^2 \subseteq \varrho$);
2) ϱ is antisymmetric (i.e., $\varrho \cap \varrho^{-1} = \iota_k^2$);
3) ϱ is transitive (i.e., $\varrho \circ \varrho = \varrho$) and
4) there exist elements o_ϱ ("**least element**") and e_ϱ ("**greatest element**") in E_k with $\{(o_\varrho, x), (x, e_\varrho) \mid x \in E_k\} \subseteq \varrho$.
It can easily be shown (see proof of Lemma 6.1.6) that the elements o_ϱ and e_ϱ are uniquely determined.
We write $a \leq_\varrho b$ instead of $(a, b) \in \varrho$ and $a <_\varrho b$, if $(a, b) \in \varrho \backslash \iota_k^2$. Furthermore, for $\mathbf{a}, \mathbf{b} \in E_k^n$ we put

$$\mathbf{a} \leq_\varrho \mathbf{b} :\Longleftrightarrow \forall i \in \{1, ..., n\} : a_i \leq_\varrho b_i.$$

Obviously, a function $f^n \in P_k$ preserves the relation $\varrho \in \mathfrak{M}_k$ if and only if the following holds:

$$\forall \mathbf{a}, \mathbf{b} \in E_k^n : \ \mathbf{a} \leq_\varrho \mathbf{b} \Longrightarrow f(\mathbf{a}) \leq_\varrho f(\mathbf{b}).$$

Thus the functions of $Pol\ \varrho$ are (non-decreasing) **monotone functions**. Obviously, we have:

Lemma 5.2.1.1

$$\forall \varrho, \varrho' \in \mathfrak{M}_k : \ (Pol\ \varrho = Pol\ \varrho' \Longleftrightarrow \varrho' = \tau\varrho \ \vee \ \varrho = \varrho').$$

■

For $k \in \{2, 3\}$ all classes of type $\mathfrak{M}$ are

$$M = Pol_2 \begin{pmatrix} 0 & 0 & 1 \\ 0 & 1 & 1 \end{pmatrix},$$

$$Pol_3 \begin{pmatrix} 0 & 1 & 2 & 0 & 0 & 1 \\ 0 & 1 & 2 & 1 & 2 & 2 \end{pmatrix} \qquad (0 <_\varrho 1 <_\varrho 2),$$

$$Pol_3 \begin{pmatrix} 0 & 1 & 2 & 0 & 0 & 2 \\ 0 & 1 & 2 & 2 & 1 & 1 \end{pmatrix} \qquad (0 <_\varrho 2 <_\varrho 1) \text{ and}$$

$$Pol_3 \begin{pmatrix} 0 & 1 & 2 & 2 & 2 & 0 \\ 0 & 1 & 2 & 0 & 1 & 1 \end{pmatrix} \qquad (2 <_\varrho 0 <_\varrho 1).$$

For larger k it is better to give the Hasse diagram, which characterizes the relation ϱ in unambiguous way, instead of the elements of ϱ. Figures 5.1 and 5.2 give the possible "basic diagrams" (i.e., diagrams without node names) for $k = 4$ and $k = 5$; hence $(E_k; \leq_\varrho)$ is a lattice for $k \leq 5$.

$k = 4$:

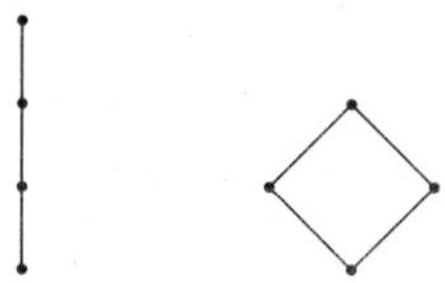

Fig. 5.1

$k = 5$:

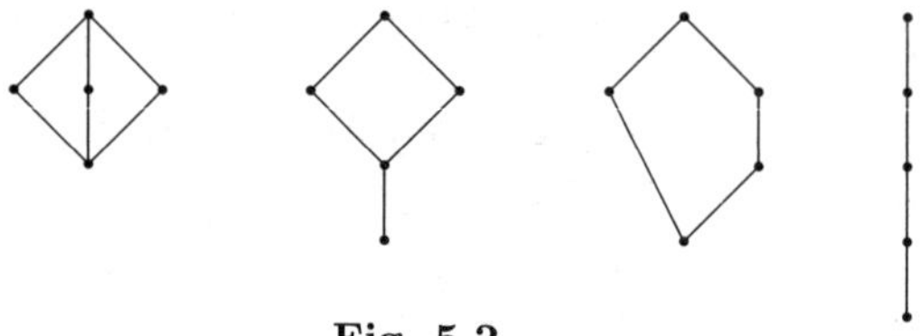

Fig. 5.2

For $k \geq 6$ there are, however, $\varrho \in \mathfrak{M}_k$, so that (E_k, ϱ) is not a lattice. Figure 5.3 is an example.

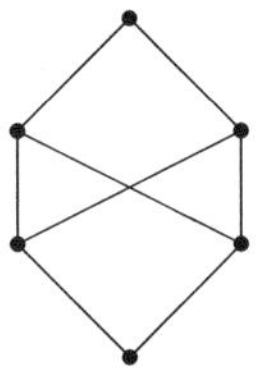

Fig. 5.3

5.2.2 Maximal Classes of Type $\mathfrak{S}$ (Maximal Classes of Autodual Functions)

For an arbitrary permutation $s \in S_k$, let

$$\varrho_s := \{(x, s(x)) \mid x \in E_k\}.$$

Functions, which preserve a relation of the form ϱ_s, are called **autodual** (in respect to s).
Let $\mathfrak{S}_k$ be the set of all relations of the form ϱ_s, where s is an arbitrary permutation on E_k with k/p cycles of the same prime length p.
Some examples for maximal classes of type $\mathfrak{S}$ are the following:

$$Pol_2 \begin{pmatrix} 0 & 1 \\ 1 & 0 \end{pmatrix},$$

$$Pol_3 \begin{pmatrix} 0 & 1 & 2 \\ 1 & 2 & 0 \end{pmatrix},$$

$$Pol_6 \begin{pmatrix} 0 & 1 & 2 & 3 & 4 & 5 \\ 1 & 0 & 3 & 2 & 5 & 4 \end{pmatrix},$$

$$Pol_6 \begin{pmatrix} 0 & 1 & 2 & 3 & 4 & 5 \\ 1 & 2 & 0 & 4 & 5 & 3 \end{pmatrix}.$$

Subsequently, we assume that an arbitrary permutation s with $\varrho_s \in \mathfrak{S}_k$ is defined as follows:
If $k = l \cdot p$ (p a prime number, $l \geq 1$) and

$$\{a_{1,0}, ..., a_{1,p-1}, a_{2,0}, ..., a_{2,p-1}, ..., a_{l,0}, ..., a_{l,p-1}\} = E_k,$$

we put

$$s(a_{r,i}) := a_{r,i+1 \ (mod\ p)} \ (r = 1, 2, ..., l; i = 0, 1, ..., p-1).$$

Then, it holds

$$s^a(a_{r,i}) = a_{r,i+a\ (mod\ p)} \text{ if } a = 0, 1, ...$$

By definition of ϱ_s, we have for arbitrary function $f^n \in Pol\ \varrho_s$:

$$s(f(x_1, x_2, ..., x_n)) = f(s(x_1), s(x_2), ..., s(x_n)) \tag{5.1}$$

or

$$f(x_1, x_2, ..., x_n) = s^i(f(s^{p-i}(x_1), s^{p-i}(x_2), ..., s^{p-i}(x_n))) \ (i = 0, 1, ..., p-1). \tag{5.2}$$

This implies that

$$\begin{aligned} &f(a_{r,i}, x_2, ..., x_n) = s^i(f(a_{r,0}, s^{p-i}(x_2), ..., s^{p-i}(x_n))) \\ &(i = 0, 1, ..., p-1; r = 1, 2, ..., l). \end{aligned} \tag{5.3}$$

Let $\widehat{S}_k^n$ be the set of all n-ary functions $f \in P_k$, for which there exists some $(n-1)$-ary functions $F_1, F_2, ..., F_l \in P_k \cup E_k$ such that

$$f(\mathbf{x}) = \sum_{r=1}^{l} \sum_{i=0}^{p-1} j_{a_{r,i}}(x_1) \cdot s^i(F_r(s^{p-i}(x_2), ..., s^{p-i}(x_n))), \tag{5.4}$$

where + is the addition *modulo* k, $\cdot$ is the multiplication *modulo* k and j_a is defined by

$$j_a(x) := \begin{cases} 1 \text{ if } & x = a, \\ 0 \text{ otherwise}, \end{cases}$$

for $a \in E_2$.
We call the function F_r **r-th component** of f.

Theorem 5.2.2.1 *It holds: Pol* $\varrho_s = \bigcup_{n \geq 1} \widehat{S}_k^n$.

Proof. For every function $f \in P_k^n$ there exists a description of the form

$$f(\mathbf{x}) = \sum_{i=0}^{k-1} j_i(x_1) \cdot f(i, x_2, ..., x_n) \ (mod\ k). \tag{5.5}$$

If $f \in Pol\ \varrho_s$ then it follows from (5.5) with the aid of (5.3):

$$f(\mathbf{x}) = \sum_{r=1}^{l} \sum_{i=0}^{p-1} j_{a_{r,i}}(x_1) \cdot s^i(f(a_{r,0}, s^{p-i}(x_2), ..., s^{p-i}(x_n))). \tag{5.6}$$

Consequently, $Pol\ \varrho_s \subseteq \bigcup_{n \geq 1} \widehat{S}_k^n$.
Let now $g \in P_k^n$ with

$$g(\mathbf{x}) = \sum_{r=1}^{l} \sum_{i=0}^{p-1} j_{a_{r,i}}(x_1) \cdot s^i(G_r(s^{p-i}(x_2), ..., s^{p-i}(x_n))) \in \widehat{S}_k^n,$$

where $G_1, G_2, ..., G_l$ are some $(n-1)$-ary functions of P_k. We have to show that $g \in Pol_k \varrho_s$.
For $i \in E_p$ and $1 \le r \le l$ it holds:

$$\begin{aligned} &g(s(a_{r,i}), s(x_2), ..., s(x_n)) = g(a_{r,i+1 \ mod \ p}, s(x_2), ..., s(x_n)) \\ &= s^{i+1}(G_r(s^{p-(i+1)}(s(x_2)), ..., s^{p-(i+1)}(s(x_n)))) \\ &= s(s^i(G_r(s^{p-i}(x_2), ..., s^{p-i}(x_n)))) \\ &= s(g(a_{r,i}, x_2, ..., x_n)), \end{aligned}$$

i.e., $g \in Pol_k \varrho_s$.
Thus, $\bigcup_{n \ge 1} \widehat{S_k^n} = Pol_k \varrho_s$. ■

We will need the following lemma to determine the number of maximal classes of P_k later.

Lemma 5.2.2.2 $\forall \varrho_s, \varrho_{s'} \in \mathfrak{S}_k$:

$Pol_k \varrho_s = Pol_k \varrho_{s'} \iff \exists i \in \{1, ..., k\} : s^i = s'$.

Proof. "$\Longrightarrow$": Let $s, s' \in \mathfrak{S}_k$ and $Pol_k \varrho_s = Pol_k \varrho_{s'}$. Then we have $Im(s) = Im(s') = E_k$ and $s, s' \in Pol_k \varrho_s$. Therefore, by Lemma 5.2.2.1, we can describe s' in the form (5.4), where $S'_0, ..., S'_{p-1}$ are nullary functions, i.e., elements of E_k. Thus, s' is unique determined by $s'(a_{r,0})$ $(r = 1, 2, ..., l)$. For every $r \in \{1, 2, ..., l\}$, $s'(a_{r,0})$ belongs to $\{a_{r,0}, a_{r,1}, ..., a_{r,p-1}\}$, since one receives a contradiction in choosing s in the opposite case, as follows:Suppose we have $s'(a_{r,0}) = a_{s,i}$ with $r \ne s$ for certain r, s, i. Set $V := \{a_{s,i}, s(a_{r,i}), ..., s^{p-1}(a_{r,i})\}$. Then, the permutation t with $t(x) := s(x)$ for all $x \in V$ and $t(x) := x$ for $x \in E_k \backslash V$ belongs to $Pol\ \varrho_s$ and, therefore, also to $Pol\ \varrho_{s'}$. Because of $s'(a_{r,0}) = a_{s,i}$ we have $(a_{r,0}, a_{s,i}) \in \varrho_{s'}$ and, by definition of t, this implies

$$t\begin{pmatrix} a_{r,0} \\ a_{s,i} \end{pmatrix} = \begin{pmatrix} a_{r,0} \\ s(a_{s,i}) \end{pmatrix}.$$

Thus, (because of $t \in Pol\ \varrho_{s'}$) $a_{s,i} = s(a_{s,i})$, in contradiction to $s \in \mathfrak{S}_k$.
Let now $s'(a_{1,0}) = a_{1,i}$ for certain $i \in E_p$. We have to show that $s'(a_{r,0}) = a_{r,i}$ for all $r \in \{2, 3, ..., l\}$ holds. Suppose $s'(a_{r,0}) = a_{r,j}$ for certain $j \in E_p \backslash \{i\}$ and $r \in \{2, 3, ..., l\}$. Then, a function g^2 with $g(a_{1,0}, a_{r,0}) = g(a_{1,0}, a_{r,j-i\ (mod\ p)}) = a_{r,0}$ belongs to $Pol_k \varrho_s$, for which we have, by (5.3),

$$g\begin{pmatrix} a_{1,0} & a_{r,0} \\ a_{1,i} & a_{r,j} \end{pmatrix} = \begin{pmatrix} a_{r,0} \\ a_{r,i} \end{pmatrix} \notin \varrho'_s.$$

Hence, the function g does not preserve the relation ϱ'_s, contrary to the assumption $Pol_k \varrho_s = Pol_k \varrho'_s$. Thus we have $s'(a_{r,0}) = a_{r,i}$ for all $r \in \{1, 2, ..., l\}$ and $s^i = s'$ holds.
"$\Longleftarrow$": If $s, s' \in \mathfrak{S}_k$ and $s^i = s'$ then $s \vdash s'$ and $s' \vdash s$. Hence $Pol_k \varrho_s \subseteq Pol_k \varrho'_s$ and $Pol_k \varrho'_s \subseteq Pol_k \varrho_s$ by Theorem 2.6.3. Thus, $Pol_k \varrho_s = Pol_k \varrho'_s$. ■

5.2.3 Maximal Classes of Type $\mathfrak{U}$ (Maximal Classes of Functions, Which Preserve Non-Trivial Equivalence Relations)

Let $\mathfrak{U}_k$ be the set of all non-trivial equivalence relations on E_k, i.e., for an arbitrary binary relation $\varrho \in R_k$ we have:

$$\varrho \in \mathfrak{U}_k \iff (\iota_k^2 \subseteq \varrho \wedge \varrho^{-1} = \varrho \wedge \varrho \circ \varrho = \varrho \wedge \varrho \notin \{\iota_k^2, E_k^2\}).$$

For $k = 2$, classes of the type $\mathfrak{U}$ do not occur. For $k = 3$ there are exactly three classes of the form $Pol\ \varrho$ with $\varrho \in \mathfrak{U}_3$:

$$Pol_3 \begin{pmatrix} 0 & 1 & 2 & 0 & 1 \\ 0 & 1 & 2 & 1 & 0 \end{pmatrix},$$

$$Pol_3 \begin{pmatrix} 0 & 1 & 2 & 0 & 2 \\ 0 & 1 & 2 & 2 & 0 \end{pmatrix} \text{ and}$$

$$Pol_3 \begin{pmatrix} 0 & 1 & 2 & 1 & 2 \\ 0 & 1 & 2 & 2 & 1 \end{pmatrix}.$$

In the following, let ϱ be an arbitrary relation of $\mathfrak{U}_k$. Further, let $\mathfrak{A}_i$ $(i \in E_{k'})$ be the equivalence classes (blocks) of this relation. By Lemma 1.4.6, an arbitrary function $f^n \in P_k$ is representable in the form (1.12). Since, in this case, the used functions f_i $(i \in E_{k'})$ and the function z belong to $Pol\ \varrho$, the function f preserves the relation ϱ if and only if $g_f \in Pol\ \varrho$ is valid. One can easily show that the functions of the form g_f $(\in Pol\ \varrho)$ are completely determined through their values on the set $\{a_i \mid i \in E_{k'}\}^n$. Hence,

$$(\{g_f \mid f \in Pol\ \varrho\}; \zeta, \tau, \Delta, \nabla, \star)$$

is isomorphic to

$$(P_{k'}; \zeta, \tau, \Delta, \nabla, \star).$$

Let now q a unary function of P_k defined by

$$\forall i \in E_{k'} : q(x) = a_i \ :\iff \ x \in \mathfrak{A}_i.$$

Then, the following lemma results from our above considerations.

Lemma 5.2.3.1 *$\forall f^n \in P_k$:*

$$f \in Pol\ \varrho \iff \exists h^n \in P_{\{a_0 \dots, a_{k'-1}\}} : g_f(\mathbf{x}) = h(q(x_1), \dots, q(x_n)).$$

■

In addition, the following is obviously valid:

Lemma 5.2.3.2 $\forall \varrho, \varrho' \in \mathfrak{U}_k : \ Pol\ \varrho = Pol\ \varrho' \iff \varrho = \varrho'.$ ■

5.2.4 Maximal Classes of Type $\mathfrak{L}$ (Maximal Classes of Quasi-Linear Functions)

The maximal classes, which are treated in this section, only occur if k is a prime number power p^m. Therefore, we assume $k = p^m$ with $p \in \mathbb{P}$ in this section.
Let $\mathfrak{G}$ be the set of all p-elementary Abelean groups of the form $(E_k, \oplus)$. All elements of such groups, which are different from the neutral element, have the order p, i.e., if o denotes the neutral element of the group $(E_k, \oplus) \in \mathfrak{G}$, then p is the smallest number with

$$\underbrace{x \oplus x \oplus ... \oplus x}_{p \text{ times}} = o$$

for all $x \in E_k \setminus \{o\}$.
Let $\mathfrak{L}_k$ be the set of all relations λ_G with $G := (E_k; \oplus) \in \mathfrak{G}$ and

$$\lambda_G := \{(a, b, c, d) \in E_k^4 \mid a \oplus b = c \oplus d\}.$$

If k is not a prime number power, we set $\mathfrak{L}_k = \emptyset$.
One can also describe the maximal classes of the form $Pol\ \lambda_G$ as follows:
Since k is a prime number power, it is well-known that one can define certain operations $+$ and $\cdot$ on E_k, so that the algebra $GF(p^m) := (E_k; +, \cdot)$ is a field, where the additive group of this field belongs to $\mathfrak{G}$. If one now chooses a suitable field with the operations $+ := \oplus$ and $\cdot$ to a group $G := (E_k, \oplus) \in \mathfrak{G}$, then (by Theorem 1.4.3) every function $f^n \in P_k$ is uniquely definable through a formula of the form

$$f(\mathbf{x}) = \sum_{(i_1 ..., i_n) \in E_k^n} a_{i_1 ... i_n} \cdot x_1^{i_1} \cdot ... \cdot x_n^{i_n}$$

with the aid of the chosen field operations.
A function g^n, which is defined through a formula of the specific form

$$g(\mathbf{x}) = a_0 + \sum_{i=1}^{n} \sum_{j=0}^{m-1} a_{ij} \cdot x_i^{p^j}, \tag{5.7}$$

is called **quasi-linear**. In particular, if $m = 1$ (i.e., k is a prime number), then g is called a **linear** function as usual.
The set of all quasi-linear functions g^n $(n = 1, 2, ...)$ of the form (5.7) is a closed set, as one can easily check with aid of the well-known fact

$$\forall i \in \{0, 1 ..., m\} : (x + y)^{p^i} = x^{p^i} + y^{p^i},$$

where $x, y \in GF(p^m)$.
Further, it holds that

Lemma 5.2.4.1 *The following statements are equivalent:*

(a) $f^n \in Pol\ \lambda_G$;
(b) f^n *is quasi-linear;*
(c) $f(x_1+y_1, x_2+y_2, ..., x_n+y_n) = f(x_1, ..., x_n) + f(y_1, ..., y_n) + f(o, ..., o)$.

Proof. (a) $\Longleftrightarrow$ (c): Because of $(x, y, x+y, o) \in \lambda_G$ every $f^n \in Pol\ \lambda_G$ satisfies the condition (c). Conversely, if (c) holds and $(x_i, y_i, u_i, v_i) \in \lambda_G$ $(i = 1, 2, ..., n)$, then $f(x_1, ..., x_n) + f(y_1, ..., y_n) = f(x_1+y_1, ..., x_n+y_n) + f(o, ..., o) = f(u_1+v_1, ..., u_n+v_n) + f(o, ..., o) = f(u_1, ..., u_n) + f(v_1, ..., v_n)$. Thus $f \in Pol\ \lambda_G$.
$(a) \Longleftrightarrow (b)$: Since every quasi-linear function is obviously a superposition over $B := \{x+y, x^{p^i}, a \cdot x, c_a \mid i \in \{1, .., m-1\} \wedge a \in E_k\}$, and by (5.8) all functions of B preserve the relation λ_G, the implication $(b) \Longrightarrow (a)$ is valid. In Theorem 5.3.6, we show that the set of all quasi-linear functions is maximal in P_k. Therefore, $(a) \Longrightarrow (b)$ is right. ■

Lemma 5.2.4.2 *Let* $G := (E_k; +)$ *and* $G' := (E_k, +')$ *be elements of* $\mathfrak{G}$. *Then, the following statements are equivalent:*

(a) $Pol\ \lambda_G = Pol\ \lambda'_G$.

(b) There exists an element $c \in E_k$ *such that die mapping*

$$\alpha : E_k \longrightarrow E_k, x \mapsto x + c$$

is an isomorphism from G' *onto* G.

(c) $\lambda_G = \lambda'_G$.

Proof. Denote o and o' the neutral elements of G and G', respectively.
$(a) \Longrightarrow (b)$: Let $Pol\ \lambda_G = Pol\ \lambda'_G$. Then, the function g^2 with $g(x, y) = x +' y$ belongs to $Pol\ \lambda_G$. Because of $(x, o', o', x), (o', y, o', y) \in \lambda'_G$ this implies

$$g(x, y) = g(x, o') + g(o', y) - g(o', o')$$

or

$$x +' y = x + y - o'. \tag{5.8}$$

With the aid of (5.8), one can verify that $\alpha(x) := x - o'$ is an isomorphic mapping from G' onto G.
$(b) \Longrightarrow (a)$: Let α be an isomorphism from G' onto G with $\alpha(x) := x + c$ for certain $c \in E_k$. Then, for arbitrary $f^n \in Pol\ \lambda_G$ by Lemma 5.2.4.1, we have

$$f(\mathbf{o}) = f(\mathbf{o}' + \mathbf{c}) = f(\mathbf{o}') + f(\mathbf{c}) - f(\mathbf{o}) \tag{5.9}$$

and

$$f(\mathbf{x} +' \mathbf{y}) = f(\mathbf{x}) + f(\mathbf{y}) + f(\mathbf{c}) - f(\mathbf{c}) - f(\mathbf{o}) - f(\mathbf{o}). \tag{5.10}$$

It results from (5.9) and (5.10):

$$f(\mathbf{x} +' \mathbf{y}) = f(\mathbf{x}) + f(\mathbf{y}) - f(\mathbf{o}').$$

Thus

$$\begin{aligned}\alpha(f(\mathbf{x} +' \mathbf{y})) &= (f(\mathbf{x}) + c) + (f(\mathbf{y}) + c) - (f(\mathbf{o}') + c)\\ &= \alpha(f(\mathbf{x})) + \alpha(f(\mathbf{y})) - \alpha(f(\mathbf{o}'))\\ &= \alpha(f(\mathbf{x}) +' f(\mathbf{y}) -' f(\mathbf{o}'))\end{aligned}$$

holds because α is an isomorphism. Consequently, by Lemma 5.2.4.1, f preserves the relation λ; hence $Pol\ \lambda_G \subseteq Pol\ \lambda'_G$. Since α^{-1} is also an isomorphism (from G onto G'), we have also $Pol\ \lambda'_G \subseteq Pol\ \lambda_G$. Thus (a) holds.
$(a) \Longleftrightarrow (c)$ follows easily by means of $(a) \Longleftrightarrow (b)$. ■

The following representation of quasi-linear functions is better suited for some purposes (for example, determining subclasses of $Pol\ \lambda_G$) than the above definition.
It is well-known that every group $G = (E_k; +)$ of $\mathfrak{G}$ is isomorphic to $(W; \oplus)$, where $W := E_p^m$ and

$$(x_1, ..., x_m) \oplus (y_1, ..., y_m) := (x_1 + y_1\ (mod\ p), ..., x_m + y_m\ (mod\ p))$$

for arbitrary $(x_1, ..., x_m), (y_1, ..., y_m) \in E_p^m$.
One can assign a function $F^n \in P_W$ to every function $f^n \in P_k$ $(k = p^m)$ by means of the isomorphism α, as follows:

$$F^n(\alpha(x_1), ..., \alpha(x_n)) := \alpha(f(x_1, ..., x_n)).$$

Let L_W^n be the set of all n-ary functions of P_W, for which there are an $\mathbf{a} \in W$ and some matrices $\mathbf{A_1}, ..., \mathbf{A_n} \in E_p^{m\times m}$ such that

$$F(X_1, ..., X_n) = \mathbf{a} + X_1 \cdot \mathbf{A_1} + ... + X_n \cdot \mathbf{A_n}, \tag{5.11}$$

where $+$ and $\cdot$ are the usual matrix operations and $X_i := (x_{i1}, ..., x_{in}) \in E_p^m$ $(i = 1, ..., n)$. Then, we have:

Lemma 5.2.4.3 *([Ros 70;a], 7.3.3)*

$L_W := \bigcup_{n\geq 1} L_W^n$ *is isomorphic to* $Pol\ \lambda_G$ *for all* $G \in \mathfrak{G}$.

Proof. It is easy to check that each function of the form (5.11) fulfills the condition (c) of Lemma 5.2.4.1, whereby L_W is isomorphic to a subset of $Pol\ \lambda_G$. $L_W \cong Pol\ \lambda_G$ follows from the fact that $|L_W^n| = p^{m+n\cdot m^2} = (p^m)^{1+n\cdot m} = k^{1+n\cdot m} = |(Pol\ \lambda_G)^n|$. ■

5.2.5 Maximal Classes of Type $\mathfrak{C}$ (Maximal Classes of Functions, Which Preserve Central Relations)

An h-ary relation γ $(1 \leq h \leq k-1)$ auf E_k is called **central**, iff γ has the three following properties:

1) γ is **totally reflexive** and non-diagonal, i.e., it holds $\iota_k^h \subseteq \gamma \neq E_k^h$ ($\iota_k^1 := \emptyset$);

2) γ is **totally symmetric**, i.e., for every permutation s on $\{1, 2, ..., h\}$ it holds:

$$(a_1, ..., a_h) \in \gamma \Longrightarrow (a_{s(1)}, ..., a_{s(h)}) \in \gamma;$$

3) there exists at least a **central element** $c \in E_k$, i.e., $(a_1, ..., a_{h-1}, c) \in \gamma$ for all $a_1, ..., a_{h-1} \in E_k$.

Let $\mathfrak{C}_k^h$ be the set of all central h-ary relations on E_k and

$$\mathfrak{C}_k := \bigcup_{h \geq 1} \mathfrak{C}_k^h.$$

Notice that

$$\mathfrak{C}_k^1 = \{\varrho \mid \emptyset \subset \varrho \subset E_k\}.$$

For $k \in \{2, 3\}$, only the following classes are defined by central relations:

$$Pol_2(0),\ Pol_2(1),$$

$$Pol_3(a)\ (a \in E_3),\ Pol_3(a\ b) \text{ if } \{a, b\} \subseteq E_3 \text{ and } a \neq b,$$

$$Pol_3\begin{pmatrix} 0 & 1 & 2 & a & b & a & c \\ 0 & 1 & 2 & b & a & c & a \end{pmatrix} \text{ if } \{a, b, c\} = E_3.$$

Because of total reflexivity and total symmetry of the relations of $\mathfrak{C}_k$, $P_{k,\{a_1,...,a_h\}}$ is a subset of $Pol\ \gamma$ for every $\gamma \in \mathfrak{C}_k$ and $(a_1, ..., a_h) \in \gamma$. With the aid of this property, one can easily prove that the following lemma holds:

Lemma 5.2.5.1 $\forall \gamma, \gamma' \in \mathfrak{C}_k :\ Pol\ \gamma = Pol\ \gamma' \Longleftrightarrow \gamma = \gamma'$. ■

5.2.6 Maximal Classes of Type $\mathfrak{B}$ (Maximal Classes of Functions, Which Preserve h-Universal Relations)

In the literature, one finds two possibilities for describing the defining relations of the maximal classes of the type $\mathfrak{B}$. Both possibilities go back to papers from I. G. Rosenberg. We begin with the definition from [Ros 65]. At the end of this section, the reader finds the second description for classes of the type $\mathfrak{B}$, which uses so-called h-regular relation sets defined in [Ros 70c].
One can represent every $a \in E_{h^m}$ ($h \geq 3$, $m \geq 1$) unique in the form

$$a = a^{(m-1)} \cdot h^{m-1} + a^{(m-2)} \cdot h^{m-2} + ... + a^{(1)} \cdot h + a^{(0)}, \tag{5.12}$$

where $a^{(m-1)}, a^{(m-2)}, ..., a^{(0)} \in E_h$ are suitably chosen.
The h-ary relation $\xi_m^h \subseteq E_{h^m}^h$ is called **h-ary elementary**, if it holds:

$$(a_0, ..., a_{h-1}) \in \xi_m^h \ :\Longleftrightarrow\ \forall i \in E_m : (a_0^{(i)}, a_1^{(i)}, ..., a_{h-1}^{(i)}) \in \iota_k^h.$$

In particular, $\xi_1^h = \iota_k^h$ for $m = 1$.
An h-ary relation ϱ on E_k is a **homomorphic inverse image** of an h-ary

relation ϱ' on $E_{k'}$, if there exists a mapping q from E_k onto $E_{k'}$, such that, for all $a_1, ..., a_h \in E_k$, it holds:

$$(a_1, a_2, ..., a_h) \in \varrho \iff (q(a_1), q(a_2), ..., q(a_h)) \in \varrho'.$$

Let $\mathfrak{B}_k^h$ be the set of all homomorphic inverse images of the h-ary elementary relation ξ_m^h and put

$$\mathfrak{B}_k := \bigcup_{h=3}^{k} \mathfrak{B}_k^h.$$

The elements of $\mathfrak{B}_k^h$ are also called **h-universal relations.**
For $k = 3$ we have $\mathfrak{B} = \{\iota_3^3\}$.
To derive a description of the functions from the clone $Pol\ \varrho$ with $\varrho \in \mathfrak{B}_k$ without using relations, we assume that ϱ is an h-ary relation ϱ on E_k which is a homomorphic inverse image of the h-ary elementary relation $\xi_m^h \subseteq E_{h^m}$ in the following. In addition, let q be the mapping from E_k onto E_{h^m} which belongs to this relation ϱ because of definition. This mapping defines a partition on E_k in the blocks

$$\mathfrak{A}_i := \{x \in E_k \mid q(x) = i\},\ i \in E_{h^m}.$$

We select a representative a_i from every one of these blocks $\mathfrak{A}_i$. Further, let r be a unary function with $r(i) = a_i$ for all $i \in E_{h^m}$. As already given in Lemma 1.4.6, an arbitrary function f^n from P_k is as a superposition over the functions z, g_f, f_i ($i \in E_{h^m}$), defined by

$$z(x, y) := x\Diamond y\ (\ x\Diamond y = y \text{ if } \exists i \in E_k : x = a_i \text{ and } y \in \mathfrak{A}_i; x\Diamond y = x \text{ otherwise}),$$

$$g_f(x_1, ..., x_n) := a_i \iff f(x_1, ..., x_n) \in \mathfrak{A}_i,$$

$$f_i(x_1, ..., x_n) := \begin{cases} f(x_1, ..., x_n) & \text{if} \quad f(x_1, ..., x_n) \in \mathfrak{A}_i, \\ a_i & \text{otherwise} \end{cases}$$
$(i \in E_{k'})$,

represents as follows:

$$f(\mathbf{x}) = ((...((g_f(\mathbf{x})\Diamond f_0(\mathbf{x}))\Diamond f_1(\mathbf{x}))\Diamond ...)\Diamond f_{h^m-1}(\mathbf{x})). \tag{5.13}$$

Since an arbitrary function $t^n \in P_{h^m}$ is describable in the form

$$t(\mathbf{x}) = (t(\mathbf{x}))^{(m-1)} \cdot h^{m-1} + (t(\mathbf{x}))^{(m-2)} \cdot h^{m-2} + ... + ((t(\mathbf{x}))^{(0)} \tag{5.14}$$

(see also (5.12)), we can describe the function from (5.13) with the aid of (5.14) as follows:

$$g_f(\mathbf{x}) = r(f'_{m-1}(\mathbf{x}) \cdot h^{m-1} + f'_{m-2}(\mathbf{x}) \cdot h^{m-2} + ... + f'_1(\mathbf{x}) \cdot h + f'_0(\mathbf{x})), \tag{5.15}$$

where $f'_i(x_1, ..., x_n) := (q(f(x_1, ..., x_n)))^{(i)}$ for $i \in E_{h^m}$. Now, one can describe the functions of $Pol\ \varrho$ without using relations, as follows:

Theorem 5.2.6.1 *([Lau 78;a])*
An n-ary function $f \in P_k$ belongs to the class Pol ϱ if and only if for the function g_f (see (5.15)) is valid:

$$\begin{aligned}&\forall\, i \in \{0,1,\ldots,m-1\}:\\ &\textit{either } (|Im(f_i)| \le h-1)\\ &\textit{or}\\ &\textit{there are } j \in \{1,\ldots,n\},\ v \in E_m,\ \textit{a permutation } s \textit{ on } E_h\\ &\textit{such that } f_i'(x_1,\ldots,x_n) = s((q(x_j))^{(v)})\,.\end{aligned} \tag{5.16}$$

Proof. Obviously, every function with the properties of the above theorem belongs to the class *Pol* ϱ. We must prove that an arbitrary n-ary function f, which does not fulfill the conditions of the theorem, does not belong to *Pol* ϱ.
Case 1: In the description (5.15), for g_f, there is an i so that f_i' is dependent at least of two variables essentially and it holds $Im(f_i') = E_h$.
Then, by Theorem 1.4.4, (b) there exists some h–tuples $r_1, ..., r_n \in \iota_k^h$ with $f_i'(r_1, ..., r_n) \in E_k^h \backslash \iota_k^h$. Consequently, $f(r_1, ..., r_n) \notin \varrho$.
Case 2: In the description (5.15), for g_f, there is an i, so that f_i' is dependent essentially only of the variable x_j $(1 \le j \le n)$, $|Im(f_i')| = E_h$ and f_i' can not be described in the form $f_i'(x_1, \ldots, x_n) = s((q(x_j))^{(v)})$.
Case 2.1: No function $F \in (Pol\ \xi_m^h)^1$ with $f_i'(x_1, ..., x_n) = F(q(x_j))$ exists.
In this case, there are tuples $\mathbf{a} = (a_1, ..., a_n)$, $\mathbf{b} = (b_1, ..., b_n) \in E_k^n$ with $f_i'(\mathbf{a}) \ne f_i'(\mathbf{b})$ and $(q(a_1), q(a_2), ..., q(a_n)) = (q(b_1), ..., q(b_n))$. By $Im(f_i') = E_h$ there exist tuples $\mathbf{c_2}, \mathbf{c_3}, ..., \mathbf{c_{h-1}} \in E_k^n$ with $\{f_i'(\mathbf{a}), f_i'(\mathbf{b}), f_i'(\mathbf{c_2}), ..., f_i'(\mathbf{c_{h-1}})\} = E_h$. Consequently, $(f(\mathbf{a}), f(\mathbf{b}), f(\mathbf{c_2}), ..., f(\mathbf{c_{h-1}})) \notin \varrho$. This and the fact that the columns of the matrix

$$\begin{pmatrix} \mathbf{a} \\ \mathbf{b} \\ \mathbf{c_2} \\ \ldots \\ \mathbf{c_{h-1}} \end{pmatrix}$$

belong to ϱ imply $f \notin Pol\ \varrho$.
Case 2.2: There exists a function $F \in (Pol\ \xi_m^h)^1$ with $f_i'(x_1, ..., x_n) = F(q(x_j))$.
With the aid of function F, one can define an m-ary function $F^* \in P_h$ as follows:

$$F^*(a^{(m-1)}, a^{(m-2)}, ..., a^{(0)}) := F(a^{(m-1)} \cdot h^{m-1} + a^{(m-2)} \cdot h^{m-2} + ... + a^{(0)}).$$

If one can not describe the function f_i' as it is given in the theorem, then the function F^* must depend on at least two variables, essentially. With the aid of Theorem 1.4.4, (b) this implies that f does not preserve ϱ. ■

We illustrate the above theorem with the following

Examples Let $h = 3$, $m = 2$, $k = 11$, $q : E_{11} \longrightarrow E_9$ be defined by $q(x) := x + 1\ (mod\ 9)$ for $x \in E_9$, $q(9) = 4$ and $q(10) = 1$. Further, let the two permutations s_1 and s_2 be defined by

x	$s_1(x)$	$s_2(x)$
0	1	2
1	0	0
2	2	1

For $x = x^{(1)} \cdot 3 + x^{(0)} \in E_9$, let $g(x) := s_1(x^{(0)})$ and $g'(x) := s_2(x^{(1)})$, i.e.,

x	$x^{(1)}$	$x^{(0)}$	$g(x)$	$g'(x)$
0	0	0	1	2
1	0	1	0	2
2	0	2	2	2
3	1	0	1	0
4	1	1	0	0
5	1	2	2	0
6	2	0	1	1
7	2	1	0	1
8	2	2	2	1

Then, the ternary functions $f, h \in P_9$ defined by

$$f(x_1, x_2, x_3) := g(x_1) \cdot 3 + g'(x_3)$$

(here $f_0(x_1, x_2, x_3) = g'(x_3) = s_2(x_3^{(1)})$ and $f_1(x_1, x_2, x_3) = g(x_1) = s_1(x_1^{(0)})$) and

$$h(x_1, x_2, x_3) := g(x_2) \cdot 3 + f'(x_1, x_2, x_3),$$

where $Im(f') \subset \{0, 1, 2\}$ and $|Im(f)| \leq 2$, both belong to Pol ζ_2.

The following is an example of a unary function $f \in P_{11}$ (see last column below) that preserves the relation ϱ:

x	$q(x)$	$(q(x))^{(1)}$	$(q(x))^{(0)}$	$s_1((q(x))^{(1)})$	$r(x)$	$q(f(x))$	$f(x)$
0	1	0	1	1	1	4	3
1	2	0	2	1	1	4	9
2	3	1	0	0	1	1	10
3	4	1	1	0	1	1	0
4	5	1	2	0	0	0	8
5	6	2	0	2	0	6	5
6	7	2	1	2	0	6	5
7	8	2	2	2	1	7	6
8	0	0	0	1	1	4	3
9	4	1	1	0	1	1	0
10	1	0	1	1	1	4	9

since

$$q(f(x)) = s_1((q(x))^{[1]}) \cdot 3 + r(x).$$

Lemma 5.2.6.2 follows from the definition of the classes of the form $Pol\ \varrho$ with $\varrho \in \mathfrak{B}_k$:

Lemma 5.2.6.2 $\forall \varrho, \varrho' \in \mathfrak{B}_k : \ Pol\ \varrho = Pol\ \varrho' \iff \varrho = \varrho'$. ■

Finally, we declare another possibility for describing of h-universal relations. We need the following concepts and notations:
Let $3 \leq h \leq k$, $m \geq 1$, $h^m \leq k$, let ϑ_i for $i \in E_m$ be an equivalence relation on E_k and $T := \{\vartheta_0, ..., \vartheta_{m-1}\}$. Then, T is called h-**regular** if T fulfills the following two conditions:

1) For every $i \in E_m$, the equivalence relation ϑ_i has exactly h equivalence classes.
2) Choosing for every $i \in E_m$ an equivalence class ε_i of the relation ϑ_i arbitrary, then $\bigcap_{i=0}^{m-1} \varepsilon_i \neq \emptyset$.

An h-ary relation $\varrho \subseteq E_k^h$ is called h-**regular** if there exists an h-regular set $T := \{\vartheta_0, ..., \vartheta_{m-1}\}$ such that for arbitrary $a_0, ..., a_{h-1} \in E_k$ it holds:

$$(a_0, ..., a_{h-1}) \in \varrho \iff (\forall i \in E_m \ \exists \{r, s\} \subset E_h : \ r \neq s \ \wedge \ (a_r, a_s) \in \vartheta_i).$$

Lemma 5.2.6.3 *([Ros 70c]) Let $3 \leq h \leq k$, $m \geq 1$, $h^m \leq k$ and $\varrho \in E_k^h$. Then, ϱ is h-universal iff ϱ is h-regular.*

Proof. "$\Longrightarrow$": Let ϱ be h-universal; i.e., there exists a surjective mapping $q : E_k \longrightarrow E_{h^m}$ with

$$(a_0, ..., a_{h-1}) \in \varrho \iff (q(a_0), ..., q(a_{h-1})) \in \xi_m^h.$$

With the aid of ϱ, one can define equivalence relations $\vartheta_i \subseteq E_k^2$ with $i \in E_m$ as follows:

$$(a, b) \in \vartheta_i \ :\Longleftrightarrow \ (q(a))^{(i)} = (q(b))^{(i)}.$$

It is easy to check that $T := \{\vartheta_0, ..., \vartheta_{m-1}\}$ is h-regular and the h-regular relation ϱ_T defined with the aid of T is identical with ϱ.
"$\Longleftarrow$": Let ϱ be h-regular and $T := \{\vartheta_0, ..., \vartheta_{m-1}\}$ denotes the corresponding h-regular set. Let

$$\varepsilon_{i,0}, \varepsilon_{i,1}, ..., \varepsilon_{i,h-1}$$

be the equivalence classes of the equivalence relation ϑ_i for $i \in E_m$. Every $x \in E_k$ belongs to exactly one equivalence class of ϑ_i, which we describe with

$$\varepsilon_{i,a_i(x)}.$$

With the help of the numbers $a_i(x) \in E_h$ one can define a mapping q as follows:

$$q : E_k \longrightarrow E_{h^m},\ x \mapsto \sum_{i=0}^{m-1} a_i(x) \cdot h^i.$$

Obviously, q is surjective. Furthermore, we have $(x_0, ..., x_{h-1}) \in \varrho$ if and only if $(\{a_i(x_0), ..., a_i(x_{h-1})) \in \iota_h^h$ for all $i \in E_m$. Consequently, q is a mapping from E_k onto E_{h^m} with

$$(x_0, ..., x_{h-1}) \in \varrho \iff (q(x_0)), ..., q(x_{h-1}) \in \xi_m^h,$$

i.e., ϱ is h-universal. ■

5.3 Proof of the Maximality of the Classes Defined in Section 5.2

An h-ary relation ϱ is called a **strong relation** iff it fulfills the following two conditions:

$$\varrho \backslash \iota_k^h \neq \emptyset \text{ and } \forall r \in \varrho \backslash \iota_k^h\ \forall s \in \varrho\ \exists f_{r,s}^1 \in Pol\varrho :\ f_{r,s}(r) = s \tag{5.17}$$

($\iota_k^h := \emptyset$ for $h = 1$).
A set $T \subseteq E_k^q$ is called **ϱ-independent** if for arbitrary pairwise different tuples $\mathbf{a_1}, \mathbf{a_2}, ..., \mathbf{a_h}$ of T ($\mathbf{a_i} := (a_{i1}, a_{i2}, ..., a_{iq})$ for $i = 1, ..., h$) there is an $j \in \{1, 2, ..., q\}$ with $(a_{1j}, a_{2j}, ..., a_{hj}) \notin \varrho$.

Lemma 5.3.1 *Let ϱ be an h-ary relation with the properties:*

(a) ϱ is a strong relation.
(b) For every $q \in \mathbb{N}$ and for every ϱ-independent set

$$T := \{\mathbf{a_1}, ..., \mathbf{a_{|T|}}\} \subseteq E_k^q$$

it holds

$$\{(f(\mathbf{a_1}), ..., f(\mathbf{a_{|T|}})) \mid f \in (Pol\varrho)^q\} = E_k^{|T|}. \tag{5.18}$$

(c) $(h = 1) \vee (h = 2 \wedge \varrho \cap \iota_k^h \in \{\emptyset, \iota_k^h\}) \vee$
$(h \geq 3 \wedge (\exists A \subseteq E_k :\ |A| = h \wedge P_{E_k,A} \subseteq Pol_k\varrho)$.

Then $Pol\varrho$ is maximal in P_k.

Proof. Let $g^m \in P_k \backslash Pol\varrho$ be arbitrary; i.e., there are some $s_1, ..., s_m \in \varrho$ with $g(s_1, ..., s_m) \notin \varrho$. Because of (a) for every $r \in \varrho \backslash \iota_k^h$ there exists a function f_{r,s_i} with $f_{r,s_i}(r) = s_i$ $(i = 1, 2, ..., m)$. Consequently, the function h_r defined by $h_r(x) := g(f_{r,s_1}(x), ..., f_{r,s_m}(x))$ has the following properties: $h_r(r) \notin \varrho$ and $h_r \in [Pol\varrho \cup \{g\}]$. Let $\{h_1, h_2, ..., h_t\} := \{h_r \mid r \in \varrho \backslash \iota_k^h\}$. It is easy to check that for $h \in \{1, 2\}$ and for every $n \in \mathbb{N}$ the set

$$\begin{aligned} T_n := &\{(x_1, ..., x_n, h_1(x_1), h_1(x_2), ..., h_1(x_n), h_2(x_1), h_2(x_2), ..., \\ &h_2(x_n), ..., h_t(x_1), h_t(x_2), ..., h_t(x_n)) \mid x_1, ..., x_n \in E_k\} \subseteq E_k^{n\cdot(t+1)} \end{aligned}$$

is ϱ-independent. Then, by (b), for every n-ary function $f \in P_k$ one can find an $n \cdot (t+1)$-ary function f' with

$$f(x_1, ..., x_n) = f'(x_1, ..., x_n, h_1(x_1), ..., h_1(x_n), ..., h_t(x_1), ..., h_t(x_n)).$$

Consequently, we have $[Pol\varrho \cup \{g\}] = P_k$ in the case $h \in \{1,2\}$. One can handle the case $h \geq 3$ in analog mode when one uses the ϱ-independent set

$$T'_n := \{(x_1, ..., x_n, h_1(g_1), ..., h_1(g_u)) \mid x_1, ..., x_n \in E_k\}$$

with $\{g_1, ..., g_u\} := P^n_{E_k,A}$ and suitably chosen h_1 instead of T_n. ■

Theorem 5.3.2 *([Jab 58], [Mart 60], [Ros 70;a])*
For every $\varrho \in \mathfrak{M}_k \cup \mathfrak{U}_k \cup \mathfrak{S}_k \cup \mathfrak{C}_k$ the clone $Pol\varrho$ is maximal in P_k.

Proof. Obviously, every $\varrho \in \mathfrak{M}_k \cup \mathfrak{U}_k \cup \mathfrak{S}_k \cup \mathfrak{C}_k$ satisfies the condition (c) of Lemma 5.3.1. In the following, we show that such relation ϱ also fulfills the conditions (a) and (b). For this purpose, $h^{n_1}_\varrho$ ($n_1 \in \{1, n\}$) denotes a function of P_k, which either fulfills $h_\varrho(r) = s$ for arbitrary r ($\in \varrho \backslash \iota^h_k$) and $s \in \varrho$ (see (5.17)), or which can take arbitrary values on an arbitrary ϱ-independent set $T \subseteq E^n_k$ (see (5.18)). If $\varrho \in \mathfrak{U}_k \cup \mathfrak{S}_k \cup \mathfrak{C}_k$, one can consider easily determination for h_ϱ, so that $h_\varrho \in Pol\varrho$ is valid. (For example, if $\varrho \in \mathfrak{C}_k$ and $h_\varrho(\mathbf{x})$ is not yet defined, one sets $h_\varrho(\mathbf{x}) = c$, where c is a central element of ϱ. For $\varrho \in \mathfrak{U}_k \cup \mathfrak{S}_k$ one can use Lemma 5.2.3.1 and Theorem 5.2.2.1.)
Let $\varrho \in \mathfrak{M}_k$ in the following. If h_ϱ is a function of the form $f_{r,s}$ with $r = \begin{pmatrix} \alpha \\ \beta \end{pmatrix} \in \varrho$ and $s = \begin{pmatrix} a \\ b \end{pmatrix} \in \varrho$, then

$$h_\varrho := \begin{cases} a \text{ if } & x <_\varrho \alpha, \\ b \text{ otherwise,} \end{cases}$$

is a function of $Pol\varrho$. If h_ϱ is supposed to be determined in the above manner through a ϱ-independent set T and (5.18), we put for all $\mathbf{x} \in E^n_k \backslash T$:

$$h_\varrho := \begin{cases} o \text{ if } & \exists \mathbf{a} \in T : \mathbf{x} <_\varrho \mathbf{a}, \\ e \text{ otherwise,} \end{cases}$$

where o is the least element and e is the greatest element of E_k in respect to ϱ. The function h_ϱ defined in this way belongs to $Pol\varrho$, as one can prove as follows:
Suppose $h \notin Pol\varrho$. Then, there exists some $\mathbf{a}, \mathbf{b} \in E^n_k$ with $\{\mathbf{a}, \mathbf{b}\} \not\subseteq T$, $\mathbf{a} <_\varrho \mathbf{b}$ and $t := (h_\varrho(\mathbf{a}), h_\varrho(\mathbf{b})) \in \{(\alpha, o), (e, \alpha), (e, o)\} \subseteq E^2_k \backslash \varrho$ for a certain $\alpha \in E_k \backslash \{o, e\}$. If $t = (\alpha, o)$, we have $\mathbf{a} \in T$ and thus $\mathbf{b} \notin T$. Because of $h_\varrho(\mathbf{b}) = o$, this implies (by the definition of h_ϱ) the existence of a $\mathbf{c} \in T$ with $\mathbf{b} <_\varrho \mathbf{c}$, whereby $\mathbf{a} <_\varrho \mathbf{c}$ for $\mathbf{a}, \mathbf{c} \in T$, in contradiction to the construction of

T. In similar way one can show that $t \in \{(e, \alpha), (e, o)\}$ is also not possible: If $t = (e, \alpha)$, then $\mathbf{b} \in T$ and $\mathbf{a} \notin T$; hence $h_\varrho(\mathbf{a}) = o$ by the definition of h_ϱ, in contradiction to $h_\varrho(\mathbf{a}) = e$. If $t = (e, o)$ then the following three cases are possible:

Case 1: $\mathbf{a} \in T$ and $\mathbf{b} \notin T$.
Because of $\mathbf{b} \notin T$ and $h_\varrho(\mathbf{b}) = o$ there exists a $\mathbf{c} \in T$ with $\mathbf{b} <_\varrho \mathbf{c}$. By $\mathbf{a} <_\varrho \mathbf{b}$; however, this implies $\mathbf{a} <_\varrho \mathbf{c}$, what was impossible for tuples $\mathbf{a}, \mathbf{c} \in T$.

Case 2: $\mathbf{a} \notin T$ and $\mathbf{b} \in T$.
This case contradicts the definition of the function h_ϱ: $h_\varrho(\mathbf{a}) = o$, since $\mathbf{a} \notin T$ and $\mathbf{a} <_\varrho \mathbf{b} \in T$.

Case 3: $\mathbf{a} \notin T$ and $\mathbf{b} \notin T$.
Because of $h_\varrho(\mathbf{b}) = o$ and $\mathbf{b} \notin T$ there exists a $\mathbf{c} \in T$ with $\mathbf{b} <_\varrho \mathbf{c}$. Then $\mathbf{a} <_\varrho \mathbf{c}$ is also valid, however; hence $h_\varrho(\mathbf{a}) = o$, in contradiction to the assumption $h_\varrho(\mathbf{a}) = e$. Thus our assertion results from Lemma 5.3.1. ■

Theorem 5.3.3 *([Ros 70;a])*
For every $\varrho \in \mathfrak{B}_k$ the clone $Pol\varrho$ is a maximal class of P_k.

Proof. Let ϱ be defined as in Section 5.2.6. Then, one can represent an arbitrary function $f^n \in P_k \backslash Pol\varrho$ in the form (4.14), (5.15) (see also Lemma 1.4.6), where, by Theorem 5.2.6.1, there exists a function f_i' that takes all values of $\{0, 1, ..., h-1\}$ on some tuples $r_1, ..., r_n \in \varrho$ (see also the proof of Theorem 5.2.6.1). Since $g(x) := (q(x))^{(i)} \in Pol\varrho$ and $P_{k,B} \subseteq Pol\varrho$ for all $B := \{b_1, .., b_h\}$ with $(b_1, ..., b_h) \in \varrho$, the function $g \star f = f_i'$ and a function of $P_{k,h}$ are superpositions over f and some functions of $P_{k,B} \subset [Pol\varrho \cup \{f\}]$. Therefore, by Lemma 1.4.5, $P_{k,h} \subseteq [Pol\varrho \cup \{f\}]$; hence all functions of P_k are superpositions over $Pol\varrho \cup \{f\}$ (see Lemma 1.4.6 and Theorem 5.2.6.1). ■

Now, we come to the proof of the maximality of the classes of type $\mathfrak{L}$.

For this purpose, let $k = p^m$ (p prime, $m \geq 1$) and let $G := (E_k; +, \cdot)$ be a field with (w.l.o.g.) the zero element 0 *and the unit element* 1. *Let Q_G be the set of all quasi-linear functions (in respect to G) of P_k.*

We need two lemmas for proof of the P_k-maximality of Q_G.

Lemma 5.3.4 *([Ros 70;a], 7.2.19)*
Let $f^n \in P_k \backslash Q_G$ ($k = p^m$). Then $[Q_G \cup \{f\}]$ contains a function of the form

$$h(x, y) = \sum_{i,j=0}^{p^m-1} a_{ij} \cdot x^i \cdot y^j,$$

which has at least a coefficient $a_{st} \neq 0$ with $0 < s, t \leq p^{m-1}$.

Proof. By Theorem 1.4.3, one can represent f in the form (1.9). Because of $f \in P_k \backslash Q_G$, the following two cases are possible:

Case 1: There exists at least a coefficient $a_{i_1 \dots i_n}$ with $|\{i_1, i_2, ..., i_n\} \setminus \{0\}| \geq 2$. Obviously, in this case, one receives the function h from the function f by identifying some variables and by substituting variables through c_1.
Case 2: Each addend different of zero in the formula (1.9) has the form $a_{0 \dots 0 i_j 0 \dots 0} \cdot x_j^{i_j}$ and there exists a q with

$$g(x) := f(c_0, ..., c_0, \underbrace{x}_{q\text{-th place}}, c_0, ..., c_0) \in P_k \setminus Q_G.$$

Since $c_0 \in Q_G$, we have $g \in [Q_G \cup \{f\}]$. Now let $g(x) := \sum_{i=0}^{p^m-1} b_i \cdot x^i$ and let d be the greatest index in such kind that d is not a power of p and $b_d \neq 0$ holds. Then one can also write d in the form $d = d_1 \cdot p^t$, where $t \geq 0$ and $d_1 > 0$ is relatively prime to p.
When one forms the function $w(x, y) := g(x+y) = \sum_{i,j=0}^{p^m-1} a_{ij} x^i y^j$, so surrenders with the aid of the binomial theorem that the coefficient of $x^{d-p^t} y^{p^t}$ is identical with $a_{d-p^t, p^t} = \binom{d}{p^t} \cdot b_d$. Next, we show that $\binom{d}{p^t}$ is relatively prime to p. Obviously, we have:

$$\binom{d}{p^t} = \binom{d_1 p^t}{p^t} = \frac{d_1 p^t (d_1 p^t - 1)..(d_1 p^t - p)..(d_1 p^t - p^2)..(d_1 p^t - p^t + 1)}{p^t (p^t - 1)......(p^t - p)....(p^t - p^2)...(p^t - p^t + 1)}.$$

In the numerator and denominator of the above number, some factors, which one can divide by a power of p, stand. When one divides the above fraction by corresponding powers of p, one receives a fraction whose factors have the form $a + b$ with $p | a$ and $p \not| b$. Therefore, $\binom{d}{p^t}$ is relatively prime to p. Hence (by $b_d \neq 0$) $a_{d-p^t, p^t} \neq 0$. ■

Lemma 5.3.5 *([Ros 70;a], 7.2.11)*
Let $h(x, y) = \sum_{i,j=0}^{p^m-1} a_{ij} x^i y^j$ with $a_{st} \neq 0$ for some $0 < s, t \leq p^m - 1$. Then the function $r(x, y) = x^s \cdot y^t$ belongs to $[Q_G \cup \{f\}]$.

Proof. Let $(u, v) \in E_{p^m}^2$ with $(u, v) \neq (s, t)$. W.l.o.g. we can assume $u \neq s$. For a primitive element α of E_k (i.e. for an $\alpha \in E_k$ with $\{\alpha^i \mid 1 \leq i \leq p^m - 1\} = E_k \setminus \{0\}$, see e.g. [Lid-N 87]) we form the superposition

$$g(x, y) := \alpha^u \cdot h(x, y) - h(\alpha \cdot x, y)$$

over $\{h(x, y), \alpha^u \cdot x - y, \alpha \cdot x\} \subseteq Q_G \cup \{f\}$. Then it holds

$$g(x, y) := \sum_{i,j=0}^{p^m-1} (\alpha^u - \alpha^i) \cdot a_{ij} \cdot x^i \cdot y^j,$$

and we put $a'_{ij} := (\alpha^u - \alpha^i) \cdot a_{ij}$. Obviously, $a'_{uj} = 0$. By the properties of a primitive element and because of $u \neq s$ we have $\alpha^u \neq \alpha^s$ and thus $a'_{st} \neq 0$.

Furthermore, $a'_{ij} = 0$, if $a_{ij} = 0$. Through repeated use of this construction for every $(u, v) \neq (s, t)$, one obtains a function d with $d(x, y) = d_{st}x^s y^t$ and $d_{st} \neq 0$. Let $b \in E_k \backslash \{0\}$ be the inverse element of d_{st}. Since $b \cdot x \in Q_G$, we have $b \cdot d_{st} \cdot x^s \cdot y^t = x^s \cdot y^t \in [Q_G \cup \{h\}]$. ■

Theorem 5.3.6 *([Ros 70;a], 7.2.12)*
Let $k = p^m$ (p prime, $m \geq 1$). Then, for every $\lambda_G \in \mathfrak{L}_k$ ($G := (E_k; +, \cdot)$ is a field) the clone $Pol_k\lambda_G$ is maximal in P_k.

Proof. W.l.o.g. let 0 be the zero element and let 1 be the unit element of G. In the proof of Lemma 5.2.4.1 we saw that $Q_G \subseteq Pol\lambda_G$. Since $Pol\lambda_G \neq P_k$ and $Q_G = [Q_G]$ are also valid, we have only to show the P_k-maximality of Q_G. Let $f \in P_k \backslash Q_G$ be arbitrary. Then, by Lemmas 5.3.4 and 5.3.5, a function $r(x, y) = x^s \cdot y^t$ belongs to $[Q_G \cup \{f\}]$, where $0 < s, t \leq p^m - 1$. One can also give s and t in the form $s = h \cdot p^u$ and $t = l \cdot p^v$, where $u, v \geq 0$ and h, l are relatively prime to p. Since $x^{p^i} \in Q_G$ $(i \in E_m)$ and $x^{p^m} = x$ holds, we have $w(x, y) := r(x^{p^{m-u}}, y^{p^{m-v}}) = x^h \cdot y^l \in [Q_G \cup \{f\}]$. If one forms now the superposition $q(x, y) := w(x + 1, y + 1)$ over $Q_G \cup \{f\}$, then one gets the following with the aid of the binomial theorem:

$$(x + 1)^h \cdot (y + 1)^l = \sum_{i,j=0}^{p^m-1} a_{ij} \cdot x^i \cdot y^j.$$

It is easy to check that

$$a_{11} = \binom{h}{1} \cdot \binom{l}{1} = h \cdot l.$$

Since h and l are relatively prime to p, this implies $a_{11} \neq 0$ and, therefore, $x \cdot y$ belongs to $[Pol_G \cup \{f\}]$ because of Lemma 5.3.5, Thus, $\{x \cdot y, x + y, a \cdot x, c_a(x) \mid a \in E_k\} \subseteq [Q_G \cup \{f\}]$ holds. Obviously, $[Q_G \cup \{f\}] = P_k$ results from that then because of Theorem 1.4.3. ■

5.4 The Number of the Maximal Classes of P_k

Let

$$R_{max}(P_k) := \mathfrak{M}_k \cup \mathfrak{S}_k \cup \mathfrak{U}_k \cup \mathfrak{L}_k \cup \mathfrak{C}_k \cup \mathfrak{B}_k.$$

Because of Lemmas 5.2.1.1 and 5.2.2.2 the following relation $\sim$ is an equivalence relations on $\mathfrak{M}_k \cup \mathfrak{S}_k$:

$$\varrho \sim \varrho' \;:\Longleftrightarrow\; \exists \mathfrak{X} \in \{\mathfrak{M}, \mathfrak{S}\} : \; \{\varrho, \varrho'\} \subseteq \mathfrak{X}_k \;\wedge\; Pol_k\varrho = Pol_k\varrho'.$$

When one selects exactly a representative from every equivalence class of $\sim$, one receives the relation sets $\mathfrak{M}_k^{\sim} \subseteq \mathfrak{M}_k$ and $\mathfrak{S}_k^{\sim} \subseteq \mathfrak{S}_k$.

Theorem 5.4.1 *([Ros 69])*
Let

$$R^{\sim}_{max}(P_k) := \mathfrak{M}^{\sim}_k \cup \mathfrak{S}^{\sim}_k \cup \mathfrak{U}_k \cup \mathfrak{L}_k \cup \mathfrak{C}_k \cup \mathfrak{B}_k.$$

Then for all $\varrho, \varrho' \in R^{\sim}_{max}$ it holds:

$$Pol_k\varrho = Pol_k\varrho' \iff \varrho = \varrho'. \tag{5.19}$$

Proof. Let $\varrho \in \mathfrak{X}_k$ and $\varrho' \in \mathfrak{Y}_k$ with $\mathfrak{X}, \mathfrak{Y} \in \{\mathfrak{M}^{\sim}, \mathfrak{S}^{\sim}, \mathfrak{U}, \mathfrak{L}, \mathfrak{C}, \mathfrak{B}\}$. If $\mathfrak{X} \neq \mathfrak{Y}$, then one can easily prove the assertion (by constructing functions f, g with $f \in Pol_k\varrho \setminus Pol_k\varrho'$ and $f \in Pol_k\varrho' \setminus Pol_k\varrho$). In the case $\mathfrak{X} = \mathfrak{Y}$ the assertion (5.20) follows from Lemmas 5.2.1.1, 5.2.2.2, 5.2.3.2, 5.2.4.3, 5.2.5.1 and 5.2.6.2. ■

Theorem 5.4.2 *([Ros 73])*
For $k \geq 2$ there is exactly

$$|R^{\sim}_{max}| = |\mathfrak{M}^{\sim}_k| + |\mathfrak{S}^{\sim}_k| + |\mathfrak{U}_k| + |\mathfrak{L}_k| + |\mathfrak{C}_k| + |\mathfrak{B}_k| \tag{5.20}$$

maximal classes of P_k and it is valid:

- $|\mathfrak{M}^{\sim}_k| = \frac{k\cdot(k-1)}{2} \cdot o(k-2)$,

  where $o(i)$ denotes the number of all different partial order relations on an i-element set ($o(0) := 1$);
- $|\mathfrak{S}^{\sim}_k| = \sum_{k=m\cdot p,\ p\ prime} \frac{1}{p-1} \cdot \frac{k!}{m!\cdot p^m}$;
- $|\mathfrak{U}_k| = z(k) - 2$,

  where $z(i)$ denotes the number of all partitions on a k-element set, which one can calculate as follows:

  $$\sum_{i=0}^{k}(-1)^i \cdot \sum_{t=0}^{i}(-1)^t \cdot \frac{t^k}{t! \cdot (i-t)!};$$
- $|\mathfrak{L}_k| = \begin{cases} \frac{(k-1)!}{p^{\binom{m}{2}}\cdot(p-1)\cdot(p^2-1)\cdot\ldots\cdot(p^m-1)}, & \text{if } k = p^m,\ p\ prime, \\ 0 & \text{otherwise;} \end{cases}$
- $|\mathfrak{C}_k| = \sum_{1\leq h<k} c_h(k) - 1$,

  where $c_h(k) := \binom{k}{1}\cdot 2^{\binom{k-1}{h}} - \binom{k}{2}\cdot 2^{\binom{k-2}{h}} + \ldots + (-1)^{k-1}\cdot\binom{k}{k}\cdot 2^{\binom{0}{h}}$ and $\binom{x}{h} := 0$, if $x < h$);
- $|\mathfrak{B}_k| = \sum_{3\leq h\leq k} b(k,h)$,

  where

  $$b(k,h) := \sum_{m\geq 0, h^m \leq k} \frac{1}{m!\cdot(h!)^m} \cdot \left(\sum_{t=0}^{h^m}(-1)^t \cdot \binom{h^m}{t} \cdot (h^m - t)^k\right)$$

  for $h \geq 3$.

(Specific numbers are given in the following tables.[3]*)*

k	2	3	4	5	6	7	8
$\|R^{\sim}_{max}\|$	5	18	82	643	15 182	7 848 984	549 761 933 169

k	2	3	4	5	6	7	8	9	10
$\|\mathfrak{M}^{\sim}_k\|$	1	3	18	190	3285	88 851	3 640 644		
$\|\mathfrak{S}^{\sim}_k\|$	1	1	3	6	35	120	105	1120	19 089
$\|\mathfrak{U}_k\|$	0	3	13	50	201	875	4398		
$\|\mathfrak{L}_k\|$	1	1	1	6	0	120	30	840	0
$\|\mathfrak{C}_k\|$	2	9	40	355	11490	7 758 205	549 758 283 980		
$\|\mathfrak{B}_k\|$	0	1	7	36	171	813	4012		

Proof. In Chapter 6, we prove that $\{Pol_k\varrho \mid \varrho \in R^{\sim}_{max}\}$ is the set of all maximal classes of P_k. Because of Theorem 5.4.1, we have to determine the numbers $|\mathfrak{X}_k|$ with $\mathfrak{X} \in \{\mathfrak{M}^{\sim}, \mathfrak{S}^{\sim}, \mathfrak{U}, \mathfrak{L}, \mathfrak{C}, \mathfrak{B}\}$.

Case 1: $\mathfrak{X} = \mathfrak{M}^{\sim}$.

Obviously, there are $k \cdot (k-1)$ possibilities to choose a greatest element and a smallest element of a relation $\varrho \in \mathfrak{M}_k$. Consequently, we have $|\mathfrak{M}_k| = k \cdot (k-1) \cdot o(k-2)$; with the aid of Lemma 5.2.1.1: $|\mathfrak{M}^{\sim}_k| = \frac{k\cdot(k-1)}{2} \cdot o(k-2)$.

Case 2: $\mathfrak{X} = \mathfrak{S}^{\sim}$.

Let $k = p \cdot m$, p prime, $m \geq 1$. Then

$$\binom{k}{p} \cdot \binom{k-p}{p} \cdot \binom{k-2p}{p} \cdot \ldots \cdot \binom{p}{p} = \frac{k!}{m! \cdot p^m}$$

is the number of the possible permutations s on E_k, which have exactly m cycles of the length p. Because of Lemma 5.2.2.2 the $p-1$ permutations s, s^2, ..., s^{p-1} determine exactly the same maximal class $Pol_k\varrho_s$. Therefore, we have $|\mathfrak{S}^{\sim}_k| = \sum_{k=m\cdot p,\ p\ prime} \frac{1}{p-1} \cdot \frac{k!}{m!\cdot p^m}$.

Case 3: $\mathfrak{X} = \mathfrak{U}$.

Obviously, $|\mathfrak{U}_k| = z(k) - 2$, where $z(k)$ denotes the number of the possible partitions of a k-element set.

Case 4: $\mathfrak{X} = \mathfrak{L}$.

Let $k = p^m$, p prime, $m \geq 1$. First, we determine $|\mathfrak{G}|$, where (as in Section 5.2.4) $\mathfrak{G}$ is the set of all elementar Abelean p-groups of the form $(E_k; \oplus)$. As is generally known, every group $G \in \mathfrak{G}$ is isomorphic to an m-fold direct product $V := (Z_p^m; +)$ of the group $(Z_p; + \,(mod\, p))$. If $s : Z_p^m \longrightarrow E_k$ is a bijective mapping, then one can define $G_s := (E_k; +_s)$ as follows:

---

[3] To $|\mathfrak{M}^{\sim}_8|$ see also [Com 66], [Rad 83] and [Roz 85].

$$\forall x,y\in E_k:\ x+_s y := s(s^{-1}(x)+s^{-1}(y)).$$

It is easy to check that

$$G_s = G_t \iff s^{-1}t \text{ is an automorphism}$$

holds for arbitrary bijective mappings s, t from Z_p^m onto E_k. Consequently, we have

$$|\mathfrak{G}| = \frac{k!}{|A|},$$

where A denotes the set of all automorphisms on V. Since $|A|$ is identical with the number of the bases of V, we obtain

$$\begin{aligned}|A| &= (p^m-1)\cdot(p^m-p)\cdot(p^m-p^2)\cdot\ldots\cdot(p^m-p^{m-1})\\ &= p\cdot p^2\cdot\ldots\cdot p^{m-1}\cdot(p^m-1)\cdot(p^{m-1}-1)\cdot\ldots(p^2-1)\cdot(p-1)\\ &= p^{\binom{m}{2}}\cdot(p^m-1)\cdot(p^{m-1}-1)\cdot\ldots\cdot(p-1).\end{aligned}$$

Furthermore, by Lemma 5.2.4.2, we have $|\mathfrak{L}_k| = \frac{|\mathfrak{G}|}{k}$. This implies our assertion about $|\mathfrak{L}_k|$.

Case 5: $\mathfrak{X} = \mathfrak{C}$.

Let A_h be the set of all h-element subsets of E_k, $1\le h\le k$. Then one can assign an element a_ϱ of $\mathfrak{P}(A_h)$ as follows to every totally reflexive, totally symmetric, h-ary relation ϱ in unique manner:

$$\{a_0,a_1,\ldots,a_{h-1}\}\in a_\varrho \iff (a_0,a_1,\ldots,a_{h-1})\in\varrho\backslash\iota_k^h. \tag{5.21}$$

Consequently, if ϱ belongs to $\mathfrak{C}_k^h$ and the set of all central elements of ϱ contains the set C, then by means of (5.21), the relation ϱ is unambiguously definite through a subset a_ϱ of $T_{h,C} := \{A\in A_h \mid A\cap C=\emptyset\}$. Because of $|T_{h,C}| = \binom{k-|C|}{h}$ and $|\mathfrak{P}(T_{h,C})| = 2^{\binom{k-|C|}{h}}$ one obtains the number of elements of $\mathfrak{C}_k^h \cup \{E_k^h\}$ through use of the principle of the inclusion and exclusion:

$$c_h(k) := \binom{k}{1}\cdot 2^{\binom{k-1}{h}} - \binom{k}{2}\cdot 2^{\binom{k-2}{h}} + \ldots + (-1)^{k-1}\cdot\binom{k}{k}\cdot 2^{\binom{0}{h}},$$

where $\binom{x}{h} := 0$ for $x<h$. Consequently, we have $|\mathfrak{C}_k| = \sum_{h=1}^{k-1} c_h(k) - 1$.

Case 6: $\mathfrak{X} = \mathfrak{B}$.

Let $h\ge 3$ and $m\ge 1$.

A bijective mapping s from E_{h^m} onto E_{h^m} is called an automorphism of ξ_m, if

$$\forall (a_0,a_1,\ldots,a_{h-1})\in\xi_m:\ (s(a_0),s(a_1),\ldots,s(a_0))\in\xi_m.$$

Let ϱ and ϱ' be relations of $\mathfrak{B}_k^h$; i.e., there exists certain mappings q or q' from E_k onto E_{h^m} or $E_{h^{m'}}$ with $\varrho = q^{-1}(\xi_m)$ and $\varrho' = q'^{\,-1}(\xi_{m'})$, respectively (see Section 5.2.6). Next, we show that the equivalence

$$\varrho = \varrho' \iff m = m' \wedge \\ (\exists \text{ automorphism } s \text{ on } \xi_m : \forall x \in E_k : q'(x) = s(q(x)) \,) \tag{5.22}$$

holds.
"$\Longrightarrow$": If $\varrho = \varrho'$, then it follows from the definition of the relations $\varrho = q^{-1}(\xi_m)$ or $\varrho' = q'^{\,-1}(\xi_{m'})$ that the mapping equivalences $\kappa_\alpha := \{(x,y) \in E_k^2 \mid \alpha(x) = \alpha(y)\}$ for $\alpha \in \{q, q'\}$ are identical. Then, $\kappa_q = \kappa_{q'}$ implies the right side of the equivalence (5.22).
"$\Longleftarrow$" follows from

$$\begin{aligned}(a_0, a_1, ..., a_{h-1}) \in \varrho &\iff (q(a_0), q(a_1), ..., q(a_{h-1})) \in \xi_m \\ &\iff (s(q(a_0)), s(q(a_1)), ..., s(q(a_{h-1}))) \in \xi_m \\ &\iff (q'(a_0), q(a_1), ..., q'(a_{h-1})) \in \xi_m \\ &\iff (a_0, a_1, ..., a_{h-1}) \in \varrho'.\end{aligned}$$

Denote $b_1(k,h)$ the number of the mappings from E_k onto E_{h^m} and denote $b_2(h,m)$ the number of the automorphisms on ξ_m. Because of (5.22), we have

$$|\mathfrak{B}_k^h| = \frac{b_1(k,h)}{b_2(h,m)}. \tag{5.23}$$

With the aid of the principle of the inclusion and exclusion, one can calculate the number $b_1(k,h)$ as follows:

$$b_1(k,h) = \sum_{t=0}^{h^m} (-1)^t \cdot \binom{h^m}{t} \cdot (h^m - t)^k. \tag{5.24}$$

For the purpose of determining $b_2(h,m)$ the following considerations follows directly from the definition of the automorphisms of ξ_m that these are some functions of $Pol_k\xi_m$. Then, we have (because of the bijectivity of the automorphisms and by Theorem 5.2.6.1) that each automorphism s of ξ_m has a representation of the form

$$s(x) = \sum_{i=0}^{m-1} s_i(x^{(\,t_x\,)}) \cdot h^i, \tag{5.25}$$

where s_0, s_1, ..., s_{m-1} are some permutations on E_h and $\{t_0, t_1, ..., t_{m-1}\} = E_m$ holds. Therefore,

$$b_2(h,m) = m! \cdot (h!)^m. \tag{5.26}$$

Then, (5.23)–(5.26) imply our assertion. ∎

We notice that there are also asymptotic estimates for the number of all maximal classes of P_k. One can take further information on that from the book [Pös- K 79] (p. 111) or the original papers [Kuz 59] and [Zac K J 64].

5.5 Remarks to the Maximal Classes of $P_k(l)$

For $A \subseteq P_k$, $l \in \mathbb{N}$, $l \leq k$, $\varrho \in R_k$ put

$$\begin{aligned} A(l) &:= \{f \in A \mid |Im(f)| \leq l\}, \\ Pol\ell\varrho &:= (Pol_k\varrho)(l), \\ \lambda_k &:= \{(a,a,b,b),(a,b,a,b),(a,b,b,a) \mid a,b \in E_k\}. \end{aligned}$$

Let $\mathfrak{M}_k(l)$ be the set of all partial orders ϱ on E_k with the property that every l-element subset of E_k has a least and a greatest element with respect to ϱ. Notice that $\mathfrak{M}_k(k) = \mathfrak{M}_k$.

We say that the h-ary relation ϱ on E_k is an l-**central** relation if $\varrho \neq E_k^h$ is totally reflexive, totally symmetric, and for every l-element subset E of E_k there exists an element $c_E \in E$ with $(a_1, ..., a_{h-1}, c_E) \in \varrho$ for all $a_1, ..., a_{h-1} \in E$. The element c_E is called a **central element** of $\varrho \cap E^h$.

Let $\mathfrak{C}_k^h(l)$ be the set of all h-ary l-central relations on E_k and let $\mathfrak{C}_k(l) := \bigcup_{h=1}^{l-1} \mathfrak{C}_k^h(l)$. Notice that $\mathfrak{C}_k(k) = \mathfrak{C}_k$.

Since any unary relation is totally reflexive and totally symmetric, we have

$$\mathfrak{C}_k^1(l) = \{\varrho \subset E_k \mid k - l + 1 \leq |\varrho| < k\}.$$

For the next definition, we need some notations of Section 5.2.6 and

$$\chi(\varrho) := \{a_1, ..., a_h\} \subseteq E_k \mid (a_1, ..., a_h) \in \varrho\}.$$

Notice that a totally reflexive and totally symmetric h-ary relation ϱ is completely determined by the set $\chi(\varrho)$.

$\mathfrak{B}_k^h(l)$ denotes the set of all h-ary non-diagonal relations with the following three properties:

(1) $\forall E \subseteq E_k:\ |E| = l \Longrightarrow\ \varrho \cap E^h \in \mathfrak{B}_k^h \cup \{E^h\}$,
(2) $\forall E \subseteq E_k:\ |E| = l\ \Longrightarrow\ (\exists id_E \in (Pol\ell\varrho)^1:\ \forall x \in E:\ id_E(x) = x)$;
(3) $\forall A, B \in \chi(\varrho):\ (A \neq B \Longrightarrow\ \exists g_{A,B} \in (Pol\ell\varrho)^1:\ g_{A,B}(A) = B)$.

Example: Let be $k = 5$, $l = 4$ and $h = 3$. Then, the totally reflexive and totally symmetric h-ary relation with

$$\chi(\varrho) := \{A_1, A_2, A_3\},\ A_1 := \{0,1,2\},\ A_2 := \{1,2,3\},\ A_3 := \{1,2,4\}$$

is an element of $\mathfrak{B}_k^h(l)$, since

$$\varrho \cap \{0,2,3,4\}^4 = \{0,2,3,4\}^4,\ \varrho \cap \{0,1,3,4\}^4 = \{0,1,3,4\}^4$$

and $\varrho \cap \{1,2,3,4\}^4$, $\varrho \cap \{0,1,2,4\}^4$ and $\varrho \cap \{0,1,2,3\}^4$ are homomorphic inverse images of the relation ι_3^3 and the functions $id_t := id_{E_5 \setminus \{t\}}$ $(t \in E_5)$ and $g_{pq} := g_{A_p,A_q}$ $(p \neq q,\ p,q \in \{1,2,3\})$ satisfy the conditions (2) and (3) (see the below indicated tables).

x	id_0	id_1	id_2	id_3	id_4
0	3	0	0	0	0
1	1	0	1	1	1
2	2	2	0	2	2
3	3	3	3	4	3
4	4	4	4	4	3

x	g_{12}	g_{13}	g_{21}	g_{23}	g_{31}	g_{32}
0	1	1	2	4	2	3
1	2	2	0	1	0	1
2	3	4	1	2	1	2
3	1	1	2	4	2	3
4	1	1	2	4	2	3

Put

$$\mathfrak{B}_k(l) := \begin{cases} \bigcup_{h=3}^{l-1} \mathfrak{B}_k^h(l) & \text{if} \quad l > 3, \\ \emptyset & \text{otherwise.} \end{cases}$$

Let $\mathfrak{H}_k(l)$ be the set of all l-ary totally reflexive and totally symmetric relations $\varrho \neq E_k^h$ that satisfy the above condition (3). Examples for such relations are, as follows:
1) ι_k^l;
2) $\varrho_E := E_k^l \setminus \{(a_{s(0)}, a_{s(1)}, ..., a_{s(l-1)}) \mid s \text{ permutation on } E_l\}$, where $E := \{a_0, ..., a_{h-1}\}$ and $|E| = l$;
3) every relation of $\mathfrak{B}_k^l$ (see [Lau 88d]). For $l = 2$ we define

$$\mathfrak{H}_k(2) := \{E_k^2 \setminus \{(a, b), (b, a)\} \mid (a, b) \in E_k^2 \setminus \iota_k^2\}.$$

Theorem 5.5.1 *([Lau 92a]; without proof) Let $A \subseteq P_k(l)$, $2 \leq l < k$ and*

$$\mathfrak{R}_k(l) := \mathfrak{M}_k(l) \cup \mathfrak{U}_k \cup \mathfrak{C}_k(l) \cup \mathfrak{B}_k(l) \cup \mathfrak{H}_k(l).$$

Then

(a) $[A] = P_k(l) \iff ((\forall \varrho \in \mathfrak{R}_k(l) : A \not\subseteq Pol\ell\varrho) \wedge (l = 2 \Longrightarrow A \not\subseteq Pol\ell\lambda_k))$.

(b) $Pol\ell\varrho$ is $P_k(l)$-maximal for every $\varrho \in \mathfrak{R}_k(l)$ and $(Pol\lambda_k)(2)$ is $P_k(2)$-maximal.

(c) The maximal classes of $P_k(2)$ are sets of the form $(Pol\ \varrho)(2)$ with

$$\varrho \in \mathfrak{C}_k^1(2) \cup \mathfrak{M}_k(2) \cup \mathfrak{U}_k \cup \mathfrak{H}_k(2) \cup \{\lambda_k\};$$

i.e., every maximal class of $Pol_k(l)$, which is not equal $(Pol_k\lambda)(2)$, is a intersection of $P_k(2)$ with a certain maximal class of P_k.

We remark that for $l \geq 3$ not every $P_k(l)$-maximal class is an intersection of $P_k(l)$ with a certain maximal class of P_k. An example for this fact is the relation

$$\{(a,b,c) \in E_5^3 \mid \{a,b,c\} \notin \{\{0,3,4\},\{1,2,3\}\}\} \in \mathfrak{C}_5^3(4).$$

Notice that the $P_k(l)$-maximality of the class $Pol_l \varrho$ for $\varrho \in \mathfrak{M}_k(l)$ was proven by V. Kolpakov in [Kol 74].

6

Rosenberg's Completeness Criterion for P_k

Since P_k is finitely generated, every proper subclass of P_k belongs to a certain maximal class of P_k; therefore, by Theorem 1.5.3.1, it follows that a subset T of P_k is complete (i.e., it holds $[T] = P_k$) if and only if T is not a subset of a maximal subclass of P_k. The aim of this chapter is to prove that the classes of Chapter 5 are all maximal classes of P_k:

Theorem 6.1 (Rosenberg's Completeness Criterion for P_k; [Ros 65], [Ros 70;a]**)**

For an arbitrary subset T of P_k it holds:

$$[T] = P_k \iff \forall \varrho \in \mathfrak{M}_k \cup \mathfrak{U}_k \cup \mathfrak{S}_k \cup \mathfrak{L}_k \cup \mathfrak{C}_k \cup \mathfrak{B}_k : T \not\subseteq Pol\varrho. \tag{6.1}$$

6.1 Proof of Completeness Criterion

By Section 4.9, the equivalence (6.1) follows from (6.2), which we still must prove:

$$\begin{aligned} &A \text{ is maximal in } P_k \implies \\ &\exists \gamma \in \mathfrak{M}_k \cup \mathfrak{U}_k \cup \mathfrak{S}_k \cup \mathfrak{L}_k \cup \mathfrak{C}_k \cup \mathfrak{B}_k : \ A \subseteq Pol\gamma. \end{aligned} \tag{6.2}$$

Denote A an arbitrary maximal class of P_k in the following. To simplify the proof of (6.2), some assumptions over A are given. These assumptions result from the following considerations:

- We show in Chapter 4 that there is exactly a P_k-maximal class that contains P_k^1. In addition, were shown that every other P_k-maximal class has the form $Pol\varrho$, where $\varrho \in \bigcup_{h=1}^{k} R_k^h$.

- For every ϱ-derivable relation ϱ', which is not diagonal, we have by Theorem 2.6.3 that $Pol\,\varrho \subseteq Pol\,\varrho' \subset P_k$. Obviously, if $Pol\,\varrho$ is P_k-maximal then for an arbitrary ϱ-derivable non-diagonal relation ϱ' it holds: $Pol\,\varrho = Pol\,\varrho'$.
- Every unary non-diagonal relation $\varrho \in R_k$ (i.e., $\emptyset \subset \varrho \subset E_k$) describes a maximal class of P_k (see Theorem 5.3.2).

Thus, w.l.o.g. we can assume that our maximal class A of P_k can be described by a certain h-ary relation ϱ: $A = Pol\,\varrho$ $(1 \leq h \leq k)$. Further, we assume w.l.o.g. that ϱ fulfills the three conditions

(*I*) $Pol\,\varrho$ is P_k-maximal and $(Pol\,\varrho)^1 \neq P_k^1$;

(*II*) For every ϱ-derivable h'-ary relation $\varrho' \in \bigcup_{h=1}^{k} R_k^h$ it holds:
(a) $\varrho' \notin Inv P_k \Longrightarrow h' \geq h$
(i.e., ϱ is a relation with minimal arity which describes the maximal class $Pol\,\varrho$);
(b) $(h' = h = 2 \wedge \varrho' = \varrho \cap (\tau\varrho)) \Longrightarrow \varrho' \in D_k^2 \cup \{\varrho\}$
(i.e., by $\varrho \cap (\tau\varrho)$ no other relation which describes $Pol\,\varrho$ either is derivable. With other words: If possible, we choose a symmetrical relation ϱ for the description of $Pol\,\varrho$;
(c) $(h' = h \geq 2 \wedge \iota_k^h \subseteq \varrho \subseteq \varrho' \wedge (\varrho$ is totally symmetric $\vee$ $(h = 2 \wedge \varrho$ is antisymmetric$)\,)) \Longrightarrow \varrho' \in \{\varrho, E_k^h\}$
(i.e., if ϱ is totally reflexive and totally symmetric or reflexive and antisymmetric, then for the description of the class $Pol\,\varrho$ we choose a binary relation ϱ which has the greatest cardinality.);

(*III*) ϱ is not a unary relation, i.e., $h \geq 2$.

Now, we want to prove:

If ϱ fulfills the obove conditions (*I*)–(*III*), then ϱ or a ϱ-derivable relation belongs to $\mathfrak{M}_k \cup \mathfrak{U}_k \cup \mathfrak{S}_k \cup \mathfrak{L}_k \cup \mathfrak{C}_k \cup \mathfrak{B}_k$. (6.3)

(6.3) implies (6.2) and our Theorem 6.1 follows from (6.2), which was already mentioned.
For $k = 2$ it is easy to check that only the relations

$$\begin{pmatrix} 0 & 1 \\ 1 & 0 \end{pmatrix} \text{ and } \begin{pmatrix} 0 & 0 & 1 \\ 0 & 1 & 1 \end{pmatrix}$$

satisfy (*I*)–(*III*); hence (6.1) is clearly for $k = 2$. Therefore, let

$$k \geq 3$$

be in the following.
In the proof of (6.1) for $k \geq 3$ we use the notations for relations and relation

operations from Sections 2.1–2.5. In particular, $\circ$ denotes the relation product and let

$$\varrho^i := \underbrace{\varrho \circ \varrho \circ ... \circ \varrho}_{i \text{ times}}.[1]$$

An h-tuple **a** from E_k^h with $h \leq k$ we often give in the form $(a_0, a_1, ..., a_{h-1})$, i.e., we also choose the indices from E_k.
Remember that a relation ϱ' is called **ϱ-derivable** iff ϱ' can be obtained by a finite number of applications of the relation operations (see 2.3) from ϱ and the diagonal relations.
The following lemma is easy to check (see also Theorem 2.11.2):

Lemma 6.1.1 *([Ros 70;a], 4.1.2)*
Each relation of the form

$$\varrho_C := \{(a_0, ..., a_{t-1}) \mid \exists a_t, ..., a_{m \cdot h-1} \in E_k : (\forall i : (a_{c_{i1}}, ..., a_{c_{ih}}) \in \varrho)\},$$

where $C := (c_{ij})_{m,h}$ is a matrix of type (m, h) with elements from $E_{m \cdot h}$, is a ϱ-derivable relation. ■

We remark that every one of the following claims on ϱ-derivability of a certain relation ϱ' almost follows from the above lemma.

For the proof of (6.1), we distinguish two cases:

Case 1: $h = 2$.

Figure 6.1 gives a survey on the possible subcases and on the following lemmas. For $2 \leq i \leq k$, the relation

$$\sigma_i(\varrho) := \{(a_1, ..., a_i) \mid \exists u \in E_k : \{(a_1, u), ..., (a_i, u)\} \subseteq \varrho\}$$

is a ϱ-derivable relation and it holds:

Lemma 6.1.2

$$\sigma_2(\varrho) = E_k^2 \implies (\forall i \in \{2, 3, ..., k\} : \sigma_i(\varrho) = E_k^i).$$

Proof. We prove the statement by induction on $i \geq 2$. For $i = 2$ the statement holds by assumption. Assume, $\sigma_{t-1} = E_k^{t-1}$ is valid for certain $t-1 \in \{2, ..., k-1\}$. Then, by definition of $\sigma_t(\varrho)$, we have $\iota_k^t \subseteq \sigma_t(\varrho)$. If $\iota_k^t \subseteq \sigma_t(\varrho) \subset E_k^t$, then $\sigma_t(\varrho)$ is no invariant of P_k and hence (by (I)) $Pol \varrho = Pol \sigma_t(\varrho)$ holds. This is, however, a contradiction of the fact that $P_k(2) \subseteq Pol \sigma_t(\varrho)$, whereby $Pol \sigma_t(\varrho)$ does not have any description by a binary relation. Therefore, $\sigma_t(\varrho) = E_k^t$. ■

Lemma 6.1.3 *Either ϱ is reflexive (i.e., $\iota_k^2 \subseteq \varrho$) or there exists a ϱ-derivable relation ϱ' which belongs to $\mathfrak{S}_k$.*

[1] For content-related reasons, it is not possible to confuse the index i with the arity of ϱ.

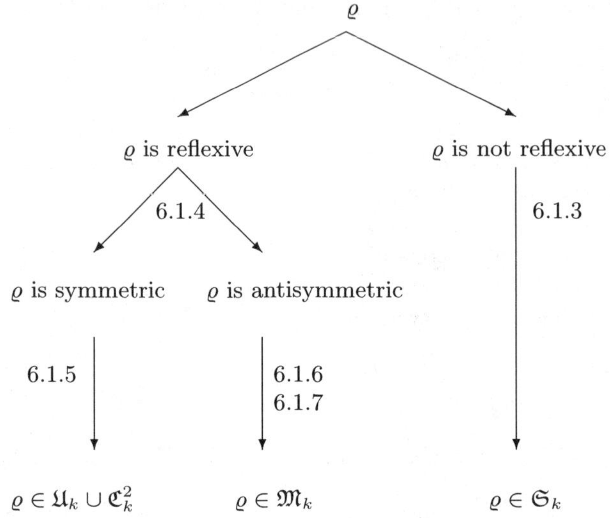

Fig. 6.1. A survey about the proof in Case 1

Proof. The relation $\varrho_1 := \varrho \cap \iota_k^2$ is ϱ-derivable. If we have $\emptyset \subset \varrho_1 \subset \iota_k^2$, then $pr\,\varrho_1$ is a non-diagonal ϱ-derivable unary relation, in contradiction to (IIa). Thus, ϱ is reflexive or areflexive.
Now, let ϱ be areflexive. Since $pr_0\varrho = pr_1\varrho = E_k$ holds by (II), we have $\iota_k^2 \subseteq \varrho \circ \zeta\varrho =: \varrho'$. Because of this, (I) and the fact that $Pol\varrho$ does not contain any constants, it follows that $\varrho' \in \{\iota_k^2, E_k^2\}$. If $\varrho' = E_k^2$, then by $\varrho \circ \zeta\varrho = \sigma_2(\varrho)$ and by Lemma 6.1.2, we get a contradiction to the areflexivity of ϱ. Therefore, $\varrho' = \iota_k^2$ holds, i.e., ϱ is a fixed-point free permutation on E_k. If ϱ has not k/t cycles of the same length, then there is an i with $\emptyset \subset \varrho^i \cap \iota_k^2 \subset E_k^2$ and we have $pr(\varrho^i \cap \iota_k^2) \in \mathfrak{C}_k^1$, in contradiction to (IIa). Thus, the permutation ϱ has cycles of the same length t. The length t is a prime number, since for $t = p \cdot q$ (p prime, $q \geq 2$), the ϱ-derivable relation $\varrho' := \varrho^q$ has only cycles of prime length p. Therefore ϱ' belongs to $\mathfrak{S}_k$. ∎

Lemma 6.1.4 *Let ϱ be reflexive. Then ϱ is either symmetric (i.e., $\varrho \cap \zeta\varrho = \varrho$) or antisymmetric (i.e., $\varrho \cap \zeta\varrho = \iota_k^2$).*

Proof. Since $\iota_k^2 \subseteq \varrho$, $\iota_k^2 \subseteq \varrho \cap \zeta\varrho \subseteq \varrho$ holds, and by (IIb) we have $\varrho \cap \zeta\varrho = \iota_k^2$ or $\varrho \cap \zeta\varrho = \varrho$. ∎

Lemma 6.1.5 *Let ϱ be reflexive and symmetric. Then $\varrho \in \mathfrak{U}_k \cup \mathfrak{C}_k^2$.*

Proof. It is easy to check that $\iota_k^2 \subset \varrho \subseteq \varrho \circ \varrho$. Then by (IIc) the following three cases are possible:
Case 1: $\varrho \circ \varrho = \varrho$.
Obviously, in this case, ϱ is transitive and $\varrho \in \mathfrak{U}_k$.

Case 2: $\varrho \circ \varrho = E_k^2$.
Since $\varrho \circ \varrho = \sigma_2(\varrho)$, $\sigma_k(\varrho) = E_k^k$ follows from Lemma 6.1.2. Hence, for E_k there is a $u \in E_k$ with $\{(e,u) \mid e \in E_k\} \subseteq \varrho$. Thus the relation ϱ belongs to $\mathfrak{C}_k^2$. ∎

Lemma 6.1.6 *Let ϱ be reflexive and antisymmetric. Then, for E_k, there exists (with respect to ϱ) a greatest element $e_\varrho \in E_k$ and (with respect to ϱ) a least element $o_\varrho \in E_k$, i.e., it holds:*

$$\{(x, e_\varrho), (o_\varrho, x) \mid x \in E_k\} \subseteq \varrho,$$

and these elements are uniquely determined.

Proof. Since ϱ is reflexive, we have $\iota_k^2 \subseteq \varrho \subseteq \sigma_2(\varrho)$. It is obvious that the relation $\sigma_2(\varrho)$ is symmetric and thus $\varrho \subset \sigma_2(\varrho)$. This and (IIc) imply $\sigma_2(\varrho) = E_k^2$. Then, because of $k \geq 3$ and with the aid of Lemma 6.1.2, we have $\sigma_k(\varrho) = E_k^k$. Hence we find an $e_\varrho \in E_k$ with $\{(x, e_\varrho) \mid x \in E_k\} \subseteq \varrho$. Then e_ϱ is a greatest element of E_k. The uniqueness of e_ϱ is to see as follows:
Assume there are two greatest elements $a, a' \in E_k$. Then we have $(a, a') \in \varrho$ and $(a', a) \in \varrho$ and thus $a = a'$ by the antisymmetry of ϱ.
The existence of a unique least element can be shown analogously starting with the ϱ-derivable relation $\sigma_i(\zeta\varrho)$ instead of $\sigma_i(\varrho)$. ∎

Lemma 6.1.7 *Let ϱ be a reflexive and antisymmetric relation with the property that E_k has a (unique) least and a (unique) greatest element with respect to ϱ. Then ϱ is transitive, i.e., $\varrho \in \mathfrak{M}_k$.*

Proof. We consider the ϱ-derivable relation ϱ^2. Since $\varrho \subseteq \varrho^2$, by (IIc) only the cases $\varrho^2 = \varrho$ (i.e., ϱ is transitive) or $\varrho^2 = E_k^2$ are possible.
Let's assume $\varrho \circ \varrho = E_k^2$ and let o_ϱ be least element of E_k with respect to ϱ. Then we have $(a, o_\varrho) \notin \varrho$ for $a \in E_k \backslash \{o_\varrho\}$, $(a, o_\varrho) \in \varrho \circ \varrho$ and there is a $b \in E_k$ with $(a,b) \in \varrho$ and $(b, o_\varrho) \in \varrho$. However, this cannot hold by the antisymmetry of ϱ and by the choose of o_ϱ. Thus the case $\varrho^2 = E_k^2$ cannot occur and $\varrho^2 = \varrho$ holds. ∎

Summing up, we have in Case 1:

$$\varrho \in \mathfrak{U}_k \cup \mathfrak{M}_k \cup \mathfrak{C}_k^2 \cup \mathfrak{S}_k.$$

Case 2: $h \geq 3$.

Figure 6.2 gives a survey about the proof in this case.

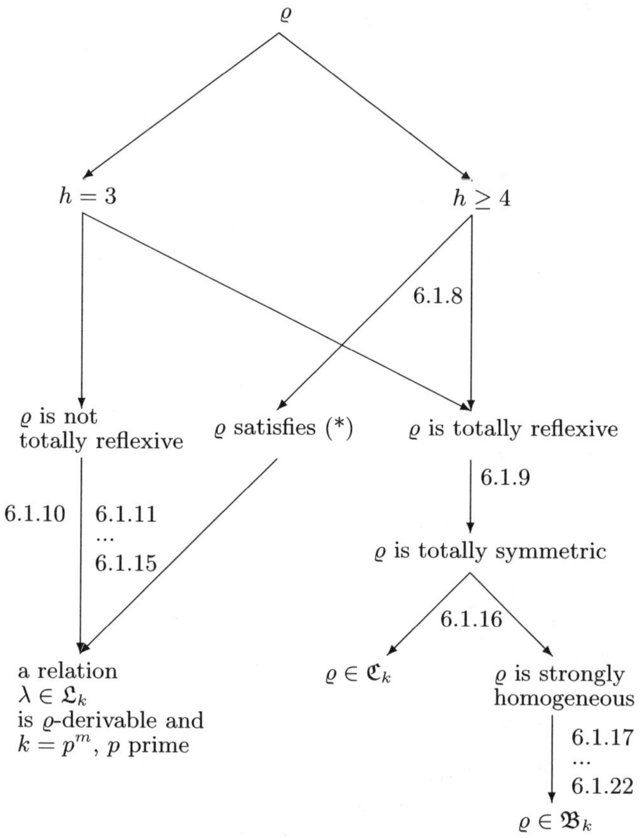

Fig. 6.2. A survey about the proof in Case 2

First, we give some conclusions of the assumptions (I)–(III) over ϱ.

$$(IV) \quad \forall r \leq h-1 \, \forall i_1, ..., i_r \in E_h : pr_{i_1,...,i_r}\varrho = E_k^r.$$

Since, if $\varrho \cap \iota_k^h = \emptyset$, $pr_{1,2}\varrho$ is not an invariant of P_k, we have by (IIa)

$$(V) \quad \varrho \cap \iota_k^h \neq \emptyset.$$

Obviously, it follows from (IIa):

$$(VI) \quad \forall \delta \in D_k^h \backslash \{E_k^h\} : \varrho \cap \delta \in D_k^h.$$

If $\varrho \backslash \iota_k^h = \emptyset$, then, by (VI), the relation ϱ is an **primitive relation** (i.e., a union of certain diagonal relations), which is preserved by all unary functions of P_k. Hence it follows from (I):

$$(VII) \quad \varrho \backslash \iota_k^h \neq \emptyset.$$

Lemma 6.1.8 *If* $4 \leq h \leq k$*, then either* ϱ *is totally reflexive or* ϱ *satisfies the following condition:*

$$(*): h = 4 \ \wedge \ |\varrho| = k^3 \ \wedge \ \delta^4_{\{0,1\},\{2,3\}} \cup \delta^4_{\{0,2\},\{1,3\}} \cup \delta^4_{\{0,3\},\{1,2\}} \subseteq \varrho.$$

Proof. First, note that for every $(a,b),(b,c),(c,d) \in E_h^2 \backslash \iota_h^2$ with $|\{a,b,c,d\}| = 4$, the implications

$$(h \geq 4 \wedge (x_0, ..., x_{h-1}) \in \varrho \wedge x_a = x_b) \implies x_b = x_c \tag{6.4}$$

and

$$(h \geq 5 \wedge (x_0, ..., x_{h-1}) \in \varrho \wedge x_a = x_b) \implies x_c = x_d \tag{6.5}$$

do **not** hold by (IV).
Next we will prove

$$(\exists (a,b) \in E_h^2 \backslash \iota_h^2 : \delta_{\{a,b\}} \subseteq \varrho) \implies (\forall (c,d) \in E_h^2 \backslash \iota_h^2 : \ \delta_{\{c,d\}} \subseteq \varrho). \tag{6.6}$$

For this, it is sufficient to show that

$$\delta_{\{0,1\}} \subseteq \varrho \implies \delta_{\{1,2\}} \subseteq \varrho \tag{6.7}$$

holds. Let $\delta_{\{0,1\}} \subseteq \varrho$. Then we have $\delta_{\{0,1,2\}} \subseteq \varrho \cap \delta_{\{1,2\}}$. Furthermore, by (VI) $\varrho \cap \delta_{\{1,2\}}$ is diagonal. Thus $\varrho \cap \delta_{\{1,2\}} \in \{\delta_{\{0,1,2\}}, \delta_{\{1,2\}}\}$. Since an implication of form (6.4) follows from $\varrho \cap \delta_{\{1,2\}} = \delta_{\{0,1,2\}}$, $\delta_{\{1,2\}} \subseteq \varrho$ holds and therefore (6.7) and (6.6).
For the proof of the total reflexivity of ϱ, we consider the relation $\varrho_1 := \varrho \cap \delta_{\{0,1\}}$, which by (V) and (VI) is a nonempty diagonal relation. Since the implications (6.4) and (6.5) are not true for ϱ, we have

$$\varrho_1 \begin{cases} = \delta_{\{0,1\}} & \text{if } h \geq 5, \\ \in \{\delta_{\{0,1\},\{2,3\}}, \delta_{\{0,1\}}\} & \text{if } h = 4. \end{cases}$$

By (6.6) we have only to examine the case $h = 4$ and

$$\delta_{\{0,1\},\{2,3\}} \cup \delta_{\{0,2\},\{1,3\}} \cup \varrho_{\{0,3\},\{1,2\}} \subseteq \varrho.$$

Obviously, either ϱ satisfies the condition (*) and our proof is closed or it holds $|\varrho| > k^3$. Let

$$|\varrho| > k^3.$$

Therefore w.l.o.g. there exists certain $a_i \in E_k$ $(i = 0, 1, 2, 3)$ with

$$(a_0, a_1, a_2, a_3) \in \varrho,\ (a_0, a_1, a_2, a_3') \in \varrho,\ a_3 \neq a_3'. \tag{6.8}$$

Then the ϱ-derivable relation

$$\varrho_2 := \{(a_1, a_2, a_3, a_3') \mid \exists a_0 : (a_0, a_1, a_2, a_3) \in \varrho \wedge (a_0, a_1, a_2, a_3') \in \varrho\}$$

has the properties

$$\varrho_2 \cap \delta_{\{2,3\}} = \delta_{\{2,3\}} \tag{6.9}$$

(since $pr_{1,2,3}\varrho = E_k^3$),

$$\delta_{\{2,3\}} \subset \varrho_2 \text{ (by (6.8)) and } \varrho_2 \subset E_k^4.$$

Thus ϱ_2 is not an invariant of P_k. Further, ϱ is totally reflexive by (6.9) and (6.6). Then, by definition of ϱ_2, we have now $\varrho \cap \delta_{\{0,1\}} \neq \delta_{\{0,1\},\{2,3\}}$, a contradiction to our assumptions on ϱ. ■

Lemma 6.1.9 *If ϱ is totally reflexive, then ϱ is totally symmetric.*

Proof. The total symmetry of ϱ follows from (IIb) via $\iota_k^h \subseteq \bigcap_{s \in S_h} \varrho_s \subseteq \varrho$, where S_h denotes the set of all permutations on E_h and ϱ_s denotes the relation $\{(a_{s(0)}, ..., a_{s(h-1)}) \mid (a_0, ..., a_{h-1}) \in \varrho\}$. ■

Consequently, for the proof of (6.1), it is sufficient to examine the following cases:

2.1. ϱ satisfies (*) or it holds: $h = 3$ and ϱ is not totally reflexive

and

2.2. $h \geq 3$ and ϱ is totally symmetric and totally reflexive.

Case 2.1: ϱ satisfies (*) or $h = 3$ and $\varrho \cap \iota_k^h \neq \iota_k^h$.

W.l.o.g. we can assume that there does not exist a ϱ-derivable relation which is non-diagonal and totally reflexive.

Lemma 6.1.10 *The ternary relation ϱ is either of the form*

(a) $\varrho = \sigma \cup \delta_{\{0,1,2\}}$ *with the properties*
$\emptyset \subset \sigma \subseteq E_k^3 \backslash \iota_k^3$ *and*
for any $a, b \in E_k$ *there are exactly a* c_1*, exactly a* c_2 *and exactly a* c_3 *with* $(c_1, a, b), (a, c_2, b), (a, b, c_3) \in \varrho$*;*

or of the form

(b) $\varrho = \sigma \cup \delta_{\{a,b\}} \cup \delta_{\{a,c\}}$ *with* $\varrho \cap \varrho_s = \delta_{\{a,b\}}$*, where* $\emptyset \subset \sigma \subseteq E_k^3 \backslash \iota_k^3$*,* $\{a, b, c\} = \{0, 1, 2\}$*,* $s(a) := b$*,* $s(b) := a$*,* $s(c) := c$ *and* $\varrho_s := \{(a_{s(0)}, a_{s(1)}, a_{s(2)}) \mid (a_0, a_1, a_2) \in \varrho\}$.

Proof. By (VI) and (VII) it is sufficient to examine w.l.o.g. the following cases:

$$\varrho = \sigma \cup \delta_{\{0,1,2\}} \tag{6.10}$$

$$\varrho = \sigma \cup \delta_{\{0,1\}} \tag{6.11}$$

$$\varrho = \sigma \cup \delta_{\{0,1\}} \cup \varrho_{\{0,2\}} \tag{6.12}$$

where $\emptyset \subset \sigma \subseteq E_k^3 \backslash \iota_k^3$.
First let ϱ be of the form (6.11). A ϱ-derivable relation in this case is the totally symmetric relation

$$\varrho_i := \{(a_1, ..., a_i) \mid \exists a, b : (\forall i : (a_i, a, b) \in \varrho)\}$$

$(i = 2, 3, ..., k)$. Obviously, this relation has the property

$$\forall i \in \{2, ..., k-1\} : \varrho_i = E_k^i \implies \varrho_{i+1} \text{ totally reflexive.} \tag{6.13}$$

By our assumption in Case 2.1 and by (I) it follows from the total symmetry and total reflexivity of ϱ_{i+1} that $\varrho_{i+1} = E_k^{i+1}$. However, this cannot occur, since for $i = k-1$ it holds

$$(0, 1, ..., k-1) \in \varrho_k \implies \exists a, b \in E_k : \ \{(0, a, b), ..., (b, a, b), ..., (k-1, a, b)\} \subseteq \varrho\}$$

and $\varrho \cap \iota_k^3 = \delta_{\{0,1\}}$ is not possible. Thus our relation ϱ is not of the form (6.11), since we can show that $\varrho_2 = E_k^2$:

> $\delta_{\{0,1\}} \subseteq \varrho$ implies $\iota_k^2 \subseteq \varrho_2$. By $pr_{0,1}\varrho = E_k^2$ there are certain a, b $(a \neq b)$ and c with $(a, b, c) \in \varrho$. Furthermore we have $(b, b, c) \in \varrho$. Thus $(a, b) \in \varrho_2$ and $\varrho_2 = E_k^2$ by (IIa).

If $\varrho = \sigma \cup \delta_{\{0,1,2\}}$ and assumed ϱ does not fulfill the condition of (a) (w.l.o.g. for c_3), then the ϱ-derivable relation

$$\varrho' := \{(b, c, c') \mid \exists a : (a, b, c) \in \varrho \ \wedge \ (a, b, c') \in \varrho\}$$

is a non-diagonal relation of the form (6.11), of which we have shown that it cannot describe a maximal class of P_k. If ϱ has the form (6.12), it must hold $\varrho \cap (\tau\varrho) = \delta_{\{0,1\}}$, since, in the opposite case, the ϱ-derivable relation $\varrho \cap (\tau\varrho)$ has the form (6.11). ■

Thus by Lemma 6.1.10, we can assume w.l.o.g. that ϱ is either of the form (6.10) or (6.12) and ϱ fulfills the conditions of Lemma 6.1.10.

Lemma 6.1.11 *It is possible to derive from ϱ a quaternary non-diagonal relation λ with the following 8 properties:*

1) $\lambda \cap \delta_{\{0,2\}} = \delta_{\{0,2\},\{1,3\}} = \lambda \cap \delta_{\{1,3\}}$

2) $\lambda \cap \delta_{\{0,3\}} = \delta_{\{0,3\},\{1,2\}} = \lambda \cap \delta_{\{1,2\}}$

3) $\varrho \cap \iota_k^3 = \delta_{\{0,1\}} \cup \delta_{\{0,2\}} \Rightarrow \delta_{\{0,1\},\{2,3\}} \subseteq \lambda$

4) $\lambda \cap (\tau\lambda) = \lambda$, *where* $\tau\lambda := \{(b, a, c, d) \mid (a, b, c, d) \in \lambda\}$

5) $\lambda \cap (\lambda)_s = \lambda$, *where* $(\lambda)_s := \{(d, c, b, a) \mid (a, b, c, d) \in \lambda\}$

6) $\forall x, y, z, z', o \in E_k : (x, y, z, o) \in \varrho \wedge (x, y, z', o) \in \varrho \implies z = z'$

7) $pr_{0,1,3}\lambda = E_k^3$

8) $pr_{0,2,3}\lambda = E_k^3$.

Proof. If ϱ the condition (*) fulfills, it is easy to check that the ϱ-derivable relation

$$\lambda := \varrho \cap \tau(\varrho) \cap \{(d, c, b, a) \mid (a, b, c, d) \in \varrho\}$$

has the properties 1)–8), since it is possible in the opposite case to derive a non-diagonal totally reflexive, ternary relation from λ, in contradiction to the assumptions in Case 2.1.

Let now $h = 3$ and let ϱ be a relation of the form (6.10) or (6.12) in the following.

A ϱ-derivable relation is

$$\lambda_1 := \begin{cases} \lambda_{11} & \text{if } \varrho \text{ has the form (6.10)}, \\ \lambda_{12} & \text{if } \varrho \text{ has the form (6.12)}, \end{cases}$$

where

$$\lambda_{11} := \{(a, b, c, d) \mid \exists u, v : \{(u, d, a), (d, v, b), (u, v, c)\} \subseteq \varrho\}$$

and

$$\begin{aligned} \lambda_{12} := \{(a, b, c, d) \mid \exists u_1, ..., u_4, v_1, ..., v_4 : \\ &\{(u_1, a, c), (u_2, b, d), (u_3, a, d), (u_4, b, c), \\ &(v_1, a, c), (v_2, b, d), (v_3, a, d), (v_4, b, c), \\ &(b, d, u_1), (a, c, u_2), (b, c, u_3), (a, d, u_4), \\ &(d, b, v_1), (c, a, v_2), (c, b, v_3), (d, a, v_4)\} \subseteq \varrho\}. \end{aligned}$$

With the help of the properties of ϱ, which are given in Lemma 6.10, it is easy to see that λ_1 fulfills the conditions 1)–3).

Since

$$\delta_{\{0,2\},\{1,3\}} \cup \delta_{\{0,3\},\{1,2\}} \subseteq \lambda_{11} \cap (\tau\lambda_{11}) \cap (\lambda_{11})_s$$

and

$$\delta_{\{0,2\},\{1,3\}} \cup \delta_{\{0,3\},\{1,2\}} \cup \delta_{\{0,1\},\{2,3\}} \subseteq \lambda_{12} \cap (\tau\lambda_{12}), \cap(\lambda_{12})_s$$

the non-diagonal relation $\lambda := \lambda_1 \cap (\tau\lambda_1) \cap (\lambda_1)_s$ has the properties 1)–5). Furthermore, for any $i, j \in E_4$, $i \neq j$, it holds

$$pr_i\lambda = E_k \text{ and } pr_{i,j}\lambda = E_k^2. \tag{6.14}$$

If λ does not fulfill 6), we can derive the non-diagonal relation

$$\lambda' := \{(c, c', a) \mid \exists b, d : (a, b, c, d) \in \lambda \wedge (a, b, c', d) \in \lambda\}$$

from λ. However λ' is of the form $\sigma \cup \delta_{\{0,1\}}$ with $\emptyset \subset \sigma \subseteq E_k^3 \backslash \iota_k^3$ (by 1) and 2)), what is not possible by the proof of Lemma 6.1.10.
Thus λ fulfills 1)–6).
If $\varrho = \sigma \cup \delta_{\{0,1,2\}}$, the properties 7) and 8) follow from (6.14), from the property (a) of Lemma 6.1.10 and from the definition of λ_1.
If ϱ has the form $\sigma \cup \delta_{\{0,1\}} \cup \delta_{\{0,2\}}$, then we have $\iota_k^3 \subseteq pr_{0,\alpha,3}\lambda$ ($\alpha \in \{1, 2\}$) by 1)–3) and hence $pr_{0,\alpha,3}\lambda = E_k^3$, since the case $pr_{0,\alpha,3}\lambda \subset E_k^3$ is a contradiction to our assumptions over ϱ. ■

By Lemma 6.1.11(6), 7)) with the aid of λ, we can define for a fixed $o \in E_k$ an operation $+_o$ on E_k as follows:

$$x +_o y = z \;:\Longleftrightarrow\; (x, y, z, o) \in \lambda. \tag{6.15}$$

Lemma 6.1.12 *The operation $+_o$ has the following properties:*

1) *$(E_k; +_o)$ is an Abelean group with the identity element o.*
2) *There exists a prime number p so that every element of $E_k \setminus \{o\}$ has the order p and it holds $k = p^m$ for certain $m \geq 1$.*

Proof. 1): All properties of $+_o$ (except the associativity), which we have to show, follow directly from Lemma 6.1.11. By 1) of Lemma 6.1.11 o is the identity element of $(E_k, +_o)$. Because of 4)–8) for every $a \in E_k$ there exists an inverse element $a' \in E_k$ with $a +_o a' = o$. The operation $+_o$ is kommutative by 4).
For proof of the associativity of $+_o$, we consider the λ-derivable relation

$$\begin{aligned}\lambda' &:= \{(a, b, c) \mid (a +_o b) +_o c = a +_o (b +_o c)\} \\ &= \{(a, b, c) \mid \exists u, v, w : \{(a, b, u, o), (u, c, v, o), (b, c, w, o), (a, w, v, o)\} \subseteq \lambda\},\end{aligned}$$

which is totally reflexive by $\{(o, a, a), (a, o, a), (a, a, o) \mid a \in E_k\} \subseteq \lambda'$ and (II). However, this is only possible if $\lambda' = E_k^3$ by our assumptions over ϱ. Thus $+_o$ is associative.
2): A λ-derivable relation is

$$\begin{aligned}\mu_n &:= \{(x, o) \mid \underbrace{x +_o x +_o \ldots +_o x}_{n \text{ times}} = o\} \\ &= \{(x, o) | \exists u_i : \{(x, x, u_1, o), (u_1, x, u_2, o), \ldots, (u_{n-2}, x, o, o)\} \subseteq \lambda\}.\end{aligned}$$

By $o +_o o = o$ for all $o \in E_k$ this relation is reflexive. Assume there exists two elements of $E_k \backslash \{o\}$ with the orders n and m ($\neq n$) respectively. Then we have $\iota_k^2 \subset \mu_n \subset E_k^2$, in contradiction to our assumptions over ϱ. Thus all elements ($\neq o$) of $(E_k, +_o)$ have the same order. Now it is easy to show that this order is a certain prime number p: If $n > 1$ is not a prime number, there exist $r, s \in \mathbb{N}$ with $n = r \cdot s$, $1 < r < n$ and $1 < s < n$. For a $b \in E_k \setminus \{0\}$ we set $c := s \cdot b := \underbrace{b +_o b +_o \ldots +_o b}_{s \text{ times}} \neq o$. Then we have $c \neq o$ and $r \cdot c = n \cdot b = o$, a contradiction to the above-proven fact that the order of b is n.
$k = p^m$ for a certain m is a conclusion from well-known theorems about Abelian groups. ■

Lemma 6.1.13 *For every $x, y, o, o_1 \in E_k$ it holds:*

1) $x +_{o_1} x = x +_o x -_o o_1$

2) $x +_{o_1} o = x -_o o_1$

3) $k \geq 4 \implies x +_{o_1} y = x +_o y -_o o_1$.

Proof. To prove $i)$ ($i \in \{1,2,3\}$) we consider certain λ-derivable relations λ_i:

$$\begin{aligned}\lambda_1 &:= \{(x, o_1, o) \mid x +_{o_1} x = x +_o x -_o o_1\}\\ &= \{(x, o_1, o) \mid \exists u, v, w : \{(x, x, u, o_1), (x, x, v, o), (o_1, w, o, o), (v, w, u, o)\}\\ &\qquad \subseteq \lambda\},\\ \lambda_2 &:= \{(x, o, o_1) \mid x +_{o_1} o = x -_o o_1\}\\ &= \{(x, o, o_1) \mid \exists u, v : \{(x, o, u, o_1), (o_1, v, o, o), (x, v, u, o)\} \subseteq \lambda\}\end{aligned}$$

and

$$\begin{aligned}\lambda_3 &:= \{(x, y, o, o_1) \mid x +_{o_1} y = x +_o y -_o o_1\}\\ &= \{(x, y, o, o_1) \mid \exists u, v, w :\\ &\qquad \{(x, y, u, o_1), (o_1, v, o, o), (x, y, w, o), (w, v, u, o)\} \subseteq \lambda\}.\end{aligned}$$

If we can show that λ_1, λ_2 and λ_3 are totally reflexive relations, then we have $\lambda_1 = \lambda_2 = E_k^3$ and $\lambda_3 = E_k^4$ for $k \geq 4$ by our assumptions in Case 2.1 and the statements of Lemma 6.1.13 are proven.
For all $x, o \in E_k$ we have:
- $(x, x, o) \in \lambda_1$, since $x +_x x = x = x +_o x -_o x$.
- $(x, o_1, x) \in \lambda_1$, because for $v = x$, $w = u$ and Lemma 6.1.11, 5) the conditions of definition of λ_1 hold.
- $(x, o, o) \in \lambda_1$, since $x +_o x = x +_o x -_o o$.

Thus 1) is proven.
For all $x, o, o_1 \in E_k$ it holds:
- $(o, o, o_1) \in \lambda_2$, because $o +_{o_1} o = o +_o o -_o o_1 = o -_o o_1$ by 1).
- $(o_1, o, o_1) \in \lambda_2$, since $o_1 +_{o_1} o = o = o_1 -_o o_1$.
- $(x, o, o) \in \lambda_2$ by $x +_o o = x = x -_o o$.

Thus b) is proven).
3) follows from
- $(x, x, o, o_1) \in \lambda_3$ by 1);
- $(o, y, o, o_1) \in \lambda_3$, since $o +_{o_1} y = y +_{o_1} o = y -_o o_1 = o +_o y -_o o_1$ by 2);
- $(o_1, y, o, o_1) \in \lambda_3$ by $o_1 +_{o_1} y = y = o_1 +_o y -_o o_1$;
- $(x, o, o, o_1) \in \lambda_3$, since (by 2)) $x +_{o_1} o = x -_o o_1 = x +_o o -_o o_1$;
- $(x, o_1, o, o_1) \in \lambda_3$ since $x +_{o_1} o_1 = x = x +_o o_1 -_o o_1$;
- $(x, y, o, o) \in \lambda_3$, since $x +_o y = x +_o y -_o o$. ■

Lemma 6.1.14 *For every $o \in E_k$ and all $a, b, c, d \in E_k$ the following equivalence holds:*

$$(a, b, c, d) \in \lambda \iff a +_o b = c +_o d.$$

Proof. If $k \geq 4$ then we have by definition of $+_o$ and by Lemmas 6.1.12 and 6.1.13:

$$\begin{aligned}(a, b, c, d) \in \lambda &\iff a +_d b = c \qquad \iff a +_d b = a +_o b -_o d = c \\ &\iff a +_o b = c +_o d.\end{aligned}$$

For $k = 3$ our statement follows from Lemma 6.1.11, 1), 2), 5) and Lemma 6.1.13, 1). ■

Then the following lemma is a conclusion from the Lemmas 6.1.11, 6.1.12, and 6.1.14:

Lemma 6.1.15 *The ϱ-derivable relation λ belongs to $\mathfrak{L}_k$.*

Proof. If $k \geq 4$, then $\lambda \in \mathfrak{L}_k$ ($k = p^m$, p prime, $m \geq 1$) follows from the Lemmas 6.1.9, 6.1.10 and 6.1.12. For $k = p = 3$ the relation

$$\lambda' := \{(a, b, c, d) \mid \exists \alpha : a +_o b = \alpha +_o \alpha \ \wedge \ c +_o d = \alpha +_o \alpha\}$$

is ϱ-derivable and belongs to $\mathfrak{L}_k$. ■

Summing up:
Our Case 2.1 is only possible if k is a prime power. Furthermore, we have for $k = p^m$:

$$Pol\varrho = Pol\lambda \ \text{ for certain } \lambda \in \mathfrak{L}_k.$$

Case 2.2:
ϱ is totally reflexive and totally symmetric.

Since $P_k(h' - 1) \subseteq Pol\varrho'$ for every totally reflexive h'-ary relation ϱ' ($h' \geq 2$), it is not possible to find a totally reflexive h'-ary relation ϱ' with $Pol\varrho = Pol\varrho'$ and $h' > h$. Thus we can assume the following property for ϱ:

$$(VIII) \quad \forall \varrho' \in [\{\varrho\} \cup InvP_k]^{h'} : \ h < h' \leq k \ \wedge \ \iota_k^{h'} \subseteq \varrho' \implies \varrho' = E_k^{h'}.$$

An h-ary, totally reflexive, and totally symmetric relation γ we will call **strongly homogeneous**, if the following holds:

$$\begin{aligned}&(\exists v_0,...,v_{h-1}\in E_k:\\&(v_0,...,v_{h-1})\in\gamma\wedge\\&\forall i\in E_h\ \forall j\in E_h\backslash\{i\}:\ (a_0,...,a_{j-1},v_i,a_{j+1},...,a_{h-1})\in\gamma)\\&\Longrightarrow(a_0,a_1,...,a_{h-1})\in\gamma.\end{aligned}\tag{6.16}$$

Lemma 6.1.16 *Either ϱ is central or strongly homogeneous.*

Proof. First we consider the ϱ-derivable relation

$$\begin{aligned}\varrho_t:=\{(a_0,...,a_{t-1})\mid\exists c:\\&(\forall i_0,...,i_{h-2}\in E_t:(a_{i_0},...,a_{i_{h-2}},c)\in\varrho))\},\end{aligned}$$

for $t\in\{h,h+1,...,k\}$. The total symmetry of ϱ implies that ϱ_t is totally symmetric. By choosing $c=a_0$ for $t=h$, we see that $\varrho\subseteq\varrho_h$. Thus by our assumption (IIc), only the following two cases are possible:
Case 1: $\varrho_h=E_k^h$.
Then we have $h<k$ since, in the opposite case, $\varrho=E_k^k$ holds by definition of ϱ. Obviously, for $h<k$, the ϱ-derivable relation ϱ_{h+1} is totally reflexive and thus, by (VIII), $\varrho_{h+1}=E_k^{h+1}$. By induction, it is easy to see that $\varrho_i=E_k^i$ for all $i\in\{h+1,...,k\}$. Then, $\varrho\in\mathfrak{C}_k$ follows from $\varrho_k=E_k^k$.
Case 2: $\varrho=\varrho_h$.
In this case, ϱ is homogeneous, i.e., ϱ has the property

$$\begin{aligned}&(\exists v:\forall i\in E_h:\\&(a_0,...,a_{i-1},v,a_{i+1},...,a_{h-1})\in\varrho)\implies(a_0,a_1,...,a_{h-1})\in\varrho.\end{aligned}$$

for arbitrary $a_0,a_1,...,a_{h-1}\in E_k$.
We consider the ϱ-derivable relation

$$\begin{aligned}\gamma_t:=\{(a_0,...,a_{t-1})\mid\exists v_0,v_1,...,v_{t-1}:\\&(\forall i_1,...,i_h\in E_t:(v_{i_1},...,v_{i_h})\in\varrho)\wedge\\&(\forall n\in E_t\forall j_1,...,j_{h-2}\in E_t\backslash\{n\}:\\&(a_{j_1},...,a_{j_{h-2}},a_n,v_n)\in\varrho)\}\end{aligned}$$

$(t\in\{h,h+1,...,k\})$. Obviously, γ_t is totally symmetric and, if $\gamma_{t-1}=E_k^{t-1}$, is totally reflexive. Furthermore, $\varrho\subseteq\gamma_h$ for $h=t$. (This follows from the definition of γ_h by setting $v_\alpha=a_\alpha$ for all $\alpha\in E_h$.) $\varrho=\gamma_h$ implies that ϱ is strongly homogeneous. Now by (IIc) we have $\gamma_h\in\{\varrho,E_k^h\}$. Thus our lemma is proven if we can show that $\gamma_h=E_k^h$ cannot occur.
Assume $\gamma_h=E_k^h$ and $h<k$. Then γ_{h+1} is totally reflexive. Thus, $\gamma_{h+1}=E_k^{h+1}$ by (VIII) and $\gamma_k=E_k^k$ via induction. Hence, for every $a_0,...,a_{h-1}\in E_k$ there exists certain $v_0,...,v_{h-1}\in E_k$ with

$$(v_0,v_1,...,v_{h-1})\in\varrho\tag{6.17}$$

and

$$\forall n \in E_h \; \forall \alpha_1, ..., \alpha_{h-2} \in E_k \backslash \{a_n\} : \; (\alpha_1, ..., \alpha_{h-2}, a_n, v_n) \in \varrho. \qquad (6.18)$$

With the help of this and by induction, we can show that

$$\forall t \geq 0 \; \forall a_0, ..., a_{t-1} \in E_k : \; (a_0, ..., a_{t-1}, v_t, v_{t+1}, ..., v_{h-1}) \in \varrho \qquad (6.19)$$

holds:
For $t = 0$, (6.19) follows from (6.17). Assume (6.19) holds for $t = n$. Then (6.19) for $t = n + 1$ is a conclusion from the induction hypothesis, the homogeneity of ϱ, the totally symmetry of ϱ and from (6.18) putting $v = v_n$ in (6.16) for the tuple $(a_0, ..., a_n, v_{n+1}, ..., v_{h-1})$. By (6.19) and $t = h$, we get a contradiction to $\varrho \subset E_k^h$:

$$\forall a_0, ..., a_{h-1} \in E_k : \; (a_0, ..., a_{h-1}) \in \varrho.$$

Hence $\gamma_h = E_k^h$ cannot occur and thus $\gamma_h = \varrho$, i.e., ϱ is strongly homogeneous. ■

By Lemma 6.1.16, we have to show for the proof of our theorem that, if ϱ is strongly homogeneous, the relation ϱ or a ϱ-derivable relation belongs to $\mathfrak{B}_k$.

Thus we can assume that ϱ is subsequently a strongly homogeneous relation.

For this relation, we will prove first that ϱ is a homomorphic inverse image of a certain relation ξ and then that the relation ξ is isomorphic to a certain elementary h-ary relation ξ_m.[2]

Lemma 6.1.17 *The relation*

$$\varepsilon := \{(a,b) \in E_k^2 \mid \forall a_0, ..., a_{h-3} \in E_k : \; (a_0, a_1, ..., a_{h-3}, a, b) \in \varrho\}$$

is an equivalence relation on E_k.

Proof. By the total reflexivity and total symmetry of ϱ, it follows directly the reflexivity and symmetry of ε. To prove the transitivity of ε, let $\{(a,b), (b,c)\} \subseteq \varepsilon$. Choosing $v_0 = \alpha_0$, $v_1 = \alpha_1$, ..., $v_{h-3} = \alpha_{h-3}$, $v_{h-2} = b$ and $v_{h-1} = b$ in (6.16) for the tuple $(\alpha_0, \alpha_1, ..., \alpha_{h-3}, a, c)$, we get $(\alpha_0, ..., \alpha_{h-3}, a, c) \in \varrho$ for arbitrary $\alpha_0, ..., \alpha_{h-3} \in E_k$. Hence, $(a,c) \in \varepsilon$. ■

By the equivalence relation ε, the set E_k is partitioned into certain (nonempty) equivalence classes A_i $(i = 1, 2, ..., r)$, from which we choose a representative α_i. With the help of the representative set

$$V := \{\alpha_1, ..., \alpha_r\}$$

we can define a mapping $F : \; E_k \longrightarrow V$ by

$$F(a) = \alpha_i \; :\Longleftrightarrow \; \{a, \alpha_i\} \subseteq A_i.$$

[2] See Lemma 6.1.22 to the concept "isomorphic".

Lemma 6.1.18 *Let ϱ be strongly homogeneous. Then*

(a) $(a_0, ..., a_{h-1}) \in \varrho \iff (F(a_0), ..., F(a_{h-1})) \in \varrho$

(b) $((\forall a_0, ..., a_{h-3} \in V : (a_0, ..., a_{h-3}, a, b) \in \varrho) \wedge \{a, b\} \subseteq V) \implies a = b.$

Proof. (a): First, we show that the equivalence

$$(a, a_1, ..., a_{h-1}) \in \varrho \iff (b, a_1, ..., a_{h-1}) \in \varrho \tag{6.20}$$

holds for $(a, b) \in \varepsilon$ and for all $a_0, ..., a_{h-1} \in E_k$.
If $(a, a_1, ..., a_{h-1}) \in \varrho$ then $(b, a_1, ..., a_{h-1}) \in \varrho$ follows from the strong homogeneity of ϱ, choosing $v_0 = v_1 = ... = v_{h-1} = a$. Since ε is symmetric, we have also proven "$\Longleftarrow$" of (6.20).
Now (a) follows from (6.20) because of $(a_i, F(a_i)) \in \varepsilon$ $(i \in E_h)$, Lemma 6.1.17 and of the total symmetry of ϱ:

$$\begin{aligned}(a_0, ..., a_{h-1}) \in \varrho &\iff (F(a_0), a_1, ..., a_{h-1}) \in \varrho \\ &\iff (F(a_0), F(a_1), a_2, ..., a_{h-1}) \in \varrho \\ &\iff \dots \\ &\iff (F(a_0), ..., F(a_{h-1})) \in \varrho.\end{aligned}$$

(b) is a conclusion from the definitions of ε and V and from (a). ■

Now put

$$\xi := F(\varrho) := \{(F(a_0), ..., F(a_{h-1})) \mid (a_0, ..., a_{h-1}) \in \varrho\}.$$

By Lemma 6.1.18, (a) ϱ is a homomorphic inverse image of this relation, i.e.,

$$\varrho = \{(a_0, ..., a_{h-1}) \in E_k^h \mid (F(a_0), ..., F(a_{h-1})) \in \xi\}$$

holds. Since $\varrho \neq E_k^h$, $\zeta \neq V^h$. Thus w.l.o.g. we can assume

$$V = E_r \text{ and } (0, 1, ..., h-1) \in V^h \backslash \xi.$$

To show that ξ is isomorphic to a certain h-ary elementary relation ξ_m ($\subseteq E_{h^m}^h$), i.e., that there is a bijective mapping φ of E_{h^m} onto V with the property

$$\xi = \{(\varphi(a_0), ..., \varphi(a_{h-1})) \mid (a_0, ..., a_{h-1}) \in \xi_m\}$$

we consider the ϱ-derivable h^i-ary graphic

$$G_{h^i}(Pol\,\varrho) := \chi_{h^i} \cup \{f(\kappa_1, ..., \kappa_{h^i}) \mid f \in (Pol\,\varrho)^{h^i}\},$$

where $\chi_{h^i} := (\kappa_1, ..., \kappa_{h^i})$ is the h^i-ary abscissa over E_k (in matrix form; see Section 2.7), $i \in \{h, h+1, ..., k\}$. Let $j_1, ..., j_i$ be the numbers of those rows of χ_{h^i} for which

$$\mathfrak{A}_i := pr_{j_1,...,j_i} \chi_{h^i}$$

is a matrix form of the relation E_h^i. Further, let

$$\mu_i := pr_{j_1,...,j_i} G_{h^i}(Pol\,\varrho).$$

Lemma 6.1.19 *If ϱ is strongly homogeneous and $(0,1,2,...,h-1) \notin \varrho$, then $\mu_i = E_k^i$ for every $i \in \{h, h+1, ..., k\}$, i.e., the functions of $Pol\varrho$ can take every value of E_k on every row of the matrix $\mathfrak{A}_i$.*

Proof. First, we remark that

$$\varrho \neq E_k^h \backslash \{(a_0, ..., a_{h-1}) \mid \{a_0, ..., a_{h-1}\} = E_h\}$$

because of $\varrho \notin \mathfrak{C}_k$. Further, if $(b_0, ..., b_{h-1}) \in \varrho$, all functions g with $Im(g) = \{b_0, ..., b_{h-1}\}$ belong to $Pol_k\varrho$. Suppose the lemma is false for $i = h$. Then, we can form an h-ary relation ϱ' with $\varrho \subset \varrho' \subset E_k^h$ with the help of the h^i-th graphic of $Pol\varrho$ by projection, contradicting (IIc).
Since all functions of $Pol\varrho$, which have at most h^i essential variables, belong to $(Pol\varrho)^{h^{i+1}}$ it is easy to show by induction that it holds:

$$\forall i \in \{h, ..., k-1\} : \ \mu_i = E_k^i \implies \iota_k^{i+1} \subseteq \mu_{i+1}.$$

Consequently, our lemma follows from (VIII). ∎

By Lemma 6.1.19, we can find an h^k-ary function $f \in Pol\varrho$ with the properties

$$Im(f) = V$$

and

$$f(\mathfrak{A}_k) = \begin{pmatrix} F(0) \\ F(1) \\ \cdots \\ F(k-1) \end{pmatrix}.$$

Next we will give some properties of this function f, from which will follows $\varrho \in \mathfrak{B}_k$.
Recall, $f \in Pol\varrho$ means on tuples of $E_h^{h^k}$ that f preserves the relation ι_h^h.
The elements z of $E_h^{h^k}$ we also give in the form

$$(z[1], z[2], ..., z[h^k])$$

and, for every $z \in E_h^{h^k}$, denote

$$z^{t,a} \ (t \in \{1, 2, ..., h^k\}, \ a \in E_h)$$

an element of $E_h^{h^k}$, which is defined by

$$z^{t,a}[i] := \begin{cases} a, & \text{if} \quad i = t, \\ z[i] & \text{otherwise} \end{cases}$$

$(i \in \{1, 2, ..., h^k\})$.

Lemma 6.1.20 *Let $t \in \{1, 2, ..., h^k\}$, $z \in E_h^{h^k}$ and $(f(z^{t,0}), ..., f(z^{t,h-1})) \in \xi$. Then*

(a) $f(z^{t,0}) = ... = f(z^{t,h-1})$,

(b) $\forall w \in E_h^{h^k} : \ f(w^{t,0}) = ... = f(w^{t,h-1})$.

Proof. (a): For proof, it is sufficient to show that (w.l.o.g.) $f(z^{t,h-2}) = f(z^{t,h-1})$. First, we prove

$$\forall b_0, ..., b_{h-3} \in E_h^{h^k} : (f(b_0), ..., f(b_{h-3}), f(z^{t,h-2}), f(z^{t,h-1})) \in \xi. \quad (6.21)$$

If $\{b_0[t], b_1[t], ..., b_{h-3}[t]\} \neq \{0, 1, ..., h-3\}$, then (6.21) holds, since in this case, all columns of the matrix

$$B := \begin{pmatrix} b_0 \\ b_1 \\ \dots \\ b_{h-3} \\ z^{t,h-2} \\ z^{t,h-1} \end{pmatrix}$$

belong to ι_h^h and f preserves ϱ. Let w.l.o.g. $b_i[t] = i$ for $i \in E_{h-2}$. Substituting the i-th row of B by $z^{t,j}$, where $i \neq j$, we obtain a matrix $B_{i,j}$ whose columns belong to ι_h^h. Hence $f(B_{i,j}) \in \xi$ holds. Consequently, by the strong homogeneity of ϱ, choosing $v_j = f(z^{t,j})$ ($j \in E_h$) for the tuple $(f(b_0), ..., f(b_{h-3}), f(z^{t,h-2}), f(z^{t,h-1}))$, we get (6.21).
Since $\{f(\alpha) \mid \alpha \in E_h^{h^k}\} = V$, for all $a_0, ..., a_{h-3} \in V$ there are certain $b_0, ..., b_{h-3} \in E_h^{h^k}$ with $f(b_i) = a_i$ ($i \in E_{h-2}$). Thus by (6.21) we have

$$\forall a_0, ..., a_{h-2} \in V : (a_0, ..., a_{h-3}, f(z^{t,h-2}), f(z^{t,h-1})) \in \xi.$$

Then, with the help of Lemma 6.1.18, (b) it follows $f(z^{t,h-2}) = f(z^{t,h-1})$.
(b): Because of (a) it is sufficient to show that

$$(f(w^{t,0}), ..., f(w^{t,h-1})) \in \xi \quad (6.22)$$

holds. By (6.21), it follows $(f(w^{t,0}), ..., f(w^{t,h-3}), f(z^{t,h-2}), f(z^{t,h-1})) \in \xi$. It is easy to check that every exchange of an i-th row in

$$\begin{pmatrix} w^{t,0} \\ w^{t,1} \\ \dots \\ w^{t,h-1} \end{pmatrix}$$

by $w^{t,j}$ for $i \in E_{h-3}$ or by $z^{t,j}$ for $i \in \{h-2, h-1\}$ and $i \neq j$, gives a matrix $\mathcal{M}$, whose columns belong to ι_h^h, and it holds $f(\mathcal{M}) \in \xi$. Hence by the strong homogeneity of ϱ, choosing $v_0 = f(w^{t,0}), ..., v_{h-3} = f(w^{t,h-3})$), $v_{h-2} = f(z^{t,h-2})$ and $v_{h-1} = f(z^{t,h-1})$, we get (6.22). ■

Lemma 6.1.21 *For the function f, there exists certain digits $t_1, ..., t_m$, which are exactly the essential digits of $f_{|E_h^{h^k}}$ (restriction of f to $E_h^{h^k}$), and f has different values on any two different tuples of $E_h^{h^k}$, i.e., it holds*

$$\forall z, w \in E_h^{h^k} : f(z) = f(w) \iff \forall i \in \{t_1, ..., t_m\} : z[i] = w[i]. \qquad (6.23)$$

Further, f has the properties

$$|Im(f)| = h^m$$

and

$$\forall r_1, ..., r_{h^k} \in E_h^h : (\{r_{t_1}, ..., r_{t_m}\} \not\subseteq \iota_h^h \implies f(r_1, ..., r_{h^k}) \notin \varrho) \qquad (6.24)$$

Proof. For every $t \in \{1, 2, ..., h^k\}$, we have either $f(z^{t,0}) = f(z^{t,1}) = ... = f(z^{t,h-1})$ for every $z \in E_h^{h^k}$ or $(f(z^{t,0}), f(z^{t,1}), ..., f(z^{t,h-1})) \notin \varrho$ for all $z \in E_h^{h^k}$ by Lemma 6.1.20. Let $T := \{t_1, ..., t_m\}$ be the set of all $t \in \{1, 2, ..., h^k\}$, for which $(f(z^{t,0}), ..., f(z^{t,h-1})) \notin \varrho$ holds. Now, we will show that the digits $t_i \in T$ have the properties of Lemma 6.1.21. First, let $f(z) = f(w)$ for certain $z, w \in E_h^{h^k}$ and assume there exists an $i \in T$ with $\alpha := z[i] \neq w[i]$. Then, the columns of the matrix

$$A := \begin{pmatrix} z^{i,0} \\ ... \\ z^{i,\alpha-1} \\ w \\ z^{i,\alpha+1} \\ ... \\ z^{i,h-1} \end{pmatrix}$$

belong to ι_h^h. Thus $f(A) \in \xi$ and by $f(w) = f(z) = f(z^{i,\alpha})$ it follows $(f(z^{i,0}), ..., f(z^{i,h-1})) \in \xi$, a contradiction to $i \in T$ and to the definition of T. Therefore "$\Longrightarrow$" in (6.23) holds. Let now $z[i] = w[i]$ for every $i \in T$ and w.l.o.g. $T = \{1, 2, ..., m\}$. $f(w) = f(z)$ is proven, if we can show that

$$f(u_{n-1}) = f(u_n) \qquad (6.25)$$

holds for every tuple $u_n := (z[1], ..., z[n], w[n+1], w[n+2], ..., w[h^k])$ and all $n \in \{m+1, m+2, ..., h^k\}$, since $f(u_m) = f(w)$ and $f(u_{h^k}) = f(z)$. By $n > m$ and the definition of T, we have $f(u_{n-1}^{n,i}) = f(u_{n-1}^{n,j})$ for every $i, j \in E_h$. As $u_{n-1}^{n,w[n]} = u_{n-1}$ and $u_{n-1}^{n,z[n]} = u_n$ we have (6.25) and (6.23) is proven.
(6.23) and $\{f(\mathbf{x}) \mid \mathbf{x} \in E_h^{h^k}\} = V$ imply $|V| = h^m$.
Finally, we prove (6.24). Assume (6.24) is false. Then there exists certain $z_0, z_1, ..., z_{h-1} \in E_h^{h^k}$ with $(f(z_0), f(z_1), ..., f(z_{h-1})) \in \xi$ and (w.l.o.g.) $z_i[1] = i$ for all $i \in E_h$ and $1 \in T$. Let $w \in E_h^{h^k}$. Then, all columns of the matrix

$$C_{i,j} := \begin{pmatrix} w^{1,0} \\ ... \\ w^{1,j-1} \\ z_j^{1,i} \\ w^{1,j+1} \\ ... \\ w^{1,h-1} \end{pmatrix}$$

belong to ι_h^h for $i \neq j$. Thus $f(C_{i,j}) \in \xi$. By choosing $v_i = f(z_i)$, for the tuple $(f(w^{1,0}), ..., f(w^{1,h-1}))$ and $i \in E_h$ from this and the strong homogeneity of ϱ it follows $(f(w^{1,0}), ..., f(w^{1,h-1})) \in \xi$. But this contradicts the definition of T and $1 \in T$. ∎

Lemma 6.1.22 *If ϱ is strongly homogeneous, then there exists for certain m, a bijective mapping φ from E_{h^m} onto V with*

$$\xi = \varphi(\xi_m) := \{(\varphi(a_0), ..., \varphi(a_{h-1})) \mid (a_0, ..., a_{h-1}) \in \xi_m\}$$

and it holds that

$$\varrho = \{(a_0, ..., a_{h-1}) \in E_k^h \mid (\varphi^{-1}(F(a_0)), ..., \varphi^{-1}(F(a_{h-1}))) \in \xi_m\},$$

i.e., $\varrho \in \mathfrak{B}_k^h$.

Proof. With the function f in Lemma 6.1.21, we can define a bijective mapping from E_{h^m} onto V as follows:

$$\begin{aligned} &\varphi(a^{(m-1)}h^{m-1} + a^{(m-2)}h^{m-2} + ... + a^{(1)}h + a^{(0)}) = f(z) \\ &:\Longleftrightarrow\ (z[t_1], z[t_2], ..., z[t_m]) = (a^{(m-1)}, a^{(m-2)}, ..., a^{(0)})), \end{aligned}$$

where $a \in E_{h^m}$. Let $(a_0, ..., a_{h-1}) \in \xi_m$. Then there are certain $z_0, ..., z_{h-1} \in E_h^{h^k}$, where $(z_0[i], ..., z_{h-1}[i]) \in \iota_h^h$ for all $i \in \{1, 2, ..., h^k\}$, with the property

$$(\varphi(a_0), \varphi(a_1), ..., \varphi(a_{h-1})) = (f(z_0), f(z_1), ..., f(z_{h-1})).$$

Since $f \in Pol\varrho$, this implies $(\varphi(a_0), ..., \varphi(a_{h-1})) \in \xi$. Hence $\varphi(\xi_m) \subseteq \xi$. Assume, there exists a tuple $(a_0, ..., a_{h-1}) \in \xi \backslash \varphi(\xi_m)$. By definition of φ, we find certain $z_0, ..., z_{h-1} \in E_h^{h^k}$ with $(f(z_0), ..., f(z_{h-1})) = (a_0, ..., a_{h-1})$, where $(z_0[i], ..., z_{h-1}[i]) \notin \iota_h^h$ for certain $i \in \{t_1, ..., t_m\}$. But this is contrary to (6.24). Thus $\xi = \varphi(\xi_m)$. The other statements follow from Lemma 6.1.18, (a) and the definition of ξ. ∎

To sum up, we have in Case 2.2:

$$\varrho \in \mathfrak{C}_k \cup \mathfrak{B}_k,$$

and Theorem 6.1 is proven.

7

Further Completeness Criteria

In the literature, one finds many articles which deal with complete sets. With the aid of complete sets, it is easy to obtain completeness criteria. In this chapter, we discuss only three types of such criteria. First we handle a criterion for Sheffer functions, which was found by G. Rousseau. Then, we show how one can reduce the conditions from Theorem 6.1 if one considers only surjective functions. Finally, we deal with criteria that indicate under which conditions a set ($\subseteq P_k$) which consists of certain unary functions and a Słupecki-function is complete in P_k.

7.1 A Criterion for Sheffer-Functions

Definitions A function $f \in P_k$ is called a **Słupecki-function**, if $f \notin Pol_k \iota_k^k$, i.e., if f depends on at least two variables essentially and takes k different values.
A function f is called a **Sheffer-function** of P_k, if every $g \in P_k$ is a superposition over f, i.e., if

$$[f] = P_k \tag{7.1}$$

holds.
Obviously, every Sheffer-function is a Słupecki-function.
The function

$$x \mid y := \overline{x \vee y}$$

which fulfills (7.1) for $k = 2$ was published by H. M. Sheffer (in [She 13]). This function and the function

$$\overline{x \wedge y},$$

which is dual to this function, are the only binary functions of P_2, which generate P_2 (cf. [Zyl 25]). One can easily prove this with the aid of the following theorem.

Theorem 7.1.1 *Let $f^n \in P_2$. Then*

$$[f] = P_2 \iff (f \notin T_0 \wedge f \notin T_1 \wedge f \notin S). \tag{7.2}$$

Proof. "$\Longrightarrow$" is trivial.
"$\Longleftarrow$": Because of $f(0,...,0) = 1$ and $f(1,...,1) = 0$, we have $f \notin M$. Since f does not preserve $\begin{pmatrix} 0 & 1 \\ 1 & 0 \end{pmatrix}$, there are some $a_1,...,a_n \in E_2$ with $f(a_1,...,a_n) = f(\overline{a}_1,...,\overline{a}_n)$. Consequently,

$$(f(0,...,0), f(a_1,...,a_n), f(\overline{a}_1,...,\overline{a}_n), f(1,...,1))$$

does not belong to

$$\alpha := \{(a,b,c,d) \in E_2^4 \mid a + b = c + d \ (mod\ 2)\},$$

although $(0, a_i, \overline{a}_i, 1) \in \alpha$ for every $i \in \{1,...,n\}$ is valid. Thus, we have also $f \notin Pol_2\alpha$. With the aid of Theorem 3.3.1, this implies $[f] = P_2$. ■

Lemma 7.1.2 *Let $f^n \in P_k$ be a Słupecki-function which preserves a certain h-ary relation $\varrho \in R_k$. Then*

(a) $(\exists s \in S_k \backslash \{e_1^1\} : \ \varrho = \{(a, s(a)) \mid a \in E_k\}) \Longrightarrow (\exists \gamma \in \mathfrak{S}_k \cup \mathfrak{C}_k^1 : \ f \in Pol\gamma)$;
(or:
$(\exists s \in S_k \backslash \{e_1^1\} : \ f(x_1,...,x_n) = s^{-1}(f(s(x_1),...,s(x_n))) \ ) \Longrightarrow$
$\exists \gamma \in \mathfrak{S}_k \cup \mathfrak{C}_k^1 : \ f \in Pol\gamma$;

(b) $\varrho \in \mathfrak{M}_k \Longrightarrow \exists \gamma \in \mathfrak{C}_k^1 : \ f \in Pol\gamma$;

(c) $(2 \le h \le k-1 \wedge \varrho \in \mathfrak{C}_k^h) \Longrightarrow \exists \gamma \in \mathfrak{C}_k^1 : f \in Pol\gamma$;

(d) $(k = p^m \ (p \ prime, \ m \ge 1) \wedge \varrho \in \mathfrak{L}_k) \Longrightarrow \exists \gamma \in \mathfrak{U}_k \cup \mathfrak{S}_k : \ f \in Pol\gamma$;

(e) $\varrho \in \mathfrak{B}_k \Longrightarrow \exists \gamma \in \mathfrak{U}_k \cup \mathfrak{S}_k : \ f \in Pol\gamma$.

Proof. (a): In the proof of Lemma 6.1.3, we had shown that, from the relation

$$\varrho_s := \{(a, s(a)) \mid a \in E_k\} \ (s \in S_k \backslash \{e_1^1\}),$$

either a relation of $\mathfrak{C}_k^1$ (in the case $\varrho_s \cap \iota_k^2 \neq \emptyset$) or a relation of $\mathfrak{S}_k$ (in the case $\varrho_s \cap \iota_k^2 = \emptyset$) are derivable, whereby we have $f \in Pol\gamma$ for a certain $\gamma \in \mathfrak{C}_k^1 \cup \mathfrak{S}_k$.
(b): Let $o \in E_k$ be the least element in respect to $\varrho \in \mathfrak{M}_k$. Because of $Im(f) = E_k$ and $f \in Pol\varrho$, $f(o,...,o) = o$ holds. Thus we have $f \in Pol\{o\}$.
(c): Let C be the set of all central elements of the relation $\varrho \in \mathfrak{C}_k^h$, $2 \le h \le k-1$. Then, because of $(a_0,...,a_{h-2}, c) \in \varrho$ for every $c \in C$ and arbitrary $a_0,...,a_{h-2} \in E_k$, it results from $f \in Pol\varrho$ and $Im(f) = E_k$ that $f(c_1,...,c_n) \in$

C is valid for all $c_1, ..., c_n \in C$. Consequently, f preserves a relation of $\mathfrak{C}_k^1$.
(d): Let $k = p^m$ (p prime, $m \geq 1$) and let f^n be a quasi-linear function. By Lemma 5.2.4.3, f is isomorphic to a certain function $\hat{f}^n \in P_A$, where $A := E_p^m$. One can describe this function $\hat{f}^n$ as follows:

$$\hat{f}(x_1, ..., x_n) = a + x_1 \cdot A_1 + ... + x_n \cdot A_n,$$

where $a, x_1, ..., x_m \in E_p^m$, $A_1, ..., A_n$ are some matrices of the type (m, m) over E_p and $+$ and $\cdot$ are the usual matrix operations over the field $(E_p; + \ (mod\ p), \cdot \ (mod\ p))$.
We form $\hat{f}_1(x) := \hat{f}(x, x, ..., x) = a + x \cdot A$, where $A := A_1 + ... + A_n$, and distinguish two cases for A:
Case 1: $det(A - I) \neq 0$. [1]
Then there exists $b \in E_p^m$ with $b \cdot (A - I) = -a$ or $a + b \cdot A = b$. Consequently, we have $\hat{f}(b, ..., b) = b$ and f preserves a relation of $\mathfrak{C}_k^1$.
Case 2: $det(A - I) = 0$.
In this case, there is a $d \in E_p^m \backslash \{o\}$, $o := (0, ..., 0)$, with $d \cdot (A - I) = o$ and it holds:

$$\begin{aligned} &\hat{f}(x_1 + d, x_2 + d, ..., x_n + d) - d \\ &= a + (x_1 + d) \cdot A_1 + ... + (x_n + d) \cdot A_n - d \\ &= a + \underbrace{d \cdot \underbrace{(A_1 + ... + A_n)}_{=A} - d}_{o} + x_1 \cdot A_1 + ... x_n \cdot A_n \\ &= \hat{f}(x_1, ..., x_n). \end{aligned}$$

Thus, $\hat{f}$ preserves the relation

$$\{(x, s(x)) \mid x \in E_p^m\} \text{ with } s(x) := x + d.$$

With the help of (a), this implies $f \in Pol\gamma$ for certain $\gamma \in \mathfrak{S}_k$.
(e): Let $\varrho \in \mathfrak{B}_k$. Then, by definition, there exists a mapping q from E_k onto E_{h^m} ($m \geq 1$) with

$$(a_0, ..., a_{h-1}) \in \varrho \implies (q(a_0), ..., q(a_{h-1})) \in \xi_m,$$

where ξ_m is the h-ary elementary relation on E_{h^m}. Because of Theorem 5.2.6.1, one can describe $f \in Pol\varrho$ with $Im(f) = E_k$ in the form (5.14), where g_f has the form

$$g_f(x_1, ..., x_n) = \sum_{i=0}^{m-1} s_i((q(x_{j_i}))^{\alpha(i)}) \cdot h^i \tag{7.3}$$

with $\{j_0, ..., j_{m-1}\} \subseteq \{1, 2, ..., n\}$, $s_0, s_1, ..., s_{m-1}$ are some permutations on E_k and α is a certain mapping from E_m into E_m. By means of (7.3), one can prove that the function f preserves the equivalence relation

[1] I denotes the identity matrix of the type (m, m).

$$\{(a,b) \in E_k^2 \mid q(a) = q(b)\}. \tag{7.4}$$

Consequently, f belongs to $Pol\gamma$ for a certain $\gamma \in \mathfrak{U}_k$, if q is not bijective. We still have to examine the case that q is bijective. W.l.o.g. let q be the identical mapping in the following. Thus our function f has the form

$$f(x_1, ..., x_n) = \sum_{i=0}^{m-1} s_i((x_{j_i})^{(\alpha(i))}) \cdot h^i \tag{7.5}$$

and we have $k = h^m$ and $m \geq 2$ (because of $f \notin Pol_k \iota_k^k$). For the mapping α, two cases are possible:

Case 1: α is not a permutation.

In this case, there is a proper subset A of E_k, which is preserved from α. Then, by means of (7.5), it is easy to check that f preserves the non-trivial equivalence relation

$$\{(a,b) \in E_k^2 \mid \forall i \in A: \; a^{(i)} = b^{(i)}\};$$

hence $f \in Pol\gamma$ holds for a certain $\gamma \in \mathfrak{U}_k$.

Case 2: α is a permutation.

We form $f_1(x) := f(x, x, ..., x)$ and set

$$\alpha^1 x := \alpha(x), \; \alpha^i x := \alpha(\alpha^{i-1} x), \; i \geq 2.$$

Then we have by assumption and (7.5):

$$g(x) := \underbrace{(f_1 \star f_1 \star ... \star f_1)}_{m \text{ times}} = \sum_{i=0}^{m-1} s_i(s_{\alpha i}(...(s_{\alpha^{m-1} i}(x^{(\alpha^m i)}))...)) \cdot h^i,$$

where $x^{(\alpha^m i)} = x^{(i)}$. The function g has the properties

$$g^{-1}(x) = \sum_{i=1}^{m-1} s^{-1}_{\alpha^{m-1} i}(s^{-1}_{\alpha^{m-2} i}(...(s_i^{-1}(x^{(i)}))...)) \cdot h^i$$

and

$$g^{-1}(f(g(x_1), g(x_2), ..., g(x_n))) = f(x_1, x_2, ..., x_n).$$

Thus, if $g \neq e_1^1$ then by (a) f preserves a certain relation of $\mathfrak{S}_k$ and (e) was proven. If $g = e_1^1$ then for every $a \in E_k$ the function f preserves the element

$$\sum_{i=0}^{m-1} s_{\alpha i}(s_{\alpha^2 i}(...(s_{\alpha^{m-1} i}(a^{(i)}))...)) \cdot h^i$$

of E_k, whereby f preserves a relation of $\mathfrak{C}_k^1$. ■

The following generalization of Theorem 7.1.1 was found by G. Rousseau.

Theorem 7.1.3 (Characterization Theorem for Sheffer-Functions; *[Rou 67]*)
Let $f^n \in P_k$. Then

$$[f] = P_k \iff \forall \varrho \in \mathfrak{C}_k^1 \cup \mathfrak{U}_k \cup \mathfrak{S}_k^{\sim} : f \notin Pol\varrho. \quad (7.6)$$

Proof. Obviously, "$\Longrightarrow$" is valid.
"$\Longleftarrow$": Let $f \in P_k$ be a function that does not preserve any relation of the set $\mathfrak{C}_k^1 \cup \mathfrak{S}_k^{\sim} \cup \mathfrak{U}_k$. Then f is a Słupecki-function and, because of Theorem 6.1, it is sufficient to show for the proof of "$\Longleftarrow$" of (7.6) that

$$\forall \varrho \in \mathfrak{M}_k \cup \mathfrak{L}_k \cup \mathfrak{B}_k \cup \bigcup_{h=2}^{k-1} \mathfrak{C}_k^h : f \notin Pol_k\varrho \quad (7.7)$$

holds. From the assumption that (7.7) is false, contradictions to the assumption result with the aid of the statements (b) - (e) from Lemma 7.1.2 in all cases to be scrutinized for ϱ. ■

Theorem 7.1.4 (*[Sch 69]*) *One can not reduce the conditions from Theorem 7.1.3; that is, these conditions are independent of each other.*

Proof. It was shown in [Sch 69] by P. Schofield that every class of the form $Pol_k\varrho$ with $\varrho \in \mathfrak{C}_k^1 \cup \mathfrak{S}_k \cup \mathfrak{U}_k$ can be generated by one function (see Section 11.2). With the help of this property, one can easily prove that our assertion is valid:
Assume there is a relation $\alpha \in \mathfrak{C}_k^1 \cup \mathfrak{S}_k^{\sim} \cup \mathfrak{U}_k$ with

$$[f] = P_k \iff \forall \varrho \in (\mathfrak{C}_k^1 \cup \mathfrak{U}_k \cup \mathfrak{S}_k^{\sim}) \setminus \{\alpha\} : f \notin Pol\varrho.$$

Furthermore, let f be a function of A with $[f] = A$. Since A is not complete in P_k, there exists a maximal class $B \neq A$ of P_k with $f \in B$. Then, f belongs to $A \cap B$ and we get $A = [f] \subseteq A \cap B \subseteq B$. Because of $A \neq B$, however, this is contradictory to the P_k-maximality of A. ■

The literary about concrete Sheffer-functions is very extensive. One finds a bibliography for this purpose for example in [Pös-K 79], p. 135, and in [Ros 77] (or [Ros 84]). The following theorem gives only one of the classical examples for Sheffer-functions, which was published in 1936.

Theorem 7.1.5 (Theorem of Webb; *[Web 36]*)
The function

$$f(x, y) := min(x, y) + 1 \ (mod\ k)$$

is a Sheffer-function for P_k.

Proof. A possible proof is the construction of a known generating system for P_k over the function f, which one can look up [Pös-K 79]. We prove the assertion here with the aid of (7.6). Superpositions over f are the functions $f_1 := \Delta f$ with $f_1(x) = x + 1 \ (mod\ k)$ and $min := \underbrace{f_1 \star f_1 \star ... \star f_1}_{k-1 \text{ times}} \star f$.

Obviously, the function f_1 does not preserve any relation $\varrho \in \mathfrak{C}_k^1$. Thus we have $f \notin Pol_k\varrho$ for all $\varrho \in \mathfrak{C}_k^1$.

Let $\varrho \in \mathfrak{S}_k$. Then there are $a, b \in E_k$ with $\{(0,a),(b,0)\} \subseteq \varrho$ and $f\begin{pmatrix} 0 & b \\ a & 0 \end{pmatrix} = \begin{pmatrix} 1 \\ 1 \end{pmatrix}$. Consequently, f does not preserve any relation $\varrho \in \mathfrak{S}_k$.

Since $f_1 \notin Pol_k\varrho$ is false for every $\varrho \in \mathfrak{U}_k$, we assume that $f_1 \in Pol_k\varrho$ for certain $\varrho \in \mathfrak{U}_k$ is valid. Then, the equivalence classes of ϱ are equipotent, and there exist two equivalence classes $[a]_\varrho$ and $[a+1]_\varrho$ of ϱ with the following properties:

$$[a]_\varrho \neq [a+1]_\varrho,\ b \in [a]_\varrho,\ a < b,\ k-1 \notin [a]_\varrho.$$

Then (a,b), $(a+1, b+1) \in \varrho$ and $min\begin{pmatrix} a & b+1 \\ b & a+1 \end{pmatrix} = \begin{pmatrix} a \\ a+1 \end{pmatrix} \notin \varrho$. Consequently, the function f does not preserve any relation of $\mathfrak{U}_k$. ■

By means of Theorems 7.1.3, it is already recognizable that the number of Sheffer-functions is quite great. The next theorem summarizes some statements about concrete numbers and certain convergence.

Theorem 7.1.6 *Let $\varphi_n(k)$ be the number of all n-ary Sheffer-functions of P_k. Then*

(a) $\varphi_n(2) = 2^{2^n-2} - 2^{2^{n-1}-1}$;

(b) $\lim_{n\to\infty} \frac{\varphi_n(2)}{2^{2^n}} = \frac{1}{4}$;

(c) $\varphi_2(3) = 3774$ *([Mar 54]);*

(d) $\lim_{k\to\infty} \frac{\varphi_n(k)}{k^{k^n}} = \lim_{n\to\infty} \frac{\varphi_n(k)}{k^{k^n}} = \frac{1}{e}$ *for* $k \geq 3$, *[Bai 67a].*

(e) For a large k, given a randomly selected f one can decide whether $f \in P_k^n$, $n \geq 2$, is a Sheffer-function with probability almost 1 by testing the condition $f(x, x, ..., x) \neq x$ for every $x \in E_k$ only; [Dav 68].

Proof. (a) and (b) are easy to check. One can find proofs for (c) –(e) in [Mar 54] and [Dav 68]. ■

7.2 A Completeness Criterion for Surjective Functions

For the case that one only considers subsets $A \subseteq P_k$ of functions with the range of values E_k (i.e., surjective functions), our general completeness criterion (7.1) can be simplified. It holds:

Theorem 7.2.1 (*[Ros 70c], [Ros 75]*)
For $k = h^m$ with $h \geq 3$ and $m \geq 1$ let

$$\mathfrak{B}_k^* := \{\varrho \in \mathfrak{B}_k \mid \exists n : \ \varrho \text{ is isomorphic to } \xi_n\}.$$

Then, for arbitrary $A \subseteq P_k[k]$ we have:

$$[A] = P_k \iff \forall \varrho \in \mathfrak{C}_k^1 \cup \mathfrak{S}_k \cup \mathfrak{U}_k \cup \mathfrak{L}_k \cup \mathfrak{B}_k^* : \ A \not\subseteq Pol_k \varrho. \tag{7.8}$$

Proof. "$\Longrightarrow$" is trivial.
"$\Longleftarrow$": Our theorem follows from Theorem 6.1 if we can show that A is not a subset of clones $Pol_k \varrho$ for every $\varrho \in \mathfrak{M}_k \cup (\bigcup_{h=2}^{k-1} \mathfrak{C}_k^h) \cup (\mathfrak{B}_k \backslash \mathfrak{B}_k^*)$. Suppose it holds:

$$\exists \varrho \in \mathfrak{M}_k \cup (\bigcup_{h=2}^{k-1} \mathfrak{C}_k^h) \cup (\mathfrak{B}_k \backslash \mathfrak{B}_k^*) : \ A \subseteq Pol_k \varrho. \tag{7.9}$$

We distinguish three cases:
Case 1: $\varrho \in \mathfrak{M}_k$.
Let o be the least element of E_k in respect to ϱ. Since each function of A has the range E_k, we have then that A is a subset of $Pol_k\{o\}$, in contradiction to our assumption. Thus Case 1 is not possible.
Case 2: $\varrho \in \mathfrak{C}_k^h$, $2 \leq h \leq k-1$.
In analog mode to the proof of Lemma 7.1.2 (c), one can show that every function of A preserves the set of the central elements of the relation ϱ in this case. However, this also contradicts our assumption.
Case 3: $\varrho \in \mathfrak{B}_k \backslash \mathfrak{B}_k^*$.
One can describe the functions of A, then, by means of formulas of the form (5.14), where g_f has the form (7.3); hence every function of A preserves the nontrivial equivalence relation (7.4). Consequently, Case 3 is not possible and, with that, the assumption (7.9) is wrong. ■

7.3 Fundamental Sets

By Chapter 4, the set $P_k^1 \cup \{f\}$, where f is a Słupecki-function, is complete in P_k. Consequently, a subset A of P_k is complete in P_k iff $P_k^1 \cup \{f\} \subseteq [A]$ holds. One can now attempt to improve this completeness criterion in such a way that the set P_k^1 is replaced by certain subsets of P_k^1.
Definitions A subset A of P_k^1 with

$$\forall f \in P_k \backslash ([P_k^1] \cup P_k(k-1)) :{}^2 \ [A \cup \{f\}] = P_k$$

[2] E.i., f is a Słupecki-function.

is called **fundamental set** in P_k. If in particular, A is a group (or a semigroup) in respect to the operation $\star$, then A is also called **fundamental group** (or **fundamental semigroup**), respectively.

We need the following definition to describe some fundamental sets:

Definition A semigroup $H \subseteq P_k^1$ is called t**-fold transitive**, $t \in \{1, 2, ..., k\}$, if for arbitrary pairwise distinct elements $a_1, ..., a_t \in E_k$ and for arbitrary pairwise distinct elements $b_1, ..., b_t \in E_k$ there exists an $h \in H$ with $h(a_i) = b_i$ for all $i \in \{1, 2, ..., t\}$.

One finds some **examples** of fundamental sets in the following theorem.

Theorem 7.3.1

(1) The following sets are fundamental semigroups:
   (a) $P_k^1 \backslash P_k^1[k]$;
   (b) a set $H \subseteq P_k^1$ *with* $k \geq 5$, *which is* $(k-1)$**-fold transitive** *([Mal 67]);*
   (c) a set $H \subseteq P_k^1$ *with* $k \in \{3, 4\}$ *and the property that* H *and also* $H \cap P_k^1(k-1)$ *are* $(k-1)$*-fold transitive ([Mal 67]).*

(2) There are fundamental groups in P_k *only for* $k \geq 5$ *and examples for fundamental groups are*
   (a) $P_k^1[k]$;
   (b) the set of all functions of P_k^1, *which are even permutations on* E_k.[3]

Proof. The statements of the theorem are easy consequences of Rosenberg's completeness criterion 6.1 (see [Pös-K 79], p. 132 and p. 135).
One can find proof that does not use Theorem 6.1, for the above statements in [Jab-L 80], 1.2.4–1.2.6. ■

[3] A permutation $s \in S_n$ is called **even**, if the inversion number $I(s)$, i.e., the number of all pairs (i, j) with $i < j$ and $s(i) > s(j)$, is even.

8

Some Properties of the Lattice $\mathbb{L}_k$

It is the aim of this first section above $\mathbb{L}_k$ to basically elaborate on the differences between $\mathbb{L}_2$, which we already completely determined in Chapter 3, and $\mathbb{L}_k$ for $k \geq 3$. In particular, we deal with cardinality statements (both about $\mathbb{L}_k$, than also about chains and antichains in $\mathbb{L}_k$) and with the embedding of $\mathbb{L}_k$ into $\mathbb{L}_{k'}$. Furthermore, the question, which cardinalities have the lattices $\mathbb{L}_k^{\downarrow}(A)$ for maximal classes A of P_k, is clarified. In the last section of chapter, there are remarks about "strategies" during the determination of "manageable" sublattices of $\mathbb{L}_k$. In addition, two examples of theorems are proven as one can determine certain subsets of $\mathbb{L}_k$.
Notice that one can find further properties of $\mathbb{L}_k$, for instance, in [Bul-K-S-S 95], [Bul-K-S-S-S 2001], [Bul 99a,b], and [Bul 2001].

8.1 Cardinality Statements

We begin with proving the existence of a subclass $C \subseteq P_k$ for $k \geq 3$, which has an infinite basis. This example was published by J. I. Janov and A. A. Mucnik in 1959.
Let

$$g^n_{I,J}(x_1, ..., x_n) := \begin{cases} 1 \text{ if} & \exists i \in I : (x_i \in \{1,2\} \wedge \\ & (\forall j \in J \cup (I \backslash \{i\}) : x_j = 2)), \\ 0 \text{ otherwise,} & \end{cases} \tag{8.1}$$

where I and J are certain disjoint subsets of $\{1, 2, ..., n\}$.
We also denote the functions $g^n_{\{1,2,...,n\},\emptyset}$ with f_n. Except the subsequently indicated tuples, the function takes only the value 0:

$$f_n \begin{pmatrix} 1 & 2 & 2 & \dots & 2 & 2 \\ 2 & 1 & 2 & \dots & 2 & 2 \\ \dots & \dots & \dots & \dots & \dots & \dots \\ 2 & 2 & 2 & \dots & 1 & 2 \\ 2 & 2 & 2 & \dots & 2 & 1 \\ 2 & 2 & 2 & \dots & 2 & 2 \end{pmatrix} = \begin{pmatrix} 1 \\ 1 \\ \dots \\ 1 \\ 1 \\ 1 \end{pmatrix}. \tag{8.2}$$

Let

$$B := \{f_i \mid i \in \mathbb{N}\} \tag{8.3}$$

and let C be the set

$$\bigcup_{n \geq 1} \{g^n_{I,J} \mid I, J \subseteq \{1, 2, ..., n\} \ \wedge \ I \cap J = \emptyset\}. \tag{8.4}$$

Lemma 8.1.1 *For the above-defined sets C and B is valid:*
(a) C is a subclass of P_k.
(b) $\forall i \in \mathbb{N}: \ f_i \notin [B \backslash \{f_i\}]$.
(c) B is an infinite basis of C.

Proof. Obviously, set C is closed in respect to the operations $\zeta, \tau, \delta, \nabla$. The completeness of C in respect to $\star$ follows from

$$g^n_{I,J} \star g^{n'}_{I',J'} = \begin{cases} g^{n+n'-1}_{\{a+n'-1 \mid a \in I\}, \{a+n'-1 \mid a \in J\}} & \text{if } 1 \notin I \cup J \ \wedge \ n \geq 2, \\ c_0^{n+n'-1} = g^{n+n'-1}_{\emptyset,\emptyset} & \text{if } 1 \in J \ \vee \ (n = 1 \ \wedge \ I = \emptyset), \\ g^{n+n'-1}_{I', J' \cup \{a+n'-1 \mid a \in J \cup (I \backslash \{1\})\}} & \text{if } 1 \in I \ \wedge \ n \geq 1. \end{cases}$$

In addition, it results from the above formulas that set B is a generating system for C that does not continue being reducible, because of

$$f_n = g^n_{I,\emptyset} \notin [\{g^{n'}_{I',J'} \in C \mid |I'| \neq |I|\}].$$

■

Theorem 8.1.2 *Let N be a countable set and let $\mathbf{L}(\mathbf{N}) := (\mathfrak{P}(N); \subseteq)$. Then, for $k \geq 3$, one can embed the lattice $\mathbf{L}(\mathbf{N})$ into the lattice $\mathbb{L}_k$; i.e., there exists a bijective mapping α from $\mathfrak{P}(N)$ in P_k, for which it holds:*

$$\forall A, B \in \mathfrak{P}(N): \ A \subseteq B \Longrightarrow \alpha(A) \subseteq \alpha(B).$$

In particular, for all $k \geq k' \geq 2$, the lattice $\mathbb{L}_{k'}$ one can embed into the lattice in $\mathbb{L}_k$.

Proof. W.l.o.g. let $N := \mathbb{N}$. Then, for every subset T of $\mathbb{N}$, one can form a subclass $C_T := [\{f_i \mid i \in T\}]$ of the class C defined in (8.4). Because of Lemma 8.1.1, we have: $T \neq T' \Longleftrightarrow C_T \neq C_{T'}$. Since $T \subseteq T' \Longrightarrow C_T \subseteq C_{T'}$, we have that $\alpha : \mathfrak{P}(\mathbb{N}) \longrightarrow P_k$, $T \longmapsto C_T$ is an embedding of $L(\mathbb{N})$ into L_k. ■

With the aid of the above theorem we can prove the statements of the following two theorems easily. In particular, the theorem shows that the crucial importance, which has the 2 in many fields of the algebra, is also provable for the function algebras. So, P_2 has countable-many subclasses, however, the set P_k for $k \geq 3$ already continuum-many subclasses:

Theorem 8.1.3 (Theorem About the Cardinality of $\mathbb{L}_k(P_k)$) *It holds:*

(a) $|\mathbb{L}_k(P_2)| = \aleph_0$. *([Pos 41])*

(b) $|\mathbb{L}_k(P_k)| = \mathfrak{c}$ *for* $k \geq 3$. *([Jan-M 59])*

Proof. (a) was proven in Chapter 3.
(b): Obviously, the set P_k is countable infinite. As is generally known, the set of all subsets of P_k has the cardinality of the continuum. Consequently, there are at least continuum-many subclasses of P_k. Thus, (b) follows from Theorem 8.1.2, since $|\mathfrak{P}(\mathbb{N})| = \mathfrak{c}$ (see e.g. [Lau 2004], volume 1). ∎

Theorem 8.1.4 *([Pös-K 79])*
For $k \geq 3$ *there are in* $\mathbb{L}_k$ *both chains and antichains of the cardinality of the continuum.*

Proof. Because of Theorem 8.1.2, it suffices to prove that there are a chain and an antichain with the cardinality of the continuum in $\mathbf{L}(\mathbf{N})$, $|N| = \aleph_0$. Let $\mathbb{Q}$ be the set of all rational numbers and $\mathbb{R}$ be the set of all real numbers. For arbitrary $s \in \mathbb{R}$ we put $M_s := \{q \in \mathbb{Q} \mid q \leq s\}$. Then, the set $\{M_s \mid s \in \mathbb{R}\}$ forms a chain of the cardinality $\mathfrak{c}$ in the lattice $\mathbf{L}(\mathbb{Q})$. For the purpose of constructing an uncountable antichain, let $H := \{(m,i) \mid m \in \mathbb{N} \wedge i \in \{0,1\}\}$ and $H_I := \{(m,0) \mid m \in I\} \cup \{(m,1) \mid m \notin I\}$. Obviously, $\{H_I \mid I \subseteq \mathbb{N}\}$ is an antichain of the cardinality $\mathfrak{c}$ in the lattice $\mathbf{L}(\mathbf{H})$ ($|H| = \aleph_0$). ∎

Because of the above theorems, the concrete determination of the elements of the lattice $\mathbb{L}_k$ seems for $k \geq 3$ to be a hopeless task. Nevertheless, one can attempt to determine certain "manageable parts" of $\mathbb{L}_k$ or one can try to get a general idea of the position of the "not manageable parts" of the lattice $\mathbb{L}_k$. The following theorem is useful for proving that certain sublattices of $\mathbb{L}_k$ are countable, which is a generalization of a theorem that was proven by I. A. Mal'tsev (see [Mal 73]).

Theorem 8.1.5 (Countability Criterion; *[Lau 86]***)**
Let A be a subclass of P_k (or an algebra with countable many elements). Furthermore, there exists an order relation $\leq$ on A with the following three properties:

(a) $f \leq g \implies [f] \subseteq [g]$,
(b) every chain is well-ordered (in respect to $\leq$*), i.e., every chain has a minimal element (in respect to* $\leq$*),*
(c) every antichain (in respect to $\leq$*) has only finitely many elements.*

Then, A has at most only countable many different subclasses (or the algebra A has at most only countable many different subalgebras).

Proof. Let B be a closed subset of A and let G be the set of all minimal elements of $A \backslash B$ (in respect to $\leq$). First, we prove that G determines the set B unambiguously. Let f be an arbitrary function of A. The following cases are possible:
Case 1: There exists a function $g \in G$ with $f \leq g$ and $f \neq g$.
Then $f \in B$, since for $f \notin B$ the function g is not a minimal element of $A \backslash B$.
Case 2: There exists a function $g \in G$ with $g \leq f$.
Because of $g \leq f$ and by assumption (a), we have that g is a superposition over f. Thus $f \in A \backslash B$, since $g \in A \backslash B$.
Case 3: Every function of G is not comparable with the f (in respect to $\leq$), or $G = \emptyset$.
Because of the assumption (b) and by Zorn's Lemma, this case is only possible for $f \in B$.
Consequently, G determines the class B unambiguously.
The set G is an antichain (in respect to $\leq$) and a finite set because of assumption (c). Every subclass B of A is also definable by a finite subset G of A. Obviously, A contains countable many elements. Consequently, there are only countable many possibilities for G and, therefore, there are only countable many closed subsets of A. ■

Let $\circ$ be an operation on E_k with the properties

$$
\begin{aligned}
&a) \quad \forall\, a, b, c \in E_k : \; a \circ (b \circ c) = (a \circ b) \circ c, \\
&b) \quad \forall\, a, b \in E_k : \; a \circ b = b \circ a, \\
&c) \quad \exists\, r \geq 1 : \; (\forall\, a \in E_k : \; \underbrace{a \circ a \circ \ldots \circ a}_{(1+r) times} = a).
\end{aligned}
$$

The properties a) - c) are e.g. fulfills, if $(E_k; \circ)$ is an Abelean group or $x \circ y := \max_\omega(x, y)$ or $x \circ y := \min_\omega(x, y)$ (ω is a partial order relation, for which exists max or min).
With the help of $\circ$ one can define a subset $K(m, \circ)$ of P_k for an $m \in \mathbb{N}$ as follows:

$$K(m, \circ) := \bigcup_{n \geq 1} K^n(m, \circ), \tag{8.5}$$

where

$$f^n \in K^n(m,\circ) :\Longleftrightarrow$$
$$\exists\ f_0^1, f_1^{n_1}, \ldots, f_t^{n_t} \in \bigcup_{i=1}^m P_k^i :$$
$$f(x_1,\ldots,x_n) =$$
$$f_0(f_1(x_1,...,x_{n_1}) \circ f_2(x_{n_1+1},...,x_{n_1+n_2}) \circ \ldots \circ f_n(x_{n_1+\ldots+n_{t-1}+1},...,x_n)) \tag{8.6}$$

In general, the set K_0 is not closed. K_0 contains, however, through suitable choice of m and $\circ$ a line of interesting subclasses of P_k (as for example the set of linear or quasi-linear functions) and it is valid:

Theorem 8.1.6 *A subclass A of P_k, for which exist an $m \in \mathbb{N}$ and an operation $\circ$ which the above properties a) - c) fulfills, with $A \subseteq K(m,\circ)$, has only countable many subclasses.*

Proof. We will define a partial order relation $\leq$ on $K(m,\circ)$, which fulfills the conditions (a)–(c) from the Theorem 8.1.5, whereby our theorem follows from Theorem 8.1.5.
Every function of $f^n \in K(m,\circ)$ with the form (8.6) is well-defined by tuples $\varphi(f) := (f_0, f_1, ..., f_t)$. Two functions f_1, f_2 with $\varphi_i(f_i) := (f_{i,0}, f_{i,1}, ..., f_{i,t_i})$, $i = 1,2$, are called **ϱ-equivalent**, iff these functions fulfill the following three conditions:
1) $f_{1,0} = f_{2,0}$;
2) $\{f_{1,1}, ..., f_{1,t_1}\} = \{f_{2,1}, ..., f_{2,t_2}\}$;
3) If f occurs in $(f_{1,1}, ..., f_{1,t_1})$ exactly q times, then there exists a certain integer s, so that f occurs in $(f_{2,1}, ..., f_{2,t_2})$ exactly $(s + r \cdot q)$ times, where r denotes the smallest integer for which $\circ$ fulfills die condition c).
Obviously, the factor set $K(m,\circ)_{/\varrho}$ is finite. Set $\{K_1, K_2, ..., K_u\} := K(m,\circ)_{/\varrho}$.
On the sets K_1, K_2, ..., K_u one can define a partial order relation $\leq$ as follows: For functions f_1, $f_2 \in K_i$, $i \in \{1,2,...,u\}$, with $\varphi(f_i) = (f_{i,0}, ..., f_{i,t_i})$ we write $f_1 \leq f_2$ if and only if for every f, which occurs in $(f_{1,1}, ..., f_{1,t_1})$ $j-1$ times and occurs in $(f_{2,1}, ..., f_{2,t_2})$ j_2 times, it holds $j_1 \leq j_2$.
This partial order relation satisfies the condition (a) of Theorem 8.1.5, since one can receive the function f from g with the aid of the operations ζ, τ, Δ by the changing of the denotation of certain r variables and by the identifying of groups of variables with $r \cdot q + 1$ $(q \geq 1)$ elements because of the property (c) of $\circ$.
Obviously, the property (b) of Theorem 8.1.5 is also valid. Thus it remains to be seen that every antichain in the set $K(m,\circ)$ in respect to $\leq$ has only finite-many elements.
On $\bigcup_{i=1}^m P_k^1$ one can define a total order. For $\varphi(f) = (f_0, ..., f_t)$, we arrange (concerning this order) that f_{i_0} is the smallest element of the set $\{f_1, ..., f_t\}$; f_{i_1} is the smallest element of the set $\{f_1, ..., f_t\} \backslash \{f_{i_0}\}$; ... and f_{i_s} is the greatest element of $\{f_1, ..., f_t\}$. If f_{i_j} occurs in $(f_1, ..., f_t)$ exactly b_j times, $j = 0,1,...,s$, then let $\alpha(f) := (b_0, b_1, ..., b_s)$. Further, let $D_i := \{\alpha(f) \mid f \in K_i\}$, $i = 1,2,...,u$. One can define a partial order relation on D_i. We set $\alpha(f_i) \leq \alpha(f_2)$ iff $f_1 \leq f_2$.
The partial ordered sets K_i and D_i are isomorphic, $i = 1,2,...,u$. Conse-

quently, for the end of our proof, it suffices to show that an arbitrary antichain C of D_i is a finite set:
Obviously, if C consists only of tuples of the length 1, then C is a finite set. Suppose our assertion is proven for tuples of the length s and all tuples of C have the length $s+1$. We select a certain tuple $(b_0..., b_s)$ in C and decompose the set C to this tuple in finite many classes as follows: Let the first class be the set of all tuples whose first coordinate is 1. All functions with the first coordinate 2 belong into the second class, ..., all functions with the first coordinate $b_0 - 1$ belong into the $(b_0 - 1)$-th class. All the tuples whose second coordinate 1 is and which are not tuples of the already defined classes, belong in the b_0-th class, and so forth. Since the tuples of C are not comparable (i.e., if $(b_1^1, ..., b_s^1) \neq (b_1^2, ..., b_s^2)$ are elements of C then there exist some $i \neq j$ with $b_i^1 < b_i^2$ and $b_j^1 > b_j^2$), all elements of $C\backslash\{(b_0..., b_s)\}$ are in the above-described classes contained. By assumption, every class has only finite many elements. Hence C is a finite set. ■

8.2 On the Cardinalities of Maximal Sublattices of $\mathbb{L}_k$

In this section, we determine the cardinality of the lattice $\mathbb{L}_k^{\downarrow}(A)$, where A is an arbitrary maximal class of P_k.
We start with a lemma that is a consequence of [Dem-H 83], Lemma 1.

Lemma 8.2.1 *Let $k = 2 \cdot l$ with $l \geq 2$, let*

$$s := (0\ 1)(2\ 3)(4\ 5)....(2l-2\ 2l-1)$$

be the cycle representation of a permutation s on E_{2l}, $\varrho_s := \{(x, s(x)) \mid x \in E_{2l}\}$ and $\widetilde{y}_i^{(n)} := (0, 0, ..., 0, \underbrace{1}_{i}, 0, ..., 0)$ for $1 \leq i \leq n$. Furthermore, let t_n $(n \geq 4)$ be an n-ary function of $Pol_k \varrho_s$ with the following properties:

(α) $(\forall i \geq 1 : \ t_n(\widetilde{y}_i^{(n)}) = 1) \ \wedge \ t_n(0, 1, 1, ..., 1) = 0$;

(β) $\forall \mathbf{x} \in E_4^n \backslash \{\widetilde{y}_1^{(n)}, ..., \widetilde{y}_n^{(n)}\} : \ x_1 \in \{0, 2\} \implies t_n(\mathbf{x}) = 2$;

(γ) $\forall \mathbf{x} \in E_k^n \backslash E_4^n : \ x_1 \in \{0, 2, 4, ..., 2l-2\} \implies t_n(x_1, ..., x_n) = x_1$.

Let ϱ_n be the n-ary relation

$$\{\widetilde{y}_i^{(n)} \mid 1 \leq i \leq n\} \cup (E_4^n \backslash \{0, 1\}^n).$$

Then

(a) The functions t_n with $n \geq 4$ are unique determined by the conditions (α)–(γ) and belong to $Pol_k \varrho_s$.
(b) $\forall n \geq 4 : \ t_n \notin Pol_k \varrho_n$.
(c) $\forall m \neq n : \ t_m \in Pol_k \varrho_n$.

Proof. (a): Because of Theorem 5.2.2.1, an arbitrary n-ary function $f^n \in Pol_k\varrho_s$ is uniquely determined by the function $f(a, x_2, ..., x_n)$ with $a \in \{0, 2, 4, ..., 2l - 2\}$. It is easy to check that the conditions (α)–(γ) determine the functions $t_n(a, x_2, ..., x_n)$ for arbitrary $a \in \{2, 4, ..., 2l - 2\}$.
(b) follows from

$$t_n \begin{pmatrix} 1 & 0 & \dots & 0 \\ 0 & 1 & \dots & 0 \\ \dots & \dots & \dots & \dots \\ 0 & 0 & \dots & 1 \end{pmatrix} = \begin{pmatrix} 1 \\ 1 \\ \dots \\ 1 \end{pmatrix}.$$

(c): Let $n \neq m$ and let $r_1, ..., r_n \in \varrho_n$ be arbitrary. If an r_i belong to $E_4^n \backslash \{0, 1\}^n$ then $t_m(r_1, ..., r_m) \in E_4^n \backslash \{0, 1\}^n$. If $\{r_1, ..., r_m\} \subseteq \{\widetilde{y_1}^{(n)}, ..., \widetilde{y_n}^{(n)}\}$, then one can show (when one distinguishes the cases $m < n$ and $m > n$) that $t_m(r_1, ..., r_n) \in \varrho_n$. ■

For the following lemma, we need the description of a subset T of the set

$$S_3 := Pol_3 \begin{pmatrix} 0 & 1 & 2 \\ 1 & 2 & 0 \end{pmatrix},$$

which is the only maximal class of type $\mathfrak{S}$ for $k = 3$.
A function f^n belongs to T if and only if f^n fulfills the following four conditions:

1) $f \in S_3 \cap Pol_3\{0, 1\}$.

2) The restriction of f onto E_2^n (we write $pr\ f$) is a function of $T_{1,\infty} \cap M \cap T_0$.

In order to be able to formulate the third and fourth condition, we need one property of the functions $g^m \in T_{1,\infty} \cap M \cap T_0$, which is a conclusion of Chapter 3: Every such function g^m can be described through a formula above a variable alphabet and through the signs $\vee$ (disjunction) and $\wedge$ (conjunction), as follows:

$$g(x_1, ..., x_m) = x_i \vee \mathfrak{A}_1 \vee ... \vee \mathfrak{A}_t,$$

where $i \in \{1, 2, ..., m\}$ and $\mathfrak{A}_j$ ($1 \leq j \leq t$) are formulas of the form $x_{j_1} \wedge x_{j_2} \wedge ... \wedge x_{j_{l_j}}$ ("conjunctions", $l_j \geq 1$). Since $x \vee (x \wedge y) = x$ as is generally known, we can assume that the sets of the indexes of the variables of the conjunctions $\mathfrak{A}_j$ in the formula for g are not contained pairwise in each other (we speak in this case from an abbreviated DNF for g).

3) If variables, which take values of E_2 in the tuple $\mathbf{a}$, exist in every conjunction of the abbreviated DNF of $pr\ f$, then $f(\mathbf{a}) \in E_2$.

4) The abbreviated DNF of $pr\ f$ has the form

$$(pr\ f)(x_1, ..., x_n) = x_{i_1} \vee x_{i_2} \vee ... x_{i_j} \vee \mathfrak{A}_1 \vee ... \vee \mathfrak{A}_p,$$

where $\mathfrak{A}_1$, ..., $\mathfrak{A}_p$ are conjunctions, which consist of at least two variables in each case, over the variable set $\{x_1, ..., x_n\} \backslash \{x_{i_1}, ..., x_{i_j}\}$. If there are tuples

$\mathbf{a_1} := (a_{11}, ..., a_{1n})$, ..., $\mathbf{a_r} := (a_{r1}, ..., a_{rn})$ of E_3^n with $f(\mathbf{x_1}) = ... = f(\mathbf{x_r}) = 0$ and every of the conjunctions $\mathfrak{A}_1$, ..., $\mathfrak{A}_p$ contains variables $x_{i_1}, x_{i_2}, ..., x_{i_j}$, which only take values of $\{0, 1\}$ in every tuple $\mathbf{a_1}$, ..., $\mathbf{a_r}$, then there exists an $s \in \{1, ..., j\}$ with $a_{1i_s} = ... = a_{ri_s} = 0$.

The following theorem was proven by S. S. Marcenkov:

Theorem 8.2.2 *([Mar 83])*

The set T is a closed set and contains a subclass, which has an infinite basis.

Proof. One can find the proof for $[T] = T$ in [Mar 83].
In the following, we construct a subclass of the form $[\Phi]$ with

$$\Phi := \{\varphi_2, \varphi_3, ...\} \subseteq T,$$

where the functions φ_n have the property $\varphi_n \notin [\Phi \backslash \{\varphi_n\}]$ for every $n \geq 2$; i.e., $[\Phi]$ is a subclass of T with infinite basis.
For every $n \geq 2$ let φ_n be a function of T^{2n}, which has the following three properties:

$$(pr\ \varphi_n)(x_1, ..., x_n, y_1, ..., y_n) := x_1 \vee ... \vee x_n \vee y_1 \cdot y_2 \cdot ... \cdot y_n \tag{8.7}$$

$(\cdot := \wedge)$,

$$\begin{aligned} &\varphi_n(\underbrace{1, 0, ..., 0}_{n}, 2, 0, ..., 0) = \varphi_n(\underbrace{0, 1, 0, ..., 0}_{n}, 0, 2, 0, ..., 0) = ... = \\ &\varphi_n(\underbrace{0, 0, ..., 1, 0}_{n}, 0, 0, ..., 2, 0) = \varphi_n(\underbrace{0, 0, ..., 0, 1}_{n}, 0, 0, 0, ..., 2) = 0, \end{aligned} \tag{8.8}$$

and $\varphi_n(\mathbf{a}) = 1$ holds for every tuple $\mathbf{a} := (a_1, ..., a_n, a_{n+1}, ..., a_{2n})$, which fulfills the condition

$$1 \in \{a_1, ..., a_n\} \implies (a_1, ..., a_n) \in \{0, 1\}^n$$

and which is different from the tuples that occur in (8.8).
Because of condition $\varphi_n \in T \subseteq S$, many more values of the function are given through the above determination. It is easy to check that one can define the function, so on remaining tuples, that $\varphi_n \in T$ is valid. How this happens is not important for the below-indicated considerations.
Some remarks on the definition of the functions φ_n:
If the elements a_i of the tuple $\mathbf{a} := (a_1, ..., a_n, a_{n+1}, ..., a_{2n})$ take only, at most, two different values, then $\varphi_n(\mathbf{a})$ is already determines by (8.7).
Let A_n be the set of all tuples of E_3^{2n} which occur in the condition (8.8).
Let now $\mathbf{a} \in E_3^{2n} \setminus A_n$ with $|\{a_1, ..., a_{2n}\}| = 3$. Suppose $a_1 = ... = a_n$. By $\varphi_n \in S$ we can assume w.l.o.g. $a_1 = ... = a_n = 1$. Then by 3) we have $\varphi_n(\mathbf{a}) \in E_2$. Because of 4) the case $\varphi_n(\mathbf{a}) = 0$ is not possible, so that $\varphi_n(\mathbf{a}) = 1$ holds.
Let $|\{a_1, ..., a_n\}| = 2$. W.l.o.g. $\{a_1, ..., a_n\} = E_2$. Then the conditions $\mathbf{a} \notin A_n$, 3) and 4) imply $\varphi_n(\mathbf{a}) = 1$.

If $|\{a_1, ..., a_n\}| = 3$, then the definition of the set T does not supply any determination for $\varphi_n(\mathbf{a})$.
Further, we remark that $(a_1, ..., a_n) \in E_2^n$ and $\varphi_n(\mathbf{a}) = 0$ are only possible if either $\mathbf{a} \in A_n$ or $a_1 = ... = a_n = 0$ holds. From that, the correctness of the fourth property results for the functions φ_n. One easily checks that the functions φ_n also fulfill the remaining three properties from the definition of set T.

Suppose φ_n is a superposition over $\{\varphi_m \mid m \neq n\}$. We consider a formula $V(x_1, ..., x_n, y_1, ..., y_n)$ with function symbols of $\Phi \backslash \{\varphi_n\}$ and with minimal complexity (related to the arity of their functional symbols), which describes the function $v(x_1, ..., x_n, y_1, ..., y_n)$, where $v(\mathbf{x}) = 0$ for all $\mathbf{x} \in A_n$ and v fulfills the equation

$$(pr\ v)(x_1, ..., x_n, y_1, ..., y_n) = x_1 \vee ... \vee x_n \vee y_1 \cdot ... \cdot y_n.$$

One can find a such formula, since by assumption $\varphi_n(\mathbf{x}) = 0$ for all $\mathbf{x} \in A_n$ and φ_n fulfills (8.7). We remark that $v \in T$. Let

$$V(x_1, ..., x_n, y_1, ..., y_n) =$$
$$\varphi_m(f_1(x_1, ..., x_n, y_1, ..., y_n), ..., f_{2m}(x_1, ..., x_n, y_1, ..., y_n)),$$

where $\{f_1, ..., f_{2m}\} \subset [\Phi \backslash \{\varphi_n\}]$.
Suppose there exists a function $f_\alpha \in \{f_1, ..., f_{2m}\}$ with $f_\alpha(\mathbf{x}) = 0$ for every $\mathbf{x} \in A_n$. Obviously,

$$pr\ v = pr\ f_1\ \vee\ pr\ f_2\ \vee\ ... pr\ f_m\ \vee\ ...\ \vee pr\ f_{2m}. \tag{8.9}$$

Consequently, the abbreviated DNF of the function $pr\ f_t$ has the form

$$x_{i_1} \vee ... \vee x_{i_s} \vee x_{j_1} \cdot ... x_{j_t} \cdot y_{l_1} \cdot ... \cdot y_{l_u}, \tag{8.10}$$

where $\{i_1, ..., i_s\} \cap \{j_1, ..., j_t\} = \emptyset$ and $u = n$ for $t = 0$.
Let $s < n$. Let B be the set of all tuple $\mathbf{x} := (x_1, ..., x_n, y_1, ..., y_n) \in A_n$ with $x_{i_1} = x_{i_2} = ... = x_{i_s} = 1$. If $t > 0$ then $x_{j_t} = 0$ for each $\mathbf{x} \in B$. If $t = 0$ then we have $u = n$, $\{l_1, ..., l_u\} = \{1, ..., n\}$ and for every y_i of an arbitrary tuple $(x_1, ..., x_n, y_1, ..., y_n) \in B$ with $i \notin \{i_1, ..., i_s\}$ we have $y_i = 0$. Consequently, in this case, every conjunction of the abbreviated DNF, which defines the function $pr\ f_\alpha$, contains a variable that has the value 0 in every tuple of B. Since $f_\alpha(\mathbf{x}) = 0$ for every $\mathbf{x} \in B$ and $f_\alpha \in T$, this is a contradiction to the property 3), because each tuple of the set B has to stand no 0 at the i_d-th place for $1 \leq d \leq s$.
Consequently, we have: $s = n$, $\{i_1, ..., i_s\} = \{1, ..., n\}$, $t = 0$, $u = n$, $\{l_1, ..., l_u\} = \{1, ..., n\}$ and

$$pr\ f_\alpha = x_1 \vee ... \vee x_n \vee y_1 \cdot ... \cdot y_n,$$

whereby the function f_α has the properties of v, which contradict the presupposed minimality of the formula V. Consequently, there is no $i \in \{1, ..., n\}$

with $f_i(\mathbf{x}) = 0$ for every $\mathbf{x} \in A_n$. Moreover, the abbreviated DNF of the functions $pr\ f_1, ..., pr\ f_m$ do not have the form (8.10). Therefore, and by property 3), $\{f_i(\mathbf{x}) \mid \mathbf{x} \in A_n\} = E_2$ for every $i \in \{1, ..., n\}$.
For each $i \in \{1, ..., m\}$ we choose a tuple $\mathbf{c_i} \in A_n$ with $f(\mathbf{c_i}) = 1$. Set

$$C := \{\mathbf{c_1}, ..., \mathbf{c_m}\}.$$

Then, for arbitrary $\mathbf{c} \in C$ is valid:

$$(f_1(\mathbf{c}), ..., f_m(\mathbf{c})) \in E_2^m, \quad 1 \in \{f_1(\mathbf{c}), ..., f_m(\mathbf{c})\}. \tag{8.11}$$

Assume there exists an $i \in \{m+1, ..., 2n\}$ with $\{f_i(\mathbf{c}) \mid \mathbf{c} \in C\} \subseteq E_2$. Then one obtains a contradiction to the property 4) of the functions φ_m, if one considers tuples of the form $(f_1(\mathbf{a}), ..., f_{2n}(\mathbf{a}))$ with $\mathbf{a} \in C$. Consequently, $2 \in \{f_i(\mathbf{c}) \mid \mathbf{c} \in C\}$ for every $i \in \{m+1, ..., 2n\}$.
Now let $m < n$. Because of the property 3) we have for each $f \in T$: $f(\mathbf{a}) = 2$ if and only if every variable of f which occurs in a conjunction of the abbreviated DNF of $pr\ f$ has the value 2 in the tuple $\mathbf{a}$. Every tuple of C has, however, the value 2 only at exactly a place. Therefore the abbreviated DNF of every function of $\{pr\ f_{m+1}, ..., pr\ f_{2m}\}$ contains a certain element of $\{y_1, ..., y_n\}$ as a unary conjunction. By (8.9) the product of these variables induces a conjunction of the form $y_{j_1} \cdot ... y_{j_t}$, where $t < n$ (because of $m < n$), in the abbreviated DNF of the function $pr\ v$. Consequently, the case $m < n$ is not possible.
Let $m > n$. Since $C \subseteq A_n$, there exists a tuple $\mathbf{a}$ with $f_i(\mathbf{a}) = 2$ for at least two $i \in \{m+1, ..., 2m\}$. By (8.11) and by the definition of the function φ_m this implies $\varphi_m(f_1(\mathbf{a}), ..., f_{2m}(\mathbf{a})) = 1$, which is a contradiction to $v(\mathbf{a}) = 0$, however. ■

After these preparations, the main result of this section can be formulated and proven:

Theorem 8.2.3 *([Dem-H 83], [Mar 83])*
Let $k \geq 3$ and let A be an arbitrary maximal class of P_k. Then $|\mathbb{L}_k^{\downarrow}(A)| = \mathfrak{c}$ if and only if A is not a maximal class of the type $\mathfrak{L}$.
If A of type $\mathfrak{L}$, then we have $k = p^m$, where $p \in \mathbb{P}$ and $m \in \mathbb{N}$, and it holds

$$|\mathbb{L}_k^{\downarrow}(A)| \begin{cases} < \aleph_0 & \text{if} \quad m = 1, \\ = \aleph_0 & \text{otherwise.} \end{cases}$$

Proof. Because of Theorem 6.1, we can assume that $A = Pol_k \varrho$ with $\varrho \in \mathfrak{M}_k \cup \mathfrak{U}_k \cup \mathfrak{S}_k \cup \mathfrak{L}_k \cup \mathfrak{C}_k \cup \mathfrak{B}_k$. For proof of $|\mathbb{L}_k^{\downarrow}(A)| = \mathfrak{c}$ for $\varrho \in \mathfrak{M}_k \cup \mathfrak{U}_k \cup \mathfrak{S}_k \cup \mathfrak{C}_k \cup \mathfrak{B}_k$, showing that the class A has a subclass with infinite basis suffices. If $\varrho \notin \mathfrak{M}_k \cup \mathfrak{S}_k \cup \mathfrak{L}_k$ there exists a 2-element subset $E \subset E_k$ with $P_{k,E} \subset A$. Then, with the aid of the Lemma 8.1.1 one can find a subclass of A with an

infinite basis easily. If A has the type $\mathfrak{M}$, then one can change the example from the Lemma 8.1.1 in the following manner: Instead of 0, 1, 2 one chooses a, b, c of E_k with $a <_\varrho b <_\varrho c$ and with the property $\forall x \in E_k : b <_\varrho x \leq_\varrho c \implies x = c$.
Consequently, A has also for $\varrho \in \mathfrak{M}_k$ a subclass with an infinite basis.
If $\varrho \in \mathfrak{S}_k$, the construction of a corresponding example is more complicated. Let $\varrho := \varrho_s \in \mathfrak{S}_k$, where s is defined as in Section 5.2.2. If s has only cycles of the length 2, then $Pol_k \varrho_s$ has a subclass with a infinite basis because of Lemma 8.2.1. In the following we assume that the cycles of s have a length ≥ 3. If $k = 3$ then $|\mathbb{L}_3^{\downarrow}(Pol_3 \varrho_s)| = \mathfrak{c}$ results from Theorem 8.2.2. If $k > 3$ then we can assume w.l.o.g. that the cycle representation of s contains a cycle of the form $(0\ 1\ 2\ ...)$. In generalization of our example from the proof of the Theorem 8.2.2 one can define then $2n$-ary functions ψ_n, which agree with the above functions φ_n on the set E_3^{2n} and which are otherwise so defined that $\psi_n \in Pol_k \varrho_s$ holds and that $[\{\psi_2, \psi_3, ...\}]$ has an infinite basis. We notice that one can find another proof for $|\mathbb{L}_k^{\downarrow}(Pol_k \varrho_s)| = \mathfrak{c}$ in the case of $k \geq 4$ in [Dem-H 83].
The remaining statements of our theorem result from Chapter 5, Theorem 8.1.6 and Chapter 13. In Chapter 13 one can find all subclasses from A for the case that $\varrho \in \mathfrak{L}_k$ and k is a prime number. Further, one finds in Chapter 13 an example of a class from linear functions, which has no basis if k is not square-free. One can generalize this example easily then for an example of a class from quasi-linear functions without basis, if $k = p^m$ and $m \geq 2$. ∎

For $k = 3$ we will show in Chapter 15 that there are exactly 7 submaximal classes B with $B \neq L_3$ and $|\mathbb{L}_3^{\downarrow}(B)| = \aleph_0$.

8.3 Some Strategies for the Determination of Sublattices of $\mathbb{L}_k$

Because of Theorem 8.1.3, there is little hope that one can ever describe all elements of $\mathbb{L}_k$ for $k \geq 3$. Nevertheless, one can attempt to get a certain idea of the construction of lattice $\mathbb{L}_k$ through the restriction to certain "manageable" sublattices. For example, one can restrict oneself to finitely generated subclasses of P_k. Subsequently, an enumeration is declared by sets for which there is a line of article in the literature and which characterize certain "sections" of $\mathbb{L}_k$ quite well:

(a) The lattice of the subclasses from linear or quasi-linear functions (see Chapter 13, 15.3 and e.g. [Sze 79], [Sze 80], [Sza-S 80], [Bag-D 82], [Sze 86], [Bul 98a,b].
(b) The set of all commutator sets or subclasses of autodual functions (e.g. [Har 74-76], [Mar 79], [Mar-D-H 80], [Csa-G 80], [Csa 80], [Mac-R 2004])
(c) subclasses of

$$U_\omega := \bigcup_{n\geq 1}\{f^n \in P_k \mid (\exists i:\ x_i = \omega) \implies f(x_1,...,x_i,...,x_n) = \omega\}$$

(this set is isomorphic to the set of all partial functions of P_{k-1}) (see e.g. [Mal 66], [Ros 88], [Ros-H 87,89,91] and Chapter 20)

(d) subclasses of P_k, which contain certain subsemigroups of $(P_k^1;\star)$ (see e.g. Chapter 4, [Had-R 94], [Kro 99])

(e) subclasses of P_k, which contain a "near unanimity function" or the discriminator

$$t(x,y,z) := \begin{cases} z, \text{ if } & x = y, \\ x, \text{ otherwise} & \end{cases}$$

(see e.g. [Bak-P 75], [Wer 78], [Csa-G 81]).

(f) The set of the inverse images of subclasses of P_l in respect to certain mappings (see Chapter 12).

(g) The set of all intersections from already determined classes (e.g. the maximal classes). (In this connection one can use for example the results on basis classifications from [Miy 71], [Miy-S-L-R 87], [Miy 88], and [Sto 87].)

(h) clones with certain properties (e.g. minimal clones, solidifyable clones etc.), see e.g. Chapter 19 and [Den-W 2000].

For some further chapters we still need two theorems, which are the basis for the determination of certain sublattices of P_k.

The following theorem is an insignificant generalization of the Theorem of Baker-Pixley which is proven in analog mode to [Wer 78] here.

Theorem 8.3.1 *Let A be a subclass of P_k, which contains a "near unanimity function" d_m for certain $m \geq 2$, i.e., an $m+1$-ary function with the property*

$$d_m(x_1,...,x_{m+1}) = x,$$
$$\text{if there is an } i \in \{1,...,m+1\} \text{ with}$$
$$x_1 = ... = x_{i-1} = x_{i+1} = ... = x_{m+1} = x.$$

Then, $A = Pol_k Inv^m A$ holds, i.e., for fixed numbers m and k there are only finite-many subclasses $A \in \mathbb{L}_k$ with $d_m \in A$.

Proof. Obviously,

$$A \subseteq Polk(Inv_k^m A).$$

Let $f^n \in Pol_k(Inv^m A)$. Further, for $T \subseteq E_k^n$ denote f_T a function of P_k, which agrees with the function f on tuples from T. We prove by induction on $|T| =: t \geq m$ that there exists a certain function $f_T \in A$ to every $T \subseteq E_k^n$; through that $f \in A$ for $t = k^m$ would be shown and $A = Pol_k(Inv_k^m A)$ would be proven.

I) $t = m$: Let $T = \{(a_{i1}, a_{i2}, ..., a_{in}) \mid i = 1, 2, ..., m\}$,

$$\varrho := \begin{pmatrix} a_{11} & a_{12} & \dots & a_{1n} \\ a_{21} & a_{22} & \dots & a_{2n} \\ \dots & \dots & \dots & \dots \\ a_{m1} & a_{m2} & \dots & a_{mn} \end{pmatrix}$$

and let ϱ' be the least relation of $Inv^m A$ with $\varrho \subseteq \varrho'$, i.e., it holds

$$\begin{pmatrix} a_1 \\ a_2 \\ \dots \\ a_m \end{pmatrix} \in \varrho' \backslash \varrho \iff \exists q \in A^n : q \begin{pmatrix} a_{11} & a_{12} & \dots & a_{1n} \\ a_{21} & a_{22} & \dots & a_{2n} \\ \dots & \dots & \dots & \dots \\ a_{m1} & a_{m2} & \dots & a_{mn} \end{pmatrix} = \begin{pmatrix} a_1 \\ a_2 \\ \dots \\ a_m \end{pmatrix}. \tag{8.12}$$

Since f^n preserves the m-ary invariants of A, we have $f(\varrho) \in \varrho'$. If $f(\varrho)$ belongs to ϱ, there is a certain function in A which agrees with the function f on tuples from T. If also on the other hand, $f(\varrho) \in \varrho' \backslash \varrho$ is valid, then there is because of (8.12) also a certain function f_T in A. Thus the above assertion was proven for $t = m$.
II) $t \longrightarrow t+1$: Suppose for every $T \subseteq E_k^n$ with $|T| = t$, $t \geq m$, there is a function $f_T \in A$ with $f_T(\mathbf{a}) = f(\mathbf{a})$ for all $\mathbf{a} \in T$. Let now $T = \{\mathbf{a_1}, \mathbf{a_2}, ..., \mathbf{a_{t+1}}\}$ and $|T| = t+1$. Because of our assumption, there exist functions $f_i \in A^n$ with $f_i(\mathbf{a}) = f(\mathbf{a})$ for all $\mathbf{a} \in T \backslash \{\mathbf{a_i}\}$, $i = 1, 2, ..., m+1$. Since $d_m \in A$, we have through $d_m(f_1(\mathbf{x}), f_2(\mathbf{x}), ..., f_{m+1}(\mathbf{x}))$ a function in A, which agrees with f on tuples aus T. ■

In Chapter 12, we generalize this theorem for $P_{k,l}$.

The last theorem of this chapter, which is a slight generalization of Lemma 3.2.3.1, shows that there are sublattices of the form $\mathbb{L}_k^{\downarrow}(A)$ of P_k whose elements one can receive by means of intersection formations of A with classes that are not subsets of A. We will use this theorem in Chapters 14 and 18.

In preparation for the announced theorem, some notations and properties are given:
Let s be a permutation of P_k defined by $s(x) := x + 1 \ (mod\, k)$, let $\varrho_s := \{(x, s(x)) \mid x \in E_k\}$ and put $S_k := Pol_k \varrho_s$. Analogously to the proof of Theorem 5.2.2.1, one can prove that a function $f^n \in P_k$ belongs to S_k if and only if there exists an $(n-1)$-ary function F of $P_k' := P_k \cup E_k$ [1] with the following property:

$$f(x_1, ..., x_n) = \sum_{i=0}^{k-1} j_i(x_1) \cdot s^i(F(s^{k-1}(x_2), ..., s^{k-i}(x_n))) \ (mod\, k) \tag{8.13}$$

($s^i(x) := x+i \ (mod\, k)$; $F(x_1, ..., x_{n-1}) := f(0, x_1, ..., x_{n-1})$). Because of (8.13) it is possible to define a bijective mapping α of S_k ($\subseteq P_k$) onto P_k' as follows:

$$\alpha : f \longrightarrow F.$$

[1] In this case, the elements of E_k are the nullary functions of P_k'.

Theorem 8.3.2 *The mapping α has the following properties:*

(a) For the operations $\widehat{\zeta}, \widehat{\tau}, \widehat{\Delta}, \widehat{\nabla}$ and $\widehat{\star}$ defined by

$$(\widehat{\zeta}f)(x_1, ..., x_n) = f(x_1, x_3, x_4, ..., x_n, x_2),$$
$$(\widehat{\tau}f)(x_1, ..., x_n) = f(x_1, x_3, x_2x_4, ..., x_n),$$
$$(\widehat{\Delta}f)(x_1, ..., x_{n-1}) = f(x_1, x_2, x_2, x_3, ..., x_{n-1}),$$
$$(\widehat{\nabla}f)(x_1, ..., x_{n+1}) = f(x_1, x_3, x_4, ..., x_{n+1}) \text{ und}$$
$$(f\widehat{\star}g)(x_1, ..., x_{m+n-2}) = f(x_1, g(x_1, ..., x_m), x_{m+1}, ..., x_{m+n-2})$$
$$(n, m \geq 2),$$

it holds $\alpha(\widehat{\gamma}f) = \gamma(\alpha(f))$ for every $\gamma \in \{\zeta, \tau, \Delta, \nabla\}$ and $\alpha(f\widehat{\star}g) = \alpha(f) \star \alpha(g)$; i.e., the algebra $(S_k; \widehat{\zeta}, \widehat{\tau}, \widehat{\Delta}, \widehat{\nabla}, \widehat{\star})$ is isomorphic to the algebra $(P_k'; \zeta, \tau, \Delta, \nabla, \star)$.

(b) For every subclass A $(\neq \emptyset)$ of S_k, $\alpha(A)$ is a subclass of P_k', and it holds $\alpha(A) \not\subseteq S_k$, $A \subseteq \alpha(A)$ and $\alpha(A) \cap S_k = A$.

Proof. (a) is easy to check.
(b): Let A be a subclass of S_k. By (a) we have that $\alpha(A)$ is also a closed set. Assume $\alpha(A) \subseteq S_k$. Then we have

$$F(x_2, ..., x_n) = s^i(F(s^{k-i}(x_2), ..., s^{k-i}(x_n)))$$

for every $i \in \{0, 1, ..., k-1\}$ and for every $f^n \in A$. Thus by (8.13) we get that the variable x_1 is fictitious for every function $f^n \in A$. However, this is not possible. Hence $\alpha(A) \not\subseteq S_k$ holds.
Let $f^n \in A$. Then $\nabla f \in A$ and therefore $\alpha(\nabla f) = f \in \alpha(A)$, i.e., $A \subseteq \alpha(A)$. If $f^n \in S_k \cap \alpha(A)$, we have $\Delta(\alpha^{-1}f) = f \in A$ and thus $S_k \cap \alpha(A) \subseteq A$. From this it follows that $A = S_k \cap \alpha(A)$, since $A \subseteq \alpha(A)$ and $A \subseteq S_k$. ■

9

Congruences and Automorphisms on Function Algebras

As generally known, one can characterize the homomorphic mappings from the algebra $\mathbf{A} := (A; F)$ into the algebra $\mathbf{B}$ through the congruences of $\mathbf{A}$ (see Chapter 4 of Part I). Therefore, we will not deal here with the homomorphisms of function algebras but only with the congruences on the subclasses on P_k.
After some basic concepts are defined, all congruences on the subclasses of P_2 are determined in this chapter. It is the aim of the following sections, then, to specify the general homomorphism theorem for function algebras and to find statements on the number of the congruences on a subclass of P_k by determining some general properties of the congruences on subclasses of P_k. Then, all congruences are determined for selected classes (among other things these are the maximal classes of P_k and certain classes from linear functions). Criteria with which one can discuss whether, on a subclass of P_k, only trivial congruences exist are also derived.
A later section deals with the connection between clone congruences and the fully invariant congruences on free algebras. The theorem found in this case has interesting inferences and is a bridge between certain investigations in the Universal Algebra and certain investigations in the theory of the function algebras.
At the end of this chapter, are some results on automorphisms. It is proven that P_k, the subclasses of P_2, and the maximal classes of P_k have only inner automorphisms.
The starting point and the basis of the subsequently compiled results was an article of A. I. Mal'tsev from the year 1966 (see [Mal 66]). Important contributions to the topic of Chapter 9 are also performed by V. V. Gorlov and I. A. Mal'tsev.

9.1 Some Basic Concepts and First Properties

Definition A **congruence** on a subclass A of P_k is an equivalence relation κ on A, which fulfills the following condition:

$$\forall (f,g),(s,t) \in \kappa \; \forall \alpha \in \{\zeta, \tau, \Delta, \nabla\} : \; (\alpha f, \alpha g) \in \kappa) \wedge (f * s, g * t) \in \kappa.$$

Let

$$Con\ A$$

be the set of all congruences on the class $A \in \mathbb{L}_k$.
In the following, we write

$$f \sim g \; (\kappa)$$

instead of

$$(f,g) \in \kappa$$

and we call such functions f and g **κ-congruent**.
Furthermore, let

$$\kappa^{(n)} := \kappa \cap (A^n \times A^n).$$

If one wants to realize whether an equivalence relation on a subclass of P_k is compatible with the operation $\star$, it is helpful to use the equivalence given in the following lemma.

Lemma 9.1.1 *Let A be a subclass of P_k and let κ be an equivalence relation on A. Then, the following two conditions are equivalent:*
(1) $\forall (f,g),(u,v) \in \kappa : \; (f \star u, g \star v) \in \kappa$
(2) $\forall (f,g) \in \kappa \; \forall t \in A : \; ((f \star t, g \star t) \in \kappa \; \wedge \; (t \star f, t \star g) \in \kappa).$

Proof. "(1) $\Longrightarrow$ (2)" follows from the reflexivity of κ.
"(2) $\Longrightarrow$ (1)": Let $(f,g),(u,v) \in \kappa$. Choosing $t = u$ in (2) and then $t = g$, one receives: $(f \star u, g \star u) \in \kappa$ and $(g \star u, g \star v) \in \kappa$. Since κ is transitive, this implies (1). ■

Next to the trivial congruences $\kappa_0 := \{(f,f) \mid f \in A\}$ and $\kappa_1 := A \times A$, the following congruence, which we want also to call **trivial congruence**, exists on every subclass of P_k:

$$\kappa_a := \{(f,g) \in A^2 \mid af = ag\}.$$

Later we obtain the following theorem as a consequence of Lemmas 9.7.1 and 9.2.1:

Theorem 9.1.2 *([Mal 66])*
The (trivial) congruences κ_0, κ_a and κ_1 are the only congruences on P_k. ■

It is easy to check that

$$\forall A \in \mathbb{L}_k \; \forall \kappa \in Con\ A : \; (\kappa_a \subset \kappa \subseteq \kappa_1 \implies \kappa = \kappa_1).$$

Consequently, the following partition of $Con\ A$, where A is a subclass of P_k, is reasonable:

$$Con_1 A := \{\kappa \in ConA \mid \kappa \not\subseteq \kappa_a\}$$

(congruences of the first kind, **non-arity congruences**)

and

$$Con_a A := \{\kappa \in ConA \mid \kappa \subseteq \kappa_a\}$$

(congruences of the second kind, **arity congruences**).

9.2 Congruences on the Subclasses of P_2

We take over the notations from Chapter 3. Further, for an arbitrary subclass A of P_2 set:

$$\kappa_c := \{(f, g) \in \kappa_a \cap (A \times A) \mid \exists a \in E_2 : f(\mathbf{x}) = g(\mathbf{x}) + a\},$$
$$\mu := \{(f, g) \in \kappa_a \cap (A \times A) \mid f \neq g \Longrightarrow \{f, g\} \subseteq C\},$$
$$\kappa_{\overline{a}} := (C_0 \times C_0) \cup (C_1 \times C_1)$$

(More generally, the relation $\kappa_{\overline{a}}$ is defined in (9.1)). It can easily be shown that κ_c is a congruence on the class A, if $A \subseteq L$; μ is a congruence on A, if $A = [A] \subseteq [P_2^1]$; and $\kappa_{\overline{a}}$ is a congruence on the set C.

Lemma 9.2.1 *Let $A \subseteq P_k$ be a clone. Then κ_1 is the only non-arity congruence on A.*

Proof. Let κ be a congruence on A with $\kappa \not\subseteq \kappa_a$. Then, there are certain f^n, $g^m \in A$ with $(f, g) \in \kappa$ and $n > m$. Consequently, we have

$$(\Delta(e_2^2 * \Delta^{n-2} f), \Delta(e_2^2 * \Delta^{n-2} g)) = (e_2^2, e_1^1) \in \kappa$$

and therefore

$$(e_2^2 * e_{t-1}^{t-1}, e_1^1 * e_{t-1}^{t-1}) = (e_t^t, e_{t-1}^{t-1}) \in \kappa$$

for $t = 2, 3, \ldots$. Then, since κ is an equivalence relation, it follows that

$$\forall s, t \in \mathbb{N} : (e_s^s, e_t^t) \in \kappa.$$

This implies $\kappa = \kappa_1$, since for every function $h^q \in A$

$$(e_2^2 * h, e_1^1 * h) = (e_{q+1}^{q+1}, h) \in \kappa$$

holds. ∎

Lemma 9.2.2 *Let $A \subseteq P_k$ be a clone and $\kappa \in Con\ A$ with*

$$\kappa \cap \{(e_1^1, c_0^1), (e_1^1, c_1^1), (e_1^2, e_2^2)\} \neq \emptyset.$$

Then κ_a is the only arity congruence on A.

Proof. Assume κ is an arity congruence on the clone A and f^n is an arbitrary function of A. Put $e := e_1^1$ and $c_a := c_a^1$ for $a \in \{0,1\}$. Then, $(e, c_a) \in \kappa$ implies $(e * f, c_a * f) = (f^n, c_a^n) \in \kappa$ and $(e_1^2, e_2^2) \in \kappa$ implies $(e_1^2(f(\mathbf{x}), x_1), e_2^2(f(\mathbf{x}), x_1)) = (f(\mathbf{x}), e_1^n(\mathbf{x})) \in \kappa$. Since κ is an equivalence relation, $\kappa = \kappa_a$ follows from that. ■

With the aid of Lemmas 9.2.1 and 9.2.2, one can easily prove the following lemma.

Lemma 9.2.3 *The congruences $\kappa_{\overline{a}}$, μ and κ_c are the only nontrivial congruences on a subclass of $[P_2^1]$.* ■

Lemma 9.2.4 *Let A be a subclass of P_2 that contains $\wedge$ or $\vee$. Then A has only trivial congruences.*

Proof. W.l.o.g. let $\wedge \in A$. Suppose there is a nontrivial congruence κ on A. Then, because of Lemma 9.2.1, we have $\kappa_0 \subset \kappa \subset \kappa_a$ and there exists κ-congruent functions f^n, $g^n \in A$ with $f \neq g$. For f and g we distinguish two cases:

Case 1: $\{\Delta^{n-1} f, \Delta^{n-1} g\} \cap \{c_0, c_1, \overline{e}\} \neq \emptyset$.

Because of $(f, g) \in \kappa \setminus \kappa_0$ one can form different unary κ-congruent functions f', g' when one identifies variables and replaces certain variables of f, g by the function $h \in \{c_0, c_1, \overline{e}\} \cap A$. W.l.o.g. $(f', g') \in \{(e, c_0), (e, c_1), (e, \overline{e}), (c_0, c_1)\}$. If $(f', g') = (e, \overline{e}) \in \kappa$, then, by $\wedge \in A$, the tuple $(e, c_0) = (e \wedge e, e \wedge \overline{e})$ belongs to κ. If $(c_0, c_1) \in \kappa$ then $(e \wedge c_0, e \wedge c_1) = (c_0, e) \in \kappa$. Consequently, we have $\{(e, c_0), (e, c_1)\} \cap \kappa \neq \emptyset$ and by Lemma 9.2.2 $\kappa = \kappa_a$, in contradiction to the assumption.

Case 2: $\Delta^{n-1} f = \Delta^{n-1} g = e$.

Since $f \neq g$, there is an $\mathbf{a} \in E_2^n \backslash \{\mathbf{0}, \mathbf{1}\}$ with (w.l.o.g.) $f(\mathbf{a}) = 0$ and $g(\mathbf{a}) = 1$. When one identifies certain variables of the functions f and g, one receives from $(f, g) \in \kappa$ the existence of two binary κ-congruent functions f'', g'' with

$$f''(0,0) = f''(0,1) = 0, f''(1,1) = 1$$

and

$$g''(0,0) = 0, g''(0,1) = g''(1,1) = 1,$$

i.e., $(f'', g'') \in \{(\wedge, e_2^2), (\wedge, \vee), (e_1^2, e_2^2), (e_1^2, \vee)\}$. If $(\wedge, e_2^2) \in \kappa$ then $(\tau\wedge, \tau e_2^2) = (\wedge, e_1^2)$ and thus (by the transitivity and symmetry of κ) we have $(e_2^2, e_1^2) \in \kappa$. Analogously, $(e_1^2, \vee) \in \kappa$ implies $(e_2^2, e_1^2) \in \kappa$. If $(\wedge, \vee) \in \kappa$ then $x = (x \wedge y) \vee x \sim (x \wedge y) \wedge x = x \wedge y\ (\kappa)$. Therefore, $(e_1^2, e_2^2) \in \kappa$ in Case 2. A contradiction from this results with the aid of Lemma 9.2.2. ■

Lemma 9.2.5 *Let A be a subclass of S, which contains the function h_2 defined by*

$$h_2(x, y, z) := xy \vee xz \vee yz.$$

Then A has only trivial congruences.

Proof. Assume κ is a nontrivial congruence on A. Then, by Lemma 9.2.1, there exists two κ-congruent functions f^n, $g^n \in A$ with $f(\mathbf{a}) = 0$ and $g(\mathbf{a}) = 1$ for a certain $\mathbf{a} \in E_2^n$. If $\mathbf{a} \in \{\mathbf{0}, \mathbf{1}\}$ then $\overline{e}$ belongs to A and $\{\Delta^{n-1}f, \Delta^{n-1}g\} = \{e, \overline{e}\}$ holds, i.e., $(e, \overline{e}) \in \kappa$. This implies $x = h_2(x, y, \overline{y}) \sim h_2(x, y, y) = y$ (κ) and then $(e_1^2, e_2^2) \in \kappa$. Therefore, by Lemma 9.2.2, $\kappa = \kappa_a$ holds, if $\mathbf{a} \in \{\mathbf{0}, \mathbf{1}\}$. If $\mathbf{a}$ does not belong to $\{\mathbf{0}, \mathbf{1}\}$, one can form two binary κ-congruent functions f' and g' when one identifies certain variables of f and of g. Since $S^2 = \{e_1^2, e_2^2, \overline{e}_1^2, \overline{e}_2^2\}$, this implies $(e_1^2, e_2^2) \in \kappa$ and thus $\kappa = \kappa_a$ by Lemma 9.2.2. This is, however, a contradiction to the assumption. ■

Theorem 9.2.6 *Let $A \subseteq L$ be a subclass, which contains the function r defined by*

$$r(x, y, z) = x + y + z.$$

Then, κ_c is the only nontrivial congruence on A.

Proof. Let κ be an arity congruence on A. The following two cases are possible:
Case 1: $\kappa_0 \subset \kappa \subseteq \kappa_c$.
In this case, there exists κ-congruent functions f and g with $f(\mathbf{x}) = g(\mathbf{x}) + 1$. Consequently, because of $r \in A$, we have :

$$x_1 = f(\mathbf{x}) + f(\mathbf{x}) + x_1 \sim g(\mathbf{x}) + f(\mathbf{x}) + x_1 = x_1 + 1 \ (\kappa),$$

i.e., $(e, \overline{e}) \in \kappa$. Therefore, $\kappa = \kappa_c$.
Case 2: $\kappa \nsubseteq \kappa_c$.
Then, there are two κ-congruent functions f^n, g^n with $f(x) = a_0 + a_1x_1 + ... + a_nx_n$, $g(x) = b_0 + b_1x_1 + ... + b_nx_n$ and (w.l.o.g.) $a_1 = 0$ and $b_1 = 1$. Thus

$$r(y, f(x, y, ..., y), f(y, y, ..., y)) = y \ \sim \ r(y, g(x, y, ..., y), g(y, y, ..., y)) = x \ (\kappa).$$

Consequently, $\kappa = \kappa_a$ by Lemma 9.2.2. ■

The following theorem is a consequence of the Post's graph and of the six above lemmas:

Theorem 9.2.7 **(Congruence Theorem for P_2,** *[Gor 73]***)**
The only nontrivial congruences on a subclass of P_2 are μ (on classes A with $A \subseteq [P_2^1]$), κ_c (on classes B with $B \subseteq L$) and $\kappa_{\overline{a}}$ (on C).
An arbitrary subclass ($\neq \emptyset$) of P_2 has at least two and at most five different congruences. The congruence lattices are given in Figure 9.1. ■

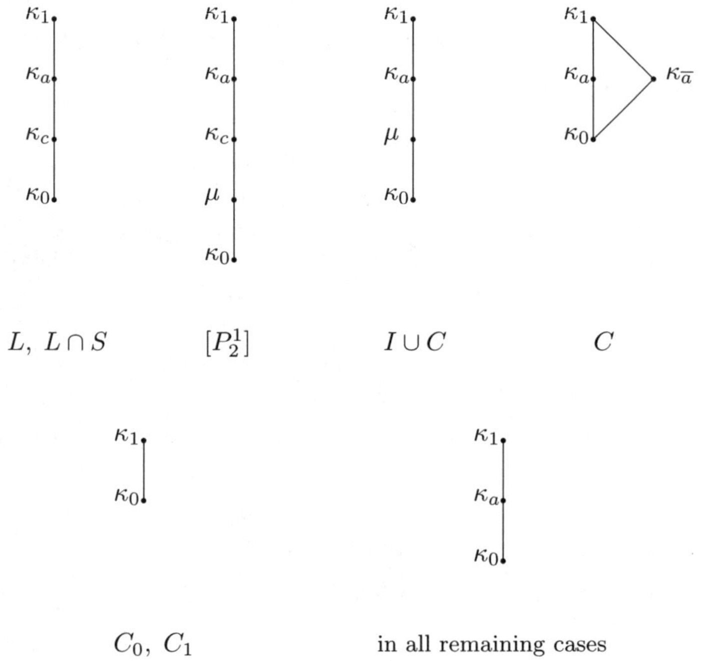

Fig. 9.1. Congruence lattices of the subclasses of P_2

9.3 Characterization of the Non-Arity Congruences

In this section let A be a subclass of P_k.

Definition We say that **two functions** f^m and g^n of $A \subseteq P_k$ **are associated** (in A), or we say that f **is associated with** g (in A), written

$$f \bowtie g,$$

iff there exist functions $u_1, u_2, ..., u_{m+n} \in A^1$ with

$$f(u_1(x), u_2(x), ..., u_m(x)) = g(u_{n+1}(x), u_{n+2}(x), ..., u_{m+n}(x)).$$

Obviously, each function $f \in A$ is associated with each function αf for $\alpha \in \{\zeta, \tau, \Delta, \nabla\}$) and with $f \star g$, where $g \in A$.

Let $p_0, p_{t+1} \in A$. We define an equivalence relation $\kappa_{\overline{a}}$ on A as follows:

$$\begin{aligned} &p_0 \sim p_{t+1} \ (\kappa_{\overline{a}}) \ :\Longleftrightarrow \\ &\exists p_1, p_2, ..., p_t \in A: \ p_0 \bowtie p_1 \bowtie p_2 \bowtie ... \bowtie p_t \bowtie p_{t+1}. \end{aligned} \tag{9.1}$$

Lemma 9.3.1 *The relation $\kappa_{\overline{a}}$ defined by (9.1) is a congruence on $A = [A] \subseteq P_k$.*

Proof. Obviously, $\kappa_{\overline{a}}$ is an equivalence relation on A. To prove the compatibility of $\kappa_{\overline{a}}$ with the superposition operations, let $p_0 \sim p_{t+1}$ $(\kappa_{\overline{a}})$ and $q_0 \sim q_{s+1}$ $(\kappa_{\overline{a}})$ be arbitrary, i.e., there exist functions p_i $(i = 1, 2, ..., t)$, q_j $(j = 1, 2, ..., s)$ of A with

$$p_0 \bowtie p_1 \bowtie p_2 \bowtie \ldots \bowtie p_t \bowtie p_{t+1}$$

and

$$q_0 \bowtie q_1 \bowtie q_2 \bowtie \ldots \bowtie q_s \bowtie q_{s+1}.$$

Because of

$$\alpha p_0 \bowtie p_0 \bowtie p_1 \bowtie p_2 \bowtie \ldots \bowtie p_t \bowtie p_{t+1} \bowtie \alpha p_{t+1}$$

$(\alpha \in \{\zeta, \tau, \Delta, \nabla\})$ we have $\alpha p_0 \sim \alpha p_{t+1}$ $(\kappa_{\overline{a}})$. Furthermore, $p_0 \star q_0 \sim p_{t+1} \star q_{s+1}$ $(\kappa_{\overline{a}})$, since

$$p_0 \star q_0 \bowtie p_0 \bowtie p_1 \bowtie p_2 \bowtie \ldots \bowtie p_{t+1} \bowtie p_{t+1} \star q_{s+1}.$$

Thus $\kappa_{\overline{a}}$ is a congruence on A. ■

Lemma 9.3.2 *Every equivalence relation κ with*

$$\kappa_{\overline{a}} \subseteq \kappa$$

is a congruence on A.

Proof. Let $f \sim g$ (κ) and $u \sim v$ (κ) be arbitrary. Then

$$\begin{array}{ll} f & \sim \alpha f \ (\kappa_{\overline{a}}) \\ \wr & \\ g & \sim \alpha g \ (\kappa_{\overline{a}}) \\ (\kappa) & \end{array}$$

$(\alpha \in \{\zeta, \tau, \Delta, \nabla\})$ and

$$\begin{array}{rcl} f \star u \sim & f & (\kappa_{\overline{a}}) \\ & \wr & \\ g \star v \sim & g & (\kappa_{\overline{a}}) \\ & (\kappa) & \end{array}.$$

Because of $\kappa_{\overline{a}} \subseteq \kappa$ we have thus $\alpha f \sim \alpha g$ (κ) and $f \star u \sim g \star v$ (κ). ■

Lemma 9.3.3 *Let κ be a non-arity congruence on a subclass A of P_k. Then $\kappa_{\overline{a}} \subseteq \kappa$.*

Proof. Let $h_1 \in A^1$ be arbitrary and let h_i be the function $\nabla^{i-1} h_1$ with $n \in \mathbb{N} \backslash \{1\}$. We prove

$$\forall i, j \in \mathbb{N} : \ h_i \sim h_j \ (\kappa). \tag{9.2}$$

Since κ is a non-arity congruence, there exist κ-congruent functions f^n and g^m of A with $n > m$. Consequently, we have

$$h_2 \star (\Delta^{n-2} f) = h_3 \sim h_2 \star (\Delta^{n-2} g) = h_2 \ (\kappa),$$

$$h_2 = \Delta h_3 \sim \Delta h_2 = h_1 \ (\kappa),$$

$$\nabla h_3 = h_4 \sim \nabla h_2 = h_3 \ (\kappa),$$

etc. Thus, (9.2) holds.
Next we prove

$$\forall f^n \in A \, \forall g_1, ..., g_n \in A^1 : \ f(x_1, ..., x_n) \sim f(g_1(x_1), ..., g_n(x_1)) \ (\kappa). \quad (9.3)$$

Let $f^n \in A$ be arbitrary and set $f_1 := \nabla f$. Then:

$$f_2 := f_1 \star h_1 \sim f_1 \star h_{n+1} =: f_3 \ (\kappa)$$

and

$$f_4 := \Delta f_2 \sim \Delta f_3 =: f_5 \ (\kappa),$$

where

$$f_4(x_1, ..., x_n) = f(x_1, ..., x_n)$$

and

$$f_5(x_1, ..., x_{2 \cdot n}) = f(x_{n+1}, ..., x_{2 \cdot n}).$$

Consequently, we have for all $g_1, ..., g_n \in A^1$:

$$\begin{aligned} &(\zeta(...((\zeta((\zeta((\zeta^{n+1} f_5) \star g_n)) \star g_{n-1})) \star g_{n-2})...) \star g_1 \\ \sim\ &(\zeta(...((\zeta((\zeta((\zeta^{n+1} f_4) \star g_n)) \star g_{n-1})) \star g_{n-2})...) \star g_1 \ (\kappa) \end{aligned}$$

which can be written more briefly in the form

$$\begin{aligned} &f_4(g_1(x_1), ..., g_n(x_1)) = f(g_1(x_1), ..., g_n(x_1)) \\ \sim\ &f_5(g_1(x_1), ..., g_n(x_1), x_1, ..., x_n) = f(x_1, ..., x_n) \ (\kappa). \end{aligned}$$

Thus (9.3) is right.
Finally, we prove the following fact:

$$\forall f, g \in A : \ (f \sim g \ (\kappa_{\overline{a}}) \implies f \sim g \ (\kappa)). \quad (9.4)$$

If $f \sim g \ (\kappa_{\overline{a}})$, there are functions $p_1, ..., p_{t+1} \in A$ with

$$p_0 := f \ \bowtie \ p_1 \ \bowtie \ p_2 \ \bowtie \ ... \ \bowtie \ p_t \ \bowtie \ p_{t+1} := g.$$

Then, by definition (9.1) and by property (9.3), we obtain

$$p_0 := f \sim p_1 \sim p_2 \sim ... \sim p_t \sim p_{t+1} := g \ (\kappa),$$

i.e., $f \sim g \ (\kappa)$. Thus $\kappa_{\overline{a}} \subseteq \kappa$. ■

Theorem 9.3.4 *(**I. A. Mal'tsev's Theorem**, [Mal 76])*
Let A be a subclass of P_k. Then A has only finite many non-arity congruences. These congruences are equivalence relations on A with $\kappa_{\overline{a}} \subseteq \kappa$.
In particular, for arbitrary subclasses A of P_k it holds:

$$Con_1 A = \{\kappa_1\} \iff \kappa_{\overline{a}} = \kappa_1$$

and

$$1 \leq |Con_1 A| \leq \mu(|A^1|),$$

where $\mu(n)$ denotes the number of possible equivalence relations on an n-element set.

Proof. The statements of our theorem result directly from Lemmas 9.3.1–9.3.3 and the fact that every function $f \in A$ is $\kappa_{\overline{a}}$–congruent to a certain unary function of A. ∎

We still notice that the above theorem is not valid if the operation ∇ is renounced. One finds an example for this purpose in [Gor 73].

The relation κ_1 is the only non-arity relation on many classes. In Lemma 9.2.1, it was shown that this is valid for all clones. In generalizing Lemma 9.2.1 we get:

Lemma 9.3.5 *([Mal 76])*
Let A be a subclass of P_k. If there is a unary function $u \in A$ with
*(a) $\forall f \in A: \ u \star f = f$ (i.e., "u is a **left unit**")*
or
*(b) $\forall g \in A^1: \ g \star u = u$ (i.e., "u is a **right zero**"),*
then κ_1 is the only non-arity congruence on A.

Proof. By Theorem 9.3.4 it is sufficient to show that $\kappa_{\overline{a}} = \kappa_1$.
(a) : If u is a left unit then for arbitrary $f^n \in A$ we have

$$f(u, u, ..., u) := u_1 \ \bowtie \ u \star u_1 = u_1.$$

Consequently, $f \bowtie u$ and therefore $\kappa_{\overline{a}} = \kappa_1$.
(b) : If u is a right zero of A then

$$\forall\ f^n \in A: \ f(u, u, ..., u) = u \ \text{ and } \ u \star u = u.$$

This implies $f \bowtie u$. Therefore $\kappa_{\overline{a}} = \kappa_1$. ∎

When one proves $\kappa_{\overline{a}} = \kappa_1$, one receives the next lemma as a conclusion of Theorem 9.3.4.

Lemma 9.3.6 *If the subclass $A \subseteq P_k$ contains all constant functions on E_k and*

$$\forall\, f_1, f_{t+1} \in A\; \exists f_2, ..., f_t \in A:\; \forall i \in \{1, 2, ..., t\}:\; Im(f_i) \cap Im(f_{i+1}) \neq \emptyset,$$

holds, then κ_1 is the only non-arity congruence on A. ■

Finally, we solve the problem how the structure of classes is which ones have the maximum number of possible non-arity congruences (see Theorem 9.3.4).

Lemma 9.3.7 *Let A be a subclass of P_k and let σ be a relation on A defined by*

$$(f^n, g^m) \in \sigma \;:\Longleftrightarrow\; \{f, g\} \subseteq A \;\wedge\; \Delta^{n-1} f = \Delta^{m-1} g. \tag{9.5}$$

Then:
(a) The relation σ is a congruence on A if and only if A has the property

$$\forall\, r, s \in A^1:\; r \star s = r. \tag{9.6}$$

(b) $\kappa_{\overline{a}} = \sigma \iff \kappa_{\overline{a}}^{(1)} = \kappa_0^{(1)}$.
(c) There are exactly $\mu(|A^1|)$ [1] non-arity congruences on A if and only if A fulfills the condition (9.6).

Proof. Obviously,

$$\sigma \subseteq \kappa_{\overline{a}} \tag{9.7}$$

holds for the relation defined by (9.5).
(a): "$\Longrightarrow$": Let σ be a congruence on A. Then, by Theorem 9.3.4 and (9.7), we have $\sigma = \kappa_{\overline{a}}$. In particular, it holds

$$(f^1, g^1) \in \sigma \iff f = g \in A^1. \tag{9.8}$$

Suppose (9.6) is false, i.e., there are certain $r, s \in A^1$ with $r \star s =: t \neq r$. Because of $r \star (s \star h) = t \star h$ for arbitrary $h \in A^1$, we have that r and t are associated in A. Because of $r \neq t$ and $\sigma = \kappa_{\overline{a}}$, this is, however, a contradiction to the equivalence (9.8). Therefore (9.6) holds.
"$\Longleftarrow$": Fulfill A the condition (9.6). Obviously, σ is an equivalence relation on A, which is compatible with all unary superposition operations. We have to show therefore only still the compatibility from σ with $\star$. Let $(f^n, g^m), (s^p, t^q) \in \sigma$ be arbitrary. Then

$$\Delta^{n-1} f = \Delta^{m-1} g =: f_1 \;\text{ and }\; \Delta^{p-1} s = \Delta^{q-1} t =: s_1.$$

This implies

$$\begin{aligned}
(\Delta^{n+p-1}(f \star s))(x) &= f(s_1(x), x, ..., x)\\
&\overset{(9.6)}{=} f(s_1((s_1(x)), s_1(x), ..., s_1(x))\\
&\overset{(9.6)}{=} f(s_1(x), s_1(x), ..., s_1(x))\\
&= g(s_1(x), s_1(x), ..., s_1(x))\\
&\overset{(9.6)}{=} g(s_1(s_1(x)), s_1(x), ..., s_1(x))\\
&\overset{(9.6)}{=} g(s_1(x), x, ..., x)\\
&= (\Delta^{m+q-1}(g \star t))(x).
\end{aligned}$$

[1] See Theorem 9.3.4.

Thus $(f \star s, g \star t) \in \sigma$.

(b): Obviously, "$\Longrightarrow$" holds.
"$\Longleftarrow$": Let $\kappa_{\overline{a}}^{(1)} = \kappa_0^{(1)}$. The following two cases are possible:
Case 1: A fulfills (9.6).
Because of (a) the relation σ in this case is a congruence, which is identical with $\kappa_{\overline{a}}$ because of (9.7) and Theorem 9.3.4.
Case 2: A does not fulfill (9.6).
In the proof of (a), we have shown that a contradiction results from $\kappa_{\overline{a}}^{(1)} \neq \kappa_0^{(1)}$.
(c) follows from Theorem 9.3.4 and (b). ■

Lemma 9.3.8 *Let κ be a non-arity congruence on $A = [A] \subseteq P_k$. Further, let $\mu := \kappa \cap (A^1 \times A^1)$. ($\mu$ is a congruence of the semigroup $(A^1; \star)$.) Then, for every equivalence class F of μ it holds:*

$$\forall g \in A^1 : \; F \star g \subseteq F.$$

Proof. Suppose μ has a certain equivalence class F, for which there are $f_1, g_1 \in F$ with $f_1 \star g_1 \notin F$. Let now $f_2^{m+1} := (\nabla f_1) \star h^m$ and $f_3^{p+1} := (\nabla f_1) \star t^p$, where $(h^m, t^p) \in \kappa$ and $m < p$. (Such h and t exist, since $\kappa \not\subseteq \kappa_a$.) Then, $(f_2, f_3) \in \kappa$, $(\Delta^{p-1} f_2, \Delta^{p-1} f_3) = (f_1, \nabla f_1) \in \kappa$ and $(f_1 \star g_1, (\nabla f_1) \star g_1) \in \kappa$. This implies $(\Delta(f_1 \star g_1), \Delta((\nabla f_1) \star g_1)) = (f_1 \star g_1, f_1) \in \kappa$, contrary to the assumption. ■

We will see in Theorem 9.8.8 that, to every congruence on $(A^1; \star)$, which fulfills the condition of Lemma 9.3.8, there is exactly a non-arity congruence on $[A^1]$.

9.4 About the Number of the Congruences on a Subclass of P_k

Denote $\mathfrak{k}_i(A)$ $(i = 1, 2)$ the cardinality of the set of all congruences i-th kind on the subclass A of P_k. Further let $\mathfrak{k}(A) := \mathfrak{k}_1(A) + \mathfrak{k}_2(A)$.
The next theorem is a direct conclusion from the results of Section 9.2:

Theorem 9.4.1 *For every closed subset A of P_2 it holds: $\mathfrak{k}_1(A) \in \{1, 2\}$, $\mathfrak{k}_2(A) \in \{1, 2, 3, 4\}$ and $\mathfrak{k}(A) \in \{2, 3, 4, 5\}$.* ■

For $k \geq 3$ the determination of all congruences is more complicated on an arbitrary class $A \subseteq P_k$. The following theorem mediates a first idea of that.

Theorem 9.4.2 *Let $k \geq 3$. For every integer $n \geq 2$ there is a subclass A_n of P_k with $\mathfrak{k}(A_n) = n$. Furthermore there exist certain closed sets B and C of P_k with $\mathfrak{k}(B) = \aleph_0$ and $\mathfrak{k}(C) = \mathfrak{c}$.*

Proof. Obviously, a closed set of P_k has at least two different congruences and at most continuum-many. Next, we give a class C with continuum-many different congruences.

Let the set C be defined as in Section 8.1. In Lemma 8.1.1 it was proven that C is closed and that C has an infinite basis. With the aid of properties of functions of C (given in the proof of Lemma 8.1.1) one can easily prove that the equivalence relation defined by

$$s^n \sim t^m \ (\kappa^N) \ :\Longleftrightarrow$$

$$\{s,t\} \subseteq C \ \wedge \ n = m \ \wedge$$
$$(s = t \ \vee \ \{s,t\} \subseteq \{g_{I,J} \in C \mid |I| \in N \ \vee \ J \neq \emptyset \ \vee \ I = J = \emptyset\})$$

is a congruence on C for every $N \subseteq \mathbb{N}$. Furthermore, it holds that $\kappa^N \neq \kappa^{N'}$ for $N \neq N'$. Therefore, by $|\mathfrak{P}(\mathbb{N})| = \mathfrak{c}$, there are continuum-many congruences on C.

Next we determine the congruences on the closed set A_t, $t \geq 0$, which is the set of all functions

$$g_J^n := g_{\emptyset,J}^n, \ |J| \leq t, \ n \geq 1.$$

For all r with $0 \leq r \leq t$, the equivalence relation π_r defined by

$$g_J^n \sim g_{J'}^{n'} \ :\Longleftrightarrow \ n = n' \ \wedge \ (J = J' \ \vee \ (|J| \leq r \ \wedge \ |J'| \leq r)\ )$$

is obvious a congruence on the class A_t. In particular, we have $\pi_0 = \kappa_0$ and $\pi_t = \kappa_a$ on A_t.

Obviously, $\kappa_0 = \kappa_a$ and κ_1 are the only congruences on A_0.

If $t \geq 1$ and κ is a congruence on A_t with $\pi_s \subset \kappa \subseteq \kappa_a$, $0 \leq s \leq t-1$, then there exist functions $g_{J_1}^n, g_{J_2}^n \in A_t$ with $g_{J_1}^n \sim g_{J_2}^n \ (\kappa)$, $J_1 \neq J_2$ and $|J_1 \geq s+1$. If $|J_2| \leq s$, we have $g_{J_1}^n \sim c_0^n \ (\kappa)$ and thus $g_J^m \sim c_0^m \ (\kappa)$ with $|J| \leq |J_1|$ and $m \geq 1$, since one can form all functions g_j^m as superpositions on g_{J_1}. Therefore, in this case $\pi_{s+1} \subseteq \kappa$. If $|J_2| \geq s+1$ then w.l.o.g. one can assume $1 \notin J_1$ and $1 \in J_2$. Hence, we obtain $g_{J_1}^n \star c_0^1 = g_{J_1}^n \ \sim \ c_0^n = g_{J_2}^n \star c_0^1 \ (\kappa)$, i.e., we have also in this case $\pi_{s+1} \subseteq \kappa$.

In summarizing we see that κ_0, π_1, π_2,..., π_{t-1} and κ_a are the only arity congruences on A_t. Because of Lemma 9.2.1, κ_1 is the only non-arity congruence on A_t. Consequently, for $t \geq 0$, A_t is a class with exactly $t+2$ different congruences.

It is easy to see that the set of all congruences on the closed set $B := \bigcup_{t \geq 0} A_t$ has the cardinality $\aleph_0$. ∎

Each class A with $|Con(A)| \in \{\aleph_0, \mathfrak{c}\}$ defined in the above proof has no basis or an infinite basis, respectively. A connection between $ord\ A = \infty$ and $|Con(A)| \in \{\aleph_0, \mathfrak{c}\}$ does not exist, however, as the examples show from the following two lemmas.

Lemma 9.4.3 *For each $k \geq 3$ there is a subclass A of P_k with $ord\ A = \infty$ and $Con\ A = 3$.*

Proof. Proving the lemma for $k = 3$ suffices. We consider the class

$$A := \bigcup_{n \geq 1} \{f \in P_{3,2} \mid \forall \mathbf{a} \in E_2^n : f(\mathbf{a}) = 0\},$$

for which we prove in Chapter 12 that it does not have any finite generating system. Subsequently, we will show that A has only trivial congruences.
Let κ be a congruence on A with $\kappa \neq \kappa_a$. Then, the following two cases are possible:
Case 1: $\kappa_0 \subset \kappa \subseteq \kappa_a$.
In this case, there are n-ary κ-congruent functions $f, g \in A$ with $f \neq g$. W.l.o.g. we can assume that $n = 3$, $f(0,1,2) = 0$ and $g(0,1,2) = 1$. Consequently, we have

$$f(c_0(x), j_2(x), x) = c_0^1(x) \sim j_2(x) = g(c_0(x), j_2(x), x)\ (\kappa).$$

Let t^m be an arbitrary m-ary function of A. Then the $(2 \cdot m)$-ary function h_t defined by

$$h_t(x_1, x_2, ..., x_{2\cdot m}) := \begin{cases} t(x_1, ..., x_m) \text{ if } & x_{m+1} = ... = x_{2\cdot m} = 0, \\ 0 & \text{otherwise} \end{cases}$$

belongs to A. From that we receive

$$\begin{aligned} & h_t(x_1, ..., x_m, c_0(x_1), ..., c_0(x_m)) = t(x_1, ..., x_m) \\ \sim\ & c_0^m(x_1, ..., x_m) = h_t(x_1, ..., x_m, j_2(x_1), ..., j_2(x_m))\ (\kappa). \end{aligned}$$

This implies $\kappa = \kappa_a$ immediately.
Case 2: $\kappa \not\subseteq \kappa_a$.
Since A has a right zero, $\kappa = \kappa_1$ results in this case from Lemma 9.3.5. Without use of Lemma 9.3.5, this can be shown as follows:
If $\kappa \not\subseteq \kappa_a$ then there are κ-congruent functions $f^n, g^m \in A$ with $n > m$. Therefore,

$$c_0^1 \star (\Delta^{n-1} f) = c_0^2 \sim c_0^1 = c_0^1 \star (\Delta^{n-1} g)\ (\kappa).$$

This implies that all constant functions of A are κ-congruent to each other. Furthermore, for a binary function $h \in A$ with $h(0,2) = 1$ we have:

$$\Delta((\tau(h \star c_0^1)) \star c_0^1) = c_0^1(x) \sim r(x,y) := \Delta((\tau(h \star c_0^2)) \star c_0^1)\ (\kappa),$$

where $r(0,2) = 1$. Because of $(c_0^1, c_0^2) \in \kappa$, this implies $(c_0^2, r) \in \kappa$ and thus $\kappa \cap \kappa_a \neq \kappa_0$. Then $\kappa = \kappa_1$ follows from our considerations for Case 1 and from the κ-congruence of the functions c_0^n for arbitrary n.
Hence A has only trivial congruences. ■

The class defined in the following lemma was published by V. L. Murskij in [Mur 65] as an example for a class without a finite basis of identities.

Lemma 9.4.4 *Let m be defined by*

$$m(x,y) := \begin{cases} x, \text{ if} & (x,y) \in \{(1,2),(2,1),(2,2)\}, \\ 0 \text{ otherwise} & \end{cases}$$

a binary function of P_3. [2] *Then, the subclass $M := [m]$ has infinite-many congruences.*

Proof. For $f^n \in M$ we denote with $\chi(f)$ the set of all tuples $\mathbf{a}$ of E_3^n with $f(\mathbf{a}) \in \{1,2\}$. Further, let

$$K_r^n := \{f \in P_3^n \mid \forall N \subseteq \chi(f) : \ (|N| \leq r \Longrightarrow (\exists i : \forall (a_1, ..., a_n) \in N : a_i = 2\}$$

and

$$K_r := \bigcup_{n \geq 1} K_r^n.$$

One can easily prove the following statement:

$$\forall f^n \in M \ \exists i : \ \mathbf{a} \in \chi(f) \implies f(\mathbf{a}) = a_i. \tag{9.9}$$

For $r \geq 1$, we consider the equivalence relation μ_r on M defined by

$$f^n \sim g^m \ (\mu_r) \ :\Longleftrightarrow \ n = m \ \wedge \ (f = g \ \vee \ \{f,g\} \subseteq K_r^n)$$

and show that it is a congruence on M. For this purpose, we choose arbitrary functions f^n, g^n, p^m, q^m of M with $(f,g),(p,q) \in \mu_r$. Then we have $(\alpha f, \alpha g) \in \mu_r$ for every $\alpha \in \{\zeta, \tau, \Delta, \nabla\}$. For the proof of $(f \star p, g \star q) \in \mu_r$, it suffices to show that

$$(f \in K_r \ \vee \ p \in K_r) \wedge (f \text{ depends essentially of } x_1) \implies f \star p \in K_r \tag{9.10}$$

holds for arbitrary $f, p \in M$. Let the first place of the function f be essential. We distinguish two cases:
Case 1: $f \in K_r$.
Let $\mathbf{a_1}, ..., \mathbf{a_s} \in \chi(f \star p)$, $1 \leq s \leq r$, $\mathbf{a_j} := (a_{j,1}, ..., a_{j,n+m-1})$, $j = 1, 2, ..., s$, $p(a_{j,1}, ..., a_{j,m}) =: b_j$. Because of $f \in K_r$ we have either $b_1 = ... = b_s = 2$ or there exists a $u \geq m+1$ with $a_{1,u} = ... = a_{s,u} = 2$. If $b_1 = ... = b_s = 2$, there is by (9.9) a v with $1 \leq v \leq m$ and $a_{v,1} = ... = a_{v,s} = 2$. Consequently, $f \star p$ belongs to K_r.
Case 2: $p \in K_r$.
In this case, since by assumption the first place of f is essential, $(f^n \star p^m)(\mathbf{a}) \in \{1,2\}$ for $\mathbf{a} \in E_3^{m+n-1}$ is only possible, if $p(a_1, ..., a_m) \in \{1,2\}$. Consequently, we have $\chi(f \star p) \subseteq \{\mathbf{a} \in E_3^{m+n-1} \mid (a_1, ..., a_m) \in \chi(p)\}$ and this implies $f \star p \in K_r$. Thus (9.10) was proven and μ_r is a congruence on M. One can

[2] We notice that the function m can be interpreted as a binary partial projection. In this case $m(a,b) = 0$ stands instead of "$m(a,b)$ is not defined".

prove that the congruences of the form μ_r are different pairwise as follows: Let $xy := m(x,y)$. The functions t^n and s^n defined by

$$t(x_1,...,x_n)$$
$$:= ((((((x_1((x_2x_1))(x_3x_1))(x_3x_2))(x_4x_1))...(x_4x_3))...)(x_nx_{n-1})(x_nx_1))(x_3x_2)) \\ ...(x_nx_{n-1}))$$
$$= \begin{cases} x_1 \text{ if} & \mathbf{x} \in \{(1,2,...,2),(2,1,2,...,2),...,(2,...,2,1),(2,...,2)\}, \\ 0 \text{ otherwise} \end{cases}$$

and

$$s(x_1,...,x_n) := t(m(x_1,x_1),...,m(x_n,x_n))$$
$$= \begin{cases} 2 \text{ if} & x_1 = ... = x_n = 2, \\ 0 \text{ otherwise} \end{cases}$$

belong to M, and it holds $(t^n, s^n) \in \mu_{n-1}$ and $(t^n, s^n) \notin \mu_n$ for arbitrary $n \geq 1$.
Consequently, there are infinite-many congruences on M. ■

With the aid of Lemma 9.10.1 and [Oat-W 80], we will later prove that the above-defined class M has continuum-many congruences.

The following theorem gives examples of classes A whose congruences are completely determined by the congruences on certain sets B with $A \subseteq B$. We use this theorem in Section 9.6.

Theorem 9.4.5 *Let S be the set of all functions $f^n \in P_k$ for which*

$$f(x_1,x_2,...,x_n) = s^{-1}(f(s(x_1),s(x_2),...,s(x_n))$$

holds, i.e., $S = Pol_k\{(x,s(x)) \mid x \in E_k\}$. Further, let α be the isomorphism defined in Theorem 8.3.2.
Then, $|Con\ A| \leq |Con\ \alpha(A)|$.
More precisely:
Let A ($\neq \emptyset$) be a subclass of S, let κ be a congruence on A and let $\alpha(\kappa)$ be a relation on $\alpha(A)$ defined by

$$(F,G) \in \alpha(\kappa) :\Longleftrightarrow (\alpha^{-1}F, \alpha^{-1}G) \in \kappa.$$

Then

(a) $\alpha(\kappa)$ is a congruence on A;

(b) $\alpha(\kappa)_{/A} = \kappa$, i.e., by means of restriction on A one can get the congruences on A from the congruences on $\alpha(A)$.

Proof. Since κ is a congruence on A, κ is compatible with the operations $\hat{\zeta}, \hat{\tau}, \hat{\Delta}, \hat{\nabla}$ and $\hat{*}$. This and Theorem 8.3.2, (a) imply our statement (a).

By Theorem 8.3.2, (b) we have $A \subset \alpha(A)$. Therefore, $\alpha(\kappa)_{/A}$ is a congruence on A. Now let f and g be arbitrary functions of A. If $(f,g) \in \kappa$ then

$(\nabla f, \nabla g) \in \kappa$ and by definition of $\alpha(\kappa)$; further, we have $(\alpha(\nabla f), \alpha(\nabla g)) = (f, g) \in \alpha(\kappa)_{/A}$, i.e., $\kappa \subseteq \alpha(\kappa)_{/A}$. If $(f, g) \in \alpha(\kappa)_{/A}$ then $(\alpha^{-1}f, \alpha^{-1}g) \in \kappa$. Since $f, g \in S$, it holds $\alpha^{-1}f = \nabla f$ and $\alpha^{-1}g = g$. Consequently, $(\nabla f, \nabla g) \in \kappa$ and $(\Delta(\nabla f), \Delta(\nabla g)) = (f, g) \in \kappa$, i.e., $\alpha(\kappa)_{/A} \subseteq \kappa$. This implies (b). ■

9.5 A Criterion for the Proof of the Countability of *Con A* for Certain $A \subseteq P_k$

It is possible to choose a congruence relation on an algebra (in particular, a function algebra) as a universe of another algebra, as follows:
Let A be the universe of a subalgebra of $\mathbf{P_k} = (P_k; \zeta, \tau, \Delta, \nabla, *)$. Then, one can define for arbitrary $(f, g), (p, q) \in A \times A$:

$$\forall\ \alpha \in \{\zeta, \tau, \Delta, \nabla\}:\ \alpha(f, g) := (\alpha f, \alpha g),$$

$$(f, g) * (p, q) := (f * p, g * q),$$

$$\sigma(f, g) := (g, f) \text{ and}$$

$$\varrho((f, g), (p, q)) := \begin{cases} (f, q) & \text{if } g = p, \\ (f, g) & \text{otherwise.} \end{cases}$$

Then each congruence on A is a universe of a subalgebra of the algebra $(A \times A; \zeta, \tau, \Delta, \nabla, *, \sigma, \varrho, (f, f)_{f \in A})$; i.e., it holds

$$Con\ A = Sub(A \times A; \zeta, \tau, \Delta, \nabla, *, \sigma, \varrho, (f, f)_{f \in A}).$$

(The tupel (f, f), where $f \in A$ is arbitrary, are certain nullary operations of the above defined algebra.) In particular, we have for the arity congruences:

$$Con_a A = Sub(\bigcup_{n \geq 1} A^n \times A^n; \zeta, \tau, \Delta, \nabla, *, \sigma, \varrho, (f, f)_{f \in A}).$$

In order to be able to prove the countability of $Con\ A$ for some $A \subseteq P_k$ later with the aid of Theorem 8.1.5, we show first that class A fulfills the conditions of Theorem 8.1.5.
Denote $\circ$ an operation on E_k with the properties

$$a)\ \forall\ a, b, c \in E_k:\ a \circ (b \circ c) = (a \circ b) \circ c,$$

$$b)\ \forall\ a, b \in E_k:\ a \circ b = b \circ a,$$

$$c)\ \exists\ r \geq 1:\ (\forall\ a \in E_k:\ \underbrace{a \circ a \circ \ldots \circ a}_{(1+r)-times} = a).$$

The properties a)–c) e.g. are fulfilled, if $(E_k, \circ)$ is an Abelean group or $x \circ y := \max_\omega(x, y)$ or $x \circ y := \min_\omega(x, y)$ (ω is a partial order relation, for which exists max or min, respectively).

With the help of $\circ$ one can define the following subset of P_k:

$$K_0 := \bigcup_{n\geq 1} \{f^n \in P_k \mid \exists\ f_0, f_1, \ldots, f_n \in P_E^1 : \ f(x_1, \ldots, x_n) = f_0(f_1(x_1) \circ f_2(x_2) \circ \ldots \circ f_n(x_n))\}.$$

In general, the set K_0 is not closed; it contains, however, some closed sets in the case of suitable choice of $\circ$ (see Theorem 9.5.2).
In analog manner to the proof of $|Sub\ A| \leq \aleph_0$ for all $A = [A] \subseteq K_0$ in Theorem 8.1.6 one can show:

Theorem 9.5.1 *([Lau 90])*
Each subclass A of P_k, for which exists an operation $\circ$ with the properties a)–c) and $A \subseteq K_0$, has at most countable-many congruences.

Proof. Since, by Theorem 9.3.4, every subclass of P_k has only finite-many non-arity congruences, it is sufficient to show that the set $Con\ (\bigcup_{n\geq 1} A^n \times A^n; \zeta, \tau, \Delta, \nabla, *, \sigma, \varrho)$ is finite or countable. For this purpose we define an order relation $\leq$, which fulfills the conditions 1)–3) from Theorem 8.1.5, on $K_0^* := \bigcup_{n\geq 1} K_0^n \times K_0^n$. Then, our theorem follows from Theorem 8.1.5.
Each pair $(f, g) \in K_0^n \times K_0^n$ with $f(\tilde{x}) = f_0(f_1(x_1) \circ \ldots \circ f_n(x_n))$ and $g(\tilde{x}) = g_0(g_1(x_1) \circ \ldots \circ g_n(x_n))$ is well-defined by the tuples $\varphi(f, g) := ((f_0, g_0), (f_1, g_1), \ldots, (f_n, g_n))$. Two pairs (f, g) and (p, q) with $\varphi(f, g) = ((f_0, g_0), \ldots, (f_n, g_n))$ and $\varphi(p, q) = ((p_0, q_0), \ldots, (p_m, q_m))$ are called $\sim$–equivalent, if they fulfill the three following conditions:

(1) $(f_0, g_0) = (p_0, q_0)$,
(2) $\{(f_1, g_1), \ldots, (f_n, g_n)\} = \{(p_1, q_1), \ldots, (p_m, q_m)\}$,
(3) If (f_i, g_i) occurs in $((f_1, g_1), \ldots, (f_n, g_n))$ exactly u times, then there exists a certain integer v so that (f_i, g_i) occurs in $((p_1, q_1), \ldots, (p_m, q_m))$ exactly $(u + v \cdot r)$ times, where r denotes the smallest integer for which $\circ$ fulfills the condition c).

Obviously, the factor set $K^*_{0/\sim}$ is finite. Set $\{K_1, ..., K_t\} := K^*_{0/\sim}$. On the sets $K_1, \ldots, K_t$ one can define an order relation $\leq$ as follows:

For $(f^n, g^n), (p^m, q^m) \in K_i$ $(i = 1, \ldots, t)$ with $\varphi(f, g) = ((f_0, g_0), \ldots, (f_n, g_n))$ and $\varphi(p, q) = ((p_0, q_0), \ldots, (p_m, q_m))$ we write $(f, g) \leq (p, q)$ if and only if for every (f_i, g_i), which occurs in $((f_1, g_1), \ldots, (f_n, g_n))$ j_1 times and occurs in $((p_1, q_1), \ldots, (p_m, q_m))$ j_2 times, it holds $j_1 \leq j_2$. This order relation $\leq$ fulfills the condition 1) of Theorem 8.1.5, since one can receive (f, g) from (p, q) with the aid of the operations ζ, τ, Δ, if $(f, g) \leq (p, q)$ is valid (because of property c) of $\circ$). Obviously the property 2) of Theorem 8.1.5 is also valid. Thus, it remains to show that every antichain in the set K_0^* in respect to $\leq$ has only finite-many elements. On $P_E^1 \times P_E^1$ one can define a total order. For $\varphi(f, g) = ((f_0, g_0), \ldots, (f_n, g_n))$, we arrange (concerning this order) that

(f_{i_0}, g_{i_0}) is the smallest element of the set $\{(f_1, g_1), \ldots, (f_n, g_n)\}$; (f_{i_1}, g_{i_1}) is the smallest element of the set $\{(f_1, g_1), \ldots, (f_n, g_n)\} \setminus \{(f_{i_0}, g_{i_0})\}, \ldots$ and (f_{i_s}, g_{i_s}) is the the greatest element of $\{(f_1, g_1), \ldots, (f_n, g_n)\}$. If (f_{i_j}, g_{i_j}) occurs in $\{(f_1, g_1), \ldots, (f_n, g_n)\}$ exactly b_j times, $j = 0, 1, \ldots, s$, then let $\alpha(f, g) := (b_0, b_1, \ldots, b_s)$. Further, let $D_i := \{\alpha(f, g) | \ (f, g) \in K_i\}$, $i = 1, 2, \ldots, t$. We set $\alpha(f, g) \leq \alpha(p, q)$ iff $(f, g) \preceq (p, q)$. The ordered sets K_i and D_i are isomorphic. In the proof of Theorem 8.1.6, it was shown that every antichain $C \subseteq D_i$ is a finite set. Therefore, $\preceq$ also fulfills the condition 3) of Theorem 8.1.5. ∎

Some consequences from Theorem 9.5.1 are summarized in the following:

Theorem 9.5.2 *If A is a subclass of one of the sets*
(a) L_k (see Chapter 4),
(b) $Pol_k \lambda$, $\lambda \in \mathfrak{L}_k$ (see Section 5.2.4),
(c) L_n (see Chapter 14)
or
(d) $[O^1 \cup \{max\}]$ (see Chapter 15),
then $|Con\ A| \leq \aleph_0$. ∎

More precise statements about $Con\ A$ for the above cases (a)–(c) are treated in the next section.

9.6 Congruences on Some Classes of Linear Functions

Let $R = (R; +, \cdot)$ be a unitary ring and let $M = (M; +, \cdot)$ be a left module on R. Then one can define a closed subset L_M of P_M as follows:

$$L_M := \bigcup_{n \geq 1} \{f^n \in P_M \mid \exists\ a_0 \in M\ \exists\ a_1, \ldots, a_n \in R: \ f(\mathbf{x}) = a_0 + \sum_{i=1}^{n} a_i \cdot x_i\}.$$

By [Sze 80], for every subclass A of L_M that contains the function r defined by

$$r(x, y, z) := x + y - z,$$

there exists a uniquely defined subring T_A of R and an uniquely defined submodule N_A of $R \times M$ on T_A with

$$A = \bigcup_{n \geq 1} \{f^n \in L_M \mid \exists\ a_0 \in M\ \exists\ a_1, \ldots, a_n \in T_A: \ f(\mathbf{x}) = a_0 + \sum_{i=1}^{n} a_i x_i \ \wedge$$
$$(1 - a_1 - a_2 - \cdots - a_n, a_0) \in N_A\}.$$

The sets T_A and N_A are defined by

$$T_A := \{a \in R \mid \ ax + (1 - a)y \in A\},$$
$$N_A := \{(1 - a, b) \in R \times M \mid \ ax + b \in A^1\}.$$

For the purpose of determining of the congruences on A, let I be an arbitrary ideal of T_A, and let U be a suitably elected submodule of N_A with

$$\forall \alpha, \beta \in T_A \; \forall u, v \in N_A : \; (\alpha - \beta \in I \wedge u - v \in U \Longrightarrow \alpha \cdot u - \beta \cdot v \in U). \quad (9.11)$$

With the help of I and U, one can define a relation $\kappa(I, U)$ ($\subseteq \kappa_a$) as follows:

$$\begin{aligned} &(a_0 + \textstyle\sum_{i=1}^n a_i x_i, \; b_0 + \sum_{i=1}^n b_i x_i) \in \kappa(I, U) \; :\Longleftrightarrow \\ &\forall i \in \{1, \ldots, n\} : \\ &a_i - b_i \in I \; \wedge \; (1 - a_1 - \cdots - a_n, a_0) - (1 - b_1 - \cdots - b_n, b_0) \in U. \end{aligned}$$

Lemma 9.6.1 *$\kappa(I, U)$ is a congruence on A.*

Proof. It can easily be shown that $\kappa(I, U)$ is an equivalence relation and that $\kappa(I, U)$ is compatible with the operations ζ, τ, Δ and ∇. Let now (f^n, g^n), $(s^m, t^m) \in \kappa(I, U)$ be arbitrary with $f(\mathbf{x}) = a_0 + a_1 x_1 + \cdots + a_n x_n$, $g(\mathbf{x}) = b_0 + b_1 x_1 + \cdots + b_n x_n$, $s(\mathbf{x}) = c_o + c_1 x_1 + \cdots + c_m x_m$ and $t(\mathbf{x}) = d_0 + d_1 x_1 + \cdots + d_m x_m$. Then, $b_i = a_i + \alpha_i$, $d_j = c_j + \beta_j$ for certain α_i, $\beta_j \in I$ $(i = 1, \ldots, n, \; j = 1, \ldots, m)$,

$$\begin{aligned} &(f * s)(\mathbf{x}) = a_0 + a_1 c_0 + a_1 c_1 x_1 + \ldots + a_1 c_m x_m + a_2 x_{m+1} + \ldots + a_n x_{m+n-1}, \\ &(g * t)(\mathbf{x}) = b_0 + b_1 d_0 + b_1 d_1 x_1 + \ldots + b_1 d_m x_m + b_2 x_{m+1} + \ldots + b_n x_{m+n-1}, \\ &a_1 c_i - b_1 d_i = -\alpha_1 \beta_i - \alpha_1 c_i - a_1 \beta_i \in I \; (i = 1, \ldots, m) \text{ and} \end{aligned}$$

$$\begin{aligned} &(1 - a_1 c_1 - \ldots - a_1 c_m - a_2 - \ldots - a_m, a_0 + a_1 c_0) \\ &-(1 - b_1 d_1 - \ldots - b_1 d_m - b_2 - \ldots - b_m, b_0 + b_1 d_0) \\ &= a_1 (1 - c_1 - \ldots - c_m, c_0) - b_1 (1 - d_1 - \ldots - d_m, d_0) \\ &+(1 - a_1 - \ldots - a_n, a_0) - (1 - b_1 - \ldots - b_n, b_0) \in U \end{aligned}$$

by (9.11) and $(f, g), (s, t) \in \kappa$. Thus $\kappa(I, U)$ is a congruence on A. ■

We need some denotation and the following three lemmas to prove that on A next to the trivial congruences only congruences of the type $\kappa(I, U)$ still exist.
The function q_a (for $a \in \mathbf{R}$) is defined by

$$q_a(x, y) := a \cdot x + (1 - a) \cdot y.$$

Lemma 9.6.2 *Let $\kappa \subseteq \kappa_a$ be a congruence on A, let $f(\mathbf{x}) = a_0 + a_1 x_1 + \ldots + a_n x_n \in A$ and let $g(\mathbf{x}) = b_0 + b_1 x_1 + \ldots + b_n x_n \in A$. Then*

$$(f, g) \in \kappa \Longleftrightarrow \forall i \in \{1, \ldots, n\} : \; (q_{a_i}, q_{b_i}) \in \kappa \; \wedge \; (\Delta^{n-1} f, \Delta^{n-1} g) \in \kappa$$

with $\Delta^{i+1} f := \Delta(\Delta^i f)$, $\Delta^1 f := \Delta f$.

Proof. Let $h \in \{f, g\}$, $h(x_1, \ldots, x_n) = \alpha_0 + \alpha_1 x_1 + \ldots + \alpha_n x_n$, and denote r_n the $(n+2)$–ary function of $[\{r\}]$ which is defined by $r_n(x_1, \ldots, x_{n+2}) :=$

$x_1+x_2+\ldots+x_{n+1}-n\cdot x_{n+2}$. Then, the statement of our lemma easily follows from the following identities:

$$q_{\alpha_i}(x,y) = r(h(y,\ldots,y,\underbrace{x}_{i},y,\ldots,y),y,h(y,\ldots,y)) \text{ and}$$

$$h(\mathbf{x}) = r_n(q_{\alpha_1}(x_1,x_1),q_{\alpha_2}(x_2,x_1),\ldots,q_{\alpha_n}(x_n,x_1),(\Delta^{n-1}h)(x_1),x_1).$$

■

Let $Q_A := \{q_a \in A^2 |\ a \in T_A\}$. One can define two operations $\oplus$, $\odot$ on Q_A as follows:

$$(q_a \oplus q_b)(x,y) := r(q_a(x,y),q_b(x,y),y) \text{ and}$$
$$(q_a \odot q_b)(x,y) := q_a(q_b(x,y),y).$$

Because of $q_a \oplus q_b = q_{a+b}$ and $q_a \odot q_b = q_{a\cdot b}$, one can easily prove that the following holds:

Lemma 9.6.3 *The algebra $(Q_A;\oplus,\odot)$ is isomorphic to $(T_A;+,\cdot)$.* ■

For arbitrary $g,h \in A^1$ and arbitrary $\alpha \in T_A$ we set now

$$(g \oplus h)(x) := r(g(x),h(x),x) \text{ and}$$
$$(\alpha \odot g)(x) := q_a(g(x),x).$$

With the help of the mapping

$$\varphi:\ A^1 \longrightarrow N_A,\quad ax+b \longmapsto (1-a,b)$$

one can easily prove the following lemma:

Lemma 9.6.4 *The left module $(A^1;\oplus,\odot)$ and $(N_A;+,\cdot)$ over T_A are isomorph.* ■

Theorem 9.6.5 *Let A be a subclass of L_M with $r \in A$. Then A has only the congruences $\kappa_0,\kappa_a,\kappa_1$ and such of the type $\kappa(I,U)$.*

Proof. Let κ be a congruence on A with $\kappa \neq \kappa_1$. By $r \in A$, A is a clone. Therefore, $\kappa \subseteq \kappa_a$ (see Lemma 9.2.1 or Lemma 9.3.5). With the aid of Lemmas 9.6.2 and 9.6.3 (or 9.6.4) one can see that κ determines a congruence μ_κ (or ν_κ) on T_A (or N_A), respectively:

$$(q_a,q_b) \in \kappa \Longleftrightarrow (a,b) \in \mu_\kappa \quad (\text{or} \tag{9.12}$$
$$(ax+b,cx+d) \in \kappa \Longleftrightarrow ((1-a,b),(1-c,d)) \in \nu_\kappa). \tag{9.13}$$

Then there is an ideal I of T_A and a submodule U of N_A with the properties

$$(a, b) \in \mu_\kappa \Longleftrightarrow a - b \in I, \tag{9.14}$$

$$(u, v) \in \nu_\kappa \Longleftrightarrow u - v \in U. \tag{9.15}$$

To prove (9.11) let $a, b \in T_A$, let $u = (1-\alpha, \beta)$, $v = (1-\gamma, \delta) \in N_A$ with $a - b \in I$ and let $u - v \in U$. Then, it holds $\{(q_a, q_b), (\alpha x + \beta, \gamma x + \delta)\} \subseteq \kappa$. Consequently, we have

$$(q_a(\alpha x + \beta, x), q_b(\gamma x + \delta, x)) = ((a\alpha + 1 - a)x + a\beta, (b\gamma + 1 - b)x + b\delta) \in \kappa.$$

Therefore,

$$((1-(a\alpha+1-a), a\beta), (1-(b\gamma+1-b), b\delta)) = ((a - a\alpha, a\beta), (b - b\gamma, b\delta)) \in \nu_\kappa$$

and thus $a \cdot (1-\alpha, \beta) - b \cdot (1-\gamma, \delta)) \in U$, i.e., (9.11) holds. By Lemma 9.6.2 and (9.12)–(9.15), we have $\kappa = \kappa(I, U)$. ∎

Next, we determine a property of the congruences on the closed subset

$$L_{M;id} := \bigcup_{n\geq 1}\{f^n \in L_M \mid \exists a_0, \ldots, a_n : \\ f(\mathbf{x}) = a_0 + \sum_{i=1}^n a_i x_i \ \wedge \ a_1 + a_2 + \ldots + a_n = 1\}$$

(a set of so-called idempotent functions) of L_M. It is easy to check that

$$L_{M;id} = L_M \cap Pol_M\{(x, x+1) \mid x \in M\}$$

holds. By Theorem 8.3.2 and with the aid of the bijective mapping

$$\alpha : \ f^n \longrightarrow F^{n-1}$$

from $L_{M;id} := L_{M;id} \backslash L_M^1$ onto L_M, where

$$F(x_1, \ldots, x_{n-1}) := f(0, x_1, \ldots, x_{n-1}),$$

we can characterize the subclasses of $L_{M;id}$ and receive as a consequence of Theorem 9.4.5:

Theorem 9.6.6 *Let A $(\neq \emptyset)$ be a subclass of $L_{M;id}$, let κ be a congruence on A and let $\alpha(\kappa)$ be a relation on $\alpha(A)$ defined by*

$$(F, G) \in \alpha(\kappa) :\Longleftrightarrow (\alpha^{-1}F, \alpha^{-1}G) \in \kappa.$$

Then:

(a) $\alpha(\kappa)$ is a congruence on A;

(b) $\alpha(\kappa)_{/A} = \kappa$, i.e., by means of restriction on A one can get the congruences on A from the congruences on $\alpha(A)$. ∎

From Theorems 9.6.5 and 9.6.6 some consequences can be drawn (with the aid of known properties of the groups, fields and vector spaces) for the classes from Chapter 13 and from [Sze 80] and [Sza-S 81]. Here only a few examples are given without proof.

Corollary 9.6.7

(1) *Let $R = M$ and let R be a finite field. Then, by [Sze 80], each subclass A of L_M with $r \in A$ and $A \not\subseteq L_{M;id}$ has the form (up to isomorphic)*

$$\begin{array}{l} L_W \cap Pol\ V \\ (\ = \bigcup_{n\geq 1}\{f^n \mid \exists\ a_0 \in V\ \exists\ a_1,\ldots,a_n \in W: \\ \qquad\qquad\qquad f(\mathbf{x}) = a_0 + a_1x_1 + \ldots + a_nx_n\}), \end{array}$$

where W is a subfield of R and V is a subspace of the vector space R on W. Then, for every nontrivial congruence κ on A there are an $I \in \{\{0\}, W\}$ and a subspace U of V with the property

$$\forall\ \alpha - \beta \in I\ \ \forall\ \ a - b \in U:\ a\cdot\alpha - b\cdot\beta \in U,$$

and it holds that

$$\begin{array}{l} \kappa = \\ \bigcup_{n\geq 1}\{(a_0 + \sum_{i=1}^n a_ix_i, b_0 + \sum_{i=1}^n b_ix_i) \mid \\ \qquad a_0 - b_0 \in U\ \wedge\ \forall\ i \in \{1,\ldots,n\}:\ a_i - b_i \in I\}. \end{array}$$

In particular,

$$\kappa_c := \{(f, g) \in \kappa_a \mid \exists\ a \in M:\ f = g + a\}$$

is the only nontrivial congruence on L_M.

(2) *Let L be a maximal class of quasi-linear functions of P_k with $k = p^m$, p prime, $m \geq 1$. Then κ_c is the only nontrivial congruence on L (see Section 5.2.4 and [Lau 81]). By Lemma 5.2.4.3, L is isomorphic to a certain set L_M, where R is the ring of all $m\times m$–matrices on E_p and M is the vector space of all $m \times 1$–matrices on E_p.*

(3) *Choosing $R = M = (E_k; + \bmod k, \cdot \bmod k)$, so there are only the following arity congruences on L_M:*

$$\begin{array}{l} \kappa_{s,t} := \bigcup_{n\geq 1}\{(a_0 + \sum_{i=1}^n a_ix_i, b_0 + \sum_{i=1}^n b_ix_i) \mid \\ \qquad\qquad s \mid a_0 - b_0\ \wedge\ \forall\ i \in \{1,\ldots,n\}:\ t \mid a_i - b_i\}, \end{array}$$

where s and t are arbitrary divisors of k with $s \mid t$. ■

Notice that *Con A* is not a finite set for every subclass A of L_M. If, for example, one finds an element z in $R\backslash\{0\}$ with $z^2 = 0$, then the closed set

$$Z := \bigcup_{n\geq 1}\{f^n \in L_M \mid \exists\ a_1,\ldots,a_n \in [\{0,z\}]_+:\ f(\mathbf{x}) = a_1x_1 + \ldots + a_nx_n\}$$

has infinite-many congruences. For the purpose of describing of some of these relations, let $var_e(f)$ be the number of the essential variables of $f \in Z$ and let

$$(f^n, g^n) \in \chi_i :\Longleftrightarrow f = g \vee (n = m \wedge var_e(f) \leq i \wedge var_e(g) \leq i)$$
$$(f, g \in Z, i \in \mathbb{N}).$$

It is easy to check that, for every $i \in \mathbb{N}$, the relation χ_i is a congruence on Z. With the aid of Theorem 9.3.4 and Theorem 9.5.1, $|Con\ Z| = \aleph_0$ results from that.

Finally, we determine all congruences on the subclasses of L_k (see Chapter 13) for $k \in \mathbb{P}$.
We start with the determining of the congruences on certain subclasses of $[P_k^1]$ for arbitrary k. The following lemma is easily provable:

Lemma 9.6.8 *Let* $C \subseteq \{c_0^1, c_1^1, ..., c_{k-1}^1\}$, *let* G *be a subgroup of* $S_k := (P_k^1[k]; \star)$, *whose elements preserve* C, *let* U *be a normal subgroup of the group* G *and let* μ *be an equivalence relation on* C, *which is preserved from all functions of* G. *Then the relation* $\kappa^{U,\mu}$ *on* $[G \cup C]$ *defined by*

$$\begin{aligned} f^n \sim g^m\ (\kappa^{U,\mu}) :\Longleftrightarrow\ & n = m\ \wedge\ (\exists i\ \exists f', g' \in C \cup G : \\ & f(\mathbf{x}) = f'(x_i)\ \wedge\ g(\mathbf{x}) = g'(x_i)\ \wedge \\ & (f' \star U = g' \star U\ \vee\ (f', g') \in \mu)) \end{aligned}$$

is a congruence on $[G \cup C]$. ■

Theorem 9.6.9 *Let* $C \subseteq \{c_0^1, c_1^1, ..., c_{k-1}^1\}$, *let* G *be a subgroup of* $S_k := (P_k^1[k]; \star)$ *and let* $G \subseteq Pol_k\{a \mid c_a \in C\}$. *Then, there are except for* κ_1 *and* κ_a *only the congruences of the type* $\kappa^{U,\mu}$ *on* $[C \cup G]$, *where* U *is an arbitrary normal subgroup of* G *and* μ *is an arbitrary equivalence relation on* C, *which is preserved from the functions of* G.

Proof. Let κ be a nontrivial congruence on $[C \cup G]$. Because of Lemma 9.2.1, $\kappa \subseteq \kappa_a$. Discussing the following three cases suffices:
Case 1: There exist κ-congruent functions f^n and g^n with $\Delta^{n-1}f \in G$ and $\Delta^{n-1}g \in C$.
Then, $\Delta^{n-1}f \sim \Delta^{n-1}g\ (\kappa)$ and thus $e_1^1 \sim c_a^1 := \Delta^{n-1}g\ (\kappa)$. Consequently, we have for every function $h^m \in [C \cup G]$: $e_1^1 \star h = h \sim c_a^m = c_a^1 \star h\ (\kappa)$. Hence, $\kappa = \kappa_a$, in contradiction to the assumption.
Case 2: There exist κ-congruent functions f^n and g^n, which essentially depend on different variables.
W.l.o.g. let $f(x_1, ..., x_n) = f_1(x_1)$, $g(x_1, ..., x_n) = g_1(x_2)$, $\{f_1, g_1\} \subseteq G$ and $(f, g) \in \kappa$. For the inverse functions f_1^{-1} and g_1^{-1} appertaining to the functions f_1 and g_1 is valid then:

$$\begin{aligned} & f(f_1^{-1}(x_1), g_1^{-1}(x_2), x_2, ..., x_2) = e_1^2(x_1, x_2) \\ \sim\ & g(f_1^{-1}(x_1), g_1^{-1}(x_2), x_2, ..., x_2) = e_2^2(x_1, x_2)\ (\kappa). \end{aligned}$$

Therefore, for arbitrary m-ary functions s and $t \in [C \cup G]$: $e_1^2(s(\mathbf{x}), t(\mathbf{x})) = s(\mathbf{x}) \sim t(\mathbf{x}) = e_2^2(s(\mathbf{x}), t(\mathbf{x}))$ (κ); i.e., in this case, we have also $\kappa = \kappa_a$, in contradiction to the assumption.

Case 3: Two arbitrary κ-congruent functions of $[C \cup G]$ are both either constant functions or the functions depend of the same variables essentially.

In this case, the congruence κ is uniquely determined by $\kappa_{/G}$ and $\kappa_{/C}$. As is generally known, a congruence of a group G causes a partition of this group in cosets, which can be formed with the aid of a normal subgroup U of this group, i.e., $f, g \in G$ are congruent iff $f \star U = g \star U$. Obviously, $\kappa_{/C}$ is an equivalence relation on C, which is preserved by G. Consequently, we have $\kappa = \kappa^{U,\kappa_{/C}}$. ■

Theorem 9.6.10 *Let $k \in \mathbb{P}$ and let $A \subseteq L_k$ (see Chapter 13). Then there are only finite-many congruences on A. The nontrivial congruences on A are*

(1) κ_c, if $A \in \{\, L_k,\ L_k \cap Pol_k\{(x, x+1) \mid x \in E_k\}\,\}$,

(2) $\kappa^\mu := \{(f^n, g^m) \in A \times A \mid (\Delta^{n-1} f, \Delta^{m-1} g) \in \mu\}$ and
$\kappa_a^\mu := \{(f^n, g^m) \in A \times A \mid n = m \ \wedge\ (\Delta^{n-1} f, \Delta^{n-1} g) \in \mu\}$
for each equivalence relation μ on A^1, if $A \subseteq [\{c_0, c_1, ..., c_{k-1}\}]$;

(3) congruences of the type $\kappa^{U,\mu}$ (see Lemma 9.6.8), if $A \subseteq [L_k^1]$ and $A \nsubseteq [\{c_0, c_1, ..., c_{k-1}\}]$.

Proof. Our theorem results from the description of the subclasses of L_k (see Chapter 13) and from Theorems 9.6.5, 9.6.6 and 9.6.9. ■

9.7 Congruences on the Maximal Classes of P_k

The following two lemmas are auxiliary statements to the proof of Theorem 9.7.3, in which all maximal classes are given that have only trivial congruences. Not only do results on the congruences of certain maximal classes follow from Lemma 9.7.1, but also those on further classes (see for example Theorem 9.10.8). The statement (a) of the following lemma was already proven by A. I. Mal'tsev in [Mal 66].

Lemma 9.7.1 *For $\omega, u \in E_k$ and $\varrho \in \mathfrak{M}_k$, where $\mathfrak{o}_\varrho$ is the smallest element of E_k in respect to ϱ and $\mathfrak{e}_\varrho$ is the greatest element of E_k in respect to ϱ, let $c^1_{\omega,u}$, $q^1_{\omega,u}$, $m^1_{\varrho,u}$ and t^2_ω be functions defined by*

$$c_{\omega,u}(x) := \begin{cases} \omega \text{ if } & x = \omega, \\ u \text{ otherwise}, & \end{cases} \qquad q_{\omega,u}(x) := \begin{cases} u \text{ if } & x = u, \\ \omega \text{ otherwise}, & \end{cases}$$

$$m_{\varrho,u}(x) := \begin{cases} \mathfrak{o}_\varrho \text{ if } & x \leq_\varrho u, \\ \mathfrak{e}_\varrho \text{ otherwise}, & \end{cases} \qquad t_\omega(x,y) := \begin{cases} \omega \text{ if } & x = \omega, \\ y \text{ otherwise}. & \end{cases}$$

Furthermore, let A be a subclass of P_k, which fulfills the following two conditions:
(a) $\exists\omega \in E_k : \ (\bigcup_{u\in E_k}\{c_{\omega,u}, q_{\omega,u}\}) \cup \{t_\omega\} \subseteq A;$
(b) $\exists\varrho \in \mathfrak{M}_k\ \exists\omega \in \{\mathfrak{o}_\varrho, \mathfrak{e}_\varrho\} : \ (\bigcup_{u\in E_k}\{c_u, m_{\varrho,u}\}) \cup \{t_\omega\} \subseteq A.$
Then, A has only trivial arity congruences.

Proof. Denote κ an arbitrary congruence on A with $\kappa_0 \subset \kappa \subseteq \kappa_a$. Then there exist two different functions r^n, $s^n \in A$ and a tuple $\mathbf{a} := (a_1, ..., a_n) \in E_k^n$ with $r(\mathbf{a}) \neq s(\mathbf{a})$ and $(r, s) \in \kappa$.
First, assume A fulfills (a). We distinguish two cases for A:
Case 1: $A \not\subseteq Pol_k\{\omega\}$.
Then there is a function $f^1 \in A$ with $f(\omega) := \alpha \neq \omega$ and we have $f \star c_{\omega,\omega} = c_\alpha \in A$. Thus because of $c_{\omega,u} \star c_\alpha = c_u$ $(u \in E_k)$, all constant functions of P_k belong to A. Consequently, $(r, s) \in \kappa$ implies the κ-congruence of certain constant functions c_a^1, c_b^1 with $a \neq b$ and $a \neq \omega$. Thus we obtain

$$c_a = q_{\omega,a} \star c_a \ \sim q_{\omega,a} \star c_b = c_\omega \ (\kappa)$$

and

$$\begin{aligned} &\forall g^n \in A : \\ &t_\omega(c_a(x_1), g(x_1, ..., x_n)) = g(x_1, ..., x_n) \\ \sim\ &t_\omega(c_\omega(x_1), g(x_1, ..., x_n)) = c_\omega^n(x_1, ..., x_n) \ (\kappa). \end{aligned} \tag{9.16}$$

Hence $\kappa = \kappa_a$.
Case 2: $A \subseteq Pol_k\{\omega\}$.
In this case, we obtain

$$\begin{aligned} &r(c_{\omega,a_1}(x_1), ..., c_{\omega,a_n}(x_n)) = c_{\omega,a} \\ \sim\ &s(c_{\omega,a_1}(x_1), ..., c_{\omega,a_n}(x_n)) = c_{\omega,b} \ (\kappa), \end{aligned}$$

$$\begin{aligned} &q_{\omega,a} \star c_{\omega,a} = c_{\omega,a} \\ \sim\ &q_{\omega,a} \star c_{\omega,b} = c_\omega \ (\kappa) \end{aligned}$$

and

$$\begin{aligned} &t_\omega(c_{\omega,a}(x), x) = x \\ \sim\ &t_\omega(c_\omega(x), x) = c_\omega(x) \ (\kappa). \end{aligned}$$

Hence $\kappa = \kappa_a$ is also valid in Case 2.

Finally, assume A fulfills the condition (b) for a certain $\varrho \in \mathfrak{M}_k$. Since the constant functions belong to A, $(r, s) \in \kappa$ implies $(c_a^1, c_b^1) \in \kappa$ for certain $a \neq b$. If $b = \omega \in \{\mathfrak{o}_\varrho, \mathfrak{e}_\varrho\}$, one obtains $\kappa = \kappa_a$ by means of (9.16). Then one can reduce the remaining cases to the case already handled through $m_{\varrho,\alpha} \star c_a \sim m_{\varrho,\alpha} \star c_b$ (κ) with suitably chosen α. ■

Lemma 9.7.2 *Let $\varrho_s \in \mathfrak{S}_k$ and let κ be a congruence on $Pol_k\varrho_s$. If there are certain κ-congruent functions $f^n, g^n \in Pol_k\varrho_s$ with $f \neq g$, then $\kappa_a \subseteq \kappa$.*

Proof. Define the relation ϱ_s as in Section 5.2.2. Then the functions f, g have representations of the form (5.4) (see Chapter 5). Since $f \neq g$, we can assume w.l.o.g. that $F_1 \neq G_1$ holds for the components F_1, G_1, i.e., there exists an $(n-1)$-tuple $\mathbf{a} := (a_2, ..., a_n)$ with $F_1(\mathbf{a}) =: c \neq d := G_1(\mathbf{a})$. Let

$$q_a(x) := \sum_{r=1}^{l} \sum_{i=0}^{p-1} j_{a_{r,i}}(x) \cdot s^i(a), \quad u(x) := \sum_{r=1}^{l} \sum_{i=0}^{p-1} j_{a_{r,i}}(x) \cdot s^i(a_{1,i})$$

and

$$h(x_1, ..., x_{m+1}) := \sum_{r=1}^{l} \sum_{i=0}^{p-1} j_{a_{r,i}}(x_i) \cdot s^i(H_r(s^{p-i}(x_2), ..., s^{p-i}(x_{m+1}))),$$

where $H_r(c, x_1, ..., x_m) = H_{1,r}(x_1, ..., x_m)$, $H_r(d, x_1, ..., x_m) = H_{2,r}(x_1, ..., x_m)$, $r = 1, 2, ..., l$, and $H_{1,1}, H_{1,1}, ..., H_{1,l}, H_{2,1}, ..., H_{2,l}$ are arbitrary functions of P_k. Since $(f, g) \in \kappa$, we have

$$\begin{aligned} f(u(x), q_{a_2}(x), ..., q_{a_n})) &= q_c(x) \\ \sim g(u(x), q_{a_2}(x), ..., q_{a_n})) &= q_d(x) \ (\kappa) \end{aligned}$$

so that

$$\begin{aligned} &h(x_1, q_c(x_1), x_2, ..., x_m) = \\ &\qquad \textstyle\sum_{r=1}^{l} \sum_{i=0}^{p-1} j_{r,i}(x_1) \cdot s^i(H_{1,r}(s^{p-i}(x_2), ..., s^{p-i}(x_m))) \\ &\sim h(x_1, q_d(x_1), x_2, ..., x_m) = \\ &\qquad \textstyle\sum_{r=1}^{l} \sum_{i=0}^{p-1} j_{r,i}(x_1) \cdot s^i(H_{2,r}(s^{p-i}(x_2), ..., s^{p-i}(x_m))) \ (\kappa). \end{aligned}$$

Therefore $\kappa = \kappa_a$. ∎

Theorem 9.7.3 *([Mal 66], [Kol 74], [Lau 79a;81])*

Let $\varrho \in (\mathfrak{C}_k \setminus \{\varrho \in \mathfrak{C}_k^1 \mid |\varrho| \geq 2\}) \cup \mathfrak{M}_k \cup \mathfrak{S}_k$. Then, the maximal class $Pol_k\varrho$ has only trivial congruences.

Proof. Because of Lemma 9.2.1, every nontrivial congruence of a maximal class is an arity congruence. With that, our theorem results for $\varrho \in \mathfrak{C}_k \setminus \{\varrho \in \mathfrak{C}_k^1 \mid |\varrho| \geq 2\}$ from Lemma 9.7.1, when one chooses ω as a central element of ϱ and one proves that $A := Pol_k\varrho$ fulfills the condition (a) of Lemma 9.7.1. If $\varrho \in \mathfrak{M}_k$ then our assertion also results of Lemma 9.7.1, since $Pol_k\varrho$ fulfills the condition (b) of Lemma 9.7.1. If $\varrho \in \mathfrak{S}_k$ then we obtain $|Con\ Pol_k\varrho| = 3$ from Lemma 9.7.2. ∎

Subsequently, we will prove that, in Theorem 9.7.4, all maximal classes with only trivial congruences were determined. We begin with determining the congruences on the remaining maximal classes of the type $\mathfrak{C}$.

Theorem 9.7.4 *Let ϱ be a unary relation on E_k with $2 \le |\varrho| \le k-1$. Then the equivalence relation $\kappa_{0,\varrho}$ defined by*

$$f^n \sim g^m \ (\kappa_{0,\varrho}) \iff n = m \ \wedge \ \{f, g\} \subset Pol_k\varrho \ \wedge \ (\ \forall \mathbf{x} \in \varrho^n : f(\mathbf{x}) = g(\mathbf{x})\ ),$$

is the only nontrivial congruence on $Pol_k\varrho$.

Proof. W.l.o.g. let $\varrho := E_l$, where $2 \le l \le k-1$.
Since $A := Pol_k E_l$ is a clone, the nontrivial congruences on A are arity congruences. Let $\kappa \in Con_a A \setminus \{\kappa_0\}$. Then the following two cases are possible:
Case 1: $\kappa_0 \subset \kappa \subseteq \kappa_{0,E_l}$.
Then there exist κ-congruent functions f^n, $g^n \in Pol_k E_l$, which are identical on tuples of E_l^n and which have different values on a certain tuple $\mathbf{a} := (a_1, a_2, ..., a_n) \in E_k^n \setminus E_l^n$.
Elements of $Pol_k E_l$ are certain unary functions $p_{a,b}$ with $p_{a,b}(a) = b$, $a \in E_k \setminus E_l$, $b \in E_k$. Also elements of $Pol_k E_l$ are the functions

$$h_i(x) := \begin{cases} c_{a_i}(x) & \text{if } a_i \in E_l, \\ p_{a,a_i}(x) & \text{if } a_i \in E_k \setminus E_l \end{cases}$$

($i = 1, 2, ..., n$; $a \in E_k \setminus E_l$). Consequently, it holds

$$f_a(x) := f(h_1(x), ..., h_n(x)) \sim g_a(x) := g(h_1(x), ..., h_n(x)) \ (\kappa),$$

where $f_a(a) \ne g_a(a)$ for all $a \in E_k \setminus E_l$.
For arbitrary κ_{0,E_l}-congruent functions r^m and s^m of $Pol_k E_l$, there exists an $(m + (k-l) \cdot m)$-ary function t with

$$t(x_1, ..., x_m, f_l(x_1), f_{l+1}(x_1), ..., f_{k-1}(x_1), ..., f_l(x_m), f_{l+1}(x_m), ..., f_{k-1}(x_m)) \\ = r(x_1, ..., x_m)$$

and

$$t(x_1, ..., x_m, g_l(x_1), g_{l+1}(x_1), ..., g_{k-1}(x_1), ..., g_l(x_m), g_{l+1}(x_m), ..., g_{k-1}(x_m)) \\ = s(x_1, ..., x_m).$$

Consequently, we have $(r, s) \in \kappa$ and hence $\kappa = \kappa_{0,E_l}$.
Case 2: $\kappa \not\subseteq \kappa_{0,E_l}$ and $\kappa \subseteq \kappa_a$.
In this case, there are certain κ-congruent functions that have different values on a certain tuple of E_l^n. It is easy to show that this implies two different constants c_a and c_b of $Pol_k E_l$ are κ-congruent. Further, for arbitrary m-ary functions u and v of $Pol_k E_l$ one can find an $(m+1)$-ary function $w \in Pol_k E_l$, for which $w(x_1, ..., x_m, a) = u(x_1, ..., x_m)$ and $w(x_1, ..., x_m, b) = v(x_1, ..., x_m)$ hold. Consequently, we have $u(\mathbf{x}) = w(\mathbf{x}, c_a(x_1)) \sim v(\mathbf{x}) = w(\mathbf{x}, c_b(x_1))$ (κ), i.e., $\kappa = \kappa_a$ holds.
Thus $Pol_k E_l$ has only four different congruences. ∎

Theorem 9.7.5 *Let $\varrho \in \mathfrak{U}_k$ and let $Z := \{\varepsilon_1, ..., \varepsilon_r\}$ be the set of all equivalence classes of the equivalence relation ϱ. Then $Pol_k\varrho$ has exactly four different congruences. The only nontrivial congruence on $Pol_k\varrho$ is the congruence κ_Z defined by*

$$f^n \sim g^m \; \kappa_Z :\Longleftrightarrow n = m \; \wedge \; \{f, g\} \subset Pol_k\varrho \; \wedge$$
$$(\forall \mathbf{a} \in E_k^n \; \exists i \in \{1, 2, ..., r\} : \; \{f(\mathbf{a}), g(\mathbf{a})\} \subseteq \varepsilon_i).$$

Proof. Denote κ be an arbitrary nontrivial congruence on $Pol_k\varrho$. Then, by Lemma 9.2.1, $\kappa_0 \subset \kappa \subseteq \kappa_a$. Our theorem is proven if we can show the following facts:

$$(\exists f^n, g^n \in Pol_k\varrho \; \exists \mathbf{a} \in E_k^n : (f, g) \in \kappa \; \wedge \; (f(\mathbf{a}), g(\mathbf{a})) \in \varrho \backslash \iota_k^2) \implies \kappa_Z \subseteq \kappa,$$
$$(\exists f^n, g^n \in Pol_k\varrho \; \exists \mathbf{a} \in E_k^n : (f, g) \in \kappa \; \wedge \; (f(\mathbf{a}), g(\mathbf{a})) \notin \varrho) \implies \kappa = \kappa_a. \tag{9.17}$$

If two different functions f^n, g^n of $Pol_k\varrho$ are κ-congruent, then it is easy to see that two different constant functions c_a, c_b are also κ-congruent. For these constant functions, the following two cases are possible:
Case 1: $(a, b) \in \varrho \backslash \iota_k^2$.
Denote s and t two arbitrary m-ary functions of $Pol_k\varrho$ with the property

$$\forall \mathbf{a} \in E_k^m \; \exists i \in \{1, 2, ..., r\} : \; \{s(\mathbf{a}), t(\mathbf{a})\} \subseteq \varepsilon.$$

The following $(m+1)$-ary function h with

$$h(x_1, ..., x_{m+1}) := \begin{cases} s(x_1, ..., x_m) \text{ if } & x_{m+1} = a, \\ t(x_1, ..., x_m) \text{ otherwise} & \end{cases}$$

belongs to $Pol_k\varrho$. Thus

$$h(x_1, ..., x_m, c_a(x_1)) = s(\mathbf{x}) \sim t(\mathbf{x}) = h(x_1, ..., x_m, c_b(x_1)) \; (\kappa) \tag{9.18}$$

is valid. Hence, $\kappa_Z \subseteq \kappa$.
Case 2: $(a, b) \notin \varrho$.
In this case, let s and t be two arbitrary m-ary functions of $Pol_k\varrho$. Then, a certain $(m+1)$-ary function h with

$$h(x_1, ..., x_m, a) = s(x_1, ..., x_m) \text{ and}$$
$$h(x_1, ..., x_m, b) = t(x_1, ..., x_m)$$

belongs to $Pol_k\varrho$. Consequently, we obtain $(s, t) \in \kappa$ as in (9.18). Thus $\kappa = \kappa_a$ and (9.17) hold. ■

The following theorem was already proven in Section 9.6 (see Corollary 9.6.7, (2).):

Theorem 9.7.6 *Let k be a prime number power and $\lambda \in \mathfrak{L}_k$. Then*

$$\kappa_c := \bigcup_{n\geq 1}\{(f^n, g^n) \mid \exists a \in E_k :\ f(\mathbf{x}) = g(\mathbf{x}) + a\}$$

is the only nontrivial congruence on $Pol_k\lambda$. ■

Finally, denote ϱ an h-ary relation on E_k, which is a homomorphic inverse image of an elementary h-ary relation $\xi_m \subseteq E_{h^m}$ ($h \geq 3$, $m \geq 1$, $k \geq h^m$). We use the same notations and representations for the functions of $Pol_k\varrho$ as in Section 5.2.6. In addition, one can use Lemma 5.2.6.1 to define the following equivalence relation κ_N on $Pol_k\varrho$:

Definition Let N be a normal subgroup of the symmetric group $(S_h; \star)$. For arbitrary $f^n, g^m \in Pol_k\varrho$ let $(f^n, g^m) \in \kappa_N$ if and only if $n = m$ and for each $i \in E_m$ at least one of the following conditions is fulfilled:
(a) $|Im(f'_i)| \leq h-1$ and $|Im(g'_i)| \leq h-1$;
(b) there exist $s_1, s_2 \in S_h$, $j \in \{1, 2, ..., n\}$ and $t \in E_m$ so that

$$f'_i(x_1, ..., x_n) = s_1((q(x_j))^{(t)}),\quad g'_i(x_1, ..., x_n) = s_2((q(x_j))^{(t)})$$
and
$$s_1 \star N = s_2 \star N.$$

Lemma 9.7.7 *The above-defined equivalence relation κ_N is a congruence on $Pol_k\varrho$.*

Proof. Let (f^n, g^n), $(r^m, s^m) \in \kappa_N$ be arbitrary. Obviously, $(\alpha f, \alpha g) \in \kappa_N$ holds for every $\alpha \in \{\zeta, \tau, \Delta, \nabla\}$. It remains to show that $u := f \star s \sim g \star t =: v$ (κ_N), i.e., $u'_i = f'_i \star r$ and $v'_i := g'_i \star s$ satisfy (a) or (b) from the definition of κ_N for each $i \in E_m$. By Lemma 5.2.6.1 the following cases are possible:
Case 1: $|Im(f'_i)| \leq h-1$ and $|Im(g'_i)| \leq h-1$.
Then $|Im(u'_i)| \leq h-1$ and $|Im(v'_i)| \leq h-1$, i.e., (a) is satisfied.
Case 2: $f'_i(x_1, ..., x_n) = s_1((q(x_j))^{(t)})$, $g'_i(x_1, ..., x_n) = s_2((q(x_j))^{(t)})$, where $s_1, s_2 \in S_h$, $s_1 \star N = s_2 \star N$, $j \in \{1, 2, ..., n\}$ and $t \in E_m$.
Case 2.1: $j \neq 1$.
Then $u'_i(x_1, ..., x_{n+m-1}) = s_1((q(x_{j+m-1}))^{(t)})$ and $v'_i(x_1, ..., x_{m+n-1}) = s_2((q(x_{j+m-1}))^{(t)})$, i.e., (b) is satisfied.
Case 2.2: $j = 1$.
Then $u'_i(x_1, ..., x_{m+n-1}) = s_1(r_t(x_1, ..., x_m)$ and $v'_i(x_1, ..., x_{m+n-1}) = s_2(s_t(x_1, ..., x_m))$.
Case 2.2.1: $|Im(r'_t)| \leq h-1$ and $|Im(s'_t)| \leq h-1$.
Then $|Im(u'_t)| \leq h-1$ and $|Im(v'_t)| \leq h-1$, i.e., (a) is satisfied.
Case 2.2.2: $r'_t(x_1, ..., x_m) = s_3((q(x_l))^{(p)}$, $s'_t(x_1, ..., x_m) = s_4((q(x_l))^{(p)}$, $\{s_3, s_4\} \in S_h$, $s_3 \star N = s_4 \star N$, $l \in \{1, 2, ..., m\}$ and $p \in E_m$.
Then we have $u'_i(x_1, ..., x_{m+n-1}) = s_1(s_3(q(x_l))^{(p)}))$ and $v'_i(x_1, ..., x_{m+n-1}) =$

$s_2(s_4(q(x_l))^{(p)}))$. Moreover, from the assumptions and from the well-known properties of normal subgroups we know that

$$(s_1 \star s_3) \star N = (s_1 \star (s_3 \star N) = (s_1 \star (s_4 \star N) = s_1 \star (N \star s_4) = \\ (s_1 \star N) \star s_4 = (s_2 \star N) \star s_4 = (s_2 \star s_4) \star N.$$

Hence (b) is satisfied. ■

Lemma 9.7.8 *Let κ be a congruence on $Pol_k\varrho$ and let e denote the unit of the symmetric group S_h. Then*

$$\kappa_0 \subset \kappa \subseteq \kappa_a \implies \kappa_{\{e\}} \subseteq \kappa.$$

Proof. Let $\kappa_0 \subset \kappa \subseteq \kappa_a$. Clearly, $T := \{f \in P_k \mid |Im(q(f))| \leq h-1\} \subseteq Pol_k\varrho$. It is easy to see that we have only trivial congruences on T. From $\kappa \subset \kappa \subseteq \kappa_a$ we get the existence of two different unary constants that are κ-congruent. Consequently, all functions of the same arity of T are mutually κ-congruent. Then, with the help of Section 5.2.6, we have $\kappa_{\{e\}} \subseteq \kappa$. ■

Lemma 9.7.9 *Let κ be a congruence on $Pol_k\varrho$, let N be a normal subgroup of the symmetric group $\mathbf{S_h}$ and let $\kappa_N \subset \kappa \subseteq \kappa_{S_h}$. Then there exists a normal subgroup N' of $\mathbf{S_h}$ with $N \subset N' \subseteq S_h$ so that $\kappa_{N'} \subseteq \kappa$.*

Proof. W.l.o.g. we can assume that $\varrho = \xi_m$, i.e., $k = h^m$ and q is the identity mapping.
Because of $\kappa_N \subset \kappa \subseteq \kappa_{S_h}$ there exist certain κ-congruent n-ary functions f, g with

$$f_i'(x_1, ..., x_n) = s_1(x_j^{(t)}),\ g_i'(x_1, ..., x_n) = s_2(x_j^{(t)})$$

for certain $i, t \in E_m$, $j \in \{1, 2, ..., n\}$ and $s_1, s_2 \in S_h$ with $s_1 \star N = s_2 \star N$. The functions $u(x) := x^{(i)}$ and $v(x) := x^{(0)} \cdot h^t$ belong to $Pol\xi_m$. Consequently, since $(f, g) \in \kappa$, we have:

$$((\Delta^{n-1}(u \star f)) \star v)(x) = s_1(x^{(0)}) \\ \sim ((\Delta^{n-1}(u \star g)) \star v)(x) = s_2(x^{(0)})\ (\kappa). \tag{9.19}$$

It is well-known that the symmetric group $\mathbf{S_h}$ has only congruences of type $\kappa_{N''}$, where N'' is a normal subgroup of $\mathbf{S_h}$. Therefore (9.19) together with $s_1 \star N \neq s_2 \star N$ imply the existence of a normal subgroup N' so that $N \subset N' \subseteq S_h$ and

$$\forall s, s' \in S_h : \ (s \star N' = s' \star N' \implies s(x^{(0)}) \ \sim \ s'(x^{(0)})\ (\kappa)\ ).$$

Then, by replacing the variable x by the function $x^{(t)} \in Pol\ \xi_m$ and with the aid of the operations ∇, ζ, τ, we get:

$$s \star N' = s' \star N' \implies a(x_1, ..., x_r) := s(x_j^{(t)}) \ \sim \ s'(x_j^{(t)}) =: b(x_1, ..., x_r)\ (\kappa)$$

for every $r \geq 1$, $j \in \{1, 2, ..., r\}$, $t \in E_m$ and $s, s' \in S_h$. Hence $\kappa_{N'} \subseteq \kappa$. ■

Lemma 9.7.10 *Let κ be a congruence on $Pol_k\varrho$ with the properties $\kappa \not\subseteq \kappa_{S_h}$ and $\kappa \subseteq \kappa_a$. Then $\kappa = \kappa_a$.*

Proof. W.l.o.g. let $\varrho = \xi_m$. Since $\kappa \not\subseteq \kappa_{S_h}$ and $\kappa \subseteq \kappa_a$, there exist κ-congruent functions f^n, g^n with $(f,g) \notin \kappa_{S_h}$, i.e., there exists an $i \in E_m$ with $(f_i', g_i') \notin \kappa_{S_h}$. We distinguish the following cases:
Case 1: $|Im(f_i')| \leq h-1$ and $|Im(g_i')| = h$.
Then by Lemma 9.7.8 we have $(f_i', c_a^n) \in \kappa$ for every $a \in E_k$. Hence $g_i'(x,...,x) =: s_1(x^{(t)}) \sim c_a(x)\ (\kappa)$ and

$$\begin{aligned} &s_2(s_1(s_1^{-1}((x^{(r)} \cdot h^t)^{(t)}))) = s_2(x^{(r)}) \\ &\sim c_a(s_1(s_1^{-1}((x^{(r)} \cdot h^t)^{(t)}))) = c_a(x)\ (\kappa) \end{aligned}$$

for every $s_2 \in S_h$, $r \in E_m$ and $a \in E_k$.
Using the operations ∇, τ, ζ we get

$$\begin{aligned} &\forall s \in S_h\ \forall j \in \{1,...,n\}\ \forall t \in E_m \forall a \in E_k : \\ &p(x_1,...,x_n) := s(x_j^{(t)}) \ \sim\ c_a^n(x_1,...,x_n)\ (\kappa). \end{aligned} \tag{9.20}$$

Moreover, since all n-ary functions f with $|Im(f)| \leq h-1$ are mutually κ-congruent, $\kappa = \kappa_a$ follows from (9.20) in the first case.
Case 2: $f_i'(x_1,...,x_n) = s_1(x_j^{(t)})$, $g_i'(x_1,...,x_n) = s_2(x_l^{(t)})$ and $(j,t) \neq (l,r)$ ($j,l \in \{1,2,...,n\}$; $s_1,s_2 \in S_h$; $t,r \in E_m$).
Case 2.1: $j \neq l$.
Then, because of $(f_i', g_i') \in \kappa$,

$$\begin{aligned} &f_i(\underbrace{x,...,x}_{l-1}, c_0(x), x, ..., x) = s_1(x^{(t)}) \\ &\sim f_i(\underbrace{x,...,x}_{l-1}, c_0(x), x, ..., x) = c_{s_2(0)}(x)\ (\kappa) \end{aligned}$$

and thus by Case 1: $\kappa = \kappa_a$.
Case 2.2: $t \neq r$.
Since $(f_i', g_i') \in \kappa$, we have $(s_1(x^{(t)}), s_2(x^{(r)})) \in \kappa$. Therefore

$$\begin{aligned} &s_1((x^{(0)} \cdot h^t)^{(t)}) = s_1(x^{(0)}) \\ &\sim s_2((x^{(0)} \cdot h^r)^{(t)}) = c_{s_2(0)}(x)\ (\kappa). \end{aligned}$$

Hence, again by Case 1, $\kappa = \kappa_a$. ■

Theorem 9.7.11 *Let V be the four group in $\mathbf{S_4}$ and let A_h be the alternating group of $\mathbf{S_h}$. Furthermore, let e be the unit of $\mathbf{S_h}$.*
For $h = 3$ and $h \geq 5$ the maximal class $Pol_k\varrho$ with $\varrho \in \mathfrak{B}_k^h$ has only the nontrivial congruences $\kappa_{\{e\}}$, κ_{A_h} and κ_{S_h}. A maximal class $Pol_k\varrho$ with $\varrho \in \mathfrak{B}_k^4$ has the nontrivial congruences $\kappa_{\{e\}}$, κ_{A_4}, κ_{S_4} and, in addition, the nontrivial congruence κ_V on $Pol_k\varrho$.

Proof. Our claim follows from Lemmas 9.7.7–9.7.10 and from well-known theorems in group theory. ∎

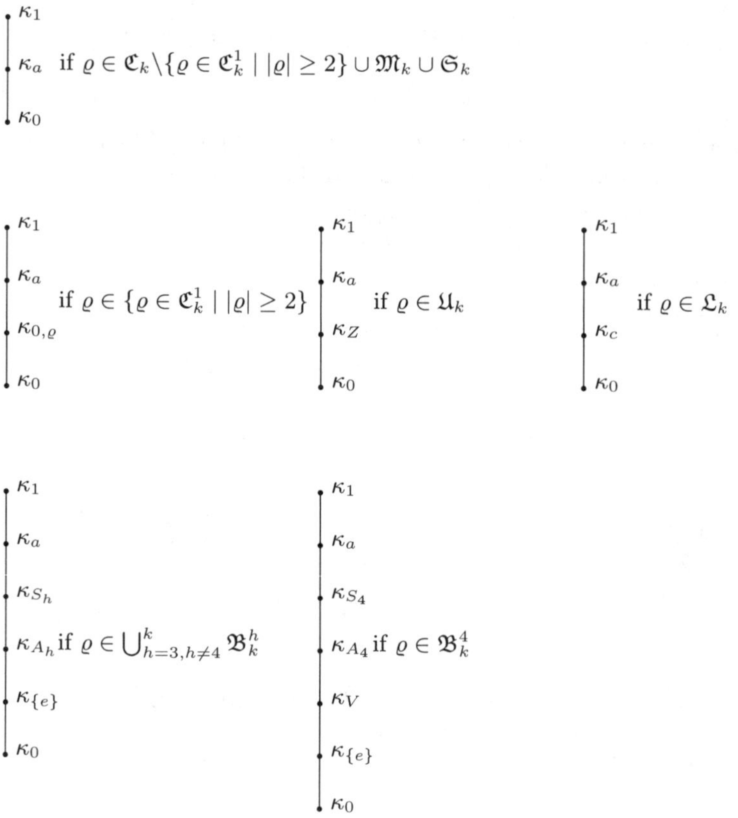

Fig. 9.2. Congruence lattices of the maximal classes $Pol_k\varrho$ of P_k

When one summarizes the above Theorems 9.7.3–9.7.6 and 9.7.11, one sees that

Theorem 9.7.12 **(Congruence Theorem for Maximal Clones;** *[Lau 79a;81])*

For a maximal class $A := Pol_k\varrho$, $\varrho \in R_{max}(P_k)$ (see Chapter 5 and 6) of P_k it holds $|Con\ A| \in \{3,4,6,7\}$ and

$$|Con\ A| = 3 \Longleftrightarrow \varrho \in (\mathfrak{C}_k \backslash \{\varrho \in \mathfrak{C}_k^1 \mid |\varrho| \geq 2\}) \cup \mathfrak{M}_k \cup \mathfrak{S}_k$$

$$|Con\ A| = 4 \Longleftrightarrow \varrho \in \{\varrho \in \mathfrak{C}_k^1 \mid |\varrho| \geq 2\} \cup \mathfrak{U}_k \cup \mathfrak{L}_k$$

$$|Con\ A| = 6 \Longleftrightarrow \varrho \in \bigcup_{h=3,\ h\neq 4}^{k} \mathfrak{B}_k^h$$

$$|Con\ A| = 7 \Longleftrightarrow \varrho \in \mathfrak{B}_k^4.$$

The possible congruence lattices of a maximal class $Pol_k\varrho$ of P_k are given in Figure 9.2. ■

9.8 Congruences on Subclasses of $[P_k^1]$

In this section, we describe the unary functions $f \in P_k$ as follows:
Let

$$\kappa_f := \{(a,b) \in E_k^2 \mid f(a) = f(b)\} \tag{9.21}$$

be the mapping equivalence of f. Denote F_1, ..., F_r the equivalence classes of this equivalence relation. Then there are some $a_1, ..., a_r \in E_k$ with

$$\forall i \in \{1,2,...,r\} : \ \forall x \in F_i : \ f(x) = a_i,$$

and we write:

$$f = \begin{pmatrix} F_1 & F_2 & ... & F_r \\ a_1 & a_2 & ... & a_r \end{pmatrix}.$$

Before we consider some general properties of the congruences on subclasses of $[P_k^1]$, we deal first with the congruences of the semigroup $(P_k^1; \star)$. We will see later that these congruences induce the congruences on the class $[P_k^1]$ (see Theorem 9.8.4).

Definition Let $r \in \{2,3,...,k\}$ and

$$\mu_r(k) := \{(f,g) \in P_k^1 \times P_k^1 \mid f = g \ \vee \ (|Im(f)| < r \ \wedge \ |Im(g)| < r \}. \tag{9.22}$$

Furthermore, let N be a normal subgroup of the symmetric group $\mathbf{S}_{\{1,2,...,r\}}$. For arbitrary $f,g \in P_k^1$ let $(f,g) \in \mu_{r,N}(k)$ if and only if the functions f and g fulfill the following conditions:

(1) $(f,g) \in \mu_r(k)$.
(2) $|Im(f)| = |Im(g)| = r$ and f, g are defined by

$$f = \begin{pmatrix} F_1 \; F_2 \; ... \; F_r \\ a_1 \; a_2 \; ... \; a_r \end{pmatrix}, \; g = \begin{pmatrix} F_1 \; F_2 \; ... \; F_r \\ a_{i_1} \; a_{i_2} \; ... \; a_{i_r} \end{pmatrix},$$

where

$$\begin{pmatrix} 1 \; 2 \; ... \; r \\ i_1 \; i_2 \; ... \; i_r \end{pmatrix} \in N$$

holds.
If k results from the context, we only write

$$\mu_r \text{ (or } \mu_{r,N})$$

instead of $\mu_r(k)$ (or $\mu_{r,N}(k)$). We remark that

$$\mu_r = \mu_{r,\{e\}}$$

holds, where e is the unit of the group $\mathbf{S}_{\{1,2,...,r\}}$.

Lemma 9.8.1 *The above-defined relation $\mu_{r,N}(k)$ is a congruence on the semigroup $(P_k^1; \star)$.*

Proof. With the help of the group properties of N one can see that $\mu_{r,N}$ is an equivalence relation.
To prove the compatibility of $\mu_{r,N}$ with $\star$, it suffices to prove the statement

$$\forall (f,g) \in \mu_{r,N} \; \forall h \in P_k^1 : \; (h \star f, h \star g) \in \mu_{r,N} \; \wedge \; (f \star h, g \star h) \in \mu_{r,N} \quad (9.23)$$

(see Lemma 9.1.1). If $(f,g) \in \mu_r$ or $|Im(h)| < r$, then (9.23) is obviously valid. Let $(f,g) \in \mu_{r,N} \backslash \mu_r$ and $|Im(h)| \geq r$ in the following. Consequently, we can assume

$$f = \begin{pmatrix} F_1 \; F_2 \; ... \; F_r \\ a_1 \; a_2 \; ... \; a_r \end{pmatrix}, \; g = \begin{pmatrix} F_1 \; F_2 \; ... \; F_r \\ a_{i_1} \; a_{i_2} \; ... \; a_{i_r} \end{pmatrix}, \; \begin{pmatrix} 1 \; 2 \; ... \; r \\ i_1 \; i_2 \; ... \; i_r \end{pmatrix} \in N.$$

If $b_i := h(a_i)$, $i = 1, 2, ..., r$, then

$$h \star f = \begin{pmatrix} F_1 \; F_2 \; ... \; F_r \\ b_1 \; b_2 \; ... \; b_r \end{pmatrix} \text{ and } h \star g = \begin{pmatrix} F_1 \; F_2 \; ... \; F_r \\ b_{i_1} \; b_{i_2} \; ... \; b_{i_r} \end{pmatrix}.$$

Thus $(h \star f, h \star g) \in \mu_{r,N}$.
Let $H_i := \{x \in E_k \mid f(h(x)) = a_i\}$, $i = 1, 2, ..., r$. If there exists i with $H_i = \emptyset$, then we have $(f \star h, g \star h) \in \mu_r$. If $H_i \neq \emptyset$ for all i, then

$$f \star h = \begin{pmatrix} H_1 \; H_2 \; ... \; H_r \\ a_1 \; a_2 \; ... \; a_r \end{pmatrix} \text{ and } g \star h = \begin{pmatrix} H_1 \; H_2 \; ... \; H_r \\ a_{i_1} \; a_{i_2} \; ... \; a_{i_r} \end{pmatrix},$$

whereby $(f \star h, g \star h) \in \mu_{r,N}$ holds. ■

The next lemma gives some properties of congruences of $(P_k^1;\star)$ that we need, to prove Theorem 9.8.3.

Lemma 9.8.2 *Let κ be a congruence of the semigroup $(P_k^1;\star)$. Then:*

(a) $\kappa \neq \kappa_0 \implies (\forall a,b \in E_k : (c_a,c_b) \in \kappa)$;

(b) $(\,(f,c_a) \in \kappa \wedge |Im(f)| \geq 2\,) \implies \mu_{|Im(f)|+1} \subseteq \kappa$;

(c) $(\,(f,g) \in \kappa \wedge |Im(f)| > |Im(g)|\,) \implies \mu_{|Im(f)|+1} \subseteq \kappa$;

(d) $(\,(f,g) \in \kappa \wedge |Im(f)| = |Im(g)| \geq 2 \wedge Im(f) \neq Im(g)\,) \implies \mu_{|Im(f)|+1} \subseteq \kappa$;

(e) $(\,(f,g) \in \kappa \wedge Im(f) = Im(g) \wedge \kappa_f \neq \kappa_g\,) \implies \mu_{|Im(f)|+1} \subseteq \kappa$; [3]

(f) $(\,(f,g) \in \kappa \wedge Im(f) = Im(g) \geq 2 \wedge \kappa_f = \kappa_g \wedge f \neq g\,) \implies \mu_{|Im(f)|} \subseteq \kappa$;

Proof. (a) and (b) are easy to check.
(c): Let $(f,g) \in \kappa$ and $|Im(f)| > |Im(g)|$. If $|Im(g)| = 1$, (c) follows from (b). Therefore we can assume $|Im(g)| \geq 2$. In addition, consider only the case $Im(g) \subset Im(f)$ suffices, since one can find an $h \in P_k^1$ with $h \star f = f$, $h \star g =: g'$ and $Im(f) \subset Im(g')$. Therefore, let $Im(g) = \{a_1,...,a_u\}$ and $Im(f) = \{a_1,...,a_u,...,a_r\}$. Then, with the help of the functions

$$p(x) := \begin{cases} a_{u+1} \text{ if } & x \in \{a_{u+1},...,a_r\}, \\ x \quad \text{otherwise} & \end{cases}$$

and

$$t_j(x) := \begin{cases} a_1 \text{ if } & x \in \{a_1,...,a_j\}, \\ x \text{ otherwise,} & \end{cases}$$

we obtain

$$f_{u+1} := p \star f \sim p \star g = g \; (\kappa)$$
$$f_{u+2-j} := t_j \star f_{u+1} \sim t_j \star g =: g_{u+1-j} \; (\kappa),$$

for $j = 2,3,...,u$, where the indices of the functions agree with the cardinalities of their ranges of values. We can rewrite the above as follows:

$$g_1 \sim f_2,\; g_2 \sim f_3,\; g_3 \sim f_4,\; ...\; g_u \sim f_{u+1}, f_{u+1} \sim f \; (\kappa).$$

This, (b) and the transitivity of κ imply our assertion (c).
(d): Let $a \in Im(g)$, $b \in Im(g)\backslash Im(f)$ and

$$q(x) := \begin{cases} a \text{ if } & x = b, \\ x \text{ otherwise.} & \end{cases}$$

Then $(f,g) \in \kappa$ implies $q \star f = f \sim q \star g =: g' \; (\kappa)$, where $|Im(g')| < |Im(f)|$. Thus, (d) follows from (c).
(e): If $Im(f) = Im(g)$ and $\kappa_f \neq \kappa_g$, there exist $a,b \in E_k$ with $(a,b) \in \kappa_f$ and $(a,b) \notin \kappa_g$. Then, with the help of the above defined function q, we have

[3] κ_f, κ_g denote the mapping equivalences of f or g, respectively (see (9.21)).

$f \star q = f \sim g \star q =: g''$ (κ) with $|Im(g'')| < |Im(f)|$. Hence (e) follows from (c).

(f): Let $(f, g) \in \kappa$, $f \neq g$, $r \geq 2$ and

$$f = \begin{pmatrix} F_1 & F_2 & \dots & F_r \\ a_1 & a_2 & \dots & a_r \end{pmatrix} \text{ and } g = \begin{pmatrix} F_{i_1} & F_{i_2} & \dots & F_{i_r} \\ a_1 & a_2 & \dots & a_r \end{pmatrix},$$

where $\begin{pmatrix} 1 & 2 & \dots & r \\ i_1 & i_2 & \dots & i_r \end{pmatrix}$ denotes a certain permutation. If $r = 2$ then (f) follows from that because of (a). If $r > 2$ then there exist indices u, v with $\{u, v\} \neq \{i_u, i_v\}$ so that $F_u \cup F_v \neq F_{i_u} \cup F_{i_v}$ holds. Then, with the help of the function

$$w(x) := \begin{cases} a_u & \text{if } x \in \{a_u, a_v\}, \\ x & \text{otherwise.} \end{cases}$$

and through $f_1 := w \star f \sim w \star g =: g_1$ (κ), one gets that two functions f_1, g_1 are κ-congruent, where $|Im(f_1)| = |Im(g_1)| = r - 1$ and the mapping equivalences κ_{f_1} and κ_{g_1} are different. Consequently, (f) follows from (e). ■

Theorem 9.8.3 *(**A. I. Mal'tsev's Theorem**; [Mal 52])*
(a) The only nontrivial congruences on $(P_k^1; \star)$ are congruences of the form $\kappa_{r,N}$, where $r \in \{2, 3, ..., k\}$ and N is an arbitrary normal subgroup of the symmetric group $\mathbf{S}_{\{1,2,...,r\}}$.
(b) Let S_r be the group of all permutations on the set $\{1, 2, ..., r\}$, and let A_r $(\subset S_r,\ r \geq 2)$ be the alternating group and let V $(\subset S_4)$ be the four group. Furthermore, denote $Con(k)$ the set of all congruences of the semigroup $(P_k^1; \star)$. Then:
$Con(2) = \{\kappa_0, \mu_2, \mu_{2,S_2}, \kappa_1\}$;
$Con(3) = \{\kappa_0, \mu_2, \mu_{2,S_2}, \mu_3, \mu_{3,A_3}, \mu_{3,S_3}, \kappa_1\}$;
$Con(4) = \{\kappa_0, \mu_2, \mu_{2,S_2}, \mu_3, \mu_{3,A_3}, \mu_{3,S_3}, \mu_4, \mu_{4,V}, \mu_{4,A_4}, \mu_{3,S_4}, \kappa_1\}$;
$Con(k) = \{\kappa_0, \mu_2, \mu_{2,S_2}, \mu_{4,V}, \kappa_1\} \cup \bigcup_{r \in \{3,4,5,...,k\}} \{\mu_r, \mu_{r,A_r}, \mu_{r,S_r}\}$ *for* $k \geq 5$.

Proof. (a): Let κ $(\neq \kappa_0)$ be a congruence of the semigroup $(P_k^1; \star)$. Then, by Lemma 9.8.2, (b)–(f), we have an $r \in \{2, ..., k-1\}$ with $\mu_r \subseteq \kappa \subset \mu_{r+1}$. To prove that a normal subgroup N with $\kappa = \mu_{r,N}$ exists, we consider the functions

$$f := \begin{pmatrix} F_1 & F_2 & \dots & F_r \\ a_1 & a_2 & \dots & a_r \end{pmatrix} \; f' := \begin{pmatrix} F_1 & F_2 & \dots & F_r \\ a_{\alpha(1)} & a_{\alpha(2)} & \dots & a_{\alpha(r)} \end{pmatrix}$$

$$g := \begin{pmatrix} F_1 & F_2 & \dots & F_r \\ b_1 & b_2 & \dots & b_r \end{pmatrix} \; g' := \begin{pmatrix} F_1 & F_2 & \dots & F_r \\ b_{\beta(1)} & b_{\beta(2)} & \dots & b_{\beta(r)} \end{pmatrix}$$

of P_k^1. Further, let

$$h_1(x) := \begin{cases} b_i \text{ if} & \exists i : x = a_i, \\ x \text{ otherwise,} \end{cases} \quad h_2(x) := \begin{cases} a_{\gamma(i)} \text{ if} & \exists i : x = a_i, \\ x \quad \text{otherwise,} \end{cases}$$

where γ is an arbitrary permutation from S_r. Now, let $F_1, ..., F_r$ be a partition of E_k and let N be the set of all permutations $\alpha \in S_r$ for which $(f, f') \in \kappa$ holds. If $\alpha, \beta \in N$, then we have $(f, f'), (g, g') \in \kappa$ and therefore also $g = h_1 \star f \sim h_1 \star f'$ (κ) and (because of transitivity) $h_1 \star f' \sim g'$ (κ); thus, according to the definition of N the following is valid:

$$\begin{pmatrix} \alpha(1)\ \alpha(2)\ ...\ \alpha(r) \\ \beta(1)\ \beta(2)\ ...\ \beta(r) \end{pmatrix} = \beta \star \alpha^{-1} \in N.$$

Therefore N is a group. The fact that N is a normal group results from

$$(f, f') \in \kappa \implies h_2 \star f \sim h_2 \star f'\ (\kappa)$$

$$\implies \begin{pmatrix} \gamma(1) & ... & \gamma(r) \\ \gamma(\alpha(1)) & ... & \gamma(\alpha(r)) \end{pmatrix} = \gamma \star \alpha \star \gamma^{-1} \in N.$$

For another partition $Q_1, ..., Q_r$ of E_k and for arbitrary functions

$$g := \begin{pmatrix} Q_1\ Q_2\ ...\ Q_r \\ d_1\ d_2\ ...\ d_r \end{pmatrix},\ q' := \begin{pmatrix} Q_1 & Q_2 & ... & Q_r \\ d_{\delta(1)} & d_{\delta(2)} & ... & d_{\delta(r)} \end{pmatrix}$$

with $\delta \in S_r$ we must still show that $(q, q') \in \kappa$ if and only if $\delta \in N$. If $(q, q') \in \kappa$ then $\delta \in N$ follows from $(q \star h_3, q' \star h_3) \in \kappa$, where $h_3(F_i) \in Q_i$ $(i = 1, ..., r)$. If $\delta \in N$ then there are κ-congruent functions h and h' with

$$h := \begin{pmatrix} F_1\ F_2\ ...\ F_r \\ d_1\ d_2\ ...\ d_r \end{pmatrix},\ h' := \begin{pmatrix} F_1 & F_2 & ... & F_r \\ d_{\delta(1)} & d_{\delta(2)} & ... & d_{\delta(r)} \end{pmatrix}$$

and $(q, q') \in \kappa$ follows from $(h \star h_4, h' \star h_4) \in \kappa$, where $h_4(Q_i) \in F_i$ $(i = 1, 2, ..., r)$.
(b) follows from well-known results over the normal subgroups of the symmetric group. ■

The following concept comes from the paper [Mal 76] by I. A. Mal'tsev.

Definitions Let A be a subclass of $[P_k^1]$, which contains a certain constant c_a. Furthermore, let α be a congruence on the semigroup $(A^1; \star)$. Then, the **α-induced relation** κ_α is defined as follows:
For arbitrary $(f^n, g^m) \in A \times A$ let $(f^n, g^m) \in \kappa_\alpha$ iff $n = m$ and the functions f and g fulfill one of the two following conditions:

(1) $(\Delta^{n-1} f, c_a) \in \alpha$ (or $(\Delta^{n-1} g, c_a) \in \alpha$);
(2) $\exists i \in \{1, ..., n\}\ \exists f_1, g_1 \in A^1 :\ (\ f(x_1, ..., x_n) = f_1(x_i)\ \wedge\ g(x_1, ..., x_n) = g_1(x_i)\ \wedge\ (f_1, g_1) \in \alpha\ )$.

We say that **the congruence κ on A is induced from the congruence** α on P_k^1 iff $\kappa = \kappa_\alpha$.

The following theorem is a special case of Theorem 9.8.8:

Theorem 9.8.4 *([Mal 76])*
Let $c_a \in A = [A] \subseteq [P_k^1]$. Then

(a) Each relation, induced from a congruence on the semigroup $(A^1; \star)$, is a congruence on A.
(b) All nontrivial arity congruences on A are given from the nontrivial congruences on A^1. ∎

The next theorem is a consequence from the above theorem and Theorem 9.8.3:

Theorem 9.8.5 *([Mal 76])*
On $[P_k^1]$ there are exactly

$$\begin{cases} 5 & \text{if } k = 2, \\ 8 & \text{if } k = 3, \\ 3 \cdot k & \text{if } k \geq 4 \end{cases}$$

different congruences. The nontrivial congruences on $[P_k^1]$ are induced from the congruences of the type $\mu_{r,N}(k)$ (see Lemma 9.8.1) on P_k^1. ∎

The rest of this section is the summary of an unpublished manuscript by B. Strauch.

Definition and Declarations

Let H be a subsemigroup of $(P_k^1; \star)$ and let μ be a congruence on H. Further, let $\{H_1, ..., H_l\}$ be the partition correlated to the relation μ, where the indexes of the blocks H_i are so chosen that the following is valid:

$$\forall i \in \{1, 2, ..., s\}\ \forall h \in H : \ H_i \star h \subseteq H_i \tag{9.24}$$

and

$$\forall i \in \{s+1, ..., l\}\ \exists h \in H : \ H_i \star h \nsubseteq H_i. \tag{9.25}$$

We set $s = 0$ if there is not any block with the property (9.24).
For functions $f^n \in [P_k^1]$, we also write f' instead of $\Delta^{n-1} f$ and put

$$ess(f) := i,$$

if $f(x_1, ..., x_n) = f'(x_i)$ with $f' \notin P_k^1[1]$. If f^n is a constant, then let $ess(f) := 0$.
With the help of μ, one can define the following relations on $[H]$:

$$\alpha_\mu := \mu \cup \{(f^n, g^m) \in [H] \times [H] \mid n = m \ \wedge \ (f', g') \in \mu \wedge (ess(f) = ess(g) \ \vee \ \exists i \in \{1, ..., s\} : \{f', g'\} \subseteq H_i) \},$$

$$\beta_\mu := \mu \cup \{(f^n, g^m) \in [H] \times [H] \mid n = m \ \wedge \ (f', g') \in \mu \ \wedge \ ess(f) = ess(g)\},$$

$$\gamma_\mu := \{(f^n, g^m) \in [H] \times [H] \mid n = m \ \wedge \ (f', g') \in \mu \}.$$

Lemma 9.8.6 *Let H and μ be defined as above. Then*

(a) $\forall f \in H_{s+1} \cup ... \cup H_l : \ \forall g \in H_1 \cup ... \cup H_s : \ f \star g \in H_1 \cup ... \cup H_s$;

(b) $(\exists t \in \{1, ..., l\} \ \exists a \in E_k : \ c_a \in H_t) \implies t \in \{1, 2, ..., s\}$.

Proof. (a): Let f and g be given as in (a) and let $(q, f \star g) \in \mu$. (a) is proven if we can show

$$\forall h \in H : \ (q \star h, q) \in \mu. \tag{9.26}$$

(9.26) results, however, from $q \star h \sim f \star (g \star h) \sim f \star g \sim q \ (\mu)$ because of $g \in H_1 \cup ... \cup H_s$.
(b): If $\{c_a, f\} \subseteq H_i$ then for every $h \in H$: $c_a \star h = c_a \sim f \star h \ (\mu)$. Therefore, $f \star h \in H_i$ and thus $i \in \{1, 2, ..., s\}$. ■

Lemma 9.8.7 *Let H and μ be defined as above. Then*

(a) α_μ is a congruence on $[H]$ with $\alpha_{|H} = \mu$.

(b) β_μ is a congruence on $[H]$ with $\beta_{|H} = \mu$, iff H does not contain any constant function or the blocks H_1, ..., H_s of μ consist only of constant functions.

(c) $\alpha_\mu = \beta_\mu$ iff $s = 0$ or the blocks H_1, ..., H_s of μ consist only of constant functions.

(d) γ_μ is a congruence on $[H]$ with $\gamma_{|H} = \mu$, iff $s = l$ holds, i.e., iff every block H_i of μ fulfills the condition (9.24).

Proof. (a): We start with proof that α_μ is an equivalence relation on $[H]$. The reflexivity and the symmetry of α_μ result immediately from the definition of α_μ. One can prove the transitivity of α_μ as follows: Let $(f^n, g^n), (g^n, h^n) \in \alpha_\mu$ be arbitrary. Then $(f', g'), (g', h') \in \mu$ and (by the transitivity of μ) $(f', h') \in \mu$. In the case $ess(f) = ess(g) = ess(h)$ we have $(f, h) \in \alpha_\mu$. Otherwise, we have $s \geq 1$ and there exists an $i \in \{1, ..., s\}$ with $\{f', g'\} \subseteq H_i$ or $\{g', h'\} \subseteq H_i$. As already was proven, the functions f', g', h' are μ-equivalent. Therefore $\{f', g', h'\} \subseteq H_i$ and thus $(f^n, h^n) \in \alpha_\mu$.
Next we show the compatibility of α_μ with the superposition operations. For the unary operations, this is clear. Let $(f, g) \in \alpha_\mu$ and $t \in [H]$ be arbitrary. To prove the compatibility of α_μ with $\star$, it is sufficient to show that $(t \star f, t \star g), (f \star t, g \star t) \in \alpha_\mu$ holds. One can check $(f \star t, g \star t) \in \alpha_\mu$ easily by defining α_μ. If t is a constant function or $ess(t) \geq 2$ holds, then also $(t \star f, t \star g) \in \alpha_\mu$. Therefore, we can assume $ess(t) = 1$. Then $(t \star f)' = t' \star f' \sim (t \star g)' = t' \star g' \ (\mu)$. If $ess(f) = ess(g)$ then also $ess(t \star f) = ess(t \star g)$. This implies $(t \star f, t \star g) \in \alpha_\mu$. In the case $ess(f) \neq ess(g)$, there is an $i \in \{1, ..., s\}$ with $\{f', g'\} \subseteq H_i$.

Let H_r be the block to which the functions $t' \star f'$ and $t' \star g'$ belong. Then, $r \leq s$ and thus $(t \star f, t \star g) \in \alpha_\mu$ follows from the following: Let $q \in H_r$ and $h \in H$ be arbitrary. Because of $\{q, t' \star f'\} \subseteq H_r$ and $\{f' \star h, f'\} \subseteq H_i$ it holds: $q \star h \sim (t' \star f') \star h = t' \star (f' \star h) \sim t' \star f' \sim q \ (\mu)$. Consequently, we have $H_r \star h \subseteq H_r$ for arbitrary $h \in H$.
Hence, α_μ is a congruence on $[H]$.
One checks the remaining statements of the lemma by means of Lemmas 9.8.6 and 9.3.8. ■

Theorem 9.8.8 *Let* $\mathbf{H} := (H; \star)$ *be a semigroup with* $H \subseteq P_k^1$. *Then, one can describe the congruences on* $[H]$ *as follows:*

(1) α_μ *with* $\mu \in Con\, \mathbf{H}$;
(2) β_μ *with* $\mu \in Con\, \mathbf{H}$, *if* H *does not contain any constant function or the equivalence classes of* μ, *which (9.24) fulfills, consist only of constant functions;*
(3) γ_μ *with* $\mu \in Con\, \mathbf{H}$ *and every equivalence class of* μ *fulfills the condition (9.24).*

Proof. First, let κ be an arbitrary arity congruence on $[H]$ and let $\mu := \kappa \cap (H \times H)$. Obviously, μ is a congruence of the semigroup $(H; \star)$ and we have (by the compatibility of κ with the operations ∇, ζ, τ) $\beta_\mu \subseteq \kappa$. If $\beta_\mu = \kappa$, then, by Lemma 9.8.7, μ and H fulfill the conditions given in (2). Therefore, let $\beta_\mu \subset \kappa$. Then there are some κ-congruent functions f^n, g^n with $f(x_1, ..., x_n) = f'(x_i)$, $g(x_1, ..., x_n) = g'(x_j)$ and $i < j$. Consequently, it holds

$$f_1(x, y) := f(\underbrace{x, ..., x}_{i}, y, y, ..., y) = f'(x) \sim$$
$$g_1(x, y) := g(\underbrace{x, ..., x}_{i}, y, y, ..., y) = g'(y) \qquad (\kappa)$$

Then, for arbitrary $q \in H$ with $(q, f') \in \kappa$, we have

$$q_1(x, y) := q(x) \sim f_1(x, y) \sim g_1(x, y) \ (\kappa).$$

Thus

$$q \star h = \Delta(q_1 \star h) \sim \Delta(g_1 \star h) = g' \ (\kappa)$$

for every $h \in H$. Consequently, the equivalence class (in respect to μ), to which the functions f' and g' belong, fulfills the condition (9.24). Thus $\kappa \subseteq \alpha_\mu$ holds. To prove $\alpha_\mu \subseteq \kappa$ let $(r^m, s^m) \in \alpha_\mu$. If $ess(r) = ess(s)$ then obviously, $(r, s) \in \kappa$. In the case $ess(r) \neq ess(s)$, it is sufficient to consider (w.l.o.g.) $m = 2$, $r(x, y) = r'(x)$ and $s(x, y) = s'(y)$. By definition of the relation α_μ the equivalence class F of μ, to which r' and s' belong, has the property

$$F \star H \subseteq F. \qquad (9.27)$$

With the help of the κ-congruent functions f_1, g_1 we have

$$r' \star f_1 \sim s' \star g_1 \ (\kappa).$$

Because of (9.27), $\{(r' \star f', r'), (s' \star g', s')\} \subseteq \mu$ is valid. Thus $\{(r' \star f_1, r), (s' \star g_1, s)\} \subseteq \kappa$. Then, the transitivity of κ implies $(r, s) \in \kappa$. Therefore $\alpha_\mu \subseteq \kappa$ and our statement on the arity congruences on $[H]$ was proven.
Finally, let κ be an arbitrary non-arity congruence on $[H]$ and let μ be the restriction of κ onto H. Then, because of Lemma 9.3.8, the inclusion $\kappa \subseteq \gamma_\mu$ is valid and the relation μ has the properties mentioned in our theorem. To prove $\gamma_\mu \subseteq \kappa$, it is sufficient to show that an arbitrary function $h_n^n \in [H]$ with $h_n^n(x_1, ..., x_n) = h'(x_n)$ is κ-congruent to h' $(= h_1^1)$. If $\kappa \not\subseteq \kappa_a$ there are κ-congruent functions f^2, g^1 $\in [H]$. Consequently, we have $h_2 = \Delta((\nabla h') \star f) \sim \Delta((\nabla h') \star g) = h_1 \ (\kappa)$, $\nabla h_2 = h_3 \sim \nabla h_1 = h_2 \ (\kappa)$, etc. Thus $h_n \sim h' \ (\kappa)$ and our theorem was proven. ∎

9.9 Congruences on Some Subclasses of $P_{k,l}$

In this section, we determine all congruences on a class A of $P_{k,l}$, which can be described with the aid of the homomorphism pr^{-1} (see Chapter 12):

$$\exists B \in \mathbb{L}_l^{\uparrow}(J_l) : \ A = pr^{-1}B, \tag{9.28}$$

i.e., A is an inverse image of a clone of P_l. Denote

$$\mathfrak{A}$$

the set of the subclasses of $P_{k,l}$ of the form (9.28).
Definition Let U_1, U_2, ..., U_t be some nonempty subsets of $E_k \backslash E_l$ and

$$\mathfrak{T}^n(U_1, U_2, ..., U_t) :=$$
$$\{(a_1, a_2, ..., a_n) \in E_k^n \mid \exists\, i \in \{1, 2, ..., t\} : \ U_i \subseteq \{a_1, ..., a_n\}\}.$$

Let $\kappa_{U_1,...,U_t}$ be a relation on a subclass of $P_{k,l}$ defined by

$$(f^n, g^m) \in \kappa_{U_1,...,U_t} \ :\Longleftrightarrow$$
$$n = m \ \wedge \ (\forall \mathbf{a} \in E_k^n \backslash \mathfrak{T}^n(U_1, ..., U_t) : f(\mathbf{a}) = g(\mathbf{a}) \ ).$$

If $\mathfrak{U} = \{U_1, ..., U_t\}$ we write briefly

$$\kappa_{\mathfrak{U}}$$

instead of $\kappa_{U_1,...,U_t}$.

One can prove the following lemma easily.

Lemma 9.9.1
(a) The above-defined relation $\kappa_{\mathfrak{U}}$ is a congruence on an arbitrary subclass of $P_{k,l}$.
(b) Let U_1, U_2 and U_3 subsets of $E_k \backslash E_l$. Then
(1) $\kappa_{U_1} \subseteq \kappa_{U_1,U_2}$,
(2) $\kappa_{U_1} \subseteq \kappa_{U_2} \iff U_2 \subseteq U_1$,
(3) $\kappa_{U_1,U_2} \subseteq \kappa_{U_3} \iff U_3 \subseteq U_1 \cap U_2$,
(4) $(\kappa_{U_1} \nsubseteq \kappa_{U_2} \wedge \kappa_{U_2} \nsubseteq \kappa_{U_1}) \iff (U_1 \nsubseteq U_2 \wedge U_2 \nsubseteq U_1)$,
(5) $\kappa_{U_1,U_2} = \kappa_{U_1} \iff U_1 \subseteq U_2$. ■

Theorem 9.9.2 *([Mal 79])*
Let $A \in \mathfrak{A}$ and let L be the set of all congruences of the form $\kappa_{\mathfrak{U}}$ on A (see above). Then, $(L; \subseteq)$ is a free distributive lattice with $k-l$ generators.

Proof. For $i = 1, 2$ let $\mathfrak{U}_i := \{U_{i1}, U_{i2}, ..., U_{it_i}\}$ and let $\kappa_{\mathfrak{U}_i}$ be a congruence on $A \in \mathfrak{A}$. We write $\kappa_{\mathfrak{U}_1} \leq \kappa_{\mathfrak{U}_2}$, if $\mathfrak{U}_1 \subseteq \mathfrak{U}_2$ and if for every $U_{2i} \in \mathfrak{U}_2$ there exists a certain $U_{1j} \in \mathfrak{U}_1$ with $U_{1j} \subseteq U_{2i}$. It results from the definition of the congruences of the type $\kappa_{\mathfrak{U}}$ that $\mathfrak{U}_1 \leq \mathfrak{U}_2$ implies $\kappa_{\mathfrak{U}_1} = \kappa_{\mathfrak{U}_2}$. Consequently, during the description of the congruences of the form $\kappa_{\mathfrak{U}}$, we can restrict ourselves to such sets $\mathfrak{U}$ with the following property:

$$\forall U \, \forall U' \, ((U \in \mathfrak{U} \wedge U \subseteq U' \subseteq E_k \backslash E_l) \implies U' \in \mathfrak{U}).$$

We call such sets J-**closed**. The set of all J-closed subsets of $\mathfrak{P}(E_k \backslash E_l)$ forms a lattice. Consequently, because of

$$(\mathfrak{U}_1 \nleq \mathfrak{U}_2 \wedge \mathfrak{U}_2 \nleq \mathfrak{U}_1) \implies \kappa_{\mathfrak{U}_1} \neq \kappa_{\mathfrak{U}_2},$$

and

$$(\mathfrak{U}_1, \mathfrak{U}_2 \text{ are } J\text{-closed} \wedge \mathfrak{U}_1 \subseteq \mathfrak{U}_2) \implies \kappa_{\mathfrak{U}_1} \subseteq \kappa_{\mathfrak{U}_2}$$

there is an isomorphism from the lattice of the congruences of the type $\kappa_{\mathfrak{U}}$ onto the lattice of the J-closed subsets of $E_k \backslash E_l$. Hence, our theorem follows from [Bir 48] (p. 146, Theorem 13). ■

As a direct consequence of Lemma 9.3.5, we get:

Lemma 9.9.3 *Let $A \in \mathfrak{A}$. Then, congruence κ_1 is the only non-arity congruence on A.* ■

Lemma 9.9.4 *Let $A \in \mathfrak{A}$, $\emptyset \neq U \subseteq E_k \backslash E_l$, κ a congruence on A and let f^n, g^n be κ-congruent functions of A, for which there exists an $\mathbf{a} \in (U \cup E_l)^n \cap \mathfrak{T}^n(U)$ with $f(\mathbf{a}) \neq g(\mathbf{a})$. Then, $\kappa_U \subseteq \kappa$.*

Proof. Let $\mathbf{a} := (a_1, a_2, ..., a_n)$. We can assume w.l.o.g. $\{a_1, a_2, ..., a_r\} = U$, $|U| = r$ and $\{a_{r+1}, ..., a_n\} \subseteq E_l$.
The functions

$$t_a(x) := \begin{cases} x \text{ if} & x \in E_l, \\ a \text{ otherwise} & \end{cases}$$

$(a \in E_l)$ belong to A. Consequently, we have

$$f(x_1, ..., x_r, t_{a_{r+1}}(x_1), ..., t_{a_n}(x_1)) =: f'(x_1, ..., x_r)$$
$$\sim g(x_1, ..., x_r, t_{a_{r+1}}(x_1), ..., t_{a_n}(x_1)) =: g'(x_1, ..., x_r) \ \ (\kappa)$$

and

$$f'(a_1, ..., a_r) \neq g'(a_1, ..., a_r).$$

Let f_1 and f_2 be arbitrary m-ary functions of A, which are identical on tuples of $E_k^m \backslash \mathfrak{T}^m(U)$, $m \geq r$. We have to show that $(f_1, f_2) \in \kappa$ holds. We start with same notations:
Let $I = (i_1, ..., i_r)$ be a variation[4] of the numbers 1, 2, ..., m, $x_i := (x_{i_1}, ..., x_{i_r})$ and let I_1, I_2, ..., I_t be all possible variations of the numbers 1, 2, ..., m taken r at a time (i.e., $t = \frac{m!}{(m-r)!}$).
The function

$$h(x_1, ..., x_m, z_1, ..., z_t) :=$$

$$\begin{cases} f_1(x_1, ..., x_m) \text{ if } & (x_1, ..., x_m) \in \mathfrak{T}^m(U) \ \wedge \ (\forall i : z_i = f'(x_{I_i})), \\ f_2(x_1, ..., x_m) \text{ otherwise.} & \end{cases}$$

belongs to A. Consequently, we have

$$h(x_1, ..., x_m, f'(x_{I_1}), ..., f'(x_{I_t})) = f_1(x_1, ..., x_m)$$
$$\sim h(x_1, ..., x_m, g'(x_{I_1}), ..., g'(x_{I_t})) = f_2(x_1, ..., x_m) \ \ (\kappa).$$

Thus $\kappa_U \subseteq \kappa$. ■

Lemma 9.9.5 *Let $A \in \mathfrak{A}$, let κ be a congruence on A and let U_1, U_2, ..., U_t, $U_{t+1} \subseteq E_k \backslash E_l$. Then*

$$(\kappa_{U_1,U_2,...,U_t} \subseteq \kappa \ \wedge \ \kappa_{U_{t+1}} \subseteq \kappa) \implies \kappa_{U_1,U_2,...,U_t,U_{t+1}} \subseteq \kappa.$$

Proof. If there is an $i \in \{1, ..., t\}$ with $A_i \subseteq A_{t+1}$, then by Lemma 9.9.1, (b) we have: $\kappa_{U_1,U_2,...,U_t} = \kappa_{U_1,U_2,...,U_t,U_{t+1}} \subseteq \kappa$. Therefore, we can assume that no $U \in \{U_1, ..., U_t\}$ is a subset of A_{t+1}. Furthermore, let f^n and g^n be arbitrary functions of A, which are identical on all tuples of $E_k^n \backslash \mathfrak{T}^n(U_1, ..., U_t, U_{t+1})$. Because of $\kappa_{U_1,...,U_t} \subseteq \kappa$ we have

$$f(x_1, ..., x_n)$$
$$\sim f'(x_1, ..., x_n) := \begin{cases} g(x_1, ..., x_n) \text{ if } & (x_1, ..., x_n) \in \mathfrak{T}^n(U_1, ..., U_t), \\ f(x_1, ..., x_n) \text{ otherwise} & \end{cases}$$
$$(\kappa).$$

Since $\kappa_{U_{t+1}} \subseteq \kappa$, it holds further that

$$f' \ \sim \ g \ (\kappa).$$

[4] Each arrangement of r different objects of a set with n different elements into a particular order is called a **variation** of n elements taken r at a time.

Thus $(f, g) \in \kappa$. Consequently, $\kappa_{U_1,...,U_t,U_{t+1}} \subseteq \kappa$. ■

We come now to the congruences on $A \in \mathfrak{A}$ which are completely determined by the congruences on $pr_l A$. So that we can better distinguish the congruences on $pr_l A$ from the congruences on A, we denote the congruences on $pr_l A$ with large Greek alphabetic characters. In particular, let K_0, K_a and K_1 be the trivial congruences on $pr_l A$.

One can check the following lemma easily.

Lemma 9.9.6 *Let $A \in \mathfrak{A}$ and let K be a congruence on pr_l A. Then the relation defined by*

$$(f, g) \in pr_l^{-1} K :\Longleftrightarrow (pr_l f, pr_l g) \in K$$

is a congruence on A.
In particular, it holds: $pr_l^{-1} K_1 = \kappa_1$ and $pr_l^{-1} K_a = \kappa_a$. ■

Lemma 9.9.7 *Let $A \in \mathfrak{A}$ and let κ be a congruence on A. If there exist κ-congruent functions f^n, $g^n \in A$, for which there is a tuple $\mathbf{a} := (a_1, a_2, ..., a_n) \in E_l^n$ with $f(\mathbf{a}) \neq g(\mathbf{a})$, then $pr_l^{-1} K_0 \subseteq \kappa$.*

Proof. The functions t_a, $a \in E_l$, which we have defined in the proof of Lemma 9.9.4, belong to A. Consequently, $(f, g) \in \kappa$ implies

$$f'(x) := f(t_{a_1}(x), ..., t_{a_n}(x)) \sim g(t_{a_1}(x), ..., t_{a_n}(x)) =: g'(x) \ (\kappa).$$

Because of $f(\mathbf{a}) \neq g(\mathbf{a})$ we have $f'(c) \neq g'(c)$ for all $c \in E_k \backslash E_l$. With the aid of Lemma 9.9.4 it follows that $\kappa_{\{l\}}, \kappa_{\{l+1\}}, ..., \kappa_{\{k-1\}} \subseteq \kappa$. Then, by Theorem 9.8.4, we have: $\kappa_{\{l\},\{l+1\},...,\{k-1\}} = pr_l^{-1} K_0 \subseteq \kappa$. ■

Lemma 9.9.8 *Let κ be a congruence on $A \in \mathfrak{A}$. Then*

$$pr_l \kappa \neq K_0 \Longrightarrow \kappa = pr_l^{-1}(pr_l \kappa).$$

Proof. Obviously, $K := pr_l \kappa$ is a congruence on $pr_l A$ and we have $\kappa \subseteq pr_l^{-1} K$. If $pr_l \kappa = K_1$ then there exists some functions f^n, $g^m \in A$ with $(f, g) \in \kappa$ and $n \neq m$. Thus, by Lemma 9.9.3, $\kappa = \kappa_1$ $(= pr_l^{-1} K_1)$. If $pr_l \kappa \neq K_1$ then $\kappa \subseteq \kappa_a$ and because of $pr_l \kappa \neq K_0$ there exist some κ-congruent functions r^t, s^t, which are not identical to a certain tuple of E_l^t. Then, by Lemma 9.9.7, we have $pr_l^{-1} K_0 \subseteq \kappa$. One can deduce $pr_l^{-1} K \subseteq \kappa$ from that and from $pr_l \kappa = K$ easy. Hence $pr_l^{-1} K = \kappa$. ■

The above lemmas can be summarized as follows:

Theorem 9.9.9 *([Lau 77])*
Let $A \in \mathfrak{A}$. Then, a nontrivial congruence on A is either of the type $pr_l^{-1} K$, where K is a congruence on $pr_l A$, or it has the form $\kappa_{U_1,U_2,...,U_t}$, where U_1, ..., U_t are some nonempty subsets of $E_k \backslash E_l$, which are pairwise incomparable (in respect to the inclusion). [5] ■

Since we determined all congruences on subclasses of P_2 in Section 9.2, can also characterize all congruences on $A \in \mathfrak{A}$ for $l = 2$ with the aid of the above theorem. As examples, we give only the congruence lattices of $P_{k,2}$ for $k \in \{3, 4, 5\}$, subsequently (in the Figures 9.3–9.5).

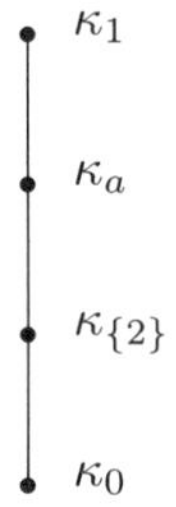

Fig. 9.3. Congruence lattice of $P_{3,2}$

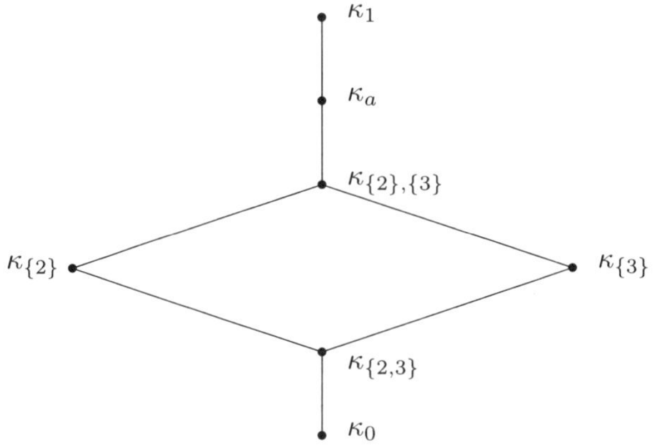

Fig. 9.4. Congruence lattice of $P_{4,2}$

---

[5] See also Theorem 9.9.2.

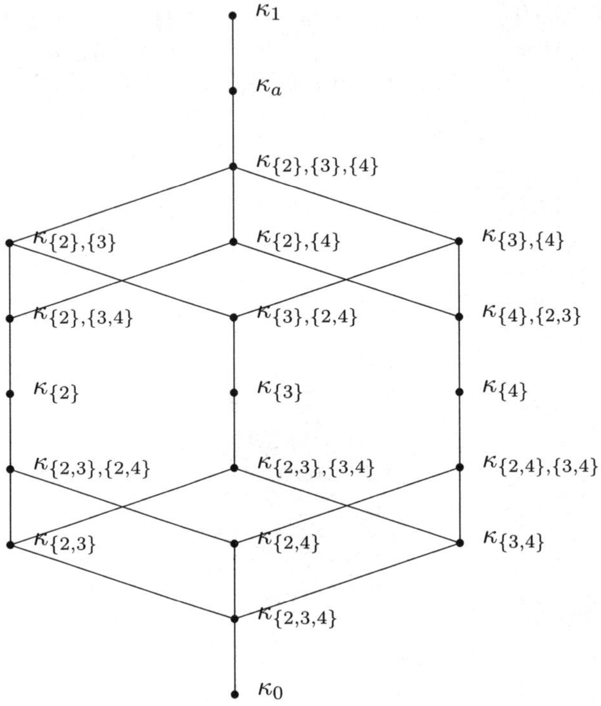

Fig. 9.5. Congruence lattice of $P_{5,2}$

9.10 Some Further General Properties of the Congruences and the l-Classes

In this section, we derive conditions with which one can realize whether a subclass of P_k has only trivial congruences. During the derivation of these conditions, we receive some general properties of the congruences as by-products. Most of the following lemmas and theorems were published by V. V. Gorlov in [Gor 77;1,2] and [Gor 79]. In the quoted papers by Gorlov proofs can be found that are excluded subsequently.

Lemma 9.10.1 *Let κ be a nontrivial congruence of the class A of P_k. Then*
(a) $e_1^1 \in A \implies \kappa_0 \subset \kappa \subset \kappa_a$,
(b) $\kappa \in Con_a A \implies \kappa^{(k)} \neq \kappa_0^{(k)}$,
(c) $(\kappa \in Con_1 A \wedge (\exists f, g \in A^1 : f \star g \neq f)) \implies \kappa^{(k)} \neq \kappa_0^{(k)}$,
(d) $e_1^1 \in A \implies (\forall n \geq 2 : \kappa^{(n)} \neq \kappa_a^{(n)})$.

Proof. (a) follows from Lemma 9.2.1.
(b): Because of (a) there are two different κ-congruent functions f^n, $g^n \in A$.

By using of the superposition operations ζ, τ, Δ, ∇ one receives obviously k-ary functions f', g' with $(f', g') \in \kappa$ and $f' \neq g'$. Consequently, (b) holds.
(c) follows from Theorem 9.3.4 and Lemma 9.3.7.
(d) Proving the contention for $n = 2$ suffices. Suppose $\kappa_a^{(2)} = \kappa^{(2)}$. Then we have $(e_1^2, e_2^2) \in \kappa$. Therefore, by Lemma 9.2.2, $\kappa = \kappa_a$. This is, however, a contradiction to our assumption. ■

With the help of an arbitrary arity congruence κ of a subclass $A \subseteq P_k$ one can define two further arity congruences:

$$\sigma_{n,\kappa} := \bigcap_{\mu \in Con_a A;\ \kappa^{(n)} = \mu^{(n)}} \mu \tag{9.29}$$

and

$$\begin{aligned}&\pi_{n,\kappa} := \\ &\{(f^p, g^q) \in A^2 \mid p = q \wedge \\ &\qquad \forall (f_1, g_1), ..., (f_p, g_p) \in \kappa^{(n)} \cup \{(e_i^n, e_i^n) \mid i \in \{1, 2, ..., n\}\} : \\ &\qquad (f(f_1, ..., f_p), g(g_1, ..., g_p)) \in \kappa^{(n)} \},\end{aligned} \tag{9.30}$$

which are completely defined through $\kappa^{(n)}$, $n \in \mathbb{N}$. Some properties of these relations that are easily proven are summarized in the following lemma.

Lemma 9.10.2 *Let A be a subclass of P_k and let $\kappa \in Con_a A$. Then*
(a) $\sigma_{n,\kappa}^{(n)} = \pi_{n,\kappa}^{(n)} = \kappa^{(n)}$,
(b) $\sigma_{n,\kappa} \subseteq \sigma_{n+1,\kappa} \subseteq \kappa \subseteq \pi_{n+1,\kappa} \subseteq \pi_{n,\kappa}$,
(c) $\kappa = \bigcap_{r \geq 1} \pi_{r,\kappa}$. ■

The following lemma is a consequence of the above and of Lemma 9.10.1.

Lemma 9.10.3 *For every nontrivial arity congruence κ of a subclass A of P_k, it holds:*

$$\kappa_0 \subset \sigma_{k,\kappa} \subseteq \kappa \subseteq \pi_{k,\kappa} \subset \kappa_a.$$

If A is in addition a clone, then

$$\kappa_0 \subset \sigma_{k,\kappa} \subseteq \kappa \subseteq \pi_{2,\kappa} \subset \kappa_a.$$

■

Since there are only finitely many possibilities for $\sigma_{k,\kappa}$ and $\pi_{k,\kappa}$, we obtain the following theorem as a consequence of Lemma 9.10.3 and Theorem 9.3.4:

Theorem 9.10.4 *The congruence lattice (or the lattice of the arity congruences) of a subclass of P_k has only finitely many atoms and dual atoms.* ■

Definition Let A be a subclass of P_k. An equivalence relation μ on A^n, $n \in \mathbb{N}$, is called an n-**congruence**, iff there exists a congruence relation $\kappa \in Con_a A$ with $\kappa^{(n)} = \mu$.

Lemma 9.10.5 *Let A be a subclass of P_k and let μ an equivalence relation on A^n. Then μ is an n-congruence if and only if μ fulfills the following condition:*

$$\forall\, f^m \in A\, \forall\, (u_0, v_0), ..., (u_n, v_n), .., (u_m, v_m) \in \mu \cup \{(e_i^n, e_i^n) \mid i \in \{1, 2, ..., n\}\} :$$
$$(f(u_1, ..., u_m), f(v_1, ..., v_m)) \in \mu \ \wedge\ (u_0(u_1, ..., u_n), v_0(v_1, ..., v_n)) \in \mu. \tag{9.31}$$

Proof. If μ is a congruence, then (9.31) obviously holds.
Assume μ fulfills (9.31). The equivalence relation π_μ defined by

$$(f^r, g^s) \in \pi_\mu :\Longleftrightarrow r = s \wedge \forall(u_1, v_1), ..., (u_r, v_r) \in \mu \cup \{(e_i^n, e_i^n) \mid i \in \{1, 2, ..., n\}\} : (f(u_1, ..., u_r), g(v_1, ..., v_r)) \in \mu)$$

is a congruence on A with $\pi_\mu^{(n)} = \mu$ (see [Gor 77a]). Thus, μ is an n-congruence on A. ■

With the help of Theorem 9.3.4, Lemma 9.3.7, and Lemma 9.10.5, easily one can prove the following theorem.

Theorem 9.10.6

(a) A subclass A of P_k has only trivial congruences if and only if $\kappa_0^{(k)}$ and $\kappa_a^{(k)}$ are the only k-congruences on A and there exist some $f, g \in A^1$ with $f \star g \neq f$.
(b) A clone A of P_k has only trivial congruences if and only if $\kappa_0^{(k)}$ and $\kappa_a^{(k)}$ are the only k-congruences on A. ■

Next, we introduce a notation for classes that are minimal classes concerning the following property: Each class B, which fulfills $A \subset B$, has only trivial congruences.

Definition A subclass A of P_k is called a **limit class** (briefly: **l-class**[6]), if it fulfills the following two conditions:
(1) $\forall B \in \mathbb{L}_k : (A \subset B \implies Con\ B = \{\kappa_0, \kappa_a, \kappa_1\})$;
(2) $\forall C \in \mathbb{L}_k : (C \subset A \implies \exists C' \in \mathbb{L}_k^{\uparrow}(C) : Con\ C' \neq \{\kappa_0, \kappa_a, \kappa_1\})$.

Lemma 9.10.7 *([Gor 77])*

(a) Let A be a subclass of P_k, which is not a clone. Then the clone $A \cup J_k$ has some nontrivial congruences.
(b) Each l-class is a clone.

Proof. (a): Since there are maximal classes (clones) with nontrivial congruences, which contain all constant functions, we can assume that $|A^n| \geq 2$ for $n \geq 2$ holds. Then, one can check that the equivalence relation

[6] In [Gor 77] the notation M-class was used instead of l-class.

$$\kappa := \{(f,f) \mid f \in J_k\} \cup \{(f^n, g^n) \mid \{f,g\} \subset A\}$$

is a nontrivial congruence on A.
(b) follows directly from (a). ■

The following theorem gives some examples of l-classes.

Theorem 9.10.8

(a) For $k = 2$ there are exactly four l-classes. These classes are L, K, D and $M \cap S$.
(b) A maximal class $Pol_k \varrho$ of P_k is an l-class if and only if $\varrho \in \{\varrho \in \mathfrak{C}_k^1 \mid |\varrho| \geq 2\} \cup \mathfrak{U}_k \cup \mathfrak{L}_k \cup \mathfrak{B}_k$.
(c) The class $[\mathfrak{U}_\omega \cup \{e_1^1\}]$ with $\omega \in E_k$ and

$$\mathfrak{U}_\omega := \bigcup_{n \geq 1} \{f^n \in P_k \mid \forall \mathbf{a} \in E_k^n \backslash (E_k \backslash \{\omega\})^n) : \; f(\mathbf{a}) = \omega\}$$

is an l-class.
(d) Among the submaximal classes of P_3 no l-classes exist.

Proof. (a) follows from Section 9.2. The statements (b) and (c) can be proven with the aid of the Theorems of Section 9.7, the Lemmas 9.7.1 and 9.10.1, and Theorem 9.10.9. One can find a proof for (d) in [Gor-L 82]. ■

Theorem 9.10.9 (**Gorlov's Theorem**; *[Gor 77]*)

(1) Let $A \subseteq P_k$ be an l-class. Then A fulfills one of the following conditions:
  (a) $|Con\ A| \geq 4$ and there exists a $2 \cdot k^k$-ary relation ϱ on E_k with $A = Pol_k \varrho$.
  (b) $|Con\ A| = 3$ and $ord\ A \leq k^{2 \cdot k^k}$.
(2) For every $k \geq 2$ there are only finitely many l-classes in $\mathbb{L}_k$.

Proof. (1): For A, the following two cases are possible:
Case 1: A has a nontrivial congruence κ.
Then, by Theorem 9.10.6, $\kappa^{(k)}$ is a nontrivial equivalence relation on A^k. If one now forms the $2 \cdot k^k$-ary relation

$$\varrho := \{(G_k(f), G_k(g)) \mid (f,g) \in \kappa^{(k)}\} \tag{9.32}$$

(see Chapter 2), then one can see that $A \subseteq Pol_k \varrho$ holds and that there exists the nontrivial congruence

$$\{(f^r, g^s) \in (Pol\ \varrho) \times (Pol\ \varrho) \mid r = s \ \wedge \ \forall (u_1, v_1), ..., (u_r, v_r) \in \kappa^{(k)} : (f(u_1, ...u_r), g(v_1, ..., v_r)) \in \kappa^{(n)}\}$$

on $Pol_k \varrho$ (see also Lemma 9.10.5). Since A is an l-class, this is only possible for $A = Pol_k \varrho$. Thus A fulfills the condition (a).

Case 2: A has only trivial congruences.
Suppose, we have $A' := [A^{2\cdot k^k}] \subset A$. Since A is an l-class, there exists a class $B \subseteq P_k$ with $A' \subset B$ and $|Con\ B| \geq 4$. Consequently, there is a nontrivial equivalence relation of the form $\kappa^{(k)}$ on B^k. This relation defines a $2 \cdot k^k$-ary relation ϱ, given by (9.32), with $B \subseteq Pol_k\varrho$. In addition, because of $A' \subset B$, one can prove $A \subseteq Pol_k\varrho$ as follows: Suppose, there is a function $f^n \in A$ that does not preserve ϱ ($n > k^{(2k^k)}$). Then there is some $r_1, ..., r_n \in \varrho$ with $f(r_1, ..., r_n) \notin \varrho$. We form the function $f'(\mathbf{x}) := f(x_{i_1}, ..., x_{i_n}) \in [f]$ with

$$\forall j, l \in \{1, ..., n\} : \ (i_j = i_l \iff r_j = r_l).$$

Then, because $|\varrho| \leq k^{2k^k}$, f' belongs indeed to A', however, it does not preserve the relation ϱ, in contradiction to $A' \subset B \subseteq Pol_k\varrho$. Consequently, $A \subseteq Pol_k\varrho$. But this is a contradiction to our assumption "A is an l-class and has only trivial congruences". Thus A fulfills the condition (b) and, therefore, (1) is proven.
(2) is a consequence of (1). ■

9.11 The Connection Between Clone Congruences and Fully Invariant Congruences

Let $\mathbf{A} = (A; W)$, $W \subseteq P_A$, be an algebra, let $V_\mathbf{A}$ be the variety generated by $\mathbf{A}$ and let $\mathbf{F}(\mathbf{X})$ $(:= (F(X); W))$ be the free algebra with countable ranges [7] of $V_\mathbf{A}$ (see Part I). For $F(X)$ we use the following model:

$$X := \{x_1, x_2, ...\},$$

$$F(X) := \bigcup_{n\geq 1}\{f(x_{i_1}, ..., x_{i_n}) \mid f \in T^{(n)}(\mathbf{A}) \ \wedge \ \{x_{i_1}, ..., x_{i_n}\} \subseteq X\}$$

(Therefore, $F(X)$ consists of all terms that can be formed from X and the functional symbols of ($\mathbf{T}(\mathbf{A}) = [W \cup J_A]$) and for arbitrary $w^n \in W$ and arbitrary $f_i(x_{i,1}, ..., x_{i,m_i}) \in F(X)$, $i = 1, ..., n+1$, it holds

$$\begin{aligned} &f_1(x_{1,1}, ..., x_{1,m_1}) = f_2(x_{2,1}, ..., x_{2,m_2}) \\ :\iff &\forall\, a_{1,1}, ..., a_{2,m_2} \in A : \ f_1(a_{1,1}, ..., a_{1,m_1}) = f_2(a_{2,1}, ..., a_{2,m_2}) \end{aligned} \tag{9.33}$$

and

$$\begin{aligned} &w(f_1(x_{1,1}, ..., x_{1,m_1}), ..., f_n(x_{n,1}, ..., x_{n,m_n})) = f_{n+1}(x_{n+1,1}, ..., x_{n+1,m_{n+1}}) \\ &:\iff \\ &\forall a_{1,1}, ..., a_{n+1,m_{n+1}} \in A : \\ &w(f_1(a_{1,1}, ..., a_{1,m_1}), ..., f_n(a_{n,1}, ..., a_{n,m_n})) = f_{n+1}(a_{n+1,1}, ..., a_{n+1,m_{n+1}}), \end{aligned}$$

[7] For example, X is countably infinite.

where

$$a_{i,j} = a_{l,k} \iff x_{i,j} = x_{l,k}$$

(see Chapter 6 of Part I).

The fully invariant congruences[8] of $F(X)$ are of interest to us here, since the lattice of these congruences is anti-isomorphic to the lattice of the subvarieties of $V_{\mathbf{A}}$. The following theorem is a well-known result of the category theory, which follows from the representation of a variety by an algebraic theory (a certain category). One finds complete information on this category-theory background in [Vog 85].

Theorem 9.11.1 *([Schw 84])*
Through

$$(t(x_1, ..., x_n), u(x_1, ..., x_n)) \in \kappa_F \iff (t, u) \in \kappa \tag{9.34}$$

every binary relation κ_F on $\mathbf{F}(\mathbf{X})$ ($\in V_{\mathbf{A}}$) is assigned to a relation $\kappa \subseteq \kappa_a$ on $T(\mathbf{A})$ and vice versa.
A binary relation κ_F is a fully invariant congruence $\mathbf{F}(\mathbf{X})$ if and only if the relation κ on $T(\mathbf{A})$ is an arity congruence of $T(\mathbf{A})$; i.e., the lattice of the fully invariant congruences of $\mathbf{F}(\mathbf{X})$ is isomorphic to the lattice of the arity congruences of $T(\mathbf{A})$ or antiisomorphic to the subvariety lattice of $V_{\mathbf{A}}$.

Proof. Since $f \in T(\mathbf{A})$ implies $\nabla f \in T(\mathbf{A})$ and because of (9.34), every relation on $F(X)$ is completely characterized through the tuples $(t(x_1, ..., x_n), u(x_1, ..., x_n))$ $(t, u \in T^{(n)}(\mathbf{A}), n = 1, 2, ...)$. Thus (9.34) describes a bijection between binary relations on $F(X)$ and arity relations on $T(\mathbf{A})$.
Let $\kappa \subseteq \kappa_a$ be a congruence relation on $T(\mathbf{A})$ and let κ_F be the relation on $F(X)$ defined by (9.34). Because of $(w(t_1, ..., t_r), w(u_1, ..., u_r)) \in \kappa$ for $(t_i, u_i) \in \kappa$, $i = 1, 2, ..., r$, and arbitrary $w \in W$, κ_F is compatible with the operations of W, i.e., κ_F is a congruence relation on $F(X)$. κ_F would be fully invariant if in addition $(t(x_1, ..., x_n), u(x_1, ..., x_n)) \in \kappa_F$ implies $(\varphi(t(x_1, ..., x_n)), \varphi(u(x_1, ..., x_n))) \in \kappa_F$ for every endomorphism φ on $F(X)$. Let φ be an endomorphism on $F(X)$ with $\varphi(x_i) = t_i(x_1, ..., x_{r_i})$, $i = 1, 2, ...,$ and $(t(x_1, ..., x_n), u(x_1, ..., x_n)) \in \kappa_F$. Then

$$\begin{aligned}
&(\varphi(t(x_1, ..., x_n), \varphi(u(x_1, ..., x_n))) \\
&= (t(\varphi(x_1), ..., \varphi(x_n)), u(\varphi(x_1), ..., \varphi(x_n))) \\
&= (t(t_1(x_1, ..., x_{r_1}), ..., t_n(x_1, ..., x_{r_n})), u(t_1(x_1, ..., x_{r_1}), ..., t_n(x_1, ..., x_{r_n}))) \\
&\in \kappa_F,
\end{aligned}$$

since $(t, u) \in \kappa$ implies $(t(t_1, ..., t_n), u(t_1, ..., t_n)) \in \kappa$. Thus κ_F is a fully invariant congruence on $\mathbf{F}(\mathbf{X})$ if κ is a congruence on $T(\mathbf{A})$.

8 These are equivalence relations on $\mathbf{F}(\mathbf{X})$, which are compatible not only with the operations of the algebra $\mathbf{F}(\mathbf{X})$ but also with every endomorphism of $\mathbf{F}(\mathbf{X})$.

Let κ_F be a fully invariant congruence on $\mathbf{F}(\mathbf{X})$ and let (f^n, g^n), $(u^m, v^m) \in \kappa$ in the following. We have to prove that $(f \star u, g \star v) \in \kappa$ and $(\alpha f, \alpha g) \in \kappa$ for $\alpha \in \{\zeta, \tau, \Delta\}$. Obviously, one can continue every mapping from X into $F(X)$ to an endomorphism on $F(X)$. We consider the endomorphisms $\varphi_1, ..., \varphi_4$ with

$$\begin{aligned}
&\varphi_1(x_1) = x_2,\ \varphi_1(x_2) = x_3,\ ...,\ \varphi_1(x_{n-1}) = x_n,\ \varphi_1(x_n) = x_1\\
&\varphi_2(x_1) = x_2,\ \varphi_2(x_2) = x_1,\ \varphi_2(x_3) = x_3,\ ...,\ \varphi_2(x_n) = x_n,\\
&\varphi_3(x_1) = \varphi_3(x_2) = x_1,\ \varphi_1(x_3) = x_2,\ ...,\ \varphi_3(x_n) = x_{n-1},\\
&\varphi_4(x_1) = u(x_1, ..., x_m),\ \varphi_4(x_2) = x_{m+1},\ ...,\ \varphi_4(x_n) = x_{m+n-1}.
\end{aligned}$$

Since, by assumption, $(f(x_1, ..., x_n), g(x_1, ..., x_n)) \in \kappa_F$ and κ_F is fully invariant, for $i = 1, 2, 3, 4$ the tuples

$$\begin{aligned}
&(\varphi_i(f(x_1, ..., x_n)), \varphi_i(g(x_1, ..., x_n)))\\
&= (f(\varphi_i(x_1), ..., \varphi_i(x_n), g(\varphi_i(x_1), ..., \varphi_i(x_n)))\\
&= \begin{cases}
(f(x_2, ..., x_n, x_1), g(x_2, ..., x_n, x_1)) & \text{if } i = 1,\\
(f(x_2, x_1, x_3, ..., x_n), g(x_2, x_1, x_3, ..., x_n)) & \text{if } i = 2,\\
(f(x_1, x_1, x_2, ..., x_{n-1}), g(x_1, x_1, x_2, ..., x_{n-1})) & \text{if } i = 3,\\
(f(u(x_1, ..., x_m), x_{m+1}, ..., x_{m+n-1}), & \\
g(u(x_1, ..., x_m), x_{m+1}, ..., x_{m+n-1})) & \text{if } i = 4,
\end{cases}
\end{aligned}$$

belong to κ_F and, therefore, $\{(\zeta f, \zeta g), (\tau f, \tau g), (\Delta f, \Delta g), (f \star u, g \star u)\} \subseteq \kappa$. In addition, since κ_F is compatible with the operations of W, we have (because of $(u(x_1, ..., x_m), v(x_1, ..., x_m)) \in \kappa_F$)

$$(g(u(x_1, ..., x_m), x_{m+1}, ..., x_{m+n-1}), g(v(x_1, ..., x_m), x_{m+1}, ..., x_{m+n-1})) \in \kappa_F$$

and (by transitivity)

$$(f(u(x_1, ..., x_m), x_{m+1}, ..., x_{m+n-1}), g(v(x_1, ..., x_m), x_{m+1}, ..., x_{m+n-1})) \in \kappa_F.$$

Thus $(f \star u, g \star v) \in \kappa$. From this, together with the above considerations, we get that κ is a congruence on $T(\mathbf{A})$.
The remaining statements of the theorem are obvious or result from Part I. ■

Analogous to the above proof and with the aid of Lemma 9.10.5, the following theorem can be proven.

Lemma 9.11.2 *Let $F(n)$ be the free algebra with n generators of $V_{\mathbf{A}}$. Through*

$$(f(x_1, ..., x_n), g(x_1, ..., x_n)) \in \mu_{F(n)} \Longleftrightarrow (f, g) \in \mu$$
$$(f,\ g\ \in T^{(n)}(\mathbf{A})\ )$$

every fully invariant congruence on $F(n)$ corresponds unambiguously with an n-congruence on $T(\mathbf{A})$ and conversely. ■

9.12 Automorphisms of Function Algebras

An automorphism of an algebra **A** is an isomorphic mapping from **A** onto **A**. In this section, we denote the image $\alpha(f)$ of a function $f \in A$, where α denotes a mapping on A, with f^α.
An automorphism α of a subclass $A \subseteq P_k$ is called an **inner automorphism** onto A, if there exists a bijective mapping φ from E_k onto E_k with the following property:

$$\forall f^n \in A: \; f^\alpha(x_1, ..., x_n) = \varphi(f(\varphi^{-1}(x_1), ..., \varphi^{-1}(x_n))$$

or

$$f^\alpha(\varphi(x_1), ..., \varphi(x_n)) = \varphi(f(x_1, ..., x_n)).$$

In the following, we put together some properties of automorphisms of subclasses of P_k, and we prove, for some classes, that they have only inner automorphisms. In particular, we prove that all automorphisms of subclasses of P_2 are inner automorphisms. This is not valid anymore for classes P_k, however, for $k \geq 3$.
The basis of these proofs is [Mal 66], from which we use many details here. In addition, we generalize some results of [Mal 66].

Lemma 9.12.1 *Let A be a subclass of P_k and let α be an automorphism onto A. Then for arbitrary $f \in A$ it holds:*

(a) $a\ f = a\ (f^\alpha)$.
(b) x_i is a fictitious variable of f iff x_i is a fictitious variable of f^α.
(c) f is a constant function iff f^α is a constant function.
(d) f is a permutation iff f^α is a permutation.

Proof. (a): Suppose there exists a function $f \in A$ with $a(f) \neq a(f^\alpha)$. We distinguish two cases.
Case 1: $n := a\ f < a(f^\alpha) =: m$.
Then $\Delta^{n-1} f = \Delta^n f$. Consequently, $\Delta^{n-1} f^\alpha = \Delta^n f^\alpha$. The last equation is not, however, possible because of $a(\Delta^{n-1} f^\alpha) = m - n + 1$ and $a(\Delta^n f^\alpha) = m - n$.
Case 2: $a\ f > a(f^\alpha) =: m$.
In this case, we have $\Delta^{m-1} f^\alpha = \Delta^m f^\alpha$ and thus $(\Delta^{m-1} f)^\alpha = (\Delta^m f)^\alpha$, in contradiction to the bijectivity of the mapping α.
Therefore, $a(f) = a(f^\alpha)$ for all $f \in A$.
(b): W.l.o.g. let $i = 1$. Obviously, x_1 is a fictitious variable of f iff $\nabla(\Delta f) = f$. The equation $\nabla(\Delta f) = f$ holds iff $\nabla(\Delta f^\alpha) = f^\alpha$. Hence (b) is right.
(c) follows directly from (b).
(d): If A contains a permutation, then all projections belong to A. The function $e := e_1^1$ is the only function of A, for which $f \star e = e \star f$ holds for all $f \in A^1$. Consequently, we have also $f^\alpha \star e^\alpha = e^\alpha \star f^\alpha$ for all $f \in A^1$. Thus $e^\alpha = e$.

Since for every permutation $s \in A$ there exists a t with $\underbrace{s \star s \star ... \star s}_{t \text{ times}} = e$, it holds $\underbrace{s^\alpha \star s^\alpha \star ... \star s^\alpha}_{t \text{ times}} = e$, whereby s^α is a permutation. ■

Theorem 9.12.2 *Let A be a subclass of P_k and let T be a subset of A^r, $r \geq 1$, which has the following three properties:*

(1) Every function of A preserves the functions of T.
(2) There exists a $\mathbf{c} \in E_k^r$ and some functions $g_0, g_1, ..., g_{k-1} \in T$ with $g_i(\mathbf{c}) = i$ for every $i \in E_k$.
(3) For each automorphism α onto A it holds $T^\alpha = T$ and $\alpha_{|T}$ is an inner automorphism onto T.

Then each automorphism onto A is an inner automorphism.

Proof. Let α be an automorphism onto A. By assumption (3), we have $T^\alpha = T$ and that $\alpha_{|T}$ is an inner automorphism onto T. The permutation φ appertaining to this inner automorphism defines an isomorphism β_φ from A onto A^{β_φ}:

$$\forall f^n \in A: \ f^{\beta_\varphi}(\mathbf{x}) = \varphi^{-1}(f(\varphi(x_1), ..., \varphi(x_n))).$$

We show that the mapping $\gamma := \alpha \Box \beta_\varphi^{-1}$ is the identical mapping, whereby $\beta_\varphi = \alpha$ is valid and the claim "α is an inner automorphism" follows:
Let $f^n \in A$ and let $\mathbf{a} = (a_1, ..., a_n) \in E_k^n$ be arbitrary. By assumption (2), for every $a_i \in E_k$, there exists a function $g_{a_i} \in T$ with $g_{a_i}(\mathbf{c}) = a_i$, $i = 1, ..., n$. Because of (1) there is a function $g \in T$ with the properties: $f(g_{a_1}(\mathbf{x}), ..., g_{a_n}(\mathbf{x})) = g(\mathbf{x})$ and $g(\mathbf{c}) = f(\mathbf{a})$. Consequently, we have $g^\gamma = f^\gamma(g_{a_1}^\gamma, ..., g_{a_n}^\gamma)$. Because the functions of T are fixed points of the mapping γ, this implies $g = f^\gamma(g_{a_1}, ..., g_{a_n})$. Thus we have $f^\gamma(\mathbf{a}) = g(\mathbf{c}) = f(\mathbf{a})$ for arbitrary $\mathbf{a} \in E_k^n$, i.e., $f = f^\gamma$ holds for every $f \in A$. ■

With the aid of the above theorem, we can prove the three following theorems.

Theorem 9.12.3 *([Mal 72b])*
If a subclass $A \subseteq P_k$ contains all constants of P_k, then each automorphism onto A is an inner automorphism.

Proof. The claim follows from Theorem 9.12.2, if one chooses $T = \{c_0, c_1, ..., c_{k-1}\}$ and uses Lemma 9.12.1, (c). ■

The following theorem generalizes Theorem 2 from [Mal 66].

Theorem 9.12.4 *Let $\emptyset \neq E \subseteq E_k$,*

$$t_{a,b}(x) := \begin{cases} a \text{ if} & x \in E, \\ b \text{ otherwise}, \end{cases}$$

$T := \{t_{a,b} \mid a \in E \wedge b \in E_k\}$, $S := P_k^1[k] \cap Pol_k E$ and let A be a subclass of P_k with $S \cup T \subseteq A$. Then, each automorphism onto A is an inner automorphism.

Proof. If $|E| = 1$ then Theorem 9.12.4 follows from the proof of Theorem 2 in [Mal 66]. If $A \not\subseteq Pol_k E$, then by $S \cup T \subseteq A$, we have that A contains all constant functions of P_k. Consequently, by Theorem 9.12.3, A has only inner automorphisms in this case. Thus we can assume $A \subseteq Pol_k E$ in the following. Let α be an automorphism onto A. We show that T fulfills the assumptions (1)–(3) of Theorem 9.12.2. Because of $A \subseteq Pol_k E$, it holds (1). (2) is also valid. Further, we have for arbitrary $g \in A^1$

$$(\forall s \in S : \ g \star s = g) \iff g \in T. \tag{9.35}$$

Thus $g \in T$ implies $g \star s = g$ for all $s \in S$ and therefore also $g^\alpha \star s^\alpha = g^\alpha$. Because the function s^α belongs to S because of Lemma 9.12.1, (d), we obtain $g^\alpha \in T$ from (9.35) Consequently, $T = T^\alpha$.
Next we show that, for the set

$$T_1 := \{t_{a,b} \mid a \in E \ \wedge \ b \in E_k \backslash E\},$$

the equation $T_1^\alpha = T_1$ holds. Obviously, we have for all $t_{a,b} \in T$:

$$t_{a,b} \star t_{a,b} = t_{c,c} \iff a = c \ \wedge \ \{a, b\} \subseteq E. \tag{9.36}$$

Therefore, $t_{a,b} \in T \backslash T_1$ implies $t_{a,b} \star t_{a,b} = t_{a,a}$ and then $t_{a,b}^\alpha \star t_{a,b}^\alpha = t_{a,a}^\alpha$. Since $t_{a,a}^\alpha$ is a constant by Lemma 9.12.1, we have $t_{a,b} \in T \backslash T_1$ because of (9.36). Consequently, $T_1^\alpha = T_1$.
If one restricts the functions of $T_1 \cup S$ to E or $E_k \backslash E$, one obtains the sets $(T_1 \cup S)_{|E} \subseteq P_E$ or $(T_1 \cup S)_{|E_k \backslash E} \subseteq P_{E_k \backslash E}$, respectively. Since the set $(T_1 \cup S)_{|E}$ or $(T_1 \cup S)_{|E_k \backslash E}$ contains all constant functions of P_E or $P_{E_k \backslash E}$, respectively, $\alpha_{|(T_1 \cup S)_{|E}}$ or $\alpha_{|(T_1 \cup S)_{|(E_k \backslash E)}}$ is an inner automorphism onto the semigroup $(T_1 \cup S; \star)$, respectively. Thus there exists a permutation φ onto E_k with

$$\forall h \in T_1 \cup S : \ h^\alpha = \varphi^{-1} \star h \star \varphi. \tag{9.37}$$

Finally we show that (9.37) also holds for every function $h \in T \backslash T_1$, whereby the condition (3) of Theorem 9.12.2 and then our theorem are proven:
Let $t_{a,b} \in T \backslash T_1$, $t_{a,b}^\alpha = t_{a',b'}$ and let $s \in S$, where s has only the fixed points a and b. Then $s \star t_{a,b} = t_{a,b}$ and

$$s^\alpha \star t_{a',b'} = t_{a',b'}. \tag{9.38}$$

(9.38) implies that a' and b' are fixed points of the permutation s^α. Because of $s^\alpha = \varphi^{-1} \star s \star \varphi$ the permutation s^α has only the fixed points $\varphi^{-1}(a)$ and $\varphi^{-1}(b)$. Consequently, $\{a', b'\} = \{\varphi^{-1}(a), \varphi^{-1}(b)\}$. Thus for $a = b$, we have shown that $t^\alpha_{a,a} = \varphi^{-1} \star t_{a,a} \star \varphi$ is valid. From that and from $t_{a,b} \star t_{a,a} = t_{a,a}$ and $t_{a',b'} \star t^\alpha_{a,a} = t^\alpha_{a,a}$, then, $a' = \varphi^{-1}(a)$ and $b' = \varphi^{-1}(b)$ hold. Consequently, we have $t_{a',b'} = \varphi^{-1} \star t_{a,b} \star \varphi$. ■

Theorem 9.12.5 *([Lau 79;b])*
All automorphisms on a maximal class of P_k are inner automorphisms.

Proof. With the aid of Theorems 9.12.3 and 9.12.4, it is easy to see that all maximal classes of P_k have only inner automorphisms, except for the maximal classes of the type $\mathfrak{S}$.
Let $A \subseteq P_k$ be a maximal class of type $\mathfrak{S}$, i.e., it holds $A = Pol_k \varrho_s$, where s, ϱ_s and the functions of A are defined as in Section 5.2.2 in the following. Denote T the set of all functions of the form

$$f_a(x) := \sum_{r=1}^{l} \sum_{i=0}^{p-1} j_{a_{r,i}}(x) \cdot s^i(a), \ (a \in E_k).$$

Our theorem is proven if we can show that T fulfills the assumptions of Theorem 9.12.2. Obviously, T fulfills the first two conditions of Theorem 9.12.2. To prove (3) from Theorem 9.12.2 denote α an automorphism onto $Pol_k \varrho_s$. First we prove $T^\alpha = T$. Obviously, for every function $g \in (Pol_k \varrho_s)^1$ and every function $f_a \in T$ there exists a function $h \in (Pol_k \varrho_s)^1$ with $f_a = h \star g$. Consequently, we have

$$f_a^\alpha = h^\alpha \star g^\alpha. \tag{9.39}$$

If the function also fulfills the condition $g^\alpha \in T$, then we have $h^\alpha \star g^\alpha \in T$ and (by (9.39)) $f_a^\alpha \in T$ for arbitrary $a \in E_k$. Thus $T^\alpha = T$.
Next we derive some properties of the function f_a:
For all $r \in \{1, 2, ..., l\}$ and $i \in E_p$ we have $f_{a_{r,0}} \star f_{a_{r,i}} = f_{a_{r,i}}$ and thus $f^\alpha_{a_{r,0}} \star f^\alpha_{a_{r,i}} = f^\alpha_{a_{r,i}}$, i.e., for every r there exists an $\widehat{r}$ with $f^\alpha_{a_{r,0}} = f_{a_{\widehat{r},0}}$. Further, it holds

$$f_{a_{r,i}} = \underbrace{(...((f_{a_{r,1}} \star f_{a_{r,1}}) \star f_{a_{r,1}})... \star f_{a_{r,1}})}_{i \text{ times}}.$$

For $f^\alpha_{a_{r,1}} = f_{a_{\widehat{r},b_r}}$ ($b_r \in E_p \backslash \{0\}$) this implies

$$f^\alpha_{a_{r,i}} = f_{a_{\widehat{r},i \cdot b_r \ (mod\ p)}}.$$

Because of $f_{a_{r,0}} \star f_{a_{1,1}} = f_{a_{r,1}}$ we have $f^\alpha_{a_{r,0}} \star f^\alpha_{a_{1,1}} = f^\alpha_{a_{r,1}} = f_{a_{\widehat{r},b_r}} = f_{a_{\widehat{r},0}} \star f_{a_{\widehat{1},b_1}} = f_{a_{\widehat{r},b_1}}$, i.e., $b_r = b_1 =: b$ holds for arbitrary $r \in \{1, 2, ..., l\}$. In summary we get:

$$f^\alpha_{a_{r,i}} = f_{a_{\widehat{r},i \cdot b \ (mod\ p)}}, \tag{9.40}$$

where $r, \widehat{r} \in \{1, 2, ..., l\}$, $i \in E_p$ and $b \in E_p \backslash \{0\}$.
Because of (9.40), one can define a permutation φ onto E_k as follows: $\varphi(a_{r,i}) := a_{\widehat{r}, i \cdot b \ (mod\ p)}$. This permutation has the property

$$\forall a_{u,v} \in E_k : \ f^{\alpha}_{a_{u,v}}(\varphi(x)) = \varphi(f_{a_{u,v}}(x)),$$

which one can prove as follows: Let $x = a_{r,i}$ ($r \in \{1, 2, ..., l\}$, $i \in E_p$). Then we have

$$f^{\alpha}_{a_{u,v}}(\varphi(a_{r,i})) = f_{a_{\widehat{u}, v \cdot b \ (mod\ p)}}(a_{\widehat{r}, i \cdot b \ (mod\ p)}) = s^{i \cdot b}(a_{\widehat{u}, v \cdot b \ (mod\ p)}) = a_{\widehat{u}, (v+i) \cdot b \ (mod\ p)}$$

and

$$\varphi(f_{a_{u,v}}(a_{r,i})) = \varphi(s^i(a_{u,v})) = \varphi(a_{u, v+i \ (mod\ p)}) = a_{\widehat{u}, (v+i) \cdot b \ (mod\ p)}.$$

Hence α is an inner automorphism onto T, and Theorem 9.12.2 implies that $Pol_k \varrho_s$ has only inner automorphisms. ■

Theorem 9.12.6 *([Gor-L 83a])*
Each automorphism on a subclass of P_2 is an inner automorphism.

Proof. Let $A = [A] \subseteq P_2$. By Chapter 4, the following cases are possible:
Case 1: $A^1 = \{c_a\}$.
Then A is the class $[c_a]$, which has obviously only inner automorphisms.
Case 2: $\{c_0, c_1\} \subseteq A^1$.
In this case, our theorem follows from Theorem 9.12.3.
Case 3: $A^1 \in \{\{c_0, e^1_1\}, \{c_1, e^1_1\}, \{e^1_1, \overline{e^1_1}\}\}$.
In this case, we obtain our claim from Theorem 9.12.2, when we elect $T = A^1$ and when we use Lemma 9.12.1, (d).
Case 4: $A^1 = \{e^1_1\}$.
By Chapter 3, considering only the following possibilities for A suffices:
Case 4.1: $A \in \{I, K, L \cap T_0 \cap S, S \cap M\}$.
In this case, there exists a function $f^n \in A$ with $[f] = A$ and $n = ord\ A$ (see Chapter 3). Then, with the help of Lemma 9.12.1, (b) and by the properties of the functions of A, one can prove that $f^{\alpha} = f$ holds for each automorphism α onto A. Because of $[f] = A$ this implies that α is the identical mapping, which is obviously an inner automorphism.
Case 4.2: $A \in \{S \cap T_0, M \cap T_0 \cap T_1, T_0 \cap T_1\} \cup \bigcup_{m \in \{2,3,...,\infty\}} \{T_{0,m} \cap T_1 \cap M, T_{0,m} \cap T_1\}$.
Choosing $T = A^2$, A fulfills the conditions (1) and (2) of Theorem 9.12.2. We prove that $T = A^2$ also fulfills condition (3):
If $A \in \{M \cap T_0 \cap T_1, T_0 \cap T_1\}$ then $A^2 = \{\wedge, \vee, e^2_1, e^2_2\}$; if $A = S \cap T_0$ then $A^2 = \{e^2_1, e^2_2\}$ and if $A \in \{T_{0,m} \cap T_1 \cap M, T_{0,m} \cap T_1\}$ then $A^2 = \{\wedge, e^2_1, e^2_2\}$.
By Lemma 9.12.1, (b) we have $(e^2_i)^{\alpha} = e^2_i$, $i \in \{1, 2\}$, for each automorphism

α onto A. When one considers this, one convinces oneself that $\alpha_{|A^2}$ is an inner automorphism onto T for each automorphism α onto A. Consequently, A fulfills every condition of Theorem 9.12.2 in Case 4.2. Thus A has only inner automorphisms. ■

Theorem 9.12.7 *([Gor-L 83a])*
If $k \geq 3$ then there are subclasses of P_k, whose automorphisms are not all inner automorphisms.

Proof. Here is an example:
Let

$$u(x) := \begin{cases} x \text{ if } & x \in E_2, \\ 0 \text{ otherwise}, \end{cases} \quad \text{and} \quad v(x) := \begin{cases} x \text{ if } & x \in E_2, \\ 1 \text{ otherwise}. \end{cases}$$

By Lemma 9.12.1, (a), (b) and Theorem 9.8.4, all automorphisms onto the set $A := [u, v, c_0]$ are induced by the automorphisms of the semigroup $(A^1; \star)$. It is easy to check that the mapping $\alpha: \ u \mapsto v, \ v \mapsto u, \ c_0 \mapsto c_0$ is an automorphism of $(A^1; \star)$. Suppose α is an inner automorphism. Then there is a permutation φ onto E_k with

$$\forall g \in \{u, v, c_0\}: \ g^\alpha = \varphi^{-1} \star g \star \varphi. \tag{9.41}$$

Because of $c_0^\alpha = c_0$ we have $\varphi(0) = 0$. (9.41) implies $\varphi \star u = v \star \varphi$ and then $\varphi(0) = v(\varphi(0))$ and $\varphi(0) = v(\varphi(2))$. Because of definitions of φ and v this is only possible if $\varphi(0) = 1$, in contrary to the above statement $\varphi(0) = 0$. Thus α is not an inner automorphism. ■

10

The Relation Degree and the Dimension of Subclasses of P_k

This chapter deals with the relation degree and the dimension of subclasses A of P_k. The *relation degree* of A is the smallest number h so that class A is unambiguously described by a relation set whose elements have the arity h at most. The *dimension* of A is the smallest arity of a relation that characterizes class A unambiguously. We begin with investigating the connections between these two complexity measures of the relational description of classes. Then we prove the relation degrees and dimensions of Post's classes, which were found by G. N. Blochina in [Blo 70]. In Section 10.3, there are further classes for which is known the relation degree or the dimension.

10.1 The Definition of the Relation Degree and of the Dimension of a Subclass of P_k

In this section we deal with two possibilities for evaluating the complexity of relational descriptions of subclasses of P_k.
Denote $ar(\varrho)$, $\varrho \in R_k^h \backslash \{\emptyset\}$, the arity h of the relation ϱ. Further, for $Q \subseteq R_k$, let

$$ar_{max}(Q) := \begin{cases} 0, & \text{if } Q = \emptyset, \\ \max\{ar(\varrho) \mid \varrho \in Q\}, & \text{if the maximum exists,} \\ \infty & \text{otherwise.} \end{cases}$$

The **relation degree** $d(A)$ of a subclass A of P_k is the smallest of the possible numbers $ar_{max}(Q)$ with $A = Pol_k Q$, i.e.,

$$d(A) := \begin{cases} \min\{ar_{max}(Q) \mid Q \subseteq R_k \ \wedge \ Pol\ Q = A\} & \text{if } \exists h \in \mathbb{N}\ \exists Q \subseteq R_k^h : \\ & \qquad Pol\ Q = A, \\ \infty & \text{otherwise.} \end{cases}$$

If one complements the above definition with the demand $|Q| = 1$, the result is the definition of the **dimension** of a subclass A of P_k:

$$dim\ A := \begin{cases} min\{ar(\varrho) \mid \varrho \in R_k \wedge Pol\ \varrho = A\} & \text{if } \exists \varrho \in R_k : Pol\ \varrho = A, \\ \infty & \text{otherwise.} \end{cases}$$

The following lemma gives some elementary connections from d and dim.

Lemma 10.1.1 *Let A be a subclass of P_k. Then*

(a) $d(A) = dim\ A = 0 \iff A = P_k$.

(b) $d(A) \leq dim\ A$.

(c) $d(A) < \infty \iff dim\ A < \infty$.

(d) If $\infty \neq d(A) < dim\ A$, *then there are certain* **proper predecessors** $A_1, ..., A_r$ *of* A, *which are pairwise not contained in each other (i.e.,* $A \subset A_i$, $A_i = [A_i]$ *and* $A_i \not\subseteq A_j$ *for* $i \neq j$ *and* $i, j = 1, 2, ..., r$*), with the following properties:*

$$A = A_1 \cap A_2 \cap ... \cap A_r, r \geq 2 \text{ and}$$
$$d(A) = max\{d(A_i) \mid i = 1, ..., r\}.$$

(e) If A has exactly a direct predecessor B (i.e., $A \subset B = [B]$ *and no subclass* B' *exists with* $A \subset B' \subset B$*), then* $d(A) = dim\ A$.

Proof. (a) follows from $Pol\ \emptyset = P_k$ and (b) results immediately from the definitions of d and dim.
(c) follows from

$$Pol\{\varrho_1, \varrho_2, ..., \varrho_t\} = Pol\ \ \varrho_1 \times \varrho_2 \times ... \times \varrho_t.$$

(d): If $d(A) \neq \infty$ and $d(A) < dim\ A$, there are certain relations $\varrho_1, ..., \varrho_r$ with the properties

$$A = Pol\ \{\varrho_1, ..., \varrho_r\} = \bigcap_{i=1}^r Pol\ \varrho_i,$$
$$r \geq 2,$$
$$\forall i \in \{1, .., r\} : A \neq Pol\ \{\varrho_1, ..., \varrho_r\} \backslash \{\varrho_i\}$$
$$\text{and } d(A) = ar_{max}\{\varrho_1, ..., \varrho_r\}.$$

If one now sets $A_i := Pol\ \varrho_i$ $(i = 1, .., r)$, most of the properties of the sets A_i mentioned in the lemma are immediately clear. The rest results from $A \subset A_i$ for all $i = 1, ..., r$ because of $d(A) < dim\ A$.
(e) follows from (b) and (d). ∎

10.2 The Dimensions and Relation Degrees of Post's Classes

The basic results of this section come from the paper [Blo 70] by G. N. Blochina.

Our aim is to prove the values given in Table 10.1. In the first column, all subclasses A of P_2 are listed (except for the isomorphic classes). Next to the values $d(A)$ and $dim(A)$, which we will subsequently prove, one can also find some descriptive relations for subclasses A. The correctness of these statements can easily be proven in the following manner:

First,

$$A \subseteq Pol\ \varrho$$

holds for the corresponding ϱ of the third column. (For most classes this is clear according to definition. For every remaining class, the statement is a conclusion from the given construction of ϱ with the aid of a certain n-ary graphic $G_n(A)$ [1].)

Then, one proves that for every direct predecessor B of A there exists an $f \in B$ [2] with $f \notin Pol\ \varrho$. The equation $Pol\ \varrho = A$ and an upper bound for $dim\ A$ ($\geq d(A)$) result from the proven facts. After Table 10.1, one finds the lower bounds for $dim\ A$, which are identical to the upper bounds and the proofs for the remaining statements of this table.

Table 10.1

A	$dim\ A$	ϱ with $A = Pol\ \varrho$	$d(A)$
P_2	0	$\emptyset$	0
T_0	1	$\{0\}$	1
S	2	$\begin{pmatrix} 0 & 1 \\ 1 & 0 \end{pmatrix}$	2
M	2	$\begin{pmatrix} 0 & 0 & 1 \\ 0 & 1 & 1 \end{pmatrix}$	2
$T_{0,2}$	2	$\begin{pmatrix} 0 & 0 & 1 \\ 0 & 1 & 0 \end{pmatrix}$	2
$T_0 \cap T_1$	2	$\begin{pmatrix} 0 \\ 1 \end{pmatrix}$	1
$T_{0,3}$	3	$E_2^3 \setminus \{\mathbf{1}\}$	3
$M \cap T_0$	3	$\begin{pmatrix} 0 & 0 & 0 \\ 0 & 0 & 1 \\ 0 & 1 & 1 \end{pmatrix}$	2
$S \cap T_0$	3	$\begin{pmatrix} 0 & 0 \\ 0 & 1 \\ 1 & 0 \end{pmatrix}$	2
$S \cap M$	3	$\begin{pmatrix} 0 & 0 & 1 \\ 0 & 1 & 1 \\ 1 & 0 & 0 \end{pmatrix}$	2

[1] See Section 2.7.

[2] It is sufficient to consider a basis of B instead of B.

A	$dim\ A$	ϱ with $A = Pol\ \varrho$	$d(A)$
$M \cap T_{0,2}$	3	$\begin{pmatrix} 0&0&0&1 \\ 0&0&1&0 \\ 0&1&1&1 \end{pmatrix}$	2
$T_1 \cap T_{0,2}$	3	$\begin{pmatrix} 0&0&1 \\ 0&1&0 \\ 1&1&1 \end{pmatrix}$	2
$L \cap T_0$	3	$\begin{pmatrix} 0&0&1&1 \\ 0&1&0&1 \\ 0&1&1&0 \end{pmatrix} = pr_{2,3,4}G_2(A)$	3
$K \cup C$	3	$\begin{pmatrix} 0&1&0&1 \\ 0&0&0&1 \\ 0&0&1&1 \end{pmatrix} = pr_{2,1,3}G_2(A)$	3
$M \cap T_{0,2}$	3	$\begin{pmatrix} 0&0&0&1 \\ 0&0&1&0 \\ 0&1&1&1 \end{pmatrix}$	2
$T_1 \cap T_{0,2}$	3	$\begin{pmatrix} 0&0&1 \\ 0&1&0 \\ 1&1&1 \end{pmatrix}$	2
$L \cap T_0$	3	$\begin{pmatrix} 0&0&1&1 \\ 0&1&0&1 \\ 0&1&1&0 \end{pmatrix} = pr_{2,3,4}G_2(A)$	3
$K \cup C$	3	$\begin{pmatrix} 0&1&0&1 \\ 0&0&0&1 \\ 0&0&1&1 \end{pmatrix} = pr_{2,1,3}G_2(A)$	3
I	3	$\begin{pmatrix} 0&1&1 \\ 1&0&1 \\ 1&1&0 \end{pmatrix} = pr_{4,6,7}G_3(A)$	3
$I \cap C_0$	3	$\begin{pmatrix} 0&0&0&1&1&1 \\ 0&0&1&0&1&0 \\ 0&1&1&0&0&1 \end{pmatrix}$	3
$\overline{I}$	3	$\begin{pmatrix} 0&1&1&0&0&1 \\ 1&0&1&0&1&0 \\ 1&1&0&1&0&0 \end{pmatrix} = pr_{4,6,7}G_3(A)$	3
$[P_2^1]$	3	$\begin{pmatrix} 0&0&0&1&1&1 \\ 0&0&1&1&0&1 \\ 0&1&0&0&1&1 \end{pmatrix} = pr_{1,2,3}G_2(A)$	3
$I \cup C$	4	$\begin{pmatrix} 0&0&0&1 \\ 0&1&0&1 \\ 1&0&0&1 \\ 1&1&0&1 \end{pmatrix} = G_2(A)$	3
$M \cap T_0 \cap T_1$	4	$\begin{pmatrix} 0&0&0 \\ 0&0&1 \\ 0&1&1 \\ 1&1&1 \end{pmatrix}$	2
L	4	$\begin{pmatrix} 0&0&0&1&1&0&1&1 \\ 0&0&1&1&0&1&0&1 \\ 0&1&0&0&1&1&0&1 \\ 0&1&1&0&0&0&1&1 \end{pmatrix} = G_2(A)$	4
$L \cap S$	4	$\begin{pmatrix} 1&0&0&0&0&1&1&1 \\ 0&1&0&0&1&0&1&1 \\ 0&0&1&0&1&1&0&1 \\ 0&0&0&1&1&1&1&0 \end{pmatrix} = pr_{1,2,3,5}G_3(A)$	4
$L \cap S \cap T_0$	4	$\begin{pmatrix} 0&0&0&0 \\ 0&0&1&1 \\ 0&1&1&0 \\ 1&0&1&0 \end{pmatrix} = pr_{1,2,4,6}G_3(A)$	3
$K \cup C_0$	4	$\begin{pmatrix} 0&0&0&0 \\ 0&1&0&0 \\ 0&1&1&0 \\ 0&1&0&1 \end{pmatrix} = pr_{1,5,6,7}G_3(A)$	3

A	$dim\ A$	ϱ with $A = Pol\ \varrho$	$d(A)$
$K \cup C_1$	4	$\begin{pmatrix} 0&0&0&1 \\ 0&1&0&1 \\ 1&0&0&1 \\ 1&1&1&1 \end{pmatrix} = G_2(A)$	3
$T_{0,4}$	4	$E_2^4 \setminus \{(1,1,1,1)\}$	4
$T_{0,3} \cap T_1$	4	$\begin{pmatrix} 0&0&0&1&0&1&1 \\ 0&0&1&0&1&0&1 \\ 0&1&0&0&1&1&0 \\ 1&1&1&1&1&1&1 \end{pmatrix} = pr_{4,6,7,8}G_3(A)$	3
$T_{0,3} \cap M$	4	$\begin{pmatrix} 0&0&0&0&1&0&1&1 \\ 0&0&0&1&0&1&0&1 \\ 0&0&1&0&0&1&1&0 \\ 0&1&1&1&1&1&1&1 \end{pmatrix} = pr_{4,6,7,8}G_3(A)$	3
$T_{0,2} \cap M \cap T_1$	4	$\begin{pmatrix} 0&0&0&1 \\ 0&0&1&0 \\ 0&1&1&1 \\ 1&1&1&1 \end{pmatrix} = pr_{3,5,7,8}G_3(A)$	2
K	5	$\begin{pmatrix} 0&0&0&0 \\ 1&0&0&0 \\ 1&0&1&0 \\ 1&1&0&0 \\ 1&1&1&1 \end{pmatrix} = pr_{1,5,6,7,8}G_3(A)$	3
$T_{0,t}$ $(t \geq 5)$	t	$E_2^t \setminus \{(1,1,...,1)\} =: \alpha_t$	t
$T_{0,t-1} \cap T_1$	t	$\alpha_{t-1} \times (1) =: \beta_t$	$t-1$
$T_{0,t-1} \cap M$	t	$\beta_t \cup \{(0,0,...,0)\} =: \gamma_t$	$t-1$
$T_{0,t-2} \cap M \cap T_1$ $(t \geq 5)$	t	$\gamma_{t-1} \times (1)$	$t-2$
$T_{0,\infty}$	∞		∞
$T_{0,\infty} \cap T_1$	∞		∞
$T_{0,\infty} \cap M$	∞		∞
$T_{0,\infty} \cap M \cap T_1$	∞		∞
C_0	∞		∞
C	∞		∞

Lemma 10.2.1 *Let*

$$\mathfrak{M} := \{C_a, C, T_{a,\infty}, T_{a,\infty} \cap T_{\overline{a}}, T_{a,\infty} \cap M, T_{a,\infty} \cap M \cap T_{\overline{a}} \mid a \in E_2\}.$$

Then $dim\ A = \infty$ *for every* $A \in \mathfrak{M}$.

Proof. Since $e_1^1 \in Pol\ \varrho$ holds for arbitrary ϱ, the contention is valid apparently for sets C_a and C. Assume for $A \in \mathfrak{M} \setminus \{C_0, C_1, C\}$ there is an h-ary relation ϱ with r columns, for which $A = Pol\ \varrho$ holds.
If one identifies the variable x_i with the variable x_j $(i \neq j)$ in the function h_r^{r+1}, one receives a function of the form

$$x_i \wedge g(x_1, ..., x_{i-1}, x_{i+1}, ..., x_r) \tag{10.1}$$

with $g \in M \cap T_0 \cap T_1$. Since all functions of the form (10.1) belong to A, h_r preserves the relation ϱ, in contradiction to $h_r \notin A$. ■

Lemma 10.2.2 *Let A be a subclass of P_2. Then*
(a) dim $A = 0 \iff A = P_2$;
(b) dim $A = 1 \iff A \in \{T_0, T_1\}$;
(c) dim $A = 2 \iff A \in \{M, S, T_0 \cap T_1, T_{0.2}, T_{1,2}\}$.

Proof. The statements (a) and (b) are trivial. One easily checks that only classes of the set $\{P_2, T_a, M, S, T_0 \cap T_1, T_{a,2} | a \in E_2\}$ are describable through binary relations on E_2. ■

As a direct consequence from Lemma 10.2.2 and the remarks before Table 10.1, we see that

Lemma 10.2.3 *If $A \in \{M \cap T_a, S \cap T_a, S \cap M, L \cap T_a, K \cup C, I, I \cup \overline{I}, I \cup C_a, [P_2^1], M \cap T_{a,2}, T_{\overline{a}} \cap T_{a,2} | a \in E_2\}$, then dim $A = 3$.* ■

Lemma 10.2.4 *If $b \geq 2$ then dim $T_{0,b} \geq b$.*

Proof. Let $b \geq 2$. Assume there is an h-ary relation ϱ with $T_{0,b} = Pol\ \varrho$, $1 \leq h < b$ and ϱ does not contain any double rows. Because of $c_0 \in T_{0,b}$, ϱ has not an 1-column. Since

$$[\{x \wedge \overline{y}\}] = \bigcup_{n \geq 1} \{f^n \in P_2 \mid \exists i\ \exists g^{n-1} \in P_2 : \\ f(x_1, ..., x_n) = x_i \wedge g(x_1, ..., x_{i-1}, x_{i-1}, ..., x_n)\} \subseteq T_{0,b},$$

is valid in addition apparently:

$$\begin{pmatrix} a_1 \\ a_2 \\ \cdots \\ a_h \end{pmatrix} \in \varrho \Longrightarrow (\forall b_1, ..., b_h \in E_2 : \begin{pmatrix} b_1 \\ b_2 \\ \cdots \\ b_h \end{pmatrix} \leq \begin{pmatrix} a_1 \\ a_2 \\ \cdots \\ a_h \end{pmatrix} \Longrightarrow \begin{pmatrix} b_1 \\ b_2 \\ \cdots \\ b_h \end{pmatrix} \in \varrho) \tag{10.2}$$

Since $T_{0,b-1} \supseteq T_{0,b} = Pol\ \varrho$ and $h_{b-1} \in T_{0,b-1} \backslash T_{0,b}$, there is certain $r_1, ..., r_b \in \varrho$ with $h_{b-1}(r_1, ..., r_b) \notin \varrho$.
Denote z_1, ..., z_t exactly the rows of $(r_1..., r_b)$ on which h_{b-1} takes the value 1. Because of definition of the class $T_{0,b-1}$, the matrix

$$\begin{pmatrix} z_1 \\ z_2 \\ \cdots \\ z_t \end{pmatrix}$$

contains an 1-column. Thus $(r_1, ..., r_b)$ has a column $\geq h_{b-1}(r_1, ..., r_b)$, in contradiction to (10.2) and $h_{b-1}(r_1, ..., r_b) \notin \varrho$. ■

Lemma 10.2.5 *If $b \geq 2$ and $a \in \{0,1\}$ then dim $M \cap T_{a,b} \geq b+1$.*

Proof. Let $b \geq 2$ and w.l.o.g. let $a = 0$.
Further, let

$$\varrho := \{(x_1, ..., x_{b+1}) \in E_2^{b+1} \backslash \{(1,1,...1)\} \mid x_{b+1} = 1\} \cup \{(0,0,...,0)\}$$

Then $M \cap T_{0,b} = Pol\ \varrho$. Assume there is an r-ary relation ϱ' with $1 \leq r \leq b$ and $Pol\varrho' = M \cap T_{0,b}$. Since $h_{b-1} \notin M \cap T_{0,b}$, there exists some $a_{ij} \in E_2$ with

$$h_{b-1}(\mathfrak{A}) := h_{b-1}\begin{pmatrix} a_{11} & a_{12} & ... & a_{1b} \\ a_{21} & a_{22} & ... & a_{2b} \\ ... & ... & ... & ... \\ a_{r1} & a_{r2} & ... & a_{rb} \end{pmatrix} = \begin{pmatrix} d_1 \\ d_2 \\ ... \\ d_r \end{pmatrix} \notin \varrho'.$$

W.l.o.g. let $d_1 = ... = d_s = 0$ and $d_{s+1} = ... = d_b = 1$. Since $(0,...,0) \in \varrho'$, only the following two cases are possible:
Case 1: $s = 0$.
Since

$$h_{b-1}(x_1, ..., x_b) = 1 \iff (\exists i: \ x_i \in E_2 \wedge (\forall j \neq i : x_j = 1)) \tag{10.3}$$

and $(1,...,1) \notin \varrho'$ (because of $c_1 \notin M \cap T_{0,b}$), we have

$$\begin{pmatrix} 0 & 1 & ... & 1 & 1 \\ 1 & 0 & ... & 1 & 1 \\ \multicolumn{5}{c}{\dots\dots\dots\dots} \\ 1 & 1 & ... & 0 & 1 \\ 1 & 1 & ... & 1 & 0 \end{pmatrix} \subseteq \mathfrak{A} \subseteq \varrho'.$$

With the help of $xy \in M \cap T_{0,b}$ this implies $\varrho' = E_2^r \backslash \{(1,...,1)\}$, in contradiction to $Pol\ \varrho' = M \cap T_{0,b}$.
Case 2: $1 \leq s \leq b-1$.
Because of (10.3), w.l.o.g., we can assume

$$\mathfrak{A} = \begin{pmatrix} a_{11} & a_{12} & ... & a_{1s} & a_{1,s+1} & ... & a_{1b} \\ \multicolumn{7}{c}{\dots\dots\dots\dots\dots\dots\dots\dots} \\ a_{s1} & a_{s2} & ... & a_{ss} & a_{s,s+1} & ... & a_{sb} \\ 0 & 1 & ... & 1 & 1 & ... & 1 \\ 1 & 0 & ... & 1 & 1 & ... & 1 \\ \multicolumn{7}{c}{\dots\dots\dots\dots\dots\dots\dots\dots} \\ 1 & 1 & ... & 0 & 1 & ... & 1 \end{pmatrix}.$$

This yields

$$h_b(\mathfrak{A}, \begin{pmatrix} a_{1b} \\ ... \\ a_{sb} \\ 1 \\ ... \\ 1 \end{pmatrix}) = \begin{pmatrix} d_1 \\ ... \\ d_s \\ d_{s+1} \\ ... \\ d_r \end{pmatrix},$$

in contradiction to $h_b \in Pol\ \varrho'$ and $(d_1, ..., d_r) \notin \varrho'$. ■

Lemma 10.2.6 *Let $A = [A] \subset T_1$ and $x \wedge y \in A$. If $A = Pol\ \varrho$, then ϱ contains an 1-row.*

Proof. The relation ϱ cannot contain a 0-column because of $A \subset T_1$. If ϱ does not contain a 1-row then there are certain $a_{ij} \in E_2$ with

$$\begin{pmatrix} 0 \\ a_{21} \\ a_{31} \\ \cdots \\ a_{h1} \end{pmatrix}, \begin{pmatrix} a_{12} \\ 0 \\ a_{32} \\ \cdots \\ a_{h2} \end{pmatrix}, ..., \begin{pmatrix} a_{1h} \\ a_{2h} \\ a_{3h} \\ \cdots \\ 0 \end{pmatrix} \in \varrho.$$

Since $x \wedge y \in A$, ϱ contains then a 0-column, however, contrary to the above remark. ■

Lemma 10.2.7 *$dim\ T_{0,b} \cap T_1 \geq b + 1$ for $b \geq 2$.*

Proof. Denote ϱ a relation of smallest arity with $Pol\ \varrho = T_{0,b} \cap T_1$. W.l.o.g., by Lemma 10.2.6, we can assume $\varrho = (1) \times \varrho'$, where obviously $T_{0,b} \cap T_1 \subset Pol\ \varrho'$ and $Pol\ \varrho' \subseteq T_1$ hold. It results, then, from the structure of Post's graph that $Pol\ \varrho' = T_{0,b}$ holds. This and Lemma 8.2.4 imply $dim\ T_{0,b} \cap T_1 \geq b + 1$. ■

Analogously, one can prove the following lemma.

Lemma 10.2.8 *$dim\ T_{0,b} \cap M \cap T_1 \geq b + 2$ for $b \geq 2$.* ■

Lemma 10.2.9 *$dim\ K \cup C_1 \geq 4$.*

Proof. By Lemma 10.2.6, every relation ϱ with $Pol\ \varrho = K \cup C_1$ contains an 1-row. If one scrutinizes the possibilities for the h-ary relations ϱ with $h \leq 3$, one receives $Pol\ \varrho \neq K \cup C_1$. ■

Lemma 10.2.10 *$dim\ K \cup C_0 \geq 4$.*

Proof. $dim\ K \cup C_0 \geq 3$ follows from Lemma 10.2.2. Assume that $Pol\ \varrho' = K \cup C_0$ holds for a certain ternary relation ϱ'. Obviously, ϱ' cannot contain an 1-column. Hence, one also does not find any $r_1, r_2, r_3 \in \varrho'$ with $r_1(r_2 \vee r_3) \notin \varrho'$. Thus, $T_{0,\infty} \cap M \cap T_1 \subseteq Pol\ \varrho'$ holds, in contradiction to $K \cup C_0 \subset T_{0,\infty} \cap M \cap T_1$. ■

Lemma 10.2.11 *$dim\ K \geq 5$.*

Proof. Obviously, $dim\ K \geq 3$. Assume an h-ary relation ϱ ($h \in \{3, 4\}$) exists without double rows and with $Pol\ \varrho = K$. The relation ϱ does not contain certainly a 0-column nor an 1-column. Furthermore, w.l.o.g. by Lemma 10.2.6, we have $\varrho = (1) \times \varrho'$, where $Pol\ \varrho' \supset K$. If one scrutinizes the possibilities for $Pol\ \varrho'$ with the aid of the Post's graph and of Lemma 10.1.1, one receives a contradiction. ■

Lemma 10.2.12 *dim* $L \geq 4$.

Proof. Because of Lemma 10.2.2, class L cannot be described with the aid of a unary or binary relation. Assume there is a ternary relation ϱ with $L = Pol\ \varrho$. Because of c_0, c_1, $\overline{x} \in L$ we have

$$\begin{pmatrix} 0 & 1 \\ 0 & 1 \\ 0 & 1 \end{pmatrix} \subseteq \varrho$$

and

$$\begin{pmatrix} a \\ b \\ c \end{pmatrix} \in \varrho \implies \begin{pmatrix} \overline{a} \\ \overline{b} \\ \overline{c} \end{pmatrix} \in \varrho.$$

Furthermore, ϱ can not contain any double rows. This implies that the relation ϱ has three linearly independent columns. Hence, because of $x + y \in L$, we have $\varrho = E_2^3$, in contradiction to $L = Pol\ \varrho$. ■

Lemma 10.2.13 *dim* $L \cap S \geq 4$.

Proof. Proving *dim* $L \cap S \neq 3$ suffices. Assume that $L \cap S = Pol\ \varrho$ holds for a certain ternary relation ϱ. Because of $c_0, c_1 \notin L \cap S$ and $\overline{x} \in L \cap S$ we have $(a, a, a) \notin \varrho$ for $a \in E_2$ and

$$\begin{pmatrix} a \\ b \\ c \end{pmatrix} \in \varrho \implies \begin{pmatrix} \overline{a} \\ \overline{b} \\ \overline{c} \end{pmatrix} \in \varrho.$$

Then, w.l.o.g., ϱ has the form

$$\varrho_1 = \begin{pmatrix} 1 & 0 \\ 0 & 1 \\ 0 & 1 \end{pmatrix}, \quad \varrho_2 = \begin{pmatrix} 1 & 0 & 0 & 1 \\ 0 & 1 & 1 & 0 \\ 0 & 1 & 0 & 1 \end{pmatrix} \quad \text{or} \quad \varrho_3 = \begin{pmatrix} 1 & 0 & 0 & 1 & 0 & 1 \\ 0 & 1 & 1 & 0 & 0 & 1 \\ 0 & 1 & 0 & 1 & 1 & 0 \end{pmatrix}.$$

Since $Pol\ \varrho_1 = Pol\ \varrho_2 = S$ and $Pol\ \varrho_3 = [\{\overline{x}\}]$, we obtain a contradiction to the supposition. ■

Lemma 10.2.14 *dim* $L \cap S \cap T_1 \geq 4$.

Proof. Since $x + y + z \in L \cap S \cap T_1$, for all columns s of the relation ϱ with $Pol\ \varrho = L \cap S \cap T_1$ we have:

$$s, s + \mathbf{1} \in \varrho \implies \forall t \in \varrho : \ s + t + (s + \mathbf{1}) = t + \mathbf{1} \in \varrho$$

$(s + \mathbf{1} := (s_1 + 1, ..., s_h + 1)$ if $s := (s_1, ..., s_h))$. Because of $x + 1 \notin L \cap S \cap T_1$ this implies

$$s \in \varrho \implies s + \mathbf{1} \notin \varrho.$$

By this and by $(a, a, a) \notin \varrho$ $(a \in E_2)$, if ϱ is ternary, then ϱ can have at most three different columns. Now one easily checks that the clone $L \cap S \cap T_1$ cannot be described by a such ternary relation. ■

Lemma 10.2.15 *dim* $I \cup C \geq 4$.

Proof. Because of Lemma 10.2.2, it is sufficient to show that *dim* $I \cup C \neq 3$ holds.
Since the classes $[P_2^1]$, $K \cup C$ and $D \cup C$ are the direct predecessors of $I \cup C$, every ternary relation ϱ with $Pol\ \varrho = I \cup C$ has w.l.o.g. the following four properties:

$$\begin{pmatrix} 0 & 1 \\ 0 & 1 \\ 0 & 1 \end{pmatrix} \subseteq \varrho \subset E_2^3,$$

$$\exists a, b, c \in E_2 : \begin{pmatrix} a \\ b \\ c \end{pmatrix} \in \varrho \wedge \begin{pmatrix} \overline{a} \\ \overline{b} \\ \overline{c} \end{pmatrix} \notin \varrho \text{ (by } \overline{x} \notin Pol\ \varrho),$$

$$\begin{pmatrix} 0 & 1 \\ 1 & 0 \\ 1 & 1 \end{pmatrix} \subseteq \varrho \wedge \begin{pmatrix} 0 \\ 0 \\ 1 \end{pmatrix} \notin \varrho \text{ (by } x \wedge y \notin Pol\ \varrho),$$

$$\begin{pmatrix} 1 & 0 \\ 0 & 1 \\ 0 & 0 \end{pmatrix} \subseteq \varrho \wedge \begin{pmatrix} 1 \\ 1 \\ 0 \end{pmatrix} \notin \varrho \text{ (by } x \vee y \notin Pol\ \varrho).$$

Apparently there is no relation ϱ that fulfills all four of these conditions. Thus *dim* $I \cup C \geq 4$. ■

Now, in summarizing, we obtain the statements of the following theorem as consequences of Lemmas 10.2.1–10.2.15 and from the above remarks (before Table 10.1).

Theorem 10.2.16 *For an arbitrary subclass A of P_2 and $t \geq 6$ the following holds:*

$dim(A) = 0 \iff A = P_2$,
$dim(A) = 1 \iff A \in \{T_0, T_1\}$,
$dim(A) = 2 \iff A \in \{M, S, T_{a,2}, T_0 \cap T_1 \mid a \in E_2\}$,
$dim(A) = 3 \iff A \in \{M \cap T_a, S \cap T_0, S \cap M, M \cap T_{a,2}, T_{\overline{a}} \cap T_{a,2}, L \cap T_a, K \cap C, I, I \cup C_a, \overline{I}, [P_2^1]\}$,
$dim(A) = 4 \iff A \in \{I \cup C, M \cap T_0 \cap T_1, L, L \cap S, L \cap S \cap T_0, K \cup C_a, T_{a,4}, T_{\overline{a}} \cap T_{a,3}, M \cap T_{a,4}, T_{\overline{a}} \cap M \cap T_{a,2} \mid a \in E_2\}$,
$dim(A) = 5 \iff A \in \{K, T_{a,5}, T_{\overline{a}} \cap T_{a,4}, M \cap T_{a,4}, T_{\overline{a}} \cap M \cap T_{a,3} \mid a \in E_2\}$,
$dim(A) = t \iff A \in \{T_{a,t}, T_{a,t-1} \cap T_{\overline{a}}, T_{a,t-1} \cap M, T_{a,t-2} \cap M \cap T_{\overline{a}} \mid a \in E_2\}$,
$dim(A) = \infty \iff A \in \{C, C_a, T_{a,\infty}, T_{a,\infty} \cap T_{\overline{a}}, T_{a,\infty} \cap M, T_{a,\infty} \cap M \cap T_{\overline{a}} \mid a \in E_2\}$. ■

Theorem 10.2.17 *For an arbitrary subclass A of P_2 and $t \geq 5$ the following holds:*

(a) $d(A) = 0 \iff A = P_2$,
$d(A) = 1 \iff A \in \{T_0, T_1, T_0 \cap T_1\}$,
$d(A) = 2 \iff A \in \{ M, S, T_{a,2}, S \cap T_0, M \cap T_a, S \cap M, T_{a,2} \cap T_{\overline{a}}, T_{a,2} \cap M, M \cap T_0 \cap T_1, T_{a,2} \cap M \cap T_{\overline{a}} \mid a \in E_2\}$,

(b) $d(A) = 3 \iff A \in \{ L \cap T_a, K \cup C, D \cup C, I, I \cup \overline{I}, I \cup C_a, [P_2^1], T_{a,3}, I \cup C, K \cup C_a, D \cup C_a, T_{a,3} \cap M, T_{a,3} \cap T_{\overline{a}}, T_{a,3} \cap M \cap T_{\overline{a}}, L \cap S \cap T_0, K, D \mid a \in E_2\}$,

(c) $d(A) = 4 \iff A \in \{L, L \cap S, T_{a,4}, T_{a,4} \cap M, T_{a,4} \cap T_{\overline{a}}, T_{a,4} \cap M \cap T_{\overline{a}} \mid a \in E_2\}$,

(d) $d(A) = t \iff A \in \{T_{a,t}, T_{a,t} \cap T_{\overline{a}}, T_{a,t} \cap M, T_{a,t} \cap M \cap T_{\overline{a}} \mid a \in E_2\}$,

(e) $d(A) = \infty \iff A \in \{C, C_a, T_{a,\infty}, T_{a,\infty} \cap T_{\overline{a}}, T_{a,\infty} \cap M, T_{a,\infty} \cap M \cap T_{\overline{a}} \mid a \in E_2\}$.

Proof. It is easy to check that the functions of $M \cap S$ preserve every unary and binary relation on E_2. Consequently, we have for an arbitrary subclass A of P_2:

$$d(A) \leq 2 \iff S \cap M \subseteq A.$$

Thus, (a) follows from the definitions of the given classes, and $d(A) \geq 3$ holds for the remaining classes A.
(b) results from the descriptive relations of the classes given in Table 10.1 or from their definitions, with the aid of

$$Pol\ \varrho \cap Pol\ \varrho' = Pol\ \varrho \times \varrho' = Pol\ \{\varrho, \varrho'\}.$$

(c) and (d) one can easy prove by means of the Post' graph and with the aid of (a), (b) and Lemma 10.1.1.
(e) follows from Lemma 10.1.1, (c) and Theorem 10.2.16. ∎

10.3 Further Examples of the Dimension and Relation Degree of Classes

We come now to the dimensions and relation degrees of some subclasses of P_k. The following theorem supplies simple examples first.

Theorem 10.3.1 *([Ros 70;a])*
Let A be a maximal class of P_k ($k \geq 2$), i.e., there exists a $\varrho \in \mathfrak{M}_k \cup \mathfrak{U}_k \cup \mathfrak{S}_k \cup \mathfrak{L}_k \cup \mathfrak{C}_k \cup \mathfrak{B}_k$ with $A = Pol \varrho$ (see Chapter 5). Then
(a) $d(A) = dim\ A$;

(b) $d(A) = 1$, *if* $\varrho \in \mathfrak{C}_k^1$;
$d(A) = 2$, *if* $\varrho \in \mathfrak{C}_k^2 \cup \mathfrak{M}_k \cup \mathfrak{S}_k \cup \mathfrak{U}_k$;
(c) $d(A) = h$, *if* $h \in \{3, 4, ..., k-1\}$ *and* $\varrho \in \mathfrak{C}_k^h \cup \mathfrak{B}_k^h$;
$d(A) = k$, *if* $A = Pol\iota_k^k$ *and* $k \geq 3$;
(d) $d(A) = 3$, *if* $\varrho \in \mathfrak{L}_k$, $k = p^m$ *and* $p \neq 2$;
$d(A) = 4$, *if* $\varrho \in \mathfrak{L}_k$, $k = p^m$ *and* $p = 2$.

Proof. (a) follows from Lemma 10.1.1, (e).
(b) results immediately from the definitions of the corresponding classes $Pol\ \varrho$.
(c): Since every h-ary relation ϱ with $\varrho \in \mathfrak{C}_k^h \cup \mathfrak{B}_k^h$ for $h \leq k-1$ and $\varrho = \iota_k^k$ for $h = k$ contains the set ι_k^h, all functions, which take at most $h-1$ different values, belong to $Pol\varrho$. Consequently, one cannot describe the class $Pol\varrho$ in this case with the aid of $(h-1)$-ary relations.
(d): Let $k = p^m$ (p prime, $m \geq 1$), $(E_k, +, \cdot)$ a field with the zero element 0 and the unit element 1,

$$\varrho = \{(a, b, c, d) \in E_k^4 \mid a + b = c + d\} \text{ (i.e. } \varrho \in \mathfrak{L}_k),$$

$A := Pol\varrho$, $dim\ A = q$ and denote λ a certain q-ary relation with $Pol\lambda = A$. Let be given the relation λ in the form of a matrix, whose rows are denoted with z_1, ..., z_q. Obviously, λ does not have any double rows and $q \geq 2$ holds. Further, we can interpret the rows (and/or the columns) of λ in natural manner as elements of a vector space $V_{|\lambda|}$ (or V_q) over the field $(E_k, +, \cdot)$, where the operations are defined as usual in matrix vector spaces. First we prove that $dimA \geq 3$ for $p \neq 2$ and $dim\ A = 4$ for $p = 2$.
Obviously, all q-ary functions f of the form $f(x_1, ..., x_q) = a_1x_1 + ... + a_qx_q$ belong to A for arbitrary a_1, a_2, ..., $a_q \in E_k$. If the rows z_1, ..., z_q are linear independent (i.e., there exist certain q linear independent columns s_1, ..., s_q $\in \lambda \subseteq E_k^q$), then for every $s \in E_k^q$ there is a certain function $f \in A$ with $f(x_1, ..., x_n) = a_1x_1 + ... + a_qx_q$ and $f(s_1, ..., s_q) = s$. Hence, $\lambda = E_k^q$ holds, in contradiction to $A = Pol\lambda \subset P_k$.
For the proof of $dim\ A \geq 3$, if $p \neq 2$, and $dim\ A = 4$, if $p = 2$, we have only to show that the rows z_1, ..., z_q ($q = 2$ for $p \neq 2$, $q \in \{2, 3\}$ for $p = 2$) are linear independent.
If $q = 2$ and if we assume that

$$z_1 = \alpha \cdot z_2 \text{ (for certain } \alpha \in E_k \backslash \{0\}),$$

then a contradiction follows from $g(x) := x + 1 \in A$ and from

$$\forall \begin{pmatrix} a \\ \alpha \cdot a \end{pmatrix} \in \lambda \backslash \left\{ \begin{pmatrix} 0 \\ 0 \end{pmatrix} \right\} : g \begin{pmatrix} a \\ \alpha \cdot a \end{pmatrix} = \begin{pmatrix} a + 1 \\ \alpha \cdot a + 1 \end{pmatrix} \notin \lambda.$$

Consequently, for arbitrary p, we have $dim\ A \geq 3$ and $pr_{i,j}\lambda = E_k^2$ for all $i, j \in \{1, 2, ..., q\}$ with $i \neq j$.
Now let $q = 3$ and $p = 2$. Obviously, the columns

$$s_1 := \begin{pmatrix} 1 \\ 0 \\ a \end{pmatrix} \quad s_2 := \begin{pmatrix} 1 \\ b \\ 0 \end{pmatrix} \quad s_3 := \begin{pmatrix} c \\ 1 \\ 0 \end{pmatrix}$$

belong to λ for certain $a, b, c \in E_k$. Then, because of $x+1$, $x^2 \in A$, we have

$$s_4 := \begin{pmatrix} 0 \\ 1 \\ a+1 \end{pmatrix} \in \lambda, \; s_5 := \begin{pmatrix} 1 \\ 0 \\ a^2 \end{pmatrix} \in \lambda.$$

Now one easily checks that

$$\begin{aligned} &s_1, s_4, s_3 \text{ if } a = 0, \\ &s_1, s_4, s_2 \text{ if } a = 1 \text{ and} \\ &s_1, s_5, s_3 \text{ if } a \notin \{0, 1\} \end{aligned}$$

are linear independent. Consequently, $dim\ A = 4$ holds in the case $p = 2$. Since

$$\Delta\lambda = \{(a, b, c) \in E_k^3 \mid a + a = b + c\}$$

obviously is not a diagonal relation, for $p \neq 2$, $A \subseteq Pol\Delta\lambda$ and A is maximal in P_k, we have $dimA = 3$ in the case $p \neq 2$. ■

It is surely a hopeless venture to determine the relation degree for all subclasses of P_k. Nevertheless, there are rather precise estimates for $d(A)$ or $dim(A)$ if A is an element of a certain sublattice of $\mathbb{L}_k$. Subsequently, two examples are given for this fact.

The following theorem is a direct consequence of generalizing the Baker-Pixley Theorem (see Theorem 8.3.1).

Theorem 10.3.2 *If a subclass A of P_k contains a certain n-ary function h with the properties*

$$h(x, y, ..., y) = h(y, x, y, ..., y) = ... = h(y, ..., y, x) = y$$

for arbitrary $x, y \in E_k$ and $n \geq 3$, then $d(A) \leq n - 1$ holds. ■

As a consequence of theorems on linear and quasi-linear functions from Chapter 13 we get:

Theorem 10.3.3 *Let $k = p^m$ (p prime, $m \geq 1$), $(E_k, +, \cdot)$ a field and $\lambda := \{(a, b, c, d) \in E_k^4 \mid a + b = c + d\}$. Then, for every subclass $A \subseteq Pol_k\lambda$, which contains the function r with*

$$r(x, y, z) = x + y - z,$$

the following holds: $d(A) = 3$ for $p \neq 2$, $3 \leq d(A) \leq 4$ for $p = 2$, $3 \leq dimA \leq 6$ for $p \neq 2$ and $3 \leq dimA \leq 7$ for $p = 2$. ■

In preparation for Theorem 10.3.7 with statements about subclasses of $[P_k^1]$, we subsequently prove three lemmas.

Lemma 10.3.4 *Let* $\mu := \delta^3_{\{1,2\}} \cup \delta^3_{\{2,3\}}$. *Then* $Pol\mu = [P_k^1]$.

Proof. Obviously, $Pol\mu \subseteq [P_k^1]$. "=" follows from Theorem 1.4.4, (a), (c). ■

Lemma 10.3.5 *Let*

$$\nu := \{(x,y) \in E_k^2 \mid x \neq y\}.$$

Then $Pol\nu = [P_k^1[k]]$.

Proof. Obviously, $(Pol\nu)^1 = P_k^1[k]$.
Assume there exists an n-ary function $f \in Pol = \nu$ that depends on at least two variables essentially. Then f takes exactly k different values and, by Theorem 1.4.4, (c) one can find certain $\mathbf{r_i} := (r_{1i},...,r_{ki}) \in \iota_k^k$ $(i = 1,...,n)$ with $f(\mathbf{r_1},...,\mathbf{r_n}) \in E_k^k \backslash \iota_k^k$. If one chooses $a_i \in E_k \backslash \{r_{1i},...,r_{ki}\}$ $(i = 1,...,n)$ now, there is so a $j \in \{1,...,k\}$ with $f(r_{j1},...,r_{jn}) = f(a_1,...,a_n)$. Since $(r_{ji}, a_i) \in \nu$ for all $i \in \{1,...,n\}$, f does not preserve the relation ν, contrary to the assumption. Thus $Pol\nu = [P_k^1[k]]$. ■

Lemma 10.3.6 *Let* $k \geq 3$ *and denote* ω *the binary relation*

$$\{(0,1),(1,0),(0,x),(x,1),(y,y+1) \mid x \in E_k \backslash \{0,1\} \wedge y \in E_k \backslash \{0,1,k-1\}\}.$$

Then for an arbitrary function $f^n \in P_k$ *the following holds:*

$$f^n \in Pol\omega \implies \exists i:\ f^n = e_i^n. \tag{10.4}$$

(I.e., the projections are the only functions of P_k*, which preserve* ω*.)*

Proof. We write $a \prec b$ instead of $(a,b) \in \omega$.
Furthermore, for arbitrary $\mathbf{a}$, $\mathbf{b} \in E_k^n$ let be:

$$\mathbf{a} \prec \mathbf{b} :\Longleftrightarrow \forall i \in \{1,...,n\}:\ a_i \prec b_i.$$

Denote f an arbitrary n-ary function of $Pol\omega$ in the following. Then the function $f_{|\{0,1\}}$ belongs to $M \cap S$, since one can derive the relations

$$\begin{pmatrix} 0 & 1 \\ 1 & 0 \end{pmatrix} \text{ and } \begin{pmatrix} 0 & 0 & 1 \\ 0 & 1 & 1 \end{pmatrix}$$

from ω as follows:

$$\omega \cap (\tau\omega) = \begin{pmatrix} 0 & 1 \\ 1 & 0 \end{pmatrix},\ pr_1(\omega \cap (\tau\omega)) = E_2,\ (\omega \circ \omega) \cap E_2^2 = \begin{pmatrix} 0 & 0 & 1 \\ 0 & 1 & 1 \end{pmatrix}$$

Because of $f_{|\{0,1\}} \in M \cap S$ one can find (w.l.o.g.) two tuples

$$\mathbf{a} := (0,0,...,0,\underbrace{1}_{i},1,...,1) \text{ and}$$

$$\mathbf{b} := (1,0,...,0,\underbrace{1}_{i},1,...,1)$$

with $f(\mathbf{a}) = 0$ and $f(\mathbf{b}) = 1$. Now we want to prove the following statements:

$$f_{|\{0,1\}} = e_1^n, \tag{10.5}$$

$$\forall \alpha \in \{0,1\} : \ f(\alpha, x_2, ..., x_n) = \alpha, \tag{10.6}$$

$$\forall \beta \in E_k \backslash \{0,1\} : \ f(\beta, x_2, ..., x_n) = \beta. \tag{10.7}$$

For this purpose we consider the tuples

$$\begin{aligned} \mathbf{u} &:= (d, 1, ..., 1, \underbrace{0}_{i}, 0, ..., 0), \\ \mathbf{v} &:= (1, 0, ..., 0, \underbrace{d}_{i}, d, ..., d), \\ \mathbf{w} &:= (0, d, ..., d, \underbrace{1}_{i}, 1, ..., 1) \end{aligned}$$

for a certain $d \in E_k \backslash \{0,1\}$ and

$$\mathbf{z} := (0, 1, 1, 1, ..., 1).$$

Then $\mathbf{a} \prec \mathbf{u} \prec \mathbf{b}$ and $\mathbf{u} \prec \mathbf{v} \prec \mathbf{w} \prec \mathbf{u}$. Consequently, we have $f(\mathbf{u}) \in E_k \backslash \{0,1\}$, $f(\mathbf{v}) = 1$ and $f(\mathbf{w}) = 0$. Then, by $\mathbf{v} \prec \mathbf{z}$, $f(\mathbf{z}) = 0$ holds. Since $f_{|\{0,1\}} \in S \cap M$, (10.5) follows obviously from $f(\mathbf{z}) = 0$.
Furthermore, for an arbitrary tuple $\mathbf{x} \in E_k^n$ there are tuples $\mathbf{y}$, $\mathbf{y}'$ with $\mathbf{y} \prec \mathbf{x} \prec \mathbf{y}'$ and

$$(y_1, y_1') = \begin{cases} (1,1) & \text{if } x_1 = 0, \\ (0,0) & \text{if } x_1 = 1, \\ (0,1) & \text{if } x_1 \in E_k \backslash \{0,1\}. \end{cases}$$

Because of $f(\mathbf{y}) \prec f(\mathbf{x}) \prec f(\mathbf{y}')$ and by (10.5) the statement (10.6) and

$$f(\mathbf{x}) \in E_k \backslash \{0,1\} \text{ if } x_1 \in E_k \backslash \{0,1\} \tag{10.8}$$

result from that.
For every tuple $\mathbf{x} \in E_k^n$ with $x_1 \in E_k \backslash \{0,1\}$ one can find $k-2$ tuples $\mathbf{x_2}$, $\mathbf{x_3}$, ..., $\mathbf{x_{k-1}}$ with the properties

$$\begin{aligned} &\mathbf{x_i} = (i, x_{i2}, ..., x_{in}) \ (i = 2, 3, ..., k-1), \\ &\mathbf{x_{x_1}} = \mathbf{x} \text{ and} \\ &\mathbf{x_2} \prec \mathbf{x_3} \prec ... \prec \mathbf{x_{k-1}}. \end{aligned}$$

(To a given tuple $\mathbf{x}$ one can construct the tuples $\mathbf{x_i}$ as follows, for example, $x_{x_1-1,j} := \overline{x}_j =: x_{x_1+1,j}$ for $x_j \in \{0,1\}$, $x_{x_1-1,j} = 0$ and $x_{x_1+1,j} = 1$ for $x_j \in E_k \backslash \{0,1\}$ $(j = 2, ..., n)$, etc.)
Then $f(\mathbf{x_2}) \prec f(\mathbf{x_3}) \prec ... \prec f(\mathbf{x_{k-1}})$ and $f(\mathbf{x_i}) \in E_k \backslash \{0,1\}$ $(i = 2, ..., k-1)$ because of (8). This is, however, possible only for $f(\mathbf{x_i}) = i$ for all $i \in \{2, ..., k-1\}$. Thus (10.4) is right and our lemma was proven. ■

Theorem 10.3.7

(1) Let A be a subclass of $[P_k^1]$ with $e_1^1 \in A$. Then

(a) $d(A) = 3$ and $3 \le dimA \le 4$ for $k = 2$;

(b) $2 \le d(A) \le k$ and $2 \le dimA \le k+3$ for $k \ge 3$.

(2) $dim[P_k^1] = d([P_k^1]) = 3$ and

$$d(A) = dimA = \begin{cases} 3 & if\ k = 2, \\ 2 & if\ k \ge 3 \end{cases}$$

for every $A \in \{[P_k^1[k]], [e_1^1]\}$.

Proof. For $k = 2$, we have the statements of the theorem already proven (see Theorems 10.2.16 and 10.2.17).
Let $k \ge 3$ in the following. Then, obviously, $d(A) \ge 2$ for all classes $A \subseteq [P_k^1]$ with $e_1^1 \in A$. By Lemma 10.3.4 and because of $A^1 = (PolG_1(A))^1$ we have, furthermore, for every $A \in V_k(e_1^1, [P_k^1])$:

$$A = Pol\{G_1(A), \mu\} = PolG_1(A) \times \mu.$$

Thus (b) holds.
The statements of (2) follow from Lemmas 10.3.5 and 10.3.6 and from the fact that only the diagonal relations of the binary relations preserve all functions of P_k^1. ■

11

On Generating Systems and Orders of the Subclasses of P_k

If $A = [A] \subseteq P_k$ is finitely generated, we call the smallest number r with $[A^r] = A$ the **order** of A. In the case that the subclass $A \subseteq P_k$ is not finitely generated, we write $ord\ A = \infty$.
During the study of subclasses of P_k or during the solution of problems in the multi-valued logics, one often deals with constructing generating systems for the considered classes. Statements about the order of the considered classes often result from these investigations. Therefore, one finds results about the order of subclasses of P_k not only in this chapter, but also in other chapters of this book.
First, some general statements about the orders of subclasses of P_k are declared in this chapter. Then we occupy ourselves with the orders of the maximal classes of P_k. It turns out, in this case, that only certain maximal classes of the type $\mathfrak{M}$ do not have any finite order. If $\varrho \in \mathfrak{M}_k$, we can prove $ord\ Pol_k\varrho = 2$ only for $k \leq 7$ and for the case where $(E_k; \varrho)$ is a lattice. By means of an example proven by G. Tardos in [Tar 86], we will be able to show, however, that for $k \geq 8$ there are classes of the type $\mathfrak{M}$ with the order ∞.

If one has a finite generating class A, it is an interesting problem to clear which cardinalities are possible for the bases of this class A. In Section 11.2, we show only that the proof for $ord\ Pol_k\varrho \leq 3$, where $\varrho \in \mathfrak{C}_k^1 \cup \mathfrak{U}_k \cup \mathfrak{S}_k$, implies the existence of functions f_ϱ with $[f_\varrho] = Pol_k\varrho$. Following corresponding notation for functions from P_k, such function f_ϱ is called a **Sheffer-function for** $Pol_k\varrho$. We notice, through the theorems of Section 12.2, Schofield's Theorem 7.1.4 is also proven.
The last section explains shortly, as one can determine the cardinalities of the possible bases of the class A, if $ord\ A < \infty$. Further, some basic ideas of [Miy 71], [Miy-S-L-R 86], [Miy 88] and [Sto 87] are explained about basis classifications. For more information on the topic, refer to the books [Miy 88] and [Sto 87] by M. Miyakawa and I. Stojmenovic.

11.1 Some General Properties of Generating Systems and Bases

The following theorem is a consequence of [Coh 65, Th. 5.4]:

Theorem 11.1.1 *([Jab 74])*
Let A be a subclass of P_k. Then the following conditions are equivalent:

(1) A is finitely generated.

(2) A has only finitely many maximal classes (with other words: The lattice $\mathbb{L}_k^{\downarrow}(A)$ is dualatomar).

(3) For every chain of the form

$$A_1 \subseteq A_2 \subseteq A_3 \subseteq ... \subseteq A_n \subseteq ... \subseteq A$$

with $\bigcup_{n \geq 1} A_n = A$ there exists an $n \in \mathbb{N}$ with $A_n = A$. ∎

Theorem 11.1.2

(a) Each subclass of P_2 is finitely generated. For every $n \in \mathbb{N}$ exists a subclass $A \subseteq P_2$ with ord $A = n$.

(b) For $k \geq 3$ there are subclasses of P_k that do not have any basis or have an infinite basis ([Jan-M 59]).

Proof. (a) follows from Chapter 3.
(b): Let $k \geq 3$. One can find a subclass of P_k with an infinite basis in Lemma 8.1.1. The set $A := [\{f_1, f_2, f_3, ...\}]$, where

$$f_n(x_1, ..., x_n) := \begin{cases} 1 \text{ if} & x_1 = ... = x_n = 2, \\ 0 \text{ otherwise,} & \end{cases}$$

is an example of a class without basis, since $f_n \star f_m = c_0^{m+n-1}$ and $\Delta f_{n+1} = f_n$ for arbitrary $m, n \in \mathbb{N}$. ∎

The following theorem is easy to check.

Theorem 11.1.3
(a) If A is a maximal class of $B \subseteq P_k$, then

$$ord\ A < \infty \implies ord\ B < \infty.$$

(b) For arbitrary subclasses A, B of P_k it holds:

$$ord[A \cup B] \leq max\{ord\ A, ord\ B\}.$$

∎

As the following theorem shows, no connection exists between the existence of a finite relation degree (see Chapter 10) and the existence of a finite order of a subclass of P_k.

Theorem 11.1.4 *([Pös-K 79])*

There are subclasses A, B, C of P_k with the properties:

(a) $d(A) = \infty$ and ord $A < \infty$.

(b) $d(B) < \infty$ and ord $B = \infty$.

(c) $d(C) = \infty$ and ord $C = \infty$.

Proof. An example for a class A is the set $T_{0,\infty}$ (see Chapter 3). (b) follows from Theorem 11.5.5. One can find an example of a class C in the book [Pös-K 79], p. 97. ■

Despite the above theorem, the juxtaposition of the orders and the relation degrees of classes is quite interesting (see [Pös-K 79], p. 94–97).

The main aim of the following sections is to prove the following theorem.

Theorem 11.1.5 (Theorem on the Orders of the Maximal Classes)

(1) Let $\varrho \in \mathfrak{M}_k$. If $2 \leq k \leq 7$ and $(E_k; \varrho)$ is a lattice, then ord $Pol_k\varrho = 2$. For $k \geq 8$ there is a certain $\varrho \in \mathfrak{M}_k$ with ord $Pol_k\varrho = \infty$.

(2) For every $\varrho \in \mathfrak{S}_k$ it holds:

$$ord\ Pol_k\varrho := \begin{cases} 3 & \text{if} \quad k = 2, \\ 2 & \text{otherwise.} \end{cases}$$

(3) If $\varrho \in \mathfrak{U}_k \cup \mathfrak{L}_k \cup \mathfrak{B}_k$, then $Pol_k\varrho = 2$.

(4) Let γ be an h-ary relation of $\mathfrak{C}_k$. Then

$$2 \leq ord\ Pol_k\gamma \leq max(2, h).$$

For $k = 3$, this theorem, already, was proven in 1962 by M. N. Gnidenko (see e.g. [Gni 65]).

We need the following theorem for proof of Theorem 11.1.5.

Theorem 11.1.6 *Let $T \subseteq E_k$ with $|T| \geq 2$, $P_{k,T} := \{f \in P_k \mid W(f) \subseteq T\}$ and $2 \leq l \leq k$. Then*

(a) ord $P_{k,T} = 2$,

(b) $\exists f \in P_{k,T} : [f] = P_{k,T}$,

(c) ord $P_k(l) = 2$.

Proof. In the case that $T = E_k$ is valid, the above theorem is, already, proven(see Theorems 1.4.2 and 7.1.6). If $|T| \le k-1$, then we can assume $T = E_l$ w.l.o.g., through which the statement (a) follows from Lemma 12.2.1 and (b) from Theorems 12.2.7 and 7.1.6. (c) follows from (a). ■

11.2 The Orders and Sheffer-Functions of the Classes of Type $\mathfrak{C}^1$, $\mathfrak{S}$ or $\mathfrak{U}$

The aim of this section is to prove that all maximal classes of type $\mathfrak{X} \in \{\mathfrak{C}^1, \mathfrak{S}, \mathfrak{U}\}$ have the order 2 for $k \ge 3$ and that for every one of these classes, Sheffer-functions exist. We remark that all other maximal classes do not have any Sheffer-functions (see Theorem 7.1.4).

Theorem 11.2.1 *([Sch 69], [Kud 70]) Let* $\varrho \in \mathfrak{C}^1_k$. *Then*
(a) ord $Pol_k\varrho = 2$
(b) $\exists f_\varrho : \; [f_\varrho] = Pol_k\varrho$.

Proof. W.l.o.g. let $\varrho = E_l$ with $1 \le l \le k-1$. Further, let $A := Pol_k E_l$. We distinguish two cases:
Case 1: $l = 1$.
In this case, A is the set of all 0-preserving functions of P_k, and it is possible to describe an arbitrary function $f^n \in A$ with the help of functions $x \vee y := max(x,y)$, $x \cdot y := min(x,y)$ (in respect to the order relation $0 < 1 < ... < k-1$),

$$j_{a;b}(x) := \begin{cases} b \text{ if} & x = a, \\ 0 \text{ otherwise} & \end{cases}$$

$(a, b \in E_k)$ as follows:

$$f(x_1, x_2, ..., x_n) = \bigvee_{\substack{\mathbf{a} = (a_1,...,a_n) \\ \in E_k^n \setminus \{(0,0,...,0)\} \\ f(\mathbf{a}) \ne 0}} j_{a_1;f(\mathbf{a})}(x_1) \cdot j_{a_2;f(\mathbf{a})}(x_2) \cdot ... \cdot j_{a_n;f(\mathbf{a})}(x_n).$$

Consequently,

$$B := \{\vee, \cdot\} \cup \bigcup_{a,b \in \{1,2,...,k-1\},\; c \in E_k} \{j_{a;b}(x) \cdot j_{c;b}(y)\} \; (\subseteq A)$$

is a generating system for A and hence *ord* $A = 2$.
To prove (b) for the first case let

$$j_{a,b;t}(x,y) := j_{a;t}(x) \cdot j_{b;t}(y) = \begin{cases} t & \text{if } (x,y)=(a,b), \\ 0 & \text{otherwise,} \end{cases}$$

and

$$l_{a;t}(x,y,z) := j_{0;k-1}(x) \cdot j_{a;t}(y) \cdot j_{a;t}(z) = \begin{cases} t & \text{if } (x,y,z)=(0,a,a), \\ 0 & \text{otherwise.} \end{cases}$$

With the help of these functions, a function $g \in A$ can be described as follows:

$$g(..., x_{a,b;t}, y_{a,b;t}, ..., u_{a;t}, v_{a;t}, w_{a;t}, ...) :=$$

$$\bigvee_{\substack{(a,b,t) \\ \in E_k^2 \times \{1,2,...,k-1\} \\ a \neq b}} j_{a,b;t}(x_{a,b;t}, y_{a,b;t}) \quad \vee$$

$$\bigvee_{\substack{(a,t) \\ \in \{1,2,...,k-1\}^2}} l_{a;t}(u_{a;t}, v_{a;t}, w_{a;t}).$$

If one identifies all variables in the function g, then one obtains the constant function c_0. One obtains an arbitrary binary function of A if one replaces certain variables of the form $u_{a;t}$ by c_0 in g and identifies certain other variables of g. Thus *ord* A in Case 1.

Case 2: $l \geq 2$.
Let f^n be an arbitrary function of A and let $f_1^n \in P_{k,l}$ with $f(\mathbf{a}) = f_1(\mathbf{a})$ for all $\mathbf{a} \in E_l^n$. With the help of f_1, the above-defined functions $\vee$, $\cdot$, $j_{a,b}$ and the function

$$l(x) := \begin{cases} l-1 & \text{if } x \in E_l, \\ 0 & \text{otherwise,} \end{cases}$$

one can describe f as follows:

$$f(x_1, x_2, ..., x_n) = f_1(x_1, ..., x_n) \cdot l(x_1) \cdot ... \cdot l(x_n) \quad \vee$$

$$\bigvee_{\substack{\mathbf{a}=(a_1,...,a_n) \\ \in E_k^n \setminus E_l^n \\ f(\mathbf{a}) \neq 0}} j_{a_1;f(\mathbf{a})}(x_1) \cdot j_{a_2;f(\mathbf{a})}(x_2) \cdot ... \cdot j_{a_n;f(\mathbf{a})}(x_n).$$

Consequently, for every set $B_1 \subseteq P_{k,l} \subset A$ with $[pr_l B_1] = P_l$, the set

$$B := B_1 \cup \{\vee, \cdot, l\} \cup \bigcup_{a \in E_k \setminus E_l,\ b,c \in E_k} \{j_{a;b}(x) \cdot j_{c;b}(y)\} \ (\subseteq A)$$

is a generating system for A. From this, Lemma 12.2.1 and Theorem 1.4.2 it follows that *ord* $A = 2$.
Since $P_{k,l}$ has a Sheffer-function t^m (see Theorems 12.2.7 and 7.1.6), one can prove analogously to the first case that the function h with

$$h(x_1, ..., x_m, x_{m+1}, ..., x_{a,b;t}, y_{a,b;t}, ..., u_{a;t}, v_{a;t}, w_{a;t}, ...) :=$$

$$t(x_1, ..., x_m) \cdot l(x_{m+1}) \ \vee \ g(..., x_{a,b;t}, y_{a,b;t}, ..., u_{a;t}, v_{a;t}, w_{a;t}, ...)$$

is a Sheffer-function for A. ∎

Theorem 11.2.2 *([Sch 69]) Let $k \geq 3$ and $\varrho \in \mathfrak{U}_k$. Then*
(a) $ord\ Pol_k \varrho = 2$
(b) $\exists f_\varrho :\ [f_\varrho] = Pol_k \varrho$.

Proof. The nontrivial equivalence relation ϱ on E_k decomposes the set E_k in the (pairwise distinct) equivalence classes

$$\mathfrak{A}_0,\ \mathfrak{A}_1,\ ...,\ \mathfrak{A}_{k'-1} \quad (k' < k).$$

We select a representative a_i from every equivalence class $\mathfrak{A}_i$ and define the function $q \in P_k^1$ by

$$\forall i \in E_{k'} :\ q(x) = a_i \ :\Longleftrightarrow\ x \in \mathfrak{A}_i.$$

Let now f^n be an arbitrary function of $Pol_k \varrho$. With the help of functions $z,\ f_i,\ g_f \in Pol_k \varrho$ defined in Lemma 1.4.6, one can describe f as follows:

$$f(\mathbf{x}) = ((...((g_f(\mathbf{x}) \diamond f_0(\mathbf{x})) \diamond f_1(\mathbf{x}))... \diamond f_{k'-1}(\mathbf{x})) \tag{11.1}$$

Furthermore, by Section 5.2.3, there exists a function $h \in P_{\{a_0,a_1,...,a_{k'}\}}$ with

$$g_f(\mathbf{x}) = h(q(x_1), ..., q(x_n)). \tag{11.2}$$

It is easy to check that the set of all functions of the form (11.2) is isomorphic to $P_{k'}$; thus the functions of this set are superpositions over the binary functions of this set. Consequently, (11.1) and $ord\ P_{k,\mathfrak{A}_i} = 2$ (see Theorem 11.1.6) implies $ord\ Pol_k \varrho = 2$.
Let p^2 be a Sheffer-function for $P_{\{a_0,a_1,...,a_{k'-1}\}}$ and let r_i^m be a Sheffer-funktion for $P_{k,\mathfrak{A}_i}$ $(i = 0, 1, ..., k' - 1)$ [1]. Then one can form the functions $p(q(x), q(y))$, $r_i(x_1, ..., x_m)$ $(i = 0, 1, ..., k' - 1)$ and z as superpositions over the function

$$\begin{aligned} &s(x, y, x_{0,1}, x_{0,2}, ..., x_{0,m}, x_{1,1}, ..., x_{1,m}, ..., x_{k'-1,1}, ..., x_{k'-1,m}) := \\ &(...((p(q(x), q(y)) \diamond r_0(x_{0,1}, ..., x_{0,m})) \diamond r_0(x_{1,1}, ..., x_{1,m}))... \\ &\qquad \diamond r_{k'-1}(x_{k'-1,1}, ..., x_{k'-1,m})) \end{aligned} \tag{11.3}$$

as follows:

[1] See Theorem 12.2.7 for the existence of such Sheffer-functions.

When one carries out only replacements of the variables x or y, one can copy the formation of superpositions in $P_{\{a_0,a_1,...,a_{k'-1}\}}$ over s. Therefore one obtains functions of the form (11.3) as superpositions over s, where p is an arbitrary function of $P^2_{\{a_0,a_1...,a_{k'-1}\}}$. Because of

$$(...((c_{a_i}(q(x_1), q(x_2)) \diamond r_0(x_1, ..., x_m)) \diamond r_0(x_1, ..., x_m))... \diamond r_{k'-1}(x_1, ..., x_m))$$
$$= r_i(x_1, ..., x_m)$$

in particular $P_{k,\mathfrak{A}_i} \subseteq [s]$ results. If one inserts functions of $\bigcup_{i=0}^{k'-1} P_{k,\mathfrak{A}_i}$ into the function s, one also obtains the functions z and $p(q(x), q(y))$ as superpositions over s. Thus we have $[s] = Pol_k\varrho$ and (b) holds. ■

Theorem 11.2.3 *([Sch 69], [Lau 78a]) For arbitrary $\varrho_s \in \mathfrak{S}_k$ is valid:*
(a)

$$ord\ Pol_k\varrho_s = \begin{cases} 3 \text{ if } & k = 2, \\ 2 \text{ otherwise.} & \end{cases}$$

(b) $\exists f : [f] = Pol_k\varrho_s$.

Proof. Let $\varrho_s \in \mathfrak{S}_k$ be defined as in Section 5.2.2. Since our assertion was prove in Chapter 3 for $k = 2$, we can assume $k \geq 3$ in the following.
(a): Because of Chapter 7, all functions

$$g(x_1, x_2) := \sum_{r=1}^{l} \sum_{i=0}^{p-1} j_{a_{r,i}}(x_1) \cdot s^i(G(s^{p-i}(x_2)) \ (mod\ k),$$

with $G \in P_k^1$ and a function

$$h(x_1, x_2, x_3) := \sum_{r=1}^{l} \sum_{i=0}^{p-1} j_{a_{r,i}}(x_1) \cdot s^i(H(s^{p-i}(x_2), s^{p-i}(x_3))) \ (mod\ k),$$

where $H \in P_k \backslash [P_k^1]$ and $|Im(H)| = k$, form a generating system for the functions of the form

$$u(x_1, x_2, ..., x_m) := \sum_{r=1}^{l} \sum_{i=0}^{p-1} j_{a_{r,i}}(x_1) \cdot s^i(U(s^{p-i}(x_2), ..., s^{p-i}(x_m))) \ (mod\ k) \tag{11.4}$$

($U \in P_k^{m-1}$, $m \geq 2$). Let now f^n be an arbitrary function of $Pol_k\varrho_s$ ($n \geq 3$). We show that $f \in [(Pol_k\varrho_3)^2]$ is valid. For this purpose, we choose in (11.4) the function U as follows:

$$U(x_1, ..., x_m) := \begin{cases} F_r(x_2, ..., x_n) \text{ if } & x_1 = a_{r,0}, r = 1, 2, ..., l, \\ 0 \quad\quad \text{otherwise,} & \end{cases}$$

where F_1, ..., F_l are the components of f from (5.4). Furthermore, let the function v be defined by

$$v(x) := \sum_{r=1}^{l}\sum_{i=0}^{p-1} j_{a_{r,i}}(x) \cdot s^i(a_{r,0}) \in (Pol_k \varrho_s)^1.$$

Then

$$u(x_1, v(x_1), x_2, ..., x_n) = f(x_1, ..., x_n)$$

holds. Consequently, we have $[(Pol_k \varrho_s)^2 \cup \{h\}] = Pol_k \varrho_s$.
Next we give a function that is a superposition over $(Pol_k \varrho_s)^2$ and which fulfills the conditions listed above for the function h. For this purpose, we choose a binary function w with the components

$$W_1(x) = ... = W_l(x) = W(x) := \begin{cases} a_{1,1} & \text{if} \quad x = a_{1,0} \wedge p > 2, \\ a_{2,0} & \text{if} \quad x = a_{1,0} \wedge p = 2, \\ a_{1,0} & \text{if} \quad x = a_{1,1}, \\ x & \text{otherwise.} \end{cases}$$

One now easily checks that $W \star w \in P_k \setminus [P_k^1]$ and $|Im(W \star w)| = k$ are valid.

(b): With the aid of the above considerations, we can easily prove that the function

$$t(x, x_1, y_1, x_2, y_2, ...x_l, y_l) := \sum_{r=1}^{l}\sum_{i=0}^{p-1} j_{a_{r,i}}(x) \cdot s^i(T(s^{p-i}(x_r), s^{p-i}(y_r))) \ (mod\ k),$$

where $[T] = P_k$, is a Sheffer-function for $Pol_k \varrho_s$. ■

11.3 Orders of the Classes of Type $\mathfrak{L}$, $\mathfrak{C}$, $\mathfrak{B}$

The following theorem results immediately from Section 5.2.4:

Theorem 11.3.1 *([Ros 70a]) For every $\varrho \in \mathfrak{L}_k$ is valid $ord\ Pol_k \varrho = 2$.* ■

Theorem 11.3.2 *([Lau 78a]) For every $\varrho \in \mathfrak{B}_k$ is valid $ord\ Pol_k \varrho = 2$.*

Proof. Subsequently, we describe the functions from the clone $Pol_k \varrho$ ($\varrho \in \mathfrak{B}_k$) as in Section 5.2.6. Because of (5.14) and (5.16), an arbitrary function $f^n \in Pol_k \varrho$ is a superposition over the functions $r(f_i')$, z, f_j ($i = 0, 1, ..., m-1$, $j = 0, 1, 2, ..., h^m - 1$) and

$$H(x_1, ..., x_m) := r(\sum_{i=1}^{m} q(x_i))^{(0)} \cdot h^{m-i}).$$

Every function of the form $r(f_i')$ or f_j belongs either to $[(Pol_k\varrho)^1]$ or to $[(Pol_k\varrho)^2]$ because of Theorem 11.1.6. Consequently, we must still prove $H \in [(Pol_k\varrho)^2]$. The following functions belong to $Pol_k\varrho$:

$$H_t(x_1, ..., x_t) := r(\textstyle\sum_{i=1}^t q(x_i))^{(0)} \cdot h^{m-i})$$

$$H_t'(x_1, x_2)) := r(\textstyle\sum_{i=1}^t q(x_1))^{(m-i)} \cdot h^{m-i} + (q(x_2))^{(0)} \cdot h^{m-t-1})$$

$(1 \le t \le m-1)$. Because of $H_t' \star H_t = H_{t+1}$ the function H $(= H_m)$ is a superposition over $(Pol_k\varrho)^2$. ■

For the rest of this section, let γ be an h-ary central relation on E_k with $2 \le h < k$, and let C be the set of the central elements of γ.

The short proof of the following theorem comes from R. Pöschel.

Theorem 11.3.3 *The clone $Pol_k\gamma$ has a finite generating system.*

Proof. Let $c \in C$. The following $(\binom{|\gamma|+1}{2} + |\gamma| + 1)$-ary function belongs to $Pol_k\gamma$:

$$h(x_1, x_2, ..., x_{|\gamma|+1}, (y_{ij})_{i,j=1,2...,|\gamma|+1;i<j})$$

$$= \begin{cases} y_{ij} \text{ if} & \exists i,j: ((i<j) \wedge (x_i = x_j) \wedge \\ & (\forall s,t: ((s<t \wedge x_s = x_t) \Rightarrow y_{st} = y_{ij}))), \\ c \quad \text{otherwise.} & \end{cases}$$

We prove by induction on the arity n of an arbitrary function $f^n \in Pol_k\gamma$ that

$$[(Pol_k\gamma)^{|\gamma|} \cup \{h\}] = Pol_k\gamma$$

holds. Obviously, our assertion is valid for all functions f^n with $n \le |\gamma|$. Assume the assertion holds for all m-ary function of $Pol_k\gamma$, where $m \ge |\gamma|$. Let $f^{m+1} \in Pol_k\gamma$ be arbitrary. Denote f_{ij} the function, which can be formed from f by identifying of the i-th variable with the j-th variable, $i \ne j$, $i = 1, 2, .., |\gamma|+1$. By induction assumption, we have $f_{ij} \in [(Pol_k\gamma)^{|\gamma|} \cup \{h\}]$. It is easy to check that

$$f(x_1, x_2, ..., x_{m+1}) = h(x_1, ..., x_{|\gamma|+1}, (f_{ij}(\mathbf{x}))_{i,j=1,2,...,|\gamma|+1;i|j}).$$

Thus $f \in [(Pol_k\gamma)^{|\gamma|} \cup \{h\}]$. ■

Theorem 11.3.4 *([Lau 78a]) It holds*

$$ord\ Pol_k\gamma \begin{cases} = 2 \ \textit{if} & k - |C| \le h, \\ \le h \ \textit{otherwise.} & \end{cases}$$

If γ fulfills the conditions $k - |C| > h$ and

$$(a_0, ..., a_{h-1}) \in \gamma \implies ((a_0, ..., a_{h-1}) \in \iota_k^h \vee (\exists i \in E_h : a_i \in C)), \quad (11.5)$$

then $ord\ Pol_k\gamma = \lceil \frac{h}{2} \rceil + 1$. [2]

Proof. Below is a procedure comprising three steps. This procedure shows that an arbitrary function, which takes $q \geq 2$ different values, is a superposition over the set $(Pol_k\gamma)^h \cup (Pol_k\gamma \cap (P_k(q-1)))$. The iteration of this procedure takes back the construction of an arbitrary function to the construction of functions that take at most $h-1$ different values. Since all such functions preserve the relation γ, we have $ord\ Pol_k\gamma \leq h$ by Theorem 11.1.6, (c).

Let f^n be an arbitrary function of $Pol_k\gamma$ with $n > 2$ and let c be any fixed element of the set C, i.e., c is a central element of the relation γ.

1.) To dismantle o f, we need the n-ary functions f_1, f_2 and the binary function g of $Pol_k\gamma$:

$$f_1(x_1, ..., x_n) := \begin{cases} f(x_1, ..., x_n) & \text{if} \quad f(x_1, ..., x_n) \notin C, \\ c & \text{otherwise}, \end{cases}$$

$$f_2(x_1, ..., x_n) := \begin{cases} c & \text{if} \quad f(x_1, ..., x_n) \notin C, \\ f(x_1, ..., x_n) & \text{otherwise}, \end{cases}$$

$$g(x_1, x_2) := \begin{cases} x_1 & \text{if} \quad x_1 \notin C \wedge x_2 = c, \\ x_2 & \text{if} \quad x_1 = c \wedge x_2 \in C, \\ c & \text{otherwise}. \end{cases}$$

We get: $f(\mathbf{x}) = g(f_1(\mathbf{x}), f_2(\mathbf{x}))$, where $f_2 \in P_{k,C} \subset [(Pol_k\gamma)^2]$ by Theorem 11.1.6, (a).

2.) In the second step the function f_1 is dismantled, where we use functions $h_a \in (Pol_k\gamma)^2$, $a \in E_k \backslash C$, defined as follows:

$$h_a(x_1, x_2) := \begin{cases} a & \text{if} \quad (x_1 = c \wedge x_2 = a) \vee (x_1 = a \wedge x_2 = c), \\ x_1 & \text{if} \quad x_1 = x_2, \\ c & \text{otherwise}. \end{cases}$$

Let $\mathfrak{E}_a := \{\mathbf{a} \in E_k^n \mid f_1(\mathbf{a}) = a\}$. Further, for $\mathbf{a} \in \mathfrak{E}_a$ we put:

$$f_{\mathbf{a}}(x_1, ..., x_n) := \begin{cases} a & \text{if} \quad \mathbf{x} = \mathbf{a}, \\ c & \text{if} \quad \mathbf{x} \in \mathfrak{E}_a \backslash \{\mathbf{a}\}, \\ f_1(\mathbf{x}) & \text{otherwise}. \end{cases}$$

If $\mathfrak{E}_a = \{\mathbf{a}_1, ..., \mathbf{a}_s\}$, then there exists the following representation for f_1:

$$f_1(\mathbf{x}) = h_a(...h_a(h_a(f_{\mathbf{a}_1}(\mathbf{x}), f_{\mathbf{a}_2}(\mathbf{x})), f_{\mathbf{a}_2}(\mathbf{x}))..., f_{\mathbf{a}_s}(\mathbf{x})).$$

Similar to f_1, one can dismantle the functions $f_{\mathbf{a}}$. In this case, we choose a $b \in E_k \backslash (C \cup \{a\})$, put $\mathfrak{E}_b := \{\mathbf{b}_1, ..., \mathbf{b}_t\}$ and for $\mathbf{b} \in \mathfrak{E}_b$ we define

$$f_{\mathbf{a},\mathbf{b}}(x_1, ..., x_n) := \begin{cases} b & \text{if} \quad \mathbf{x} = \mathbf{b}, \\ c & \text{if} \quad \mathbf{x} \in \mathfrak{E}_b \backslash \{\mathbf{b}\}, \\ f_{\mathbf{a}}(\mathbf{x}) & \text{otherwise}. \end{cases}$$

[2] $\lceil x \rceil$ denotes the greatest integer z with $z \leq x$.

Then the function $f_{\mathbf{a}}$ has the representation

$$f_{\mathbf{a}}(\mathbf{x}) = h_b(...h_b(h_b(f_{\mathbf{a},\mathbf{b}_1}(\mathbf{x}), f_{\mathbf{a},\mathbf{b}_2}(\mathbf{x})), f_{\mathbf{a},\mathbf{b}_3}(\mathbf{x}))...f_{\mathbf{a},\mathbf{b}_t}(\mathbf{x})).$$

Then, one dismantles the functions in analogous manner $f_{\mathbf{a},\mathbf{b}}$, and so forth. As a result, the function f_1 is dismantled in certain functions $h_{\mathbf{a};\alpha} \in Pol_k\gamma$ defined by

$$h_{\mathbf{a};\alpha}(x_1, ..., x_n) := \begin{cases} \alpha \text{ if} & \mathbf{x} = \mathbf{a}, \\ c \text{ otherwise}, \end{cases}$$

where $\alpha \in (E_k \backslash C) \cup \{c\}$ and $\mathbf{a} \in E_k^n$.

3.) In this step, we construct representations for the n-ary functions $u \in Pol_k\gamma$, which are defined by

$$u(x_1, ..., x_n) := \begin{cases} d_i \text{ if} & \mathbf{x} = \mathbf{d}_i, \ i = 1, 2, ..., q, \\ c \text{ otherwise}, \end{cases}$$

where $\{d_1, ..., d_q\} \subseteq E_k \backslash C$, $d_i \neq d_j$ for $i \neq j$ and $\mathbf{d}_i = (d_{i1}, ..., d_{im})$; $i, j = 1, 2, ..., q$.
If $q \leq h - 1$ and for the case that $q = h$ and $(d_1, ..., d_q) \in \gamma$, we have $u \in [(Pol_k\gamma)^2]$. Consequently, the following three cases must still be examined:
Case 1: $q = h$ and $(d_1, ..., d_q) \notin \gamma$.
In this case, there exists a j $(1 \leq j \leq n)$ with $(d_{1j}, d_{2j}, ..., d_{hj}) \notin \gamma$, since $u \in Pol_k\gamma$. Put

$$u_1(x_1, .., x_n) := \begin{cases} d_1 \text{ if} & \mathbf{x} \in \{\mathbf{d}_1, \mathbf{d}_2, ..., \mathbf{d}_h\}, \\ c \text{ otherwise}, \end{cases}$$

and

$$u_2(x_1, x_2) := \begin{cases} d_i \text{ if} & x_1 = d_{ij} \wedge x_2 = d_1, \ i = 1, 2, ..., h, \\ c \text{ otherwise}. \end{cases}$$

Because of $u_1, u_2 \in Pol_k k\gamma$ and $u(\mathbf{x}) = u_2(x_j, u_1(\mathbf{x}))$ we have $u \in [(Pol_k\gamma)^2]$.
Case 2: There is a set $\{e_1, e_2, ..., e_h\} \subseteq \{d_1, ..., d_q\}$ with $(e_1, ..., e_h) \in \gamma \backslash \iota_k^h$.
The functions

$$v(x_1, ..., x_h) := \begin{cases} e_i \text{ if} & x_1 = ... = x_{i-1} = x_{i+1} = ... = x_h = e_i, \\ & x_i = c, \ i = 1, 2, ..., h, \\ x_1 \text{ if} & x_1 = ... = x_h \in \{d_1, .., d_q\} \backslash \{e_1, ..., e_h\} \\ c \text{ otherwise}, \end{cases}$$

and

$$v_i(x_1, ..., x_n) := \begin{cases} c & \text{if} \quad u(x_1, ..., x_n) = e_i, \\ u(x_1, ..., x_n) & \text{otherwise} \end{cases}$$

$(i = 1, 2, ..., h)$ belong to $Pol_k\gamma$. It is easy to check that $u(\mathbf{x}) = v(v_1(\mathbf{x}), v_2(\mathbf{x}), ..., v_h(\mathbf{x}))$ holds, where $|Im(v_i)| < |Im(u)|$.

Case 3: $q > h$ and for an arbitrary set $\{e_1, e_2, ..., e_h\} \subseteq \{d_1, ..., d_q\}$ it holds: $(e_1, e_2, ..., e_h) \notin \gamma \setminus \iota_k^h$.
We need the functions

$$w(x_1, ..., x_{\lceil \frac{h}{2} \rceil + 1}) := \begin{cases} d_{2i-1} & \text{if } x_1 = ... = x_{\lceil \frac{h}{2} \rceil + 1} = d_{2i-1},\ i = 1, 2, .., \lceil \frac{h}{2} \rceil, \\ d_{2i} & \text{if } x_1 = ... x_{i-1} = x_{i+1} = ... = x_{\lceil \frac{h}{2} \rceil + 1} = d_{2i}, \\ & \quad x_i = d_{2i-1}, i = 1, 2, .., \lceil \frac{h}{2} \rceil, \\ d_h & \text{if } h \text{ odd } \wedge\ x_1 = ... = x_{\lceil \frac{h}{2} \rceil + 1} = d_h, \\ d_{h+1} & \text{if } x_1 = ... = x_{\lceil \frac{h}{2} \rceil} = d_{h+1} \wedge x_{\lceil \frac{h}{2} \rceil + 1} = d_h, \\ x_1 & \text{if } x_1 = ... = x_{\lceil \frac{h}{2} \rceil + 1} \in \{d_{h+2}, ..., d_q\}, \\ c & \text{otherwise,} \end{cases}$$

$$w_i(x_1, ..., x_n) := \begin{cases} d_{2i-1} & \text{if} \quad u(\mathbf{x}) \in \{d_{2i-1}, d_{2i}\}, \\ u(\mathbf{x}) & \text{otherwise} \end{cases}$$

$(i = 1, 2, ..., \lceil \frac{h}{2} \rceil)$ and

$$w_{\lceil \frac{h}{2} \rceil + 1}(x_1, ..., x_n) := \begin{cases} d_h & \text{if} \quad u(\mathbf{x}) \in \{d_h, d_{h+1}\}, \\ u(\mathbf{x}) & \text{otherwise,} \end{cases}$$

of $Pol_k\gamma$ for the dismantling of the function u. It holds:

$$u(\mathbf{x}) = w(w_1(\mathbf{x}), w_2(\mathbf{x}), ..., w_{\lceil \frac{h}{2} \rceil + 1}(\mathbf{x})),$$

where $|Im(w_i)| < |Im(u)|$.
The iteration of the above procedure proves the contention

$$ord\ Pol_k\gamma = \begin{cases} 2 & \text{if} \quad k - |C| \leq k, \\ \leq h & \text{otherwise.} \end{cases}$$

If $k - |C| > h$ and the second case is not possible for any function $f \in Pol_k\gamma$,i.e., γ fulfills the condition (11.1), it results from the above procedure then that $ord\ Pol_k\gamma \leq \lceil \frac{h}{2} \rceil + 1$ is valid. We still have to show that $w \notin [(Pol_k\gamma)^{\lceil \frac{h}{2} \rceil}$ holds, if γ fulfills the condition (11.1).
Denote $\mathfrak{A}$ a matrix whose rows are just the tuples of $E_k^{\lceil \frac{h}{2} \rceil + 1}$ on which the function w takes the values d_1, d_2, ..., d_{h+1}. It is easy to check that certain rows $\mathbf{a}_1, ..., \mathbf{a}_h$ ($\mathbf{a}_i := (a_{i1}, ..., a_{ir})$, $i = 1, 2, ..., h$, $r \leq \lceil \frac{h}{2} \rceil$) are found in every matrix, which can be formed from the matrix $\mathfrak{A}$ by deleting at least a column, so that is valid for every $i \in \{1, 2, ..., r\}$: $(a_{i1}, a_{2i}, ..., a_{ir}) \in \iota_k^h$. Consequently, an arbitrary $(\frac{h}{2} + 1)$-ary function of $Pol_k\gamma$, which depends on at most $\lceil \frac{h}{2} \rceil$ places essentially, takes either a value from the set C on at least a tuple of $\mathfrak{A}$ or two different rows of $\mathfrak{A}$, on which the function takes the very same value, exist. Further, every $\lceil \frac{h}{2} \rceil$-ary function of $Pol_k\gamma$ can take only then $h+1$ different values from the set $E_k \backslash C$ on different tuples $\mathbf{b}_1$, ...,$\mathbf{b}_{h+1}$, if there is at least a column that contains $h+1$ different values of the set $E_k \backslash C$ among the columns of the matrix

$$\begin{pmatrix} \mathbf{b}_1 \\ \mathbf{b}_2 \\ \dots \\ \mathbf{b}_{h+1} \end{pmatrix}.$$

Consequently, we have $w \notin (Pol_k\gamma)^{\lceil \frac{h}{2} \rceil}$ and $ord\ Pol_k\gamma = \lceil \frac{h}{2} \rceil + 1$, if γ fulfills the condition (11.1). ■

11.4 The Order of $Pol_k\varrho$ for $\varrho \in \mathfrak{M}_k$ and $k \le 7$

In this section let ϱ be a relation of $\mathfrak{M}_k$. The following lemma generalizes a theorem from [Jab 58], p. 83.

Lemma 11.4.1 *Let $E \subseteq E_k$ with $|E| \ge 2$. Further, let ϱ' be a partial order relation on E with the following properties: $\varrho' \subseteq \varrho$ and $(E; \varrho')$ is a lattice. Then*

$$\{f \in Pol_k\varrho \mid Im(f) \subseteq E\} \subseteq [(Pol_k\varrho)^2].$$

Proof. Obviously, ϱ' has a least element o and has a greatest element e. It is easy to check that there exists a unary function i with $Im(i) = E$ and $i(a) = a$ for all $a \in E$. We show that an arbitrary n-ary function $f \in P_{k,E} \cap Pol_k\varrho$ is a superposition over $(sup_{\varrho'}(i(x_1), i(x_2)))$, $(inf_{\varrho'}(i(x_1), i(x_2)))$ and the functions

$$m_{b,a}(x) := \begin{cases} a \text{ if } & x \ge_\varrho b, \\ o \text{ otherwise} & \end{cases}$$

$(a \in E,\ b \in E_k)$. For all $\mathbf{a} := (a_1, a_2, ..., a_n) \in E_k^n$ is valid:

$$f_{\mathbf{a}}(x_1, ..., x_n) := \begin{cases} f(\mathbf{a}) \text{ if } & \mathbf{x} \ge_\varrho \mathbf{a}, \\ o \quad \text{otherwise,} & \end{cases}$$

$= inf_{\varrho'}(m_{a_1,f(\mathbf{a})}(\mathbf{x}), m_{a_2,f(\mathbf{a})}(\mathbf{x}), ..., m_{a_n,f(\mathbf{a})}(\mathbf{x})) \in [(Pol_k\varrho)^2]$. Then one can represent the function f^n as follows:

$$f(\mathbf{x}) = sup_{\varrho'}(f_{\mathbf{a}_1}(\mathbf{x}), f_{\mathbf{a}_2}(\mathbf{x}), ..., f_{\mathbf{a}_{k^n}}(\mathbf{x})),$$

where $\{\mathbf{a}_1, ..., \mathbf{a}_{k^n}\} = E_k^n$. Consequently, $f \in [(Pol_k\varrho)^2]$. ■

A consequence of Lemma 11.4.1 is

Theorem 11.4.2 *If $(E;\varrho)$ a lattice then $ord\ Pol_k\varrho = 2$.* ■

Theorem 11.4.3 *If $k \le 7$ then $ord\ Pol_k\varrho = 2$.*

Proof. Let $k = 6$ and ϱ the partial order relation defined by Figure 11.1. We will prove the theorem only for this relation. In the remaining cases, the relation fulfills the conditions of Theorem 11.4.2 or one can lead the proof similarly to the proof which is subsequently given.

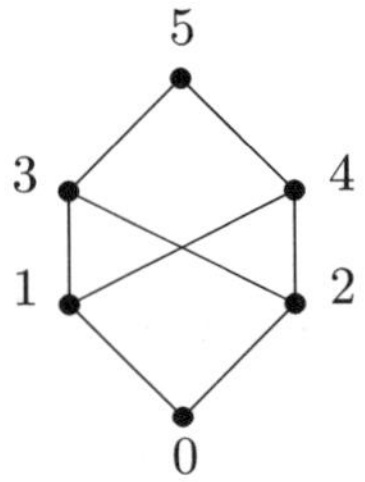

Fig. 11.1

We start with the definitions of some functions of $(Pol_k\varrho)^2$:

$$i_1(x) := \begin{cases} 0 \text{ if } & x \in E_3, \\ x \text{ otherwise}, \end{cases} \qquad i_2(x) := \begin{cases} x \text{ if } & x \in E_3, \\ 5 \text{ otherwise}, \end{cases}$$

$$e_1(x) := \begin{cases} 2 \text{ if } & x = 1, \\ 1 \text{ if } & x = 2, \\ \quad \text{otherwise}, \end{cases} \qquad e_2(x) := \begin{cases} 4 \text{ if } & x = 3, \\ 3 \text{ if } & x = 4, \\ x \text{ otherwise}, \end{cases}$$

$sup_{\varrho_i}(x_1, x_2)$, $i = 0, 1, 2$ and $inf_{\varrho_j}(x_1, x_2)$, $j = 3, 4, 5$, where the relations ϱ_i and ϱ_j are defined in Figure 11.2.

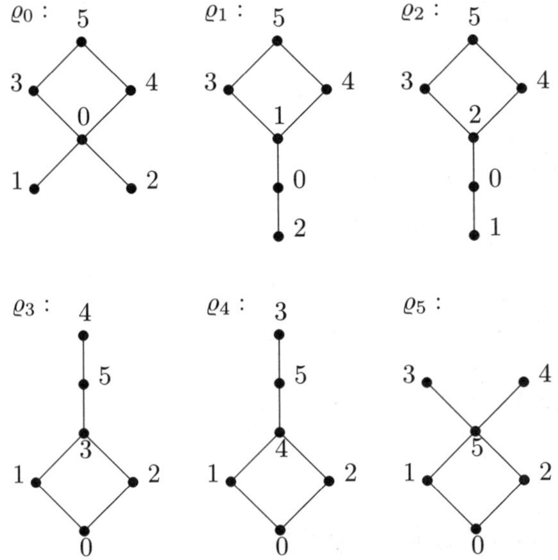

Fig. 11.2

Let $f^n \in Pol_6\varrho$ be arbitrary. We prove by induction on n that $f^n \in [(Pol_6\varrho)^2]$ holds. For $n = 1, 2$ our assertion is trivial. Assume the assertion is valid for $n - 1 \geq 2$ and we show that it is valid for n. If $|Im(f)| \leq 5$, then, by Lemma 11.4.1, $f \in [(Pol_6\varrho)^2]$. Let $|Im(f)| = 6$ in the following. We show through a dismantling procedure that f^n is a superposition over certain binary functions and certain functions g^m with $|Im(g)| \leq 5$ or $m \leq n-1$. Because of induction assumption and Lemma 11.4.1, our theorem would then be proven.

1.) During the first dismantling of function f, we use, the following functions of $Pol_6\varrho$: $sup_\varrho(x_1, i_1(x_2))$,

$$f_1^n(\mathbf{x}) \begin{cases} = f(\mathbf{x}) & \text{if} \quad f(\mathbf{x}) \in E_3, \\ \leq_\varrho f(\mathbf{x}) & \text{otherwise,} \end{cases}$$

where we assume the validity of

$$\neg\exists f_1' \in Pol_6\varrho :$$
$$\left(\exists \mathbf{a} : \; f_1'(\mathbf{a}) <_\varrho f_1(\mathbf{a}) \;\wedge\; f_1'((\mathbf{x}) \begin{cases} = f(\mathbf{x}) & \text{if} \quad f(\mathbf{x}) \in E_3, \\ \leq_\varrho f(\mathbf{x}) & \text{otherwise,} \end{cases} \right),$$

and

$$f_2^n(\mathbf{x}) := \begin{cases} 0 & \text{if} \quad f(\mathbf{x}) \in E_3, \\ f(\mathbf{x}) & \text{otherwise.} \end{cases}$$

We get

$$f(\mathbf{x}) = sup_\varrho(f_1(\mathbf{x}), i_1(f_2(\mathbf{x}))).$$

2.) Because of $f_2 \in [(Pol_6\varrho)^2]$ we must still find only a dismantling of the kind described above for the function f_1. Let

$$N_i := \{\mathbf{a} \in E_6^n \mid f_1(\mathbf{a}) = i \;\wedge\; \neg\exists \mathbf{a}' \in E_6^n : \; \mathbf{a}' <_\varrho \mathbf{a} \;\wedge\; f_1(\mathbf{a}') = i\}$$

$(i = 1, 2)$,

$$N_1 := \{\mathbf{a}_1, \mathbf{a}_2, ..., \mathbf{a}_s\}, \qquad N_2 := \{\mathbf{b}_1, \mathbf{b}_2, ..., \mathbf{b}_t\}, \qquad N := N_1 \cup N_2,$$

and for $N' \subset N$ let $f_{N'}$ be an n-ary function of $Pol_6\varrho$ with

$$f_{N'}(\mathbf{x}) : \begin{cases} = 1 & \text{if } (\exists \mathbf{a} \in N' \cap N_1 : \mathbf{x} \geq_\varrho \mathbf{a}) \wedge (\neg\exists \mathbf{b} \in N' \cap N_2 : \mathbf{x} \geq_\varrho \mathbf{b}), \\ = 2 & \text{if } (\neg\exists \mathbf{a} \in N' \cap N_1 : \mathbf{x} \geq_\varrho \mathbf{a}) \wedge (\exists \mathbf{b} \in N' \cap N_2 : \mathbf{x} \geq_\varrho \mathbf{b}), \\ \leq_\varrho f_1(\mathbf{x}) & \text{if } (\exists \mathbf{a} \in N' \cap N_1 : \mathbf{x} \geq_\varrho \mathbf{a}) \wedge (\exists \mathbf{b} \in N' \cap N_2 : \mathbf{x} \geq_\varrho \mathbf{b}), \\ = 0 & \text{otherwise,} \end{cases}$$

where for an arbitrary $\mathbf{a} \in E_6^n$ we assume that $f_N(\mathbf{a}) = 5$ is valid if and only if there is not any function $f_{N'} \in Pol_6\varrho$ with $f_{N'}(\mathbf{a}) \in \{3, 4\}$. We get the following description for f_1:

$$f_1(\mathbf{x}) = sup_{\varrho_1}(f_{N^{(1)}}(\mathbf{x}), f_{N^{(2)}}(\mathbf{x}), ..., f_{N^{(1)}}(\mathbf{x}))$$

and

$$f_{N^{(i)}}(\mathbf{x}) = sup_{\varrho 2}(f_{N^{(i,1)}}(\mathbf{x}), f_{N^{(i,2)}}(\mathbf{x}), ..., f_{N^{(i,t)}}(\mathbf{x}))$$

$(i = 1, 2, ..., s)$.

3.) We search dismantling for the functions $f_{i,j} := f_{N^{(i,j)}}, 1 \leq i \leq s, 1 \leq j \leq t$. Let

$$S_{i,j} := \{\mathbf{c} \in E_6^n \mid \mathbf{a}_i <_\varrho \mathbf{c} \wedge \mathbf{b}_j <_\varrho \mathbf{c} \wedge (\neg\exists \mathbf{c}' : \mathbf{a}_i <_\varrho \mathbf{c} <_\varrho \mathbf{c} \wedge \mathbf{b}_j <_\varrho \mathbf{c}' <_\varrho \mathbf{c})\}.$$

Case 1: $|S_{i,j}| = 1$ or $|\{f_{i,j}(\mathbf{a}) \mid \mathbf{a} \in S_{i,j}| = 1$.
In this case, we have $|Im(f_{i,j})| \leq 5$. Then by Lemma 11.4.1: $f_{i,j} \in [(Pol_k\varrho)^2]$.

Case 2: $|S_{i,j}| > 1$ and $|\{f_{i,j}(\mathbf{a}) \mid \mathbf{a} \in S_{i,j}\}| > 1$.
Let $\mathbf{a}_i = (a_{i1}, a_{i2}, ..., a_{in})$ and $\mathbf{b}_j := (b_{j1}, b_{j2}, ..., b_{jn})$. W.l.o.g. we can assume that

$$(a_{ip}, b_{jp}) \begin{cases} = (1,2) & \text{if } p = 1, 2, ..., u, \\ = (2,1) & \text{if } p = u, u+1, ..., u+v, \\ \notin \{(1,2),(2,1)\} & \text{if } p = u+v+1, ..., n, \end{cases}$$

where $u \geq 1$ or $v \geq 1$ because of $|S_{i,j}| > 1$.

Case 2.1: $u + v =: m < n$.
Let

$$\mathbf{c}^\star := (sup_\varrho(a_{i,m+1}, b_{j,m+1}), sup_\varrho(a_{i,m+2}, b_{j,m+2}), ..., sup_\varrho(a_{i,n}, b_{j,n})),$$

$$g(x_1, ..., x_m) := \begin{cases} 1 & \text{if} & \mathbf{x} \in \{1,3,4,5\}^m \backslash \{3,4,5\}^m, \\ 2 & \text{if} & \mathbf{x} \in \{2,3,4,5\}^m \backslash \{3,4,5\}^m, \\ f_{i,j}(x_1, ..., x_m, \mathbf{c}^\star) & \text{if} & \mathbf{x} \in \{1,3,4,5\}^m, \\ 0 & \text{otherwise,} \end{cases}$$

and

$$h(x_1, ..., x_n) := \begin{cases} 1 \text{ if} & \mathbf{x} \geq_\varrho \mathbf{a}_i \wedge \mathbf{x} \not\geq_\varrho \mathbf{b}_j, \\ 2 \text{ if} & \mathbf{x} \not\geq_\varrho \mathbf{a}_i \wedge \mathbf{x} \geq_\varrho \mathbf{b}_j, \\ 4 \text{ if} & \mathbf{x} \geq_\varrho \mathbf{a}_i \wedge \mathbf{x} \geq_\varrho \mathbf{b}_j, \\ 0 \text{ otherwise.} \end{cases}$$

We get

$$f_{i,j}(\mathbf{x}) = inf_{\varrho 3}(g(x_1, ..., x_u, e_1(x_{u+1}, ..., e_1(x_m), h(x_1, ..., x_n)),$$

i.e., $f_{i,j} \in [(Pol_6\varrho)^2]$ by induction assumption and Lemma 11.4.1.

Case 2.2: $u + v = n$.
Let

$$T_{i,j} := \{\mathbf{a} \in E_6^n \mid f_{i,j}(\mathbf{a}) \in \{3,4\} \wedge (\neg\exists \mathbf{a}' \in E_6^n : \mathbf{a}' >_\varrho \mathbf{a} \wedge f_{i,j}(\mathbf{a}' \in \{3,4\}\}.$$

Case 2.2.1: $T_{i,j} \subseteq \{3,4\}^n$.
For

$$\{\mathbf{d} \in T_{i,j} \mid f_{i,j}(\mathbf{d}) = 4\} =: \{\mathbf{d}_1, ..., \mathbf{d}_l\},$$

with $\mathbf{d}_r := (d_{r1}, ..., d_{rn})$, $r = 1, 2, ..., l$ and

$$q_a(x) := \begin{cases} x & \text{if } a = 4, \\ e_2(x) & \text{if } a = 3, \end{cases}$$

we obtain

$$\begin{aligned} & f_{i,j}(\mathbf{x}) = \\ & inf_{\varrho 4}(inf_{\varrho 3}(q_{d_{11}}(x_1), ..., q_{d_{1u}}(x_u), q_{d_{1,u+1}}(e_1(x_{u+1}), ..., q_{d_{1n}}(e_1(x_n))), \\ & \qquad inf_{\varrho 3}(q_{d_{21}}(x_1), ..., q_{d_{2u}}(x_u), q_{d_{2,u+1}}(e_1(x_{u+1}), ..., q_{d_{2n}}(e_1(x_n))), ..., \\ & \qquad \cdots\cdots\cdots \\ & \qquad inf_{\varrho 3}(q_{d_{l1}}(x_1), ..., q_{d_{lu}}(x_u), q_{d_{l,u+1}}(e_1(x_{u+1}), ..., q_{d_{ln}}(e_1(x_n)))), \end{aligned}$$

i.e., $f_{i,j} \in [(Pol_6\varrho)^2]$.

Case 2.2.2: $T_{i,j} \not\subseteq \{3,4\}^n$.

Case 2.2.2.1: $\neg(\exists \mathbf{a} := (a_1, ..., a_n), (b_1, ..., b_n) \in T_{i,j} : f_{i,j}(\mathbf{a}) = 3 \wedge f_{i,j}(\mathbf{b}) = 4 \wedge (\forall p \in \{1, ..., n\} : (a_p, b_q) \in \{(3,4), (4,3)\})$.

Here one can show $f_{i,j} \in [Pol\ \varrho]^{n-1}$ when one uses a procedure that results from the construction steps 1.), 2.) and 3.) (including Case 2.1) listed above, when one the following replacements are carried out: f through $f_{i,j}$, 0 through 5, 1 through 3, 2 through 4, 3 through 2, 4 through 2, 5 through 0, $\leq_\varrho$ through $\geq_\varrho$, i_1 through i_2, e_1 through e_2, inf through sup ... .

Case 2.2.2.2: $\exists \mathbf{a} := (a_1, ..., a_n), (b_1, ..., b_n) \in T_{i,j} : f_{i,j}(\mathbf{a}) = 3 \wedge f_{i,j}(\mathbf{b}) = 4 \wedge (\forall p \in \{1, ..., n\} : (a_p, b_q) \in \{(3,4), (4,3)\}$.

W.l.o.g. let $u = n$ and

$$(a_p, b_q) = \begin{cases} (3,4) & \text{if } p = 1, 2, ..., y, \\ (4,3) & \text{if } p = y+1, ..., n. \end{cases}$$

We need the following functions of $Pol_6\varrho$:

$$z_1(x_1, ..., x_n) := \begin{cases} 5 & \text{if} \qquad \mathbf{x} = \mathbf{a}, \\ f_{i,j}(\mathbf{x}) & \text{otherwise}, \end{cases}$$

$$z_2(x_1, ..., x_n) := \begin{cases} 5 & \text{if} \qquad \mathbf{x} \neq \mathbf{a} \wedge f_{i,j}(\mathbf{x}) = 3, \\ f_{i,j}(\mathbf{x}) & \text{otherwise}, \end{cases}$$

$$z_3(x_1, ..., x_n) := \begin{cases} 5 & \text{if} \qquad \mathbf{x} = \mathbf{b}, \\ z_2(\mathbf{x}) & \text{otherwise}, \end{cases}$$

and

$$\begin{aligned} z_4(x_1, ..., x_n) &:= \begin{cases} 5 & \text{if} \qquad \mathbf{x} \neq \mathbf{b} \wedge z_2(\mathbf{x}) = 4, \\ z_2(\mathbf{x}) & \text{otherwise}, \end{cases} \\ &= inf_{\varrho 5}(x_1, ..., x_y, e_2(x_{y+1}), ..., e_2(x_n)). \end{aligned}$$

We get

$$f_{i,j}(\mathbf{x}) = inf_{\varrho 3}(z_1(\mathbf{x}), z_2\mathbf{x}))$$

and

$$z_2(\mathbf{x}) = inf_{\varrho_4}(z_3(\mathbf{x}), z_4(\mathbf{x})).$$

Then, for the functions z_1 and z_3, one can distinguish cases as above and repeat the constructions where appropriate. Therefore, one can reduce the Case 2.2.2.2 to the previous cases.

Thus $f \in [(Pol_6\varrho)^2]$ holds. ■

We still notice that one cannot transfer the methods from the proof of Theorem 11.4.3 for arbitrary relations $\varrho^\star \in \mathfrak{M}_k$, since there are no binary functions in $Pol_8\varrho^\star$ with the properties used of function sup_{ϱ_1} for $k = 8$ and for the relation $\varrho^\star$, which is defined in Figure 11.3. Suppose there exists a function $f^2 \in Pol_8\varrho^\star$ with $f(1,0) = f(0,1) = 1$ and $f(x,x) = x$ for all $x \in E_8$. Then, $f(1,0) = 1$, $f(2,2) = 2$ and $f(3,3) = 3$ implies $f(3,2) = 3$. Further, $f(0,1) = 1$, $f(2,2) = 2$ and $f(4,4) = 4$ implies $f(2,4) = 4$. Consequently, we have $f(3,4) \in \{5,6,7\}$. This is, however, a contradiction to $f(5,5) = 5$, $f(6,6) = 6$ and $f \in Pol_8\varrho^\star$.

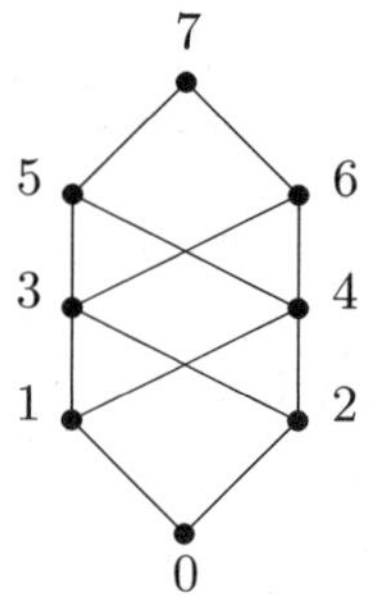

Fig. 11.3

In the next section, we show that the above-described class does not have a finite order.

11.5 A Maximal Clone of Monotone Functions That Is Not Finitely Generated

This section basically displays the content of the paper [Tar 86] by G. Tardos.

The set E_8 is written down in this section in the form:

$$\{0, \alpha, \alpha', \beta, \beta', \gamma, \gamma', 1\},$$

On E_8 we consider the partial order relation ϱ which is defined by Figure 11.4. If from monotone functions the speech is subsequently, only functions are with that meant from $Pol_8\varrho$.

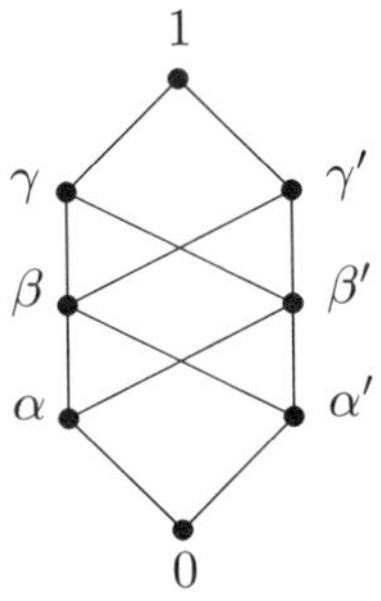

Fig. 11.4

The signs $<, >, \geq, \leq$ refer to this order relation or are explained for tuples from E_8^n, on which the relation works coordinate wisely.

The following definition is clarified by means of Figure 11.5 which indicates a Hasse-diagram for certain n-tuples. A denotation of the form $x; a$ at a knot of this diagram means that $f(x) = a$ is valid for a corresponding n-ary function f.

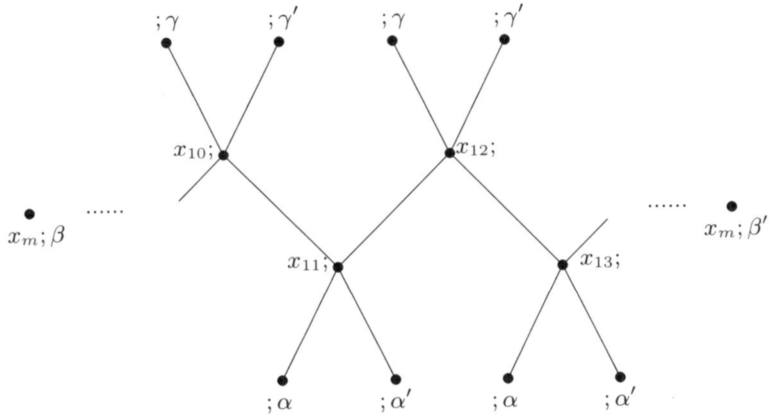

Fig. 11.5

Definition Let $\emptyset \neq Q \subseteq E_8^n$ and $f_1 : Q \longrightarrow E_8$. We call a sequence $x_m, ..., x_{m'}$ of pairwise distinct elements from the set E_8^n a **zigzag** for f_1 if the following seven conditions hold:

(a) $m = 0$ or $m = 1$ and $m' > m + 2$.

(b) $x_m, x_{m'} \in Q$ and $x_i \in E_8^n \backslash Q$ for all $i \in \{m+1, ..., m'-1\}$.

(c) $f_1(x_m) = \beta$ and $f_1(x_{m'}) = \beta'$.

(d) $\forall\ 2i \in \{m+1, ..., m'\} : x_{2i} > x_{2i-1}$.

(e) $\forall\, 2i \in \{m, m+1, ..., m'-1\} : x_{2i} > x_{2i+1}$.

(f) $\forall\, 2i+1 \in \{m+1, ..., m'-1\}\ \exists y, y' \in Q :\ y, y' > x_{2i+1} \wedge f_1(y) = \alpha \wedge f_1(y') = \alpha'$.

(g) $\forall\, 2i \in \{m+1, ..., m'-1\}\ \exists z, z' \in Q :\ z, z' > x_{2i} \wedge f_1(z) = \gamma \wedge f_1(z') = \gamma'$.

Lemma 11.5.1 *Let $\emptyset \neq Q \subseteq E_8^n$ and let $f_1 : Q \longrightarrow E_8$ be a mapping. Then there is a monotone function $f : E_8^n \longrightarrow E_8$ with $f_{|Q} = f_1$ if and only if the following two conditions hold:*

(1) f_1 is monotone;

(2) there is no zigzag in Q for f_1.

Proof. "$\Longrightarrow$": The condition (1) is trivial. To prove (2) we assume by way of contradiction that $x_m, ..., x_{m'} \in Q$ is a zigzag for f_1 according to the above definition and that $f : E_8^n \longrightarrow E_8$ is a monotone extension of f_1. First, let $m = 0$ and $m' = 2q+1$, $q \in \mathbb{N}$ in the zigzag (see Figure 11.6).

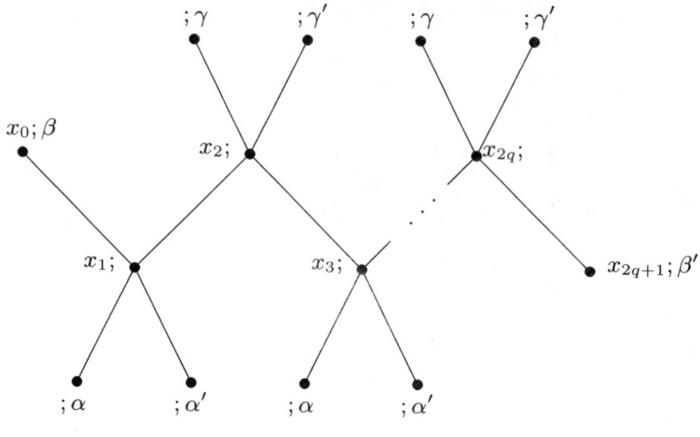

Fig. 11.6

Since $x_1, ..., x_{2q} \in E_8^n$ and f_1 preserves ϱ on these tuples, we have $f_1(x_1) = \beta$. Continuation of this consideration supplies $f_1(x_{2q}) = \beta$, which is not possible because of $x_{2q+1} < x_{2q}$ and $f_1(x_{2q+1}) = \beta'$. For the other possibilities for m and m' one shows in analog mode that the assumption of the existence of a zigzag for f_1 supplies a contradiction.
"$\Longleftarrow$": Assume that the conditions (1) and (2) are satisfied. We construct a monotone function $f : E_8^n \longrightarrow E_8$ extending the function f_1 by defining of the following sets $H_x := f^{-1}(x)$ for each $x \in E_8$. For $q \in E_8^n$ let

$$f_\star(q) := \{f_1(y) \mid y \in Q \wedge y \leq q\},$$
$$f^\star(q) := \{f_1(z) \mid z \in Q \wedge z \geq q\}.$$

Furthermore, let

$$\begin{aligned}
H_0 &:= \{q \in E_8^n \mid f_\star(q) \subseteq \{0\}\},\\
H_1 &:= \{q \in E_8^n \mid f^\star(q) \subseteq \{1\}\}\backslash H_0,\\
H_\alpha &:= \{q \in E_8^n \mid f_\star(q) \subseteq \{0,\alpha\}\}\backslash (H_0 \cup H_1),\\
H_{\alpha'} &:= \{q \in E_8^n \mid f_\star(q) \subseteq \{0,\alpha'\}\}\backslash (H_0 \cup H_1 \cup H_\alpha),\\
H_\gamma &:= \{q \in E_8^n \mid f^\star(q) \subseteq \{1,\gamma\}\}\backslash (H_0 \cup H_1 \cup H_\alpha \cup H_{\alpha'}),\\
H_{\gamma'} &:= \{q \in E_8^n \mid f^\star(q) \subseteq \{1,\gamma'\}\}\backslash (H_0 \cup H_1 \cup H_\alpha \cup H_{\alpha'} \cup H_\gamma)
\end{aligned}$$

and

$$H := E_8^n \backslash (H_0 \cup H_1 \cup H_\alpha \cup H_{\alpha'} \cup H_\gamma \cup H_{\gamma'}).$$

To partition the set H, we consider the elements of H as vertices of a non-directional graph in which an edge joins the vertices $x, y \in H$ if and only if $x < y$ or $y < x$ in E_8^n is valid. Then, graph G consists of certain connected maximal subgraphs (the so-called components) $K_1 \ldots, K_t$, with which we define the sets H_β and $H_{\beta'}$ as follows:

$$\begin{aligned}
H_\beta &:= \{x \in H \mid \exists i \in \{1, \ldots, t\} : \; x \in K_i \;\wedge\; \beta \in \{f_1(a) \mid a \in K_i\}\},\\
H_{\beta'} &:= H \backslash H_\beta.
\end{aligned}$$

The subsets H_x for $x \in E_8$ now form a partition of E_8^n, so they really determine a function $f : E_8^n \longrightarrow E_8$ satisfying

$$\forall x \in E_8^n \; \forall e \in E_8 : \; f(x) = e \; :\Longleftrightarrow \; x \in H_e.$$

We prove that this function preserves the relation ϱ and that it agrees with the function f_1 on E_8^n.
One can check the monotony from f easily, using the following implications, which result from the definitions of the sets H_e, $e \in E_8$ and which are valid for every $a \in E_8^n$:

$$\begin{aligned}
\forall e \in \{\alpha, \alpha'\} : \; f(a) = e \;&\Longrightarrow\; e \in f_\star(a)\\
\forall e \in \{\gamma, \gamma'\} : \; f(a) = e \;&\Longrightarrow\; e \in f^\star(a).
\end{aligned}$$

Finally we show $f_{|Q} = f_1$.
Obviously, for arbitrary $z \in Q$ with $f_1(z) \neq \beta'$ we have $f(z) = f_1(z)$. To show that $f(z) = f_1(z)$ holds for elements $z \in Q$ with $f_1(z) = \beta'$ as well, we must prove that no component G contains elements from both $f_1^{-1}(\beta)$ and $f_1^{-1}(\beta')$. We use here the complicated condition (2). By the definition of H for every $z \in H$, we have

$$\begin{aligned}
f_\star(z) \not\subseteq \{0, \alpha\}, \; f_\star(z) \not\subseteq \{0, \alpha'\},\\
f^\star(z) \not\subseteq \{1, \gamma\}, \; f^\star(z) \not\subseteq \{1, \gamma'\}.
\end{aligned}$$

By condition (1) these conditions are equivalent to

$$\beta \in f_\star(z) \text{ or } \beta' \in f_\star(z) \text{ or } \{\alpha, \alpha'\} \subseteq f_\star(z) \tag{11.6}$$

and

$$\beta \in f^\star(z) \text{ or } \beta' \in f^\star(z) \text{ or } \{\gamma, \gamma'\} \subseteq f^\star(z). \tag{11.7}$$

Suppose a component K of the graph G contains a $x_m \in f_1^{-1}(\beta)$ and a $x_{m'} \in f^{-1}(\beta')$. We choose x_m and $x_{m'}$ so that are the least distance among the possible vertices. So, there is a path between x_m and $x_{m'}$. Take the shortest path. If x, y and z are three consecutive elements of this path then neither $x < y < z$ nor $x > y > z$ can hold since y would then be redundant. Thus, we can choose the indices so that the path be $x_m, ..., x_{m'}$ satisfies the conditions (a), (c), (d), and (e) in the definition of a zigzag. If $m < 2i < m'$ then there is no $z \in Q$ with $f_1(z) = \beta$ or $f_1(z) = \beta'$ and $z \geq x_{2i}$, since otherwise either $z, x_{2i+1}, ..., x_{m'}$ or $x_m, ..., x_{2i-1}, z$ would be a shorter path in K from a element of $f^{-1}(\beta)$ to an element $f^{-1}(\beta')$. Consequently, (b) holds. Condition (f) also holds because for x_{2i} only $\{\gamma, \gamma'\} \subseteq f^\star(x_{2i})$ can occur among the possibilities in (11.3). Similarly, condition (g) holds. Hence $x_m, ..., x_{m'}$ is a zigzag for f_1 contrary to (2). ■

Next, for $n \geq 3$, we shall define $(n+5)$-ary relations $\mu_0, \mu_1, ..., \mu_n$ and μ on E_8, which we need to prove the following lemmas.

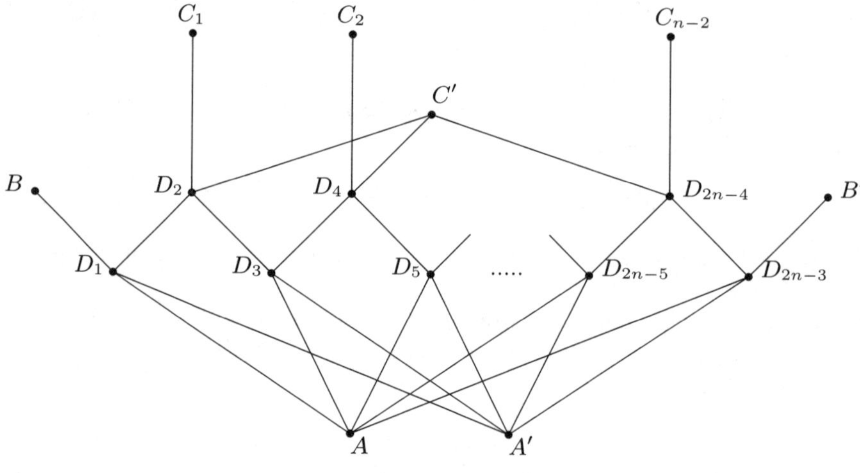

Fig. 11.7. Poset Q_0

Definitions Let $n \geq 3$. Let Q_0 ($\subseteq E_8^t$) [3] the poset defined by Figure 11.7. Further, we take

$$Q_0' := \{A, A', B, B', C_1, C_2, ..., C_{n-2}, C'\},$$
$$Q_1' := Q_0' \backslash \{B\} \text{ and } Q_2' := Q_0' \backslash \{B'\}.$$

Let

[3] Later we choose $t = 2 \cdot n$.

$$T := (a, a', b, b', c_1, c_2, ..., c_n, c') \in E_8^{n+5}.$$

The index of c is always understood modulo n. Let $f'_i : Q'_0 \longrightarrow E_8$ $(i = 1, ..., n)$ be the following mapping:

$$\begin{aligned} &f'_i(A) := a, \quad f'_i(A') := a', \\ &f'_i(B) := b, \quad f'_i(B') := b', \\ &f'_i(C_j) := c_{i+j} \ (j = 1, ..., n-2) \ \text{ and } \ f'_i(C') := c'. \end{aligned}$$

(Clearly, the mapping f'_i depends on the choice of the tuple T.)
We declare $T \in \mu_0$ iff for each $i \in \{1, ..., n\}$ both $(f'_i)_{|Q'_1}$ and $(f'_i)_{|Q'_2}$ can be extended monotonously to Q_0.
Furthermore, for $j \in \{1, ..., n\}$ let $T \in \mu_j$ iff $T \in \mu_0$ and f'_j can be extended monotonously to a monotone mapping on Q_0.
Finally, let

$$\mu := \bigcup_{j=1}^{n} \mu_j. \tag{11.8}$$

Lemma 11.5.2 *If $T \in \mu$ then T belongs to at least $n-2$ of the relations μ_j with $1 \leq j \leq n$.*

Proof. Let $T := (a, a', b, b', c_1, ..., c_n, c')$ and $l := |\{i \in \{1, ..., n\} \mid T \in \mu_i\}|$. $T \in \mu$ implies $T \in \mu_0$. If $T \in \mu_i$ for all $i \in \{1, ..., n\}$ then $l = n$. Thus, it is enough to consider the case $T \in \mu_0$ but $T \notin \mu_i$ for some $i \in \{1, ..., n\}$. By the cyclical nature of the relations μ_i, we may assume $T \notin \mu_1$. It means that f'_1 cannot be extended monotonously to Q_0. Because of $T \in \mu_0$ both $(f'_1)_{|Q'_1}$ and $(f'_1)_{|Q'_2}$ have monotone extensions to Q_0. Therefore, f'_1 is monotone. So by Lemma 11.5.1, there is a zigzag $x_m, ..., x_{m'}$ in Q_0 for f'_1. Let H be the set of all elements of Q_0 comparable with some x_i $(m < i < m')$. Since $x_m, ...x_{m'}$ is a zigzag for $(f'_1)_{|(Q'_0 \cap H)}$ as well, it follows that $(f'_1)_{|(Q'_0 \cap H)}$ cannot be extended monotonously to Q_0. This implies that both B and B' are in H. But $x_{m+1}, ..., x_{m'-1}$ is a series of distinct elements in $Q_0 \setminus Q'_0$ such that every two consecutive elements are comparable. These two properties imply that

$$\{x_i \mid m < i < m'\} = \{D_j \mid 1 \leq j \leq 2n-3\},$$

where the correspondence is either in direct or reverse order depending on whether $f'_1(B) = \beta$ or $f'_1(B) = \beta'$. So, by condition (g) in the definition of a zigzag

$$\{a, a'\} = \{f'_i(A), f'_i(A')\} = \{\alpha, \alpha'\},$$

and by condition (f) we have

$$\{c_i, c'\} = \{f'_i(C_{i-1}), f'_i(C')\} = \{\gamma, \gamma'\}$$

for $i = 2, ..., n-1$. W.l.o.g. we may assume that $c' = \gamma'$ and $c_i = \gamma$ for all $i = 2, ..., n-1$. Because of condition (c) of the definition of a zigzag we have

$$\{b, b'\} = \{f_i'(B), f_i'(B')\} = \{\beta, \beta'\}.$$

Using the fact that $T \in \mu_0$, one can show that $\{c_1, c_n\} \subseteq \{\gamma, \gamma', 1\}$. So, we have determined all tuples $T \in \mu_0 \backslash \mu_1$. By symmetry, we get similar characterizations for $\mu_0 \backslash \mu_i$ for $i = 2, ..., n$. Applying them, we obtain:

$$c_1 = c_2 = \gamma \implies l = 0 \implies T \notin \mu,$$

$$((c_1 = \gamma \wedge c_n \neq \gamma) \vee (c_1 \neq \gamma \wedge c_n = \gamma)) \implies l = n - 2,$$

$$(c_1 \neq \gamma \wedge c_n \neq \gamma) \implies l = n - 1.$$

■

Lemma 11.5.3 *Let $n \geq 3$. For every $l < \frac{n}{2}$ the l-ary monotone functions on E_8 preserve the relation μ.*

Proof. Let g be an l-ary function of $Pol_8\varrho$ and let $T_i \in \mu$ for $i = 1, 2, ..., l$. Then by Lemma 11.5.2 we have

$$\forall i \in \{1, 2, ..., l\} : |\{j \in \{1, ..., n\} \mid T_i \notin \mu_j\}| \leq 2.$$

Thus there is an index $j \in \{1, ..., n\}$ with $T_i \in \mu_j$ for all $i = 1, 2, ..., l$. Since one can derive the relations μ_i $(1 \leq i \leq n)$ from the t-th graphic $G_t(Pol_8\varrho)$ by means of the operation *pr* (projection), the relations m_i belong to $Inv_8(Pol_8\varrho)$ (see Chapter 2). Therefore, function g preserves the relation μ_j, from which we receive $g(T_1..., T_l) \in \mu_j \subseteq \mu$. ■

Lemma 11.5.4 *For $n \geq 3$ there exist $2n$-ary monotone functions on E_8, which do not preserve the relation μ.*

Proof. We consider a matrix $\mathfrak{A}$ of the type $(n + 5, 2n)$, whose construction is given in the following table, where $a, a', b, b', c_1, ..., c_n, c'$ are the rows of $\mathfrak{A}$ and $T_1, T_2, ..., T_{2n}$ are the columns of $\mathfrak{A}$. Furthermore, let $T := (\alpha, \alpha', \beta, \beta', \gamma, \gamma, ..., \gamma, \gamma')^T$ be a column matrix of length $n + 5$, which is also defined in the following table.

	T_1	T_2	T_3	T_4	...	T_n	T_{n+1}	T_{n+2}	T_{n+3}	T_{n+4}	...	T_{2n}	T
a	α	α	α	α	...	α	α	α	α	α	...	α	α
a'	α'	α'	α'	α'	...	α'	α'	α'	α'	α'	...	α'	α'
b	β	β	β	β	...	β	1	1	1	1	...	1	β
b'	β'	β'	β'	β'	...	β'	1	1	1	1	...	1	β'
c_1	1	γ	γ	γ	...	γ	1	β	γ	γ	...	β'	γ
c_2	γ	1	γ	γ	...	γ	β'	1	β	γ	...	γ	γ
c_3	γ	γ	1	γ	...	γ	γ	β'	1	β	...	γ	γ
c_4	γ	γ	γ	1	...	γ	γ	γ	β'	1	...	γ	γ
.	.	.	.	.	...	.	.	.	.	.	...	.	.
.	.	.	.	.	...	.	.	.	.	.	...	.	.
.	.	.	.	.	...	.	.	.	.	.	...	.	.
c_n	γ	γ	γ	γ	...	1	β	γ	γ	γ	...	1	γ
c'	γ'	γ'	γ'	γ'	...	γ'	γ'	γ'	γ'	γ'	...	γ'	γ'

Let

$$Q' := \{a, a', b, b', c_1, ..., c_n, c'\} \subseteq E_8^{2n}$$

and $f_1 : Q' \longrightarrow E_8$ be a function defined by

$$f_1(\mathfrak{A}) = f_1(T_1, T_2, ..., T_{2n}) := T.$$

Clearly, f_1 is monotone. Suppose that there is a zigzag $x_m, ..., x_{m'} \in E_8^{2n}$ for f_1. Thus we have

$$x_m = b,\ x_{m'} = b',$$
$$m < 2i + 1 < m' \implies a < x_{2i+1} \ \wedge\ a' < x_{2n+1}$$
$$m < 2i < m' \implies c' > x_{2i} \ \wedge\ (\exists q_i \in \{1, 2, ..., n\} : c_{q_i} > x_{2i}).$$

Let $p := \lfloor \frac{1}{2}(m' - 1) \rfloor$.[4] Each $j \in \{1, ..., n\}$ appears among the numbers $q_1, ..., q_p$, since we could not otherwise choose the j-th coordinates of $x_m, ..., x_{m'}$ to satisfy the above conditions. Let l $(1 \leq l \leq n)$ be the number that appears last from the beginning in the sequence $q_1, ..., q_p$. Then there is an interval in this sequence with first last element $l-1$ and $l+1$ or the reverse and containing no other element equal to $l - 1$, l or $l + 1$ considered again modulo n. However, in this case, we cannot choose the $(n + l)$-th coordinates of $x_{m+1}, ..., x_{m'-1}$ to satisfy the above conditions. The contradiction proves that there is no zigzag in E_8^{2n} for f_1.
So, by Lemma 11.5.1, we can extend f_1 to a $2n$-ary monotone function h with

$$h(T_1, ..., T_n) = T.$$

Using that $\{T_1, ..., T_n, T\} \subseteq \mu_0$ and the characterization of $\mu_0 \backslash \mu_i$ $(1 \leq i \leq n)$ one can easily check that $T_1, ..., T_n$ belong to μ but $T \notin \mu$. Thus the $2n$-ary monotone function h does not preserve μ. ■

Theorem 11.5.5 *([Tar 86]) For $k \geq 8$ there is a $\varrho \in \mathfrak{M}_k$, so that $Pol_k \varrho$ does not have a finite order.*

Proof. Let $\varrho \in \mathfrak{M}_k$ defined by the Hasse-diagram of Figure 11.4. Choosing a finite set $A \subset Pol_8 \varrho$ we have a number l, such that every element of A is at most l-ary. Taking $n \geq 2 \cdot l$ and defining μ corresponding to this n, we get that $A \subseteq Pol_8\{\mu, \varrho\} \subseteq Pol_8 \varrho$ by Lemma 11.5.3, but $(Pol_8 \varrho)^n \nsubseteq Pol_8\{\mu, \varrho\}$ by Lemma 11.5.4. So the finite set A does not generate the whole clone. ■

Presumably $ord\ Pol_k \varrho = \infty$ is valid for every relation ϱ, in whose diagram one can embed the diagram given in Figure 11.4.

[4] $\lfloor x \rfloor$ denotes the **floor** of $x \in \mathbb{R}$, i.e., the largest integer which is $\leq x$.

11.6 Classifications and Basis Enumerations in P_k

Let A be a class of P_k that is finitely generating. Then, A has only finite many A-maximal classes $M_1, M_2, ..., M_t$ and it is valid for all $B \subseteq A$:

$$[B] = A \iff (\forall i \in \{1, ..., t\} : B \not\subseteq M_i).$$

A generating system $B := \{f_1, ..., f_r\}$ of A with the following property is especially interesting: $[B \setminus \{f_i\}] \neq A$ for every $i \in \{1, ..., r\}$. Such a set B is called **basis** of A and the number r is the **rank** of B. It is briefly explained in this section, as one can classify the bases of A and one can also determine, then, the ranks of the bases.
With all maximal classes $M_1, ..., M_t$ of the class A, can be classified the functions, by their membership in the A-maximal sets as follows:
For $f \in A$ we put

$$\chi_i(f) := \begin{cases} 0 \text{ if } f \in M_i, \\ 1 \text{ if } f \notin M_i \end{cases}$$

$(i = 1, 2, ..., t)$ and

$$\chi(f) := (\chi_1(f), \chi_2(f), ..., \chi_t(f)).$$

We call $\chi(f)$ the **characteristic vector** of f.
We put $f \equiv g$ iff $f, g \in A$ and $\chi(f) = \chi(g)$. Obviously, $\equiv$ is an equivalence relation on A, and so it partitions A into pairwise disjoint nonempty sets (called equivalence classes or blocks). Let $f \in A$ with $\chi(f) = (a_1, ..., a_t)$. Then, it is easy to see that the block $[f]_\equiv$ of $\equiv$ has the form

$$T_{a_1} \cap T_{a_2} \cap .... \cap T_{a_t},$$

where, for $i \in \{1, ..., t\}$,

$$T_{a_i} := \begin{cases} M_i & \text{if } \ a_i = 0, \\ A \setminus M_i & \text{for } \ a_i = 1. \end{cases}$$

If $f \in B \subseteq A$ and $\chi(f) = \chi(g)$, then we have

$$[B] = A \iff [\{g\} \cup (B \setminus \{f\})] = A.$$

In other words, it suffices to study the completeness in A up to the equivalence relation $\equiv$.
Further, it is easy to see that

$$\forall B \subseteq A : ([B] = A \iff (\bigvee_{f \in B} \chi(f)) = (1, 1, 1, ..., 1)),$$

where $\bigvee$ is the usual componentwise logical $\vee$ of the tuples ($\in E_2^t$).
A set $B := \{f_1, ..., f_r\} \subseteq A$ is a basis of A iff $[B] = A$ and

$$\forall j \in \{1, ..., r\} : (\bigvee_{f \in B \setminus \{f_j\}} \chi(f)) \neq (1, 1, 1, ..., 1)) \qquad (11.9)$$

Once we know all the characteristic vectors, we can find all complete sets in A and all bases by a direct combinatorial check (which may be done by a simple computer program, provided t is not large (see [Sto 87])). If to $\alpha := (\alpha_1, ..., \alpha_t) \in E_2^t$ we associate the set $I_\alpha := \{i \mid \alpha_i = 1\}$ and if $I_1, ..., I_m$ are the subsets of $\{1, 2, ..., t\}$ corresponding to the characteristic vectors, the completeness problem is reduced to listing irredundant coverings (i.e., no proper subset covers $\{1, 2, ..., t\}$).
The study of classes also provides information on the classes of P_k, which are the intersections of families of A-maximal sets, which is of independent interest.[5]
The characteristic vectors can also be applied to seek the set of classes of functions that make an incomplete set complete.

One finds many results above characteristic vectors and basis enumerations in the books [Mas 88] and [Sto 87]; e.g., classification of $A \in \{P_2, P_3, P_{k,2}, L_p\}$, $p \in \mathbb{P}$, and classifications of the maximal classes of P_3 and of $P_{k,2}$.
Subsequently, the concepts and ideas introduced above are only explained as an example.
Let $A = P_2$. Then, by Theorem 3.2.4.2, we can say

$$M_1 := T_0,\ M_2 := T_1,\ M_3 := S,\ M_4 := L,\ M_5 := M.$$

Then there are exactly 15 characteristic vectors $(a_1, ..., a_5)$ for P_2 ([Jab 52], [Ibu-N-N 63], [Kri 65]):

number of $(a_1, ..., a_5)$	a_1	a_2	a_3	a_4	a_5	example for f with $\chi(f) = (a_1, ..., a_5)$
1	1	1	1	1	1	$\overline{x \wedge y}$
2	1	1	0	1	1	$(x \wedge y) \vee (x \wedge z) \vee (y \wedge z)$
3	0	1	1	1	1	$x \wedge \overline{y}$
4	1	0	1	1	1	$x \vee \overline{y}$
5	1	1	0	0	1	$\overline{x}$
6	1	0	1	0	1	$x + y + 1$
7	0	1	1	0	1	$x + y$
8	0	0	1	1	1	$x \wedge (y \vee \overline{z})$
9	1	0	1	0	0	c_1
10	0	1	1	0	0	c_0
11	0	0	1	1	0	$x \vee y$
12	0	0	0	1	1	$(x \wedge (y \vee \overline{z})) \vee (\overline{x} \wedge y \wedge \overline{z})$
13	0	0	0	1	0	$(x \wedge y) \vee (x \wedge z) \vee (y \wedge z)$
14	0	0	0	0	1	$x + y + z$
15	0	0	0	0	0	e_1^1

[5] E.g., for $A = P_3$ with one exception the least nontrivial intersections are all minimal clones.

With the aid of the above, one can classify the bases of P_2 as follows:

Theorem 11.6.1 *([Ibu-N-N 63], [Kri 65]; without proof)*
Let $B := \{f_1, ..., f_r\}$ *be an arbitrary basis of* P_2. *Then* $r \in \{1, 2, 3, 4\}$ *and for* $\chi(B) := \{\chi(f_1), ..., \chi(f_r)\}$ *is valid:*

$\|B\|$	*the sets of the numbers of the lements of* $\chi(B)$
1	$\{1\}$
2	$\{2, x\}$, *where* $x \in \{3, 4, 6, 7, 8, 9, 10, 11\}$ $\{3, x\}$, *where* $x \in \{4, 5, 6, 9\}$ $\{4, x\}$, *where* $x \in \{5, 7, 10\}$ $\{5, x\}$, *where* $x \in \{8, 11\}$
3	$\{5, x, y\}$, *where* $x \in \{6, 7, 9, 10\}$ *and* $y \in \{12, 13\}$ $\{6, 7, x\}$, *where* $x \in \{8, 11, 12, 13\}$ $\{6, 6, 10\}$ $\{6, 10, x\}$, *where* $x \in \{11, 12, 13\}$ $\{7, 8, 9\}$ $\{7, 9, x\}$, *where* $x \in \{11, 12, 13\}$ $\{9, 10, x\}$, *where* $x \in \{8, 12\}$
4	$\{9, 10, 11, 14\}$, $\{9, 10, 13, 14\}$

In other words: There are exactly 42 aggregates[6] *for* P_2. *1 aggregate has the rank 1, 17 aggregates have the rank 2, 22 aggregates have the rank 3 and 2 aggregates have the rank 4.*

[6] An **aggregate** is the set of all bases having the the same set of characteristic vectors.

12

Subclasses of $P_{k,2}$

In this chapter, we will to deal with subclasses of the class

$$P_{k,2} := \{f \in P_k \mid W(f) \subseteq \{0,1\}\}.$$

When one restricts the domain of a function $f^n \in P_{k,2}$ to the set E_2^n, a homomorphic mapping pr ("projection") from $P_{k,2}$ onto P_2 can be defined.
Since the image prA of a subclass A of $P_{k,2}$ is a subclass of P_2 and the subclasses of P_2 are known, one can hope +to find certain properties of the inverse images ($\subseteq P_{k,2}$) through the known properties of the images ($\subseteq P_2$). This hope confirmed itself in a certain sense (see e.g. Theorem 12.2.5 and Theorem 9.7.6).
On the other hand, $P_{k,2}$, $k \geq 3$ also reflects the negative properties of P_k because the examples of classes from Section 8.1.1 with infinite and without bases are subclasses of $P_{k,2}$.
Further, the functions of $P_{k,2}$ are important since they can be interpreted as predicates.
Some further applications are subsequently mentioned:
Functions of $P_{3,2}$ permit the description of a decision (values 0, 1) with abstention from voting (value 2).
Special functions of $P_{3,2}$ are of interest in the theory of noncorrect algorithms (see e.g. [Sch 78]). In [Eps-F-R 74] G. Epstein, G. Frieder and D. C. Rine mention that functions of $P_{k,2}$ are useful for describing logico-arithmetical branchings in programs where the arithmetical constants (mostly $k > 2$) are arguments and the two logical constants form the range. They also give a function of $P_{3,2}$ which is used in control of real-time processes and in aeronautics.

In 1973/1974 G. Burosch dealt with the set $P_{k,2}$. Some years later he, J. Dassow, W. Harnau and the authoress continued the study of the subclasses of $P_{k,2}$.
The investigations of $P_{k,2}$ concerned the following problems:

(1) For a given subclass B of P_2 determine the set of all subclasses of $P_{k,2}$, which can be projected to B, i.e., the restrictions of the functions to arguments of E_2 gives a function of B.
(2) As completely as possible, construct the lattice of the subclasses of $P_{3,2}$.
(3) For a given subclass of $P_{k,2}$, decide whether it is finitely generated and construct a system of generators (if possible).

A summary of the achieved results on the closed subsets of $P_{k,2}$ was published in [Bur-D-H-L 85]. Unfortunately, a proof could not be given in [Bur-D-H-L 85] for every theorem, since the extent of this proof was too great.
After B. Csákány showed the author how the Theorem of Baker and Pixley can be generalized (see Theorems 8.3.1 and 12.3.1), she found new and shorter proofs.

The subsequently proofs to the theorems and lemmas without reference come mostly from the papers [Lau 88;a;b] (or [Lau 84c]) but also from [Bur-D-H-L 85] and [Lau 77;a;b].
In addition, the results of N. Grünwald from [Grü 83;a;b]which continued the investigations of the team Burosch-Dassow-Harnau-Lau.

The chapter is organized as follows:
Section 12.1 contains the basic concepts and notations. In Section 12.2, one can find results on inverse images of subclasses of P_2 (with respect to the above projection). The remaining sections deal with the determination of the cardinality and with the determination of the elements of the set

$$\mathfrak{N}_k(B) := \{A \in \mathbb{L}_k^{\downarrow}(P_{k,2}) \mid pr\ A = B\}$$

for subclasses B of P_2. In Section 12.3, one can find some structure statements about $\mathbb{L}_k^{\downarrow}(P_{k,2})$. In addition, this section clarifies whether the set $\mathfrak{N}_k(B)$ is finite or infinite or has the cardinality of continuum. In Section 12.4, one can find all subclasses A of $P_{k,l}$ whose projection is the class P_l or $Pol_l\{\alpha\}$ ($\alpha \in E_l$, $2 \leq l < k$). After that, in Section 12.5, the maximal and the submaximal classes of $P_{k,2}$ are determined. Then, the investigations are continued from Section 12.3 for $k = 3$, i.e., for many classes $B \subseteq P_2$ with $|\mathfrak{N}_3(B)| \leq \aleph_0$ the elements of the set $\mathfrak{N}_3(B)$ are determined.

12.1 Notations

In this chapter, let $2 \leq l < k$ and $P_{k,l}$ as defined in Section 1.1.
In general, we define functions of $P_{k,l}$ by formulae over an alphabet $X := \{x, x_1, x_2, ...\}$. In contrast, we define the Boolean functions over the alphabet $Y := \{y, y_1, y_2, ...\}$, and we use the notations for functions and closed sets of P_2 from Chapter 3. By k_1 and j_a, where $a \in E_k$, we denote the unary functions of $P_{k,l}$ given by

$$k_1(x) := \begin{cases} 1 \text{ if} & x \in E_2, \\ 0 \text{ otherwise}, \end{cases} \qquad j_a(x) := \begin{cases} 1 \text{ if} & x = a, \\ 0 \text{ otherwise}, \end{cases}$$

respectively.
To characterize the subclasses of $P_{k,l}$ we need the following homomorphism of $P_{k,l}$ onto P_l, which we denote by pr_l or only with

$$pr$$

and which we call "**projection**" [1]):
For $f^n \in P_{k,l}$ and $g^m \in P_l$ let $pr\ f^n = g^m$ if and only if $n = m$ and $f(\mathbf{a}) = g(\mathbf{a})$ for all $\mathbf{a} \in E_l^n$.
If $B \subseteq P_l$, we call the subset

$$pr^{-1}B := \{f \in P_{k,l} \mid pr\ f \in B\}$$

of $P_{k,l}$ **inverse image of** B.
We say a subclass A of $P_{k,l}$ is B-**projectable** iff $pr_l A = B$. Denote

$$\mathfrak{N}_k(B)$$

the set of all B-projectable subclasses of $P_{k,l}$.
By

$$Pol_{P_{k,l}}\varrho := P_{k,2} \cap Pol_k \varrho$$

one can describe subclasses of $P_{k,l}$. If the index $P_{k,l}$ can be seen from the context, we write only $Pol\ \varrho$ instead of $Pol_{P_{k,l}}\varrho$. .
If $B = Pol_l \varrho\ (\subseteq P_l)$, $\varrho \subseteq E_l^h$ and $\mathbf{a_1}, ..., \mathbf{a_r}$ are some tuples of $E_k^h \backslash E_l^h$, then we also denote the closed set $Pol_{P_{k,l}}(\varrho \cup \{\mathbf{a_1}, ..., \mathbf{a_r})\}$ with

$$B^{\mathbf{a_1},\ldots,\mathbf{a_r}}.$$

In particular, these notations will be used in Section 12.6 for $l = 2$ and $B \in \{T_0, T_1, M\}$.
Furthermore, let

$$Z_{a,b} := Pol_{P_{k,l}} \begin{pmatrix} 0 & 1 & \ldots & l-1 & a \\ 0 & 1 & \ldots & l-1 & b \end{pmatrix} \qquad \{a, b\} \subseteq E_k.$$

12.2 Some Properties of the Inverse Images

With the mapping pr, we can prove that some properties of the subclasses of P_l are transmitted to their inverse images in $P_{k,l}$.

---

[1] To avoid confusion with the notation for functions e_i^n, the functions e_i^n are called **selectors** at some places of this chapter.

Lemma 12.2.1 *(a) Let $A \subseteq P_{k,l}$, $[pr\ A] = P_l$ and $b \in E_k$. Then, the set*

$$A \cup \{j_a \mid a \in E_k \backslash \{b\}\}$$

is a generating system for $P_{k,l}$.

(b) ord $P_{k,l} = 2$.

Proof. (a): Since $pr\ A$ is a generating system for P_l, there are binary functions h and g of $[A]$ with $(pr\ h)(y_1, y_2) = y_1 + y_2\ (mod\ l)$ and $(pr\ g)(y_1, y_2) = y_1 \cdot y_2\ (mod\ l)$. Furthermore, the constant functions $c_0, c_1, ..., c_{l-1}$ and the function

$$j_b(x) = 1 + (l-1) \cdot \sum_{a=0,\ a \neq b}^{k-1} j_a(x)\ (mod\ l)$$

belong to $[A]$. Further, any function f^n of $P_{k,l}$ has a representation of the form

$$f(x_1, ..., x_n) = \sum_{(a_1,...,a_n) \in E_k^n} f(a_1, ..., a_n) \cdot j_{a_1}(x_1) \cdot j_{a_2}(x_2) \cdot ... \cdot j_{a_n}(x_n)\ (mod\ l), \tag{12.1}$$

i.e., f belongs to $[\{g, h, c_0, ..., c_{l-1}, j_0, ..., j_{k-1}\}] \subseteq [A]$.
(b) follows from (a) and Theorem 1.4.2, (b). ■

Lemma 12.2.2 *Let B be a subclass of P_l, which contains the set J_l of all projections (selectors) of P_l. Then, for every set $A \subseteq P_{k,l}$ with $[pr\ A] = pr\ B$, the set $A \cup pr^{-1} J_l$ is a generating system for $pr^{-1}B$.*

Proof. Let $f^n \in pr^{-1}B$. Since $[pr\ A] = B$, there exists a function $f_1^n \in [A]$ with $pr\ f_1 = pr\ f$. In $pr^{-1}J_l$ one can find the $(n+1)$-ary function

$$h(\mathbf{x}) := \begin{cases} x_{n+1} & \text{if} \quad \mathbf{x} \in E_l^{n+1}, \\ f(x_1, ..., x_n) & \text{otherwise.} \end{cases}$$

Consequently, $f(x_1, ..., x_n) = h(x_1, ..., x_n, f_1(x_1, ..., x_n))$ is a superposition over $A \cup pr^{-1} J_l$. ■

By the above lemma, we obtain the following theorem.

Theorem 12.2.3 *If $J_l \subseteq B' \subset B \subseteq P_l$ and B' is a maximal class of the class B then is also $pr^{-1}B'$ maximal in $pr^{-1}B$.* ■

It is easy to see that the above theorem does not hold if B does not contain J_l (see also Theorem 12.2.6).

Lemma 12.2.4 *The order of $pr^{-1}J_l$ is 3.*

Proof. Let

$$i(x) := \begin{cases} x \text{ if} & x \in E_l, \\ 0 \text{ otherwise.} \end{cases}$$

Every n-ary function f of $pr^{-1}J_l$ with $pr\ f = e_i^n$ can be represented in the following manner:

$$\begin{aligned} f(x_1, ..., x_n) &= i(x_i) \cdot k_1(x_1) \cdot ... \cdot k_1(x_n) + \\ &\textstyle\sum_{\mathbf{a} \in E_k^n \setminus E_l^n} f(a_1, ..., a_n) \cdot j_{a_1}(x_1) \cdot ... \cdot j_{a_n}(x_n)\ (mod\ l). \end{aligned} \tag{12.2}$$

We prove that (12.2) is a superposition over

$$\begin{aligned} &i(x_1) \cdot k_1(x_2)\ (mod\ l), \\ &i(x_1) + a \cdot j_q(x_2)\ (mod\ l), \\ &i(x_1) + i(x_2) \cdot j_q(x_3)\ (mod\ l), \\ &i(x_3) + i(x_1) \cdot j_p(x_2) \cdot j_q(x_3)\ (mod\ l), \\ &(p, q \in E_k;\ q > 1,\ a \in E_l). \end{aligned} \tag{12.3}$$

First, we construct the function

$$i(x) + a \cdot j_{a_1}(x_1) \cdot j_{a_2}(x_2) \cdot ... \cdot j_{a_n}(x_n)\ (mod\ l) \tag{12.4}$$

for $(a_1, ..., a_n) \in E_k^n \setminus E_l^n$, $a \in E_l$.
W.l.o.g. let $a_1 \in E_k \setminus E_l$. The function $i(x)+(i(x_1)+i(x_3))j_{a_2}(x_2)j_{a_1}(x_1)j_{a_1}(x_1)$ $= i(x)+i(x_3)j_{a_1}(x_1)j_{a_2}(x_2)\ (mod\ l)$ is obtained from $i(x)+i(x_2)j_{a_1}(x_1)$ if we substitute the variable x_2 by $i(x_1)+i(x_3)j_{a_2}(x_2)j_{a_1}(x_1)$. By substitution of x_3 by $i(x_1) + i(x_4)j_{a_3}(x_3)j_{a_1}(x_1)$ we generate the function $i(x) + i(x_4)j_{a_1}(x_1)j_{a_2}(x_2)j_{a_3}(x_3)\ (mod\ l)$. By iterated application of these constructions we obtain the function

$$i(x) + i(x_{n+1})j_{a_1}(x_1)j_{a_2}(x_2)...j_{a_n}(x_n)\ (mod\ l). \tag{12.5}$$

The function (12.4) is generated from (12.5) by substituting x_{n+1} by $i(x_1) + a \cdot j_{a_1}(x_1)$.
Obviously,

$$i(x) + \sum_{\mathbf{a} \in E_k^n \setminus E_l^n} f(a_1, ..., a_n)j_{a_1}(x_1)j_{a_2}(x_2)...j_{a_n}(x_n)\ (mod\ l) \tag{12.6}$$

is a superposition of functions of type (12.4).
If we substitute x in (12.6) by

$$i(x_i) \cdot k_1(x_1) \cdot ... \cdot k_1(x_n) \in [i(x) \cdot k_1(x_1)]$$

we obtain the function (12.2); i.e., (12.3) is a generating system for $pr^{-1}J_l$ and $ord\ pr^{-1}J_l \leq 3$. Since the functions of $(pr^{-1}J_l)^2$ preserve the relation

$\lambda_l \cup \{(2,2,2,2)\}$ (see Lemma 4.1), the function $i(x_1) + j_1(x_1)j_1(x_2)j_2(x_3) \in pr^{-1}J_l$ does not preserve this relation, however, we have $ord\, pr^{-1}J_l = 3$. ∎

Theorem 12.2.5 *Let $B \subseteq P_l$ be a clone. Then*

$$ord\, B \leq ord\, pr^{-1}B \leq max(3, ord\, B).$$

Every inverse image ($\subseteq P_{k,2}$) of a clone $B \subseteq P_2$ is finitely generated.

Proof. The statements of the theorem are consequences of Theorems 12.3.2 and 12.3.4 and of Chapter 3. ∎

The following theorem gives information on the order of the remaining inverse images ($\subseteq P_{k,2}$) of subclasses ($\subseteq P_2$).

Theorem 12.2.6 *The inverse images $pr^{-1}C_0$, $pr^{-1}C_1$ and $pr^{-1}C$ have no finite basis.*

Proof. Let A be an inverse image of $P_{k,2}$ whose projection is generated by $\{c_a, c_b\}$, $\{a,b\} \subseteq E_2$.
Then A contains the functions

$$f^n(x_1,...,x_n) := \begin{cases} \alpha \text{ if} & \exists i, \alpha : x_i = 2 \wedge \\ & x_1 = ... = x_{i-1} = x_{i+1} = ... = x_n = \alpha \in E_2, \\ a \text{ otherwise} & \end{cases}$$

for all $n \in \mathbb{N}$.
A has no finite basis if we can prove that, for each n, the function f^n is not a superposition of the $(n-1)$-ary functions of A. It suffices to prove that f^n has no representation by a formula

$$h(x_1,...,x_n) := g_0(g_1(x_1,...,x_n),...,g_{n-1}(x_1,...,x_n)), \tag{12.7}$$

where g_i $(0 \leq i \leq n)$ is a function of A or g_i $(1 \leq i \leq n)$ is defined by

$$g_i(x_1,...,x_n) = x_j \tag{12.8}$$

for some $j \in \{1,2,...,n\}$. W.l.o.g. we can assume that g_1 is not a function of form (12.8). Then, for $\alpha \in E_2$

$$\beta_\alpha := (g_1(2,\alpha,...,\alpha),...,g_{n-1}(2,\alpha,...,\alpha))$$

is a tuple of E_2^n and thus $g_0(\beta_0) = g_0(\beta_1)$. However by definition $f^n(2,\alpha,...,\alpha) = \alpha$ for every $\alpha \in E_2$. Therefore, formula (12.7) does not define the function f^n. ∎

Theorem 12.2.7

(a) Let $A \subseteq P_{k,l}$ be an inverse image whose projection $pr_l A$ is a clone. Then A has a basis with exactly r elements if and only if $pr\ A$ has such a basis.

(b) $\exists f \in P_{k,l} :\ [f] = P_{k,l}$.

Proof. (a): If $[\{f_1, ..., f_r\}] = A$ then $[\{pr\ f_1, pr\ f_2, ..., pr\ f_r\}] = pr\ A$ holds. Let $pr\ A = [\{g_1, ..., g_r\}]$, where g_1 denotes a function that stands in the construction formula of e_1^1 "outside"; i.e., we have $e_1^1(x) = g_1(t_1(x), ..., t_n(x))$ for certain $t_1, ..., t_n$ of $[\{g_1, ..., g_r\}]$. Further, denote $f_1, ..., f_r$ certain functions of A with $pr\ f_i = g_i$, $i = 1, ..., r$ and let

$$\begin{aligned} & f(x_1, ..., x_n, x, ..., x_{p,q}, ..., x'_{p,q}, ..., x''_{p,q}, ..., x'''_{p,q}, ..., x^{IV}_{p,q}, ...) \\ & := f_1(x_1, ..., x_n) \cdot k_1(x) + \sum_{\substack{p,q \\ q \geq l \\ 0 \leq p < q \leq k-1}} i(x_{p,q}) \cdot j_p(x'_{p,q}) \cdot j_q(x''_{p,q}) \\ & \quad + \sum_{\substack{p,q \\ q \geq l \\ 0 \leq p < q \leq k-1}} j_p(x'''_{p,q}) \cdot j_q(x^{IV}_{p,q}) \ (mod\ l). \end{aligned}$$

We prove that $\{f, f_2, ..., f_r\}$ is a generating system for A. By Lemmas 12.2.2 and 12.2.4, it is sufficient to show that the function system (12.3) belongs to $[\{f, f_2, ..., f_r\}]$.
Due to the choice of the function f_1 and by $e_1^1 \in pr_l A$, the function

$$i'(x_1)k_1(x) + \sum_{\substack{p,q \\ q \geq l \\ 0 \leq p < q \leq k-1}} (i(x_{p,q})j_p(x'_{p,q})j_q(x''_{p,q}) + j_p(x'''_{p,q})j_q(x^{IV}_{p,q})) \tag{12.9}$$

with $pr\ i' = pr\ i$ is a superposition over the functions $f, f_2, ..., f_n$. If one identifies all variables $\neq x_1$ and $\neq x$ in (12.9) with x, one receives the function $i'(x_1)k_1(x)$. Therefore, the functions $i(x) = i'(x)k_1(x)$ and $i(x_1)k_1(x_2) = i'(i(x_1))k_1(x_2)$ are also superpositions over $f, f_2, ..., f_n$. By identifying certain variables and by substituting x_1 by $i(x)$ in (12.9), we obtain the functions $i(x) + i(x_1)j_p(x_2)j_q(x_3)$ and $i(x) + j_p(x_1)j_q(x_2)$, where $q \geq l$ and $0 \leq p < q \leq k-1$. This implies that the functions

$$i(x_3) + i(x_1)j_p(x_2)j_q(x_3),$$
$$i(x_1) + j_0(i(x_2))j_q(x_2) = i(x_1) + j_q(x_2) =: g(x_1, x_2),$$
$$\underbrace{g(...g(g(g}_{a \text{ times}}(x_1, x_2), x_2), x_2)..., x_2) = i(x_1) + a \cdot j_q(x_2) \quad \text{and}$$
$$i(x_1) + i(x_2)j_0(i(x_3))j_q(x_3) = i(x_1) + i(x_2)j_q(x_3)$$

for $q \geq l$, $0 \leq p < q \leq k-1$ and $a \in E_l$ are superpositions over $\{f, f_2, ..., f_r\}$. Hence all functions of the system (12.3) belong to $[\{f, f_2, ..., f_r\}]$, and our theorem was proven.
(b) follows from (a) with the aid of Theorem 7.1.6. ■

12.3 On the Number of the B-projectable Subclasses of $P_{k,2}$, $B \subseteq P_2$

In this section, d_m denotes an $(m+1)$-ary function with the property

$$d_m(x_2, x_1, ..., x_1) = d_m(x_1, x_2, x_1, ..., x_1) = ... = d_m(x_1, ..., x_1, x_2) = x_1,$$

where $m \geq 2$. We call such a function a **near unanimity function**. Examples of P_2 are the functions h_m with $m \geq 2$ (see Chapter 3).

To generalize Theorem 8.3.1, we get:

Theorem 12.3.1 *Let A be a subclass of $P_{k,l}$ whose projection $pr_l A$ contains a "near unanimity function" d_m for a certain $m \geq 2$. Then $A = Pol_{P_{k,l}} Inv^m A$, i.e., for a fixed m and for a fixed k there exist only finite many subclasses $A \subseteq P_{k,l}$ with $d_m \in pr_l A$.*

Proof.[2] Obviously,

$$A \subseteq Pol(Inv^m A).$$

Let $f^n \in Pol(Inv^m\ A)$ be arbitrary. Further, denote f_T for $T \subseteq E_k^n$ a function of $P_{k,l}$ which is identical to f on tuples of T. We prove through induction on $|T| =: t \geq m$ that there is a certain function f_T in A for every $T \subseteq E_k^n$. $f \in A$ for $t = k^m$, with which $A = Pol(Inv^m A)$ would be proven.
I) $t = m$: Let $T = \{(a_{i1}, a_{i2}, ..., a_{in}) \mid i = 1, 2, ..., m\}$,

$$\varrho := \begin{pmatrix} a_{11} & a_{12} & ... & a_{1n} \\ a_{21} & a_{22} & ... & a_{2n} \\ \multicolumn{4}{c}{\dotfill} \\ a_{m1} & a_{m2} & ... & a_{mn} \end{pmatrix}$$

and let ϱ' be the smallest relation of $Inv^m A$ with $\varrho \subseteq \varrho'$, i.e., we have

[2] The proof goes back to a manuscript by B. Csákány.

$$\begin{pmatrix} a_1 \\ a_2 \\ \dots \\ a_m \end{pmatrix} \in \varrho' \backslash \varrho \iff \exists q \in A^n : q \begin{pmatrix} a_{11} & a_{12} & \dots & a_{1n} \\ a_{21} & a_{22} & \dots & a_{2n} \\ \dots & \dots & \dots & \dots \\ a_{m1} & a_{m2} & \dots & a_{mn} \end{pmatrix} = \begin{pmatrix} a_1 \\ a_2 \\ \dots \\ a_m \end{pmatrix}. \tag{12.10}$$

Since f^n preserves the m-ary invariants of A, $f(\varrho) \in \varrho'$ holds. If $f(\varrho)$ belongs to ϱ, a certain function of $A \cap pr^{-1}[e_1^1]$ agrees with the function f on tuples of T. On the other hand, if $f(\varrho) \in \varrho' \backslash \varrho$, then there is also a function f_T in A.Consequently, our assertion is proven for $t = m$.

II) $t \longrightarrow t+1$: Assume for every $T \subseteq E_k^n$ with $|T| = t$, $t \geq m$, one can find a function $f_T \in A$ with $f_T(\mathbf{a}) = f(\mathbf{a})$ for all $\mathbf{a} \in T$. Now, let $T = \{\mathbf{a_1}, \mathbf{a_2}, ..., \mathbf{a_{t+1}}\}$ and $|T| = t+1$. Because of our assumption, there are functions $f_i \in A^n$ with $f_i(\mathbf{a}) = f(\mathbf{a})$ for all $\mathbf{a} \in T \backslash \{\mathbf{a_i}\}$, $i = 1, 2, ..., m+1$. Since there is also a function $d'_m \in A$ with $pr_l d'_m = d_m$, the function $d'_m(f_1(\mathbf{x}), f_2(\mathbf{x}), ..., f_{m+1}(\mathbf{x}))$ belongs to A. It is obvious that this function agrees with f on all tuples of T. ■

As a direct consequence of the above proven theorem and Chapter 3, we get:

Theorem 12.3.2 *Let A be a clone of P_2 with $S \cap M \subseteq A$ or $T_{a,m} \cap T_{\overline{a}} \cap M \subseteq A$ for certain $m \in \mathbb{N} \setminus \{1\}$ and $a \in E_2$. Then there are only finite many A-projectable subclasses of $P_{k,2}$.* ■

One can find the exact description of the A-projectable classes from Theorem 12.3.2 in Sections 12.6 and 12.7. To prepare for the proofs in these sections, we derive some properties of the subclasses of $P_{k,2}$, whose projections contain a certain function h_m.

Lemma 12.3.3 *Let $A = \bigcap_{i=1}^m Pol_{P_{k,l}} \varrho_i$ and $\varrho_i \subseteq E_k^{h_i}$, $i = 1, 2, ..., m$. Then the following statements are equivalent:*

(1) $pr_l A$ contains the selectors of P_l.
(2) There exists a mapping t from E_k onto E_l with the fixed points $0, 1, ..., l-1$ and the property

$$\begin{aligned} t(\varrho_i) &:= \{(t(a_1), t(a_2), ..., t(a_{h_t})) \mid (a_1, ..., a_{h_i}) \in \varrho_i\} \\ &= \varrho_i \cap E_l^{h_i} \text{ for every } i \in \{1, 2, ..., m\}. \end{aligned}$$

(3) $pr_l A = \bigcap_{i=1}^m Pol_l(\varrho_i \cap E_l^{h_i})$.

Proof. Obviously, if $pr_l A$ contains the selectors, there is an e_1^1-projectable function $t \in A$ ($e_1^1 \in P_l$) that has the properties given in (2). Thus (1) implies (2). The implication "(3) $\Longrightarrow$ (1)" is trivial. It remains to be proven "(2) $\Longrightarrow$ (3)": Obviously, the relations $\varrho_i \cap E_l^{h_i}$ ($i = 1, 2, ..., m$) can be derived from the relations $\varrho_1, ..., \varrho_m$ and the invariants of $P_{k,l}$. This implies $A \subseteq \bigcap_{i=1}^m Pol_{P_{k,l}}(\varrho_i \cap E_l^{h_i})$. Thus, $pr_l A \subseteq \bigcap_{i=1}^m Pol_l(\varrho_i \cap E_l^{h_i})$.

Now let $g^n \in \bigcap_{i=1}^{m}(Pol_l \varrho_i \cap E_l^{h_i})$ be arbitrary. The function $f(x_1, ..., x_n) := g(t(x_1), ..., t(x_n))$, where t fulfills the conditions from (2), preserves all relations $\varrho_1, ..., \varrho_m$ and has the projection g. Consequently, $\bigcap_{i=1}^{m} Pol_l(\varrho_i \cap E_l^{h_i}) \subseteq pr_l A$. Hence "(2) $\Longrightarrow$ (3)" is proven. ■

Theorem 12.3.4 *Let A be a subclass of $P_{k,2}$, whose projection $pr_2 A$ ($\subseteq P_2$) is an intersection of at least two classes that are incomparable (with respect to inclusion) and for which there exists a certain $m \in \mathbb{N} \setminus \{1\}$ with $h_m \in pr_2 A$. Then, A can be described as an intersection of at least two incomparable classes of $P_{k,2}$ whose projections are different from $pr\ A$.*

Proof. Let $r \geq 2$ be the smallest number for which the function h_r belongs to $pr\ A$. By Theorem 12.3.1, there are certain r-ary relations $\varrho_1, ..., \varrho_m$ with $A = \bigcap_{i=1}^{m} Pol_{P_{k,2}} \varrho_i$. For proof of the theorem, it suffices that $pr\ A \subset Pol_{P_{k,2}} \varrho_i$ is correct for every $i \in \{1, 2, ..., m\}$. Obviously, A contains an e_1^1-projectable function. Consequently, by Lemma 12.3.3, $pr\ Pol\ \varrho_i = Pol_2(\varrho_i \cap E_2^r)$. Further, one receives by scrutinizing the possibilities for $pr\ A$, where $h_r \in pr\ A$, and considering the results (about the smallest arities of relations ϱ with $pr\ A = Pol\ \varrho$) from Section 11.2 that $pr\ A$ and $Pol_2(\varrho_i \cap E_2^r)$ are different. Hence, it holds $pr\ A \subset Pol \varrho_i$, $i = 1, 2, ..., m$. ■

The following **procedure for the determination of all subclasses $A \subseteq P_{k,2}$ with $M \cap S \subseteq pr_2 A$ or $T_{a,m} \cap T_{\overline{a}} \cap M \subseteq pr_2 A$ ($a \in E_2$, $m \in \mathbb{N}\backslash\{1\}$)** results from Theorems 3.1, 12.3.1, and 12.3.4 :

(1) Determination of all A-projectable subclasses of $P_{k,2}$, where A is P_2, T_0, T_1, M, S or $T_{a,m}$ ($a \in E_2$, $m \in \mathbb{N}\backslash\{1\}$).
(2) Formation of "permissible" intersections that is, only formation of such intersections whose projections contain a function h_m.

In the first step of this procedure, because of Theorem 12.3.1, one has to find all h_i-ary relations ϱ_i, $i = 1, 2, ..., r$, with $pr \bigcap_{i=1}^{r} Pol\ \varrho_i \supseteq B$, $B \in \{T_0, T_1, M, S, T_{0,m}, T_{1,m}\}$. The following Lemma 12.3.5 and Theorem 12.3.6 give some helpful considerations for this purpose.

Lemma 12.3.5 *Let B be a clone of P_l and let ϱ_i be an h_i-ary relation ($\subseteq E_k^{h_i}$) for all $i \in \{1, ..., r\}$. Then $B \subseteq pr \bigcap_{i=1}^{r} Pol_{P_{k,l}} \varrho_i$ if and only if there is a mapping t from E_k onto E_l with the fixed points $0, 1, ..., l-1$ and with the properties $t(\varrho_i) = \varrho_i \cap E_l^{h_i}$ and $\varrho_i \cap E_l^{h_i} \in Inv\ B$ for all $i \in \{1, ..., r\}$.*

Proof. If $B \subseteq pr \bigcap_{i=1}^{r} Pol\ \varrho_i$, then there is an e_1^1-projectable function $t \in \bigcap_{i=1}^{r} Pol\ \varrho_i$ with the properties $t(a) = a$ for all $a \in E_l$, $t(\varrho_i) = \varrho_i \cap E_l^{h_i}$ and $\varrho_i \cap E_l^{h_i} \in Inv\ B$ for all $i \in \{1, 2, ..., r\}$. Conversely, if a mapping t exists with these properties, then by Lemma 12.3.3, we have $pr \bigcap_{i=1}^{r} Pol\ \varrho_i = \bigcap_{i=1}^{r} Pol\ (\varrho_i \cap E_l^{h_i})$. Thus, $B \subseteq \bigcap_{i=1}^{r} Pol\ (\varrho_i \cap E_l^{h_i}) = pr \bigcap_{i=1}^{r} Pol\ \varrho_i$ holds because of $\varrho_i \cap E_l^{h_i} \in Inv\ B$, $i = 1, ..., r$. ■

Theorem 12.3.6 *Let $\varrho_1, ..., \varrho_r$ be subsets of $E_k^2 \backslash E_2^2$, $\varrho \subseteq E_2^2$, $\varrho \neq \emptyset$ and Pol $\varrho \in \{P_2, T_0, T_1, S, M, T_{0,2}, T_{1,2}\}$, i.e., w.l.o.g. let*

$$\varrho \in \left\{ \begin{pmatrix} 0\ 0\ 1\ 1 \\ 0\ 1\ 0\ 1 \end{pmatrix}, \begin{pmatrix} 0\ 1 \\ 0\ 1 \end{pmatrix}, \begin{pmatrix} 0\ 0 \\ 0\ 1 \end{pmatrix}, \begin{pmatrix} 1\ 0 \\ 1\ 1 \end{pmatrix}, \begin{pmatrix} 0 \\ 0 \end{pmatrix}, \begin{pmatrix} 1 \\ 1 \end{pmatrix}, \begin{pmatrix} 0\ 1 \\ 1\ 0 \end{pmatrix}, \begin{pmatrix} 0\ 0\ 1 \\ 0\ 1\ 1 \end{pmatrix}, \begin{pmatrix} 0\ 1\ 0 \\ 0\ 0\ 1 \end{pmatrix}, \begin{pmatrix} 1\ 0\ 1 \\ 1\ 1\ 0 \end{pmatrix} \right\}.$$

Then $pr(\bigcap_{i=1}^r Pol\ (\varrho \cup \varrho_i)) = Pol_2 \varrho$ iff ϱ fulfills one of the following conditions:

(1) $\varrho \in \left\{ \begin{pmatrix} 0\ 0\ 1\ 1 \\ 0\ 1\ 0\ 1 \end{pmatrix}, \begin{pmatrix} 0\ 1\ 0 \\ 0\ 0\ 1 \end{pmatrix} \begin{pmatrix} 1\ 0\ 1 \\ 1\ 1\ 0 \end{pmatrix} \right\}$;

(2) $\varrho = \begin{pmatrix} 0\ 0 \\ 1\ 1 \end{pmatrix}$ *and there are no elements $a_1, ..., a_q, b_1, .., b_q$ $(q \geq 2)$ with $0 \in \{a_1, b_1\}$, $1 \in \{a_q, b_q\}$, $\{a_i, b_i\} \cap \{a_{i+1}, b_{i+1}\} \neq \emptyset$ for all $i \in \{1, 2, ..., q-1\}$ and*
$\begin{pmatrix} a_1 & a_2 & ... & a_q \\ b_1 & b_2 & ... & b_q \end{pmatrix} \subseteq \varrho_1 \cup \varrho_2 \cup ... \cup \varrho_r$;

(3) $\varrho = \begin{pmatrix} a \\ a \end{pmatrix}$, *$a \in E_2$ and $\varrho_1 \cup \varrho_2 \cup ... \cup \varrho_r \subseteq (E_k \backslash \{\overline{a}\})^2$;*

(4) $\varrho = \begin{pmatrix} a & a \\ a & \overline{a} \end{pmatrix}$, *$a \in E_2$ and $\varrho_1 \cup \varrho_2 \cup ... \cup \varrho_r \subseteq E_k^2 \backslash \begin{pmatrix} \overline{a} & \overline{a} & ... & \overline{a} \\ 2 & 3 & ... & k-1 \end{pmatrix}$;*

(5) $\varrho = \begin{pmatrix} 0\ 1 \\ 1\ 0 \end{pmatrix}$, *$(\varrho_1 \cup \varrho_2 \cup ... \cup \varrho_r) \cap \{(a, a) \mid a \in E_k\} = \emptyset$ and there are no elements $a_1, ..., a_q, b_1, ..., b_q$, $q \geq 2$, with the properties*

$\begin{pmatrix} a_1 & ... & a_q \\ b_1 & ... & b_q \end{pmatrix} \subseteq \varrho_1 \cup \varrho_2 \cup ... \cup \varrho_r \quad \wedge$
$(\forall i \in \{1, ..., q-1\} : \ \{a_i, b_i\} \neq \{a_{i+1}, b_{i+1}\} \ \wedge \ \{a_i, b_i\} \cap \{a_{i+1}, b_{i+1}\} \neq \emptyset)$
$\wedge\ ((0 \in \{a_1, b_1\}\ \wedge 1 \in \{a_q, b_q\}\ \wedge$ *q is even* $\vee\ (\{a_1, b_1\} \cap \{a_q, b_q\} \neq \emptyset$
$\wedge$ *q is odd*));

(6) $\varrho = \begin{pmatrix} 0\ 0\ 1 \\ 0\ 1\ 1 \end{pmatrix}$ *and there are no elements $a_1, ..., a_q$ of $E_k \backslash E_2$ with*

$$\begin{pmatrix} 1 & a_1 & a_2 & ... & a_{q-2} & a_{q-1} & a_q \\ a_1 & a_2 & a_3 & ... & a_{q-1} & a_q & 0 \end{pmatrix} \subseteq \varrho_1 \cup ... \cup \varrho_r.$$

Proof. One easily proves the above statements with the aid of Lemma 12.3.5. ∎

Subsequently, we deal with subclasses of $P_{k,2}$, whose projection classes comprise linear functions.

Theorem 12.3.7 *Let $L \cap T_0 \cap S \subseteq B \subseteq L$. Then there exists at least countable-infinite-many B-projectable subclasses of $P_{k,2}$ for every $k \geq 3$, i.e., $|\mathfrak{N}_k(B)| \geq \aleph_0$. Further, for every B-projectable class A of $P_{k,2}$ and for every $n \geq 1$, A^n has a generating system that consists of at most $|A^1| + 2$ functions of A.*

Proof. $|\mathfrak{N}_k(B)| \geq \aleph_0$ for $L \cap T_0 \cap S \subseteq B \subseteq L$ follows from the fact that the sets

$$L_{B,r} := \bigcup_{n\geq 1}\{f^n \in P_{k,2} \mid (\exists a_i, a_I : f(\mathbf{x}) = a_0 + \sum_{i=1}^{n} a_i \cdot j_1(x_i) + \sum_{\substack{I \subseteq \{1,...,n\} \\ |I| \leq r}} a_I \cdot \bigwedge_{i\in I} j_2(x_i)) \wedge pr\ f \in B\}$$

for $r = 1, 2, ...$ are closed and $L_{B,r} \neq L_{B,r'}$ holds for $r \neq r'$.
Now, let $A \subseteq P_{k,2}$ and $L \cap T_0 \cap S \subseteq pr\ A$. Then there is a function $g \in A$ with $(pr\ g)(y_1, y_2, y_3) = y_1 + y_2 + y_3$. Thus the function $r(x_1, x_2, x_3) := i(x_1) + i(x_2) + i(x_3)$ with $i(x) := g(x, x, x)$ and $pr\ i = e_1^1$ belongs to A.
We will show that for every $n \geq 1$ that A^n has a generating system of at most $|A^1| + 2$ functions of A.
Let $A^n = \{f_1, f_2, ..., f_t\}$. Then, by $r \in A$, the functions

$$q(x_1, x_2, ..., x_{n\cdot t}, x) := \sum_{j=1}^{t} f_t(x_{(j-1)\cdot n+1}, ..., x_{j\cdot n}) + a \cdot i(x)$$

and

$$p_s(x) := \sum_{\substack{j=1 \\ j\neq s}}^{t} f_j(x, x, ..., x) + (a+1) \cdot i(x),$$

belong to A, where

$$a = \begin{cases} 0, & \text{if } t \text{ odd}, \\ 1, & \text{if } t \text{ even}. \end{cases}$$

Then we have

$$f_s(x_1, ..., x_n) =$$

$$r(q(x_1, ..., x_1, \underbrace{x_1}_{\substack{((s-1)\cdot n+1)\text{-th} \\ \text{place}}}, x_2, ..., \underbrace{x_n}_{\substack{((s\cdot n)\text{-th} \\ \text{place}}}, x_1, ..., x_1, p_s(x_1), x_1)$$

for $s = 1, 2, ..., t$. Hence, every function of A^n is a superposition over $A^1 \cup \{r, q\}$. ∎

Theorem 12.3.7 will be specified for $k = 3$ through Theorem 12.6.9.
The following theorem states something about the cardinality $\mathfrak{N}_k(B)$ of the remaining classes $B \subseteq L$.

Theorem 12.3.8 *Let $\emptyset \neq B \subseteq [P_2^1]$ and $k \geq 3$. Then there are continuum-many B-projectable subclasses of $P_{k,2}$.*

Proof. Denote N a subset of $\mathbb{N}\backslash\{1\}$. For every N, one can define a certain $[P_2^1]$-projectable subset Q_N of $P_{k,2}$, as follows inductively:
Let $Q_N^1 = P_{k,2}^1$ and Q_N^n $(n \geq 2)$ be the set of all functions $f^n \in P_{k,2}$, which fulfill the following conditions:

(1) $f \in [Q_N^{n-1}]_{\zeta,\tau,\nabla,\Delta}$.
(2) There exists a nonempty subset I of $\{1, 2, ..., n\}$, an $a \in E_2$, an n-ary function $f_1 \in [Q_N^{n-1}]_{\zeta,\tau,\nabla,\Delta}$ which depends on at least $|I|$ variables fictitiously, and a function $t \in P_{k,2}^n$ with

$$f(\mathbf{x}) = (f_1(\mathbf{x}) \cdot \bigwedge_{i \in I} k_1(x_i) \ \vee \ t(\mathbf{x}) \cdot \bigwedge_{i \in I} j_2(x_i))^a,$$

where

$$g^a := \begin{cases} g & \text{if } a = 1, \\ \overline{g} & \text{if } a = 0. \end{cases}$$

(3) If $n \in N$, there are an $h \in Q_N^1$, an i of $\{1, ..., n\}$ and an $a \in E_2$ with

$$f(\mathbf{x}) = (h(x_i) \cdot \bigwedge_{q=1}^{n} k_1(x_q) \vee \bigvee_{q=1}^{n} j_2(x_1)...j_2(x_{q-1}) j_1(x_q) j_2(x_{q+1})...f_2(x_n))^a.$$

By (1), the set Q_N is a closed set with respect to the unary operations $\zeta, \tau, \nabla, \Delta$. One can show that the set Q_N is also closed with respect to the operation $\star$ as follows:
Let f^n and g^m be arbitrary functions of Q_N. Through induction on $r := n+m$, $r \geq 2$, we will show that $f \star g \in Q_N$.
If $r = 2$ $(n = m = 1)$ then $f \star g \in Q_N$ is obvious.
Suppose for all $n + m \leq r - 1$, $f^n \star g^m \in Q_N$ holds for arbitrary f and g of Q_N. Now let $n + m = r$. $f \star g \in Q_N$ is surely valid for $n = 1$. If $n \geq 2$, then, by (1), it is sufficient to study the following cases:
Case 1: f fulfills (2).
In this case, the function $f \star g$ is not essentially dependent from all their variables or it also fulfills the condition (2). Consequently, $f \star g$ belongs to Q_N.
Case 2: f fulfills (3) and w.l.o.g. let be $a = 1$.
In this case, we have for $i = 1$:

$$\begin{aligned} (f \star g)(\mathbf{x}) &= (h \star g)(x_1, ..., x_m) k_1(x_{m+1})...k_1(x_{m+n-1}) \vee \\ &\quad g(x_1, ..., x_m) j_2(x_{m+1})...j_2(x_{m+n-1}). \end{aligned}$$

For $i \neq 1$ it holds:

$$(f \star g)(\mathbf{x}) = h(x_{i+m-1}) k_1(x_{m+1})...k_1(x_{m+n-1}) \ \vee g(\mathbf{x}) j_2(x_{m+1})...j_2(x_{m+n-1}).$$

Hence, the function $f \star g$ has the property (2) both for $i = 1$ and for $i \neq 1$. Consequently, Q_N is closed.

It follows from the definition of the sets of type Q_N and from the above proof (for $Q_N = [Q_N]$) that the sets Q_N, $Q_{N'}$ are pairwise different for $N \neq N'$. Therefore, there are continuum-many $[P_2^1]$-projectable subclasses of $P_{k,2}$.

The following facts are easy to check:
For every B with $\emptyset \neq B \subseteq [P_2^1]$ and for arbitrary $N \subseteq \mathbb{N}\backslash\{1\}$, the set $Q_N \cap pr^{-1}B$ is a B-projectable subclass of $P_{k,2}$ and it holds $Q_N \cap pr^{-1}B \neq Q_{N'} \cap pr^{-1}B$ for all $N \neq N'$ with $N, N' \subseteq \mathbb{N}\backslash\{1\}$.

This implies Theorem 12.3.8. ∎

One can deal with the infinite-digit relations defined in the following lemma as with the finite-digit relations. Further, analogously to Chapter 2, one can introduce the same concepts as in Chapter 2 for the infinite-digit relations.

Lemma 12.3.9 *([Grü 84]) Let be*

$$E_2^\infty\backslash\{\mathbf{1}\} := \{(a_1, a_2, a_3, ...) \mid \forall i \in \mathbb{N}:\ a_i \in \{0,1\}\ \wedge\ \exists j \in \mathbb{N}:\ a_j = 0\},$$

$$\varrho_t := \{(a_1, a_2, ..., a_{t+1}, 2, 2, 2, ...) \mid \exists i \in \{1, 2, ..., t+1\} : (a_i = 2\ \wedge\ \forall j \in \{1, 2, ..., t+1\}\backslash\{i\} :\ a_j = 0)\},$$

$(t \in \mathbb{N})$,

$f_q^{(q+1)}$*, $q \in \mathbb{N}$, defined by*

$$f_q(\mathbf{x}) := \begin{cases} 1 & \text{if } \mathbf{x} \in \{(2,0,...,0),(0,2,0,...,0),...,(0,0,...,0,2)\}, \\ 0 & \text{otherwise;} \end{cases}$$

$$A_t := Pol_{P_{k,2}}(E_2^\infty\backslash\{\mathbf{1}\}) \cup \varrho_t,\quad (t \in \mathbb{N})$$

and

$$A_T := \bigcap_{i \in T} A_i \quad \text{if } T \subseteq \mathbb{N}.$$

Then

(a) $\forall i \in \mathbb{N}:\ f_i \notin A_i$;

(b) $\forall i \in \mathbb{N}\ \forall j \in \mathbb{N}\backslash\{i\}:\ f_j \in A_i$;

(c) $\forall T, T' \subseteq \mathbb{N}:\ T \neq T' \implies A_T \neq A_{T'}$.

Proof. (a) follows directly from the definition of the functions f_i and (c) is a conclusion from (b). Therefore, we only have to prove (b).
Suppose for certain $i, j \in \mathbb{N}$ with $i \neq j$ it holds $f_j \in A_i$. Then there are some columns $\mathbf{s_1}$, $\mathbf{s_2}$, ..., $\mathbf{s_{j+1}} \in E_2^\infty\backslash\{\mathbf{1}\} \cup \varrho_i$ with

$$f_j(\mathbf{s_1}, \mathbf{s_2}, ..., \mathbf{s_{j+1}}) = \mathbf{1}. \tag{12.11}$$

Obviously, this is only valid if $\{\mathbf{s_1}, ..., \mathbf{s_{j+1}}\} \not\subseteq E_2^\infty\backslash\{\mathbf{1}\}$. The case $\varrho_i \not\subseteq \{\mathbf{s_1}, ..., \mathbf{s_{j+1}}\}$ is not possible, since in this case, the matrix $(\mathbf{s_1}, ..., \mathbf{s_{j+1}})$ would have a row with at least a zero and no two, what (12.11) contradicts. Therefore we can presuppose $\varrho_i \subseteq \{\mathbf{s_1}, ..., \mathbf{s_{j+1}}\}$. Then, because of $i \neq j$ we have either

$\{\mathbf{s_1}, ..., \mathbf{s_{j+1}}\} \cap (E_2^\infty \backslash \{\mathbf{1}\}) \neq \emptyset$ or $\mathbf{s_q} = \mathbf{s_r}$ for a certain $q \neq r$. In the first case, one finds in $(\mathbf{s_1}..., \mathbf{s_{j+1}})$ (infinite many) rows with the element 2 in at least two places and with an element of E_2. In the second case, there is a row with the element 2 in at least two places and with an element 0. In every one of these cases, a contradiction to (12.11) results with that from the definition of the functions f_j. ■

Theorem 12.3.10 *([Grü 84])*

For every $a \in E_2$ and every $B \in \{K, K \cup C_0, D, D \cup C_1, T_{a,\infty} \cap T_{\overline{a}} \cap M, T_{a,\infty} \cap M, T_{a,\infty} \cap T_{\overline{a}}, T_{a,\infty}\}$ it holds $|\mathfrak{N}_k(B)| = \mathfrak{c}$.

Proof. W.l.o.g. let $a = 0$ and $K \subseteq B$. Obviously, the sets A_T, $T \subseteq \mathbb{N}$, defined in Lemma 12.3.9, belong to $\mathfrak{N}_k(T_{0,\infty})$. $|\mathfrak{N}_k(T_{0,\infty})| = \mathfrak{c}$ results from that then because of Lemma 12.3.9, (c). Since $pr\ f_q \in K \subseteq B$, statements analogous to (c) of Lemma 12.3.9 are valid for the sets $A_T \cap pr^{-1}B'$ with $B' \in \{K, K \cup C_0, T_1, M, M \cap T_1\}$. This implies the remaining assertions of our theorem result. ■

Theorem 12.3.11 *For every $B \in \{K \cup C, K \cup C_1, D \cup C, D \cup C_0\}$ it holds $|\mathfrak{N}_k(B)| = \mathfrak{c}$.*

Proof. Let

$$f_n(x_1, ..., x_n) = \bigvee_{i=1}^n j_2(x_1)...j_2(x_{i-1})j_1(x_i)j_2(x_{i+1})...j_2(x_n),$$
$$g_n(x_1, ..., x_n) = j_2(x_1)j_2(x_2)...j_2(x_n),$$
$$N \subseteq \mathbb{N}\backslash\{1\}$$
$$A_N := [\{j_1(x_1)j_1(x_2)\} \cup \{f_i, g_j \mid i \in \mathbb{N}\backslash(N \cup \{1\}) \wedge j \in \mathbb{N}\backslash\{1\}\}]$$

and

$$\varrho_i := \begin{pmatrix} 1 & 2 & \dots & 2 \\ 2 & 1 & \dots & 2 \\ \dots & \dots & \dots & \dots \\ 2 & 2 & \dots & 1 \end{pmatrix} \cup E_k^i \backslash \{1, 2, ...k-1\}^i, \qquad i \in \mathbb{N}\backslash\{1\}.$$

It is easy to check that the functions of $A_{\{j\}}$ preserve the relations ϱ_i for all $i \in \mathbb{N}\backslash\{1, j\}$. On the other hand, the function f_i does not preserve the relation ϱ_j. Due to this property, there are continuum-many subclasses of the type A_N with the projection $K \cup C_0$. Since $[A_N \cup \{c_1\}] = [\{c_1\}] \cup A_N$ is obviously valid, we have $|\mathfrak{N}_k(K \cup C)| = \mathfrak{c}$ too.
Since the sets $K \cup C$, $D \cup C$ and $K \cup C_1$, $D \cup C_0$ are isomorphic, respectively, only the proof for $|\mathfrak{N}_k(K \cup C_1)| = \mathfrak{c}$ is still missing.
Denote f'_n and g'_n functions defined by

$$f'_n(\mathbf{x}) := f_n(\mathbf{x}) \vee j_1(x_1)j_1(x_2)...j_1(x_n)$$

or

$$g'_n(\mathbf{x}) := g_n(\mathbf{x}) \vee j_1(x_1)j_1(x_2)...j_1(x_n)$$

and let

$$A'_N := [\{f'_i, g'_j \mid i \in \mathbb{N}\backslash(N \cup \{1\}) \ \wedge \ j \in \mathbb{N}\}] \quad \text{if } N \subseteq \mathbb{N}.$$

Obviously, all functions f'_n for $n \neq i$ preserve the relation

$$\varrho'_i := \{1\} \times \varrho_i, \quad i \in \mathbb{N}\backslash\{1\}.$$

Furthermore, we have $[A'_N \cup \{c_1\}] = [\{c_1\}] \cup A'_N$ and $pr([\{c_1\}] \cup A'_N) = K \cup C_1$. Consequently, there are continuum-many $K \cup C_1$-projectable subclasses of $P_{k,2}$. ■

With the above theorem, we are ready with a first coarse survey over $|\mathfrak{N}_k(B)|$ for all subclasses B of P_2. One finds more precise statements over $|\mathfrak{N}_k(B)|$ for certain B in the following section and for $k = 3$ in Sections 12.6 and 12.7.

12.4 The P_l-projectable and the $Pol_l\{\alpha\}$-projectable Subclasses of $P_{k,l}$

Let R be the set of all relations of the form

$$\begin{pmatrix} 0 & 1 & 2 & ... & l-1 \\ 0 & 1 & 2 & ... & l-1 \end{pmatrix} \cup \varrho$$

with $\varrho \subseteq E_k^2 \backslash E_l^2$ and

$$R(\alpha) := \{\{\alpha\} \cup \varrho \mid \varrho \subseteq E_k \backslash E_l\}.$$

A subset Q of R is called **permissible**, iff no elements $a_1, ..., a_r, b_1, ..., b_r$ with $\begin{pmatrix} a_1 & ... & a_r \\ b_1 & ... & b_r \end{pmatrix} \subseteq \bigcup_{\sigma \in Q} \sigma$, $\{a_1, b_r\} \subseteq E_l$, $a_1 \neq b_r$ and $\{a_i, b_i\} \cap \{a_{i+1}, b_{i+1}\} \neq \emptyset$ for all $i \in \{1, 2, ..., r\}$ exist.
Furthermore, a subset Q of $R \cup R(\alpha)$ is called α-**permissible**, iff $Q \cap R(\alpha) \neq \emptyset$, $Q \cap R$ is permissible and no elements $a_1, ..., a_r, b_1, ..., b_r$ and no relation $\varrho \in Q \cap R(\alpha)$ with $\begin{pmatrix} a_1 & ... & a_r \\ b_1 & ... & b_r \end{pmatrix} \subseteq \bigcup_{\sigma \in R \cap Q} \sigma$, $a_1 \in E_l \backslash \{\alpha\}$, $b_r \in \varrho \backslash \{\alpha\}$ and $\{a_i, b_i\} \cap \{a_{i+1}, b_{i+1}\} \neq \emptyset$ for all $i \in \{1, 2, ..., r\}$ exist.
With the help of Lemma 12.3.5, it is easy to prove that $Pol_{P_{k,l}} Q$ for $Q \subseteq R$ is P_l-projectable iff Q is permissible.
Obviously, for $Q \subseteq R \cup R(\alpha)$ we have $pr(Pol_{P_{k,l}} Q) = Pol_l\{\alpha\}$ iff Q is α-permissible.
Denote $T_n(Q)$ the set of all tuples of E_k^n, on which all functions of $Pol\ Q$ have the value α.

With the help of $Q \subseteq R \cup R(\alpha)$, where Q is permissible or α-permissible, one can define an equivalence relation $Eq(Q)$ on the set of all tuples with elements of E_k as follows:

$$(\mathbf{a}, \mathbf{b}) \in Eq(Q) \;:\Longleftrightarrow\; \exists n : \; \{\mathbf{a}, \mathbf{b}\} \subseteq E_k^n \;\wedge\; (\{\mathbf{a}, \mathbf{b}\} \subseteq T_n(Q) \vee \\ (\{\mathbf{a}, \mathbf{b}\} \not\subseteq T_n(Q) \;\wedge\; \forall f \in Pol\, Q : \; f(\mathbf{a}) = f(\mathbf{b}))).$$

Let σ_1, σ_2 be two equivalence relations. Then σ_1 is **finer** than σ_2, if $(\mathbf{a}, \mathbf{b}) \in \sigma_1$ implies $(\mathbf{a}, \mathbf{b}) \in \sigma_2$ for all $\mathbf{a}, \mathbf{b}$. We write for that

$$\sigma_1 \sqsupset \sigma_2$$

For two permissible (or α-permissible) sets Q_1, Q_2, we call Q_2 a **minimal coarsening** of Q_1, if $Eq(Q_1) \sqsupset Eq(Q_2)$ holds and no permissible (or α-permissible) relation set Q' exists with $Eq(Q_1) \sqsupset Eq(Q') \sqsupset Eq(Q_2)$.

Theorem 12.4.1 *Let Q be a permissible relation set. Then*

(1) The only maximal classes of $Pol_{P_{k,l}}Q$ are the following classes:
(a) $Pol_{P_{k,l}}Q'$, where Q' is an arbitrary minimal coarsening of Q,
(b) $pr_l^{-1}B \cap Pol_{P_{k,l}}Q$, where B is an arbitrary maximal class of P_l.
(2) The classes, defined in (a) and (b), with different definitions are pairwise incomparable (with respect to inclusion).
(3) A set $A \subseteq Pol_{P_{k,l}}Q$ is Pol Q-complete if and only if $A \not\subseteq M$ holds for every class M defined above in (a) or (b).

Proof. The second statement is easy to check and (1) follows from (2) and (3). Consequently, it is sufficient to prove the statement (3).
A *Pol Q*-complete set A is obviously no subset of an arbitrary class defined in (a) or (b). Let $A \subseteq Pol\, Q$ be an arbitrary set with $A \not\subseteq M$ for every class M, which is defined in (a) or (b). We consider two n-tuples $\mathbf{a} := (a_1, a_2, ..., a_n)$, $\mathbf{b} := (b_1, b_2, ..., b_n)$ with $(\mathbf{a}, \mathbf{b}) \notin Eq(Q)$ and we choose a minimal j, for which there exists $i_1, ..., i_j \in \{1, 2, ..., n\}$ with $((a_{i_1}, ..., a_{i_j}), (b_{i_1}, ..., b_{i_j})) \notin Eq(Q)$. Now,

$$\varrho := \begin{pmatrix} 0 & 1 & ... & l-1 & a_{i_1} & ... & a_{i_j} \\ 0 & 1 & ... & l-1 & b_{i_1} & ... & b_{i_j} \end{pmatrix}$$

and $Q' := Q \cup \{\varrho\}$. If Q' is permissible, then Q' is a minimal coarsening of Q, and thus there is a function $f \in A$ with $f \notin Pol\, Q'$. If Q' is not permissible, $pr\; Pol\, Q' \neq P_l$ holds. Hence, one can find a maximal class B of P_l with $Pol\, Q' \subseteq Pol\, Q \cap pr^{-1}B$. Therefore, there exists a function $f \in A$ with $f \notin Pol\, Q'$. Obviously, for all $a \in E_l$, there is a c_a-projectable function $p_a \in [A]$. Because of $p_a \star p_a = c_a$, all constant functions $c_0, c_1, ..., c_{l-1}$ belong to $[A]$. Hence there is a j-ary function $f_1 \in [A]$ with $f_1(a_{i_1}, ..., a_{i_j}) \neq f_1(b_{i_1}, ..., b_{i_j})$ and, by this, there is a function $f_2 \in [A]$ with $f_2(a_1, ..., a_n) \neq f_2(b_1, ..., b_n)$.
Now let $n \geq 1$. We showed that there are functions $g_1^n ..., g_w^n \in [A]$ with the

following property: for every pair $(\mathbf{a}, \mathbf{b}) \notin Eq(Q)$ of tuples, there is a function g_i with $g_i(\mathbf{a}) \neq g_i(\mathbf{b})$. Hence, for an arbitrary function $f^n \in Pol\ Q$ one can find a function $g \in P_l$ with $g(g_1(\mathbf{x}), ..., g_w(\mathbf{x})) = f(\mathbf{x})$. Since $A \nsubseteq pr_l^{-1} B \cap Pol_{P_{k,l}} Q$, where B is an arbitrary maximal class of P_l, there is a g-projectable function in $[A]$. Consequently, $f \in [A]$. Thus A is $Pol\ Q$-complete. ■

As a direct consequence of Theorem 12.4.1, we get:

Theorem 12.4.2 *The set of the closed sets $A \subseteq P_{k,l}$ with $pr_l A = P_l$ is identical with the set of all classes of the form $Pol_{P_{k,l}} Q$ with permissible Q.* ■

Theorem 12.4.3 (Completeness Criterion for $P_{k,l}$, [Lau 75])
For all l with $2 \leq l \leq k-1$ it holds:

(1) The maximal classes of $P_{k,l}$ are exactly the following sets:
(a) $pr_l^{-1} B$, where B is an arbitrary maximal class of P_l;
(b) $Pol \begin{pmatrix} 0 & 1 & .. & l-1 & i \\ 0 & 1 & .. & l-1 & t \end{pmatrix}$, $0 \leq t \leq k-2$, $l \leq i \leq k-1$, $t < i$.
(2) A set $T \subseteq P_{k,l}$ is $P_{k,l}$-complete if and only if $T \nsubseteq B$ for all classes B defined in (1).

Proof. The statements (1) and (2) are consequences of Theorem 12.4.1. A proof that does not use Theorem 12.4.1 can be found in [Lau 75]. ■

As we will see below, one can prove the following theorem similar to the above theorem.

Theorem 12.4.4 *Let Q be an α-permissible subset of $R \cup R(\alpha)$. Then:*

(1) The only maximal classes of $Pol_{P_{k,l}} Q$ are sets of the type
(a) $Pol_{P_{k,l}} Q'$, where Q' is a minimal α-permissible coarsening of Q, and
(b) $pr^{-1} B \cap Pol_{P_{k,l}} Q$, where B is a maximal class of $Pol_l\{\alpha\}$.
(2) The sets, defined in (a) and (b), with different definitions are pairwise incomparable (with respect to inclusion).
(3) A set $A \subseteq Pol_{P_{k,l}} Q$ is $Pol\ Q$-complete if and only if $A \nsubseteq M$ holds for every class M which is defined above in (a) or (b).

Proof. We prove only (3), "$\Longleftarrow$" here. The remaining statements of the theorem are either obvious or easily provable.
Let $A \subseteq Pol_{P_{k,l}} Q$ and $A \nsubseteq M$ for every class M which is defined above in (a) or (b).
Denote $\mathbf{a} := (a_1, ..., a_n)$ an n-tuple, which does not belong to $T_n(Q)$. Then there is a minimal j for $i_1, ..., i_j \in \{1, 2, ..., n\}$ with $(a_{i_1}, ..., a_{i_j}) \notin T_j(Q)$. Now

we put $\varrho := \{\alpha, a_{i_1}, ..., a_{i_j}\}$ and $Q' := Q \cup \{\varrho\}$. If Q' is α-permissible, then Q' is a minimal coarsening of Q; thus in A there is a function f with $f \notin Pol\ Q'$. If Q' is not permissible, then $pr\ Pol\ Q' \neq Pol_l\{\alpha\}$. Therefore, there exists a maximal class B of $Pol\ \{\alpha\}$ with $Pol\ Q' \subseteq Pol\ Q \cap pr^{-1}B$ and, hence, there exists a function $f \in A$ with $f \notin Pol\ Q'$ in this case.
Since $c_\alpha \in [A]$, there is in $[A]$ a j-ary function f_1 with $f_1(a_{i_1}, ..., a_{i_j}) \neq \alpha$; thus there is also an n-ary function $f_2 \in [A]$ with $f_2(a_1, ..., a_n) \neq \alpha$.
Next we consider two n-tuples $\mathbf{a} := (a_1, ..., a_n)$, $\mathbf{b} := (b_1, ..., b_n)$ with $(\mathbf{a}, \mathbf{b}) \notin Eq(Q)$ and $\{\mathbf{a}, \mathbf{b}\} \nsubseteq T_n(Q)$. Analogous to the considerations from proof of Theorem 12.3.1, one can prove that there is a function $r \in [A]$ and that there are certain $i_1, ..., i_j \in \{1, ..., n\}$ with

$$r\begin{pmatrix} 0 & 1 & \dots & l-1 & a_{i_1} & \dots & a_{i_j} \\ 0 & 1 & \dots & l-1 & b_{i_1} & \dots & b_{i_j} \end{pmatrix} \notin \begin{pmatrix} 0 & 1 & \dots & l-1 \\ 0 & 1 & \dots & l-1 \end{pmatrix}.$$

Because – as already was proven – functions s and t with $s(a_{i_1}, ..., a_{i_j}) \neq \alpha$ and $t(b_{i_1}, ..., b_{i_j}) \neq \alpha$ exist and because certain functions u_β with

$$u_\beta\left(s\begin{pmatrix} a_{i_1} & \dots & a_{i_j} \\ b_{i_1} & \dots & a_{i_j} \end{pmatrix}, t\begin{pmatrix} a_{i_1} & \dots & a_{i_j} \\ b_{i_1} & \dots & a_{i_j} \end{pmatrix}\right) = \begin{pmatrix} \beta \\ \beta \end{pmatrix}, \quad \beta \in E_l,$$

belong to $[A]$, a certain function r' with

$$r'\begin{pmatrix} a_{i_1} & \dots & a_{i_j} \\ b_{i_1} & \dots & b_{i_j} \end{pmatrix} \notin \begin{pmatrix} 0 & 1 & \dots & l-1 \\ 0 & 1 & \dots & l-1 \end{pmatrix}$$

is a superposition over $r, u_0, ..., u_{l-1}$. Consequently, there exists an n-ary function $r'' \in [A]$ with $r''(\mathbf{a}) \neq r''(\mathbf{b})$.
Now, let $n \geq 1$. As proven above, there are functions $q_1, ..., q_v, g_1, ..., g_w \in [A]$ with the following property: for every $\mathbf{a} \in E_k^n \backslash T_n(Q)$ there exists a q_i with $q_i(\mathbf{a}) \neq \alpha$ and for every pair $\mathbf{c}, \mathbf{d} \in E_k^n \backslash T_n(Q)$ with $(\mathbf{c}, \mathbf{d}) \notin Eq(Q)$ there exists a function g_j with $g_j(\mathbf{c}) \neq g_j(\mathbf{d})$.
Let $f^n \in Pol\ Q$ be arbitrary. Then one can find a function $g \in Pol_l\{\alpha\}$ with

$$g(q_1(\mathbf{x}), ..., q_v(\mathbf{x}), g_1(\mathbf{x}), ..., g_w(\mathbf{x})) = f(\mathbf{x}).$$

Since $pr[A] = Pol_l\{\alpha\}$, there exists in $[A]$ a g-projectable function. Consequently, $f \in [A]$ holds. Hence the set A is $Pol\ Q$-complete. ■

As a consequence of Theorem 12.4.4, we get:

Theorem 12.4.5 *The only $Pol_l\{\alpha\}$-projectable subclasses of $P_{k,l}$ with $\alpha \in E_l$ are sets of the form $Pol_{P_{k,l}}Q$, where Q are α-permissible subsets Q of $R \cup R(\alpha)$.* ■

12.5 The Maximal and the Submaximal Classes of $P_{k,2}$

A direct consequence of Theorem 12.4.1 is the following theorem, already proven in [Bur 74].

Theorem 12.5.1 *The maximal classes of $P_{k,2}$ are exactly the following sets:*

(1) $pr^{-1}B$, *where B is maximal in P_2;*

(2) $Z_{i,t} := Pol\begin{pmatrix} 0 & 1 & i \\ 0 & 1 & t \end{pmatrix}$, $0 \le t \le k-2$, $2 \le i \le k-1$, $t < i$.

The above sets are all pairwise incomparable (with respect to $\subseteq$), so that there are exactly

$$\frac{1}{2} \cdot (k-2) \cdot (k+1) + 5$$

maximal classes of $P_{k,2}$. ■

Next we determine the submaximal classes of $P_{k,2}$, i.e., we determine the maximal classes of the classes of Theorem 12.4.1.
The following theorem is a special case of Theorem 12.4.1. This is easy to see considering the following facts:

$$Pol\begin{pmatrix} 0 & 1 & i & r \\ 0 & 1 & t & s \end{pmatrix} \subseteq Pol\begin{pmatrix} 0 & 1 & i \\ 0 & 1 & t \end{pmatrix} \cap Pol\begin{pmatrix} 0 & 1 & r \\ 0 & 1 & s \end{pmatrix}$$

and

$$Pol\begin{pmatrix} 0 & 1 & i & i \\ 0 & 1 & t & i \end{pmatrix} = Pol\begin{pmatrix} 0 & 1 & i & t \\ 0 & 1 & t & t \end{pmatrix}$$

for $0 \le t \le k-2$, $2 \le i \le k-1$, $t < i$.

Theorem 12.5.2

(1) The maximal classes of $Z_{i,t}$ ($0 \le t \le k-2$, $2 \le i \le k-1$, $t < i$) are exactly the following sets

(a) $Z_{i,t} \cap Z_{j,l}$, $0 \le l \le k-2$, $2 \le j \le k-1$, $l < j$, $l \ne t$ *or* $i \ne j$, $\{0,1\} \ne \{t,l\}$ *for* $i \ne j$;

(b) $Pol\begin{pmatrix} 0 & 1 & i & j \\ 0 & 1 & t & j \end{pmatrix}$, $2 \le j \le k-1$, $j \ne i$ *for* $t \in E_2$;

(c) $Z_{i,t} \cap pr^{-1}B$, *where B is a maximal class of P_2.*

(2) $Z_{i,t}$ *has exactly* $\frac{1}{2} \cdot (k-1) \cdot k + 2$ *maximal classes.* ■

A consequence from Theorem 12.4.4 is

Theorem 12.5.3

(1) $pr^{-1}T_\alpha$ ($\alpha \in E_2$) has exactly the following maximal classes:
(a) $pr^{-1}T_\alpha \cap pr^{-1}B$, where B is T_α-maximal;
(b) $Z_{i,t} \cap pr^{-1}T_\alpha$, $0 \le t \le k-2$, $2 \le i \le k-1$, $t < i$, $t \ne \alpha$;
(c) $Pol_{P_{k,2}}\{\alpha, i\}$, $2 \le i \le k-1$.

(2) The classes listed in (a)– (c) are pairwise incomparable with respect to the inclusion, so that there are exactly $4 + \frac{1}{2} \cdot (k-2) \cdot (k+1)$ in $pr^{-1}T_\alpha$ maximal classes.

(3) A set $A \subseteq pr^{-1}T_\alpha$ is $pr^{-1}T_\alpha$-complete if and only if $A \not\subseteq C$ holds for all classes C listed in (a)–(c). ■

Theorem 12.5.4

(1) $pr^{-1}S$ has exactly the following maximal classes:
(a) $pr^{-1}B$, where B is a maximal of S;
(b) $Z_{i,t} \cap pr^{-1}S$, $0 \le t < i < k$, $i \ge 2$;
(c) $S^{(i,t)} := Pol\begin{pmatrix} 0 & 1 & i \\ 1 & 0 & t \end{pmatrix}$, $2 \le t < i < k$.

(2) The classes listed in (a)– (c) are pairwise incomparable with respect to the inclusion, so that there are exactly $2 + (k-1) \cdot (k-2)$ in $pr^{-1}S$ maximal classes.

(3) A set $A \subseteq pr^{-1}S$ is $pr^{-1}S$-complete if and only if $A \not\subseteq C$ holds for all classes C listed in (a)–(c).

Proof. Let A be an arbitrary subset of $pr^{-1}S$ with $A \not\subseteq C$ for all classes C listed in (a)–(c). Then $pr[A] = S$ holds. Now let $f^n \in pr^{-1}S$ be arbitrary. We show that $f \in [A]$.

Let $h_{i,t} \in A \backslash S^{(i,t)}$ and $f_{i,t} \in A \backslash Z_{i,t}$. Then, superpositions over these functions and over (suitably chosen) inverse images ($\in [A]$) of functions of S are the binary functions $h'_{i,t}$ and $f'_{i,t}$ of $[A]$ with

$$\begin{pmatrix} h'_{i,t}(1,i) \\ h'_{i,t}(0,t) \end{pmatrix} = \begin{pmatrix} 1 \\ 1 \end{pmatrix} \text{ and } \begin{pmatrix} f'_{i,t}(1,i) \\ f'_{i,t}(1,t) \end{pmatrix} = \begin{pmatrix} 1 \\ 0 \end{pmatrix}.$$

In $[A]$ there is a inverse image $l(x)$ of $e_1^1 \in S$. It holds

$$\begin{pmatrix} l(i) \\ l(t) \end{pmatrix} = \begin{pmatrix} c \\ c \end{pmatrix} \text{ or } \begin{pmatrix} l(i) \\ l(t) \end{pmatrix} = \begin{pmatrix} c \\ \overline{c} \end{pmatrix},$$

W.l.o.g. we can assume that $c = 1$. In the first case $\begin{pmatrix} l(i) \\ l(t) \end{pmatrix} = \begin{pmatrix} 1 \\ 1 \end{pmatrix}$ we have

$$f''_{i,t}(x) := f'_{i,t}(l(x), x) \in [A] \text{ and } \begin{pmatrix} f''_{i,t}(i) \\ f''_{i,t}(t) \end{pmatrix} = \begin{pmatrix} 1 \\ 0 \end{pmatrix}.$$

This implies

$$h''_{i,t}(x) := h'_{i,t}(f'_{i,t}(x), x) \in [A] \text{ and } \begin{pmatrix} h''_{i,t}(i) \\ h''_{i,t}(t) \end{pmatrix} = \begin{pmatrix} 1 \\ 1 \end{pmatrix}.$$

In the second case $\begin{pmatrix} l(i) \\ l(t) \end{pmatrix} = \begin{pmatrix} 1 \\ 0 \end{pmatrix}$ we put

$$h''_{i,t}(x) := h'_{i,t}(l(x), x) \text{ and } f''_{i,t}(x) := f'_{i,t}(h'_{i,t}(x), x),$$

where the above functions have the same properties as the functions in the first case. Therefore, there are certain functions of $[A]$, which have on $\begin{pmatrix} i \\ t \end{pmatrix}$ the values $\begin{pmatrix} 1 \\ 0 \end{pmatrix}$ or $\begin{pmatrix} 1 \\ 1 \end{pmatrix}$.

Now we assign a tuple of the form

$$\mathbf{w} := (..., h''_{i,t}(a_1), ..., h''_{i,t}(a_n), ..., f''_{i,t}(a_1), ..., f''_{i,t}(a_n), ...)$$

to every n-tuple $\mathbf{a} := (a_1, ..., a_n) \in E_k^n$, where i, t take all possible values with $0 \le t \le k-2$, $2 \le i \le k-1$, $i > t$ (in a fixed order). All tuples of the form $\mathbf{w}$ belong to a certain Cartesian product E_2^σ. Denote

$$W$$

the set of all tuples $\mathbf{w}$ defined above.

Next we prove: If $\mathbf{w}, \mathbf{w}' \in W$ were assigned to $\mathbf{a}, \mathbf{a}' \in E_k^n$ ($\mathbf{a}' := (a'_1, ..., a'_n)$), respectively, then it holds:

$$\mathbf{a} \ne \mathbf{a}' \implies \mathbf{w} \ne \mathbf{w}' \tag{12.12}$$

and

$$\mathbf{a} = \overline{\mathbf{a}'} \iff \mathbf{w} = \overline{\mathbf{w}'}, \tag{12.13}$$

where $\overline{\mathbf{a}} := (\overline{a_1}, ..., \overline{a_n})$, $\overline{0} := 1$, $\overline{1} := 0$, $\overline{a} := a$ for all $a \in E_k \backslash E_2$.

Let $\mathbf{a} \ne \mathbf{a}'$, where $\mathbf{a} := (a_1, ..., a_n)$ and $\mathbf{a}' := (a'_1, ..., a'_n)$. Then there exists a j with $c := a_j \ne a'_j =: d$. If $\begin{pmatrix} c \\ d \end{pmatrix} = \begin{pmatrix} 0 \\ 1 \end{pmatrix}$ or $\begin{pmatrix} c \\ d \end{pmatrix} = \begin{pmatrix} 1 \\ 0 \end{pmatrix}$, (12.12) is valid certainly. If $\{c, d\} \ne \{0, 1\}$ and w.l.o.g. $c > d$, then by definition of $f''_{c,d}$ the statement (12.12) holds. If $\mathbf{a} = \overline{\mathbf{a}'}$ then $\mathbf{w} = \overline{\mathbf{w}'}$, since all considered functions belong to $pr^{-1}S$. For $\mathbf{a} \ne \overline{\mathbf{a}'}$ the two cases are possible:

Case 1: The matrix $\begin{pmatrix} \mathbf{a} \\ \mathbf{a}' \end{pmatrix}$ has a column $\begin{pmatrix} p \\ q \end{pmatrix}$ with

$$\begin{pmatrix} p \\ q \end{pmatrix} \notin \left\{ \begin{pmatrix} 0 \\ 0 \end{pmatrix}, \begin{pmatrix} 0 \\ 1 \end{pmatrix}, \begin{pmatrix} 1 \\ 0 \end{pmatrix}, \begin{pmatrix} 1 \\ 1 \end{pmatrix} \right\}.$$

Case 2: The matrix $\begin{pmatrix} \mathbf{a} \\ \mathbf{a}' \end{pmatrix}$ has a column $\begin{pmatrix} q' \\ q' \end{pmatrix}$ with $q' \in E_2$.

In the second case, $\mathbf{w} \neq \mathbf{w}'$ holds. In the first case, one can prove $\mathbf{w} \neq \mathbf{w}'$ easily when one one assumes $p > q$ considering the condition $\begin{pmatrix} h''_{i,t}(i) \\ h''_{i,t}(t) \end{pmatrix} = \begin{pmatrix} 1 \\ 1 \end{pmatrix}$.

Let h be a mapping from W onto E_2 defined by

$$h(\mathbf{w}) = h(w_1, ..., w_\sigma) := f(\mathbf{a}).$$

Because of (12.12) and (12.13), an α-ary function $h_1 \in S$ can be found, whose restriction onto W is identical with h. Hence there is an inverse image h' of h in $[A]$, and, by construction, f is a superposition over h', $f''_{i,t}$ and $h''_{i,t}$. Thus $f \in [A]$ and $[A] = pr^{-1}S$; i.e., the third statement of Theorem 12.5.4 was proven.

One checks the remaining statements of the theorem easily. ■

Lemma 12.5.5 *Let $A \subseteq pr^{-1}M$, $[pr\ M] = M$ and $\{k_1, j_0(x_1) \cdot j_i(x_2) \mid i = 2, 3, ..., k-1\} \subseteq [A]$. Then $[A] = pr^{-1}M$.*

Proof. Obviously, for every function $g \in M$ there exists a function $g' \in [A]$ with $pr\ g' = g$. Consequently, certain inverse images of the conjunction, disjunction, $n \star n = c_0$ ($n \in [A]$, $pr\ n = c_0$), $i(x) \cdot k_1(x) = j_1(x)$ ($i \in [A]$, $pr\ i = pr\ j_1$) and $j_0(c_0(x)) \cdot j_p(x) = j_p(x)$, $2 \leq p \leq k-1$ are superpositions over A.

One can describe every function $f^n \neq c_0^n$ of $pr^{-1}M$ by

$$\begin{aligned} f(x_1, .., x_n) &= f_1(x_1, ..., x_n) \cdot k_1(x_1) \cdot ... \cdot k_1(x_n) \\ &\quad \vee \bigvee_{\substack{(a_1, ..., a_n) \\ \in E_k^n \setminus E_2^n \\ f(a_1, ..., a_n) = 1}} j_{a_1}(x_1) \cdot j_{a_2}(x_2) \cdot ... \cdot j_{a_n}(x_n), \end{aligned} \tag{12.14}$$

where $f_1 \in [A]$ and $pr\ f_1 = pr\ f$. (12.14) is obvious a superposition over A and thus $[A] = pr^{-1}M$. ■

Theorem 12.5.6

(1) $pr^{-1}M$ has exactly the following maximal classes:
(a) $pr^{-1}B$, where B is maximal in M;
(b) $Z_{i,t} \cap pr^{-1}M$, $2 \leq t < i < k$;

(c) $M^{(0,r)}$, $M^{(r,s)}$, $1 \le s < k$, $2 \le r < k$;

$$\left(M^{(a,b)} := Pol \begin{pmatrix} 0 & 1 & 0 & a \\ 0 & 1 & 1 & b \end{pmatrix} \right).$$

(2) The sets, defined in a)–(c), with different definitions are pairwise incomparable (with respect to inclusion) so that $pr^{-1}M$ *has exactly* $4 + \frac{3}{2}(k-1) \cdot (k-2)$ *maximal classes.*

(3) A set $A \subseteq pr^{-1}M$ *is* $pr^{-1}M$*-complete if and only if* $A \not\subseteq T$ *holds for every class* T *which is defined in (a)–(c).*

Proof. Obviously, if $[A] = pr^{-1}M$, then A is not a subset of the sets (a)–(c). Now let A be subset of $pr^{-1}M$ with $A \not\subseteq T$ for every class T defined in (a)–(c). Then we have $[pr\ A] = M$ and $c_0, c_1 \in [A]$. Further, $[A]$ contains functions $f_{i,t}, g_{r,s}, q_r$ and p_r with the properties:

$$f_{i,t} \notin Z_{i,t}, \quad g_{r,s} \notin M^{(r,s)}, \quad q_r \notin M^{(r,r)}, \quad p_r \notin M^{(0.r)}.$$

Then certain unary or binary functions $f'_{i,t}, g'_{r,s}, q'_r$ and p'_r with

$$f'_{i,t}(i) \neq f'_{i,t}(t),$$

$$g'_{r,s}\begin{pmatrix} 0 & r \\ 1 & s \end{pmatrix} = \begin{pmatrix} 1 \\ 0 \end{pmatrix}, \quad q'_r\begin{pmatrix} 0 & r \\ 1 & r \end{pmatrix} = \begin{pmatrix} 1 \\ 0 \end{pmatrix} \quad \text{and} \quad p'_r\begin{pmatrix} 0 & 0 \\ 1 & r \end{pmatrix} = \begin{pmatrix} 1 \\ 0 \end{pmatrix},$$

$2 \le t < i \le k-1$, $1 \le s \le k-1$, $2 \le r \le k-1$ are obvious superpositions over $\{c_0, c_1, f_{i,t}, g_{r,s}, q_r, p_r\}$. For the function

$$f''_{i,t}(x) := \begin{cases} g'_{i,t}(f'_{i,t}(x), x), & \text{if } f'_{i,t}(i) = 0, \\ g'_{t,i}(f'_{i,t}(x), x), & \text{if } f'_{i,t}(i) = 1, \end{cases}$$

it holds $f'_{i,t}(a) = \overline{f''_{i,t}(a)}$ for $a \in \{i,t\}$. Thus w.l.o.g. we can assume for all permissible i, t in the following: $f'_{i,t}(i) = 1$ and $f''_{i,t}(i) = 0$.
Let $i \in [A]$, $pr\ i = pr\ j_1$ and let E be the set of all $A \in E_k \backslash E_2$ with $i(a) = 0$. Then, for $r \in E$ and $g'_{r,1}(i(x), x) =: g''_{r,1}(x)$, we have $g''_{r,1}(r) = 1$ and $g''_{r,1}(1) = 0$. Consequently,

$$i(x) \vee \bigvee_{r \in E} g''_{r,1}(x) = \overline{j_0(x)}$$

and

$$q'_r(i(x_1), x_2) \cdot g'_{r,1}(i(x_1), x_2) \cdot (\bigwedge_{t=2}^{r-1} f'_{r,t}(x_2)) \cdot (\bigwedge_{i=r+1}^{k-1} f''_{i,r}(x_2)) = j_0(x_1) \cdot j_r(x_2) \in [A]$$

for $r \geq 2$. Furthermore, it holds

$$\bigwedge_{r=2}^{k-1} p'_r(j_0(c_0(x)) \cdot j_r(x), x) = k_1(x) \in [A].$$

Thus, we showed that the generating system for $pr^{-1}M$ of Lemma 12.5.5 belongs to $[A]$. Therefore, the set A is $pr^{-1}M$-complete.
One checks the remaining statements of the theorem easily. ■

We come now to the maximal classes of $pr^{-1}L$.
First we give some remarks on a possible description of functions of $pr^{-1}L$. Because of $j_0 = 1+j_1(x)+...+j_{k-1}(x)$ it follows from Section 12.1 that every function $f^n \in P_{k,2}$ has an unambiguous description (up to the order of the summands) of the form

$$f(\mathbf{x}) = a + \sum_{\substack{I_1,...,I_{k-1} \subseteq \\ \{1,2,...,n\}}} a_{I_1,...,I_{k-1}} \cdot k_{I_1,...,I_{k-1}}(\mathbf{x}), \tag{12.15}$$

where $a_{I_1,...,I_{k-1}} \in E_2$,

$$k_{I_1,...,I_{k-1}}(x_1,...,x_n) := \bigwedge_{i=1}^{k-1} \bigwedge_{q \in I_i} j_i(x_q) \tag{12.16}$$

and the sets $I_1,...,I_{k-1}$ in (12.15) are pairwise disjunct.
The functions $k_{I_1,...,I_{k-1}}$ with $a_{I_1,...,I_{k-1}} = 1$ in (12.15) are called **components** of f . Let K_f be the set of all components of f. Denote $k_{f,I}$ an n-ary function of the form

$$k_{f,I}(\mathbf{x}) := \sum_{\substack{I_1,...,I_{k-1} \\ I_1 \cup ... \cup I_{k-1} = I \\ k_{I_1,...,I_{k-1}} \in K_f}} k_{I_1,...,I_{k-1}}(\mathbf{x}),$$

where $I \subseteq \{1,2,...,n\}$.

Lemma 12.5.7 *Let $\{f,g,h\} \subseteq pr^{-1}L$, $(pr\ g)(y_1,y_2) = y_1 + y_2$ and $(pr\ h)(y) = \overline{y}$. Then every function of the form $k_{f,I}$ is a superposition over $\{c_0,f,g,h\}$.*

Proof. It is easy to see that the function $f'(\mathbf{x}) := f(\mathbf{x}) + f(0,0,...,0)$ is a superposition over the functions f and h. Denote r the smallest number, for which f' has a component with r essential variables. W.l.o.g. let

$$f'(\mathbf{x}) = k_{f,\{1,2,...,r\}} + f''(\mathbf{x}).$$

Then it holds

$$f'(x_1,...,x_r,c_0,...,c_0) = k_{f,\{1,2,...,r\}}(\mathbf{x}) \in [\{c_0,f,g,h\}]$$

and

$$f''(\mathbf{x}) = g(f'(\mathbf{x}), k_{f,\{1,2,...,r\}}(\mathbf{x})) \in [\{c_0,f,g,h\}],$$

where $K_{f''} \subset K_{f'} \subseteq K_f$ and every component of f'' has at least r essential variables. Through repetition of this construction, one receives the statement of the lemma. ■

Theorem 12.5.8

(1) $pr^{-1}L$ *has exactly the following maximal classes:*
  (a) $pr^{-1}B$, *where* B *is a maximal class of* L;
  (b) $Z_{i,t} \cap pr^{-1}L$, $2 \leq t < i < k$;
  (c) $L_q := Pol_{P_{k,l}}\{(q,q,q,q),(a,b,c,d) \mid \{a,b,c,d\} \subseteq E_2 \;\wedge\; a+b = c+d \; (mod\ 2)\}$, $2 \leq q < k$.

(2) The classes listed in (a)–(c) are pairwise incomparable (with respect to the inclusion) so that there are exactly $4+\frac{1}{2}(k-1)\cdot(k-2)$ $pr^{-1}L$*-maximal classes.*

(3) A set $A \subseteq pr^{-1}L$ *is* $pr^{-1}L$*-complete if and only if* $A \not\subseteq Q$ *holds for all classes* Q *listed in (a)–(c).*

Proof. The statements (1) and (2) are consequences from (3). Since "$\Longrightarrow$" of (3) is trivial, we prove only (3), "$\Longleftarrow$".
Let A be a subset of $pr^{-1}L$ with $A \not\subseteq Q$ for all classes Q which are defined in (a)–(c). Because of $A \not\subseteq pr^{-1}B$, B maximal in L, $pr\,[A] = L$ holds. Consequently, $c_0, c_1 \in [A]$ and one can find functions g, $h \in [A]$ with $(pr\ g)(y_1,y_2) = y_1+y_2$ and $(pr\ h)(y) = \overline{y}$. Since A is, in addition, not a subset of the sets $Z_{i,t}$, $2 \leq t < i \leq k-1$, there are unary functions $g_{i,t} \in [A]$ with $g_{i,t}(i) = 1$ and $g_{i,t}(t) = 0$, $t \neq i$, $2 \leq i \leq k-1$, $2 \leq t \leq k-1$.
Let now $q \in \{2,3,...,k-1\}$. Since the functions of A not all preserve the relation

$$\begin{pmatrix} 0 & 0 & 0 & 1 & 1 & 0 & 1 & 1 & q \\ 0 & 0 & 1 & 1 & 0 & 1 & 0 & 1 & q \\ 0 & 1 & 0 & 0 & 1 & 1 & 0 & 1 & q \\ 0 & 1 & 1 & 0 & 0 & 0 & 1 & 1 & q \end{pmatrix},$$

a function f with $pr(f(x_1,...,x_n,q)) \notin L$ belongs to $[A]$. Then by the maximality of L in P_2 and by $pr[A] = L$, for every function $u^m \in P_2$ there is a $(m+1)$-ary function $u' \in [A]$ with

$$u'(y_1,...,y_m,q) = u(y_1,...,y_m).$$

Consequently, a function of the form

$$\begin{aligned} v(x_1,...,x_k) \;&=\; a + \textstyle\sum_{i=1}^{k} a_i j_i(x_i) + j_1(x_1)\cdot ... \cdot j_1(x_{k-1})\cdot j_q(x_k) + \\ &\textstyle\sum_{p=2,p\neq q}^{k-1} v_p(x_1,...,x_{k-1})\cdot j_p(x_k) \end{aligned}$$

belongs to $[A]$ for some $a, a_i \in E_2$ and some functions $v_p \in P_{k,2}$. By Lemma 12.5.7 the functions of the type $k_{v,I}$, $I \subseteq \{1,2,...,k\}$, are superpositions over A. If one adds all functions of the form $k_{v,I}$ with $|I| \leq k-1$ to v, one receives so the function

$$v(x_1, ..., x_k) := j_1(x_1) \cdot ... \cdot j_1(x_{k-1}) \cdot j_q(x_k) + \sum_{\substack{I_1, ..., I_{k-1}, p \\ I_1\cup...\cup I_{k-1} = \{1, ..., k-1\} \\ p \in \{2,3,...,k-1\}\setminus\{q\}}} a_{I_1,...,I_{k-1}} \cdot k_{I_1,...,I_{k-1}}(\mathbf{x}) \cdot j_p(x_k),$$

which also belongs to $[A]$ and for which

$$v'(x_1, x_2, g_{q,2}(x_3), g_{q,3}(x_3), ..., g_{q,q-1}(x_3), g_{q,q+1}(x_3), ..., g_{q,k-1}(x_3), x_3)$$
$$= j_1(x_1) \cdot j_1(x_2) \cdot j_q(x_3) \in [A]$$

holds. With the help of (12.15) it is easy to see that every function of $pr^{-1}L$ is a superposition over

$$\{c_0, c_1, g, h\} \cup \{j_1(x_1)j_1(x_2)j_q(x_3) \mid q \in \{2,3,...,k-1\}\}.$$

Thus, $[A] = pr^{-1}L$. ■

12.6 The Classes A with $M \cap T_0 \cap T_1 \subseteq prA$ or $L \cap T_0 \cap S \subseteq prA$ or $prA = M \cap S$

In this and in the following section, we specify the cardinality statements of Section 12.3 about $\mathfrak{N}_k(A)$ ($A \in \mathbb{L}_2$) for $k = 3$. In addition, we provide a concrete description of the elements of $\mathfrak{N}_3(A)$ if $|\mathfrak{N}_3(A)| \leq \aleph_0$ is valid. Table 12.1 gives a first survey of these statements. The statements of the last three rows of Table 12.1 were proven already in Theorems 12.3.8, 12.3.10, and 12.3.11. The cardinality statements of the eighth row are a consequence of Theorem 12.3.1. The remaining statements in the table result from the following theorems.

Theorem 12.6.1 *It holds:*

(1) $\mathfrak{N}_3(P_2) = \{P_{3,2}, Z_{2,0}, Z_{2,1}\}$;
(2) $\mathfrak{N}_3(S) = \{pr^{-1}S, pr^{-1}S \cap Z_{2,0}, pr^{-1}S \cap Z_{2,1}\}$;
(3) $\mathfrak{N}_3(T_a) = \{pr^{-1}T_a, T_a^{(2)}, pr^{-1}T_a \cap Z_{2,0}, pr^{-1}T_a \cap Z_{2,1}\}$ ($a \in E_2$);
(4) $\mathfrak{N}_3(T_0 \cap T_1) = \{pr^{-1}(T_0 \cap T_1), T_0^{(2)} \cap pr^{-1}(T_0 \cap T_1), T_1^{(2)} \cap pr^{-1}(T_0 \cap T_1), Z_{2,0} \cap pr^{-1}(T_0 \cap T_1), Z_{2,1} \cap pr^{-1}(T_0 \cap T_1)\}$;
(5) $\mathfrak{N}_3(T_0 \cap S) = \{pr^{-1}(T_0 \cap S), T_0^{(2)} \cap pr^{-1}(T_0 \cap S), T_1^{(2)} \cap pr^{-1}(T_0 \cap S), Z_{2,0} \cap pr^{-1}(T_0 \cap S), Z_{2,1} \cap pr^{-1}(T_0 \cap S)\}$.

(One finds Hasse diagrams of some of the sets listed above in the Figure 12.1.)

Table 12.1

B	$\vert\mathfrak{N}_3(B)\vert$
P_2, S	3
T_a $(a \in E_2)$	4
$T_0 \cap T_1$, $T_0 \cap S$	5
M	23
$M \cap T_a$ $(a \in E_2)$	36
$M \cap T_0 \cap T_1$	49
$T_{a,2}$ $(a \in E_2)$	148
$T_{a,m}, T_{a,m} \cap T_{\overline{a}}$, $T_{a,m} \cap M$, $T_{a,m} \cap T_{\overline{a}} \cap M$, $S \cap M$ $(a \in E_2,\ m \in \{2,3,...\})$	$< \aleph_0$
L, $L \cap T_0$, $L \cap T_1$, $L \cap S$, $L \cap T_0 \cap S$	$\aleph_0$
$T_{a,\infty}, T_{a,\infty} \cap T_{\overline{a}}$, $T_{a,\infty} \cap M$, $T_{a,\infty} \cap T_{\overline{a}} \cap M$ $(a \in E_2)$	$\mathfrak{c}$
$K \cup C$, $K \cup C_a$, K, $D \cup C$, $D \cup C_a$, D $(a \in E_2)$	$\mathfrak{c}$
$[P_2^1]$, $I \cup C$, $\overline{I}$, $I \cup C_a$, I, C, C_a $(a \in E_2)$	$\mathfrak{c}$

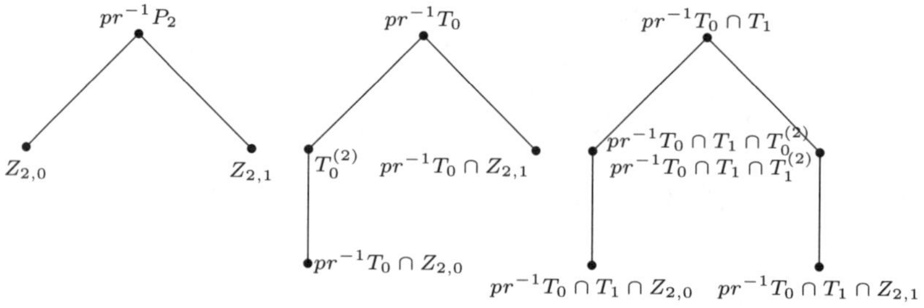

Fig. 12.1

Proof. The intersections of $pr^{-1}B$, $B \subseteq P_2$, with $Z_{2,0}$ or $Z_{2,1}$ are always the smallest B-projectable subclasses of $P_{3,2}$. Thus, (1) is a consequence from Theorem 12.5.1 and (2) is a consequence from Theorem 12.5.4. By Theorem 12.5.3, the maximal classes of $pr^{-1}T_a$ belonging to $\mathfrak{N}_3(T_a)$ are just the sets $Z_{2,\overline{a}} \cap pr^{-1}T_a$ and $T_a^{(2)} = Pol_{P_{3,2}}\{a, 2\}$. With the aid of Theorem 12.4.4, it is easy to see that $Z_{2,a} \cap pr^{-1}T_a$ is the only T_a-projectable maximal class of $T_a^{(2)}$. From that and from the above remarks, (3) follows.
By Theorem 12.3.4, the statements (4) and (5) are consequences from (1)–(3).
■

We come now to the M-projectable subclasses A of $P_{3,2}$. Because of $h_2 \in M$, Theorem 12.3.1, and Lemma 12.3.5 one can find some $\varrho_1, ..., \varrho_m \subseteq E_3^2\backslash E_2^2$ and an A' with $pr\ A' = P_2$ for such class A so that

$$A = \left(\bigcap_{i=1}^{m} Pol\left(\begin{pmatrix} 0 & 0 & 1 \\ 0 & 1 & 1 \end{pmatrix}\right) \cup \varrho_i\right) \cap A' \tag{12.17}$$

holds. Because of Theorem 12.3.6, (6) must be valid for the relations $\varrho_1, ..., \varrho_m$ in addition:

$$\begin{pmatrix} 2 & 1 \\ 0 & 2 \end{pmatrix} \not\subseteq \varrho_1 \cup \varrho_2 \cup ... \cup \varrho_m. \tag{12.18}$$

We call relations $\varrho_1, ..., \varrho_m$ with the above property (12.18) **M-permissible**. The following lemma gives some properties of M-projectable subclasses of $P_{3,2}$, which we need to prove Theorem 12.6.3.

Lemma 12.6.2 *Let be the relations $\varrho_1, ..., \varrho_m \subseteq E_3^2\backslash E_2^2$ M-permissible and let $a \in E_2$. Then:*

(1) $\varrho_1 \subseteq \varrho_2 \implies Pol\begin{pmatrix} 0 & 0 & 1 \\ 0 & 1 & 1 \end{pmatrix} \cup \varrho_2 \subseteq Pol\begin{pmatrix} 0 & 0 & 1 \\ 0 & 1 & 1 \end{pmatrix} \cup \varrho_1;$

(2) $\begin{pmatrix} a & 2 \\ 2 & a \end{pmatrix} \subseteq \varrho_1 \cup ... \cup \varrho_m \implies \bigcap_{i=1}^m Pol\begin{pmatrix} 0 & 0 & 1 \\ 0 & 1 & 1 \end{pmatrix} \cup \varrho_i \subseteq pr^{-1}M \cap Z_{2,a};$

(3) $Pol\begin{pmatrix} 0 & 0 & 1 & 0 & 1 \\ 0 & 1 & 1 & 2 & 2 \end{pmatrix} = Pol\begin{pmatrix} 0 & 0 & 1 & 1 \\ 0 & 1 & 1 & 2 \end{pmatrix};$

(4) $Pol\begin{pmatrix} 0 & 0 & 1 & 0 & 2 \\ 0 & 1 & 1 & 2 & 1 \end{pmatrix} = Pol\begin{pmatrix} 0 & 0 & 1 & 0 \\ 0 & 1 & 1 & 2 \end{pmatrix} \cap Pol\begin{pmatrix} 0 & 0 & 1 & 2 \\ 0 & 1 & 1 & 1 \end{pmatrix};$

(5) $Pol\begin{pmatrix} 0 & 0 & 1 & 1 & 2 \\ 0 & 1 & 1 & 2 & 2 \end{pmatrix} = Pol\begin{pmatrix} 0 & 0 & 1 & 0 & 1 & 2 \\ 0 & 1 & 1 & 2 & 2 & 2 \end{pmatrix};$

(6) $Pol\begin{pmatrix} 0 & 0 & 1 & 2 & 2 \\ 0 & 1 & 1 & 0 & 1 \end{pmatrix} = Pol\begin{pmatrix} 0 & 0 & 1 & 2 \\ 0 & 1 & 1 & 0 \end{pmatrix} \cap Pol\begin{pmatrix} 0 & 0 & 1 & 2 \\ 0 & 1 & 1 & 1 \end{pmatrix};$

(7) $Pol\begin{pmatrix} 0 & 0 & 1 & 2 & 2 \\ 0 & 1 & 1 & 0 & 2 \end{pmatrix} = Pol\begin{pmatrix} 0 & 0 & 1 & 2 & 2 & 2 \\ 0 & 1 & 1 & 0 & 1 & 2 \end{pmatrix};$

(8) $Pol\begin{pmatrix} 0 & 0 & 1 & 0 & 2 & 2 \\ 0 & 1 & 1 & 2 & 1 & 2 \end{pmatrix} = Pol\begin{pmatrix} 0 & 0 & 1 & 0 & 2 \\ 0 & 1 & 1 & 2 & 2 \end{pmatrix} \cap Pol\begin{pmatrix} 0 & 0 & 1 & 2 & 2 \\ 0 & 1 & 1 & 1 & 2 \end{pmatrix}.$

Proof. The statement (1) is trivial. Since

$$\begin{pmatrix} 0 & 0 & 1 & 2 \\ 0 & 1 & 1 & a \end{pmatrix} \circ \begin{pmatrix} 0 & 0 & 1 & a \\ 0 & 1 & 1 & 2 \end{pmatrix} =: \varrho = \begin{cases} \begin{pmatrix} 0 & 0 & 1 & 0 & 2 & 2 & 2 \\ 0 & 1 & 1 & 2 & 0 & 2 & 1 \end{pmatrix} & \text{if } a = 0, \\ \begin{pmatrix} 0 & 0 & 1 & 1 & 2 & 2 & 0 \\ 0 & 1 & 1 & 2 & 1 & 2 & 2 \end{pmatrix} & \text{if } a = 1 \end{cases}$$

and $(\varrho \cap \tau\varrho) \circ \begin{pmatrix} 0 & 1 \\ 0 & 1 \end{pmatrix} = \begin{pmatrix} 0 & 1 & 2 \\ 0 & 1 & a \end{pmatrix}$, we have

$$Pol\begin{pmatrix} 0 & 0 & 1 & 2 \\ 0 & 1 & 1 & a \end{pmatrix} \cap \begin{pmatrix} 0 & 0 & 1 & a \\ 0 & 1 & 1 & 2 \end{pmatrix} \subseteq pr^{-1}M \cap Z_{2,a}.$$

With the aid of (1), statement (2) results.
(3) follows from (1) and

$$\begin{pmatrix} 0&0&1 \\ 0&1&1 \end{pmatrix} \circ \begin{pmatrix} 0&0&1&1 \\ 0&1&1&2 \end{pmatrix} = \begin{pmatrix} 0&0&1&0&1 \\ 0&1&1&2&2 \end{pmatrix}.$$

Because of (1) and

$$\begin{pmatrix} 0&0&1&2 \\ 0&1&1&1 \end{pmatrix} \circ \begin{pmatrix} 0&0&1&0 \\ 0&1&1&2 \end{pmatrix} = \begin{pmatrix} 0&0&1&2&0 \\ 0&1&1&1&2 \end{pmatrix}$$

(4) holds.
(5) follows from (1) and

$$\begin{pmatrix} 0&0&1&1&2 \\ 0&1&1&2&2 \end{pmatrix} \circ \begin{pmatrix} 0&0&1&1&2 \\ 0&1&1&2&2 \end{pmatrix} = \begin{pmatrix} 0&0&1&0&1&2 \\ 0&1&1&2&2&2 \end{pmatrix}.$$

The classes from (6) or (7) are isomorphic to those from (3) or (4) ($0 \to 1$, $1 \to 0$, $2 \to 2$).
The last statement of the lemma follows from (1) and

$$\begin{pmatrix} 0&0&1&0&2 \\ 0&1&1&2&2 \end{pmatrix} \circ \begin{pmatrix} 0&0&1&2&2 \\ 0&1&1&1&2 \end{pmatrix} = \begin{pmatrix} 0&0&1&0&2&2 \\ 0&1&1&2&1&2 \end{pmatrix}.$$

■

Theorem 12.6.3 *The M-projectable subclasses of $P_{3,2}$ are exactly the following:*

(1) $pr^{-1}M$,
(2) $M^{(02)} := Pol\begin{pmatrix} 0&0&1&0 \\ 0&1&1&2 \end{pmatrix}$,
(3) $M^{(22)} := Pol\begin{pmatrix} 0&0&1&2 \\ 0&1&1&2 \end{pmatrix}$,
(4) $M^{(21)} := Pol\begin{pmatrix} 0&0&1&2 \\ 0&1&1&1 \end{pmatrix}$,
(5) $M^{(12)} := Pol\begin{pmatrix} 0&0&1&1 \\ 0&1&1&2 \end{pmatrix}$,
(6) $M^{(02)} \cap M^{(22)}$,
(7) $M^{(02)} \cap M^{(21)}$,
(8) $M^{(21)} \cap M^{(22)}$,
(9) $M^{(20)} := Pol\begin{pmatrix} 0&0&1&2 \\ 0&1&1&0 \end{pmatrix}$,
(10) $M^{(12)} \cap M^{(22)}$,
(11) $M^{(02)(22)} := Pol\begin{pmatrix} 0&0&1&0&2 \\ 0&1&1&2&2 \end{pmatrix}$,
(12) $M^{(02)} \cap M^{(22)} \cap M^{(21)}$,

(13) $M^{(21)(22)} := Pol\begin{pmatrix} 0 & 0 & 1 & 2 & 2 \\ 0 & 1 & 1 & 1 & 2 \end{pmatrix}$,
(14) $M^{(20)} \cap M^{(22)}$,
(15) $M^{(12)} \cap M^{(02)(22)}$,
(16) $M^{(21)} \cap M^{(02)(22)}$,
(17) $M^{(02)} \cap M^{(21)(22)}$,
(18) $M^{(20)} \cap M^{(21)(22)}$,
(19) $M^{(02)(12)(22)} := Pol\begin{pmatrix} 0 & 0 & 1 & 0 & 1 & 2 \\ 0 & 1 & 1 & 2 & 2 & 2 \end{pmatrix}$,
(20) $M^{(02)(21)(22)} := Pol\begin{pmatrix} 0 & 0 & 1 & 0 & 2 & 2 \\ 0 & 1 & 1 & 2 & 1 & 2 \end{pmatrix}$,
(21) $M^{(20)(21)(22)} := Pol\begin{pmatrix} 0 & 0 & 1 & 2 & 2 & 2 \\ 0 & 1 & 1 & 0 & 1 & 2 \end{pmatrix}$,
(22) $pr^{-1}M \cap Z_{2,1}$,
(23) $pr^{-1}M \cap Z_{2,0}$

(see Figure 12.2).

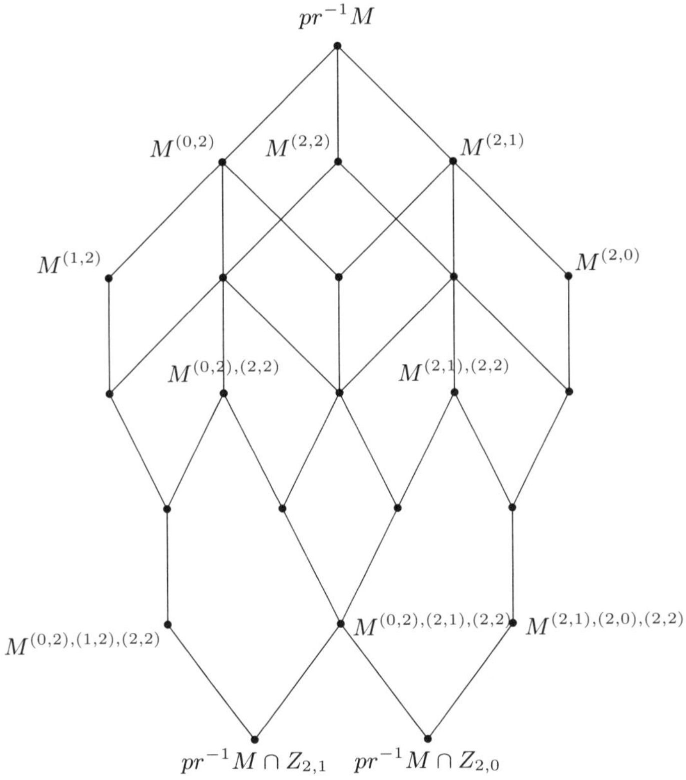

Fig. 12.2

Proof. Because of Lemma 12.6.2 (1),

$$\begin{pmatrix} 0 & 0 & 1 \\ 0 & 1 & 1 \end{pmatrix} \circ \begin{pmatrix} 0 & 0 & 1 & 1 \\ 0 & 1 & 1 & 2 \end{pmatrix} = \begin{pmatrix} 0 & 0 & 1 & 0 & 1 \\ 0 & 1 & 1 & 2 & 2 \end{pmatrix}$$

or

$$\begin{pmatrix} 0 & 0 & 1 & 2 \\ 0 & 1 & 1 & 0 \end{pmatrix} \circ \begin{pmatrix} 0 & 0 & 1 \\ 0 & 1 & 1 \end{pmatrix} = \begin{pmatrix} 0 & 0 & 1 & 2 & 2 \\ 0 & 1 & 1 & 0 & 1 \end{pmatrix}$$

we have $M^{(12)} \subseteq M^{(02)}$ or $M^{(20)} \subseteq M^{(21)}$. With the help of Table 12.2 one can prove the other inclusions (non-inclusions) of the classes (1) - (23). In Table 12.2, the sign $+$ (or $-$) shows whether a function f_i ($i = 1, ..., 10$) belongs (or does not belong) to a class of the left column of the table, respectively.

Table 12.2

	j_1	$\overline{j_0}$	j_2	k_1	f_1	f_2	f_3	f_4	f_5	f_6	f_7	f_8	f_9	f_{10}
1. $pr^{-1}M$	+	+	+	+	+	+	+	+	+	+	+	+	+	+
2. $M^{(02)}$	+	+	+	−	+	−	+	+	+	+	+	+	+	−
3. $M^{(22)}$	+	+	+	+	−	−	+	−	+	+	+	+	+	+
4. $M^{(21)}$	+	+	−	+	−	+	−	+	+	+	+	+	−	+
5. $M^{(12)}$	−	+	+	−	+	−	+	−	−	+	−	−	+	−
6. $M^{(02)} \cap M^{(22)}$	+	+	+	−	−	−	+	−	+	+	+	+	+	−
7. $M^{(02)} \cap M^{(21)}$	+	+	−	−	−	−	−	+	+	+	+	+	−	−
8. $M^{(21)} \cap M^{(22)}$	+	+	−	+	−	−	−	−	+	+	+	+	−	+
9. $M^{(20)}$	+	−	−	+	−	+	−	−	−	−	+	−	−	+
10. $M^{(12)} \cap M^{(22)}$	−	+	+	−	−	−	+	−	−	+	−	−	+	−
11. $M^{(02)(22)}$	+	+	+	−	−	−	−	−	+	−	−	−	+	−
12. $M^{(02)} \cap M^{(22)} \cap M^{(21)}$	+	+	−	−	−	−	−	−	+	+	+	+	−	−
13. $M^{(21)(22)}$	+	+	−	+	−	−	−	−	−	+	−	+	−	+
14. $M^{(20)} \cap M^{(22)}$	+	−	−	+	−	−	−	−	−	−	+	−	−	+
15. $M^{(12)} \cap M^{(02)(22)}$	−	+	+	−	−	−	−	−	−	−	−	−	+	−
16. $M^{(21)} \cap M^{(02)(22)}$	+	+	−	−	−	−	−	−	+	−	−	−	−	−
17. $M^{(02)} \cap M^{(21)(22)}$	+	+	−	−	−	−	−	−	−	+	−	+	−	−
18. $M^{(20)} \cap M^{(21)(22)}$	+	−	−	+	−	−	−	−	−	−	−	−	−	+
19. $M^{(02)(12)(22)}$	−	+	+	−	−	−	−	−	−	−	−	−	−	−
20. $M^{(02)(21)(22)}$	+	+	−	−	−	−	−	−	−	−	−	−	−	−
21. $M^{(20)(21)(22)}$	+	−	−	+	−	−	−	−	−	−	−	−	−	−
22. $pr^{-1}M \cap Z_{21}$	−	+	−	−	−	−	−	−	−	−	−	−	−	−
23. $pr^{-1}M \cap Z_{20}$	+	−	−	−	−	−	−	−	−	−	−	−	−	−

The functions f_1–f_{10} from the above table are defined as follows:

$$f_1(x_1, x_2) := j_0(x_1)j_2(x_2),$$
$$f_2(x_1, x_2) := \overline{j_1(x_1)j_2(x_2)},$$
$$f_3(x_1, x_2) := k_1(x_1)j_2(x_2),$$
$$f_4(x_1, x_2) := j_0(x_1)j_2(x_2) \vee j_1(x_2),$$

$$\begin{aligned}
f_5(x_1,x_2) &:= j_1(x_1)j_1(x_2) \vee j_2(x_1)j_2(x_2),\\
f_6(x_1,x_2) &:= k_1(x_1)\overline{j_0(x_2)} \vee j_2(x_1)j_1(x_2),\\
f_7(x_1,x_2) &:= k_1(x_1)k_1(x_2) \vee j_2(x_1)j_2(x_2),\\
f_8(x_1,x_2) &:= k_1(x_1)j_2(x_2) \vee j_1(x_2),\\
f_9(x_1,x_2) &:= j_1(x_1)j_2(x_2),\\
f_{10}(x_1,x_2) &:= \overline{j_0(x_1)j_2(x_2)}.
\end{aligned}$$

Thus it remains to show that no further classes are described by (12.17) than the ones listed above.

By Theorem 12.6.1, only the sets $P_{3,2}$, $Z_{2,0}$, and $Z_{2,1}$ are possible for A' in (12.17). Since $pr^{-1}M \cap Z_{2,0}$ and $pr^{-1}M \cap Z_{2,1}$ are minimal M-projectable classes of $P_{3,2}$, the M-projectable classes different from these classes are described as follows

$$\bigcap_{i=1}^{m} Pol\begin{pmatrix} 0 & 0 & 1 \\ 0 & 1 & 1 \end{pmatrix} \cup \varrho_i, \tag{12.19}$$

where the relations $\varrho_1, ..., \varrho_m$ are M-permissible subsets of $E_3^2\backslash E_2^2$. With the aid of the statement (2) of Lemma 12.6.2 and considering (12.18), one can be convinced that the relations ϱ_i, $i = 1, ..., m$, must be from the set

$$\left\{\begin{pmatrix} 0 \\ 2 \end{pmatrix}, \begin{pmatrix} 1 \\ 2 \end{pmatrix}, \begin{pmatrix} 2 \\ 0 \end{pmatrix}, \begin{pmatrix} 2 \\ 1 \end{pmatrix}, \begin{pmatrix} 2 \\ 2 \end{pmatrix}, \begin{pmatrix} 0 & 1 \\ 2 & 2 \end{pmatrix}, \begin{pmatrix} 0 & 2 \\ 2 & 1 \end{pmatrix}, \begin{pmatrix} 0 & 2 \\ 2 & 2 \end{pmatrix}, \begin{pmatrix} 1 & 2 \\ 2 & 2 \end{pmatrix},\right.$$
$$\left.\begin{pmatrix} 2 & 2 \\ 0 & 1 \end{pmatrix}, \begin{pmatrix} 2 & 2 \\ 0 & 2 \end{pmatrix}, \begin{pmatrix} 2 & 2 \\ 1 & 2 \end{pmatrix}, \begin{pmatrix} 0 & 1 & 2 \\ 2 & 2 & 2 \end{pmatrix}, \begin{pmatrix} 0 & 2 & 2 \\ 2 & 1 & 2 \end{pmatrix}, \begin{pmatrix} 2 & 2 & 2 \\ 0 & 1 & 2 \end{pmatrix}\right\}.$$

Because of Lemma 12.6.2, (3)–(8) we can assume that the relations $\varrho_1, ..., \varrho_m$ belong to

$$\left\{\begin{pmatrix} 0 \\ 2 \end{pmatrix}, \begin{pmatrix} 1 \\ 2 \end{pmatrix}, \begin{pmatrix} 2 \\ 0 \end{pmatrix}, \begin{pmatrix} 2 \\ 1 \end{pmatrix}, \begin{pmatrix} 2 \\ 2 \end{pmatrix}, \begin{pmatrix} 0 & 2 \\ 2 & 2 \end{pmatrix}, \begin{pmatrix} 2 & 2 \\ 1 & 2 \end{pmatrix}, \begin{pmatrix} 0 & 1 & 2 \\ 2 & 2 & 2 \end{pmatrix}, \begin{pmatrix} 2 & 2 & 2 \\ 0 & 1 & 2 \end{pmatrix}\right\}.$$

Now it is not difficult to find all classes, which are describable through (12.19), in a step-by-step way. Because of Lemma 12.6.2, (1) or Theorem 12.5.6, the largest M-projectable classes lying below $pr^{-1}M$ are the sets $M^{(02)}$, $M^{(22)}$ and $M^{(21)}$. Then (with the help of Lemma 12.6.2, (2)) possible intersections of these sets and the next largest classes of the form (12.19) with $m = 1$ belong to layers 1–4 in Figure 12.2. As intersections of classes of the fourth layer (i.e., as next classes of the form (3) with $m = 1$), only the classes of the fifth layer are possible, and so forth. ■

Theorem 12.6.4 *Let $a \in E_2$. Then*

$$\begin{aligned}
\mathfrak{N}_3(M \cap T_a) \;=\; & \{A \cap pr^{-1}M \cap T_a \mid A \in \mathfrak{N}_3(M)\}\cup \\
& \cup\{A \cap T_a^{(2)} \mid A \in \mathfrak{N}_3(M) \wedge A \not\subseteq M^{(12)} \wedge A \not\subseteq M^{(20)}\}
\end{aligned}$$

and

$$|\mathfrak{N}_3(M \cap T_a)| = 36.$$

Proof. W.l.o.g. let $a = 0$. Put $A_3 := M \cap T_0$. It is easy to check that the 36 sets given in the theorem are all A_3-projectable and pairwise different (see Figure 12.3).
By Theorem 12.3.4, for every class A of $\mathfrak{N}_3(A_3)$ there are certain sets A_1, A_2 with $A = A_1 \cap A_2$, $A_1 \in \mathfrak{N}_3(M)$, and $A_2 \in \mathfrak{N}_3(T_0)$. Because of Theorem 12.3.6, (3) the sets of $\mathfrak{N}_3(A_3)$ different from $Z_{2,0} \cap pr^{-1}A_3$ and $Z_{2,1} \cap pr^{-1}A_3$ belong to the set

$$\{A \cap pr^{-1}A_3, A \cap T_0^{(2)} \mid A \in \mathfrak{N}_3(M) \backslash \{Z_{2,0} \cap pr^{-1}M, Z_{2,1} \cap pr^{-1}M\}\}$$

Since $pr(T_0^{(2)} \cap M^{(12)}) \neq M \cap T_0$ and $pr(M_0^{(20)} \cap pr^{-1}(M \cap T_0)) \subseteq T_0^{(2)}$ is obviously valid, the set $\mathfrak{N}_3(A_3)$ agrees with the set given in Theorem 12.6.4, and it holds $|\mathfrak{N}_3(A_3)| = 36$ by Theorem 12.6.3. ∎

Theorem 12.6.5 *Exactly 49 subclasses of $P_{3,2}$ are $M \cap T_0 \cap T_1$-projectable, and it holds*

$$\begin{aligned}&\mathfrak{N}_3(M \cap T_0 \cap T_1)\\ =\ &\{A \cap pr^{-1}(M \cap T_0 \cap T_1) \mid A \in \mathfrak{N}_3(M)\} \cup \\ &\{A \cap T_a^{(2)} \mid a \in E_2 \wedge A \in \mathfrak{N}_3(M) \wedge A \not\subseteq M^{(12)} \wedge A \not\subseteq M^{(20)}\}\end{aligned}$$

(see Figure 12.4, where $A_4 := M \cap T_0 \cap T_1$).

Proof. Analogous to the considerations in the proof of Theorem 12.6.4 the statements from Theorem 12.6.5 are a consequence of Theorems 12.3.4, 12.6.1, and 12.6.3, where one must consider that

$$pr(T_0^{(2)} \cap M^{(12)} \cap pr^{-1}T_1) \neq M \cap T_0 \cap T_1$$

and

$$pr(T_1^{(2)} \cap M^{(20)} \cap pr^{-1}T_0) \neq M \cap T_0 \cap T_1.$$

∎

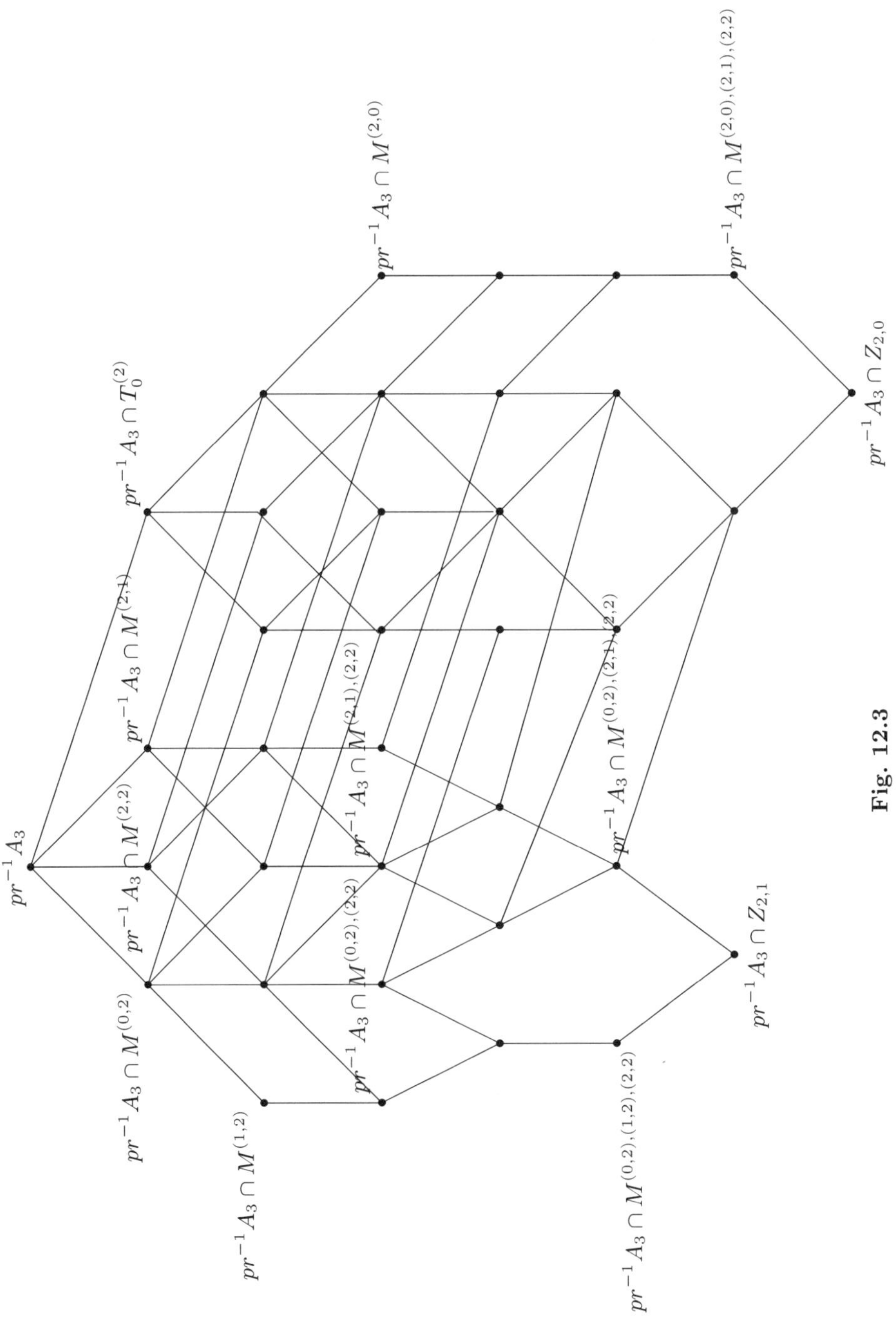
$pr^{-1} A_3$
$pr^{-1} A_3 \cap T_0^{(2)}$
$pr^{-1} A_3 \cap M^{(2,1)}$
$pr^{-1} A_3 \cap M^{(2,2)}$
$pr^{-1} A_3 \cap M^{(0,2)}$
$pr^{-1} A_3 \cap M^{(1,2)}$
$pr^{-1} A_3 \cap M^{(2,0)}$
$pr^{-1} A_3 \cap M^{(2,1),(2,2)}$
$pr^{-1} A_3 \cap M^{(0,2),(2,2)}$
$pr^{-1} A_3 \cap M^{(0,2),(2,1),(2,2)}$
$pr^{-1} A_3 \cap M^{(0,2),(1,2),(2,2)}$
$pr^{-1} A_3 \cap M^{(2,0),(2,1),(2,2)}$
$pr^{-1} A_3 \cap Z_{2,0}$
$pr^{-1} A_3 \cap Z_{2,1}$

Fig. 12.3

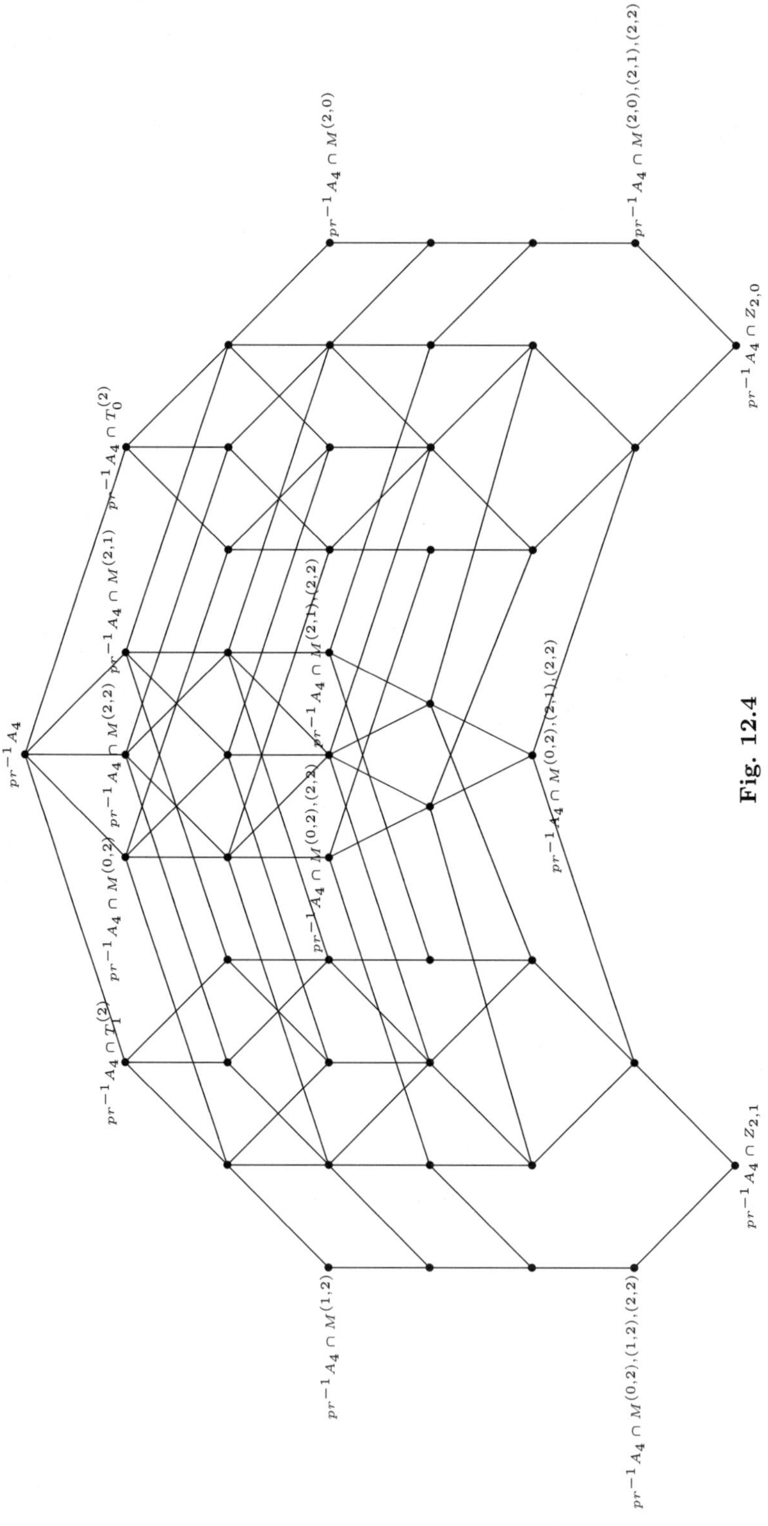
$pr^{-1}A_4$
$pr^{-1}A_4 \cap T_0^{(2)}$
$pr^{-1}A_4 \cap M^{(2,1)}$
$pr^{-1}A_4 \cap M^{(2,2)}$
$pr^{-1}A_4 \cap M^{(0,2)}$
$pr^{-1}A_4 \cap T_1^{(2)}$
$pr^{-1}A_4 \cap M^{(2,0)}$
$pr^{-1}A_4 \cap M^{(2,1),(2,2)}$
$pr^{-1}A_4 \cap M^{(0,2),(2,2)}$
$pr^{-1}A_4 \cap M^{(1,2)}$
$pr^{-1}A_4 \cap M^{(0,2),(2,1),(2,2)}$
$pr^{-1}A_4 \cap M^{(2,0),(2,1),(2,2)}$
$pr^{-1}A_4 \cap M^{(0,2),(1,2),(2,2)}$
$pr^{-1}A_4 \cap Z_{2,0}$
$pr^{-1}A_4 \cap Z_{2,1}$

Fig. 12.4

Now we come to the determining of sets $\mathfrak{N}_3(A)$ with

$$A \in \{L, L\cap T_0, L\cap T_1, L\cap S, L\cap T_0\cap S\}.$$

As already shown in Section 12.5 while preparing proof of Theorem 12.5.8, there are some $a, a_i, a_{I,J}$ of E_2 with

$$f(\mathbf{x}) = a + \sum_{i=1}^{n} a_i j_1(x_i) + \sum_{\substack{I,J \\ I\cup J\subseteq\{1,...,n\} \\ I\cap J=\emptyset \\ J\neq\emptyset}} a_{I,J}\cdot k_{I,J}(\mathbf{x}) \qquad (12.20)$$

for every function $f^n \in pr^{-1}L$. The functions

$$k_{I,J}(\mathbf{x}) := (\bigwedge_{i\in I} j_1(x_i))\cdot(\bigwedge_{j\in J} j_2(x_j))$$

with $a_{I,J} = 1$ in (12.20) are called **components** of f . Let K_f be the set of all components of f. Further, let $K := \bigcup_{f\in pr^{-1}L} K_f$ and $K_{\mathfrak{M}} := \bigcup_{f\in\mathfrak{M}} K_f$, $\mathfrak{M} \subseteq P_{3,2}$.

It is easy to check that the subsequently defined subsets of $pr^{-1}L$ are closed and are finitely generated by the given subsets:

$$\begin{aligned}
L_2 \quad &:= Pol\begin{pmatrix} 0&0&0&1&1&0&1&1&2\\ 0&0&1&1&0&1&0&1&2\\ 0&1&0&0&1&1&0&1&2\\ 0&1&1&0&0&0&1&1&2 \end{pmatrix}\\
&= \{f\in pr^{-1}L \mid K_f \subseteq \{k_{I,J}\in K \mid |I|\leq 1\}\}\\
&= [j_1(x_1)+j_1(x_2), c_1, j_1(x_1)\cdot j_2(x_2)],\\
L_{2,r} &:= \{f\in pr^{-1}L \mid K_f \subseteq \{k_{I,J}\in K \mid I=\emptyset \wedge |J|\leq 4\}\}\\
&= \begin{cases} [j_1(x_1)+j_1(x_2), c_1, j_2(x_1)\cdot\ldots\cdot j_2(x_r)] & \text{if } r<\infty,\\ \left[\{j_1(x_1)+j_1(x_2), c_1\}\cup\bigcup_{q\geq 1} j_2(x_1)\cdot\ldots\cdot j_2(x_q)\}\right] & \text{if } r=\infty \end{cases}\\
&\quad (1\leq r\leq\infty),
\end{aligned}$$

$$\begin{aligned}
&Z_{2,0}\cap pr^{-1}L\\
&= \textstyle\bigcup_{n\geq 1}\{f^n\in P_{3,2} \mid \exists a, a_1, ..., a_n \in E_2 :\ f(\mathbf{x}) = a + \sum_{i=1}^{n} a_i\cdot j_1(x_i)\},\\
&= [j_1(x_1)+j_1(x_2), c_1],
\end{aligned}$$

$$\begin{aligned}
&Z_{2,1}\cap pr^{-1}L\\
&= \textstyle\bigcup_{n\geq 1}\{f^n\in P_{3,2} \mid \exists a, a_1, ..., a_n \in E_2 :\\
&\qquad\qquad f(\mathbf{x}) = a + \textstyle\sum_{i=1}^{n} a_i\cdot(j_1(x_i)+j_2(x_i))\},\\
&= [j_0(x_1)+j_0(x_2), c_1].
\end{aligned}$$

Further put

$$\begin{aligned} K' &:= \{k_{I,J} \in K \mid J \neq \emptyset \wedge I \cap J = \emptyset\}, \\ K_1 &:= \{k_{I,J} \in K' \mid |I| \geq 1\}, \\ K_2 &:= \{k_{I,J} \in K' \mid |I| \leq 1\}, \\ K_3 &:= K_1 \cap K_2 = \{k_{I,J} \in K \mid |I| = 1\}, \\ K_{0,r} &:= \{k_{\emptyset,J} \in K' \mid |J| \leq r\}, \qquad 1 \leq r \leq \infty. \end{aligned}$$

Lemma 12.6.6 *Let $f^n \in P_{3,2}$, pr $f^n = c_0^n$, $f^n \neq c_0^n$ and $K_f \subseteq K_{0,\infty}$. Further, let r be the smallest number for which there is a function $k_{\emptyset,J} \in K_f$ with $|J| = r$. Then*

$$k_{\emptyset,\{1,\dots,r\}}(\mathbf{x}) = j_2(x_1) \cdot \dots \cdot j_2(x_r) \in [f, j_1].$$

Proof. By assumption, in K_f there is a function $k_{\emptyset,J}$ with $J' \not\subseteq J$ for all $k_{\emptyset,J'} \in K_f \backslash \{k_{\emptyset,J}\}$. Consequently, we have

$$f(h_1(x_1), \dots, h_n(x_n)) = k_{\emptyset,J}(\mathbf{x}),$$

if

$$h_i(x) := \begin{cases} j_1(x) & \text{if } i \notin J, \\ x & \text{if } i \in J \end{cases}$$

and $i = 1, \dots, n$. ∎

Lemma 12.6.7 *Let $f^n \in P_{3,2}$, pr $f^n = c_0^n$ and $K_f \not\subseteq K_{0,\infty}$. Then there is an $a \in E_2$ with*

(1) $j_1(x_1)j_2(x_2) + a \cdot j_2(x_2) \in [f, j_1]$;

(2) $j_1(x_1)j_1(x_2)j_2(x_3) + a \cdot j_2(x_3) \in [f, j_1]$, if furthermore $K_f \cap (K_1 \backslash K_2) \neq \emptyset$ holds.

Proof. First we prove (1) through induction on the arity n of f.
$n = 2$: Since $K_f \not\subseteq K_{0,\infty}$ we can assume w.l.o.g.

$$\begin{aligned} f^2(x_1, x_2) &= j_1(x_1)j_2(x_2) + \alpha \cdot j_2(x_1)j_1(x_2) + \beta \cdot j_2(x_1) + \gamma \cdot j_2(x_2) + \\ &\quad + \delta \cdot j_2(x_1)j_2(x_2) \qquad (\alpha, \beta, \gamma, \delta \in E_2). \end{aligned}$$

Thus we have $f(j_1(x_1), x_2) = j_1(x_1)j_2(x_2) + \gamma \cdot j_2(x_2)$, i.e., (1) is right for $n = 2$.
$n-1 \longrightarrow n$: Suppose (1) holds for all $(n-1)$-ary functions f with pr $f = c_0^{n-1}$ and $K_f \not\subseteq K_{0,\infty}$, $n > 2$.
Now let f be an arbitrary n-ary function with pr $f = c_0^n$ and $K_f \not\subseteq K_{0,\infty}$. Then there are $(n-1)$-ary functions g, h and q with (w.l.o.g.)

$$f(\mathbf{x}) = j_1(x_1) \cdot g(x_2, \dots, x_n) + j_2(x_1) \cdot h(x_2, \dots, x_n) + q(x_2, \dots, x_n),$$

where $g \neq c_0^{n-1}$. Consequently,

$$f(j_1(x_1), x_2, ..., x_n) = j_1(x_1) \cdot g(x_2, ..., x_n) + q(x_2, ..., x_n) =: f'(\mathbf{x}).$$

Case 1: $K_g \subseteq K_{0,\infty}$.
With the aid of the proof of Lemma 12.6.6, it is easy to see that $j_1(x_1)j_2(x_2)+ a \cdot j_2(x_2)$ is a superposition over $\{f', j_1\}$ for certain $a \in E_2$.
Case 2: $K_g \not\subseteq K_{0,\infty}$.
By induction assumption, the function $j_1(x_2)j_2(x_3) + b \cdot j_2(x_3)$ is a superposition on g and j_1 for certain $b \in E_2$. Thus we can construct the function

$$f''(x_1, x_2, x_3) := j_1(x_1)(j_1(x_2)j_2(x_3) + b \cdot j_2(x_3)) + q'(x_2, x_3),$$

where q' denotes a c_0-projectable function, as a superposition over f and j_1 (under the given conditions). For $b = 1$ we have

$$f''(x_1, x_2, x_2) = j_1(x_1)j_2(x_2) + a \cdot j_2(x_2), \qquad a := q'(2, 2).$$

If $b = 0$ then we construct the function

$$\begin{aligned} f'''(x_1, x_2, x_3) &:= f''(x_1, j_1(x_2), x_3) \\ &= j_1(x_1)j_1(x_2)j_2(x_3) + \alpha \cdot j_1(x_2)j_2(x_3) + \beta \cdot j_2(x_3) \end{aligned}$$

first. Consequently,

$$f'''(x_2, x_1, x_2) = j_1(x_1)j_2(x_2) + \beta \cdot j_2(x_2)$$

if $\alpha = 1$, and

$$f'''(x_1, x_1, x_2) = j_1(x_1)j_2(x_2) + \beta \cdot j_2(x_2)$$

in the case $\alpha = 0$. Hence (1) holds for each n-ary function $f \in P_{3,2}$ with $pr\ f \neq c_0^n$ and $K_f \not\subseteq K_{0,\infty}$.
Proof for (2) is obtained through the transfer of proof for (1). ■

Lemma 12.6.8 *Let* $B \in \{L, L\cap T_0, L\cap S, L\cap T_0\cap S\}$, $A \subseteq Z_{2,0}\cap pr^{-1}B$ *and* $[pr\ A] = B$. *Then*

(1) $pr^{-1}B = [A \cup \{j_1(x) + j_1(x_1)j_1(x_2)j_2(x_3), j_1(x) + j_2(x_1)\}]$,
(2) $L_2 \cap pr^{-1}B = [A \cup \{j_1(x) + j_1(x_1)j_2(x_2), j_1(x) + j_2(x_1)\}]$,
(3) $T_0^{(2)} \cap pr^{-1}B = [A \cup \{j_1(x) + j_1(x_1)j_1(x_2)j_2(x_3)\}]$,
(4) $T_0^{(2)} \cap L_2 \cap pr^{-1}B = [A \cup \{j_1(x) + j_1(x_1)j_2(x_2)\}]$,
(5) $L_{2,r} \cap pr^{-1}B = [A \cup \{j_1(x) + j_2(x_1)j_2(x_2)...j_2(x_r)\}]$, $r \geq 1$,
(6) $L_{2,\infty} \cap pr^{-1}B = [A \cup \bigcup_{q\geq 1}\{j_1(x) + j_2(x_1)j_2(x_2)...j_2(x_q)\}]$,
(7) $Z_{2,0} \cap pr^{-1}B = [A]$.

Proof. The proof results from definitions of the considered classes. ■

Theorem 12.6.9 *Let $B \in \{L, L\cap T_0, L\cap S, L\cap T_0\cap S\}$. Then the following B-projectable classes ($\subseteq P_{3,2}$) only exist:*

(1) $pr^{-1}B$,
(2) $L_2 \cap pr^{-1}B$,
(3) $L_{2,r} \cap pr^{-1}B$, $r = 1, 2, ...,$
(4) $L_{2,\infty} \cap pr^{-1}B$,
(5) $Z_{2,a} \cap pr^{-1}B$, $a \in E_2$,
(6) $T_a^{(2)} \cap pr^{-1}B$, $a \in E_2$, if $B \cap T_a = B$,
(7) $T_a^{(2)} \cap L_2 \cap pr^{-1}B$, $a \in E_2$, if $B \cap T_a = B$

(see Figure 12.5).

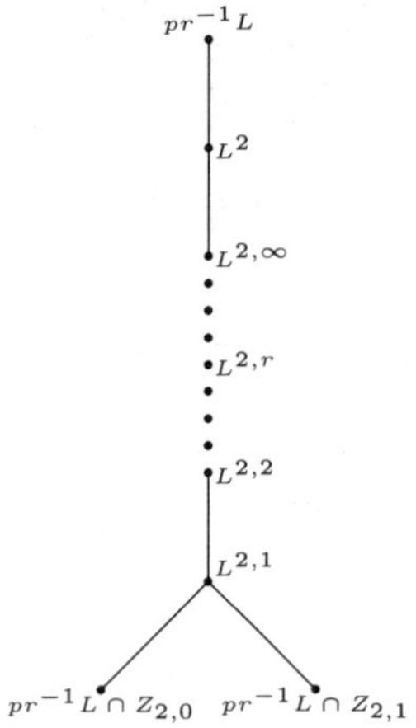

Fig. 12.5

Proof. Let A be a B-projectable subclass of $P_{3,2}$. Then, j_1 or $j_1 + j_2$ belongs to A. W.l.o.g. we can assume $j_1 \in A$, since $s \star j_1 \star s = j_1 + j_2$ holds for $s(x) := 2x+1 \ (mod\ 3)$ and thus all considerations are in the case $j_1 \in A$ isomorphically to those of the case $j_1 + j_2 \in A$. Consequently, we have $Z_{2,0} \cap pr^{-1}B \subseteq A$ and $j_1(x_1) + j_1(x_2) + j_1(x_3) \in A$.

For K_A ($= \bigcup_{f\in A} K_f$) the following cases are possible:

Case 1: $K_A \subseteq K_{0,\infty}$ (i.e., $A \subseteq L_{2,\infty}$).

Case 1.1: $K_A = \emptyset$.

In this case we have: $A = Z_{2,0} \cap pr^{-1}B$.

Case 1.2: It exists an $r \geq 1$ with $K_A \not\subseteq K_{0,r-1}$ and $K_A \subseteq K_{0,r}$.

Then, there is a function $f^n \in A$ with $K_f \cap K_{0,r} \neq \emptyset$. Consequently, the function

$$f_1(x, \mathbf{x}) := j_1(x) + f(j_1(x_1), ..., j_1(x_n)) + f(\mathbf{x}) = j_1(x) + f_1'(\mathbf{x})$$

with

$$f_1'(\mathbf{x}) := \sum_{k_{\emptyset,J}\in K_f} k_{\emptyset,J}(\mathbf{x}).$$

belongs to A. By Lemma 12.6.6, one can construct the function $j_2(x_1)...j_2(x_t)$, where

$$t := \min_{k_{\emptyset,J}\in K_f} |J|,$$

as a superposition over f_1' and j_1. Consequently, $j_1(x) + j_2(x_1)....j_2(x_{t'}) \in A$ for $t' \leq t$.
A superposition over these functions and f_1 is also

$$\begin{aligned} f_2(x,\mathbf{x}) &:= j_1(x) + f_1'(\mathbf{x}) + \sum_{\substack{k_{\emptyset,J} \in K_f \\ |J| \leq t}} k_{\emptyset,J}(\mathbf{x}) \\ &= j_1(x) + f_2'(\mathbf{x}) \quad \text{with} \quad f_2'(\mathbf{x}) := \sum_{\substack{k_{\emptyset,J} \in K_f \\ |J| > t}} k_{\emptyset,J}(\mathbf{x}). \end{aligned}$$

If $f_2' \neq c_0^n$, then we have the function $j_2(x_1)...j_2(x_s)$, where

$$s := \min_{k_{\emptyset,J}\in K_f,\ |J|>t} |J|,$$

by substituting certain variables by j_1 and by possibly changing the numbering of the variables from f_2'. Consequently, the function

$$f_3'(x,\mathbf{x}) := j_1(x) + \sum_{\substack{k_{\emptyset,J} \in K_f \\ |J| > s}} k_{\emptyset,J}(\mathbf{x})$$

belongs to A, etc. By iterated application of these constructions we obtain the function

$$j_1(x) + j_2(x_1)...j_2(x_r) \in A.$$

Hence, by Lemma 12.6.8, (5), we have $A = L_{2,r} \cap pr^{-1}B$.

Case 1.3: $K_A \cap K_{0,r} \neq \emptyset$ for all $r \geq 1$.
Because of the considerations from Case 1.2 and Lemma 12.6.8, (6) is valid: $A = L_{2,\infty} \cap^{-1} B$. $A = L_{2,\infty} \cap^{-1} B$.

Case 2: $K_A \not\subseteq K_{2,\infty}$.
In this case, there is a function $f^n \in A$ with $K_f \cap K_1 \neq \emptyset$. Then, the function

$$f_1(x,\mathbf{x}) := j_1(x) + f(\mathbf{x}) + f(j_1(x_1),...,j_1(x_n)) = j_1(x) + f_1'(\mathbf{x})$$

with

$$f_1'(\mathbf{x}) := j_1(x) + \sum_{k_{I,J}\in K_f} k_{I,J}(\mathbf{x})$$

belongs to A.
With the aid of Lemma 12.6.6, it is easy to see that the function

$$f_1''(x_1,x_2) := j_1(x_1)j_2(x_2) + a\cdot j_2(x_2),\ a\in E_2$$

is a superposition over $\{j_1, f_1'\}$. Therefore, the function

$$g(x,x_1,x_2) := j_1(x) + j_1(x_1)j_2(x_2) + a\cdot j_2(x_2)$$

is a superposition on $\{f_1, j_1\}$, i.e., $g \in A$. Consequently,

$$g(g(x,x_1,x_2),x_2,x_2) = j_1(x) + j_1(x_1)j_2(x_2)$$

belongs to A. Then, by Lemma 12.6.8, we have $T_0^{(2)} \cap L_2 \cap pr^{-1}B \subseteq A$. If $A \neq T_0^{(2)} \cap L_2 \cap pr^{-1}B$ then the following three cases are possible:
Case 2.1: $A \subseteq L_2$ and $pr\ A \not\subseteq T_0$.
In this case, there is a function $p \in A$ with $p(0) = 1$. For $(pr\ p)(y) = \overline{y}$ it holds

$$j_1(x) + \overline{j_1(x_1)}j_2(x_1) = j_1(x) + j_2(x_1) \in A.$$

If $pr\ p = c_1$ then we also obtain

$$j_1(x) + j_2(x_1) = j_1(x) + p(p(x_1))j_2(x_1) \in A.$$

Consequently, by Lemma 12.6.8, $A = L_2 \cap pr^{-1}B$.
Case 2.2: $A \subseteq L_2$, $pr\ A \subseteq T_0$ and $pr\ A \not\subseteq T_0^{(2)}$.
In this case, there is a function $r^2 \in A$ with $r(0,0) = 0$ and $r(0,2) = 1$, and it holds that

$$j_1(x) + r(j_1(x_1),x_1) + r(j_1(x_1),j_1(x_1)) = j_1(x) + j_2(x_1) \in A.$$

Therefore, by Lemma 12.6.8, $A = L_2 \cap pr^{-1}B$.
Case 2.3: $A \not\subseteq L_2$.
Denote q an n-ary function of A, which does not belong to L_2. Then,

$$q_1(x,\mathbf{x}) := j_1(x) + q(\mathbf{x}) + q(j_1(x_1),...,j_1(x_n)) = j_1(x) + q_1'(\mathbf{x}),$$

where

$$q_1'(\mathbf{x}) := \sum_{k_{I,J}\in K_q} k_{I,j}(\mathbf{x}),$$

is a function of A. By Lemma 12.6.6, a function of the form

$$j_1(x_1)j_1(x_2)j_2(x_3) + a\cdot j_2(x_3)$$

is a superposition over $\{q_1', j_1\}$. Thus,

$$q_2(x,x_1,x_2,x_3) := j_1(x) + j_1(x_1)j_1(x_2)j_2(x_3) + a\cdot j_2(x_3) \in A$$

and therefore

$$q_2(q_2(x,x_1,x_2,x_3),x_3,x_3) := j_1(x) + j_1(x_1)j_1(x_2)j_2(x_3) \in A.$$

Thus by Lemma 12.6.8, we have $T_0^{(2)} \cap pr^{-1}B \subseteq A$. If $a \neq T_0^{(2)} \cap pr^{-1}B$, then there is either a function $p \in A$ with $p(0) = 1$, or a function $r \in A$ with $r(0,0) = 0$ and $r(0,2) = 1$. As already shown in the above Case 2.1 or 2.2, $j_1(x) + j_2(x_1) \in A$ results. Thus by Lemma 12.6.8, (1), $A = pr^{-1}B$ holds. ■

Next we study the set $\mathfrak{N}_3(T_{0,m})$ with $m \in \mathbb{N}\backslash\{1\}$. By Theorem 12.3.1 (or 12.3.2), for every $m \in \mathbb{N}\backslash\{1\}$, we know that $\mathfrak{N}_3(T_{0,m})$ is a finite set. In the following, we determine the elements of $\mathfrak{N}_3(T_{0,m})$ and we show that $|\mathfrak{N}_3(T_{0,2})| = 148$ holds. Since the description of the elements of the other sets $\mathfrak{N}_3(A)$ with $S \cap M \subseteq A$ or $T_{a,m} \cap T_{\overline{a}} \cap M \subseteq A$ requires a lot of place, we must renounce this description here and refer the reader to the dissertation by N. Grünwald (see [Grü 84]). In the dissertation by N. Grünwald, one also finds the proofs left out in the following.

Lemma 12.6.10 *For every set*

$$T := A \cap \left(\bigcap_{j \in J} Pol_{P_{3,2}}(E_2^m \backslash \{\mathbf{1}\} \cup \varrho_j)\right)$$

with $A \in \{T_0^{(2)}, P_{3,2}\}$, $\varrho_j \subseteq E_3^m \backslash E_2^m$, *and every set*

$$B := \{g \in K \mid g \text{ has the value 1 on at most } m \text{ tuples}\} \cup \{g_1^{m+1}\}$$

with $pr\ g_1^{m+1} = h_m$ *($\in P_2$; see Chapter 3) it holds* $[B] = T$.

Proof. Let $f^n \in T$ and let $\mathbf{a_1}, ..., \mathbf{a_r}$ be the tuples of E_3^n, on which the function f^n has the value 1. By induction on r, we prove $f^n \in [B]$: For $r \leq m$ the assertion is obviously right. For $r > m$ we assume that all functions of T, which have the value 1 on less than r tuples, belong to $[B]$. By definition of T, the functions

$$f_i^n(\mathbf{x}) := \begin{cases} 0 & \text{if} \quad \mathbf{x} = \mathbf{a_i}, \\ f^n(\mathbf{x}) & \text{otherwise} \end{cases}$$

$(i = 1, ..., m+1)$ belong to T. Thus, by assumption, we have $f_1, ..., f_n \in [B]$. Then our assertion follows from

$$f^n(\mathbf{x}) = g_1(f_1(\mathbf{x}), f_2(\mathbf{x}), ..., f_{m+1}(\mathbf{x})) \in [B]. \quad ■$$

Lemma 12.6.11 *Let*

$$W_{1,m} := Pol_{P_{3,2}}(E_3^m \backslash \{\mathbf{1}\}),$$
$$W_{2,m} := Pol_{P_{3,2}}(E_3^m \backslash \{1,2\}^m).$$

Then

(a) For arbitrary sets $A_i \subseteq W_{i,m}$ $(i \in \{1,2\})$ *with* $[pr\ A_i] = T_{0,m}$ *the set*

$$B_1 := A_1 \cup \{j_1(x_1) \cdot j_2(x_2), j_1(x), j_1(x_1) \cdot j_0(x_2)\}$$

is a generating system for $W_{1,m}$ *and*

$$B_2 := A_2 \cup \{j_1(x_1) \cdot j_2(x_2), j_1(x), j_1(x_1) \cdot j_0(x_2)\}$$

is a generating system for $W_{2,m}$.
(b) The set $pr^{-1}T_{0,2} \cap Z_{2,0}$ *is maximal in* $W_{1,m}$.
(c) The set $pr^{-1}T_{0,2} \cap Z_{2,1}$ *is maximal in* $W_{2,m}$.
(d) For all $A \in \mathfrak{N}_3(T_{0,m})\backslash\{pr^{-1}T_{0,2}\cap Z_{2,0}, pr^{-1}T_{0,2}\cap Z_{2,1}\}$ *it holds* $W_{1,n} \subseteq A$.

Proof. See [Grü 84]. ■

Lemma 12.6.12 *Let A be a set of the following form*

$$\bigcap_{i\in I} Pol\ ((E_2^m\backslash\{\mathbf{1}\}) \cup \varrho_i) \quad mit\ \varrho_i \subseteq (E_3^m\backslash(E_2^m \cup \{\mathbf{2}\})),\ m \geq 2. \qquad (12.21)$$

Then

(1) A is $T_{0,m}$*-projectable and the only maximal classes of A that belong to* $\mathfrak{N}_3(T_{0,m})$ *are the following classes:*
(a) $A \cap T_0^{(2)}$;
(b) $A \cap Pol\ ((E_2^m\backslash\{\mathbf{1}\} \cup \delta)$, *where* δ *fulfills the following conditions:*

$$\delta \subseteq (E_3^m\backslash(E_2^m \cup \{\mathbf{2}\},$$
$$\nexists i \in I:\ \delta \subseteq \varrho_i,$$
$$\forall \varepsilon \subset \delta \exists i \in I:\ \varepsilon\varrho_i.$$

(2) If $B \subseteq A$, $[pr\ B] = T_{0,m}$, $B \nsubseteq Z_{2,i} \cap pr^{-1}T_{0,m}$, $i = 0, 1$, *and if B is not contained in the sets of the above type (a) or (b), then* $[B] = A$ *holds.*

(3) If special $A = W_{2,m}$, *then* $Z_{2,1} \cap pr^{-1}T_{0,m}$ *is the only* $T_{0,m}$*-projectable maximal class of* $W_{2,m}$.

Proof. See [Grü 84]. ■

Lemma 12.6.13 *Denote A a set of the form*

$$T_0^{(2)} \cap \bigcap_{i\in I} Pol\ ((E_2^m\backslash\{\mathbf{1}\}) \cup \varrho_i) \quad mit\ \varrho_i \subseteq (E_3^m\backslash E_2^m,\ m \geq 2. \qquad (12.22)$$

Then:

(1) A is $T_{0,m}$*-projectable, and the only maximal classes of A, which belong to* $\mathfrak{N}_3(T_{0,m})$, *are classes of the following form:*

$$A \cap Pol\ ((E_2^m\backslash\{\mathbf{1}\} \cup \delta), \qquad (12.23)$$

where δ *fulfills the following conditions:*

$$\delta \subseteq E_3^m\backslash E_2^m,$$
$$\nexists i \in I:\ \delta \subseteq \varrho_i,$$
$$\forall \varepsilon \subset \delta \exists i \in I:\ \varepsilon\varrho_i.$$

(2) If $B \subseteq A$, $[pr\ B] = T_{0,m}$, $B \not\subseteq Z_{2,i} \cap pr^{-1} T_{0,m}$, $i = 0, 1$, and $B \not\subseteq C$ for every class C that was described in (1), then it holds $[B] = A$.

(3) If in particular $A = W_{1,m}$ then $Z_{2,0} \cap pr^{-1} T_{0,m}$ is the only $T_{0,m}$-projectable maximal class of $W_{1,m}$.

Proof. See [Grü 84]. ■

Theorem 12.6.14 *([Grü 84])*

There are only finitely many subclasses of $P_{3,2}$, which are $T_{0,m}$-projectable. One receives a concrete description of these subclasses with the aid of Lemmas 12.6.12 and 12.6.13.

Proof. See [Grü 84]. ■

In the following table are explanations to notations that we need to describe the $T_{0,2}$-projectable subclasses of $P_{3,2}$.

Table 12.3

i	τ_i'	i	τ_i'
1	$\begin{pmatrix} 1 \\ 2 \end{pmatrix}$	10	$\begin{pmatrix} 1 & 2 & 0 \\ 2 & 1 & 2 \end{pmatrix}$
2	$\begin{pmatrix} 0 \\ 2 \end{pmatrix}$	11	$\begin{pmatrix} 1 & 2 & 2 \\ 2 & 1 & 2 \end{pmatrix}$
3	$\begin{pmatrix} 2 \\ 2 \end{pmatrix}$	12	$\begin{pmatrix} 2 & 2 & 2 \\ 1 & 0 & 2 \end{pmatrix}$
4	$\begin{pmatrix} 1 & 2 \\ 2 & 1 \end{pmatrix}$	13	$\begin{pmatrix} 2 & 0 & 2 \\ 1 & 2 & 2 \end{pmatrix}$
5	$\begin{pmatrix} 1 & 2 \\ 2 & 2 \end{pmatrix}$	14	$\begin{pmatrix} 2 & 2 & 0 \\ 1 & 0 & 2 \end{pmatrix}$
6	$\begin{pmatrix} 1 & 0 \\ 2 & 2 \end{pmatrix}$	15	$\begin{pmatrix} 0 & 2 & 2 \\ 2 & 0 & 2 \end{pmatrix}$
7	$\begin{pmatrix} 1 & 2 \\ 2 & 0 \end{pmatrix}$	16	$\begin{pmatrix} 1 & 2 & 0 & 2 \\ 2 & 1 & 2 & 0 \end{pmatrix}$
8	$\begin{pmatrix} 2 & 2 \\ 0 & 2 \end{pmatrix}$	17	$\begin{pmatrix} 1 & 2 & 0 & 2 \\ 2 & 1 & 2 & 2 \end{pmatrix}$
9	$\begin{pmatrix} 2 & 0 \\ 0 & 2 \end{pmatrix}$	18	$\begin{pmatrix} 2 & 2 & 0 & 2 \\ 1 & 0 & 2 & 2 \end{pmatrix}$
		19	$\begin{pmatrix} 1 & 2 & 0 & 2 & 2 \\ 2 & 1 & 2 & 0 & 2 \end{pmatrix}$

Theorem 12.6.15 *([Grü 83;a;b])*

Let $\tau_0 := \{0, 2\}$ and $\tau_i := \begin{pmatrix} 0 & 0 & 1 \\ 0 & 1 & 0 \end{pmatrix} \cup \tau_i'$ for $i = 1, 2, ..., 19$.

There are exactly 148 $T_{0,2}$-projectable subclasses of $P_{3,2}$ that can be described as follows:

1): $pr^{-1}T_{0,2}$,
2)–5): *Pol* τ_i *with* $i \in \{0,1,2,3\}$,
6)–10): *Pol* $\{\tau_i,\tau_j\}$ *with* $(i,j) \in \{(1,3),(1,2),(2,3),(1,0),(2,0)\}$;
11)–12): *Pol* $\{\tau_i,\tau_j,\tau_k\}$ *with* $(i,j,k) \in \{(1,2,3),(1,2,0)\}$;
13)–18): *Pol* τ_i *with* $i \in \{4,5,6,7,8,9\}$;
19)–30): *Pol* $\{\tau_i,\tau_j\}$ *with* $(i,j) \in \{(4,3),(4,2),(4,0),(5,2),(6,3),(6,0),(7,3),(7,0),(8,1),(9,0),(9,1),(9,3)\}$;
31)–34): *Pol* $\{\tau_i,\tau_j,\tau_k\}$ *with* $(i,j,k) \in \{(4,2,3),(4,2,0),(9,1,3),(9,1,0)\}$;
35)–49): *Pol* $\{\tau_i,\tau_j\}$ *with* $(i,j) \in \{(4,5),((4,6),(4,7),(4,8),(4,9),(5,6),(5,7),(5,8),(5,9),((6,7),(6,8),(6,9),(7,8),(7,9),(8,9)\}$;
50)–63): *Pol* $\{\tau_i,\tau_j,\tau_k\}$ *with* $(i,j,k) \in \{(4,5,2),(4,6,0),(4,7,0),(4,6,3),(4,7,3),(4,9,0),(4,9,3),(6,7,0),(6,7,3),(6,9,4),(6,9,3),(7,9,0),(7,9,3),(8,9,1)\}$;
64)–69): *Pol* τ_i *with* $i \in \{10,11,12,13,14,15\}$;
70)–75): *Pol* $\{\tau_i,\tau_j\}$ *with* $(i,j) \in \{(10,0),(10,3),(11,2),(14,0),(14,3),(15,1)\}$;
76)–95): *Pol* $\{\tau_i,\tau_j\}$ *with* $(i,j) \in \{(10,5),(10,8),(10,9),(14,0),(11,6),(11,7),(11,9),(12,4),(12,7),(12,9),(13,4),(13,7),(13,9),(14,4),(14,5),(14,8),(15,4),(15,5),(15,7)\}$;
96)–99): *Pol* $\{\tau_i,\tau_j,\tau_k\}$ *with* $(i,j,k) \in \{((10,9,0),(10,9,3),(14,4,0),(14,4,3)\}$;
100)–114): *Pol* $\{\tau_i,\tau_j\}$ *with* $(i,j) \in \{(10,11),(10,12),(10,13),(10,14),(10,15),(11,12),(11,13),(11,14),(11,15),(12,13),(12,14),(12,15),(13,14),(13,15),(14,15)\}$;
115)–117): *Pol* τ_i *with* $i \in \{16,17,18\}$;
118)–119): *Pol* $\{\tau_i,\tau_j\}$ *with* $(i,j) \in \{(16,0),(16,3)\}$;
120)–123): *Pol* $\{\tau_i,\tau_j\}$ *with* $(i,j) \in \{(16,8),(16,5),(17,9),(18,4)\}$;
124): *Pol* $\{\tau_{16},\tau_8,\tau_5\}$;
125)–132): *Pol* $\{\tau_i,\tau_j\}$ *with* $(i,j) \in \{(16,15),(16,12),(16,13),(16,11),(17,15),(17,14),(18,11),(18,10)\}$;
133)–134): *Pol* $\{\tau_i,\tau_j,\tau_k\}$ *with* $(i,j,k) \in \{(16,15,5),(16,11,8)\}$;
135)–142): *Pol* $\{\tau_i,\tau_j,\tau_k\}$ *with* $(i,j,k) \in \{(16,15,12),(16,15,14),(16,15,11),(16,12,13),(16,12,11),(16,13,11),(17,15,14),(18,11,10)\}$;
143)–144): *Pol* $\{\tau_i,\tau_j\}$ *with* $(i,j) \in \{(16,17),(16,18),(17,18)\}$;
146): *Pol* τ_{19};
147)–148): $Z_{2,1} \cap pr^{-1}T_{0,2}$, $Z_{2,0} \cap pr^{-1}T_{0,2}$.

Proof. The following five statements are direct consequences from Lemmas 12.6.12 and 12.6.13:

(1) $T_{0,2}^{(02)(20)} := Pol\begin{pmatrix} 0 & 1 & 0 & 0 & 2 \\ 0 & 0 & 1 & 2 & 0 \end{pmatrix}$ has exactly the following $T_{0,2}$-projectable maximal classes:
a) $T_0^{(2)} \cap T_{0,2}^{(02)(20)}$;
b) $T_{0,2}^{(02)(20)} \cap Pol\begin{pmatrix} 0 & 1 & 0 & 1 \\ 0 & 0 & 1 & 2 \end{pmatrix}$;
c) $Z_{2,1} \cap pr^{-1}T_{0,2}$.

(2) If $A := Pol\left(\begin{pmatrix} 0 & 1 & 0 \\ 0 & 0 & 1 \end{pmatrix} \cup \varrho_i\right)$ with $\varrho_i \subseteq \begin{pmatrix} 2 & 0 & 2 & 1 \\ 0 & 2 & 1 & 2 \end{pmatrix}$ and $A \neq Pol\begin{pmatrix} 0 & 1 & 0 & 0 & 2 \\ 0 & 0 & 1 & 2 & 0 \end{pmatrix}$, then A has exactly the following $T_{0,2}$-projectable maximal classes:
a) $T_0^{(2)} \cap A$;
b) $A \cap Pol\left(\begin{pmatrix} 0 & 1 & 0 \\ 0 & 0 & 1 \end{pmatrix} \cup \delta\right)$, where δ fulfills the conditions

$$\begin{array}{l} \delta \subseteq \begin{pmatrix} 2 & 0 & 2 & 1 \\ 0 & 2 & 1 & 2 \end{pmatrix} \\ \nexists i \in I : \delta \subseteq \varrho_i \\ \forall \varepsilon \subset \delta\ \exists i \in I :\ \varepsilon \subseteq \varrho_i. \end{array}$$

(3) If $A = T_0^{(2)} \cap Pol\begin{pmatrix} 0 & 1 & 0 & 2 & 0 & 2 & 1 & 2 \\ 0 & 0 & 1 & 0 & 2 & 1 & 2 & 2 \end{pmatrix}$ then $Z_{2,0} \cap pr^{-1}T_{0,2}$ is the only maximal class of A, which is $T_{0,2}$-projectable.

(4) If $A := T_0^{(2)} \cap Pol\left(\begin{pmatrix} 0 & 1 & 0 \\ 0 & 0 & 1 \end{pmatrix} \cup \varrho_i\right)$ with $\varrho_i \subseteq \begin{pmatrix} 2 & 0 & 2 & 1 & 2 \\ 0 & 2 & 1 & 2 & 2 \end{pmatrix}$ and $A \neq Pol\begin{pmatrix} 0 & 1 & 0 & 2 & 0 & 2 & 1 & 2 \\ 0 & 0 & 1 & 0 & 2 & 1 & 2 & 2 \end{pmatrix}$, then A has exactly the following $T_{0,2}$-projectable maximal classes:
$A \cap Pol\left(\begin{pmatrix} 0 & 1 & 0 \\ 0 & 0 & 1 \end{pmatrix} \cup \delta\right)$, where δ fulfills the following conditions:

$$\begin{array}{l} \delta \subseteq \begin{pmatrix} 2 & 0 & 2 & 1 & 2 \\ 0 & 2 & 1 & 2 & 2 \end{pmatrix} \\ \nexists i \in I : \delta \subseteq \varrho_i \\ \forall \varepsilon \subset \delta\ \exists i \in I :\ \varepsilon \subseteq \varrho_i. \end{array}$$

(5) There is not any proper subset of $Z_{2,i} \cap pr^{-1}T_{0,2}$, $i = 0, 1$, whose projection agrees with $T_{0,2}$.

With the aid of the above statements, our theorem can be proven as follows: One obtains the $T_{0,2}$-projectable maximal classes of $pr^{-1}T_{0,2} = Pol\begin{pmatrix} 0 & 0 & 1 \\ 0 & 1 & 0 \end{pmatrix}$ by means of (2). Then, one obtains the $T_{0,2}$-projectable maximal classes of these classes by means of (1)–(4), etc. ■

13

Classes of Linear Functions

In this chapter, the elements of the clone

$$L_k := \bigcup_{n\geq 1}\{f^n \in P_k \mid \exists a_0, ..., a_n \in E_k : \ f(\mathbf{x}) = a_0 + \sum_{i=1}^{n} a_i \cdot x_i \ (mod\ k)\}.$$

and the elements of a generalization of the set L_k are called **linear functions**. The lattice of the subclasses of L_k belongs to the earliest and best investigated sublattices of $\mathbb{L}_k$. For the case that k is a prime number, all subclasses, which are no subsets of $[L_k^1]$, were determined by A. A. Salomaa in [Sal 64]. The results of [Sal 64] were also proven by J. Bagyinszki and J. Demetrovics in [Bag-D 82] and complemented with the remaining subclasses of $[L_k^1]$, $p \in \mathbb{P}$. Many results about linear functions were obtained by Á. Szendrei. For example, she proven in [Sze 78] that L_k has only finitely many subclasses, if k is square-free.[1] In addition, she showed in [Sze 78] that an arbitrary class has at most the order 2 (or 3), if k is square-free and k is an odd number (or an even number), respectively. One finds in [Sze 80] a short and fine determination of the subclasses not contained in $[L_p^1]$ of L_p (p prime). Similarly Theorem 13.2.1 is proven here.
For the case that k is not square-free one easily find a class that does not have any finite basis (see Lemma 13.3.7).
In Section 13.1, we start with properties of certain subclasses of the set

$$\begin{aligned} U_d \ := \ & \bigcup_{n\geq 1}\{f^n \in L_k \mid \exists a_0, ..., a_n \in E_k \ \exists j \in \{1, ..., n\} : \\ & f(\mathbf{x}) = a_0 + a_j \cdot x_j + d \cdot \textstyle\sum_{i=1, i\neq j} a_i \cdot x_i\}. \end{aligned}$$

This set is closed, if d is a divisor of k (notation: $d \mid k$). Then, with the aid of the results from the first section new proofs are given for the theorems of [Sal 64] and [Bag-D 82]. Section 13.3 gives a survey of further results, which were found by A. A. Bulatov, Á. Szendrei and the author, to classes from linear functions.

[1] Through that, a presumption of A. A. Salomaa from [Sal 64] was confirmed.

13.1 Some Properties of the Subclasses of U_d That Contain r_d

In this section, let $d \in E_k$ be a divisor of k. Further, the ternary function r_d is defined by

$$r_d(x, y, z) := x + d \cdot y - d \cdot z \ (mod\ k).$$

Lemma 13.1.1 *For every subclass T of U_d, which contains the function r_d, it holds:*
(a) $T = [T^1 \cup \{r_d\}]$,
(b) $T = L_k \cap Pol_k\{(a, a+b) \mid a + bx \in T^1\}$.

Proof. (a): Let $f^n \in T$ be arbitrary and w.l.o.g.

$$f(x_1, ..., x_n) = a_0 + a_1x_1 + d \cdot \sum_{i=2}^{n} a_ix_i.$$

Obviously, one can obtain every function of the type

$$g(x_1, ..., x_m) := x_1 + d \cdot \sum_{i=2}^{m} b_ix_i \text{ with } b_2 + ... + b_m = 0 \ (mod\ k)$$

by identifying variables of a function of the form $r_d \star r_d \star ... \star r_d$. In particular, the function

$$f'(x, y, x_1, ..., x_n) := x + d(-(a_2 + ... + a_m)y + \sum_{i=1}^{n} a_ix_i)$$

is a superposition over r_d. Consequently, by

$$f(x_1, ..., x_n) = f'(f(x_1, ..., x_1), x_1, x_2, ..., x_n),$$

we have $f \in [T^1 \cup \{r_d\}]$.
(b): Because of (a), $T = L_k \cap Pol_k G_1(T)$ holds. The matrix $G_1(T)$ is, however, unambiguously determined by the first two rows ($x = 0$, $x = 1$). Thus we have $T = L_k \cap pr_{0,1}G_1(T)$, where $pr_{0,1}G_1(T) = \{(a, a+b) \mid a + bx \in T^1\}$. ■

Lemma 13.1.2 *Let $f(x_1, ..., x_n) = a_0 + a_1x_1 + d \cdot \sum_{i=2}^{n} a_ix_i$ and $a_1 \sqcap k = a_2 \sqcap k = 1$. Then, the function r_d is a superposition over f.*

Proof. Let $f_1(x, y, z) := f(x_1, x_2, x_3, ..., x_3)$. Then, the functions

$$f_i(x, y, z) := f_{i-1}(f_1(x, z, z), y, z), \ i = 2, 3, ...$$

are superpositions over f. Since $a_2 \sqcap k = 1$, there exists a t with $a_1^t = 1 \ (mod\ k)$. Consequently, $f_t(x, y, z) = a + x + a_2dy + bdz$ for certain $a, b \in E_k$.

The functions

$$f_{t,1} := f_t \text{ and}$$
$$f_{t,i}(x,y,z) := f_{t,i-1}(f_t(x,y,z),y,z),\ i = 2,3,...$$

belong to $[\{f\}]$. Since $a_2 \sqcap k = 1$, there exists an s with $a_2 s = 1 \ (mod\ k)$. Consequently, $f_{t,s}(x,y,z) = as + x + dy + sbdz$ is valid. Next we form the superpositions

$$f_{t,s,1} := f_{t,s} \text{ and}$$
$$f_{t,s,i}(x,y,z) := f_{t,s,i-1}(f_{t,s}(x,y,z),y,z),\ i = 2,3,...$$

For $i = k$ we obtain $f_{t,s,k}(x,y,z) = x+dy+(k-1)dz$, i.e., r_d is a superposition over f. ∎

Subsequently, the subclasses of L_k that contain the function r_1 are determined. Obviously, a subset $A \subseteq L_k^1$ determines a subclass T of the form $T = [T^1 \cup \{r_1\}]$ if and only if $[A \cup \{r_1\}]^1 = A$ holds. Further we have:

Lemma 13.1.3

(a) If $[A \cup \{r_1\}]^1 = A$ for a subset A of L_k^1, then the binary relation

$$\varrho_A := \{(a, a+b-1) \mid a + bx \in A\}$$

is a subgroup of the direct product $(E_k;+)^2$.

(b) Conversely, if γ is a subgroup of $(E_k;+)^2$, then

$$[A_\gamma \cup \{r_1\}]^1 = A,$$

where $A_\gamma := \{a + bx \in L_k^1 \mid (a, a+b-1) \in \gamma\}$.

Proof. (a): Let $(a, a+b-1), (a', a'+b'-1) \in \varrho_A$ be arbitrary, i.e., the functions $a + bx$ and $a' + b'x$ belong to A. Then, by assumption, r_1 preserves the functions x, $a + bx$ and $a' + b'x$ of A. Consequently,

$$r_1(a + bx, a' + b'x, x) = a + a' + (b + b' - 1)x \in A$$

holds and therefore $(a + a', a + a' + b + b' - 2) \in \varrho_A$. Thus, by definition of ϱ, we have

$$(a, a+b-1) + (a', a'+b'-1) \in \varrho_A$$

for arbitrary $(a, a+b-1)$, $(a', a'+b'-1) \in \varrho_A$. Hence, ϱ_A is a subgroup of $(E_k;+)^2$.

(b): We have to show that A_γ is closed in respect to the operation $\star$ and that the function r_1 preserves the functions of A_γ.

Let $f(x) := a + bx$, $g(x) := a' + b'x$ and $h(x) := a'' + b''x$ be arbitrary of A_γ. Then the function $(f \star g)(x) = a + a'b + bb'x$ belongs to A_γ, since $(a, a+b-1) \in \gamma$, $(a', a'+b'-1) \in \gamma$ and

$$(a, a+b-1) + b \cdot (a', a'+b'-1) = (a + a'b, a + a'b + bb' - 1) \in \gamma.$$

Further

$$r_1(a+bx, a'+b'x, a''+b''x) = (a+a'-a'')+(b+b'-b'')x$$

holds. Since

$$\begin{aligned}&(a+a'-a'', a+a'-a''+b+b'-b''-1)\\&=(a,a+b-1)+(a',a'+b'-1)-(a'',a''+b''-1)\end{aligned}$$

and γ is a group, the function $(a+a'-a'')+(b+b'-b'')x$ belongs to A_γ. Consequently, we have $[A_\gamma \cup \{r_1\}]^1 = A_\gamma$. ■

When we summarize Lemmas 13.1.1–13.1.3, or as a consequence of [Sze 80], Lemma 4.3, we obtain:

Theorem 13.1.4 *The lattice of the closed subsets of U_d $(d|k)$, which contain the function r_d, is finite for arbitrary $k \in \mathbb{N}$ and is isomorphic to the subgroup lattice of the group $(E_k;+)^2$ for $d=1$.* ■

One can describe the subgroups of the group $(E_k;+)^2$ easily. The subsequently given description is a special case of Theorem 4.3.1 from [Sco 64], p. 71:

Theorem 13.1.5

(a) *Let γ be a subgroup of $(E_k;+)^2$. Then*

$$\begin{aligned}&\gamma_1 := \{a \mid \exists b \in E_k :\ (a,b) \in \gamma\}, && \gamma_2 := \{b \mid \exists a \in E_k :\ (a,b) \in \gamma\},\\&N_1 := \{a \mid (a,0) \in \gamma\}, && N_2 := \{b \mid (0,b) \in \gamma\}\end{aligned}$$

are subgroups of $(E_k;+)$, $N_i \subseteq \gamma_i$ $(i=1,2)$ and the mapping α from the factor set γ_1/N_1 onto the factor set γ_2/N_2 with

$$\alpha(a+N_1) = b+N_2 \ :\Longleftrightarrow\ (a,b) \in \gamma$$

is an isomorphism from γ_1/N_1 onto γ_2/N_2.

(b) *Conversely: Let γ_1, γ_2, N_1, N_2 be subgroups of $(E_k;+)$ with $N_i \subseteq \gamma_i$ $(i=1,2)$ and let α be an isomorphism from γ_1/N_1 onto γ_2/N_2. Then*

$$\gamma := \{(a,b) \mid a \in \gamma_1 \ \wedge\ b \in \gamma_2 \ \wedge\ \alpha(a+N_1) = b+N_2\}$$

is a universe of a subgroup of $(E_k;+)^2$. ■

With the help of Theorem 13.1.5 and with well-known theorems about cyclic groups, one can easily determine the cardinality of $\mathbb{L}_k([\{r_1\}], L_k)$. For this purpose, let φ be the **Euler function** (i.e., $\varphi(n)$ is the number of all $q \in \{1,2,...,n-1\}$ with $n \sqcap q = 1$) and let $t(q,k)$ be the number of all $n \in \mathbb{N}$ with $n|k$ and $q|n$.

Theorem 13.1.6 *It holds:*

$$|\mathbb{L}_k([\{r_1\}], L_k)| = \sum_{\substack{q \\ q \in \mathbb{N},\ q|k}} \varphi(q) \cdot (t(q,k))^2.$$

Proof. Because of Theorem 13.1.4, we have to determine only the number of the subgroups of $(E_k;+)^2$ to the proof.
Let $\gamma \subseteq E_k^2$ be a subgroup of $(E_k;+)^2$. This holds (by Theorem 13.1.5) if and only if there exist subgroups γ_i, N_i ($N_i \subseteq \gamma_i$, $i = 1,2$) of $(E_k;+)$ and an isomorphism α from γ_1/N_1 onto γ_2/N_2 with $\gamma = \{(a,b) \mid \alpha(a+N_1) = b+N_2\}$. Moreover, it holds $|\gamma_1/N_1| = |\gamma_2/N_2| =: q$ and $q|k$. Since $(E_k;+)$ is a cyclic group, the groups γ_i, N_i and γ_i/N_i ($i = 1,2$) are cyclic. Therefore, for fixed q and k, there are exactly $t(q,k)$ possibilities for the choice of γ_i/N_i ($i \in \{1,2\}$). Further, if g is a generating element of γ_1/N_1, then the mapping α is completely determined by $\alpha(g)$, where $\alpha(g)$ is a generating element of γ_2/N_2.
As is generally known, a cyclic group of the order q has exactly $\varphi(q)$ generating elements. Thus, for determining the mapping α there are exactly $\varphi(q)$ possibilities.
Consequently, for γ with $|\gamma_1/N_1| = |\gamma_2/N_2| = q$ there are exactly $\varphi(q) \cdot (t(q,k))^2$ possibilities, whereby our assertion is proven. ■
We notice that one can prove further properties of the elements of $\mathbb{L}_k([\{r_1\}], L_k)$ with the aid of Lemma 13.1.1. For example, every subclass T of L_k with $r_1 \in T$ is finitely axiomatizable (see [McK 78]) and such a class has only finite many congruences, which are determined through the congruences on $[\{r_1\}]$ and of $(T^1;\star)$ (see Chapter 9).

13.2 The Subclasses of Linear Functions of P_k with $k \in \mathbb{P}$

In this section, let p be an arbitrary prime number. Because of Lemma 13.1.2, every class $A \subseteq L_p$ with $A \not\subseteq [L_p^1]$ contains the function r_1. Therefore, by Lemma 13.1.3 and Theorem 13.1.4, such a class A is determined by a certain subgroup of the group $(E_k;+)^2$. Obviously, by Theorem 13.1.5, there are only the following $p+3$ subgroups of $(E_k;+)^2$:

$$E_p^2,\ \{(0,0)\},\ \{(0,x) \mid x \in E_p\},\ \text{and}\ \{(x, t\cdot x) \mid x \in E_p\},\ \text{if}\ t \in E_p.$$

Consequently, by Section 13.1, we have

Theorem 13.2.1 *([Sal 64], [Bag-D 82], [Sze 80])*
L_p has exactly $p+3$ subclasses which are not subsets of $[L_p^1]$:

L_p
$L_p \cap Pol_p\{(0,1)\}$
$(= \bigcup_{n\geq 1}\{f^n \in L_p \mid f(x_1,...,x_n) = \sum_{i=1}^n a_i x_i \ \wedge \ a_1 + ... + a_n = 1\} = [r_1])$,
$L_p \cap Pol_p\{(x, x+1) \mid x \in E_p\}$
$(= \bigcup_{n\geq 1}\{f^n \in L_p \mid f(x_1,...,x_n) = a_0 + \sum_{i=1}^n a_i x_i \ \wedge \ a_1 + ... + a_n = 1\})$,
$L_p \cap Pol_p\{0\}$ *and*
$L_p \cap Pol_p\{(x, tx+1) \mid x \in E_p\}$
$(= L_p \cap Pol_p\{(1-t)^{-1}\})$, *where* $t \in E_p\backslash\{1\}$. ■

Since one obtains an arbitrary subclass A of $[L_p^1]$ from a subsemigroup A^1 of $(L_p^1;\star)$ by means of $[A^1]_{\zeta,\tau,\nabla}$, it is sufficient to determine the subsemigroups of $(L_p^1;\star)$ for the description of the remaining subclasses of L_p.
Let A be an arbitrary subsemigroup of $(L_p^1;\star)$. Then $A = A' \cup C_A$, where A' is either the empty set or a subgroup of the group $S := (\{ax+b \mid a \in E_p\backslash\{0\}\};\star)$ and $C_A \subseteq \{c_a \mid a \in E_p\} =: C$. If $A' \neq \emptyset$ then the set $U_a := \{a \mid \exists b: \ ax+b \in A'\}$ is a subgroup of the group $G := (E_p\backslash\{0\};\cdot)$. Now, one can easily see that for an arbitrary subgroup U of G, $G/U =: \{N_1, N_2, ..., N_{p-1}\}$, $\alpha \in E_p$ and $I \subseteq \{1, 2, ..., \frac{p-1}{|U|}\}$ the following subsets of L_p^1 are closed in respect to $\star$:

$T_{\alpha,U} := \{ax + (1-a)\alpha \mid a \in U\}$ $(\subseteq L_p^1 \cap Pol_p\{\alpha\})$,
$S_U := \{ax + b \mid a \in U \ \wedge \ b \in E_p\}$,
$T_{\alpha,U} \cup C_{\alpha,U,I}$ with $C_{\alpha,U,I} := \{c_{\alpha+n \ (mod\ p)} \mid n \in \bigcup_{i\in N} N_i\}$ and $C_{\alpha,U,\emptyset} := \{c_\alpha\}$,
$T_{\alpha,U} \cup C_{\alpha,U,I} \cup \{c_\alpha\}$ and
$S_U \cup C$.

Theorem 13.2.2 *([Bag-D 82])*
An arbitrary subsemigroup of $(L_p^1;\star)$ *is either a subset of* C *or has the form* $T_{\alpha,U}$, S_U, $T_{\alpha,U} \cup C_{\alpha,U,I}$, $T_{\alpha,U} \cup C_{\alpha,U,I} \cup \{c_\alpha\}$ *or* $S_U \cup C$.

Proof. Let A be an arbitrary subsemigroup of $(L_p^1;\star)$ with $A = A' \cup C_A$ $(A' \subseteq S,\ C_A \subseteq C)$. The following cases are possible:
Case 1: $C_A = \emptyset$.
1.1.: $A' \subseteq L_p^1 \cap Pol\{\alpha\}$ for a certain $\alpha \in E_p$.
In this case, A' has the form $T_{\alpha,U}$, since $L_p^1 \cap Pol\{\alpha\}$ is a cyclic group.
1.2.: For all $\alpha \in E_p$ it holds $A' \not\subseteq L_p^1 \cap \{\alpha\}$.
Then $A' \neq \{x\}$ and A' contains a function $x + a$ $(a \neq 0)$ or at least two functions of the form $f(x) := bx + (1-b)\beta$, $g(x) := cx + (1-c)\gamma$ with $c, b \in E_p\backslash\{0,1\}$ and $\beta \neq \gamma$, since a function $ax + b$ with $a \in E_p\backslash\{0,1\}$ has exactly a fixed point. Then, in the first case, all functions $x, x+1, ..., x+p-1$ belong to A', and in the second case, $(f^{-1} \star g \star f \star g^{-1})(x) = x + d \in A'$ with

$d \neq 0$ (because of $\beta \neq \gamma$). Thus, $A' = S_U$ for certain $U \leq G$.
Case 2: $C_A \neq \emptyset$.
Because of considerations for Case 1, set A is either a subset of C or of the form $S_U \cup C$ or $T_{\alpha,U} \cup C_A$, where the functions of $T_{\alpha,U}$ preserve C_A. The latter is valid if and only if C_A has the form $C_{\alpha,U;I}$ or $C_{\alpha,U;I} \cup \{c_\alpha\}$ for certain I. ∎

Theorem 13.2.3 *([Bag-D 82])*
L_p has exactly

$$2 \cdot t(p-1) + 2^p \cdot (2-p) + p + 3 + 2 \cdot p \cdot \sum_{\substack{h \\ h \mid (p-1)}} 2^h$$

closed subsets (including $\emptyset$), where $t(p-1)$ is the number of all divisors $n \in \mathbb{N}$ of $p-1$.

Proof. Because of Theorem 13.2.1, we only have to determine the number of all subsemigroups which were described in Theorem 13.2.2. Since $T_{\alpha,U} \subseteq L_p^1 \cap Pol\{\alpha\}$, $L_p^1 \cap Pol\{\alpha\} \cap Pol\{\beta\} = \{x\}$ for $\alpha \neq \beta$ and U is a subgroup of the cyclic group $G = (E_p \backslash \{0\}; \cdot)$, there is exactly $p \cdot t(p-1) - p + 1$ subsemigroups of the form $T_{\alpha,U}$.
Obviously, there exist $2 \cdot t(p-1)$ sets of the form S_U and $S_U \cup C$ $(U \leq G)$. Since there are exactly 2^h possibilities for the choice of I $(\subseteq \{1, 2, ..., h\},\ h := \frac{p-1}{|U|})$, one receives (considering the equivalence

$$\begin{aligned}
&T_{\alpha,U} \cup C_{\alpha,U,I} = T_{\beta,V,J} \cup C_{\beta,V,J} \\
&\iff (\alpha, U, I) = (\beta, V, J) \ \vee \ (U = V = \{1\} \ \wedge \ I = J)
\end{aligned}$$

the following number of the subsemigroups of $(L_p^1; \star)$ of the form $T_{\alpha,U} \cup C_{\alpha,U,I}$ or $T_{\alpha,U} \cup C_{\alpha,U,I} \cup \{c_\alpha\}$:

$$\begin{aligned}
&p \cdot \sum_{\substack{h \\ h \mid (p-1)}} (2^{h+1} - 1) - (p+1) \cdot (2^p - 1) \\
&= 2 \cdot p \cdot ((\sum_{\substack{h \\ h \mid (p-1)}} 2^h) - p \cdot t(p-1) - (p-1) \cdot (2^p - 1).
\end{aligned}$$

By adding, one immediately obtains the number given in our theorem. ∎

Next, we provide some statements about the order of the subclasses of L_p. One finds generating systems for these classes in [Bag-D 82].

Theorem 13.2.4 *([Bag-D 82])*
For an arbitrary subclass $A \subseteq L_p$ it holds:

$$ord\ A = \begin{cases} 1, & \text{if} \quad A \subseteq L_p^1, \\ 3, & \text{if} \quad p = 2 \text{ and } A \in \{L_2, L_2 \cap Pol_2\begin{pmatrix} 0 & 1 \\ 1 & 0 \end{pmatrix}, [r_1]\}, \\ 2 & \text{otherwise.} \end{cases}$$

Proof. For $p = 2$ the above statements were proven in Chapter 3. Because of Lemmas 13.1.1 and 13.1.2, we have $[A^1 \cup \{r_1\}] = A$ and therefore $1 < ord\ A \leq 3$ for arbitrary $A \in \mathbb{L}_p^{\downarrow}(L_p)$ with $A \not\subseteq [L_p^1]$. If $p \neq 2$ then (by Lemma 13.1.2) $r_1 \in [\Delta r_1]$. This implies our theorem. ■

13.3 A Survey of Further Results on Linear Functions

In this section, $k \geq 2$ is arbitrary.
Let

$$L_{k,id} := \bigcup_{n \geq 1} \{\sum_{i=1}^{n} a_i x_i \in L_k \mid \sum_{i=1}^{n} a_i = 1\}$$

the set of all **idempotent functions** of L_k.
For $d|k$ we set

$$I_{k,d} := \bigcup_{n\geq 1}\{f^n \in L_{k,id} \mid \exists i \in \{1,...,n\}\ \exists a_1,...,a_n \in E_k : \\ f(\mathbf{x}) = a_i x_i + d \cdot \sum_{j \in \{1,...,i-1,i+1,...n\}} a_j x_j \ \}$$

Theorem 13.3.1 *([Sze 76], [Sze 82]; without proof)*
Any non-trivial closed subset A of $L_{k,id}$ has a unique representation of the form

$$A = \bigcap_{i=1}^{m} I_{k,d_i}, \tag{13.1}$$

where $m \geq 1$ and $d_1, ..., d_m$ (> 1) are pairwise relatively prime proper divisors of k.
If k is odd, then any subclass of $L_{k,id}$ has a order ≤ 2. If, in turn, k is even, then the order of any subclass is at most 3 and is equal to 3 if and only if in the representation (13.1) of A $d_1 \cdot ... \cdot d_m$ is odd.

To determine subclasses of L_k, the following lemmas are useful.
We notice that the proof (in [Sze 78]) for Lemma 13.3.2 uses Theorem 13.3.1 and that the proof (in [Lau 88a]) for Lemma 13.3.3 requires Lemma 13.3.2, so that the proof of Theorem 13.3.1 forms the basis of the proofs for most statements of this section.

Lemma 13.3.2 *([Sze 78]; without proof) Let*

$$f(\mathbf{x}) := a_0 + \sum_{i=1}^{n} a_i x_i \ (mod\ k) \in L_k$$

with $0 \notin \{a_1, ..., a_n\}$, $n \geq 2$ and $a_1 \sqcap ... \sqcap a_n \sqcap k = 1$. Then *(a)* $r_{q_1 \cdot \ldots \cdot q_n} \in [\{f\}]$, *where*

$$q_i := a_1 \sqcap a_2 \sqcap \ldots \sqcap a_{i-1} \sqcap a_{i+1} \sqcap \ldots \sqcap a_n \sqcap k,$$

$i = 1, 2, \ldots$ *([Sze 78], Basic Lemma);*

(b) $f \in [[f]_{id} \cup [f]^1]$, *where*

$$[A]_{id} := [A] \cap L_{k,id}$$

for arbitrary $A \subseteq L_k$ *([Sze 78], Corollary 3.1).*

Lemma 13.3.3 *([Lau 88a]; without proof) Let* A *be a subclass of* L_k *which is no subset of*

$$W_q := U_q \cup \bigcup_{t;t|q;t\neq 1} Q_t$$

for every divisor $q \neq 1$ *of* k*. Then the function* r_1 *belongs to* A*.*

For every $k, d \in \mathbb{N}$

$$d||k$$

is a short notation for "$d|k$ and $d \sqcap \frac{k}{d} = 1$". For lack of a better name, d will be called a **full divisor of** k.

The following lemma is easy to prove.

Lemma 13.3.4 *Let* $k, d \in \mathbb{N}$ *with* $d||k$*. Furthermore* $d\mathbb{Z}_k$ *denotes the set* $\{d \cdot z \ (mod\ k) \mid z \in \mathbb{Z}_k\}$*. Then:*

(a) $\mathbf{d\mathbb{Z}_k} := (d\mathbb{Z}_k; + \ (mod\ k), \cdot \ (mod\ k))$ *is a ring which is isomorphic to* $\mathbb{Z}_{\mathbf{k/d}}$*.*

(b) $\mathbf{d\mathbb{Z}_k}$ *has a unit which will be denoted by* e_d*.*

(c) $e_d + e_{n/d} = 1 \ (mod\ k)$*.* ∎

For $\alpha, d \in E_k$ with $d|k$ let

$$Q_{d,\alpha} := \bigcup_{n\geq 1} \{a_0 \cdot \alpha + \sum_{i=1}^{n} a_i x_i \ (mod\ k) \mid a_0 + \sum_{i=1}^{n} a_i x_i \in Q_d\}.$$

Lemma 13.3.5 *([Sze 78]) Let* $k, d \in \mathbb{N}$ *with* $d||k$*. Then:*

(a) The mapping

$$\psi_d : \ L_{k/d} \longrightarrow Q_{d,e_d}, \ a_0 + \sum_{i=1}^{n} a_i x_i \mapsto e_d \cdot (a_0 + \sum_{i=1}^{n} a_i x_i)$$

is an isomorphism.

(b) The mapping

$$\varphi_d : Q_d \longrightarrow Q_{d,e_d}, \ a_0 + \sum_{i=1}^{n} a_i x_i \mapsto e_d \cdot a_0 + \sum_{i=1}^{n} a_i x_i$$

is a homomorphism.

(c) *For any* $f \in Q_d$ *there exists a unique constant* $c_d(f)$ *such that*

$$f = \varphi_d(f) + c_d(f) \ (mod\ k).$$

Moreover, if $f, g \in Q_d$*, the constant* $c_d(f)$ *has the following properties:*

$$\forall \alpha \in \{\zeta, \tau, \Delta, \nabla\}: \ c_d(\alpha f) = c_d(f)$$
$$c_d(f \star g) = c_d(f).$$

Proof. (a), (b), and (c) are easy to check with the aid of Lemma 13.3.4. ■

Theorem 13.3.6 *([Sze 78]) Let* $k \in \mathbb{N}\backslash\{1\}$ *be square-free. Then, for arbitrary subclass of* L_k*,*

$$ord\ A \le \begin{cases} 2 \text{ if} & k \text{ is odd,} \\ 3 \text{ otherwise,} & \end{cases}$$

i.e., there are only a finite number of subclasses of L_k*.*

Proof. Let $f^n \in L_k \setminus [L_k^1]$ be arbitrary. Set $A := [f]$. To show our theorem, it suffices to prove the following statement:

$$f \in \begin{cases} [A^2] \text{ if} & k \text{ is odd,} \\ [A^3] \text{ otherwise.} & \end{cases} \tag{13.2}$$

W.l.o.g. we can assume that f is an n-ary function with $n \ge 4$ and defined by

$$f(\mathbf{x}) := a_0 + \sum_{i=1}^{n} a_i \cdot x_i \ (mod\ k),$$

where $0 \notin \{a_1, a_2, ..., a_n\}$. For $q := a_1 \sqcap a_2 \sqcap ... \sqcap a_n \sqcap k$ the following two cases are possible:

Case 1: $q = 1$.

In this case, we have $[f] = [[f]_{id} \cup [\{f\}]^1]$ by Lemma 13.3.2, (b), whereby our theorem follows from Theorem 13.3.1.

Case 2: $q \neq 1$.

With the aid of Lemma 13.3.5, one can reduce this case to the first case (see [Sze 78] for details). ■

We notice that the paper [Sze 78] contains precise representations of the subclasses of L_k, where k is square-free.

Lemma 13.3.7 *If* $k \in \mathbb{N}$ *is not square-free, then there exists a subclass of* L_k *that is not finitely generated and does not have a basis.*

Proof. If k can not be represented as a product of different prime numbers, then there is a $t \in E_k\backslash\{0\}$ with $t^2 = 0 \ (mod\ k)$. Then, the set

$$Q_t := \bigcup_{n\geq 1}\{f^n \in L_k \mid \exists a_0, ..., a_n \in E_k : \ f(\mathbf{x}) = a_0 + t \cdot \sum_{i=1}^{n} a_i \cdot x_i \ (mod\ k)\}$$

is a subclass of L_k, whose functions f^n, g^m for arbitrary n, m have the property that the number of the essential variables of $f \star g$ is at most $n-1$. Therefore, Q_t does not have a finite generating system. Since by Theorem 8.1.6 $|\mathbb{L}_k^{\downarrow}(L_k)| \leq \aleph_0$ holds, every class of linear functions that does not have a finite generating system, does not have a basis. ■

The following theorem is a consequence of Theorems 13.3.6, 3.2.2.2, 13.2.3, and 8.1.6, Lemma 13.3.7, and [Lau-S 90] (for $k = 6$):

Theorem 13.3.8 *For arbitrary $k \in \mathbb{N} \setminus \{1\}$ it holds:*

$$|\mathbb{L}_k(L_k)| \begin{cases} < \aleph_0 & \text{if } k \text{ is square-free,} \\ = \aleph_0 & \text{otherwise.} \end{cases}$$

In particular, we have:

k	2	3	4	5	6	7	8	9
$\|\mathbb{L}_k^{\downarrow}(L_k)\|$	15	38	$\aleph_0$	319	7524	470	$\aleph_0$	$\aleph_0$

■

A solution for the completeness problem for L_k, where $k \in \mathbb{N}$ is arbitrary, follows from

Theorem 13.3.9 *([Lau 88a]; without proof) Let $k = p_1^{\alpha_1} p_2^{\alpha_2} ... p_s^{\alpha_s}$ ($p_i \neq p_j$ for $i \neq j$; $p_i \in \mathbb{P}, \alpha_i \in \mathbb{N}$ for $i = 1, ..., s$). Then, L_k has exactly $s + 2^s - 1 + \sum_{i=1}^{s} p_i$ pairwise different maximal classes:*

(1) $S_p := \bigcup_{n\geq 1}\{f^n \in L_k \mid \exists a_0, ..., a_n \in E_k : \ f(\mathbf{x}) = a_0 + \sum_{i=1}^{n} a_i x_i \ (mod\ k) \ \wedge$

$\sum_{i=1}^{n} a_i = 1 \ (mod\ p)\}$, where $p \in \{p_1, ..., p_s\}$;

(2) $W_q := U_q \cup \bigcup_{t|q, t\neq 1} Q_t$, where q is a square-free divisor of k;

(3) $T_{a,p} := L_k \cap Pol_k\{x \in E_k \mid p|(x - a)\}$, where $a \in E_k$ and $p \in \{p_1, ..., p_s\}$.

As a consequence from the above theorem and from Lemma 13.3.3 we get:

Lemma 13.3.10 *Let A be an subclass of L_k. Then either $r_1 \in A$ or there exists a square-free divisor q of k with $A \subseteq W_q$.* ■

Since all subclasses of L_k that contain r_1 were determined in Section 13.1, we need all subclasses of the maximal classes of the type W_q for a complete description of $\mathbb{L}_k^{\downarrow}(L_k)$. For $k = 4$ the following theorem gives a rough description of the missing classes.

Theorem 13.3.11 *([Lau 88a]; without proof) It holds:*

(a) $|\mathbb{L}_4^{\downarrow}([L_4^1])| = 189.$
(b) $|\mathbb{L}_4([r_1], L_4)| = 15.$
(c) Exactly 20 subclasses of L_4 *contain the function* r_2*, but not the function* r_1*.*
(d) A subclass of L_4*, which contains neither* r_1 *nor* r_2*, is a subset of* $Q_2 \cup [L_4^1]$ *and there exist a subclass* B *of* $[L_4^1]$ *and some* $t_0, t_1, t_2, t_3 \in \mathbb{N} \cup \{\infty\}$ *with*

$$A = B \cup K_{0,t_0} \cup K_{1,t_1} \cup K_{2,t_2} \cup K_{3,t_3},$$

where

$$K_{a,r} := \begin{cases} \emptyset & \text{if } r = 1, \\ [\{a + 2 \cdot \sum_{i=1}^{r} x_i \ (mod\ 4)\}] & \text{if } r > 1 \end{cases}$$

$(a \in E_4,\ r \in \mathbb{N})$ *and* $K_{a,\infty} := \bigcup_{r \geq 2} K_{a,r}.$

We notice that the lattice $\mathbb{L}_k(L_k)$ for $k = p^2$ (p prime) was studied in detail in [Bul-I 2002] and [Bul 2002].

To describe further results about classes of linear function, we generalize the set L_k as follows:
Let $\mathbf{R} = (R; +, \cdot, -, 0, 1)$ be a finite unitary ring and let

$$\mathbf{M} := (M; +, -, (r)_{r \in R}, 0)$$

be a finite faithful[2] module over $\mathbf{R}$. Further, we consider the mapping $\odot$ defined by

$$\odot :\ R \times M \longrightarrow M,\ (r, x) \mapsto r \odot x := r(x),$$

which fulfills the axioms (M_1)–(M_4) (see Part I, Section 1.2.8). Then one can define a closed subset $\mathcal{L}_M$ of P_M as follows:

$$\mathcal{L}_M := \bigcup_{n \geq 1} \{f^n \in P_M \mid \exists\, a_0 \in M\ \exists\, a_1, \ldots, a_n \in R :\ f(\mathbf{x}) = a_0 + \sum_{i=1}^{n} a_i \odot x_i\}.$$

It is usual to call the elements of $\mathcal{L}_M$ also **linear functions** over the (left) module $\mathbf{M}$.
We remark that one can define analogously a set $\mathcal{L}_{M'}$ for the case that $\mathbf{M}'$ is a right module.
If one sets $\mathbf{R} = \mathbb{Z}_p^{m \times m}$ (the ring of all (m, m)-matrices over the field $\mathbb{Z}_\mathbf{p}$; p prime) and $\mathbf{M} = \mathbb{Z}_\mathbf{p}^{\mathbf{m} \times \mathbf{1}}$ (the module of all $(m, 1)$-matrices over $\mathbf{R}$), then $\mathcal{L}_M$ is isomorphic to the set of all quasi-linear functions of P_{p^m} (see Section 5.2.4). This matrix representation of the quasi-linear function was used in [Sze-S 81] to describe all subclasses that contain certain sets of unary quasi-linear functions:

[2] A module $\mathbf{M}$ over a ring $\mathbf{R}$ is **faithful** iff $\{r \in R \mid \forall m \in M :\ r(m) = 0\} = \{0\}$.

Theorem 13.3.12 *([Sze-S 81]; without proof) Let p prime, $m \in \mathbb{N}$, $p^m \geq 3$, $\mathbf{R} = \mathbb{Z}_p^{m\times m}$ and $\mathbf{M} = \mathbb{Z}_\mathbf{p}^{\mathbf{m}\times\mathbf{1}}$. Moreover, set*

$$
\begin{aligned}
I &:= \{(i,j) \in \mathbb{N}_0^2 \mid 0 \leq j \leq i \leq k-1\} \cup \{(-1,0),(k-1,k)\},\\
H_{(r,s)} &:= [\{A \cdot x \in \mathcal{L}_M^1 \mid rg(A) \leq r \ \vee \ rg(A) = k\}]\\
&\quad \cup \bigcup\nolimits_{n\geq 2}\{f^n \in \mathcal{L}_M \cap Pol_M\{\mathbf{0}\} \mid \dim(Im(f)) \leq s\},\\
&\quad \text{where } (r,s) \in I;^{3}\\
H_{(r,s)};c &:= \{f + c \in \mathcal{L}_M \mid f \in H_{(r,s)} \ \wedge \ c \in M\}, \text{where } (r,s) \in I.
\end{aligned}
$$

Then:

(a) $H_{(r,s)} \subseteq H_{(r',s')}$ *iff* $r \leq r'$ *and* $s \leq s'$.
(b) $H_{(k-1,k)} = \mathcal{L}_M \cap Pol_M\{\mathbf{0}\} = \bigcup_{n\geq 1}\{\sum_{i=1}^n A_i \cdot x_i \mid A_1,\dots,A_n \in \mathbb{Z}_p^{m\times m}\}$ *and* $H_{(k-1,k);c} = \mathcal{L}_M$.
(c) The sets $H_{(r,s)}$, $(r,s) \in I$, *are the only subclasses of* $\mathcal{L}_M \cap Pol_M\{\mathbf{0}\}$ *containing* $(\mathcal{L}_M \cap Pol_M\{\mathbf{0}\})^1$.
(d) The sets $H_{(r,s);c}$, $(r,s) \in I$, *are the only subclasses of* $\mathcal{L}_M$ *containing* $\mathcal{L}_M^1$.

For the case $p = m = 2$, one can find in [Kro-R 2003] all subclasses of quasi-linear selfdual[4] functions, which contain all unary quasi-linear selfdual functions.

In the case that $\mathbf{R} = \mathbf{M}$ is a finite field $\mathbf{F}$, one can find results on $\mathbb{L}_F^{\downarrow}(\mathcal{L}_F)$ in [Sze 80]. Among other things, the following theorem was proven in [Sze 80]:

Theorem 13.3.13 *(without proof) For any finite field* $\mathbf{F}$, *the class* $\mathcal{L}_F$ *has only finitely many subclasses.*
Every subclass A *of* $\mathcal{L}_F$ *with* $A \not\subseteq [L_F^1]$ *contains the function* $x + y - z$ *and is a class of the following form (a) or (b):*

(a) $I(E,S) := \bigcup_{n\geq 1}\{a_0 + \sum_{i=1}^n a_i \odot x_i \mid \{a_1,\dots,a_n\} \subseteq E \ \wedge \ (\exists s,s' \in S : a_0 = s - (\sum_{i=1}^n a_i) \odot s'\}$, *where* E *is a universe of a subfield of* $\mathbf{F}$ *and* $S = V + a$ *for an* $a \in F$ *and a subspace* V *of* $\mathbf{F}$, *considered as a vector space over* $\mathbf{E}$.
(b) $I(E,S_0) := \bigcup_{n\geq 1}\{a_0 + \sum_{i=1}^n a_i \odot x_i \mid \{a_1,\dots,a_n\} \subseteq E \ \wedge \ \sum_{i=1}^n a_i = 1 \ \wedge \ a_0 \in S_0\}$, *where* E *is a universe of a subfield of* $\mathbf{F}$ *and* S_0 *is a subspace of* $\mathbf{F}$, *considered as a vector space over* $\mathbf{E}$.

Theorem 13.3.14 *([Sze 80], without proof) Let* $k = p^m$ *(p prime,* $m \geq 1$*),* $\mathbf{F} := (E_k;+,\cdot)$ *a field and let* QL_F *be the set of all quasi-linear functions of* P_k *(see Section 5.2.4). Further, we set*

$$
\begin{aligned}
L_F &:= \bigcup\nolimits_{n\geq 1}\{f^n \in QL_k \mid \exists a_0,\dots,a_n \in E_k : \ f(\mathbf{x}) = a_0 + \sum\nolimits_{i=1}^n a_i \cdot x_i\},\\
L_{F;d} &:= [L_F \cup \{x^{p^d}\}] \quad \text{where } d \in \mathbb{N} \text{ is a divisor of } k.
\end{aligned}
$$

---

[3] $rg(A)$ denotes the **range** of the matrix A and $dim\ V$ is the **dimension** of the vector space V. We remark that $Im(f)$ is a vector space, if $f(\mathbf{x}) = \sum_{i=1}^n A_i \cdot x_i$.
[4] with respect to a fixed-point-free permutation π of order p

(In particular, $L_{F;1} = QL_k$ and $L_{F;k} = L_F$.) Then for any subclass $A \subset P_k$ with $L_F \subseteq A$, there exists a divisor d of k such that $A = L_{F;d}$.

Theorem 13.3.15 *([Sze 80], without proof) Let* $\mathbf{M}$ *be a faithful unitary module over a unitary ring* $\mathbf{R}$. *Then for any subclass A of $\mathcal{L}_M$ containing the operation $x+y-z$ there exists a unique subring* $\mathbf{T}$ *of* $\mathbf{R}$ *and a unique* $\mathbf{T}$*-submodule* $\mathbf{N}$ *of* $\mathbf{T} \times \mathbf{M}$ *such that*

$$A = \bigcup_{n \geq 1} \{m + \sum_{i=1}^{n} r_i \odot x_i \mid \{r_1, ..., r_n\} \subseteq T \ \wedge \ (1 - \sum_{i=1}^{n} r_i, m) \in N\}$$

(see also Section 9.6).

Next we give some results by A. A. Bulatov about the elements of $\mathbb{L}_M^{\downarrow}(\mathcal{L}_M)$, where the unitary ring $\mathbf{R}$ is commutative (i.e., $\cdot$ is kommutative). The following theorem is the basis for a classification of the elements of $\mathbb{L}_M^{\downarrow}(\mathcal{L}_M)$.

Theorem 13.3.16 *([Bul 98a]; without proof) Let* $\mathbf{R}$ *be kommutative. Then:*

(1) The minimal classes [5] of $\mathbb{L}_M^{\downarrow}(\mathcal{L}_M)$ are exactly the following classes:
  (a) J_M;
  (b) $[c_a]$, where $a \in M$;
  (c) $[\varepsilon \odot x + (1 - \varepsilon) \odot b]$, where $b \in M$, $\varepsilon \in R \setminus \{0, 1\}$ and $\varepsilon \cdot \varepsilon = \varepsilon$ (e.g., ε is an **idempotent** *of the ring R).*

(2) The minimal clones, which are elements of $\mathbb{L}_M^{\downarrow}(\mathcal{L}_M)$, are exactly the following clones:
  (a) $J_M \cup A$, where A is a minimal class of the form (b) or (c) (see (1));
  (b) $[x + b]$, where $b \in M$ is of prime additive order;
  (c) $[\alpha \odot x + b]$, where $b \in M$, $\alpha \in R \setminus \{1\}$ and there exists a prime number p with $\alpha^p = 1$ and $(1 + \alpha + \alpha^2 + ... + \alpha^{p-1}) \odot b = 0$;
  (d) $[x - r \odot y + r \odot z]$, where $r \in R \setminus \{0\}$ with $r^2 = 0$ and r is of prime additive order;
  (e) $[x + y - z]$, if char $\mathbf{R}$ *is a prime number.[6]*

In [Bul 98b], it was shown that any element of $\mathbb{L}_M^{\downarrow}(\mathcal{L}_M)$ can be assigned to one of four types, defined below. The definitions of these types use properties of coefficients of functions and, in addition, some special functions. Classes of the *first type* are those containing a function $\varepsilon \odot x + (1 - \varepsilon) \odot y$ where ε is an idempotent of $\mathbf{R}$ ($\varepsilon \neq 0, 1$). Similarily, classes of the *second type* are those not of the first type and containing the function $\varepsilon \odot x + (1 - \varepsilon) \odot b$ where ε is an idempotent of $\mathbf{R}$ ($\varepsilon \neq 0, 1$) and $b \in M$. A function $f = \sum_{i=1}^{n} \alpha_i \odot x_i + a$ is called **primitive**, if there exists $j \in \{1, ..., n\}$ such that α_j is an invertible

---

[5] See Chapter 19.

[6] If there is a least positive integer n with $\underbrace{x + x + ... + x}_{n \text{ times}} = 0$ for all $x \in R$, then the ring $\mathbf{R}$ is said to have **characteristic** n (notation *char* $\mathbf{R}$).

element and α_i is a nilpotent element[7] whenever $i \neq j$. Classes of the *third type* consist of primitive functions and functions whose coefficients are nilpotent elements. Finally, a class belongs to the *fourth type* if it contains the Mal'tsev function $x + y - z$ and is not of the first type.
The classes of the forth type were described by Á. Szendrei in [Sze 80] (see Theorem 13.3.15). Classes of the first and second type were studied in [Bul 98a]. The paper [Bul 98b] deals with the classes of primitive functions and with the lattice of classes of the third type. In particular, A. A. Bulatov completely describes the lattice of classes of primitive functions in [Bul 98b].

We need the following notations for the last theorem of this section:
In [Pös-K 79], the direct product of functions and subclasses of P_k was defined in the following way. Let a set $A := A_1 \times A_2$ be Cartesian product of the sets A_1 and A_2. The **direct product of the functions** $f_1 \in P_{A_1}^{(n)}$ and $f_2 \in P_{A_2}^{(n)}$ is the function $f \in P_A^{(n)}$ defined by the equality

$$f((b_1, c_1), ..., (b_n, c_n)) = (f(b_1, ..., b_n), f(c_1, ..., c_n))$$

for arbitrary $b_1, ..., b_n \in A_1$, $c_1, ..., c_n \in A_2$. In this case, we shall write $f = f_1 \otimes f_2$. The set

$$C = \{f_1 \otimes f_2 \mid f_1 \in C_1^{(n)}, f_2 \in C_2^{(n)},\ n \in \mathbb{N}\}$$

is the **direct product of the classes** $C_1 \subseteq P_{A_1}$, $C_2 \subseteq P_{A_2}$ and is denoted by $C_1 \otimes C_2$.
It is well known that, for any idempotent $\varepsilon \in R \setminus \{0, 1\}$, we can represent the module $\mathbf{M}$ as the direct sum of two submodules

$$(\varepsilon \odot \mathbf{M}) \oplus ((\mathbf{1} - \varepsilon) \odot \mathbf{M})$$

(this is the so-called **Peirce decomposition**). Further, each function $f(\mathbf{x}) := a + \sum_{i=1}^{n} \alpha_i \odot x_i \in \mathcal{L}_M$ can be represented as $f = f_1 \otimes f_2$ where $f_1(\mathbf{x}) = \varepsilon \odot a + \sum_{i=1}^{n} \alpha_i x_i$ and $f_2(\mathbf{x}) = (1 - \varepsilon) \odot a + \sum_{i=1}^{n} \alpha_i \odot x_i$ are functions on the sets $\varepsilon \odot M$ and $(1-\varepsilon) \odot M$, respectively. To show this take $(b_1, c_1), \ldots, (b_n, c_n) \in M = (\varepsilon \odot M) \times ((1 - \varepsilon) \odot M)$. Then

$$\begin{aligned}
& f((b_1, c_1), \ldots, (b_n, c_n)) \\
= \ & f(b_1 + c_1, \ldots, b_n + c_n) \\
= \ & a + \sum_{i=1}^{n} \alpha_i \odot (b_i + c_i) \\
= \ & \varepsilon \odot a + \sum_{i=1}^{n} \varepsilon \alpha_i \odot (b_i + c_i) + (1 - \varepsilon) \odot a + \sum_{i=1}^{n} (1 - \varepsilon) \alpha_i \odot (b_i + c_i)
\end{aligned}$$

[7] An element a of a ring $\mathbf{R} = (R; +, \cdot, -, 0)$ is called **nilpotent** iff there exists an $n \in \mathbb{N}$ with $a^n = 0$.

$$= \varepsilon \odot a + \sum_{i=1}^{n} \alpha_i \odot b_i + (1-\varepsilon) \odot a + \sum_{i=1}^{n} \alpha_i \odot c_i$$
$$= (f_1(b_1, \ldots, b_n), f_2(c_1, \ldots, c_n)).$$

Conversely, the function $f = f_1 \otimes f_2 = (b+c) + \sum_{i=1}^{n}(\varepsilon\gamma_i + (1-\varepsilon)\zeta_i) \odot x_i$ corresponds to the given pair of functions $f_1 = b + \sum_{i=1}^{n} \gamma_i \odot x_i \in \mathcal{L}_{\varepsilon \odot M}$, $f_2 = c + \sum_{i=1}^{n} \zeta_i \odot x_i \in \mathcal{L}_{(1-\varepsilon)\odot M}$. The modules $\varepsilon \odot M$, $(1-\varepsilon) \odot M$ can be considered as εR- and $(1-\varepsilon)R$-modules, respectively. So, $\mathcal{L}_M = \mathcal{L}_{M_1} \times \mathcal{L}_{M_2}$ where $M_1 = \varepsilon \odot M$ is an εR-module and $M_2 = (1-\varepsilon) \odot M$ is an $(1-\varepsilon)R$-module. Let C be a class of all functions preserving the kernels of projections from A onto A_1 and A_2. In [Pös-K 79] (3.3.4, 3.3.5, p. 85) it was noted that the interval $\mathbb{L}_{A_1 \times A_2}(J_{A_1} \times J_{A_2}, C)$ can be decomposed into the direct product $[J_{A_1}, P_{A_1}] \times [J_{A_2}, P_{A_2}]$. Using this decomposition we obtain the next statement.

Theorem 13.3.17 *([Bul 98a]; without proof) For any idempotent $\varepsilon \in R$,*

(a) if the function $\varepsilon \odot x + (1-\varepsilon) \odot y$ belongs to a subclass C of $\mathcal{L}_M$, then $C = C_1 \otimes C_2$ where C_1, C_2 are subclasses of $\mathcal{L}_{\varepsilon \odot M}$ and $\mathcal{L}_{(1-\varepsilon)\odot M}$, respectively and $\mathcal{L}_1, \mathcal{L}_2$ both contain all projections;
(b) $\mathbb{L}_M([\varepsilon \odot x + (1-\varepsilon) \odot y], \mathcal{L}_M) \cong$ $\mathbb{L}_{\varepsilon \odot M}(J_{\varepsilon \odot M}, \mathcal{L}_{\varepsilon \odot M}) \times \mathbb{L}_{(1-\varepsilon)\odot M}(J_{(1-\varepsilon)\odot M}, \mathcal{L}_{(1-\varepsilon)\odot M})$.

For a proper ideal I of $\mathbf{R}$, let $\mathcal{L}I_M$ be the set of all functions $f \in \mathcal{L}_M$ whose coefficients belong to I. Further, let $\mathcal{LP}_M$ be the set of all primitive functions of $\mathcal{L}_M$.
For a ring $\mathbf{R}$ let $idemp(R)$ be the set of all idempotents $\neq 0$. Moreover, put

$$K_M(R', M') = \{\sum_{i=1}^{n} \varrho_i x_i + a \mid n \geq 1,\ \varrho_1, \ldots, \varrho_n \in R',\ (1 - \sum_{i=1}^{n} \varrho_i, a) \in M'\}$$

where R' is a unitary subring of R and M' is a submodule of R'-module $R' \times M$.

Theorem 13.3.18 *([Bul 98a]; without proof) The set of all maximal clones of $\mathcal{L}_M$ comprise exactly the following clones:*

(a) $(\mathcal{L}I_{\varepsilon\odot M} \cup \mathcal{LP}_{\varepsilon \odot M}) \otimes \mathcal{L}_{(1-\varepsilon)\odot M}$ where $\varepsilon \in idemp(R)$;
(b) $K_{\varepsilon M}(R', R' \times \varepsilon \odot M) \otimes \mathcal{L}_{(1-\varepsilon)\odot M}$ where $\varepsilon \in idemp(R)$, R' is a maximal unitary subring of εR without proper idempotents;
(c) $K_{\varepsilon \odot M}(\varepsilon R, M') \otimes \mathcal{L}_{(1-\varepsilon)\odot M}$ where $\varepsilon \in idemp(R)$ is a minimal idempotent and M' is a maximal submodule of $\varepsilon R \times \varepsilon M$.

14

Submaximal Classes of P_3

A subclass (or a subclone) of P_k is called **submaximal** if it is covered by a maximal class (clone).[1] The concept *submaximal class* was introduced of I. G. Rosenberg in [Ros 74]. In [Ros 74] one also finds the first results about submaximal classes of P_k (see Chapter 17 for details).
In general, the submaximal classes seem to be interesting for the following reasons.[2] The largely unknown lattice $\mathbb{L}_k$ has intervals with antichains of cardinality $\mathfrak{c}$ situated far down from the top. It is not unreasonable to assume that the lattice $\mathbb{L}_k$ is nicer near the top, and therefore the submaximal clones are good candidates. The problem of determining certain submaximal clones also came up in the study of shortest maximal chains in the lattice (see [Sze 83]). Given a maximal clone M, one can ask for a completeness criterion for M: under what conditions does the clone $[F]$ generated by some $F \subseteq M$ coincide with M? In the case that M is finitely generated, a full list of clones maximal in M would provide a general criterion, because then $[F]$ if and only if F is contained in no clone maximal in M. An application could be a characterization of Sheffer functions for M. V. B. Kudrjavcev and P. Schofield proven that they exist exactly for maximal clones of the form $Pol_k\varrho$ with $\varrho \in \mathfrak{C}_k^1 \cup \mathfrak{S}_k \cup \mathfrak{U}_k$ (see Chapter 7), but the examples in the proofs have many variables. It would be interesting to have simple criteria of type Theorem 7.1.3, which, in its turn, could lead to the question: what is the minimum number of functional values

[1] It is easy to see that every submaximal class is a clone: Assume there is a maximal class $M \subset P_k$ and a submaximal class $S \subset M$ with $J_k \cap S = \emptyset$ and $[S \cup \{f\}] = M$ for all $f \in M \setminus S$. Then, $M = S \cup J_k$, since M is a clone (see Theorem 5.1.1) and $J_k = [e]$ for all $e \in J_k$. However, $M = S \cup J_k$ with $S \cap J_k = \emptyset$ is not possible, since $S \cup J_k \neq P_k(k-1) \cup [P_k^1]$ and every maximal class A $(\neq P_k(k-1) \cup [P_k^1])$ of P_k contains an idempotent function g with $g \notin J_k$. For example, if $\varrho \in \mathfrak{M}_k$ and o is the smallest element of ϱ, then the function f^2, defined by $f(o,x) = f(x,o) = o$ for all $x \in E_k$ and $f(x,y) = x$ otherwise, is idempotent and belongs to $(Pol_k\varrho) \setminus J_k$. With the help of the theorems from Chapter 5, one can easily find idempotent functions for the other possible cases.

[2] The following is an indirect quotation from [Ros-S 85].

whose knowledge can guarantee that a function is a Sheffer function?
Finally, the submaximal clones may be of interest on their own, e.g., as a source of examples and counter-examples.

For arbitrary k, the full list of the maximal classes of a maximal class A of P_k is only known, if A has the type $\mathfrak{S}$ ([Ros-S 84], see Section 18.1), $\mathfrak{C}^1$ (see Chapters 16 and 17) or $A = Pol_k\varrho$, where $\varrho := E_{k-1}^2 \cup \{(k-1, k-1)\}$ (see Section 18.3). There are, however, some papers in which one finds submaximal classes for specific k or only such submaximal classes that contain certain functions.
In Section 14.1, one finds a complete description of all submaximal classes of P_3 and some remarks about generalization of the given theorems. The papers [Mac 79], [Mar-D-H 80], [Sal 64], [Bag-D 82], and [Lau 82a] form the basis of this description. The theorems from the first section are proven then in Sections 14.2–14.9. In Section 14.10, we will prove that there are 5 submaximal classes with finitely many subclasses, 7 with countably many subclasses, and the remaining 146 with uncountably many subclasses. All elements of the lattices $\mathbb{L}_3^{\downarrow}(A)$, where A is a submaximal class with $|\mathbb{L}_3^{\downarrow}(A)| \le \aleph_0$ are determined in Chapter 15.

14.1 A Survey of the Submaximal Classes of P_3

The following theorem was proven 1958 by Jablonskij and is a special case of Rosenberg's Theorem 6.1.

Theorem 14.1.1 *([Jab 58])*
P_3 has exactly 18 maximal classes:

(1) $Pol_3\{0\}$ (2) $Pol_3\{1\}$

(3) $Pol_3\{2\}$ (4) $Pol_3\{0,1\}$

(5) $Pol_3\{0,2\}$ (6) $Pol_3\{1,2\}$

(7) $Pol_3\begin{pmatrix} 0&1&2 \\ 1&2&0 \end{pmatrix}$ (8) $Pol_3\begin{pmatrix} 0&1&2&0&1 \\ 0&1&2&1&0 \end{pmatrix}$

(9) $Pol_3\begin{pmatrix} 0&1&2&0&2 \\ 0&1&2&2&0 \end{pmatrix}$ (10) $Pol_3\begin{pmatrix} 0&1&2&1&2 \\ 0&1&2&2&1 \end{pmatrix}$

(11) $Pol_3\begin{pmatrix} 0&1&2&0&0&1 \\ 0&1&2&1&2&2 \end{pmatrix}$ (12) $Pol_3\begin{pmatrix} 0&1&2&0&0&2 \\ 0&1&2&2&1&1 \end{pmatrix}$

(13) $Pol_3\begin{pmatrix} 0&1&2&1&1&0 \\ 0&1&2&0&2&2 \end{pmatrix}$ (14) $Pol_3\begin{pmatrix} 0&1&2&0&1&0&2 \\ 0&1&2&1&0&2&0 \end{pmatrix}$

(15) $Pol_3\begin{pmatrix} 0&1&2&1&0&1&2 \\ 0&1&2&0&1&2&1 \end{pmatrix}$ (16) $Pol_3\begin{pmatrix} 0&1&2&2&0&2&1 \\ 0&1&2&0&2&1&2 \end{pmatrix}$

(17) $Pol_3\{(a,b,c,d) \in E_3^4 \mid a+b = c+d \ (mod\ 3)\}$

(18) $Pol_3\{(a,b,c) \in E_3^3 \mid |\{a,b,c\}| \le 2\}$ ■

In the following theorems, one finds all A-maximal classes for every maximal class A from Theorem 14.1.1.
The next two theorems are special cases of general statements about the maximal classes of $Pol_k E$ with $1 \leq |E| \leq k-1$ (see Chapters 16 and 17).

Theorem 14.1.2 *([Lau 82a])*
Let $\{a,b,c\} := E_3$. Then $T_a := Pol_3\{a\}$ has exactly the following 12 maximal classes:

(1) $T_a \cap Pol_3\{b\}$

(2) $T_a \cap Pol_3\{c\}$

(3) $T_a \cap Pol_3\{a,b\}$

(4) $T_a \cap Pol_3\{a,c\}$

(5) $T_a \cap Pol_3\{b,c\}$

(6) $T \cap Pol_3 \begin{pmatrix} a & b & c & b & c \\ a & b & c & c & b \end{pmatrix}$

(7) $T_a \cap Pol_3 \begin{pmatrix} a & b & c & a & b & a & c \\ a & b & c & b & a & c & a \end{pmatrix}$

(8) $T_a \cap Pol_3 \begin{pmatrix} a & b & c & a & a & b \\ a & b & c & b & c & c \end{pmatrix}$

(9) $T_a \cap Pol_3 \begin{pmatrix} a & b & c & a & a & c \\ a & b & c & c & b & b \end{pmatrix}$

(10) $Pol_3 \begin{pmatrix} a & b & c \\ a & c & b \end{pmatrix}$

(11) $Pol_3 \begin{pmatrix} a & a & b & a & c \\ a & b & a & c & a \end{pmatrix}$

(12) $Pol_3 \begin{pmatrix} a & a & b & a & c & b & c \\ a & b & a & c & a & c & b \end{pmatrix}$

Theorem 14.1.3 *([Lau 82a])*
Let $\{a,b,c\} := E_3$. Then $T_{a,b} := Pol_3\{a,b\}$ has exactly the following 15 maximal classes:

(1) $T_{a,b} \cap Pol_3\{a\}$

(2) $T_{a,b} \cap Pol_3\{b\}$

(3) $T_{a,b} \cap Pol_3 \begin{pmatrix} a & b & a \\ a & b & b \end{pmatrix}$

(4) $T_{a,b} \cap Pol_3 \begin{pmatrix} a & b \\ b & a \end{pmatrix}$

(5) $T_{a,b} \cap Pol_3 \begin{pmatrix} a & a & a & b & b & a & b & b \\ a & a & b & b & a & b & a & b \\ a & b & a & a & b & b & a & b \\ a & b & b & a & a & a & b & b \end{pmatrix}$

(6) $T_{a,b} \cap Pol_3\{c\}$

(7) $T_{a,b} \cap Pol_3 \begin{pmatrix} 0 & 1 & 2 & a & b \\ 0 & 1 & 2 & b & a \end{pmatrix}$

(8) $T_{a,b} \cap Pol_3 \begin{pmatrix} 0 & 1 & 2 & a & c & b & c \\ 0 & 1 & 2 & c & a & c & b \end{pmatrix}$

(9) $Pol_3 \begin{pmatrix} a & b & a \\ a & b & c \end{pmatrix}$

(10) $Pol_3 \begin{pmatrix} a & b & b \\ a & b & c \end{pmatrix}$

(11) $Pol_3 \begin{pmatrix} a & a & b & b & a \\ a & b & a & b & c \end{pmatrix}$

(12) $Pol_3 \begin{pmatrix} a & a & b & b & b \\ a & b & a & b & c \end{pmatrix}$

(13) $Pol_3 \begin{pmatrix} a & a & b & b & a & c & b & c \\ a & b & a & b & c & a & c & b \end{pmatrix}$

(14) $Pol_3 \begin{pmatrix} a & b & a & b & a & b & a & b \\ a & b & a & b & b & a & a & b \\ a & b & b & a & c & c & c & c \end{pmatrix}$

(15) $Pol_3 \begin{pmatrix} a & b & b & a & a & b & b & a & a & b \\ a & b & a & b & a & b & a & b & a & b \\ a & b & a & a & b & a & b & b & c & c \end{pmatrix}$

The following theorem is a special case of a theorem from [Lar 93], in which B. Larose determined all maximal classes of $Pol_k \varrho$, where ϱ is a total order on E_k and $2 \le k \le 5$.

Theorem 14.1.4 *([Mac 79])*

Let $E_3 := \{a, b, c\}$ and let min, max be defined by

x	y	$max(x, y)$	$min(x, y)$
a	a	a	a
a	b	b	a
a	c	c	a
b	a	b	a
b	b	b	b
b	c	c	b
c	a	c	a
c	b	c	b
c	c	c	c

Then $O := Pol \begin{pmatrix} 0 & 1 & 2 & a & a & b \\ 0 & 1 & 2 & b & c & c \end{pmatrix}$ *has exactly the following 13 maximal classes:*

(1) $O \cap Pol\{a\}$

(2) $O \cap Pol\{c\}$

(3) $O \cap Pol\{a, b\}$

(4) $O \cap Pol\{a, c\}$

(5) $O \cap Pol\{b, c\}$

(6) $O \cap Pol_3 \begin{pmatrix} 0 & 1 & 2 & a & b \\ 0 & 1 & 2 & b & a \end{pmatrix}$

(7) $O \cap Pol_3 \begin{pmatrix} 0 & 1 & 2 & b & c \\ 0 & 1 & 2 & c & b \end{pmatrix}$

(8) $O \cap Pol_3 \begin{pmatrix} 0 & 1 & 2 & a & b & a & c \\ 0 & 1 & 2 & b & a & c & a \end{pmatrix}$

(9) $O \cap Pol_3 \begin{pmatrix} 0 & 1 & 2 & b & b & b & c \\ 0 & 1 & 2 & a & b & c & b \end{pmatrix}$

(10) $O \cap Pol_3 \begin{pmatrix} 0 & 1 & 2 & c & a & c & b \\ 0 & 1 & 2 & a & c & b & c \end{pmatrix}$

(11) $O \cap Pol_3 \iota_3^3$

(12) $Pol_3\begin{pmatrix} 0&1&2&0&0&1&0&0&1\\ 0&1&2&0&0&1&1&2&2\\ 0&1&2&1&2&2&0&0&1 \end{pmatrix} = [\{min\}\cup O^1]$ [3]

(13) $Pol_3\begin{pmatrix} 0&1&2&0&0&1&1&2&2\\ 0&1&2&1&2&2&0&0&1\\ 0&1&2&1&2&2&1&2&2 \end{pmatrix} = [\{max\}\cup O^1]$ [4]

The following theorem is a special case of theorems from [Ros-S 84] and [Lau 84]:

Theorem 14.1.5 *([Mar-D-H 80])*

$S := Pol_3\begin{pmatrix} 0&1&2\\ 1&2&0 \end{pmatrix}$ *has exactly two maximal classes:*

(1) $S\cap Pol_3\{0\}$

(2) $S\cap Pol_3\lambda_3$.

The following theorem follows from Section 13.2:

Theorem 14.1.6 *([Bag-D 82])*

$L_3 := Pol_3\lambda_3$ *has exactly 5 maximal classes:*

(1) $L_3\cap Pol\{0\}$.

(2) $L_3\cap Pol\{1\}$.

(3) $L_3\cap Pol\{2\}$.

(4) $L_3\cap Pol_3\begin{pmatrix} 0&1&2\\ 1&2&0 \end{pmatrix}$

(5) $[(L_3)^1]$. ■

Theorem 14.1.7 *([Lau 82a])*

Let $E_3 := \{a,b,c\}$. *Then* $C := Pol_3\begin{pmatrix} 0&1&2&a&b&a&c\\ 0&1&2&b&a&c&a \end{pmatrix}$ *has exactly the following 7 maximal classes:*

(1) $C\cap Pol\{a\}$

(2) $C\cap Pol\{a,b\}$

---

[3] The equality of these two sets can be proven as follows: It is easily checked that $[\{min\}\cup O^1]\subseteq Pol_3(...)$ is valid. Then one proves that $[\{min\}\cup O^1]$ is maximal in O (see for this purpose also [Mac 79]).

[4] The equality of these two sets results from the considerations to the statement (12) and the fact that the classes (12) and (13) are isomorphic.

(3) $C \cap Pol\{a,c\}$

(4) $C \cap Pol\{b,c\}$

(5) $C \cap Pol_3 \begin{pmatrix} 0 & 1 & 2 & a & a \\ 0 & 1 & 2 & b & c \end{pmatrix}$

(6) $C \cap Pol_3 \begin{pmatrix} 0 & 1 & 2 & a & b & a & c & b \\ 0 & 1 & 2 & b & a & c & a & c \end{pmatrix}$

(7) $C \cap Pol_3 \begin{pmatrix} a & b & a & c & a & b & a & a & b & b & a & b & c & a & a & c & c & a & c \\ b & a & c & a & a & a & b & a & b & a & b & b & a & c & a & c & a & c & c \\ c & c & b & b & a & a & a & b & a & b & b & b & a & a & c & a & c & c & c \end{pmatrix}$

Theorem 14.1.8 *([Lau 82a])*
Let $E_3 := \{a,b,c\}$. *Then* $U := Pol_3 \begin{pmatrix} 0 & 1 & 2 & a & b \\ 0 & 1 & 2 & b & a \end{pmatrix}$ *has exactly the following 13 maximal classes:*

(1) $U \cap Pol\{c\}$

(2) $U \cap Pol\{a,b\}$

(3) $U \cap Pol\{a,c\}$

(4) $U \cap Pol\{b,c\}$

(5) $U \cap Pol_3 \begin{pmatrix} 0 & 1 & 2 & a & b & a & c \\ 0 & 1 & 2 & b & a & c & a \end{pmatrix}$

(6) $U \cap Pol_3 \begin{pmatrix} 0 & 1 & 2 & b & a & b & c \\ 0 & 1 & 2 & a & b & c & b \end{pmatrix}$

(7) $Pol_3 \begin{pmatrix} 0 & 1 & 2 & a \\ 0 & 1 & 2 & b \end{pmatrix}$

(8) $Pol_3 \begin{pmatrix} a & c & b & c \\ c & a & c & b \end{pmatrix}$

(9) $Pol_3 \begin{pmatrix} 0 & 1 & 2 & a & b & a & b \\ 0 & 1 & 2 & b & a & c & c \end{pmatrix}$

(10) $Pol_3 \begin{pmatrix} 0 & 1 & 2 & a & a & b & b & c & c & a & b \\ 0 & 1 & 2 & a & a & b & b & c & c & b & a \\ 0 & 1 & 2 & b & c & a & c & a & b & c & c \end{pmatrix}$

(11) $Pol_3 \begin{pmatrix} a & a & a & a & b & b & b & b & a & b & c & c & c \\ a & a & b & b & a & a & b & b & a & b & c & c & c \\ a & b & a & b & a & b & a & b & c & c & a & b & c \end{pmatrix}$

(12) $Pol_3 \begin{pmatrix} a & a & a & b & b & a & b & b & c \\ a & a & b & b & a & b & a & b & c \\ a & b & a & a & b & b & a & b & c \\ a & b & b & a & a & a & b & b & c \end{pmatrix}$

$$(13)\ Pol_3 E_2^4 \cup \begin{pmatrix} a & a & b & b & a & a & b & b & a & a & b & b & c & c & c & c & c & c & c & c & c & c & c & c & c \\ a & b & a & b & c & c & c & c & c & c & c & c & a & a & b & b & a & a & b & b & c & c & c & c & c \\ c & c & c & c & a & b & a & b & c & c & c & c & a & b & a & b & c & c & c & c & a & a & b & b & c \\ c & c & c & c & c & c & c & c & a & b & a & b & c & c & c & c & a & b & a & b & a & b & a & b & c \end{pmatrix}.$$

Theorem 14.1.9 *([Lau 82a])*
Let $P_3(2) := \{f \in P_3 \mid |Im(f)| \leq 2\}$. *The set* $Pol_3\iota_3^3$ $(= P_3(2) \cup [P_3^1])$ *has exactly 5 maximal classes:*

(1) $P_3(2) \cup [\{s_1, s_2\}]$

(2) $P_3(2) \cup [\{s_1, s_3\}]$

(3) $P_3(2) \cup [\{s_1, s_6\}]$

(4) $P_3(2) \cup [\{s_1, s_4, s_5\}]$

(5) $\bigcup_{n\geq 1}\{f^n \in P_3 | \exists f_i \in P_3^1 :$
$f(x_1, ..., x_n) = f_0(f_1(x_1) + ... + f_n(x_n))\ (mod\ 2)\} \cup [P_3^1]$.

(The definitions of the functions $s_1, ..., s_6$ *are given in Section 15.2, Table 15.1.)*

Table 14.1 gives a summary of the above-described submaximal classes, where $\{a, b, c\} := E_3$ and $\{\alpha, \beta, \gamma\} := E_3$. The submaximal classes that can be described as intersections of maximal clones are labelled with numbers 1–17. The other classes are labelled with numbers 18–43 in the order that they occur in Theorems 14.1.2–14.1.9.
With the aid of Table 14.1 and with

$$Pol_3 \begin{pmatrix} a & b & a \\ a & b & c \end{pmatrix} = Pol\{a, b\} \cap Pol_3 \begin{pmatrix} 0 & 1 & 2 & a & c \\ 0 & 1 & 2 & c & a \end{pmatrix},[5]\ \{a, b, c\} = E_3,$$

one can prove the following theorem:

Theorem 14.1.10 *P_3 has exactly 158 submaximal classes.* ■

[5] This equation can be proven as follows:
Because of $\Delta \begin{pmatrix} a & b & a \\ a & b & c \end{pmatrix} = \{a, b\}$ and
$\begin{pmatrix} a & b & c \\ a & b & a \end{pmatrix} \circ \begin{pmatrix} a & b & a \\ a & b & c \end{pmatrix} = \begin{pmatrix} 0 & 1 & 2 & a & c \\ 0 & 1 & 2 & c & a \end{pmatrix}$ it holds
$Pol_3 \begin{pmatrix} a & b & a \\ a & b & c \end{pmatrix} \subseteq Pol\{a, b\} \cap Pol_3 \begin{pmatrix} 0 & 1 & 2 & a & c \\ 0 & 1 & 2 & c & a \end{pmatrix}$.
Since, in addition, $\Delta\left(\{a, b\} \times \begin{pmatrix} 0 & 1 & 2 & a & c \\ 0 & 1 & 2 & c & a \end{pmatrix}\right) = Pol_3 \begin{pmatrix} a & b & a \\ a & b & c \end{pmatrix}$ is valid, we have also
$Pol\{a, b\} \cap Pol_3 \begin{pmatrix} 0 & 1 & 2 & a & c \\ 0 & 1 & 2 & c & a \end{pmatrix} \subseteq Pol_3 \begin{pmatrix} a & b & a \\ a & b & c \end{pmatrix}$.

Table 14.1 Submaximal classes of P_3 [6]

i	A	$n_i(A)$
1	$Pol\{a\} \cap Pol\{b\}$	3
2	$Pol\{a\} \cap Pol\{a,b\}$	6
3	$Pol\{a\} \cap Pol\{b,c\}$	3
4	$Pol\{a\} \cap Pol\begin{pmatrix} 0 & 1 & 2 & b & c \\ 0 & 1 & 2 & c & b \end{pmatrix}$	3
5	$Pol\{a\} \cap Pol\begin{pmatrix} 0 & 1 & 2 & a & b & a & c \\ 0 & 1 & 2 & b & a & c & a \end{pmatrix}$	3
6	$Pol\{a\} \cap Pol\begin{pmatrix} 0 & 1 & 2 & a & a & b \\ 0 & 1 & 2 & b & c & c \end{pmatrix}$	6
7	$Pol\{a\} \cap Pol\lambda_3$	3
8	$Pol\{0\} \cap \begin{pmatrix} 0 & 1 & 2 \\ 1 & 2 & 0 \end{pmatrix}$	1
9	$Pol\{a,b\} \cap Pol\begin{pmatrix} 0 & 1 & 2 & a & b \\ 0 & 1 & 2 & b & a \end{pmatrix}$	3
10	$Pol\{a,b\} \cap Pol\begin{pmatrix} 0 & 1 & 2 & a & c \\ 0 & 1 & 2 & c & a \end{pmatrix}$	6
11	$Pol\{a,b\} \cap Pol\begin{pmatrix} 0 & 1 & 2 & \alpha & \alpha & \beta \\ 0 & 1 & 2 & \beta & \gamma & \gamma \end{pmatrix}$	9
12	$Pol\{a,b\} \cap Pol\begin{pmatrix} 0 & 1 & 2 & \alpha & \beta & \alpha & \gamma \\ 0 & 1 & 2 & \beta & \alpha & \gamma & \alpha \end{pmatrix}$	9
13	$Pol\begin{pmatrix} 0 & 1 & 2 & a & b \\ 0 & 1 & 2 & b & a \end{pmatrix} \cap Pol\begin{pmatrix} 0 & 1 & 2 & a & a & b \\ 0 & 1 & 2 & b & c & c \end{pmatrix}$	6
14	$Pol\begin{pmatrix} 0 & 1 & 2 & a & b \\ 0 & 1 & 2 & b & a \end{pmatrix} \cap Pol\begin{pmatrix} 0 & 1 & 2 & a & b & a & c \\ 0 & 1 & 2 & b & a & c & a \end{pmatrix}$	6
15	$Pol\begin{pmatrix} 0 & 1 & 2 & a & a & b \\ 0 & 1 & 2 & b & c & c \end{pmatrix} \cap Pol\begin{pmatrix} 0 & 1 & 2 & \alpha & \beta & \alpha & \gamma \\ 0 & 1 & 2 & \beta & \alpha & \gamma & \alpha \end{pmatrix}$	9
16	$Pol\begin{pmatrix} 0 & 1 & 2 & a & a & b \\ 0 & 1 & 2 & b & c & c \end{pmatrix} \cap Pol\iota_3^3$	3
17	$Pol\begin{pmatrix} 0 & 1 & 2 \\ 1 & 2 & 0 \end{pmatrix} \cap Pol\lambda_3$	1
18	$Pol\begin{pmatrix} a & b & c \\ a & c & b \end{pmatrix}$	3
19	$Pol\begin{pmatrix} a & a & b & a & c \\ a & b & a & c & a \end{pmatrix}$	3
20	$Pol\begin{pmatrix} a & a & b & a & c & b & c \\ a & b & a & c & a & c & b \end{pmatrix}$	3
21	$Pol\begin{pmatrix} a & b & a \\ a & b & b \end{pmatrix}$	3
22	$Pol\begin{pmatrix} a & b \\ b & a \end{pmatrix}$	3
23	$Pol\begin{pmatrix} a & a & a & b & b & a & b & b \\ a & a & b & b & a & b & a & b \\ a & b & a & a & b & b & a & b \\ a & b & b & a & a & a & b & b \end{pmatrix}$	3
24	$Pol\begin{pmatrix} a & a & b & b & a \\ a & b & a & b & c \end{pmatrix}$	6
25	$Pol\begin{pmatrix} a & a & b & b & a & c & b & c \\ a & b & a & b & c & a & c & b \end{pmatrix}$	3

[6] $n_i(A)$ denotes the number of possibilities for A in case i.

i	A	$n_i(A)$
26	$Pol\begin{pmatrix} a & b & a & b & a & b & a & b \\ a & b & a & b & b & a & a & b \\ a & b & b & a & c & c & c & c \end{pmatrix}$	3
27	$Pol\begin{pmatrix} a & b & b & a & a & b & b & a & a & b \\ a & b & a & b & a & b & a & b & a & b \\ a & b & a & a & b & a & b & b & c & c \end{pmatrix}$	3
28	$[\{max\} \cup O^1]$	3
29	$[\{min\} \cup O^1]$	3
30	$[(L_3)^1]$	1
31	$Pol\begin{pmatrix} 0 & 1 & 2 & a \\ 0 & 1 & 2 & b \end{pmatrix}$	3
32	$Pol\begin{pmatrix} a & c & b & c \\ c & a & c & b \end{pmatrix}$	3
33	$Pol\begin{pmatrix} 0 & 1 & 2 & a & b & a & b \\ 0 & 1 & 2 & b & a & c & c \end{pmatrix}$	3
34	$Pol\begin{pmatrix} 0 & 1 & 2 & a & a & b & b & c & c & a & b \\ 0 & 1 & 2 & a & a & b & b & c & c & b & a \\ 0 & 1 & 2 & b & c & a & c & a & b & c & c \end{pmatrix}$	3
35	$Pol\begin{pmatrix} a & a & a & a & b & b & b & b & a & b & c & c & c \\ a & a & b & b & a & a & b & b & a & b & c & c & c \\ a & b & a & b & a & b & a & b & c & c & a & b & c \end{pmatrix}$	3
36	$Pol\begin{pmatrix} a & a & a & b & b & a & b & b & c \\ a & a & b & b & a & b & a & b & c \\ a & b & a & a & b & b & a & b & c \\ a & b & b & a & a & a & b & b & c \end{pmatrix}$	3
37	$Pol E_2^4 \cup$ $\begin{pmatrix} a & a & b & b & a & a & b & b & a & a & b & b & c & c & c & c & c & c & c & c & c & c & c & c & c \\ a & b & a & b & c & c & c & c & c & c & c & c & a & a & b & b & a & a & b & b & c & c & c & c & c \\ c & c & c & c & a & b & a & b & c & c & c & c & a & b & a & b & c & c & c & c & a & a & b & b & c \\ c & c & c & c & c & c & c & c & a & b & a & b & c & c & c & c & a & b & a & b & a & b & a & b & c \end{pmatrix}$	3
38	$Pol\begin{pmatrix} 0 & 1 & 2 & a & a \\ 0 & 1 & 2 & b & c \end{pmatrix}$	3
39	$Pol\begin{pmatrix} 0 & 1 & 2 & a & b & a & c & b \\ 0 & 1 & 2 & b & a & c & a & c \end{pmatrix}$	3
40	$Pol\begin{pmatrix} a & b & a & c & a & b & a & a & b & b & a & b & c & a & a & c & c & a & c \\ b & a & c & a & a & a & b & a & b & a & b & b & a & c & a & c & a & c & c \\ c & c & b & b & a & a & a & b & a & b & b & b & a & a & c & a & c & c & c \end{pmatrix}$	3
41	$P_3(2) \cup [X],\ X \in \{\{s_1, s_2\}, \{s_1, s_3\}, \{s_1, s_6\}\}$	3
42	$P_3(2) \cup [\{s_1, s_4, s_5\}]$	1
43	$\bigcup_{n\geq 1}\{f^n \in P_3 \mid \exists f_i \in P_3^1 :$ $f(x_1, ..., x_n) =$ $f_0(f_1(x_1) + f_2(x_2) + ... + f_n(x_n))\ (mod\ 2)\} \cup$ $[P_3^1]$	1

14.2 Some Declarations and Lemmas for Sections 14.3–14.9

In this chapter, we use the notations from Chapter 15, Table 15.1 for the unary functions of P_3.
We prove Theorems 14.1.2–14.1.9 as follows in the next sections:
Suppose in the theorem a list is indicated with subclasses $(1), (2), ..., (r_T)$ for the class T. To prove that this list indicates all submaximal classes of T, we will show that every subset $A \subseteq T$, which is not contained in any subclass of the list, is a generating system for T. Then, it remains to show that the listed classes $(1), (2), ..., (r_T)$ are proper subsets of the set T, and the set of all these classes is an antichain; i.e., in this set, no two elements are comparable in respect to $\subseteq$. For this purpose, we give some functions ($\in T$) and a table that shows whether any function of these functions belongs to the indicated class. The sign + stands for "the function belongs to the class". If the considered function does not belong to the class, we write -. We leave the readers to check the given table and, with the aid of the table, to prove that the classes $(1), (2), ..., (r_T)$ are incomparable.

If we assume in the proof that $A \subseteq T$ and A is not a subset of the class (i), then there exists a function $f_i \in A$ which does not belong to the class (i). Since we can choose $[A]$ instead of A, we can assume w.l.o.g. that the following holds:

(*): If the class (i) is given in the form $T \cap Pol_3\varrho$ or $Pol_3\varrho$, where $\varrho := (\sigma_1, \sigma_2, ..., \sigma_m)$, then $f_i(\sigma_1, \sigma_2, ..., \sigma_m) \notin \varrho$.

Now some lemmas are given that we subsequently need more often and which are consequences from some theorems of Chapters 3 and 12.
Already in Chapter 3 (see Theorem 3.3.1) the following was proven:

Lemma 14.2.1 *Let A be an arbitrary subset of P_2. Then, $[A] = P_2$ if and only if $A \not\subseteq B$ for every class B of the following list:*
(1) $Pol_2\{0\}$
(2) $Pol_2\{1\}$
(3) $Pol_2\begin{pmatrix} 0 & 1 \\ 1 & 0 \end{pmatrix}$
(4) $Pol_2\begin{pmatrix} 0 & 0 & 1 \\ 0 & 1 & 1 \end{pmatrix}$
(5) $Pol_2\begin{pmatrix} 0 & 0 & 0 & 1 & 1 & 0 & 1 & 1 \\ 0 & 0 & 1 & 1 & 0 & 1 & 0 & 1 \\ 0 & 1 & 0 & 0 & 1 & 1 & 0 & 1 \\ 0 & 1 & 1 & 0 & 0 & 0 & 1 & 1 \end{pmatrix}$. ■

Lemma 14.2.2 *Let A be an arbitrary subset of $T_0 := Pol_2\{0\} \subset P_2$. Then, $[A] = T_0$ if and only if $A \not\subseteq B$ for every class B of the following list:*
(1) $T_0 \cap Pol_2\{1\}$

(2) $T_0 \cap Pol_2 \begin{pmatrix} 0 & 0 & 1 \\ 0 & 1 & 1 \end{pmatrix}$

(3) $T_0 \cap Pol_2 \begin{pmatrix} 0 & 1 & 0 \\ 0 & 0 & 1 \end{pmatrix}$

(4) $T_0 \cap Pol_2 \begin{pmatrix} 0 & 0 & 0 & 1 & 1 & 0 & 1 & 1 \\ 0 & 0 & 1 & 1 & 0 & 1 & 0 & 1 \\ 0 & 1 & 0 & 0 & 1 & 1 & 0 & 1 \\ 0 & 1 & 1 & 0 & 0 & 0 & 1 & 1 \end{pmatrix}$. ■

As a consequence of Theorem 12.4.3, we get:

Lemma 14.2.3 *Let A be an arbitrary subset of $P_{3,2}$. Then $[A] = P_{3,2}$ if and only if $A \not\subseteq B$ for every class B of the following list:*
(1) $P_{3,2} \cap Pol_3\{0\}$
(2) $P_{3,2} \cap Pol_3\{1\}$
(3) $P_{3,2} \cap Pol_3 \begin{pmatrix} 0 & 1 \\ 1 & 0 \end{pmatrix}$
(4) $P_{3,2} \cap Pol \begin{pmatrix} 0 & 0 & 1 \\ 0 & 1 & 1 \end{pmatrix}$
(5) $P_{3,2} \cap Pol_3 \begin{pmatrix} 0 & 0 & 0 & 1 & 1 & 0 & 1 & 1 \\ 0 & 0 & 1 & 1 & 0 & 1 & 0 & 1 \\ 0 & 1 & 0 & 0 & 1 & 1 & 0 & 1 \\ 0 & 1 & 1 & 0 & 0 & 0 & 1 & 1 \end{pmatrix}$
(6) $P_{3,2} \cap Pol_3 \begin{pmatrix} 0 & 1 & 0 \\ 0 & 1 & 2 \end{pmatrix}$
(7) $P_{3,2} \cap Pol_3 \begin{pmatrix} 0 & 1 & 1 \\ 0 & 1 & 2 \end{pmatrix}$. ■

The next lemma follows from Theorem 12.5.3:

Lemma 14.2.4 *Let A be an arbitrary subset of $pr^{-1}T_0 := P_{3,2} \cap Pol_3\{0\} \subset P_3$. Then $[A] = pr^{-1}T_0$ if and only if $A \not\subseteq B$ for every class B of the following list:*
(1) $pr^{-1}T_0 \cap Pol_3\{1\}$
(2) $pr^{-1}T_0 \cap Pol_3 \begin{pmatrix} 0 & 0 & 1 \\ 0 & 1 & 1 \end{pmatrix}$
(3) $pr^{-1}T_0 \cap Pol_3 \begin{pmatrix} 0 & 1 & 0 \\ 0 & 0 & 1 \end{pmatrix}$
(4) $pr^{-1}T_0 \cap Pol_3 \begin{pmatrix} 0 & 0 & 0 & 1 & 1 & 0 & 1 & 1 \\ 0 & 0 & 1 & 1 & 0 & 1 & 0 & 1 \\ 0 & 1 & 0 & 0 & 1 & 1 & 0 & 1 \\ 0 & 1 & 1 & 0 & 0 & 0 & 1 & 1 \end{pmatrix}$
(5) $pr^{-1}T_0 \cap Pol_3\{0, 2\}$
(6) $pr^{-1}T_0 \cap Pol_3 \begin{pmatrix} 0 & 1 & 1 \\ 0 & 1 & 2 \end{pmatrix}$. ■

14.3 Proof of Theorem 14.1.2

W.l.o.g. we assume $a = 0$, $b = 1$ and $c = 2$. Further, in this proof, let A be an arbitrary subset of $Pol\{0\}$, which is not contained in any class ((1)–(12)) from the above list, which is given in Theorem 14.1.2. Then, there are some functions $f_1, f_2, ..., f_{12} \in [A]$ with the above property (*).
First we prove that $(Pol_3\{0\})^1 \subseteq [A]$.
The function f_1 belongs to $\{c_0, j_2, u_2, u_1, s_2, u_5\}$. If $f_1 \in \{j_2, u_1\}$ then $c_0 = f_1 \star f_1 \in [A]$. If $f_1 = u_2$ then $f_2 \star f_1 \in \{c_0, j_2\}$. For $f_1 \in \{s_2, u_5\}$ we have $f_5(x, f_1(x)) \in \{c_0, j_2, u_2\}$. Therefore, c_0 is a superposition over A. The function $f_3'(x) := f_3(c_o, x) \in \{u_1, s_2, u_5\}$ also belongs to $[A]$. Thus we have to distinguish three cases:
Case 1: $f_3' = u_1$.
In this case $j_1 = f_4(c_0, u_1)$ and a function $f_4'(x) = f_4(c_0, x) \in \{j_2, j_5, s_2\}$ belongs to $[A]$.
Case 1.1: $f_4' = j_2$.
Then $u_1 \star j_2 = u_2 \in [A]$, $f_{11}'(x) := f_{11}(c_0, j_2, j_1, u_2, u_1) \in [A]$ and $f_{11}''(x) := f_{11}(c_o, j_1, j_2, u_1, u_2) \in [A]$, where $\{f_{11}', f_{11}''\} \cap \{s_2, j_5, u_5\} \neq \emptyset$.
Because of $f_{12}(c_0, j_2, j_1, u_2, u_1, x, s_2) \in \{j_5, u_5\}$ and $j_2 \star u_5 = j_5$ we can assume $j_5 \in [A]$. Then $j_2 \star j_5 = u_5$ and $\{f_7(c_0, j_5, u_5, j_2, j_1, u_2, u_1), f_7(c_0, j_5, u_5, j_1, j_2, u_1, u_2)\} = \{s_1, s_2\} \subseteq [A]$. Therefore, $(Pol_3\{0\})^1 \subseteq [A]$ in Case 1.1.
Case 1.2: $f_4' = s_2$.
Because of $j_1 \star s_2 = j_2$ one can reduce this case to Case 1.1.
Case 1.3: $f_4' = j_5$.
In this case, functions $u_5 = u_1 \star j_5$ and $f_9(c_0, j_5, u_5, u_1, j_1, x) \in \{j_2, u_2, s_2\}$ are superpositions over A. Therefore, because of $j_5 \star u_2 = j_2$, we have given either Case 1.1 or Case 1.2.
Case 2: $f_3' = u_5$.
Here we can form the superpositions $j_5 = f_4(c_0, u_5)$ and

$$f_6'(x_1, x_2) := j_5(f_6(c_0, j_5(x_1), u_5(x_1), x_1, x_2))$$

over A, where $f_6'\begin{pmatrix} 1 & 2 \\ 2 & 1 \end{pmatrix} \in \begin{pmatrix} 0 & 1 \\ 1 & 0 \end{pmatrix}$. W.l.o.g. let $f_6'\begin{pmatrix} 1 & 2 \\ 2 & 1 \end{pmatrix} = \begin{pmatrix} 0 \\ 1 \end{pmatrix}$. If $f_6'(1,1) = 1$ we obtain Case 1 because of $u_5(f_6'(j_5(x), x)) = u_1$. If $f_6'(1,1) = 0$ then it holds $f_6'(x, j_5(x)) = j_2$, $u_5 \star j_2 = u_2$ and $f_8(c_0, j_5, u_5, j_2, u_2, x) \in \{j_1, u_1, s_2\}$. Since $u_5 \star j_1 = u_1$ and $u_2 \star s_2 = u_1$, we have also $u_1 \in [A]$ if $f_6'(1,1) = 0$, i.e., one can reduce Case 2 to Case 1.
Case 3: $f_3' = s_2$.
Because of $f_{10}(c_0, s_2, x) \in \{j_1, j_2, j_5, u_1, u_2, u_5\}$, $s_2 \star j_i = u_i$, $i \in \{1, 2, 5\}$, and $u_2 \star s_2 = u_1$, we obtain either Case 1 or Case 2.

Consequently, $(Pol\{0\})^1 \subseteq [A]$ is proven.

Next we prove that $P_{3,2} \cap Pol_3\{0\}$ is a subset of $[A]$. Since $(P_{3,2} \cap Pol_3\{0\})^1 \subseteq [A]$ was already shown, we have to find a subset of $[A] \cap P_{3,2}$ that is not a subset of $Pol\begin{pmatrix} 0 & 0 & 1 \\ 0 & 1 & 0 \end{pmatrix}$, $Pol\begin{pmatrix} 0 & 0 & 1 \\ 0 & 1 & 1 \end{pmatrix}$ and $Pol\lambda_2 := \{(a,b,c,d) \in E_2^4 \mid a+b = c+d \ (mod\ 2)\}$, to be able to use Lemma 14.2.4. Obviously, the function

$$\widehat{f}_{11}(x_1, x_2) := j_5(f_{11}(c_0, x_1, x_2, u_1(x_1), u_1(x_2))) \in [A] \cap P_{3,2}$$

does not preserve the relation $\begin{pmatrix} 0 & 0 & 1 \\ 0 & 1 & 0 \end{pmatrix}$. For the function

$$f_7'(x_1, x_2) := f_7(c_0, \widehat{f}_{11}(x_1, x_2), u_1(\widehat{f}_{11}(x_1, x_2)), x_1, x_2, u_1(x_1), u_1(x_2)) \in [A]$$

we can assume w.l.o.g. that $f_7'\begin{pmatrix} 0 & 1 \\ 1 & 0 \end{pmatrix} = \begin{pmatrix} 1 \\ 0 \end{pmatrix}$. If $f_7'(1,1) \in \{0,2\}$ we have $j_1 \star f_7' \notin Pol_3\begin{pmatrix} 0 & 0 & 1 \\ 0 & 1 & 1 \end{pmatrix} \cup Pol_3\lambda_2$. If $f_7'(1,1) = 1$ then this is valid for $j_2 \star f_7'$ instead of $j_1 \star f_7'$.

Thus by Lemma 14.2.4, we have $P_{3,2} \cap Pol_3\{0\} \subseteq [A]$.

After these preparations we can easily show that an arbitrary function $f^n \in Pol_3\{0\}$ is a superposition over A. For this purpose, we need the functions q_1, $q_2 \in P_{3,2}$ defined by

$$q_1(\mathbf{x}) := \begin{cases} 0 & \text{if } f(\mathbf{x}) \in \{0,1\}, \\ 1 & \text{if } f(\mathbf{x}) = 2, \end{cases}$$

$$q_2(\mathbf{x}) := \begin{cases} 0 & \text{if } f(\mathbf{x}) \in \{0,2\}, \\ 1 & \text{if } f(\mathbf{x}) = 1, \end{cases}$$

Then $f_7'(q_1(\mathbf{x}), q_2(\mathbf{x})) = f(\mathbf{x}) \in [A]$ holds.

Consequently, $[A] = Pol_3\{0\}$.

As explained in Section 14.2, our theorem results from Table 14.2, where the binary functions h_1, h_2,..., h_8 are defined in Table 14.3.

Table 14.2

	c_0	j_1	s_2	u_5	h_1	h_2	h_3	h_4	h_5	h_6	h_7	h_8
(1)	−	+	−	−	+	+	+	−	+	−	+	−
(2)	−	−	−	+	+	+	−	−	+	−	−	−
(3)	+	+	−	−	−	+	−	−	+	+	−	−
(4)	+	+	−	+	+	−	−	−	+	−	+	+
(5)	−	−	+	+	−	+	+	−	+	−	−	−
(6)	+	−	+	+	−	−	+	−	+	−	−	−
(7)	+	+	+	+	−	−	−	−	−	+	−	−
(8)	+	−	−	+	−	−	−	−	+	+	−	−
(9)	+	+	−	+	−	−	+	−	−	−	−	−
(10)	+	−	+	−	−	−	−	+	−	−	−	−
(11)	+	+	+	+	−	−	−	−	−	−	+	−
(12)	+	+	+	−	−	−	−	−	−	−	+	+

Table 14.3

x_1	x_2	h_1	h_2	h_3	h_4	h_5	h_6	h_7	h_8
0	0	0	0	0	0	0	0	0	0
0	1	1	1	1	1	1	0	0	1
0	2	2	1	1	2	2	1	0	0
1	0	2	1	2	1	1	0	2	2
1	1	1	1	1	2	1	0	1	0
1	2	0	1	1	0	2	1	0	0
2	0	2	0	1	2	2	1	0	0
2	1	0	1	1	0	2	1	0	0
2	2	2	2	1	1	2	1	0	0

14.4 Proof of Theorem 14.1.3

W.l.o.g. we assume $a = 0$, $b = 1$ and $c = 2$. Further, in this proof let A be an arbitrary subset of $Pol\{0,1\}$, which is not contained in any class ((1)–(15)) from the above list, which is given in Theorem 14.1.3. Then there are some functions $f_1, f_2, ..., f_{15} \in [A]$ with the above property (*).
With the help of Lemma 14.2.1, one can easily prove that every function of P_2 is a restriction of a function of $[A]$. Therefore, a function $g \in \{c_0, j_2, u_2\}$ and a function $h \in \{j_0, j_4, s_3\}$ are superpositions over A.
If $g \in \{c_0, j_2\}$ then the functions $g \star g = c_0$ and $h \star c_0 = c_1$ belong to $[A]$. If $g = u_2$ then $f_6' := f_6 \star g \in \{c_0, c_1, j_2, j_3\}$ and $\{f_6' \star f_6', h \star f_6' \star f_6'\} = \{c_0, c_1\}$. Thus the constant functions c_0 and c_1 are superpositions over A.
Next we prove that the remaining unary functions of $Pol_3\{0,1\}$ and all functions of $P_{3,2}$ are also superpositions over A. For this purpose, we distinguish two cases for the function h:
Case 1: $h \in \{j_0, j_4\}$.
Then $h' := h \star h \in \{j_1, j_5\}$ and the functions $\widehat{f_i} := h' \star f_i$, $i \in \{1,2,3,4,5\}$ belong to $[A]$, where $f_i' \in P_{3,2}$ and f_i' does not belong to the class (i). Further, the functions $f_9' := h(f_9(c_0, c_1, x))$ and $f_{10}' := h(f_{10}(c_0, c_1, x))$ belong to $[A]$. It is easy to check that $\{h, f_9', f_{10}'\} \subseteq P_{3,2}$ and that $\{h, f_9', f_{10}'\}$ is no subset of the classes $Pol_3 \begin{pmatrix} 0 & 1 & 0 \\ 0 & 1 & 2 \end{pmatrix}$ and $Pol_3 \begin{pmatrix} 0 & 1 & 1 \\ 1 & 1 & 2 \end{pmatrix}$. Consequently, by Lemma 14.2.3, every function of $P_{3,2}$ is a superposition over A.
A function of $[A]$ is also $f_{11}' := f_{11}(c_0, j_2, j_0, c_1, x) \in \{v_2, s_3\}$.
Case 1.1: $f_{11}' = v_2$.
In this case, we have $f_{12}' := f_{12}(c_0, j_2, j_3, c_1, v_2) = u_2 \in [A]$ and

$$\{f_{15}(c_0, c_1, j_0, j_1, j_2, j_3, j_4, j_5, u_2, v_2), f_{15}(c_0, c_1, j_1, j_0, j_2, j_3, j_5, j_4, u_2, v_2)\}$$
$$= \{s_1, s_2\}.$$

Consequently, $(Pol_3\{0,1\})^1 \subseteq [A]$.
Case 1.2: $f_{11}' = s_2$.

First we form a ternary function $\widehat{f_{15}}$ as a superposition over $P_{3,2} \cup \{f_{15}\}$, for which

$$\widehat{f_{15}} \begin{pmatrix} 0 & 0 & 1 \\ 1 & 0 & 1 \\ 0 & 2 & 2 \end{pmatrix} = \begin{pmatrix} 0 \\ 1 \\ 2 \end{pmatrix}$$

holds. If $\widehat{f_{15}}(0,1,0) = 0$ then $s_3(\widehat{f_{15}}(c_0, x, s_3)) = v_2 \in [A]$. If $\widehat{f_{15}}(0,1,0) = 1$ then $\widehat{f_{15}}(j_0, x, s_3) = v_2 \in [A]$. Consequently, we can reduce Case 1.2 to Case 1.1.

Consequently, $P_{3,2} \cup (Pol_3\{0,1\})^1 \subseteq [A]$ was proven in Case 1.

Case 2: $h = s_3$.

W.l.o.g. we can assume $f_8 \begin{pmatrix} 0 & 1 & 2 & 0 & 2 & 1 & 2 \\ 0 & 1 & 2 & 2 & 0 & 2 & 1 \end{pmatrix} = \begin{pmatrix} 0 \\ 1 \end{pmatrix}$. Consequently, $f_8'(x_1, x_2, x_3) := f_8(c_0, c_1, x_1, x_2, x_3, s_3(x_2), s_3(x_3))$ belongs to $[A]$, and we have $f_8' \begin{pmatrix} 2 & 0 & 2 \\ 2 & 2 & 0 \end{pmatrix} = \begin{pmatrix} 0 \\ 1 \end{pmatrix}$. Then the functions $f_8'' := f_8'(s_3, c_0, x)$ and $f_8''' := f_8'(x, s_3, c_0)$ are superpositions over A, where $\{f_8'', f_8'''\} \cap \{j_0, j_3, j_2, j_4\} \neq \emptyset$. Thus, we have either Case 1 or $j_2 = s_3 \star j_3$ and $j_3 = s_3 \star j_2$ belong to $[A]$. A superposition over A is also a certain binary function q with the property

$$q \begin{pmatrix} 0 & 0 \\ 0 & 1 \\ 1 & 0 \\ 1 & 1 \end{pmatrix} = \begin{pmatrix} 1 \\ 0 \\ 0 \\ 0 \end{pmatrix}.$$

If $\{q(0,2), q(2,0), q(1,2), q(2,1), q(2,2)\} \cap \{0,1\} \neq \emptyset$ then

$$\{q(c_0, x), q(x, c_0), q(j_2, x), q(x, j_2), q(x, x)\} \cap \{j_0, j_4\} \neq \emptyset,$$

i.e., one can reduce Case 2 to Case 1 here.

Therefore, we can assume $q(\mathbf{a}) = 2$ for every $\mathbf{a} \in E_3^2 \backslash E_2^2$ in the following. Then $q(j_3, x) = u_2$, $s_3 \star u_2 = v_2$ and $f_{14}' := f_{14}(c_0, c_1, j_2, j_3, x, s_3, u_2, v_2) \in \{j_0, j_1, j_4, j_5\}$ belong to $[A]$. Since $s_3 \star j_1 = j_4$ and $s_3 \star j_5 = j_0$, we have again Case 1.

Hence

$$P_{3,2} \cup (Pol_3\{0,1\})^1 \subseteq [A].$$

W.l.o.g. we can assume $f_7 \begin{pmatrix} 0 & 1 & 2 & 0 & 1 \\ 0 & 1 & 2 & 1 & 0 \end{pmatrix} = \begin{pmatrix} 0 \\ 2 \end{pmatrix}$. Then $f_7'(x_1, x_2) := u_2(f_7(c_0, c_1, x_1, x_2, j_0(x_2))) \in [A]$, where

$$f_7' \begin{pmatrix} 0 & 0 \\ 0 & 1 \\ 1 & 0 \\ 1 & 1 \\ 2 & 0 \\ 2 & 1 \end{pmatrix} = \begin{pmatrix} 0 \\ 0 \\ 0 \\ 0 \\ 0 \\ 2 \end{pmatrix}.$$

Further, $f'_{13}(x_1,x_2) := u_2(f_{13}(c_0,j_2(x_1),j_2(x_2),c_1,x_1,x_2,s_3(x_1),s_3(x_2))) \in [A]$, where

$$f'_{13}\begin{pmatrix} 0 & 0 \\ 0 & 2 \\ 2 & 0 \end{pmatrix} = \begin{pmatrix} 0 \\ 2 \\ 2 \end{pmatrix}.$$

It is easy to see that one obtains a binary function f'_{15} with

$$f'_{15}\begin{pmatrix} 0 & 0 \\ 1 & 0 \\ 0 & 2 \end{pmatrix} = \begin{pmatrix} 0 \\ 1 \\ 2 \end{pmatrix}$$

as a superposition over $P_{3,2} \cup (Pol_3\{0,1\})^1 \cup \{f_{15}\}$.
Let f^n be an arbitrary function of $Pol_3\{0,1\}$. We show $f \in [A]$. For this purpose, let $f_{\widetilde{\alpha}_1,...,\widetilde{\alpha}_r;\beta}$ ($\widetilde{\alpha}_1,...,\widetilde{\alpha}_r \in E_3^n, \beta \in E_3$) be an n-ary function defined by

$$f_{\widetilde{\alpha_1},...,\widetilde{\alpha_r};\beta}(\mathbf{x}) := \begin{cases} \beta \text{ if} & \mathbf{x} \in \{\widetilde{\alpha}_1,...,\widetilde{\alpha}_r\}, \\ 0 \text{ otherwise.} \end{cases}$$

Obviously, the functions of the type $f_{\widetilde{\alpha};1}$ are superpositions over A for every $\widetilde{\alpha} \in E_3^n$. If $\widetilde{\alpha} = (\alpha_1,...,\alpha_n) \in E_3^n$ and $\alpha_i = 2$ then $f_{\widetilde{\alpha};2} \in [A]$ follows from

$$f_{\widetilde{\alpha};2} = f'_7(x_i, f_{\widetilde{\alpha};1}(\mathbf{x})).$$

Furthermore, we have

$$f_{\widetilde{\alpha_1},...,\widetilde{\alpha_r};2}(\mathbf{x}) = f'_{13}(f_{\widetilde{\alpha_1};2}(\mathbf{x}), f_{\widetilde{\alpha_2},\widetilde{\alpha_3},...,\widetilde{\alpha_r};2}(\mathbf{x})).$$

Consequently, all functions of the type $f_{\widetilde{\alpha_1},...,\widetilde{\alpha_r};2}$ with $\{\widetilde{\alpha_1},...,\widetilde{\alpha_r}\} \subseteq E_3^n \backslash E_2^n$ are superpositions over A.
Then $f \in [A]$ follows from

$$f(\mathbf{x}) = f'_{15}(f_{\widetilde{\beta_1},...,\widetilde{\beta_s};1}(\mathbf{x}), f_{\widetilde{\gamma_1},...,\widetilde{\gamma_t};2}(\mathbf{x})),$$

where $\{\widetilde{\beta_1},...,\widetilde{\beta_s}\} := \{\mathbf{x} \in E_3^n \mid f(\mathbf{x}) = 1\}$ and $\{\widetilde{\gamma_1},...,\widetilde{\gamma_t}\} := \{\mathbf{x} \in E_3^n \mid f(\mathbf{x}) = 2\}$. Consequently, $[A] = Pol_3\{0,1\}$.
Then our theorem follows from Table 14.4. The functions $g_1,...,g_{10}$ of Table 14.4 are defined in Table 14.5. The function h^3 is defined by

$$h(\mathbf{x}) := \begin{cases} 2 \text{ if} & \mathbf{x} \in E_3^3 \backslash E_2^3, \\ 1 \text{ if} & \mathbf{x} \in \{(0,1,1),(1,0,1),(1,1,0),(1,1,1)\}, \\ 0 \text{ otherwise.} \end{cases}$$

Table 14.4

	j_0	j_4	u_2	s_3	g_1	g_2	g_3	g_4	g_5	g_6	g_7	g_8	g_9	g_{10}	h
(1)	−	−	+	−	+	−	+	−	+	−	+	+	+	+	+
(2)	−	−	−	−	−	+	−	−	+	+	+	−	+	+	+
(3)	−	−	+	−	+	+	−	−	+	−	+	+	+	+	+
(4)	+	+	−	+	−	−	−	+	+	−	−	−	+	−	+
(5)	+	+	+	+	+	+	−	+	+	−	−	+	+	−	−
(6)	−	−	+	+	−	−	+	−	+	+	+	+	+	−	+
(7)	+	+	+	+	−	−	−	−	−	+	+	+	−	+	+
(8)	−	−	+	+	+	+	−	−	−	−	+	+	−	−	+
(9)	−	+	+	−	+	−	−	+	−	−	−	+	−	+	−
(10)	+	−	−	−	−	+	−	−	+	−	+	−	−	−	−
(11)	+	+	+	−	+	−	−	−	−	−	−	+	−	+	−
(12)	+	+	−	−	−	+	−	−	−	+	+	−	−	+	−
(13)	+	+	+	+	−	+	−	+	+	+	−	−	−	+	−
(14)	−	−	+	+	+	+	−	−	−	−	+	+	+	−	+
(15)	+	+	+	+	+	+	−	−	−	−	−	−	−	+	−

Table 14.5

x_1	x_2	g_1	g_2	g_3	g_4	g_5	g_6	g_7	g_8	g_9	g_{10}
0	0	0	1	0	1	0	1	0	0	0	0
0	1	0	1	1	1	0	0	1	0	0	0
1	0	0	1	0	0	1	1	1	0	1	0
1	1	0	1	0	0	1	1	1	0	1	1
0	2	0	1	0	1	0	1	2	2	1	0
1	2	2	2	2	2	2	0	2	2	2	0
2	0	0	1	1	1	1	2	2	2	2	0
2	1	2	1	2	1	1	2	2	2	2	0
2	2	0	1	2	1	2	2	2	2	2	0

14.5 Proof of Theorem 14.1.4

W.l.o.g. we assume $a = 0$, $b = 1$ and $c = 2$. Further, in this proof, let A be an arbitrary subset of O, which is not contained in any class ((1)–(13)) from the above list, given in Theorem 14.1.4. Then, there are some functions f_1, ..., $f_{13} \in [A]$ with the above property (*).

First we prove that $O^1 \backslash \{x\} \subseteq [A]$ holds. For the unary function f_1, we have $f_1(0) \in \{1, 2\}$. Then, either $f_1 = c_2$ or we obtain c_2 by forming $f_3(x, f_1(x))$. Consequently, $\{c_0, c_1, c_2\} \subseteq [\{f_2, f_4, f_5, c_2\}] \subseteq [A]$. W.l.o.g. we can assume

$$f_8 \begin{pmatrix} 0 & 1 & 2 & 2 & 0 & 2 & 1 \\ 0 & 1 & 2 & 0 & 2 & 1 & 2 \end{pmatrix} \in \begin{pmatrix} 0 \\ 1 \end{pmatrix}.$$

Since f_8 is monotone, we have

$$f_8 \begin{pmatrix} 0 & 1 & 2 & 0 & 0 & 1 & 0 \\ 0 & 1 & 2 & 0 & 2 & 1 & 2 \end{pmatrix} \in \begin{pmatrix} 0 \\ 1 \end{pmatrix}.$$

Therefore, $f_8(c_0, c_1, c_2, c_0, x, c_1, x) \in \{j_2, j_5\}$ is valid. Since we can assume w.l.o.g.

$$f_{10}\begin{pmatrix} 0 & 1 & 2 & 1 & 2 \\ 0 & 1 & 2 & 2 & 1 \end{pmatrix} \in \begin{pmatrix} 0 & 0 \\ 1 & 2 \end{pmatrix},$$

and since $f_{10} \in O$, we have $f_{10}(0,1,2,0,2) = 0$ and $f_{10}(0,1,2,2,2) \in \{1,2\}$. Thus $f_{10}(c_0, c_1, c_2, x, c_2) \in \{j_2, u_2\}$. Since $j_5 \star u_2 = j_2$, $j_2 \in [A]$ holds. With the aid of functions f_6 and f_9 one can show analogously that $v_5 \in [A]$ also holds. Because of $j_5 = j_2 \star v_5$ and $v_2 = v_5 \star j_2$ we have $j_5,\ v_2 \in [A]$. W.l.o.g. let

$$f_7\begin{pmatrix} 0 & 1 & 2 & 1 & 0 & 1 & 2 \\ 0 & 1 & 2 & 0 & 1 & 2 & 1 \end{pmatrix} \in \begin{pmatrix} 0 \\ 2 \end{pmatrix}.$$

Thus,

$$f_7\begin{pmatrix} 0 & 1 & 2 & 1 & 0 & 1 & 1 \\ 0 & 1 & 2 & 1 & 1 & 2 & 1 \end{pmatrix} \in \begin{pmatrix} 0 \\ 2 \end{pmatrix}$$

and therefore $f_7(c_0, c_1, c_2, c_1, j_5, v_5, c_1) = u_5 \in [A]$. Consequently, by $u_5 \star j_2 = u_2$, we have $O^1\backslash\{x\} \subseteq [A]$.

A conclusion of the Fundamental Lemma of Jablonskij (see Theorem 1.4.4, (a)) is the following fact: For the function f_{11}, there are some $a, b, c \in E_3$ with the property (w.l.o.g.):

$$f_{11}\begin{pmatrix} a & c \\ b & c \\ b & d \end{pmatrix} \in \begin{pmatrix} 0 \\ 1 \\ 2 \end{pmatrix},$$

where (because of $f_{11} \in O$) $(a,c) < (b,c)$ and $(b,c) < (b,d)$. If one replaces the variables of function f_{11} by certain functions from $O^1\backslash\{x\}$, one obtains a function $f'_{11} \in [A]$ with

$$f'_{11}\begin{pmatrix} 0 & 0 \\ 2 & 0 \\ 2 & 2 \end{pmatrix} \in \begin{pmatrix} 0 \\ 1 \\ 2 \end{pmatrix}.$$

Now we can form $g_\alpha(x,y) := f'_{11}(u_5(x), u_2(y))$, where

$$g_\alpha(x,y) = \begin{cases} \alpha & \text{if} \quad (x,y) = (0,2), \\ v_2(y) & \text{if} \quad x \in \{1,2\}, \\ 0 & \text{otherwise.} \end{cases}$$

Next, with the help of the function g_α, we show that

$$[A] \cap \{min, max\} \neq \emptyset \tag{14.1}$$

is valid, where we use the notation $x \vee y := max(x,y)$ and $x \wedge y := min(x,y)$. We distinguish three cases:

Case 1: $\alpha = 0$.

In this case, we have $g_0(x,y) = u_5(x) \wedge v_2(y)$ and

$$g_0(u_5(g_0(x, v_5(y))), v_2(g_0(u_2(x), y))) = x \wedge y \in [A]$$

holds.
Case 2: $a = 1$.
Then $g_1(x, y) = (u_5(x) \wedge v_2(y)) \vee j_2(y)$, whereby

$$g_1(u_5(g_1(u_2(x), u_2(y))), u_5(j_5(g_1(x, u_5(y))))) = x \vee y \in [A]$$

holds.
Case 3: $a = 2$.
Since $g_2(x, y) = (u_5(x) \wedge v_2(y)) \vee u_2(y)$ is valid in this case, we have

$$g_2(u_5(g_2(x, u_5(y))), v_2(g_2(u_2(x), y))) = x \vee y \in [A]$$

Thus (14.1) is proven.
Since the classes (12) and (13) are isomorphic, we can assume w.l.o.g. that $min \in [A]$. The function f_{12} has the property

$$f_{12}\begin{pmatrix} 0 & 1 & 2 & 0 & 0 & 1 & 0 & 0 & 1 \\ 0 & 1 & 2 & 0 & 0 & 1 & 1 & 2 & 2 \\ 0 & 1 & 2 & 1 & 2 & 2 & 0 & 0 & 1 \end{pmatrix} \in \begin{pmatrix} 0 & 0 & 0 & 0 & 1 \\ 1 & 2 & 1 & 2 & 2 \\ 1 & 1 & 2 & 2 & 2 \end{pmatrix}.$$

If one forms $f'_{12}(x, y) := f_{12}(c_0, c_1, c_2, x, u_2(x), v_2(x), y, u_5(y), v_5(y))$, either the function $u_5(f'_{12}(x, y))$ or the function $u_2(f'_{12}(x, y))$ is identical to the function $g(x, y) := u_5(x) \vee u_5(y)$. Therefore, we get $x \vee y = g(x, y) \wedge (v_5(g(u_2(x), u_2(y)))) \in [A]$. Since $(O^1 \backslash \{x\}) \cup \{\wedge, \vee\}$ is a generating system of O (see Section 11.4), we have shown $[A] = O$.
Then our theorem follows from Table 14.6, where $g_1 := \wedge$, $g_2 := \vee$, $g_3(x, y) := (j_5(x) \wedge j_5(y)) \vee j_2(y)$ and $g_4 := u_5 \star g_3$.

Table 14.6

	c_0	c_1	c_2	j_2	j_5	u_2	u_5	v_2	v_5	g_1	g_2	g_3	g_4
(1)	+	−	−	+	+	+	+	−	−	+	+	+	+
(2)	−	−	+	−	−	+	+	+	+	+	+	−	+
(3)	+	+	−	+	+	+	−	+	−	+	+	+	−
(4)	+	−	+	−	−	+	+	−	−	+	+	−	+
(5)	−	+	+	−	+	−	+	+	+	+	+	+	+
(6)	+	+	+	+	+	+	+	−	−	+	−	+	+
(7)	+	+	+	+	+	−	−	+	+	+	+	+	−
(8)	+	+	+	−	−	+	+	+	+	−	+	−	+
(9)	+	+	+	+	+	+	−	+	−	+	+	+	−
(10)	+	+	+	−	+	−	+	+	+	+	+	−	+
(11)	+	+	+	+	+	+	+	+	+	−	−	+	+
(12)	+	+	+	+	+	+	+	+	+	+	−	−	−
(13)	+	+	+	+	+	+	+	+	+	−	+	−	−

14.6 Proof of Theorem 14.1.5

Let A be an arbitrary subset of S, which is no subset of $Pol_3\{0\}$ and no subset of L. Then there are two functions $f_1,\ f_2 \in [A]$ with the above property (*). To prove $[A] = S$ it is sufficient to show that $\alpha([A]) := \{f \star c_0 \mid f \in [A]\} = P_3$ holds (see Theorem 8.3.2). $\alpha([A]) = P_3$ is proven, if one can show that $\alpha([A]) \nsubseteq B$ for every class B from the list of Theorem 14.1.1.
Obviously, we have $f_1^1 \in \{s_4, s_5\}$, $S^1 = \{s_4, s_5, e_1^1\} \subseteq [f_1]$ and f_1 is no element of a maximal class of P_3 with the marking (1), ..., (6), (8), ..., (16) of Theorem 14.1.1. Because of $\alpha(e_1^2) = c_0^1$, $\alpha([A])$ is no subset of a maximal class of P_3 with the marking (7). Further, $\alpha(f_2) \notin B$ for each maximal class B of P_3 with the marking (17) or (18). Consequently, $\alpha([A]) = P_3$ and therefore $[A] = S$. Then, our Theorem 14.1.5 follows from $S \cap L \nsubseteq S \cap Pol_3\{0\}$ and $S \cap L \nsupseteq S \cap Pol_3\{0\}$.

14.7 Proof of Theorem 14.1.7

W.l.o.g. we assume $a = 0$, $b = 1$ and $c = 2$. Further, in this proof, let A be an arbitrary subset of C, which is not contained in any class with the marking (1)–(7) from the above list, which is given in Theorem 14.1.7. Then there are some functions $f_1,\ ...,\ f_7 \in [A]$ with the above property (*).
First we show that $\{c_0, c_1, c_2\} \subset [A]$. Obviously, $f_1 \in \{j_0, j_4, j_3, c_1, u_0, u_4, u_3, c_2\}$. We distinguish the following three cases:
Case 1: $f_1 \in \{c_1, c_2, j_3, u_4\}$.
In this case $f_1' := f_1 \star f_1 = c_\alpha,\ \alpha \in \{1, 2\}$ holds. Since $f_2, f_3 \in C$ and $f_2(0, 1) = 2$ or $f_3(0, 2) = 1$, we have $f_2(1, 1) \in \{0, 2\}$ or $f_3(2, 2) \in \{0, 1\}$. Therefore, $\{c_1, c_2\}$ or $\{c_0, c_\alpha\}$ is a subset of $[A]$. With the aid of functions f_2, f_3, f_4 $(\in [A])$ one can prove $\{c_0, c_1, c_2\} \subset [A]$ easily.
Case 2: $f_1 = j_0$.
In this case, the functions $j_0 \star j_0 = j_5$ and $f_2'(x) := f_2(j_5, j_0) \in \{c_2, u_2\}$ belong to $[A]$. If $f_2' = c_2$, then we obtain by Case 1 that $\{c_0, c_1, c_2\} \subset [A]$. If $f_2' = u_0$ then $u_5 = u_0 \star u_0 \in [A]$.
For the function f_7 is valid:

$$f_7 \begin{pmatrix} 0 & 1 & 0 & 2 & 0 & 1 & 0 & 0 & 1 & 1 & 0 & 1 & 2 & 0 & 0 & 2 & 2 & 0 & 2 \\ 1 & 0 & 2 & 0 & 0 & 0 & 1 & 0 & 1 & 0 & 0 & 1 & 0 & 2 & 0 & 2 & 0 & 2 & 2 \\ 2 & 2 & 1 & 1 & 0 & 0 & 0 & 1 & 0 & 1 & 1 & 1 & 0 & 0 & 2 & 0 & 2 & 2 & 2 \end{pmatrix} \in \begin{pmatrix} 1 & 2 \\ 1 & 2 \\ 2 & 1 \end{pmatrix}.$$

Since f_7 preserves the relation $\begin{pmatrix} 0 & 1 & 2 & 0 & 1 & 0 & 2 \\ 0 & 1 & 2 & 1 & 0 & 2 & 0 \end{pmatrix}$, we have

$$f_7 \begin{pmatrix} 0 & 2 & 0 & 1 & 0 & 0 & . & . & . & . & . & 0 & 0 & 0 & . & . & . & . & 0 \\ 2 & 0 & 1 & 0 & 1 & 1 & . & . & . & . & . & 1 & 2 & 2 & . & . & . & . & 2 \end{pmatrix} = \begin{pmatrix} 0 \\ 0 \end{pmatrix}.$$

Consequently,

$$j_0(f_7(u_5, u_0, j_5, j_0, j_5, j_5, ..., j_5, u_5, u_5, ..., u_5) = c_0 \in [A].$$

Therefore, one can reduce Case 2 to Case 1.
Case 3: $f_1 \in \{j_4, u_0, u_3\}$.
It is easy to check that one can also reduce this case to the first case analogously to Case 2.

Thus

$$\{c_0, c_1, c_2\} \subset [A]$$

is proven.

Since $f_6 \in C$ and

$$f_6 \begin{pmatrix} 0 & 1 & 2 & 0 & 1 & 0 & 2 & 1 \\ 0 & 1 & 2 & 1 & 0 & 2 & 2 & 2 \end{pmatrix} = \begin{pmatrix} 2 \\ 1 \end{pmatrix}$$

holds, we have

$$f_6 \begin{pmatrix} 0 & 1 & 2 & 0 & 1 & 0 & 2 & 0 \\ 0 & 1 & 2 & 1 & 0 & 2 & 2 & 0 \end{pmatrix} = \begin{pmatrix} 0 \\ 0 \end{pmatrix}.$$

Consequently, the functions $f_6'(x) := f_6(c_0, c_1, c_2, c_0, c_1, c_0, c_2, x)$ and $f_6''(x) := f_6(c_0, c_1, c_2, c_0, c_1, c_2, c_0, x)$ belong to $[A]$, where $f_6' \begin{pmatrix} 0 \\ 1 \end{pmatrix} = \begin{pmatrix} 0 \\ 2 \end{pmatrix}$ and $f_6'' \begin{pmatrix} 0 \\ 2 \end{pmatrix} = \begin{pmatrix} 0 \\ 1 \end{pmatrix}$. Therefore, we can assume w.l.o.g. $f_5 \begin{pmatrix} 0 & 1 & 2 & 0 & 0 \\ 0 & 1 & 2 & 1 & 2 \end{pmatrix} = \begin{pmatrix} 1 \\ 0 \end{pmatrix}$. Then

$$p(x) := f_5(c_0, c_1, c_2, x, f_6'(x)) \in \{j_0, j_4\}.$$

It is easy to check that $f_6(0, 1, 2, 1, 0, 2, 0, 1) \in \{0, 2\}$. Consequently, $f_6'''(x) := f_6(c_0, c_1, c_2, c_1, c_0, c_2, c_0, x) \in \{j_2, s_2\}$. Therefore, we have to distinguish the following two cases for f_6''':
Case 1: $f_6''' = j_2$.
Then, the functions $p \star j_2 = j_3$, $f_6' \star j_2 = u_2$, $f_6' \star j_3 = u_3$, $f_6(c_0, c_1, c_2, j_2, j_3, u_2, u_3, x) = s_2$, $j_2 \star s_2 = j_1$, $j_3 \star s_2 = j_4$, $s_2 \star j_1 = u_1$ and $s_2 \star j_4 = u_4$ are superpositions over A. Consequently, the function

$$\begin{aligned} &f_7'(x_1, x_2) := \\ &f_7(x_1, x_2, s_2(x_1), s_2(x_2), c_0, j_1(x_2), j_1(x_1), j_2(x_1), j_3(x_1), j_4(x_1), j_4(x_2), \\ &\quad c_1, u_1(x_1), u_2(x_1), u_3(x_1), j_4(x_1), j_4(x_2), c_2) \end{aligned}$$

belongs to $[A]$, where

$$f_7' \begin{pmatrix} 0 & 1 \\ 1 & 0 \\ 2 & 2 \end{pmatrix} = \begin{pmatrix} \alpha \\ \alpha \\ \beta \end{pmatrix}$$

and $\{\alpha, \beta\} = \{1, 2\}$. Since $s_2 \in [A]$, we can assume $\alpha = 1$ and $\beta = 2$. Because of $f_7' \in C$, we have $f_7'(0, 0) = f_7'(0, 2) = f_7'(2, 0) = 0$ and $\{f_7'(1, 2), f_7'(2, 1)\} \subseteq$

$\{0,1\}$. Consequently, the function $f_7''(x_1,x_2) := j_2(f_7'(u_1(x_1),u_1(x_2))) \in [A]$ has the properties

$$f_7''\begin{pmatrix} 0 & 0 \\ 0 & 1 \\ 1 & 0 \\ 1 & 1 \end{pmatrix} = \begin{pmatrix} 0 \\ 0 \\ 0 \\ 1 \end{pmatrix}$$

and $f_7'' \in P_{3,2}$. Obviously, the set $\{j_2, p, c_0, c_1, f_7''\}$ is a subset of $P_{3,2}$, however no subset of the maximal classes of $P_{3,2}$ (see Lemma 14.2.3). Consequently, the sets $P_{3,2}$ and $\{s_2 \star r \mid r \in P_{3,2}\} = \{r' \in P_3 \mid Im(r') \subseteq \{0,2\}\}$ are subsets of $[A]$.

With the aid of the functions

$$t_i(x_1,x_2) := \begin{cases} 2 \text{ if} & (2,i) = (x_1,x_2), \\ 0 \text{ otherwise,} & \end{cases}$$

$i \in \{1,2\}$, of $[A]$ we can form a function $t \in [A]$ by

$$\begin{aligned} t(x_1,x_2) &:= f_7'(s_2(f_7'(s_2(x_2),t_1(x_1,x_2))),t_2(x_1,x_2)) \\ &= \begin{cases} 1 \text{ if} & (x_1,x_2) = (2,1), \\ 2 \text{ if} & (x_1,x_2) = (2,2), \\ 0 \text{ otherwise.} & \end{cases} \end{aligned}$$

Let S and T be nonempty disjoint subsets of E_3^n with the property that for different arbitrary tuples $\sigma := (\sigma_1,...,\sigma_n) \in T$ and $\tau := (\tau_1,...,\tau_n) \in T$ there exists an i with $\{\sigma_i,\tau_i\} = \{1,2\}$.

Then the function

$$f_{S,T}(\mathbf{x}) := \begin{cases} 1 \text{ if} & \mathbf{x} \in S, \\ 2 \text{ if} & \mathbf{x} \in T, \\ 0 \text{ otherwise,} & \end{cases}$$

belongs to C. $[A] = C$ would be proven in Case 1, if we could show that every function of the form $f_{S,T}$ is an element of $[A]$.

For $\sigma := (\sigma_1,...,\sigma_n)$, $\tau := (\tau_1,...,\tau_n)$, $\{\sigma_i,\tau_i\} = \{1,2\}$ and

$$g_{\sigma,\tau}(\mathbf{x}) := \begin{cases} 2 \text{ if} & \mathbf{x} \in \{\sigma,\tau\}, \\ 0 \text{ otherwise,} & \end{cases}$$

we have

$$f_{\{\sigma\},\{\tau\}}(\mathbf{x}) := \begin{cases} t(g_{\{\sigma\},\{\tau\}}(\mathbf{x}),x_i) & \text{if } (\sigma_i,\tau_i) = (1,2), \\ t(g_{\{\sigma\},\{\tau\}}(\mathbf{x}),s_2(x_i)) & \text{if } (\sigma_i,\tau_i) = (2,1), \end{cases}$$

i.e., $f_{\{\sigma\},\{\tau\}} \in [A]$.

Furthermore,

$$f_{S,T}(\mathbf{x}) = f_7'(f_{\{\sigma\},T}(\mathbf{x}), f_{S\setminus\{\sigma\},T}(\mathbf{x})),$$
$$f_{S,T}(\mathbf{x}) = s_2(f_7'(s_2(f_{\{\sigma\},T}(\mathbf{x})), s_2(f_{S\setminus\{\sigma\},T}(\mathbf{x}))))$$

is valid. Consequently, the functions of the type $f_{S,T}$ are superpositions over A (see Section 11.2). Thus, $[A] = C$ holds in Case 1.

Case 2: $f_6''' = s_2$.

If the above function p is the function j_4, then we can reduce Case 2 to Case 1 because of $j_4 \star j_4 \star s_2 = j_2$. Consequently, we can assume $p = j_0 \in [A]$. Then we have $\{j_0 \star j_0 = j_5, s_2 \star j_0 = u_0, s_2 \star j_5 = u_5\} \subseteq A$. Further, a function of $[A]$ is also the function

$$\widehat{f_7}(x_1, x_2, x_3) :=$$
$$f_7(x_1, x_2, s_2(x_1), s_2(x_2), c_0, j_0(x_1), j_0(x_2), x_3, j_0(x_3), j_5(x_2), j_5(x_1), c_1, u_0(x_1), u_0(x_2), u_5(x_3), u_0(x_3), u_5(x_2), u_5(x_1), c_2),$$

where

$$\widehat{f_7}\begin{pmatrix} 0 & 1 & 0 \\ 1 & 0 & 0 \\ 2 & 2 & 1 \end{pmatrix} \in \begin{pmatrix} 1 & 2 \\ 1 & 2 \\ 2 & 1 \end{pmatrix}.$$

Since $\widehat{f_7} \in C$, we have $\widehat{f_7}(2,0,0) = 0$. Thus $j_0(\widehat{f_7}(x, j_0, c_0)) = j_2 \in [A]$, i.e., one can reduce Case 2 to Case 1.

We have proven $[A] = C$ with that.

Then our theorem follows from Table 14.7 and Table 14.8.

Table 14.7

	c_1	p_1	p_2	p_3	p_4	p_5	p_6
(1)	−	+	+	+	−	+	+
(2)	+	−	+	−	−	+	+
(3)	−	−	−	+	+	+	−
(4)	+	−	−	+	+	−	−
(5)	+	−	−	+	−	+	+
(6)	+	−	−	−	+	+	−
(7)	+	−	−	−	+	−	+

Table 14.8

x_1	x_2	p_1	p_2	p_3	p_4	p_5	p_6
0	0	0	0	0	2	0	0
0	1	0	1	2	0	1	0
0	2	1	0	0	0	0	0
1	0	0	0	0	2	1	0
1	1	2	0	2	2	1	0
1	2	0	2	1	2	0	0
2	0	1	1	2	0	0	0
2	1	0	0	2	2	0	2
2	2	0	0	2	2	2	1

14.8 Proof of Theorem 14.1.8

[7] W.l.o.g. we can assume $a = 0$, $b = 1$, and $c = 2$. Further, in this proof let A be an arbitrary subset of U, which is not contained in any class from the above list, which is given in Theorem 14.1.8. Then there are some functions $f_1, ..., f_{13} \in [A]$ with the above property $(^\star)$.

First we show $\{c_0, c_1, c_2\} \subset [A]$.

[7] Essential parts of this proof come from the proof of the following theorem, found by W. Harnau:
Let $A \subseteq U$ be arbitrary. Then $[A] = U$ if and only if $U^1 \backslash \{u_3, v_3\} \subseteq [A]$ and $A \not\subseteq B$ for every class B with the marking (9), (11), (12), (13) in Theorem 14.1.8.

Because of $f_2(0,1) = 2$ and $f_2 \in U$ we have $f_2(x,x) \in \{u_3, v_3, c_2\}$. Obviously, c_0, c_1 and c_2 are superpositions over $\{c_2, f_1, f_3, f_4\}$. If $f_2(x,x) \neq c_2$ then we can assume either $\{c_1, u_2\} \subseteq [A]$ or $\{c_o, v_2\} \subseteq [A]$ because of $u_3 \star u_3 = u_2$, $v_3 \star v_3 = v_2$, $f_3(u_2, u_3) \in \{j_3, c_1, v_2\}$, $f_4(v_2, v_3) \in \{c_0, j_2, u_2\}$, $j_3 \star j_3 = c_1$, $j_2 \star j_2 = c_0$, $v_2 \star u_3 = v_3$, $u_2 \star v_3 = u_3$, $f_8(u_2, u_3, v_2, v_3) \in \{c_0, c_1, c_2, j_2, j_3\}$ and $u_3 \star c_2 = c_0$. Since $u_2 \star c_1 = c_0$, $v_2 \star c_0 = c_1$ and $f_2(c_0, c_1) = c_2$, we have $\{c_0, c_1, c_2\} \subset [A]$.

Next we prove $U^1 \cup P_{3,2} \subseteq [A]$.

By definition of f_9 we have

$$f_9 \begin{pmatrix} 0 & 1 & 2 & 0 & 1 & 0 & 1 \\ 0 & 1 & 2 & 1 & 0 & 2 & 2 \end{pmatrix} \in \begin{pmatrix} 2 & 2 \\ 0 & 1 \end{pmatrix},$$

i.e.,

$$f_9 \begin{pmatrix} 0 & 1 & 2 & 1 & 0 & 1 & 1 \\ 0 & 1 & 2 & 1 & 0 & 0 & 0 \end{pmatrix} = \begin{pmatrix} 2 \\ 2 \end{pmatrix}$$

and therefore $f_9(c_0, c_1, c_2, c_1, c_0, x, x) \in \{u_3, v_3\}$. Since $v_3 \star v_3 = v_2$, we have that either $\{u_2, u_3\}$ or $\{v_2, v_3\}$ is a subset of $[A]$.

W.l.o.g. let

$$f_5 \begin{pmatrix} 0 & 1 & 2 & 0 & 1 & 0 & 2 \\ 0 & 1 & 2 & 1 & 0 & 2 & 0 \end{pmatrix} = \begin{pmatrix} 1 \\ 2 \end{pmatrix}$$

and

$$f_6 \begin{pmatrix} 0 & 1 & 2 & 1 & 0 & 1 & 2 \\ 0 & 1 & 2 & 1 & 0 & 2 & 1 \end{pmatrix} = \begin{pmatrix} 0 \\ 2 \end{pmatrix},$$

i.e., we have $f_5(c_0, c_1, c_2, c_0, c_1, u_2, u_3) = v_2$ and $f_6(c_0, c_1, c_2, c_1, c_0, v_2, v_3) = u_2$. Thus, because of $v_2 \star u_3 = v_3$ and $u_2 \star v_3 = u_3$, the set $\{u_2, u_3, v_2, v_3\}$ is a subset of $[A]$.

Further, the function f_7' with $f_7'(x) := f_7(c_0, c_1, c_2, x) \in \{j_0, j_4, s_3\}$ belongs to $[A]$. We distinguish two cases for f_7':

Case 1: $f_7' \in \{j_0, j_4\}$.

In this case, the set $\{f_7', c_0, c_1, f_7'(v_2), f_7'(f_{12}(x_1, ..., x_8, c_2))\}$ is a subset of $[A] \cap P_{3,2}$, however, no subset of a maximal class of $P_{3,2}$ (see Lemma 14.2.3). Consequently, $P_{3,2}$ is a subset of $[A]$.

Obviously, the functions s_1 and s_3 are superpositions over $\{f_{11}, c_2, u_2, u_3, v_2, v_3\} \cup P_{3,2}$.

Therefore, $P_{3,2} \cup U^1 \subseteq [A]$ in Case 1.

Case 2: $f_7' = s_3$.

W.l.o.g. we can assume

$$f_{10} \begin{pmatrix} 0 & 1 & 2 & 0 & 0 & 1 & 1 & 2 & 2 & 0 & 1 \\ 0 & 1 & 2 & 0 & 0 & 1 & 1 & 2 & 2 & 1 & 0 \\ 0 & 1 & 2 & 1 & 2 & 0 & 2 & 0 & 1 & 2 & 2 \end{pmatrix} \in \begin{pmatrix} 1 & 1 \\ 0 & 0 \\ 0 & 1 \end{pmatrix}.$$

Since $f_{10} \in U$, we have $f_{10}(0, 1, 2, 0, 2, 1, 2, 0, 1, 2, 2) \in \{0, 1\}$, i.e.,

$$f_{10}(c_0, c_1, c_2, c_0, u_2, c_1, u_2, c_1, v_2, u_3, v_3, x, s_3) \in \{j_0, j_4\}.$$

Therefore, one can reduce Case 2 to Case 1.
Thus $P_{3,2} \cup U^1 \subseteq [A]$.
Then, a binary function f'_{11} with

$$f'_{11}\begin{pmatrix} 0 & 0 \\ 1 & 0 \\ 0 & 2 \end{pmatrix} = \begin{pmatrix} 0 \\ 1 \\ 2 \end{pmatrix}$$

is a superposition over A. It is easy to prove that a 6-ary function f'_{13} with

$$f'_{13}\begin{pmatrix} 0 & 0 & 0 & 2 & 2 & 2 \\ 1 & 2 & 2 & 0 & 0 & 2 \\ 2 & 1 & 2 & 1 & 2 & 0 \\ 2 & 2 & 1 & 2 & 1 & 1 \end{pmatrix} = \begin{pmatrix} 0 \\ 2 \\ 2 \\ 2 \end{pmatrix}$$

is a superposition over $P_{3,2} \cup U^1 \cup \{f_{13}\}$. Since $f'_{13} \in U$, we have

$$f'_{13}\begin{pmatrix} 0 & 0 & 0 & 2 & 2 & 2 \\ 0 & 2 & 2 & 0 & 0 & 2 \\ 2 & 0 & 2 & 0 & 2 & 0 \\ 2 & 2 & 0 & 2 & 0 & 0 \end{pmatrix} = \begin{pmatrix} 0 \\ 2 \\ 2 \\ 2 \end{pmatrix}.$$

We form

$$f''_{13}(x_1, x_2, x_3) := u_2(f'_{13}(x_1, x_2, x_3, u_3(x_3), u_3(x_2), u_3(x_1))) \in [A].$$

Because of $\{f''_{13}(0,0,2), f''_{13}(0,2,0), f''_{13}(2,0,0)\} \subseteq \{0,2\}$ we can assume w.l.o.g. $f''_{13}(0,2,0) = f''_{13}(2,0,0)$. Then: $f''_{13}(x_1, x_2, c_0) \in \{d_1, d_2\}$, where d_1 and d_2 are defined in Table 14.9.

Table 14.9

x_1	x_2	d_1	d_2	d_3	d_4	min	max	d_5	d_6	d_7	d_8	d_9
0	0	0	0	0	0	0	0	0	0	2	1	0
0	1	0	0	0	1	0	1	1	1	2	0	0
1	0	0	0	0	1	0	1	1	0	2	0	0
1	1	0	0	1	1	1	1	0	0	2	1	1
0	2	0	2	0	0	0	2	2	0	0	1	2
1	2	0	2	1	0	1	2	2	2	1	1	2
2	0	0	2	0	0	0	2	2	1	2	1	2
2	1	0	2	1	0	1	2	2	1	2	1	2
2	1	2	2	0	0	2	2	0	2	2	1	2

Thus because of

$$u_5(d_2(u_3(x_1), u_3(x_2))) = d_1(x_1, x_2)$$

and

$$u_3(d_1(u_3(x_1), u_3(x_2))) = d_2(x_1, x_2),$$

the functions d_1 and d_2 are superpositions over A. Further, we have

$$min(x_1, x_2) = f'_{11}(d_3(x_1, x_2), d_1(x_1, x_2)) \in [A]$$

and

$$max(x_1, x_2) = f'_{11}(d_4(x_1, x_2), d_2(x_1, x_2)) \in [A]$$

(see Table 14.9). With that, it was shown that the basis functions, determined in [Gni 65] [8] for the class U, belong to $[A]$. Therefore, $[A] = U$.

Then, our theorem follows from Tables 14.9 and 14.10, where the ternary function $d \in U$ is defined by

$$d(x_1, x_2, x_3) := \begin{cases} 2 \text{ if} & x_1 = x_2 = 2 \ \vee \ x_1 = x_3 = 2 \ \vee \ x_2 = x_3 = 2, \\ 0 \text{ otherwise.} & \end{cases}$$

Table 14.10

	c_1	c_2	j_2	u_3	v_3	s_3	d_5	d_6	d_7	d_8	d_9	d	max	min
(1)	−	+	−	−	−	+	−	+	+	−	+	+	+	+
(2)	+	−	+	−	−	+	+	+	−	+	+	+	+	+
(3)	−	+	−	+	−	−	+	−	+	−	+	+	+	+
(4)	+	+	−	−	+	−	−	−	+	+	+	−	+	+
(5)	+	+	+	+	−	−	−	−	−	+	+	+	−	+
(6)	+	+	+	−	+	−	−	−	−	+	−	−	+	+
(7)	+	+	+	+	+	−	−	−	+	−	+	+	+	+
(8)	−	−	−	+	+	+	−	+	−	−	−	+	−	−
(9)	+	+	+	−	−	+	−	+	−	+	+	+	+	+
(10)	+	+	+	+	+	+	−	−	+	−	+	+	+	−
(11)	+	+	+	+	+	+	−	−	−	+	−	+	−	+
(12)	+	+	+	+	+	+	+	−	+	+	−	+	−	−
(13)	+	+	+	+	+	+	+	+	−	+	−	−	−	−

14.9 Proof of Theorem 14.1.9

Let A be an arbitrary subset of $B := Pol\iota_3^3$, which is not contained in any class from the list, which is given in Theorem 14.1.9. Then, for each $i \in \{1, 2, ..., 5\}$ there is a function $f_i \in A$ which does not belong to the class with marking (i). Obviously, $S := \{s_1, s_2, ..., s_6\}$ is a subset of $[\{f_1, f_2, f_3, f_4\}]$.
For the function $f'_5(x) := f(x, x, ..., x) \in [A]$, the following cases are possible:
Case 1: $f'_5 \in B^1 \backslash \{c_0, c_1, c_2\}$.
In this case, one can easily prove $(P_3(2))^1 \subseteq [A]$ when one forms the superpositions of the form $s \star f'_5 \star s'$ and $s \star f'_5 \star f'_5 \star s'$, where $\{s, s'\} \subset S$. Consequently, $P_3^1 \subseteq [A]$.
Then, by Theorem 4.3, we have $[P_3^1 \cup \{f_5\}] = Pol\iota_3^3 = [A]$.

[8] See Chapter 11.

Case 2: $f_5' \in \{c_0, c_1, c_2\}$.
Obviously, we have $\{c_o, c_1, c_2\} \subseteq [S \cup \{f_5'\}]$ in this case. Since $f_5 \in P_3(2)\backslash[P_3^1]$, there exist some tuples $\mathbf{a} := (a_1, ..., a_n)$ and $\mathbf{a}' := (a_1, ..., a_{i-1}, a_i', a_{i+1}, ..., a_n)$ with $a_i \neq a_i'$ and $f_5(\mathbf{a}) \neq f_5(\mathbf{a}')$. Therefore, a unary function of $B^1\backslash\{c_0, c_1, c_2\}$ belongs to $[A]$. Hence, one can reduce Case 2 to Case 1.
Consequently, $[A] = Pol\iota_3^3$ is proven.
Since one can easily prove that the classes with the marking (1)–(5) are B-maximal, our theorem holds.

14.10 On the Cardinality of $\mathbb{L}_3^{\downarrow}(A)$ for Submaximal Clones A

The aim of this section is to prove

Theorem 14.10.1 *([Bul-L-S 96])*
There are 5 submaximal clones ($\subset P_3$) with finitely many subclasses and 7 submaximal clones ($\subset P_3$) whose subclass lattice is infinite but countable. The subclass lattices of the remaining 146 submaximal clones of P_3 have the cardinality of continuum.
More precise:
Let $F_1, ..., F_{43}$ be the 43 families of submaximal clones of P_3 given in Table 14.1 and let $A \in F_i$ with $i \in \{1, 2, ..., 43\}$. Then

$$|\mathbb{L}_3^{\downarrow}(A)| = \begin{cases} 5 \;\; \textit{if} & i = 17, \\ 8 \;\; \textit{if} & i = 7, \\ 32 \;\; \textit{if} & i = 30, \\ \aleph_0 \;\; \textit{if} & i \in \{28, 29, 43\}, \\ \mathfrak{c} \;\; \textit{otherwise.} & \end{cases}$$

We need the following lemmas for proof of the above theorem.

Lemma 14.10.2 *Let $n \in \mathbb{N} \setminus \{1, 2\}$ and $p_n \in P_3^n$ be defined by*

$$p_n(x_1, ..., x_n) := \begin{cases} 1 \;\; \textit{if} & \exists i : (x_i = 1 \, \wedge \, \forall i \neq j : x_i = 2), \\ 0 \;\; \textit{otherwise.} & \end{cases}$$

Let π_n be the n-ary relation

$$\alpha_n := \{(1, 2, 2, ..., 2), (2, 1, 2, ..., 2), ..., (2, 2, , ..., 2, 1)\} \cup (E_3^n \backslash \{1, 2\}^n).$$

Then

(a) $\forall n \geq 3 : \; p_n \notin Pol_3\alpha_n$.

(b) $\forall m \neq n: \; p_m \in Pol_3\alpha_n$.

(c) $(p_i)_{i\geq 3}$ *is an infinite basis for the clone* $[\{p_i \mid i \geq 3\}]$.

Proof. (a) follows from

$$p_n \begin{pmatrix} 1 & 2 & 2 & \dots & 2 \\ 2 & 1 & 2 & \dots & 2 \\ \dots & \dots & \dots & \dots & \dots \\ 2 & 2 & 2 & \dots & 1 \end{pmatrix} = \begin{pmatrix} 1 \\ 1 \\ \dots \\ 1 \end{pmatrix} \notin \alpha_n.$$

(b) is easy to check by considering the two cases $m < n$ and $m > n$.
(c) Let $j \geq 3$. By (b), we have that $p_i \in Pol_3\alpha_j$ for all $i \geq 3$, $i \neq j$, and thus $[\{p_3, p_4, ...,\} \backslash \{p_j\}] \subseteq Pol_3\alpha_j$. Since $p_j \notin Pol_3\alpha_j$, we deduce that $p_j \notin [\{p_3, p_4, ...,\} \backslash \{p_j\}]$. Consequently, $\{p_3, p_4, ...\}$ is an independent set of operations, proving (c). ■

Lemma 14.10.3 *Let* $n \in \mathbb{N} \setminus \{1\}$ *and* $q_n \in P_3^n$ *be defined by*

$$q_n(x_1, ..., x_n) := \begin{cases} 1 & \text{if} \quad \forall i : x_i = 2, \\ p_n(\mathbf{x}) & \text{otherwise}. \end{cases}$$

Then $\{q_i \mid i \geq 2\}$ *is an infinite basis for the clone* $[\{q_i \mid i \geq 2\}]$.

Proof. The proof can be found in [Jan-M 59] (see also Lemma 8.1.1). ■

Lemma 14.10.4 *Let* $n \in \mathbb{N} \setminus \{1\}$ *and* $r_n \in P_3^n$ *be defined by*

$$r_n(x_1, ..., x_n) := \begin{cases} 1 \text{ if} & \exists i : (x_i = 2 \wedge \forall j \neq i : x_j = 1), \\ 2 \text{ if} & x_1 = ... = x_n = 2, \\ 0 \text{ otherwise}. \end{cases}$$

Let ϱ_n *be the* n*-ary relation*

$$\varrho_n := \{(a_1, ..., a_n) \in E_3^n \mid (\exists i : a_i = 2 \wedge \forall j \neq i : a_j = 1) \vee (\exists i : a_i = 0)\}.$$

Then

(a) $\forall n \geq 2: \; r_n \notin Pol_3\varrho_n$.

(b) $\forall n \neq m: \; r_m \in Pol_3\varrho_n$.

(c) $\{r_i \mid i \geq 2\}$ *is an infinite basis for* $[\{r_i \mid i \geq 2\}]$.

Proof. (a) follows from

$$r_n \begin{pmatrix} 2 & 1 & 1 & \dots & 1 \\ 1 & 2 & 1 & \dots & 1 \\ \dots & \dots & \dots & \dots & \dots \\ 1 & 1 & 1 & \dots & 2 \end{pmatrix} = \begin{pmatrix} 1 \\ 1 \\ \dots \\ 1 \end{pmatrix}.$$

(b): Let $m \neq n$, $m, n \geq 2$, $\mathfrak{A} = (a_{ij})_{n,m}$ be an $n \times m$ matrix on E_3 with columns $a_1, ..., a_m \in \varrho_n$, where $a_i = (a_{1i}, ..., a_{ni})$ $(i = 1, ..., n)$. If $a_{ij} = 0$ for some i, j

then $r_m(a_{i1}, ..., a_{im}) = 0$ and by definition of ϱ_n, $r_n(\mathfrak{A}) = r_n(a_1, ..., a_n) \in \varrho$. So assume that $a_1, ..., a_m \in \{1,2\}^n$. We distinguish two cases:
Case 1: $m < n$.
At least one row of the matrix $\mathfrak{A}$ consists of 1s, hence $r_m(\mathfrak{A}) \in \varrho_n$.
Case 2: $m > n$.
In this case, there are $i \neq j$ with $a_i = a_j$. Hence, one row of the matrix $\mathfrak{A}$ consists of 1s or contains at least two 2s, and so $r_m(\mathfrak{A}) \in \varrho_n$. Therefore (b) holds.
(c) follows straightforward from (a) and (b). ■

Lemma 14.10.5 *Let $n \in \mathbb{N} \setminus \{1,2\}$ and $s_n \in P_3^n$ be defined by*

$$s_n(x_1, ..., x_n) :=$$
$$\begin{cases} x_1 & \text{if } \exists i : (x_i = 1 \ \wedge \ \forall j \neq i : x_j = 0) \ \vee (x_i = 0 \ \wedge \forall j \neq i : \ x_j = 1), \\ 2 & \text{otherwise.} \end{cases}$$

Let σ_n be the n-ary relation

$$\sigma_n := \{(a_1, ..., a_n, a_{n+1}) \in E_3^n \mid (\exists i \in \{1,2,...,n\} : (x_i = 1 \ \wedge \ \forall j \neq i : \ x_i = 0) \ \vee a_i = 2\}.$$

Then

(a) $\forall n \geq 3 : \ s_n \notin Pol_3 \sigma_n$.

(b) $\forall n \neq m : \ s_m \in Pol_3 \sigma_n$.

(c) $\{s_i \mid i \geq 3\}$ *is an infinite basis for* $[\{s_i \mid i \geq 3\}]$.

Proof. (a) follows from

$$s_n \begin{pmatrix} 1 & 0 & 0 & ... & 0 \\ 0 & 1 & 0 & ... & 0 \\ \multicolumn{5}{c}{\dotfill} \\ 0 & 0 & 0 & ... & 1 \\ 0 & 0 & 0 & ... & 0 \end{pmatrix} = \begin{pmatrix} 1 \\ 0 \\ ... \\ 0 \\ 2 \end{pmatrix}.$$

(b) and (c) are easy to verify. ■

The following lemma was found by B. Strauch.

Lemma 14.10.6 *Let $n \in \mathbb{N} \setminus \{1,2\}$ and let the n-ary function $t_n \in P_3^n$ be defined by*

$$t_n(x_1, ..., x_n) := \begin{cases} 0 & \text{if} \quad \exists i : \ (x_i \in E_2 \ \wedge \ \forall j \neq i : \ x_j = 0), \\ 2 & \text{if} \quad \mathbf{x} = \mathbf{1} \vee (\exists i : \ x_i = 2), \\ 1 & \text{otherwise.} \end{cases}$$

Moreover, let

$$\alpha_m := \{(a_1, ..., a_m, \overline{a_1}, ..., \overline{a_m}) \in E_2^{2\cdot m} \mid (\exists i : a_i = 1) \wedge \forall j \neq i : \ a_j = 0)\},$$
$$\overline{x} := x + 1 \ (mod\ 2),$$

$$\beta_m := (E_2^m \backslash \{0\}^m) \times \{1\}^m,$$

$$\gamma_m := E_3^{2\cdot m} \backslash E_2^{2\cdot m},$$

$$\tau_m := \alpha_m \cup \beta_m \cup \gamma_m.$$

Then

(a) $\forall n \geq 3 : \ t_n \notin Pol_3\tau_n$.

(b) $\forall m \neq n : \ t_n \in Pol_3\tau_m$.

(c) $\{t_n \mid n \geq 3\}$ *is an infinite basis for* $[\{t_n \mid s \geq 3\}]$.

Proof. (a) follows from

$$s_n \begin{pmatrix} 1 & 0 & 0 & ... & 0 & 0 \\ 0 & 1 & 0 & ... & 0 & 0 \\ \cdots & \cdots & \cdots & \cdots & \cdots & \cdots \\ 0 & 0 & 0 & ... & 0 & 1 \\ 0 & 1 & 1 & ... & 1 & 1 \\ 1 & 0 & 1 & ... & 1 & 1 \\ \cdots & \cdots & \cdots & \cdots & \cdots & \cdots \\ 1 & 1 & 1 & ... & 1 & 0 \end{pmatrix} = \begin{pmatrix} 0 \\ 0 \\ \cdots \\ 0 \\ 1 \\ 1 \\ \cdots \\ 1 \end{pmatrix} \notin \tau_n.$$

(b): Let $m, n \geq 3$, $m \neq n$ and $r_1, ..., r_n \in \tau_m$. If $\{r_1, ..., r_n\} \not\subseteq \alpha_m \cup \beta_m$ or $\{r_1, ..., r_n\} \subseteq \beta_m$, we have $t_n(r_1, ..., r_n) \in \tau_m$. So assume $\{r_1, ..., r_n\} \subseteq \alpha_m \cup \beta_m$ and $\{r_1, ..., r_n\} \not\subseteq \beta_m$. It is easy to see that $r_i = r_j$ for some $i \neq j$ implies $t_n(r_1, ..., r_n) \in \tau_m$. Further, by the definition of t_n

$$t_n(r_1, ..., r_n) = \{o\}^m \times \{1\}^m \tag{14.2}$$

is possible only if $t_n(r_1, ..., r_n) \notin \tau_m$. For pairwise distinct $r_1, ..., r_n \in \alpha_m \cup \beta_m$, the condition (14.2) can be valid only if $m = n$. Since we assume $m \neq n$, $t_n(r_1, ..., r_n) \in \tau_m$ holds.
(c) follows from (a) and (b). ∎

The following two lemmas were found by A. Bulatov. For these lemmas, we need some notations. For every $n \in \mathbb{N}$ let

$$\underline{n} := \{1, 2, ..., n\}.$$

Furthermore, let

$$\varrho := \begin{pmatrix} 0 & 1 & 0 & 2 \\ 1 & 0 & 2 & 0 \end{pmatrix} \tag{14.3}$$

and for $n \geq 3$

$$\varrho_n := \{(a_1, a_2, ..., a_n) \mid \exists i : (a_i \in \{1,2\} \wedge (\forall j \neq i : a_j = 0)\} \cup (\{1,2\}^n \setminus \{2\}^n). \quad (14.4)$$

We want to prove that the co-clone

$$[\{\varrho\} \cup \{\varrho_m \mid m \geq 3\}] \quad (14.5)$$

has an infinite basis.

Lemma 14.10.7

For a finite index-set I and every $i \in I$ let π_i be an m_i-ary relation of $\{\varrho\} \cup \{\varrho_m \mid m \geq 3\}$. Moreover, let φ_i be a mapping of $\underline{m_i}$ in $\underline{n}$, $i \in I$. If the relation ϱ_l, $l \geq 3$ has the form

$$\begin{aligned} \varrho_l &= \{(a_1, ..., a_l) \in E_3^l \mid \exists a_{l+1}, ..., a_n \in E_3 : \forall i \in I : \\ & (a_{\varphi_i(1)}, a_{\varphi_i(2)}, ..., a_{\varphi_i(m_i)}) \in \pi_i\} \end{aligned} \quad (14.6)$$

for a some n and if there exists an $i \in I$ with $\pi_i \neq \varrho$, then:
(a) If a tuple $(a_1, ..., a_n) \in E_3^n$ fulfills the condition

$$\forall i \in I : (a_{\varphi_i(1)}, a_{\varphi_i(2)}, ..., a_{\varphi_i(m_i)}) \in \pi_i \quad (14.7)$$

(we say : "$(a_1, ..., a_n)$ is permissible") and if $a_j = 2$ for a some $j \in \underline{n}$, then the n-tuple $(a_1, ..., a_{j-1}, 1, a_{j+1}, .., a_n)$ is also permissible.

(b) $\exists \alpha \in I : \varphi_\alpha(\underline{m_\alpha}) \subseteq \underline{l} \wedge \pi_\alpha \neq \varrho$.

(c) Let $\varphi_\alpha(\underline{m_\alpha}) \subseteq \underline{l}$ and $\pi_\alpha \neq \varrho$. Then φ_α is bijective and $\pi_\alpha = \varrho_l$ holds.

Proof. (a) follows straightforwad from the observation that, if we replace a 2 with a 1 in a tuple of ϱ_l or ϱ, we obtain again a tuple from ϱ_l or ϱ, respectively.
(b): Suppose that

$$\forall i \in I : (\pi_i \neq \varrho \implies \varphi_i(\underline{m_i}) \not\subseteq \underline{l}). \quad (14.8)$$

Because of (14.4) and (14.6), there exists a permissible n-tuple of the form $\mathbf{a} := (a_1, ..., a_l, a_{l+1}, ..., a_n)$ with $a_1 = 1$ and $a_2 = ... = a_l = 2$; i.e.,

$$\forall i \in I : (a_{\varphi_i(1)}, ..., a_{\varphi_i(m_i)}) \in \pi_i. \quad (14.9)$$

By (a) we can suppose w.l.o.g. that

$$\forall i \in \{l+1, l+2, ..., n\} : a_i \in \{0, 1\}.$$

We now show that (14.8) and (14.9) imply the permissibility of the tuple $\mathbf{b} := (b_1, ..., b_n)$, where $b_1 = ... = b_l = 2$ and $b_i = a_i$ for all $i \in \{l+1, l+2, ..., n\}$, which leads to the contradiction $(2, 2, ..., 2) \in \varrho_l$.
The following cases are possible for $\alpha \in I$:

Case 1: $1 \notin \varphi_\alpha(\underline{m_\alpha})$.
In this case, the permissibility of $\mathbf{b}$ follows from (14.9).
Case 2: $1 \in \varphi_\alpha(\underline{m_\alpha})$; i.e., there exists an $i \leq m_\alpha$ with $\varphi_\alpha(i) = 1$.
Case 2.1: $\pi_\alpha = \varrho$.
W.l.o.g. let $\varphi_\alpha(1) = 1$. By (14.8) we have $\varphi_\alpha(2) \in \{3, ..., n\}$, hence $(b_{\varphi_\alpha(1)}, b_{\varphi_\alpha(2)}) \in \{(2,1),(2,0)\} \subseteq \varrho$.
Case 2.2: $\pi_\alpha = \varrho_s$.
Case 2.2.1: $0 \in \{a_{\varphi_\alpha(1)}, ..., a_{\varphi_\alpha(1)}\}$.
By (14.8) and the definition of ϱ_s we obtain

$$(a_{\varphi_\alpha(1)}, ..., a_{\varphi_\alpha(i-1)}, a_1, a_{\varphi_\alpha(i+1)}, ..., a_{\varphi_\alpha(s)}) = (0, ..., 0, 1, 0, ..., 0). \quad (14.10)$$

Since $a_k = b_k$ for all $k \geq 2$ and $b_1 = 2$, the inclusion $(b_{\varphi_\alpha(1)}, ..., b_{\varphi_\alpha(s)}) \in \varrho_s$ follows from (14.10).
Case 2.2.2: $\{a_{\varphi_\alpha(1)}, ..., a_{\varphi_\alpha(s)}\} \subseteq \{1, 2\}$.
Since $\varphi_\alpha(\underline{m_\alpha}) \nsubseteq \underline{m}$ there exists a j such that $\varphi_\alpha(j) \notin \underline{m}$, $b_{\varphi_\alpha(j)} = a_{\varphi_\alpha(j)} = 1$ and $\{b_{\varphi_\alpha(1)}, ..., b_{\varphi_\alpha(s)}\} \subseteq \{1, 2\}$, hence $(b_{\varphi_\alpha(1)}, ..., b_{\varphi_\alpha(s)})$ belongs to π_α and therefore $\mathbf{b}$ is permissible as required.

(c): By assumption, there is an $s \geq 3$ with $\pi_\alpha = \varrho_s$. Then $m_\alpha = s$. First we prove that φ_α is injective. Suppose φ_α is not injective. Then the following two cases are possible:
Case 1: $|\varphi_\alpha(\underline{m_\alpha})| = 1$, i.e., $\varphi_\alpha(\underline{s}) = \{j\}$ for a some $j \leq m$.
Then for every permissible tuple $\mathbf{a} := (a_1, ..., a_s)$ we have $(a_{\varphi_\alpha(1)}, a_{\varphi_\alpha(2)}, ..., a_{\varphi_\alpha(s)}) = (a_j, a_j, ..., a_j) = (1, 1, ..., 1)$ which contradicts (14.6) because all elements of $\{0, 1, 2\}$ occur in every row of the relation ϱ_l.
Case 2: $\exists i_1, i_2, i_3 \in \underline{s} : i_1 \neq i_2 \wedge \varphi_\alpha(i_1) = \varphi_\alpha(i_2) \wedge \varphi_\alpha(i_3) \neq \varphi_\alpha(i_1)$.
W.l.o.g. let $\varphi(i_1) = 1$. Then for the tuple $(a_1, ..., a_l) := (1, 0, 0, ..., 0) \in \varrho_l$ we obtain $(a_{\varphi_\alpha(1)}, a_{\varphi_\alpha(2)}, ..., a_{\varphi_\alpha(s)}) \in \varrho_s$, which is not possible, because of $a_{\varphi(i_1)} = a_{\varphi_\alpha(i_2)} = 1$ and $a_{\varphi_\alpha(i_3)} = 0$.
Consequently, φ_α is injective.
We now show that φ_α is surjective.
Suppose that $\varphi_\alpha(\underline{m_\alpha}) \subset \underline{m}$. We may assume w.l.o.g. that $1 \notin \varphi_\alpha(\underline{m_\alpha})$. Consider the tuple $\mathbf{a} := (a_1, a_2, ..., a_l) = (1, 2, 2, ..., 2) \in \varrho_l$. Then $(a_{\varphi_\alpha(1)}, ..., a_{\varphi_\alpha(s)}) = (2, 2, 2, ..., 2) \notin \varrho_s$, in contradiction with (14.6).
We have shown that φ_α is a bijection. Combining this with the properties of ϱ_l, we deduce that $\pi_\alpha = \varrho_l$. ■

Lemma 14.10.8 *Let $n \in \mathbb{N} \setminus \{1, 2\}$ and let ϱ and ϱ_n be as in (14.3) and (14.4). Then:*

(a) $\varrho \notin [\{\varrho_n \mid n \geq 3\}]$.

(b) $\forall l \geq 3 : \varrho_l \notin [\{\varrho\} \cup \{\varrho_n \mid n \in \mathbb{N} \setminus \{1, 2, l\}\}]$.

(c) The co-clone $[\{\varrho\} \cup \{\varrho_n \mid n \geq 3\}]$ has the infinite basis $\{\varrho, \varrho_3, \varrho_4, ...\}$.

Proof. (a) follows from the fact that the constant function c_1 preserves the relation ϱ_n for all $n \geq 3$ but does not preserve the relation ϱ.

(b): Suppose for an $l \geq 3$ we have $\varrho_l \in [\{\varrho\} \cup \{\varrho_n \mid n \in \mathbb{N} \setminus \{1, 2, l\}\}]$. Then, by Theorem 2.11.2,(a),(b)

$$\begin{aligned} \varrho_l \ = \ & \{(a_{\varphi(1)}, , ..., a_{\varphi(l)}) \in E_3^l \mid \exists a_{l+1}, ..., a_n \in E_3 : \ \forall i \in I : \\ & (a_{\varphi_i(1)}, a_{\varphi_i(2)}, ..., a_{\varphi_i(m_i)}) \in \pi_i\}, \end{aligned} \tag{14.11}$$

holds, where I is an index-set, π_i are some m_i-ary relations of $\{\varrho\} \cup \{\varrho_m \mid m \geq 3, m \neq l\}$ for all $i \in I$ and $\varphi : \underline{l} \longrightarrow \underline{n}$ and $\varphi_i : \underline{m_i} \longrightarrow \underline{n}$ are mappings. There must be at least a relation under the relations π_i, which is different from the relation ϱ, since in the opposite case, $\varrho_l \in [\{\varrho\}]$ holds and this implies $Pol_3\varrho \cup \{c_1\} \subseteq Pol_3\varrho_l$, which by means of Theorem 14.1.8 $[Pol_3\varrho \cup \{c_1\}] = Pol_3 \begin{pmatrix} 0 & 1 & 2 & 1 & 2 \\ 0 & 1 & 2 & 2 & 1 \end{pmatrix} \subseteq Pol_3\varrho_l$ results from, which contradicts $c_2 \notin Pol_3\varrho_l$ and $c_2 \in Pol_3 \begin{pmatrix} 0 & 1 & 2 & 1 & 2 \\ 0 & 1 & 2 & 2 & 1 \end{pmatrix}$.

It results now from the definition of the relation ϱ_l that φ is an injective mapping. W.l.o.g. we can assume that $\varphi(x) = x$ for all $x \in \underline{l}$. Therefore (14.11) agrees with (14.6) and, with the help of Lemma 14.10.7, we have a contradiction to our assumption.

(c) follows directly from (a) and (b). ■

Proof of Theorem 14.10.1:

The statements for the classes with the numbers 7, 17, or 30 of Table 14.1 are consequences from Chapter 13 (see also Theorem 15.1.1).

For the classes with the number 28, 29, or 43, our assertion follows from Theorem 8.1.6 and Sections 15.3 and 15.4.

Since the class of Theorem 8.2.2 is a subclass of $A_8 = Pol_3\{0\} \cap Pol_3 \begin{pmatrix} 0 & 1 & 2 \\ 1 & 2 & 0 \end{pmatrix}$ we have $|\mathbb{L}^{\downarrow}(A_8)| = \mathfrak{c}$.

The clones $A \in F_j$ with

$$j \in J := \{9, 14, 24, 25, 27, 33, 35, 37, 39, 40, 41, 42\}$$

obviously satisfy the condition

$$\exists a, b \in E_3 : \ a \neq b \ \wedge \ P_{3,\{a,b\}} \subseteq A_i.$$

Hence for every $j \in J$ there exists a subclass of A with infinite basis such that the range of its functions is a 2-element set (see for instance Lemma 14.10.2). An example of a subclass of $A \in F_{22}$ with infinite basis is given in the proof of Theorem 12.3.8.

Let $A \in F_{32}$ with $a = 1$, $b = 2$ and $c = 0$. We have shown in Lemma 14.10.8 that there are uncountable many co-clones containing Inv_3A. This means that $|\mathbb{L}_3(J_3, A_{32})| = \mathfrak{c}$ (see Chapter 2).

Finally, we use constructions given in Lemmas 14.10.2–14.10.6 to show that $|\mathbb{L}_3^{\downarrow}(A)| = \mathfrak{c}$ with $A \in F_i$ for the remaining i. All results are presented in Table 14.11.

Table 14.11

Let $A \in F_i$, where $i =$	Considered cases for $a, b, c, \alpha, \beta, \gamma$ (w.l.o.g.) a b c α β γ	Use the construction given in Lemma 14.10.j, where $j =$
1	0 2 1	4
2,4	0 1 2	4
3	2 0 1	4
5,19,20,21,23,26,31	0 1 2	2
6,13,16,36,38	0 1 2	3
10	0 2 1	2
11	0 1 0 1 2	3
11	0 2 0 1 2	8
11	0 1 2 1 0	3
12	0 1 2 0 1 2	2
12	0 2 1 1 0 2	2
15	0 1 2 0 1 2	3
18	2 0 1	5
34	0 1 2	5

■

15

Finite and Countably Infinite Sublattices of Depth 1 or 2 of $\mathbb{L}_3$

We say that the class $A \in \mathbb{L}_k$ (or the lattice $\mathbb{L}_k^{\downarrow}(A)$ of subclasses of $A = [A] \subseteq P_k$) has the **depth** t, if t is the least integer for which there are some classes $A_1, ..., A_{t-1} \in \mathbb{L}_k$ with

$$A \subset A_1 \subset A_2 \subset ... \subset A_{t-1} \subset P_k.$$

In particular, it holds that the maximal classes of P_k have the depth 1 and the submaximal classes of P_k have the depth 2.
For $k = 3$ by Theorems 13.2.3 and 8.1.6, there are finite and countably infinite sublattices of depth 1 or 2. In this chapter, these sublattices will be determined.
The finite lattice $\mathbb{L}_3^{\downarrow}(L_3)$ of depth 1 can be found in Section 15.1. This lattice is a conclusion from Chapter 13. In addition, by Section 14.10, this lattice is the only finite lattice of depth 1, and furthermore, this lattice contains all finite sublattices of $\mathbb{L}_3$ of depth 2.
In Section 15.2 is a description of all subsemigroups (or subclasses) of $(P_s^1; \star)$ (or $[P_3^1]$), respectively. These subclasses are also subclasses of the submaximal class $\mathfrak{L}$ of all quasilinear functions of P_3. The list of all subsemigroups of $(P_s^1; \star)$ is then an important aid in Section 15.3 during the determination of the remaining elements of lattice $\mathbb{L}_3^{\downarrow}(\mathfrak{L})$.
Because of Theorem 8.1.6, a further countable sublattice of depth 2 is $\mathbb{L}_3^{\downarrow}([O^1 \cup \{max\}])$. This sublattice is given in Section 15.4. Except for isomorphic lattices, all countable sublattices of $\mathbb{L}_3$ of depth 2 are traced with that (see Section 14.10).

15.1 The Lattice of Subclasses of P_3 of Linear Functions

Let L_3 be the set of all linear functions of P_3. As a consequence of Theorem 13.2.1, one can see that $\mathbb{L}_3^{\downarrow}(L_3)$ has exactly 6 elements, which are not subsets of $[L_3^1]$. The subclasses of $[L_3^1]$ is obtained from Theorem 13.2.2 or from Section 15.2. Thus it holds:

Theorem 15.1.1 *The class L_3 of all linear functions of P_3 has exactly 38 subclasses:*

$$L_3, L_3 \cap Pol_3\{a\}(a \in E_3), L_3 \cap \begin{pmatrix} 0\,1\,2 \\ 1\,2\,0 \end{pmatrix}, L_3 \cap \begin{pmatrix} 0 \\ 1 \end{pmatrix}, [H_i],$$

where
$i \in \{1, 2, 3, 4, 5, 6, 7, 8, 601, 602, 603, 604, 605, 606, 607, 608, 1201, 1202, 1203, 1204, 1232, 1233, 1234, 1235, 1263, 1264, 1265, 1266, 1294, 1295, 1297, 1298\}$
(see Table 15.10 of Section 15.2). ■

15.2 The Subsemigroups of $(P_3^1; \star)$

In this section, we determine the subclasses of $[P_3^1]$. It suffices to describe all subsemigroups of $(P_3^1; \star)$ for this.
Since the subsemigroups of $(P_3^1; \star)$ play a role in many investigations of P_3, it is particularly a question of clarifying the construction of these subsemigroups. To facilitate checking the following considerations without a lot of expenditure also by hand, some tables are given at the end of this section.
The functions of P_3^1 and their notations are given in Table 15.1. As usual the operation $\star$ is defined by

$$(f \star g)(x) \;:=\; f(g(x))$$

for all $f, g \in P_3^1$ (see Table 15.2).
The number statement of the following theorem was published by G. Wilde and Sh. Raney in 1972 without proof (as a result of a computer calculation).

Theorem 15.2.1 *([Wil-R 72], [Lau 84a])*
$(P_3^1; *)$ has exactly

1299

subsemigroups (including $\emptyset$), which are listed in Table 15.10.

Proof.[1] In preparation for the proof of the above theorem, some notations, which we use during the description of the subsemigroups of $(P_3^1, *)$, are given. Let

$$\begin{aligned} &C := \{c_0, c_1, c_2\}, \\ &J := \{c_0, c_1, j_0, j_1, j_2, j_3, j_4, j_5\}, \quad U := \{c_0, c_2, u_0, u_1, u_2, u_3, u_4, u_5\}, \\ &V := \{c_1, c_2, v_0, v_1, v_2, v_3, v_4, v_5\}, \; S := \{s_1, s_2, s_3, s_4, s_5, s_6\}. \end{aligned}$$

Starting from the possible subsemigroups J_i ($i \in \{-3, -2, -1, 0, 1, 2, ..., 41\}$) of J ($= J_{41}$), given in Table 15.4, one can construct, with the aid of the following mappings,

[1] Basically, the following proof comes from [Lau 84a], which required correction in some places. I owe K. Todorov (Sofia) and Anne Fearnley (Montreal) indications of the mistakes.

$$\varphi_i : f \to s_i^{-1} * f * s_i \ (i \in \{2,3,...,6\})$$

and the notation

$$\varphi_i(A) := \{ \ s_i^{-1} * f * s_i \mid f \in A \ \} \ (A \subseteq P_3^1)$$

the classes isomorphic to J_i. These isomorphic classes are given in Tables 15.5–15.9, where we use the notations

$$U_i := \varphi_2(J_i) \text{ and } V_i := \varphi_6(J_i)$$
$$(i \in \{-3,-2,-1,0,1,2,...,41\}).$$

Furthermore we use the notations:

$$S_0 := \emptyset,\ S_1 := \{s_1\},$$
$$S_2 := \{s_1,s_3\},\ S_3 := \{s_1,s_2\},\ S_4 := \{s_1,s_6\},$$
$$S_5 := \{s_1,s_4,s_5\},\ S_6 := S.$$

To make the structure of the subsemigroups recognizable short, we assign a tuple $\tau(H)$ as follows to every subsemigroup H:

$$\tau(H) := (a,b,c,d) :\Longleftrightarrow H \cap J = J_a \wedge H \cap U = U_b \wedge H \cap V = V_c \wedge$$
$$H \cap S = S_d \ .$$

Obviously, it holds

$$H \neq H' \Longleftrightarrow \tau(H) \neq \tau(H').$$

H always denotes an arbitrary subsemigroup of $(P_3^1;*)$.
The number of possibilities for H in one of the below cases i for H is denoted with

$$n_i.$$

In Cases 4.1 and 4.2 some possibilities are already included in other cases for H, so that we must change the number n_i into the number n_i', i.e., n_i denotes the number of possibilities for H in Case i, which were not included in the preceded cases.
In most of the following cases i for H, the numbers n_i or/and n_i' and the possibilities for H are given. To check the following proof is an arduous matter that should be carried out only with the aid of a computer.
A summary of all possibilities for H together with the corresponding characteristic tuples $\tau(H)$ can be found in Table 15.10.

Case 1: $H \subseteq C$.
Obviously, $n_1 = 8$ and $H = H_i$ ($i = 1,...,8$; see Table 15.10) is an arbitrary subset of C.
Case 2: $H \not\subseteq C$ and $H \subseteq A \in \{J,U,V\}$.
Case 2.1: $H \subseteq J$.
It is not hard to check that $n_{2.1} = 41$ and $H \in \{J_1,J_2,...,J_{41}\}$ (or $H = H_t$, $t \in \{9,10,...,49\}$, see Table 15.10).

Case 2.2: $H \subseteq U$.
By $\varphi_2(J) = U$ and by Case 2.1, we have $n_{2.2} = 41$, and H is one of the following sets: $U_i := \varphi_2(J_i)$ with $i \in \{1, 2, ..., 41\}$ (or $H = H_t$ for $t \in \{50, 51, ..., 90\}$).
Case 2.3: $H \subseteq V$.
By $\varphi_6(J) = V$ and by Case 2.1, we have $n_{2.3} = 41$, and for H only, the following sets are possible: $V_i := \varphi_6(J_i)$ with $i \in \{1, 2, ..., 41\}$ (or $H = H_i$, $i \in \{91, 92, ..., 131\}$).
Case 3: $H \subseteq J \cup U$, $H \not\subseteq J$ and $H \not\subseteq U$.
Case 3.1: $H \subseteq C \cup \{j_1, j_4, u_2, u_3\} = J_{25} \cup U_{25}$.
The possibilities for H are:
$J_3 \cup \{c_0, c_2\}$, $\{c_0, c_1\} \cup U_3$, $J_{12} \cup \{c_0, c_2\}$, $\{c_0, c_1\} \cup U_{12}$, $J_{25} \cup \{c_0, c_2\}$, $\{c_0, c_1\} \cup U_{25}$ and $J_p \cup U_q$, where $(p, q) \in \{(3, 3), (3, 12), (12, 3), (12, 12), (3, 25), (25, 3), (12, 25), (25, 12), (25, 25)\}$ (or $H = H_i$ with $i \in \{132, 133, ..., 146\}$); i.e., we have $n_{3.1} = 15$.
Case 3.2: $H \cap J \not\subseteq \{c_0, c_1, j_1, j_4\}$ and $H \cap U \subseteq \{c_0, c_2, u_2, u_3\}$.
For $H \cap U$ only the following sets are possible: $\{c_2\}$, $\{c_0, c_2\}$, $\{c_0, u_2\} = U_3$, $\{c_0, c_2, u_2\} = U_{12}$ and $\{c_0, c_2, u_2, u_3\} = U_{25}$.
Case 3.2.1: $H \cap U = \{c_2\}$.
Obviously, $H = J_8 \cup \{c_2\}$ ($= H_{147}$).
Case 3.2.2: $H \cap U = \{c_0, c_2\}$.
In this case, $H \cap J$ is an arbitrary subsemigroup of J, which contains $\{c_0, c_1\}$, but no subset is of J_{25}, i.e., it holds $n_{3.2.2} = 19$ and $H = \{c_0, c_2\} \cup J_i$ with $i \in \{13, 14, 15, 22, 23, 24, 26, 27, 28, 29, 33, 34, 35, 36, 37, 38, 39, 40, 41\}$ (or $H = H_t$ with $t \in \{148, 149, ..., 166\}$).
Case 3.2.3: $H \cap U = U_3$.
In this case, we have $n_{3.2.3} = 17$ and $H = U_3 \cup J_i$ with $i \in \{4, 13, 14, 16, 18, 23, 24, 27, 28, 30, 33, 34, 36, 38, 39, 40, 41\}$ (or $H = H_t$ with $t \in \{167, 168, ..., 183\}$).
Case 3.2.4: $H \cap U = U_{12}$.
It is easy to check that $n_{3.2.4} = 13$ and $H = U_{12} \cup J_i$ with $i \in \{13, 14, 23, 24, 27, 28, 33, 34, 36, 38, 39, 40, 41\}$ (or $H = H_t$ with $t \in \{184, 185, ..., 196\}$).
Case 3.2.5: $H \cap U = U_{25}$.
In this case, $H \cap J$ is an arbitrary subsemigroup of J, which contains $\{j_2, j_3\}$. Thus, we have $n_{3.2.5} = 7$ and $H = U_{25} \cup J_i$ with $i \in \{27, 33, 36, 38, 39, 40, 41\}$ (or $H = H_t$ with $t \in \{197, 198, ..., 203\}$).
In summary, we get: $n_{3.2} = 57$.
Case 3.3: $H \cap J \subseteq \{c_0, c_1, j_1, j_4\}$ and $H \cap U \not\subseteq \{c_0, c_1, u_2, u_3\}$.
In this case, H is isomorphic to a subsemigroup of $J \cup U$ which fulfills the conditions of Case 3.2. Consequently, we have $n_{3.3} = 57$ and $H = \varphi_2(H_i)$ with $i \in \{147, 148, ..., 203\}$ (or $H = H_t$ with $t \in \{204, 205, ..., 260\}$). We remark that $\varphi_2(J_a \cup U_b) = J_b \cup U_a$.
Case 3.4: $H \cap J \not\subseteq \{c_0, c_1, j_1, j_4\}$ and $H \cap U \not\subseteq \{c_0, c_1, u_2, u_3\}$.
Then there are two functions f and g of H with

$$f\begin{pmatrix}0\\1\end{pmatrix} \in \begin{pmatrix}0\,2\\2\,0\end{pmatrix} \text{ and } g\begin{pmatrix}0\\2\end{pmatrix} \in \begin{pmatrix}0\,1\\1\,0\end{pmatrix}.$$

This implies $|H \cap J| = |H \cap U|$ and, if $(H \backslash C) \cap J = \{j_i \mid i \in I\}$ with $I \subseteq \{0, 1, ..., 5\}$, $(H \backslash C) \cap U \in \{\{u_i \mid i \in I\}, \{u_{5-i} \mid i \in I\}\}$. Examining the possible cases yields $n_{3.4} = 19$ and $H = J_p \cup U_q$ with $(p, q) \in \{(2, 2), (5, 5), (8, 8), (9, 9), (15, 15), (16, 16), (17, 18, (18, 17), (22, 22), (23, 23), (24, 24), (26, 28), (28, 26), (30, 30), (34, 34), (37, 38), (38, 37), (39, 39), (41, 41)\}$.
Summing up, we get $n_3 = 148$.
Case 4: $H \subseteq A \cup B \in \{J \cup V, U \cup V\}$, $H \nsubseteq A$ and $H \nsubseteq B$.
Case 4.1: $A = J$ and $B = V$.
By $\varphi_3(J \cup U) = J \cup V$ we have $n_{4.1} = 148$, and H is isomorphic to a subsemigroup which we have already determined in Case 3. Hence, one receives a list of the subsemigroups with the aid of the results from the third case, where one has to consider $\varphi_3(J_a \cup U_b) = \varphi_3(J_a) \cup \varphi_5(J_b)$ (see Table 15.7 and 15.9). Twenty-three of the 148 subsemigroups were already determined, so that we have $n'_{4.1} = 125$. [2] In Table 15.10 (for the Case 4.1), 148 sets are given, where every already listed set, is characterized by their first number; this first number is given in bold point.
Case 4.2: $A = U$ and $B = V$.
By $\varphi_4(J \cup U) = U \cup V$, we have $n_{4.2} = 148$ and H is isomorphic to a subsemigroup that we have already determined in Case 3. While listing the subsemigroups, one notices that $\varphi_4(J_a \cup U_b) = \varphi_4(J_a) \cup V_b$ holds. With the aid of Table 15.10 one sees that 48 of the subsemigroups were determined in previous cases. Thus $n'_{4.2} = 102$.
Case 5: $H \subseteq J \cup U \cup V$, $H \nsubseteq J \cup U$, $H \nsubseteq J \cup V$ and $H \nsubseteq U \cup V$.
Case 5.1: $H \cap (U \cup V) \subseteq C \cup \{u_2, u_3, v_2, v_3\}$.
For $H \cap (U \cup V)$, only the sets $U_3 \cup V_8 = \{c_0, c_1, u_2, v_2\}$, $U_{12} \cup V_{15} = \{c_0, c_1, c_2, u_2, v_2\}$ and $U_{25} \cup V_{22} = \{c_0, c_1, c_2, u_2, u_3, v_2, v_3\}$ come into consideration.
Case 5.1.1: $H \cap (U \cup V) = U_3 \cup V_8$.
Then, all the subsets J_i of J which contain the two functions c_0, c_1 and which have the properties

$$\begin{aligned} j_0 \in J_i \vee j_4 \in J_i &\Longrightarrow \{j_2, j_3\} \subseteq J_i, \\ j_1 \in J_i &\Longrightarrow j_3 \in J_i, \\ j_5 \in J_i &\Longrightarrow j_2 \in J_i. \end{aligned}$$

are possible for $H \cap J$. Thus $n_{5.1.1} = 11$ and $H = J_i \cup U_3 \cup V_8$ with $i \in \{13, 14, 24, 27, 28, 33, 36, 38, 39, 40, 41\}$.
Case 5.1.2: $H \cap (U \cup V) = U_{12} \cup V_{15}$.
With the aid of Case 5.1.1, we obtain $n_{5.1.2} = 11$ and $H = J_i \cup U_{12} \cup V_{15}$, $i \in \{13, 14, 24, 27, 28, 33, 36, 38, 39, 40, 41\}$.
Case 5.1.3: $H \cap (U \cup V) = U_{25} \cup V_{22}$.
Then $H \cap J$ is a subsemigroup of J, which contains j_2 and j_3. Consequently, $n_{5.1.3} = 7$ and $H = J_i \cup U_{25} \cup V_{22}$, $i \in \{27, 33, 36, 38, 39, 40, 41\}$.
In summary, we obtain $n_{5.1} = 29$.

[2] This was not taken into account in [Lau 84a]!

Case 5.2: $H \cap (J \cup V) \subseteq C \cup \{j_1, j_4, v_1, v_4\}$.
Because of $\varphi_2(C \cup \{j_1, j_4, v_1, v_4\}) = U_{25} \cup V_{22}$, we have $n_{5.2} = 29$ and $H = \varphi_2(A)$, where A is a set which we have already determined in Case 5.1. More exactly: If $A = J_a \cup U_b \cup V_c$ then $\varphi_2(A) = J_b \cup U_a \cup \varphi_5(J_c)$.
Case 5.3: $H \cap (J \cup U) \subseteq C \cup \{j_0, j_5, u_0, u_5\}$.
This case is also isomorphic to Case 5.1. One can obtain all 29 possibilities for H with the aid of the mapping φ_6 from the sets determined in Case 5.1, where $\varphi_6(J_a \cup U_b \cup V_c) = J_c \cup \varphi_4(J_b) \cup V_a$ must be pointed out.
Case 5.4: H fulfills none of the conditions from Cases 5.1, 5.2, or 5.3.
In this case, there are functions $f, g, h \in H$ with

$$f\begin{pmatrix}0\\1\end{pmatrix} \in \begin{pmatrix}0\,2\,1\,2\\2\,0\,2\,1\end{pmatrix}, g\begin{pmatrix}0\\2\end{pmatrix} \in \begin{pmatrix}0\,1\,1\,2\\1\,0\,2\,1\end{pmatrix} \text{ and } h\begin{pmatrix}1\\2\end{pmatrix} \in \begin{pmatrix}0\,1\,0\,2\\1\,0\,2\,0\end{pmatrix}.$$

Since certain functions j_a, u_b, v_c for certain a, b, c belong to H, one easily checks that $C \subset H$ and, if $(H \backslash C) \cap J = \{j_i \mid i \in I\}$, it holds $(H\backslash C) \cap U \in \{\{u_i \mid i \in I\}, \{u_{5-i} \mid i \in I\}\}$ and $(H\backslash C) \cap V \in \{\{v_i \mid i \in I\}, \{v_{5-i} \mid i \in I\}\}$. Examining the possible cases using the results of Case 2 yields $n_{5.4} = 7$ and $H = J_p \cup U_q \cup V_r$ with $(p, q, r) \in \{(24, 24, 26), (26, 28, 24), (28, 26, 28), (37, 38, 39), (38, 37, 38), (39, 39, 37), (41, 41, 41)\}$. Consequently, $n_5 = 94$.
Case 6: $S \cap H \neq \emptyset$.
Since S as is well-known has 6 subgroups, the following cases are possible:
Case 6.1: $S \cap H = \{s_1\}$.
Then we have $H = A \cup \{s_1\}$, where A is one of the 600 subsemigroups of $J \cup U \cup V$ determined above.
Case 6.2: $S \cap H = \{s_1, s_3\}$.
In this case, $H \backslash S$ is one of the following 31 sets: $\emptyset$, $\{c_2\}$, $\{c_0, c_1\}$, C, J_{27}, J_{32}, J_{37}, J, $J_{27} \cup \{c_2\}$, $J_{37} \cup \{c_2\}$, $C \cup J$, $U_1 \cup V_2$, $U_3 \cup V_8$, $U_6 \cup V_5$, $U_{12} \cup V_{15}$, $U_{10} \cup V_9$, $U_{25} \cup V_{22}$, $U_{21} \cup V_{16}$, $U_{29} \cup V_{23}$, $U_{31} \cup V_{30}$, $U_{35} \cup V_{34}$, $U_{38} \cup V_{39}$, $U \cup V$, $J_{27} \cup U_3 \cup V_8$, $J_{27} \cup U_{12} \cup V_{15}$, $J_{27} \cup U_{25} \cup V_{22}$, $J_{37} \cup U_{38} \cup V_{39}$, $J \cup U_{12} \cup V_{15}$, $J \cup U_3 \cup V_8$, $J \cup U_{25} \cup V_{22}$, $J \cup U \cup V$.
Case 6.3: $S \cap H = \{s_1, s_2\}$.
In this case, $n_{6.3} = 31$ and the possibilities for H be determined with the results from Case 6.2 and with the mapping φ_6.
Case 6.4: $S \cap H = \{s_1, s_6\}$.
In this case, $n_{6.4} = 31$ and one can determine the possibilities for H, using Case 6.2 and the mapping φ_2.
Case 6.5: $S \cap H \in \{\{s_1, s_4, s_5\}, S\}$.
For $H \backslash S$ only the sets $\emptyset$, C and $J \cup U \cup V$ are possible, whereby $n_{6.5} = 6$.
Consequently, we have $n_6 = 699$.

In summary, we get that $(P_3^1; *)$ has exactly 1299 different subsemigroups. ■

We remark that one finds further information about the given subsemigroups in [Bij-T 91].

Table 15.1

x	$j_0(x)$	$j_1(x)$	$j_2(x)$	$j_3(x)$	$j_4(x)$	$j_5(x)$	$u_0(x)$	$u_1(x)$	$u_2(x)$	$u_3(x)$	$u_4(x)$	$u_5(x)$
0	1	0	0	1	1	0	2	0	0	2	2	0
1	0	1	0	1	0	1	0	2	0	2	0	2
2	0	0	1	0	1	1	0	0	2	0	2	2

x	$v_0(x)$	$v_1(x)$	$v_2(x)$	$v_3(x)$	$v_4(x)$	$v_5(x)$	$s_1(x)$	$s_2(x)$	$s_3(x)$	$s_4(x)$	$s_5(x)$	$s_6(x)$
0	2	1	1	2	2	1	0	0	1	1	2	2
1	1	2	1	2	1	2	1	2	0	2	0	1
2	1	1	2	1	2	2	2	1	2	0	1	0

Table 15.2

$f * g$	j_i	u_i	v_i	s_2	s_3	s_4	s_5	s_6
j_0	j_{5-i}	j_{5-i}	c_0	j_0	j_1	j_2	j_1	j_2
j_1	j_i	c_0	j_{5-i}	j_2	j_0	j_0	j_2	j_1
j_2	c_0	j_i	j_i	j_1	j_2	j_1	j_0	j_0
j_3	c_1	j_{5-i}	j_{5-i}	j_4	j_3	j_4	j_5	j_5
j_4	j_{5-i}	c_1	j_i	j_3	j_5	j_5	j_3	j_4
j_5	j_i	j_i	c_1	j_5	j_4	j_3	j_4	j_3
u_0	u_{5-i}	u_{5-i}	c_0	u_0	u_1	u_2	u_1	u_2
u_1	u_i	c_0	u_{5-i}	u_2	u_0	u_0	u_2	u_1
u_2	c_0	u_i	u_i	u_1	u_2	u_1	u_0	u_0
u_3	c_2	u_{5-i}	u_{5-i}	u_4	u_3	u_4	u_5	u_5
u_4	u_{5-i}	c_2	u_i	u_3	u_5	u_5	u_3	u_4
u_5	u_i	u_i	c_2	u_5	u_4	u_3	u_4	u_3
v_0	v_{5-i}	v_{5-i}	c_1	v_0	v_1	v_2	v_1	v_2
v_1	v_i	c_1	v_{5-i}	v_2	v_0	v_0	v_2	v_1
v_2	c_1	v_i	v_i	v_1	v_2	v_1	v_0	v_0
v_3	c_2	v_{5-i}	v_{5-i}	v_4	v_3	v_4	v_5	v_5
v_4	v_{5-i}	c_2	v_i	v_3	v_5	v_5	v_3	v_4
v_5	v_i	v_i	c_2	v_5	v_4	v_3	v_4	v_3
s_2	u_i	j_i	v_{5-i}	s_1	s_5	s_6	s_3	s_4
s_3	j_{5-i}	v_i	u_i	s_4	s_1	s_2	s_6	s_5
s_4	v_i	j_{5-i}	u_{5-i}	s_3	s_6	s_5	s_1	s_2
s_5	u_{5-i}	v_{5-i}	j_i	s_6	s_2	s_1	s_4	s_3
s_6	v_{5-i}	u_{5-i}	j_{5-i}	s_5	s_4	s_3	s_2	s_1

Table 15.3

f	$\varphi_2(f) = s_2 * f * s_2$	$\varphi_3(f) = s_3 * f * s_3$	$\varphi_5(f) = s_4 * f * s_5$	$\varphi_4(f) = s_5 * f * s_4$	$\varphi_6(f) = s_6 * f * s_6$
c_0	c_0	c_1	c_1	c_2	c_2
c_1	c_2	c_0	c_2	c_0	c_1
c_2	c_1	c_2	c_0	c_1	c_0
j_0	u_0	j_4	v_1	u_3	v_3
j_1	u_2	j_5	v_2	u_5	v_4
j_2	u_1	j_3	v_0	u_4	v_5
j_3	u_4	j_2	v_5	u_1	v_0
j_4	u_3	j_0	v_3	u_0	v_1
j_5	u_5	j_1	v_4	u_2	v_2
u_0	j_0	v_1	j_4	v_3	u_3
u_1	j_2	v_0	j_3	v_5	u_4
u_2	j_1	v_2	j_5	v_4	u_5
u_3	j_4	v_3	j_0	v_1	u_0
u_4	j_3	v_5	j_2	v_0	u_1
u_5	j_5	v_4	j_1	v_2	u_2
v_0	v_5	u_1	u_4	j_2	j_3
v_1	v_3	u_0	u_3	j_0	j_4
v_2	v_4	u_2	u_5	j_1	j_5
v_3	v_1	u_3	u_0	j_4	j_0
v_4	v_2	u_5	u_2	j_5	j_1
v_5	v_0	u_4	u_1	j_3	j_2
s_1	s_1	s_1	s_1	s_1	s_1
s_2	s_2	s_6	s_6	s_3	s_3
s_3	s_6	s_3	s_2	s_6	s_2
s_4	s_5	s_5	s_4	s_4	s_5
s_5	s_4	s_4	s_5	s_5	s_4
s_6	s_3	s_2	s_3	s_2	s_6

Table 15.4

i	J_i
-3	$\{c_0, c_1\}$
-2	$\{c_1\}$
-1	$\{c_0\}$
0	$\emptyset$
1	$\{j_1\}$
2	$\{j_5\}$
3	$\{c_0, j_1\}$
4	$\{c_0, j_2\}$
5	$\{c_0, j_5\}$
6	$\{c_1, j_1\}$
7	$\{c_1, j_3\}$
8	$\{c_1, j_5\}$
9	$\{j_0, j_5\}$
10	$\{j_1, j_4\}$
11	$\{j_1, j_5\}$
12	$\{c_0, c_1, j_1\}$
13	$\{c_0, c_1, j_2\}$
14	$\{c_0, c_1, j_3\}$
15	$\{c_0, c_1, j_5\}$
16	$\{c_0, j_1, j_2\}$
17	$\{c_0, j_1, j_5\}$
18	$\{c_0, j_2, j_5\}$
19	$\{c_1, j_1, j_3\}$
20	$\{c_1, j_1, j_5\}$
21	$\{c_1, j_3, j_5\}$
22	$\{c_0, c_1, j_0, j_5\}$
23	$\{c_0, c_1, j_1, j_2\}$
24	$\{c_0, c_1, j_1, j_3\}$
25	$\{c_0, c_1, j_1, j_4\}$
26	$\{c_0, c_1, j_1, j_5\}$
27	$\{c_0, c_1, j_2, j_3\}$
28	$\{c_0, c_1, j_2, j_5\}$
29	$\{c_0, c_1, j_3, j_5\}$
30	$\{c_0, j_1, j_2, j_5\}$
31	$\{c_1, j_1, j_3, j_5\}$
32	$\{j_0, j_1, j_4, j_5\}$
33	$\{c_0, c_1, j_1, j_2, j_3\}$
34	$\{c_0, c_1, j_1, j_2, j_5\}$
35	$\{c_0, c_1, j_1, j_3, j_5\}$
36	$\{c_0, c_1, j_2, j_3, j_5\}$
37	$\{c_0, c_1, j_0, j_1, j_4, j_5\}$
38	$\{c_0, c_1, j_0, j_2, j_3, j_5\}$
39	$\{c_0, c_1, j_1, j_2, j_3, j_4\}$
40	$\{c_0, c_1, j_1, j_2, j_3, j_5\}$
41	$\{c_0, c_1, j_0, j_1, j_2, j_3, j_4, j_5\}$

Table 15.5

i	$U_i := \varphi_2(J_i)$
-3	$\{c_0, c_2\}$
-2	$\{c_2\}$
-1	$\{c_0\}$
0	$\emptyset$
1	$\{u_2\}$
2	$\{u_5\}$
3	$\{c_0, u_2\}$
4	$\{c_0, u_1\}$
5	$\{c_0, u_5\}$
6	$\{c_2, u_2\}$
7	$\{c_2, u_4\}$
8	$\{c_2, u_5\}$
9	$\{u_0, u_5\}$
10	$\{u_2, u_3\}$
11	$\{u_2, u_5\}$
12	$\{c_0, c_2, u_2\}$
13	$\{c_0, c_2, u_1\}$
14	$\{c_0, c_2, u_4\}$
15	$\{c_0, c_2, u_5\}$
16	$\{c_0, u_2, u_1\}$
17	$\{c_0, u_2, u_5\}$
18	$\{c_0, u_1, u_5\}$
19	$\{c_2, u_2, u_4\}$
20	$\{c_2, u_2, u_5\}$
21	$\{c_2, u_4, u_5\}$
22	$\{c_0, c_2, u_0, u_5\}$
23	$\{c_0, c_2, u_2, u_1\}$
24	$\{c_0, c_2, u_2, u_4\}$
25	$\{c_0, c_2, u_2, u_3\}$
26	$\{c_0, c_2, u_2, u_5\}$
27	$\{c_0, c_2, u_1, u_4\}$
28	$\{c_0, c_2, u_1, u_5\}$
29	$\{c_0, c_2, u_4, u_5\}$
30	$\{c_0, u_2, u_1, u_5\}$
31	$\{c_2, u_2, u_4, u_5\}$
32	$\{u_0, u_2, u_3, u_5\}$
33	$\{c_0, c_2, u_2, u_1, u_4\}$
34	$\{c_0, c_2, u_2, u_1, u_5\}$
35	$\{c_0, c_2, u_2, u_4, u_5\}$
36	$\{c_0, c_2, u_1, u_4, u_5\}$
37	$\{c_0, c_2, u_0, u_2, u_3, u_5\}$
38	$\{c_0, c_2, u_0, u_1, u_4, u_5\}$
39	$\{c_0, c_2, u_2, u_1, u_4, u_3\}$
40	$\{c_0, c_2, u_2, u_1, u_4, u_5\}$
41	$\{c_0, c_2, u_0, u_2, u_1, u_4, u_3, u_5\}$

Table 15.6

i	$V_i := \varphi_6(J_i)$
-3	$\{c_2, c_1\}$
-2	$\{c_1\}$
-1	$\{c_2\}$
0	$\emptyset$
1	$\{v_4\}$
2	$\{v_2\}$
3	$\{c_2, v_4\}$
4	$\{c_2, v_5\}$
5	$\{c_2, v_2\}$
6	$\{c_1, v_4\}$
7	$\{c_1, v_0\}$
8	$\{c_1, v_2\}$
9	$\{v_3, v_2\}$
10	$\{v_4, v_1\}$
11	$\{v_4, v_2\}$
12	$\{c_2, c_1, v_4\}$
13	$\{c_2, c_1, v_5\}$
14	$\{c_2, c_1, v_0\}$
15	$\{c_2, c_1, v_2\}$
16	$\{c_2, v_4, v_5\}$
17	$\{c_2, v_4, v_2\}$
18	$\{c_2, v_5, v_2\}$
19	$\{c_1, v_4, v_0\}$
20	$\{c_1, v_4, v_2\}$
21	$\{c_1, v_0, v_2\}$
22	$\{c_2, c_1, v_3, v_2\}$
23	$\{c_2, c_1, v_4, v_5\}$
24	$\{c_2, c_1, v_4, v_0\}$
25	$\{c_2, c_1, v_4, v_1\}$
26	$\{c_2, c_1, v_4, v_2\}$
27	$\{c_2, c_1, v_5, v_0\}$
28	$\{c_2, c_1, v_5, v_2\}$
29	$\{c_2, c_1, v_0, v_2\}$
30	$\{c_2, v_4, v_5, v_2\}$
31	$\{c_1, v_4, v_0, v_2\}$
32	$\{v_3, v_4, v_1, v_2\}$
33	$\{c_2, c_1, v_4, v_5, v_0\}$
34	$\{c_2, c_1, v_4, v_5, v_2\}$
35	$\{c_2, c_1, v_4, v_0, v_2\}$
36	$\{c_2, c_1, v_5, v_0, v_2\}$
37	$\{c_2, c_1, v_3, v_4, v_1, v_2\}$
38	$\{c_2, c_1, v_3, v_5, v_0, v_2\}$
39	$\{c_2, c_1, v_4, v_5, v_0, v_1\}$
40	$\{c_2, c_1, v_4, v_5, v_0, v_2\}$
41	$\{c_2, c_1, v_3, v_4, v_5, v_0, v_1, v_2\}$

Table 15.7

i	$\varphi_3(J_i) = J_t$	t
-3	$\{c_1, c_0\}$	-3
-2	$\{c_0\}$	-1
-1	$\{c_1\}$	-2
0	$\emptyset$	0
1	$\{j_5\}$	2
2	$\{j_1\}$	1
3	$\{c_1, j_5\}$	8
4	$\{c_1, j_3\}$	7
5	$\{c_1, j_1\}$	6
6	$\{c_0, j_5\}$	5
7	$\{c_0, j_2\}$	4
8	$\{c_0, j_1\}$	3
9	$\{j_4, j_1\}$	10
10	$\{j_5, j_0\}$	9
11	$\{j_5, j_1\}$	11
12	$\{c_1, c_0, j_5\}$	15
13	$\{c_1, c_0, j_3\}$	14
14	$\{c_1, c_0, j_2\}$	13
15	$\{c_1, c_0, j_1\}$	12
16	$\{c_1, j_5, j_3\}$	21
17	$\{c_1, j_5, j_1\}$	20
18	$\{c_1, j_3, j_1\}$	19
19	$\{c_0, j_5, j_2\}$	18
20	$\{c_0, j_5, j_1\}$	17
21	$\{c_0, j_2, j_1\}$	16
22	$\{c_1, c_0, j_4, j_1\}$	25
23	$\{c_1, c_0, j_5, j_3\}$	29
24	$\{c_1, c_0, j_5, j_2\}$	28
25	$\{c_1, c_0, j_5, j_0\}$	22
26	$\{c_1, c_0, j_5, j_1\}$	26
27	$\{c_1, c_0, j_3, j_2\}$	27
28	$\{c_1, c_0, j_3, j_1\}$	24
29	$\{c_1, c_0, j_2, j_1\}$	23
30	$\{c_1, j_5, j_3, j_1\}$	31
31	$\{c_0, j_5, j_2, j_1\}$	30
32	$\{j_4, j_5, j_0, j_1\}$	32
33	$\{c_1, c_0, j_5, j_3, j_2\}$	36
34	$\{c_1, c_0, j_5, j_3, j_1\}$	35
35	$\{c_1, c_0, j_5, j_2, j_1\}$	34
36	$\{c_1, c_0, j_3, j_2, j_1\}$	33
37	$\{c_1, c_0, j_4, j_5, j_0, j_1\}$	37
38	$\{c_1, c_0, j_4, j_3, j_2, j_1\}$	39
39	$\{c_1, c_0, j_5, j_3, j_2, j_0\}$	38
40	$\{c_1, c_0, j_5, j_3, j_2, j_1\}$	40
41	$\{c_1, c_0, j_4, j_5, j_3, j_2, j_0, j_1\}$	41

Table 15.8

i	$\varphi_4(J_i) = U_t$	t
-3	$\{c_2, c_0\}$	-3
-2	$\{c_0\}$	-1
-1	$\{c_2\}$	-2
0	$\emptyset$	0
1	$\{u_5\}$	2
2	$\{u_2\}$	1
3	$\{c_2, u_5\}$	8
4	$\{c_2, u_4\}$	7
5	$\{c_2, u_2\}$	6
6	$\{c_0, u_5\}$	5
7	$\{c_0, u_1\}$	4
8	$\{c_0, u_2\}$	3
9	$\{u_3, u_2\}$	10
10	$\{u_5, u_0\}$	9
11	$\{u_5, u_2\}$	11
12	$\{c_2, c_0, u_5\}$	15
13	$\{c_2, c_0, u_4\}$	14
14	$\{c_2, c_0, u_1\}$	13
15	$\{c_2, c_0, u_2\}$	12
16	$\{c_2, u_5, u_4\}$	21
17	$\{c_2, u_5, u_2\}$	20
18	$\{c_2, u_4, u_2\}$	19
19	$\{c_0, u_5, u_1\}$	18
20	$\{c_0, u_5, u_2\}$	17
21	$\{c_0, u_1, u_2\}$	16
22	$\{c_2, c_0, u_3, u_2\}$	25
23	$\{c_2, c_0, u_5, u_4\}$	29
24	$\{c_2, c_0, u_5, u_1\}$	28
25	$\{c_2, c_0, u_5, u_0\}$	22
26	$\{c_2, c_0, u_5, u_2\}$	26
27	$\{c_2, c_0, u_4, u_1\}$	27
28	$\{c_2, c_0, u_4, u_2\}$	24
29	$\{c_2, c_0, u_1, u_2\}$	23
30	$\{c_2, u_5, u_4, u_2\}$	31
31	$\{c_0, u_5, u_1, u_2\}$	30
32	$\{u_3, u_5, u_0, u_2\}$	32
33	$\{c_2, c_0, u_5, u_4, u_1\}$	36
34	$\{c_2, c_0, u_5, u_4, u_2\}$	35
35	$\{c_2, c_0, u_5, u_1, u_2\}$	34
36	$\{c_2, c_0, u_4, u_1, u_2\}$	33
37	$\{c_2, c_0, u_3, u_5, u_0, u_2\}$	37
38	$\{c_2, c_0, u_3, u_4, u_1, u_2\}$	39
39	$\{c_2, c_0, u_5, u_4, u_1, u_0\}$	38
40	$\{c_2, c_0, u_5, u_4, u_1, u_2\}$	40
41	$\{c_2, c_0, u_3, u_5, u_4, u_1, u_0, u_2\}$	41

Table 15.9

i	$\varphi_5(J_i) = V_t$	t
-3	$\{c_1, c_2\}$	-3
-2	$\{c_2\}$	-1
-1	$\{c_1\}$	-2
0	$\emptyset$	0
1	$\{v_2\}$	2
2	$\{v_4\}$	1
3	$\{c_1, v_2\}$	8
4	$\{c_1, v_0\}$	7
5	$\{c_1, v_4\}$	6
6	$\{c_2, v_2\}$	5
7	$\{c_2, v_5\}$	4
8	$\{c_2, v_4\}$	3
9	$\{v_1, v_4\}$	10
10	$\{v_2, v_3\}$	9
11	$\{v_2, v_4\}$	11
12	$\{c_1, c_2, v_2\}$	15
13	$\{c_1, c_2, v_0\}$	14
14	$\{c_1, c_2, v_5\}$	13
15	$\{c_1, c_2, v_4\}$	12
16	$\{c_1, v_2, v_0\}$	21
17	$\{c_1, v_2, v_4\}$	20
18	$\{c_1, v_0, v_4\}$	19
19	$\{c_2, v_2, v_5\}$	18
20	$\{c_2, v_2, v_4\}$	17
21	$\{c_2, v_5, v_4\}$	16
22	$\{c_1, c_2, v_1, v_4\}$	25
23	$\{c_1, c_2, v_2, v_0\}$	29
24	$\{c_1, c_2, v_2, v_5\}$	28
25	$\{c_1, c_2, v_2, v_3\}$	22
26	$\{c_1, c_2, v_2, v_4\}$	26
27	$\{c_1, c_2, v_0, v_5\}$	27
28	$\{c_1, c_2, v_0, v_4\}$	24
29	$\{c_1, c_2, v_5, v_4\}$	23
30	$\{c_1, v_2, v_0, v_4\}$	31
31	$\{c_2, v_2, v_5, v_4\}$	30
32	$\{v_1, v_2, v_3, v_4\}$	32
33	$\{c_1, c_2, v_2, v_0, v_5\}$	36
34	$\{c_1, c_2, v_2, v_0, v_4\}$	35
35	$\{c_1, c_2, v_2, v_5, v_4\}$	34
36	$\{c_1, c_2, v_0, v_5, v_4\}$	33
37	$\{c_1, c_2, v_1, v_2, v_3, v_4\}$	37
38	$\{c_1, c_2, v_1, v_0, v_5, v_4\}$	39
39	$\{c_1, c_2, v_2, v_0, v_5, v_3\}$	38
40	$\{c_1, c_2, v_2, v_0, v_5, v_4\}$	40
41	$\{c_1, c_2, v_1, v_2, v_0, v_5, v_3, v_4\}$	41

Table 15.10[3]

i	$\tau(H_i)$	H_i	Case
1	$(0,0,0,0)$	$\emptyset$	1
2	$(-1,-1,0,0)$	$\{c_0\}$	
3	$(-1,0,-1,0)$	$\{c_1\}$	
4	$(0,-2,-2,0)$	$\{c_2\}$	
5	$(-3,-1,-1,0)$	$\{c_0,c_1\}$	
6	$(-1,-3,-2,0)$	$\{c_0,c_2\}$	
7	$(-2,-2,-3,0)$	$\{c_1,c_2\}$	
8	$(-3,-3,-3,0)$	$\{c_0,c_1,c_2\}$	
9	$(1,0,0,0)$	$\{j_1\}$	2.1
10	$(2,0,0,0)$	$\{j_5\}$	
11	$(3,-1,0,0)$	$\{c_0,j_1\}$	
12	$(4,-1,0,0)$	$\{c_0,j_2\}$	
13	$(5,-1,0,0)$	$\{c_0,j_5\}$	
14	$(6,0,-2,0)$	$\{c_1,j_1\}$	
15	$(7,0,-2,0)$	$\{c_1,j_3\}$	
16	$(8,0,-2,0)$	$\{c_1,j_5\}$	
17	$(9,0,0,0)$	$\{j_0,j_5\}$	
18	$(10,0,0,0)$	$\{j_1,j_4\}$	
19	$(11,0,0,0)$	$\{j_1,j_5\}$	
20	$(12,-1,-2,0)$	$\{c_0,c_1,j_1\}$	
21	$(13,-1,-2,0)$	$\{c_0,c_1,j_2\}$	
22	$(14,-1,-2,0)$	$\{c_0,c_1,j_3\}$	
23	$(15,-1,-2,0)$	$\{c_0,c_1,j_5\}$	
24	$(16,-1,-2,0)$	$\{c_0,j_1,j_2\}$	
25	$(17,-1,-2,0)$	$\{c_0,j_1,j_5\}$	
26	$(18,-1,0,0)$	$\{c_0,j_2,j_5\}$	
27	$(19,0,-2,0)$	$\{c_1,j_1,j_3\}$	
28	$(20,0,-2,0)$	$\{c_1,j_1,j_5\}$	
29	$(21,0,-2,0)$	$\{c_1,j_3,j_5\}$	
30	$(22,-1,-2,0)$	$\{c_0,c_1,j_0,j_5\}$	
31	$(23,-1,-2,0)$	$\{c_0,c_1,j_1,j_2\}$	
32	$(24,-1,-2,0)$	$\{c_0,c_1,j_1,j_3\}$	
33	$(25,-1,-2,0)$	$\{c_0,c_1,j_1,j_4\}$	
34	$(26,-1,-2,0)$	$\{c_0,c_1,j_1,j_5\}$	
35	$(27,-1,-2,0)$	$\{c_0,c_1,j_2,j_3\}$	
36	$(28,-1,-2,0)$	$\{c_0,c_1,j_2,j_5\}$	
37	$(29,-1,-2,0)$	$\{c_0,c_1,j_3,j_5\}$	
38	$(30,-1,-2,0)$	$\{c_0,j_1,j_2,j_5\}$	
39	$(31,0,-2,0)$	$\{c_1,j_1,j_3,j_5\}$	
40	$(32,0,0,0)$	$\{j_0,j_1,j_4,j_5\}$	
41	$(33,-1,-2,0)$	$\{c_0,c_1,j_1,j_2,j_3\}$	
42	$(34,-1,-2,0)$	$\{c_0,c_1,j_1,j_2,j_5\}$	
43	$(35,-1,-2,0)$	$\{c_0,c_1,j_1,j_3,j_5\}$	
44	$(36,-1,-2,0)$	$\{c_0,c_1,j_2,j_3,j_5\}$	
45	$(37,-1,-2,0)$	$\{c_0,c_1,j_0,j_1,j_4,j_5\}$	
46	$(38,-1,-2,0)$	$\{c_0,c_1,j_0,j_2,j_3,j_5\}$	
47	$(39,-1,-2,0)$	$\{c_0,c_1,j_1,j_2,j_3,j_4\}$	
48	$(40,-1,-2,0)$	$\{c_0,c_1,j_1,j_2,j_3,j_5\}$	
49	$(41,-1,-2,0)$	$\{c_0,c_1,j_0,j_1,j_2,j_3,j_4,j_5\}$	
50	$(0,1,0,0)$	$\{u_2\}$	2.2
51	$(0,2,0,0)$	$\{u_5\}$	
52	$(-1,3,0,0)$	$\{c_0,u_2\}$	
53	$(-1,4,0,0)$	$\{c_0,u_1\}$	
54	$(-1,5,0,0)$	$\{c_0,u_5\}$	
55	$(0,6,-1,0)$	$\{c_2,u_2\}$	
56	$(0,7,-1,0)$	$\{c_2,u_4\}$	
57	$(0,8,-1,0)$	$\{c_2,u_5\}$	
58	$(0,9,0,0)$	$\{u_0,u_5\}$	
59	$(0,10,0,0)$	$\{u_2,u_3\}$	
60	$(0,11,0,0)$	$\{u_2,u_5\}$	

[3] If the sets H_i are unions of certain sets, repeated constants are not removed in the descriptions of the sets to make the construction of the sets H_i recognizable.

i	$\tau(H_i)$	H_i	Case
61	$(-1,12,-1,0)$	$\{c_0,c_2,u_2\}$	
62	$(-1,13,-1,0)$	$\{c_0,c_2,u_1\}$	
63	$(-1,14,-1,0)$	$\{c_0,c_2,u_4\}$	
64	$(-1,15,-1,0)$	$\{c_0,c_2,u_5\}$	
65	$(-1,16,-1,0)$	$\{c_0,u_2,u_1\}$	
66	$(-1,17,-1,0)$	$\{c_0,u_2,u_5\}$	
67	$(-1,18,0,0)$	$\{c_0,u_1,u_5\}$	
68	$(0,19,-1,0)$	$\{c_2,u_2,u_4\}$	
69	$(0,20,-1,0)$	$\{c_2,u_2,u_5\}$	
70	$(0,21,-1,0)$	$\{c_2,u_4,u_5\}$	
71	$(-1,22,-1,0)$	$\{c_0,c_2,u_0,u_5\}$	
72	$(-1,23,-1,0)$	$\{c_0,c_2,u_2,u_1\}$	
73	$(-1,24,-1,0)$	$\{c_0,c_2,u_2,u_4\}$	
74	$(-1,25,-1,0)$	$\{c_0,c_2,u_2,u_3\}$	
75	$(-1,26,-1,0)$	$\{c_0,c_2,u_2,u_5\}$	
76	$(-1,27,-1,0)$	$\{c_0,c_2,u_1,u_4\}$	
77	$(-1,28,-1,0)$	$\{c_0,c_2,u_1,u_5\}$	
78	$(-1,29,-1,0)$	$\{c_0,c_2,u_4,u_5\}$	
79	$(-1,30,-1,0)$	$\{c_0,u_2,u_1,u_5\}$	
80	$(0,31,-1,0)$	$\{c_2,u_2,u_4,u_5\}$	
81	$(0,32,0,0)$	$\{u_0,u_2,u_3,u_5\}$	
82	$(-1,33,-1,0)$	$\{c_0,c_2,u_2,u_1,u_4\}$	
83	$(-1,34,-1,0)$	$\{c_0,c_2,u_2,u_1,u_5\}$	
84	$(-1,35,-1,0)$	$\{c_0,c_2,u_2,u_4,u_5\}$	
85	$(-1,36,-1,0)$	$\{c_0,c_2,u_1,u_4,u_5\}$	
86	$(-1,37,-1,0)$	$\{c_0,c_2,u_0,u_2,u_3,u_5\}$	
87	$(-1,38,-1,0)$	$\{c_0,c_2,u_0,u_1,u_4,u_5\}$	
88	$(-1,39,-1,0)$	$\{c_0,c_2,u_2,u_1,u_4,u_3\}$	
89	$(-1,40,-1,0)$	$\{c_0,c_2,u_2,u_1,u_4,u_5\}$	
90	$(-1,41,-1,0)$	$\{c_0,c_2,u_0,u_2,u_1,u_4,u_3,u_5\}$	
91	$(0,0,1,0)$	$\{v_4\}$	2.3
92	$(0,0,2,0)$	$\{v_2\}$	
93	$(0,-2,3,0)$	$\{c_2,v_4\}$	
94	$(0,-2,4,0)$	$\{c_2,v_5\}$	
95	$(0,-2,5,0)$	$\{c_2,v_2\}$	
96	$(-2,0,6,0)$	$\{c_1,v_4\}$	
97	$(-2,0,7,0)$	$\{c_1,v_0\}$	
98	$(-2,0,8,0)$	$\{c_1,v_2\}$	
99	$(0,0,9,0)$	$\{v_3,v_2\}$	
100	$(0,0,10,0)$	$\{v_4,v_1\}$	
101	$(0,0,11,0)$	$\{v_4,v_2\}$	
102	$(-2,-2,12,0)$	$\{c_2,c_1,v_4\}$	
103	$(-2,-2,13,0)$	$\{c_2,c_1,v_5\}$	
104	$(-2,-2,14,0)$	$\{c_2,c_1,v_0\}$	
105	$(-2,-2,15,0)$	$\{c_2,c_1,v_2\}$	
106	$(0,-2,16,0)$	$\{c_2,v_4,v_5\}$	
107	$(0,-2,17,0)$	$\{c_2,v_4,v_2\}$	
108	$(0,-2,18,0)$	$\{c_2,v_5,v_2\}$	
109	$(-2,0,19,0)$	$\{c_1,v_4,v_0\}$	
110	$(-2,0,20,0)$	$\{c_1,v_4,v_2\}$	
111	$(-2,0,21,0)$	$\{c_1,v_0,v_2\}$	
112	$(-2,-2,22,0)$	$\{c_2,c_1,v_3,v_2\}$	
113	$(-2,-2,23,0)$	$\{c_2,c_1,v_4,v_5\}$	
114	$(-2,-2,24,0)$	$\{c_2,c_1,v_4,v_0\}$	
115	$(-2,-2,25,0)$	$\{c_2,c_1,v_4,v_1\}$	
116	$(-2,-2,26,0)$	$\{c_2,c_1,v_4,v_2\}$	
117	$(-2,-2,27,0)$	$\{c_2,c_1,v_5,v_0\}$	
118	$(-2,-2,28,0)$	$\{c_2,c_1,v_5,v_2\}$	
119	$(-2,-2,29,0)$	$\{c_2,c_1,v_0,v_2\}$	
120	$(0,-2,30,0)$	$\{c_2,v_4,v_5,v_2\}$	
121	$(-2,0,31,0)$	$\{c_1,v_4,v_0,v_2\}$	
122	$(0,0,32,0)$	$\{v_3,v_4,v_1,v_2\}$	
123	$(-2,-2,33,0)$	$\{c_2,c_1,v_4,v_5,v_0\}$	
124	$(-2,-2,34,0)$	$\{c_2,c_1,v_4,v_5,v_2\}$	
125	$(-2,-2,35,0)$	$\{c_2,c_1,v_4,v_0,v_2\}$	

i	$\tau(H_i)$	H_i	Case
126	$(-2,-2,36,0)$	$\{c_2, c_1, v_5, v_0, v_2\}$	
127	$(-2,-2,37,0)$	$\{c_2, c_1, v_3, v_4, v_1, v_2\}$	
128	$(-2,-2,38,0)$	$\{c_2, c_1, v_3, v_5, v_0, v_2\}$	
129	$(-2,-2,39,0)$	$\{c_2, c_1, v_4, v_5, v_0, v_1\}$	
130	$(-2,-2,40,0)$	$\{c_2, c_1, v_4, v_5, v_0, v_2\}$	
131	$(-2,-2,41,0)$	$\{c_2, c_1, v_3, v_4, v_5, v_0, v_1, v_2\}$	
132	$(3,-3,-1,0)$	$\{c_0, j_1, c_2\}$	3.1
133	$(-3,3,-2,0)$	$\{c_0, c_1, c_0, u_2\}$	
134	$(12,-3,-3,0)$	$\{c_0, c_1, j_1, c_0, c_2\}$	
135	$(-3,12,-3,0)$	$\{c_0, c_1, c_0, c_2, u_2\}$	
136	$(25,-3,-3,0)$	$\{c_0, c_1, j_1, j_4, c_0, c_2\}$	
137	$(-3,25,-3,0)$	$\{c_0, c_1, c_0, c_2, u_2, u_3\}$	
138	$(3,3,0,0)$	$\{c_0, j_1, c_0, u_2\}$	
139	$(3,12,-1,0)$	$\{c_0, j_1, c_0, c_2, u_2\}$	
140	$(12,3,-2,0)$	$\{c_0, c_1, j_1, c_0, u_2\}$	
141	$(12,12,-3,0)$	$\{c_0, c_1, j_1, c_0, c_2, u_2\}$	
142	$(3,25,-1,0)$	$\{c_0, j_1, c_0, c_2, u_2, u_3\}$	
143	$(25,3,-2,0)$	$\{c_0, c_1, j_1, j_4, c_0, u_2\}$	
144	$(12,25,-3,0)$	$\{c_0, c_1, j_1, c_0, c_2, u_2, u_3\}$	
145	$(25,12,-3,0)$	$\{c_0, c_1, j_1, j_4, c_0, c_2, u_2\}$	
146	$(25,25,-3,0)$	$\{c_0, c_1, j_1, j_4, c_0, c_2, u_2, u_3\}$	
147	$(8,-2,-3,0)$	$\{c_1, j_5, c_2\}$	3.2.1
148	$(13,-3,-3,0)$	$\{c_0, c_1, j_2, c_0, c_2\}$	3.2.2
149	$(14,-3,-3,0)$	$\{c_0, c_1, j_3, c_0, c_2\}$	
150	$(15,-3,-3,0)$	$\{c_0, c_1, j_5, c_0, c_2\}$	
151	$(22,-3,-3,0)$	$\{c_0, c_1, j_0, j_5, c_0, c_2\}$	
152	$(23,-3,-3,0)$	$\{c_0, c_1, j_1, j_2, c_0, c_2\}$	
153	$(24,-3,-3,0)$	$\{c_0, c_1, j_1, j_3, c_0, c_2\}$	
154	$(26,-3,-3,0)$	$\{c_0, c_1, j_1, j_5, c_0, c_2\}$	
155	$(27,-3,-3,0)$	$\{c_0, c_1, j_2, j_3, c_0, c_2\}$	
156	$(28,-3,-3,0)$	$\{c_0, c_1, j_2, j_5, c_0, c_2\}$	
157	$(29,-3,-3,0)$	$\{c_0, c_1, j_3, j_5, c_0, c_2\}$	
158	$(33,-3,-3,0)$	$\{c_0, c_1, j_1, j_2, j_3, c_0, c_2\}$	
159	$(34,-3,-3,0)$	$\{c_0, c_1, j_1, j_2, j_5, c_0, c_2\}$	
160	$(35,-3,-3,0)$	$\{c_0, c_1, j_1, j_3, j_5, c_0, c_2\}$	
161	$(36,-3,-3,0)$	$\{c_0, c_1, j_2, j_3, j_5, c_0, c_2\}$	
162	$(37,-3,-3,0)$	$\{c_0, c_1, j_0, j_1, j_4, j_5, c_0, c_2\}$	
163	$(38,-3,-3,0)$	$\{c_0, c_1, j_0, j_2, j_3, j_5, c_0, c_2\}$	
164	$(39,-3,-3,0)$	$\{c_0, c_1, j_1, j_2, j_3, j_4, c_0, c_2\}$	
165	$(40,-3,-3,0)$	$\{c_0, c_1, j_1, j_2, j_3, j_5, c_0, c_2\}$	
166	$(41,-3,-3,0)$	$J \cup \{c_0, c_2\}$	
167	$(4,3,0,0)$	$\{c_0, j_2, c_0, u_2\}$	3.2.3
168	$(13,3,-2,0)$	$\{c_0, c_1, j_2, c_0, u_2\}$	
169	$(14,3,-2,0)$	$\{c_0, c_1, j_3, c_0, u_2\}$	
170	$(16,3,0,0)$	$\{c_0, j_1, j_2, c_0, u_2\}$	
171	$(18,3,0,0)$	$\{c_0, j_2, j_5, c_0, u_2\}$	
172	$(23,3,-2,0)$	$\{c_0, c_1, j_1, j_2, c_0, u_2\}$	
173	$(24,3,-2,0)$	$\{c_0, c_1, j_1, j_3, c_0, u_2\}$	
174	$(27,3,-2,0)$	$\{c_0, c_1, j_2, j_3, c_0, u_2\}$	
175	$(28,3,-2,0)$	$\{c_0, c_1, j_2, j_5, c_0, u_2\}$	
176	$(30,3,0,0)$	$\{c_0, j_1, j_2, j_5, c_0, u_2\}$	
177	$(33,3,-2,0)$	$\{c_0, c_1, j_1, j_2, j_3, c_0, u_2\}$	
178	$(34,3,-2,0)$	$\{c_0, c_1, j_1, j_2, j_5, c_0, u_2\}$	
179	$(36,3,-2,0)$	$\{c_0, c_1, j_2, j_3, j_5, c_0, u_2\}$	
180	$(38,3,-2,0)$	$\{c_0, c_1, j_0, j_2, j_3, j_5, c_0, u_2\}$	
181	$(39,3,-2,0)$	$\{c_0, c_1, j_1, j_2, j_3, j_4, c_0, u_2\}$	
182	$(40,3,-2,0)$	$\{c_0, c_1, j_1, j_2, j_3, j_5, c_0, u_2\}$	
183	$(41,3,-2,0)$	$J \cup \{c_0, u_2\}$	
184	$(13,12,-3,0)$	$\{c_0, c_1, j_2, c_0, c_2, u_2\}$	3.2.4
185	$(14,12,-3,0)$	$\{c_0, c_1, j_3, c_0, c_2, u_2\}$	
186	$(23,12,-3,0)$	$\{c_0, c_1, j_1, j_2, c_0, c_2, u_2\}$	
187	$(24,12,-3,0)$	$\{c_0, c_1, j_1, j_3, c_0, c_2, u_2\}$	
188	$(27,12,-3,0)$	$\{c_0, c_1, j_2, j_3, c_0, c_2, u_2\}$	
189	$(28,12,-3,0)$	$\{c_0, c_1, j_2, j_5, c_0, c_2, u_2\}$	
190	$(33,12,-3,0)$	$\{c_0, c_1, j_1, j_2, j_3, c_0, c_2, u_2\}$	

i	$\tau(H_i)$	H_i	Case
191	$(34, 12, -3, 0)$	$\{c_0, c_1, j_1, j_2, j_5, c_0, c_2, u_2\}$	
192	$(36, 12, -3, 0)$	$\{c_0, c_1, j_2, j_3, j_5, c_0, c_2, u_2\}$	
193	$(38, 12, -3, 0)$	$\{c_0, c_1, j_0, j_2, j_3, j_5, c_0, c_2, u_2\}$	
194	$(39, 12, -3, 0)$	$\{c_0, c_1, j_1, j_2, j_3, j_4, c_0, c_2, u_2\}$	
195	$(40, 12, -3, 0)$	$\{c_0, c_1, j_1, j_2, j_3, j_5, c_0, c_2, u_2\}$	
196	$(41, 12, -3, 0)$	$J \cup \{c_0, c_2, u_2\}$	
197	$(27, 25, -3, 0)$	$\{c_0, c_1, j_2, j_3, c_0, c_2, u_2, u_3\}$	3.2.5
198	$(33, 25, -3, 0)$	$\{c_0, c_1, j_1, j_2, j_3, c_0, c_2, u_2, u_3\}$	
199	$(36, 25, -3, 0)$	$\{c_0, c_1, j_2, j_3, j_5, c_0, c_2, u_2, u_3\}$	
200	$(38, 25, -3, 0)$	$\{c_0, c_1, j_0, j_2, j_3, j_5, c_0, c_2, u_2, u_3\}$	
201	$(39, 25, -3, 0)$	$\{c_0, c_1, j_1, j_2, j_3, j_4, c_0, c_2, u_2, u_3\}$	
202	$(40, 25, -3, 0)$	$\{c_0, c_1, j_1, j_2, j_3, j_5, c_0, c_2, u_2, u_3\}$	
203	$(41, 25, -3, 0)$	$J \cup \{c_0, c_2, u_2, u_3\}$	
204	$(-2, 8, -3, 0)$	$\{c_1, c_2, u_5\}$	3.3
205	$(-3, 13, -3, 0)$	$\{c_0, c_1, c_0, c_2, u_1\}$	
206	$(-3, 14, -3, 0)$	$\{c_0, c_1, c_0, c_2, u_4\}$	
207	$(-3, 15, -3, 0)$	$\{c_0, c_1, c_0, c_2, u_5\}$	
208	$(-3, 22, -3, 0)$	$\{c_0, c_1, c_0, c_2, u_0, u_5\}$	
209	$(-3, 23, -3, 0)$	$\{c_0, c_1, c_0, c_2, u_2, u_1\}$	
210	$(-3, 24, -3, 0)$	$\{c_0, c_1, c_0, c_2, u_2, u_4\}$	
211	$(-3, 26, -3, 0)$	$\{c_0, c_1, c_0, c_2, u_2, u_5\}$	
212	$(-3, 27, -3, 0)$	$\{c_0, c_1, c_0, c_2, u_1, u_4\}$	
213	$(-3, 28, -3, 0)$	$\{c_0, c_1, c_0, c_2, u_1, u_5\}$	
214	$(-3, 29, -3, 0)$	$\{c_0, c_1, c_0, c_2, u_4, u_5\}$	
215	$(-3, 33, -3, 0)$	$\{c_0, c_1, c_0, c_2, u_2, u_1, u_4\}$	
216	$(-3, 34, -3, 0)$	$\{c_0, c_1, c_0, c_2, u_2, u_1, u_5\}$	
217	$(-3, 35, -3, 0)$	$\{c_0, c_1, c_0, c_2, u_2, u_4, u_5\}$	
218	$(-3, 36, -3, 0)$	$\{c_0, c_1, c_0, c_2, u_1, u_4, u_5\}$	
219	$(-3, 37, -3, 0)$	$\{c_0, c_1, c_0, c_2, u_0, u_2, u_3, u_5\}$	
220	$(-3, 38, -3, 0)$	$\{c_0, c_1, c_0, c_2, u_0, u_1, u_4, u_5\}$	
221	$(-3, 39, -3, 0)$	$\{c_0, c_1, c_0, c_2, u_2, u_1, u_4, u_3\}$	
222	$(-3, 40, -3, 0)$	$\{c_0, c_1, c_0, c_2, u_2, u_1, u_4, u_5\}$	
223	$(-3, 41, -3, 0)$	$\{c_0, c_1\} \cup U$	
224	$(3, 4, 0, 0)$	$\{c_0, j_1, c_0, u_1\}$	
225	$(3, 13, -1, 0)$	$\{c_0, j_1, c_0, c_2, u_1\}$	
226	$(3, 14, -1, 0)$	$\{c_0, j_1, c_0, c_2, u_4\}$	
227	$(3, 16, 0, 0)$	$\{c_0, j_1, c_0, u_2, u_1\}$	
228	$(3, 18, 0, 0)$	$\{c_0, j_1, c_0, u_1, u_5\}$	
229	$(3, 23, -1, 0)$	$\{c_0, j_1, c_0, c_2, u_2, u_1\}$	
230	$(3, 24, -1, 0)$	$\{c_0, j_1, c_0, c_2, u_2, u_4\}$	
231	$(3, 27, -1, 0)$	$\{c_0, j_1, c_0, c_2, u_1, u_4\}$	
232	$(3, 28, -1, 0)$	$\{c_0, j_1, c_0, c_2, u_1, u_5\}$	
233	$(3, 30, 0, 0)$	$\{c_0, j_1, c_0, u_2, u_1, u_5\}$	
234	$(3, 33, -1, 0)$	$\{c_0, j_1, c_0, c_2, u_2, u_1, u_4\}$	
235	$(3, 34, -1, 0)$	$\{c_0, j_1, c_0, c_2, u_2, u_1, u_5\}$	
236	$(3, 36, -1, 0)$	$\{c_0, j_1, c_0, c_2, u_1, u_4, u_5\}$	
237	$(3, 38, -1, 0)$	$\{c_0, j_1, c_0, c_2, u_0, u_1, u_4, u_5\}$	
238	$(3, 39, -1, 0)$	$\{c_0, j_1, c_0, c_2, u_2, u_1, u_4, u_3\}$	
239	$(3, 40, -1, 0)$	$\{c_0, j_1, c_0, c_2, u_2, u_1, u_4, u_5\}$	
240	$(3, 41, -1, 0)$	$\{c_0, j_1\} \cup U$	
241	$(12, 13, -3, 0)$	$\{c_0, c_1, j_1, c_0, c_2, u_1\}$	
242	$(12, 14, -3, 0)$	$\{c_0, c_1, j_1, c_0, c_2, u_4\}$	
243	$(12, 23, -3, 0)$	$\{c_0, c_1, j_1, c_0, c_2, u_2, u_1\}$	
244	$(12, 24, -3, 0)$	$\{c_0, c_1, j_1, c_0, c_2, u_2, u_4\}$	
245	$(12, 27, -3, 0)$	$\{c_0, c_1, j_1, c_0, c_2, u_1, u_4\}$	
246	$(12, 28, -3, 0)$	$\{c_0, c_1, j_1, c_0, c_2, u_1, u_5\}$	
247	$(12, 33, -3, 0)$	$\{c_0, c_1, j_1, c_0, c_2, u_2, u_1, u_4\}$	
248	$(12, 34, -3, 0)$	$\{c_0, c_1, j_1, c_0, c_2, u_2, u_1, u_5\}$	
249	$(12, 36, -3, 0)$	$\{c_0, c_1, j_1, c_0, c_2, u_1, u_4, u_5\}$	
250	$(12, 38, -3, 0)$	$\{c_0, c_1, j_1, c_0, c_2, u_0, u_1, u_4, u_5\}$	
251	$(12, 39, -3, 0)$	$\{c_0, c_1, j_1, c_0, c_2, u_2, u_1, u_4, u_3\}$	
252	$(12, 40, -3, 0)$	$\{c_0, c_1, j_1, c_0, c_2, u_2, u_1, u_4, u_5\}$	
253	$(12, 41, -3, 0)$	$\{c_0, c_1, j_1\} \cup U$	
254	$(25, 27, -3, 0)$	$\{c_0, c_1, j_1, j_4, c_0, c_2, u_1, u_4\}$	
255	$(25, 33, -3, 0)$	$\{c_0, c_1, j_1, j_4, c_0, c_2, u_2, u_1, u_4\}$	

i	$\tau(H_i)$	H_i	Case
256	$(25, 36, -3, 0)$	$\{c_0, c_1, j_1, j_4, c_0, c_2, u_1, u_4, u_5\}$	
257	$(25, 38, -3, 0)$	$\{c_0, c_1, j_1, j_4, c_0, c_2, u_0, u_1, u_4, u_5\}$	
258	$(25, 39, -3, 0)$	$\{c_0, c_1, j_1, j_4, c_0, c_2, u_2, u_1, u_4, u_3\}$	
259	$(25, 40, -3, 0)$	$\{c_0, c_1, j_1, j_4, c_0, c_2, u_2, u_1, u_4, u_5\}$	
260	$(25, 41, -3, 0)$	$\{c_0, c_1, j_1, j_4\} \cup U$	
261	$(2, 2, 0, 0)$	$\{j_5, u_5\}$	3.4
262	$(5, 5, 0, 0)$	$\{c_0, j_5, c_0, u_5\}$	
263	$(8, 8, -3, 0)$	$\{c_1, j_5, c_2, u_5\}$	
264	$(9, 9, 0, 0)$	$\{j_0, j_5, u_0, u_5\}$	
265	$(15, 15, -2, 0)$	$\{c_0, c_1, j_5, c_0, c_2, u_5\}$	
266	$(16, 16, 0, 0)$	$\{c_0, j_1, j_2, c_0, u_2, u_1\}$	
267	$(17, 18, 0, 0)$	$\{c_0, j_1, j_5, c_0, u_1, u_5\}$	
268	$(18, 17, 0, 0)$	$\{c_0, j_2, j_5, c_0, u_2, u_5\}$	
269	$(22, 22, -3, 0)$	$\{c_0, c_1, j_0, j_5, c_0, c_2, u_0, u_5\}$	
270	$(23, 23, -3, 0)$	$\{c_0, c_1, j_1, j_2, c_0, c_2, u_2, u_1\}$	
271	$(24, 24, -3, 0)$	$\{c_0, c_1, j_1, j_3, c_0, c_2, u_2, u_4\}$	
272	$(26, 28, -3, 0)$	$\{c_0, c_1, j_1, j_5, c_0, c_2, u_1, u_5\}$	
273	$(28, 26, -3, 0)$	$\{c_0, c_1, j_2, j_5, c_0, c_2, u_2, u_5\}$	
274	$(30, 30, 0, 0)$	$\{c_0, j_1, j_2, j_5, c_0, u_2, u_1, u_5\}$	
275	$(34, 34, -3, 0)$	$\{c_0, c_1, j_1, j_2, j_5, c_0, c_2, u_2, u_1, u_5\}$	
276	$(37, 38, -3, 0)$	$\{c_0, c_1, j_0, j_1, j_4, j_5, c_0, c_2, u_0, u_1, u_4, u_5\}$	
277	$(38, 37, -3, 0)$	$\{c_0, c_1, j_0, j_2, j_3, j_5, c_0, c_2, u_0, u_2, u_3, u_5\}$	
278	$(39, 39, -3, 0)$	$\{c_0, c_1, j_1, j_2, j_3, j_4, c_0, c_2, u_2, u_1, u_4, u_3\}$	
279	$(41, 41, -3, 0)$	$J \cup U$	
147	$(8, -2, -3, 0)$	$\{c_1, j_5, c_2, c_1\}$	4.1
280	$(-3, -1, 8, 0)$	$\{c_0, c_1, c_1, v_2\}$	
150	$(15, -3, -3, 0)$	$\{c_0, c_1, j_5, c_2, c_1\}$	
281	$(-3, -3, 15, 0)$	$\{c_0, c_1, c_2, c_1, v_2\}$	
151	$(22, -3, -3, 0)$	$\{c_0, c_1, j_0, j_5, c_2, c_1\}$	
282	$(-3, -3, 22, 0)$	$\{c_0, c_1, c_2, c_1, v_3, v_2\}$	
283	$(8, 0, 8, 0)$	$\{c_1, j_5, c_1, v_2\}$	
284	$(8, -1, 15, 0)$	$\{c_1, j_5, c_2, c_1, v_2\}$	
285	$(15, -2, 8, 0)$	$\{c_0, c_1, j_5, c_1, v_2\}$	
286	$(15, -3, 15, 0)$	$\{c_0, c_1, j_5, c_2, c_1, v_2\}$	
287	$(8, -1, 22, 0)$	$\{c_1, j_5, c_2, c_1, v_3, v_2\}$	
288	$(22, -2, 8, 0)$	$\{c_0, c_1, j_0, j_5, c_1, v_2\}$	
289	$(15, -3, 22, 0)$	$\{c_0, c_1, j_5, c_2, c_1, v_3, v_2\}$	
290	$(22, -3, 15, 0)$	$\{c_0, c_1, j_0, j_5, c_2, c_1, v_2\}$	
291	$(22, -3, 22, 0)$	$\{c_0, c_1, j_0, j_5, c_2, c_1, v_3, v_2\}$	
132	$(3, -3, -1, 0)$	$\{c_0, j_1, c_2\}$	
149	$(14, -3, -3, 0)$	$\{c_0, c_1, j_3, c_2, c_1\}$	
148	$(13, -3, -3, 0)$	$\{c_0, c_1, j_2, c_2, c_1\}$	
134	$(12, -3, -3, 0)$	$\{c_0, c_1, j_1, c_2, c_1\}$	
136	$(25, -3, -3, 0)$	$\{c_0, c_1, j_1, j_4, c_2, c_1\}$	
157	$(29, -3, -3, 0)$	$\{c_0, c_1, j_3, j_5, c_2, c_1\}$	
156	$(28, -3, -3, 0)$	$\{c_0, c_1, j_2, j_5, c_2, c_1\}$	
154	$(26, -3, -3, 0)$	$\{c_0, c_1, j_1, j_5, c_2, c_1\}$	
155	$(27, -3, -3, 0)$	$\{c_0, c_1, j_2, j_3, c_2, c_1\}$	
153	$(24, -3, -3, 0)$	$\{c_0, c_1, j_1, j_3, c_2, c_1\}$	
152	$(23, -3, -3, 0)$	$\{c_0, c_1, j_1, j_2, c_2, c_1\}$	
161	$(36, -3, -3, 0)$	$\{c_0, c_1, j_2, j_3, j_5, c_2, c_1\}$	
160	$(35, -3, -3, 0)$	$\{c_0, c_1, j_1, j_3, j_5, c_2, c_1\}$	
159	$(34, -3, -3, 0)$	$\{c_0, c_1, j_1, j_2, j_5, c_2, c_1\}$	
158	$(33, -3, -3, 0)$	$\{c_0, c_1, j_1, j_2, j_3, c_2, c_1\}$	
162	$(37, -3, -3, 0)$	$\{c_0, c_1, j_0, j_1, j_4, j_5, c_2, c_1\}$	
164	$(39, -3, -3, 0)$	$\{c_0, c_1, j_1, j_2, j_3, j_4, c_2, c_1\}$	
163	$(38, -3, -3, 0)$	$\{c_0, c_1, j_0, j_2, j_3, j_5, c_2, c_1\}$	
165	$(40, -3, -3, 0)$	$\{c_0, c_1, j_1, j_2, j_3, j_5, c_2, c_1\}$	
166	$(41, -3, -3, 0)$	$J \cup \{c_2, c_1\}$	
292	$(7, 0, 8, 0)$	$\{c_1, j_3, c_1, v_2\}$	
293	$(14, -2, 8, 0)$	$\{c_0, c_1, j_3, c_1, v_2\}$	
294	$(13, -2, 8, 0)$	$\{c_0, c_1, j_2, c_1, v_2\}$	
295	$(21, 0, 8, 0)$	$\{c_1, j_3, j_5, c_1, v_2\}$	
296	$(19, 0, 8, 0)$	$\{c_1, j_1, j_3, c_1, v_2\}$	
297	$(29, -2, 8, 0)$	$\{c_0, c_1, j_3, j_5, c_1, v_2\}$	

i	$\tau(H_i)$	H_i	Case
298	$(28,-2,8,0)$	$\{c_0,c_1,j_2,j_5,c_1,v_2\}$	
299	$(27,-2,8,0)$	$\{c_0,c_1,j_2,j_3,c_1,v_2\}$	
300	$(24,-2,8,0)$	$\{c_0,c_1,j_1,j_3,c_1,v_2\}$	
301	$(31,0,8,0)$	$\{c_1,j_1,j_3,j_5,c_1,v_2\}$	
302	$(36,-2,8,0)$	$\{c_0,c_1,j_2,j_3,j_5,c_1,v_2\}$	
303	$(35,-2,8,0)$	$\{c_0,c_1,j_1,j_3,j_5,c_1,v_2\}$	
304	$(33,-2,8,0)$	$\{c_0,c_1,j_1,j_2,j_3,c_1,v_2\}$	
305	$(39,-2,8,0)$	$\{c_0,c_1,j_1,j_2,j_3,j_4,c_1,v_2\}$	
306	$(38,-2,8,0)$	$\{c_0,c_1,j_0,j_2,j_3,j_5,c_1,v_2\}$	
307	$(40,-2,8,0)$	$\{c_0,c_1,j_1,j_2,j_3,j_5,c_1,v_2\}$	
308	$(41,-2,8,0)$	$J\cup\{c_1,v_2\}$	
309	$(14,-3,15,0)$	$\{c_0,c_1,j_3,c_2,c_1,v_2\}$	
310	$(13,-3,15,0)$	$\{c_0,c_1,j_2,c_2,c_1,v_2\}$	
311	$(29,-3,15,0)$	$\{c_0,c_1,j_3,j_5,c_2,c_1,v_2\}$	
312	$(28,-3,15,0)$	$\{c_0,c_1,j_2,j_5,c_2,c_1,v_2\}$	
313	$(27,-3,15,0)$	$\{c_0,c_1,j_2,j_3,c_2,c_1,v_2\}$	
314	$(24,-3,15,0)$	$\{c_0,c_1,j_1,j_3,c_2,c_1,v_2\}$	
315	$(36,-3,15,0)$	$\{c_0,c_1,j_2,j_3,j_5,c_2,c_1,v_2\}$	
316	$(35,-3,15,0)$	$\{c_0,c_1,j_1,j_3,j_5,c_2,c_1,v_2\}$	
317	$(33,-3,15,0)$	$\{c_0,c_1,j_1,j_2,j_3,c_2,c_1,v_2\}$	
318	$(39,-3,15,0)$	$\{c_0,c_1,j_1,j_2,j_3,j_4,c_2,c_1,v_2\}$	
319	$(38,-3,15,0)$	$\{c_0,c_1,j_0,j_2,j_3,j_5,c_2,c_1,v_2\}$	
320	$(40,-3,15,0)$	$\{c_0,c_1,j_1,j_2,j_3,j_5,c_2,c_1,v_2\}$	
321	$(41,-3,15,0)$	$J\cup\{c_2,c_1,v_2\}$	
322	$(27,-3,22,0)$	$\{c_0,c_1,j_2,j_3,c_2,c_1,v_3,v_2\}$	
323	$(36,-3,22,0)$	$\{c_0,c_1,j_2,j_3,j_5,c_2,c_1,v_3,v_2\}$	
324	$(33,-3,22,0)$	$\{c_0,c_1,j_1,j_2,j_3,c_2,c_1,v_3,v_2\}$	
325	$(39,-3,22,0)$	$\{c_0,c_1,j_1,j_2,j_3,j_4,c_2,c_1,v_3,v_2\}$	
326	$(38,-3,22,0)$	$\{c_0,c_1,j_0,j_2,j_3,j_5,c_2,c_1,v_3,v_2\}$	
327	$(40,-3,22,0)$	$\{c_0,c_1,j_1,j_2,j_3,j_5,c_2,c_1,v_3,v_2\}$	
328	$(41,-3,22,0)$	$J\cup\{c_2,c_1,v_3,v_2\}$	
329	$(-1,-3,3,0)$	$\{c_0,c_2,v_4\}$	
330	$(-3,-3,14,0)$	$\{c_0,c_1,c_2,c_1,v_0\}$	
331	$(-3,-3,13,0)$	$\{c_0,c_1,c_2,c_1,v_5\}$	
332	$(-3,-3,12,0)$	$\{c_0,c_1,c_2,c_1,v_4\}$	
333	$(-3,-3,25,0)$	$\{c_0,c_1,c_2,c_1,v_4,v_1\}$	
334	$(-3,-3,29,0)$	$\{c_0,c_1,c_2,c_1,v_0,v_2\}$	
335	$(-3,-3,28,0)$	$\{c_0,c_1,c_2,c_1,v_5,v_2\}$	
336	$(-3,-3,26,0)$	$\{c_0,c_1,c_2,c_1,v_4,v_2\}$	
337	$(-3,-3,27,0)$	$\{c_0,c_1,c_2,c_1,v_5,v_0\}$	
338	$(-3,-3,24,0)$	$\{c_0,c_1,c_2,c_1,v_4,v_0\}$	
339	$(-3,-3,23,0)$	$\{c_0,c_1,c_2,c_1,v_4,v_5\}$	
340	$(-3,-3,36,0)$	$\{c_0,c_1,c_2,c_1,v_5,v_0,v_2\}$	
341	$(-3,-3,35,0)$	$\{c_0,c_1,c_2,c_1,v_4,v_0,v_2\}$	
342	$(-3,-3,34,0)$	$\{c_0,c_1,c_2,c_1,v_4,v_5,v_2\}$	
343	$(-3,-3,33,0)$	$\{c_0,c_1,c_2,c_1,v_4,v_5,v_0\}$	
344	$(-3,-3,37,0)$	$\{c_0,c_1,c_2,c_1,v_3,v_4,v_1,v_2\}$	
345	$(-3,-3,39,0)$	$\{c_0,c_1,c_2,c_1,v_4,v_5,v_0,v_1\}$	
346	$(-3,-3,38,0)$	$\{c_0,c_1,c_2,c_1,v_3,v_5,v_0,v_2\}$	
347	$(-3,-3,40,0)$	$\{c_0,c_1,c_2,c_1,v_4,v_5,v_0,v_2\}$	
348	$(-3,-3,41,0)$	$\{c_0,c_1\}\cup V$	
349	$(8,0,7,0)$	$\{c_1,j_5,c_1,v_0\}$	
350	$(8,-1,14,0)$	$\{c_1,j_5,c_2,c_1,v_0\}$	
351	$(8,-1,13,0)$	$\{c_1,j_5,c_2,c_1,v_5\}$	
352	$(8,0,21,0)$	$\{c_1,j_5,c_1,v_0,v_2\}$	
353	$(8,0,19,0)$	$\{c_1,j_5,c_1,v_4,v_0\}$	
354	$(8,-1,29,0)$	$\{c_1,j_5,c_2,c_1,v_0,v_2\}$	
355	$(8,-1,28,0)$	$\{c_1,j_5,c_2,c_1,v_5,v_2\}$	
356	$(8,-1,27,0)$	$\{c_1,j_5,c_2,c_1,v_5,v_0\}$	
357	$(8,-1,24,0)$	$\{c_1,j_5,c_2,c_1,v_4,v_0\}$	
358	$(8,0,31,0)$	$\{c_1,j_5,c_1,v_4,v_0,v_2\}$	
359	$(8,-1,36,0)$	$\{c_1,j_5,c_2,c_1,v_5,v_0,v_2\}$	
360	$(8,-1,35,0)$	$\{c_1,j_5,c_2,c_1,v_4,v_0,v_2\}$	
361	$(8,-1,33,0)$	$\{c_1,j_5,c_2,c_1,v_4,v_5,v_0\}$	
362	$(8,-1,39,0)$	$\{c_1,j_5,c_2,c_1,v_4,v_5,v_0,v_1\}$	

i	$\tau(H_i)$	H_i	Case
363	$(8,-1,38,0)$	$\{c_1, j_5, c_2, c_1, v_3, v_5, v_0, v_2\}$	
364	$(8,-1,40,0)$	$\{c_1, j_5, c_2, c_1, v_4, v_5, v_0, v_2\}$	
365	$(8,-1,41,0)$	$\{c_1, j_5\} \cup V$	
366	$(15,-3,14,0)$	$\{c_0, c_1, j_5, c_2, c_1, v_0\}$	
367	$(15,-3,13,0)$	$\{c_0, c_1, j_5, c_2, c_1, v_5\}$	
368	$(15,-3,29,0)$	$\{c_0, c_1, j_5, c_2, c_1, v_0, v_2\}$	
369	$(15,-3,28,0)$	$\{c_0, c_1, j_5, c_2, c_1, v_5, v_2\}$	
370	$(15,-3,27,0)$	$\{c_0, c_1, j_5, c_2, c_1, v_5, v_0\}$	
371	$(15,-3,24,0)$	$\{c_0, c_1, j_5, c_2, c_1, v_4, v_0\}$	
372	$(15,-3,36,0)$	$\{c_0, c_1, j_5, c_2, c_1, v_5, v_0, v_2\}$	
373	$(15,-3,35,0)$	$\{c_0, c_1, j_5, c_2, c_1, v_4, v_0, v_2\}$	
374	$(15,-3,33,0)$	$\{c_0, c_1, j_5, c_2, c_1, v_4, v_5, v_0\}$	
375	$(15,-3,39,0)$	$\{c_0, c_1, j_5, c_2, c_1, v_4, v_5, v_0, v_1\}$	
376	$(15,-3,38,0)$	$\{c_0, c_1, j_5, c_2, c_1, v_3, v_5, v_0, v_2\}$	
377	$(15,-3,40,0)$	$\{c_0, c_1, j_5, c_2, c_1, v_4, v_5, v_0, v_2\}$	
378	$(15,-3,41,0)$	$\{c_0, c_1, j_5\} \cup V$	
379	$(22,-3,27,0)$	$\{c_0, c_1, j_0, j_5, c_2, c_1, v_5, v_0\}$	
380	$(22,-3,36,0)$	$\{c_0, c_1, j_0, j_5, c_2, c_1, v_5, v_0, v_2\}$	
381	$(22,-3,33,0)$	$\{c_0, c_1, j_0, j_5, c_2, c_1, v_4, v_5, v_0\}$	
382	$(22,-3,39,0)$	$\{c_0, c_1, j_0, j_5, c_2, c_1, v_4, v_5, v_0, v_1\}$	
383	$(22,-3,38,0)$	$\{c_0, c_1, j_0, j_5, c_2, c_1, v_3, v_5, v_0, v_2\}$	
384	$(22,-3,40,0)$	$\{c_0, c_1, j_0, j_5, c_2, c_1, v_4, v_5, v_0, v_2\}$	
385	$(22,-3,41,0)$	$\{c_0, c_1, j_0, j_5\} \cup V$	
386	$(1,0,1,0)$	$\{j_1, v_4\}$	
387	$(6,0,6,0)$	$\{c_1, j_1, c_1, v_4\}$	
388	$(3,-3,3,0)$	$\{c_0, j_1, c_2, v_4\}$	
389	$(10,0,10,0)$	$\{j_1, j_4, v_4, v_1\}$	
390	$(12,-3,12,0)$	$\{c_0, c_1, j_1, c_2, c_1, v_4\}$	
391	$(21,0,21,0)$	$\{c_1, j_3, j_5, c_1, v_0, v_2\}$	
392	$(20,0,19,0)$	$\{c_1, j_1, j_5, c_1, v_4, v_0\}$	
393	$(19,0,20,0)$	$\{c_1, j_1, j_3, c_1, v_4, v_2\}$	
394	$(25,-3,25,0)$	$\{c_0, c_1, j_1, j_4, c_2, c_1, v_4, v_1\}$	
395	$(29,-3,29,0)$	$\{c_0, c_1, j_3, j_5, c_2, c_1, v_0, v_2\}$	
396	$(28,-3,28,0)$	$\{c_0, c_1, j_2, j_5, c_2, c_1, v_5, v_2\}$	
397	$(26,-3,24,0)$	$\{c_0, c_1, j_1, j_5, c_2, c_1, v_4, v_0\}$	
398	$(24,-3,26,0)$	$\{c_0, c_1, j_1, j_3, c_2, c_1, v_4, v_2\}$	
399	$(31,0,31,0)$	$\{c_1, j_1, j_3, j_5, c_1, v_4, v_0, v_2\}$	
400	$(35,-3,35,0)$	$\{c_0, c_1, j_1, j_3, j_5, c_2, c_1, v_4, v_0, v_2\}$	
401	$(37,-3,39,0)$	$\{c_0, c_1, j_0, j_1, j_4, j_5, c_2, c_1, v_4, v_5, v_0, v_1\}$	
402	$(39,-3,37,0)$	$\{c_0, c_1, j_1, j_2, j_3, j_4, c_2, c_1, v_3, v_4, v_1, v_2\}$	
403	$(38,-3,38,0)$	$\{c_0, c_1, j_0, j_2, j_3, j_5, c_2, c_1, v_3, v_5, v_0, v_2\}$	
404	$(41,-3,41,0)$	$J \cup V$	
204	$(-2,8,-3,0)$	$\{c_2, u_5, c_2, c_1\}$	4.2
352	$(-1,-3,3,0)$	$\{c_0, c_2, c_2, v_4\}$	
207	$(-3,15,-3,0)$	$\{c_0, c_2, u_5, c_2, c_1\}$	
355	$(-3,-3,12,0)$	$\{c_0, c_2, c_2, c_1, v_4\}$	
208	$(-3,22,-3,0)$	$\{c_0, c_2, u_0, u_5, c_2, c_1\}$	
356	$(-3,-3,25,0)$	$\{c_0, c_2, c_2, c_1, v_4, v_1\}$	
405	$(0,8,3,0)$	$\{c_2, u_5, c_2, v_4\}$	
406	$(-2,8,12,0)$	$\{c_2, u_5, c_2, c_1, v_4\}$	
407	$(-1,15,3,0)$	$\{c_0, c_2, u_5, c_2, v_4\}$	
408	$(-3,15,12,0)$	$\{c_0, c_2, u_5, c_2, c_1, v_4\}$	
409	$(-2,8,25,0)$	$\{c_2, u_5, c_2, c_1, v_4, v_1\}$	
410	$(-1,22,3,0)$	$\{c_0, c_2, u_0, u_5, c_2, v_4\}$	
411	$(-3,15,25,0)$	$\{c_0, c_2, u_5, c_2, c_1, v_4, v_1\}$	
412	$(-3,22,12,0)$	$\{c_0, c_2, u_0, u_5, c_2, c_1, v_4\}$	
413	$(-3,22,25,0)$	$\{c_0, c_2, u_0, u_5, c_2, c_1, v_4, v_1\}$	
414	$(-2,3,-2,0)$	$\{c_2, u_2, c_1\}$	
206	$(-3,14,-3,0)$	$\{c_0, c_2, u_4, c_2, c_1\}$	
205	$(-3,13,-3,0)$	$\{c_0, c_2, u_1, c_2, c_1\}$	
135	$(-3,12,-3,0)$	$\{c_0, c_2, u_2, c_2, c_1\}$	
137	$(-3,25,-3,0)$	$\{c_0, c_2, u_2, u_3, c_2, c_1\}$	
214	$(-3,29,-3,0)$	$\{c_0, c_2, u_4, u_5, c_2, c_1\}$	
213	$(-3,28,-3,0)$	$\{c_0, c_2, u_1, u_5, c_2, c_1\}$	
211	$(-3,26,-3,0)$	$\{c_0, c_2, u_2, u_5, c_2, c_1\}$	

i	$\tau(H_i)$	H_i	Case
212	$(-3, 27, -3, 0)$	$\{c_0, c_2, u_1, u_4, c_2, c_1\}$	
210	$(-3, 24, -3, 0)$	$\{c_0, c_2, u_2, u_4, c_2, c_1\}$	
209	$(-3, 23, -3, 0)$	$\{c_0, c_2, u_2, u_1, c_2, c_1\}$	
218	$(-3, 36, -3, 0)$	$\{c_0, c_2, u_1, u_4, u_5, c_2, c_1\}$	
217	$(-3, 35, -3, 0)$	$\{c_0, c_2, u_2, u_4, u_5, c_2, c_1\}$	
216	$(-3, 34, -3, 0)$	$\{c_0, c_2, u_2, u_1, u_5, c_2, c_1\}$	
215	$(-3, 33, -3, 0)$	$\{c_0, c_2, u_2, u_1, u_4, c_2, c_1\}$	
219	$(-3, 37, -3, 0)$	$\{c_2, c_1, v_3, v_4, v_1, v_2, c_2, c_1\}$	
221	$(-3, 39, -3, 0)$	$\{c_0, c_2, u_2, u_1, u_4, u_3, c_2, c_1\}$	
220	$(-3, 38, -3, 0)$	$\{c_0, c_2, u_0, u_1, u_4, u_5, c_2, c_1\}$	
222	$(-3, 40, -3, 0)$	$\{c_0, c_2, u_2, u_1, u_4, u_5, c_2, c_1\}$	
223	$(-3, 41, -3, 0)$	$U \cup \{c_2, c_1\}$	
415	$(0, 7, 3, 0)$	$\{c_2, u_4, c_2, v_4\}$	
416	$(-1, 14, 3, 0)$	$\{c_0, c_2, u_4, c_2, v_4\}$	
417	$(-1, 13, 3, 0)$	$\{c_0, c_2, u_1, c_2, v_4\}$	
418	$(0, 21, 3, 0)$	$\{c_2, u_4, u_5, c_2, v_4\}$	
419	$(0, 19, 3, 0)$	$\{c_2, u_2, u_4, c_2, v_4\}$	
420	$(-1, 29, 3, 0)$	$\{c_0, c_2, u_4, u_5, c_2, v_4\}$	
421	$(-1, 28, 3, 0)$	$\{c_0, c_2, u_1, u_5, c_2, v_4\}$	
422	$(-1, 27, 3, 0)$	$\{c_0, c_2, u_1, u_4, c_2, v_4\}$	
423	$(-1, 24, 3, 0)$	$\{c_0, c_2, u_2, u_4, c_2, v_4\}$	
424	$(0, 31, 3, 0)$	$\{c_2, u_2, u_4, u_5, c_2, v_4\}$	
425	$(-1, 36, 3, 0)$	$\{c_0, c_2, u_1, u_4, u_5, c_2, v_4\}$	
426	$(-1, 35, 3, 0)$	$\{c_0, c_2, u_2, u_4, u_5, c_2, v_4\}$	
427	$(-1, 33, 3, 0)$	$\{c_0, c_2, u_2, u_1, u_4, c_2, v_4\}$	
428	$(-1, 39, 3, 0)$	$\{c_0, c_2, u_2, u_1, u_4, u_3, c_2, v_4\}$	
429	$(-1, 38, 3, 0)$	$\{c_0, c_2, u_0, u_1, u_4, u_5, c_2, v_4\}$	
430	$(-1, 40, 3, 0)$	$\{c_0, c_2, u_2, u_1, u_4, u_5, c_2, v_4\}$	
431	$(-1, 41, 3, 0)$	$U \cup \{c_2, v_4\}$	
432	$(-3, 14, 12, 0)$	$\{c_0, c_2, u_4, c_2, c_1, v_4\}$	
433	$(-3, 13, 12, 0)$	$\{c_0, c_2, u_1, c_2, c_1, v_4\}$	
434	$(-3, 29, 12, 0)$	$\{c_0, c_2, u_4, u_5, c_2, c_1, v_4\}$	
435	$(-3, 28, 12, 0)$	$\{c_0, c_2, u_1, u_5, c_2, c_1, v_4\}$	
436	$(-3, 27, 12, 0)$	$\{c_0, c_2, u_1, u_4, c_2, c_1, v_4\}$	
437	$(-3, 24, 12, 0)$	$\{c_0, c_2, u_2, u_4, c_2, c_1, v_4\}$	
438	$(-3, 36, 12, 0)$	$\{c_0, c_2, u_1, u_4, u_5, c_2, c_1, v_4\}$	
439	$(-3, 35, 12, 0)$	$\{c_0, c_2, u_2, u_4, u_5, c_2, c_1, v_4\}$	
440	$(-3, 33, 12, 0)$	$\{c_0, c_2, u_2, u_1, u_4, c_2, c_1, v_4\}$	
441	$(-3, 39, 12, 0)$	$\{c_0, c_2, u_2, u_1, u_4, u_3, c_2, c_1, v_4\}$	
442	$(-3, 38, 12, 0)$	$\{c_0, c_2, u_0, u_1, u_4, u_5, c_2, c_1, v_4\}$	
443	$(-3, 40, 12, 0)$	$\{c_0, c_2, u_2, u_1, u_4, u_5, c_2, c_1, v_4\}$	
444	$(-3, 41, 12, 0)$	$U \cup \{c_2, c_1, v_4\}$	
445	$(-3, 27, 25, 0)$	$\{c_0, c_2, u_1, u_4, c_2, c_1, v_4, v_1\}$	
446	$(-3, 36, 25, 0)$	$\{c_0, c_2, u_1, u_4, u_5, c_2, c_1, v_4, v_1\}$	
447	$(-3, 33, 25, 0)$	$\{c_0, c_2, u_2, u_1, u_4, c_2, c_1, v_4, v_1\}$	
448	$(-3, 39, 25, 0)$	$\{c_0, c_2, u_2, u_1, u_4, u_3, c_2, c_1, v_4, v_1\}$	
449	$(-3, 38, 25, 0)$	$\{c_0, c_2, u_0, u_1, u_4, u_5, c_2, c_1, v_4, v_1\}$	
450	$(-3, 40, 25, 0)$	$\{c_0, c_2, u_2, u_1, u_4, u_5, c_2, c_1, v_4, v_1\}$	
451	$(-3, 41, 25, 0)$	$U \cup \{c_2, c_1, v_4, v_1\}$	
281	$(-3, -1, 8, 0)$	$\{c_0, c_1, v_2\}$	
354	$(-3, -3, 13, 0)$	$\{c_0, c_2, c_2, c_1, v_5\}$	
353	$(-3, -3, 14, 0)$	$\{c_0, c_2, c_2, c_1, v_0\}$	
283	$(-3, -3, 15, 0)$	$\{c_0, c_2, c_2, c_1, v_2\}$	
285	$(-3, -3, 22, 0)$	$\{c_0, c_2, c_2, c_1, v_3, v_2\}$	
362	$(-3, -3, 23, 0)$	$\{c_0, c_2, c_2, c_1, v_4, v_5\}$	
361	$(-3, -3, 24, 0)$	$\{c_0, c_2, c_2, c_1, v_4, v_0\}$	
359	$(-3, -3, 26, 0)$	$\{c_0, c_2, c_2, c_1, v_4, v_2\}$	
360	$(-3, -3, 27, 0)$	$\{c_0, c_2, c_2, c_1, v_5, v_0\}$	
358	$(-3, -3, 28, 0)$	$\{c_0, c_2, c_2, c_1, v_5, v_2\}$	
357	$(-3, -3, 29, 0)$	$\{c_0, c_2, c_2, c_1, v_0, v_2\}$	
366	$(-3, -3, 33, 0)$	$\{c_0, c_2, c_2, c_1, v_4, v_5, v_0\}$	
365	$(-3, -3, 34, 0)$	$\{c_0, c_2, c_2, c_1, v_4, v_5, v_2\}$	
364	$(-3, -3, 35, 0)$	$\{c_0, c_2, c_2, c_1, v_4, v_0, v_2\}$	
363	$(-3, -3, 36, 0)$	$\{c_0, c_2, c_2, c_1, v_5, v_0, v_2\}$	
367	$(-3, -3, 37, 0)$	$\{c_0, c_2, c_2, c_1, v_3, v_4, v_1, v_2\}$	

i	$\tau(H_i)$	H_i	Case
369	$(-3,-3,38,0)$	$\{c_0,c_2,c_2,c_1,v_3,v_5,v_0,v_2\}$	
368	$(-3,-3,39,0)$	$\{c_0,c_2,c_2,c_1,v_4,v_5,v_0,v_1\}$	
370	$(-3,-3,40,0)$	$\{c_0,c_2,c_2,c_1,v_4,v_5,v_0,v_2\}$	
371	$(-3,-3,41,0)$	$\{c_0,c_2\}\cup V$	
452	$(0,8,4,0)$	$\{c_2,u_5,c_2,v_5\}$	
453	$(-2,8,13,0)$	$\{c_2,u_5,c_2,c_1,v_5\}$	
454	$(-2,8,14,0)$	$\{c_2,u_5,c_2,c_1,v_0\}$	
455	$(0,8,16,0)$	$\{c_2,u_5,c_2,v_4,v_5\}$	
456	$(0,8,18,0)$	$\{c_2,u_5,c_2,v_5,v_2\}$	
457	$(-2,8,23,0)$	$\{c_2,u_5,c_2,c_1,v_4,v_5\}$	
458	$(-2,8,24,0)$	$\{c_2,u_5,c_2,c_1,v_4,v_0\}$	
459	$(-2,8,27,0)$	$\{c_2,u_5,c_2,c_1,v_5,v_0\}$	
450	$(-2,8,28,0)$	$\{c_2,u_5,c_2,c_1,v_5,v_2\}$	
461	$(0,8,30,0)$	$\{c_2,u_5,c_2,v_4,v_5,v_2\}$	
462	$(-2,8,33,0)$	$\{c_2,u_5,c_2,c_1,v_4,v_5,v_0\}$	
463	$(-2,8,34,0)$	$\{c_2,u_5,c_2,c_1,v_4,v_5,v_2\}$	
464	$(-2,8,36,0)$	$\{c_2,u_5,c_2,c_1,v_5,v_0,v_2\}$	
465	$(-2,8,38,0)$	$\{c_2,u_5,c_2,c_1,v_3,v_5,v_0,v_2\}$	
466	$(-2,8,39,0)$	$\{c_2,u_5,c_2,c_1,v_4,v_5,v_0,v_1\}$	
467	$(-2,8,40,0)$	$\{c_2,u_5,c_2,c_1,v_4,v_5,v_0,v_2\}$	
468	$(-2,8,41,0)$	$\{c_2,u_5\}\cup V$	
469	$(-3,15,13,0)$	$\{c_0,c_2,u_5,c_2,c_1,v_5\}$	
470	$(-3,15,14,0)$	$\{c_0,c_2,u_5,c_2,c_1,v_0\}$	
471	$(-3,15,23,0)$	$\{c_0,c_2,u_5,c_2,c_1,v_4,v_5\}$	
472	$(-3,15,24,0)$	$\{c_0,c_2,u_5,c_2,c_1,v_4,v_0\}$	
473	$(-3,15,27,0)$	$\{c_0,c_2,u_5,c_2,c_1,v_5,v_0\}$	
474	$(-3,15,28,0)$	$\{c_0,c_2,u_5,c_2,c_1,v_5,v_2\}$	
475	$(-3,15,33,0)$	$\{c_0,c_2,u_5,c_2,c_1,v_4,v_5,v_0\}$	
476	$(-3,15,34,0)$	$\{c_0,c_2,u_5,c_2,c_1,v_4,v_5,v_2\}$	
477	$(-3,15,36,0)$	$\{c_0,c_2,u_5,c_2,c_1,v_5,v_0,v_2\}$	
478	$(-3,15,38,0)$	$\{c_0,c_2,u_5,c_2,c_1,v_3,v_5,v_0,v_2\}$	
479	$(-3,15,39,0)$	$\{c_0,c_2,u_5,c_2,c_1,v_4,v_5,v_0,v_1\}$	
480	$(-3,15,40,0)$	$\{c_0,c_2,u_5,c_2,c_1,v_4,v_5,v_0,v_2\}$	
481	$(-3,15,41,0)$	$\{c_0,c_2,u_5\}\cup V$	
482	$(-3,22,27,0)$	$\{c_0,c_2,u_0,u_5,c_2,c_1,v_5,v_0\}$	
483	$(-3,22,33,0)$	$\{c_0,c_2,u_0,u_5,c_2,c_1,v_4,v_5,v_0\}$	
484	$(-3,22,36,0)$	$\{c_0,c_2,u_0,u_5,c_2,c_1,v_5,v_0,v_2\}$	
485	$(-3,22,38,0)$	$\{c_0,c_2,u_0,u_5,c_2,c_1,v_3,v_5,v_0,v_2\}$	
486	$(-3,22,39,0)$	$\{c_0,c_2,u_0,u_5,c_2,c_1,v_4,v_5,v_0,v_1\}$	
487	$(-3,22,40,0)$	$\{c_0,c_2,u_0,u_5,c_2,c_1,v_4,v_5,v_0,v_2\}$	
488	$(-3,22,41,0)$	$\{c_0,c_2,u_0,u_5\}\cup V$	
489	$(0,1,2,0)$	$\{u_2,v_2\}$	
490	$(0,6,5,0)$	$\{c_2,u_2,c_2,v_2\}$	
491	$(-3,3,8,0)$	$\{c_0,u_2,c_1,v_2\}$	
492	$(0,10,9,0)$	$\{u_2,u_3,v_3,v_2\}$	
493	$(-3,12,15,0)$	$\{c_0,c_2,u_2,c_2,c_1,v_2\}$	
494	$(0,21,16,0)$	$\{c_2,u_4,u_5,c_2,v_4,v_5\}$	
495	$(0,20,18,0)$	$\{c_2,u_2,u_5,c_2,v_5,v_2\}$	
496	$(0,19,17,0)$	$\{c_2,u_2,u_4,c_2,v_4,v_2\}$	
497	$(-3,25,22,0)$	$\{c_0,c_2,u_2,u_3,c_2,c_1,v_3,v_2\}$	
498	$(-3,29,23,0)$	$\{c_0,c_2,u_4,u_5,c_2,c_1,v_4,v_5\}$	
499	$(-3,28,24,0)$	$\{c_0,c_2,u_1,u_5,c_2,c_1,v_4,v_0\}$	
500	$(-3,26,28,0)$	$\{c_0,c_2,u_2,u_5,c_2,c_1,v_5,v_2\}$	
501	$(-3,24,26,0)$	$\{c_0,c_2,u_2,u_4,c_2,c_1,v_4,v_2\}$	
502	$(0,31,30,0)$	$\{c_2,u_2,u_4,u_5,c_2,v_4,v_5,v_2\}$	
503	$(-3,35,34,0)$	$\{c_0,c_2,u_2,u_4,u_5,c_2,c_1,v_4,v_5,v_2\}$	
504	$(-3,37,38,0)$	$\{c_2,c_1,v_3,v_4,v_1,v_2,c_2,c_1,v_3,v_5,v_0,v_2\}$	
505	$(-3,39,37,0)$	$\{c_0,c_2,u_2,u_1,u_4,u_3,c_2,c_1,v_3,v_4,v_1,v_2\}$	
506	$(-3,38,39,0)$	$\{c_0,c_2,u_0,u_1,u_4,u_5,c_2,c_1,v_4,v_5,v_0,v_1\}$	
507	$(-3,41,41,0)$	$U\cup V$	
508	$(13,3,8,0)$	$\{c_0,c_1,j_2,c_0,u_2,c_1,v_2\}$	5.1.1
509	$(14,3,8,0)$	$\{c_0,c_1,j_3,c_0,u_2,c_1,v_2\}$	
510	$(24,3,8,0)$	$\{c_0,c_1,j_1,j_3,c_0,u_2,c_1,v_2\}$	
511	$(27,3,8,0)$	$\{c_0,c_1,j_2,j_3,c_0,u_2,c_1,v_2\}$	
512	$(28,3,8,0)$	$\{c_0,c_1,j_2,j_5,c_0,u_2,c_1,v_2\}$	

i	$\tau(H_i)$	H_i	Case
513	$(33,3,8,0)$	$\{c_0,c_1,j_1,j_2,j_3,c_0,u_2,c_1,v_2\}$	
514	$(36,3,8,0)$	$\{c_0,c_1,j_2,j_3,j_5,c_0,u_2,c_1,v_2\}$	
515	$(38,3,8,0)$	$\{c_0,c_1,j_0,j_2,j_3,j_5,c_0,u_2,c_1,v_2\}$	
516	$(39,3,8,0)$	$\{c_0,c_1,j_1,j_2,j_3,j_4,c_0,u_2,c_1,v_2\}$	
517	$(40,3,8,0)$	$\{c_0,c_1,j_1,j_2,j_3,j_5,c_0,u_2,c_1,v_2\}$	
518	$(41,3,8,0)$	$J\cup\{c_0,u_2,c_1,v_2\}$	
519	$(13,12,15,0)$	$\{c_0,c_1,j_2,c_0,c_2,u_2,c_2,c_1,v_2\}$	5.1.2
520	$(14,12,15,0)$	$\{c_0,c_1,j_3,c_0,c_2,u_2,c_2,c_1,v_2\}$	
521	$(24,12,15,0)$	$\{c_0,c_1,j_1,j_3,c_0,c_2,u_2,c_2,c_1,v_2\}$	
522	$(27,12,15,0)$	$\{c_0,c_1,j_2,j_3,c_0,c_2,u_2,c_2,c_1,v_2\}$	
523	$(28,12,15,0)$	$\{c_0,c_1,j_2,j_5,c_0,c_2,u_2,c_2,c_1,v_2\}$	
524	$(33,12,15,0)$	$\{c_0,c_1,j_1,j_2,j_3,c_0,c_2,u_2,c_2,c_1,v_2\}$	
525	$(36,12,15,0)$	$\{c_0,c_1,j_2,j_3,j_5,c_0,c_2,u_2,c_2,c_1,v_2\}$	
526	$(38,12,15,0)$	$\{c_0,c_1,j_0,j_2,j_3,j_5,c_0,c_2,u_2,c_2,c_1,v_2\}$	
527	$(39,12,15,0)$	$\{c_0,c_1,j_1,j_2,j_3,j_4,c_0,c_2,u_2,c_2,c_1,v_2\}$	
528	$(40,12,15,0)$	$\{c_0,c_1,j_1,j_2,j_3,j_5,c_0,c_2,u_2,c_2,c_1,v_2\}$	
529	$(41,12,15,0)$	$J\cup\{c_0,c_2,u_2,c_2,c_1,v_2\}$	
530	$(27,25,22,0)$	$\{c_0,c_1,j_2,j_3,c_0,c_2,u_2,u_3,c_2,c_1,v_3,v_2\}$	5.1.3
531	$(33,25,22,0)$	$\{c_0,c_1,j_1,j_2,j_3,c_0,c_2,u_2,u_3,c_2,c_1,v_3,v_2\}$	
532	$(36,25,22,0)$	$\{c_0,c_1,j_2,j_3,j_5,c_0,c_2,u_2,u_3,c_2,c_1,v_3,v_2\}$	
532	$(38,25,22,0)$	$\{c_0,c_1,j_0,j_2,j_3,j_5,c_0,c_2,u_2,u_3,c_2,c_1,v_3,v_2\}$	
533	$(39,25,22,0)$	$\{c_0,c_1,j_1,j_2,j_3,j_4,c_0,c_2,u_2,u_3,c_2,c_1,v_3,v_2\}$	
534	$(40,25,22,0)$	$\{c_0,c_1,j_1,j_2,j_3,j_5,c_0,c_2,u_2,u_3,c_2,c_1,v_3,v_2\}$	
535	$(41,25,22,0)$	$J\cup\{c_0,c_2,u_2,u_3,c_2,c_1,v_3,v_2\}$	
536	$(3,13,3,0)$	$\{c_0,j_1,c_0,c_2,u_1,c_2,v_4\}$	5.2
537	$(3,14,3,0)$	$\{c_0,j_1,c_0,c_2,u_4,c_2,v_4\}$	
538	$(3,24,3,0)$	$\{c_0,j_1,c_0,c_2,u_2,u_4,c_2,v_4\}$	
539	$(3,27,3,0)$	$\{c_0,j_1,c_0,c_2,u_1,u_4,c_2,v_4\}$	
540	$(3,28,3,0)$	$\{c_0,j_1,c_0,c_2,u_1,u_5,c_2,v_4\}$	
541	$(3,33,3,0)$	$\{c_0,j_1,c_0,c_2,u_2,u_1,u_4,c_2,v_4\}$	
542	$(3,36,3,0)$	$\{c_0,j_1,c_0,c_2,u_1,u_4,u_5,c_2,v_4\}$	
543	$(3,38,3,0)$	$\{c_0,j_1,c_0,c_2,u_0,u_1,u_4,u_5,c_2,v_4\}$	
544	$(3,39,3,0)$	$\{c_0,j_1,c_0,c_2,u_2,u_1,u_4,u_3,c_2,v_4\}$	
545	$(3,40,3,0)$	$\{c_0,j_1,c_0,c_2,u_2,u_1,u_4,u_5,c_2,v_4\}$	
546	$(3,41,3,0)$	$\{c_0,j_1\}\cup U\cup\{c_2,v_4\}$	
547	$(12,13,12,0)$	$\{c_0,c_1,j_1,c_0,c_2,u_1,c_2,c_1,v_4\}$	
548	$(12,14,12,0)$	$\{c_0,c_1,j_1,c_0,c_2,u_4,c_2,c_1,v_4\}$	
549	$(12,24,12,0)$	$\{c_0,c_1,j_1,c_0,c_2,u_2,u_4,c_2,c_1,v_4\}$	
550	$(12,27,12,0)$	$\{c_0,c_1,j_1,c_0,c_2,u_1,u_4,c_2,c_1,v_4\}$	
551	$(12,28,12,0)$	$\{c_0,c_1,j_1,c_0,c_2,u_1,u_5,c_2,c_1,v_4\}$	
552	$(12,33,12,0)$	$\{c_0,c_1,j_1,c_0,c_2,u_2,u_1,u_4,c_2,c_1,v_4\}$	
553	$(12,36,12,0)$	$\{c_0,c_1,j_1,c_0,c_2,u_1,u_4,u_5,c_2,c_1,v_4\}$	
554	$(12,38,12,0)$	$\{c_0,c_1,j_1,c_0,c_2,u_0,u_1,u_4,u_5,c_2,c_1,v_4\}$	
555	$(12,39,12,0)$	$\{c_0,c_1,j_1,c_0,c_2,u_2,u_1,u_4,u_3,c_2,c_1,v_4\}$	
556	$(12,40,12,0)$	$\{c_0,c_1,j_1,c_0,c_2,u_2,u_1,u_4,u_5,c_2,c_1,v_4\}$	
557	$(12,41,12,0)$	$\{c_0,c_1,j_1\}\cup U\cup\{c_2,c_1,v_4\}$	
558	$(25,27,25,0)$	$\{c_0,c_1,j_1,j_4,c_0,c_2,u_1,u_4,c_2,c_1,v_4,v_1\}$	
559	$(25,33,25,0)$	$\{c_0,c_1,j_1,j_4,c_0,c_2,u_2,u_1,u_4,c_2,c_1,v_4,v_1\}$	
560	$(25,36,25,0)$	$\{c_0,c_1,j_1,j_4,c_0,c_2,u_1,u_4,u_5,c_2,c_1,v_4,v_1\}$	
561	$(25,38,25,0)$	$\{c_0,c_1,j_1,j_4,c_0,c_2,u_0,u_1,u_4,u_5,c_2,c_1,v_4,v_1\}$	
562	$(25,39,25,0)$	$\{c_0,c_1,j_1,j_4,c_0,c_2,u_2,u_1,u_4,u_3,c_2,c_1,v_4,v_1\}$	
563	$(25,40,25,0)$	$\{c_0,c_1,j_1,j_4,c_0,c_2,u_2,u_1,u_4,u_5,c_2,c_1,v_4,v_1\}$	
564	$(25,41,25,0)$	$\{c_0,c_1,j_1,j_4\}\cup U\cup\{c_2,c_1,v_4,v_1\}$	
565	$(8,8,13,0)$	$\{c_1,j_5,c_2,u_5,c_2,c_1,v_5\}$	5.3
566	$(8,8,14,0)$	$\{c_1,j_5,c_2,u_5,c_2,c_1,v_0\}$	
567	$(8,8,24,0)$	$\{c_1,j_5,c_2,u_5,c_2,c_1,v_4,v_0\}$	
568	$(8,8,27,0)$	$\{c_1,j_5,c_2,u_5,c_2,c_1,v_5,v_0\}$	
569	$(8,8,28,0)$	$\{c_1,j_5,c_2,u_5,c_2,c_1,v_5,v_2\}$	
570	$(8,8,33,0)$	$\{c_1,j_5,c_2,u_5,c_2,c_1,v_4,v_5,v_0\}$	
571	$(8,8,36,0)$	$\{c_1,j_5,c_2,u_5,c_2,c_1,v_5,v_0,v_2\}$	
572	$(8,8,38,0)$	$\{c_1,j_5,c_2,u_5,c_2,c_1,v_3,v_5,v_0,v_2\}$	
573	$(8,8,39,0)$	$\{c_1,j_5,c_2,u_5,c_2,c_1,v_4,v_5,v_0,v_1\}$	
574	$(8,8,40,0)$	$\{c_1,j_5,c_2,u_5,c_2,c_1,v_4,v_5,v_0,v_2\}$	
575	$(8,8,41,0)$	$\{c_1,j_5,c_2,u_5\}\cup V$	
576	$(15,15,13,0)$	$\{c_0,c_1,j_5,c_0,c_2,u_5,c_2,c_1,v_5\}$	

i	$\tau(H_i)$	H_i	Case
577	$(15,15,14,0)$	$\{c_0,c_1,j_5,c_0,c_2,u_5,c_2,c_1,v_0\}$	
578	$(15,15,24,0)$	$\{c_0,c_1,j_5,c_0,c_2,u_5,c_2,c_1,v_4,v_0\}$	
579	$(15,15,27,0)$	$\{c_0,c_1,j_5,c_0,c_2,u_5,c_2,c_1,v_5,v_0\}$	
580	$(15,15,28,0)$	$\{c_0,c_1,j_5,c_0,c_2,u_5,c_2,c_1,v_5,v_2\}$	
581	$(15,15,33,0)$	$\{c_0,c_1,j_5,c_0,c_2,u_5,c_2,c_1,v_4,v_5,v_0\}$	
582	$(15,15,36,0)$	$\{c_0,c_1,j_5,c_0,c_2,u_5,c_2,c_1,v_5,v_0,v_2\}$	
583	$(15,15,38,0)$	$\{c_0,c_1,j_5,c_0,c_2,u_5,c_2,c_1,v_3,v_5,v_0,v_2\}$	
584	$(15,15,39,0)$	$\{c_0,c_1,j_5,c_0,c_2,u_5,c_2,c_1,v_4,v_5,v_0,v_1\}$	
585	$(15,15,40,0)$	$\{c_0,c_1,j_5,c_0,c_2,u_5,c_2,c_1,v_4,v_5,v_0,v_2\}$	
586	$(15,15,41,0)$	$\{c_0,c_1,j_5,c_0,c_2,u_5\}\cup V$	
587	$(22,22,27,0)$	$\{c_0,c_1,j_0,j_5,c_0,c_2,u_0,u_5,c_2,c_1,v_5,v_0\}$	
588	$(22,22,33,0)$	$\{c_0,c_1,j_0,j_5,c_0,c_2,u_0,u_5,c_2,c_1,v_4,v_5,v_0\}$	
589	$(22,22,36,0)$	$\{c_0,c_1,j_0,j_5,c_0,c_2,u_0,u_5,c_2,c_1,v_5,v_0,v_2\}$	
590	$(22,22,38,0)$	$\{c_0,c_1,j_0,j_5,c_0,c_2,u_0,u_5,c_2,c_1,v_3,v_5,v_0,v_2\}$	
591	$(22,22,39,0)$	$\{c_0,c_1,j_0,j_5,c_0,c_2,u_0,u_5,c_2,c_1,v_4,v_5,v_0,v_1\}$	
592	$(22,22,40,0)$	$\{c_0,c_1,j_0,j_5,c_0,c_2,u_0,u_5,c_2,c_1,v_4,v_5,v_0,v_2\}$	
593	$(22,22,41,0)$	$\{c_0,c_1,j_0,j_5,c_0,c_2,u_0,u_5\}\cup V$	
594	$(24,24,26,0)$	$\{c_0,c_1,j_1,j_3,c_0,c_2,u_2,u_4,c_2,c_1,v_4,v_2\}$	5.4
595	$(26,28,24,0)$	$\{c_0,c_1,j_1,j_5,c_0,c_2,u_1,u_5,c_2,c_1,v_4,v_0\}$	
596	$(28,26,28,0)$	$\{c_0,c_1,j_2,j_5,c_0,c_2,u_2,u_5,c_2,c_1,v_5,v_2\}$	
597	$(37,38,39,0)$	$\{c_0,c_1,j_0,j_1,j_4,j_5,c_0,c_2,u_0,u_1,u_4,u_5,c_2,c_1,v_4,v_5,v_0,v_1\}$	
598	$(38,37,38,0)$	$\{c_0,c_1,j_0,j_2,j_3,j_5,c_0,c_2,u_0,u_2,u_3,u_5,c_2,c_1,v_3,v_5,v_0,v_2\}$	
599	$(39,39,37,0)$	$\{c_0,c_1,j_1,j_2,j_3,j_4,c_0,c_2,u_2,u_1,u_4,u_3,c_2,c_1,v_3,v_4,v_1,v_2\}$	
600	$(41,41,41,0)$	$J\cup U\cup V$	
$600+t$	$(...,1)$	$H_t\cup\{s_1\}$ $(t\in\{1,2,...,600\})$	6.1
1201	$(0,0,0,2)$	$\{s_1,s_3\}$	6.2
1202	$(0,-2,-1,2)$	$\{c_2,s_1,s_3\}$	
1203	$(-3-1,-2,2)$	$\{c_0,c_1,s_1,s_3\}$	
1204	$(-3,-3,-3,2)$	$\{c_0,c_1,c_2,s_1,s_3\}$	
1205	$(27,-1,-2,2)$	$\{c_0,c_1,j_2,j_3,s_1,s_3\}$	
1206	$(32,0,0,2)$	$\{j_0,j_1,j_4,j_5,s_1,s_3\}$	
1207	$(37,-1,-2,2)$	$\{c_0,c_1,j_0,j_1,j_4,j_5,s_1,s_3\}$	
1208	$(41,-1,-2,2)$	$J\cup\{s_1,s_3\}$	
1209	$(27,-3,-3,2)$	$\{c_0,c_1,j_2,j_3,c_2,s_1,s_3\}$	
1210	$(37,-3,-3,2)$	$\{c_0,c_1,j_0,j_1,j_4,j_5,c_2,s_1,s_3\}$	
1211	$(41,-3,-3,2)$	$J\cup\{c_0,c_2,s_1,s_3\}$	
1212	$(0,1,2,2)$	$\{u_2,v_2,s_1,s_3\}$	
1213	$(-3,3,8,2)$	$\{c_0,c_1,u_2,v_2,s_1,s_3\}$	
1214	$(0,6,5,2)$	$\{c_2,u_2,c_2,v_2,s_1,s_3\}$	
1215	$(-3,12,15,2)$	$\{c_0,c_2,u_2,c_2,c_1,v_2,s_1,s_3\}$	
1216	$(0,10,9,2)$	$\{u_2,u_3,v_3,v_2,s_1,s_3\}$	
1217	$(-3,25,22,2)$	$\{c_0,c_2,u_2,u_3,c_2,c_1,v_3,v_2,s_1,s_3\}$	
1218	$(0,21,16,2)$	$\{c_2,u_4,u_5,c_2,v_4,v_5,s_1,s_3\}$	
1219	$(-3,29,23,2)$	$\{c_0,c_2,u_4,u_5,c_2,c_1,v_4,v_5,s_1,s_3\}$	
1220	$(0,31,30,2)$	$\{c_2,u_2,u_4,u_5,c_2,v_4,v_5,v_2,s_1,s_3\}$	
1221	$(-3,35,34,2)$	$\{c_0,c_2,u_2,u_4,u_5,c_2,c_1,v_4,v_5,v_2,s_1,s_3\}$	
1222	$(-3,38,39,2)$	$\{c_0,c_2,u_0,u_1,u_4,u_5,c_2,c_1,v_4,v_5,v_0,v_1,s_1,s_3\}$	
1223	$(-3,41,41,2)$	$U\cup V\cup\{s_1,s_3\}$	
1224	$(27,3,8,2)$	$\{c_0,c_1,j_2,j_3,c_0,u_2,c_1,v_2,s_1,s_3\}$	
1225	$(27,12,15,2)$	$\{c_0,c_1,j_2,j_3,c_0,c_2,u_2,c_2,c_1,v_2,s_1,s_3\}$	
1226	$(27,25,22,2)$	$\{c_0,c_1,j_2,j_3,c_0,c_2,u_2,u_3,c_2,c_1,v_3,v_2,s_1,s_3\}$	
1227	$(37,38,39,2)$	$\{c_0,c_1,j_0,j_1,j_4,j_5,c_0,c_2,u_0,u_1,u_4,u_5,c_2,c_1,v_4,$ $v_5,v_0,v_1,s_1,s_3\}$	
1228	$(41,12,15,2)$	$J\cup\{c_0,c_2,u_2,c_2,c_1,v_2,s_1,s_3\}$	
1229	$(41,3,8,2)$	$J\cup\{c_0,u_2,c_1,v_2,s_1,s_3\}$	
1230	$(41,25,22,2)$	$J\cup\{c_0,c_2,u_2,u_3,c_2,c_1,v_2,v_3,s_1,s_3\}$	
1231	$(41,41,41,2)$	$J\cup U\cup V\cup\{s_1,s_3\}$	
1232	$(0,0,0,3)$	$\{s_1,s_2\}$	6.3
1233	$(-1,-1,0,3)$	$\{c_0,s_1,s_2\}$	
1234	$(-2,-2,-3,3)$	$\{c_2,c_1,s_1,s_2\}$	
1235	$(-3,-3,-3,3)$	$\{c_0,c_1,c_2,s_1,s_2\}$	
1236	$(-2,-2,27,3)$	$\{c_2,c_1,v_5,v_0,s_1,s_2\}$	
1237	$(0,0,32,3)$	$\{v_3,v_4,v_1,v_2,s_1,s_2\}$	
1238	$(-2,-2,37,3)$	$\{c_2,c_1,v_3,v_4,v_1,v_2,s_1,s_2\}$	

i	$\tau(H_i)$	H_i	Case
1239	$(-2,-2,41,3)$	$V\cup\{s_1,s_2\}$	
1240	$(-3,-1,27,3)$	$\{c_0,c_2,c_1,v_5,v_0,s_1,s_2\}$	
1241	$(-3,-1,37,3)$	$\{c_0,c_2,c_1,v_3,v_4,v_1,v_2,s_1,s_2\}$	
1242	$(-3,-3,41,3)$	$\{c_0,c_2\}\cup V\cup\{s_1,s_2\}$	
1243	$(2,2,0,3)$	$\{j_5,u_5,s_1,s_2\}$	
1244	$(8,8,-3,3)$	$\{c_1,j_5,c_2,u_5,s_1,s_2\}$	
1245	$(5,5,0,3)$	$\{c_0,j_5,c_0,u_5,s_1,s_2\}$	
1246	$(15,15,-3,3)$	$\{c_0,c_1,j_5,c_0,c_2,u_5,s_1,s_2\}$	
1247	$(9,9,0,3)$	$\{j_0,j_5,u_0,u_5,s_1,s_2\}$	
1248	$(22,22,-3,3)$	$\{c_0,c_1,j_0,j_5,c_0,c_2,u_0,u_5,s_1,s_2\}$	
1249	$(16,16,0,3)$	$\{c_0,j_1,j_2,c_0,u_2,u_1,s_1,s_2\}$	
1250	$(23,23,-3,3)$	$\{c_0,c_1,j_1,j_2,c_0,c_2,u_2,u_1,s_1,s_2\}$	
1251	$(30,30,0,3)$	$\{c_0,j_1,j_2,j_5,c_0,u_2,u_1,u_5,s_1,s_2\}$	
1252	$(34,34,-3,3)$	$\{c_0,c_1,j_1,j_2,j_5,c_0,c_2,u_2,u_1,u_5,s_1,s_2\}$	
1253	$(39,39,-3,3)$	$\{c_0,c_1,j_1,j_2,j_3,j_4,c_0,c_2,u_2,u_1,u_4,u_3,s_1,s_2\}$	
1254	$(41,41,-3,3)$	$J\cup U\cup\{s_1,s_2\}$	
1255	$(8,8,27,3)$	$\{c_1,j_5,c_2,u_5,c_2,c_1,v_5,v_0,s_1,s_2\}$	
1256	$(15,15,27,3)$	$\{c_0,c_1,j_5,c_0,c_2,u_5,c_2,c_1,v_5,v_0,s_1,s_2\}$	
1257	$(22,22,27,3)$	$\{c_0,c_1,j_0,j_5,c_0,c_2,u_0,u_5,c_2,c_1,v_5,v_0,s_1,s_2\}$	
1258	$(39,39,37,3)$	$\{c_0,c_1,j_1,j_2,j_3,j_4,c_0,c_2,u_2,u_1,u_4,u_3,c_2,c_1,v_3,$ $v_4,v_1,v_2,s_1,s_2\}$	
1259	$(15,15,41,3)$	$\{c_0,c_1,j_5,c_0,c_2,u_5\}\cup V\cup\{s_1,s_2\}$	
1260	$(8,8,41,3)$	$\{c_1,j_5,c_2,u_5\}\cup V\cup\{s_1,s_2\}$	
1261	$(22,22,41,3)$	$\{c_0,c_1,j_0,j_5,c_0,c_2,u_0,u_5\}\cup V\cup\{s_1,s_2\}$	
1262	$(41,41,41,3)$	$J\cup U\cup V\cup\{s_1,s_2\}$	
1263	$(0,0,0,4)$	$\{s_1,s_6\}$	6.4
1264	$(-2,0,-2,4)$	$\{c_1,s_1,s_6\}$	
1265	$(-1,-3,-1,4)$	$\{c_0,c_2,s_1,s_6\}$	
1266	$(-3,-3,-3,4)$	$\{c_0,c_1,c_2,s_1,s_6\}$	
1267	$(-1,27,-1,4)$	$\{c_0,c_2,u_1,u_4,s_1,s_6\}$	
1268	$(0,32,0,4)$	$\{u_0,u_2,u_3,u_5,s_1,s_6\}$	
1269	$(-1,37,-1,4)$	$\{c_0,c_2,u_0,u_2,u_3,u_5,s_1,s_6\}$	
1270	$(-1,41,-1,4)$	$U\cup\{s_1,s_6\}$	
1271	$(-2,27,-3,4)$	$\{c_1,c_0,c_2,u_1,u_4,s_1,s_6\}$	
1272	$(-2,37,-3,4)$	$\{c_1,c_0,c_2,u_0,u_2,u_3,u_5,s_1,s_6\}$	
1273	$(-3,41,-3,4)$	$\{c_0,c_1\}\cup U\cup\{s_1,s_6\}$	
1274	$(1,0,1,4)$	$\{j_1,v_4,s_1,s_6\}$	
1275	$(6,0,6,4)$	$\{c_1,j_1,c_1,v_4,s_1,s_6\}$	
1276	$(3,-3,3,4)$	$\{c_0,j_1,c_2,v_4,s_1,s_6\}$	
1277	$(12,-3,12,4)$	$\{c_0,c_1,j_1,c_2,c_1,v_4,s_1,s_6\}$	
1278	$(10,0,10,4)$	$\{j_1,j_4,v_4,v_1,s_1,s_6\}$	
1279	$(25,-3,25,4)$	$\{c_0,c_1,j_1,j_4,c_2,c_1,v_4,v_1,s_1,s_6\}$	
1280	$(21,0,21,4)$	$\{c_1,j_3,j_5,c_1,v_0,v_2,s_1,s_6\}$	
1281	$(29,-3,29,4)$	$\{c_0,c_1,j_3,j_5,c_2,c_1,v_0,v_2,s_1,s_6\}$	
1282	$(31,0,31,4)$	$\{c_1,j_1,j_3,j_5,c_1,v_4,v_0,v_2,s_1,s_6\}$	
1283	$(35,-3,35,4)$	$\{c_0,c_1,j_1,j_3,j_5,c_2,c_1,v_4,v_0,v_2,s_1,s_6\}$	
1284	$(38,-3,38,4)$	$\{c_0,c_1,j_0,j_2,j_3,j_5,c_2,c_1,v_3,v_5,v_0,v_2,s_1,s_6\}$	
1285	$(41,-3,41,4)$	$J\cup V\cup\{s_1,s_6\}$	
1286	$(3,27,3,4)$	$\{c_0,j_1,c_0,c_2,u_1,u_4,c_2,v_4,s_1,s_6\}$	
1287	$(12,27,12,4)$	$\{c_0,c_1,j_1,c_0,c_2,u_1,u_4,c_2,c_1,v_4,s_1,s_6\}$	
1288	$(25,27,25,4)$	$\{c_0,c_1,j_1,j_4,c_0,c_2,u_1,u_4,c_2,c_1,v_4,v_1,s_1,s_6\}$	
1289	$(38,37,38,4)$	$\{c_0,c_1,j_0,j_2,j_3,j_5,c_o,c_2,u_0,u_2,u_3,u_5,c_2,c_1,v_3,$ $v_5,v_0,v_2,s_1,s_6\}$	
1290	$(12,41,12,4)$	$\{c_0,c_1,j_1\}\cup U\cup\{c_2,c_1,v_4,s_1,s_6\}$	
1291	$(3,41,3,4)$	$\{c_0,j_1\}\cup U\cup\{c_2,v_4,s_1,s_6\}$	
1292	$(25,41,25,4)$	$\{c_0,c_1,j_1,j_4\}\cup U\cup\{c_2,c_1,v_4,v_1,s_1,s_6\}$	
1293	$(41,41,41,4)$	$J\cup U\cup V\cup\{s_1,s_6\}$	
1294	$(0,0,0,5)$	$\{s_1,s_4,s_5\}$	6.5
1295	$(-3,-3,-3,5)$	$\{c_0,c_1,c_2,s_1,s_4,s_5\}$	
1296	$(41,41,41,5)$	$J\cup U\cup V\cup\{s_1,s_4,s_5\}$	
1297	$(0,0,0,6)$	$\{s_1,s_2,s_3,s_4,s_5,s_6\}$	6.6
1298	$(-3,-3,-3,6)$	$\{c_0,c_1,c_2,s_1,s_2,s_3,s_4,s_5,s_6\}$	
1299	$(41,41,41,6)$	$J\cup U\cup V\cup\{s_1,s_2,s_3,s_4,s_5,s_6\}$	

15.3 Classes of Quasilinear Functions of P_3 that are not Subsets of $[P_3^1]$

The elements of the set

$$\mathfrak{L} :=$$
$$[P_3^1] \cup \bigcup_{n\geq 1}\{f^n \in P_3 \mid \exists a \in E_2\ \exists f_0 \in P_k^1\ \exists f_1, ..., f_n \in P_{3,2}^1 : \\ f(\mathbf{x}) = f_0(a+f_1(x_1)+f_2(x_2)+...+f_n(x_n)\ (mod\ 2))\},$$

are called **quasilinear** functions in this section. The set $\mathfrak{L}$ agrees with the set L_3 from Chapter 4. In Lemma 4.1, it was proven that

$$\mathfrak{L} = Pol_3\lambda_3 \quad \text{with}$$
$$\lambda_3 := \{(a,a,b,b), (a,b,a,b), (a,b,b,a) \mid a,b \in E_3\}$$

holds and that $\mathfrak{L}$ is a submaximal class of P_3.
$\mathfrak{L}$ was examined already by I. A. Mal'tsev in [Mal 72] and [Mal 73b]. In [Mal 72], one also finds proof that $\mathfrak{L}$ has exactly countable-infinite-many subclasses (see Theorem 8.1.5). As in [Lau 85] we will determine all subclasses of $\mathfrak{L}$ with the aid of the 1299 subsemigroups of $(P_3^1; \star)$. Further, we determine the order of theses subclasses.
After introducing some notations in Section 15.3.1, we determine in Section 15.3.2 the subclasses of $\mathfrak{L}$ whose functions take only values from the set $\{0,1\}$. With the aid of an isomorphic mapping, we obtain all subclasses of $\mathfrak{L}$ whose functions take only values from the set $\{0,2\}$. In Section 15.3.3, all subclasses of $\mathfrak{L}$, whose functions take only values from $\{0,1\}$ or $\{0,2\}$, are determined then as follows: We begin with the proof of certain (necessary and sufficient) criteria with which one can easily determine whether the union of certain classes, which are given in Section 15.3.3, is again a class. Then we determine the remaining classes, which do not fulfill the conditions of the criteria.
With the aid of the subclasses of $\{f \in \mathfrak{L} \mid Im(f) \subseteq \{0,1\} \vee Im(f) \subseteq \{0,2\}\}$, the remaining classes are easily described, as shown in Section 15.3.4.

Notably, another kind of derivation of the subclasses of $\{f \in \mathfrak{L} \mid Im(f) \subseteq \{0,1\} \vee Im(f) \subseteq \{0,2\}\}$ in [Dem-M 89] can be found. In [Dem-M 89], one can also find rough drafts of lattices from such classes.

15.3.1 Some Notations

We arrange to write only $+$ instead of $+$ $(mod\ 2)$. Denote L the set of all linear functions of P_2, and let

$$\mathfrak{L}_{a,b} := \{f \in \mathfrak{L} \mid Im(f) \subseteq \{a,b\}\},\ \{a,b\} \subseteq E_3,\ a \neq b.$$

Let $pr_{a,b}$ be a mapping from $\mathfrak{L}_{a,b}$ onto L $(\subseteq P_2)$, which is defined by

$$pr_{a,b}f^n = F^n \;:\Longleftrightarrow\; \forall \mathbf{x} \in \{a,b\}^n : \; g(f(x_1,...,x_n)) = F(g(x_1),...,g(x_n)),$$

where

$$g\begin{pmatrix} a \\ b \end{pmatrix} = \begin{pmatrix} 0 \\ 1 \end{pmatrix}, \; f^n \in \mathfrak{L}_{a,b}, \text{ and } F \in L.$$

With the aid of an arbitrary subclass A of L (see Theorem 3.2.2) one can describe a subclass of $\mathfrak{L}_{a,b}$ by

$$pr_{a,b}^{-1}A := \{f \in \mathfrak{L}_{a,b} \mid pr_{a,b}f \in A\}.$$

Further, let $Z_{a,b}$ be the notation for the set

$$Pol_3 \begin{pmatrix} b\; c\; b \\ b\; c\; a \end{pmatrix}, \; \{a,b,c\} = E_3.$$

For the description of certain isomorphic classes, we use the automorphisms

$$\varphi_i : \mathfrak{L} \longrightarrow \mathfrak{L}, \; f^n \mapsto s_i^{-1}(f(s_i(x_1),...,s_i(x_n))), \; (i = 1,2,...,6)$$

from Section 15.2.
Obviously, one can describes every subclass A of $\mathfrak{L}$ in the form

$$A = (A \cap \mathfrak{L}_{0,1}) \cup (A \cap \mathfrak{L}_{0,2}) \cup (A \cap \mathfrak{L}_{1,2}) \cup (A \cap [P_3^1[3]]).$$

We determine, therefore, the subclasses of $\mathfrak{L}_{0,1}$ in Section 15.3.2 and then in Section 15.3.3 the subclasses of $\mathfrak{L}_{0,1} \cup \mathfrak{L}_{0,2}$ that are not contained in $\mathfrak{L}_{0,1}$ or $\mathfrak{L}_{0,2}$. With the aid of these results, it is easy to determine the remaining subclasses of $\mathfrak{L}$ in Section 15.3.4.

15.3.2 Subclasses of $\mathfrak{L}_{0,1}$

The following lemma results from the definition of the class $\mathfrak{L}_{0,1}$ and the facts

$$j_0 = 1 + j_1 + j_2, \; j_3 = 1 + j_2, \; j_4 = 1 + j_1, \; j_5 = j_1 + j_2$$

immediately.

Lemma 15.3.2.1 *It holds:*

$$\mathfrak{L}_{0,1} = \bigcup_{n \geq 1}\{f^n \in P_3 \mid \exists a_0,...,a_n,b_1,...,b_n \in E_2 : \\ f(\mathbf{x}) = a_0 + \textstyle\sum_{i=1}^n (a_i j_1(x_i) + b_i j_2(x_i))\}.$$

■

With the aid of this lemma, one can see that the identities

$$pr_{0,1}^{-1}A = \bigcup_{n \geq 1}\{f^n \in \mathfrak{L}_{0,1} \mid \exists a_0,...,a_n,b_1,...,b_n \in E_2 : \\ f(\mathbf{x}) = a_0 + \textstyle\sum_{i=1}^n (a_i j_1(x_i) + b_i j_2(x_i)) \;\wedge \\ (pr_{0,1}f)(\mathbf{y}) = a_0 + \textstyle\sum_{i=1}^n a_i y_i \;\in A\}$$

and

$$Z_{2,c} \cap pr_{0,1}^{-1} A = \bigcup_{n\geq 1} \{f^n \in \mathfrak{L}_{0,1} \mid \exists a_0, ..., a_n, c \in E_2 : \\ f(\mathbf{x}) = a_0 + \sum_{i=1}^n (a_i(j_1(x_i) + c j_2(x_i))) \wedge \\ (pr_{0,1} f)(\mathbf{y}) = a_0 + \sum_{i=1}^n a_i y_i \in A\}$$

hold, where $c \in E_2$ and A denotes a closed subset of L.
The sets

$$B_{c,r} := \bigcup_{n\geq 1} \{f^n \in \mathfrak{L}_{0,1} \mid \exists a_1, ..., a_n \in E_2 : \\ f(\mathbf{x}) = c + \sum_{i=1}^n a_i j_2(x_i) \wedge \\ (\text{at most } r \text{ of the } a_i \text{ are } 1)\ \}$$

and

$$B_r := B_{0,r} \cup B_{1,r},$$

where $r \in \{\infty, 1, 2, ...\}$, are also subclasses of $\mathfrak{L}_{0,1}$. Obviously, $B_{c,\infty}$ and B_∞ do not have any basis. Therefore, they are also not finitely generated.

Lemma 15.3.2.2 *Let A be a subclass of $\mathfrak{L}_{0,1}$, which contains a function $i(x_1) + j_2(x_2)$ with $i \in \{j_1, j_5\}$. Then, it holds that $A = pr_{0,1}^{-1}(pr_{0,1}A)$ and $A = [A' \cup \{j_1(x_1) + j_2(x_2)\}]$ for every A' with $A' \subseteq A$ and $[pr_{0,1}A'] = pr_{0,1}A$.*

Proof. Obviously, $A \subseteq pr_{0,1}^{-1}(pr_{0,1}A)$. Let $f^n \in pr_{0,1}^{-1}(pr_{0,1}A)$ and $f(\mathbf{x}) = a_0 + \sum_{i=1}^n (a_i j_1(x_i) + b_i j_2(x_i))$. We want to show that f belongs to A. Because of $pr_{0,1}f \in pr_{0,1}A$ there is a function $f' \in A$ with $f'(\mathbf{x}) = a_0 + \sum_{i=1}^n (a_i j_1(x_i) + b_i' j_2(x_i))$. A superposition over $i(x_1) + j_2(x_2)$ is obviously the function $q(x, x_1, ..., x_n) = i(x) + \sum_{i=1}^n (b_i + b_i') j_2(x_i)$. Consequently, we have $f(\mathbf{x}) = q(f'(\mathbf{x}), x_1, ..., x_n) \in A$ and therefore $A = pr_{0,1}^{-1}(pr_{0,1}A)$.
Since $f' \in [A']$ holds for every A' with $[pr_{0,1}A'] = pr_{0,1}A$, the equation $A = [A' \cup \{j_1(x_1) + j_2(x_2)\}]$ also results from the one shown above. ■

Lemma 15.3.2.3 *Let A be a subclass of $\mathfrak{L}_{0,1}$ and $pr_{0,1}A \not\subseteq [L^1]$. Then, A is either the set $pr_{0,1}^{-1}(pr_{0,1}A)$ or $Z_{2,0} \cap pr_{0,1}^{-1}(pr_{0,1}A)$ or the set $Z_{2,1} \cap pr_{0,1}^{-1}(pr_{0,1}A)$.*

Proof. For $f(\mathbf{x}) = a_0 + \sum_{i=1}^n (a_i j_1(x_i) + b_i j_2(x_i))$ denote $Ch(f)$ the set $\{(a_i, b_i) \mid i = 1, 2, ..., n\}$. We distinguish the following three cases for A:
Case 1: There exists an $a \in E_2$ so that $Ch(f) \subseteq \{(1, a), (0, 0)\}$ for every $f \in A$ holds.

In this case, $f \in A$ preserves the relation $\begin{pmatrix} 0 & 1 & a \\ 0 & 1 & 2 \end{pmatrix}$, and we have $A = pr_{0,1}^{-1}(pr_{0,1}A) \cap Z_{2,a}$.

Case 2: A contains a function f with $(0,1) \in Ch(f)$.
W.l.o.g. let $(a_1, b_1) = (0, 1)$. Then, $f'(x_1, x_2) := f(x_1, x_2, ..., x_2) = a_0 + j_2(x_1) + r(x_2)$, where $r \in \{c_0, j_1, j_2, j_5\}$. If $r \in \{j_1, j_5\}$ then we have $f'(x_1, f'(x_2, x_2)) = j_2(x_1) + i(x_2) \in A$, $i \in \{j_1, j_5\}$; thus Lemma 15.3.2.3 follows from Lemma 15.3.2.2. If $r \in \{c_0, j_2\}$ then we obtain $f''(x) :=$

$f'(x, f'(x,x)) \in \{j_2, j_3\}$. Since $pr_{0,1}A \not\subseteq [L^1]$, an inverse image h of the function H with $H(y_1, y_2, y_3) = y_1 + y_2 + y_3$ belongs to A because of Lemma 3.2.2.1. Consequently, $h(x_1, f''(x_2), f''(f''(x_2))) = i(x_1) + j_2(x_2)$, $i \in \{j_1, j_5\}$, is a function of A. Therefore, Lemma 15.3.2.3 also follows from Lemma 15.3.2.2 for $r \in \{c_0, j_2\}$.

Case 3: A contains a function f with $\{(1,0),(1,1)\} \subseteq Ch(f)$.
W.l.o.g. let $(a_1, b_1) = (1,0)$ and $(a_2, b_2) = (1,1)$. Then the function $f(x_1, x_1, x_2, ..., x_2)$ fulfills the condition of Case 2. ■

Theorem 15.3.2.4 *Let A be a subclass of $\mathfrak{L}_{0,1}$ with $A \not\subseteq [\mathfrak{L}^1_{0,1}] \cup B_\infty$. Then*

$$A \in \{pr_{0,1}^{-1}(pr_{0,1}A), Z_{2,0} \cap pr_{0,1}^{-1}(pr_{0,1}A), Z_{2,1} \cap pr_{0,1}^{-1}(pr_{0,1}A)\}.$$

Proof. If $A \not\subseteq [\mathfrak{L}^1_{0,1}] \cup B_\infty$, there is a function $f \in A$ with $pr_{0,1}f \notin [P_2^1]$. Let w.l.o.g.

$$f(\mathbf{x}) = a_0 + \sum_{i=1}^{n}(a_i j_1(x_i) + b_i j_2(x_i)) \in A,$$

where $a_1 = 1$ and $(a_2, b_2) \neq (0,0)$. The following two cases are possible:

Case 1: $(a_2, b_2) = (0,1)$.
One can obtain the function $f'(x_1, x_2, x_3) = a_0 + j_1(x_1) + b_1 j_2(x_1) + j_2(x_2) + p(x_3)$ with $p \in \{c_0, j_1, j_2, j_5\}$ by identifying the variables $x_3, ..., x_n$ from f. Then

$$f'(f'(x_1, x_2, x_2), x_1, x_2) = i(x_1) + j_2(x_2),\ i \in \{j_1, j_5\}.$$

Thus by Lemma 15.3.2.2, $A = pr_{0,1}^{-1}(pr_{0,1}A)$.

Case 1: $a_2 = 1$.
In this case, we have $pr_{0,1}A \not\subseteq [L^1]$ and our theorem follows from Lemma 15.3.2.3. ■

Theorem 15.3.2.5 *The subclasses ($\neq \emptyset$) of $[L^1_{0,1}] \cup B_\infty$ are exactly* $[c_0]$, $[c_1]$, $[c_0, c_1]$, $[J_i]$, $[J_a] \cup B_r$, $[J_b] \cup B_{0,r} \cup B_{1,s}$, $[J_c] \cup B_{0,r}$, $[J_d] \cup B_{1,r}$, *where* $\{r, s\} \subset \{\infty, 2, 3, ...\}$, $1 \leq i \leq 41$, $a \in \{27, 33, 36, 38, 39, 40, 41\}$, $b \in \{27, 33, 36, 40\}$, $c \in \{4, 13, 16, 18, 23, 27, 28, 30, 33, 34, 36, 40\}$ *and* $d \in \{7, 14, 19, 21, 24, 27, 29, 31, 33, 35, 36, 40\}$.

Proof. One obtains the statements of the theorem easily by means of the properties of the functions of $[L^1_{0,1}] \cup B_\infty$ and the results from Section 15.2 about the subsemigroups of $(P_3^1; \star)$. ■

Lemma 15.3.2.6 *Let A be a subclass of $\mathfrak{L}_{0,1}$ and $pr_{0,1}A \not\subseteq [L^1]$. Then there is a function $h^3 \in A$ with $h(x_1, x_2, x_3) := i(x_1) + i(x_2) + i(x_3)$, $i \in \{j_1, j_5\}$, and it holds $A = [A^1 \cup \{h\}]$.*

Proof. By Lemma 3.2.2.1 we have $pr_{0,1}A = [(pr_{0,1}A)^1 \cup \{pr_{0,1}h\}]$. If $A = Z_{2,a} \cap pr_{0,1}^{-1}(pr_{0,1}A)$, $a \in E_2$, then Lemma 15.3.2.6 holds. If $A = pr_{0,1}^{-1}(pr_{0,1}A)$ then $j_1(x_1) + j_2(x_2) \in A$. By Lemma 15.3.2.2 and the above remark, we have $A = [A^1 \cup \{h, j_1(x_1) + j_2(x_2)\}]$, where $h \in A$ is an inverse image of the function H^3 with $H(y_1, y_2, y_3) = y_1 + y_2 + y_3$. Because of $\{j_1, j_5\} \subseteq A$, we can assume $h(x_1, x_2, x_3) = j_1(x_1) + j_1(x_2) + j_1(x_3)$, and it holds $h(x_1, x_2, j_5(x_2)) = j_1(x_1) + j_1(x_2) \in [A^1 \cup \{h\}]$; i.e., we have $A = [A^1 \cup \{h\}]$ in the case $A = pr_{0,1}^{-1}(pr_{0,1}A)$. Since there are no other possibilities because of Theorem 15.3.2.4, our Lemma 15.3.2.6 is valid. ■

Theorem 15.3.2.7 *Let A be a subclass of $\mathfrak{L}_{0,1}$. Then*

$$ord\ A = \begin{cases} 2, \text{ if } A \not\subseteq [\mathfrak{L}_{0,1}^1] \cup B_\infty \wedge (C \cap A^1 \neq \emptyset \vee pr_{0,1}A \subseteq [L^1]), \\ 3, \text{ if } pr_{0,1}A \not\subseteq [L^1] \wedge C \cap A^1 = \emptyset, \\ t, \text{ if } A \subseteq [\mathfrak{L}_{0,1}^1] \cup B_\infty \wedge A \cap (B_t \backslash [A^1]) \neq \emptyset \wedge \\ \qquad (\forall\, r \geq t:\ A \cap (B_r \backslash [A^1]) = \emptyset). \end{cases}$$

Proof. By Theorem 15.3.2.4, the following three cases are possible for A:
Case 1: $A = Z_{2,a} \cap pr_{0,1}^{-1}(pr_{0,1}A)$, $a \in E_2$.
In this case, $ord\ A = ord\ pr_{0,1}A$. Then, the statements of the theorem follow from Theorem 3.1.1.
Case 2: $A = pr_{0,1}^{-1}(pr_{0,1}A)$ and $A \not\subseteq [\mathfrak{L}_{0,1}^1] \cup B_\infty$.
By Lemma 15.3.2.6 and 15.3.2.2, it holds in this case:

$$ord\ pr_{0,1}A \leq ord\ A \begin{cases} \leq 3 \text{ if } pr_{0,1}A \not\subseteq [L^1], \\ = 2 \text{ if } pr_{0,1}A \subseteq [L^1]. \end{cases}$$

As is generally known, only the classes of L, which do not contain any constant functions and which are not subsets of $[L^1]$, have the order 3. If a constant function belongs to $pr_{0,1}A$ then the function h from Lemma 15.3.2.6 is obviously a superposition over binary functions of A. Thus the statements of the theorem hold in Case 2.
Case 3: $A \subseteq [\mathfrak{L}_{0,1}^1 \cup B_\infty$.
In this case, the order of A follows from Theorem 15.3.2.5 and from $[\mathfrak{L}_{0,1}^1] \cup B_r \subset [\mathfrak{L}_{0,1}^1] \cup B_{r+1}$, $2 \leq r \leq \infty$. ■

One can obtain the following lemma as a direct consequence from Theorem 15.3.2.7:

Theorem 15.3.2.8 *The only finitely generated subclasses of $\mathfrak{L}_{0,1}$ are the classes*
$[J_a] \cup B_\infty$, $[J_b] \cup B_{0,\infty}$, $[J_c] \cup B_{1,\infty}$, $[J_d] \cup B_{0,\infty} \cup B_{1,s}$ and $[J_d] \cup B_{0,s} \cup B_{1,\infty}$, where $a \in \{27, 33, 36, 38, 39, 40, 41\}$, $b \in \{4, 13, 16, 18, 23, 27, 28, 30, 33, 34, 36, 40\}$, $c \in \{7, 14, 19, 21, 24, 27, 29, 31, 33, 35, 36, 40\}$, $d \in \{27, 33, 36, 40\}$ and $2 \leq r, s \leq \infty$. ■

15.3.3 The Subclasses of $\mathfrak{L}_{0,1} \cup \mathfrak{L}_{0,2}$ That Are Not Subclasses of $\mathfrak{L}_{0,1}$ or $\mathfrak{L}_{0,2}$

The aim of this section is first of all the derivation of a necessary and sufficient criterion with which one can find out whether a set $A_1 \cup A_2$ ($A_1 \subseteq \mathfrak{L}_{0,1}$, $A_2 \subseteq \mathfrak{L}_{0,2}$) is closed.

Put

$$A \star A' := \{f \star g \mid f \in A \ \wedge \ g \in A'\}.$$

Obviously, it holds:

Lemma 15.3.3.1 *Let A_1 and A_2 be subclasses of $\mathfrak{L}$ with $A_1 \subseteq \mathfrak{L}_{0,1}$ and $A_2 \subseteq \mathfrak{L}_{0,2}$. Then, $A_1 \cup A_2$ is closed if and only if $A_1 \star A_2 \subseteq A_1$ and $A_2 \star A_1 \subseteq A_2$.* ∎

Lemma 15.3.3.2 *Let A_1 and A_2 be subclasses of $\mathfrak{L}$, $A_1 \subseteq \mathfrak{L}_{0,1}$, $A_2 \subseteq \mathfrak{L}_{0,2}$, $(A_1 \cup A_2)^1$ a subsemigroup of $(P_3^1; \star)$ and $\{i, j\} = \{1, 2\}$. Then*
(a) $A_1^1 \subseteq \{c_0, c_1, j_1, j_4\} \Longrightarrow A_1 \star A_2 \subseteq A_1$,
(b) $A_2^1 \subseteq \{c_0, c_2, u_2, u_3\} \Longrightarrow A_2 \star A_1 \subseteq A_2$,
(c) $pr_{0,i}A_i \not\subseteq [L^1] \Longrightarrow A_i \star A_j \subseteq A_i$,
(d) $(pr_{0,1}A_1 \subseteq [L^1] \ \wedge \ j_1(x_1) + j_2(x_2) \in A_1 \ \wedge \ A_2^1 \subseteq \{c_0, c_2, u_2, u_3\}) \Longrightarrow A_1 \star A_2 \subseteq A_1$.

Proof. (a): By Section 15.3.2 and by the assumptions over A_1, A_1 is a subclass of $Z_{2,0} \cap \mathfrak{L}_{0,1} = [c_1, j_1(x_1) + j_1(x_2)]$. Thus, because of $j_1 \begin{pmatrix} 0 \\ 2 \end{pmatrix} = \begin{pmatrix} 0 \\ 0 \end{pmatrix}$, we have $A_1 \star A_2 = A_1$.
(b): By assumption and Section 15.3.2, A_2 is a subset of $\varphi_2(Z_{2,0} \cap \mathfrak{L}_{0,1})$. Thus, because of $u_2 \begin{pmatrix} 0 \\ 1 \end{pmatrix} = \begin{pmatrix} 0 \\ 0 \end{pmatrix}$, we have $A_2 \star A_1 \subseteq A_2$.
(c): Let w.l.o.g. $pr_{0,1}A_1 \subseteq [L^1]$. Then, because of Lemma 15.3.2.6, $A_1 = [A_1^1 \cup \{h\}]$, where $h \in A_1$ is an arbitrary inverse image of the function $H(y_1, y_2, y_3) = y_1 + y_2 + y_3$. Suppose, $A_1 \star A_2 \not\subseteq A_1$ holds. Then, $A_1' := [A_1 \cup (A_1 \star A_2)]$ is a closed set with $A_1' \supset A_1$ and $A_1' = [(A_1')^1 \cup \{h\}]$. Consequently, we have $(A_1')^1 \supset A_1^1$, i.e., there is a function $g \in A_1$ and there are certain functions $p_1, ..., p_n \in (A_1 \cup A_2)^1$ with $g'(x) := g(p_1(x), p_2(x), ..., p_n(x)) \notin A_1^1$. However, we have also $g(x_1, ..., x_n) = g_0(g_1(x_1), g_2(x_2), ..., g_n(x_n))$ for certain functions $g_0 \in [\{h\}]$ and $g_1, ..., g_n \in A_1^1$. Therefore, we obtain $g' = g_0(g_1 \star p_1, g_2 \star p_2, ..., g_n \star p_n)$. Moreover, by assumption, $g_i \star p_i \in (A_1 \cup A_2)^1$, $i = 1, 2, ..., n$, and $A_1 = [A_1]$. Therefore, we have $g' \in A_1$, contrary to the assumption.
(d): Let $pr_{0,1}A_1 \subseteq [L^1]$, $j_1(x_1) + j_2(x_2) \in A_1$ and $A_2^1 \subseteq \{c_0, c_2, u_2, u_3\}$. Then, by Section 15.3.2, $A_1 = [A_1^1 \cup \{j_1(x_1) + j_2(x_2)\}]$ and $A_2 \subseteq \varphi_2(Z_{2,0} \cap \mathfrak{L}_{0,1}) = [c_2, u_2(x_1) \oplus u_2(x_2)]$ with

x	y	$x \oplus y$
0	0	0
0	2	2
2	0	2
2	2	0

.

Since $j_1\begin{pmatrix}0\\2\end{pmatrix} = \begin{pmatrix}0\\0\end{pmatrix}$, $j_2(a \oplus u_2(x_1) \oplus u_2(x_2) \oplus ... \oplus u_m(x_m)) = j_2(a) + j_2(x_1) + j_2(x_2) + ... + j_2(x_m)$ and $(A_1 \cup A_2)^1$ is closed in respect to $\star$, it follows $A_1 \star A_2 \subseteq A_1$. ■

Theorem 15.3.3.3 *Let A_1 and A_2 be subclasses of $\mathfrak{L}$, $A_1 \subseteq \mathfrak{L}_{0,1}$, $A_2 \subseteq \mathfrak{L}_{0,2}$, $A_1 \not\subseteq [\mathfrak{L}^1_{0,1}] \cup B_\infty \cup \varphi_2(Z_{2,0} \cap \mathfrak{L}_{0,1})$, $A_2 \not\subseteq \varphi_2([\mathfrak{L}^1_{0,1}] \cup B_\infty) \cup (Z_{2,0} \cap \mathfrak{L}_{0,1})$ and $A_1^1 \subseteq \{c_0, c_1, j_1, j_4\}$ or $A_2^1 \subseteq \{c_0, c_1, u_2, u_3\}$. Then, $A_1 \cup A_2$ is closed if and only if $(A_1 \cup A_2)^1$ is a subsemigroup of $(P_3^1; \star)$.*

Proof. If $A_1 \cup A_2$ is closed, then $(A_1 \cup A_2)^1$ is obviously a subsemigroup of $(P_3^1; \star)$.
Let $(A_1 \cup A_2)^1$ be a closed set in respect to $\star$. Then the following three cases are possible:
Case 1: $(A_1 \cup A_2)^1 \subseteq \{c_0, c_1, c_2, j_1, j_4, u_2, u_3\}$.
In this case, by Lemma 15.3.3.2, (a), (b), and Lemma 15.3.3.1, $A_1 \cup A_2$ is closed.
Case 2: $A_1^1 \not\subseteq \{c_0, c_1, j_1, j_4\}$ and $A_2^1 \subseteq \{c_0, c_2, u_2, u_3\}$.
Because of Lemma 15.3.3.2, (b) it holds that $A_2 \star A_1 \subseteq A_2$. Since A_1 is not a subset of $[\mathfrak{L}^1_{0,1}] \cup B_\infty$, we have $pr_{0,1}A_1 \not\subseteq [L^1]$ or $pr_{0,1}A_1 \subseteq [L^1]$ and $j_1(x_1) + j_2(x_2) \in A_1$. With the aid of Lemma 15.3.3.2, (c), (d): $A_1 \star A_2 \subseteq A_1$. Thus, because of Lemma 15.3.3.1, $A_1 \star A_2$ is closed.
Case 3: $(A_1^1 \subseteq \{c_0, c_1, j_1, j_4\}$ and $(A_2^1 \not\subseteq \{c_0, c_2, u_2, u_3\}$.
Because of $\varphi_2(\{c_0, c_1, j_1, j_4\}) = \{c_0, c_2, u_2, u_3\}$ the classes that fulfill the conditions of the third case are isomorphic to such classes that fulfill the conditions of the second case. Therefore, Theorem 15.3.3.3 is also valid in the third case. ■

Theorem 15.3.3.4 *Let A_1 and A_2 be subclasses of $\mathfrak{L}$, $A_1 \subseteq \mathfrak{L}_{0,1}$, $A_2 \subseteq \mathfrak{L}_{0,2}$, $A_1 \cup A_2 \not\subseteq [\mathfrak{L}^1]$, $A_1^1 \not\subseteq \{c_0, c_1, j_1, j_4\}$ and $A_2^1 \not\subseteq \{c_0, c_2, u_2, u_3\}$. Then, $A_1 \cup A_2$ is closed if and only if $pr_{0,1}A_1 \not\subseteq [L^1]$ and $pr_{0,2}A_2 \not\subseteq [L^1]$ hold and $(A_1 \cup A_2)^1$ is closed in respect to $\star$.*

Proof. If $pr_{0,i}A_i \not\subseteq [L^1]$ for $i = 1, 2$ and $(A_1 \cup A_2)^1$ is a subsemigroup of $(P_3^1; \star)$, then Theorem 15.3.3.2, (c) and Lemma 15.3.3.1 imply that $A_1 \cup A_2$ is closed.
Let $A_1 \cup A_2$ be a subclass of $\mathfrak{L}$. Since, by assumption, $A_1^1 \not\subseteq \{c_0, c_1, j_1, j_4\}$ and $A_2^1 \not\subseteq \{c_0, c_2, u_2, u_3\}$, there is a function $p^1 \in A_1^1$ with $p\begin{pmatrix}0\\2\end{pmatrix} \in \begin{pmatrix}0\,1\\1\,0\end{pmatrix}$ and a function $q \in A_2^1$ with $p\begin{pmatrix}0\\1\end{pmatrix} \in \begin{pmatrix}0\,2\\2\,0\end{pmatrix}$. Because $A_1 \cup A_2$ is closed, we have $p \star A_2 \subseteq A_1$ and $q \star A_1 \subseteq A_2$. Obviously, from this and the assumption $A_1 \cup A_2 \not\subseteq [\mathfrak{L}^1]$, it follows that $A_1 \not\subseteq [\mathfrak{L}^1]$ and $A_2 \not\subseteq [\mathfrak{L}^1]$. Suppose $pr_{0,1}A_1 \subseteq$

$[L^1]$. Then, by Section 15.3.2, a function $t(x_1, x_2) = g(x_1) + j_2(x_2)$ with $g \in A_1^1$ belongs to A_1. Therefore the function

$$t'(x_1, x_2) := \begin{cases} t(x_1, q(x_2)) & \text{if } g \in \{j_0, j_1, j_4, j_5\}, \\ t(q(x_1), q(x_2)) & \text{if } g \in \{j_2, j_3\}, \end{cases}$$

belongs to A_1 and it holds $pr_{0,1}t' \notin [L^1]$ obviously, contrary to the assumption. Consequently, $pr_{0,1}A_1 \not\subseteq [L^1]$. Similarly, one can show that $pr_{0,2}A_2 \not\subseteq [L^1]$. ■

To complete the description of all subclasses of $\mathfrak{L}_{0,1} \cup \mathfrak{L}_{0,2}$, which are not subclasses of $\mathfrak{L}_{0,1}$, $\mathfrak{L}_{0,2}$ or $[\mathfrak{L}^1]$, we are only still missing the subclasses of $[\mathfrak{L}^1_{0,1}] \cup B_\infty \cup \varphi_2(Z_{2,0} \cap \mathfrak{L}_{0,1})$ and of $\varphi_2(\mathfrak{L}^1_{0,1} \cup B_\infty) \cup (Z_{2,0} \cap \mathfrak{L}_{0,1})$ because of Theorems 15.3.3.3 and 15.3.3.4. Since the two sets to be examined are isomorphic, determining only the subclasses of one suffices, as in the following theorem.

Theorem 15.3.3.5 *The subclasses of* $[\mathfrak{L}^1_{0,1}] \cup B_\infty \cup \varphi_2(Z_{2,0} \cap \mathfrak{L}_{0,1})$*, which are not subclasses of* $[\mathfrak{L}^1]$, $\mathfrak{L}_{0,1}$ *or* $\mathfrak{L}_{0,2}$*, are the following classes:*
$A \cup A'$, *where* $A \in \{[c_0, c_1], [J_a] \cup B_\infty, [J_b] \mid a \in \{27, 33, 36, 38, 39, 40, 41\}, b \in \{3, 12, 25\}\}$ *and* $A' \in \{\varphi_2(Z_{2,0} \cap \mathfrak{L}_{0,1}), \varphi_2([c_0, j_1(x_1) + j_1(x_2)])\}$,
$[J_c] \cup B_{0,\infty} \cup \varphi([c_0, j_1(x_1) + j_1(x_2)])$, $c \in \{4, 13, 16, 18, 23, 28, 30, 34\}$,
$[J_d] \cup B_{1,\infty} \cup \varphi_2([c_0, j_1(x_1) + j_1(x_2)])$, $d \in \{14, 24\}$, *and*
$[A_1] \cup B_r$,
$[J_e] \cup B_{0,r} \cup [c_0, u_2]$,
$[J_f] \cup B_{1,r} \cup [c_0, u_2]$,
$[J_g] \cup B_{\alpha,r} \cup B_{\beta,r-1} \cup [A_2]$,
$[J_g] \cup B_{0,r} \cup B_{1,s} \cup [c_0, u_2]$, *where*
$e \in \{4, 13, 16, 18, 23, 27, 28, 30, 33, 34, 36, 40\}$, $f \in \{7, 14, 19, 21, 24, 27, 29, 31, 33, 35, 36, 40\}$, $g \in \{27, 33, 36, 40\}$, $2 \leq r, s \leq \infty$, $\{\alpha, \beta\} = E_2$, A_1 *is a subsemigroup of* $J \cup U_{25}$ *with* $(A_1 \setminus \{c_0\}) \cap U_{25} \neq \emptyset$*, which contains* $\{j_2, j_3\}$*, and* $A_2 \in \{[c_2], [c_0, c_2, u_2]\}$.

Proof. The theorem follows from Sections 15.2 and 15.3.2. ■

15.3.4 The Remaining Subclasses of $\mathfrak{L}$

Obviously, Lemma 15.3.3.1 implies the following

Lemma 15.3.4.1 *Let* A_1, A_2 *and* A_3 *be subclasses of* $\mathfrak{L}$ *with* $A_1 \subseteq \mathfrak{L}_{0,1}$, $A_2 \subseteq \mathfrak{L}_{0,2}$ *and* $A_3 \subseteq \mathfrak{L}_{1,2}$*. Then* $A_1 \cup A_2 \cup A_3$ *is closed if and only if* $A_1 \cup A_2$, $A_1 \cup A_3$ *and* $A_2 \cup A_3$ *are closed sets and* $(A_1 \cup A_2 \cup A_3)^1$ *is a semigroup.* ■

Since the sets $\mathfrak{L}_{0,1} \cup \mathfrak{L}_{0,2}$, $\mathfrak{L}_{0,1} \cup \mathfrak{L}_{1,2}$, $\mathfrak{L}_{0,2} \cup \mathfrak{L}_{1,2}$ are isomorphic, one has a complete description of the subclasses of $\mathfrak{L}_{0,1} \cup \mathfrak{L}_{0,2} \cup \mathfrak{L}_{1,2}$ through Lemma 15.3.4.1

and Sections 15.3.2 and 15.3.3. The following theorem now gives information about the missing subclasses of $\mathfrak{L}$.

Theorem 15.3.4.2 *The subclasses of $\mathfrak{L}$, which are not subsets of $[\mathfrak{L}]^1$ and which contain at least a permutation, are*
$A_1 \cup [s_1]$ *(A_1 is a subclass of $\mathfrak{L}_{0,1} \cup \mathfrak{L}_{0,2} \cup \mathfrak{L}_{1,2}$),*
$A_2 \cup [s_1, s_i]$ *($i \in \{2,3,6\}$, A_2 is a subclass of $\mathfrak{L}_{0,1} \cup \mathfrak{L}_{0,2} \cup \mathfrak{L}_{1,2}$ with $s_i \star A_2 \subseteq A_2$, $A_2^1 \cup \{s_1, s_i\}$ is a subsemigroup of $(P_3^1; \star)$,*
$\mathfrak{L}_{0,1} \cup \mathfrak{L}_{0,2} \cup \mathfrak{L}_{1,2} \cup [s_1, s_4, s_5]$ *and* $\mathfrak{L}$.

Proof. The theorem follows from Section 15.2 and from the properties of the functions of $\mathfrak{L}$. ∎

15.4 The Subclasses of $[O^1 \cup \{max\}]$

15.4.1 Some Descriptions of the Class M

Some functions of P_3, which are used in Sections 15.4.2–15.4.6, are defined in the following two tables (see also Table 15.1).

Table 15.11

x	j_2	j_5	u_2	u_5	v_2	v_5	s_1
0	0	0	0	0	1	1	0
1	0	1	0	2	1	2	1
2	1	1	2	2	2	2	2

Table 15.12

x y	$x \vee y := max(x,y)$
0 0	0
0 1	1
0 2	2
1 0	1
1 1	1
1 2	2
2 0	2
2 1	2
2 2	2

Our object of investigation is the set

$$M := [\{c_0, c_1, c_2, j_2, j_5, u_2, u_5, v_2, v_5, max\}],$$

which, by [Mac 79] (see Theorem 14.1.4), is a maximal class of the class

$$O := Pol_3 \begin{pmatrix} 0\,1\,2\,0\,0\,1 \\ 0\,1\,2\,1\,2\,2 \end{pmatrix}.$$

M can also be described in the form

$$[O^1 \cup \{max\}]$$

or

$$\bigcup_{n \geq 1} \{f^n \in P_3 \mid \exists f_1, ..., f_n \in O^1 : \ f(x_1, ..., x_n) = f_1(x_1) \vee ... \vee f_n(x_n)\}.$$

15.4.2 Some Lemmas and a Rough Partition of the Subclasses of M

One checks the following lemma easily:

Lemma 15.4.2.1
(a) For every function $f^n \in M$ there is exactly a representation of the following form:

$$f(x_1, ..., x_n) = f_1(x_1) \vee ... \vee f_n(x_n) \tag{15.1}$$

with

$$f_i(x) := f(\underbrace{0, ..., 0}_{i-1}, x, 0, ..., 0) \tag{15.2}$$

$(i = 1, ..., n)$.
(b) If the function f is given in the form (15.1) with (15.2), for every $g \in M^1$ it holds:

$$(g \star f)(x_1, ..., x_n) = (g \star f_1)(x_1) \vee ... \vee (g \star f_n)(x_n).$$

■

We agree to represent functions f^n of M in the form (15.1) in which the functions f_i are given by (15.2).

Further, denote $F(f)$ the set of all functions f_i of (15.2) that describe the function f. Let

$$num_f(f_i)$$

be the number of occurrence of function f_i in (15.1) for f, where f_i is defined by (15.2). If f arises from the context, instead of $F(f)$, we will write only F, and instead of $num_f(f_i)$, we will write only $num(f_i)$.

Lemma 15.4.2.2 is the basis for the following theorems about bases and generating systems for subclasses of M.

Lemma 15.4.2.2 *Let $f^n \in M$, $n \geq 2$,*

$$f(x_1, ..., x_n) := f_1(x_1) \vee ... \vee f_n(x_n), \tag{15.3}$$

$F := \{f_1, ..., f_n\}$, $\{c_0, c_1, c_2\} \cap F = \emptyset$ (i.e., all variables of f are essential) and denote $num(g)$ the number of functions $g \in F$ that occur in (15.3). Then
(a) $f \in [[\{f\}]^2]$ if and only if the function f fulfills at least one of the following conditions:

1) $s_1 \in F$ *and* $u_5 \notin F$;
2) $num(s_1) = 1$ *and* $F \in \{\{s_1, u_5\}, \{s_1, u_2, u_5\}\}$;
3) $F \subseteq \{u_2, u_5\}$;
4) $F \in \{\{j_5\}, \{j_2, u_5\}, \{j_5, u_2\}, \{j_5, u_5\}, \{j_2, j_5, u_2\}, \{j_2, j_5, u_5\}, \{j_2, u_2, u_5\}, \{j_5, u_2, u_5\}, \{j_2, j_5, u_2, u_5\}\}$;
5) $num(j_5) = 1$ *and* $F = \{j_2, j_5\}$;
6) $num(v_2) = 1$ *and* $F = \{v_2, v_5\}$;
7) $num(u_2) \geq 2$ *and* $F = \{j_2, u_2\}$;
8) $n = 2$ *and* $F \in \{\{j_2\}, \{v_5\}, \{j_2, u_2\}\}$.

(b) $f \in [[\{f\}]^3]$ *and* $f \notin [[\{f\}]^2]$ *if and only if the function* f *fulfills at least one of the following conditions:*
1) $num(s_1) \geq 2$ *and* $u_5 \in F$;
2) $num(s_1) = 1, u_5 \in F$ *and* $F \notin \{\{s_1, u_5\}, \{s_1, u_2, u_5\}\}$;
3) $num(j_5) \geq 2$ *and* $F = \{j_2, j_5\}$;
4) $num(u_2) = 1$ *and* $F = \{j_2, u_2\}$;
5) $num(v_2) \geq 2$ *and* $F = \{v_2, v_5\}$;
6) $n = 3$ *and* $F \in \{\{j_2\}, \{v_5\}\}$.

(c) $f \notin [[\{f\}]^{n-1}]$ *and* $n \geq 4$ *if and only if* $F \in \{\{j_2\}, \{v_5\}, \{j_2, u_2\}\}$ *and* $num(u_2) \leq 1$.

Proof. Since, by assumption, all variables of f are essential, we have $F \subseteq \{j_2, j_5, u_2, u_5, v_2, v_5\}$. If $\{v_2, v_5\} \cap F \neq \emptyset$ then $F \subseteq \{v_2, v_5\}$. Thus for f the following cases are possible:
Case 1: $s_1 \in F$.
Case 1.1: $u_5 \notin F$.
Let w.l.o.g. $f_1 = s_1$. The functions

$$h_t(x, y) := f(\underbrace{x, ..., x}_{t-1}, y, x, ..., x) = x \vee f_t(y)$$

are superpositions over f for every $t \in \{2, 3, ..., n\}$. Then one can obtain function f as a superposition over these functions as follows:

$$(...((h_n \star h_{n-1}) \star h_{n-2})... \star h_3) \star h_2.$$

Therefore, $f \in [[\{f\}]^2]$ in Case 1.
Case 1.2: $u_5 \in F$.
Let w.l.o.g. $f_1 = s_1$ and $f_2 = u_5$. Then the ternary functions $x \vee u_5(y) \vee f_i(z)$ ($i = 3, ..., n$) are superpositions over f, and these functions form a generating system for f. Therefore, $f \in [[\{f\}]^3]$. This is reducible to $f \in [[\{f\}]^2]$ iff

$$num(s_1) = 1 \text{ and } F \in \{\{s_1, u_5\}, \{s_1, u_2, u_5\}\}.$$

Case 2: $F \subseteq \{j_2, j_5, u_2, u_5\}$.
Case 2.1: $F = \{j_2\}$.
Because of $j_2 \star j_2 = c_0$, we have $j_2(x_1) \vee ... \vee j_2(x_r) \notin [\{j_2(x_1) \vee ... \vee j_2(x_{r-1})\}]$ for all $r \geq 2$.
Case 2.2: $F = \{j_5\}$.
In this case, we have $f \in [\{j_5(x) \vee j_5(x_2)\}]$, i.e., $f \in [[\{f\}]^2]$.
Case 2.3: $F = \{j_2, j_5\}$.
By $(...((g \star g) \star g)...) \star g$, where $g(x, y) := j_5(x) \vee j_2(y)$, one can obtain all functions of the form $j_5(x_1) \vee j_2(x_2) \vee ... \vee j_2(x_t)$ for $t \geq 2$. Thus $f \in [[\{f\}]^2]$, if $num(j_5) = 1$. If $num(j_5) \geq 2$ then $j_5(x) \vee j_5(y) \vee j_2(z)$ belongs to $[\{f\}]^3$ and we have $f \in [[\{f\}]^3]$ and $f \notin [[\{f\}]^2]$.
Case 2.4: $F \subseteq \{u_2, u_5\}$.

Since $u_p \star u_q = u_q$ for all $p, q \in \{2, 5\}$, it follows from Lemma 15.4.2.1, (b) that $f \in [[\{f\}]^2]$.

Case 2.5: $F = \{j_2, u_2\}$.

Because of $j_2 \star u_2 = j_2$, $u_2 \star j_2 = c_0$, $u_2 \star u_2 = u_2$ and $u_2(x) \vee j_2(x) = u_2$ we have in this case:

$$num(u_2) \geq 2 \implies \{j_2(x) \vee u_2(y), u_2(x) \vee u_2(y)\} \subseteq [\{f\}]^2 \implies f \in [[\{f\}]^2]$$

and

$$num(u_2) = 1 \implies f \notin [[\{f\}]^{n-1}].$$

Case 2.6: $F = \{j_2, u_5\}$.

Since $\Delta^{n-1} f = u_5$, the functions u_5 and $j_2(u_5(x)) \vee u_5(y) = j_5(x) \vee u_5(y)$ belong to $[\{f\}]^2$. Thus by $j_5 \star u_5 = j_5$ and $u_5 \star j_5 = u_5$, functions of the form $j_5(x_1) \vee ... \vee j_5(x_r) \vee u_5(x_{r+1}) \vee ... \vee u_5(x_m)$ are superpositions over $[\{f\}]^2$ for arbitrary $m > r \geq 1$. From this and by $j_5(j_2(x_1) \vee u_5(x_2)) \vee u_5(x_2) = j_2(x_1) \vee u_5(x_2)$) follows then that $f \in [[\{f\}]^2]$.

Case 2.7: $F = \{j_5, u_2\}$.

If $num(u_2) = 1$ or $num(j_5) \geq 2$, we have $n = 2$ or $h(x, y) := j_5(x) \vee j_5(y) \vee u_2(z) \in [\{f\}]^3$. Thus, $h(x, y, y) = j_5(x) \vee y \in [\{f\}]^2$. Then functions of the form $x_1 \vee j_5(x_2) \vee ... \vee j_5(x_t)$ $(t \geq 2)$ are superpositions over h. Replacing x_1 by $j_5(x_1) \vee u_2(x_2) \in [\{f\}]^2$ and then identifying variables shows that $f \in [[\{f\}]^2]$ in the case $num(u_2) = 1$.

If we have $num(u_2) \geq 2$ and $num(j_5) = 1$, then the function $j_5(x) \vee u_2(y) \vee u_2(z)$ belongs to $[\{f\}]^3$ and, therefore, the function $x \vee u_2(y)$ also belongs to $[\{f\}]^3$. From this, $f \in [[\{f\}]^2]$.

Case 2.8: $F = \{j_5, u_5\}$.

Because of $j_5(j_5(x_1) \vee u_5(x_2)) \vee u_5(j_5(x_3) \vee u_5(x_4)) = j_5(x_1) \vee j_5(x_2) \vee u_5(x_3) \vee u_5(x_4)$ we have $f \in [[\{f\}]^2]$.

Case 2.9: $F = \{j_2, j_5, u_2\}$.

In this case, $j_2(x) \vee y$ and $j_5(x) \vee u_2(y)$ are some superpositions over f. Thus $f \in [[\{f\}]^2]$ is an easy conclusion from our considerations of Cases 1 and 2.7.

Case 2.10: $F = \{j_2, j_5, u_5\}$.

Some superpositions over f are $u_5 = \Delta^{n-1} f$, $j_2(x) \vee u_5(y)$ and $j_5(x) \vee u_5(y)$. Hence, and by Case 2.8, we have $f \in [[\{f\}]^2]$.

Case 2.11: $F = \{j_2, u_2, u_5\}$.

Then, the functions $u_5 = \Delta^{n-1} f$, $u_2(x) \vee u_5(y)$, $j_2(x) \vee u_5(y)$ and $j_2(u_5(x)) \vee u_5(y) = j_5(x) \vee u_5(y)$ are superpositions over f. Using considerations from Cases 2.4, 2.6, and 2.8, we obtain $f \in [[\{f\}]^2]$.

Case 2.12: $F = \{j_5, u_2, u_5\}$.

Then the functions $x \vee u_5(y)$, $u_2(x) \vee u_5(y)$ and $j_5(x) \vee u_5(y)$ are superpositions over f. By Cases 2.4 and 2.8, we get $f \in [[\{f\}]^2]$.

Case 2.13: $F = \{j_2, j_5, u_2, u_5\}$.

Since $j_2(x) \vee u_5(y)$, $j_5(x) \vee u_5(y)$, $u_2(x) \vee u_5(y) \in [\{f\}]^2$, one can obtain $f \in [[\{f\}]^2]$ by proceeding to the above cases analogously.

Case 3: $F \subseteq \{v_2, v_5\}$.

Case 3.1: $F = \{v_2\}$.
Obviously, $f \in [\{v_2(x) \vee v_2(y)\}]$, i.e., $f \in [[\{f\}]^2]$.
Case 3.2: $F = \{v_5\}$.
Because of $v_5 \star v_5 = c_2$ we have $f \notin [[\{f\}]^{n-1}]$.
Case 3.3: $F = \{v_2, v_5\}$.
This case resembles Case 2.3. Thus we have:

$$num(v_2) = 1 \Longrightarrow f \in [[\{f\}]^2]$$

and

$$num(v_2) \geq 2 \implies f \in [[\{f\}]^3] \wedge f \notin [[\{f\}]^2].$$

■

Next, we want to consider a rough partition of the lattice $\mathbb{L}_3(M)$. Let

$$R := \{f \in M \mid |Im(f)| \leq 2\}.$$

Then every subclass T of M has the form

$$(T \cap R) \cup (T \cap Pol_3 \begin{pmatrix} 0 \\ 2 \end{pmatrix}), \tag{15.4}$$

since all functions of M with $|Im(f)| \geq 3$ have the property $f(0, ..., 0) = 0$ and $f(2, ..., 2) = 2$. Thus one can obtain all subclasses of M if every subclass T that fulfills exactly one of the following conditions (I)–(IV) is determined:

(I) $T \subseteq [M^1]$;
(II) $T \subseteq R$ and $T \not\subseteq [M^1]$;
(III) $T \subseteq M \cap Pol_3 \begin{pmatrix} 0 \\ 2 \end{pmatrix}$, $T \not\subseteq R$ and $T \not\subseteq [M^1]$;
(IV) $\exists T_1 \in \mathbb{L}_3(R)\backslash\{\emptyset\}\ \exists T_2 \in \mathbb{L}_3(M \cap Pol_3 \begin{pmatrix} 0 \\ 2 \end{pmatrix})\backslash\mathbb{L}_3(R)$:
$T = T_1 \cup T_2 \wedge T \not\subseteq [M^1]$.

We determine the subclasses of M in Sections 15.4.3–15.4.6, in compliance with the above partition *(I)–(IV)* of the subclasses.

First, we determine the maximal classes of M.

Theorem 15.4.2.3 *M has exactly 8 maximal classes. These classes are:*

(1) $M \cap Pol\{0\} = \{f \in M \mid F(f) \subseteq \{c_0, j_2, j_5, u_2, u_5, s_1\}\}$;

(2) $M \cap Pol\{2\} = \{f \in M \mid F(f) \cap \{c_2, u_2, u_5, v_2, v_5, s_1\} \neq \emptyset\}$;

(3) $M \cap Pol\{1, 2\} = \{f \in M \mid F(f) \cap \{c_1, c_2, j_5, u_5, v_2, v_5, s_1\} \neq \emptyset\}$;

(4) $M \cap Pol\begin{pmatrix} 0\,1\,2\,0 \\ 0\,1\,2\,1 \end{pmatrix}$

$= \{f \in M \mid F(f) \subseteq \{c_0, c_1, j_2, j_5, u_2, s_1\} \vee F(f) \subseteq \{c_1, c_2, v_2\}\};$

(5) $M \cap Pol\begin{pmatrix} 0\,1\,2\,1 \\ 0\,1\,2\,2 \end{pmatrix}$

$= \{f \in M | F(f) \subseteq \{c_0, c_1, j_5, u_5, s_1\} \vee F(f) \subseteq \{c_1, c_2, v_2, v_5\}\};$

(6) $M \cap Pol\begin{pmatrix} 0\,1\,2\,0\,1 \\ 0\,1\,2\,2\,2 \end{pmatrix}$

$= \{f \in M | F(f) \subseteq \{c_0, c_2, u_2, u_5, s_1\} \vee F(f) \subseteq \{c_1, v_2, v_5\}\};$

(7) $M \cap Pol\begin{pmatrix} 0\,1\,2\,0\,1 \\ 0\,1\,2\,1\,2 \end{pmatrix}$

$= \{f \in M \mid F(f) \subseteq \{c_0, c_1, j_2, j_5, s_1\} \vee F(f) \subseteq \{c_1, c_2, v_2, v_5\}\};$

(8) $[\{s_1\}] \cup R := [\{s_1\}] \cup \{f \in M \mid |Im(f)| \leq 2\}$.

Proof. Since $M = [M^1 \cup \{max\}]$ and $max \in T$ for all classes T, which are defined by (1)–(7), it is easy to prove that the classes (1)–(7) are M-maximal. One can show the M-maximality of $[\{s_1\}] \cup R$ as follows:
Let $f \in M \backslash ([\{s_1\}] \cup R)$. Since $c_0 \in R$, one of the following 9 functions is a superposition over $R \cup \{f\}$: $x \vee y$, $x \vee g(y)$ $(g \in \{j_2, j_5, u_2, u_5\})$, $h_1(x) \vee h_2(y)$ $(h_1 \in \{j_2, j_5\}, h_2 \in \{u_2, u_3\})$. Through substitution of x, y by certain functions of $\{j_2, j_5, u_2, u_5\}$, one can reduce these 9 cases to the case: $t(x, y) := j_5(x) \vee u_2(y) \in [R \cup \{f\}]$. Because of $j_5(t(x, u_5(y))) = j_5(x) \vee j_5(y)$ and $u_5(j_2(x) \vee j_2(y)) = u_2(x) \vee u_2(y)$ it holds $x \vee y = t(j_5(x) \vee j_5(y), u_2(x) \vee u_2(y))$. Hence $[R \cup \{f\}] = M$ and $[\{s_1\}] \cup R$ is M-maximal.
We still have to show that the class M does not have any further maximal classes. Suppose $T \subset M$ is an M-maximal class that is different from the 8 M-maximal classes listed above. Then, there is for every $i \in \{1, 2, ..., 8\}$, a certain function $f_i \in T$ that does not belong to the class (i). Consequently, it holds $f'_2 \in \{c_0, c_1, j_2, j_5\}$ for $f'_2(x) := f_2(x, x, ..., x)$, and $f_3(a_1, a_2, ..., a_n) = 0$ for certain $a_1, a_2, ..., a_n \in \{1, 2\}$. Hence, $c_0 \in T$, since $j_2 \star j_2 = c_0$ and $f_3(g_1(x), ..., g_n(x)) = c_0$, if $f'_2 \in \{c_1, j_5\}$ and

$$g_i(x) := \begin{cases} x & \text{if } a_i = 2, \\ f'_2(x) & \text{if } a_i = 1 \end{cases}$$

$(i = 1, 2, ..., n)$. Some unary functions f'_t with $f'_4 \in \{u_5, v_5\}$, $f'_5 \in \{j_2, u_2\}$, $f'_6 \in \{j_2, j_5\}$ and $f'_7 \in \{u_2, u_5\}$ are superpositions over $\{f_t, c_0\}$ $(t = 4, 5, 6, 7)$. It is easy to check that $\{j_2, j_5, u_2, u_5\} \subset [\{f'_4, f'_5, f'_6, f'_7\}]$ holds. In the above proof of the M-maximality of $[\{s_1\}] \cup R$, we have shown already that $max \in [\{j_2, j_5, u_2, u_5, f_8\}]$ holds. Since in addition $u_i \vee c_1 = v_i$ $(i = 2, 5)$, $v_2(c_0) = c_1$ and $v_5(c_1) = c_2$, we have $M^1 \cup \{max\} \subseteq T$. Consequently, $T = M$, in contradiction to $T \subset M$. ■

15.4.3 The Subclasses of $[M^1]$

Since all subclasses of $[P_3^1]$ were already determined in Section 15.2, one obtains the following theorem as a consequence of Theorem 15.2.1:

Theorem 15.4.3.1 *$[M^1]$ has exactly 190 pairwise distinct subclasses H_i. These classes H_i are defined in Table 15.13 through H_i^1 for $i \in \{1,2,...,95\}$, and through $H_{95+i} := H_i \cup [\{s_1\}]$ for $i = 1,2,...,95$.* ■

Table 15.13

i	H_i^1	i	H_i^1	i	H_i^1
1	$\emptyset$	2	$\{c_0\}$	3	$\{c_1\}$
4	$\{c_2\}$	5	$\{c_0,c_1\}$	6	$\{c_0,c_2\}$
7	$\{c_1,c_2\}$	8	$\{c_0,c_1,c_2\}$	9	$\{j_5\}$
10	$\{j_2,c_0\}$	11	$\{j_5,c_0\}$	12	$\{j_5,c_1\}$
13	$\{j_2,c_0,c_1\}$	14	$\{j_5,c_0,c_1\}$	15	$\{j_2,j_5,c_0\}$
16	$\{j_2,j_5,c_0,c_1\}$	17	$\{u_2\}$	18	$\{u_5\}$
19	$\{u_2,c_0\}$	20	$\{u_5,c_0\}$	21	$\{u_2,c_2\}$
22	$\{u_5,c_2\}$	23	$\{u_2,u_5\}$	24	$\{u_2,c_0,c_2\}$
25	$\{u_5,c_0,c_2\}$	26	$\{u_2,u_5,c_0\}$	27	$\{u_2,u_5,c_2\}$
28	$\{u_2,u_5,c_0,c_2\}$	29	$\{v_2\}$	30	$\{v_5,c_2\}$
31	$\{v_2,c_2\}$	32	$\{v_2,c_1\}$	33	$\{v_5,c_1,c_2\}$
34	$\{v_2,c_1,c_2\}$	35	$\{v_2,v_5,c_2\}$	36	$\{v_2,v_5,c_1,c_2\}$
37	$\{j_5,c_1,c_2\}$	38	$\{j_2,c_0,c_1,c_2\}$	39	$\{j_5,c_0,c_1,c_2\}$
40	$\{j_2,j_5,c_0,c_1,c_2\}$	41	$\{u_2,c_0,c_1\}$	42	$\{u_5,c_1,c_2\}$
43	$\{u_2,c_0,c_1,c_2\}$	44	$\{u_5,c_0,c_1,c_2\}$	45	$\{u_2,u_5,c_0,c_1,c_2\}$
46	$\{v_2,c_0,c_1\}$	47	$\{v_5,c_0,c_1,c_2\}$	48	$\{v_2,c_0,c_1,c_2\}$
49	$\{v_2,v_5,c_0,c_1,c_2\}$	50	$\{j_5,u_5\}$	51	$\{j_2,u_2,c_0\}$
52	$\{j_5,u_5,c_0\}$	53	$\{j_2,u_2,c_0,c_1\}$	54	$\{j_5,u_5,c_1,c_2\}$
55	$\{j_5,u_5,c_0,c_1,c_2\}$	56	$\{j_2,u_2,c_0,c_1,c_2\}$	57	$\{j_2,j_5,u_2,c_0\}$
58	$\{j_2,j_5,u_2,c_0,c_1\}$	59	$\{j_2,j_5,u_2,c_0,c_1,c_2\}$	60	$\{j_2,j_5,u_2,u_5,c_0\}$
61	$\{j_2,j_5,u_2,u_5,c_0,c_1,c_2\}$	62	$\{j_2,v_2,c_0,c_1\}$	63	$\{j_2,v_2,c_0,c_1,c_2\}$
64	$\{j_5,v_2,c_1\}$	65	$\{j_5,v_2,c_0,c_1\}$	66	$\{j_5,v_2,c_1,c_2\}$
67	$\{j_5,v_2,c_0,c_1,c_2\}$	68	$\{j_5,v_5,c_1,c_2\}$	69	$\{j_5,v_5,c_0,c_1,c_2\}$
70	$\{j_2,j_5,v_2,c_0,c_1\}$	71	$\{j_2,j_5,v_2,c_0,c_1,c_2\}$	72	$\{j_5,v_2,v_5,c_1,c_2\}$
73	$\{j_5,v_2,v_5,c_0,c_1,c_2\}$	74	$\{j_2,j_5,v_2,v_5,c_0,c_1,c_2\}$	75	$\{u_2,v_2\}$
76	$\{u_2,v_2,c_0,c_1\}$	77	$\{u_2,v_2,c_2\}$	78	$\{u_2,v_2,c_0,c_1,c_2\}$
79	$\{u_5,v_5,c_2\}$	80	$\{u_5,v_5,c_1,c_2\}$	81	$\{u_5,v_5,c_0,c_1,c_2\}$
82	$\{u_5,v_2,v_5,c_2\}$	83	$\{u_5,v_2,v_5,c_1,c_2\}$	84	$\{u_5,v_2,v_5,c_0,c_1,c_2\}$
85	$\{u_2,u_5,v_2,v_5,c_2\}$	86	$\{u_2,u_5,v_2,v_5,c_0,c_1,c_2\}$	87	$\{j_2,u_2,v_2,c_0,c_1\}$
88	$\{j_2,u_2,v_2,c_0,c_1,c_2\}$	89	$\{j_5,u_5,v_5,c_1,c_2\}$	90	$\{j_5,u_5,v_5,c_0,c_1,c_2\}$
91	$\{j_2,j_5,u_2,v_2,c_0,c_1\}$	92	$\{j_2,j_5,u_2,v_2,c_0,c_1,c_2\}$	93	$\{j_5,u_5,v_2,v_5,c_1,c_2\}$
94	$\{j_5,u_5,v_2,v_5,c_0,c_1,c_2\}$	95	$\{j_2,j_5,u_2,u_5,v_2,v_5\} \cup H_8^1$		

15.4.4 The Subclasses of *R*

Theorem 15.4.4.1 *The class*

$$J_1 := \{f \in M \mid Im(f) \subseteq \{0,1\}\}$$

has exactly the following (countable infinite-many) subclasses, which are not subclasses of $[M^1]$:

$$\begin{aligned}
&J_1,\\
&J_2 := J_1 \cap Pol\{0\} = \{f \in J_1 \mid f \notin [\{c_1\}]\},\\
&J_3 := J_1 \cap Pol\{1\} = \{f \in J_1 \mid f \notin [\{c_0\}]\},\\
&J_4 := J_1 \cap Pol\{0\} \cap Pol\{1\},\\
&J_5 := \{f \in J_1 \mid num_f(j_5) \le 1\},\\
&J_6 := J_2 \cap J_5,\\
&J_7 := J_3 \cap J_5,\\
&J_8 := J_4 \cap J_5,\\
&J_9 := \{f \in J_1 \mid f \in [\{j_5\}] \vee num_f(j_5) = 0\},\\
&J_{10} := \{f \in J_9 \mid f \notin [\{c_1\}]\},\\
&J_{11} := \{f \in J_9 \mid f \notin [\{j_5\}]\},\\
&J_{12} := \{f \in J_9 \mid f \notin [\{c_1, j_5\}],\\
&J_{9,r} := \{f \in J_9 \mid num_f(j_2) \le r\},\\
&J_{10,r} := J_{10} \cap J_{9,r},\\
&J_{11,r} := J_{11} \cap J_{9,r},\\
&J_{12,r} := J_{12} \cap J_{9,r},\\
&J_{13} := \{f \in J_1 \mid num_f(j_2) = 0\},\\
&J_{14} := J_{13} \cap J_2,\\
&J_{15} := J_{13} \cap J_3,\\
&J_{16} := J_{13} \cap J_4,
\end{aligned}$$

where $r = 1, 2, 3, 4, \ldots.$.

Proof. With the aid of the Hasse diagram of the above classes (see Figure 15.1) and the generating systems of these classes (see Table 15.14), one can prove the theorem. ■

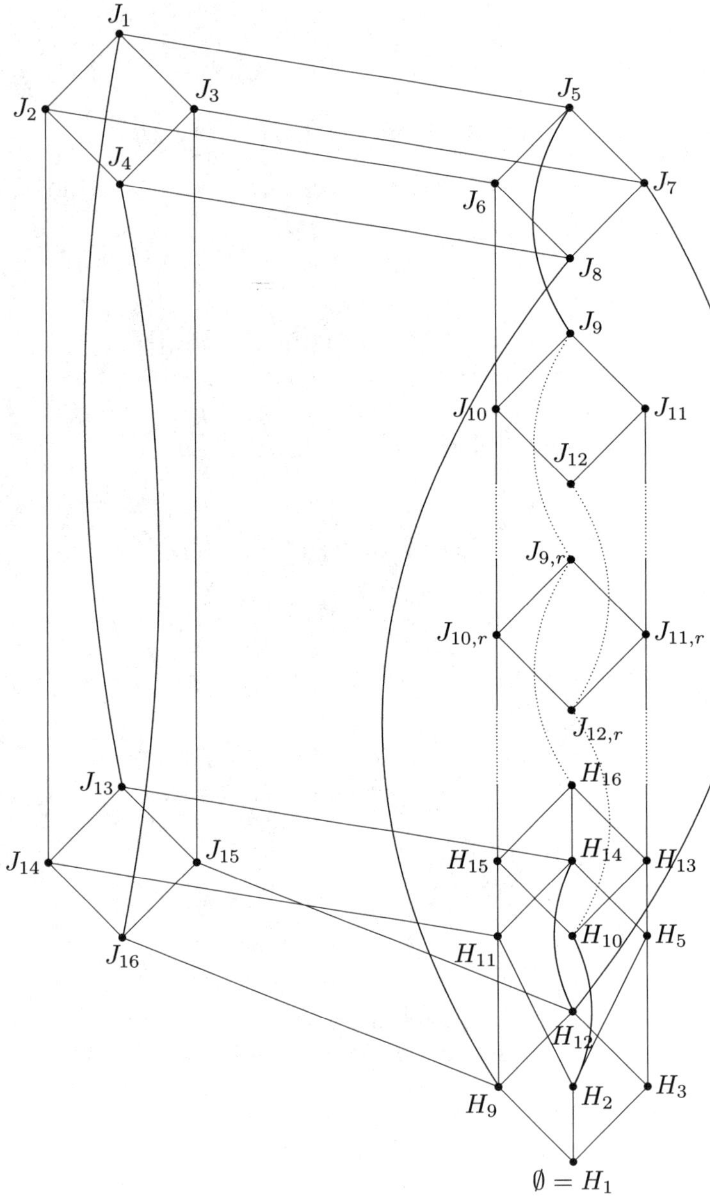

Fig. 15.1. The subclasses of J_1

Table 15.14

A	generating system (or basis, if it exists) for A	A^1
J_1	$\{j_5(x) \vee j_5(y), j_2, c_1\}$	$\{j_2, j_5, c_0, c_1\}$
J_2	$\{j_5(x) \vee j_5(y), j_2\}$	$\{j_2, j_5, c_0\}$
J_3	$\{j_5(x) \vee j_5(y), j_2(x) \vee j_5(y), c_1\}$	$\{j_5, c_1\}$
J_4	$\{j_5(x) \vee j_5(y), j_2(x) \vee j_5(y)\}$	$\{j_5\}$
J_5	$\{j_5(x) \vee j_2(y), c_0, c_1\}$	$\{j_2, j_5, c_0, c_1\}$
J_6	$\{j_5(x) \vee j_2(y), c_0\}$	$\{j_2, j_5, c_0\}$
J_7	$\{j_5(x) \vee j_2(y), c_1\}$	$\{j_5, c_1\}$
J_8	$\{j_5(x) \vee j_2(y)\}$	$\{j_5\}$
J_9	$\{j_5, c_1, j_2(x_1) \vee ... \vee j_2(x_r) \mid r \in \mathbb{N}\}$	$\{j_2, j_5, c_0, c_1\}$
J_{10}	$\{j_5, j_2(x_1) \vee ... \vee j_2(x_r) \mid r \in \mathbb{N}\}$	$\{j_2, j_5, c_0\}$
J_{11}	$\{c_1, j_2(x_1) \vee ... \vee j_2(x_r) \mid r \in \mathbb{N}\}$	$\{j_2, c_0, c_1\}$
J_{12}	$\{j_2(x_1) \vee ... \vee j_2(x_r) \mid r \in \mathbb{N}\}$	$\{j_2, c_0\}$
$J_{9,r}$	$\{j_5, c_1, j_2(x_1) \vee ... \vee j_2(x_r)\}$	$\{j_2, j_5, c_0, c_1\}$
$J_{10,r}$	$\{j_5, j_2(x_1) \vee ... \vee j_2(x_r)\}$	$\{j_2, j_5, c_0\}$
$J_{11,r}$	$\{c_1, j_2(x_1) \vee ... \vee j_2(x_r)\}$	$\{j_2, c_0, c_1\}$
$J_{12,r}$	$\{j_2(x_1) \vee ... \vee j_2(x_r)\}$	$\{j_2, c_0\}$
J_{13}	$\{j_5(x) \vee j_5(y), c_0, c_1\}$	$\{j_5, c_0, c_1\}$
J_{14}	$\{j_5(x) \vee j_5(y), c_0\}$	$\{j_5, c_0\}$
J_{15}	$\{j_5(x) \vee j_5(y), c_1\}$	$\{j_5, c_1\}$
J_{16}	$\{j_5(x) \vee j_5(y)\}$	$\{j_5\}$

In analog mode, the following two theorems can be proven when one uses Figures 15.2 and 15.3 and Tables 15.15 and 15.16.

Theorem 15.4.4.2 *The class*

$$U_1 := \{f \in M \mid Im(f) \subseteq \{0,2\}\}$$

has exactly the following 14 pairwise different subclasses, which are not subclasses of $[M^1]$*:*

U_1,
$U_2 := U_1 \cap Pol\{0\} = \{f \in U_1 \mid f \notin [\{c_2\}]\}$,
$U_3 := \{f \in U_1 \mid F(f) \subseteq \{c_0, c_2, u_2\}\}$,
$U_4 := \{f \in U_1 \mid F(f) \subseteq \{c_0, c_2, u_5\}\}$,
$U_5 := U_1 \cap Pol\{2\} = \{f \in U_1 \mid f \notin [\{c_0\}]\}$,
$U_6 := \{f \in U_3 \mid f \notin [\{c_2\}]\}$,
$U_7 := \{f \in U_4 \mid f \notin [\{c_2\}]\}$,
$U_8 := \{f \in U_3 \mid f \notin [\{c_0\}]\}$,
$U_9 := \{f \in U_1 \mid f \in [\{c_2\}] \vee num_f(u_5) \geq 1\}$,
$U_{10} := \{f \in U_4 \mid f \notin [\{c_0\}]\}$,
$U_{11} := \{f \in U_1 \mid f \notin [\{c_0, c_2\}]\}$,
$U_{12} := \{f \in U_3 \mid f \notin [\{c_0, c_2\}]\}$,
$U_{13} := \{f \in U_9 \mid f \notin [\{c_2\}]\}$,
$U_{14} := \{f \in U_4 \mid f \notin [\{c_0, c_2\}]\}$. ■

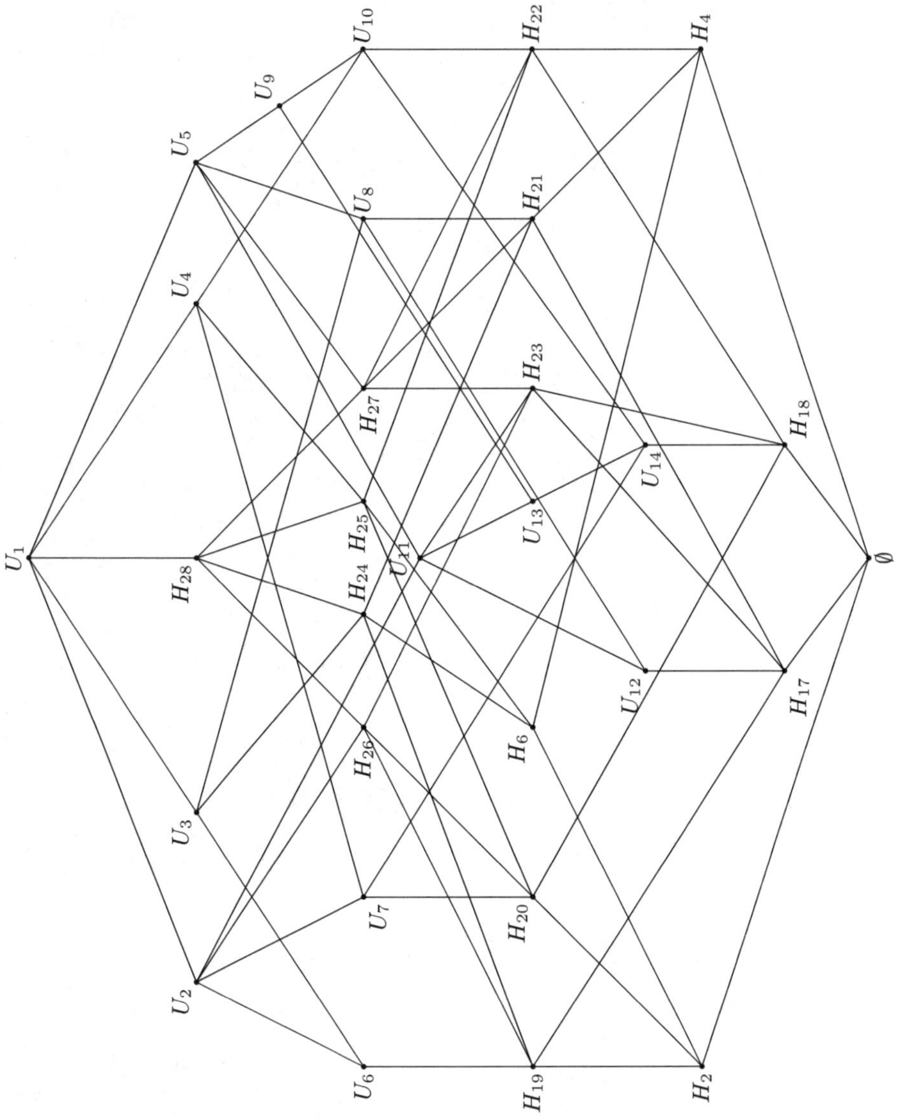

Fig. 15.2. The subclasses of U_1

Table 15.15

A	Basis for A	A^1
U_1	$\{u_2(x) \vee u_2(y), u_5, c_0, c_1\}$	$\{u_2, u_5, c_0, c_2\}$
U_2	$\{u_2(x) \vee u_2(y), u_5, c_0\}$	$\{u_2, u_5, c_0\}$
U_3	$\{u_2(x) \vee u_2(y), c_0, c_2\}$	$\{u_2, c_0, c_2\}$
U_4	$\{u_5(x) \vee u_5(y), c_0, c_2\}$	$\{u_5, c_0, c_2\}$
U_5	$\{u_2(x) \vee u_2(y), u_5, c_2\}$	$\{u_2, u_5, c_2\}$
U_6	$\{u_2(x) \vee u_2(y), c_0\}$	$\{u_2, c_0\}$
U_7	$\{u_5(x) \vee u_5(y), c_0\}$	$\{u_5, c_0\}$
U_8	$\{u_2(x) \vee u_2(y), c_2\}$	$\{u_2, c_2\}$
U_9	$\{u_2(x) \vee u_5(y), c_2\}$	$\{u_5, c_2\}$
U_{10}	$\{u_5(x) \vee u_5(y), c_2\}$	$\{u_5, c_2\}$
U_{11}	$\{u_2(x) \vee u_2(y), u_5\}$	$\{u_2, u_5\}$
U_{12}	$\{u_2(x) \vee u_2(y)\}$	$\{u_2\}$
U_{13}	$\{u_2(x) \vee u_5(y)\}$	$\{u_5\}$
U_{14}	$\{u_5(x) \vee u_5(y)\}$	$\{u_5\}$

Theorem 15.4.4.3 *The class*

$$V_1 := \{f \in M \mid Im(f) \subseteq \{1,2\}\}$$

has exactly the following (countable infinite-many) subclasses, which are not subclasses of $[M^1]$ *:*

V_1,
$V_2 := V_1 \cap Pol(2) = \{f \in V_1 \mid f \notin [\{c_1\}]\}$,
$V_3 := \{f \in V_1 \mid f \in [\{c_1, c_2, v_2\}] \vee num_f(v_5) \geq 1\}$,
$V_4 := \{f \in V_3 \mid f \notin [\{c_1\}]\}$,
$V_5 := \{f \in V_1 \mid f \in [\{c_1, c_2, v_2\}] \vee num_f(v_2) \leq 1\}$,
$V_6 := \{f \in V_5 \mid f \in [\{c_1\}]\}$,
$V_7 := \{f \in V_1 \mid f \in [\{c_2, v_2\}] \vee F(f) \subseteq \{c_1, v_5\}\}$,
$V_8 := \{f \in V_7 \mid f \notin [\{c_1\}]\}$,
$V_9 := \{f \in V_7 \mid f \notin [\{v_2\}]\}$,
$V_{10} := \{f \in V_7 \mid f \notin [\{c_1, v_2\}]\}$,
$V_{7,r} := \{f \in V_7 \mid num_f(v_5) \leq r\}$,
$V_{8,r} := \{f \in V_{7,r} \mid f \notin [\{c_1\}]\}$,
$V_{9,r} := \{f \in V_{7,r} \mid f \notin [\{v_2\}]\}$,
$V_{10,r} := \{f \in V_{7,r} \mid f \notin [\{c_1, v_2\}]\}$,
$V_{11} := \{f \in V_3 \mid f \notin [\{v_2\}]\}$,
$V_{12} := V_5 \cap V_{11}$,
$V_{13} := \{f \in V_{11} \mid f \notin [\{c_1\}]\}$,
$V_{14} := \{f \in V_{12} \mid f \notin [\{c_1\}]\}$,
$V_{15} := \{f \in V_1 \mid F(f) \subseteq \{c_1, c_2, v_2\}\}$,
$V_{16} := \{f \in V_{15} \mid f \notin [\{c_1\}]\}$,
$V_{17} := \{f \in V_{15} \mid f \notin [\{c_2\}]\}$,
$V_{18} := \{f \in V_{15} \mid f \notin [\{c_1, c_2\}]\}$,
where $r = 1, 2, 3, 4, \ldots$. ■

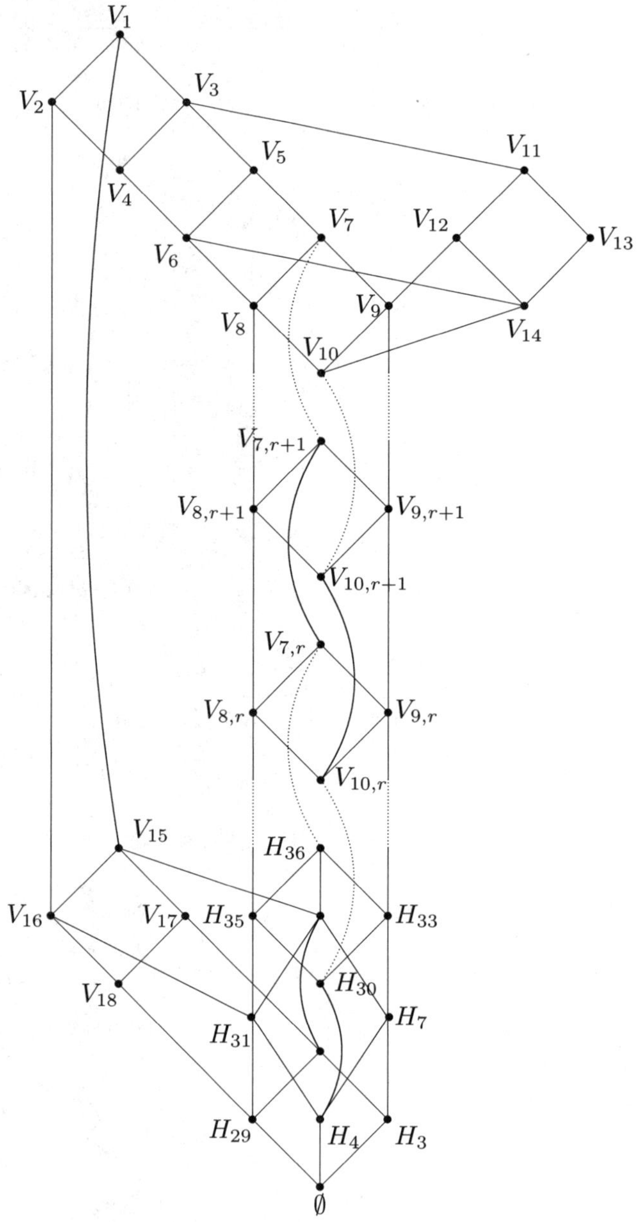

Fig. 15.3. The subclasses of V_1

Table 15.16

A	Generating system (or basis, if it exists) for A	A^1
V_1	$\{v_2(x) \vee v_2(y), v_5, c_1\}$	$\{v_2, v_5, c_1, c_2\}$
V_2	$\{v_2(x) \vee v_2(y), v_5\}$	$\{v_2, v_5, c_2\}$
V_3	$\{v_2(x) \vee v_2(y) \vee v_5(z), v_2, c_1\}$	$\{v_2, v_5, c_1, c_2\}$
V_4	$\{v_2(x) \vee v_2(y) \vee v_5(z), v_5\}$	$\{v_2, v_5, c_2\}$
V_5	$\{v_2(x) \vee v_5(y), v_2, c_1\}$	$\{v_2, v_5, c_1, c_2\}$
V_6	$\{v_2(x) \vee v_5(y), v_2\}$	$\{v_2, v_5, c_2\}$
V_7	$\{v_2, c_1, v_5(x_1) \vee ... \vee v_5(x_r) \mid r \in \mathbb{N}\}$	$\{v_2, v_5, c_1, c_2\}$
V_8	$\{v_2, v_5(x_1) \vee ... \vee v_5(x_r) \mid r \in \mathbb{N}\}$	$\{v_2, v_5, c_2\}$
V_9	$\{c_1, v_5(x_1) \vee ... \vee v_5(x_r) \mid r \in \mathbb{N}\}$	$\{v_5, c_1, c_2\}$
V_{10}	$\{v_5(x_1) \vee ... \vee v_5(x_r) \mid r \in \mathbb{N}\}$	$\{v_5, c_2\}$
$V_{7,r}$	$\{v_2, c_1, v_5(x_1) \vee ... \vee v_5(x_r)\}$	$\{v_2, v_5, c_1, c_2\}$
$V_{8,r}$	$\{v_2, v_5(x_1) \vee ... \vee v_5(x_r)\}$	$\{v_2, v_5, c_2\}$
$V_{9,r}$	$\{c_1, v_5(x_1) \vee ... \vee v_5(x_r)\}$	$\{v_5, c_1, c_2\}$
$V_{10,r}$	$\{v_5(x_1) \vee ... \vee v_5(x_r)\}$	$\{v_5, c_2\}$
V_{11}	$\{v_2(x) \vee v_2(y) \vee v_5(z), c_1\}$	$\{v_5, c_1, c_2\}$
V_{12}	$\{v_2(x) \vee v_5(y), c_1\}$	$\{v_5, c_1, c_2\}$
V_{13}	$\{v_2(x) \vee v_2(y) \vee v_5(z)\}$	$\{v_5, c_2\}$
V_{14}	$\{v_2(x) \vee v_5(y)\}$	$\{v_5, c_2\}$
V_{15}	$\{v_2(x) \vee v_2(y), c_1, c_2\}$	$\{v_2, c_1, c_2\}$
V_{16}	$\{v_2(x) \vee v_2(y), c_2\}$	$\{v_2, c_2\}$
V_{17}	$\{v_2(x) \vee v_2(y), c_1\}$	$\{v_2, c_1\}$
V_{18}	$\{v_2(x) \vee v_2(y)\}$	$\{v_2\}$

Theorem 15.4.4.4 *Class $J_1 \cup U_1$ has exactly the following subclasses (ordered with respect to equality of the unary functions), which are not subclasses of J_1 or U_1 or $[(J_1 \cup U_1)^1]$:*

$J_3 \cup H_4, J_7 \cup H_4, J_{15} \cup H_4$;
$J_{11} \cup H_6, J_{11,r} \cup H_6$;
$J_{13} \cup H_6$;
$J_1 \cup H_6, J_5 \cup H_6, J_9 \cup H_6, J_{9,r} \cup H_6$;
$H_5 \cup U_6$;
$H_3 \cup U_9, H_3 \cup U_{10}$;
$H_5 \cup U_3$;
$H_5 \cup U_4$;
$H_5 \cup U_1$;
$J_4 \cup U_{13}, J_{16} \cup U_{14}$;
$J_{12} \cup H_{19}, J_{12} \cup U_6, J_{12,r} \cup H_{19}$;
$J_{14} \cup U_7$;
$J_{11} \cup H_{19}, J_{11} \cup U_6, J_{11,r} \cup H_{19}$;
$J_3 \cup U_9$;
$J_{13} \cup U_4$;
$J_{11} \cup H_{24}, J_{11,r} \cup H_{24}, J_{11} \cup U_3$;
$J_{10} \cup H_{19}, J_{10,r} \cup H_{19}, J_{10} \cup U_6, J_6 \cup H_{19}, J_6 \cup U_6, J_2 \cup H_{19}, J_2 \cup U_6$;

$J_9 \cup H_{19}, J_{9,r} \cup H_{19}, J_5 \cup H_{19}, J_1 \cup H_{19}, J_9 \cup U_6, J_5 \cup U_6, J_1 \cup U_6$;
$J_9 \cup H_{24}, J_{9,r} \cup H_{24}, J_5 \cup H_{24}, J_1 \cup H_{24}, J_5 \cup U_3, J_9 \cup U_3, J_1 \cup U_3$;
$J_2 \cup U_2$;
$J_1 \cup U_1$;
where $r = 1, 2, 3, \ldots$.

Proof. Let T be an arbitrary subclass of $J_1 \cup U_1$, which was not described in Theorems 15.4.3.1, 15.4.4.1, or 15.4.4.2. Then, obviously $T_1 := T \cap J_1$ and $T_2 := T \cap U_1$ are subclasses of J_1 or U_1 with $T = T_1 \cup T_2$. Further, $[T^1]$ is a subclass H of $[M^1]$ with $H^1 = T^1$. The possibilities for H follow from Theorems 15.4.3.1, 15.4.4.1, and 15.4.4.2. These possibilities imply the possibilities for T_1 and T_2, which are given in Table 15.17. When one selects the sets $T_1 \cup T_2 \not\subseteq [M^1]$, which are closed, one receives the statement of the theorem. ∎

Table 15.17

$[T^1]$	Possibilities for	
	$T \cap J_1$	$T \cap U_1$
H_{37}	J_3, J_7, J_{15}	H_4
H_{38}	$J_{11}, J_{11,r}$	H_6
H_{39}	J_{13}	H_6
H_{40}	$J_1, J_5, J_9, J_{9,r}$	H_6
H_{41}	H_5	U_6
H_{42}	H_3	U_9, U_{10}
H_{43}	H_5	U_3
H_{44}	H_5	U_4
H_{45}	H_5	U_1
H_{50}	H_9, J_4, J_8, J_{16}	H_{18}, U_{13}, U_{14}
H_{51}	$H_{10}, J_{12}, J_{12,r}$	H_{19}, U_6
H_{52}	H_{11}, J_{14}	H_{20}, U_7
H_{53}	$H_{13}, J_{11}, J_{11,r}$	H_{19}, U_6
H_{54}	H_{12}, J_3, J_7, J_{15}	H_{22}, U_9
H_{55}	H_{14}, J_{13}	H_{25}, U_4
H_{56}	$H_{13}, J_{11}, J_{11,r}$	H_{24}, U_3
H_{57}	$H_{15}, J_2, J_6, J_{10}, J_{10,r}$	H_{19}, U_6
H_{58}	$H_{16}, J_1, J_5, J_9, J_{9,r}$	H_{19}, U_6
H_{59}	$H_{16}, J_1, J_5, J_9, J_{9,r}$	H_{24}, U_3
H_{60}	$H_{15}, J_2, J_6, J_{10}, J_{10,r}$	H_{26}, U_2
H_{61}	$H_{16}, J_1, J_5, J_9, J_{9,r}$	H_{28}, U_1

Theorem 15.4.4.5 *Class* $J_1 \cup V_1$ *has exactly the following subclasses, which are not subclasses of* J_1 *or* V_1 *or* $[(J_1 \cup V_1)^1]$:
$J_3 \cup H_7, J_7 \cup H_7, J_{15} \cup H_7$;
$J_{11} \cup H_7, J_{11,r} \cup H_7$;
$J_{13} \cup H_7$;
$J_1 \cup H_7, J_5 \cup H_7, J_9 \cup H_7, J_{9,r} \cup H_7$;

$H_5 \cup V_{17}$;
$H_5 \cup V_9, H_5 \cup V_{9,r}, H_5 \cup V_{11}, H_5 \cup V_{12}$;
$H_5 \cup V_{15}$;
$H_5 \cup V_1, H_5 \cup V_3, H_5 \cup V_5, H_5 \cup V_7, H_5 \cup V_{7,r}$;
$J_{11} \cup H_{32}, J_{11} \cup V_{17}, J_{11,r} \cup H_{32}$;
$J_{11} \cup H_{34}, J_{11} \cup V_{15}, J_{11,r} \cup H_{34}$;
$H_{12} \cup V_{17}, J_3 \cup H_{32}, J_3 \cup V_{17}, J_7 \cup H_{32}, J_7 \cup V_{17}, J_{15} \cup H_{32}, J_{15} \cup V_{17}$;
$H_{14} \cup V_{17}, J_{13} \cup H_{32}, J_{13} \cup V_{17}$;
$H_{12} \cup V_{15}, J_3 \cup H_{34}, J_3 \cup V_{15}, J_7 \cup H_{34}, J_7 \cup V_{15}, J_{15} \cup H_{34}, J_{15} \cup V_{15}$;
$H_{14} \cup V_{15}, J_{13} \cup H_{34}, J_{13} \cup V_{15}$;
$H_{12} \cup V_9, H_{12} \cup V_{11}, H_{12} \cup V_{12}, H_{12} \cup V_{9,r}, J_{15} \cup V_9, J_{15} \cup V_{12}, J_{15} \cup V_{11}, J_7 \cup V_{12}, J_7 \cup V_{11}, J_3 \cup V_{11}$;
$H_{14} \cup V_9, H_{14} \cup V_{11}, H_{14} \cup V_{12}, H_{14} \cup V_{9,r}, J_{13} \cup V_9, J_{13} \cup V_{11}, J_{13} \cup V_{12}$;
$J_1 \cup H_{32}, J_1 \cup V_{17}, J_5 \cup H_{32}, J_5 \cup V_{17}, J_9 \cup H_{32}, J_9 \cup V_{17}, J_{9,r} \cup H_{32}$;
$J_1 \cup H_{34}, J_1 \cup V_{15}, J_5 \cup H_{34}, J_5 \cup V_{15}, J_9 \cup H_{34}, J_9 \cup V_{15}, J_{9,r} \cup H_{34}$;
$H_{12} \cup V_1, H_{12} \cup V_3, H_{12} \cup V_5, H_{12} \cup V_7, H_{12} \cup V_{7,r}, J_{15} \cup V_7, J_{15} \cup V_5, J_{15} \cup V_3, J_{15} \cup V_1, J_7 \cup V_5, J_7 \cup V_3, J_7 \cup V_1, J_3 \cup V_3, J_3 \cup V_1$;
$H_{14} \cup V_1, H_{14} \cup V_3, H_{14} \cup V_5, H_{14} \cup V_7, H_{14} \cup V_{7,r}, J_{13} \cup V_1, J_{13} \cup V_3, J_{13} \cup V_5, J_{13} \cup V_7$;
$J_1 \cup V_1$;

where $r = 1, 2, 3, \ldots$.

Table 15.18

$[T^1]$	Possibilities for $T \cap J_1$	Possibilities for $T \cap V_1$
H_{37}	H_{12}, J_3, J_7, J_{15}	H_7
H_{38}	$H_{13}, J_{11}, J_{11,r}$	H_7
H_{39}	H_{14}, J_{13}	H_7
H_{40}	$H_{16}, J_1, J_5, J_9, J_{9,r}$	H_7
H_{46}	H_5	H_{32}, V_{17}
H_{47}	H_5	$H_{33}, V_9, V_{9,r}, V_{11}, V_{12}$
H_{48}	H_5	H_{34}, V_{15}
H_{49}	H_5	$H_{36}, V_1, V_3, V_5, V_7, V_{7,r}$
H_{62}	$H_{13}, J_{11}, J_{11,r}$	H_{32}, V_{17}
H_{63}	$H_{13}, J_{11}, J_{11,r}$	H_{34}, V_{15}
H_{64}	H_{12}, J_3, J_7, J_{15}	H_{32}, V_{17}
H_{65}	H_{14}, J_{13}	H_{32}, V_{17}
H_{66}	H_{12}, J_3, J_7, J_{15}	H_{34}, V_{15}
H_{67}	H_{14}, J_{13}	H_{34}, V_{15}
H_{68}	H_{12}, J_3, J_7, J_{15}	$H_{33}, V_9, V_{9,r}, V_{11}, V_{12}$
H_{69}	H_{14}, J_{13}	$H_{33}, V_9, V_{9,r}, V_{11}, V_{12}$
H_{70}	$H_{16}, J_1, J_5, J_9, J_{9,r}$	H_{32}, V_{17}
H_{71}	$H_{16}, J_1, J_5, J_9, J_{9,r}$	H_{34}, V_{15}
H_{72}	H_{12}, J_3, J_7, J_{15}	$H_{36}, V_1, V_3, V_5, V_7, V_{7,r}$
H_{73}	H_{14}, J_{13}	$H_{36}, V_1, V_3, V_5, V_7, V_{7,r}$
H_{74}	$H_{16}, J_1, J_5, J_9, J_{9,r}$	$H_{36}, V_1, V_3, V_5, V_7, V_{7,r}$

Proof. Let T be an arbitrary subclass of $J_1 \cup V_1$, which was not described in Theorems 15.4.3.1, 15.4.4.1, or 15.4.4.3. Then obviously $T_1 := T \cap J_1$ and $T_2 := T \cap V_1$ are subclasses of J_1 or V_1 with $T = T_1 \cup T_2$. Further, $[T^1]$ is a subclass H of $[M^1]$ with $H^1 = T^1$. The possibilities for H follow from Theorems 15.4.3.1, 15.4.4.1, and 15.4.4.3. These possibilities imply the possibilities for T_1 and T_2, which are given in Table 15.18. When one selects the sets $T_1 \cup T_2 \not\subseteq [M^1]$, which are closed, one receives the statement of the theorem. ■

Theorem 15.4.4.6 *Class $U_1 \cup V_1$ has exactly the following subclasses, which are not subclasses of U_1 or V_1 or $[(U_1 \cup V_1)^1]$:*
$U_6 \cup H_3$;
$U_9 \cup H_7$;
$U_3 \cup H_7$;
$U_4 \cup H_7$;
$U_1 \cup H_7$;
$H_2 \cup V_{17}$;
$H_6 \cup V_9, H_6 \cup V_{9,r}, H_6 \cup V_{11}, H_6 \cup V_{12}$;
$H_6 \cup V_{15}$;
$H_6 \cup V_1, H_6 \cup V_3, H_6 \cup V_5, H_6 \cup V_7, H_6 \cup V_{7,r}$;
$U_{12} \cup V_{18}$;
$U_6 \cup V_{17}$;
$U_8 \cup V_{16}$;
$U_3 \cup V_{15}$;
$H_{22} \cup V_{10}, H_{22} \cup V_{10,r}, H_{22} \cup V_{13}, H_{22} \cup V_{14}, U_9 \cup V_{13}$;
$H_{22} \cup V_9, H_{22} \cup V_{9,r}, H_{22} \cup V_{11}, H_{22} \cup V_{12}, U_9 \cup V_{11}$;
$H_{25} \cup V_9, H_{25} \cup V_{9,r}, H_{25} \cup V_{11}, H_{25} \cup V_{12}, U_4 \cup V_9$;
$H_{22} \cup V_2, H_{22} \cup V_4, H_{22} \cup V_6, H_{22} \cup V_8, H_{22} \cup V_{8,r}, U_9 \cup V_2, U_9 \cup V_4$;
$H_{22} \cup V_1, H_{22} \cup V_3, H_{22} \cup V_5, H_{22} \cup V_7, H_{22} \cup V_{7,r}, U_9 \cup V_1, U_9 \cup V_3$;
$H_{25} \cup V_1, H_{25} \cup V_3, H_{25} \cup V_5, H_{25} \cup V_7, H_{25} \cup V_{7,r}, U_4 \cup V_1, U_4 \cup V_3, U_4 \cup V_5, U_4 \cup V_7$;
$U_5 \cup V_2$;
$U_1 \cup V_1$;
where $r = 1, 2, 3, \ldots$.

Proof. Let T be an arbitrary subclass of $U_1 \cup V_1$, which was not described in Theorems 15.4.3.1, 15.4.4.2, or 15.4.4.3. Then obviously $T_1 := T \cap U_1$ and $T_2 := T \cap V_1$ are subclasses of U_1 or V_1 with $T = T_1 \cup T_2$. Further, $[T^1]$ is a subclass H of $[M^1]$ with $H^1 = T^1$. The possibilities for H follow from Theorems 15.4.3.1, 15.4.4.2, and 15.4.4.5, which are given in Table 15.19. When one selects the sets $T_1 \cup T_2 \not\subseteq [M^1]$, which are closed, one receives the statement of the theorem. ■

Table 15.19

$[T^1]$	Possibilities for $T \cap U_1$	$T \cap V_1$
H_{41}	H_{19}, U_6	H_3
H_{42}	H_{22}, U_9	H_7
H_{43}	H_{24}, U_3	H_7
H_{44}	H_{25}, U_4	H_7
H_{45}	H_{28}, U_1	H_7
H_{46}	H_2	H_{32}, V_{17}
H_{47}	H_6	$H_{33}, V_9, V_{9,r}, V_{11}, V_{12}$
H_{48}	H_6	H_{34}, V_{15}
H_{49}	H_6	$H_{36}, V_1, V_3, V_5, V_7, V_{7,r}$
H_{75}	H_{17}, U_{12}	H_{29}, V_{18}
H_{76}	H_{19}, U_6	H_{32}, V_{17}
H_{77}	H_{21}, U_8	H_{31}, V_{16}
H_{78}	H_{24}, U_3	H_{34}, V_{15}
H_{79}	H_{22}, U_9	$H_{30}, V_{10}, V_{10,r}, V_{13}, V_{14}$
H_{80}	H_{22}, U_9	$H_{33}, V_9, V_{9,r}, V_{11}, V_{12}$
H_{81}	H_{25}, U_4	$H_{33}, V_9, V_{9,r}, V_{11}, V_{12}$
H_{82}	H_{22}, U_9	$H_{35}, V_2, V_4, V_6, V_8, V_{8,r}$
H_{83}	H_{22}, U_9	$H_{36}, V_1, V_3, V_5, V_7, V_{7,r}$
H_{84}	H_{25}, U_4	$H_{36}, V_1, V_3, V_5, V_7, V_{7,r}$
H_{85}	H_{27}, U_5	$H_{35}, V_2, V_4, V_6, V_8, V_{8,r}$
H_{86}	H_{28}, U_1	$H_{36}, V_1, V_3, V_5, V_7, V_{7,r}$

Theorem 15.4.4.7 *Class $J_1 \cup U_1 \cup V_1$ has the following subclasses, which are not subclasses of $J_1 \cup U_1$, $J_1 \cup V_1$, $U_1 \cup V_1$ or $[M^1]$:*
$J_{11} \cup H_{19} \cup H_{32}$, $J_{11,r} \cup H_{19} \cup H_{32}$, $J_{11,r} \cup U_6 \cup V_{17}$;
$J_{11} \cup H_{24} \cup H_{34}$, $J_{11,r} \cup H_{24} \cup H_{34}$, $J_{11} \cup U_3 \cup V_{15}$;
$H_{12} \cup H_{22} \cup V_9$, $H_{12} \cup H_{22} \cup V_{9,r}$, $H_{12} \cup H_{22} \cup V_{11}$, $H_{12} \cup H_{22} \cup V_{12}$, $J_3 \cup U_9 \cup V_{11}$;
$H_{14} \cup H_{25} \cup V_9$, $H_{14} \cup H_{25} \cup V_{9,r}$, $H_{14} \cup H_{25} \cup V_{11}$, $H_{14} \cup H_{25} \cup V_{12}$, $J_{13} \cup U_4 \cup V_9$;
$J_9 \cup H_{19} \cup H_{32}$, $J_{9,r} \cup H_{19} \cup H_{32}$, $J_5 \cup H_{19} \cup H_{32}$, $J_1 \cup H_{19} \cup H_{32}$, $J_5 \cup U_6 \cup V_{17}$, $J_1 \cup U_6 \cup V_{17}$, $J_9 \cup U_6 \cup V_{17}$;
$J_9 \cup H_{24} \cup H_{34}$, $J_{9,r} \cup H_{24} \cup H_{34}$, $J_5 \cup H_{24} \cup H_{34}$, $J_1 \cup H_{24} \cup H_{34}$, $J_1 \cup U_3 \cup V_{15}$, $J_5 \cup U_3 \cup V_{15}$, $J_9 \cup U_3 \cup V_{15}$;
$H_{12} \cup H_{22} \cup V_1$, $H_{12} \cup H_{22} \cup V_3$, $H_{12} \cup H_{22} \cup V_5$, $H_{12} \cup H_{22} \cup V_7$, $H_{12} \cup H_{22} \cup V_{7,r}$, $J_3 \cup U_9 \cup V_3$, $J_3 \cup U_9 \cup V_1$;
$H_{14} \cup H_{25} \cup V_1$, $H_{14} \cup H_{25} \cup V_3$, $H_{14} \cup H_{25} \cup V_5$, $H_{14} \cup H_{25} \cup V_7$, $H_{14} \cup H_{25} \cup V_{7,r}$, $J_{13} \cup U_4 \cup V_1$, $J_{13} \cup U_4 \cup V_3$, $J_{13} \cup U_4 \cup V_5$, $J_{13} \cup U_4 \cup V_7$;
$J_1 \cup U_1 \cup V_1$;
where $r = 1, 2, 3, \ldots$.

Proof. It is easy to check that a class T of the form $T_1 \cup T_2 \cup T_3$ with $T_1 \in \mathbb{L}_3(J_1)\backslash\{\emptyset\}$, $T_2 \in \mathbb{L}_3(U_1)\backslash\{\emptyset\}$ and $T_3 \in \mathbb{L}_3(V_1)\backslash\{\emptyset\}$ is closed if and only if the sets $T_i \cup T_j$ are closed for all $i, j \in \{1, 2, 3\}$ and $i \neq j$. Thus, our theorem follows from Theorems 15.4.4.4–15.4.4.6 with the aid of Theorem 15.4.3.1, in which one can find the possibilities for T^1. ∎

15.4.5 The Subclasses of $M \cap Pol_3\{(0,2)\}$

To receive a coarse partition of the lattice of the subclasses of

$$\begin{aligned} A_1 &:= M \cap Pol_3 \binom{0}{2} \\ &= \{f \in M \mid f \notin [\{c_0, c_1, c_2\}] \ \wedge \ F(f) \cap \{v_2, v_5\} = \emptyset \ \wedge \\ &\qquad (\{j_2, j_5\} \cap F(f) \neq \emptyset \Longrightarrow \{s_1, u_2, u_5\} \cap F(f) \neq \emptyset)\}, \end{aligned}$$

we determine the maximal classes of A_1 first.

Lemma 15.4.5.1 *A_1 has exactly three maximal classes:*

(1) $A_2 := A_1 \cap Pol_3\{0,1\} = \{f \in A_1 \mid u_5 \notin F(f)\}$;

(2) $A_3 := A_1 \cap Pol_3\{1,2\} = \{f \in A_1 \mid \{j_2, u_2\} \cap F(f) \neq \emptyset \Longrightarrow \{s_1, j_5, u_5\} \cap F(f) \neq \emptyset\}$;

(3) $B_1 := \{f \in A_1 \mid F(f) \subseteq \{c_0, u_2, u_5, s_1\}\}$.

Proof. One can conclude from Theorem 15.4.2.2 that

$$\begin{aligned} A_1 &= [\{max, x \vee j_2(y), u_2, u_5\}], \\ A_2 &= [\{max, x \vee j_5(y), u_2\}], \\ A_3 &= [\{max, x \vee j_2(y), x \vee u_2(y), u_5\}], \\ B_1 &= [\{max, u_2, u_5\}]. \end{aligned}$$

With the aid of the above statements, it is easy to prove the A_1-maximality of A_2, A_3, and B.
Denote now T an arbitrary subset of A_1, which is not a subset of X for all $X \in \{A_2, A_3, B\}$. Then there are some functions q_i $(i = 1, 2, 3)$ with $q_1 \in T \setminus B_1$, $q_2 \in T \setminus A_2$ and $q_3 \in T \setminus A_3$. By identifying the variables in the functions q_2 and q_3, one obtains the functions u_5 and u_2. The function q_1 has at least two essential variables and it holds that $F(q_1) \cap \{j_2, j_5\} \neq \emptyset$. By identifying certain variables of q_1 and substituting certain variables of q_1 through the functions u_2, u_5 we obtain the functions $j_5(x) \vee u_2(y)$ and $j_5(x) \vee u_5(y)$. In the proof of Lemma 15.4.2.2, we showed that an arbitrary function $t \in M$ with $F(t) \in \{\{j_5, u_2\}, \{j_5, u_5\}\}$ is a superposition over the above-constructed functions. Then, by identifying variables in a function $t^4 \in M$ with $F(t) = \{j_5, u_2\}$ and $num_t(j_5) = num_t(u_2) = 2$, we obtain $max \in [\{q_1, q_2, q_3\}]$. Since in addition $x \vee (j_5(u_2(y)) \vee u_2(x)) = x \vee j_2(y)$ holds, we have $[\{q_1, q_2, q_3\}] = A_3$, whereby our lemma is proven. ∎

Obviously, it holds

$$I_1 := A_2 \cap A_3 = A_1 \cap Pol_3\{1\},$$

and the functions f of I_1 are idempotent; i.e., $f(x,x,...,x) = x$. Subsequently, we determine the elements of

$$\mathbb{L}_3(A_1)\backslash(\mathbb{L}_3(R) \cup \mathbb{L}_3([M^1]),$$

which belong to B_1 and I_1, and then the remaining elements of

$$\mathbb{L}_3(A_1)\backslash(\mathbb{L}_3(R) \cup \mathbb{L}_3([M^1])$$

(see Figure 15.4). With the aid of Theorems 15.4.3.1 and 15.4.4.2, one obtains a complete description of $\mathbb{L}_3(A_1)$.

Theorem 15.4.5.2 *B_1 has exactly the following 20 subclasses, which are not subsets of $[M^1]$ or U_1:*
$B_1 = [\{max, u_2, u_5\}]$,
$B_2 := \{f \in B_1 \mid F(f) \cap \{s_1, u_5\} \neq \emptyset\} = [\{max, x \vee u_2(y), u_5\}]$,
$B_3 := \{f \in B_1 \mid num_f(s_1) \leq 1\} = [\{x \vee u_2(y), x \vee u_5(y), u_2\}]$,
$B_4 := \{f \in B_1 \mid u_5 \notin F(f)\} = [\{max, u_2\}]$,
$B_5 := \{f \in B_2 \mid num_f(s_1) \geq 2 \Longrightarrow u_5 \in F(f)\} = [\{x \vee y \vee u_5(z), x \vee u_2(y)\}]$,
$B_6 := \{f \in B_2 \mid u_2 \in F(f) \Longrightarrow u_5 \in F(f)\} = [\{max, u_2(x) \vee u_5(y)\}]$,
$B_7 := B_5 \cap B_6 = [\{x \vee y \vee u_5(z), u_2(x) \vee u_5(y), s_1\}]$,
$B_8 := \{f \in B_1 \mid u_2 \notin F(f)\} = [\{max, u_5\}]$,
$B_9 := B_2 \cap B_3 = [\{x \vee u_2(y), u_5\}]$,
$B_{10} := B_2 \cap B_4 = [\{max, x \vee u_2(y)\}]$,
$B_{11} := B_7\backslash[\{s_1\}]$,
$B_{12} := B_7 \cap B_8 = [\{x \vee y \vee u_5(z), s_1\}]$,
$B_{13} := B_7 \cap B_3 = [\{x \vee u_5(y), u_2(x) \vee u_5(y), s_1\}]$,
$B_{14} := B_3 \cap B_4 = [\{x \vee u_2(y), u_2\}]$,
$B_{15} := B_{11} \cap B_{12} = [\{x \vee y \vee u_5(z)\}]$,
$B_{16} := B_{12} \cap B_{13} = [\{x \vee u_5(y), s_1\}]$,
$B_{17} := B_9 \cap B_{13} = [\{x \vee u_5(y), u_2(x) \vee u_5(y)\}]$,
$B_{18} := B_9 \cap B_{14} = [\{x \vee u_2(y)\}]$,
$B_{19} := [\{max\}]$,
$B_{20} := [\{x \vee u_5(y)\}]$
(See Figure 15.5).

Proof. Except for the functions f with $u_5 \in F(f)$ and $num_f(s_1) \geq 2$, for which $f \in [[\{f\}]^3]$ is valid, we have by Lemma 15.4.2.2 $f \in [\{f\}]^2$ for all other functions $f \in B_1\backslash[M^1]$. Consequently, one can describe an arbitrary subclass B of B_1 in the form of a closure of a certain subset of

$$\{x \vee y \vee u_5(z), max, x \vee u_2(y), x \vee u_5, u_2(y) \vee u_5(y), u_2, u_5, s_1\}.$$

When one examines the possible cases for B ($\not\subseteq U_1$ or $\not\subseteq [M^1]$) with the aid of Figure 15.5, one obtains our theorem. ■

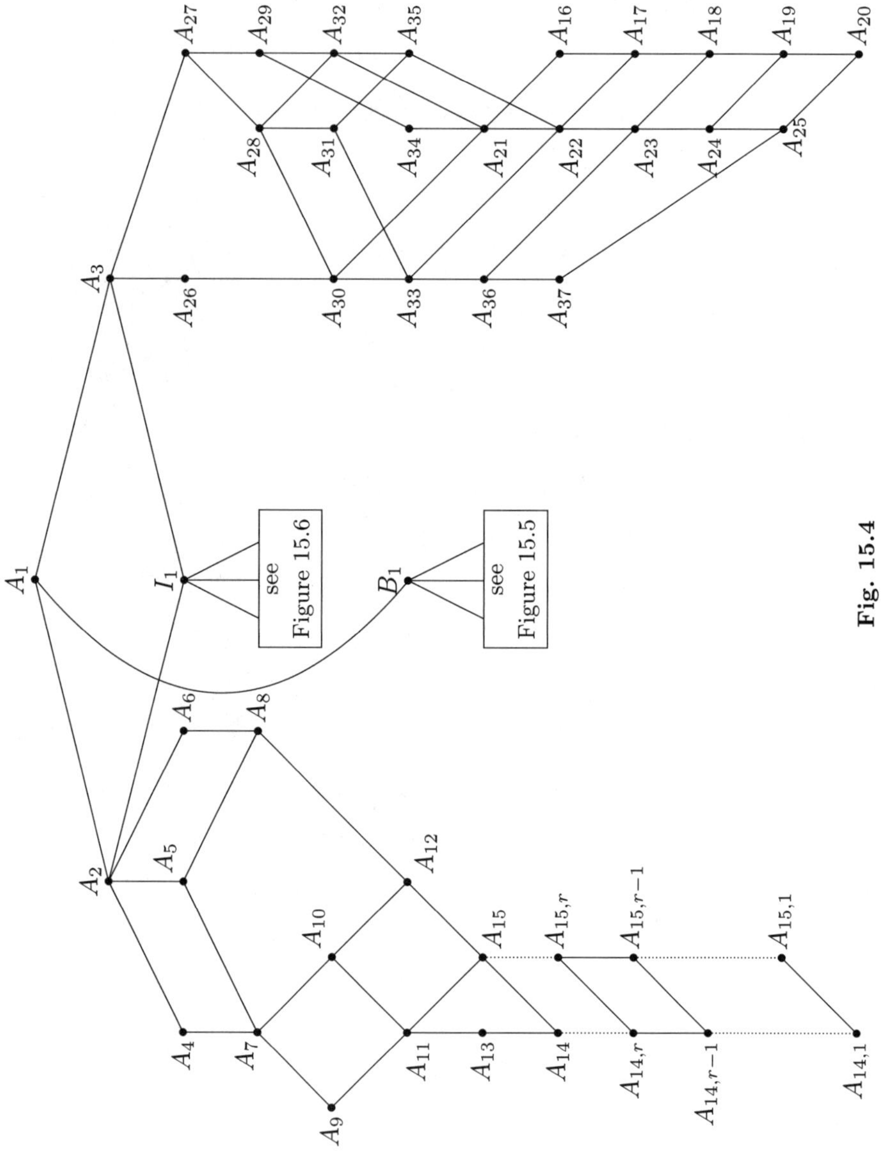
A1
A2
A3
I1
B1
see Figure 15.6
see Figure 15.5
A4
A5
A6
A7
A8
A9
A10
A11
A12
A13
A14
A15
A14,r
A15,r
A14,r−1
A15,r−1
A14,1
A15,1
A16
A17
A18
A19
A20
A21
A22
A23
A24
A25
A26
A27
A28
A29
A30
A31
A32
A33
A34
A35
A36
A37

Fig. 15.4

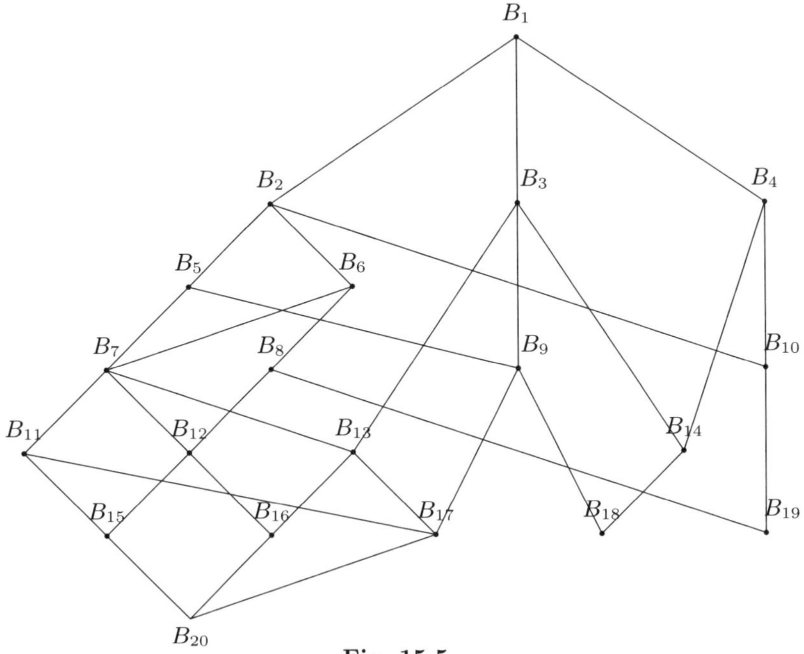

Fig. 15.5

Theorem 15.4.5.3 *I_1 has exactly 17 subclasses, which are not subsets of $[M^1]$:*
$I_1 = [\{max, u_2(x) \vee j_5(y)\}]$,
$I_2 := [\{max, x \vee j_5(y), x \vee u_2(y)\}]$,
$I_3 := [\{x \vee j_5(y), j_5(x) \vee u_2(y)\}]$,
$I_4 := [\{x \vee u_2(y), j_5(x) \vee u_2(y)\}]$,
$I_5 := [\{max, x \vee j_2(y), x \vee j_5(y)\}]$,
$I_6 := [\{max, x \vee j_2(y), x \vee u_2(y)\}]$,
$I_7 := [\{x \vee j_2(y), j_5(y) \vee u_2(y)\}]$,
$I_8 := [\{x \vee j_5(y), x \vee j_2(y)\}]$,
$I_9 := [\{x \vee j_2(y), x \vee u_2(y)\}]$,
$I_{10} := [\{max, x \vee j_5(y)\}]$,
$I_{11} := [\{max, x \vee j_2(y)\}]$,
$B_{10} = [\{max, x \vee u_2(y)\}]$,
$I_{12} := [\{j_5(x) \vee u_2(y)\}]$,
$I_{13} := [\{x \vee j_5(y)\}]$,
$I_{14} := [\{x \vee j_2(y)\}]$,
$B_{18} = [\{x \vee u_2(y)\}]$,
$B_{19} = [\{max\}]$.

Proof. Because of Theorem 15.4.2.2, every subclass of I_1 has a generating system from binary functions of I_1. Consequently, one obtains the subclasses of I_1, which are not subclasses of $[M^1]$, through closure of the subsets of

$$\{max, x \vee j_2(y), x \vee j_5(y), x \vee u_2(y), j_5(x) \vee u_2(y)\}.$$

(During the forming of these classes one notices that the equations

$$x \vee j_5(j_5(x) \vee u_2(y)) = x \vee j_2(y)$$

and

$$u_2(x) \vee j_5(x \vee u_2(y)) = x \vee j_2(y)$$

are valid.)
The Figure 15.6 gives the Hasse diagram of the classes constructed in this manner. ■

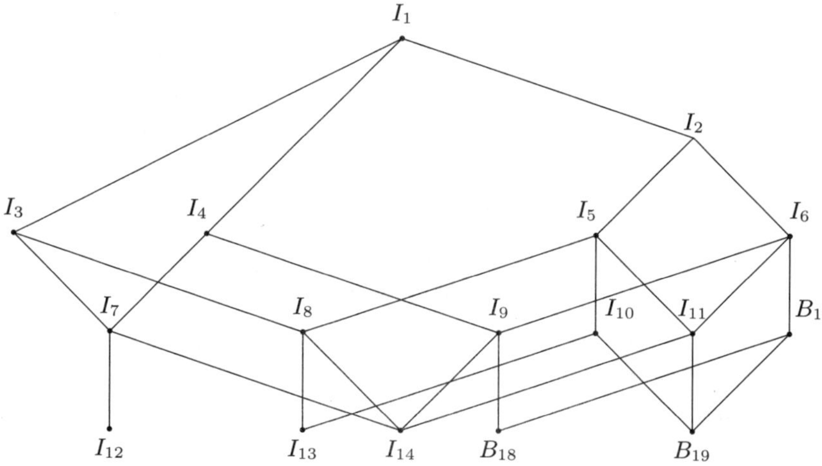

Fig. 15.6

Theorem 15.4.5.4 *The following classes are the only subclasses of A_2 that are not contained in I_1, B_1 or $[M^1]$:*
$A_2 = [\{max, x \vee j_5(y), u_2\}]$,
$A_4 := [\{max, j_2(x) \vee u_2(y)\}]$,
$A_5 := [\{x \vee u_2(y), j_5(x) \vee u_2(y), u_2\}]$,
$A_6 := [\{x \vee j_5(y), u_2\}]$,
$A_7 := [\{x \vee j_2(y), x \vee u_2(y), u_2\}]$,
$A_8 := [\{j_5(x) \vee u_2(y), u_2\}]$,
$A_9 := [\{x \vee u_2(y), j_2(x) \vee u_2(y)\}]$,
$A_{10} := [\{x \vee j_2(y), u_2(x) \vee u_2(y)\}]$,
$A_{11} := [\{s_1, j_2(x) \vee u_2(y), u_2(x) \vee u_2(y)\}]$,
$A_{12} := [\{x \vee j_2(y), u_2\}]$,
$A_{13} := A_{11} \backslash [\{s_1\}]$,
$A_{14} := [\{u_2(x_1) \vee j_2(x_2) \vee ... \vee j_2(x_n) \mid n \in \mathbb{N} \backslash \{1\}\}]$,

$A_{14,r} := [\{f \in A_{10} \mid num_f(j_2) \le r\}]$,
$A_{15} := A_{14} \cup [\{s_1\}]$,
$A_{15,r} := A_{14,r} \cup [\{s_1\}]$,
where $r = 1, 2, \ldots$.

Proof. Let A be a subclass of A_2, which is not a subset of I_1, B_1 or $[M^1]$. Because of $A \not\subseteq I_1 = A_2 \cap A_3$, u_2 belongs to A. Then, by $A \not\subseteq [M^1]$ and $A \not\subseteq B_1$, we have $j_2(x) \vee u_2(y) \in A$. Thus A contains the class $A_{14,1}$. One can verify the remaining statements of our theorem easily with the aid of the above-noted generating systems of the classes, Figure 15.4, and Theorem 15.4.2.2. ■

Theorem 15.4.5.5 *The following classes are the only subclasses of* A_3 *that are not contained in* I_1, B_1 *or* $[M^1]$:
$A_3 = [\{max, j_5(x) \vee u_2(y), u_5\}]$,
$A_{16} := \{f \in A_1 \mid u_5 \in F(f)\} = [\{j_5(x) \vee u_5(y), j_2(x) \vee u_5(y)\}]$,
$A_{17} := \{f \in A_{16} \mid \{j_2, u_2\} \cap F(f) \neq \emptyset\} = [\{u_5(x) \vee j_5(y), x \vee y \vee u_5(z)\}]$,
$A_{18} := \{f \in A_{17} \mid num_f(s_1) \le 1\} = [\{x \vee j_5(y) \vee u_5(z)\}]$,
$A_{19} := \{f \in A_{18} \mid num_f(s_1) = 1 \Longrightarrow j_5 \notin F(f)\} = [\{x \vee u_5(y), j_5(x) \vee u_5(y)\}]$,
$A_{20} := \{f \in A_{19} \mid num_f(s_1) = 0\} = [\{j_5(x) \vee u_5(y)\}]$,
$A_{21} := A_{16} \cup [\{s_1\}]$,
$A_{22} := A_{17} \cup [\{s_1\}]$,
$A_{23} := A_{18} \cup [\{s_1\}]$,
$A_{24} := A_{19} \cup [\{s_1\}]$,
$A_{25} := A_{20} \cup [\{s_1\}]$,
$A_{26} := I_3 \cup A_{16}$,
$A_{27} := I_2 \cup A_{16}$,
$A_{28} := I_5 \cup A_{16}$,
$A_{29} := B_{10} \cup A_{16}$,
$A_{30} := I_8 \cup A_{16}$,
$A_{31} := I_{10} \cup A_{17}$,
$A_{32} := B_{19} \cup A_{16}$,
$A_{33} := I_{13} \cup A_{17}$,
$A_{34} := B_{18} \cup A_{16}$,
$A_{35} := B_{19} \cup A_{17}$,
$A_{36} := I_{13} \cup A_{18}$,
$A_{37} := I_{13} \cup A_{20}$.

Proof. Let T be a subclass of A_3 with $T \not\subseteq I_1$, $T \not\subseteq B_1$ and $T \not\subseteq [M^1]$.
It follows from $T \not\subseteq I_1$, $T \not\subseteq B_1$, $T \not\subseteq [M^1]$ and $j_2 \star u_5 = j_5$ that the functions u_5 and $j_5(x) \vee u_5(y)$ belong to T. Hence we have $A_{20} \subseteq T$.
Since for an arbitrary function $f \in A_3$ either $f(1, \ldots, 1) = 1$ or $f(1, \ldots, 1) = 2$ holds, and the case $f(1, \ldots, 1) = 2$ and $f \in A_3$ is only possible, if $u_5 \in F(f)$,

the class T has the form

$$(T \cap I_1) \cup (T \cap A_{16}). \tag{15.5}$$

First we show the possibilities for $T \cap A_{16}$.
Because of

$$j_2(u_5(y)) \vee u_5(j_2(x) \vee u_5(y)) = u_2(x) \vee u_5(y)$$

and

$$j_5(u_2(x) \vee u_5(y)) \vee u_5(y) = j_2(x) \vee u_5(y)$$

we have

$$j_2(x) \vee u_5(y) \in T \iff u_2(x) \vee u_5(y) \in T. \tag{15.6}$$

Further, it holds that:

$$j_5(x_1) \vee j_5(x_2) \vee u_5(x_3) \vee u_5(x_4) \in [\{j_5(x) \vee u_5(y)\}]$$

This implies

$$\begin{aligned} & x \vee y \vee u_5(z) = j_5(g(z,x)) \vee j_5(g(z,y)) \vee u_5(g(x,z)) \vee u_5(g(y,z)) \\ & \in [\{j_5(x) \vee u_5(y), g\}] \end{aligned} \tag{15.7}$$

for every $g(x,y) \in \{j_2(x) \vee u_5(y), u_2(x) \vee u_5(y)\}$.
With the aid of (15.6) and (15.7), it is easy to check that the classes A_{16}, ..., A_{20} are the only possibilities for T if $T \subseteq A_{16}$ holds. The remaining possibilities for T can be obtained with the help of (15.5) and Theorem 15.4.5.3, when one determines the classes $I \in \mathbb{L}_3(I_1)\backslash\{\emptyset\}$ and $A \in \{A_{16}, ..., A_{20}\}$ with $I \cup A = [I \cup A]$. ■

15.4.6 The Remaining Subclasses of M

The following lemma is the basis for determining the subclasses of M that are still missing.

Lemma 15.4.6.1 *Let T be a subclass of M, which is not a subset of $[M^1]$, R or A_1. Furthermore, let*

$$T_1 := T \cap R$$

and

$$T_2 := T \cap Pol \begin{pmatrix} 0 \\ 2 \end{pmatrix}.$$

Then T fulfills one of the two following conditions:

(a) $T \in \{T_1 \cup [\{s_1\}] \mid T_1 \in \mathbb{L}_3(R)\backslash\mathbb{L}_3([M^1])\}$.

(b) $T_2 \cap \{max, x \vee j_2(y), x \vee j_5(y), x \vee u_2(y), x \vee u_5(y), j_2(x) \vee u_2(y), j_5(x) \vee u_2(y), j_5(x) \vee u_5(y)\} \neq \emptyset$.

Proof. For T the following two cases are possible:
Case 1: $T_2 \subseteq [M^1]$.
Because of $T \not\subseteq R$ and $(T \cap Pol \begin{pmatrix} 0 \\ 2 \end{pmatrix})^1 \subseteq \{u_2, u_5, s_1\}$, this case is only possible for $s_1 \in T_2$. Since $T_2 \backslash [\{s_1\}] \subseteq R$ and classes of the form $T_1 \cup [\{s_1\}]$ are closed for all $T_1 \in \mathbb{L}_3(R)$, T fulfills the condition (a).
Case 2: $T_2 \not\subseteq [M^1]$.
By Section 15.4.5, T fulfills the condition (b). ∎

In the following, denote T a subclass of M with the properties:

$$T \cap R \neq \emptyset$$

and

$$T \cap \{max, x \vee j_2(y), x \vee j_5(y), x \vee u_2(y), x \vee u_5(y), j_2(x) \vee u_2(y), j_5(x) \vee u_2(y), j_5(x) \vee u_5(y)\} \neq \emptyset.$$

Furthermore, we use the notations:

$$T_1 := T \cap R$$

and

$$T_2 := T \cap Pol \begin{pmatrix} 0 \\ 2 \end{pmatrix} \quad (= T \cap A_1).$$

Theorem 15.4.6.2 *If $T \subseteq J_1 \cup A_1$ then T is one of the following classes:*
1) $H_2 \cup B_{19}, J_{12} \cup I_{11}, J_{14} \cup I_{10}, J_2 \cup I_5$
(classes that contain the set $\{c_0, max\}$);
2) $J_2 \cup I_8, J_{14} \cup I_{13}, J_i \cup I_{14} (i \in \{1, 2, 5, 6, 11, 12\})$
(classes T with $c_0 \in T$, $max \notin T$ and $\{x \vee j_5(y), x \vee j_2(y)\} \cap T \neq \emptyset$);
3) $H_3 \cup I_{14}, J_7 \cup I_{14}, J_3 \cup I_{14}$
(classes T with $c_1 \in T$ and $c_0 \notin T$);
4) $J_{16} \cup I_{13}, J_{16} \cup I_{10}, J_8 \cup I_4, J_8 \cup I_7, J_8 \cup I_{14}, J_4 \cup I_5, J_4 \cup I_1, J_4 \cup I_8, J_4 \cup I_3, J_4 \cup I_{14}$
(classes that do not contain constant functions).

Proof. Because of $x \vee c_1(x) = v_2(x)$ the set $J_1 \cup A_1$ is not closed. To determine all closed subsets T of $J_1 \cup A_1$, we distinguish the following cases:
Case 1: $c_0 \in T_1$.
Then for every function $f \in T$, we have $F(f) \subseteq T^1$. Thus T_2 is a subset of I_1, and we have $\{max, x \vee j_5(y), x \vee j_2(y)\} \cap T \neq \emptyset$ and $\{j_5(x) \vee u_2(y), x \vee u_2(y)\} \cap T = \emptyset$. If $\{max, x \vee j_5(y)\} \cap T \neq \emptyset$, T does not contain c_1, since $x \vee c_1(x) = v_2(x) \notin T$.
Case 1.1: $max \in T_2$.
In this case, the class T is also describable in the form $T = [T^1 \cup [max\}]$. With

the aid of Theorems 15.4.3.1, 15.4.4.1, 15.4.5.3 and Table 15.14, one obtains the classes given in *1)*.
Case 1.2: $max \notin T_2$.
Case 1.2.1: $x \vee j_5(y) \in T_2$.
If one examines the possibilities that arise from Theorems 15.4.4.1 and 15.4.5.3 for T, then one sees that only the sets $J_{14} \cup I_{13}$ and $J_2 \cup I_8$ are closed.
Case 1.2.2: $x \vee j_5(y) \notin T_2$ and $x \vee j_2(y) \in T_2$.
In this case, we have $T_2 = I_{14}$ and T_1 contains J_{12}. Then, the possibilities for T are: $J_i \cup I_{14}$, where $i \in \{1, 2, 5, 6, 11, 12\}$).
Case 2: $c_0 \notin T_1$ and $c_1 \in T_1$.
Because of $x \vee c_1(x) = v_2(x)$, $c_1 \vee u_i = v_i$ $(i = 2, 5)$ and $j_2(c_1) \vee u_2 = u_2$, only classes T are possible with $\{max, x \vee j_5(y), x \vee u_2(y), x \vee u_5(y), j_2(x) \vee u_2(y), j_5(x) \vee u_2(y), j_5(x) \vee u_5(y)\} \cap T = \emptyset$. Hence, by Theorem 15.4.4.1 and Section 15.4.5, $T_1 \in \{H_3, J_7, J_3\}$ and $T_2 = I_{14}$, where every possibility supplies a closed class, which is given in *3)*.
Case 3: $\{c_0, c_1\} \cap T = \emptyset$.
In this case, by Theorem 15.4.4.1, $T_1 \in \{H_9, J_{16}, J_8, J_4\}$. Further, we have

$$\{max, x \vee j_2(y), x \vee j_5(y), x \vee u_2(y), j_5(x) \vee u_2(y)\} \cap T \neq \emptyset$$

and

$$\{x \vee u_5(y), j_2(x) \vee u_2(y), j_5(x) \vee u_5(y)\} \cap T = \emptyset.$$

Thus $T_1 = H_9$ is not possible. We must only continue to examine, therefore, the following three cases:
Case 3.1: $T_1 = J_{16}$.
Then, T can not contain the functions $x \vee j_2(y)$, $x \vee u_2(y)$ and $j_5(x) \vee u_2(y)$; thus $T_2 \in \{I_{13}, B_{19}, I_{10}\}$ and $T \in \{J_{16} \cup I_{13}, J_{16} \cup I_{10}\}$.
Case 3.2: $T_1 = J_8$.
In this case, we have $T \cap \{max, x \vee j_5(y)\} = \emptyset$ and $T_2 \subseteq I_4$. By scrutinizing the possibilities that result, one receives $T \in \{J_8 \cup I_4, J_8 \cup I_7, J_8 \cup I_{14}\}$.
Case 3.3: $T_1 = J_4$.
The following cases are still possible, then:
Case 3.3.1: $max \in T_2$.
Then T contains the class $J_4 \cup I_5$. Because of $j_5(x) \vee u_2(y) \in [\{j_5, x \vee u_2(y)\}]$ the set $J_4 \cup I_2$ is not closed, and only the classes T with $T = J_4 \cup I_5$ or $T = J_4 \cup I_1$ fulfill the conditions of this case.
Case 3.3.2: $max \notin T_2$.
Case 3.3.2.1: $x \vee j_5(y) \in T$.
Since $x \vee j_5(j_5(x) \vee j_2(y)) = x \vee j_2(y)$, T contains the class I_8 and it holds that $T \in \{J_4 \cup I_8, J_4 \cup I_3\}$.
Case 3.3.2.2: $\{j_5(x) \vee u_2(y), x \vee u_2(y)\} \cap T \neq \emptyset$ and $x \vee j_5(y) \notin T$.
Because of $(j_5(x) \vee j_5(y)) \vee u_2(x) = x \vee j_5(y)$, this case is not possible.
Case 3.3.2.3: $x \vee j_2(y) \in T$ and $\{x \vee j_5(y), x \vee u_2(y), j_5(x) \vee u_2(y)\} \cap T = \emptyset$.
Only the class $J_4 \cup I_{14}$ can be T in this case. ■

Theorem 15.4.6.3 *If $T \subseteq U_1 \cup A_1$, then T is exactly one of the following classes:*
1) $H_2 \cup B_{19}, H_6 \cup B_{19}, U_6 \cup B_4, U_7 \cup B_8, U_3 \cup B_4, U_4 \cup B_8, U_2 \cup B_1, U_1 \cup B_1$ *(classes that contain $\{c_0, max\}$;*
2) $U_1 \cup B_3, U_2 \cup B_3, U_3 \cup B_{14}, U_6 \cup B_{14}, U_4 \cup B_{16}, U_7 \cup B_{16}$ *(classes T with $c_0 \in T$ and $max \notin T$);*
3) $H_4 \cup B_{19}, H_4 \cup B_{10}, U_5 \cup B_1, U_8 \cup B_4, U_9 \cup B_2, U_9 \cup B_6, U_{10} \cup B_8$ *(classes T with $\{c_2, max\} \subseteq T$ and $c_0 \notin T$);*
4) $H_4 \cup B_{18}, U_5 \cup B_3, U_8 \cup B_{14}, U_9 \cup B_5, U_9 \cup B_9$ *(classes T with $\{c_2, x \vee u_2(y)\}$, $c_0 \notin T$ and* max $\notin T$*);*
5) $U_9 \cup B_i$ *($i \in \{7, 11, 13, 17\}$),* $U_{10} \cup B_t$ *($t \in \{12, 15, 16, 20\}$) (classes T with $\{c_2, x \vee u_5(y)\} \subseteq T$ and $\{c_0, max, x \vee u_2(y)\} \cap T = \emptyset$).*

Proof. Since $U_1 \backslash [\{c_0, c_1\}] \subseteq A_1$, we have $T \cap \{c_0, c_2\} \neq \emptyset$.
Case 1: $c_0 \in T_1$.
In this case, it holds that $F(f) \subseteq T^1$ for all $f \in T$. Because of Section 15.4.5, then, we have either B_{18} or B_{19} or B_{20} as a subset of T.
Case 1.1: $max \in T_2 (B_{19} \subseteq T)$.
Then $[T^1 \cup \{max\}] = T$, and because of Theorems 15.4.3.1, 15.4.4.2, and 15.4.5.2, for T there are only the possibilities given in *1)*.
Case 1.2: $max \notin T_2$.
If $x \vee u_2(y) \in T_2$ (i.e., $B_{18} \subseteq T_2$) then u_2 belongs to T_1, and by Theorem 15.4.5.2, we have $T_2 \in \{B_3, B_{14}\}$. Thus

$$T \in \{U_6 \cup B_{14}, U_3 \cup B_{14}, U_2 \cup B_3, U_1 \cup B_3\}.$$

In the case $x \vee u_5(y) \in T_2$ (i.e., $B_{20} \subseteq T_2$) and $x \vee u_2(y) \notin T$, only $T_1 \in \{U_4, U_7\}$ and $T_2 = B_{16}$ are possible (because of $c_0 \in T$). Therefore, in Case 1.2, only the classes given in *2)* are possible.
Case 2: $c_0 \notin T_1$ and $c_2 \in T_1$.
Since $j_r(c_2) \vee u_s = v_s$ and $x \vee j_r(c_2) = v_2$ $(r, s \in \{2, 5\})$, we have

$$T \cap \{j_r(x) \vee u_s(y), x \vee j_r(y) \mid r, s \in \{2, 5\}\} = \emptyset$$

and, by Section 15.4.5, T contains either B_{18}, B_{19} or B_{20}.
Case 2.1: $max \in T_2$.
Then T_1 belongs to $\{H_4, U_5, U_8, U_9, U_{10}\}$ and T_2 belongs to $\{B_1, B_2, B_4, B_6, B_8, B_{10}, B_{19}\}$. In *3)* those classes of the form $T_1 \cup T_2$ are given, which are closed.
Case 2.2: $max \notin T$ and $x \vee u_2(y) \in T$.
In this case, by Theorem 15.4.5.2, we have $T_2 \in \{B_3, B_5, B_9, B_{14}, B_{18}\}$. Then T_1 belongs to the set $\{H_4, U_5, U_8, U_9\}$. When one considers the possible cases, one receives the classes given in *4)*.
Case 2.3: $\{max, x \vee u_2(y)\} \cap T = \emptyset$ and $x \vee u_5(y) \in T$.

Then it holds that $T_2 \in \{U_9, U_{10}\}$ and $T_2 \in \{B_7, B_{11}, B_{12}, B_{13}, B_{15}, B_{16}, B_{17}, B_{20}\}$. The closed sets of the form $T_1 \cup T_2$ are given in *5)*. ■

Theorem 15.4.6.4 *If* $T \subseteq V_1 \cup A_1$, *then* T *is exactly one of the following classes:*

1) $V_1 \cup I_5, V_1 \cup I_{12}, V_1 \cup B_{19}$,
$V_{15} \cup I_s$ $(s \in \{2,5,6,10,11\})$,
$V_{15} \cup B_{10}, V_{15} \cup B_{19}$,
$V_{17} \cup I_t$ $(t \in \{2,5,6,10,11\})$,
$V_{17} \cup B_{10}, V_{17} \cup B_{19}$
(classes that belong $\{c_1, max\}$*);*

2) $V_{15} \cup B_{18}, V_{15} \cup I_9, V_{17} \cup B_{18}, V_{17} \cup I_9$
(classes T *with* $\{c_1, x \vee u_2(y)\} \subset T$ *and* $\{max, x \vee j_5(y)\} \cap T = \emptyset$*);*

3) $V \cup I_8 (V \in \{H_{32}, H_{34}, V_1, V_3, V_{15}, V_{17}\})$,
$V \cup I_{13} (V \in \{H_{32}, H_{34}, V_1, V_3, V_5, V_7, V_{15}, V_{17}\})$
(classes T *with* $\{c_1, x \vee j_5(y)\} \subset T$ *and* $\{max, x \vee u_2(y)\} \cap T = \emptyset$*);*

4) $V \cup I_{14} (V \in \{H_3, H_{32}, H_{34}, V_{15}, V_{17}\})$
(classes T *with* $x \vee j_2(y) \in T$ *and* $\{max, x \vee u_2(y), x \vee j_5(y)\} \cap T = \emptyset$*);*

5) $V_2 \cup B_{19}, V_{10} \cup B_{19}, V_{13} \cup B_{19}$,
$V_2 \cup I_{13}, V_4 \cup I_{13}, V_8 \cup I_{13}$,
$V_2 \cup I_8, V_4 \cup I_8$,
$V_2 \cup I_{10}$,
$V_2 \cup I_5$
(classes T *with* $\{c_2, v_5\} \subset T$ *and* $c_1 \notin T$*);*

6) $H_4 \cup B_{10}, H_4 \cup B_{19}$,
$H_{31} \cup I_r$ $(r \in \{3,7,8,12,13,14\})$,
$V_{16} \cup B_s$ $(s \in \{10,18,19\})$,
$V_{16} \cup I_t$ $(t \in \{1,2,4,5,6,9,10,11\})$
(classes T *with* $c_2 \in T$ *and* $T \cap \{c_1, v_5\} = \emptyset$*);*

7) $H_{29} \cup I_r$ $(r \in \{3,7,8,12,13,14\})$,
$V_{18} \cup B_s (s \in \{10,18,19\})$,
$V_{18} \cup I_t (t \in \{1,2,4,5,6,9,10,11\})$
(classes T *with* $T \cap \{c_1, c_2\} = \emptyset$*).*

Proof. Because of $U_1 \cap A_1 \neq \emptyset$, the set $A_1 \cup V_1$ is not closed and, therefore, $T \subset V_1 \cup A_1$. Since $u_5 \in A$ for every $A \in \{B_{20}, A_{20}\}$, $u_2 \in A_{14,1}$ and $u_2 \in B_{14}$, T_2 is a certain subclass of I_1.

Case 1: $c_1 \in T_1$.

Then, because of $j_5 \vee u_2(c_1) = j_5$, we have $j_5(x) \vee u_2(y) \notin T_2$.

Case 1.1: $max \in T_2$.

Since $x \vee c_1 = v_2(x)$, it holds that $V_{17} \subseteq T_1$. With the aid of Theorems 15.4.4.3 and 15.4.5.3, one obtains only the possibilities given in *1)* for T.

Case 1.2: $max \notin T_2$.
By Theorem 15.4.5.3 and because of $j_5(x) \vee u_2(y) \notin T$, only the following three cases are possible:
Case 1.2.1: $x \vee u_2(y) \in T_2$.
Because of $c_1 \vee u_2 = v_2$ it holds $V_{17} \subseteq T_1$. Further, because $x \vee u_2(v_5) = u_5(x)$, $v_5 \notin T$. Hence,

$$T \in \{V_{15} \cup B_{18}, V_{15} \cup I_9, V_{17} \cup B_{18}, V_{17} \cup I_9\}.$$

Case 1.2.2: $x \vee u_2(y) \notin T_2$ and $x \vee j_5(y) \in T_2$.
Because of Theorem 15.4.5.3, this case is only possible for $T_2 \in \{I_8, I_{13}\}$. With the aid of Theorem 15.4.4.3, this implies that T is a class given in *3)*.
Case 1.2.3: $T_2 = I_{14}(= [\{x \vee j_2(y)\}])$.
Because of $x \vee j_2(v_5(y)) = x \vee j_5(y)$, the function v_5 does not belong to T_1. With the aid of Theorem 15.4.4.3 this implies that T is a class which is given in *4)*.
Case 2: $c_1 \notin T_1$ and $c_2 \in T_1$.
We distinguish two cases:
Case 2.1: $v_5 \in T_1$.
Because of $j_5(x) \vee u_2(v_5(x)) = u_5(x) \notin T$ and $x \vee u_2(v_5(x)) = u_5(x) \notin T$, T_2 cannot contain $j_5(x) \vee u_2(y)$ or $x \vee u_2(y)$. Further, we have:

$$x \vee j_2(y) \in T \implies x \vee j_2(v_5(y)) = x \vee j_5(y) \in T.$$

Thus $T_2 \in \{B_{19}, I_5, I_8, I_{10}, I_{13}\}$ and $V_{10} \subseteq T_1$. Because of $c_1 \notin T$, this implies $T_1 \in \{V_2, V_4, V_6, V_8, V_{10}, V_{13}, V_{14}\}$. The possibilities resulting for T are given in *5)*.
Case 2.2: $v_5 \notin T_1$.
Then, because of Theorem 15.4.4.3, $T_1 \in \{H_4, H_{31}, V_{16}\}$. The possibilities resulting for T are given in *6)*.
Case 3: $\{c_1, c_2\} \cap T = \emptyset$.
Because of $v_5 \star v_5 = c_2$, $v_5 \notin T$. Thus, either $T_1 = H_{29} = [\{v_2\}]$ or $T_1 = V_{18} = [\{v_2(x) \vee v_2(y)\}]$. With the aid of Theorem 15.4.5.3, this implies that T is a class which is given in *7)*. ■

Theorem 15.4.6.5 *If $T \subseteq (J_1 \cup U_1) \cup A_1$, $T_1 \not\subseteq J_1 \cup A_1$ and $T_1 \not\subseteq U_1 \cup A_1$, then T is exactly one of the following classes:*

1) $(J_4 \cup U_{13}) \cup A_i$ $(i \in \{3, 16, 21, 26, 30\})$,
$(J_{16} \cup U_{14}) \cup A_m$ $(m \in \{17, 18, 19, 20, 22, 23, 24, 25, 33, 35, 36, 37\})$,
$(J_{12} \cup U_6) \cup A_p$ $(p \in \{4, 7, 9, 10, 11, 13\})$,
$(J_{12} \cup H_{19}) \cup A_n$ $(n \in \{12, 14, 15\})$,
$(J_{12} \cup H_{19}) \cup A_{s,r}$ $(s \in \{14, 15\}),\ r \in \mathbb{N})$,
$(J_{12,r} \cup H_{19}) \cup A_{s,r'}$ $(s \in \{14, 15\};\ r, r' \in \mathbb{N}; r' \leq r)$,
$(J_{14} \cup U_7) \cup A_q$ $(q \in \{17, 18, 19, 20, 22, 23, 24, 25, 33, 35, 36, 37\})$,

$(J_{10,r} \cup H_{19}) \cup A_{s,r'}$ $(s \in \{14, 15\};\ r, r' \in \mathbb{N}; r' \leq r)$,
$(J_{10} \cup H_{19}) \cup A_t$ $(t \in \{12, 14, 15\})$,
$(J_{10} \cup H_{19}) \cup A_{s,r}$ $(s \in \{14, 15\}, r \in \mathbb{N})$,
$(J_{16} \cup H_{19}) \cup A_i$ $(i \in \{8, 12, 14, 15\})$,
$(J_{16} \cup H_{19}) \cup A_{s,r}$ $(s \in \{14, 15\}, r \in \mathbb{N})$,
$(J_{10} \cup U_6) \cup A_m$ $(m \in \{11, 13\})$,
$(J_6 \cup U_6) \cup A_n$ $(n \in \{5, 10, 11, 13\})$,
$(J_2 \cup H_{19}) \cup A_p$ $(p \in \{6, 12, 14, 15\})$,
$(J_2 \cup H_{19}) \cup A_{s,r}$ $(s \in \{14, 15\}, r \in \mathbb{N})$,
$(J_2 \cup U_6) \cup A_q$ $(q \in \{2, 10, 11, 13\})$,
$(J_2 \cup U_2) \cup A_1$
(classes T with $\{c_0, c_1\} \cap T = \emptyset$);

2) $(J_{11} \cup H_{19}) \cup A_i$ $(i \in \{12, 14, 15\})$,
$(J_{11} \cup H_{19}) \cup A_{s,r}$ $(s \in \{14, 15\},\ r \in \mathbb{N})$,
$(J_{11} \cup U_6) \cup A_m$ $(m \in \{10, 11, 13\})$,
$(J_{11,r} \cup H_{19}) \cup A_{s,r'}$ $(s \in \{14, 15\};\ r, r' \in \mathbb{N};\ r' \leq r)$,
$(J_9 \cup H_{19}) \cup A_{s,r}$ $(s \in \{14, 15\},\ r \in \mathbb{N})$,
$(J_{9,r} \cup H_{19}) \cup A_{s,r'}$ $(s \in \{14, 15\};\ r, r' \in \mathbb{N};\ r' \leq r)$,
$(J_5 \cup H_{19}) \cup A_n$ $(n \in \{12, 14, 15\})$,
$(J_5 \cup H_{19}) \cup A_{s,r'}$ $(s \in \{14, 15\};\ r \in \mathbb{N})$,
$(J_1 \cup H_{19}) \cup A_p$ $(p \in \{12, 14, 15\})$,
$(J_1 \cup H_{19}) \cup A_{s,r}$ $(s \in \{14, 15\}, r \in \mathbb{N})$,
$(J_9 \cup U_6) \cup A_q$ $(q \in \{11, 13\})$,
$(J_i \cup U_6) \cup A_t$ $(i \in \{1, 5\},\ t \in \{10, 11, 13\})$
(classes T with $c_1 \in T$ and $c_2 \notin T$).

Proof.
Case 1: $c_2 \notin T_1$.
Case 1.1: $c_1 \notin T_1$.
In this case, T is a subset of $(J_2 \cup U_2) \cup A_1$. Table 15.20 indicates the possibilities for T_1 that result from Theorem 15.4.4.4:

Table 15.20

T_1^1	T_1
H_{50}^1	$J_4 \cup U_{13}, J_{16} \cup U_{14}$
H_{51}^1	$J_{12} \cup H_{19}, J_{12} \cup U_6, J_{12,r} \cup H_{19}$
H_{52}^1	$J_{14} \cup U_7$
H_{57}^1	$J_{10} \cup H_{19}, J_{10,r} \cup H_{19}, J_{10} \cup U_6, J_6 \cup H_{19}, J_6 \cup U_6, J_2 \cup H_{19}, J_2 \cup U_6$
H_{60}^1	$J_2 \cup U_2$

Then, by Section 14.4.5, we have

$$T \cap \{j_2(x) \vee u_2(y), j_5(x) \vee u_5(y)\} \neq \emptyset.$$

Consequently, with the aid of Theorems 15.4.5.4 and 15.4.5.5, one obtains the classes of *1)* as possibilities for T, where these classes are sorted after the cases that result from Table 15.20.

Case 1.2: $c_1 \in T_1$.

Because of $T \cap R \not\subseteq V_1$, this case is only possible for

$$T \cap \{max, x \vee u_2(y), x \vee u_5(y), x \vee j_5(y), j_5(x) \vee u_2(y), j_5(x) \vee u_5(y)\} = \emptyset.$$

Consequently, we have $A_{14,1} \subseteq T_2 \subseteq A_{10}$ or $T_2 = A_{16}$.

Because of Theorem 15.4.4.4, T_1 can be only a class from Table 15.21.

Table 15.21

T_2^1	T_2
H_{14}^1	$H_5 \cup U_6$
H_{53}^1	$J_{11} \cup H_{19}, J_{11} \cup U_6, J_{11,r} \cup H_{19}$
H_{58}^1	$J_9 \cup H_{19}, J_{9,r} \cup H_{19}, J_5 \cup H_{19}, J_1 \cup H_{19}, J_9 \cup U_6, J_5 \cup U_6, J_1 \cup U_6$

When one scrutinizes the cases resulting from that, one receives the classes of *2)*.

Case 2: $c_2 \in T_1$.

Then, by $T \cap R \not\subseteq V_1$, we have

$$\{j_2(x) \vee u_2(y), j_5(x) \vee u_2(y), j_5(x) \vee u_5(y), x \vee j_5(y)\} \cap T = \emptyset$$

and $T_2 \subseteq B_1$. Thus T contains $x \vee u_2(y)$ or $x \vee u_5(y)$.

If $T \cap J_1 \neq [\{c_1\}]$, then we have $T \cap \{j_2, j_5\} \neq \emptyset$. However, by $T_2 \subseteq B_1$, this cannot be possible. Equally, the case $T \cap J_1 = [\{c_1\}]$ is not possible, since $v_2 \in [\{c_1, x \vee u_2(y)\}]$ and $v_5 \in [\{c_1, x \vee u_5(y)\}]$. Thus the second case cannot occur. ■

Theorem 15.4.6.6 *If $T \subseteq (J_1 \cup V_1) \cup A_1$, $T_1 \not\subseteq J_1 \cup A_1$ and $T_1 \not\subseteq V_1 \cup A_1$, then T is exactly one of the following classes:*

1) $(H_{14} \cup V_i) \cup B_{19}$ $(i \in \{1, 15, 17\})$,
$(J_{11} \cup V_m) \cup I_{11}$ $(m \in \{15, 17\})$,
$(J_{13} \cup V_n) \cup I_{10}$ $(n \in \{1, 15, 17\})$,
$(J_1 \cup V_p) \cup I_5$ $(p \in \{1, 15, 17\})$
(classes that contain $\{c_0, max\}$);

2) $(J_{13} \cup V) \cup I_{13}$ $(V \in \{H_{32}, H_{34}, V_1, V_3, V_5, V_7, V_{17}\})$,
$(J_{11} \cup V) \cup I_{14}$ $(V \in \{H_{32}, V_{17}, H_{34}, V_{15}\})$,
$(J_i \cup V) \cup I_{14}$ $(i \in \{1, 5\}, V \in \{H_{32}, V_{17}, H_{34}, V_{15}\})$,
$(J_1 \cup V) \cup I_8$ $(V \in \{H_{32}, V_{17}, H_{34}, V_{15}, V_1\})$
(classes that contain c_0 but not max);

3) $(J_{15} \cup V_{15}) \cup I_{10}, (J_3 \cup V_{15}) \cup I_5, J_3 \cup V_{15}) \cup I_1$,
$(J_7 \cup V_{15}) \cup I_4$,

$(J_{15} \cup H_{34}) \cup I_{15}, (J_{15} \cup V_{15}) \cup I_{15}, (J_3 \cup H_{34}) \cup I_{10}, (J_3 \cup V_{15}) \cup I_{10},$
$(J_7 \cup H_{34}) \cup I_7, (J_7 \cup V_{15}) \cup I_7,$
$(J_3 \cup H_{34}) \cup I_{14}, (J_3 \cup V_{15}) \cup I_{14}, (J_7 \cup H_{34}) \cup I_{14}, (J_7 \cup V_{15}) \cup I_{14},$
$(J_{15} \cup V_1) \cup I_{10}, (J_3 \cup V_1) \cup I_5, (J_3 \cup V_1) \cup I_1,$
$(J_3 \cup V_1) \cup I_8,$
$(J_{15} \cup V_7) \cup I_{13}, (J_{15} \cup V_5) \cup I_{13}, (J_{15} \cup V_3) \cup I_{13}, (J_{15} \cup V_1) \cup I_{13}$
(classes that contain c_2 but not c_0);

4) $(J_{15} \cup V_{17}) \cup I_{10}, (J_{15} \cup V_{17}) \cup I_5, (J_{15} \cup V_{17}) \cup I_1,$
$(J_7 \cup V_{17}) \cup I_4, (J_7 \cup V_{17}) \cup I_1,$
$(J_{15} \cup H_{32}) \cup I_{13}, (J_{15} \cup V_{17}) \cup I_{13}, (J_3 \cup H_{32}) \cup I_8, (J_3 \cup V_{17}) \cup I_3,$
$(J_7 \cup H_{32}) \cup I_{12}, (J_7 \cup H_{32}) \cup I_7,$
$(J_7 \cup H_{32}) \cup I_{14}, (J_7 \cup V_{17}) \cup I_{14}, (J_3 \cup H_{32}) \cup I_{14}, (J_3 \cup V_{17}) \cup I_{14}$
(classes T with $\{c_0, c_1, c_2\} \cap T = \{c_1\}$).

Proof. Because of $T \cap U_1 = \emptyset$, we have

$$\{j_2(x) \vee u_2(y), j_5(x) \vee u_5(y), x \vee u_5(y)\} \cap T = \emptyset,$$

Thus $T_2 \subseteq I_1$. Then the following cases are possible:

Case 1: $c_0 \in T_1$.

Then $\{j_5(x) \vee u_2(y), x \vee u_2(y)\} \cap T = \emptyset$.

Case 1.1: $max \in T_2$.

Because of $T = [T^1 \cup \{max\}]$, T is one of the classes given in *1)* (see also Theorems 15.4.3.1 and 15.4.5.3).

Case 1.2: $max \notin T_2$.

In this case, by Theorem 15.4.5.3, T_2 belongs to $\{I_{13}, I_8, I_{14}\}$ and it holds that

$$\begin{aligned} T_2 = I_{13} &\implies T_1^1 \in \{H_{65}^1, H_{67}^1, H_{73}^1\}, \\ T_2 = I_{14} &\implies T_1^1 \in \{H_{62}^1, H_{63}^1, H_{70}^1, H_{71}^1\}, \\ T_2 = I_8 &\implies T_1^1 \in \{H_{70}^1, H_{71}^1, H_{74}^1\}. \end{aligned}$$

From this and from Theorem 15.4.4.5, we get the possibilities given in *2)* for T.

Case 2: $c_0 \notin T_1$.

Then, because of $T \cap J_1 \neq \emptyset$ and $T \cap V_1 \neq \emptyset$, the function c_1 belongs to T.

Case 2.1: $c_2 \in T_1$.

Then $T_1^1 \in \{H_{66}^1, H_{72}^1\}$. Because of Theorems 15.4.4.5 and 15.4.5.3, the possibilities *3)* for T follow.

Case 2.2: $c_2 \notin T_1$.

In this case, we have $T_1^1 = H_{64}^1$. Then, with the help of Theorems 15.4.4.5 and 15.4.5.3, one obtains the possibilities given in *4)* for T. ■

Theorem 15.4.6.7 *If $T \subseteq (U_1 \cup V_1) \cup A_1$, $T_1 \nsubseteq U_1 \cup A_1$ and $T_1 \nsubseteq V_1 \cup A_1$, then T is exactly one of the following classes:*

1) $(H_2 \cup V_{17}) \cup B_{19}, (H_6 \cup V_{15}) \cup B_{19}, (H_6 \cup V_1) \cup B_{19}, (U_6 \cup V_{17}) \cup B_4, (U_3 \cup V_{15}) \cup B_4, (U_4 \cup V_1) \cup B_8, (U_1 \cup V_1) \cup B_1$
(classes that contain max);

2) $(U_6 \cup V_{17}) \cup B_{14}, (U_3 \cup V_{15}) \cup B_{14}$,
$(U_1 \cup V_1) \cup B_3$
(classes that contain $x \vee u_2(y)$ but not max);

3) $A \cup B_i$ $(A \in \{U_4 \cup V_9, U_4 \cup V_1, U_4 \cup V_3, U_4 \cup V_5, U_4 \cup V_7\}, i \in \{16, 20\})$
(classes T with $T \cap \{max, x \vee u_2(y)\} = \emptyset$ and $x \vee j_5(y) \in T$).

Proof. Because of $T \nsubseteq V_1 \cup A_1$, we have $c_0 \in T_1$. $T \cap V_1 \neq [\{c_2\}]$ implies $c_1 \in T_1$. Hence, $T_2 \subseteq B_1$. By Theorem 15.4.5.2, only the following cases are possible:

Case 1: $max \in T_2$.
Then $T = [T^1 \cup \{max\}]$ and one obtains the possibilities given in *1)* of T with the aid of Theorems 15.4.3.1, 15.4.4.6, and 15.4.5.2.

Case 2: $max \notin T_2$.
Then $\{x \vee u_2(y), x \vee u_5(y)\} \cap T \neq \emptyset$.

Case 2.1: $x \vee u_2(y) \in T_2$.
Every class T that satisfies this condition is given in *2)*.

Case 2.2: $x \vee u_5(y) \in T$ and $x \vee u_2(y) \notin T$.
Then T_1^1 is an element of $\{H_{81}^1, H_{84}^1, H_{86}^1\}$ and the possibilities for T are given in *3)*. ■

Theorem 15.4.6.8 *If $T \subseteq R \cup A_1$, $T_1 \nsubseteq J_1 \cup U_1 \cup A_1$, $T_1 \nsubseteq J_1 \cup V_1 \cup A_1$ and $T_1 \nsubseteq U_1 \cup V_1 \cup A_1$, then T is exactly one of the following classes:*

1) $(J_{11} \cup U_6 \cup V_{17}) \cup A_4, (J_{11} \cup U_3 \cup V_5) \cup A_4$,
$(J_1 \cup U_6 \cup V_{17}) \cup A_2, (J_1 \cup U_3 \cup V_{15}) \cup A_2$,
$(J_{13} \cup U_4 \cup V_1) \cup A_{31}$,
$(J_1 \cup U_1 \cup V_1) \cup A_1$
(classes that contain $\{c_0, max\}$);

2) $(J_{11} \cup U_6 \cup V_{17}) \cup A_7$,
$(J_{11} \cup H_{19} \cup H_{32}) \cup A_{12}, (J_{11} \cup U_6 \cup V_{17}) \cup A_{10}$,
$(J_{11} \cup H_{19} \cup H_{32}) \cup A$ $(A \in \{A_{14}, A_{15}, A_{14,r}, A_{15,r} \mid r \in \mathbb{N}\})$,
$(J_{11} \cup U_6 \cup V_{17}) \cup A_i$ $(i \in \{11, 13\})$,
$(J_{11,r} \cup H_{19} \cup H_{32}) \cup A_{s,r'}$ $(s \in \{14, 15\};\ r, r' \in \mathbb{N};\ r' \leq r)$
(classes T with $max \notin T$ and $T_1^1 = H_{87}^1$);

3) $(J_{11} \cup U_3 \cup V_{15}) \cup A_7$,
$(J_{11} \cup H_{24} \cup H_{34}) \cup A_{12}, (J_{11} \cup U_3 \cup V_{17}) \cup A_{10}$,
$(J_{11} \cup H_{24} \cup H_{34}) \cup A$ $(A \in \{A_{14}, A_{15}, A_{14,r}, A_{15,r} \mid r \in \mathbb{N}\})$,
$(J_{11} \cup U_3 \cup V_{15}) \cup A_i$ $(i \in \{11, 13\})$,

$(J_{11,r} \cup H_{24} \cup H_{34}) \cup A_{s,r'}$ ($s \in \{14,15\}$; $r, r' \in \mathbb{N}$; $r' \leq r$)
(classes T with $max \notin T$ and $T_1^1 = H_{88}^1$);

4) $(J_{13} \cup U_4 \cup V_9) \cup A_i$ ($i \in \{19, 20, 24, 25, 36, 37\}$)
(classes T with $max \notin T$ and $T_1^1 = H_{90}^1$);

5) $(J_5 \cup U_6 \cup V_{17}) \cup A_5$,
$(J_1 \cup H_{19} \cup H_{32}) \cup A_6$,
$(J_5 \cup H_{19} \cup H_{32}) \cup A_{12}$, $(J_1 \cup H_{19} \cup H_{32}) \cup A_{12}$,
$(J_5 \cup U_6 \cup V_{17}) \cup A_{10}$, $(J_1 \cup U_6 \cup V_{17}) \cup A_{10}$,
$(J_5 \cup H_{19} \cup H_{32}) \cup A_8$,
$(J_i \cup U_6 \cup V_{17}) \cup A_t$ ($i \in \{1,5,9\}$, $t \in \{11,13\}$),
$(J_i \cup H_{19} \cup H_{32}) \cup A_t$ ($i \in \{1,5,9\}$, $t \in \{14,15\}$),
$(J_{9,r} \cup H_{19} \cup H_{32}) \cup A_{s,r'}$ ($s \in \{14,15\}$; $r, r' \in \mathbb{N}$; $r' \leq r$),
$(J_i \cup H_{19} \cup H_{32}) \cup A_{s,r}$ ($i \in \{1,5,9\}$, $s \in \{14,15\}, r \in \mathbb{N}$)
(classes T with $max \notin T$ and $T_1^1 = H_{91}^1$);

6) $(J_5 \cup U_3 \cup V_{15}) \cup A_5$,
$(J_1 \cup H_{24} \cup H_{34}) \cup A_6$,
$(J_i \cup H_{24} \cup H_{34}) \cup A_{12}$ ($i \in \{1,5\}$),
$(J_i \cup U_3 \cup V_{15}) \cup A_{10}$ ($i \in \{1,5\}$),
$(J_5 \cup H_{24} \cup H_{34}) \cup A_8$,
$(J_i \cup U_3 \cup V_{15}) \cup A_t$ ($i \in \{1,5,9\}$, $t \in \{11,13\}$),
$(J_i \cup H_{24} \cup H_{34}) \cup A_t$ ($i \in \{1,5,9\}, t \in \{14,15\}$),
$(J_{9,r} \cup H_{24} \cup H_{34}) \cup A_{s,r'}$ ($s \in \{14,15\}$; $r, r' \in \mathbb{N}$; $r' \leq r$),
$(J_i \cup H_{24} \cup H_{34}) \cup A_{s,r}$ ($i \in \{1,5,9\}$, $s \in \{14,15\}$, $r \in \mathbb{N}$)
(classes T with $max \notin T$ and $T_1^1 = H_{94}^1$);

7) $(J_{13} \cup U_4 \cup V_i) \cup A_t$ ($i \in \{1,3,5\}$, $t \in \{18,19,23,24,36\}$),
$(J_{13} \cup U_4 \cup V_i) \cup A_t$ ($i \in \{1,3,5,7\}$, $t \in \{20,25,37\}$)
(classes T with $max \notin T$ and $T_1^1 = H_{94}^1$);

8) M
(class T with $max \notin T$ and $T_1^1 = H_{91}^1$).

Proof. Because of $T \not\subseteq (J_1 \cup V_1) \cup A_1$, we have $c_0 \in T_1$.

Case 1: $max \in T_2$.

Then $T = [T^1 \cup \{max\}]$ and one obtains the possibilities for T given in *1)* with the aid of Theorem 15.4.3.1.

Case 2: $max \notin T_2$.

Since $c_0 \in T$, we have $T_1^1 = H_i^1$ with $i \in \{87, 88, 90, 91, 92, 94, 95\}$ by Theorems 15.4.3.1 and 15.4.4.7. With the aid of Theorems 15.4..4.7, 15.4.5.2–15.4.5.5, one obtains the possibilities given in *2)–8)* for T. ∎

16

The Maximal Classes of $\bigcap_{a\in Q} Pol_k\{a\}$ for $Q \subseteq E_k$

In this section we describe all maximal classes of the subclass

$$T_Q := \bigcap_{a\in Q} Pol_k\{a\}$$

of P_k for arbitrary Q with $\emptyset \neq Q \subseteq E_k$, $k \geq 2$. With the aid of these classes, a completeness criterion for T_Q can easily be formulated. This criterion implies necessary and sufficient conditions regarding whether a finite algebra is semi-primal and has only trivial subalgebras.
Moreover, if $|Q| \geq 2$, we prove that every maximal class of T_Q is an intersection of T_Q with a certain maximal classes of P_k or $Pol_k\{a\}$ $(a \in Q)$. Presumably, something similar is valid for the maximal classes of $T_{Q'} := \bigcap_{\varrho\in Q'} Pol_k\varrho$, where $Q' \subseteq \mathfrak{P}(E_k)$ and $|Q'| \geq 2$. For $k = 3$ this presumption was proven in [Lau 95b].

16.1 Notations

We say that a relation ϱ' is derivable from the relation ϱ **with the aid of** Inv_kT_Q (or briefly ϱ' is **ϱ-derivable**), if $\varrho' \in [\{\varrho\} \cup Inv_kT_Q]$ (see Section 2.4). In this case we also write

$$\{\varrho\} \cup T_Q \vdash \varrho'$$

or briefly

$$\varrho \vdash \varrho'.$$

The following lemma provides the basis of later proofs and summarizes some well-known statements that can easily be checked.

Lemma 16.1.1

(a) $\forall \varrho, \varrho' \in R_k : (Pol\varrho \subseteq T_Q \;\wedge\; (\{\varrho\} \cup Inv_kT_Q \vdash \varrho') \Longrightarrow Pol\varrho \subseteq T_Q \cap Pol\varrho');$
(b) $Inv_kP_k = D_k$ *(see Chapter 2);*

(c) For every relation $\varrho \in Inv_k T_Q$ there are some relations $\varrho_1, ..., \varrho_r \in \{\{a\} \mid a \in Q\} \cup D_k$ and a relation $\varrho' \in D_k$ with $\varrho = (\varrho_1 \times \varrho_2 \times ... \times \varrho_r) \cap \varrho'$.

■

Next we define some relation sets that we need to describe of the maximal classes of T_Q $(Q \subseteq E_k)$. In this case, we also use the notations from Chapter 5.

$$\mathfrak{M}_{k;Q} := \begin{cases} \{\varrho \in \mathfrak{M}_k \mid a \text{ is greatest or smallest element of } E_k \text{ in respect to } \varrho\ \}, \\ \quad \text{if } Q = \{a\}, \\ \{\varrho \in \mathfrak{M}_k \mid a \text{ is greatest (smallest) and } b \text{ is smallest (greatest) element of } E_k \text{ in respect to } \varrho\ \}, \text{ if } Q = \{a\} \text{ and } a \neq b, \\ \emptyset \text{ otherwise;} \end{cases}$$

$$\mathfrak{U}_{k;Q} := \begin{cases} \{\varrho \in \mathfrak{U}_k \mid \forall x \in E_k\ \forall q \in Q:\ (x,q) \in \varrho \implies x = q\}, \\ \quad \text{if } |Q| \le k-2, \\ \emptyset \text{ otherwise;} \end{cases}$$

$$\mathfrak{S}_{k;Q} := \begin{cases} \{\varrho \in \mathfrak{S}_k \mid \forall x, y \in E_k : (x,y) \in \varrho\ \wedge x \in Q \implies y \in Q\}, \\ \quad \text{if } |Q| = s \cdot p, k = t \cdot p, p \text{ prime}, \\ \emptyset \text{ otherwise;} \end{cases}$$

$\mathfrak{P}_{k,Q} := \{\ \{(x, s(x)) \mid x \in E_k\} \mid s$ is a permutation on E_k with exactly one fixed point $(\in Q)$; all proper cycles of s have the same prime number length, and s preserves Q };

$$\mathfrak{L}_{k;Q} := \begin{cases} \{\ \{(a,b,c) \in E_k^3 \mid a +_G b = c\} \mid (E_k; +_G) \text{ is an elementar Abelean 2-group with the neutral element } q\ \}, \\ \quad \text{if } k = 2^m,\ m \ge 1 \text{ and } Q = \{q\}, \\ \emptyset \text{ otherwise;} \end{cases}$$

$\mathfrak{C}_{k;Q} :=$
$(\mathfrak{C}_k^1 \backslash \{\{q\} \mid q \in Q\}) \cup \bigcup_{h=2}^{k-1} \{\varrho \in \mathfrak{C}_k^h \mid \forall q \in Q : q \text{ is a central element of } \varrho\}$

(in particular, we have $\mathfrak{C}_{k;E_k} = \mathfrak{C}_k^1 \backslash \{\{q\} \mid q \in E_k\}$)
and

$$\mathfrak{N}_{k;Q} := \begin{cases} \{\ (\varrho \backslash \iota_k^2) \cup \{(q,q)\} \mid \varrho \in \mathfrak{C}_{k;Q}^2 \cup \{E_k^2\}\}, \text{ if } Q = \{q\}, \\ \{\varrho \subseteq E_k^2 \mid \exists q \in Q : \{(x,q),(q,x) \mid x \in E_k\} \subseteq \varrho\ \wedge\ \tau\rho = \rho\ \wedge \\ \varrho \cap \iota_k^2 = \{(q,q)\} \wedge\ \varrho \cap ((Q \backslash \{q\}) \times (E_k \backslash \{q\})) = \emptyset\}\ , \text{ if } 2 \le |Q|. \end{cases}$$

Obviously, $\mathfrak{N}_{k;Q} \subseteq \bigcup_{a\in Q} \mathfrak{N}_{k;\{a\}}$.
An element $c \in Q$ with $\{(x,c),(c,x) \mid x \in E_k \backslash \{c\}\} \subseteq \varrho$ $(\in \mathfrak{N}_{k;Q})$ is called **central element** of ϱ.
If one considers an arbitrary other finite set A instead of E_k, then the relation

sets $\mathfrak{M}_A$, $\mathfrak{M}_{A;Q}$, $\mathfrak{U}_A$, $\mathfrak{U}_{A;Q}$, $\mathfrak{S}_A$, $\mathfrak{S}_{A;Q}$, $\mathfrak{P}_{A;Q}$, $\mathfrak{L}_A$, $\mathfrak{L}_{A;Q}$, $\mathfrak{C}_A$, $\mathfrak{C}_{A;Q}$, $\mathfrak{N}_{A;Q}$ and $\mathfrak{B}_A$ for $Q \subseteq A$ can be defined when one replaces the set E_k by A in the above definitions or in the definitions of Chapter 5.

16.2 Results of Chapter 16

Our aim is to prove the following

Theorem 16.2.1 *([Sze 91], [Lau 82b], [[Lau 95a])*
Let $R_{max}(T_Q) := \mathfrak{M}_{k;Q} \cup \mathfrak{U}_{k;Q} \cup \mathfrak{S}_{k;Q} \cup \mathfrak{L}_{k;Q} \cup \mathfrak{C}_{k;Q} \cup \mathfrak{N}_{k;Q} \cup \mathfrak{P}_{k;Q}$. *Then*

$$\{T_Q \cap Pol_k\varrho \mid \varrho \in R_{max}(T_Q)\}$$

is the set of all maximal classes of T_Q *for* $\emptyset \neq Q \subseteq E_k$.

The following theorem is a direct conclusion from the above theorem and from the fact that T_Q is finitely generating (see Lemma 16.3.1):

Theorem 16.2.2 (Completeness Criterion for T_Q)

For an arbitrary subset M *of* T_Q *is valid:*

$$[M] = T_Q \iff \forall \varrho \in R_{max}(T_Q) : M \not\subseteq T_Q \cap Pol_k\varrho.$$

■

The next theorem is a special case of Theorem 16.2.2:

Theorem 16.2.3 *(***Completeness Criterion for the Class of all Idempotent Functions of** P_k*)*

For an arbitrary subset M *of* T_{E_k} *(* $= \bigcup_{n\geq 1}\{f^n \in P_k \mid f(x,...,x) = x\}$*) with* $k \geq 3$ *is valid:*

$$[M] = T_{E_k} \iff \forall \varrho \in \mathfrak{C}^1_{k;E_k} \cup \mathfrak{S}_{k;E_k} \cup \mathfrak{P}_{k;E_k} \cup \mathfrak{N}_{k;E_k} : M \not\subseteq T_{E_k} \cap Pol_k\varrho.$$ ■

Theorem 16.2.1 can also be formulated in the language of the Universal Algebra:
A finite algebra $(A; F)$ $(F \subseteq P_A)$ is called **semi-primal**, if $[F] = Pol_A Inv^1_A F$ holds (see [Fos-P 64], [Den 82], [Pös-K 79], p. 143) or [Den-W 2002]. Let $Sub(\mathfrak{A})$ be the set of all subalgebras of $\mathfrak{A}$. Then $Inv^1_A F$ is the set of all universes of algebras of $Sub(\mathfrak{A})$. If $Sub(\mathfrak{A})\backslash\{\mathfrak{A}\}$ contains only 1-element algebras and $Q := \{a \mid (\{a\}; F) \in Sub(\mathfrak{A})\}$ holds, then every $\varrho \in Inv\ F \cap (\mathfrak{P}_{A;Q} \cup \mathfrak{S}_{k;Q})$

defines a non-trivial automorphism s ($s(a) = b :\Longleftrightarrow (a,b) \in \varrho$) of the algebra $\mathfrak{A}$ and the relations of $\mathfrak{U}_{k;Q}$ are some non-trivial congruences of $\mathfrak{A}$. Then the following theorem follows from Theorem 16.2.1:

Theorem 16.2.4 *([Sze 91])*

Let $\mathfrak{A} = (A; F)$ be a finite algebra with the property that $(Sub\mathfrak{A})\backslash\{\mathfrak{A}\}$ contains only 1-element algebras. Then the following conditions are equivalent:

(a) $(A; F)$ is semi-primal with $\{a \in A \mid (\{a\}; F) \in Sub\mathfrak{A}\} = Q$.

(b) $\mathfrak{A}$ has no proper automorphisms, is simple (i.e., $\mathfrak{A}$ has only trivial congruences) and the direct product $\mathfrak{A}^h$ of $\mathfrak{A}$ for $h = 2, 3, ..., |A|-1$ has no subalgebra whose universe is an element of the set $\mathfrak{M}_{A;Q} \cup \mathfrak{L}_{A;Q} \cup \mathfrak{N}_{A;Q} \cup \mathfrak{C}_{A;Q}$.

■

16.3 Some Lemmas

Lemma 16.3.1 $T_Q = [T_Q^3]$ *for all Q with $\emptyset \neq Q \subseteq E_k$.*

Proof. For $k = 2$, our assertion is valid by Chapter 3. Let $k \geq 3$. Then the below-defined functions $\vee$, $\cdot$, w, $r_{a;b}$, if $\{a,b\} \subset E_k$, $q_{a,b;c}$, if $a \neq b$ or $\{a,b\} \not\subseteq Q$ and $\{a,b,c\} \subseteq E_k$, belong to T_Q:

$$x \vee y := max(x,y),\ x \cdot y := min(x,y) \text{ in respect to the total order } 0 < 1 < 2 < \ldots < k-1;$$

$$w(x,y) := \begin{cases} x & \text{if } x = y \in Q, \\ 0 & \text{otherwise;} \end{cases}$$

$$j_{a;b}(x) := \begin{cases} b & \text{if } x = a, \\ 0 & \text{otherwise;} \end{cases}$$

$$r_{a;b}(x,y,z) := x \vee j_{a;b}(y) \cdot z;$$
$$q_{a,b;c}(x,y,z) := x \vee j_{a;c}(y) \cdot j_{b;c}(z).$$

We show that the above-defined functions form a generating system for T_Q. The function $w_n := \underbrace{w \star w \star ... \star w}_{n-1 \text{ times}}$ with

$$w_n(x_1, ..., x_n) = \begin{cases} x & \text{if } x_1 = x_2 = ... = x_n = x \in Q, \\ 0 & \text{otherwise} \end{cases}$$

($n \geq 1$) is a superposition over w.
Let f^n be an arbitrary function of T_Q, which is different from w_n. Then one can represent f as follows:

$$f(x_1,...,x_n) = w_n(x_1,...,x_n) \vee \bigvee_{\substack{\boldsymbol{a}=(a_1,...,a_n)\\ \in E_k^n\backslash\{(q,q,...,q)\mid q\in Q\}\\ f(\boldsymbol{a})\neq 0}} j_{a_1;f(\boldsymbol{a})}(x_1)\cdot ... \cdot j_{a_n;f(\boldsymbol{a})}(x_n).$$

Therefore, f is a superposition over the set
$B := \{w_n\} \cup$
$\{x \vee j_{a_1;b}(x_1)\cdot ... \cdot j_{a_n;b}(x_n) \mid (a_1,...,a_n) \in E_k^n\backslash\{(q,q,...,q) \mid q \in Q\} \wedge b \in E_k\}$.
An arbitrary function of $B\backslash\{w_n\}$ is generated from functions of the type $r_{a;b}$ and $q_{a,b;c}$, since

$$\begin{aligned}&r_{a_1,...,a_i,a_{i+1};b}(x,x_1,...,x_i,x_{i+1},y) :=\\ &r_{a_1,...,a_i;b}(x,x_1,...,x_i,r_{a_{i+1};b}(x,x_{i+1},y)) =\\ &x \vee j_{a_1;b}(x_1)\cdot ... \cdot j_{a_n;b}(x_n)\cdot y\end{aligned}$$

$(i = 1,...,n)$ and

$$\begin{aligned}&r_{a_1,...,a_n;b}(x,x_1,...,x_n,q_{a_i,a_j;b}(x,x_i,x_j)) =\\ &x \vee j_{a_1;b}(x_1)\cdot ... \cdot j_{a_n;b}(x_n).\end{aligned}$$

Consequently, $[T^3] = T$. ■

The next lemma is a conclusion of the above lemma:

Lemma 16.3.2 *For every Q with $\emptyset \neq Q \subseteq E_k$, the lattice of the subclasses of T_Q is dual atomar and T_Q has only finite-many maximal classes.* ■

Since the maximal classes of T_Q are clones[1], one can easily show the following

Lemma 16.3.3 *For every maximal class M of T_Q there exists an h-ary relation ϱ_M with $M = Pol_k\varrho_M$ and $1 \leq h \leq k^3$.* ■

Lemma 16.3.4 *Let $\emptyset \neq Q \subseteq E_k$. Then*

(a) $\forall \varrho \in \mathfrak{M}_k \cup \mathfrak{U}_k \cup \mathfrak{S}_k \cup \mathfrak{L}_k \cup \mathfrak{B}_k \cup (\mathfrak{C}_k\backslash\{\{q\} \mid q \in Q\}) : T_Q \not\subseteq Pol_k\varrho$;
(b) $V_k^{\uparrow}(T_Q) = \{T_{Q'} \mid Q' \subseteq Q\}$ ($T_\emptyset := P_k$);
(c) $\forall \varrho \in \bigcup_{a\in Q}(\mathfrak{P}_{k;\{a\}} \cup \mathfrak{N}_{k;\{a\}}) : T_Q \cap Pol_k\varrho \subset T_Q$.

Proof. (a) and (c) are easy to check.
(b): It is sufficient to prove the following statement for $\emptyset \neq Q \subseteq E_k$ and $f \in P_k$:

$$(a \in Q \ \wedge f(a,a,...,a) \neq a) \implies T_{Q\backslash\{a\}} \subseteq [T_Q \cup \{f\}].$$

Let $f(a,a,...,a) \neq a$ for a certain $a \in Q$, $Q \subseteq E_k$ and $f'(x) := f(x,x,...,x)$. To prove $T_{Q\backslash\{a\}} \subseteq [T_Q \cup \{f\}]$, let $g^m \in T_{Q\backslash\{a\}}$ be arbitrary. Then the function h_g^{m+1} defined by

[1] One can prove this analogous to the proof of footnote 1 of Chapter 14.

$$h_g(x_1,..,x_{m+1}) := \begin{cases} x, & \text{if } x_1 = ... = x_{m+1} = x \in Q, \\ u, & \text{if } x_1 = ... = x_m = u \in Q\backslash\{a\} \wedge \\ & \qquad x_{m+1} = f'(u), \\ g(x_1,...,x_m) & \text{otherwise,} \end{cases}$$

belongs to T_Q and $g(x_1,...,x_m) = h_g(x_1,...,x_m,f'(x_1))$ is valid. Therefore, $g \in [T_Q \cup \{f\}]$ holds. ∎

Lemma 16.3.5 *For every relation* $\gamma \in R := \mathfrak{M}_k \cup \mathfrak{U}_k \cup \mathfrak{S}_k \cup \mathfrak{L}_k \cup \mathfrak{B}_k \cup (\mathfrak{C}_k\backslash\{\{q\} \mid q \in Q\})$ *there exists a* $\gamma' \in [\{\gamma\} \cup Inv_k T_Q]$, *which belongs to*

$$\mathfrak{M}_{k;Q} \cup \mathfrak{U}_{k;Q} \cup \mathfrak{S}_{k;Q} \cup \mathfrak{L}_{k;Q} \cup \mathfrak{C}_{k;Q} \cup \bigcup_{q\in Q} (\mathfrak{P}_{k;\{q\}} \cup \mathfrak{N}_{k;\{q\}}).$$

Proof. Examining the case

$$\gamma \in R\backslash(\mathfrak{M}_{k;Q} \cup \mathfrak{U}_{k;Q} \cup \mathfrak{S}_{k;Q} \cup \mathfrak{C}_{k;Q}) \tag{16.1}$$

suffices. Further, we can assume that γ is h-ary, where $h \geq 2$ because of (16.1). First we form the $(h-1)$-ary relation

$$\gamma_a := pr_{1,2,...,h-1}(\Delta(\{a\} \times \gamma)) \tag{16.2}$$

for $a \in Q$. If $h = 2$ (i.e., $\gamma \in (\mathfrak{C}_k^2\backslash\mathfrak{C}_{k;Q}^2) \cup (\mathfrak{M}_k\backslash\mathfrak{M}_{k;Q}) \cup (\mathfrak{U}_k\backslash\mathfrak{U}_{k;Q}) \cup (\mathfrak{S}_k\backslash\mathfrak{S}_{k;Q})$), then we have

$$\gamma_a = \{x \in E_k \mid (a,x) \in \gamma\}. \tag{16.3}$$

Because of (16.1), it is easy to check that there is an $a \in Q$ such that the relation γ_a belongs to $\mathfrak{C}_{k;Q}^1 = \{E \subset E_k \mid |E| \geq 2 \vee \exists b \in E_k\backslash Q : E = \{b\}\}$:
If $\gamma \in \mathfrak{M}_k\backslash\mathfrak{M}_{k;Q}$, then for this purpose, one can choose $a \in Q \setminus \{o, e\}$, where o is the smallest element and e is the greatest element of E_k in respect to γ. If $\gamma \in \mathfrak{U}_k\backslash\mathfrak{U}_{k;Q}$, then an $a \in Q$ of an at least 2-element equivalence class of γ fulfills the above condition. If $\gamma \in \mathfrak{S}_k\backslash\mathfrak{S}_{k;Q}$, then there exists an $a \in Q$ and a $b \in E_k\backslash Q$ with $(a,b) \in \gamma$; thus $\gamma_a = \{b\} \in \mathfrak{C}_{k;Q}^1$. If $\gamma \in \mathfrak{C}_k^2\backslash\mathfrak{C}_{k;Q}^2$, there is an $a \in Q$, which is no central element of γ. Consequently, we also have $\gamma_a \in \mathfrak{C}_{k;Q}^1$ in this case.
Now, let $h \geq 3$ and $\gamma \notin \mathfrak{L}_k$, i.e., $\gamma \in \mathfrak{C}_k^h \cup \mathfrak{B}_k^h$. Choose $a \in Q \setminus C$, if $\gamma \in \mathfrak{C}_k$ and C is the set of all central elements of γ. For $\gamma \in \mathfrak{B}_k^h$ let $a \in Q$ be arbitrary. Then the relation γ_a defined by (16.2) is an $(h-1)$-ary reflexive, totally symmetric relation with the same central elements as γ and the new central element a. Since a is no central element of γ, there exist $d_1,...,d_{h-1} \in E_k$ with $(a,d_1,...,d_{h-1}) \notin \gamma$. Consequently, we have $\gamma_a \neq E_k^{h-1}$ and γ_a is a central relation. Through repetitions of the above construction, one obtains a γ-derivable relation of $\mathfrak{C}_{k;Q}$.
Finally, let $\gamma \in \mathfrak{L}_k$, where $k = p^m$, p prime and $m \geq 1$. Then there is an elementar Abelean p-group $(E_k;+)$ with $\gamma = \{(x,y,u,v) \in E_k^4 \mid x+y = u+v\}$. By Lemma 5.2.4.2, we can assume w.l.o.g. that the neutral element o of the p-group $(E_k;+)$ belongs to Q. For $p \neq 2$ we now form the γ-derivable relation

$$\gamma' := pr_{3,4}(\delta^5_{\{0,1,2\}} \cap (\{o\} \times \gamma)) = \{(x,y) \in E_k^2 \mid x + y = o\}.$$

Obviously, $\gamma' \in \mathfrak{P}_{k;\{o\}}$, since for $p \neq 2$ the equation $x + x = o$ is valid only if $x = o$.
If $p = 2$, then one can form the relation

$$\gamma'' := pr_{1,2,3}(\Delta(\{o\} \times \gamma)) = \{(x,y,z) \in E_k^3 \mid x + y = z\} \in [\{\gamma\} \cup Inv_k T_Q].$$

Thus $\gamma'' \in \mathfrak{L}_{k;\{o\}}$ in the case $\{o\} = Q$. If $|Q| \geq 2$ there is an $a \in Q \backslash \{o\}$ and the γ-derivable relation

$$\gamma''' := pr_{0,1}((\gamma'' \times \{a\}) \cap \delta^4_{\{2,3\}}) = \{(x, a - x) \mid x \in E_k\}$$

belongs to $\mathfrak{S}_k$. Further, we have that either $\gamma''' \in \mathfrak{S}_{k;Q}$ is valid or (see above) it is possible to derive from γ''' a relation of $\mathfrak{C}^1_{k;Q}$. ■

Lemma 16.3.6 *Let* $|Q| \geq 2$ *and* $\gamma \in \bigcup_{a \in Q} \mathfrak{P}_{k;\{a\}} \cup \mathfrak{N}_{k;\{a\}}$. *Then there exists a* $\varrho \in [\{\gamma\} \cup Inv_k T_Q]$ *with* $\varrho \in \mathfrak{P}_{k;Q} \cup \mathfrak{N}_{k;Q} \cup \mathfrak{C}^1_{k;Q}$.

Proof. First let $\gamma \in (\bigcup_{a \in Q} \mathfrak{P}_{k;\{a\}}) \backslash \mathfrak{P}_{k;Q}$. Then there is $(b,c) \in \gamma$ with $b \in Q$ and $c \in E_k \backslash Q$. The γ-derivable relation

$$\gamma_b := pr_1(\Delta(\{b\} \times \gamma)) = \{x \in E_k \mid (b,x) \in \gamma\} \tag{16.4}$$

belong to $\mathfrak{C}_{k;Q}$ (because of $\gamma_b = \{c\}$).
Finally, let $\gamma \in (\bigcup_{a \in Q} \mathfrak{N}_{k;\{a\}}) \backslash \mathfrak{N}_{k;Q}$; i.e., γ has the following properties:

- $\exists q \in Q : \iota^2_k \cap \gamma = \{(q,q)\} \wedge \{(x,q),(q,x) \mid x \in E_k\} \subseteq \gamma$;
- γ is symmetric and
- $\exists b \in Q \setminus \{q\}\ \exists c \in E_k \backslash \{b,q\} : (b,c) \in \gamma$.

We form the γ-derivable relation γ_b (see (16.4)). Then we have $\{q,c\} \subseteq \gamma_b$ and $b \notin \gamma_b$. Thus $\gamma_b \in \mathfrak{C}^1_{k;Q}$. ■

Lemma 16.3.7 *Let* $A := Pol_k \varrho$ *be* T_Q*-maximal, where* $\varrho \in R^t_k$. *Moreover,* $V^{\uparrow}_k(A) \backslash \{A\} = V^{\uparrow}_k(T_Q)$. *Then* $|Q| = 1$ *is valid and there exists a relation*

$$\gamma \in \mathfrak{P}_{k;Q} \cup \mathfrak{N}_{k;Q}$$

with $A \subseteq Pol_k \gamma$.

Proof. Because of Lemma 16.3.4, the relation ϱ has the following properties:

(I) If $|Q| \geq 2$, then there is a relation $\gamma \in [\{\varrho\} \cup Inv_k T_Q]$ with

$$\gamma \in (R_{max}(P_k) \backslash \{\{q\} \mid q \in Q\}) \cup \bigcup_{a \in Q} R_{max}(T_{\{a\}}).$$

In particular, it follows from (I) that $t \geq 2$.

W.l.o.g. we can assume the following three properties of ϱ:

(II) ϱ does not have any double rows.
(III) Every ϱ-derivable $(t-1)$-ary relation belongs to $Inv_k T_Q$ (see 16.1.1,(c)).
(IV) No ϱ-derivable t-ary relation ϱ' with the properties $Pol_k\varrho' \subset T_Q$ and $|\varrho'| < |\varrho|$ exists.

From assumptions (II) - (IV), some further properties of the relation ϱ follow:

(V) ϱ does not have a constant row. (Suppose, ϱ has a constant row $(a, a, ..., a)$. Because of (I) we have $a \in Q$ and it is valid $\varrho = pr_{0,...,i-1}\varrho \times \{a\} \times pr_{i+1,...,t-1}\varrho$ for certain $i \in E_t$. Then, with the help of (III) and 16.1.1,(c) one can prove that ϱ is an invariant of T_Q, contrary to our assumptions about ϱ.)

A direct consequence from (V) and (III) is:

(VI) $\forall i \in E_t : pr_{0,...,i-1,i+1,...,t-1}\varrho = E_k^{t-1}$.

Next we show that

(VII) $(t \geq 3 \vee |Q| = 1) \implies \varrho \cap \delta^t_{\{0,1,...,t-1\}} \in \{\{(q,q,...,q)\} \mid q \in Q\}$;
$(t = 2) \implies \varrho \cap \delta^t_{\{0,1\}} \in \{\emptyset, \{(q,q)\} \mid q \in Q\}$

holds. By (III) we have

$$\varrho \cap \delta^t_{\{0,1,...,t-1\}} \in \{\emptyset, \{(q,q,...,q)\}, \delta^t_{\{0,1,...,t-1\}} \mid q \in Q\}.$$

$\varrho \cap \delta^t_{\{0,1,...,t-1\}} = \delta^t_{\{0,1,...,t-1\}}$ is not possible, since $A = Pol_k\varrho \subset T_Q$ and T_Q contains at most a constant.
If $Q = \{a\}$ and $\varrho \cap \delta^t_{\{0,1,...,t-1\}} = \emptyset$, the equation $pr_{0,...,t-2}\varrho = E_k^{t-1}$ (see (VI)) implies $pr_t((\{a\} \times \varrho) \cap \delta^{t+1}_{\{0,...,t-1\}} \in \mathfrak{C}^1_{k;\{a\}}$, a contradiction to (I).
In the case that $t \geq 3$, it follows from $\varrho \cap \delta^t_{\{0,1,...,t-1\}} = \emptyset$ that $pr_{t-2,t-1}(\varrho \cap \delta^t_{\{0,...,t-2\}}) \notin Inv_k T_Q$, which contradicts (III). Therefore (VII) holds.

As generally known (see Chapter 6), the relation

$$\sigma_i(\varrho) := \{(a_1, ..., a_i) \in E_k^i \mid \exists u \in E_k : \{(a_1,u), ..., (a_i,u)\} \subseteq \varrho\}$$

is derivable from the relation ϱ for $t = 2$, and it is valid that

(VIII) $(t = 2 \wedge \varrho \circ (\tau\varrho) = E_k^2) \implies \forall i \geq 2 : \sigma_i(\varrho) = E_k^i$.

For the relation ϱ, the following three cases are possible:
Case 1: t = 2.
We consider the ϱ-derivable relation $\varrho \cap (\tau\varrho)$.
Case 1.1: $\varrho \cap (\tau\varrho) \in \{\emptyset, \{(q,q)\}\}$ for a certain $q \in Q$.
Then ϱ is antisymmetric. Because of (VI) the relation $\varrho \circ (\tau\varrho)$ has the property $\iota_k^2 \subseteq \varrho \circ (\tau\varrho)$.
Case 1.1.1: $\varrho \circ (\tau\varrho) = \iota_k^2$.
Because of $pr_0\varrho = pr_1\varrho = E_k$ (see (VI)) we have $|\varrho| \geq k$. If $|\varrho| > k$, there exists $a, b, c \in E_k$ with $(a,c), (b,c) \in \varrho$ and $a \neq b$. Thus $(a,b) \in \varrho \circ (\tau\varrho)$,

which is not possible because of $\varrho \circ (\tau\varrho) = \iota_k^2$. Therefore $|\varrho| = k$, and ϱ has the form $\{(x, s(x)) \mid x \in E_k\}$, where $s \neq e_1^1$ is a permutation on E_k, which has at most a fixed point (namely q). Assume the permutation has proper cycles of different length. Let r (≥ 2) be the length of a smallest proper cycle. Then we have

$$pr_0((\underbrace{\varrho \circ \varrho \circ \ldots \circ \varrho}_{r \text{ times}}) \cap \iota_k^2) \in \mathfrak{C}_{k;Q}^1,$$

a contradiction to (I). Therefore, all proper cycles of s have the very same length l. Suppose $l = p \cdot m$, p prime and $m \geq 2$. Then $\underbrace{\varrho \circ \varrho \circ \ldots \circ \varrho}_{m \text{ times}}$ has proper cycles of the length p; i.e., a relation of the set $\mathfrak{S}_k \cup \mathfrak{P}_{k;\{q\}}$ is derivable from the relation ϱ. But, this contradicts the condition (I) for $|Q| \geq 2$ or $\varrho \in \mathfrak{S}_k$. Therefore, $Q = \{q\}$ and $A \subseteq Pol_k\gamma$ for a certain $\gamma \in \mathfrak{P}_{k;\{q\}}$ in Case 1.1.1.

Case 1.1.2: $\iota_k^2 \subset \varrho \circ (\tau\varrho) \subset E_k^2$.

In this case, we see that the relation $\varrho' = \varrho \circ (\tau\varrho)$ is not an invariant of T_Q and that $Pol_k\varrho' \not\subseteq T_Q$ holds, which is not possible because of $V_k^{\uparrow}(A)\backslash\{A\} = V_k^{\uparrow}(T_Q)$.

Case 1.1.3: $\varrho \circ (\tau\varrho) = E_k^2$.

By (VIII) we have $\sigma_k(\varrho) = E_k^k$. Consequently, there exists a $u \in E_k$ with $(x, u) \in \varrho$ for all $x \in E_k$. Because of $\varrho \cap \iota_k^2 \in \{\emptyset, \{(q, q)\}\}$ ($q \in Q$), this is only possible for $\varrho \cap \iota_k^2 = \{(q, q)\}$ and $u = q$.

If $(\tau\varrho) \circ \varrho = E_k^2$ is valid, we can prove $\{(x, q) \mid x \in E_k\} \subseteq \tau\varrho$ in analog mode too, through which we receive a contradiction to the antisymmetry of the relation ϱ. Therefore $\iota_k^2 \subseteq (\tau\varrho) \circ \varrho \subset E_k^2$. Consequently, Case 1.1.3 is reducible to Cases 1.1.1 and 1.1.2.

Case 1.2: $\{(q, q)\} \subset \varrho \cap (\tau\varrho) \subset \varrho$.

Because of condition (III) this case is not possible.

Case 1.3: $\varrho \cap (\tau\varrho) = \varrho$.

In this case, ϱ is symmetric. Further, we have $\iota_k^2 \subseteq \varrho \circ \varrho$. We distinguish three cases:

Case 1.3.1: $\varrho \circ \varrho = \iota_k^2$.

Then ϱ is a permutation with at most a fixed point q (if $\varrho \cap \iota_k^2 = \{(q, q)\}$), and every proper cycle of ϱ has the length 2 because of symmetry of ϱ, i.e., $\varrho \in \mathfrak{S}_k \cup \mathfrak{P}_{k;\{q\}}$. Thus, as in Case 1.1.1, we obtain a contradiction to the condition (I) either or $|Q| = 1$ and $A \subseteq Pol_k\gamma$ for certain $\gamma \in \mathfrak{P}_{k;Q}$ are valid.

Case 1.3.2: $\iota_k^2 \subset \varrho \circ \varrho \subset E_k^2$.

This case can be excluded as Case 1.1.2.

Case 1.3.3: $\varrho \circ \varrho = E_k^2$.

Because of $\varrho \circ \varrho = \varrho \circ (\tau\varrho) = E_k^2$ and by (VIII), we have $\sigma_k(\varrho) = E_k^k$,; i.e., there is $u \in E_k$ with $\{(x, u) \mid x \in E_k\} \subseteq \varrho$. Consequently, $\varrho \cap \iota_k^2 = \{(q, q)\}$ and $u = q$. This and the symmetry of ϱ implies that q is a central element of ϱ. Therefore, ϱ belongs to $\mathfrak{N}_{k;\{q\}}$. Because of (I), this is only possible for $|Q| = 1$.

Case 2: $t = 3$.
Since $pr_{0,1}\varrho = E_k^2$ (by (VI)), $\varrho \cap \delta^3_{\{0,1,2\}} = \{(q,q,q)\}$ for certain $q \in Q$ (see (VII)) and (III) is valid, we have $\Delta\varrho = E_k \times \{q\}$, i.e., $(a,a,q) \in \varrho$ for all $a \in E_k$. Analogously, one can prove that the tuples (a,q,a) and (q,a,a) belong to ϱ for every $a \in E_k$. Next we prove that ϱ is totally symmetric. Assume the relation is not totally symmetric. Then the ϱ-derivable relation

$$\varrho' := \bigcap_{\substack{s \\ s \text{ is permutation on } E_3}} \{(a_{s(0)}, a_{s(1)}, a_{s(2)}) \mid (a_0,a_1,a_2) \in \varrho\}$$

is totally symmetric. Because of

$$\begin{pmatrix} a & a & q \\ a & q & a \\ q & a & a \end{pmatrix} \subseteq \varrho' \subset \varrho \ (a \neq q)$$

the relation ϱ' is not, however, an invariant of T_Q. This is a contradiction to the condition (IV).

Next we prove

(IX) $\forall a,b \in E_k : \{(a,b,c),(a,b,c')\} \subseteq \varrho \implies c = c'$.

Suppose there are a,b,c,c' with $\{(a,b,c),(a,b,c')\} \subseteq \varrho$, and $c \neq c'$. Then the ϱ-derivable relation

$$\begin{aligned} \varrho_1 &:= pr_{0,2,5}((\varrho \times \varrho) \cap \delta^6_{\{0,3\},\{1,4\}}) \\ &= \{(x,y,z) \mid \exists u \in E_k : \{(x,u,y),(x,u,z)\} \subseteq \varrho\} \end{aligned}$$

has the property $Pol_k\varrho_1 \not\subseteq T_Q$ because of $\delta^3_{\{0,1,2\}} \subseteq \varrho_1$. Consequently, ϱ_1 is a diagonal relation. Further, it is easy to check that $(a,c,c') \in \varrho_1$ and $(d,q,q) \in \varrho_1$ for all $d \in E_k$ hold. Therefore, ϱ_1 is the diagonal relation E_k^3. For our relation ϱ, this means that for arbitrary $(x,y,z) \in E_k^3$ there is a u with $(x,u,y) \in \varrho$ and $(x,u,z) \in \varrho$. Then, when one chooses $x = y = q$ and $z \neq q$ and considers the total symmetry of the relation ϱ, there exists an r with $r \neq q$ and $(q,q,r) \in \varrho$. Above $\Delta\varrho = E_k \times \{q\}$ was, however, proven. Therefore, our assumption $c \neq c'$ was false. With that, (IX) is valid.
Now we consider the ϱ-derivable relation

$$\varrho_2 := \varrho \circ \varrho = \{(a,b,c,d) \mid \exists u \in E_k : \{(a,b,u),(u,c,d)\} \subseteq \varrho\}.$$

Because of $\{(a,a,q),(q,a,a) \mid a \in E_k\} \subseteq \varrho$ we have $\delta^4_{\{0,1,2,3\}} \subseteq \varrho_2$ and thus $Pol_k\varrho_2 \not\subseteq T_Q$. Obviously, the relation ϱ_2 does not have any double rows. Because of our assumptions about the relation ϱ, however, this is possible only for $\varrho_2 = E_k^4$. Then, by definition of ϱ_2, we have $\{(a,b,u),(u,c,a)\} \subseteq \varrho$ for arbitrary $a,b,c \in E_k$, and certain u. Since ϱ is totally symmetric, however,

$\{(a,u,b),(a,u,c)\} \subseteq \varrho$, where $b \neq c$ is possible, contrary to (IX). Therefore, Case 2 is not possible.

Case 3: $t \geq 4$.

Because of $pr_{0,1,\ldots,t-2}\varrho = E_k^{t-1}$ (see (VI)), (II), (III) and (VII) we have $\varrho \cap \delta^t_{\{0,\ldots,h-3\}} = \delta^{t-1}_{\{0,\ldots,h-3\}} \times \{q\}$ for certain $q \in Q$. Since $pr_{0,\ldots,h-3,h-1}\varrho = E_k^{t-1}$ is also valid, one can analogously prove $\varrho \cap \delta^t_{\{0,\ldots,h-3\}} = \delta^{t-2}_{\{0,..,h-3\}} \times \{q\} \times E_k$, contrary to that just shown. Therefore, there is no t-ary relation ϱ with the above-demanded properties and $t \geq 4$. ∎

An equivalence relation $\sim$ is defined by

$$\varrho \sim \varrho' \;:\Longleftrightarrow\; T_Q \cap Pol_k\varrho = T_Q \cap Pol_k\varrho'$$

on a set $\mathfrak{A} \subseteq R_k$. We select a representative from every equivalence class of $\sim$ now and obtain a certain subset of $\mathfrak{A}$, which we denote with $\mathfrak{A}^{\sim}$, where $\mathfrak{A} \in \{\mathfrak{P}_{k;Q}, \mathfrak{M}_{k;Q}, \mathfrak{S}_{k;Q}\}$ in the following.

Lemma 16.3.8 *Let* $\emptyset \neq Q \subseteq E_k$. *Then*

(a) $\forall \varrho, \varrho \in \mathfrak{M}_{k;Q} : \varrho \sim \varrho' \iff (\varrho' = \tau\varrho \,\vee\, \varrho = \varrho')$;

(b) $\forall \varrho, \varrho' \in \mathfrak{P}_{k;Q} \cup \mathfrak{S}_{k;Q} : \varrho \sim \varrho' \iff (\exists t : \varrho' = \underbrace{\varrho \circ \varrho \circ \ldots \circ \varrho}_{t \text{ times}})$;

(c) $\forall \varrho, \varrho' \in \mathfrak{M}^{\sim}_{k;Q} \cup \mathfrak{S}^{\sim}_{k;Q} \cup \mathfrak{P}^{\sim}_{k;Q} \cup \mathfrak{U}_{k;Q} \cup \mathfrak{N}_{k;Q} \cup \mathfrak{L}_{k;Q} \cup \mathfrak{C}_{k;Q} : \varrho \sim \varrho' \iff \varrho \neq \varrho'$.

Proof. The statements (a) and (b) are direct conclusions from the proof of corresponding statements about relations of $\mathfrak{M}_k \cup \mathfrak{S}_k$ (see Chapter 5). To prove (c), we agree that o_ϱ denotes the smallest element of E_k and that e_ϱ denotes the greatest element of E_k (in respect to $\varrho \in \mathfrak{M}_{k;Q}$). Because of (a), we can assume w.l.o.g. that $o_\varrho \in Q$. Now, let ϱ and ϱ' be two different relations of $\mathfrak{M}^{\sim}_{k;Q} \cup \mathfrak{S}^{\sim}_{k;Q} \cup \mathfrak{P}^{\sim}_{k;Q} \cup \mathfrak{U}_{k;Q} \cup \mathfrak{N}_{k;Q} \cup \mathfrak{L}_{k;Q} \cup \mathfrak{C}_{k;Q}$. To prove

$$T_Q \cap Pol_k\varrho \not\subseteq T_Q \cap Pol_k\varrho' \tag{16.5}$$

we distinguish the following 9 cases:

Case 1: $\{\varrho, \varrho'\} \subseteq \mathfrak{C}^1_{k;Q} \cup \mathfrak{U}_{k;Q} \cup \mathfrak{P}^{\sim}_{k;Q} \cup \mathfrak{S}^{\sim}_{k;Q}$.

If $|Q| \geq k-1$ then $\mathfrak{U}_{k;Q} = \mathfrak{P}_{k;Q} = \mathfrak{S}_{k;Q} = \emptyset$ and (16.5) is obviously valid for the relations ϱ, ϱ'. Let now $|Q| \leq k-2$. Denote ω the relation $\{(q,q) \mid q \in Q\} \cup (E_k \backslash Q)^2$ of $\mathfrak{U}_{k;Q}$. It is easy to check that the set

$$A(\gamma) := \bigcup_{n \geq 1} \{f^n \in P_{E_k \backslash Q} \mid \exists f' \in T_Q \cap Pol_k\gamma : (\forall \mathbf{a} \in (E_k \backslash Q)^n : f'(\mathbf{a}) = f(\mathbf{a}))\}$$

is a maximal class of $P_{E_k\backslash Q}$ for $\gamma \in (\mathfrak{C}^1_{k;Q} \backslash \{\gamma \mid \gamma \subseteq Q \vee \gamma = E_k \backslash Q\}) \cup (\mathfrak{U}_{k;Q} \backslash \{\omega\}) \cup \mathfrak{P}^{\sim}_{k;Q} \cup \mathfrak{S}^{\sim}_{k;Q}$ and that $A(\gamma) = P_{E_k\backslash Q}$ holds for $\gamma \in \{\gamma \in \mathfrak{C}^1_k \mid \gamma \subseteq Q \vee \gamma = E_k \backslash Q\} \cup \{\omega\}$. With the aid of Chapters 5 and 6, our assertion (16.5) results from that for $\varrho' \in \{\gamma \in \mathfrak{C}^1_k \mid \gamma \subseteq Q \vee \gamma = E_k \backslash Q\} \cup \{\omega\}$. If $\varrho' \in \{\gamma \in \mathfrak{C}^1_k \mid \gamma \subseteq Q \vee \gamma = E_k \backslash Q\} \cup \{\omega\}$, then it is also easy to prove that

there is a $|\varrho'|$-ary function $f_1 \in T_Q \cap Pol_k\varrho$ with $f_1(\varrho') \notin \varrho'$. Consequently, (16.5) is also valid for the remaining relations ϱ' of the Case 1.
Case 2: $\varrho \in \mathfrak{C}^1_{k;Q}$ and $\varrho' \in \mathfrak{M}^{\sim}_{k;Q} \cup (\mathfrak{C}_{k;Q} \backslash \mathfrak{C}^1_{k;Q}) \cup \mathfrak{N}_{k;Q} \cup \mathfrak{L}_{k;Q}$.
The following $|\varrho'|$-ary function f_2 with the properties

$$f_2(\varrho') \notin \varrho', f_2(q, ..., q) = q \text{ for all } q \in Q$$

and

$$f_2(\boldsymbol{a}) = c \text{ for the remaining tuples } \boldsymbol{a}, \text{ where } c \in \varrho,$$

belongs to $T_Q \cap Pol_k\varrho$. Thus (16.5) holds.
Case 3: $\varrho \in \mathfrak{M}^{\sim}_{k;Q} \cup \mathfrak{U}_{k;Q} \cup \mathfrak{P}^{\sim}_{k;Q} \cup \mathfrak{S}^{\sim}_{k;Q}$ and $\varrho' \in (\mathfrak{C}_{k;Q} \backslash \mathfrak{C}^1_{k;Q}) \cup \mathfrak{N}_{k;Q} \cup \mathfrak{L}_{k;Q}$.
Let ϱ' be an h'-ary relation. Further, let $q \in Q$ and $a \in E_k \backslash Q$. In addition, if $\varrho' \in \mathfrak{N}_{k;Q}$, then we choose $q \in Q$ so that $(q, q) \in \varrho'$ holds. Then the h'-tuples $(q, a, a, \ldots, a)$, $(a, q, a, \ldots, a)$, $\ldots$, $(a, a, \ldots, a, q)$ belong to ϱ', though not all belong to ϱ. Consequently, there exists an h'-ary function $f_3 \in T_Q \cap Pol_k\varrho$ with

$$f_3 \begin{pmatrix} q & a & a & . & . & . & a \\ a & q & a & . & . & . & a \\ . & . & . & . & . & . & . \\ a & a & a & . & . & . & q \end{pmatrix} \notin \varrho'.$$

Hence (16.5) holds in Case 3.
Case 4: $\varrho \in \mathfrak{P}^{\sim}_{k;Q} \cup \mathfrak{S}^{\sim}_{k;Q}$ and $\varrho' \in \mathfrak{M}^{\sim}_{k;Q}$.
This case can occur only for $|Q| \leq 2$. Further, we have $\mathfrak{S}^{\sim}_{k;Q} = \emptyset$, if $|Q| = 1$ or $k = 3$. It is easy to check that $(T_Q \cap Pol_k\varrho) \not\subseteq (T_Q \cap Pol_k\varrho')^1$ is valid for $k \geq 3$.
Case 5: $\varrho \in \mathfrak{U}_{k;Q}$ and $\varrho' \in \mathfrak{M}^{\sim}_{k;Q}$.
For $o_{\varrho'} \in Q$ the binary function f_4 defined by

$$f_4(x, y) := \begin{cases} y \text{ if } x = o_{\varrho'}, \\ o_{\varrho'} \text{ otherwise,} \end{cases}$$

belongs to $T_Q \cap Pol_k\varrho$. However, f_4 does not preserve the relation ϱ' with the smallest element $o_{\varrho'}$, since

$$(f_4(o_{\varrho'}, e_{\varrho'}), f(\alpha, e_{\varrho'})) = (e_{\varrho'}, o_{\varrho'}) \text{ if } \alpha \in E_k \backslash \{o_{\varrho'}, e_{\varrho'}\} .$$

Case 6: $\varrho \in \mathfrak{M}^{\sim}_{k;Q}$ and $\varrho' \in \mathfrak{C}^1_{k;Q}$.
Let $a \in E_k \backslash \varrho'$ and

$$t_a(x, y) := \begin{cases} x \text{ if } x = y \in Q, \\ a \text{ otherwise.} \end{cases}$$

The function $t_a \in T_Q$ preserves ϱ and does not preserve ϱ'. Therefore, (16.5) holds.

Case 7: $\varrho \in \mathfrak{M}^{\sim}_{k;Q}$ and $\varrho' \in \mathfrak{M}^{\sim}_{k;Q} \cup \mathfrak{U}_{k;Q} \cup \mathfrak{P}^{\sim}_{k;Q} \cup \mathfrak{S}^{\sim}_{k;Q}$.
Let $\varrho' = \begin{pmatrix} \mathbf{r} \\ \mathbf{s} \end{pmatrix}$, $\{\mathbf{r}, \mathbf{s}\} \subset E_k^{|\varrho'|}$ and f_5 a $|\varrho'|$-ary function defined by

$$f_5(\varrho') := \begin{pmatrix} \alpha \\ \beta \end{pmatrix} \begin{cases} \notin \varrho' & \text{if} \quad \varrho' \not\subseteq \varrho, \\ \in \varrho \backslash \varrho' & \text{if} \quad \varrho' \subset \varrho \end{cases}$$

and, if $\mathbf{r} <_\varrho \mathbf{s}$ and hence $\alpha <_\varrho \beta$,

$$f_5(\boldsymbol{a}) := \begin{cases} o_\varrho & \text{if} \qquad \boldsymbol{a} = (o_\varrho, o_\varrho, ..., o_\varrho), \\ \alpha & \text{if} \qquad \boldsymbol{o_\varrho} <_\varrho \boldsymbol{a} <_\varrho \boldsymbol{s}, \\ \beta & \text{if} \qquad \boldsymbol{a} = \boldsymbol{s}, \\ e_\varrho & \text{otherwise,} \end{cases}$$

and, if $\mathbf{s}$ and $\mathbf{r}$ are incomparable in respect to ϱ,

$$f_5(\mathbf{a}) := \begin{cases} o_\varrho & \text{if} \qquad \boldsymbol{a} <_\varrho \boldsymbol{r} \vee \boldsymbol{a} <_\varrho \boldsymbol{s}, \\ \alpha & \text{if} \qquad \boldsymbol{a} = \boldsymbol{r}, \\ \beta & \text{if} \qquad \boldsymbol{a} = \boldsymbol{s}, \\ e_\varrho & \text{otherwise.} \end{cases}$$

It is easy to check that $f_5 \in T_Q \cap Pol_k \varrho$ and $f_5 \notin T_Q \cap Pol_k \varrho'$.
Case 8: $\varrho \in (\mathfrak{C}_{k;Q} \backslash \mathfrak{C}^1_{k;Q}) \cup \mathfrak{N}_{k;Q}$ and $\varrho' \in \mathfrak{U}_{k;Q} \cup \mathfrak{M}^{\sim}_{k;Q} \cup \mathfrak{C}_{k;Q} \cup \mathfrak{P}^{\sim}_{k;Q} \cup \mathfrak{S}_{k;Q} \cup \mathfrak{N}_{k;Q} \cup \mathfrak{L}_{k;Q}$.
Let c be a central element of ϱ. Further, let f_6 be an $|\varrho'|$-ary function defined by

$$f_6(\varrho') \begin{cases} \notin \varrho' & \text{if} \quad \varrho' \not\subseteq \varrho, \\ \in \varrho \backslash \varrho' & \text{if} \quad \varrho' \subset \varrho \end{cases}$$

and $f_6(q, q, ..., q) = q$ for all $q \in Q$ and $f_6(\boldsymbol{a}) = c$ for the remaining tuples $\boldsymbol{a} \in E_k^{|\varrho'|}$. Then $f_6 \in T_Q \cap Pol_k \varrho$ and $f_6 \notin T_Q \cap Pol_k \varrho'$. Therefore, (16.5) holds.
Case 9: $\varrho \in \mathfrak{L}_{k;Q}$.
Then $|Q| = 1$ and $k = 2^m, m \geq 1$. W.l.o.g. let $Q = \{0\}$. As shown in Chapter 5, one can define the relation ϱ with the aid of an elementary Abelean 2-group G, where 0 is the neutral element of G. It follows from the proof of the Lemma 16.3.5 that the relation ϱ is derivable from a relation $\{0\} \times \lambda$, where $\lambda \in \mathfrak{L}_{2^m}$. Since $\varrho \circ \varrho = \lambda$ is in addition valid, we have $T_{\{0\}} \cap Pol_k \lambda = Pol_k \varrho$. Therefore, by Chapter 5, every n-ary function $f \in Pol_k \varrho$ is a quasi-linear function of the form

$$f(x_1, ..., x_n) = \sum_{i=1}^{n} \sum_{j=0}^{m-1} a_{i,j} \cdot x_i^{2^j},$$

where $a_{1,0}, ..., a_{n,m-1} \in E_{2^m}$ and $+(= +_G)$ and $\cdot$ are the operations of a certain field $(E_{2^m}; +, \cdot)$ with $G = (E_{2^m}; +)$.

Case 9.1: $\varrho' \in \mathfrak{C}^1_{k;Q}$.
Let $a \notin \varrho'$ and $b \in \varrho'$. Then the quasi-linear unary function f_7 defined by $f_7(x) := (a \cdot b^{-1}) \cdot x$ does not preserve ϱ' because of $f_7(b) = a$, i.e., (16.5) holds.
Case 9.2: $\varrho' \in \mathfrak{P}^{\sim}_{k;Q}$.
Let 1 the unit of the multiplicative group of the field $(E_{2^m}; +, \cdot)$. Since the order of this group is odd, we have $a^2 \neq 1$ for all $a \in E_{2^m}\backslash\{0,1\}$. The unary function f_8 defined by $f_8(x) := x^2$ belongs to $Pol_k\varrho$, but it does not preserve ϱ', since (by definition of ϱ') there is an $a \in E_k\backslash\{1\}$ with $(1, a) \in \varrho'$ and $(f_8(1), f_8(a)) = (1, a^2) \notin \varrho'$ because of $a^2 \neq a$ and because of the above.
Case 9.3: $\varrho' \in \mathfrak{U}_{k;Q}$.
The binary function f_9 with $f_9(x, y) : \ = x + y$ belongs to $Pol_k\varrho$. One can prove $f_9 \notin Pol_k\varrho'$ as follows: Let $(a, b) \in \varrho'$ and $a \neq b$. Then

$$f_9\begin{pmatrix} a & a \\ b & a \end{pmatrix} = \begin{pmatrix} 0 \\ a+b \end{pmatrix} \notin \varrho',$$

since $a \neq b$ implies $a + b \neq 0$.
Case 9.4: $\varrho' \in \mathfrak{M}^{\sim}_{k;Q}$.
In this case, the above-defined function f_9 does not preserve the relation ϱ', since

$$f_9\begin{pmatrix} 0 & a \\ e_{\varrho'} & e_{\varrho'} \end{pmatrix} = \begin{pmatrix} a \\ 0 \end{pmatrix} \notin \varrho'$$

holds for $a \in E_k\backslash\{0\}$.
Case 9.5: $\varrho' \in \mathfrak{L}_{k;Q}$.
For arbitrary $\lambda \in \mathfrak{L}_{2^2,0}$ we have $\{(0, x, x), (x, 0, x), (x, x, 0) \mid x \in E_k\} \subseteq \lambda$. Because of $\varrho \neq \varrho'$, there are certain a and b of E_k with $(a, b, a +_G b) \in \varrho$ and $(a, b, a +_G b) \notin \varrho'$. Consequently, the ternary function $f_{10} \in Pol_k\varrho$ defined by $f_{10}(x, y, z) := x +_G y +_G z$ does not preserve the relation ϱ', since

$$f_{10}\begin{pmatrix} 0 & a & 0 \\ a & a & b \\ a & 0 & b \end{pmatrix} = \begin{pmatrix} a \\ b \\ a +_G b \end{pmatrix}$$

holds.
Case 9.6: $\varrho' \in (\mathfrak{C}_{k;Q}\backslash\mathfrak{C}^1_{k;Q}) \cup \mathfrak{N}_{k;Q}$.
Let ϱ' be an h'-ary relation and $(a_1, \ldots, a_{h'}) \notin \varrho'$. Then the h'-ary function f_{11} defined by

$$f_{11}(x_1, \ldots, x_{h'}) := a_1 \cdot x_1 +_G a_2 \cdot x_2 +_G \cdots +_G a_{h'} \cdot x_{h'}$$

does not preserve the relation ϱ', since, if 1 is the unit of the field $(E_k; +_G, \cdot)$,

$$f_{11}\begin{pmatrix} 1 & 0 & 0 & . & . & . & 0 \\ 0 & 1 & 0 & . & . & . & 0 \\ . & . & . & . & . & . & . \\ 0 & 0 & 0 & . & . & . & 1 \end{pmatrix} = \begin{pmatrix} a_1 \\ a_2 \\ . \\ a_{h'} \end{pmatrix}$$

holds and because 0 is a central element of ϱ'. Therefore, (16.5) holds in Case 9.6.
Since $T_Q \cap Pol_k\varrho \not\subseteq T_Q \cap Pol_k\varrho'$ implies $\varrho \neq \varrho'$, (c) is proven. ■

16.4 Proof of Theorem 16.2.1

Put

$$R_{max}(T_Q) := \mathfrak{M}_{k;Q} \cup \mathfrak{U}_{k;Q} \cup \mathfrak{S}_{k;Q} \cup \mathfrak{L}_{k;Q} \cup \mathfrak{C}_{k;Q} \cup \mathfrak{P}_{k;Q} \cup \mathfrak{N}_{k;Q}.$$

Let A be an arbitrary maximal class of T_Q with $\emptyset \neq Q \subseteq E_k$. We will prove by induction on $r := |Q|$ that

$$\exists \varrho \in R_{max}(T_Q) : A \subseteq T_Q \cap Pol_k\varrho \tag{16.6}$$

holds.
r = 1 : If $V_k^{\uparrow}(A) \backslash \{A\} = V_k^{\uparrow}(T_Q)$, then (16.6) follows from Lemma 16.3.7. If $V_k^{\uparrow}(A)\backslash\{A\} \neq V_k^{\uparrow}(T_Q)$, then there exists a maximal class B_1 of P_k with $B_1 \neq Pol_k\varrho$ and $A = T_Q \cap B_1$ because of Lemma 16.3.4, (b). Then, by Chapter 6, there is a relation $\gamma \in \mathfrak{M}_k \cup \mathfrak{U}_k \cup \mathfrak{S}_k \cup \mathfrak{L}_k \cup \mathfrak{B}_k \cup \mathfrak{C}_k \backslash \{Q\}$ with $\gamma \in Inv_k A$. Therefore, (16.6) follows from Lemma 16.3.5.
r − 1 ⟶ r : Suppose, (16.6) holds for all $Q' \subseteq E_k$ with $1 \leq |Q'| \leq r-1$. Let Q ($\subseteq E_k$) be an r-element set. Because of $r = |Q| \geq 2$ and Lemma 16.3.7, we have $V_k^{\uparrow}(A)\backslash\{A\} \neq V_k^{\uparrow}(T_Q)$. Consequently, there exists a subclass B of P_k with

$$A = B \cap T_Q \text{ and } B \not\subseteq T_Q. \tag{16.7}$$

We distinguish two cases for B:
Case 1: $\forall a \in Q : B \not\subseteq Pol_k\{a\}$.
Since $A \neq P_k$ and because of Chapter 6, there is a $\gamma \in R_{max}(P_k)\backslash\{\{a\} \mid a \in Q\}$ with $B \subseteq Pol_k\gamma$. Then this, (16.7), and Lemma 16.3.4, (a) imply $A = B \cap T_Q \subseteq T_Q \cap Pol_k\gamma \subset T_Q$. Because $\gamma \in Inv_k A$, the statement (16.6) results through Lemmas 16.3.5 and 16.3.6.
Case 2: $\exists Q' \subset Q : B \subseteq T_{Q'} \wedge (\forall b \in Q\backslash Q' : B \not\subseteq Pol_k\{b\})$.
Because of $B \notin V_k^{\uparrow}(T_Q)$, we have $B \subset T_{Q'}$. Thus there exists a maximal class B_2 of $T_{Q'}$ with $B \subseteq B_2$ (see Lemma 16.3.2). By induction assumption, there is a $\varrho \in R_{max}(T_{Q'})$ with $B_2 \subseteq T_{Q'} \cap Pol_k\varrho$. Because of Lemma 16.3.4, (a), (c) we have $T_{Q'} \cap Pol_k\varrho \neq T_{Q'}$. This implies $B_2 = T_{Q'} \cap Pol_k\varrho$. Furthermore, the following is valid: $\varrho \in \mathfrak{M}_k \cup \mathfrak{S}_k \cup \mathfrak{U}_k \cup (\mathfrak{C}_k\backslash\{\{q\} \mid q \in Q\}) \cup \bigcup_{a \in Q'}(\mathfrak{P}_{k;\{a\}} \cup \mathfrak{N}_{k;\{a\}})$ (notice $\forall b \in Q\backslash Q' : B \not\subseteq Pol_k\{b\}$). This and Lemmas 16.3.5 and 16.3.6 imply (16.6).
Consequently, (16.6) is proven.
It is easy to see that $T_Q \cap Pol_k\varrho \neq T_Q$ holds for every $\varrho \in R_{max}(T_Q)$.
Therefore, Theorem 16.2.1 follows from (16.6) and Lemma 16.3.7. ■

17

Maximal Classes of $Pol_k E_l$ for $2 \leq l < k$

This chapter continues the investigations of Chapter 16 and generalizes Theorem 14.1.3. For arbitrary $k, l \in \mathbb{N}$ with

$$2 \leq l \leq k-1$$

all maximal classes of $Pol_k E_l$ are determined. With the help of these maximal classes, one can easily give a completeness criterion for $Pol_k E_l$.
The proofs given in this chapter resemble those ones from Chapter 6; that is, the results of this chapter were achieved with the means which were developed by I. G. Rosenberg in [Ros 70a].

17.1 Notations, Definitions, and Some Lemmas

A relation ϱ, for which there are certain $t_1, ..., t_n \in \{l, k\}$ and a relation $\delta \in D_k$ with

$$\varrho = (E_{t_1} \times E_{t_2} \times ... \times E_{t_h}) \cap \delta,$$

is called E_l**-diagonal**.
For relations $\varrho \in D_k \cup \{\iota_k^h\}$, we agree with the following brief notations:

$$\varrho(t_1, t_2, ..., t_n) := \varrho \cap (E_{t_1} \times E_{t_2} \times ... \times E_{t_h}) \quad (t_1, ..., t_h \in \{l, k\})$$

and

$$\varrho[t] := \varrho(\underbrace{l, l, ..., l}_{t \text{ times}}, k, k, ..., k).$$

In this chapter, we say that a relation ϱ' is ϱ**-derivable** if one can receive ϱ' by a finite number of applications of the relation operations (see Section 2.3) from the relations of $\{\varrho\} \cup Inv_k Pol_k E_l$. In this case, we also write

$$\{\varrho\} \cup Inv_k Pol_k E_l \vdash \varrho'$$

or

$$\varrho' \in [\{\varrho\} \cup Inv_k Pol_k E_l].$$

In the following proofs of this chapter, we assume that the reader knows the statements of the following lemma. The lemma contains statements that were already proven or can easily be proven.

Lemma 17.1.1 *It holds:*

(a) $\forall \varrho \in R_k : \; Pol_k E_l \cap Pol_k \varrho = Pol_k E_l \times \varrho$*;*

(b) $(\exists i \in E_l : \; pr_i \varrho = E_l) \implies Pol_k E_l \times \varrho = Pol_k \varrho$*;*

(c) $\forall \varrho, \varrho' \in R_k :$
$((Pol_k \varrho \subseteq Pol_k E_l \;\wedge\; (\{\varrho\} \cup Inv_k Pol_k E_l \;\vdash\; \varrho')) \implies$
$Pol_k \varrho \subseteq Pol_k E_l \cap Pol_k \varrho')$*;*

(d) $Inv_k P_k = D_k$ *(see Theorem 2.6.2);*

(e) A relation ϱ *belongs to* $Inv_k Pol_k E_l$ *if and only if* ϱ *is* E_l*-diagonal.* ■

In addition to those notations introduced in Chapter 5, we need the following for relations and relation sets:

$\mathfrak{U}_{k;E_l} := \{\varrho \in \mathfrak{U}_k \mid \varrho \cap E_l^2 \in \{\iota_l^2, E_l^2\} \wedge \bigcup_{a\in E_l}\{x \mid (a,x) \in \varrho\} \in \{E_l, E_k\}\}.$

$S_{k;E_l} := \{\{(x, s(x)) \mid x \in E_k\} \mid$ s is a permutation on E_k with exactly l fixed points ($\in E_l$), and all proper cycles of s have the same prime length $\}$;

Put

$\mathfrak{C}^{(1)}_{k;E_l} := \{\varrho \in \mathfrak{C}^1_k \mid \varrho \cap E_l \in \{\emptyset, E_l\} \wedge \varrho \neq E_l\},$

$\mathfrak{C}^h_{k;E_l} := \{\varrho \in \mathfrak{C}^h_k \mid$ each element of E_l is a central element of ϱ $\}$
$(2 \le h \le k-1)$;

$$\begin{aligned} Z\mathfrak{C}^2_{k;E_l} := \{\varrho \in \mathfrak{C}^2_k \mid\; & \varrho \cap E_l^2 = \iota_l^2 \;\wedge \\ & (\forall x \in E_k \;\exists \alpha_x \in E_l : \; (x,\alpha_x) \in \varrho) \;\wedge \\ & (\forall a,b \in E_k \backslash E_l : ((\exists u \in E_l : \{(a,u),(u,b)\} \subseteq \varrho) \implies \\ & (a,b) \in \varrho))\} \end{aligned}$$

(For example: If $k = 3$ and $l = 2$ then $\begin{pmatrix} 0 & 1 & 2 & 2 & 2 & 0 & 1 \\ 0 & 1 & 2 & 0 & 1 & 2 & 2 \end{pmatrix}$ belongs to $Z\mathfrak{C}_{k;E_l}$.)

and

$\mathfrak{C}_{k;E_l} := Z\mathfrak{C}^2_{k;E_l} \cup \bigcup_{h=1}^{k-1} \mathfrak{C}^h_{k;E_l}.$

Let $Z_{k;E_l}$ be the set of all binary relations that fulfill the following four conditions:

1) $\varrho \subset E_l \times E_k$;
2) $\varrho \cap E_l^2 = \iota_l^2$;

3) $\{x \mid \exists y: (x,y) \in \varrho\} = E_l \ \wedge \ \{y \mid \exists x: (x,y) \in \varrho\} = E_k$;
4) $\exists c \in E_k \backslash E_l : E_l \times \{c\} \subseteq \varrho$.

One finds connections between the sets $Z\mathfrak{C}_{k;E_l}$ and $Z_{k;E_l}$ explained in Lemma 17.3.5.

Let $N_{k;E_l}$ be the set of all binary relations of the form

$$(\varrho \backslash \iota_k^2) \cup E_l^2,$$

where either $\varrho = E_k^2$ or $\varrho \in \mathfrak{C}_{k;E_l}^2$ [1] holds.
An h-ary relation ϱ is called **$(l;r)$-central** if ϱ fulfills the following five conditions:

1) $\varrho \subset E_l^r \times E_k^{h-r}$, $2 \leq h \leq k-1$, $1 \leq r \leq h-1$;
2) ϱ is **totally $(l;r)$-reflexive**, i.e., $\iota_k^h[r] \subseteq \varrho$;
3) ϱ is **totally $(l;r)$-symmetric**, i.e., for every permutation $s \in S_h[r] := \{f \mid f \text{ is permutation on } E_h \ \wedge \forall a \in E_r : f(a) \in E_r\}$ it holds:

$$(a_0, a_1, ..., a_{h-1}) \in \varrho \implies (a_{s(0)}, a_{s(1)}, ..., a_{s(h-1)}) \in \varrho.$$

4) ϱ is a **weakly $(l;r)$-central** relation, i.e.,

$$\{(a_1, ..., a_r, b_1, ..., b_{h-r}) \in E_l^r \times E_k^{h-r} \mid \exists i \in \{1, ..., h-r\} : b_i \in E_l\} \subseteq \varrho;$$

5) ϱ has a **central element** $c \in E_l$, i.e., for arbitrary $a_1, ..., a_r \in E_l$ and arbitrary $b_1, ..., b_{h-r} \in E_k$ it holds:

$$c \in \{a_1, ..., a_r\} \implies (a_1, ..., a_r, b_1, ..., b_{h-r}) \in \varrho.$$

We denote the set of all h-ary $(l;r)$-central relations with $C_{k;E_l}^h[r]$.
Further, let

$$C_{k;E_l} := \bigcup_{h=2}^{k-1} \bigcup_{r=1}^{h-1} C_{k;E_l}^h[r].$$

Let $H_{k;E_l}$ be the set of all ternary relations ϱ with the following six properties:

1) $\varrho \subseteq E_l^2 \times E_k$;
2) $\varrho \cap E_l^3 = \{(x,x,y) \mid x, y \in E_l\}$;
3) $\varrho \cap \iota_k^3 = \{(x,x,z) \mid x \in E_l \wedge z \in E_k\}$;
4) $\exists u \in E_k \backslash E_l : E_l^2 \times \{u\} \subseteq \varrho$;
5) ϱ is totally $(l;2)$-symmetric;
6) ϱ is **strongly $(l;2)$-homogeneous**, i.e.,

$$\forall x, y \in E_l \ \forall z \in E_k :$$
$$(\exists \alpha, \beta \in E_l : \{(x,\alpha,z), (\beta,y,z), (\alpha,\beta,z)\} \subseteq \varrho) \Longrightarrow (x,y,z) \in \varrho.$$

[1] I.e., ϱ is symmetric and $\iota_k^2 \cup (E_l \times E_k) \cup (E_k \times E_l) \subseteq \varrho$.

With the help of the notations given in the definition of an $(l;r)$-central relation and with

$$\gamma_{r,s} := \{(a_1, ..., a_r, b_1, ..., b_s) \in E_l^r \times E_k^s \mid (\exists i \in \{1, ..., s\} : b_i \in E_l) \vee (a_1, ..., a_r, b_1, ..., b_s) \in \iota_k^h[r]\}$$

one can define the following relations:

An h-ary relation ϱ is called **$(l;2)$-universal** relation $(3 \leq h \leq k-l+2)$ if ϱ fulfills the following six conditions:

1) $\varrho \subset E_l^2 \times E_k^s$, $s := h-2$, $3 \leq h \leq k-l+2$.
2) ϱ is totally $(l;2)$-reflexive.
3) ϱ is totally $(l;2)$-symmetric.
4) ϱ is a weakly $(l;2)$-central relation.
5) The relation

$$\varrho_{\mathbf{b}} := \{(a_1, a_2) \mid (a_1, a_2, b_1, ..., b_s) \in \varrho\}$$

   is an equivalence relation for every $\mathbf{b} := (b_1, ..., b_s) \in (E_k \backslash E_l)^s \backslash \iota_k^s$.
6) $s \in \{1, k-l\}$ and $\varrho = \gamma_{2;s}$ or $\gamma_{2;s} \subset \varrho$.

Let

$$B_{k;E_l}^h[2]$$

be the set of all h-ary $(l;2)$-universal relations.

In the following definitions, we use notations introduced in Section 5.2.6.

An h-ary relation ϱ is called an **$(l;r)$-universal** if ϱ fulfills the following six conditions:

1) $\varrho \subset E_l^r \times E_k^s$, $s := h-r$, $3 \leq r \leq l$, $1 \leq s \leq k-l$.
2) ϱ is totally $(l;r)$-reflexive.
3) ϱ is totally $(l;r)$-symmetric.
4) ϱ is a weakly $(l;r)$-central relation.
5) For every $\mathbf{b} := (b_1, ..., b_s) \in (E_k \backslash E_l)^s \backslash \iota_k^s$ with $E_l^r \times \{\mathbf{b}\} \not\subseteq \varrho$ it holds

$$\varrho_{\mathbf{b}} := \{(a_1, ..., a_r) \mid (a_1, ..., a_r, b_1, ..., b_s) \in \varrho\} \in \mathfrak{B}_l,$$

   i.e., there exists an $m[\mathbf{b}] \in \mathbb{N}$ and a mapping

$$q_{\mathbf{b}} : E_l \longrightarrow E_{r^{m[\mathbf{b}]}}$$

   with the property

$$\begin{array}{l}\forall a_1, ..., a_r \in E_l : \\ ((a_1, ..., a_r, b_1, ..., b_s) \in \varrho \\ \Longleftrightarrow \\ \forall i \in E_{m[\mathbf{b}]} : ((q_{\mathbf{b}}(a_1))^{(i)}, ..., (q_{\mathbf{b}}(a_r))^{(i)}) \in \iota_r^r).\end{array}$$

6) If $s \geq 2$ then:
For every tuple $(\alpha, \beta) := (\alpha_1, ..., \alpha_r, \beta_1, ..., \beta_s) \in (E_l^r \times E_k^s) \backslash \varrho$ there is an n-ary function $f_{(\alpha,\beta)}$ that preserves the relation ϱ, and there is a matrix $A_{(\alpha,\beta)}$ of the type $k \times n$, whose columns belong to $\{\alpha_1, ..., \alpha_r\}^l \times \{\beta_1, ..., \beta_s\}^{k-l}$ with the property

$$f_{(\alpha,\beta)}(A_{(\alpha,\beta)}) := \begin{pmatrix} q_\beta(0) \\ q_\beta(1) \\ \dots \\ q_\beta(l-1) \\ l \\ l+1 \\ \dots \\ k-1 \end{pmatrix}.$$

The simplest relations, which fulfill the above six conditions, are the above-defined relations $\gamma_{r;s}$ with $(r, s) \in \{(l, 1), (l, k-l)\}$ (see also the Lemmas 17.5.13 and 17.5.14).

Let $B_{k;E_l}[r]$ be the set of all h-ary $(l; r)$-universal relations for $r \geq 3$ and put

$$B_{k;E_l} := \bigcup_{\substack{3 \leq h \leq k,\ 2 \leq r \leq l \\ 1 \leq h-r \leq k-l}} B^h_{k;E_l}[r].$$

17.2 Results of Chapter 17

The aim of Chapter 17 is to prove the following theorem:

Theorem 17.2.1 *Let* $2 \leq l \leq k-1$,

$R_{max}(P_l) := \mathfrak{M}_l \cup \mathfrak{S}_l \cup \mathfrak{U}_l \cup \mathfrak{L}_l \cup \mathfrak{C}_l \cup \mathfrak{B}_l$

and

$R_{max}(Pol_k E_l) := R_{max}(P_l) \cup \mathfrak{U}_{k;E_l} \cup \mathfrak{C}_{k;E_l} \cup S_{k;E_l} \cup N_{k;E_l} \cup Z_{k;E_l} \cup C_{k;E_k} \cup H_{k;E_l} \cup B_{k;E_l}.$

Then

$$\{Pol_k E_l \times \varrho \mid \varrho \in R_{max}(Pol_k E_l)\}$$

is the set of all maximal classes of $Pol_k E_l$.

The next theorem follows from the above theorem and from the fact that $Pol_k E_l$ is finitely generated (see Lemma 17.4.1).

Theorem 17.2.2 (Completeness Criterion for $Pol_k E_l$)
For an arbitrary subset M of $Pol_k E_l$ it holds:

$$[M] = Pol_k E_l \iff \forall \varrho \in R_{max}(Pol_k E_l) : M \not\subseteq Pol_k \varrho \times E_l.$$

■

The following theorem (basically given in [Ros 74]) is also a conclusion from Theorem 17.2.1.

Theorem 17.2.3 *Let M be an arbitrary subset of $Pol_k E_l$ with $(Pol_k E_l)^1 \subseteq M$. Then, $[M] = Pol_k E_l$ if and only if $M \not\subseteq Pol_k \varrho$ for all ϱ of the following list:*

$$\lambda_l := \begin{cases} \{(a,b,c,d) \in E_2^4 \mid a+b = c+d \, (mod\ 2)\} & if\ l = 2, \\ \iota_l^l & if\ l \geq 3, \end{cases}$$

$$\mu := \{(x,y) \in E_k^2 \mid \{x,y\} \subseteq E_l \vee x = y\} \ (\in \mathfrak{U}_{k;E_l}),$$

$$\kappa_h := \iota_k^h \cup \{(a_1,...,a_h) \in E_k^h \mid \exists i \in \{1,...,h\} : a_i \in E_l\}$$
$$(\in \mathfrak{C}_{k;E_l}) \ if\, 2 \leq h \leq k-l \ and\ \kappa_h \neq E_k^h,$$

$$\kappa_* := (E_l \times E_k) \cup (E_k \times E_l),$$
$$\gamma_{r,s} \ if\, (s,t) \in \{(2,1),(l,1),(2,k-l),(l,k-l)\}.$$

In other words: $Pol_k E_l$ has exactly

$$n_{k,l} := \begin{cases} 4 & if & (k,l) = (3,2), \\ k-l+4 & if & (k,l) \in \{(k,2),(k,k-1)\}\ and\ k \geq 4, \\ k-l+6 & otherwise, & \end{cases}$$

maximal classes that contain $(Pol_k E_l)^{(1)}$:
$Pol_k \varrho_1$ $(\varrho_1 \in \{\lambda_l, \kappa_*, \gamma_{2,1}, \gamma_{l,1}, \gamma_{2,k-l}, \gamma_{l,k-l}\})$,
$Pol_k \varrho_2 \times E_l$ $(\varrho_2 \in \{\mu, \kappa_h \mid \kappa_h \neq E_k^h \wedge 2 \leq h \leq k-l\})$. ■

17.3 Maximality Proofs

The following lemma is a conclusion of Chapter 6:

Lemma 17.3.1 *For every $\varrho \in R_{max}(P_l)$ the clone $Pol_k \varrho$ is a maximal clone of $Pol_k E_l$, $2 \leq l \leq k-1$.* ■

Lemma 17.3.2 *Let $\varrho \in \mathfrak{C}^1_{k;E_l} \backslash \mathfrak{C}^1_l$, i.e., $\emptyset \subset \varrho \subset E_k$, $\varrho \backslash E_l \neq \emptyset$ and $\varrho \cap E_l \in \{\emptyset, E_l\}$. Then, $Pol_k \varrho \times E_l$ is maximal in $Pol_k E_l$.*

Proof. The unary functions f_a $(a \in \varrho)$ with

$$f_a(x) := \begin{cases} 0 & \text{if} \quad x \in E_l, \\ a & \text{otherwise,} \end{cases}$$

belong to $Pol_k\varrho \times E_l$. Let now $f^n \in Pol_kE_l \backslash Pol_k\varrho$ be arbitrary. Then there exist certain $a_1, ..., a_n \in \varrho$ with $f(a_1, ..., a_n) =: \alpha \in E_k \backslash \varrho$. Consequently, we have $f_*(x) := f(f_{a_1}(x), ..., f_{a_n}(x)) \in [Pol_k\varrho \times E_l \cup \{f\}]$, where $f_*(\beta) \notin \varrho$ for every $\beta \in \varrho \backslash E_l$.
For every function $g^m \in Pol_kE_l$, one finds a $2 \cdot m$-ary function h_g with

$$h_g(x_1, ..., x_{2\cdot m}) := \begin{cases} g(x_1, ..., x_m) & \text{if } (x_{m+1}, ..., x_{2\cdot m}) = (f_*(x_1), ..., f_*(x_m)), \\ 0 & \text{if } (x_{m+1}, ..., x_{2\cdot m}) \in E_l^m \text{ and} \\ & \quad (x_{m+1}, ..., x_{2\cdot m}) \neq (f_*(x_1), ..., f_*(x_m)), \\ a & \text{otherwise} \end{cases}$$

$(a \in \varrho)$ in $Pol_k\varrho \times E_l$. Consequently,

$$g(x_1, ..., x_m) = h_g(x_1, ..., x_m, f_*(x_1), ..., f_*(x_m))$$

is a superposition over $Pol_k\varrho \times E_l \cup \{f\}$. ■

Lemma 17.3.3 *Let ϱ be a relation of $\mathfrak{U}_{k;E_l}$ and let $[a]_\varrho$ be the equivalence class $\{x \in E_k \mid (a, x) \in \varrho\}$, $a \in E_k$. Then is valid:*
(1) ϱ fulfills exactly one of the following conditions:
(a) $\varrho \cap E_l^2 = \iota_l^2 \;\wedge\; \bigcup_{a \in E_l} [a]_\varrho = E_l$;
(b) $\varrho \cap E_l^2 = \iota_l^2 \;\wedge\; \bigcup_{a \in E_l} [a]_\varrho = E_k$;
(c) $\varrho \cap E_l^2 = E_l^2 \;\wedge\; \bigcup_{a \in E_l} [a]_\varrho = E_l$.
(2) $Pol_k\varrho \times E_l$ is maximal in Pol_kE_l.

Proof. (1) follows from the definition of the set $\mathfrak{U}_{k;E_l}$ and from the fact that the conditions $\varrho \cap E_l^2 = \iota_l^2$ and $\bigcup_{a \in E_l} [a]_\varrho = E_k$ only of the relation E_k^2, which does not belong to $\mathfrak{U}_k$, are fulfilled.
(2): Let $\varrho' := \varrho \cap (E_k \backslash E_l)^2$ and $\varrho'' := \varrho \cap (E_l \times (E_k \backslash E_l))$. If ϱ fulfills (a) or (b) of (1), then $\varrho'' = \emptyset$ and for arbitrary $\mathbf{a} := (a_1, a_2) \in \varrho'$ and $\mathbf{b} := (b_1, b_2) \in \varrho$ we have that the function

$$f_{\mathbf{a};\mathbf{b}}(x) := \begin{cases} 0 & \text{if} \quad x \in E_l, \\ b_1 & \text{if} \quad x = a_1, \\ b_2 & \text{otherwise,} \end{cases}$$

belongs to $Pol_k\varrho \times E_l$. If ϱ fulfills the condition (c), then one can find exactly an element of E_l in each equivalence class of ϱ, through which, for every $\mathbf{a} := (a_1, a_2) \in \varrho' \cup \varrho''$ and all $\mathbf{b} \in \varrho$, the function

$$f'_{\mathbf{a};\mathbf{b}}(x) := \begin{cases} \alpha \text{ if} & x \in E_l \wedge [\alpha]_\varrho = [a_1]_\varrho \wedge \alpha \in E_l, \\ b_1 \text{ if} & x = a_1, \\ b_2 \text{ otherwise,} & \end{cases}$$

preserves the relation ϱ.
Let $f \in Pol_k E_l \backslash Pol_k \varrho$ be arbitrary. Then, certain unary functions $h_{\mathbf{a}}$ with $h_{\mathbf{a}}(\mathbf{a}) \notin \varrho$ for all $a \in \varrho' \cup \varrho''$ are superpositions over f and functions of the form $f_{\mathbf{a};\mathbf{b}}$ or $f'_{\mathbf{a};\mathbf{b}}$. We briefly denote all functions of this form with h_1, ..., h_t. If one considers a matrix of the type $(k^m, t \cdot (m+1))$, whose rows are (in arbitrary sequence) of the form

$$\mathbf{y} := (x_1..., x_m, h_1(x_1)..., h_1(x_m)..., h_t(x_1)..., h_t(x_m))$$

$(x_1..., x_m \in E_k)$, then one sees that each two rows that are different and do not consist only of elements of the set E_l, contain at least a column of $E_k^2 \backslash \varrho$. In analogous mode to the proof of the maximality of $Pol_k \varrho$ in P_k (see Theorem 5.3.2) one can easily prove that, for every m-ary function $g \in Pol_k E_l$, there is a function $h_g \in Pol_k \varrho \times E_l$ with $h_g(\mathbf{y}) = g(x_1..., x_m)$, whereby $[Pol_k \varrho \times E_l \cup \{f\}] = Pol_k E_l$ is valid. ■

Analogous to the proof of Lemma 17.3.3 or analogous to the proof of similar statements about maximal classes (see Section 5.3) one can prove the following lemma.

Lemma 17.3.4 *If $\varrho \in S_{k;E_l} \cup N_{k;E_l} \cup C^2_{k;E_l}[1] \cup \mathfrak{C}^2_{k;E_l}$ then $Pol_k \varrho \times E_l$ is maximal in $Pol_k E_l$.* ■

Lemma 17.3.5
(1) Let ϱ be a relation of $Z\mathfrak{C}^2_{k;E_l}$; i.e., it holds that
(a) $\iota_k^2 \subseteq \varrho \subset E_k^2$ and ϱ is symmetric;
(b) $\varrho \cap E_l^2 = \iota_l^2$;
(c) $\exists c \in E_k \backslash E_l : \; E_k \times \{c\} \subseteq \varrho$;
(d) $\forall x, y, z \in E_k : \; \{(x,z),(y,z)\} \subseteq \varrho \implies (x,y) \in \varrho$.
Then $\varrho' := \varrho \cap (E_l \times E_k)$ belongs to $Z_{k;E_l}$, i.e., ϱ' has the following properties:
(e) $\varrho' \subset E_l \times E_k$, $pr_0 \varrho' = E_l$ and $pr_1 \varrho' = E_k$;
(f) $\varrho' \cap E_l^2 = \iota_l^2$;
(g) $\exists c \in E_k \backslash E_l : \; E_l \times \{c\} \subseteq \varrho'$.

(2) $Pol_k \varrho \times E_l = Pol_k \varrho \cap (E_l \times E_k)$ is valid for every $\varrho \in Z\mathfrak{C}^2_{k;E_l}$.

Proof. (1) one can easily check.
(2): Because of (1), we have $Pol_k \varrho \times E_l \subseteq Pol_k \varrho \cap (E_l \times E_k)$ for every $\varrho \in Z\mathfrak{C}_{k;E_l l}$. To prove the reverse inclusion, we show $\varrho = (\tau\varrho') \circ \varrho'$, where $\varrho' := \varrho \cap (E_l \times E_k)$:
Let $(a,b) \in (\tau\varrho') \circ \varrho'$. Then there exists a $c \in E_l$ with $(a,c) \in \tau\varrho'$ and $(c,b) \in \varrho'$. Therefore, $(a,b) \in \varrho$ because of (d) and the symmetry of ϱ. Thus we have $(\tau\varrho') \circ \varrho' \subseteq \varrho$.

Let $(a,b) \in \varrho$. Because of (e) there exist certain $x_a, x_b \in E_l$ with $(x_a, a) \in \varrho'$ and $(x_b, b) \in \varrho'$. Moreover, because of (g) we have $(x_a, c) \in \varrho'$ and $(x_b, c) \in \varrho'$, through what $(x_a, x_b) \in \varrho$ holds because of (d). However, this is only possible for $x_a = x_b$ because of (b). Therefore we have $(a, x_a) \in \tau\varrho'$ and $(x_a, b) \in \varrho'$ through what $(a,b) \in (\tau\varrho') \circ \varrho'$ holds. ■

One can describe classes of the form $Pol_k\varrho \times E_l$ with $\varrho \in Z\mathfrak{C}_{k;E_l}$ through relations of the set $Z_{k;E_l}$. Consequently, the Pol_kE_l-maximality of these classes surrenders through the following lemma.

Lemma 17.3.6 *Let $\varrho \in Z_{k;E_l}$. Then $Pol_k\varrho$ is maximal in Pol_kE_l.*

Proof. The following unary functions belong to $Pol_k\varrho$:

$$f_{(\alpha,\beta)} := \begin{cases} \alpha \text{ if } & x \in E_l, \\ \beta \text{ otherwise} & \end{cases}$$

$((\alpha,\beta) \in \varrho)$,

$$g_{(c,d)}(x) := \begin{cases} 0 \text{ if } & x = c, \\ 1 \text{ if } & x \in (E_l\backslash\{c\}) \cup \{d\}, \\ c \text{ otherwise} & \end{cases}$$

$((c,d) \in (E_l \times E_k)\backslash\varrho)$ and

$$h_{(u,v)}(x) := \begin{cases} 0 \text{ if } & u \in E_l \cup \{u\}, \\ w \text{ otherwise} & \end{cases}$$

$((u,v) \in E_k^2;\ u \neq v;\ v, w \notin E_l$ and $(0,w) \in \varrho)$.
Further, $Pol_k\varrho$ has the following property:
For every set T of t-tuples with

$$\begin{array}{l} \forall \mathbf{a} := (a_1, ..., a_t), \mathbf{b} := (b_1, ..., b_t) \in T: \\ \mathbf{a} \neq \mathbf{b} \implies \exists i \in \{1, ..., n\}: \ (a_i, b_i) \in \{(0,1),(1,0)\} \end{array} \tag{17.1}$$

and for every determination $f(\mathbf{a}) \in E_k$ for the t-tuples $\mathbf{a}$ of T, where $\mathbf{a} \in T \cap E_l^n$ implies $f(\mathbf{a}) \in E_l$, there is a certain determination for the remaining tuples $\mathbf{x}$ of E_k^n, so that f belongs to $Pol_k\varrho$. If $\mathbf{x} \notin E_l^n$ then one can set $f(\mathbf{x}) = w$, where $E_l \times \{w\} \subseteq \varrho$, with that $f \in Pol_k\varrho$. If $\mathbf{x} \in E_l^t$, then (because of $E_l^2 \cap \varrho = \iota_k^2$) maximally a tuple of T can influence the value $f(\mathbf{x})$, with which the value $f(\mathbf{x})$ is also easily defined in this case.
Let now $f^n \in Pol_kE_l\backslash Pol_k\varrho$ be arbitrary. Then there are certain $(a_1, ..., a_n)$, $(b_1, ..., b_n)$ with $(a_i, b_i) \in \varrho$ for all $i \in \{1, ..., n\}$ and $(f(a_1, ..., a_n), f(b_1, ..., b_n)) \notin \varrho$. Thus a unary function f_1 with

$$\begin{pmatrix} c \\ d \end{pmatrix} := f_1 \begin{pmatrix} \alpha \\ \beta \end{pmatrix} \in (E_l \times E_k)\backslash\varrho,$$

where $(\alpha,\beta) \in \varrho\backslash\iota_l^2$ is arbitrary, is a superposition over the functions $f_{(\alpha_i,\beta_i)}$ and f.

Further, if one forms $f_2(x) := g_{(c,d)}(f_1(x))$, then the function f_2 has the property

$$f_2\begin{pmatrix}\alpha\\ \beta\end{pmatrix} = \begin{pmatrix}0\\ 1\end{pmatrix}.$$

Consequently, the unary functions of the form $f_2 \star h_{(u,v)}$ belong to $[Pol_k\varrho \cup \{f\}]$, where $\{(f_2 \star h_{(u,v)}(u), (f_2 \star h_{(u,v)}(v)\} = E_2$ for arbitrary $u, v \in E_k$ with $u \neq v$ and $v \in E_l$.
Let $\{r_1, ..., r_q\} := \{(u,v) \in E_k^2 \mid u \neq v \ \wedge \ v \notin E_l\}$ and denote $s_{r_1}, ..., s_{r_q}$ certain functions of $[Pol_k\varrho \cup \{f\}]$ with $s_{r_i}(r_i) = \begin{pmatrix}0\\ 1\end{pmatrix}$. Then

$$T := \{(x_1, ..., x_m, s_{r_1}(x_1), ..., s_{r_1}(x_m), ..., s_{r_q}(x_1), ..., s_{r_q}(x_m)) \mid x_1, ..., x_m \in E_k\}$$

is a set of tuples with the property (17.1). Thus, by our above considerations, for each m-ary function $g \in Pol_k E_l$ there is a certain function $h_g \in Pol_k\varrho$ with

$$g(x_1, ..., x_m) = h_g(x_1, ..., x_m, s_{r_1}(x_1), ..., s_{r_q}(x_m)),$$

whereby $[Pol_k\varrho \cup \{f\}] = Pol_k E_l$ is proven. ■

Lemma 17.3.7 *For every $\varrho \in H_{k;E_l}$, $Pol_k\varrho$ is a maximal class of $Pol_k E_l$.*

Proof. Let $\varrho \in H_{k;E_l}$. Then ϱ fulfills the conditions 1)–6) of the definition of the elements of $H_{k;E_l}$, which we refer more often to in the following considerations. We begin with some easily proven auxiliary statements:
1) Let $(a,b,c) \in (E_l^2 \times E_k)\backslash\varrho$ with $c \notin E_l$. Then, because of 6), there exists a unary function $f_{a,b,c} \in Pol_k\varrho$ with

$$f_{a,b,c}\begin{pmatrix}a\\ b\\ c\end{pmatrix} = \begin{pmatrix}0\\ 1\\ 0\end{pmatrix}$$

(for $x \in E_k\backslash(E_l \cup \{c\})$ one can set $f(x) = u$, where $E_l \times \{u\} \subseteq \varrho$ (see 4)); for $x \in E_l$, the value $f(x)$ is determined either by $f\begin{pmatrix}a\\ b\end{pmatrix} = \begin{pmatrix}0\\ 1\end{pmatrix}$ or is arbitrary of E_l).
2) For every $\mathbf{a} \in E_k^n$, the n-ary function $g_{\mathbf{a}}$ defined by

$$g_{\mathbf{a}}(x_1, ..., x_n) := \begin{cases} 1 \text{ if} & \mathbf{x} = \mathbf{a},\\ 0 \text{ if} & \mathbf{x} \in E_l^n\backslash\{\mathbf{a}\},\\ u \text{ otherwise,} & \end{cases}$$

where u fulfills 4), belongs to $Pol_k\varrho$.
3) For each $\alpha := (a,b,c) \in \varrho\backslash\delta^3_{k;\{0,1\}}$ with $c \in E_k\backslash E_l$ and for each $\alpha' := (a', b', c') \in \varrho$ there is a unary function $t_{\alpha;\alpha'} \in Pol_k\varrho$ with

$$t_{\alpha;\alpha'}\begin{pmatrix}a\\ b\\ c\end{pmatrix} = \begin{pmatrix}a'\\ b'\\ c'\end{pmatrix}.$$

Let now $f^n \in Pol_k E_l \backslash Pol_k \varrho$ be arbitrary. Then there are certain $\mathbf{a} := (a_1, ..., a_n)$, $\mathbf{b} := (b_1, ..., b_n)$, $\mathbf{c} := (c_1, ..., c_n) \in E_k^n$ with $(a_i, b_i, c_i) \in \varrho$ for $i = 1, ..., n$ and

$$f\begin{pmatrix}\mathbf{a}\\ \mathbf{b}\\ \mathbf{c}\end{pmatrix} = \begin{pmatrix}a'\\ b'\\ c'\end{pmatrix} \notin \varrho.$$

By replacing certain variables of f through functions defined in 3), one obtains a unary function f' with

$$f'\begin{pmatrix}a\\ b\\ c\end{pmatrix} = \begin{pmatrix}a'\\ b'\\ c'\end{pmatrix}$$

for arbitrary $(a, b, c) \in \varrho\backslash\delta^3_{k;\{0,1\}}$ with $c \notin E_l$.
If $c' \in E_k\backslash E_l$, then the function $f'' := f_{(a',b',c')} \star f'$ has the property

$$f''\begin{pmatrix}a\\ b\\ c\end{pmatrix} = \begin{pmatrix}0\\ 1\\ 0\end{pmatrix}.$$

If $c' \in E_l$, then w.l.o.g. we can assume $(a', b', c') = (0, 1, 0)$.
Thus with the help of functions defined in 2), we obtain functions of the form

$$q^n_{\mathbf{a}}(\mathbf{x}) := \begin{cases} 1 \text{ if} & \mathbf{x} = \mathbf{a}, \\ 0 \text{ otherwise,} & \end{cases}$$

for every $\mathbf{a} \in E_k^n$, $n \in \mathbb{N}$.
Let now $g^m \in Pol_k E_l$ be arbitrary. To prove $g_m \in [Pol_k \varrho \cup \{f\}]$, we set $E_k^m = \{\mathbf{a_1}, ..., \mathbf{a_{k^m}}\}$ and consider the $(m + k^m)$-tuples of the form

$$\mathbf{y_i} := (\mathbf{a_i}, 0, ..., 0, \underbrace{1}_{(m+1)\text{-th place}}, 0, ..., 0)$$

for $i = 1, ..., k^m$. Then, because of 4) and 6), there is an $(m+k^m)$-ary function $h_g \in Pol_k \varrho$ with $h_g(\mathbf{y_i}) = g(\mathbf{a_i})$ for each $i \in \{1, ..., k^m\}$. Consequently, g is a superposition over $\{h_g\} \cup \{q_{\mathbf{a}} \mid \mathbf{a} \in E_k^m\}$ and therefore also over $Pol_k \varrho \cup \{f\}$. ■

Lemma 17.3.8 *For every $\varrho \in C^h_{k;E_l}[r] \cup \mathfrak{C}^h_{k;E_l}$ $(3 \leq h \leq k-1,\ 1 \leq r \leq h-1)$ the clone $Pol_k \varrho \times E_l$ is maximal in $Pol_k E_l$.*

Proof. First let $\varrho \in C^h_{k;E_l}[r]$ and let $f \in Pol_k E_l \backslash Pol_k \varrho$ be arbitrary.
Every function f_1 of $Pol_k E_l$ with $Im(f_1) = \{a_1, ..., a_h\}$ and $(a_1, ..., a_h) \in \varrho$ belongs to $Pol_k \varrho$. Hence, for a certain $\mathbf{b} := (b_1, ..., b_h) \in (E_l^r \times E_k^s)\backslash\varrho$, where $s := h-r$, each function of $Pol_k E_l$ with the image $\{b_1, ..., b_h\}$ is a superposition over $Pol_k \varrho \cup \{f\}$. Let $\{h_1, h_2, ..., h_q\} := P^m_{k;\{b_1,...,b_h\}} \cap Pol_k E_l$ and let $c \in E_l$ be a central element of ϱ. We choose an m-ary function $g \in Pol_k E_l$ arbitrary

and put $\mathbf{y} := (x_1, ..., x_m, h_1(x_1, ..., x_m), ..., h_q(x_1, ..., x_m))$. Obviously, the $(m \cdot (q+1))$-ary function h_g defined by

$$h_g(\mathbf{y}) := g(x_1, ..., x_m) \qquad \text{and}$$
$$h_g(\mathbf{y}') := c \text{ for the remaining tuples } \mathbf{y}'$$

preserves the relation ϱ. Thus the function g is a superposition over $Pol_k \varrho \cup \{f\}$, whereby our Lemma is proven for $(l; r)$-central relations ϱ. Analogously, one can lead the proof for the central relations of $\mathfrak{C}^h_{k;E_l}$. ■

Lemma 17.3.9 *For every $\varrho \in B^h_{k;E_l}[2]$ with $3 \le h \le k-l+2$ the clone $Pol_k \varrho$ is maximal in $Pol_k E_l$.*

Proof. Let $f \in Pol_k E_l \backslash Pol_k \varrho$ be arbitrary. To prove $[Pol_k \varrho \cup \{f\}] = Pol_k E_l$, we distinguish the following cases:
Case 1: $\varrho = \gamma_{2,1}$.
Each unary function of $Pol_k E_l$ and each function f_1 with $Im(f_1) = \{a, b, c\}$, where $(a, b, c) \in \gamma_{2,1}$, preserves the relation $\gamma_{2,1}$. Thus each function of $Pol_k E_l$ with the image $\{0, 1, l\}$ is a superposition over $Pol_k \varrho \cup \{f\}$.
Let now $g^m \in Pol_k E_l$ be arbitrary. Set $E_l^m = \{\mathbf{a_1}, ..., \mathbf{a_{l^m}}\}$. The $(m + l^m)$-ary function h_g with

$$h_g(x_1, ..., x_m, 0, ..., 0, \underbrace{1}_{(i+m)\text{-th place}}, 0, ..., 0) := g(\mathbf{a_i})$$

for all $x_1, ..., x_m \in E_l$, $i = 1, 2, .., l^m$,
$h_g(y_1, ..., y_m, l, l, ..., l) := g(y_1, ..., y_m)$ for all $(y_1, ..., y_m) \in E_k^m \backslash E_l^m$,
and $h_g(\mathbf{z}) := 0$ for the remaining tuples $\mathbf{z} \in E_k^{m+l^m}$

belongs to $Pol_k \varrho$.
By replacing certain variables of the function f_g through certain functions with the image $\{0, 1, l\}$, one obtains the function g; that is, $g \in [Pol_k \varrho \cup \{f\}]$ is proven.
Case 2: $\varrho = \gamma_{2,k-l}$ and $k - l \ge 2$.
Clearly, each unary function of $Pol_k E_l$ and each function $h \in Pol_k E_l$ with $|Im(h)| \le k - l + 1$ preserves ϱ. Thus $\{g \in Pol_k E_l \mid |Im(g)| \le h - l + 2\} \subseteq [Pol_k \varrho \cup \{f\}]$. Consequently, $[Pol_k \varrho \cup \{f\}] = Pol_k E_l$ for $l = 2$. If $l \ge 3$, one can prove the $Pol_k E_l$-maximality of $Pol_k \varrho$ with the help of the l-ary function $t \in Pol_k \varrho$ defined by

$$t\begin{pmatrix} 1 & 0 & 0 & \dots & 0 \\ 0 & 1 & 0 & \dots & 0 \\ \dots & \dots & \dots & \dots & \dots \\ 0 & 0 & 0 & \dots & 1 \\ l & l & l & \dots & l \\ \dots & \dots & \dots & \dots & \dots \\ k-1 & k-1 & k-1 & \dots & k-1 \end{pmatrix} = \begin{pmatrix} 0 \\ 1 \\ \dots \\ l-1 \\ l \\ \dots \\ k-1 \end{pmatrix}$$

and $t(\mathbf{x}) = 0$ otherwise.

Case 3: $\gamma_{r,s} \subseteq \varrho \subset E_l^2 \times E_k^s$.
If $s = 1$, one can similar to Case 1 prove $[Pol_k\varrho \cup \{f\}] = Pol_k E_l$, when w.l.o.g. one assumes $E_l^2 \times \{l\} \not\subseteq \varrho$ and $(0,1) \notin \varepsilon_l := \{(a,b) \mid (a,b,l) \in \varrho\}$ and if one changes the definition of the function h_g as follows:

$$\forall (\mathbf{x},\mathbf{y}) := (x_1,..,x_m,y_1,...,y_{l^m}) \in E_l^{m+l^m}\ \forall i :$$
$$((\forall j \in \{1,..,l^m\}\backslash\{i\} : (y_j, 0) \in \varepsilon_l) \wedge (y_i, 1) \in \varepsilon_l) \implies h_g(\mathbf{x},\mathbf{y}) := g(\mathbf{a_i}).$$

Let $s \geq 2$.
Because of assumption, for every $\mathbf{b} := (b_1,...,b_s) \in (E_k\backslash E_l)^s\backslash\varrho$, we have that

$$\varepsilon_{\mathbf{b}} := \{(x,y) \mid (x,y,b_1,...,b_s) \in \varrho\}$$

is an equivalence relation on E_l with at least two equivalence classes. For tuples $\mathbf{y_i} := (y_{i_1},...,y_{i_t}) \in E_l^t$ (i = 1, 2), we define:

$$\mathbf{y_1} \sim_{\varepsilon_\mathbf{b}} \mathbf{y_2} \ :\Longleftrightarrow\ \forall j :\ (y_{1_j}, y_{2_j}) \in \varepsilon_{\mathbf{b}}.$$

Since f does not preserve the relation ϱ, there are certain $r_1,...r_n \in \varrho$ and an $\alpha := (a_1,a_2,b_1,...,b_s) \in (E_l^2 \times E_k^s)\backslash\varrho$ with $f(r_1,...,r_n) = \alpha$. Hence, each function $h \in Pol_k\varrho$ with $Im(h) \subseteq \{a_1,a_2,b_1,...,b_s\}$ is a superposition over $Pol_k\varrho \cup \{f\}$.
Let the m-ary function $g \in Pol_k E_l$ be arbitrary. Then the $m \cdot (q+1)$-ary function h_g defined by

$$\begin{aligned}
&\mathbf{y} := (x_1,...,x_m), h_1(x_1,...,x_m),...,h_q(x_1,...,x_m))\\
&(\{h_1,...,h_q\} := \{h \in (Pol_k E_l)^m \mid Im(f) = \{a_1,a_2,b_1,...,b_s\}\},\\
&h_g(\mathbf{y}) := g(x_1,...,x_m),\\
&\forall \mathbf{y}' \in E_l^m : (\mathbf{y}' \sim_{\varepsilon_\mathbf{b}} \mathbf{y} \implies h_g(\mathbf{y}') = h_g(\mathbf{y}),\\
&h(\mathbf{y}'') := 0 \text{ for the remaining tuples } \mathbf{y}''
\end{aligned}$$

preserves ϱ. Therefore, $g(\mathbf{x}) = h_g(\mathbf{x}, h_1(\mathbf{x}),...,h_q(\mathbf{x})) \in [Pol_k\varrho \cup \{f\}]$ holds. ■

Lemma 17.3.10 *For every $\varrho \in B_{k;E_l}^h[r]$ with $3 \leq r \leq l$ and $1 \leq s := h - r \leq k - l$, the clone $Pol_k\varrho$ is maximal in $Pol_k E_l$.*

Proof. Let $f^n \in Pol_k E_l\backslash Pol_k\varrho$ be arbitrary. Then, there are some $r_1,...,r_n \in \varrho$ and an $\alpha := (a_1,...,a_r,b_1,...,b_s) \in (E_l^r \times E_k^s)\backslash\varrho$ with $f(r_1,...,r_n) = \alpha$. Consequently, each function $h \in Pol_k E_l$ with $Im(h) \subseteq \{a_1,...,a_r,b_1,...,b_s\}$ is a superposition over $Pol_k\varrho \cup \{f\}$.
We distinguish the following two cases:
Case 1: $s = 1$.
Because of $(a_1,...,a_r,b_1) \notin \varrho$, the relation $\varrho_{b_1} := \{(x_1,...,x_r) \mid (x_1,...,x_r,b_1) \in \varrho\}$ belongs to $\mathfrak{B}_l^r$. First let

$$(a_1,...,a_r) := (0,1,...,r-1),\ l := r^q \text{ and}$$
$$\varrho_{b_1} := \{(x_1,...,x_r) \mid \forall i \in E_q :\ (a_1^{(i)},...,a_r^{(i)}) \in \iota_r^r\}$$

(notations: see Section 5.2.6). Let the m-ary function $g \in Pol_k E_l$ be arbitrary. Then the $(q+m)$-ary function h_g defined by

$$h_g(y_1, ..., y_q, x_1, ..., x_m) := \Sigma_{i=0}^{q-1} y_{i+1}^{(0)} \cdot h^{q-i}$$
$$\text{if } (y_1, ..., y_m, x_1, ..., x_m) \in E_l^{q+m},$$
$$h_g(b_1, b_1, ..., b_1, x_1, ..., x_m) := g(x_1, ..., x_m)$$
$$\text{if } (x_1, ..., x_m) \in E_k^m \backslash E_l^m$$

and

$h_g(\mathbf{z}) := 0$ for the remaining tuples $\mathbf{z} \in E_k^{q+m}$

preserves ϱ. Moreover, the functions g_j $(j \in E_q)$ defined by

$$g_f(x_1, ..., x_m) := \begin{cases} (g(x_1, ..., x_m))^{(j)} & \text{if } (x_1, ..., x_m) \in E_l^m, \\ b_1 & \text{otherwise,} \end{cases}$$

are superpositions over $Pol_k \varrho \cup \{f\}$, since the images of these functions are subsets of $\{0, 1, ..., r-1, b_1\}$. Thus, because of $g(\mathbf{x}) = h_g(g_{q-1}(\mathbf{x}), ..., g_0(\mathbf{x}), \mathbf{x})$, $\mathbf{x} := (x_1, ..., x_m)$, g is a superposition over $Pol_k \varrho \cup \{f\}$ in the above case.
If $\varrho_{\mathbf{b}}$ is a homomorphic inverse image of an r-ary elementary relation, then this case can be reduced to the above case, using the constructions from Section 5.2.6.
Case 2: $s \geq 2$.
Since ϱ fulfills the condition 6) of the definition of the elements of $\mathfrak{B}_{k;E_l}^h[r]$ in this case, an arbitrary function of $Pol_k \varrho$ is a superposition over

$$\{f_{(a_1,...,a_s,b_1,...b_s)}\} \cup P_{k;\{a_1,...,a_r,b_1,...,b_s\}} \subseteq [Pol_k \varrho \cup \{f\}]$$

(see also Lemma 1.4.6). ■

17.4 Some Lemmas

It is subsequently shown in analog mode to corresponding statements about P_k that every maximal class A of P_k can be described with the aid of a relation ϱ_A in the form $Pol_k \varrho_A$.

Lemma 17.4.1 *$Pol_k E_l$ has a generating system of functions of $(Pol_k E_l)^2$.*

Proof. For $a, b \in E_k$ let $j_{a;b}$ be a unary function of P_k defined by

$$j_{a;b}(x) := \begin{cases} b & \text{if } x = a, \\ 0 & \text{otherwise.} \end{cases}$$

The function $j_{a;b}$ preserves E_l (i.e., $j_{a;b} \in Pol_k E_l$) iff $b \in E_l$ or $a \in E_k \backslash E_l$. Also, the following functions belong to $Pol_k E_l$:

$$x \vee y := max(x, y),$$

$$x \cdot y := min(x, y)$$

(in respect to the total order $0 < 1 < 2 < ... < k-1$)
and

$$r_{a;b}(x, y) := j_{a;b}(x) \cdot y.$$

Then one can describe an arbitrary function $f^n \in Pol_k E_l$ as follows:

$$f(x_1, ..., x_n) = \bigvee_{\mathbf{a} = (a_1, ..., a_n) \in E_k^n} j_{a_1;f(\mathbf{a})}(x_1) \cdot ... \cdot j_{a_n;f(\mathbf{a})}(x_n)$$

where $f(\mathbf{a}) \in E_l$, if $(a_1, ..., a_n) \in E_l^n$. Consequently,

$$\{\vee, \cdot\} \cup \{j_{a;b} \mid a \in E_k \backslash E_l \ \wedge \ b \in E_l\} \cup \{r_{a;b} \mid a, b \in E_k\}$$

is a generating system for $Pol_k E_l$. ■

The following lemma is a conclusion of the above lemma.

Lemma 17.4.2 *The lattice of the subclasses of $Pol_k E_l$ is dual atomar, and $Pol_k E_l$ has only finitely many maximal classes.* ■

Since one can easily prove that all maximal classes of $Pol_k E_l$ are clones, the following is also valid:

Lemma 17.4.3 *For every maximal class M of $Pol_k E_l$ there exists at most a k^2-ary relation ϱ_M with $M = Pol_k \varrho_M$.* ■

Obviously, the only maximal classes of $Pol_k E_l$ describable through relations from R_l are the classes of the form $Pol_k \varrho$ with $\varrho \in R_{max}(P_l)$.
Possible descriptive relations for the maximal classes of $Pol_k E_l$ are relations of the set $R_{max}(P_l) \cup R_{max}(P_k)$, which we will scrutinize in Section 17.6.
We begin, however, by determining those maximal classes A of $Pol_k E_l$, which are not representable in the form $Pol_k \varrho \times E_l$ with $\varrho \in R_{max}(P_l) \cup R_{max}(P_k)$.

17.5 Not Through Relations of $R_{max}(P_l) \cup R_{max}(P_k)$ Describable Classes

In this section, let A be an arbitrary maximal class of $Pol_k E_l$ with

$$pr_l A = P_l \qquad \text{and}$$

$$\mathbb{L}_k^{\uparrow}(A) = \{A, \ Pol_k E_l, \ P_k\}.$$

Then, by Lemma 17.4.2, there is a relation $\varrho \in R_k^h$ with $h \leq k^2$ and

$$A = Pol_k E_l \cap Pol_k \varrho.$$

The assumptions above A and ϱ are equivalent to

(I) $A \notin \{Pol_k\gamma \cap Pol_k E_l \mid \gamma \in R_{max}(P_l) \cup R_{max}(P_k)\}$

or

(I′) No relation from the set $R_{max}(P_l) \cup R_{max}(P_k)$ is derivable from relations of $\{\varrho\} \cup Inv_k Pol_k E_l$.

Since $R_k^1 \backslash \{E_l\} \subseteq R_{max}(P_k)$, (I′) implies

(I″) $h \geq 2$ holds and no relation of the set $\mathfrak{C}_k^1 \backslash \{E_l\}$ is derivable from ϱ.

W.l.o.g. we can assume the following properties of the relation ϱ:

(II) $\varrho \subseteq E_l^r \times E_k^s$ $(r + s = h; 0 \leq r \leq h; 1 \leq s \leq h)$.

(III) A is not describable through a relation $\varrho' \subseteq E_l^{r'} \times E_k^{h-r'}$ with $r' > r$.
In other words: Every relation $\varrho' \in [\{\varrho\} \cup Inv_k Pol_k E_l]$ with $\varrho' \subseteq E_l^{r'} \times E_k^{h-r'}$ and $r' > r$ belongs to $Inv_k Pol_k E_l$.

(IV) Every h'-ary relation $\varrho' \in [\{\varrho\} \cup Inv_k Pol_k E_l]$ with $h' < h$ belongs to $Inv_k Pol_k E_l$.

(V) If possible, choose a symmetric relation ϱ for the description of A, i.e., for every permutation $s \in S_h$ and

$$\varrho_s := \{(a_{s(0)}, a_{s(1)}, ..., a_{s(h-1)}) \mid (a_0, ..., a_{h-1}) \in \varrho\}$$

is valid: $\varrho \cap \varrho_s \in \{\varrho\} \cup Inv_k Pol_k E_l$.

(VI) From the possible $(l;r)$-symmetric, h-ary relations ϱ', which describe A, the relation ϱ is that relation with the greatest cardinality, i.e., $\forall \varrho' \subseteq E_l^r \times E_k^s$:
$(A = Pol_k E_l \times \varrho' \wedge (\varrho' \text{ is } (l;r)\text{-symmetric})) \Longrightarrow |\varrho'| \leq |\varrho|$.

The following is a conclusion from *(II)–(IV)*:

(VII) $\forall i \in \{0,1,...,h-1\}$:
$pr_{0,1,...,i-1,i+1,...,h-1}\varrho = pr_{0,1,...,i-1,i+1,...,h-1}E_l^r \times E_k^s$.
(In particular, ϱ has no double rows.)

Lemma 17.5.1 $\varrho \cap \delta^h_{k;\{0,1,...,h-1\}} = \delta^h_{l;\{0,1,...,h-1\}}$.

Proof. By (IV) we have $\varrho \cap \delta^h_{k;\{0,1,...,h-1\}} \in \{\emptyset, \delta^h_{l;\{0,1,...,h-1\}}, \delta^h_{k;\{0,1,...,h-1\}}\}$. $\varrho \cap \delta^h_{k;\{0,1,...,h-1\}} = \delta^h_{k;\{0,1,...,h-1\}}$ cannot occur, since $A \subset Pol_k E_l$ and $Pol_k E_l$ does not contain all constant functions of P_k. If $\varrho \cap \delta^h_{k;\{0,1,...,h-1\}} = \emptyset$, then the ϱ-derivable relation $\varrho' := pr_{0,h}(E_l \times \varrho) \cap \delta^{h+1}_{k;\{0,1,...,h-1\}} = \{(x,y) \in E_l \times E_k \mid (x,x,...,x,y) \in E_l \times \varrho\}$ has the following properties:
- $\varrho' \cap \iota_k^2 = \emptyset$;
- $\varrho' \cap E_l^2 \neq \emptyset \implies \varrho' \cap E_l^2 \notin Inv_k Pol_k E_l$;
- $\varrho' \cap E_l^2 = \emptyset \implies pr_1\varrho' \in \mathfrak{C}_k^1 \backslash \{E_l\}$.

Thus because of our assumptions (I') and (I''), the case just seen cannot occur for $\varrho \cap \delta^h_{k;\{0,...,h-1\}}$. Hence our assertion is valid. ∎

In the following, we distinguish two cases for ϱ, and we start with

Case 1: h = 2.

Because of (I'') we have

$$pr_0\varrho \in \{E_l, E_k\} \quad \text{and} \quad pr_1\varrho = E_k$$

in this case. As is generally accepted, the relations

$$\sigma_i(\varrho) := \{(a_1,...,a_i) \mid \exists u \in E_k : \ \{(u,a_1),...,(u,a_i)\} \subseteq \varrho\}$$

$(i = 2,3,...)$ are ϱ-derivable for $h = 2$.

Lemma 17.5.2 *Let $(\tau\varrho) \circ \varrho = E_k^2$. Then $\sigma_i(\varrho) = E_k^i$ for all $i \geq 2$.*

Proof. We prove the statement by induction on $i \geq 2$. For $i = 2$ the statement holds by assumption, since $\sigma_2(\varrho) = (\tau\varrho) \circ \varrho$. Assume, $\sigma_i(\varrho) = E_k^i$, $i \geq 2$. Then, by definition of $\sigma_{i+1}(\varrho)$, we have $\iota_k^{i+1} \subseteq \sigma_{i+1}(\varrho)$. The case $\sigma_{i+1}(\varrho) \neq E_k^{i+1}$ implies $Pol_k\varrho \cap Pol_k E_l \subset Pol_k \sigma_{i+1}(\varrho) \notin \{Pol_k E_l, P_k\}$, contrary to (I). Therefore, we have $\sigma_{i+1}(\varrho) = E_k^{i+1}$. ∎

By Lemma 17.5.1, the ϱ-derivable relation $\varrho \cap (\tau\varrho)$ has the property $\iota_l^2 = \iota_k^2 \cap \varrho \cap (\tau\varrho)$. Thus, by (V), $\varrho \in \{\iota_l^2, \varrho, E_l^2\}$ holds; i.e., we must examine three cases for ϱ, and do so in the following three lemmas.

Lemma 17.5.3 *Let $\varrho \cap (\tau\varrho) = \iota_l^2$. Then there exists a ϱ-derivable relation $\gamma \in S_{k;E_l}$ whose proper cycles have a length ≥ 3.*

Proof. First we form the ϱ-derivable relation

$$\varrho' := (\tau\varrho) \circ \varrho = \{(x, y) \mid \exists u \in E_k : \{(u, x), (u, y)\} \subseteq \varrho\}.$$

Because of $pr_1\varrho = E_k$, we have $\iota_k^2 \subseteq \varrho'$ and $|\varrho| \geq k$. The following two cases are possible:

Case 1: $\varrho' = \iota_k^2$.

If $|\varrho| > k$ or $pr_0\varrho = E_l$, there exist $a, b, c \in E_k$ with $(c, a), (c, b) \in \varrho$ and $a \neq b$, whereby $(a, b) \in \varrho'$ with $a \neq b$ follows, contrary to our assumption in the first case. Therefore $|\varrho| = k$, and it holds that

$$\varrho = \{(x, s(x)) \mid x \in E_k\}$$

for a certain permutation $s \in S_k \backslash \{id\}$ with the fixed points $0, 1, ..., l-1$ (S_k denotes the set of all permutations on E_k and id is the identity permutation). Because of $\varrho \cap (\tau\varrho) = \iota_k^2$, every proper cycle of ϱ has at least the length 3. If ϱ has proper cycles with different lengths, and if $r \geq 3$ is the smallest length of the proper cycles, then the relation $\Delta((\varrho \circ \varrho \circ ... \circ \varrho) \cap \iota_k^2)$ is a ϱ-derivable relation of $\mathfrak{C}_k^1 \backslash \{E_l\}$, contrary to (I''). Therefore, all proper cycles of ϱ have the same length $q \geq 3$. The length q is a prime number, since for $q = p \cdot m$ (p prime, $m \geq 2$), the ϱ-derivable relation

$$\underbrace{\varrho \circ \varrho \circ ... \circ \varrho}_{m \text{ times}}$$

has only proper cycles of the length p. Consequently, it is possible to derive from ϱ a relation of $S_{k;E_l}$, which has proper cycles of the length ≥ 3.

Case 2: $\iota_k^2 \subset \varrho'$.

In this case, we have $\{c_0, c_1, ..., c_{k-1}\} \subseteq Pol_k\varrho'$, whereby $\varrho' = E_k^2$ is valid because of (I'). With the aid of Lemma 17.5.2, this implies $\sigma_k(\varrho) = E_k^2$. Therefore, ϱ has a central element u, i.e., $\{(u, x) \mid x \in E_k\} \subseteq \varrho$. Because of $\varrho \cap \iota_k^2 = \iota_l^2$, u belongs to E_l. Consequently, $\iota_l^2 \subset \varrho \cap E_l^2$ and hence (by(I')) $\varrho \cap E_l^2 = E_l^2$, contrary to our assumption $\varrho \cap (\tau\varrho) = \iota_l^2$. Thus Case 2 is not possible. ∎

Lemma 17.5.4 *If $\varrho \cap (\tau\varrho) = \varrho$, then either $\varrho \in N_{k;E_l}$ or it is possible derive from ϱ a permutation of the set $S_{k;E_l}$, whose proper cycles have the same length 2.*

Proof. The symmetry of ϱ and $pr_1\varrho = E_k$ imply $pr_0\varrho = E_k$. Consequently, $\iota_k^2 \subseteq \varrho \circ \varrho$. Because of (I'), this is only possible if $\varrho \circ \varrho \in \{\iota_k^2, E_k^2\}$; i.e., we have to distinguish the following two cases.

Case 1: $\varrho \circ \varrho = \iota_k^2$.

Since $\varrho \circ \varrho = (\tau\varrho) \circ \varrho$, one can show (analogous to the proof of Lemma 17.5.3, Case 1) that there is a ϱ-derivable relation $\gamma \in S_{k;E_l}$. Because of the symmetry of ϱ, every proper cycle of ϱ has the length 2.

Case 2: $\varrho \circ \varrho = E_k^2$.
Since $\varrho \circ \varrho = (\tau\varrho) \circ \varrho$, Lemma 17.5.2 implies $\sigma_k(\varrho) = E_k^2$. Therefore, there is a $u \in E_k$ with $\{(u, x) \mid x \in E_k\} \subseteq \varrho$. Because of $\varrho \cap \iota_k^2 = \iota_l^2$ (see Lemma 17.5.1) u belongs to E_l. Hence $\iota_l^2 \subset \varrho \cap E_l^2$, whereby $E_l^2 \subseteq \varrho$ (because of our assumptions in this section). Now, if one forms $\varrho' := \varrho \cap (E_l \times E_k)$, then $E_l^2 \subset \varrho' \subseteq E_l \times E_k$. Since the case $\varrho' \subset E_l \times E_k$ is a contradiction to our assumption (III), we have $E_l \times E_k \subseteq \varrho$ and then (because of symmetry of ϱ) $\varrho \in N_{k;E_l}$. ■

Lemma 17.5.5 *If $\varrho \cap (\tau\varrho) = E_l^2$, then $r = 1$ and $\varrho \in C_{k;E_l}^2$.*

Proof. Suppose, $r = 0$. By (III) and by $\varrho \cap (\tau\varrho) = E_l^2$, for $\varrho_0 := \varrho \cap (E_l \times E_k)$ and $\varrho_1 := \varrho \cap (E_k \times E_l)$ the following cases are to study (w.l.o.g.): $\varrho_0 = \varrho_1 = E_l^2$ or $\varrho_0 = E_l \times E_k$ and $\varrho_1 = E_l^2$. In both cases, the ϱ-derivable relation $\varrho_i := \underbrace{\varrho \circ \varrho \circ \ldots \circ \varrho}_{i \text{ times}}$ for every $i \in \mathbb{N}$ has the property that $\varrho_i \cap (E_k \times (E_k \times E_l)) = E_l^2$ holds. Furthermore, there is an i with $\iota_l^2 \subset \varrho_i \cap \iota_k^2$. Consequently, if $r = 0$, then there is a ϱ-derivable relation that describes a class B with $B \not\subseteq Pol_k E_l$ and $B \neq P_k$. This is, however, excluded from this section. Therefore, we can set $r = 1$ in the following; i.e., $\varrho \subseteq E_l \times E_k$ holds. If one forms the relation $(\tau\varrho) \circ \varrho$, one can see that $E_k \times E_l \subseteq (\tau\varrho) \circ \varrho$ and $\iota_k^2 \subseteq (\tau\varrho) \circ \varrho$ are valid. By our assumptions over ϱ, this is only possible if $(\tau\varrho) \circ \varrho = E_k^2$ holds. Because of Lemma 17.5.2, this implies that there is a $u \in E_l$ with $\{(u, x) \mid x \in E_k\} \subseteq \varrho$. Consequently, $\varrho \in C_{k;E_l}^2$. ■

Let's summarize:

In Case 1 our maximal class A has the form:

$$Pol_k \varrho \text{ with } \varrho \in S_{k;E_l} \cup N_{k;E_l} \cup C_{k;E_l}^2.$$

Case 2: $h \geq 3$.

Lemma 17.5.6 *Let $s \geq 2$ and $E_l^h \subseteq \varrho$. Then*

$$\bigcup_{i=0}^{s-1} E_l^r \times E_k^i \times E_l \times E_k^{s-i-1} \subseteq \varrho.$$

Proof. First let $r = 0$, i.e., $pr_i \varrho = E_k$ for every $i \in E_h$. Because of Lemma 17.5.1 and (VII), we have $\varrho \cap \delta_{k;\{1,2,...,h-1\}} = \delta_{k;\{1,2,...,h-1\}}[1]$ and therefore $pr_i(\varrho \cap (E_l \times E_k^{h-1})) = E_k$ for every $i \in \{1, 2, ..., h-1\}$. Consequently, by (III), $\varrho \cap (E_l \times E_k^{h-1})$ can be only the relation $E_l \times E_k^{h-1}$. Analogously, one can prove $E_k^i \times E_l \times E_k^{h-i-1} \subseteq \varrho$ for $i \in \{1, ..., s-1\}$. Thus our assertion is valid for $r = 0$.
Let now $r \geq 1$. Because of (VII), we have $pr_i(\varrho \cap (E_l^{r+1} \times E_k^{h-r-1})) = E_k$ for $i \in \{r+1, ..., h-1\}$. In analog mode, one can prove that our assertion for $r \geq 1$ and $h - r \geq 2$ follows. ■

Lemma 17.5.7 *Let $h = 3$. Then for ϱ is valid either*
(a) $r = 2$ and $\varrho \cap \iota_k^3 = \delta_{k;\{0,1\}}[2]$
or
(b) $r \in \{1,2\}$ and ϱ is $(l;r)$-reflexive, i.e.,

$$\iota_k^3[r] := \iota_k^3 \cap (E_l^r \times E_k^{3-r}) \subseteq \varrho.$$

Proof. We must study the following three cases:
Case 1: $r = 0$.
(VII) and Lemma 17.5.1 imply

$$\varrho \cap \delta_{k;\{1,2\}} = \delta_{k;\{0,1\}}[1]. \qquad (17.2)$$

Analogously, one can show $\delta_{k;\{0,1\}}(k,k,l) \cup \delta_{k;\{0,2\}}(k,l,k) \subseteq \varrho$. Consequently, we have $\iota_l^3 \subseteq \varrho \cap E_l^h$, which can be possible because of (I') only for $\varrho \cap E_l^3 = E_l^3$. With the help of Lemma 17.5.6, this implies

$$E_l \times E_k^2 \cup E_k \times E_l \times E_k \cup E_k^2 \times E_l \subseteq \varrho. \qquad (17.3)$$

Now, if one forms the relation $\varrho' := \Delta\varrho$, then this relation has the properties

$$E_l \times E_k \cup E_k \times E_l \subseteq \varrho' \qquad \text{(by (17.3))}$$

and

$$\varrho' \neq E_k^2 \qquad \text{(by (17.2))},$$

whereby ϱ' is no invariant of $Pol_k E_l$, contrary to (IV). Therefore, the case $r = 0$ does not occur.
Case 2: $r = 1$.
Since by (VI) $pr_{1,2}\varrho = E_k^2$ is valid, we have

$$\varrho \cap E_3^3 \in \{E_l^3, \delta_{l;\{0,1\}}^3, \delta_{l;\{0,2\}}^3\}. \qquad (17.4)$$

In addition, as in Case 1, (17.2) is valid. Because of (17.2) and (17.4), the relation $\varrho \cap E_l^3$ does not have any double rows. Hence, by (I'), $\varrho \cap E_l^3 = E_l^3$ follows. Then with the help of Lemma 17.5.6 and (17.2), it follows that $\iota_k^3[1] \subseteq \varrho$.
Case 3: $r = 2$.
Because of $pr_{1,2}\varrho = E_l \times E_k$, we have that (17.4) is valid and, in addition, that

$$\delta_{k;\{0,1\}}[2] \subseteq \varrho \qquad (17.5)$$

holds. In analog mode to the considerations from Case 2, either $\varrho \cap E_l^3 = E_l^3$ or $\varrho \cap E_l^3 = \delta_{l;\{0,1\}}^3$ results. If $E_l^3 \subseteq \varrho$ then, because of (17.5), the relation ϱ is $(l;2)$-reflexive. If $\varrho \cap E_l^3 = \delta_{l;\{0,1\}}^3$ then $\varrho \cap \iota_k^3 = \delta_{k;\{0,1\}}^3[2]$ must be valid, whereby our Lemma is proven. ∎

Lemma 17.5.8 *For $h \geq 4$, $\iota_k^h \cap (E_l^r \times E_k^s) \subseteq \varrho$ holds, i.e., ϱ is $(l;r)$-reflexive.*

Proof. One can lead the proof analogously to the proof of Lemma 6.1.8: Because of (VII), for arbitrary $(a,b),(b,c),(c,d) \in E_h^2 \backslash \iota_h^2$ with $|\{a,b,c,d\}| = 4$ the following implications are **not** valid:

$$(h \geq 4 \ \wedge \ (x_0,...,x_{h-1}) \in \varrho \ \wedge \ x_a = x_b) \implies x_b = x_c \tag{17.6}$$

$$(h \geq 5 \ \wedge \ (x_0,...,x_{h-1}) \in \varrho \ \wedge \ x_a = x_b) \implies x_c = x_d. \tag{17.7}$$

Next we prove

$$\exists (a,b) \in E_h^2 \backslash \iota_h^2 : \ \delta_{\{a,b\}}[r] \subseteq \varrho \implies (\forall (c,d) \in E_h^2 \backslash \iota_h^2 : \ \delta_{\{c,d\}}[r] \subseteq \varrho). \tag{17.8}$$

Let $\delta_{\{a,b\}}[r] \subseteq \varrho$ ($\{a,b\} \subseteq E_k$, $a \neq b$), $\{b,c\} \subseteq E_k$ and $c \notin \{a,b\}$. To prove (17.8), it is sufficient to show that $\delta_{\{b,c\}}[r] \subseteq \varrho$ is valid. Because of $\delta_{\{a,b\}}[r] \subseteq \varrho$, we have $\delta_{\{a,b,c\}}[r] \subseteq \varrho$. The assumption (IV) implies that $\varrho \cap \delta_{\{b,c\}}[r]$ is E_l-diagonal. Consequently, we have $\varrho \cap \delta_{\{b,c\}}[r] \in \{\delta_{\{a,b,c\}}[r], \delta_{\{b,c\}}[r]\}$. The case $\varrho \cap \delta_{\{b,c\}}[r] = \delta_{\{a,b,c\}}[r]$ leads to an implication of the form (17.6). Therefore, $\varrho \cap \delta_{\{b,c\}}[r] = \delta_{\{b,c\}}[r]$ is only possible, whereby (17.8) is right. Hence, it remains to show that $\delta_{\{a,b\}}[r] \subseteq \varrho$ holds. W.l.o.g. let $(a,b) = (0,1)$.
Set $\varrho_1 := \varrho \cap \delta_{\{0,1\}}[r]$. Because of (VII) and (IV), the relation ϱ_1 is a nonempty E_l-diagonal relation. Consequently, since implications (5) and (6) are false, we have:

$$\varrho' \begin{cases} = \delta_{\{0,1\}}[r] & \text{if } h \geq 5, \\ \in \{\delta_{\{0,1\},\{2,3\}}[r], \delta_{\{0,1\}}[r]\} & \text{if } h = 4. \end{cases}$$

Thus, because of (17.8), our Lemma were proven for $h \geq 5$ or $h = 4$ and $\varrho_1 \neq \delta_{\{0,1\},\{2,3\}}[r]$. Hence only the case $h = 4$ and

$$\varrho \cap \delta_{\{a,b\}}[r] \in \{\delta_{\{0,1\},\{2,3\}}[r], \delta_{\{0,2\},\{1,3\}}[r], \delta_{\{0,3\},\{1,2\}}[r]\}. \tag{17.9}$$

must still be examined. Because of $pr_{0,1,2}\varrho = pr_{0,1,2}E_l^r \times E_k^s$ and because of (IV) we have $|\varrho| > |E_l^r \times E_k^{s-1}|$. Consequently, there are certain $a_0, a_1, a_2, a_3, a_3' \in E_k$ with

$$(a_0,a_1,a_2,a_3), (a_0,a_1,a_2,a_3') \in \varrho \text{ and } a_3 \neq a_3'. \tag{17.10}$$

Then, the relation

$$\varrho_2 := \{(x_1,x_2,x_3,x_3') \in E_l^r \times E_k^{4-s} \mid \exists x_0 \in E_k : (x_0,x_1,x_2,x_3) \in \varrho \wedge (x_0,x_1,x_2,x_3') \in \varrho\},$$

which is derivable from ϱ and E_l, has the properties:

$$\varrho_2 \cap \delta_{\{2,3\}}[r] = \delta_{\{2,3\}}[r] \quad \text{(since } pr_{1,2,3}\varrho = pr_{1,2,3}E_l^r \times E_k^{4-r}\text{)},$$

$$\delta_{\{2,3\}}[r] \subset \varrho_2 \quad \text{(because of (17.10))}$$

and

$\varrho_2 \subset E_l^r \times E_k^{4-r}$

(By assumption (IV) we have:
$(a_0, a_1, a_1, a_3) \in \varrho \Longrightarrow a_0 = a_3$.
Consequently:
$((a_1, a_2, a_3, a_3') \in \varrho_2 \wedge a_1 = a_2) \Longrightarrow a_3 = a_3'$.)

Thus ϱ is no invariant of $Pol_k E_l$. In addition, because of the above facts, ϱ_2 is a totally $(l;r)$-reflexive relation. Then, by definition of ϱ_2, we have $\varrho \cap \delta_{\{1,2\}}[r] \neq \delta_{\{0,3\},\{1,2\}}[r]$, contrary to (17.9). ■

Because of Lemmas 17.5.7 and 17.5.8, the following three cases are possible for $h \geq 3$:

Case 2.1: $h = 3$, $\varrho \cap E_l^3 = \delta^3_{l;\{0,1\}}$ and $\varrho \cap \iota_k^3 = \delta^3_{k;\{0,1\}}[2]$.

The investigation of this case concludes with minimal expenditure, with the aid of the following lemma.

Lemma 17.5.9 *In Case 2.1, relation ϱ has the following properties:*
(a) $\exists u \in E_k \backslash E_l : E_l^2 \times \{u\} \subseteq \varrho$;
(b) ϱ is $(l;2)$-symmetric;
(c) ϱ is strongly $(l;2)$-homogeneous, i.e.,

$$\forall x, y \in E_l \forall z \in E_k : (\exists \alpha, \beta \in E_l : \{(x, \alpha, z), (\beta, y, z), (\alpha, \beta, z)\} \subseteq \varrho) \Longrightarrow (x, y, z) \in \varrho.$$

Proof. (a): The relation

$$\gamma_t := \{(a_1, b_1, a_2, b_2, ..., a_t, b_t) \in E_l^{2\cdot t} \mid \exists u \in E_k : \{(a_1, b_1, u), (a_2, b_2, u), ..., (a_t, b_t, u)\} \subseteq \varrho\}$$

is a ϱ-derivable relation. Because of $\delta^3_{k;\{0,1\}}[2] \subseteq \varrho$, we have

$$\{(x_1, x_1, x_2, x_2, ..., x_t, x_t) \in E_l^{2\cdot t} \mid x_1, ..., x_t \in E_l\} \subseteq \gamma_t.$$

Further, $\{(y_1, y_2, y_1, y_2, ..., y_1, y_2) \in E_l^{2\cdot t} \mid y_1, y_2 \in E_l\} \subseteq \gamma_t$ holds, since $pr_{0,1}\varrho = E_l^2$. Consequently, the relation γ_t does not have any double rows, which is possible according to the assumption only for $\gamma_t = E_l^{2\cdot t}$. For $t = l \cdot (l-1)$, this implies that there exists a $u \in E_k$ with $E_k \times \{u\} \subseteq \varrho$. The element u belongs to $E_k \backslash E_l$, since the opposite case $u \in E_l$ supplies a contradiction to $\varrho \cap E_l^3 = \delta^3_{l;\{0,1\}}$.
(b): Let $\varrho' := \{(x, y, z) \in E_l^2 \times E_k \mid \{(x, y, z), (y, x, z)\} \subseteq \varrho\}$. Obviously, the relation $\delta^3_{l;\{0,1\}} \cup (E_l^2 \times \{u\})$, which is not E_l-diagonal, is a subset of ϱ'. By (V) this is only possible for $\varrho = \varrho'$, whereby ϱ is $(l;2)$-symmetric.
(c): If one forms the ϱ-derivable relation

$$\varrho_1 := \{(a_1, a_2, b) \in E_l^2 \times E_k \mid \exists x_1, x_2 \in E_l : \{(a_1, x_1, b), (x_2, a_2, b), (x_1, x_2, b)\} \subseteq \varrho\},$$

then one can see that ϱ_1 is an $(l;2)$-symmetric relation with $\varrho \subseteq \varrho_1$ because of (b). By (VI) this is only possible for $\varrho_1 \in \{\varrho, E_l^2 \times E_k\}$. Suppose, $\varrho_1 = E_l^2 \times E_k$. Then, in particular, $(0,1,0) \in \varrho_1 \backslash \varrho$. This implies the existence of certain $x_1, x_2 \in E_l$ with $(0,x_1,0), (x_2,1,0), (x_1,x_2,0) \in \varrho$. Because of $\varrho \cap E_l^3 = \delta^3_{l;\{0,1\}}$ the first two conditions are only fulfilled for $x_1 = 0$ and $x_2 = 1$, contrary to $(0,1,0) \notin \varrho$. Consequently, $\varrho_1 = \varrho$ and, therefore (c) is valid. ∎

Hence in Case 2.1 we have

$$\varrho \in H^3_{k;E_l}.$$

Case 2.2.: $h \geq 3$ and ϱ is $(l;r)$-reflexive.

The following lemma summarizes some consequences from our assumptions in this case and from the properties obtained from ϱ.

Lemma 17.5.10
(a) ϱ is $(l;r)$-symmetric.
(b) $E_l^h \subseteq \varrho$.
(c) For $s \geq 2$, ϱ is a weakly $(l;r)$-central relation, i.e.,

$$\bigcup_{i=0}^{s-1} E_l^r \times E_k^i \times E_l \times E_k^{s-i-1} \subseteq \varrho.$$

(d) $r \geq 1$.

Proof. (a): We consider the ϱ-derivable relation

$$\varrho' := \bigcap_{s \in S_h[r]} \varrho_s,$$

where $\varrho_s := \{(a_{s(0)}, a_{s(1)}, ..., a_{s(h-1)}) \mid (a_0, a_1, ..., a_{h-1}) \in \varrho\}$. By assumption, $\iota_k^h[r] \subseteq \varrho'$, where $\iota_k^h[r]$ is not E_l-diagonal. By (V) this is only possible if $\varrho = \varrho'$, whereby ϱ is $(l;r)$-symmetric.
(b) follows from $\iota_l^h \subseteq \varrho$ and (I').
(c) follows from (b) and Lemma 17.5.6.
(d): The case $r = 0$ (and therefore $\iota_k^h \subseteq \varrho$) cannot occur because of Lemma 17.5.1. ∎

Subsequently, we need the following property of set A more often.

Lemma 17.5.11 *It is not possible that the maximal class A $(= Pol_k E_l \times \varrho)$ is describable with the aid of an h'-ary totally $(l;r)$-reflexive and totally $(l;r)$-symmetric relation ϱ' with $h' > h$.*

Proof. Let ϱ' be an h'-ary totally $(l;r)$-reflexive and totally $(l;r)$-symmetric relation, where $h' > h$. Set $A = Pol_k E_l \times \varrho'$. Then A' contains every function $f \in Pol_k E_l$ with

$$Im(f) \subseteq T \in \{\{a_1, .., a_r, b_1, ..., b_{h'-r}\} \mid (a_1, ..., a_r, b_1, ..., b_{h'-r}) \in \varrho'\}.$$

Then, because of $h < h'$, it is not possible to describe A by a relation $\varrho \subseteq E_l^r \times E_k^{h-1}$, contrary to our assumptions. ■

Case $r = 1$ cannot occur, as shown by the following lemma:

Lemma 17.5.12 *If $r = 1$ then $\varrho \in C_{k;E_l}^h[1]$.*

Proof. Because of Lemma 17.5.10,(a),(c) we must only show that ϱ has an $(l;1)$-central element of E_l. The following relation is a ϱ-derivable relation:

$$\gamma_t := \{(a_1, ..., a_t) \in E_l \times E_k^{t-1} \mid \exists c \in E_l : \forall i_1, ..., i_{h-1} \in \{1, ..., t\} : (c, a_{i_1}, ..., a_{i_{h-1}}) \in \varrho\}.$$

First we show that $\gamma_h = E_l \times E_k^{h-1}$ holds. For this purpose, let $(a_1, ..., a_h)$ be an arbitrary element of $E_l \times E_k^{h-1}$. By (VII) for $(a_2, ..., a_h)$ there is a certain $\alpha \in E_l$ with $(\alpha, a_2, ..., a_h) \in \varrho$. Since ϱ is weakly $(l,1)$-central because of Lemma 17.5.10,(c), we have $(\alpha, a_{i-1}, ..., a_{i_{h-1}}) \in \varrho$ for all $\{i_1, ..., i_{h-1}\} \subseteq \{1, 2, ..., h\}$ and $1 \in \{i_1, ..., i_{h-1}\}$. Consequently, $(a_1, ..., a_h)$ belongs to γ_h, whereby $\gamma_h = E_l \times E_k^{h-1}$ holds. Because of $\gamma_h = E_l \times E_k^{h-1}$ it is easy to prove that the relation γ_{h+1} is $(l;1)$-reflexive, which is valid only for $\gamma_{h+1} = E_l \times E_k^h$ because of Lemma 17.5.11. Therefore, by induction, we have $\gamma_k = E_l \times E_k^{k-1}$. The existence of an $(l;1)$-central element follows immediately from that. ■

We can, therefore, always presuppose that

$$r \geq 2.$$

By Lemma 17.5.10, the relation

$$\gamma_{r,s} := \{(a_1, ..., a_r, b_1, ..., b_s) \in E_l^r \times E_k^s \mid (\exists i \in \{1, ..., s\} : b_i \in E_l) \vee (a_1, ..., a_r, b_1, ..., b_s) \in \iota_k^{r+s}[r]\}$$

is a subset of ϱ. Next we clarify for which r, s the equation $\varrho = \gamma_{r,s}$ is valid.

Lemma 17.5.13 *The relation $\gamma_{r,s}$ has the following properties:*
(a) $(4 \leq r \leq l \wedge 1 \leq s \leq k - l) \implies \gamma_{r,s} \in [\{\gamma_{r-1,s}\} \cup Inv_k Pol_k E_l]$;
(b) $(2 \leq r \leq l \wedge 3 \leq s \leq k - l) \implies \gamma_{r,s} \in [\{\gamma_{r,s-1}\} \cup Inv_k Pol_k E_l]$;
(c) $(3 \leq r \leq l \wedge s = 1) \implies \gamma_{l,1} \in [\{\gamma_{r,1}\} \cup Inv_k Pol_k E_l]$;
(d) $(r = 2 \wedge 2 \leq s \leq k - l) \implies \gamma_{2,k-l} \in [\{\gamma_{2,s}\} \cup Inv_k Pol_k E_l]$;
(e) $(3 \leq r \leq l \wedge 2 \leq s \leq k - l) \implies \gamma_{l,k-l} \in [\{\gamma_{r,s}\} \cup Inv_k Pol_k E_l]$;

Proof. (a): Obviously, the $(r + s)$-ary relation

$$\alpha_1 := \{(a_1, ..., a_r, b_1, ..., b_s) \in E_l^r \times E_k^s \mid \exists u \in E_l : (a_1, a_2, ..., a_{r-2}, u, b_1, ..., b_s) \in \gamma_{r-1,s} \wedge (\forall j_1, ..., j_{r-4} \in \{1, 2, .., r-2\} : (u, a_{j_1}, a_{j_2}, ..., a_{j_{r-4}}, a_{r-1}, a_r, b_1, ..., b_s) \in \gamma_{r-1,s})\}$$

is $\gamma_{r-1,s}$-derivable. Consequently, to prove (a), it suffices to show the equation $\alpha_1 = \gamma_{r,s}$:
Let $\mathbf{x} := (a_1, ..., a_r, b_1, ..., b_s) \in \gamma_{r,s}$ be arbitrary. The following table shows that $\gamma_{r,s} \subseteq \alpha_1$.

i	Case i for $\mathbf{x}$	$u \in E_l$ such that $\mathbf{x} \in \alpha_1$
1	$\exists i \in \{1, ..., s\} : b_i \in E_l$	$u \in E_l$
2	$\exists \{i, j\} \subseteq \{1, ..., r\} : i < j \wedge a_i = a_j$	
2.1	$\{i, j\} \subseteq \{1, 2, ..., r-2\}$	$u = a_{r-1}$
2.2	$i \in \{1, 2, ..., r-2\} \wedge j \in \{r-1, r\}$	$u = a_j$
2.3	$i = r-1 \wedge j = r$	$u = a_1$
3	$\exists \{i, j\} \subseteq \{1, ..., s\} : i < j \wedge b_i = b_j$	$u \in E_l$

Suppose there is a tuple $(x_1, ..., x_r, y_1, ..., y_s) \in \alpha_1 \backslash \gamma_{r,s}$. Since the elements of this tuple are pairwise different, an element $u \in E_l$, for which $(x_1, x_2 ..., x_{r-2}, u, y_1 ... y_s) \in \gamma_{r-1,s}$ holds, must belong to the set $\{x_1 ..., x_{r-2}\}$. Analogously, one can show that the remaining conditions of the definition of α_1 are only fulfilled if $u \in \{x_{r-1}, x_r\}$. Therefore, no element u that fulfills all conditions from the definition of α_1 exists, whereby $\alpha_1 = \gamma_{r,s}$ is proven.
(b): The relation

$$\begin{aligned}\alpha_2 := \{(a_1, ..., a_r, b_1, ..., b_s) \in E_l^r \times E_k^s \mid \exists u_1, u_2 \in E_l :\\ \forall i_1, ..., i_{r-2} \in \{1, ..., r\} :\\ (u_1, a_1, a_{i_1}, ..., a_{i_{r-2}}, b_1, ..., b_{s-2}, b_{s-1}) \in \gamma_{r,s-1} \wedge\\ (u_2, a_2, a_{1_1}, ..., a_{i_{r-2}}, b_1, ..., b_{s-2}, b_s) \in \gamma_{r,s-1} \wedge\\ (u_1, u_2, a_{i_1}, ..., a_{i_{r-2}}, b_2, b_3, ..., b_{s-1}, b_s) \in \gamma_{r,s-1}\}\end{aligned}$$

is an $\gamma_{r,s-1}$-derivable relation. Analogous to the proof of (a), one can show that $\alpha_2 = \gamma_{r,s}$. We only prove here that $\gamma_{r,s} \subseteq \alpha_2$ is valid. However, this follows from the following table, where $\mathbf{x} := (a_1, ..., a_r, b_1, ..., b_s) \in \gamma_{r,s}$ is arbitrary.

i	Case i for $\mathbf{x}$	$u_1, u_2 \in E_l$ such that $\mathbf{x} \in \alpha_1$
1	$\exists i \in \{1, ..., s-2\} : b_i \in E_l$	$u_1 = u_2$
2	$b_{s-1} \in E_l$	$u_2 = a_2$
3	$b_s \in E_l$	$u_1 = a_1$
4	$\exists \{i, j\} \subseteq \{1, ..., s\} : i < j \wedge b_i = b_j$	
4.1	$\{i, j\} \subseteq \{1, 2, ..., s-2, s-1\}$	$u_1 = u_2 = a_2$
4.2	$\{i, j\} \subseteq \{1, 2, ..., s-2, s\}$	$u_1 = u_2 = a_1$
4.3	$\{i, j\} \subseteq \{2, 3, ..., s-1, s\}$	$u_1 = a_1, u_2 = a_2$
5	$\exists \{i, j\} \subseteq \{1, ..., r\} : i < j \wedge a_i = a_j$	$u_1 = a_1, u_2 = a_2$

The statements (c)–(e) follow from (a) and (b). ■

Then Lemma 17.5.13, (c)–(e) implies

Lemma 17.5.14 *If* $\varrho = \gamma_{r,s}$ *then* $(r, s) \in \{(2, 1), (l, 1), (2, k-l), (l, k-l)\}$. ■

Note: As is generally known, every relation that is preserved from all unary functions of P_k, is representable through the union of certain diagonal relations on E_k. Then, as one can easily prove, every relation that is preserved from all unary functions of $Pol_k E_l$ is representable through the union of certain E_l-diagonal relations on E_k. Consequently, because of Lemma 17.5.14 (and Lemmas 17.3.9 and 17.3.10) all maximal classes of $Pol_k E_l$, which all functions of $(Pol_k E_l)^1$ contain and fulfill the conditions of this section and of Case 2, were determined.

We say that an h-ary, totally reflexive and totally symmetric relation γ ($\subseteq E_l^r \times E_k^s$) is **strongly** $(l;r)$**-homogeneous**, if the following is valid:

$$\forall a_0, ..., a_{r-1} \in E_l \; \forall b_0, ..., b_{s-1} \in E_k :$$
$$(\, (\exists v_0, ..., v_{r-1} \in E_l : (v_0, ..., v_{r-1}, b_0, ..., b_{s-1}) \in \varrho \, \wedge$$
$$(\forall i \in E_r \forall j \in E_r \backslash \{i\} :$$
$$(a_0, ..., a_{j-1}, v_i, a_{j+1}, ..., a_{r-1}, b_0, ..., b_{s-1}) \in \varrho))$$
$$\Longrightarrow (a_0, ..., a_{r-1}, b_0, ..., b_{s-1}) \in \varrho \,)$$

Lemma 17.5.15 *ϱ is either an $(l;r)$-central or a strong $(l;r)$-homogeneous relation.*

Proof. First, for $t \in \{r, r+1, ..., l\}$, we consider the ϱ-derivable relation

$$\varrho_t := \{(a_0, ..., a_{t-1}, b_0, ..., b_{s-1}) \in E_l^t \times E_k^s \mid \exists c \in E_l :$$
$$(\forall i_0, ..., i_{r-2} \in E_t :$$
$$(a_{i_0}, ..., a_{i_{r-2}}, c, b_0, ..., b_{s-1}) \in \varrho))\}.$$

Since ϱ is totally $(l;r)$-symmetric, this relation is also totally $(l;r)$-symmetric. If one put $c = a_0$ for $t = r$, then one can see that $\varrho \subseteq \varrho_r$ holds. Therefore, because of our assumption (VI), the following two cases are possible:

Case 1: $\varrho_r = E_l^r \times E_k^s$.

This case is only possible for $r < l$, since, in the opposite case, the definition of ϱ_t implies $\varrho = E_l^l \times E_k^s$. For $r < l$ the ϱ-derivable relation ϱ_{r+1} is obvious totally $(l;r+1)$-reflexive, whereby $\varrho_{r+1} = E_l^{r+1} \times E_k^s$ follows with the aid of Lemma 17.5.11. One easily shows by induction that $\varrho_i = E_l^i \times E_k^s$ is valid for every $i \in \{r+1, ..., l\}$. Then, $\varrho_l = E_l^l \times E_k^s$ implies $\varrho \in C_{k;E_l}^h[r]$.

Case 2: $\varrho = \varrho_r$.

In this case, the relation ϱ is $(l;r)$**-homogeneous**; that is, it fulfills the condition

$$(\exists v \in E_l : \; \forall i \in E_r :$$
$$(a_0, ..., a_{i-1}, v, a_{i+1}, ..., a_{r-1}, b_0, ..., b_{s-1}) \in \varrho)$$
$$\Longrightarrow \; (a_0, a_1, ..., a_{r-1}, b_0, ..., b_{s-1}) \in \varrho$$

for arbitrary $a_0, \, a_1, \, ..., \, a_{r-1} \in E_l$ and arbitrary $b_0, ..., b_{s-1} \in E_k$. The total $(l;r)$-symmetric relation

$$\begin{aligned}\gamma_t := \{&(a_0, ..., a_{t-1}, b_0, ..., b_{s-1}) \mid \exists v_0, v_1, ..., v_{t-1} \in E_l : \\ &(\forall i_1, ..., i_r \in E_t : (v_{i_1}, ..., v_{i_r}, b_0, ..., b_{s-1}) \in \varrho) \wedge \\ &(\forall n \in E_t \; \forall j_1, ..., j_{r-2} \in E_t \backslash \{n\} : \\ &(a_{j_1}, ..., a_{j_{r-2}}, a_n, v_n, b_0, ..., b_{s-1}) \in \varrho)\}\end{aligned}$$

($t \in \{r, r+1, ..., l\}$) is ϱ-derivable and, if $\gamma_{t-1} = E_l^{t-1} \times E_k^s$, totally $(l;t)$-reflexive. In particular, for $r = t$ we have $\varrho \subseteq \gamma_r$. (To see this, one chooses $v_\alpha = a_\alpha$ in the definition of γ_r for every $\alpha \in E_r$.) And $\varrho = \gamma_r$ means that ϱ is strongly $(l;r)$-homogeneous. Now, by assumption (VI), we have that $\gamma_r \in \{\varrho, E_l^r \times E_k^s\}$ holds. Thus our lemma would be proven, if we could show that the case $\gamma_r = E_l^r \times E_k^s$ does not occur. Suppose $\gamma_r = E_l^r \times E_k^s$ and $r < l$. Then γ_{r+1} is totally $(l;r+1)$-reflexive and therefore $\gamma_{r+1} = E_l^{r+1} \times E_k^s$ because of Lemma 17.5.11. Thus, one can prove by induction that $\gamma_l = E_l^l \times E_k^s$. Therefore, for arbitrary $a_0, ..., a_{r-1} \in E_l$ and $b_0, ..., b_{s-1} \in E_k$, there are certain $v_0, ..., v_{r-1} \in E_l$ with

$$(v_0, v_1, ..., v_{r-1}, b_0, ..., b_{s-1}) \in \varrho \tag{17.11}$$

and

$$\forall n \in E_r \; \forall \alpha_1, ..., \alpha_{h-2} \in E_l \backslash \{a_n\} : \; (\alpha_1, ..., \alpha_{r-2}, a_n, v_n, b_0, ..., b_{s-1}) \in \varrho. \tag{17.12}$$

By induction we show that for arbitrary $a_0, ..., a_{r-1} \in E_l$ and $b_0, ..., b_{s-1} \in E_k$ is valid:

$$\forall t \geq 0 : \; (a_0, ..., a_{t-1}, v_t, v_{t+1}, ..., v_{r-1}, b_0, ..., b_{s-1}) \in \varrho. \tag{17.13}$$

For $t = 0$, (17.13) holds because of (17.11). Assume (17.13) is valid for $t = n$. Then, for $t = n+1$ the statement (17.13) follows from this assumption, from the $(l;r)$-homogeneousness of ϱ (considering (17.12)) and from the total $(l;r)$-symmetry of ϱ, when one chooses $v = v_n$ for the tuple $(a_0, ..., a_n, v_{n+1}, ..., v_{r-1}, b_0, ..., b_{s-1})$.
(17.13) implies $(a_0, ..., a_{r-1}, b_0, ..., b_{s-1}) \in \varrho$ for arbitrary $a_0, ..., a_{r-1} \in E_l$ and $b_0, ..., b_{s-1} \in E_k$, contrary to $\varrho \subset E_l^r \times E_k^s$. Consequently, our assumption $\gamma_r = E_l^r \times E_k^s$ is false and therefore ϱ is strongly $(l;r)$-homogeneous. ■

To conclude our proof in this section, because of Lemma 17.5.15, it remains to show that if ϱ is strongly $(l;r)$-homogeneous the relation ϱ or a relation derivable from ϱ belongs to $B_{k;E_l}[r]$. Because of Lemma 17.5.10, we have only to prove conditions 5)–6) from the definition of an $(l;r)$-universal relation.

Thus we can assume that ϱ is a strong $(l;r)$-homogeneous relation in the following.

Lemma 17.5.16 *For every* $\mathbf{b} := (b_0, ..., b_{s-1}) \in (E_k \backslash E_l)^s$ *the relation*

$$\varepsilon_{\mathbf{b}} := \{(a, b) \in E_l^2 \mid \forall a_0, ..., a_{r-3} \in E_l : \; (a_0, a_1, ..., a_{r-3}, a, b, b_0, ..., b_{s-1}) \in \varrho\}$$

is an equivalence relation on E_l.

Proof. By the total $(l;r)$-reflexivity and total $(l;r)$-symmetry of ϱ, the reflexivity and symmetry of $\varepsilon_{\mathbf{b}}$ follows directly. To prove the transitivity of $\varepsilon_{\mathbf{b}}$ let $\{(a,b),(b,c)\} \subseteq \varepsilon_{\mathbf{b}}$. Choosing $v_0 = \alpha_0$, $v_1 = \alpha_1$, ..., $v_{r-3} = \alpha_{r-3}$, $v_{r-2} = b$ and $v_{r-1} = b$ in the definition of a strong $(l;r)$-homogeneous relation for the tuple

$$(\alpha_0, \alpha_1, ..., \alpha_{h-3}, a, c, b_0, ..., b_{s-1})$$

we get $(\alpha_0, ..., \alpha_{r-3}, a, c, b_0, ..., b_{s-1}) \in \varrho$ for arbitrary $\alpha_0, ..., \alpha_{r-3} \in E_l$, whereby $(a,c) \in \varepsilon_{\mathbf{b}}$. ■

The following lemma is a direct consequence from Lemmas 17.5.10, 17.5.14, and 17.5.16:

Lemma 17.5.17 *If $r = 2$ then the strong $(l;2)$-homogeneous relation ϱ belongs to $B^h_{k;E_l}[2]$.* ■

From the above lemma, we can assume that

$$r \geq 3$$

in our further considerations.

By the equivalence relation $\varepsilon_{\mathbf{b}}$, the set E_l is partitioned into certain (nonempty) equivalence classes $A_i[\mathbf{b}]$ $(i = 1, 2, ..., t[\mathbf{b}])$, from which we choose a representative $\alpha_i[\mathbf{b}]$. With the help of the representative set

$$V_{\mathbf{b}} := \{\alpha_1[\mathbf{b}], ..., \alpha_{t[\mathbf{b}]}[\mathbf{b}]\},$$

we can define a mapping $F_{\mathbf{b}} : E_l \longrightarrow V_{\mathbf{b}}$ by

$$F_{\mathbf{b}}(a) = \alpha_i[\mathbf{b}] \; :\Longleftrightarrow \; \{a, \alpha_i[\mathbf{b}]\} \subseteq A_i[\mathbf{b}].$$

Lemma 17.5.18 *Let ϱ be strongly $(l;r)$-homogeneous. Then for every $\mathbf{b} := (b_0, ..., b_{s-1}) \in (E_k \backslash E_l)^s \backslash \iota^s_k$ is valid:*

(a) $(a_0, ..., a_{r-1}, b_0, ..., b_{s-1}) \in \varrho \iff (F_{\mathbf{b}}(a_0), ..., F_{\mathbf{b}}(a_{r-1}), b_0, ..., b_{s-1}) \in \varrho;$

(b) $((\forall a_0, ..., a_{r-3} \in V_{\mathbf{b}} : \;\; (a_0, ..., a_{r-3}, a, b, b_0, ..., b_{s-1}) \in \varrho) \;\wedge\; \{a,b\} \subseteq V_{\mathbf{b}}) \implies a = b.$

Proof. (a): First we show that the equivalence

$$(a, a_1, ..., a_{r-1}, b_0, ..., b_{s-1}) \in \varrho \iff (b, a_1, ..., a_{r-1}, b_0, ..., b_{s-1}) \in \varrho \quad (17.14)$$

is valid for $(a,b) \in \varepsilon_{\mathbf{b}}$ and for arbitrary $a_0, ..., a_{r-1} \in E_l$.
If $(a, a_1, ..., a_{r-1}, b_0, ..., b_{s-1}) \in \varrho$, then $(b, a_1, ..., a_{r-1}, b_0, ..., b_{s-1}) \in \varrho$ follows from the strong $(l;r)$-homogeneousness of ϱ, choosing $v_0 = v_1 = ... = v_{r-1} = a$. Since $\varepsilon_{\mathbf{b}}$ is symmetric, we have also proven "$\Longleftarrow$" of (17.14).
Now (a) follows from $(a_i, F_{\mathbf{b}}(a_i)) \in \varepsilon_{\mathbf{b}}$ $(i \in E_r)$, (17.14), Lemma 17.5.16 and of the total $(l;r)$-symmetry of ϱ:

$$\begin{array}{lr}
(a_0, ..., a_{r-1}, b_0, ..., b_{s-1}) \in \varrho & \iff \\
(F_{\mathbf{b}}(a_0), a_1, ..., a_{r-1}, b_0, ..., b_{s-1}) \in \varrho & \iff \\
(F_{\mathbf{b}}(a_0), F_{\mathbf{b}}(a_1), a_2, ..., a_{r-1}, b_0, ..., b_{s-1}) \in \varrho & \iff \\
 & \vdots \\
 & \iff \\
(F_{\mathbf{b}}(a_0), ..., F_{\mathbf{b}}(a_{r-1}), b_0, ..., b_{s-1}) \in \varrho. &
\end{array}$$

(b) is a conclusion from the definitions of $\varepsilon_{\mathbf{b}}$ and $V_{\mathbf{b}}$ and from (a). ∎

Now put

$$\xi_{\mathbf{b}} := F_{\mathbf{b}}(\varrho_{\mathbf{b}}) := \{(F_{\mathbf{b}}(a_0), ..., F_{\mathbf{b}}(a_{r-1})) \mid (a_0, ..., a_{r-1}, b_0, ..., b_{s-1}) \in \varrho\}.$$

By Lemma 17.5.18, (a) $\varrho_{\mathbf{b}}$ is a homomorphic inverse image of this relation; i.e., it holds that

$$\varrho_{\mathbf{b}} = \{(a_0, ..., a_{r-1}) \in E_l^r \times E_k^s \mid (F_{\mathbf{b}}(a_0), ..., F_{\mathbf{b}}(a_{r-1})) \in \xi_{\mathbf{b}}\}.$$

Because of $\varrho \neq E_l^r \times E_k^s$ there is at least a $\mathbf{b} \in (E_k \backslash E_l)^s \backslash \iota_k^s$ with $E_l^r \times \{\mathbf{b}\} \nsubseteq \varrho$. Thus w.l.o.g. we can assume

$$V_{\mathbf{b}} = E_{t[\mathbf{b}]} \text{ and } (0, 1, ..., r-1, b_0, ..., b_{s-1}) \in (V_{\mathbf{b}}^r \times \{\mathbf{b}\}) \backslash \xi_{\mathbf{b}}.$$

To prove that ϱ has the properties 5) and (only for $s \geq 2$) 6) from the definition of an $(l;r)$-universal relation, we consider the ϱ-derivable q-ary graphic

$$G_q(Pol\,\varrho) := \chi_q \cup \{g(\kappa_1, ..., \kappa_q) \mid g \in (Pol\,\varrho)^q\},$$

where $\chi_q := (\kappa_1, ..., \kappa_q)$ is the q-ary abscissa over E_k (in matrix form, see Section 2.7) and $q \in \mathbb{N}$. Let $m_1, ..., m_i, n_1, ..., n_j$ be the numbers of those rows of $\chi_{r^i \cdot s^j}$ for which

$$\mathfrak{A}_{i,j} := pr_{m_1,...,m_i,n_1,...,n_j} \chi_{r^i \cdot s^j}$$

is a matrix form of the relation

$$E_r^i \times \{b_0, ..., b_{s-1}\}^j$$

$(r \leq i \leq l;\ s \leq j \leq k-l)$. Further, let

$$\mu_{i,j} := pr_{m_1,...,m_i,n_1,...,n_j} G_{r^i \cdot s^j}(Pol\,\varrho).$$

Lemma 17.5.19 *If ϱ is strongly $(l;r)$-homogeneous and $(0, 1, 2, ..., r-1, b_0, ..., b_{s-1}) \notin \varrho$, then $\mu_{i,j} = E_l^i \times E_k^j$ for all i, j with $r \leq i \leq l$ and $j = 1$, if $s = 1$, and $s \leq j \leq k-l$.*

Proof. First we remark that

$$\varrho \neq (E_l^r \times E_k^s) \backslash \{(a_0, ..., a_{r-1}, b_0, ..., b_{s-1}) \mid \{a_0, ..., a_{r-1}\} = E_r\}$$

because of $\varrho \notin C^h_{k;E_l}[r]$. Furthermore, if $(\beta_0, ..., \beta_{r-1}) \in \varrho$, every function g with $Im(g) = \{\beta_0, ..., \beta_{h-1}\}$ belongs to $Pol\varrho$.
Suppose the lemma is false for $(i,j) = (r,s)$. Then we can form an h-ary relation ϱ' with $\varrho \subset \varrho' \subset E_l^r \times E_k^s$ with the help of the $r^r \cdot s^s$-ary graphic of $Pol\varrho$ by projection, contradicting (VI).
Since all functions of $Pol\varrho$ with at most $r^i \cdot s^j$ essential variables belong to $(Pol\varrho)^{r^{i+1} \cdot s^j}$ or $(Pol\varrho)^{r^i \cdot s^{j+1}}$, it is easy to show by induction that the following holds:

$$\forall i \in \{r, ..., l-1\} : \mu_{i,j} = E_l^i \times E_k^j \implies \iota_k^{i+j+1}[i+1] \subseteq \mu_{i+1,j}$$

and

$$\forall i \in \{s, ..., k-l-1\} : \mu_{i,j} = E_l^i \times E_k^j \implies \iota_k^{i+j+1}[i] \subseteq \mu_{i,j+1},$$

if $s > 1$.
Consequently, our lemma follows from Lemma 17.5.11. ■

Let

$$\nu_s := \begin{cases} r^l & \text{if } s = 1, \\ r^l \cdot s^{k-l} & \text{if } s \geq 2. \end{cases}$$

Because of Lemma 17.5.19, we can find a ν_s-ary function $f_{\mathbf{b}} \in Pol\varrho$ for $s \geq 2$ with den properties

$$Im(f_{\mathbf{b}}) = V_{\mathbf{b}} \cup (E_k \backslash E_l)$$

and

$$f_{\mathbf{b}}(\mathfrak{A}_{l,k-l}) = \begin{pmatrix} F_{\mathbf{b}}(0) \\ F_{\mathbf{b}}(1) \\ ... \\ F_{\mathbf{b}}(l-1) \\ l \\ l+1 \\ ... \\ k-1 \end{pmatrix}.$$

In the case that $s = 1$, an analogous statement is valid for the matrix $\mathfrak{A}_{l,1}$:

$$f_{\mathbf{b}}(\mathfrak{A}_{l,1}) = \begin{pmatrix} F_{\mathbf{b}}(0) \\ F_{\mathbf{b}}(1) \\ ... \\ F_{\mathbf{b}}(l-1) \\ b_0 \end{pmatrix}.$$

Therefore, ϱ fulfills condition 6) of the definition of an $(l;r)$-universal relation.
Let $w_0, ..., w_{s-1}$ be certain rows of $\mathfrak{A}_{l,k-l}$ or $\mathfrak{A}_{l,1}$ with

$$f(w_i) = b_i \ (i = 0, ..., s-1).$$

Next we give some properties of the function $f_{\mathbf{b}}$ on ν_s-ary tuples of $E_r^{\nu_s}$, from which will follow that $\varrho \in B^h_{k;E_l}[r]$.
We also give elements z of $E_r^{\nu_s}$ in the form

$$(z[1], z[2], ..., z[\nu_s])$$

and, for every $z \in E_r^{\nu_s}$, denote

$$z^{t,a} \qquad (t \in \{1,2,...,\nu_s\},\ a \in E_r)$$

an element of $E_r^{\nu_s}$, which is defined by

$$z^{t,a}[i] := \begin{cases} a, & \text{if} \qquad i = t, \\ z[i] & \text{otherwise} \end{cases}$$

$(i \in \{1,2,...,\nu_s\})$.

Lemma 17.5.20 *Let* $t \in \{1,2,...,\nu_s\}$, $z \in E_r^{\nu_s}$ *and* $(f_{\mathbf{b}}(z^{t,0}), ..., f_{\mathbf{b}}(z^{t,r-1})) \in \xi_{\mathbf{b}}$. *Then*
(a) $f_{\mathbf{b}}(z^{t,0}) = ... = f_{\mathbf{b}}(z^{t,r-1})$,
(b) $\forall w \in E_r^{\nu_s} :\ f_{\mathbf{b}}(w^{t,0}) = ... = f_{\mathbf{b}}(w^{t,r-1})$.

Proof. (a): For proof, it is sufficient to show that w.l.o.g. $f_{\mathbf{b}}(z^{t,r-2}) = f_{\mathbf{b}}(z^{t,r-1})$. First we prove

$$\forall \beta_0, ..., \beta_{r-3} \in E_r^{\nu_s} :\ (f_{\mathbf{b}}(\beta_0), ..., f_{\mathbf{b}}(\beta_{r-3}), f_{\mathbf{b}}(z^{t,r-2}), f_{\mathbf{b}}(z^{t,r-1})) \in \xi_{\mathbf{b}}. \tag{17.15}$$

If $\{\beta_0[t], \beta_1[t], ..., \beta_{r-3}[t]\} \neq \{0,1,...,r-3\}$, then (17.15) holds, since in this case all columns of the matrix

$$B := \begin{pmatrix} \beta_0 \\ \beta_1 \\ \cdots \\ \beta_{r-3} \\ z^{t,r-2} \\ z^{t,r-1} \\ w_0 \\ \cdots \\ w_{s-1} \end{pmatrix}$$

belong to $\iota_k^h[r]$ and $f_{\mathbf{b}}$ preserves the relation ϱ.
Let w.l.o.g. $\beta_i[t] = i$ for $i \in E_{r-2}$. Substituting the i-th row of B by $z^{t,j}$, where $i \neq j$, we obtain a matrix $B_{i,j}$ the columns of which belong to $\iota_k^h[r]$, whereby $f_{\mathbf{b}}(B_{i,j}) \in \xi_{\mathbf{b}} \times \{\mathbf{b}\}$ holds. Consequently, by the strong $(l;r)$-homogeneousness of ϱ, choosing $v_j = f_{\mathbf{b}}(z^{t,j})$ $(j \in E_r)$ for the tuple $(f_{\mathbf{b}}(\beta_0), ..., f_{\mathbf{b}}(\beta_{r-3}), f_{\mathbf{b}}(z^{t,r-2}), f_{\mathbf{b}}(z^{t,r-1}), b_0, ..., b_{s-1})$, we get (17.15).
Since $\{f_{\mathbf{b}}(\mathbf{x}) \mid \mathbf{x} \in E_r^{\nu_s}\} = V_{\mathbf{b}}$, there are certain $\beta_0, ..., \beta_{r-3} \in E_r^{\nu_s}$ with $f_{\mathbf{b}}(\beta_i) = a_i$ $(i \in E_{r-2})$ for arbitrary $a_0, ..., a_{r-3} \in V_{\mathbf{b}}$. Consequently, by (17.15) we have

$\forall a_0, ..., a_{r-2} \in V_{\mathbf{b}}$:
$(a_0, ..., a_{r-3}, f_{\mathbf{b}}(z^{t,r-2}), f_{\mathbf{b}}(z^{t,r-1}), b_0, ..., b_{s-1}) \in \xi_{\mathbf{b}} \times \{\mathbf{b}\}$.

Because of Lemma 17.5.18, (b), this implies $f_{\mathbf{b}}(z^{t,r-2}) = f_{\mathbf{b}}(z^{t,r-1})$.
(b): Because of (a), it is sufficient to show that

$$(f_{\mathbf{b}}(w^{t,0}), ..., f_{\mathbf{b}}(w^{t,r-1})) \in \xi_{\mathbf{b}} \tag{17.16}$$

holds. (17.15) implies

$$(f_{\mathbf{b}}(w^{t,0}), ..., f_{\mathbf{b}}(w^{t,r-3}), f_{\mathbf{b}}(z^{t,r-2}), f_{\mathbf{b}}(z^{t,r-1})) \in \xi_{\mathbf{b}}.$$

Now it is easy to check that any exchange of an i-th row in

$$\begin{pmatrix} w^{t,0} \\ ... \\ w^{t,r-1} \\ w_0 \\ ... \\ w_{s-1} \end{pmatrix}$$

by $w^{t,j}$ for $i \in E_{r-3}$ or by $z^{t,j}$ for $i \in \{r-2, r-1\}$ and $i \neq j$ gives a matrix $\mathcal{M}$, whose columns belong to $\iota_k^h[r]$ and it holds $f(\mathcal{M}) \in \xi_{\mathbf{b}} \times \{\mathbf{b}\}$. Hence, by the strong $(l;r)$-homogeneousness of ϱ, choosing $v_0 = f_{\mathbf{b}}(w^{t,0})$, ..., $v_{r-3} = f_{\mathbf{b}}(w^{t,r-3})$, $v_{r-2} = f_{\mathbf{b}}(z^{t,r-2})$, $v_{r-1} = f_{\mathbf{b}}(z^{t,r-1})$, we get (17.16). ■

Lemma 17.5.21 *For the function $f_{\mathbf{b}}$, there are certain digits $t_1, ..., t_{m[\mathbf{b}]}$ that are exactly the essential digits of $(f_{\mathbf{b}})_{|E_r^{\nu_s}}$ (restriction of $f_{\mathbf{b}}$ to $E_r^{\nu_s}$); i.e., it holds that*

$$\forall z, w \in E_r^{\nu_s} : \; f_{\mathbf{b}}(z) = f_{\mathbf{b}}(w) \iff \forall i \in \{t_1, ..., t_{m[\mathbf{b}]}\} : \; z[i] = w[i]. \tag{17.17}$$

Furthermore, f has the properties

$$|Im(f_{\mathbf{b}})| = r^{m[\mathbf{b}]}$$

and

$$\forall r_1, ..., r_{\nu_s} \in E_r^r \times \{\mathbf{b}\} : \; (\{r_{t_1}, ..., r_{t_{m[\mathbf{b}]}}\} \not\subseteq \iota_k^h[r] \implies f_{\mathbf{b}}(r_1, ..., r_{\nu_s}) \notin \varrho) \tag{17.18}$$

Proof. For every $t \in \{1, 2, ..., \nu_s\}$, we have either

$$f_{\mathbf{b}}(z^{t,0}) = f_{\mathbf{b}}(z^{t,1}) = ... = f_{\mathbf{b}}(z^{t,r-1})$$

for every $z \in E_r^{\nu_s}$ or

$$(f_{\mathbf{b}}(z^{t,0}), f_{\mathbf{b}}(z^{t,1}), ..., f_{\mathbf{b}}(z^{t,r-1}), b_0, ..., b_{s-1}) \notin \varrho$$

for all $z \in E_r^{\nu_s}$ by Lemma 17.5.20. Let $T := \{t_1, ..., t_{m[\mathbf{b}]}\}$ be the set of all $t \in \{1, 2, ..., \nu_s\}$, for which $(f_{\mathbf{b}}(z^{t,0}), ..., f_{\mathbf{b}}(z^{t,r-1}), b_0, ..., b_{s-1}) \notin \varrho$ holds. Now we will show that the digits $t_i \in T$ of $f_{\mathbf{b}}$ have the properties of Lemma 17.5.21. First let $f_{\mathbf{b}}(z) = f_{\mathbf{b}}(w)$ for certain $z, w \in E_r^{\nu_s}$ and assume there exists an $i \in T$ with $\alpha := z[i] \neq w[i]$. Then the columns of the matrix

$$A := \begin{pmatrix} z^{i,0} \\ \dots \\ z^{i,\alpha-1} \\ w \\ z^{i,\alpha+1} \\ \dots \\ z^{i,r-1} \\ w_0 \\ \dots \\ w_{s-1} \end{pmatrix}$$

belong to $\iota_k^h[r]$, whereby $f_{\mathbf{b}}(A) \in \xi_{\mathbf{b}} \times \{\mathbf{b}\}$ holds. Because of $f_{\mathbf{b}}(w) = f_{\mathbf{b}}(z) = f_{\mathbf{b}}(z^{i,\alpha})$ follows

$$(f_{\mathbf{b}}(z^{i,0}), ..., f_{\mathbf{b}}(z^{i,r-1}), b_0, ..., b_{s-1}) \in \xi_{\mathbf{b}},$$

contrary to $i \in T$ and the definition of T. Therefore "$\Longrightarrow$" in (17.17) holds. Let now $z[i] = w[i]$ for every $i \in T$ and w.l.o.g. $T = \{1, 2, ..., m[\mathbf{b}]\}$. $f_{\mathbf{b}}(w) = f_{\mathbf{b}}(z)$ is proven if we can show that

$$f_{\mathbf{b}}(u_{n-1}) = f_{\mathbf{b}}(u_n) \tag{17.19}$$

holds for every tuple

$$u_n := (z[1], ..., z[n], w[n+1], w[n+2], ..., w[\nu_s])$$

and all $n \in \{m+1, m+2, ..., \nu_s\}$, since $f_{\mathbf{b}}(u_m) = f_{\mathbf{b}}(w)$ and $f_{\mathbf{b}}(u_{\nu_s}) = f_{\mathbf{b}}(z)$. By $n > m[\mathbf{b}]$ and the definition of T, we have $f_{\mathbf{b}}(u_{n-1}^{n,i}) = f_{\mathbf{b}}(u_{n-1}^{n,j})$ for arbitrary $i, j \in E_r$. As $u_{n-1}^{n,w[n]} = u_{n-1}$ and $u_{n-1}^{n,z[n]} = u_n$, we have (17.19), whereby (17.17) is proven.

From (17.17) and $\{f_{\mathbf{b}}(\mathbf{x}) \mid \mathbf{x} \in E_r^{\nu_s}\} = V_{\mathbf{b}}$ follows $|V_{\mathbf{b}}| = r^{m[\mathbf{b}]}$.

Finally we prove (17.18). Assume (17.18) is false. Then there are certain $z_0, z_1, ..., z_{r-1} \in E_r^{\nu_s}$ with $(f_{\mathbf{b}}(z_0), f_{\mathbf{b}}(z_1), ..., f_{\mathbf{b}}(z_{r-1}), b_0, ..., b_{s-1}) \in \xi_{\mathbf{b}} \times \{\mathbf{b}\}$ and (w.l.o.g.) $z_i[1] = i$ for every $i \in E_r$ and $1 \in T$. Let $w, z \in E_r^{\nu_s}$ be arbitrary. Then each column of the matrix

$$C_{i,j} := \begin{pmatrix} w^{1,0} \\ \dots \\ w^{1,j-1} \\ z^{1,i} \\ w^{1,j+1} \\ \dots \\ w^{1,r-1} \\ w_0 \\ \dots \\ w_{s-1} \end{pmatrix}$$

belongs to $\iota_k^h[r]$ for $i \neq j$. Thus $f_{\mathbf{b}}(C_{i,j}) \in \xi_{\mathbf{b}} \times \{\mathbf{b}\}$. By choosing $v_i = f_{\mathbf{b}}(z^{1,i})$ for the tuple

$$(f_{\mathbf{b}}(w^{1,0}), ..., f_{\mathbf{b}}(w^{1,r-1}), b_0, ..., b_{s-1})$$

and $i \in E_r$, $(f_{\mathbf{b}}(w^{1,0}), ..., f_{\mathbf{b}}(w^{1,r-1}), b_0, ..., b_{s-1}) \in \xi_{\mathbf{b}} \times \{\mathbf{b}\}$ follows from this and the strong $(l;r)$-homogeneousness of ϱ, contrary to the definition of T and $1 \in T$. ■

Lemma 17.5.22 *Let ϱ be strongly $(l;r)$-homogeneous, $\mathbf{b} \in (E_k \backslash E_l)^s \backslash \iota_k^s$ and $E_l^r \times \{\mathbf{b}\} \not\subseteq \varrho$. Then for certain $m[\mathbf{b}] \in \mathbb{N}$ there is a bijective mapping $\varphi_{\mathbf{b}}$ from $E_{r^{m[\mathbf{b}]}}$ onto $V_{\mathbf{b}}$ with*

$$\xi_{\mathbf{b}} = \varphi_{\mathbf{b}}(\xi_{m[\mathbf{b}]}) := \{(\varphi_{\mathbf{b}}(a_0), ..., \varphi_{\mathbf{b}}(a_{r-1})) \mid (a_0, ..., a_{r-1}) \in \xi_{m[\mathbf{b}]}\}$$

($\xi_{m[\mathbf{b}]}$ denotes an $r-$ary elementary relation)
and it holds that

$$\varrho_{\mathbf{b}} = \{(a_0, ..., a_{r-1}) \in E_l^r \mid (\varphi_{\mathbf{b}}^{-1}(F(a_0)), ..., \varphi_{\mathbf{b}}^{-1}(F(a_{r-1}))) \in \xi_{m[\mathbf{b}]}\};$$

i.e., $\varrho \in B_{k;E_l}^h[r]$.

Proof. Subsequently, we denote $m[\mathbf{b}]$ with m. Because of Lemma 17.5.21, a bijective mapping from E_{r^m} onto $V_{\mathbf{b}}$ is defined by

$$\begin{aligned} &\varphi_{\mathbf{b}}(a^{(m-1)}h^{m-1} + a^{(m-2)}h^{m-2} + ... + a^{(1)}h + a^{(0)}) = f_{\mathbf{b}}(z) \\ &:\Longleftrightarrow (z[t_1], z[t_2], ..., z[t_m]) = (a^{(m-1)}, a^{(m-2)}, ..., a^{(0)}), \end{aligned}$$

where $a \in E_{r^m}$.
Let now $(a_0, ..., a_{r-1}) \in \xi_m$. Then there are certain $z_0, ..., z_{r-1} \in E_r^{\nu_s}$ with the following properties: $(z_0[i], ..., z_{r-1}[i]) \in \iota_r^r$ for every $i \in \{1, 2, ..., \nu_s\}$ and

$$(\varphi_{\mathbf{b}}(a_0), \varphi_{\mathbf{b}}(a_1), ..., \varphi_{\mathbf{b}}(a_{r-1})) = (f_{\mathbf{b}}(z_0), f_{\mathbf{b}}(z_1), ..., f_{\mathbf{b}}(z_{r-1})).$$

Then, because of $f_{\mathbf{b}} \in Pol\,\varrho$, we have

$$(\varphi_{\mathbf{b}}(a_0), ..., \varphi_{\mathbf{b}}(a_{r-1}), b_0, ..., b_{s-1}) \in \xi_{\mathbf{b}} \times \{\mathbf{b}\}$$

and therefore $\varphi_{\mathbf{b}}(\xi_m) \subseteq \xi_{\mathbf{b}}$. Suppose there exists

$$(a_0, ..., a_{r-1}) \in \xi_{\mathbf{b}} \backslash \varphi_{\mathbf{b}}(\xi_m).$$

Then, by definition of $\varphi_{\mathbf{b}}$, there exist certain $z_0, ..., z_{r-1} \in E_r^{\nu_s}$ and an $i \in \{t_1, ..., t_m\}$ with $(f_{\mathbf{b}}(z_0), ..., f_{\mathbf{b}}(z_{r-1})) = (a_0, ..., a_{r-1})$ and $(z_0[i], ..., z_{r-1}[i]) \notin \iota_r^r$. But this is contrary to (17.18). Thus $\xi_{\mathbf{b}} = \varphi_{\mathbf{b}}(\xi_m)$. The remaining assertions follow from Lemma 17.5.18, (a) and the definition of $\xi_{\mathbf{b}}$. ■

Summing up, we obtain in Case 2.2,

$$\varrho \in C_{k;E_l}^h[r] \cup B_{k;E_l}^h.$$

17.6 Classes Describable by Relations of $R_{max}(P_l) \cup R_{max}(P_k)$

The aim of this section is to prove the following:

Theorem 17.6.1 *Let ϱ be an h-ary relation of $R_{max}(P_l) \cup R_{max}(P_k)$. Then $Pol_k\varrho \times E_l$ is a maximal class of $Pol_k E_l$ if and only if $\varrho \in R_{max}(P_l) \cup \mathfrak{U}_{k;E_l} \cup \mathfrak{C}_{k;E_l}$.*

With the help of Sections 5.2 and 5.4, one can easily prove the following lemma:

Lemma 17.6.2 *It holds:*
(a) $\forall \varrho \in R_{max}(P_k) \backslash R_{max}(P_l)$: $pr_{E_l}(Pol_k\varrho \times E_l) \neq P_l \implies Pol_k\varrho \times E_l$ is not maximal in $Pol_k E_l$.
(b) $\forall \varrho, \varrho' \in R_{max}(P_k)$: $Pol_k\varrho \neq Pol_k\varrho' \ \wedge\ pr_{E_l}Pol_k\varrho \times E_l = pr_{E_l}Pol_k\varrho' \times E_l = P_l \implies Pol_k\varrho \times E_l \neq Pol_k\varrho' \times E_l$. ■

Lemma 17.6.3 *For every relation $\varrho \in \mathfrak{M}_k \cup \mathfrak{S}_k \cup \mathfrak{L}_k$, there is a certain ϱ-derivable relation which belongs to $R_{max}(P_l) \cup \mathfrak{C}_{k;E_l}$.*

Proof. Let ϱ be an arbitrary h-ary relation of $\mathfrak{M}_k \cup \mathfrak{S}_k \cup \mathfrak{L}_k$. Then the following relations are ϱ-derivable:

$$\varrho' := \varrho \cap E_l^h$$

and

$$\varrho_{E_l} := pr_{1,...,h-1}(\Delta(E_l \times \varrho)).$$

We distinguish two cases for ϱ:
Case 1: $\varrho \in \mathfrak{M}_k \cup \mathfrak{S}_k$.
Then the following three cases are possible for the relation ϱ':
Case 1.1: $\varrho' = \emptyset$.
This case is possible only for $\varrho' \in \mathfrak{S}_k$ and the relation ϱ_{E_l} ($= \{x \in E_k \mid \exists a \in E_l : (a, x) \in \varrho\}$) belongs to $\mathfrak{C}_{k;E_l}^1$, since (by $\varrho' = \emptyset$) $(a, b) \in \varrho$ for all $a \in E_l$ implies $b \notin E_l$.

Case 1.2: $\emptyset \subset \varrho' \subset \iota_l^2$ or $\iota_l^2 \subset \varrho' \subset E_l^2$.
Then, $\varrho' \in R_l \backslash D_l$, and because of Chapter 6, one can derive a relation of $R_{max}(P_l)$ from the relations of $\{\varrho'\} \cup D_l$.
Case 1.3: $\varrho' = \iota_l^2$.
This case is possible only for $\varrho \in \mathfrak{M}_k$. Then, all elements of E_l are incomparable in respect to the partial order relation ϱ, and both the smallest element o_ϱ and the greatest element e_ϱ of E_k (in respect to ϱ) belong to the set $E_k \backslash E_l$. Since $e_\varrho \in \varrho_{E_l}$ and $o_\varrho \notin \varrho_{E_l}$, we have $E_l \subseteq \varrho_{E_l}$ and $E_l \subset \varrho_{E_l} \subset E_k$ for the above defined relation ϱ_{E_l}, obviously. Consequently, $\varrho_{E_l} \in \mathfrak{C}^1_{k;E_l}$ and our assertion also holds in Case 1.3.
Case 2: $\varrho \in \mathfrak{L}_k$.
In this case, $k = p^m$ (p prime, $m \geq 1$) and

$$\varrho = \{(a,b,c,d) \in E_k^4 \mid a + b = c + d\},$$

where $(E_k; +)$ is an elementar Abelean p-group. W.l.o.g. (see Lemma 5.2.4.2) we can assume that the element $0 \in E_l$ is the neutral element of $(E_k; +)$. Then the relation ϱ' is not E_l-diagonal because of

$$\{(a,0,0,a), (a,0,a,0), (0,a,0,a), (0,a,a,0) \mid a \in E_l\} \subseteq \varrho' \subseteq E_l^4$$

and $(0,1,1,1) \notin \varrho'$. Consequently, by Chapter 6, one can derive a relation of $R_{max}(P_l)$ from relations of $\{\varrho\} \cup D_l$. ■

Lemma 17.6.4 *For every $\varrho \in \mathfrak{B}_k^h$ $(3 \leq h \leq k)$ the clone $Pol_k \varrho \times E_l$ is not maximal in $Pol_k E_l$.*

Proof. Let $\varrho \in \mathfrak{B}_k^h$. We consider the ϱ-derivable relations

$$\varrho_r := \varrho \cap (E_l^r \times E_k^{h-r}) \; (0 \leq r \leq h).$$

Since $\iota_k^h \subseteq \varrho$ and $h \geq 3$, the sets ι_l^h and $\{(x_1, ..., x_h) \in E_k^h \mid |\{x_1, ..., x_h\}| \leq 2\}$ are subsets of ϱ. Consequently, the relations ϱ_r do not have any double rows for every $r \in \{0, 1, ..., h\}$, and it holds that

$$\forall i \in E_h : \; pr_i \varrho_r = pr_i E_l^r \times E_k^{h-r}.$$

Therefore, $\varrho_r \in Inv_k Pol_k E_l$ implies $\varrho_r = E_l^r \times E_k^{h-r}$.
Let r_* be the smallest number, for which is valid $\varrho_{r_*} \notin Inv_k Pol_k E_l$ and $\varrho_r \in Inv_k Pol_k E_l$ for all $r < r_*$. The following cases are possible:
Case 1: $r_* = 0$.
Because of the above remarks, we have $\varrho_1 = E_l \times E_k^{h-1} \subseteq \varrho$, whereby every element $c \in E_l$ is a central element of the relation ϱ, which cannot, however, be for the relation $\varrho \in \mathfrak{B}_k$.
Case 2: $r_* = 1$.
In this case, the relation ϱ_1 is totally $(l;1)$-reflexive, totally $(l;1)$-symmetric, weakly $(l;1)$-central and different from $E_l \times E_k^{h-1}$. In addition, we can assume that for every $(a_2, ..., a_h) \in E_k^{h-1}$ there exists a $c \in E_l$ with $(c, a_2, ..., a_h) \in \varrho_1$.

Namely, if this is false, we can derive a relation of $\mathfrak{C}_k$ by forming $pr_{1,...,h-1}\varrho_1$, whereby $Pol_k\varrho \times E_l$ is not a maximal class of $Pol_k E_l$ because of Lemma 17.6.2. Let

$$\gamma_t := \{(a_1, ..., a_t) \in E_l \times E_k^{t-1} \mid \exists c \in E_l : \forall i_2, ..., i_h \in \{1, 2, ..., t\} : (c, a_{i_2}, ..., a_{i_h}) \in \varrho_1\}.$$

This relation has the property

$$(\gamma_t = E_l \times E_k^{t-1} \wedge t \geq h) \implies \gamma_{t+1} = E_l \times E_k^t \vee (\exists \varrho' \in \mathfrak{C}_k : \gamma_t \vdash \varrho') \quad (17.20)$$

as one can prove: Because of $\gamma_t = E_l \times E_k^{t-1}$ and the above properties of ϱ_1, the relation γ_{t+1} is strongly $(l;1)$-reflexive, strongly $(l;1)$-symmetric and weakly $(l;1)$-central. Thus, if $\gamma'_{t+1} := pr_{1,...,t+1}\gamma_{t+1} \neq E_k^t$, then $\gamma'_{t+1} \in \mathfrak{C}_k$. If $\gamma'_{t+1} = E_k^t$, then for every $(a_2, ..., a_{t+1}) \in E_k^t$ there is an $\alpha \in E_l$ with $(\alpha, a_2, ..., a_{t+1}) \in \gamma_{t+1}$ and we can show $\gamma_{t+1} = E_l \times E_k^t$, as follows:
Let $(a_1, a_2, ..., a_{t+1}) \in E_l \times E_k^t$. Then there exists an $\alpha \in E_l$ with $(\alpha, a_2, ..., a_{t+1}) \in \gamma_{t+1}$. Because of the definition of γ_{t+1}, this implies

$$\exists c \in E_l \forall i_2, ..., i_h \in \{2, ..., t+1\} : (c, a_{i_2}, ... a_{i_h}) \in \varrho_1.$$

In addition, we have

$$\forall i_2, ..., i_h \in \{1, 2, ..., t+1\} : (\exists q \in \{2, ..., h\} : i_q = 1) \implies (c, a_{i_2}, ..., a_{i_h}) \in \varrho_1,$$

since ϱ_1 is a weakly $(l;1)$-central relation. Therefore $(a_1, a_2, ..., a_{t+1}) \in \gamma_{t+1}$ and $\gamma_{t+1} = E_l \times E_k^t$ is proven, i.e., (17.20) holds.
Analogous to the just carried out considerations or to the proof of Lemma 17.5.12, one can show $\gamma_h = E_l \times E_k^{h-1}$. From that and from (17.20), it follows that a relation of $\mathfrak{C}_k$ is ϱ-derivable. Consequently, $Pol_k\varrho \times E_l$ is not maximal in $Pol_k E_l$.

Case 3: $r_* \geq 2$.
We consider the (proper subclass) $A := Pol_k\varrho_{r_*}$ of $Pol_k E_l$, where $Pol_k\varrho \times E_l \subseteq A$ holds, since ϱ_{r_*} is ϱ-derivable. Next we show that $Pol_k\varrho \times E_l \neq A$ (and with that, our assertion) through the construction of a function $f \in A$, which ϱ does not preserve. For this purpose, let T be a matrix, which is a matrix representation of the relation $\varrho \backslash E_l^h$. Then, the $|T|$-ary function f defined by

$$f(T) = \alpha \text{ for certain } \alpha \in E_k^h \backslash \varrho \text{ and}$$
$$f(\mathbf{x}) = 0 \text{ for the remaining tuples } \mathbf{x} \in E_k^{|T|}$$

preserves ϱ_{r_*}, but does not preserve ϱ. ■

Lemma 17.6.5 *Let $\varrho \in \mathfrak{U}_k$. Then the class $Pol_k\varrho \times E_l$ is $Pol_k E_l$-maximal if and only if $\varrho \in \mathfrak{U}_{k;E_l}$.*

Proof. Let $[a]_\varrho := \{x \in E_k \mid (a, x) \in \varrho\}$. If $\varrho \notin \mathfrak{U}_{k;E_l}$, then we have either $\varrho' := \varrho \cap E_l^2 \notin \{\iota_l^2, E_l^2\}$ (and therefore $\varrho' \in \mathfrak{U}_l$) or $\bigcup_{a \in E_l} [a]_\varrho \notin \{E_l, E_k\}$. In the first case a relation of $\mathfrak{U}_l \subseteq R_{max}(P_l)$ is ϱ-derivable. In the second case, the ϱ-derivable relation $\varrho_{E_l} := pr_1(\Delta(E_l \times \varrho)) = \{x \in E_k \mid \exists a \in E_l : (a, x) \in \varrho\}$ belongs to $\mathfrak{C}^1_{k;E_l}$. Our assertion results from that and from Lemma 17.3.3. ■

Lemma 17.6.6 *Let $\varrho \in \mathfrak{C}_k$. Then, $Pol_k \varrho \times E_l$ is $Pol_k E_l$-maximal if and only if $\varrho \in \mathfrak{C}_l^1 \cup \mathfrak{C}_{k;E_l}$.*

Proof. If ϱ is a unary relation and $\varrho \notin \mathfrak{C}_l^1 \cup \mathfrak{C}_{k;E_l}^1$, then the ϱ-derivable relation $\varrho \cap E_l$ belongs to $\mathfrak{C}_l^1 \cup \mathfrak{C}_{k;E_l}^1$. Therefore, our assertion for unary relation follows from Lemmas 17.3.1 and 17.3.2. Let now $\varrho \in \mathfrak{C}_k^h$ with $2 \leq h \leq k-1$ and let $Pol_k \varrho \times E_l$ be $Pol_k E_l$-maximal, where it is not possible to describe the class $Pol_k \varrho \times E_l$ with the aid of a relation of $R_l \backslash D_l$. Because of Lemmas 17.3.5, and 17.3.6–17.3.8 it is sufficient proof to show that $\varrho \in \mathfrak{C}_{k;E_l}$ is valid. Obviously,

$$\varrho \cap E_l^2 \in \{\iota_l^2, E_l^2\} \text{ if } h = 2$$

and

$$E_l^h \subseteq \varrho \text{ if } h \geq 3$$

result from our assumptions about the relation ϱ. Then the following cases are possible:

Case 1: $h = 2$ and $\varrho \cap E_l^2 = \iota_l^2$.
Then every central element of ϱ belongs to $E_k \backslash E_l$.
We form $\varrho_1 := \varrho \cap (E_l \times E_k)$. This relation has the following properties:

$$\begin{array}{ll} \varrho_1 \cap E_l^2 = \iota_l^2,\ pr_0 \varrho_1 = E_l,\ \varrho_1 \subset E_l \times E_k, & \\ \exists c \in E_k \backslash E_l : E_l \times \{c\} \subseteq \varrho_1 & \text{and} \\ \{c\} \cup E_l \subseteq pr_1 \varrho_1. & \end{array}$$

If $pr_1 \varrho_1 \neq E_k$, then one can derive a relation of $\mathfrak{C}_{k;E_l}^1$ from ϱ_1 and therefore from ϱ. Consequently, the ϱ-derivable relation ϱ_1 belongs to $Z_{k;E_l}$. If there are $a, c \in E_k \backslash E_l$ and $b \in E_l$ with $\{(a,b),(b,c)\} \subseteq \varrho$ and $(a,c) \notin \varrho$, then the binary function f defined by

$$f(x,y) := \begin{cases} a \text{ if} & (x,y) = (a,b), \\ c \text{ if} & (x,y) = (b,c), \\ b \text{ otherwise}, & \end{cases}$$

preserves ϱ_1, but does not belong to $Pol_k \varrho \times E_l$. Since, by assumption, $Pol_k \varrho \times E_l$ is $Pol_k E_l$-maximal,

$$\forall a, c \in E_k\ \forall b \in E_l :\ \{(a,b),(b,c)\} \subseteq \varrho \implies (a,c) \in \varrho$$

results. Hence $\varrho \in Z\mathfrak{C}_{k;E_l}^2$ and our lemma follows from Lemmas 17.3.5 and 17.3.6.

Case 2: $h \geq 2$ and $E_l^h \subseteq \varrho$.
Case 2.1: ϱ has a central element c which belongs to E_l.
For the ϱ-derivable relation $\varrho' := \varrho \cap (E_l \times E_k^{h-1})$ the following two cases are possible:
Case 2.1.1: $\varrho' \neq E_l \times E_k^{h-1}$.
Then, $\varrho' \notin Inv_k Pol_k E_l$ and it holds

$$Pol_k\varrho \times E_l \subset Pol_k\varrho' \subset Pol_k E_l,$$

since one can easily check that the h-ary function f with

$$f\begin{pmatrix} c & l & l & \dots & l \\ l & c & l & \dots & l \\ \dots & \dots & \dots & \dots & \dots \\ l & l & l & \dots & c \end{pmatrix} = \begin{pmatrix} a_1 \\ a_2 \\ \dots \\ a_h \end{pmatrix} \notin \varrho$$

and

$$f(\mathbf{x}) = c \text{ for the remaining tuples } \mathbf{x} \text{ of } E_k^h$$

preserves the relation ϱ, however f does not preserve the relation ϱ'.
Case 2.1.2: $\varrho' = E_l \times E_k^{h-1}$.
In this case, every element of E_l is a central element of ϱ and ϱ belongs to $\mathfrak{C}_{k;E_l}$.
Case 2.2: Every central element of ϱ belongs to $E_k \backslash E_l$.
If $\varrho_1 \neq E_l^2 \times E_k^{h-2}$ then the ϱ-derivable relation $\varrho_1 := \varrho \cap (E_l^2 \times E_k^{h-2})$ is not an invariant of $Pol_k E_l$ and

$$Pol_k\varrho \times E_l \subset Pol_k\varrho_1 \subset Pol_k E_l,$$

holds, since the h-ary function f with

$$f\begin{pmatrix} c & 0 & 0 & \dots & 0 \\ l & c & 0 & \dots & 0 \\ \dots & \dots & \dots & \dots & \dots \\ l & 0 & 0 & \dots & c \end{pmatrix} := \begin{pmatrix} a_1 \\ a_2 \\ \dots \\ a_h \end{pmatrix} \notin \varrho$$

(c is a central element von ϱ)

$f(\mathbf{x}) := 0$ for $\mathbf{x} \in E_l^h$ and

$f(\mathbf{x}) := c$ for the remaining tuples $\mathbf{x}$

belongs to $Pol_k\varrho_1 \backslash Pol_k\varrho$.
Therefore, we can assume that $E_l^2 \times E_k^{h-2} \subseteq \varrho$ or

$$\{(a_1, ..., a_h) \in E_k^h \mid \exists i \neq j : \ \{a_i, a_j\} \subseteq E_l\} \subseteq \varrho. \tag{17.21}$$

We consider the relation

$$\varrho_h := \{(a_0, ..., a_{h-1}) \mid \exists u \in E_l : \forall i \in E_h : (a_0, ..., a_{i-1}, u, a_{i+1}, ..., a_{h-1}) \in \varrho\},$$

which is totally symmetric. W.l.o.g. we can assume that for arbitrary $\mathbf{a} := (a_1, a_3, ..., a_h) \in E_k^{h-1}$ there is a $u_{\mathbf{a}} \in E_l$ with $(u_{\mathbf{a}}, a_1, a_3, ..., a_h) \in \varrho$. Namely, if this is not valid, we have that the ϱ-derivable relation

$$\varrho' := pr_{1,...,h-1}((E_l \times E_k^{h-1}) \cap \varrho)$$

belongs to $\mathfrak{C}_k^{h-1}$; i.e., E_l is the set of all central elements of ϱ.
Because of symmetry of ϱ we have in addition $(a_1, u_{\mathbf{a}}, a_3, ..., a_h) \in \varrho$. Then,

with the total reflexivity of ϱ, $(a_1, a_1, a_3, ..., a_h) \in \varrho_h$ holds. Thus ϱ_h is totally reflexive.

Next we show that each element $u \in E_l$ is a central element of ϱ_h:
Let $(\alpha, a_2, ..., a_h) \in E_l \times E_k^{h-1}$ be arbitrary. Then, as already shown, there is a $u \in E_l$ with $(u, a_2, ..., a_h) \in \varrho$. In addition, because of (17.21), we have $(\alpha, a_2, ..., a_{i-1}, u, a_{i+1}, ..., a_h) \in \varrho$ for each $i \in \{2, ..., h\}$. By definition of ϱ_h, this implies $(\alpha, a_2, ..., a_h) \in \varrho_h$.
Therefore, ϱ_h is either a central relation with central elements of E_l or $\varrho_h = E_k^h$ holds. Because of Lemma 17.6.2, we must continue to examine the case $\varrho_h = E_k^h$.
We form the relation

$$\varrho_t := \{(a_1, ..., a_t) \mid \exists u \in E_l : \forall \{i_1, ..., i_{h-1}\} \subseteq \{1, ..., t\} : (u, a_{i_1}, ..., a_{i_{h-1}}) \in \varrho\}$$

$(h+1 \leq t \leq k)$. Then, $\varrho_{t-1} = E_k^{t-1}$ implies $\iota_k^t \subseteq \varrho_t$ $(h+1 \leq t \leq k)$. Therefore, we have either $\varrho_t = E_k^t$ or $\varrho_t \notin Inv_kPol_kE_l$. Since ϱ does not have any central elements of E_l, there is a tuple $(\alpha_1, \alpha_2, ..., \alpha_h) \in (E_l \times E_k^{h-1}) \backslash \varrho$. Then the h-ary function g with

$$g\begin{pmatrix} c & 0 & 0 & \dots & 0 \\ l & c & 0 & \dots & 0 \\ \dots & \dots & \dots & \dots & \dots \\ l & 0 & 0 & \dots & c \end{pmatrix} := \begin{pmatrix} \alpha_1 \\ \alpha_2 \\ \dots \\ \alpha_h \end{pmatrix} \notin \varrho$$

(c is a central element of ϱ)

$g(\mathbf{x}) := \alpha_1$ otherwise,

does not preserve ϱ. However, g preserves the relation ϱ_t for $t \geq h+1$. Thus $Pol_k\varrho \times E_l$ is Pol_kE_l-maximal only if $\varrho_t = E_k^t$ holds for all $t \in \{h, ..., k\}$. It results, however, from $\varrho_k = E_k^k$ that ϱ has a certain central element of E_l, which we had excluded in Case 2.2. Therefore, this case cannot occur. ■

A direct conclusion from Lemmas 17.6.2–17.6.6 is Theorem 17.6.1, whereby the theorems of Section 17.2 are also proven.

18

Further Submaximal Classes of P_k

As already indicated in Chapter 14, there are few results about submaximal classes for arbitrary k. Supplementary to Chapters 16 and 17, all submaximal classes of a maximal class of the type $\mathfrak{S}$ are described in this chapter. It is then shown how one can prove the special case $k \in \mathbb{P}$ of this general description. The rest of this chapter deals with submaximal classes of P_k which lie below a maximal class of the type $\mathfrak{U}$. In Section 18.2, one can find some maximal classes of $Pol_k\varrho$, where $\varrho \in \mathfrak{U}_k$ is arbitrary. Then, in Section 18.3, the list from Section 18.2 is completed to the list of all maximal classes of $Pol_k\varrho$ for $\varrho = E_{k-1}^2 \cup \{(k-1, k-1)\}$.

18.1 The Maximal Classes of $Pol_k\varrho_s$ for $\varrho_s \in \mathfrak{S}_k$

In this section, p always denotes a prime number. Further, let $k := p \cdot l$ with $l \in \mathbb{N}$. Denote s a fixed point free permutation on the set E_k, whose cycles have the same length p. Furthermore, let

$$\varrho_s := \{(x, s(x)) \mid x \in E_k\}.$$

By Chapter 5

$$S := Pol_k\varrho_s$$

is a maximal class of type $\mathfrak{S}$. For a description of the maximal classes of S, we need the following concepts and definitions.
Let

$$\theta_s := \{(x, y) \in E_k^2 \mid \exists i \in \{0, 1, ..., p-1\} : y = s^i(x)\}.$$

It is easy to see that θ_s is an equivalence relation on E_k.
An h-ary relation $\gamma \in R_k$ is called **θ_s-closed**, if

$$\forall (x_1, ..., x_h) \in \gamma\ \forall (y_1, ..., y_h) \in R_k^h :$$
$$(\forall i \in \{1, ..., h\} : (x_i, y_i) \in \theta_s) \Longrightarrow (y_1, ..., y_h) \in \gamma.$$

Obviously, a θ_s-closed relation $\varrho \subseteq E_k^h$ is the homomorphic inverse image of an h-ary relation ϱ' on the factor set (quotient set) $(E_k)_{/\theta_s}$ in respect to the mapping $E_k \longrightarrow (E_k)_{/\theta_s}$, $x \mapsto x/\theta_s$. It is easy to check that the following statements are valid:
– An equivalence relation ϱ is θ_s-closed if and only if $\theta_s \subseteq \varrho$.
– A central relation is θ_s-closed if and only if it is the homomorphic inverse image of a central relation defined on $(E_k)_{/\theta_s}$.
– An h-regular relation ϱ, which is defined as in Section 5.2.1 [1], is θ_s-closed if and only if $\theta_s \subseteq \vartheta_0 \cap \vartheta_1 \cap ... \cap \vartheta_{m-1}$.

An equivalence relation ε on E_k is called **transversal to** s, if $s \in Pol_k \varepsilon$ and $\varepsilon \cap \theta_s = \iota_k^2$; i.e., the permutation s maps each equivalence class (block) of ε onto another equivalence class of ε.
A unary relation μ on E_k is called **transversal to** s, if $s^i(x) \notin \mu$ whenever $i \in \{1, ..., p-1\}$ and $x \in E_k$.
Let q, r be prime numbers and let n be the least positive integer with

$$q^n = 1 \pmod r.$$

Moreover, denote $GF(q^n)$ a field with the order q^n.[2] Let

$$\mathfrak{G}(q, r)$$

be the set of all mappings of the form $GF(q^n) \longrightarrow GF(q^n)$, $x \mapsto a \cdot x + b$ with $a, b \in GF(q^n)$ and $a^r = 1$. It is easy to see that the algebra

$$\mathfrak{G}(\mathbf{q}, \mathbf{r}) := (\mathfrak{G}(q, r); \Box)$$

is a group.
P. P. Palfy proved the following fact (see [Ros-S 85]):

Lemma 18.1.1 *(without proof)*
A finite group **G** *has a maximal subgroup of order p (p prime) if and only if* **G** *is isomorphic to one of the groups listed below:*

(1) an Abelean group of order $p \cdot q$*, where* $q \in \mathbb{P}$*;*
(2) $\mathfrak{G}(\mathbf{p}, \mathbf{q})$*, where* $q \in \mathbb{P}$ *and* $p = 1 \pmod q$*;*
(3) $\mathfrak{G}(\mathbf{p}, \mathbf{q})$*, where* $q \in \mathbb{P} \setminus \{p\}$*.*

Now we define permutation groups with the aid of groups of Lemma 18.1.1. For a group $\mathbf{G} := (G; \circ)$ whose order divides $k = |G| \cdot r$, we consider a partition of E_k into $|G|$-element blocks $A_1, ..., A_r$ and select arbitrary bijections

$$\varphi_i : A_i \longrightarrow G,$$

[1] See also Lemma 5.2.6.3.
[2] As generally known, there is (up to isomorphism) exactly one finite field of order q^n.

$1 \le i \le r$. Then, with the help of the mappings φ_i, for every $g \in G$ one can define a permutation π_g on E_k as follows:

$$\forall i \in \{1,...,r\}\ \forall x \in A_i :\ \pi_g(x) := \varphi^{-1}(g \circ (\varphi_i(x))).$$

It is easy to check that the set of all permutations of the form π_g with $g \in G$ together with the operation $\square$ forms a group, which will be called a **semiregular representation** of the group **G**.

The following theorem was found by I. G. Rosenberg and Á. Szendrei.

Theorem 18.1.2 (**Rosenberg-Szendrei Theorem**; *[Ros-S 85]; without proof)*
The following list describes all maximal classes of $S := Pol_k\varrho_s$:

(1) $Pol_k\{(x, \pi(x)) \mid \pi \in G\}$, where G is a semiregular representation of a group, which belongs to the ones described in Lemma 18.1.1, (1)–(3) and for which $s \in G$ is valid;
(2) $S \cap Pol_k\lambda_G$, where $\lambda_G \in \mathfrak{L}_k$, $G := (E_k, \oplus)$ is an elementary Abelean p-group (see Section 5.4) and there exists an element $c \in E_k$ with $s(x) = x \oplus c$;
(3) $S \cap Pol_k\varepsilon$, where $\varepsilon \in \mathfrak{U}_k$ is either θ_s-closed or transversal to s.
(4) $S \cap Pol_k\gamma$, where $\gamma \in \mathfrak{C}_k$ and γ is either θ_s-closed or a nonempty unary relation transversal to s;
(5) $S \cap Pol_k\varrho$, where $\varrho \in \mathfrak{B}_k$ and ϱ is θ_s-closed.

There are also certain k for which the description of the maximal classes of S is simple, as the following theorem shows:

Theorem 18.1.3 *([Sze 84], [Lau 84b])*
Let $k = p \in \mathbb{P}$ and $\varrho_s \in \mathfrak{S}_p$. Then $S := Pol_p\varrho_s$ has exactly two maximal classes. One maximal class has the type (2) and the other maximal has the type (4) from Theorem 18.1.2.
Choosing $s(x) := x + 1 \pmod p$, one can describe these maximal classes as follows: $S \cap Pol_p\{0\}$ and $S \cap Pol_p\lambda$ with $\lambda := \{(a,b,c,d) \in E_k^4 \mid a + b = c + d \pmod p\}$.

The above theorem is a special case of the following theorem:

Theorem 18.1.4 *([Lau 84b])*
Let $k \in \mathbb{N}$, $s(x) := x + 1 \pmod k$, $S := Pol_k\{(x, s(x)) \mid x \in E_k\}$ [3] and let T_k be the set of all divisors $\in \mathbb{N}$ of k. Then the following list gives all maximal classes of S:

---
[3] S is a maximal clone of P_k iff $k \in \mathbb{P}$. For $k \notin \mathbb{P}$ the set S is a subclone of a certain clone of the type $\mathfrak{U}$.

(1) $S \cap Pol_k\gamma_r$*, where* $r \in T_k \setminus \{1\}$ *and*

$$\gamma_r := \{x \in E_k \mid r \textit{ divides } x\};$$

(2) $S \cap Pol_k\varrho_t$*, if* $k \notin \mathbb{P}$, $t \in T_k \setminus \{1, k\}$ *and*

$$\varrho_t := \{(x, y) \in E_k \mid r \textit{ divides } (x - y)\};$$

(3) $S \cap L_k$*, if* $k \in \mathbb{P}$ *and*

$$L_k := Pol_k\{(a, b, c, d) \in E_k^4 \mid a + b = c + d \ (mod\ k)\} =$$
$$\bigcup_{n\geq 1}\{f^n \in P_k \mid \exists a_0, ..., a_n \in E_k : \ f(\mathbf{x}) = a_0 + \sum_{i=1}^{n} a_i \cdot x_i \ (mod\ k)\}.$$

Proof. We prove the above theorem with the aid of Theorem 6.1 and Theorem 8.3.2. Let α be the mapping defined in Theorem 8.3.2.
First we prove that the classes described in (1)–(3) are S-maximal. For this purpose, let $r \in T_k\setminus\{1\}$, $t \in T_k\setminus\{1, k\}$ and $A \in \{Pol_k\gamma_r, Pol_k\varrho_t, L_k\}$ be arbitrary. By Chapter 5, the classes $Pol_k\gamma_r$, $Pol_k\varrho_t$ and L_k are maximal in P_k. Consequently, because of Theorem 8.3.2, (a), we have to show that $\alpha(S \cap A) = A$ holds.
Obviously, because of $c_0 \in A$, we have $\alpha(S\cap A) \subseteq A$. For the proof of $A \subseteq \alpha(S\cap A)$, we show that every function $f^n \in S$ with $\alpha(f) \in A$ belongs to A. We distinguish three cases:
Case 1: $A = Pol_k\gamma_r$.
Let $a_1, ..., a_n \in \gamma_r$ and $f^n \in S$ with $\alpha(f) \in Pol_k\gamma_r$ be arbitrary. By definition of γ_r, $b_i := a_i - a_1$ (mod k) is an element of γ_r for every $i \in \{1, ..., n\}$. Thus, because of $\alpha(f) \in Pol_k\gamma_r$, we have $b := f(0, b_2, ..., b_n) \in \gamma_r$ and therefore $b + a_1$ (mod k) $=$ $f(a_1, ..., a_n) \in \gamma_r$. Consequently, f preserves the relation γ_r, and $\alpha(S \cap Pol_k\gamma_r) = Pol_k\gamma_r$ is valid.
Case 2: $A = Pol_k\varrho_t$.
Let $(a_1, b_1), ..., (a_n, b_n) \in \varrho_t$ and $f^n \in S$ with $\alpha(f) \in Pol_k\varrho_t$ be arbitrary. Set $c_i := a_i - a_1$ (mod k) and $d_i := b_i - b_1$ (mod k) for $i = 1, ..., n$. Then $(c_i, d_i) \in \varrho_t$. Further, because of $\alpha(f) \in Pol_k\varrho_t$, we have

$$\begin{pmatrix} u \\ v \end{pmatrix} := f\begin{pmatrix} 0 & c_2 & ... & c_n \\ 0 & d_2 & ... & d_n \end{pmatrix} \in \varrho_t.$$

Since $f \in S$,

$$f\begin{pmatrix} a_1 & a_2 & ... & a_n \\ b_1 & b_2 & ... & b_n \end{pmatrix} = \begin{pmatrix} u + a_1 \text{ (mod } k) \\ v + b_1 \text{ (mod } k) \end{pmatrix} \in \varrho_t$$

results. Consequently, f preserves the relation ϱ_t and $\alpha(S\cap Pol_k\varrho_t) = Pol_k\varrho_t$ holds.
Case 3: $A = L_k$.
It is easy to check that

$$S \cap L_k = \bigcup_{n\geq 1}\{f^n \in P_k \mid \exists a_0, ..., a_n \in E_k : \ f(\mathbf{x}) = a_0 + \sum_{i=1}^{n} a_i \cdot x_i \text{ (mod } k) \wedge a_1 + ... + a_n = 1 \text{ (mod } k)\}.$$

This implies $\alpha(S \cap L_k) = L_k$.

Now we come to the completeness proof. For this purpose, we choose an arbitrary

subset M of S with $M \not\subseteq A$ for every class A defined in (1)–(3). Our theorem is proven if we can show that $[M] = S$ holds.
$S^1 = \{s, s^2, ..., s^{k-1} = e_1^1\} \subseteq [M]$ follows from the following considerations:
Because of $M \not\subseteq S \cap Pol_k\gamma_k$, there is a certain function $f^n \in M$ with $a := f(0,0,...,0) \neq 0$. Consequently, $\Delta^{n-1}f = s^a$ belongs to $[M]$. If a and k are relatively prime, then $[s^a]^1 = S^1$. If a and k are not relatively prime, there is a $t \in T_k \setminus \{1\}$ with $[s^a]^1 = \{s^x \mid x \in \gamma_t\}$. Because of $M \not\subseteq S \cap Pol_k\gamma_t$, a certain function g_t^m with $b := g_t(a_1, ..., a_m) \notin \gamma_t$ for certain $a_1, ..., a_m \in \gamma_t$ belongs to M. Consequently, we have $g_t(s^{a_1}, ..., s^{a_m}) = s^b \in [M]$. Further, there exists a $t' \in T_k \setminus \{t\}$ with $[s^a, s^b]^1 = \{s^x \mid x \in \gamma_{t'}\}$. If $t' = 1$, then $S^1 \subseteq [M]$. If $t' \neq 1$, there is a function $g_{t'} \in [M]$, which does not preserve $\gamma_{t'}$. Analogous to the above, one can form a function $s^c \in S^1 \setminus [s^a, s^b]$ as a superposition over $\{g_{t'}, s^a, s^b\}$, and two cases are possible. The iteration of the construction above shows $S^1 \subseteq [M]$.
Because of Theorem 8.3.2, $[M] = S$ is proven if we can show that $\alpha([M]) = P_k'$ holds. By Theorem 6.1, $\alpha([M]) = P_k'$ is proven if one can show that for every relation $\varrho \in \mathfrak{M}_k \cup \mathfrak{U}_k \cup \mathfrak{S}_k \cup \mathfrak{L}_k \cup \mathfrak{C}_k \cup \mathfrak{B}_k$ a function $f_\varrho \in [M]$ that does not preserve the relation ϱ exists.
Since $\alpha(s^a) = a$ and $s^a \in [M]$ for $a \in E_k$, the constants $0, 1, ..., k-1$ belong to $\alpha([M])$. Further, $[M] \subseteq \alpha([M])$ by Theorem 8.3.2, (b). Because of $\{a, s^a \mid a \in E_k\} \subseteq \alpha([M])$ we have

$$\forall \varrho \in \mathfrak{C}_k^1 \cup \mathfrak{S}_k \cup \mathfrak{M}_k : \ \alpha([M]) \not\subseteq Pol_k\varrho. \tag{18.1}$$

Let $\varrho \in \mathfrak{C}_k^h$ with $2 \le h \le k-1$, let c be a central element of ϱ and $(a_1, ..., a_h) \in E_k^h \setminus \varrho$. Then the function $s^{a_1-c} \in [M]$ does not preserve ϱ because of

$$s^{a_1-c}\begin{pmatrix} c \\ a_2 - a_1 + c \pmod{k} \\ \vdots \\ a_h - a_1 + c \pmod{k} \end{pmatrix} = \begin{pmatrix} a_1 \\ a_2 \\ \vdots \\ a_h \end{pmatrix}.$$

Consequently, we have

$$\forall \varrho \in \bigcup_{h=2}^{k-1} \mathfrak{C}_k^h : \ \alpha([M]) \not\subseteq Pol_k\varrho. \tag{18.2}$$

Next we determine all such relations $\varrho \in \mathfrak{U}_k$ that are preserved from the permutation s. In the following, let ϱ be an arbitrary equivalence relation on E_k and $U := \{x \in E_k \mid (0, x) \in \varrho\}$. Suppose, $s \in Pol_k\varrho$. Then because of $(s^{k-a}(a), s^{k-a}(b)) = (0, b-a)$ and $(s^a(0), s^a(b-a)) = (a, b)$, we have

$$\forall (a,b) \in E_k^2 : (a,b) \in \varrho \iff (0, b-a) \in \varrho \iff b - a \in U. \tag{18.3}$$

With the help of (18.3) and with the properties of equivalence relations, one can easily prove that U is a subgroup of the cyclic group $(E_k; + \pmod{k})$. As is generally known, every subgroup of a cyclic group with k elements can be described with the aid of a divider of k. Hence, there exists a t with $t \in T_k$ and $U = \{x \in E_k \mid t|x\}$. By (18.3), this implies $\varrho = \varrho_t$. Consequently, if $k \in \mathbb{P}$, s preserves only trivial equivalence relations and, if $k \notin \mathbb{P}$, s preserves only equivalence relations, defined in (2). Therefore, by our assumptions of M, we have

$$\forall \varrho \in \mathfrak{U}_k : \alpha([M]) \not\subseteq Pol_k \varrho. \tag{18.4}$$

Because of $|Im(f)| = k$ for all $f \in S$, $S^1 = L_p^1$ and $M \not\subseteq S \cap L_p$ for $p \in \mathbb{P}$ and, if $k \notin \mathbb{P}$, $S^1 = (Pol_k \varrho_t)^1$ and $M \not\subseteq S \cap Pol_k \varrho_t$ for $t \in T_k \setminus \{1, k\}$

$$\alpha([M]) \not\subseteq Pol_k \iota_k^k \tag{18.5}$$

is obviously valid.
Next, let $\varrho \in \mathfrak{B}_k^h$ be arbitrary with $3 \leq h \leq k-1$. Then by definition there is a surjective mapping $q : E_k \longrightarrow E_{h^m}$ with $m \geq 1$ and

$$(a_1, ..., a_h) \in \varrho \iff (q(a_1), ..., q(a_h)) \in \xi_m,$$

where ξ_m is an h-ary elementary relation on E_{h^m} (see Chapter 5). If q is a bijective mapping, then with the aid of Theorem 5.2.6.1, one can prove that s does not preserve the relation ϱ. If q is not bijective, then the mapping equivalence $\sigma := \{(x, y) \in E_k^2 \mid q(x) = q(y)\}$ belongs to $\mathfrak{U}_k$. We have shown that a function f_σ^n, which does not preserve the relation σ, belongs to $\alpha([M]) \cap S$. Therefore, there are tuples $(a_1, b_1), ..., (a_n, b_n) \in \sigma$ with $(c, d) := (f_\sigma(a_1, ..., a_n), f_\sigma(b_1, ..., b_n)) \notin \sigma$. Because of $\iota_h^h \cap E_{h^m}^h \subseteq \xi_m$, we have $\{(a_i, b_i, x_3, ..., x_h) \mid x_3, ..., x_h \in E_k\} \subseteq \varrho$ for all $i = 1, ..., n$. The definition of ϱ and the choice of the elements c and d imply the existence of certain $u_3, ..., u_h \in E_k$ with $(c, d, u_3, ..., u_h) \notin \varrho$. $Im(f_\sigma) = E_k$ holds because of $f_\sigma \in S$. Consequently, there exist certain $(a_i, b_i, c_{i3}, ..., c_{in}) \in \varrho$, $i = 1, ..., n$, with

$$f_\sigma \begin{pmatrix} a_1 & a_2 & \dots & a_n \\ b_1 & b_2 & \dots & b_n \\ c_{13} & c_{23} & \dots & c_{n3} \\ \dots & \dots & \dots & \dots \\ c_{1h} & c_{2h} & \dots & c_{nh} \end{pmatrix} = \begin{pmatrix} c \\ d \\ u_3 \\ \dots \\ u_h \end{pmatrix}.$$

Hence, f_σ does not preserve the relation $\varrho \in \mathfrak{B}_k^h$ and

$$\forall \varrho \in \bigcup_{h=3}^{k-1} \mathfrak{B}_k^h : \alpha([M]) \not\subseteq Pol_k \varrho \tag{18.6}$$

is valid. It still remains to be proven that

$$\forall \lambda \in \mathfrak{L}_k : \alpha([M]) \not\subseteq Pol_k \lambda. \tag{18.7}$$

For this purpose let $\lambda \in \mathfrak{L}_k$. Then there exist $p \in \mathbb{P}$, $m \in \mathbb{N}$ with $k = p^m$ and an elementary Abelean p-group $(E_k; \oplus)$ with the property

$$\lambda := \{(a, b, c, d) \in E_k^4 \mid a \oplus b = c \oplus d\}.$$

W.l.o.g.[4] let 0 be the neutral element of the group $(E_k; \oplus)$.
If $k \in \mathbb{P}$, then either $S^1 \not\subseteq Pol_k \lambda$ or $\lambda = \{(a, b, c, d) \in E_k^4 \mid a + b = c + d \pmod{p}\}$. If $k \in \mathbb{P}$, (18.6) results from that and then (with the aid of (18.1), (18.2), (18.4)–(18.6) and Theorems 8.3.2 and 6.1) our theorem.
Let $k \notin \mathbb{P}$ in the following.

[4] See Section 5.2.4

To prove (18.7), we show first that there is a relation $\gamma \in \mathfrak{U}_k \cup \mathfrak{M}_k \cup \mathfrak{C}_k \cup \mathfrak{B}_k$ with $S \cap Pol_k\lambda \subseteq S \cap Pol_k\gamma$.
It is easy to check that the following three relations are invariants of $S \cap Pol_k\lambda$:

$$\lambda_1 := \{(a_1, ..., a_p) \in E_k^p \mid a_1 \oplus a_2 \oplus ... \oplus a_p = 0\},$$

$$\lambda_2 := \{(a_1, ..., a_p, b_1, ..., b_p) \in E_k^{2\cdot p} \mid a_1 \oplus a_2 \oplus ... \oplus a_p = b_1 \oplus b_2 \oplus ... \oplus b_p\}$$

and, where + denotes the addition modulo k,

$$\varrho' := \{(i, i+1, i+2, ..., i+p-1) \mid i \in E_k\}.$$

For $i \in E_k$ we set:

$$t_i := (i \oplus (i+1) \oplus (i+2) \oplus ... \oplus (i+p-1),$$

where + is the addition modulo k. Because of $k \neq p$ the equations $t_0 = t_1 = ... = t_{k-1}$ are not possible. If there is a j with $t_j = 0$ (this is possible only for $p \neq 2$), then $pr_1(\varrho' \cap \lambda_1) \in \mathfrak{C}_k^1$. If $t_i \neq 0$ for all $i \in E_k$, there are certain r, s with $t_r = t_s$ and $r \neq s$, whereby the relation $\varrho'' := pr_{1,p+1}((\varrho' \times \varrho') \cap \lambda_2)$ is a binary reflexive non-diagonal relation. A conclusion of Chapter 6 is the fact that a relation $\gamma \in \mathfrak{U}_k \cup \mathfrak{M}_k \cup \mathfrak{C}_k \cup \mathfrak{B}_k$ is derivable from ϱ''.
Consequently, we obtain from the assumption $\alpha([M]) \subseteq Pol_k\lambda$ a contradiction to the statements (18.1), (18.2), and (18.4)–(18.6). Our assumption was therefore false, and (18.7) is valid.
Thus, the set $\alpha([M])$ is no subset of an arbitrary maximal class of P_k'. Hence, $\alpha([M]) = P_k'$ is valid. This and Theorem 8.3.2 imply $[M] = S$. ■

18.2 Some Maximal Classes of a Maximal Class of Type $\mathfrak{U}$

Let

$$\varrho$$

be an arbitrary relation of $\mathfrak{U}_k$, where

$$\mathcal{A}_i \ (i \in E_t)$$

denote the equivalence classes of this relation. In addition,

$$a/\varrho$$

denotes the equivalence class of ϱ, which contains $a \in E_k$.
One checks the following lemma easily (see Table 18.1):[5]

[5] See also Lemma 1.4.6.

Lemma 18.2.1 *Let $\mathcal{A}_i$ $(i \in E_t)$ be a partition the set E_k and $a_i \in \mathcal{A}_i$ $(i \in E_t)$. Furthermore let*

$$x \diamond y := \begin{cases} y & \text{for } \exists i \in E_t : x = a_i \text{ and } y \in \mathcal{A}_i, \\ x & \text{otherwise.} \end{cases}$$

Then, an arbitrary function $f^n \in P_k$ is a superposition over the functions z, g_f, f_i $(i \in E_t)$, defined by

$$\begin{aligned} & z(x,y) := x \diamond y, \\ & g_f(x_1,...,x_n) := a_i \iff f(x_1,...,x_n) \in \mathcal{A}_i, \\ & f_i(x_1,...,x_n) := \begin{cases} f(x_1,...,x_n) & \text{for} \quad f(x_1,...,x_n) \in \mathcal{A}_i, \\ a_i & \text{otherwise} \end{cases} \\ & (i \in E_t), \end{aligned}$$

and a representation of f is given by

$$f(\mathbf{x}) = ((...((g_f(\mathbf{x}) \diamond f_0(\mathbf{x})) \diamond f_1(\mathbf{x})) \diamond ...) \diamond f_{t-1}(\mathbf{x})). \tag{18.8}$$

■

Table 18.1

$\mathbf{x}$	$f(\mathbf{x})$	$g_f(\mathbf{x})$	$f_i(\mathbf{x})$	$g_f \diamond f_0$	$(g_f \diamond f_0) \diamond f_1$	...
⋮	$\}\in \mathcal{A}_0$	$\}= a_0$	$\}= a_i$	$\}= f(\mathbf{x})$	$\}= f(\mathbf{x})$	...
⋮	$\}\in \mathcal{A}_1$	$\}= a_1$	$\}= a_i$	$\}= a_1$	$\}= f(\mathbf{x})$	...
⋮	$\}\in \mathcal{A}_2$	$\}= a_2$	$\}= a_i$	$\}= a_2$	$\}= a_2$	...
⋮	⋮	⋮	⋮	⋮	⋮	⋮
⋮	$\}\in \mathcal{A}_i$	$\}= a_i$	$\}= f(\mathbf{x})$	$\}= a_i$	$\}= a_i$	...
⋮	⋮	⋮	⋮	⋮	⋮	⋮
⋮	$\}\in \mathcal{A}_{t-1}$	$\}= a_{t-1}$	$\}= a_i$	$\}= a_{t-1}$	$\}= a_{t-1}$	...

Remark (18.8) is also valid, if one defines

$$x \diamond y := \begin{cases} y & \text{for } \exists i \in E_t : \{x,y\} \subseteq \mathcal{A}_i, \\ x & \text{otherwise} \end{cases} \tag{18.9}$$

and if g_f fulfills the following condition:

$$\forall \mathbf{x} \in E_k^n : (g_f(\mathbf{x}), f(\mathbf{x})) \in \varrho.$$

Let $f^n \in P_k$ be arbitrary and let (18.8) be a representation of f. Then $f \in Pol_k\varrho$ if and only if $g_f \in Pol\ \varrho$, since the functions f_i $(i \in E_t)$ and the function z preserve ϱ. Obviously, every function of the form g_f ($\in Pol\ \varrho$) is unambiguously characterized through its values on $\{a_i \mid i \in E_t\}^n$. Consequently,

$$(\{g_f \mid f \in Pol\ \varrho\}; \zeta, \tau, \Delta, \nabla, \star)$$

is isomorphic to

$$(P_t; \zeta, \tau, \Delta, \nabla, \star).$$

Let q a unary function of P_k defined by

$$\forall i \in E_t : q(x) = a_i \ :\Longleftrightarrow\ x \in \mathcal{A}_i.$$

Lemma 18.2.2 follows from the above considerations:

Lemma 18.2.2 $\forall f^n \in P_k$:

$$f \in Pol\ \varrho \iff \exists\, h^n \in P_{\{a_0 \dots, a_{t-1}\}} : g_f(\mathbf{x}) = h(q(x_1), \dots, q(x_n)).$$

■

Since $[P_k^2] = P_k$ and $[P_{k,A}^2] = P_{k,A}$, the fact

$$ord\ Pol_k \varrho = 2$$

follows from Lemmas 18.2.1 and 18.2.2 (see also proof of Theorem 11.2.2). In analog mode to a corresponding statement for P_k, the following lemma results from that then:

Lemma 18.2.3 *For every proper subclone T of $Pol_k \varrho$ there is a maximal clone of $Pol_k \varrho$, which contains T. $Pol_k \varrho$ has only finite many maximal clones. For every maximal class M of $Pol_k \varrho$, there is a relation $\psi \in R_k^{k^2}$ with $M = Pol_k \psi$.* ■

W.l.o.g. let

$$|\mathcal{A}_0| \geq |\mathcal{A}_1| \geq \dots \geq |\mathcal{A}_{t-1}|$$

and

$$\mathcal{A}_0 := \{0, 1, \dots, l-1\}, \qquad l \geq 2$$

in the following. For the purpose of defining a homomorphism $\alpha_\varphi : P_k \longrightarrow P_t$ we need the mapping φ defined by

$$\varphi : E_k \longrightarrow E_t,\ \forall i \in E_t\ \forall x \in \mathcal{A}_i : \ \varphi(x) := i.$$

Obviously,

$$\alpha_\varphi : P_k \longrightarrow P_t,\ f \mapsto f^{\alpha_\varphi}$$
$$\forall x_1, \dots, x_n \in E_k : \ f^{\alpha_\varphi}(\varphi(x_1), \dots, \varphi(x_n)) := \varphi(f(x_1, \dots, x_n))$$

is a homomorphism.
Next we describe maximal clones of

$$U := Pol_k\varrho$$

which contain all unary functions of U or which have the form $Pol_k\varrho \cap Pol_k\varrho'$, where $Pol_k\varrho'$ is P_k-maximal. Some of these clones can be described in the form

$$Pol_k\beta_i$$

for certain indices i. Put

$$\lambda := \{(a,a,b,b),(a,b,a,b),(a,b,b,a) \mid a,b \in \{0,1\}\},$$

$$\iota_i^h := \{(a_1,...,a_h) \in E_i^h \mid |\{a_1,...,a_h\}| \le h-1\},$$

$$\beta_1 := \begin{cases} \varphi^{-1}(\lambda) & \text{if } t=2, \\ \varphi^{-1}(\iota_t^t) & \text{if } t \ge 3, \end{cases}$$

$$\lambda_\star := \{(a,a,b,b),(a,b,a,b),(a,b,b,a) \mid \exists i : |\mathcal{A}_i| = 2 \ \wedge \ \{a,b\} \subseteq \mathcal{A}_i\} \cup \{(x,x,x,x) \mid \exists j : \ \mathcal{A}_j = \{x\}\},$$

$$\beta_2 := \begin{cases} \lambda_\star & \text{if } l=2, \\ \iota_\star & \text{if } l \ge 3. \end{cases}$$

For the purpose of describing a third type of maximal clones of U, we first define the following relation for $i_1,...,i_r \in \mathbb{N}$:

$$\gamma_{i_1,...,i_r} := \{(x_{1,1},x_{1,2},...,x_{1,i_1},x_{2,1},x_{2,2},...,x_{2,i_2},...,x_{r,1},x_{r,2},...,x_{r,i_r}) \in E_k^{i_1+...+i_r} \mid \forall j \in \{1,...,r\}\ \exists q : \ \{x_{j,1},x_{j,2},...,x_{j,i_j}\} \subseteq \mathcal{A}_q\}.$$

Then we can define the relations $\beta_{i_1,...,i_r}$, as follows:

$$\mathbf{x} := (x_{1,1},x_{1,2},...,x_{1,i_1},x_{2,1},x_{2,2},...,x_{2,i_2},...,x_{r,1},x_{r,2},...,x_{r,i_r}) \in \beta_{i_1,...,i_r}$$

$$:\Longleftrightarrow$$

$$\mathbf{x} \in \gamma_{i_1,...,i_r} \ \wedge \ (\exists s \in \{i_1,...,i_r\} : |\{x_{s,1},x_{s,2},...,x_{s,i_j}\}| \le i_j - 1).$$

For the case $|\mathcal{A}_2| = ... = |\mathcal{A}_t| = 1$ we need the following relations:

$$\sigma_{r,s} := \{(x_1,...,x_r,y_1,...,y_s) \in E_k^{r+s} \mid x_1 = ... = x_r \ \vee \ (x_1,...,x_r) \in \iota_l^r \ \vee \ (\{x_1,...,x_r\} = E_l \Longrightarrow |\{x_1,...,x_r,y_1,...,y_s\}| \le r+s-1\}$$

$(2 \le r \le l,\ 1 \le s \le k-l)$.

Lemma 18.2.4 *Let $\emptyset \subset \sigma \subset E_k$. Then $Pol_k\sigma \cap Pol_k\varrho$ is a maximal clone of $U := Pol_k\varrho$ if and only if the following condition is valid:*

$$(\exists I \subset \{0,1,...,t-1\} : \ \sigma = \bigcup_{i\in I} \mathcal{A}_i) \ \vee \ (\forall j \in \{0,1,...,t-1\} : \sigma \cap \mathcal{A}_j \ne \emptyset). \tag{18.10}$$

Proof. "$\Longrightarrow$": Let $Pol_k\sigma \cap Pol_k\varrho$ be U-maximal and let

$$\sigma' := \{y \in E_k \mid \exists x \in E_k : \ (x, x, y) \in \sigma \times \varrho\}.$$

The following two cases are possible:
Case 1: $\emptyset \subset \sigma' \subset E_k$.
In this case, we have

$$U \cap Pol_k\sigma \subseteq U \cap Pol_k\sigma' \subset U$$

and σ' has the form

$$\exists I \subseteq \{1, 2, ..., t\} : \ \sigma = \bigcup_{i \in I} \mathcal{A}_i,$$

since

$$a \in \sigma' \implies (\forall b \in a/\varrho : b \in \sigma')$$

holds ($a/\varrho := \{x \in E_k \mid (a, x) \in \varrho\}$).
Case 2: $\sigma' = E_k$.
(Such a case occurs, for example, if $k = 3$, $\varrho := \{(0,0), (1,1), (2,2), (0,1), (1,0)\}$ and $\sigma := \{0, 2\}$.)
It is easy to check that σ fulfills the condition

$$\forall j \in \{0, 1, ..., t-1\} : \sigma \cap \mathcal{A}_j \neq \emptyset$$

in this case.
"$\Longleftarrow$": Let $f^n \in U \backslash Pol_k\sigma$, where σ has the property (18.10). Then there are $a_1, ..., a_n \in \sigma$ with $f(a_1, ..., a_n) = a \in E_k \backslash \sigma$. Since the constant c_s with $s \in \sigma$ belongs to $U \cap Pol_k\sigma$, we have $c_a \in [(U \cap Pol_k\sigma) \cup \{f\}]$.
Let now $g^m \in U$ be arbitrary. We show that $g \in [(U \cap Pol_k\sigma) \cup \{f\}]$. For this purpose, we consider a function $h_g^{m+1} \in U$ with $h_g(x_1, ..., x_m, a) = g(x_1, ..., x_m)$. Our assertion is proven, if we can show that there is a such function h_g in the set $U \cap Pol_k\sigma$. If $\sigma = \bigcup_{i \in I} \mathcal{A}_i$ for a certain $I \subset E_t$, we set

$$h_g(x_1, ..., x_{m+1}) := \begin{cases} g(x_1, ..., x_m), & \text{if} \quad x_{m+1} \in a/\varrho, \\ s & \text{otherwise,} \end{cases}$$

where $s \in \sigma$. Obviously, $h_g \in U \cap Pol_k\sigma$.
Let $\sigma \cap \mathcal{A}_j \neq \emptyset$ for all $j \in E_t$. Then one must choose $h_g(a'_1, ..., a'_m, a'_{m+1}) \in (b/\varrho) \cap \sigma$, if $h_g(a_1, ..., a_m, a_{m+1}) = b$ was chosen and $(a_i, a'_i) \in \varrho$ for every $i \in \{1, ..., m{+}1\}$ is valid. For the remaining tuples $\mathbf{x}$, one can set e.g. $h_g(\mathbf{x}) = s$, where $s \in \sigma$, so that $h_g \in U \cap Pol_k\sigma$. ■

Lemma 18.2.5

(1) Let $\gamma \subseteq E_t^h$ and let $Pol_t\gamma$ P_t-maximal. Furthermore, let

$$\alpha_\varphi^{-1}(\gamma) := \{(a_1, ..., a_h) \in E_k^h \mid (\alpha_\varphi(a_1), ..., \alpha_\varphi(a_h)) \in \gamma\}.$$

Then $Pol_k\alpha_\varphi^{-1}(\gamma)$ is U-maximal.

(2) $Pol_k \beta_1$ *is* U*-maximal.*

(3) Let A *be a subset of* U *with* $U^1 \subseteq A$ *and* $A \not\subseteq Pol_k \beta_1$. *Then*

$$\forall f \in P_t \; \exists g \in [A] : \; \alpha_\varphi(g) = f. \tag{18.11}$$

Proof. (1): Let $f \in U \backslash Pol_k \alpha_\varphi^{-1}(\gamma)$ be arbitrary. Then we have obvious $[\alpha_\varphi(Pol_k \alpha_\varphi^{-1}(\gamma) \cup \{f\})] = P_t$. Consequently,

$$\forall h \in P_t \; \exists g \in Pol_k \alpha_\varphi^{-1}(\gamma) : \; \alpha_\varphi(g) = h. \tag{18.12}$$

For arbitrary functions $q \in U$ it follows from the definition of $Pol_k \alpha_\varphi^{-1}(\gamma)$:

$$q \in Pol_k \alpha_\varphi^{-1}(\gamma) \iff \alpha_\varphi(q) \in Pol_t \gamma. \tag{18.13}$$

Case 1: $\{c_0, c_1, ..., c_{t-1}\} \subset Pol_t \gamma$.
With the help of (18.13) and the fact that the constant functions $c_0, ..., c_{t-1}$ of P_t belong to $Pol_t \gamma$, one gets:

$$\{\diamond\} \cup \bigcup_{i=0}^{t-1} P_{k,\mathcal{A}_i} \subseteq Pol_k \alpha_\varphi^{-1}(\gamma). \tag{18.14}$$

Then (with the help of Lemma 18.2.1), (18.12), and (18.14) imply $[Pol_k \alpha_\varphi^{-1}(\gamma) \cup \{f\}] = U$, i.e., $Pol_k \alpha_\varphi^{-1}(\gamma)$ is U-maximal.
Case 2: $\gamma \subset E_t$.
In this case, our assertion follows from Lemma 18.2.4.
Case 3: $\gamma \in \mathfrak{S}_t$.
Because of (18.12), one can find a unary function o with $\alpha_\varphi(o) = c_0$ in $[\alpha_\varphi(Pol_k \alpha_\varphi^{-1}(\gamma) \cup \{f\})]$. Since a unary function p with

$$\forall x \in \mathcal{A}_0 : \; p(x) = 0$$

belongs to $Pol_k \alpha_\varphi^{-1}(\gamma)$, the constant $c_0 = p \star o$ is a superposition over $Pol_k \alpha_\varphi^{-1}(\gamma) \cup \{f\}$. Let $g \in U^m$ be arbitrary. Then there is an $(m+1)$-ary function h_g with

$$h_g(0, x_1, ..., x_m) = g(x_1, ..., x_m)$$

in $Pol_k \alpha_\varphi^{-1}(\gamma)$, whereby $g = h_g \star c_0 \in [Pol_k \alpha_\varphi^{-1}(\gamma) \cup \{f\}]$ holds. Consequently, $Pol_k \alpha_\varphi^{-1}(\gamma)$ is U-maximal.

(2) is a special case of (1).

(3): A conclusion of Rosenberg's Theorem is

$$[\alpha_\varphi(A)] = P_t,$$

i.e., (18.11) holds. ∎

Lemma 18.2.6 *For some $i \in E_t$ let $|\mathcal{A}_i| \geq 2$. Furthermore, let $\gamma \subseteq \mathcal{A}_i^h$ be a relation that describes a maximal clone $Pol_{\mathcal{A}_i}\gamma$ in $P_{\mathcal{A}_i}$, where all constant functions of $P_{\mathcal{A}_i}$ belong to $Pol_{\mathcal{A}_i}\gamma$. Set*

$$\gamma' := \gamma \cup \bigcup_{j=1,\ j\neq i}^{t-1} \mathcal{A}_j^h.$$

Then the clone $Pol_k\gamma'$ is U-maximal.

Proof. W.l.o.g. let $A_i := E_r$ with $r \geq 2$ and

$$\gamma \in \mathfrak{U}_r \cup \mathfrak{M}_r \cup \mathfrak{L}_r \cup \mathfrak{B}_r \cup \bigcup_{h=2}^{r-1} \mathfrak{C}_r^h.$$

Obviously, $Pol_k\gamma' \subset U$ and

$$\{\diamond, c_0, c_1, ..., c_{k-1}\} \cup U_0 \cup \bigcup_{j=1,\ j\neq i}^{t-1} P_{k,\mathcal{A}_j} \subseteq Pol_k\gamma'.$$

Elements of $Pol_k\gamma'$ are also all n-ary functions q_g with

$$q_g(\mathbf{x}) := \begin{cases} g(\mathbf{x}) & \text{if} \qquad \mathbf{x} \in E_r^n, \\ 0 & \text{otherwise,} \end{cases}$$

for arbitrary $g \in Pol_r\gamma$. Moreover, all unary functions $p_{d_0,...,d_{r-1}}$ defined by

$$p_{d_0,...,d_{r-1}}(x) := d_j \iff x \in \mathcal{A}_j$$

$(j = 0, 1, ..., t-1)$ belong to $Pol_k\gamma'$ for all $d_0, d_1, ..., d_{t-1} \in \mathcal{A}_i = E_r$.
Let $f \in U \backslash Pol_k\gamma'$ be arbitrary. To prove $[\{f\} \cup Pol_k\gamma'] = U$, we have only to show that $P_{k,r} \subseteq [\{f\} \cup Pol_k\gamma']$ is valid because of Lemma 18.2.2 and because of the above considerations. Obviously, there are certain $(a_{1j}, a_{2j}, ..., a_{hj}) \in \gamma$ for $j = 1, ..., m$ and certain $(b_{1l}, b_{2l}, ..., b_{hl}) \in \gamma' \backslash \gamma$ for $l = 1, ...n$ with

$$f\begin{pmatrix} a_{11} & a_{12} & ... & a_{1m} & b_{11} & b_{12} & ... & b_{1n} \\ a_{21} & a_{22} & ... & a_{2m} & b_{21} & b_{22} & ... & b_{2n} \\ \dots & & & & & & & \dots \\ a_{h1} & a_{h2} & ... & a_{hm} & b_{h1} & b_{h2} & ... & b_{hn} \end{pmatrix} = \begin{pmatrix} \alpha_1 \\ \alpha_2 \\ ... \\ \alpha_h \end{pmatrix} \notin E_r^h \backslash \gamma.$$

By choosing certain m-ary functions $f_1..., f_n$ from $\bigcup_{j=1,\ j\neq i}^{t-1} P_{k,\mathcal{A}_j}$ one receives the m-ary function $f' \in P_{k,r}$ by means of

$$f'(x_1, ..., x_m) := q_{e_1^1}^1(f(x_1, ..., x_m, f_1(x_1, ..., x_m), ..., f_n(x_1, ..., x_m))),$$

where

$$f'\begin{pmatrix} a_{11} & a_{12} & \dots & a_{1m} \\ a_{21} & a_{22} & \dots & a_{2m} \\ \dots & \dots & \dots & \dots \\ a_{h1} & a_{h2} & \dots & a_{hm} \end{pmatrix} = \begin{pmatrix} \alpha_1 \\ \alpha_2 \\ \dots \\ \alpha_h \end{pmatrix} \notin E_r^h \backslash \gamma$$

holds, as a superposition over $Pol_k\gamma' \cup \{f\}$. With the help of the completeness criterion for $P_{k,r}$ (see Theorem 12.4.3 or the proof of the following lemma) one can easy prove that $\{f'\} \cup \{q_g \mid g \in P_r\} \cup \{p_{d_0,...,d_{t-1}} \mid d_0, ..., d_{r-1} \in E_r\}$ is a generating system for $P_{k,r}$, whereby $P_{k,r} \subseteq [\{f\} \cup Pol_k\gamma']$ was shown. With that, a generating system for the clone U was proven in $\{f\} \cup Pol_k\gamma'$. Therefore, $Pol_k\gamma'$ is U-maximal. ∎

Lemma 18.2.7

(1) $Pol_k\beta_2$ is U-maximal.
(2) Let A be a subset of U with $U^1 \subseteq A$ and $A \not\subseteq Pol_k\beta_2$. Furthermore, A fulfills the condition (18.11). Then

$$\forall i \in \{0, 1, ..., t-1\} : \ P_{k,\mathcal{A}_i} \subseteq [A]. \tag{18.15}$$

Proof. (1): For proof, we use the completeness criterion for $P_{k,s}$ from Chapter 12:

Let $pr : P_{k,s} \longrightarrow P_s$ be defined by

$$pr(f^n) = g^m \ :\Longleftrightarrow\ n = m \wedge \forall \mathbf{x} \in E_s^n : f(\mathbf{x}) = g(\mathbf{x}).$$

With the aid of the mapping pr, one can describe the following maximal classes of $P_{k,s}$:

$$pr^{-1}M := \{f \in P_{k,s} \mid pr(f) \in M\},$$

where M is an arbitrary maximal clone of P_s. Furthermore, let

$$\zeta_{i,t} := \{(x, x) \mid x \in E_s\} \cup \{(i, t)\}$$

for all i, t with $i < t \le k-1$ and $s \le t$. Then, for all $A \subseteq P_{k,s}$ is valid:

$[A] = P_{k,s} \iff$ *A is no subset of every set of the form $pr^{-1}M$ (where M is a maximal clone of P_s) and A does not preserve every relation of the form $\zeta_{i,t}$.*

(This means that the set $P_{k,s}^1 \cup \{f\}$, where $f \in P_{k,s}$ and $pr(f) \notin L$ for $s = 2$ and $pr(f) \notin Pol_s\iota_s^s$ for $s \ge 3$, is complete in $P_{k,s}$.)
Obviously, $U^1 \subset Pol_k\beta_2$.
Let $f \in U\backslash Pol_k\beta_2$ be arbitrary. Then a function $g \in P_{k,l}$ with $pr(g) \notin L$ for $l = 2$ and with $pr(g) \notin Pol_l\iota_l^l$ for $l \ge 3$ is a superposition over $U^1 \cup \{f\}$. Since $U^1 \subseteq Pol_k\beta_2$ in addition, we have

$$P_{k,l} \subseteq [Pol_k\beta_2 \cup \{f\}]$$

because of the completeness criterion, and also

$$\forall i \in E_t : P_{k,\mathcal{A}_i} \subseteq [Pol_k\beta_2 \cup \{f\}].$$

It is easy to check that the function $\diamond$ defined in the remark on Lemma 18.2.1 preserves the relation β_2. Moreover, we have

$$\forall h \in P_t \; \exists g \in Pol_k\beta_2 : \; \alpha_\varphi(g) = h.$$

Consequently, the clone $[Pol_k\beta_2 \cup \{f\}]$ contains a generating system for U, whereby (1) is proven.

(2) follows from the proof for (1). ■

Lemma 18.2.8 *Let $|\mathcal{A}_1| = ... = |\mathcal{A}_{t-1}| = 1$ (i.e., $\mathcal{A}_0 = E_l$ is the only equivalence class of ϱ that contains at least two elements). Furthermore, let*

$$U_0 := \{f \in U \mid |Im(f) \cap E_l| \leq 1\},$$
$$U_{1,\alpha} := \{f \in U \mid Im(f) \subseteq E_l \cup \{\alpha\}\},$$
$$U_1 := \bigcup_{\alpha \in E_k \setminus E_l} U_{1,\alpha}.$$

Then

(a) $\forall f \in U \backslash Pol_k\sigma_{l,1} : \; U_1 \subseteq [\{f\} \cup U^1 \cup P_{k,l} \cup U_0]$;

(b) $Pol_k\sigma_{l,1}$ is U-maximal.

Proof. The definition of $\sigma_{r,s}$, which one can find before Lemma 18.2.4, implies

$$\sigma_{l,1} = \{(x_1, ..., x_l, y) \in E_k^{l+1} \mid x_1 = ... = x_l \;\vee\; \{x_1, ..., x_{l+1}\} \subseteq E_l \;\vee\; (\{x_1, ..., x_l\} \in \iota_l^l\}$$

(a): Let $f \in U \backslash Pol_k\sigma_{l,1}$ be arbitrary and

$$M := \{f\} \cup U^1 \cup P_{k,l} \cup U_0.$$

Then, an n-ary function f_1 with the property

$$f_1 \begin{pmatrix} 0 & 0 & a_{13} & a_{14} & \cdots & a_{1n} \\ 0 & 1 & a_{23} & a_{24} & \cdots & a_{2n} \\ 0 & 2 & a_{33} & a_{34} & \cdots & a_{3n} \\ \cdots & \cdots & \cdots & \cdots & \cdots & \cdots \\ 0 & l-1 & a_{l3} & a_{l4} & \cdots & a_{ln} \\ l & 0 & b_3 & b_4 & \cdots & b_n \end{pmatrix} = \begin{pmatrix} 0 \\ 1 \\ 2 \\ \cdots \\ l-1 \\ l \end{pmatrix},$$

where $(a_{1i}, a_{2i}, ..., a_{li})$ $(i = 3, 4, ..., n)$ are certain tuples of ι_l^l and $b_3, ..., b_n$ are certain elements of E_k, is a superposition over M. The above property of f_1 and $f_1 \in U$ implies that f_1 has only values of E_l on tuples of E_l^n.

Next we consider the function

$$f_1' := (f_1 \star c_0)_{|E_l} \in P_l,$$

that is, we consider the restriction of $f_1(c_0(x_1), x_2..., x_n)$ on $(x_1..., x_n) \in E_l^n$. Then the following cases are possible:

Case 1: x_2 is the only essential variable of f_1'.

In this case, we have

$$f_1 \begin{pmatrix} 0 & 0 & 0 & 0 & \cdots & 0 \\ 0 & 1 & 0 & 0 & \cdots & 0 \\ 0 & 2 & 0 & 0 & \cdots & 0 \\ \cdots & \cdots & \cdots & \cdots & \cdots & \cdots \\ 0 & l-1 & 0 & 0 & \cdots & 0 \\ l & 0 & b_3 & b_4 & \cdots & b_n \end{pmatrix} = \begin{pmatrix} 0 \\ 1 \\ 2 \\ \cdots \\ l-1 \\ l \end{pmatrix},$$

and for every $\alpha \in E_k \backslash E_l$, a certain binary function f_α with the property

$$f_\alpha \begin{pmatrix} 0 & 0 \\ 0 & 1 \\ 0 & 2 \\ \cdots & \cdots \\ 0 & l-1 \\ l & 0 \end{pmatrix} = \begin{pmatrix} 0 \\ 1 \\ 2 \\ \cdots \\ l-1 \\ \alpha \end{pmatrix}$$

is a superposition over M.
Let g be an arbitrary m-ary function of $U_{1,\alpha}$ with $\alpha \in E_k \backslash E_l$. The m-ary functions $g_1 \in U_0$ and $g_2 \in P_{k,l}$ defined by

$$g_1(\mathbf{x}) := \begin{cases} 0, & \text{if} \quad g(\mathbf{x}) \in E_l, \\ l & \text{otherwise}, \end{cases}$$

and

$$g_2(\mathbf{x}) := \begin{cases} = f(\mathbf{x}), & \text{if} \quad g(\mathbf{x}) \in E_l, \\ 0 & \text{otherwise}, \end{cases}$$

belong to M, whereby $g \in [M]$ holds because of

$$g(\mathbf{x}) = f_\alpha(g_1(\mathbf{x}), g_2(\mathbf{x})).$$

Consequently, we have $U_{1,\alpha} \subseteq [M]$ and (since $U^1 \subseteq M$) also $U_1 \subseteq [M]$, i.e., our assertion (a) is proven in Case 1.

Case 2: x_2 and x_i for certain $i \in \{3, .., n\}$ are essential variables of f_1'.

In this case, the Fundamental Lemma of Jablonskij (see Theorem 1.4.4) implies the existence of some $c_i, d_i, e_{ji} \in E_l$ $(i = 2, ..., n;\ j = 4, ..., l)$ with

$$f_1\begin{pmatrix} 0 & c_2 & c_3 & c_4 & \cdots & c_n \\ 0 & c_2 & d_3 & d_4 & \cdots & d_n \\ 0 & d_2 & d_3 & d_4 & \cdots & d_n \\ 0 & e_{42} & e_{43} & e_{44} & \cdots & e_{4n} \\ \multicolumn{6}{c}{\dotfill} \\ 0 & e_{l2} & e_{l3} & e_{l4} & \cdots & e_{ln} \\ l & 0 & b_3 & b_4 & \cdots & b_n \end{pmatrix} = \begin{pmatrix} \alpha_0 \\ \alpha_1 \\ \alpha_2 \\ \cdots \\ \alpha_{l-1} \\ l \end{pmatrix}$$

and $\{\alpha_0, ..., \alpha_{l-1}\} = E_l$.
First we show that

$$U_{\alpha_1,\alpha_2,l} := \{f \in U \mid Im(f) \subseteq \{\alpha_1, \alpha_2, l\}\ \} \subseteq [M] \tag{18.16}$$

holds. For this purpose, we use the property

$$f_1\begin{pmatrix} 0 & c_2 & d_3 & d_4 & \cdots & d_n \\ 0 & d_2 & d_3 & d_4 & \cdots & d_n \\ l & 0 & b_3 & b_4 & \cdots & b_n \end{pmatrix} = \begin{pmatrix} \alpha_1 \\ \alpha_2 \\ l \end{pmatrix}.$$

Let h be an arbitrary m-ary function of $U_{\alpha_1,\alpha_2,l}$. Then the m-ary functions $h_1, h_2, ..., h_n$ defined by

$$h_1(\mathbf{x}) := \begin{cases} 0, & \text{if} \quad h(\mathbf{x}) = \alpha_1, \\ l & \text{otherwise} \end{cases} \quad (\in U_0),$$

$$h_2(\mathbf{x}) := \begin{cases} c_2, & \text{if} \quad h(\mathbf{x}) = \alpha_1, \\ d_2, & \text{if} \quad h(\mathbf{x}) = \alpha_2, \\ 0 & \text{otherwise} \end{cases} \quad (\in P_{k,l}),$$

$$h_i(\mathbf{x}) := \begin{cases} d_i, & \text{if} \quad h(\mathbf{x}) \in \{\alpha_1, \alpha_2\} \\ b_i & \text{otherwise} \end{cases} \quad (\in U_0),$$

$(i = 3, ..., n)$, belong to $[M]$, and we have:

$$h(\mathbf{x}) = f_1(h_1(\mathbf{x}), h_2(\mathbf{x}), ..., h_n(\mathbf{x})),$$

whereby (18.16) is proven.
Since $U^1 \subseteq M$, (18.16) implies

$$\bigcup_{a,b\in E_l,\ c\in E_k\setminus E_l} \{f \in U \mid Im(f) \subseteq \{a, b, c\}\ \} \subseteq [M].$$

Then, analogously to the above and with the aid of

$$f_1\begin{pmatrix} 0 & c_2 & c_3 & c_4 & \cdots & c_n \\ 0 & c_2 & d_3 & d_4 & \cdots & d_n \\ 0 & d_2 & d_3 & d_4 & \cdots & d_n \\ l & 0 & b_3 & b_4 & \cdots & b_n \end{pmatrix} = \begin{pmatrix} \alpha_0 \\ \alpha_1 \\ \alpha_2 \\ l \end{pmatrix},$$

one can show that

$$\{f \in U \mid Im(f) \subseteq \{\alpha_0, \alpha_1, \alpha_2, l\}\ \} \subseteq [M]$$

is valid. Consequently, because of $U^1 \subset M$, we have

$$\bigcup_{\beta_0,\beta_1,\beta_2 \in E_l,\ \gamma \in E_k \backslash E_l} \{f \in U \mid Im(f) \subseteq \{\beta_0, \beta_1, \beta_2, \gamma\}\ \} \subseteq [M].$$

Then (a) results by means of induction from what was shown till now.

(b): It is easy to see that the binary function z defined by

$$z(x,y) := \begin{cases} x \text{ if } & x \in E_l, \\ y \text{ otherwise}, & \end{cases}$$

preserves the relation $\sigma_{l,1}$. Moreover, we obviously have:

$$U^1 \cup P_{k,l} \cup U_0 \subseteq Pol_k \sigma_{l,1}. \tag{18.17}$$

Let $f \in U \backslash Pol_k \sigma_{l,1}$ be arbitrary. Then, because of (a) and (18.17),

$$U_1 \subseteq [\{f\} \cup Pol_k \sigma_{l,1}]$$

holds. Let q be an arbitrary m-ary function of U. Then, one can form q as a superposition over z and the m-ary functions q_1, q_2 defined by

$$q_1(\mathbf{x}) := \begin{cases} q(\mathbf{x}), \text{ if } & q(\mathbf{x}) \in E_l, \\ l \quad \text{ otherwise} & \end{cases} \quad (\in U_1),$$

and

$$q_2(\mathbf{x}) := \begin{cases} 0, \quad \text{ if } & q(\mathbf{x}) \in E_l, \\ q(\mathbf{x}) \text{ otherwise} & \end{cases} \quad (\in U_0),$$

as follows:

$$q(\mathbf{x}) = z(q_1(\mathbf{x}), q_2(\mathbf{x})).$$

From that and from what was shown till now $[\{f\} \cup Pol_k \sigma_{l,1}] = U$, through which (b) is proven. ■

Theorem 18.2.9 *Let $k = l + 1$. Then U has exactly three maximal clones that have U^1 as a subset. These clones are*

$$Pol_k \beta_1,\ Pol_k \beta_2,\ Pol_k \sigma_{l,1}. \tag{18.18}$$

Proof. The U-maximality of the given clones was proven in Lemmas 18.2.5, 18.2.7, and 18.2.8.
Let A be an arbitrary subset of U which has U^1 as a subset and which is no subset of the clones given in (18.18). To prove our theorem, we have to show that $[A] = U$.
Lemmas 18.2.5 and 18.2.7 and $U^1 \subseteq U$ imply

$$U_0 \cup P_{k,l} \subseteq A.$$

Consequently, with the aid of Lemma 18.2.8, (a), we have that $U_1 = U$ is a subset of $[A]$. ■

18.3 The Maximal Classes of $Pol_k(E^2_{k\text{-}1} \cup \{(k\text{-}1, k\text{-}1)\})$

In this section let $k \geq 3$,

$$\varrho := E^2_{k-1} \cup \{(k-1, k-1)\}$$

and

$$U := Pol_k \varrho.$$

The sets

$$U_0 := \{f \in U \mid k-1 \in Im(f) \ \wedge \ |Im(f)| \leq 2\}$$

and

$$U_{a,b,k-1} := \{f \in U \mid Im(f) \subseteq \{a, b, k-1\}\ \}.$$

are subsets of U.

18.3.1 Definitions of the U-Maximal Classes

For $k = 3$, the clone U has exactly 13 maximal classes, which were given in Chapter 13, Theorem 13.1.8. In generalization of these classes (for $a = 0$ and $b = 1$), we will define six types of classes from which in the following two sections it is shown that they are the only maximal classes of the clone U.

Type I (homomorphic inverse images of the maximal classes of P_2): With the help of the mapping

$$\varphi : E_k \longrightarrow E_2, \ \forall x \in E_{k-1} : \ \varphi(x) := 0, \ \varphi(k-1) := 1.$$

one can define a **homomorphic inverse image** $\alpha_\varphi^{-1}(\gamma)$ of a relation $\gamma \subseteq E_2^h$ as follows:

$$\alpha_\varphi^{-1}(\gamma) := \{(x_1, ..., x_h) \in E_k^h \mid (\varphi(x_1), \varphi(x_2), ..., \varphi(x_h)) \in \gamma\}.$$

In the case that γ describes a maximal class of P_2, we obtain the following maximal classes of U:

$$U \cap Pol_k \ \alpha_\varphi^{-1}(\{0\}),$$

$$U \cap Pol_k \ \alpha_\varphi^{-1}(\{1\}),$$

$$U \cap Pol_k \ \alpha_\varphi^{-1}\begin{pmatrix} 0 & 1 \\ 1 & 0 \end{pmatrix},$$

$$Pol_k \ \alpha_\varphi^{-1}\begin{pmatrix} 0 & 0 & 1 \\ 0 & 1 & 1 \end{pmatrix},$$

$$Pol_k \ \alpha_\varphi^{-1}\begin{pmatrix} 0 & 0 & 1 & 1 & 0 & 1 & 0 & 1 \\ 0 & 1 & 1 & 0 & 1 & 0 & 0 & 1 \\ 1 & 0 & 0 & 1 & 1 & 0 & 0 & 1 \\ 1 & 1 & 0 & 0 & 0 & 1 & 0 & 1 \end{pmatrix}.$$

For $k = 3$, these are the classes with the numbers (2), (1), (8), (9), and (13) of Theorem 13.1.8.

Type II (*U*-maximal classes described by certain unary relations): Let σ be a subset of E_k, which has one of the following two properties:

$$1)\quad \sigma = E_{k-1},$$
$$2)\quad \{k-1\} \subseteq \sigma \subset E_k.$$

Then $U \cap Pol_k\sigma$ is maximal in U (see Lemma 18.2.4).

For $k = 3$ there are exactly four maximal clones that are describable in this manner. Two of these classes were already recorded by means of type I.

Type III (*U*-maximal classes that are determined by the maximal classes of P_{k-1}; first possibility): Let $\gamma \subseteq E_{k-1}^h$, $2 \le h \le k-1$, be an h-ary relation, which describes by means of $Pol_{k-1}\gamma$ a maximal class of P_{k-1} and which contains all constant functions of P_{k-1}. Then

$$Pol_k(\gamma \cup \{(k-1, k-1, ..., k-1)\})$$

is U-maximal.

For $k = 3$ there are exactly two U-maximal clones of type III:

$$Pol_3 \begin{pmatrix} 0 & 1 & 2 & 0 \\ 0 & 1 & 2 & 1 \end{pmatrix},$$

$$Pol_3 \begin{pmatrix} 0 & 0 & 0 & 1 & 1 & 0 & 1 & 1 & 2 \\ 0 & 0 & 1 & 1 & 0 & 1 & 0 & 1 & 2 \\ 0 & 1 & 0 & 0 & 1 & 1 & 0 & 1 & 2 \\ 0 & 1 & 1 & 0 & 0 & 0 & 1 & 1 & 2 \end{pmatrix}.$$

The U-maximality of the classes of type III follows from Lemma 18.2.6.

Type IV (*U*-maximal classes that are determined by the maximal classes of P_{k-1}; second possibility): This type of U-maximal class occurs only for $k \ge 4$, since for $k = 3$ the conditions (a) and (b) mentioned below cannot be met. Denote $\gamma \subseteq E_{k-1}^h$, $2 \le h \le k-1$, an h-ary relation with the following properties:

(a) $Pol_{k-1}\gamma$ is a maximal class of P_{k-1};

(b) γ is totally reflexive and totally symmetric,

i.e., we can assume

$$\gamma \in \mathfrak{U}_{k-1} \cup \bigcup_{h=2}^{k-2} \mathfrak{C}_{k-1}^h \cup \bigcup_{h=3}^{k-1} \mathfrak{B}_{k-1}^h.$$

Moreover, let

$$\begin{aligned}\gamma^\star &:= E_{k-1}^{h+1} \cup \{(x, x, ..., x, y) \mid x, y \in E_k\} \\ &\quad \cup \{(x_1, ..., x_h, k-1) \mid (x_1, ..., x_h) \in \gamma\}.\end{aligned}$$

Then $Pol_k\gamma^\star$ is U-maximal by Lemma 18.3.2.4.

Example Let $k = 4$ and

$$\gamma := \begin{pmatrix} 0 & 1 & 2 & 3 & 0 & 3 \\ 0 & 1 & 2 & 3 & 3 & 0 \end{pmatrix}.$$

It is well-known that $Pol_3\gamma$ is a maximal class of P_3 and that γ is a reflexive and symmetric relation, whereby the relation

$$\gamma^\star := E_3^3 \cup \{(x, x, y) \mid x, y \in E_4\} \cup \begin{pmatrix} 0 & 1 & 2 & 3 & 0 & 3 \\ 0 & 1 & 2 & 3 & 3 & 0 \\ 3 & 3 & 3 & 3 & 3 & 3 \end{pmatrix}$$

describes a maximal class of U.

Type V (U-maximal classes that are described by certain binary central relations): Denote τ a binary central relation $\subseteq E_{k-1}^2$ (i.e., a binary reflexive and symmetric relation, which has at least a central element $c \in E_{k-1}$ with $\{(c, x) \mid x \in E_{k-1}\} \subseteq \tau$ and which is different from E_{k-1}^2.)
If τ fulfills the two following conditions

(a) $\varrho \subseteq \tau$,

(b) $\exists c \in E_{k-1} : c$ is central element of τ,

then $U \cap Pol_k\tau$ is U-maximal by Lemma 18.3.2.9.

Examples For $k = 3$ there are exactly two maximal classes of this type:

$$U \cap Pol_3 \begin{pmatrix} 0 & 1 & 2 & 0 & 1 & 0 & 2 \\ 0 & 1 & 2 & 1 & 0 & 2 & 0 \end{pmatrix}$$

$$U \cap Pol_3 \begin{pmatrix} 0 & 1 & 2 & 1 & 0 & 1 & 2 \\ 0 & 1 & 2 & 0 & 1 & 2 & 1 \end{pmatrix}.$$

Type VI (Two U-maximal classes that are described by ternary relations): For $A \in \{E_{k-1}, \{k-1\}\}$ let

$$\sigma_A := (E_{k-1}^2 \times A) \cup \{(x, x, y) \mid x, y \in E_k\}.$$

The class $Pol_k\sigma_{E_{k-1}}$ is U-maximal by Lemma 18.3.2.5 and the class $Pol_k\sigma_{\{k-1\}}$ is U-maximal by Lemma 18.3.2.7.

Example For $k = 3$ these classes have the number (10) and (11) in Theorem 13.3.8.

18.3.2 Proof of the U-Maximality of the Classes Defined in 18.3.1

The U-maximality of a class of the type I, II or III follows from Lemmas 18.2.4–18.2.6.

Lemma 18.3.2.1 *Let M be a subset of U which has the two following properties:*

(1) The set

$$\{f^n \in P_{k-1} \mid \exists g^n \in M \cap Pol_k\{k-1\} \cap Pol_k E_{k-1} : \forall \mathbf{x} \in E_{k-1}^n : \; g(\mathbf{x}) = f(\mathbf{x})\}$$

is complete in P_{k-1}.

(2) There are two different elements $a, b \in E_{k-1}$ with

$$U_{a,b,k-1} := \{f \in U \mid Im(f) \subseteq \{a, b, k-1\}\} \subseteq [M].$$

Then M is complete in U.

Proof. Because of (1) a $(k-1)$-ary function p with the property

$$p\begin{pmatrix} b & a & a & \dots & a & a & a \\ a & b & a & \dots & a & a & a \\ a & a & b & \dots & a & a & a \\ \dots & \dots & \dots & \dots & \dots & \dots & \dots \\ a & a & a & \dots & b & a & a \\ a & a & a & \dots & a & b & a \\ a & a & a & \dots & a & a & b \\ k-1 & k-1 & k-1 & \dots & k-1 & k-1 & k-1 \end{pmatrix} = \begin{pmatrix} 0 \\ 1 \\ 2 \\ \dots \\ k-4 \\ k-3 \\ k-2 \\ k-1 \end{pmatrix}$$

belongs to $[M]$. Then one can form an arbitrary n-ary function $u \in U$ as a superposition over $\{p\} \cup U_{a,b,k-1} \subseteq [M]$ as follows:
The n-ary functions $u_0, u_1, ..., u_{k-2}$ defined by

$$u_i(\mathbf{x}) := \begin{cases} b & \text{if } u(\mathbf{x}) = i, \\ a & \text{if } u(\mathbf{x}) \in E_{k-1}\backslash\{i\}, \\ k-1 & \text{if } u(\mathbf{x}) = k-1 \end{cases}$$

$(i = 0, 1, ..., k-2)$ belong to $U_{a,b,k-1}$. Then

$$u = p(u_0, u_1, ..., u_{k-2}) \in [M],$$

whereby $[M] = U$ is proven. Therefore, M is complete in U. ■

Lemma 18.3.2.2 *Let M be a subset of U, which has the two following properties:*

(1) The set

$$\{f^n \in P_{k-1} \mid \exists g^n \in M \cap Pol_k E_{k-1} : \ \forall \mathbf{x} \in E^n_{k-1} : \ g(\mathbf{x}) = f(\mathbf{x})\}$$

is complete in P_{k-1}.

(2) There are two different elements $a, b \in E_{k-1}$ *with*

$$P_{k,\{a,b\}} := \{f \in P_k \mid Im(f) \subseteq \{a, b\} \} \subseteq [M].$$

Then $P_{k,k-1} \subseteq [M]$.

Proof. The proof is similar to the proof of Lemma 18.3.2.1: Because of (1) a $(k-1)$-ary function p with the property

$$p\begin{pmatrix} b & a & a & \dots & a & a & a \\ a & b & a & \dots & a & a & a \\ a & a & b & \dots & a & a & a \\ \dots & \dots & \dots & \dots & \dots & \dots & \dots \\ a & a & a & \dots & b & a & a \\ a & a & a & \dots & a & b & a \\ a & a & a & \dots & a & a & b \end{pmatrix} = \begin{pmatrix} 0 \\ 1 \\ 2 \\ \dots \\ k-4 \\ k-3 \\ k-2 \end{pmatrix}$$

belongs to $[M]$. Then one can form an arbitrary n-ary function $u \in P_{k,k-1}$ as a superposition over $\{p\} \cup P_{k,\{a,b\}} \subseteq [M]$, as follows: The n-ary functions $u_0, u_1, ..., u_{k-2}$ defined by

$$u_i(\mathbf{x}) := \begin{cases} b & \text{if } u(\mathbf{x}) = i, \\ a & \text{if } u(\mathbf{x}) \in E_{k-1} \backslash \{i\} \end{cases}$$

$(i = 0, 1, ..., k-2)$ belong to $P_{k,\{a,b\}}$. Then

$$u = p(u_0, u_1, ..., u_{k-2}) \in [M],$$

whereby $P_{k,k-1} \subseteq [M]$ is proven. ■

Lemma 18.3.2.3 *For each* $f \in U \backslash Pol_k(\iota^{k-1}_{k-1})^\star$ *and for each* $(a, b) \in E^2_{k-1} \backslash \iota^2_{k-1}$ *the set*

$$\{f\} \cup U^1 \cup P_{k,k-1} \cup U_{a,b,k-1}$$

is complete in U.

Proof. Let $f \in U \backslash Pol_k(\iota^{k-1}_{k-1})^\star$ be arbitrary. Then, w.l.o.g. there are certain $(a_{1i}, a_{2i}, ..., a_{ki}) \in E^k_{k-1}$ $(i = 1, ..., m)$ and $(b_{1j}, b_{2j}, ..., b_{k-1,j}) \in \iota^{k-1}_{k-1}$ $(j = 1, ..., n)$ with

$$f\begin{pmatrix} a_{11} & a_{12} & \dots & a_{1m} & b_{11} & b_{12} & \dots & b_{1n} \\ a_{21} & a_{22} & \dots & a_{2m} & b_{21} & b_{22} & \dots & b_{2n} \\ \dots & \dots & \dots & \dots & \dots & \dots & \dots & \dots \\ a_{k-1,1} & a_{k-2,2} & \dots & a_{k-1,m} & b_{k-1,1} & b_{k-1,2} & \dots & b_{k-1,n} \\ a_{k1} & a_{k2} & \dots & a_{km} & k-1 & k-1 & \dots & k-1 \end{pmatrix} = \begin{pmatrix} 0 \\ 1 \\ \dots \\ k-2 \\ k-1 \end{pmatrix}.$$

By replacing of some variables of f through certain functions of $P_{k,k-1}$ one can form an n-ary function $f' \in U \backslash Pol_k(\iota_{k-1}^{k-1})^\star$ with $f'(k-1, k-1, ..., k-1) = k-1$. Consequently, the set

$$\{f'\} \cup \{g \in U^1 \mid g(k-1) = k-1\}$$

fulfills the assumption (1) of Lemma 18.3.2.1. Then with the help of Lemma 18.3.2.1, one can prove $[\{f\} \cup U^1 \cup P_{k,k-1} \cup U_{a,b,k-1}] = U$. ■

Lemma 18.3.2.4 *Let* $\gamma \in \mathfrak{U}_{k-1} \cup \mathfrak{B}_{k-1} \cup \bigcup_{k=2}^{k-2} \mathfrak{C}_{k-1}^h$ *be an* h*-ary relation. Then* $Pol_k\gamma^\star$ *is* U*-maximal.*

Proof. Clearly, $Pol_k\gamma^\star$ is a proper subset of U. For each n-ary function $g \in Pol_{k-1}\gamma$ we define an n-ary function $f_g \in P_k$ by

$$f_g(\mathbf{x}) := \begin{cases} g(\mathbf{x}) & \text{if} \quad \mathbf{x} \in E_{k-1}^n, \\ k-1 & \text{if} \quad \mathbf{x} = (k-1, k-1, ..., k-1), \\ 0 & \text{otherwise.} \end{cases}$$

It is easy to check that f_g belongs to $Pol_k\gamma^\star$. Moreover, we have $P_{k,k-1} \subseteq Pol_k\gamma^\star$ and there are two different elements $a, b \in E_{k-1}$ with $U_{a,b,k-1} \subseteq Pol_k\gamma^\star$. More precisely: If $\gamma \in \mathfrak{U}_{k-1} \cup \mathfrak{C}_{k-1}^2$ one chooses the elements a and b so that $(a,b) \in \gamma \backslash \iota_{k-1}^2$. If γ has an arity $h \geq 3$, then one can choose the elements a and b arbitrary.
Now let $f \in U \backslash Pol_k\gamma^\star$ be arbitrary. Then, w.l.o.g. there are certain tuples $(a_{1i}, a_{2i}, ..., a_{h+1,i}) \in E_{k-1}^{h+1}$ $(i = 1, ..., n)$ and $(b_{1j}, b_{2j}, ..., b_{hj}) \in \gamma$ $(j = 1, ..., m)$ with

$$f\begin{pmatrix} a_{11} & a_{12} & \cdots & a_{1n} & b_{11} & b_{12} & \cdots & b_{1m} \\ a_{21} & a_{22} & \cdots & a_{2n} & b_{21} & b_{22} & \cdots & b_{2m} \\ & & & \cdot & & & & \\ a_{h1} & a_{h2} & \cdots & a_{hn} & b_{h1} & b_{h2} & \cdots & b_{hm} \\ a_{h+1,1} & a_{h+1,2} & \cdots & a_{h+1,n} & k-1 & k-1 & \cdots & k-1 \end{pmatrix} = \begin{pmatrix} a_1 \\ a_2 \\ \cdots \\ a_h \\ k-1 \end{pmatrix}$$

and $(a_1, ..., a_h) \in E_{k-1}^h \backslash \gamma$. By replacing some variables of f through certain functions of $P_{k,k-1} \subseteq Pol_k\gamma^\star$ one can form a function $f' \in [\{f\} \cup Pol_k\gamma^\star]$, which preserves the relations $\{k-1\}$ and E_{k-1} but does not preserve the relation γ. Then, the maximality of $Pol_{k-1}\gamma$ in P_{k-1} implies that the set $\{f'\} \cup \{f_g \mid g \in Pol_k\gamma\}$ fulfills the condition (1) of Lemma 18.3.2.1. Since we have already proven that $Pol_k\gamma^\star$ fulfills condition (2) of Lemma 18.3.2.1, the set $\{f\} \cup Pol_k\gamma^\star$ is complete in U. Hence $Pol_k\gamma^\star$ is U-maximal. ■

Lemma 18.3.2.5 *Let*

$$\sigma_{E_{k-1}} := E_{k-1}^3 \cup \{(a,a,b) \mid k-1 \in \{a,b\} \subseteq E_k\}.$$

Then

(1) For every function $f \in U \backslash Pol_k \sigma_{E_{k-1}}$ *there are two different elements* $a, b \in E_{k-1}$ *with* $U_{a,b,k-1} \subseteq [\{f\} \cup P_{k,k-1} \cup U_0]$.

(2) $Pol_k \sigma_{E_{k-1}}$ *is U-maximal.*

Proof. (1): Let

$$f \in U \backslash Pol_k \sigma_{E_{k-1}}$$

be arbitrary. W.l.o.g., we can assume that there are certain $(a_{1j}, a_{2j}, a_{3j}) \in E_{k-1}^3$ $(j = 1, 2, ..., n)$ and certain $(b_i, b_i, c_i) \in E_k^3 \backslash E_{k-1}^3$ $(i = 1, 2, ..., m)$ with the properties

$$f \begin{pmatrix} a_{11} & \dots & a_{1n} & b_1 & \dots & b_m \\ a_{21} & \dots & a_{2n} & b_1 & \dots & b_m \\ a_{31} & \dots & a_{3n} & c_1 & \dots & c_m \end{pmatrix} = \begin{pmatrix} a \\ b \\ k-1 \end{pmatrix},$$

$a \neq b$ and $\{a, b\} \subseteq E_{k-1}$. We choose some functions $f_i \in P_{k,k-1}$ and $g_j \in U_0$ with the properties

$$f_i \begin{pmatrix} 0 & 0 \\ 0 & 1 \\ k-1 & 0 \end{pmatrix} = \begin{pmatrix} a_{1i} \\ a_{2i} \\ a_{3i} \end{pmatrix}$$

$(i = 1, ..., n)$ and

$$g_j \begin{pmatrix} 0 & 0 \\ 0 & 1 \\ k-1 & 0 \end{pmatrix} = \begin{pmatrix} b_j \\ b_j \\ c_j \end{pmatrix}$$

$(j = 1, ..., m)$. Thus the binary function f' with

$$f' := f(f_1, f_2, ..., f_n, g_1, ..., g_m)$$

belongs to $[\{f\} \cup P_{k,k-1} \cup U_0]$ and it holds that

$$f' \begin{pmatrix} 0 & 0 \\ 0 & 1 \\ k-1 & 0 \end{pmatrix} = \begin{pmatrix} a \\ b \\ k-1 \end{pmatrix}.$$

Next we show that every function of $U_{a,b,k-1} := \{q \in U \mid Im(q) \subseteq \{a, b, k-1\}\}$ is a superposition over $\{f'\} \cup U_0 \cup P_{k,k-1}$:
Let $q^t \in U_{a,b,k-1}$ be arbitrary. Then the t-ary function q_1 with

$$q_1(\mathbf{x}) := \begin{cases} 0 & \text{if} \quad q(\mathbf{x}) \in \{a, b\}, \\ k-1 & \text{otherwise,} \end{cases}$$

belongs to U_0 and the t-ary function q_2 with

$$q_2(\mathbf{x}) := \begin{cases} 1 & \text{if} \quad q(\mathbf{x}) = b, \\ 0 & \text{otherwise,} \end{cases}$$

belongs to $P_{k,k-1}$. Because of

$$q = f'(q_1, q_2) \in [\{f\} \cup P_{k,k-1} \cup U_0]$$

we have $U_{a,b,k-1} \subseteq [\{f\} \cup P_{k,k-1} \cup U_0]$.
(2): Clearly, $U \neq Pol_k\sigma_{E_{k-1}}$. Since from $\sigma_{E_{k-1}}$ the relation ϱ is derivable, $Pol_k\sigma_{E_{k-1}} \subset U$ holds. Let $f \in U \backslash Pol_k\sigma_{E_{k-1}}$ be arbitrary. Since $P_{k,k-1} \cup U_0 \subseteq Pol_k\sigma_{E_{k-1}}$, above statement (1) implies that $U_{a,b,k-1} \subseteq [\{f\} \cup Pol_k\sigma_{E_{k-1}}]$ is valid for two certain elements $a, b \in E_{k-1}$, whereby $\{f\} \cup Pol_k\sigma_{E_{k-1}}$ fulfills the condition (2) of Lemma 18.3.2.1.
Obviously, each n-ary function $p \in U$ with

$$\forall \mathbf{x} \in E_k^n : p(\mathbf{x}) = k-1 \iff \mathbf{x} = (k-1, k-1, ..., k-1)$$

is a function of $Pol_k\sigma_{E_{k-1}}$. Consequently, $\{f\} \cup Pol_k\sigma_{E_{k-1}}$ also fulfills the condition (1) of Lemma 18.3.2.1. Therefore, the U-maximality of $Pol_k\sigma_{E_{k-1}}$ follows from Lemma 18.3.2.1. ∎

Lemma 18.3.2.6 *Denote max $\in P_k$ the binary function that is defined with respect to the order $0 < 1 < 2 < ...k-1$ as usual. Then, the set*

$$U' := \{\max\} \cup P_{k,k-1} \cup U_0$$

is complete in U.

Proof. Obviously, $P_{k,k-1} \cup U_0 \subseteq U$. Since the function max has the property

$$\max(x, y) = k-1 \iff k-1 \in \{x, y\},$$

max belongs to U. Consequently, U' is a subset of U.
Let $f^n \in U$ be arbitrary. Then, the n-ary functions f_1 and f_2 defined by

$$f_1(\mathbf{x}) := \begin{cases} f(\mathbf{x}), & \text{if} \quad f(\mathbf{x}) \in E_{k-1}, \\ 0 & \text{otherwise,} \end{cases}$$

and

$$f_2(\mathbf{x}) := \begin{cases} 0, & \text{if} \quad f(\mathbf{x}) \in E_{k-1}, \\ k-1 & \text{otherwise,} \end{cases}$$

belong to U'. Hence, the U-completeness of U' follows from $f = \max(f_1, f_2)$. ∎

Lemma 18.3.2.7 *Let*

$$\sigma_{\{k-1\}} := (E_{k-1}^2 \times \{k-1\}) \cup \{(a, a, b) \mid \{a, b\} \subseteq E_k\}.$$

Then:

(1) For every function $f \in U \backslash Pol_k\sigma_{\{k-1\}}$ and every set $M \subseteq U$ with $[pr_{E_{k-1}} M] = P_{k-1}$, where

$$pr_{E_{k-1}} M := \{g^n \in P_{k-1} \mid \exists g_1 \in M \cap Pol_k E_{k-1} : \forall \mathbf{x} \in E_{k-1}^n : g(\mathbf{x}) = g_1(\mathbf{x})\},$$

is valid:

$$P_{k,k-1} \subseteq [\{f\} \cup U_0 \cup M \cup (U^1 \cap Pol_k\{k-1\})].$$

(2) $Pol_k\sigma_{\{k-1\}}$ *is maximal in* U.

Proof. (1): Let

$$f \in U \backslash Pol_k\sigma_{\{k-1\}}$$

be arbitrary. W.l.o.g. we can assume that there are certain $a_i, b_i \in E_{k-1}$ $(i = 1, 2, ..., r)$, $c_j \in E_k$ $(j = 1, 2, ..., s)$ and $d_l, e_l \in E_{k-1}$ $(l = 1, ..., t)$ with the property

$$f\begin{pmatrix} a_1 & \dots & a_r & k-1 & \dots & k-1 & d_1 & \dots & d_t \\ b_1 & \dots & b_r & k-1 & \dots & k-1 & d_1 & \dots & d_t \\ k-1 & \dots & k-1 & c_1 & \dots & c_s & e_1 & \dots & e_t \end{pmatrix} = \begin{pmatrix} a \\ b \\ c \end{pmatrix},$$

where $a \neq b$ and $\{a, b, c\} \subseteq E_{k-1}$. Then, because of $f \in U$, we have

$$f\begin{pmatrix} a_1 & \dots & a_r & k-1 & \dots & k-1 & d_1 & \dots & d_t \\ b_1 & \dots & b_r & k-1 & \dots & k-1 & d_1 & \dots & d_t \\ k-1 & \dots & k-1 & c_1 & \dots & c_s & d_1 & \dots & d_t \end{pmatrix} = \begin{pmatrix} a \\ b \\ c' \end{pmatrix}$$

for a certain $c' \in E_{k-1}$. W.l.o.g. let

$$c' \neq a.$$

We choose some unary functions $f_i \in U^1 \cap Pol_k\{k-1\}$ $(i = 0, 1, ..., r)$ and $g_j \in U_0$ with the properties

$$f_0(x) = \begin{cases} 0 & \text{if } x = a, \\ 1 & \text{if } x \in E_{k-1} \backslash \{a\}, \end{cases}$$

$$f_i\begin{pmatrix} 0 \\ 1 \\ k-1 \end{pmatrix} = \begin{pmatrix} a_i \\ b_i \\ k-1 \end{pmatrix}$$

$(i = 1, ..., r)$ and

$$g_j\begin{pmatrix} 0 \\ 1 \\ k-1 \end{pmatrix} = \begin{pmatrix} k-1 \\ k-1 \\ c_j \end{pmatrix}$$

$(j = 1, ..., s)$. Therefore, the unary function f' with

$$f' := f_0(f(f_1, f_2, ..., f_r, g_1, ..., g_s, c_{d_1}, ..., c_{d_t}))$$

belongs to $[\{f\} \cup (U^1 \cap Pol_k\{k-1\}) \cup U_0]$, where $f' \in P_{k,2}$ (because of $f' \in U$ and $f'(k-1) \neq k-1$) and

$$f'\begin{pmatrix} 0 \\ 1 \\ k-1 \end{pmatrix} = \begin{pmatrix} 0 \\ 1 \\ 1 \end{pmatrix}$$

are valid. Now it is possible to prove that $P_{k,2} \subseteq [\{f'\} \cup U_0 \cup M]$ holds with the help of the following **completeness criterion for** $P_{k,2}$ (see Theorem 12.4.3):

A subset $T \subseteq P_{k,2}$ is a generating system for $P_{k,2}$ if and only if T fulfills the following two conditions:

(i) The set

$$pr_{\{0,1\}}\, T := \{f^n \in P_2 \mid \exists f_1^n \in T : \ \forall \mathbf{x} \in E_2^n : \ f(\mathbf{x}) = f_1(\mathbf{x})\}$$

is a generating system for P_2.

(ii) For each $a \in E_{k-1}$ and each $b \in E_k \backslash \{0,1\}$ with $a < b$ it is valid:

$$T \not\subseteq P_{k,2} \cap Pol_k \begin{pmatrix} 0 & 1 & a \\ 0 & 1 & b \end{pmatrix}.$$

Clearly, the set $pr_{\{0,1\}}\{f' \star g \mid g \in M\}$ fulfills the above condition *(i)* because of our assumption $pr_{E_{k-1}} M = P_{k-1}$.
For arbitrary $a \in E_{k-1}$, the unary function q_a with

$$q_a(x) := \begin{cases} 0 & \text{if} \quad x = a, \\ k-1 & \text{if} \quad x = k-1, \\ 1 & \text{otherwise}, \end{cases}$$

belongs to $U^1 \cap Pol_k\{k-1\}$. Then, for arbitrary $a \in E_{k-1}$ and $b \in E_k \backslash \{0,1\}$ with $a < b$ we have

$$(f' \star q_a)\begin{pmatrix} a \\ b \end{pmatrix} = \begin{pmatrix} 0 \\ 1 \end{pmatrix},$$

whereby $\{f'\} \cup (U^1 \cap Pol_k\{k-1\})$ also fulfills the conditions of *(ii)*.
Thus by the completeness theorem for $P_{k,2}$,

$$P_{k,2} \subseteq [\{f\} \cup U_0 \cup M \cup (U^1 \cap Pol_k\{k-1\})]$$

is valid. This and Lemma 18.3.2.2 imply our assertion (1).
(2): Clearly, $Pol_k \sigma_{\{k-1\}} \neq U$. Further, it is easy to prove that ϱ is derivable from $\sigma_{\{k-1\}}$. Hence $Pol_k \sigma_{\{k-1\}} \subset U$.
Let $f \in U \backslash Pol_k \sigma_{\{k-1\}}$ be arbitrary. Obviously, $U_0 \subseteq Pol_k \sigma_{\{k-1\}}$ and each n-ary function $p \in U$ $(n \in \mathbb{N})$ with the property

$$\forall \mathbf{x} \in E_k^n : \ p(\mathbf{x}) = k-1 \iff \mathbf{x} \in E_k^n \backslash E_{k-1}^n$$

belongs to $Pol_k \sigma_{\{k-1\}}$. In particular, $\max \in Pol_k \sigma_{\{k-1\}}$. Then $[\{f\} \cup Pol_k \sigma_{\{k-1\}}] = U$ follows from statement (1) and Lemma 18.3.2.6. ■

Lemma 18.3.2.8 *For certain $t \in \{1, 2, ..., k-2\}$ we set $C := \{t, t+1, ..., k-2\}$. Furthermore let*

$$\alpha := \{(x,x) \mid x \in E_k\} \cup E_{k-1}^2 \cup \{(x,y),(y,x) \mid x \in E_k, \ y \in C\},$$

i.e., α is a binary central relation, where C is the set of all central elements of α and α has the property

$$E_k^2 \backslash \alpha = \{(u, k-1), (k-1, u) \mid u \in E_t\}.$$

Then $\max \in U \cap Pol_k \alpha$.

Proof. Because of Lemma 18.3.2.6, we have $\max \in U$. Suppose max does not preserve the relation α. Then there are certain $(a, b),\ (c, d) \in \alpha$ and a $u \in E_t$ with

$$\max \begin{pmatrix} a & c \\ b & d \end{pmatrix} = \begin{pmatrix} u \\ k-1 \end{pmatrix}.$$

W.l.o.g. we can assume that $a = u$ and $d = k - 1$, i.e., $c \leq u$ and $b \leq k - 1$. Then, because of $u \in E_t$ we have however $c \in E_t$, contrary to $(c, d) = (c, k - 1) \in \alpha$. ∎

Lemma 18.3.2.9 *Let $\alpha \in \mathfrak{C}_k^2$ be a binary central relation with $\varrho \subseteq \alpha$ and the property that at least an element of E_{k-1} is a central element of α. Then $U \cap Pol_k \alpha$ is U-maximal.*

Proof. Because of $\varrho \subseteq \alpha$ and $\varrho \notin D_k$, element $k - 1$ cannot be a central element of α. Therefore, w.l.o.g. we can assume that the set

$$C := \{t, t+1, ..., k-2\}$$

for a certain $t \in \{1, 2, ..., k-2\}$ is the set of all central elements of α, whereby we have

$$\alpha = \{(x, x) \mid x \in E_k\} \cup E_{k-1}^2 \cup \{(x, y), (y, x) \mid x \in E_k,\ y \in C\}$$

and

$$E_k^2 \backslash \alpha = \{(u, k-1), (k-1, u) \mid u \in E_t\}.$$

Notice that, with that, $k - 2$ is a central element of α.
Let $f^n \in U \backslash Pol_k \alpha$ be arbitrary. Then there are certain $(a_1, b_1), ...,$ $(a_n, b_n) \in \alpha$ with

$$f \begin{pmatrix} a_1 & a_2 & \dots & a_n \\ b_1 & b_2 & \dots & b_n \end{pmatrix} = \begin{pmatrix} a \\ k-1 \end{pmatrix}$$

and $(a, k-1) \notin \alpha$. Since the functions $t_{r,s}$ $(i = 1, ..., n)$ with $(r, s) \in \alpha$ and q_a $(a \in E_{k-1})$ with

$$t_{r,s}(x) := \begin{cases} r & \text{fü } x \in E_{k-1}, \\ s & \text{if } x = k-1, \end{cases}$$

and

$$q_{a,b}(x) := \begin{cases} b & \text{fü } x = a, \\ k-2 & \text{if } x \in E_{k-1} \backslash \{a\}, \\ k-1 & \text{if } x = k-1, \end{cases}$$

for arbitrary $b \in E_{k-1}$ belong to $U \cap Pol_k \alpha$, the functions

$$t_{b,k-1} = q_{a,b}(f(t_{a_1,b_1}, ..., t_{a_n,b_n}))$$

and

$$t_{k-1,b} = q_{a,b}(f(t_{b_1,a_1}, ..., t_{b_n,a_n}))$$

are superpositions over $\{f\} \cup (U \cap Pol_k\alpha)$ for all $b \in E_{k-1}$. Consequently, $U_0 \subseteq [\{f\} \cup (U \cap Pol_k\alpha)]$. Moreover, max $\in U \cap Pol_k\alpha$ (see Lemma 18.3.2.8) and (because of $\varrho \subseteq \alpha$) $P_{k,k-1} \subseteq Pol_k\alpha$. Then the U-completeness of $\{f\} \cup (U \cap Pol_k\alpha)$ follows from Lemma 18.3.2.6. ■

18.3.3 Proof of the Completeness Criterion for U

Let M be an arbitrary subset of U, which is not contained in any class from type I–VI. We show that $U = [M]$ results from this assumption when we prove the generating system from Lemma 18.3.2.3 in $[M]$.

Since M is no subset of a class of type I, it results from the completeness theorem for P_2 that the following is valid:

$$\forall g^m \in P_2 \; \exists g_1^m \in [M] : \; \alpha_\varphi(g_1) = g. \tag{18.19}$$

In particular, (18.19) implies:

$$c_{k-1} \in [M]. \tag{18.20}$$

With the aid of the functions of M, which do not belong to the clones of type II, one sees from (18.20) that the constant functions of P_k belong to $[M]$:

$$c_0, c_1, ..., c_{k-1} \in [M]. \tag{18.21}$$

Because of (18.19), there is a unary function in $[M]$ that is an inverse image of the function $g \in [M]$ with

x	$g(x)$
0	1
1	0

.

Let g' be this function with

x	$g'(x)$
0	$k-1$
1	$k-1$
2	$k-1$
.	.
.	.
$k-2$	$k-1$
$k-1$	a

,

where a is a certain element of E_{k-1}.

The relation

$$E_{k-1}^2 \cup \{(k-1,k-1),(a,k-1),(k-1,a)\}$$

describes a certain class of the type V. Thus there is a function $q \in M$ that does not preserve this relation. W.l.o.g. we can assume

$$q\begin{pmatrix} a_1 & a_2 & \dots & a_n & k-1 & a & k-1 \\ b_1 & b_2 & \dots & b_n & k-1 & k-1 & a \end{pmatrix} = \begin{pmatrix} b \\ k-1 \end{pmatrix},$$

where $a_1,...,a_n,b_1,...,b_n$ are certain elements of E_{k-1}. Because of $q \in U$, we have also

$$q\begin{pmatrix} a_1 & a_2 & \dots & a_n & k-1 & a & k-1 \\ a_1 & a_2 & \dots & a_n & k-1 & k-1 & a \end{pmatrix} = \begin{pmatrix} b \\ k-1 \end{pmatrix}.$$

Consequently, the functions

$$t_{b,k-1}(x) := q(c_{a_1}(x),...,c_{a_2}(x),c_{k-1}(x),g'(g'(x)),g'(x))$$

and

$$t_{k-1,b}(x) := q(c_{a_1}(x),...,c_{a_2}(x),c_{k-1}(x),g'(x),g'(g'(x)))$$

belong to $[M]$, where

x	$t_{b,k-1}(x)$	$t_{k-1,b}$
0	b	$k-1$
1	b	$k-1$
2	b	$k-1$
.	.	.
.	.	.
$k-2$	b	$k-1$
$k-1$	$k-1$	b

.

With the aid of further functions of M, which do not belong to the clones of type V, in analog mode, we see that all functions of the form

$$t_{\alpha,\beta}(x) := \begin{cases} \alpha & \text{if } x \in E_{k-1}, \\ \beta & \text{if } x = k-1, \end{cases}$$

belong to $[M]$ for all $\alpha,\beta \in E_{k-1}$ with $k-1 \in \{\alpha,\beta\}$.
Next we show that

$$U_0 := \{f \in U \mid |Im(f)| \le 2 \land k-1 \in Im(f)\} \subseteq [M] \qquad (18.22)$$

holds.
For this purpose, let f^n be an arbitrary function of U_0 with $Im(f) = \{a, k-1\}$. Because of (18.19) there is an n-ary function $f' \in [M]$ with the property

$$\forall \mathbf{x} \in E_k^n : \; f'(\mathbf{x}) = k-1 \iff f(\mathbf{x}) = k-1.$$

With the aid of the above function $t_{a,k-1} \in [M]$, one can prove that

$$t_{a,k-1} \star f' = f \in [M].$$

Therefore, and because of (18.21), (18.22) is valid.

If one forms superpositions over the constants and the functions of M, which do not belong to the clones of type III, one obtains functions $g_\gamma^m \in [M]$ with

$\mathbf{x}$	$g_\gamma(\mathbf{x})$
$\in E_{k-1}^m$	$\}= g_{\gamma'} \notin Pol_{k-1}\gamma$
otherwise	certain values

for all γ with the properties:
(a) $Pol_{k-1}\gamma$ is maximal in P_{k-1} and
(b) $c_0, ..., c_{k-1} \in Pol_{k-1}\gamma$.
Hence, because of (18.21), there are functions $g_\gamma \in [M]$ that do not fulfill (b). This and the completeness criterion for P_{k-1} imply that the following function belongs to $[M]$:

x y	$s(x,y)$
$\in E_{k-1}^2$	$\}= max(x,y) + 1 \ (mod\ k-1)$
$k-1$ $k-1$	α
otherwise	certain values

We distinguish two cases for $\alpha := s(k-1, k-1)$:

Case 1: $\alpha \in E_{k-1}$.

By $s \in [M]$ the functions

$$s' := \Delta s \quad (s'(x) = s(x,x))$$

and

$$s'' := (s')^{k-1}, \quad (s''(x) = \underbrace{s'(s'(s'(...s'(x)...)))}_{k-1}\ )$$

belong to $[M]$. Let $\beta := s''(k-1)$ and $\gamma \in E_{k-1}\backslash\{\beta\}$.
With the help of the completeness criterion for $P_{k,k-1}$, one can easily prove that the set

$$M_1 := \{s'' \star s,\ t_{\beta,\gamma} = s'' \star t_{k-1,\gamma}\} \subseteq [M]$$

is a generating system for $P_{k,k-1}$. The **completeness criterion for** $P_{k,k-1}$ is a special case of Theorem 12.4.3 and says:

An arbitrary subset $T \subseteq P_{k,k-1}$ is a generating system for $P_{k,k-1}$, if and only if T fulfills the following two conditions:

(i) The set

$$pr\ T := \{f^{n)} \in P_{k-1} \mid \exists f_1^n \in T : \ \forall \mathbf{x} \in E_{k-1}^n : \ f(\mathbf{x}) = f_1(\mathbf{x})\}$$

is a generating system for P_{k-1}.

(ii) For all $a \in E_k$:

$$T \not\subseteq P_{k,k-1} \cap Pol_k \begin{pmatrix} 0 & 1 & \dots & k-2 & a \\ 0 & 1 & \dots & k-2 & k-1 \end{pmatrix}.$$

Because of $s'' \star s \in P_{k,k-1}$,

$$\forall x, y \in E_{k-1} : (s'' \star s)(x, y) = \max(x, y) + 1 \ (mod\ k-1)$$

and the fact that $\max(x, y) + 1$ is a Sheffer function for P_{k-1} [6] the set $T := M_1$ fulfills *(i)*.
Because of

$$t_{\beta,\gamma}\begin{pmatrix} a \\ k-1 \end{pmatrix} = \begin{pmatrix} \beta \\ \gamma \end{pmatrix}$$

$T := M_1$ also fulfills the condition *(ii)*.
Thus

$$P_{k,k-1} \subseteq [M]$$

was proven in Case 1. Then Lemma 18.3.2.5 implies the existence of two different elements $a, b \in E_{k-1}$ with

$$U_{a,b,k-1} \subseteq [M]. \qquad (18.23)$$

Except for the unary function $e \in U \backslash U_0$ with

x	$e(x)$
0	$\in E_{k-1}$
1	
.	
.	
$k-2$	
$k-1$	$k-1$

we have proven that the other unary functions of U belong to $[M]$. Next, with the aid of the completeness theorem for P_{k-1}, we show that the functions of the form e are superpositions over

$$\{c_0, c_1, \dots, c_{k-1}\} \cup U_0 \cup P_{k,k-1} \subseteq [M]$$

and functions of M, which do not belong to the clones of type IV.

[6] See Theorem 7.1.5.

Let $\gamma \in \mathfrak{U}_{k-1} \cup \mathfrak{B}_{k-1} \cup \bigcup_{h=1}^{k-2} \mathfrak{C}_{k-1}$ be an h-ary relation; i.e., γ is a totally reflexive and totally symmetric relation that describes the maximal class $Pol_{k-1}\gamma$ of P_{k-1}. Then, the relation

$$\begin{aligned}\gamma^\star \;:=\; & E_{k-1}^{h+1} \cup \{(x,x,...,x,y) \mid x,y \in E_k\} \\ & \cup \{(x_1,...,x_h,k-1) \mid (x_1,...,x_h) \in \gamma\}\end{aligned}$$

describes a maximal class of U of type IV.

If one replaces the variables of a function $f_\gamma \in M \backslash Pol_k\gamma^\star$ by certain functions of the set $\{c_0, c_1, ..., c_{k-1}\} \cup U_0 \cup P_{k,k-1} \subseteq [M]$, one receives an n-ary function $f'_\gamma \in [M]$ with

$\mathbf{x}$	$f'_\gamma(\mathbf{x})$
$\in E_{k-1}^n$	$\}= f''_\gamma(\mathbf{x}) \notin Pol_{k-1}\gamma$
$k-1\ k-1\ ...\ k-1$	$k-1$
otherwise	certain values

(18.24)

for all $\gamma \in \mathfrak{U}_{k-1} \cup \mathfrak{B}_{k-1} \cup \bigcup_{h=1}^{k-2} \mathfrak{C}_{k-1}$. Since function $t_{a,k-1}$ belongs to $U_0 \subseteq [M]$ for every $a \in E_{k-1}$, there are also functions $f'_\gamma \in [M]$ with the property (18.24) for all $\gamma \in \mathfrak{S}_{k-1} \cup \mathfrak{C}_{k-1}^1$. Because of (18.23), there is a function $f'_\gamma \in [M]$ with the property (18.24) for every $\gamma \in \mathfrak{M}_{k-1} \cup \mathfrak{L}_{k-1}$. With the help of the completeness criterion for P_{k-1} and the fact that the function $f_\gamma \in [M]$ preserves the relation $\{k-1\}$ for every $\gamma \in \mathfrak{M}_{k-1} \cup \mathfrak{S}_{k-1} \cup \mathfrak{U}_{k-1} \cup \bigcup_{h=1}^{k-2} \mathfrak{C}_{k-1} \cup \bigcup_{h=3}^{k-1} \mathfrak{B}_{k-1}$, one can prove that $U^1 \cap Pol_k\{k-1\} \subseteq [M]$. Consequently, $U^1 \subseteq [M]$.

To summarize, we have proven $P_{k,k-1} \cup U_0 \cup U_{a,b,k-1} \cup U^1 \subseteq [M]$ in Case 1. Since a function of $U \backslash Pol_k(\iota_{k-1}^{k-1})^\star$ belongs to M, this and Lemma 18.3.2.3 implies $[M] = U$.

Case 2: $\alpha = k-1$.

Since function s preserves $\{k-1\}$ in this case and s is a Sheffer function for P_{k-1} on the tuples of E_{k-1}^2, every unary function e with $e(k-1) = k-1$ is a superposition over M. Furthermore, $[pr_{E_{k-1}} M] = P_{k-1}$. Since we have proven that $U_0 \subseteq [M]$ holds, the first statement of Lemma 18.3.2.7 implies $P_{k,k-1} \subseteq [M]$. Therefore, there exists a function of $[M]$ that fulfills the assumption of Case 1. Consequently, $[M] = U$ in Case 2, and we have proven the following theorem:

Theorem 18.3.3.1 *(***Completeness Theorem for** U*)*

(1) The clones defined in Section 18.3.1 are the only maximal classes of U.

(2) An arbitrary set $M \subseteq U$ is U-complete if and only if there exists no class of the type I, II, ... or VI that has M as a subset. ■

19

Minimal Classes and Minimal Clones of P_k

In this chapter we deal with classes (of $\mathbb{L}_k$), which are either direct predecessors of the empty set (so-called *minimal classes*) or which are direct predecessors of the set of all projections (so-called *minimal clones*).
It will turn out that it is not heavy to determine the minimal classes. However, for the minimal clones only partial results can be given.

19.1 Minimal Classes

A subclass A of P_k is called a **minimal class**, if no subclass $B \neq \emptyset$ with $B \subset A$ exists. Because of $[\Delta^{n-1} f] \subset [f]$ for every function $f \in P_k \backslash [P_k^1]$, every minimal classes has the order 1. If $k = 2$ then $[c_0]$, $[c_1]$ and $[e_1^1]$ are the only minimal classes. For arbitrary k one can easy check that J_k is a minimal class and, for every other minimal class A, there exists a unary function $f \in P_k^1(k-1)$ with the property $f \star f = f$ and $[f] = A$.
Then a minimal class of P_6 is e.g.

x	$f(x)$
0	1
1	1
2	1
3	3
4	4
5	4

By generalizing this example, one obtains the following theorem which is a conclusion from [Pös-K 79], Section 4.4.

Theorem 19.1.1 *P_k contains exactly*

$$1+\sum_{r=1}^{k-1}\binom{k}{r}\cdot r^{k-r}$$

minimal classes. These are class J_k and the classes of the form $[f]$ with $f \in P_k^1(k-1)$ and $f \star f = f$.
One can describe a function $f \in P_k^1(k-1)$ with the property $f \star f = f$ as follows:
Let θ be an equivalence relation on E_k, which is different from ι_k^2 $(= \kappa_0)$ and has exactly r equivalence classes. Further, denote $\mathbf{a} := (a_1, ..., a_r)$ an r-tuple with elements, which in pairs are not θ-equivalent. For such a θ und such an $\mathbf{a}$, one can define the mapping $f_{\mathbf{a}}^{\theta} \in P_k$ by

$$f_{\mathbf{a}}^{\theta}(x) = a_i \;:\Longleftrightarrow\; x \in [a_i]_\theta.$$

Then, for a function $f \in P_k^1(k-1)$, $f \star f = f$ holds, if and only if there is an equivalence relation θ and a tuple $\mathbf{a}$ of the above form with $f = f_{\mathbf{a}}^{\theta}$. ■

19.2 The Five Types of Minimal Clones

A clone $A \subseteq P_k$ is called **minimal in** P_k, iff J_k is a maximal clone of A, i.e., it holds

$$\forall f \in A \setminus J_k :\; [J_k \cup \{f\}] = A. \tag{19.1}$$

Therefore, a basis of a minimal clone has at most two elements. Furthermore, it is easy to see that the following implication holds:

$$(A \text{ is a minimal clone with ord } A \geq 2) \implies \forall f \in A \setminus J_k :\; A = [f]. \tag{19.2}$$

From Chapter 3 it follows directly:

Theorem 19.2.1 *P_2 has exactly 7 minimal clones. These are the clones $I \cup C_0 = [e_1^1, c_0]$, $I \cup C_1 = [e_1^1, c_1]$, $\overline{I} = [\overline{e_1^1}]$, $K = [\wedge]$, $D = [\vee]$, $L \cap S \cap T_0 = [r]$ and $[S \cap M \cap T_0] = [h_2^3]$.* ■

One has solved the problem of describing minimal clones of the form $[f^n] \subseteq P_k$ for arbitrary k thus far only for $n = 1$.

Theorem 19.2.2 *([Pös-K 79], [Har 74]; without proof) P_k has exactly*

$$\sum_{r=1}^{k-1}\binom{k}{r}\cdot r^{k-r}+\sum_{r=0}^{k-2}\binom{k}{r}\cdot\sum_{p\cdot t=k-r,p\in\mathbb{P},t\in\mathbb{N}}\frac{(k-r)!}{p^t\cdot t!\cdot(p-1)}$$

minimal clones of the order 1. A minimal clone A of the order 1 has either the form $A = J_k \cup [f]$, where $[f]$ is a minimal class different to J_k (see Theorem 19.1.1), or has the form $A = [s]$, where $s \in P_k^1[k] \setminus \{e_1^1\}$ is a permutation for which there is a prime number p with $s^p = e_1^1$.

The following theorem supplies the existence of at most finite many minimal clones (for a fixed k) and a coarse division of the minimal clones.

Theorem 19.2.3 (**Rosenberg's Classification of the Minimal Clones**; *[Ros 82]*)

For every $k \in \mathbb{N} \setminus \{1\}$ there is only finite many minimal clones. If $A = [J_k \cup \{f^n\}]$ is an arbitrary minimal clone of P_k, where $[A^{n-1}] = J_k$, then this clone fulfills one of the following five conditions:

(1) $n = 1$ and A is described in Theorem 19.2.2.
(2) $n = 2$ and f is idempotent, i.e., $f(x, x) = x$ holds for arbitrary $x \in E_k$.
(3) $n = 3$ and

$$\forall x, y \in E_k : \ f(x, x, y) = f(x, y, x) = f(y, x, x) = y, \tag{19.3}$$

i.e., f is a so-called ternary **minority function**.
A minority function g^3 of P_k generates a minimal clone if and only if $g(x, y, z) = x \oplus y \oplus z$ and $(E_k; \oplus)$ is an elementary 2-group.
(4) $n = 3$ and

$$\forall x, y \in E_k : \ f(x, x, y) = f(x, y, x) = f(y, x, x) = x, \tag{19.4}$$

i.e., f is a so-called ternary **majority function**.
(5) [1] $n \in \{3, 4, ..., k\}$ and f is a **semiprojection**, *i.e., there exists an $i \in \{1, ..., n\}$ with $f(a_1, ..., a_n) = a_i$ for every tuple $(a_1, ..., a_n) \in E_k$ with $|\{x_1, ..., x_n\}| \leq n - 1$.*

Proof. Let A be an arbitrary minimal clone of the order $n \in \mathbb{N}$. Because of Theorem 19.2.2 and by (19.2), we can assume w.l.o.g. that $n \geq 2$ and that $A = [f^n]$ for certain $f \in P_k$. Obviously, because of *ord* $f \geq 2$ and the minimality of the clone A, it holds:

$$\forall i \in \{1, ..., n-1\} : \Delta^i f \in J_k \tag{19.5}$$

In particular this implies $f(x, x, ..., x) = x$, i.e., f is idempotent. Thus, (2) is proven for $n = 2$. Because of (19.5) we obtain for $n = 3$ that

---

[1] The following statement is also mentioned *Świerczkowski Lemma* in the literature. S. Świerczkowski published in [Swi 60] a theorem from which statement (5) of Theorem 19.2.3 results.

$$\{f(x,x,y), f(x,y,x), f(y,x,x)\} \subseteq \{x,y\}.$$

Consequently, the following eight cases are possible:

	Case							
	1	2	3	4	5	6	7	8
$f(x,x,y) =$	x	x	x	x	y	y	y	y
$f(x,y,x) =$	x	x	y	y	x	x	y	y
$f(y,x,x) =$	x	y	x	y	x	y	x	y

In Case 1, the function f fulfills the condition (19.4). In Cases 2, 3, and 5 the function f is a semiprojection, i.e., f fulfills the condition (5). Case 8 gives the condition (19.3). The remaining cases 4, 6, and 7 cannot occur for a function f which generates a minimal clone. This can be shown as follows:
Suppose it holds Case 6; i.e., we have $f(x,x,y) = f(y,x,x) = y$ and $f(x,y,x) = x$. Then the function $g(x,y,z) := f(x, f(x,y,z), z)$ is a majority function. Since every ternary superposition $t \notin J_k$ over the majority function f is a majority function, too [2], this is a contradiction to the minimality of clone A. In analog mode, the other cases can be led to a contradiction.
Let now $n = 4$. Then, by assumption $[A^3] = J_k$. Therefore, in particular $\Delta f \in J_k$, i.e., we have $f(x_1,x_1,x_3,x_4) \in \{x_1,x_3,x_4\}$. Then the following two cases are possible:
Case 1: $f(x_1,x_1,x_3,x_4) = x_1$.
Then, because of $[A^3] = J_k$ it holds:

$$\begin{aligned}
x_a &:= f(x_1,x_2,x_1,x_4) \in \{x_1,x_2,x_4\},\\
x_b &:= f(x_1,x_2,x_3,x_1) \in \{x_1,x_2,x_3\},\\
x_c &:= f(x_1,x_2,x_2,x_4) \in \{x_1,x_2,x_4\},\\
x_d &:= f(x_1,x_2,x_3,x_2) \in \{x_1,x_2,x_3\},\\
x_e &:= f(x_1,x_2,x_3,x_3) \in \{x_1,x_2,x_3\}.
\end{aligned}$$

If one puts $x_1 = x_2$ in the above equations and if one compares this with $f(x_1,x_1,x_3,x_4) = x_1$, it follows

$$\{a,b,c,d,e\} \subseteq \{1,2\}.$$

Since there exists a $t \in \{1,2\}$ with $f(x_1,x_2,x_1,x_1) = x_t$ in addition, one obtain from the above equations:

$$a = b = c = d = e = t.$$

Thus f is a semiprojection in Case 1.

[2] Use induction on the number of occurrences of f in t; see e.g. [Sza 83b] or [Qua 95].

Case 2: There exists an $i \in \{3,4\}$ with $f(x_1,x_1,x_3,x_4) = x_i$.
If $i = 3$ then $f(x_1,x_2,x_3,x_3) = x_3$. Consequently, we can continue analogously to the first case with the proof. If $i = 4$, then we have $f(x_1,x_2,x_3,x_1) = x_1$ and one can show (as in the first case) that f is a semiprojection.

In analog mode to the case $n = 4$, one can examine the case $n \geq 5$.

It remains to show that a minority function g^3, which generates a minimal clone, fulfills condition (4). One can show this most easily with the aid of the following statement, proven by Á. Szendrei in [Sze 87]:
Let $k \geq 2$ and $d \in P_k$ be a ternary function, which fulfills the Mal'tsev-condition

$$d(x,y,y) = d(y,y,x) = x.$$

Then $[d]$ is a minimal clone of P_k if and only if there exists an elementary Abelean p-group $(E_k;\oplus)$ with $d(x,y,z) = x \ominus y \oplus z$. [3]
Since a minority function g fulfills the Mal'tsev-condition according to definition, the function g has the form given in the above statement. However, the function $g(x,y,z) = x \ominus y \oplus z$ is a minority function iff $p = 2$. ∎

Theorem 19.2.4 *([Pal 86])*
For every $t \in \{1,2,...,k\}$ there is a minimal clone of the order t.

Proof. For $t = 1$ one can find examples for minimal clones of the order 1 in Theorem 19.2.1. In [Pös-K 79] one finds proof that the binary function f with $f(x,y) = x \circ y$, where the operation $\circ$ fulfills the equations $x \circ x = x$ (idempotent law), $(x \circ y) \circ z = x \circ (y \circ z)$ (associative law), $(x \circ y) \circ x = x \circ y$ (absorption law), generates a minimal clone.
Let $t \in \{3,...,k-1\}$. Denote $a_1,...,a_{t+1}$ pairwise distinct elements of E_k. The t-ary semiprojection f is defined by

$$f(x_1,...,x_t) := \begin{cases} a_{t+1} & \text{if} \quad x_1 = a_1 \wedge \{x_2,...,x_t\} = \{a_2,...,a_t\}, \\ x_1 & \text{otherwise.} \end{cases}$$

In [Pal 86] it was proven that $[f]$ is a minimal clone of the order t.
Put

$$g(x_1,...,x_k) := \begin{cases} x_1 & \text{if} \quad |\{x_1,...,x_k\}| \leq k-1, \\ x_2 & \text{otherwise.} \end{cases}$$

Then it is easy to see that $[g]$ is a minimal clone of the order k. ∎

Of the many results found in the literature on minimal clones, only some are given without proof.

[3] One finds this statement also proven in [Qua 95].

Theorem 19.2.5 *([Csa 83b]; without proof)*
P_3 has exactly 84 minimal clones. One obtains every one of these clones by using an inner automorphism of P_3 onto exactly one of the 24 following clones (under that 4 of the order 1, 12 of the order 2 and 8 of the order 3): $[j_2]$, $[u_2]$, $[s_2]$, $[s_4]$ (see Table 15.1), $[b_i]$ and $[m_j]$, where $i \in \{1, 2, ..., 12\}$, b_i idempotent, $j \in \{1, 2, ..., 8\}$, m_j majority function and

x	y	b_1	b_2	b_3	b_4	b_5	b_6	b_7	b_8	b_9	b_{10}	b_{11}	b_{12}
0	1	0	0	0	0	0	0	0	0	0	0	0	2
1	0	0	0	0	0	0	0	0	1	1	2	0	2
0	2	0	0	0	0	0	0	0	0	0	0	2	1
2	0	0	2	0	0	2	2	2	2	2	1	2	1
1	2	0	0	1	1	1	1	2	0	0	1	1	0
2	1	0	2	1	2	1	2	2	0	2	2	1	0

x	y	z	m_1	m_2	m_3	m_4	m_5	m_6	m_7	m_8
0	1	2	0	0	2	0	0	0	0	1
0	2	1	0	1	1	0	0	0	0	2
1	0	2	0	1	2	0	0	0	2	0
1	2	0	0	0	0	0	0	2	2	2
2	0	1	0	0	1	0	2	2	1	0
2	1	0	0	1	0	0	2	2	1	1

Note that the cardinality of the set of all subclasses of P_3, which contain a fixed minimal clone of the order 1, can be found in [Pan-V 2000].

It was proven by B. Szczepara in his 210-page long Ph.D. thesis that there are exactly 2182 binary minimal clones for $k = 4$ (see [Szc 95]). All minimal clones, which are generated by majority functions of P_4, were determined by T. Waldhauser in [Wal 2000].
In [Lev-P 96] one can find all minimal clones C of the order 2 with $|C^2| \in \{3, 4, 6\}$. Furthermore, one can find examples for minimal clones C of the order 2 with $|C^2| = 2t + 2$ ($t \geq 1$) or $|C^2| = 3t + 2$ ($t \geq 2$) in this paper.
Up to isomorphic functions, all binary commutative functions, which generate minimal clones of the order 2, were determined in [Kea-S 99].
In [Pös-K 79] it was proven that P_k is the smallest clone, which contains all minimal clones of P_k. Further, it was proven that J_k is the intersection of all maximal clones of P_k. The following problem results from that: How many minimal (or maximal) clones $M_1, ..., M_t$ does one need so that at least $[\bigcup_{i=1}^t M_i] = P_k$ (or $\bigcap_{i=1}^t M_i = J_k$) is valid, respectively? In [Zsa 92] it was proven that $t \leq 3$ holds. The solution $t = 2$ for the above problem can be found in [Czé-H-K-P-S 2001]. A necessary and sufficient condition for $[f, g] = P_k$, if $[f]$ and $[g]$ are minimal clones, can be found in [Ros-M 2001]. The answer 3 for the corresponding problems during the investigation of *partial* minimal (or maximal) clones was proven in [Had-M-R 2002].

One finds supplements to this chapter in the survey articles [Qua 95] and [Csa 2002].

In the next chapter, we deal with partial functions, for which we will handle similar problems as in the previous chapters. The next theorem shows that the problem of determining minimal partial clones can be reduced to the problem of determining minimal clones.

Theorem 19.2.6 *([Bör-H-P 91]; without proof)*
Let C be a partial clone of $\widetilde{P_k}$. Then C is either a minimal clone of P_k or C is generated by a partial n-ary projection with a nontrivial totally reflexive and totally symmetric domain $D(e) \subset E_k^n$ for certain $n \leq k$.
Denote $t(k)$ the number of all minimal clones of P_k and $m(k)$ the number of all partial minimal clones of $\widetilde{P_k}$. Then

$$m(k) = t(k) + \sum_{i=1}^{k} (2^{\binom{k}{i}} - 1).$$

In particular, we have $m(2) = 11$ and $m(3) = 99$, where $t(2) = 7$ and $t(3) = 84$ (see Theorems 19.2.1 and 19.2.5).

20

Partial Function Algebras

In Part I, Chapter 1, we introduced the concept of partial operation over a set A. By choosing $A = E_k$ and replacing the concept of "operation" with the concept of "function", we get the concept "partial function" over E_k. One can then introduce certain modified Mal'tsev-operations over the set $\widetilde{P_k}$ of all partial functions on E_k. Then the set $\widetilde{P_k}$ together with these operations forms a so-called (full) partial function algebra $(\widetilde{P_k}; \tau, \zeta, \Delta, \nabla, \star)$, which can be examined similar to the function algebra $(P_k; \tau, \zeta, \Delta, \nabla, \star)$.
The choice of results on partial function in this chapter focuses on questions that were already treated for P_k in the previous chapter.
After a composition of some basic concepts in Section 20.1, Section 20.2 shows that the lattice of all partial clones of $\widetilde{P_k}$ is isomorphic to a certain sublattice of the lattice of all clones of P_{k+1}. Thus, one gets many properties of the partial clones from the properties of the clones that were already found. To find certain partial clones, however, with the aid of the above-mentioned isomorphism, is not possible, because of the absence of results on clones. One could, for example, not solve the completeness problem for $\widetilde{P_k}$ with the help of isomorphism.
In Section 20.3, we show how one can describe partial clones by relations.
Sections 20.4 and 20.5 deal with the maximal partial clones with whose aid, analogously to P_k, one can solve the completeness problem of the partial logic. We will prove that $\widetilde{P_2}$ has exactly 8 and $\widetilde{P_3}$ has exactly 58 maximal partial clones. In Section 20.5 one can find the complete list of all maximal clones of $\widetilde{P_k}$ for arbitrary $k \in \mathbb{N}$, which was found by L. Haddad and I. G. Rosenberg. The list is given without proof. In Section 20.6, we determine the descriptive relations of the maximal clones of P_k that are also descriptive relations of the maximal partial clones of $\widetilde{P_k}$. In addition, a survey those papers that deal with determining the orders of the maximal partial clones. Section 20.7 deals with determining the cardinality of the set $\mathcal{I}(A) := \{C \subseteq \widetilde{P_k} \mid C = [C] \wedge C \cap P_k = A\}$, where A is an arbitrary maximal clone of P_k. We prove, that, if A has the type $\mathfrak{U}$, $\mathfrak{S}$ or $\mathfrak{C}$, $\mathcal{I}(A)$ is a finite set.

On the other hand, the set $\mathcal{I}(A)$ has the cardinality of continuum, if A has the type $\mathfrak{L}$ or $\mathfrak{B}$. For the type $\mathfrak{M}$ we can give only partial results.
Section 20.8 gives a survey on the cardinalities of the sets $\mathcal{I}(A)$, where A is an arbitrary subclass of P_2.
In last section, we determine the congruences on the maximal partial clones. It is proven particularly that $\widetilde{P_k}$ has exactly 4 congruences, whereas a maximal partial clone has exactly 4, 8, or 10 pairwise distinct congruences.

20.1 Basic Concepts

Let A be nonempty sets and let T be a proper subset of A^n. For an arbitrary mapping f from T into A and for $(a_1, ..., a_n) \in A^n \backslash T$, one can use the notation

$$f(a_1, ..., a_n) = \infty$$

with $\infty \notin A$, to indicate that $f(a_1, ..., a_n)$ is not defined. Then

$$D(f) := \{(x_1, ..., x_n) \in A^n \mid f(a_1, ..., a_n) \neq \infty\}$$

is the domain of f.
In the following, we use these notations for $A = E_k$ and for functions which are defined over subsets of E_k^n. More exact:
For a fixated $k \in \mathbb{N} \setminus \{1\}$ let

$$\widetilde{E_k} := E_k \cup \{\infty\}.$$

An n-ary mapping f of the form

$$f : E_k^n \longrightarrow \widetilde{E_k}$$

is called (n-ary) **partial function**. and the set of all n-ary partial functions let

$$\widetilde{P_k}^n.$$

Furthermore, we put

$$\widetilde{P_k} := \bigcup_{n \geq 1} \widetilde{P_k}^n.$$

By the above definition, the functions of P_k are also partial functions. To distinguish these functions from functions of the set $\widetilde{P_k} \setminus P_k$, we call these the **total functions** of $\widetilde{P_k}$.
A function $g \in \widetilde{P_k}^n$ is called **subfunction of** $f \in \widetilde{P_k}^n$, if $g(\mathbf{a}) \in \{f(\mathbf{a}), \infty\}$ holds for all $\mathbf{a} \in E_k^n$. In the following, the notation

$$g \subseteq_p f$$

is used to indicated the fact that g is a subfunction of f.
We take the notations introduced for total functions and we set

$$c_\infty^n(x_1, ..., x_n) := \infty,$$

i.e., c_∞^n is the notation for the n-ary function with empty domain, and

$$C_\infty := \{c_\infty^n \mid n \in \mathbb{N}\}.$$

The **reduction of** a function $f^n \in P_k$ to $T \subset E_k^n$ is a function defined by

$$g(\mathbf{x}) := \begin{cases} f(\mathbf{x}), & \text{if} \quad \mathbf{x} \in T, \\ \infty & \text{otherwise}, \end{cases}$$

which is also denoted with

$$f_{|T}.$$

Now we consider the operations over $\widetilde{P_k}$.

For arbitrary $f^n, g^m \in \widetilde{P_k}$ we define the unary Mal'tsev-operations analogous to Chapter 1 and the binary operation $\star$ as follows:

$$(\zeta f)(x_1, x_2, ..., x_n) := f(x_2, x_3, ..., x_n, x_1),$$
$$(\tau f)(x_1, x_2, ..., x_n) := f(x_2, x_1, x_3, ..., x_n),$$
$$(\Delta f)(x_1, x_2, ..., x_{n-1}) := f(x_1, x_1, x_2, ..., x_{n-1}) \text{ for } n \geq 2,$$
$$\zeta f = \tau f = \Delta f = f \text{ for } n = 1,$$
$$(\nabla f)(x_1, x_2, ..., x_{n+1}) := f(x_2, x_3, ..., x_{n+1}) \text{ and}$$
$$(f * g)(x_1, ..., x_{m+n-1}) :=$$

$$\begin{cases} f(g(x_1, ..., x_m), x_{m+1}, ..., x_{m+n-1}) & \text{if } g(x_1, ..., x_m) \in E_k, \\ \infty & \text{otherwise.} \end{cases}$$

We adopt the concepts (such as closure, closed set, clone ...) and the notations (such as [...]) coupled with the Mal'tsev-operations from Chapter 1. If distinctions are required, we complete these concepts with the word "partial". For example, "partial clone" instead of "clone" if a closed set of $\widetilde{P_k}$, which contains the set J_k of all projections of P_k, is meant.

A partial clone C is called **strong** if it contains all subfunctions of its functions.

If C is a clone of P_k, then let

$$Str(C) := \{f \in \widetilde{P_k} \mid \exists g \in P_k : \ g_{|D(f)} = f\}.$$

It is easy to check that $Str(C)$ is a partial strong clone.

As we see in Section 20.3, partial clones, like the clones of P_k, are describable with the help of relations (on $\widetilde{E_k}$) suitably chosen.

At first we want to show, however, how one can embed the lattice of the subclones of $\widetilde{P_k}$ into the lattice of the subclones of P_{k+1} isomorphically.

20.2 One-Point Extension

In this section, let

$$A := E_k \text{ and } B := E_k \cup \{\infty\}.$$

The following mappings establish relations between $\widetilde{P_A} := \widetilde{P_k}$ and P_B (isomorphic to P_{k+1}):

$$+: \widetilde{P_A} \longrightarrow P_B,\ f^n \mapsto f^n_+,$$

$$-: P_B \longrightarrow \widetilde{P_A},\ g^n \mapsto g^n_-$$

where f^n_+ is the so-called **extended function** defined by

$$f_+(\mathbf{x}) := \begin{cases} f(\mathbf{x}) & \text{if} \quad \mathbf{x} \in D(f), \\ \infty & \text{otherwise,} \end{cases}$$

and g^n_-, the so-called **restricted** function, is defined by

$$g_-(\mathbf{x}) := \begin{cases} g(\mathbf{x}) & \text{if} \quad \mathbf{x} \in E^n_k \ \wedge\ g(\mathbf{x}) \in E_k, \\ \infty & \text{otherwise.} \end{cases}$$

Furthermore, for subsets $F \subseteq P_A$ and $G \subseteq P_B$ we put:

$$F_+ := \{f_+ \mid f \in F\} \text{ and } G_- := \{g_- \mid g \in G\}.$$

To distinguish the projections from P_A of the projections from P_B, we use the notation $e^n_{i,X}$ with $X \in \{A, B\}$, i.e.,

$$\forall x_1, ..., x_n \in X :\ e^n_{i,X}(x_1, ..., x_n) := x_i$$

($i \in \{1, ..., n\}$) holds. Let J_X be the set of all projections of the form $e^n_{i,X}$. One can prove that, if $f \in \widetilde{P_A}$ and $D(f) \neq \emptyset$, $\nabla f_+ \notin F_+$ holds. In the following, the clones ($\subseteq P_B$), which are formed by closing certain subsets of P_B, are important to further considerations:

$$H := [(J_k)_+] \text{ and } U := [(\widetilde{P_k})_+].$$

The next Lemma gives properties of the clones defined above.

Lemma 20.2.1 *([Ros 88])*

(1) A function $q^t \in P_B$ belongs to U, if and only if the following two statements hold:
   (a) q preserves ∞ and
   (b) for every essential place i of q and for arbitrary $b_1, ..., b_n \in B$ it holds $q(b_1, ..., b_{i-1}, \infty, b_{i+1}, ..., _t) = \infty$.

(2) H is a minimal clone of P_B.

Proof. (1): Denote U' the set of all functions of P_B, which fulfills conditions (a) and (b) above. At first we show that U' is a clone. The equation $U' = U$ from which the assertion (1) immediately follows is then proven.
Let $f^n, g^m \in U'$ be arbitrary. Then obviously $\tau f, \zeta f, \nabla f \in U'$ and $\Delta f \in U'$ for $n = 1$. Let $n > 1$. If $\Delta\, f$ depends on the first place essentially, then f depends at least on the places 1 and 2 essentially. Consequently, we have $(\Delta\, f)(\infty, x_3, ..., x_n) = f(\infty, \infty, x_3, ..., x_n) = \infty$ for all $x_3, ..., x_n \in B$. If Δf depends on the i-th place with $2 \leq i \leq n-1$ essentially, then f depends on the $(i+1)$-th place essentially and

$$(\Delta\, f)(x_1, x_3, ..., x_{i-1}, \infty, x_{i+1}, ..., x_n) = f(x_1, x_1, x_3, ..., x_{i-1}, \infty, x_{i+1}, ..., x_n)$$
$$= \infty$$

holds for all $x_1, x_3, .., ,, x_n \in B$. Consequently, the function $\Delta\, f$ also belongs to U' in the case $n > 1$.
To prove $f \star g \in U'$, we put $h^{m+n-1} := f \star g$. Obviously h fulfills the condition (a). Let h be dependent of the i-th place and let $\mathbf{x} := (x_1, ..., x_{m+n-1}) \in B^{m+n-1}$ be arbitrary, where $x_i = \infty$. Then the following two cases are possible:
Case 1: $i \in \{1, ..., m\}$.
In this case, the first place of f and the i-th place of g is essential. Thus by assumption, we have $g(x_1, ..., x_m) = \infty$ and $h(\mathbf{x}) = f(\infty, x_{m+1}, ..., x_{m+n-1}) = \infty$.
Case 2: $i \in \{m+1, ..., m+n-1\}$.
Since in this case f depends on the $(i-m+1)$-th place essentially, $h(\mathbf{x}) = \infty$ holds.
It also holds that $h = f \star g \in U'$. Thus U' is a closed set. Since the function $e^2_{1,B}$ fulfills conditions (a) and (b), U' is a clone.
It is easy to check that $(P_A)_+ \subseteq U'$ and $U \subseteq U'$ hold. For the missing proof of $U' \subseteq U$, we consider an arbitrary function $q^t \in U'$ with at least an essential place. Using the operations ζ, τ, Δ we get from q a function $q_1 \in U'$ that depends on all its places essentially and which belongs to $(P_A)_+$. Since we can get again the function q by using the operations ∇, ζ, τ from q_1, we have $q \in [(P_A)_+] = U$. Thus $U' = U$ and (1) is proven.

(2): It is easy to check that $J_B \subset [(e^2_{1,A})_+] = H$ holds. To prove the maximality of J_B in H we put $b := |B|$ and $\varrho := \{(x_1, ..., x_b) \in B^b \mid |\{x_1, ..., x_b\}| = b\}$. Since $(e^2_{1,A})_+ \in Pol_B\varrho$, we have $H \subseteq Pol_B\varrho$.
Let $q^t \in H \setminus J_B$ be arbitrary. Since no constants belong to $Pol_B\varrho$, the function q has at least an essential place i. Suppose i is the only one essential place of q. Then we have

$$q(x_1, ..., x_{i-1}, x_i, x_{i+1}, ..., x_n) = g(x_i, ..., x_i, ..., x_i) = x_i,$$

i.e., $q^t = e^t_{i,B}$, in contradiction to $q \notin J_B$. Thus the function q has also at least two essential places. Using the operations ζ, τ, Δ we get from q a binary function g, which depends on both places essentially, preserves all elements of B, and belongs to $H \subset U$; i.e., for all $x \in B$ it holds:

$$g(x,x) = x,\ g(x,\infty) = g(\infty,x) = \infty,\ g(\infty,\infty) = \infty.$$

Consequently, $g = (e^2_{1,A})_+$ and therefore $H \subseteq [g] \subseteq H$. Thus J_B is a maximal clone of H. ■

Lemma 20.2.2 *For arbitrary $f^n, g^m \in U$ holds:*

(1) $\alpha(f_-) = (\alpha\, f)_-$ for every operation $\alpha \in \{\zeta, \tau, \Delta, \nabla\}$;
(2) $f_- \star g_- = h_-$, where

$$h^{m+n-1}(x_1, ..., x_{m+n-1}) :=$$
$$(e^2_1)_+(f(g(x_1, ..., x_m), x_{m+1}, ..., x_{m+n-1}), g(x_1, ..., x_m));$$

(3) $f_- \star g_- \subseteq_p (f \star g)_-$;
(4) $f_- \star g_- = (f \star g)_-$, if the first place of f is essential;
(5) $f_- \star g_- = (f \star g)_-$, if $g_- \in P_k$.

Proof. (1): For $\alpha \in \{\zeta, \tau, \delta\}$ the assertion is obvious. For arbitrary $(x_1, ..., x_{n+1}) \in A$ is by definition $(\nabla f)(x_1, ..., x_{n+1}) = f(x_2, ..., x_{n+1})$. Furthermore $(x_1, ..., x_{n+1}) \in D((\nabla f)_-)$ holds iff $(x_2, ..., x_{n+1}) \in D(f_-)$. Consequently, we have $(\nabla f)_-(x_1, ..., x_{n+1}) = ((\nabla f)_-)(x_2, ..., x_{n+1})$ for arbitrary $(x_2, ..., x_{n+1}) \in D(f_-)$.
(b2): Let $\mathbf{x} \in A^{m+n-1}$ be arbitrary. It is easy to check that $h(\mathbf{x}) = \infty$ iff $(f \star g)(\mathbf{x}) = \infty$. If $h(\mathbf{x}) \neq \infty$, then $h(\mathbf{x}) = (f \star g)(\mathbf{x})$. Consequently, $D(h_-) = \{(x_1, ..., x_{m+n-1}) \in A^{m+n-1} \mid (x_1, ..., x_m) \in D(g_-) \wedge (g_-(x_1, ..., x_m), x_{m+1}, ..., x_{n+m-1}) \in D(f_-)\} = D(f_- \star g_-)$. Thus, for $\mathbf{x} := (x_1, ..., x_{m+n-1}) \in D(h_-)$ we have $g(x_1, ..., x_m) = g_-(x_1, ..., x_m)$ and $h(\mathbf{x}) = (f \star g)(\mathbf{x}) \in A$. Hence $h_- = f_- \star g_-$.
(3): Let $\mathbf{x} := (x_1, ..., x_{m+n-1}) \in D(f_- \star g_-)$ be arbitrary. Then $(x_1, ..., x_m) \in D(g_-)$ and $(g(x_1, ..., x_m), x_{m+1}, ..., x_{m+n-1}) \in D(f_-)$. Thus $(f \star g)(\mathbf{x}) \neq \infty$. Consequently, $\mathbf{x}$ belongs to $D((f \star g)_-)$ and we have

$$(f_- \star g_-)(\mathbf{x}) = f(g(x_1, ..., x_m), x_{m+1}, ..., x_{m+n-1}) = (f \star g)(\mathbf{x}),$$

i.e., $f_- \star g_- \subseteq_p (f \star g)_-$.
(4): Let the first place of f be essential and let $\mathbf{x} := (x_1, ..., x_{m+n-1}) \in D((f \star g)_-)$ be arbitrary. Because of $f \in U$ it follows that $g(x_1, ..., x_m) \neq \infty$. Thus $(x_1, ..., x_m) \in D(g_-)$ and $(g(x_1, ..., x_m), x_{m+1}, ..., x_{m+n-1}) \in D(f_-)$. Hence $\mathbf{x} \in D(f_- \star g_-)$ and by (3) we have $f_- \star g_- = (f \star g)_-$.
(5): Let $g \in P_A$ and let $\mathbf{x} := (x_1, ..., x_{m+n-1}) \in A^{m+n-1}$ be arbitrary. Then, $(x_1, ..., x_m) \in D(g_-)$ and, because of

$$\mathbf{x} \in D(f_- \star g_-) \iff (g(x-1, ..., x_m), x_{m+1}, ..., x_{m+n-1}) \in D(f_-)$$
$$\iff \mathbf{x} \in D((f \star g)_-),$$

the assertion follows with the help of (c). ■

Lemma 20.2.3 *For arbitrary $f, g \in \widetilde{P_A}$ it holds:*

(1) $\alpha(f_+) = (\alpha\, f)_+$ for every operation $\alpha \in \{\zeta, \tau, \Delta\}$;
(2) $\nabla(f_+) \neq (\nabla f)_+$, if $f \notin C_\infty$;
(3) $f_+ \star g_+ = (f \star g)_+$.

Proof. (1) and (3) are immediate conclusions from the definition of the mapping $+$.
(2) is easy to check. ■

Lemma 20.2.4 *Let $G \subseteq U \subseteq P_B$ be a clone and let $F \subseteq \widetilde{P_A}$ be a partial clone with $G_- \subseteq F$. Then*

$$([G])_- \subseteq F. \tag{20.1}$$

Proof. Let $f, g \in G$ be arbitrary. Then by assumption the function f_- and g_- belong to F. Because of Lemma 20.2.2, (1) it follows that $(\alpha f)_- = \alpha(f_-) \in F$ for every $\alpha \in \{\zeta, \tau, \Delta, \nabla\}$.
Let $h := f \star g \in G$. If the first place of f is essential, then $h_- = (f \star g)_- = f_- \star g_- \in F$ holds by Lemma 20.2.2, (4). If the first place of f is fictitious, then $h^{m+n-1} = f^n \star g^m = f^n \star e^m_{1,B}$. Since $(e^m_{1,B})_- = e^m_{1,A}$, we have $h_- = f_- \star e^m_{1,A}$ by Lemma 20.2.2, (5). ■

Theorem 20.2.5 *([Ros 88], [Bör-P 90], [Bör 97])*

(1) For every partial clone $F \subseteq \widetilde{P_A}$ we have $F = ([F_+])_-$.
(2) For every clone $G \subseteq P_B$ with $H \subseteq G$ the set G_- is a partial clone of $\widetilde{P_A}$ with the property $G = [(G_-)]_+$.
(3) The mapping

$$\varphi : \mathbb{L}_B(H; U) \longrightarrow \mathbb{L}_A(J_A; \widetilde{P_A}),\ G \mapsto G_- \tag{20.2}$$

is a lattice isomorphism between the lattices $\mathbb{L}_\mathbf{B}(\mathbf{H}, \mathbf{U})$ and $\mathbb{L}_\mathbf{A}(\mathbf{J_A}; \widetilde{\mathbf{P_A}})$, where $\varphi^{-1}(F) = [F_+]$ holds for every $F \in \mathbb{L}_A(J_A; \widetilde{P_A})$.

Proof. (1): Because of $f = (f_+)_-$ for every $f \in P_A$ we have $F \subseteq ([F]_+)_-$. It follows from $(F_+)_- \subseteq F$ and Lemma 20.2.4 that $([F_+])_- \subseteq F$.

(2): Let r^n and s^m be two arbitrary functions of G_-. Then there are two total functions u^n and v^m of G with $u_- = r$ and $v_- = s$. Because of Lemma 20.2.1, (a) we have $\alpha r = \alpha(u_-) = (\alpha u)_- \in G_-$ for all $\alpha \in \{\zeta, \tau, \Delta, \nabla\}$. Furthermore, by Lemma 20.2.1, (b) the following is valid:

$$\begin{aligned}
& (r \star s)(x_1, ..., x_{m+n-1}) \\
= \ & (u_- \star v)(x_1, ..., x_{m+n-1}) \\
= \ & ((e^2_{1,A})_+(u(v(x_1, ..., x_m), x_{m+1}, ..., x_{m+n-1}), v(x_1, ..., x_m)))_-.
\end{aligned}$$

Thus $r \star s \in G_-$. Consequently, G_- is a partial clone.
Let $h_1^n \in G \subseteq U$ be arbitrary. By using the operations ζ, τ, Δ one can form

from h_1 a function h_2^m, which depends on all its places essentially. Then, for the function h_2 and for arbitrary $j \in \{1, ..., m\}$ and arbitrary $x_1, ..., x_m \in B$, it holds:

$$h_2(x_1, ..., x_{j-1}, \infty, x_{j+1}, ...x_m) = \infty,$$

i.e., $h_2 = ((h_2)-)_+ \in (G_-)_+$. With the help of the operations $\zeta, \tau, \Delta, \nabla$, one can form the function h_1 from the function h_2. Consequently, $h_1^n \in [(G_-)_+]$ and $G \subseteq [(G_-)_+]$ hold.
Let $h_3^t \in G_-$ be arbitrary. Then there is a function $h_4^t \in G$ with $(h_4)_- = h_3$. Because of $H \subseteq G$ we have $(e_{1,A}^{t+1})_+ \in G$. Then, for the function h_5^t with

$$h_5(x_1, ..., x_t) := (e_{1,A}^{t+1})_+(h_4^t(x_1, ..., x_t), x_1, ..., x_t)$$

it follows: $h_5 \in G$ and $h_5 = (h_3)_+$. Thus $(G_-)_+ \subseteq G$.

(3): Let G_1 and G_2 be two different clones of P_B with $H \subseteq G_1 \subset G_2 \subseteq U$. By (1) $(G_1)_-$ and $(G_2)_-$ are partial clones of $\mathbb{L}_A[J_A, P_A]$. Obviously, $(G_1)_- \subseteq (G_2)_-$. Suppose, $(G_1)_- = (G_2)_-$. Then, by (2) it follows that $G_1 = [((G_1)_-)_+] = [((G_2)_-)_+] = G_2$, in contradiction to $G_1 \neq G_2$. Consequently, the mapping φ is an order-preserving bijective mapping from $\mathbb{L}_B(H; U)$ into $\mathbb{L}_A(J_A; P_A)$.
Let $F \in \mathbb{L}_A[J_A, P_A]$ be arbitrary. Then $G := [F_+]$ is a clone with $H \subseteq G \subseteq U$, since if $e_{1,A}^2 \in F$, then $(e_{1,A}^2)_+ \in G$ and $F_+ \subseteq (P_A)_+ \subseteq U$ also hold. Further, we have $\varphi(G) = ([F_+])_- = F$ by (1). Consequently, φ is surjective and $\varphi^{-1}(F) = [F_+]$. ■

20.3 Description of Partial Clones by Relations

To describe closed subsets of $\widetilde{P_k}$, h-ary relations (i.e., subsets of $\widetilde{E_k}^h$), $h \geq 1$, are suitable. We often write the elements of relations in the form of columns and we often give a relation in the form of a matrix, the columns of which are the elements of the relation.
The set of all h-ary relations over $\widetilde{E_k}^h$ let

$$\widetilde{R_k}^h$$

and we put

$$\widetilde{R_k} := \bigcup_{h \geq 1} \widetilde{R_k}^h.$$

We say that a **function** $f \in \widetilde{P_k}$ **preserves an** h**-ary relation** ϱ **over** $\widetilde{E_k}$, iff for all $\mathbf{r_1}, \mathbf{r_2}, ..., \mathbf{r_n} \in \varrho$ with $\mathbf{r_i} := (r_{1i}, r_{2i}, ..., r_{hi})$, $i = 1, 2, ..., n$, it holds

$$f(\mathbf{r_1}, ..., \mathbf{r_n}) := \begin{pmatrix} f(r_{11}, r_{12}, ..., r_{1n}) \\ f(r_{21}, r_{22}, ..., r_{2n}) \\ \vdots \\ f(r_{h1}, r_{h2}, ..., r_{hn}) \end{pmatrix} \in \varrho,$$

where $f(\mathbf{a}) = \infty$ is defined for all $\mathbf{a} \in \widetilde{E_k}^n \setminus E_k^h$.
Let

$$pPol_k \varrho$$

be the set of all functions of $\widetilde{P_k}$ that preserve the relation $\varrho \subseteq \widetilde{E_k}^h$. Furthermore,

$$pPOL_k \varrho := pPol_k(\varrho \cup (\widetilde{E_k}^h \setminus E_k^h)).$$

The following Lemma is easy to check.

Lemma 20.3.1 *For every relation $\varrho \in R_k^h$ is valid:*

(1) $pPol_k \rho$ and $pPOL_k \rho$ are partial clones.
(2) $Str(Pol_k \varrho) \subseteq pPOL_k \varrho$. ■

Lemma 20.3.2 *For each relation $\varrho \in \widetilde{R_k}^h$, the following conditions are equivalent:*

(1) $pPol_k \varrho$ is a partial clone;
(2) $e_1^2 \in pPol_k \varrho$;
(3) If $(a_1, \ldots, a_h), (b_1, \ldots, b_h) \in \varrho$ and $(c_1, \ldots, c_h)$ is defined by

$$c_i := \begin{cases} a_i & \text{if } b_i \in E_k, \\ \infty & \text{if } b_i = \infty, \end{cases}$$

$(i = 1, ..., n)$, then $(c_1, \ldots, c_h) \in \varrho$ holds.

Proof. Obviously, (1) $\Longleftrightarrow$ (2) holds by definition. (2) $\Longleftrightarrow$ (3) follows from

$$e_1^2 \begin{pmatrix} a_1 & b_1 \\ \vdots & \vdots \\ a_h & b_h \end{pmatrix} = \begin{pmatrix} c_1 \\ \vdots \\ c_h \end{pmatrix}.$$

■

The following two lemmas give important properties that are needed to determine the maximal partial clones. [1] We need the following concept for the wording of the lemmas:

The relation $\varrho \in R_k^h$ is called **irredundant**, iff it fulfills the following two conditions:
1) for all i, j with $1 \leq i < j \leq h$, there is a tuple $(a_1, ..., a_h) \in \varrho$ with $a_i \neq a_j$;
2) No $i \in \{1, ..., h\}$ exists, such that $(a_1, ..., a_h) \in \varrho$ implies $(a_1, ..., a_{i-1}, x, a_{i+1}, ..., a_h) \in \varrho$ for all $x \in E_k$.

We say that for $i = 1, ..., n$ the relations $\chi_i \subseteq \{1, ..., t\}^{h_i}$ **cover** the set $\{1, ..., t\}$, if for every $x \in \{1, ..., t\}$ there exists an $i \in \{1, ..., n\}$ such that $x \in \{a_1, ..., a_{h_i}\}$ holds for at least a tuple $(a_1, ..., a_{h_i}) \in \chi_i$.

[1] As one can gather from Sections 20.4 and 20.5, the maximal partial clones are all strong clones, with an exception.

Lemma 20.3.3 *([Rom 81], without proof) Let $C \subseteq \widetilde{P_k}$ be a strong partial clone. Then there is a certain nonempty set $M \subseteq R_k$ of irredundant relations with $C = \bigcap_{\varrho \in M} pPOL_k\varrho$.*

Lemma 20.3.4 *(**Representation Lemma of B. A. Romov**, [Rom 81]; without proof) Let $\alpha_i \subseteq E_k^{h_i}$ for $i = 1, ..., n$ and let $\beta \subseteq E_k^t$ be an irredundant relation. Then*

$$\bigcap_{i=1}^{n} pPOL_k\alpha_i \subseteq pPOL_k\beta$$

if and only if there are certain (help-)relations $\chi_i \subseteq \{1, ..., t\}^{h_i}$ for $i = 1, ..., n$ that cover the set $\{1, ..., t\}$ and for which

$$\beta = \{(b_1, ..., b_t) \in E_k^t \mid \forall j \in \{1, ..., n\}\ \forall (i_1^j, ..., i_{h_j}^j) \in \chi_j : (b_{i_1^j}, ..., b_{i_{h_j}^j}) \in \alpha_j\}$$

holds.

20.4 The Maximal Partial Classes of $\widetilde{P_2}$ and $\widetilde{P_3}$

Lemma 20.4.1 *Let f be an n-ary function of P_3, which essentially depends on at least two variables (w.l.o.g. let x_1 and x_2 be the essential variables). Furthermore, let $\{a, b, c\} = E_3$ and $\delta_I := \{(a_0, a_1, a_2) \in E_3^3, \mid \forall \alpha, \beta \in I : a_\alpha = a_\beta\}$, $I \subseteq E_k$. Then*
(a) $Im(f) = E_3 \Rightarrow \exists\, \mathbf{r_1}, ..., \mathbf{r_n} \in \delta^3_{\{0,1\}} \cup \delta^3_{\{1,2\}} : f(\mathbf{r_1}, ..., \mathbf{r_n}) \in E_3^3 \backslash \iota_3^3$;
(b) $|Im(f)| = 2 \Rightarrow \exists\, \mathbf{r_1}, ..., \mathbf{r_n} \in \delta^3_{\{a,b\}} \cup \delta^3_{\{b,c\}} : f(\mathbf{r_1}, ..., \mathbf{r_n}) \in \delta^3_{\{a,c\}} \backslash \delta^3_{\{0,1,2\}}$;
(c) $Im(f) = E_3 \Rightarrow [\{f\} \cup \{\, g \in P_3^1 \mid g(a) = g(b) \ \vee\ g(b) = g(c)\, \}] = P_3$.

Proof. (a) is a special case of the "fundamental lemma of Jablonskij" and (b) is an easy conclusion from this lemma (see Theorem 1.4.4).
(c): W.l.o.g. let $a = 0$, $b = 1$ and $c = 2$. Obviously, then, we have

$$\begin{aligned} &\{\, g \in P_3^1 \mid g(0) = g(1) \ \vee\ g(1) = g(2)\, \} \\ =\ &\{c_0, c_1, c_2, j_\alpha, u_\alpha, v_\alpha \mid \alpha \in \{0, 2, 3, 5\}\}. \end{aligned}$$

It is easy to check that this set of unary functions is not a subset of maximal classes of type $\mathfrak{M}$, $\mathfrak{U}$, $\mathfrak{S}$, $\mathfrak{C}$ and $\mathfrak{L}$. Since $Im(f) = E_3$ and $f \in P_3 \backslash [P_3^1]$ hold, f does not preserve by (a) the relation ι_3^3. Thus (c) follows from the completeness criterion for P_3. ■

Lemma 20.4.2 (a) *For every $g \in \widetilde{P_k} \backslash (P_k \cup [\{c_\infty\}])$ it holds $[P_k \cup \{g\}] = \widetilde{P_k}$.*
(b) *$P_k \cup [\{c_\infty\}]$ is the only maximal class of $\widetilde{P_k}$ that contains P_k.*

Proof. (a): Let $g^m \in \widetilde{P_k} \backslash (P_k \cup [\{c_\infty\}])$. Since the function $g' := e_2^2 * g$ has $k + 1$ different values, we can assume w.l.o.g. $Im(g) = \widetilde{E_k}$. Consequently, there are $k+1$ tuples $\mathbf{a_i} := (a_{i1}, a_{i2}, ..., a_{im}) \in E_k^m$ with $g(\mathbf{a_i}) = i$ ($i \in \widetilde{E_k}$). Let f^n be an arbitrary function of $\widetilde{P_k}$. Independently from f, one can define the following functions f_j ($j = 1, 2, ..., m$):

$$f_j(b_1, ..., b_n) = a_{ij} :\Longleftrightarrow f(b_1, ..., b_n) = i$$

($b_1, ..., b_n \in E_k$; $i \in \widetilde{E_k}$). Thus we have $f(\mathbf{x}) = g(f_1(\mathbf{x}), ..., f_m(\mathbf{x}))$. Hence $f \in [P_k \cup \{g\}]$.
(b) follows directly from (a). ∎

The next four lemmas deal with the maximality of certain subclasses of $\widetilde{P_3}$ (or $\widetilde{P_k}$) in $\widetilde{P_3}$ (or $\widetilde{P_k}$), respectively.

The following statement (a) was proven in [Bur 67] (see also Chapter 4) and (b) of the following lemma was proven in [Rom 80] (or in [Lau 77], [Lau 88]).

Lemma 20.4.3 *Let*

$$\begin{aligned}
\varrho_1 &:= \{(a,a,b,b),(a,b,a,b) \mid a,b \in E_k\},\\
\varrho_2 &:= \{(a,a,b,b),(a,b,a,b),(a,b,b,a) \mid a,b \in E_k\},\\
\varrho_i &:= \{(a_1,a_2,...,a_i) \in E_k^i \mid |\{a_1,...,a_i\}| \le i-1\} \ (i = 3, ..., k).
\end{aligned}$$

Then
(a) *The classes $Pol_k\varrho_i$ ($i = 1,2,...,k$) are the only proper subclasses of P_k that contain P_k^1. Furthermore, it holds that*

$$[P_k^1] = Pol_k\varrho_1 \subset Pol_k\varrho_2 \subset ... \subset Pol_k\varrho_{k-1} \subset Pol_k\varrho_k \subset P_k.$$

(b) *The classes $pPOL_k\varrho_i$ ($i = 1, 2, ..., k$) are maximal classes of $\widetilde{P_k}$ and the only maximal classes of $\widetilde{P_k}$ that contain P_k^1.* ∎

Lemma 20.4.4 *Let $\varrho := \delta \cup \sigma$, where δ denotes a certain h-ary diagonal relation, which is different from E_k^h, and σ fulfills the condition $\emptyset \ne \sigma \subseteq \{(a_1, ..., a_h) \in E_k^h \mid |\{a_1, ..., a_h\}| = h\}$.*
There exists to every $\mathbf{a} := (a_1, ..., a_h) \in \sigma$ a certain equivalence relation ε_a on E_k with the following two properties:

(1) For every $i \in \{1, ..., h\}$ there exists exactly an equivalence class of ε_a which contains a_i.
(2) To every $\mathbf{b} \in \varrho$ one can find in $pPOL_k\varrho$ a unary function $g_{\mathbf{a},\mathbf{b}}$ with $g_{\mathbf{a},\mathbf{b}}(\mathbf{a}) = \mathbf{b}$ and $g_{\mathbf{a},\mathbf{b}}(x) = g_{\mathbf{a},\mathbf{b}}(y)$ for all $(x, y) \in \varepsilon_{\mathbf{a}}$.

Then $pPOL_k\varrho$ is a maximal class of $\widetilde{P_k}$.

Proof. Obviously, $pPOL_k\varrho \ne \widetilde{P_k}$. Let $f \in \widetilde{P_k} \backslash pPOL_k\varrho$. Since ϱ has the properties (1) and (2), one gets a certain unary function h_r with $h_r(r) \in E_k^h \backslash \varrho$ as a superposition over unary functions of $pPOL_k\varrho$ and f for every $r \in \sigma$. Let $\sigma = \{r_1, ..., r_m\}$. One can find in $pPOL_k\varrho$ certain functions, which arbitrary values have of $\widetilde{E_k}$ on rows of the form

$$(x_1, x_2, g_{r_1}(x_1), g_{r_2}(x_1), ..., g_{r_m}(x_1), g_{r_1}(x_2), g_{r_2}(x_2), ..., g_{r_m}(x_2))$$

and otherwise only have the value ∞. Consequently, arbitrary functions of P_k^2 are superpositions over $\{f\} \cup pPOL_k\varrho$. Hence (by well-known properties of P_k) it follows that $P_k \subseteq [\{f\} \cup pPOL_k\varrho]$. Since $pPOL_k\varrho$ obtains functions with exactly $k+1$ different values, it follows from Lemma 20.4.2, (a) that $[\{f\} \cup pPOL_k\varrho] = \widetilde{P_k}$ holds. Hence $pPOL_k\varrho$ is a maximal class of $\widetilde{P_k}$. ■

Table 20.1

i	τ_i	i	τ_i
1	$\{0\}$	2	$\{1\}$
3	$\{2\}$	4	$\{0,1\}$
5	$\{0,2\}$	6	$\{1,2\}$
7	$\begin{pmatrix} 0 \\ 1 \end{pmatrix}$	8	$\begin{pmatrix} 0 \\ 2 \end{pmatrix}$
9	$\begin{pmatrix} 1 \\ 2 \end{pmatrix}$	10	$\begin{pmatrix} 0 & 1 \\ 1 & 0 \end{pmatrix}$
11	$\begin{pmatrix} 0 & 2 \\ 2 & 0 \end{pmatrix}$	12	$\begin{pmatrix} 1 & 2 \\ 2 & 1 \end{pmatrix}$
13	$\begin{pmatrix} 0 & 0 \\ 1 & 2 \end{pmatrix}$	14	$\begin{pmatrix} 0 & 2 \\ 1 & 1 \end{pmatrix}$
15	$\begin{pmatrix} 0 & 1 \\ 2 & 2 \end{pmatrix}$	16	$\begin{pmatrix} 0 & 1 & 0 & 2 \\ 1 & 0 & 2 & 0 \end{pmatrix}$
17	$\begin{pmatrix} 0 & 1 & 1 & 2 \\ 1 & 0 & 2 & 1 \end{pmatrix}$	18	$\begin{pmatrix} 0 & 2 & 1 & 2 \\ 2 & 0 & 2 & 1 \end{pmatrix}$
19	$\begin{pmatrix} 0 & 1 & 2 & 0 \\ 0 & 1 & 2 & 1 \end{pmatrix}$	20	$\begin{pmatrix} 0 & 1 & 2 & 0 \\ 0 & 1 & 2 & 2 \end{pmatrix}$
21	$\begin{pmatrix} 0 & 1 & 2 & 1 \\ 0 & 1 & 2 & 2 \end{pmatrix}$	22	$\begin{pmatrix} 0 & 1 & 2 & 0 & 1 \\ 0 & 1 & 2 & 1 & 0 \end{pmatrix}$
23	$\begin{pmatrix} 0 & 1 & 2 & 0 & 2 \\ 0 & 1 & 2 & 2 & 0 \end{pmatrix}$	24	$\begin{pmatrix} 0 & 1 & 2 & 1 & 2 \\ 0 & 1 & 2 & 2 & 1 \end{pmatrix}$
25	$\begin{pmatrix} 0 & 1 & 2 & 0 & 0 \\ 0 & 1 & 2 & 1 & 2 \end{pmatrix}$	26	$\begin{pmatrix} 0 & 1 & 2 & 0 & 2 \\ 0 & 1 & 2 & 1 & 1 \end{pmatrix}$
27	$\begin{pmatrix} 0 & 1 & 2 & 0 & 1 \\ 0 & 1 & 2 & 2 & 2 \end{pmatrix}$	28	$\begin{pmatrix} 0 & 1 & 2 & 0 & 2 \\ 0 & 1 & 2 & 1 & 0 \end{pmatrix}$
29	$\begin{pmatrix} 0 & 1 & 2 & 1 & 2 \\ 0 & 1 & 2 & 0 & 1 \end{pmatrix}$	30	$\begin{pmatrix} 0 & 1 & 2 & 2 & 1 \\ 0 & 1 & 2 & 0 & 2 \end{pmatrix}$
31	$\begin{pmatrix} 0 & 1 & 2 & 0 & 0 & 2 \\ 0 & 1 & 2 & 1 & 2 & 1 \end{pmatrix}$	32	$\begin{pmatrix} 0 & 1 & 2 & 0 & 2 & 2 \\ 0 & 1 & 2 & 1 & 1 & 0 \end{pmatrix}$
33	$\begin{pmatrix} 0 & 1 & 2 & 0 & 0 & 1 \\ 0 & 1 & 2 & 1 & 2 & 2 \end{pmatrix}$	34	$\begin{pmatrix} 0 & 1 & 2 & 0 & 1 & 0 & 2 \\ 0 & 1 & 2 & 1 & 0 & 2 & 0 \end{pmatrix}$
35	$\begin{pmatrix} 0 & 1 & 2 & 0 & 1 & 1 & 2 \\ 0 & 1 & 2 & 1 & 0 & 2 & 1 \end{pmatrix}$	36	$\begin{pmatrix} 0 & 1 & 2 & 0 & 2 & 1 & 2 \\ 0 & 1 & 2 & 2 & 0 & 2 & 1 \end{pmatrix}$
37	$\begin{pmatrix} 0 \\ 1 \\ 2 \end{pmatrix}$	38	$\begin{pmatrix} 0 & 0 \\ 1 & 2 \\ 2 & 1 \end{pmatrix}$
39	$\begin{pmatrix} 0 & 2 \\ 1 & 1 \\ 2 & 0 \end{pmatrix}$	40	$\begin{pmatrix} 0 & 1 \\ 1 & 0 \\ 2 & 2 \end{pmatrix}$

Table 20.2

i	τ_i	i	τ_i
41	$\begin{pmatrix} 0 & 1 & 2 \\ 1 & 2 & 0 \\ 2 & 0 & 1 \end{pmatrix}$	42	$\begin{pmatrix} 0 & 0 & 1 & 1 & 2 & 2 \\ 1 & 2 & 0 & 2 & 0 & 1 \\ 2 & 1 & 2 & 0 & 1 & 0 \end{pmatrix}$
43	$\begin{pmatrix} 0 & 1 & 2 & 0 \\ 0 & 1 & 2 & 1 \\ 0 & 1 & 2 & 2 \end{pmatrix}$	44	$\begin{pmatrix} 0 & 1 & 2 & 0 & 0 \\ 0 & 1 & 2 & 1 & 2 \\ 0 & 1 & 2 & 2 & 1 \end{pmatrix}$
45	$\begin{pmatrix} 0 & 1 & 2 & 0 & 2 \\ 0 & 1 & 2 & 1 & 1 \\ 0 & 1 & 2 & 2 & 0 \end{pmatrix}$	46	$\begin{pmatrix} 0 & 1 & 2 & 0 & 1 \\ 0 & 1 & 2 & 1 & 0 \\ 0 & 1 & 2 & 2 & 2 \end{pmatrix}$
47	$\begin{pmatrix} 0 & 1 & 2 & 0 & 1 & 2 \\ 0 & 1 & 2 & 1 & 2 & 0 \\ 0 & 1 & 2 & 2 & 0 & 1 \end{pmatrix}$	48	$\begin{pmatrix} 0 & 1 & 2 & 0 & 0 & 1 & 1 & 2 & 2 \\ 0 & 1 & 2 & 1 & 2 & 0 & 2 & 0 & 1 \\ 0 & 1 & 2 & 2 & 1 & 2 & 0 & 1 & 0 \end{pmatrix}$

Table 20.3

i	τ_i
49	$\{\ (0,1,2),(a,a,b)\mid a,b\in E_3\ \}$
50	$\{\ (0,1,2),(a,b,a)\mid a,b\in E_3\ \}$
51	$\{\ (0,1,2),(b,a,a)\mid a,b\in E_3\ \}$
52	$\{\ (0,1,2),(1,0,2),(a,a,b)\mid a,b\in E_3\ \}$
53	$\{\ (0,1,2),(2,1,0),(a,b,a)\mid a,b\in E_3\ \}$
54	$\{\ (0,1,2),(0,2,1),(b,a,a)\mid a,b\in E_3\ \}$
55	$\{\ (a,a,b,b),(a,b,a,b)\mid a,b\in E_3\ \}$
56	$\{\ (a,a,b,b),(a,b,a,b),(a,b,b,a)\mid a,b\in E_3\ \}$
57	$\{\ (a,b,c)\in E_3^3\mid \lvert\{a,b,c\}\rvert\ \le\ 2\ \}$

Lemma 20.4.5

(1) For every $\varrho\in\{\{0\},\{1\},\{(0,1)\},\{(0,1),(1,0)\},\{(0,0),(0,1),(1,1)\},\lambda_2:=\{(a,a,b,b),(a,b,a,b),(a,b,b,a)\mid a,b\in E_2\},G_2([P_2^1])\}$, the set $pPOl_2\varrho$ is a maximal class of $\widetilde{P_2}$.

(2) The classes $pPOL_3\tau_i$ ($i\in\{1,2,...,57\}$, see Tables 20.1 - 20.3) are maximal classes of $\widetilde{P_3}$.

Proof. (1): The maximality of $pPOL_2\varrho$ for $\varrho\in\{\lambda_2,G_2([P_2^1])\}$ was proven in Lemma 20.4.3.

Next we show that $pPOL_2\{(0,1)\}$ is a maximal class of $\widetilde{P_2}$: Let $f^n\in\widetilde{P_2}\setminus pPOL_2\{(0,1)\}$ be arbitrary. Then $\Delta^{n-1}f\in\{c_0,c_1,\overline{e_1^1}\}$. We distinguish two cases:

Case 1: $c_a\in[f]$ for certain $a\in E_2$.

The unary function g with $g(a)=\overline{a}$ and $g(\overline{a})=\infty$ belongs to $pPOL_2\{(0,1)\}$, whereby $\{c_0,c_1\}\subset[\{f\}\cup pPOL_2\{(0,1)\}]$. With the aid of Theorem 3.2.4.1, it is easy to prove that $P_2=[(T_0\cap T_1)\cup\{c_0,c_1\}]\subseteq[\{f\}\cup pPOL_2\{(0,1)\}]$.

Further, we have $pPOL_2\{(0,1)\} \not\subseteq P_2 \cup [c_\infty]$. Therefore, by Lemma 20.4.2, $[\{f\} \cup pPOL_2\{(0,1)\}] = \widetilde{P_2}$ holds, i.e., $pPOL_2\{(0,1)\}$ is a maximal class of $\widetilde{P_2}$.

Case 2: $\overline{e_1^1} \in [f]$.

In this case, by Theorem 3.2.4.1, $P_2 = [(T_0 \cap T_1) \cup \{\overline{e_1^1}\}] \subseteq [\{f\} \cup pPOL_2\{(0,1)\}]$, whereby Case 2 is reducible to Case 1.

The maximality of the other classes $pPOL_2\varrho$ is easy to check.

(2): For $i \in \{1,2,...,54\}$ one can easily prove the lemma with the help of Lemma 20.4.4. If $i \in \{55,56,57\}$, the above statement follows from Lemma 20.4.3. ∎

Theorem 20.4.6 *([Fre 66])*
$\widetilde{P_2}$ has exactly 8 maximal classes:
$pPOL_2\sigma_i$, where $\sigma_0 := \{0\}$, $\sigma_1 := \{1\}$, $\sigma_2 := \{(0,1)\}$, $\sigma_3 := \{(0,1),(1,0)\}$, $\sigma_4 := \{(0,0),(0,1),(1,1)\}$, $\sigma_5 := G_2([P_2^1]) = \{(a,a,b,b),(a,b,a,b) \mid a,b \in E_2\}$ and $\sigma_6 := \{(a,a,b,b),(a,b,a,b),(a,b,b,a) \mid a,b \in E_2\}$, and $P_2 \cup [c_\infty] = pPol_2(E_k^2 \cup \{(\infty,\infty)\})$.

Proof. Because of Lemmas 20.4.2 and 20.4.5, we must show that each subset A of $\widetilde{P_2}$, which fulfills $A \not\subseteq pPOL_3\tau_i$ for all $i \in \{0,1,2,...,6\}$ and $A \not\subseteq P_2 \cup [c_\infty]$, is complete in $\widetilde{P_2}$.

Let $A \subseteq \widetilde{P_2}$ be an arbitrary set, for which there are $f_0,...,f_6,f_7 \in A$ with $f_i \notin pPOL_2\sigma_i$ $(i = 0,1,...,6)$ and $f_7 \notin P_2 \cup [c_\infty]$.

Let $g \in \widetilde{P_2}$ defined by $g(x) := f_2(x,x,...,x)$. Then $g \in [A]$ and $g \in \{c_0,c_1,\overline{e_1^1}\}$. Consequently, since $c_1 \in [f_0,c_0]$, $c_0 \in [f_1,c_1]$, $\{c_0,c_1,e_1^1\} \subset [f_3,\overline{e_1^1}]$ and $\overline{e_1^1} \in [c_0,c_1,f_4]$, we have $P_2^1 \subset [A]$. Thus there exists a function $h \in [\{f_5\} \cup P_2^1] \subseteq [A]$ with $h \in P_2 \setminus [P_2^1]$. We distinguish two cases:

Case 1: h is non-linear.

In this case, by Theorem 3.2.4.1, we have $[P_2^1 \cup \{h\}] = P_2 \subseteq [A]$. Because of $A \not\subseteq P_2 \cup [c_\infty]$ and Lemma 20.4.2, this implies $[A] = \widetilde{P_2}$.

Case 2: $h \notin [P_2^1]$ is a linear function.

Then it is easy to check that all binary linear functions belong to $[A]$, whereby (by $f_6 \in A$) a non-linear function of P_2 is a superposition over A. Thus Case 2 is reducible to Case 1. ∎

Theorem 20.4.7 *([Lau 77], [Rom 80])*
$\widetilde{P_3}$ has exactly 58 maximal classes. These classes are the sets $pPOL_3\tau_i$ $(i = 1,2,3,...,57)$, where τ_i are given in Tables 20.1–20.3, and the set $P_2 \cup [c_\infty]$.

Proof.[2] Because of Lemmas 20.4.2 and 20.4.5, we must show that each subset M of $\widetilde{P_3}$, which fulfills $M \not\subseteq pPOL_3\tau_i$ for all $i \in \{1,2,...,57\}$ and $M \not\subseteq P_2 \cup [c_\infty]$, is complete in $\widetilde{P_3}$.
Let $M \subseteq \widetilde{P_3}$ be an arbitrary set that fulfills $M \not\subseteq P_2 \cup [c_\infty]$ and $M \not\subseteq pPOL_3\tau_i$ for all $i \in \{1,2,...,57\}$). Consequently, there are functions $f_i^{n_i} \in M \setminus pPOL_3\tau_i$ for all $i = 1,\ 2,\ ...,\ 57$) and $f_{58} \in M \setminus (P_2 \cup [c_\infty])$. If

$$\tau_i = (\sigma_1\ \sigma_2\ ...\ \sigma_{m_i})$$

($i \in \{1,2,...,57\}$) is an h_i-ary relation, then we can assume w.l.o.g. $n_i = m_i$ and

$$f_i(\sigma_1, \sigma_2, ..., \sigma_{m_i}) \in E_3^{h_i} \backslash \tau_i\ .$$

As already mentioned, we must show that $[M] = \widetilde{P_3}$. First we prove

$$\{c_0, c_1, c_2\} \subseteq [M]. \tag{20.3}$$

The function f_{37} is unary and belongs to $P_3^1 \backslash \{s_1\}$. Then the following three cases are possible:
Case 1: $f_{37} = c_a$ ($a \in E_3$).
Obviously, we have $\{c_0, c_1, c_2\} \subseteq [\{f_{37}, f_1, f_2, ..., f_6\}]$ in this case.
Case 2: $|Im(f_{37})| = 2$.
W.l.o.g. let $Im(f_{37}) = \{0,1\}$, i.e., $f_{37} \in \{j_0, j_1, ..., j_5\}$. Because of $j_2 * j_2 = c_0$, $j_0 * j_0 = j_5$, $j_3 * j_3 = c_1$, $j_4 * j_4 = j_1$ and Case 1 it is sufficient to assume that $f_{37} \in \{j_1, j_5\}$.
2.1: $f_{37} = j_1$.
We form $f_7' := f_7 * j_1 \in \{c_0, c_1, c_2, j_4, u_1, u_4, v_1, v_4\}$. Because of Case 1, $u_1 * u_1 = c_0$, $v_1 * v_1 = v_4$ and $u_4 * u_4 = c_2$, we can confine ourselves to $f_7' \in \{j_4, v_4\}$.
2.1.1: $f_7' = j_4$.
Putting the functions f_{37} ($= j_1$) and f_7' ($= j_4$) into f_{10} provides a unary function $f_{10}'(x) := f_{10}(j_1(x), j_4(x))$ with $f_{10}' \in \{c_0, c_1, c_2, u_1, u_4, v_1, v_4\}$. Because of $u_1 * u_1 = c_0$, $u_4 * u_4 = c_2$ and $v_1 * v_1 = v_4$ we still have to examine the case $f_{10}' = v_4$. It holds $v_4 * j_4 = v_1$. If we form $f_{17}'(x) := f_{17}(j_1(x), j_4(x), v_1(x), v_4(x))$, then $f_{17}' \in \{c_0, c_1, c_2, u_1, u_4\}$ holds and $f_{17}' * f_{17}'$ is a constant function. With that we have reduced the Case 2.1.1 to the first case.
2.1.2: $f_7' = v_4$.
The function $f_{14}'(x) := f_{14}(j_1(x), v_4(x))$ belongs to $\{c_0, c_1, c_2, j_4, u_1, u_4, v_1\}$. Because of $u_1 * u_1 = c_0$, $u_4 * u_4 = c_2$, $j_1 * v_1 = j_4$ and by Case 2.1.1, we reduce Case 2.1.2 to the first case.
2.2: $f_{37} = j_5$.
In this case, one can proceed to Case 2.1 analogously using the function f_{16} instead of f_{17} and f_{13} instead of f_{14}.
Case 3: $|Im(f_{37})| = 3$ and $f_{37} \neq s_1$.
With the help of functions f_{38}, ..., f_{42}, this case is reduced at first to Case 1 or 2 and, therefore, to Case 1.

Consequently, the constant functions belong to $[M]$.

Let $f_{43}'(x) := f_{43}(c_0(x), c_1(x), c_2(x), x)$. This function belongs to $P_3^1 \cap [M]$ and has at least two different values. If f_{43}' is a permutation, then with the help of functions

[2] The definitions of the following unary functions of P_3 are in Chapter 15, Table 15.1.

f_{44}, ..., f_{48}, we can form a function with 2-element range. Thus w.l.o.g. we can assume $Im(f'_{43}) = E_2$ in the following; i.e., $f'_{43} \in \{j_0, j_1, ..., j_5\}$.
Let

$$M_{a,b} := \{f \in P_3^1 \mid f(a) = f(b)\}.$$

Next

$$\exists\, a, b \in E_3 :\ a \neq b \ \wedge\ M_{a,b} \subset [M] \tag{20.4}$$

shall be proven. Because of $j_0 * j_0 = j_5$, $j_4 * j_4 = j_1$ and for reasons of the duality, we can assume that $f'_{43} \in \{j_1, j_2\}$.
Case 1: $f'_{43} = j_1$.
One receives $f'_{19} \in \{j_4, u_1, u_4, v_1, v_4\}$ as a superposition over the constant functions, j_1 and f_{19}. Because of $v_1 * v_1 = v_4$ and $f_{22}(c_0, c_1, c_2, j_1, j_4) \in \{u_1, u_4, v_1, v_4\}$ we can assume $f'_{19} \in \{u_1, u_4, v_4\}$.
1.1: $f'_{19} = u_1$.
In this case, we can form a function $f'_{25}(x) := f_{25}(c_0(x), c_1(x), c_2(x), j_1(x), u_1(x))$ with $f'_{25} \in \{j_4, u_4, v_1, v_4\}$. Further, we can assume w.l.o.g. $f'_{25} \in \{j_4, u_4, v_4\}$.
1.1.1: $f'_{25} = j_4$.
It holds that $u_1 * j_4 = u_4$. Consequently, the unary function $f'_{34} := f_{34}(c_0, c_1, c_2, j_1, j_4, u_1, u_4) \in \{v_1, v_4\}$ is a superposition over M. Thus by $v_1 * v_1 = v_4$ and $v_4 * j_4 = v_1$ it holds: $\{c_0, c_1, c_2, j_1, j_4, u_1, u_4, v_1, v_4\} = \{f \in P_3^1 \mid f(0) = f(2)\} \subset [M]$.
1.1.2: $f'_{25} = u_4$.
For the function $f'_{23} := f_{23}(c_0, c_1, c_2, u_1, u_4)$ we have $f'_{23} \in \{j_1, j_4, v_1, v_4\}$. Because of $v_1 * v_1 = v_4$ and $j_4 * j_4 = j_1$, we can assume $f'_{23} \in \{j_1, v_4\}$.
1.1.2.1: $f'_{23} = j_1$.
When one substitutes u_1 instead of u_4 and u_4 instead of u_1 into f_{23}, one receives $f''_{23} = j_4$ instead of f'_{23}. Now one can continue the proof as in Case 1.1.1.
1.1.2.2: $f'_{23} = v_4$.
When one substitutes u_1 instead of u_4 and u_4 instead of u_1 into f_{23}, one receives the function $f''_{23} = v_1$ instead of f'_{23}. Further, we have $f'_{36} := f_{36}(c_0, c_1, c_2, u_1, u_4, v_1, v_4) \in \{j_1, j_4\}$. If $f'_{36} = j_1$, we substitute u_1 instead of u_4, u_4 instead of u_1, v_1 instead of v_4 and v_4 instead of v_1 into f_{36} and we receive the function $f''_{36} = j_4$ instead of f'_{36}. Thus Case 1.1.2.2 was also reduced to Case 1.1.1.
1.1.3: $f'_{13} = v_4$.
Then, $f'_{31} := f_{31}(c_0, c_1, c_2, j_1, u_1, v_4) \in \{j_4, u_4, v_1\}$ is a superposition over M. Because of $j_1 * v_1 = j_4$, this case is reducible to Cases 1.1.1 and 1.1.2.
1.2: $f'_{19} = u_4$.
We can form $f'_{28} := f_{28}(c_0, c_1, c_2, j_1, u_4) \in \{j_4, u_1, v_1, v_4\}$ or w.l.o.g. (by $v_1 * v_1 = v_4$ and $u_4 * j_4 = u_1$) $f'_{28} \in \{u_1, v_4\}$.
1.2.1: $f'_{28} = u_1$.
In this case, one can continue the proof as in Case 1.1.
1.2.2: $f'_{28} = v_4$.
Here we have $f'_{32} := f_{32}(c_0, c_1, c_2, j_1, v_4, u_4) \in \{j_4, u_1, v_1\}$. Because of $v_4 * j_4 = v_1$ and $u_4 * v_1 = u_1$ we can assume $f'_{32} = u_1$. Thus Case 1.2 is reducible to Case 1.1.
1.3: $f'_{19} = v_4$.
We form $f'_{26} := f_{26}(c_0, c_1, c_2, j_1, v_4)$. Then, $f'_{26} \in \{j_4, u_1, u_4, v_1\}$ or w.l.o.g. (by $v_4 * j_4 = v_1$) $f'_{26} \in \{u_1, u_4, v_1\}$.
1.3.1: $f'_{26} \in \{u_1, u_4\}$.
Continuation of the proof as in Cases 1.1 or 1.2.

1.3.2: $f'_{26} = v_1$.
It holds that $j_1 * v_1 = j_4$. Consequently, we can form a function $f'_{35} := f_{35}(c_0, c_1, c_2, j_1, j_4, v_1, v_4) \in \{u_1, u_4\}$; i.e., Case 1.3.2 is reducible to Cases 1.1 and 1.2.
Hence, we have proven (20.4) in Case 1.

Case 2: $f'_{43} = j_2$.
Then $f'_{19} := f_{19}(c_0, c_1, c_2, j_2) \in \{j_3, u_2, u_3, v_2, v_3\}$. Because of $u_3 * u_3 = u_2$ and $v_3 * v_3 = v_2$ let w.l.o.g. $f'_{19} \in \{j_3, u_2, v_2\}$.
2.1: $f'_{19} = j_3$.
Set $f'_{22} := f_{22}(c_0, c_1, c_2, j_2, j_3)$ and $f''_{22} := f_{22}(c_0, c_1, c_2, j_3, j_2)$. Then $\{f'_{22}, f''_{22}\} \in \{\{u_2, u_3\}, \{v_2, v_3\}\}$ and $\{u_2, u_3, v_2, v_3\} \subseteq [\{f'_{22}, f''_{22}, j_2, j_3, f_{34}, f_{35}\}]$. Thus $M_{0,1} \subseteq [M]$ is proven.
2.2: $f'_{19} = u_2$.
Here we have $f'_{25} := f_{25}(c_0, c_1, c_2, j_2, u_2) \in \{j_3, u_3, v_2, v_3\}$. Because of $j_2 * u_3 = j_3$, $v_3 * v_3 = v_2$ and Case 2.1 we can assume $f'_{25} = v_2$. Since $f_{33}(c_0, c_1, c_2, j_2, u_2, v_2) \in \{j_3, u_3, v_3\}$, $j_2 * u_3 = j_3$ and $j_2 * v_3 = j_3$, Case 2.2 is reducible to Case 2.1.
2.3: $f'_{19} = v_2$.
Analogously to 2.2.

Thus (20.4) is proven.

Therefore, w.l.o.g. we can assume

$$M_{0,1} := \{\, f \in P_3^1 \mid f(0) = f(1) \,\} = \{c_0, c_1, c_2, j_2, j_3, u_2, u_3, v_2, v_3\} \subset [M]. \quad (20.5)$$

By substituting the functions of M_{01} and identifying variables in f_{49}, one can form a function $f'_{49} \in P_3^1$ with $f'_{49}(0) \neq f'_{49}(1)$ and $f'_{49} \neq s_1$.
Case 1: $|Im(f'_{49})| = 2$.
In this case it is easy to check that $M_{0,1} \cup M_{a,b} \subset [M]$ for a certain $(a,b) \in \{(0,2),(1,2)\}$. W.l.o.g. let $(a,b) = (1,2)$. Further, by (20.4) w.l.o.g. we can assume that, for certain $\alpha, \beta, \gamma, \delta \in E_3$,

$$f_{55}\begin{pmatrix} \alpha & \gamma \\ \alpha & \delta \\ \beta & \gamma \\ \beta & \delta \end{pmatrix} \in E_3^4 \backslash \tau_{55}$$

holds. Then we can form the function $f'_{55}(x,y) := f_{55}(g_1(x), g_2(y)) \in [M]$, where g_1 and g_2 are certain functions of $M_{0,1} \cup M_{1,2}$ with $g_1\begin{pmatrix} 0 \\ 1 \end{pmatrix} = \begin{pmatrix} \alpha \\ \beta \end{pmatrix}$ and $g_2\begin{pmatrix} 0 \\ 1 \end{pmatrix} = \begin{pmatrix} \gamma \\ \delta \end{pmatrix}$. Obviously, $f'_{55} \in P_3 \backslash [P_3^1]$.
If $|Im(f'_{55})| = 3$, then it follows from Lemma 20.4.1, (c) that $P_3 \subseteq [M]$. This implies $[M] = \widetilde{P_3}$ with the help of Lemma 20.4.2, (a) and $f_{58} \in \widetilde{P_3} \backslash (P_3 \cup [\{c_\infty\}])$.
Let

$$|Im(f'_{55})| = 2 \quad (20.6)$$

be in the following. Because of Lemma 20.4.1, (b) there are certain $a_1, a_2, a_3, b_1, b_2, b_3 \in E_3$ with

$$f'_{55}\begin{pmatrix} a_1 & a_2 \\ b_1 & a_2 \\ a_1 & b_2 \end{pmatrix} = \begin{pmatrix} a_3 \\ b_3 \\ b_3 \end{pmatrix},$$

where $a_3 \neq b_3$. By substituting functions of $M_{0,1} \cup M_{1,2}$ ($\subset [M]$) into f'_{55}, one can form a unary function f''_{55} with the property: $M_{0,2} \subseteq [\{f''_{55}\} \cup M_{0,1} \cup M_{1,2}]$. Then, with the help of function $f_{57} \in M$ and Lemma 20.4.1, (a), it follows $P_3^1 \subset [M]$. By Lemma 20.4.3, (a) we have then $Pol_3\tau_{56} \subseteq [M]$. With the help of functions f_{56} and f_{57} of M, it is easy to prove that all functions of P_3 are superpositions on M. Because of $f_{58} \in M$ and Lemma 20.4.2, (a), this implies $[M] = \widetilde{P}_3$.
Case 2: $|Im(f'_{49})| = 3$.
In this case, f'_{49} is a permutation $\neq s_1$. If $f_{49} \neq s_3$, then, because of (20.5), one can form a certain unary function g with $g(0) \neq g(1)$ as a superposition over M and one continues to be able to use the proof as in the first case.
If $f'_{49} = s_3$, then it is possible to form a unary function $h \in [M_{0,1} \cup \{f_{52}\}](\subset [M])$, which is either a permutation $\notin \{s_1, s_3\}$ or h is a function with $|Im(h)| = 2$ and $h \notin M_{0,1}$. Consequently, one can completely reduce the second case to the first case.

■

Next, some remarks on partial Sheffer functions:
A partial function $f \in \widetilde{P_k}$ is called **Sheffer** iff $[f] = \widetilde{P_k}$.
In [Had-R 91] all partial Sheffer functions for $k = 2$ and all binary Sheffer function for $k = 3$ were described. Further, the statement

$$[f] = \widetilde{P_2} \iff$$
$$(\forall A \in \{pPOL_2\{0\}, pPOL_2\{1\}, pPOL_2\{(0,1),(1,0)\}, P_2 \cup [c_\infty]\} : f \notin A)$$

was proven. In [Had-L 2006] one finds the proof of the following criterion:

$$[f] = \widetilde{P_3} \iff$$
$$(\forall A \in \{pPOL_3\tau_i \mid i \in \{1,2,3,4,5,6,10,11,12,16,17,18,19,20,21,22,23,$$
$$24,41,42,47,48,55,56,57\}\} \cup \{P_2 \cup [c_\infty]\} : f \notin A),$$

where the relations τ_i are defined in Tables 20.1–20.3. In [Had-L 2006] one finds also the proof that it is not possible to reduce the conditions from the above criteria; i.e., these conditions are independent of each other.

20.5 The Completeness Criterion for $\widetilde{P_k}$

In analog mode to the sixth chapter, one can show that a completeness criterion for $\widetilde{P_k}$ can be found using the maximal partial classes of $\widetilde{P_k}$. Subsequently, the maximal partial classes are described in a form found by L. Haddad and I. G. Rosenberg. The following definitions are needed:
Definitions Let Eq_h be the set of all equivalence relations over $\{1,...,h\}$. An h-ary relation $\varrho \subseteq E_k^h$ is called

- **areflexive**, if $\varrho \cap \delta_\varepsilon = \emptyset$ for every $\varepsilon \in Eq_h$, $\varepsilon \neq \iota_h^2$, i.e., for all $(x_1, \ldots, x_h) \in \varrho$ we have $x_i \neq x_j$ for all $1 \leq i < j \leq h$.

- **quasi-diagonal**, if $\varrho = \sigma \cup \delta_\varepsilon$, where σ is a nonempty areflexive relation, $\varepsilon \in Eq_h \setminus \{\iota_h^2\}$ holds, and further $\varrho \neq E_k^2$ for $h = 2$.

Furthermore

$$\begin{aligned}
\varrho_1 &:= \{(a,a,b,b),(a,b,a,b) \mid a,b \in E_k\ \} \\
&= \delta_{\{1,2\},\{3,4\}} \cup \delta_{\{1,3\},\{2,4\}}, \\
\varrho_2 &:= \{(a,a,b,b),(a,b,a,b),(a,b,b,a) \mid a,b \in E_k\ \} \\
&:= \delta_{\{1,2\},\{3,4\}} \cup \delta_{\{1,3\},\{2,4\}} \cup \delta_{\{1,4\},\{2,3\}}.
\end{aligned}$$

In the following, denote ϱ an h-ary relation of the form

$$\varrho = \sigma \cup (\bigcup_{\varepsilon \in F} (\delta_\varepsilon)),$$

where σ is an areflexive h-ary relation and $F \subset Eq_h$. Let

$$G_\sigma : = \{\pi \in S_h \mid \sigma \cap \sigma^{(\pi)} \neq \emptyset\},$$

where S_h denotes the set of all permutations over the set $\{1,...,h\}$ and $\sigma^{(\pi)} := \{(a_{\pi(1)},...,a_{\pi(h)}) \mid (a_1,...,a_h) \in \sigma\}$.
The **model** of ϱ is the h-ary relation

$$M(\varrho) : = \{(\pi(1),\ldots,\pi(h)) \mid \pi \in G_\sigma\} \bigcup (\bigcup_{\varepsilon \in F} \{(x_1,\ldots,x_h) \in \{1,\ldots,h\}^h \mid (i,j) \in \varepsilon \Rightarrow x_i = x_j\})$$

on the set $\{1,\ldots,h\}$.[3]

Suppose h, F, and σ fulfill exactly one of the following five conditions:

i) $h \geq 2$, $F = \emptyset$ and $\sigma \neq \emptyset$, i.e., ϱ is a nonempty h-ary areflexive relation;

ii) $h \geq 2$, $F = \{\varepsilon\}$, where $\varepsilon \neq \iota_h^2$, $\sigma \neq \emptyset$ and $\sigma \cup \delta_\varepsilon \neq E_k^2$, i.e., ϱ is a trivial quasi-diagonale h-ary relation;

iii) $h = 4$ and $F = \{\{\{1,2\},\{3,4\}\},\ \{\{1,3\},\{2,4\}\},\ \{\{1,4\},\{2,3\}\}\}$, i.e., $\varrho = \sigma \cup \varrho_2$, where σ is an areflexive 4-ary relation (the empty set is possible);

iv) $h = 4$ and $F = \{\{\{1,2\},\{3,4\}\},\ \{\{1,3\},\{2,4\}\}\}$, i.e., $\varrho = \sigma \cup \varrho_1$, where σ is an areflexive 4-ary relation ($= \emptyset$ is possible);

v) $h \neq 2$, $h \leq k$, $F = \bigcup_{1 \leq i \langle j \leq h} \{i,j\}$ and $\varrho \neq E_k^h$, i.e., ϱ is a totally reflexive and totally symmetric not trivial relation.[4]

We say that ϱ is **coherent**, iff

[3] In some papers the set E_h is elected instead of $\{1...,h\}$ in defining the model.
[4] For $h = 1$ we have $\emptyset \subset \varrho \subset E_k$.

(1) $G_\sigma = \{\pi \in S_h \mid \sigma^{(\pi)} = \sigma\}$ and $\pi(\varepsilon) := \{(\pi(x), \pi(y)) \mid (x, y) \in \varepsilon\} = \varepsilon$ for all $\pi \in G_\sigma$, if ϱ fulfills either the above condition i) or ii),
$G_\sigma = \{\pi \in S_h \mid \sigma^{(\pi)} = \sigma\} \cup \{\pi \in S_h \mid \pi(F) = F\}$, if ϱ fulfills iii) or iv),
$G_\sigma = \{\pi \in S_h \mid \sigma^{(\pi)} = \sigma\} = S_h$, if ϱ fulfills the condition v), and
(2) for every nonempty subset σ' of σ there exists a relational homomorphism $\gamma : E_k \to \{1, \ldots, h\}$ of σ' in $M(\varrho)$, such that $(\gamma(i_1), \ldots, \gamma(i_h)) = (1, \ldots, h)$ for at least an h-tuple $(i_1, \ldots, i_h) \in \sigma'$.

Theorem 20.5.1 (Haddad-Rosenberg Theorem; *[Had-R 89], [Had-R 92]; without proof***)**

Let $k \geq 2$. For every proper partial subclass A of $\widetilde{P_k}$ there is a maximal partial clone that contains A. If C is a maximal partial clone of $\widetilde{P_k}$, then either $C = P_k \cup \{f \in \widetilde{P_k} \mid D(f) = \emptyset\}$ $(= pPol_k E_k^2 \cup \{(\infty, \infty)\})$ or $C = pPOL_k\varrho$, where ϱ is one of the following relations:

(1) an h-ary not trivial totally reflexive and totally symmetric relation with $1 \leq h \leq k$;

(2) an h-ary areflexive or quasidiagonal relation with $h \geq 2$, which is coherent;

(3) a quaternary relation ϱ_2 or ϱ_1;

(4) a quaternary coherent relation $\sigma \cup \tau_i$, where $i \in \{1, 2\}$ and $\sigma \neq \emptyset$ is a quaternary areflexive relation.

Let $\widetilde{R}_{max}$ be the set of all relations ($\in R_k$) given in the above theorem. Then the following theorem is a consequence of the above:

Theorem 20.5.2 (Completeness Criterion for $\widetilde{P_k}$; *[Had-R 92]***)**
Let $C \subseteq \widetilde{P_k}$. Then $[C] = \widetilde{P_k}$ if and only if $C \not\subseteq pPOL_k\varrho$ for all $\varrho \in \widetilde{R}_{max}$ and $C \not\subseteq P_k \cup \{f \in \widetilde{P_k} \mid D(f) = \emptyset\}$. ■

20.6 Some Properties of the Maximal Partial Clones of $\widetilde{P_k}$

We show first that each maximal clone of P_k is a subset of exactly a maximal partial clone of $\widetilde{P_k}$. Then we specify the relations $\varrho \in R_{max} := \mathfrak{M}_k \cup \mathfrak{U}_k \cup \mathfrak{S}_k \cup \mathfrak{L}_k \cup \mathfrak{C}_k \cup \mathfrak{B}_k$, with the property that $pPOL_k\varrho$ is a maximal partial clone of $\widetilde{P_k}$.
A survey of the orders of the maximal partial clones forms the end of this section.

Next to the notations from the fifth chapter, we still need the following notations for certain relation sets:
Let $P_{k,p}$ be the set of all fixed-point-free permutations on E_k whose cycles have the same prim number length p.
Let

$$\mathfrak{S}_{k,p} := \{ \{(x, s(x)) \mid x \in E_k\} \mid s \in P_{k,p}\}$$

and let $\mathfrak{M}_k^\star$ be the set of all $\varrho \in \mathfrak{M}_k$ with the property that $(E_k; \varrho)$ is a lattice.

Theorem 20.6.1 *For every maximal clone $C \subseteq P_k$ there is exactly one maximal partial clone $C' \subseteq \widetilde{P_k}$ with $C' \cap P_k = C$.*

Proof. Let C be a partial clone of $\widetilde{P_k}$. Put

$$\widehat{C} := \bigcup_{n\geq 1}\{f \in \widetilde{P_k} \mid \forall f_1, \ldots, f_n \in C \cap P_k^2 :$$
$$(f(f_1, \ldots, f_n) \in P_k \Longrightarrow f(f_1, \ldots, f_n) \in C) \}.$$

Notice that $C \cap P_k^2 \neq \emptyset$. As C is a partial clone, $C \subseteq \widehat{C}$. It was already shown in [Fre 66] that $\widehat{C}$ is a partial clone with the following property:

$$C \neq \widetilde{P_k} \Longrightarrow \widehat{C} \neq \widetilde{P_k}.$$

Consequently, if C is a maximal partial clone, then

$$C = \widehat{C}. \tag{20.7}$$

Let M be a maximal clone of P_k and let C_1, C_2 be two maximal partial clones of P_k that contain M. Then, by (20.7) $C_1 = \widehat{C_1}$ and $C_2 = \widehat{C_2}$. From $M \neq P_k$ and $[P_k^2] = P_k$ (see Theorem 1.4.2), we have

$$M^2 = (C_1 \cap P_k)^2 = (C_2 \cap P_2)^2 \subset P_k^2.$$

By definition of $\widehat{C}$, we have $\widehat{C_1} = \widehat{C_2}$ and by (20.7), $C_1 = C_2$. ■

Theorem 20.6.2 *([Had-L 2000]; without proof)*

(1) Let $M := \mathfrak{C}_k \cup \mathfrak{M}_k \cup \mathfrak{S}_k \cup \mathfrak{U}_k \cup \mathfrak{L}_k \cup \mathfrak{B}_k$ and $M_1 := \mathfrak{S}_k \backslash \mathfrak{S}_{k,2}$. For every $\varrho \in M \setminus (\mathfrak{L}_k \cup M_1)$ the set $pPOL_k\varrho$ is a maximal partial clone of $\widetilde{P_k}$.
(2) For $s \in P_{k,p}$ with $p \geq 3$ let

$$s^o := \{(x, s(x)) \mid x \in E_k\}$$
$$s^\square := \{(x, s(x), s^2(x), \ldots, s^{p-1}(x)) \mid x \in E_k\}.$$

Then $pPOL_k s^\square$ is a maximal partial clone that properly contains the partial clone $pPOL_k s^0$.

We need the following notations for the statement still missing on classes of type $\mathfrak{L}$.
Let $p \in \mathbb{P}$, $m \in \mathbb{N}$ and $W := E_p^m$. As shown in Section 5.2.4, a maximal class of P_{p^m} of type $\mathfrak{L}$ is isomorphic to the set L_W. The set L_W one can also describe in the form $Pol_W\lambda$ with $\lambda := \{(a,b,c,d) \in (E_p^m)^4 \mid a \oplus b = c \oplus d\}$, where $(a_1,...,a_m) \oplus (b_1,...,b_m) := (a_1+b_1,...,a_m+b_m)$ and $+$ is the addition modulo p.
With the help of $\oplus$ and $\odot$ ($\alpha \odot (a_1,...,a_m) := (\alpha \cdot a_1 \pmod p),...,\alpha \cdot a_m \pmod p))$ for $\alpha \in E_p$ and $(a_1,...,a_m) \in W$) we can define the following p-ary relation over W:

$$\lambda_p := \{(a, a \oplus b, a \oplus 2 \odot b, ..., a \oplus (p-1) \odot b) \mid a, b \in W\}.$$

Theorem 20.6.3 *([Had-L 2000]; without proof)*
For $p \in \mathbb{P}$ the partial clone $pPOL_W\lambda_p$ is a maximal clone of $\widetilde{P_W}$ that properly contains the partial clone $pPOL_W\lambda$.

Now, we come to some theorems that deal with the finite generating of partial clones.
It was shown already in [Fre 66] that $\widetilde{P_2}$ has partial subclasses, which are not finitely generating. In [Lau 88], it was proven that there are maximal partial clones of $\widetilde{P_2}$ that are not finitely generating. These are exactly the partial clones $pPOL_2\varrho$ with $\varrho \in \{\lambda_2, G_2([P_2^1])\}$ (see also [Bör-H 97]). The order of each finite generating maximal partial clone C of $\widetilde{P_2}$ agrees with the order of $C \cap P_2$.

We need the following concept for the following criterion about the finite generating of strong partial clones:
Let $r > 0$ and let C be a clone of P_k. Then C is called r-**separable**, if for every $n > 0$ and every $\mathbf{b} \in E_k^n$ there are certain n-ary functions $g_1, ..., g_r \in C$ such that the mapping $g: E_k^n \longrightarrow E_k^r$ with $g(\mathbf{a}) := (g_1(\mathbf{a}), ..., g_r(\mathbf{a}))$ has the property $g^{-1}(g(\mathbf{b})) = \{\mathbf{b}\}$ for all $\mathbf{b} \in E_k^n$.

Theorem 20.6.4 *([Bör-H 97]; without proof)*
Let $C \subseteq P_k$ be a clone. Then the partial clone $Str(C)$ is finitely generated if and only if C is finitely generated and there is an $r \in \mathbb{N}$ such that C is r-separable.
If C r-separable and finitely generated, then $ord(Str(C)) \leq \max\{ord(C), r\}$.

Theorem 20.6.5 *([Noz-L 97], [Had-L 2000]; without proof)*
Let $k \geq 3$. Then

(1) For every $\varrho \in \mathfrak{M}_k \cup \mathfrak{U}_k \cup \mathfrak{C}_k^1 \cup \mathfrak{C}_k^2$ it holds $ord(pPOL_k\varrho) = 2$.

(2) For every $\varrho \in \mathfrak{C}_k^h$ *with* $3 \le h \le k-1$ *it holds* $ord(pPOL_k\varrho) \le h$.
(3) For every $\varrho \in \mathfrak{M}_k \cup \mathfrak{B}_k$ *it holds* $ord(Str(Pol_\varrho)) = 2$.
(4) $ord(pPOL_k s^\square) = 2$ *(see Theorem 20.6.2).*

20.7 Intervals of Partial Clones That Contain a Maximal Clone

Since there are many results about subclasses of P_k, it is obvious to classify the subclasses C of $\widetilde{P_k}$ after $P_k \cap C$. For an arbitrary subclass A of P_k:

$$\mathcal{I}(A) := \{F \subseteq \widetilde{P_k} \mid [F] = F \;\wedge\; F \cap P_k = A\}.$$

This section aims to obtain cardinality statements over the sets $\mathcal{I}(A)$ for the maximal clones A of P_k. First, however, some general properties of the set $\mathcal{I}(A)$:

Lemma 20.7.1 *([Str 97]; without proof)*
Let $A \subseteq P_k$ *be a clone. Then*

(1) $\mathcal{I}(A)$ *contains the partial clones* $A \cup C_\infty$ *and* $Str(A)$ *as well all partial clones of the form* $pPOL_k\varrho$ *with* $Pol_k\varrho = A$.
(2) If A *is finitely generated with* $ord\, A = n$, *then* $pPOL_kG_n(A)$ *is the greatest element in the lattice* $(\mathcal{I}(A); \subseteq)$.

Now we obtain from (2) of the above lemma:

Theorem 20.7.2 *If* A *is a finitely generated clone on* E_k *of the order* n, *then* $\mathcal{I}(A)$ *is exactly the interval* $[A, pPOL_kG_n(A)]$ *of the lattice of all partial clones of* $\widetilde{P_k}$. ■

Lemma 20.7.3 *Let* $A = [A] \subseteq P_k$. *Then*

(1) A *is a maximal subclass of* $A \cup C_\infty$.
(2) If $\{c_0^1, \ldots, c_{k-1}^1\} \subset A$ *then* $A \cup C_\infty \subseteq B$ *for all* $B \in \mathcal{I}(A)\backslash\{A\}$.

Proof. (1) is clear.
(2): Let $B \in \mathcal{I}(A)\backslash\{A\}$. If $B = A \cup C_\infty$, we do not have to prove anything. Otherwise, there is a function $f^n \in B\backslash(A \cup C_\infty)$. Let $f(a_1, \ldots, a_n) = \infty$. Now $c_\infty^1 = f(c_{a_1}^1, \ldots, c_{a_n}^1) \in B$ and from this one can easily verify that $C_\infty \subseteq B$. ■

Theorem 20.7.4 *([Had-L-R 2002; without proof) Let* $\emptyset \ne \varrho \subset E_k$, $T := Pol_k\varrho$ *and*

$$T_\infty := \bigcup_{n \ge 1} \{f^n \in \widetilde{P_k} \mid f(\varrho^n) = \{\infty\}\}.$$

Then $\mathcal{I}(T)$ *consists exactly of the partial clones:*

T, $T \cup C_\infty$, $pPol_k\varrho$, $T \cup T_\infty$, $pPol_k\varrho \cup C_\infty$, $pPol_k\varrho \cup T_\infty$, $\mathrm{pPOL}_{\,k}\varrho$.

The partial clones are pairwise distinct for $|\varrho| > 1$ whereas for $|\varrho| = 1$ $pPol_k\varrho \cup T_\infty$ and $pPOL_k\varrho$ coincide. Their inclusions are shown in Figure 20.1.

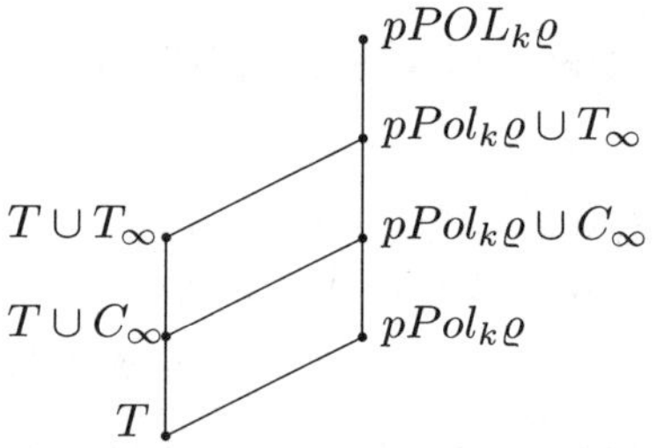

Fig. 20.1

Theorem 20.7.5 *([Had-L-R 2002]; without proof)*
Let $h \geq 2$ and $\varrho \in \mathfrak{C}_k^h$. Then the set $\mathcal{I}(Pol_k\varrho)$ is a 3-element chain:

$$Pol_k\varrho \subset (Pol_k\varrho) \cup C_\infty \subset pPol_k\varrho.$$

Theorem 20.7.6 *([Had-L-R 2002]; without proof)*
Let $\varrho \in \mathfrak{U}_k$, $X := Pol_k\varrho$ and $\varrho_1 := \varrho \cup \{(\infty, \infty)\}$. Then $\mathcal{I}(X)$ is a 4-element chain:

$$X \subset X \cup C_\infty \subset pPol_k\varrho_1 \subset POL_k\varrho.$$

For clones M of type $\mathfrak{M}$, there are only partial results over $\mathcal{I}(M)$. The following is one of these results:

Theorem 20.7.7 *([Had-L-R 2002]; without proof)*
Let $\leq\ \in \mathfrak{M}_k$. Set

$$M := Pol_k \leq,$$
$$\leq_0 := \leq \cup \{(\infty, x) \mid x \in \widetilde{E_k}\},$$
$$\leq_1 := \leq \cup \{(x, \infty) \mid x \in \widetilde{E_k}\},$$
$$\varrho_2 := \{(x,y,z) \in E_k^3 \mid x \leq y \leq z\} \cup \{(\infty, x, y), (x, y, \infty) \mid x, y \in E_k,\ x \leq y\} \cup (\{\infty\} \times E_k \times \{\infty\}),$$
$$M_i := pPol_k(\leq_i)\ (i = 0, 1) \quad \textit{and}$$
$$M_2 := pPol_k\varrho_2.$$

Then

$$\{M, M \cup C_\infty, M_0, M_1, M_2, Str(M)\}$$

is the set of all partial clones from $\mathcal{I}(M)$ included in $Str(M)$. Their inclusions are shown in Figure 20.2.

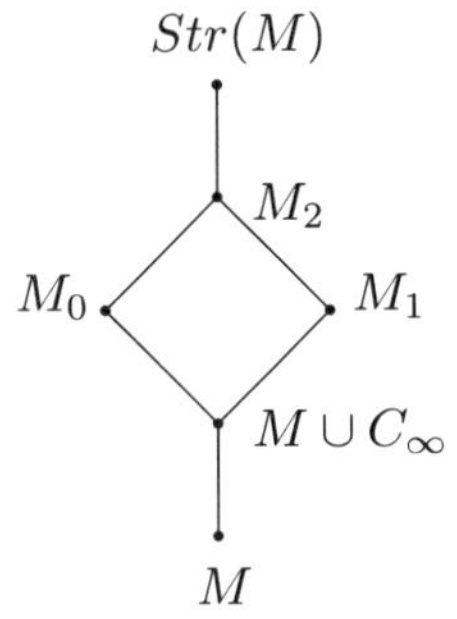

Fig. 20.2

Because of

$$Str(Pol_k \leq) = Pol_k \leq \iff \leq \in \mathfrak{M}_k^\star,$$

Theorem 20.7.8 follows.

Theorem 20.7.8 *For every $\varrho \in \mathfrak{M}_k^\star$ is $|\mathcal{I}(Pol_k\varrho)| = 6$.*

Next we prove that $\mathcal{I}(A)$ is a finite set if A is a maximal class of type $\mathfrak{S}$. For this, we need some notations:

Let p be a prime factor of k and $s \in P_{k,p}$ a fixed-point-free permutation on E_k comprising of cycles of the same length p. Set

$$s^o := \{(x, s(x)) \mid x \in E_k\},\ S := Pol_k s^o,$$
$$s^\square := \{(x, s(x), \dots, s^{p-1}(x)) \mid x \in E_k\} \text{ and } S_{max} := pPOL_k s^\square.$$

By Lemma 20.6.2, we have that S_{max} is the (unique) maximal partial clone containing the maximal clone $Pol_k s^o$. As usual, the powers of s are defined recursively by setting $s^0(x) := x$ and $s^{i+1}(x) := s(s^i(x))$ for all $x \in E_k$. To describe functions of S_{max}, we define the following relations on E_k^n:

$$\mathbf{x} =_s \mathbf{y} :\iff (\exists i \in \{0, 1, ..., p-1\} : s^i(\mathbf{x}) = \mathbf{y}),$$

where $s^i(\mathbf{x}) = s^i((x_1, \dots, x_n)) := (s^i(x_1), \dots, s^i(x_n))$.
The relation $=_s$ is an equivalence relation of E_k^n with $r_n := k^n/p$ equivalence classes (or blocks) which we denote by $U_1, \dots, U_{r_n}$. Fix $\mathbf{v}_i \in U_i$ $(i = 1, \dots, r_n)$ and set

$$V_n := \{\mathbf{v}_1, \mathbf{v}_2, \dots, \mathbf{v}_{r_n}\}.$$

Since for all $f^n \in S$, $\mathbf{b} \in E_k^n$ and $i \in E_p$

$$f(s^i(\mathbf{b})) = s^i(f(\mathbf{b})),$$

each function $f^n \in S$ is fully determined by its values on V_n.
Set

$$\gamma := \{(a_0, \ldots, a_{p-1}) \in \widetilde{E_k}^p \backslash E_k^p \mid \exists \alpha \in E_k \ \forall i \in E_p \ (a_i \neq \infty \implies a_i = s^i(\alpha)\}.$$

For $i = 1, 2, \ldots, p$ let the p-ary relation

$$\gamma_i$$

consist of all $(a_0, a_1, \ldots, a_{p-1}) \in \gamma$ with exactly i coordinates ∞. Furthermore, set $\gamma_0 := s^\square$.
For every $I \subseteq E_p$ let

$$\tau_I := \{(a_0, a_1, \ldots, a_{p-1}) \in \gamma \mid (\forall i \in I: \ a_i = \infty) \ \wedge \ (\forall j \in E_p : \backslash I \ a_j \neq \infty)\}.$$

For every function $f^n \in S_{max}$ set

$$\chi_\star(f) := \{(f(\mathbf{a}), f(s(\mathbf{a})), \ldots, f(s^{p-1}(\mathbf{a}))) \mid \mathbf{a} \in V_n\}$$

and

$$\chi(f) := \{(f(\mathbf{a}), f(s(\mathbf{a})), \ldots, f(s^{p-1}(\mathbf{a}))) \mid \mathbf{a} \in \mathbf{k}^n\}.$$

Notice that $\chi_\star(f)$ and $\chi(f)$ are p-ary relations on $\widetilde{E_k}$ with $\chi_\star(f) \subseteq \chi(f)$ and

$$\chi(f) \subseteq (\widetilde{E_k}^p \backslash E_k^p) \cup s^\square, \tag{20.8}$$

and hence, in general, $\chi(f)$ is not a subrelation of $s^\square$. It is easy to check that

$$\chi(f) = \{(a_i, a_{i+1}, \ldots, a_{p-1}, a_0, a_1, \ldots, a_{i-1}) \mid (a_0, a_1, ..., a_{p-1}) \in \chi_\star(f), \ i \in E_p\}. \tag{20.9}$$

Moreover, for $R \subseteq (\widetilde{E_k}^p \backslash E_k^p) \cup s^\square$ and $\alpha \in \{\chi_\star, \chi\}$, set

$$\alpha^{-1}(R) := \{g \in S_{max} \mid \alpha(g) \subseteq R\}.$$

Then it holds that

$$\alpha^{-1}(R) = \{g \in S_{max} \mid \forall \ r_1, \ldots, r_n \in \gamma_0 \ \ g_+(r_1, \ldots, r_n) \in R\}.$$

We start with

Lemma 20.7.9 *Let $f^n \in S_{max}$ and $I, I' \subseteq E_p$. Then*

(1) $G_f := \{g^n \in Str\ (S) \mid D(g) = D(f)\} \subseteq [S \cup \{f\}]$,

(2) $\chi_\star^{-1}(\ \chi_\star(f)\ ) \subseteq [S \cup \{f\}]$,

(3) $(\alpha_0, \alpha_1, \ldots, \alpha_{p-1}) \in \chi_\star(f) \implies \chi_\star^{-1}(\ \{(\alpha_1, \alpha_2, \ldots, \alpha_{p-1}, \alpha_0)\} \cup \chi_\star(f)) \subseteq [S \cup \{f\}]$,

(4) *there is a function* $g \in [S \cup \{f\}]$ *with* $\chi_\star(g) = \chi(f)$,

(5) $\chi^{-1}(\ \chi(f)\ ) \subseteq [S \cup \{f\}]$,

(6) $\chi(f) \cap \tau_I \neq \emptyset \implies \chi^{-1}(\tau_I \cup \chi(f)) \subseteq [S \cup \{f\}]$,

(7) $(\chi(f) \cap \tau_I \neq \emptyset\ \wedge\ \chi(f) \cap \tau_{I'} \neq \emptyset) \implies \chi^{-1}(\tau_{I \cup I'} \cup \chi(f)) \subseteq [S \cup \{f\}]$,

(8) $(p \in \{2,3\}\ \wedge\ j \in \{1, \ldots, p\}\ \wedge\ \chi(f) \cap \gamma_j \neq \emptyset) \implies \chi^{-1}(\bigcup_{i=j}^{p} \gamma_i \cup \chi(f)) \subseteq [S \cup \{f\}]$,

(9) $(p = 3 \wedge \chi(f) \cap (\widetilde{E_k}^3 \setminus (E_k^3 \cup \gamma)) \neq \emptyset) \implies \chi^{-1}(\widetilde{E_k}^3 \setminus E_k^3 \cup \chi(f)) \subseteq [S \cup \{f\}]$.

Proof. (1): Let $g^n \in G_f$ and $g_1 \in S$ with $g_{1|D(f)} = g$. Then

$$g = e_1^2(g_1, f) \in [S \cup \{f\}].$$

(2): Let $g^m \in \chi_\star^{-1}(\chi_\star(f))$ be arbitrary. Then there is for every $\mathbf{v} \in V_m$ a $\mathbf{b_v} := (b_{\mathbf{v}1}, b_{\mathbf{v}2}, \ldots, b_{\mathbf{v}n}) \in \chi_\star(f)$ with

$$(g(\mathbf{v}), g(s(\mathbf{v})), \ldots, g(s^{p-1}(\mathbf{v}))) = (f(\mathbf{b_v}), f(s(\mathbf{b_v})), \ldots, f(s^{p-1}(\mathbf{b_v}))).$$

It is easy to check that the functions $g_i^m \in S$ $(i = 1, 2, \ldots, n)$ exist with $g_i(\mathbf{v}) = b_{\mathbf{v}i}$ for all $i \in \{1, \ldots, n\}$ and $\mathbf{v} \in V_m$. Then we have that $g = f(g_1, \ldots, g_n)$ and (2) hold.
(3): Let $\mathbf{a}_1, \ldots, \mathbf{a}_q \in V_n$ with $\mathbf{a}_i = (a_{i1}, a_{i2}, \ldots, a_{in})$ $(i = 1, \ldots, q)$,

$$\chi_\star(f) = \{(f(\mathbf{a}_i), f(s(\mathbf{a}_i)), \ldots, f(s^{p-1}(\mathbf{a}_i))) \mid i = 1, \ldots, q\}$$

and

$$(f(\mathbf{a}_q), f(s(\mathbf{a}_q)), \ldots, f(s^{p-1}(\mathbf{a}_q))) = (\alpha_0, \alpha_1, \ldots, \alpha_{p-1}).$$

Then there exists a $t \in \mathbb{N}$, $\mathbf{b}_1, \ldots, \mathbf{b}_{q+1} \in V_t$ and functions $g_j^t \in S$ $(j = 1, 2, \ldots, n)$ with $g_j(\mathbf{b}_i) := a_{ij}$ $(i = 1, 2, ..., q,\ j = 1, \ldots, n)$ and $g_j(\mathbf{b}_{q+1}) := s(a_{qj})$ $(j = 1, \ldots, n)$. Let $g := f(g_1, \ldots, g_n)$. So we have $g \in [S \cup \{f\}]$ and it is easy to check that $\chi_\star(f) \cup \{(\alpha_1, \alpha_2, \ldots, \alpha_{p-1}, \alpha_0)\} \subseteq \chi(g)$. Consequently, (3) follows from (2).
(4) follows from (20.9) and (3).
(5) follows from (4) and (2).
In the following, we can assume

$$\chi_\star(f) = \chi(f).$$

(6): Let $(a_0, a_1, \ldots, a_{p-1}) \in \chi(f) \cap \tau_I$. Then there exists an $\alpha \in E_k$ with

$$\forall i \in E_p\ (a_i \neq \infty \implies a_i = s^i(\alpha)\ .$$

Obviously, there are a $t \in \mathbb{N}$, a function $q^t \in S$ with $\chi_\star(q) = \gamma_0$ and a function $h^2 \in S$ with

$$h(x,y) := \begin{cases} y & \text{if} \quad x \in \{\alpha, s(\alpha), \ldots, s^{p-1}(\alpha)\}, \\ x & \text{otherwise.} \end{cases}$$

Then

$$h_1 := h(f(e_1^{n+t}, e_2^{n+t} ..., e_n^{n+t}), q(e_{n+1}^{n+t}, \ldots, e_{n+t}^{n+t})) \in [S \cup \{f\}]$$

and it is easy to check that $\tau_I \cup \chi(f) \subseteq \chi(h_1)$. Thus by (5) we have

$$\chi^{-1}(\tau_I \cup \chi(f)) \subseteq \chi^{-1}(\chi(h_1)) \subseteq [S \cup \{f\}];$$

i.e., (6) holds.
(7): Let $\mathbf{a}_1, \ldots, \mathbf{a}_q \in V_n$ with $\mathbf{a}_i = (a_{i1}, a_{i2}, \ldots, a_{in})$ $(i = 1, \ldots, q)$,

$$\chi(f) = \{(f(\mathbf{a}_i), f(s(\mathbf{a}_i)), \ldots, f(s^{p-1}(\mathbf{a}_i))) \mid i = 1, \ldots, q\},$$

$$(f(\mathbf{a}_{q-1}), f(s(\mathbf{a}_{q-1})), \ldots, f(s^{p-1}(\mathbf{a}_{q-1}))) \in \tau_I$$

and

$$(f(\mathbf{a}_q), f(s(\mathbf{a}_q)), \ldots, f(s^{p-1}(\mathbf{a}_q))) \in \tau_{I'}.$$

Then there exists $t \in \mathbb{N}$, $\mathbf{b}_1, \ldots, \mathbf{b}_{q+1} \in V_t$ and functions $g_j^t, h_j^t \in S$ $(j = 1, 2, \ldots, n)$ with

$$g_j(\mathbf{b}_i) := h_j(\mathbf{b}_i) := a_{ij}$$

$(i = 1, 2, ..., q,\ j = 1, \ldots, n)$ and

$$g_j(\mathbf{b}_{q+1}) := a_{q-1,j},\ h_j(\mathbf{b}_{q+1}) := a_{q,j}$$

$(j = 1, \ldots, n)$. Let

$$g := e_1^2(f(g_1, \ldots, g_n), f(h_1, \ldots, h_n)).$$

Then $g \in [S \cup \{f\}]$, and it is easy to check that $\chi(f) \subseteq \chi(g)$ and $\chi(g) \cap \tau_{I \cup I'} \neq \emptyset$. Then, it follows from (6): $\chi^{-1}(\tau_{I \cup I'} \cup \chi(f)) \subseteq [S \cup \{g\}] \subseteq [S \cup \{f\}]$.
(8) follows from (3) and (5)–(7).
(9): Let $p = 3$, $(a, b, \infty) \in \chi(f)$, $a, b \in E_k$ and $s(a) \neq b$. Furthermore, let $t \in \mathbb{N}$, $k^t/3 \geq k^2$ and $\{(\alpha_i, \beta_i) \mid i = 1, 2, ..., k^2\} := E_k^2$. Then there are $\mathbf{c}_1, \ldots, \mathbf{c}_{k^2} \in V_t$ with $\mathbf{x} \neq_s \mathbf{y}$ for all $\mathbf{x}, \mathbf{y} \in V_{t+1}^\star := \{(a, \mathbf{c}_i), (b, s(\mathbf{c}_i)) \mid i = 1, 2, \ldots, k^2\}$ with $\mathbf{x} \neq \mathbf{y}$. Since we can choose V_{t+1} such that $V_{t+1}^\star \subseteq V_{t+1}$ holds, there is a function $g^{t+1} \in S$ with

$$g(x, \mathbf{x}) := \begin{cases} \alpha_i & \text{if } x = a,\ \mathbf{x} = \mathbf{c}_i, \\ \beta_i & \text{if } x = b,\ \mathbf{x} = \mathbf{c}_i, \\ x & \text{if } \mathbf{x} \in V_{t+1} \backslash V_{t+1}^\star. \end{cases}$$

Then

$$h := g(f(e_1^{n+t}, \ldots, e_n^{n+t}), e_{n+1}^{n+t}, \ldots, e_{n+t}^{n+t}) \in [S \cup \{f\}]$$

and it is easy to check that

$$E_k^2 \times \{\infty\} \cup \chi(f) \subseteq \chi(h).$$

Then by (3) and (8) it follows that (9) holds. ■

Theorem 20.7.10 *([Had-L-R 2002]) The set $\mathcal{I}(S)$ is finite.*

Proof. Obviously,

$$S_{max} = \bigcup_{R \subseteq (\widetilde{E_k}^p \setminus E_k^p) \cup s^{\square}} \chi^{-1}(R). \tag{20.10}$$

By Lemma 20.7.9, (5) we have

$$\chi^{-1}(\chi(f)) \subseteq [S \cup \{f\}] \tag{20.11}$$

for every function $f^n \in S_{max}$.
Let $G := \{\chi(f) \mid f \in S_{max}\}$. By (20.8) G is a finite set.
Let $C \in \mathcal{I}(S)$. Obviously, $H := \{\chi(f) \mid f \in C\} \subseteq G$ is finite. Thus $H = \{\chi(\ell_1), \ldots, \chi(\ell_h)\}$ for certain $\ell_1, \ldots \ell_h \in C$. Furthermore, it holds that $C \subseteq \chi^{-1}(\chi(\ell_1)) \cup \ldots \cup \chi^{-1}(\chi(\ell_h))$, where, by (20.11),

$$\chi^{-1}(\chi(\ell_1)) \cup \ldots \cup \chi^{-1}(\chi(\ell_h)) \subseteq [S \cup \{\ell_1\}] \cup \ldots \cup [S \cup \{\ell_h\}] \subseteq [S \cup \{\ell_1, \ldots, \ell_h\}] \subseteq C.$$

Thus $C = [S \cup \{\ell_1, \ldots, \ell_h\}]$. Consequently, the partial clone $C \in \mathcal{I}(S)$ is generated from S and from not more than $|G|$ functions of $S_{max} \setminus S$. ■

Now we determine exactly the set $\mathcal{I}(S)$ for the cases $p = 2$ and $p = 3$. We begin with the case $p = 2$.

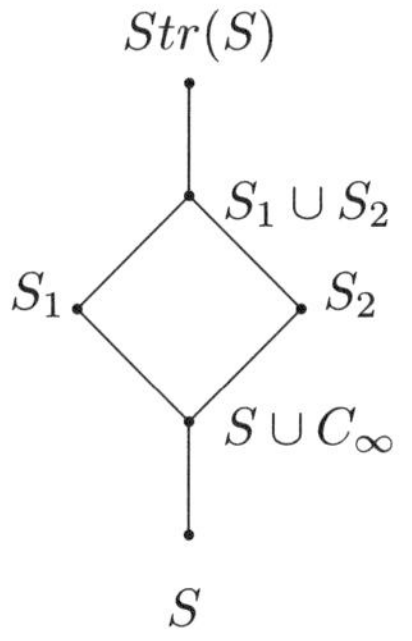

Fig. 20.3

Theorem 20.7.11 *([Had-L-R 2002]; without proof)*
Let $k \geq 2$, $s \in P_{k,2}$ and $S = Pol_k s^o$. Furthermore, let

$$S_1 := \chi^{-1}(\gamma_0 \cup \gamma_2) \text{ and } S_2 := S \cup \chi^{-1}(\gamma_1 \cup \gamma_2).$$

Then

$$\mathcal{I}(S) = \{S, S \cup C_\infty, S_1, S_2, S_1 \cup S_2, Str(S)\},$$

where $S_{max} = \text{pPOL}\,_k s^o = Str(S)$.
The lattice $(\mathcal{I}(S)); \subseteq)$ is given in Figure 20.3.

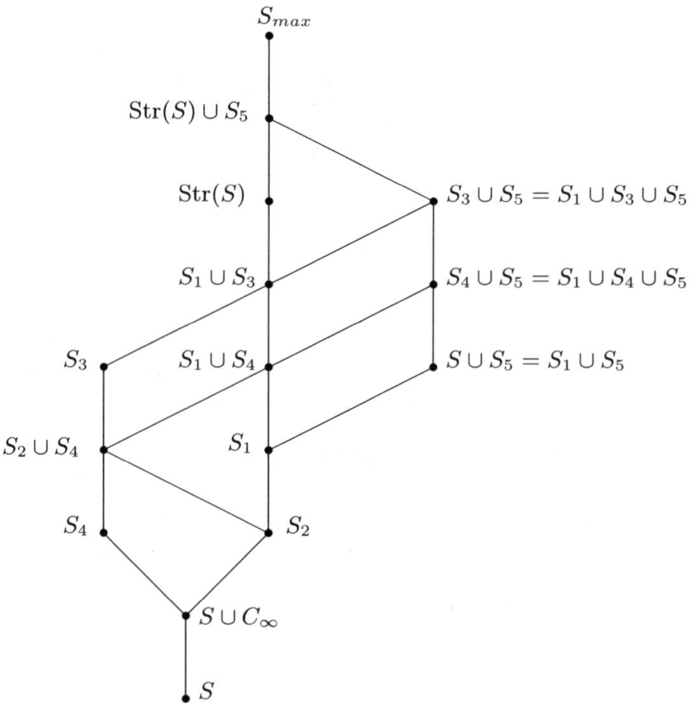

Fig. 20.4

Theorem 20.7.12 *([Had-L-R 2002]; without proof)*
Let $s \in P_{k,3}$ and $S = \text{Pol}\,_k s^o$. Furthermore, let

$$S_1 := S \cup \chi^{-1}(\gamma_1 \cup \gamma_2 \cup \gamma_3), \;\; S_2 := S \cup \chi^{-1}(\gamma_2 \cup \gamma_3),$$
$$S_3 := \chi^{-1}(\gamma_0 \cup \gamma_2 \cup \gamma_3), \;\; S_4 := \chi^{-1}(\gamma_0 \cup \gamma_3) \text{ and } S_5 := \chi^{-1}(\widetilde{E_k}^3 \setminus E_k^3),$$

where $\mathrm{Str}(S) = \chi^{-1}(\gamma_0 \cup \ldots \cup \gamma_3)$. *Then*

$$\mathcal{I}(S) = \{S, S \cup C_\infty, S_2, S_4, S_2 \cup S_4, S_1, S_3, S_1 \cup S_4, S_1 \cup S_3, S \cup S_5, S_4 \cup S_5, S_3 \cup S_5, \text{Str}\ (S), \text{Str}\ (S) \cup S_5, S_{max}\}.$$

The lattice $(\mathcal{I}(S); \subseteq)$ *is given in Figure* 20.4.

Theorem 20.7.13 *([Ale-V 94], [Bör-H 98], [Had-L 2003])*
For every $\varrho \in \mathfrak{L}_k \cup \mathfrak{B}_k$*, the set* $\mathcal{I}(Pol_k\varrho)$ *has the cardinality of continuum.*

Proof. For $k = 2$ (and therefore $\varrho = \{(a, b, c, d) \mid a + b = c + d \pmod 2\}$) the theorem was proven in [Ale-V 94].
If k is an arbitrary prime number power, then $|\mathcal{I}(Pol_k\varrho)| = \mathfrak{c}$ with $\varrho \in \mathfrak{L}_k$ was proven in [Had-L 2003], where the proof of the general result includes the proof from [Ale-V 94].
$|\mathcal{I}(Pol_k\iota_k^k)| = \mathfrak{c}$ was shown in [Bör-H 98]. This result and Theorem 5.2.6.1 were used in [Had-L 2003] to prove the remaining statements of the theorem. ■

20.8 Intervals of Boolean Partial Classes

In continuation of the examinations of Section 20.7, one can find cardinality statements over the set $\mathcal{I}(A)$ for arbitrary subclasses A of P_2 in the following. For this we use the notations from Chapter 3.

Theorem 20.8.1 *([Ale-V 94], [Str 97b])*
Let A *be a subclass of* P_2 *with* $A \subseteq L$ *or*

$$A \subseteq B \in \{\overline{I} \cup C, D \cup C, K \cup C, T_{0,\infty}, T_{1,\infty}\}.$$

Then the set $\mathcal{I}(A)$ *has the cardinality of continuum.*

Proof. For $A = L$, the theorem was proven by V. B. Alekzeev and L. L. Voronenko in [Ale-V 94]. By easy modification of the proof from [Ale-V 94], one gets the statements of the theorem for $A \in \{L, L \cap T_0, L \cap T_1, L \cap T_0 \cap T_1\}$. One finds proof of the remaining statements of the theorem in [Str 97b]. ■

Theorem 20.8.2 *([Ale-V 94], [Str 97a], [Str 95], [Str 96], [Lau 2006])*
Let A *be a subclass of* P_2 *with*

$$T_0 \cap T_1 \cap M \subseteq A \ \ or \ \ T_0 \cap S \subseteq A\ (or\ T_1 \cap S \subseteq A).$$

Then $\mathcal{I}(A)$ *is a finite set and it holds that*

A	$\lvert\mathcal{I}(A)\rvert$
P_2	3
T_a $(a \in \{0,1\})$	6
M	6
S	6
$T_0 \cap T_1$	30
$T_a \cap M$ $(a \in \{0,1\})$	15
$T_0 \cap T_1 \cap M$	101
$T_0 \cap S$	$? < 2000$

For the remaining subclasses A of P_k, i.e., $M \cap T_{\overline{a}} \cap T_{a,\infty} \subset A \subseteq T_{a,2}$ for certain $a \in E_2$ or $A = T_0 \cap M \cap S$, it holds $|\mathcal{I}(A)| \geq \aleph_0$.

Proof. $\mathcal{I}(P_2) = \{P_2, P_2 \cup C_\infty, \widetilde{P_2}\}$ and therefore $|\mathcal{I}(P_2)| = 3$ follows from Lemma 20.4.2.
The statements of the theorem over maximal classes of P_2 different from L were proven by V. B. Alekzeev and L. L. Voronenko and by B. Strauch independently of each other. (see [Ale-V 94], [Str 94], [Str 97a]). Certain elements of the sets $\mathcal{I}(A)$ with $A \in \{T_0, T_1, S, M\}$ were already determined in [Lau 88] at the determination of submaximal classes of $\widetilde{P_2}$. On can find the description of the elements of $\mathcal{I}(A)$ with $A \in \{T_0, T_1, S, M\}$ in Theorems 20.7.4, 20.7.7, and 20.7.11.
The sets $\mathcal{I}(T_0 \cap T_1)$ and $\mathcal{I}(T_0 \cap M)$ were determined in [Str 97a]. One can find $\mathcal{I}(M \cap T_0 \cap T_1)$ in [Str 95].
The finiteness of the set $\mathcal{I}(S \cap T_0 \cap T_1)$ was proven in [Str 96].
In [Lau 2006] it was proven that every set $\mathcal{I}(A)$ with $A \subseteq T_{0,2}$ has infinitely many elements. ∎

20.9 On Congruences of Partial Clones

This section is a revised version of the papers [Lau-D 90] and [Lau-D 91].
For an arbitrary partial clone C, we define the following equivalence relations which – as A. I. Mal'tsev in [Mal 66] was already proving – are congruences over C and, for $C = \widetilde{P_k}$, are the only possible congruences (see Theorem 20.9.3):

$$\kappa_0 := \{(f,f) \mid f \in C\}$$

$$\kappa_1 := C \times C$$

$$\kappa_a := \{(f^n, g^m) \in C \times C \mid n = m\}$$

$$\kappa_\infty := \kappa_0 \cup \{(f^n, g^m) \in C \times C \mid D(f) = D(g) = \emptyset\}$$

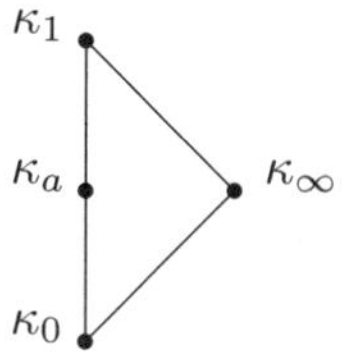

One is easily able to transfer many results from Chapter 9 (because of results of Section 20.2) to partial clones. Therefore, only the congruences on the maximal partial clones shall be determined. For this we need the following notations, which are introduced for relations $\varrho \in R_k^h$.

$$\kappa_1(\varrho) := \kappa_0 \cup \{(f^m, g^n) \in \kappa_1 \mid \forall r_1, ..., r_{max(m,n)} \in \varrho : \{f(r_1, ..., r_m), g(r_1, ..., r_n)\} \subseteq (\widetilde{E_k})^h \setminus E_k^h\},$$
$$\kappa_a(\varrho) := \kappa_1(\varrho) \cap \kappa_a,$$
$$U(\varrho) := \{\alpha \in E_k \mid \forall (a_1, ..., a_h) \in \varrho : \ \alpha \notin \{a_1, ..., a_h\}\},$$
$$\mu_0(\varrho) := \{(f^m, g^m) \in \kappa_a \mid \forall \mathbf{a} \in (E_k \setminus U(\varrho))^m : \ f(\mathbf{a}) = g(\mathbf{a})\},$$
$$\mu(\varrho) := \{(f^m, g^m) \in \kappa_a \mid \forall r_1, ..., r_m \in \varrho : \{f(r_1, ..., r_m), g(r_1, ..., r_m)\} \subseteq (\widetilde{E_k})^h \setminus E_k^h \vee f(r_1, ..., r_m) = g(r_1, ..., r_m)\}.$$

Obviously, the above relations are equivalence relations with the following Hasse-diagram:

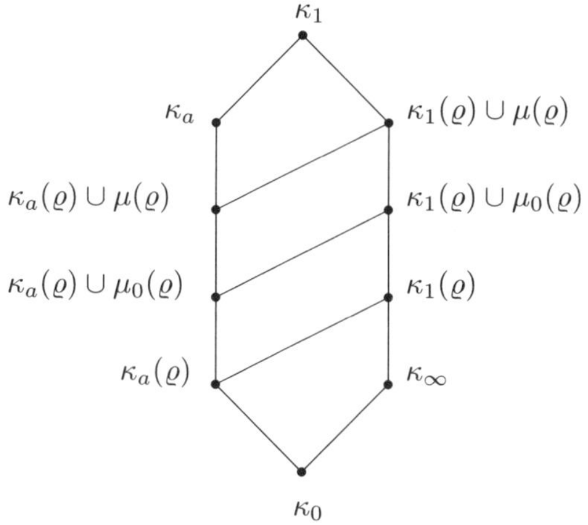

Since $E_k \setminus U(\varrho) = \varrho$, if ϱ is a unary relation, we have in this case $\mu(\varrho) = \mu_0(\varrho)$. If $U(\varrho) = \emptyset$, then $\mu_0(\varrho) = \kappa_0$. If all constant functions and a function c_∞^n belong to the partial clone $pPOL_k\varrho$, then we have $\kappa_1(\varrho) = \kappa_\infty$ and $\kappa_a(\varrho) \cup \mu(\varrho) = \kappa_0$.

The following Lemma is easy to prove.

Lemma 20.9.1

(1) Let $C \subseteq \widetilde{P_k}$. Then the relations κ_0, κ_a, κ_1, κ_∞ are congruences on C.
(2) Let $\varrho \subseteq E_k^h$ and $C := pPOL_k\varrho \subseteq \widetilde{P_k}$. Then the relations $\kappa_a(\varrho)$, $\kappa_a(\varrho) \cup \mu_0(\varrho)$, $\kappa_a(\varrho) \cup \mu(\varrho)$ $\kappa_1(\varrho)$, $\kappa_1(\varrho) \cup \mu_0(\varrho)$, $\kappa_1(\varrho) \cup \mu(\varrho)$ are congruences on C. ■

In the following, it shall be proven that a maximal partial clone of $\widetilde{P_k}$ of the form $pPOL_k\varrho$ has only those congruences given above. Then, dependently of ϱ, one has 4, 8 or 10 pairwise distinct congruences per partial clone $pPOL_k\varrho$.

Lemma 20.9.2 *Let C be a partial clone on E_k and κ a congruence of C. Then*

(a) $\kappa \cap (P_k \times P_k) \nsubseteq \kappa_a \implies \kappa = \kappa_1$;
(b) $(c_\infty^1 \in C \wedge \kappa \nsubseteq \kappa_a) \implies \kappa_\infty \subseteq \kappa$.

Proof. (a): Let $\kappa \cap (P_k \times P_k) \nsubseteq \kappa_a$. Then there are functions f^m, $g^n \in C \cap P_k$ with $m < n$ and $(f, g) \in \kappa$. Consequently, we have

$$f_1^1 := \Delta^{n-2} f \sim g_1^2 := \Delta^{n-2} g \ (\kappa)$$

and

$$e_2^2 = e_2^2 \star f_1 \sim e_3^3 = e_2^2 \star g \ (\kappa).$$

Thus $(e_n^n, e_1^1) \in \kappa$ for all $n \geq 1$ and $e_1^1 \star h = h \sim e_2^2 \star h = e_{t+1}^{t+1}$ (κ) for all $h^t \in C$, i.e., $\kappa = \kappa_1$.

(b): Let $\kappa \nsubseteq \kappa_a$ and $c_\infty \in C$. Then there are two functions f^m, g^n, $m < n$, in C with $(f, g) \in \kappa$. Thus

$$(\Delta^{n-2} g) \star c_\infty^1 = c_\infty^2 \sim (\Delta^{n-2} f) \star c_\infty^1 = c_\infty^1 \ (\kappa).$$

Consequently, $c_\infty^2 \star e_1^r = c_\infty^{r+1} \sim c_\infty^1 \star e_1^r = c_\infty^r$ (κ) for arbitrary $r \in \mathbb{N}$. Therefore, $\kappa_\infty \subseteq \kappa$. ■

Theorem 20.9.3 *([Mal 66]) Let C be a clone with*

$$Str(\{c_a \mid a \in E_k\}) \subset C.$$

Then C has exactly the following four congruences: κ_a, κ_a, κ_∞, κ_1.

Proof. Let $\kappa \neq \kappa_0$ be a congruence of C. Then the following cases are possible:

Case 1: $\kappa_0 \subset \kappa \subseteq \kappa_a$.
In this case, there are two functions f^m, $g^n \in C$ and an n-tuple $\mathbf{a} := (a_1, ..., a_n)$ with $f(c_{a_1}, ..., c_{a_n}) =: c_\alpha$, $g(c_{a_1}, ..., c_{a_n}) =: g'$, $g' \in \{c_\beta, c_\infty\}$, $\{\alpha, \beta\} \subseteq E_k$, $(c_\alpha, g') \in \kappa$ and $c_\alpha \neq g'$.

Since $Str(\{c_a \mid a \in E_k\}) \subset C$, there is a function $h^1 \in C$ with $D(h) = \{\alpha\}$ and $h(\alpha) = \alpha$. Thus we have $h \star c^1_\alpha \sim h \star g' = c^1_\infty$ (κ) and $e^1_1 = \Delta(e^2_2 \star c_\alpha) \sim \Delta(e^2_2 \star c^1_\infty) = c^1_\infty$ (κ). We obtain $e^1_1 \star t = t \sim c^n_\infty$ (κ) for all $t^n \in C$. Thus $\kappa_a \subseteq \kappa$ and $\kappa = \kappa_a$.

Case 2: $\kappa_0 \not\subset \kappa_a$.
Since $c_\infty \in C$, it follows from Lemma 20.9.2: $\kappa_\infty \subseteq \kappa$. Suppose, $\kappa \neq \kappa_\infty$. Then there are two functions f^m, $g^n \in C$ with $m \neq n$, $\{f, g\} \not\subseteq \{c^n_\infty \mid n \in \mathbb{N}\}$ and $(f, g) \in \kappa$. W.l.o.g. we can assume that $f \neq c^m_\infty$ and $m < n$. Let $\mathbf{a} := (a_1, ..., a_m) \in D(f)$ and $f(\mathbf{a}) = \alpha \in E_k$. Then

$$\begin{aligned} &(...((\Delta((\Delta((\Delta(f \star c_{a_1})) \star c_{a_2})) \star c_{a_3})) \star c_{a_4}))... \star c_{a_m} = f(c_{a_1}, ..., c_{a_m}) = c_\alpha \\ \sim\ &(...((\Delta((\Delta((\Delta(g \star c_{a_1})) \star c_{a_2})) \star c_{a_3})) \star c_{a_4}))... \star c_{a_m} =: g^r_1\ (\kappa) \end{aligned}$$

with $r := n - m + 1$. W.l.o.g. let $r = 2$. If $g_1 = c^2_\infty$, we obtain $(c^1_\alpha, c^1_\infty) \in \kappa$ and analogously to Case 1, $\kappa_a \subseteq \kappa$. $\kappa_a \subseteq \kappa$ and $\kappa_\infty \subseteq \kappa$ imply $\kappa = \kappa_1$. If $g_1 \neq c^2_\infty$, there exists a $(b_1, b_2) \in E^2_k$ with $g_1(b_1, , b_2) \in E_k$ and thus we have

$$c^1_\alpha = c^1_\alpha(\tau(c_\alpha \star c_{b_1}) \star c_{b_2}) \sim c^1_\alpha(\tau(g_1 \star c_{b_1}) \star c_{b_2}) = c^2_\alpha\ (\kappa).$$

By Lemma 20.9.2, (a) we get $\kappa = \kappa_1$. ∎

Theorem 20.9.4 *([Lau-D 90])*
Let $C = P_k \cup \{c^n_\infty \mid n \in \mathbb{N}\}$ or $C = pPOL_k\varrho$ with $\varrho \in \widetilde{R}_{max} \backslash \{\varrho \in \widetilde{R}_{max} \mid \emptyset \subset \varrho \subset E_k$ or ϱ is a coherent areflexive relation$\}$, where $\widetilde{R}_{max}$ denotes the set of all relations ϱ by which a maximal partial clone $pPOL_k\varrho$ is described (see 20.5).
Then C has exactly the following four congruences: κ_0, κ_a, κ_∞, κ_1.

Proof. If $C = pPOL_k\varrho$ with $\varrho \in \widetilde{R}_{max} \backslash \{\varrho \in \widetilde{R}_{max} \mid \emptyset \subset \varrho \subset E_k$ or ϱ is a coherent areflexive relation$\}$, our theorem follows from Theorem 20.9.3. It was proven in Chapter 9 (see Theorem 9.1.2) that P_k has only the congruences $\kappa_0, \kappa_a, \kappa_1$. From this and from Lemma 20.9.2 follows our theorem for $C = P_k \cup \{c^n_\infty \mid n \in \mathbb{N}\}$. ∎

Lemma 20.9.5 *Let κ be a congruence of $pPOL_k\varrho$. Then*

(a) $\kappa_0 \subset \kappa \subseteq \kappa_a \implies \kappa_a(\varrho) \subseteq \kappa$,

(b) $\kappa_\infty \subset \kappa \implies \kappa_1(\varrho) \subseteq \kappa$,

(c) $\kappa_a(\varrho) \subset \kappa \subseteq \kappa_1(\varrho) \implies \kappa = \kappa_1(\varrho)$,

(d) $\kappa_a(\varrho) \cup \mu_0(\varrho) \subset \kappa \subseteq \kappa_1(\varrho) \cup \mu_0(\varrho) \implies \kappa = \kappa_1(\varrho) \cup \mu_0(\varrho)$,

(e) $\kappa_a(\varrho) \cup \mu(\varrho) \subset \kappa \subseteq \kappa_1(\varrho) \cup \mu(\varrho) \implies \kappa = \kappa_1(\varrho) \cup \mu(\varrho)$.

Proof. (a): Let $\kappa_0 \subset \kappa \subseteq \kappa_a$. Then there are functions f^n, g^n with $(f^n, g^n) \in \kappa$ and certain $\mathbf{a} := (a_1, ..., a_n) \in E^n_k$, $a, b \in E_k$, $a \neq b$ with $f(\mathbf{a}) =: a \neq b := g(\mathbf{a})$. Furthermore, $pPOL_k\varrho$ contains functions t_α $(\alpha \in E_k)$ and $h_{\beta,\gamma}$ $(\beta, \gamma \in E_k)$ with

$$t_\alpha(x,y) := \begin{cases} y & \text{if} \quad x = \alpha, \\ \infty & \text{otherwise}, \end{cases}$$

and

$$h_{\beta,\gamma}(x) := \begin{cases} \gamma & \text{if} \quad x = \beta, \\ \infty & \text{otherwise}, \end{cases}$$

where, if ϱ is areflexive or $\{c_a \mid a \in E_k\} \subset pPOL_k\varrho$ holds, β and γ are arbitrary elements of E_k; otherwise we must choose, however, $\beta \notin \{a \mid (a,a,...,a) \notin \varrho\}$.
Thus we obtain

$$f(h_{\beta,a_1}(x),...,h_{\beta,a_n}(x)) = h_{\beta,a}(x) \sim g(h_{\beta,a_1}(x),...,h_{\beta,a_n}(x)) = h_{\beta,b}(x)\ (\kappa)$$

and

$$t_a(h_{\beta,a}(x),y) = t_\beta(x,y) \sim t_a(h_{\beta,b}(x),y) = c^2_\infty(x,y)\ (\kappa).$$

Let u^m be an arbitrary function from $pPOL_k\varrho$ with the property

$$\forall (r_{11},r_{21},...,r_{h1}),...,(r_{1n},r_{2n},...,r_{hn}) \in \varrho\ \exists i \in \{1,...,h\}:$$
$$(r_{i1},r_{i2},...,r_{in}) \notin D(u).$$

Then the function v^m defined by

$$v(\mathbf{x}) := \begin{cases} \beta & \text{if} \quad u(\mathbf{x}) \in E_k, \\ \infty & \text{otherwise}, \end{cases}$$

belongs to $pPOL_k\varrho$. Consequently, we have

$$t_\beta(v(\mathbf{x}),u(\mathbf{x})) = u(\mathbf{x}) \sim c^2_\infty(v(\mathbf{x}),u(\mathbf{x})) = c^m_\infty(\mathbf{x})\ (\kappa),$$

i.e., it holds $\kappa_a(\varrho) \subseteq \kappa$.
(b): For $\kappa_\infty \subset \kappa$ and $\kappa \cap \kappa_a \neq \kappa_0$, we have $\kappa_a(\varrho) \subseteq \kappa$ because of (a), since the inclusion $\kappa_a(\varrho) \subseteq \kappa'$ follows from $\kappa \cap \kappa_a = \kappa' \neq \kappa_0$ and since the inclusion $\kappa_a(\varrho) \subseteq \kappa$ follows from $\kappa' \subseteq \kappa$.
$\kappa_a(\varrho) \subseteq \kappa$ and $\kappa_\infty \subseteq \kappa$ imply $\kappa_1(\varrho) \subseteq \varrho$, since, if $(f^m, g^n) \in \kappa_1(\varrho)$ $(m \neq n)$, the inclusions $(f^m, c^m_\infty) \in \kappa$ and $(c^n_\infty, g^n) \in \kappa$ follow, because of $\kappa_a(\varrho) \subseteq \kappa$ and $(c^m_\infty, c^n_\infty) \in \kappa$ by $\kappa_\infty \subseteq \kappa$. By the transitivity of κ we get $(f^m, g^n) \in \kappa$.
If $\kappa_\infty \subset \kappa$ and $\kappa \cap \kappa_a = \kappa_0$, there are two functions s^l, t^r $(l \neq r)$ in $pPOL_k\varrho$ with $(s,t) \in \kappa$. We can assume $l > r$ and $s \neq c^l_\infty$, i.e., there exists an element $\mathbf{a} := (a_1,...,a_l)$ with $s(\mathbf{a}) \in E_k$. We distinguish the following two cases:

Case 1: For all $i \in \{1,...,l\}$ it holds $(a_i,...,a_i) \in \varrho$.
In this case we have $\{c_{a_1},...,c_{a_l}\} \subset pPOL_k\varrho$. From $(s,t) \in \kappa$ and $\kappa_\infty \subseteq \kappa$ the existence of two κ-kongruent constant functions with different arities follows. Therefore, by Lemma 20.9.2, (a) we get $(P_k \cap pPOL_k\varrho) \times (P_k \cap pPOL_k\varrho) \subseteq \kappa$, a contradiction to $\kappa \cap \kappa_a = \kappa_0$.

Case 2: There exists an $i \in \{1, ..., l\}$ with $(a_i, ..., a_i) \notin \varrho$.
W.l.o.g. we can assume $(a_1, ..., a_1) \notin \varrho$. Then the function $h^1_{a_1,a_j}$ belongs to $pPOL_k\varrho$ for certain $a_j \in E_k$, $j = 1, ..., n$. Further, we can assume $a_1 \neq a_{l-r}$. (If $a_1 = a_{l-r}$, then we can choose $s' := s \star e_2^2$ and $t' := t \star e^2$ and then use the operations ζ and τ to get functions s'' and t'' with $(s'', t'') \in \kappa$, $s''(\mathbf{a}) \in E_k$, $(a_1'', ..., a_1'') \notin \varrho$ and $a_1'' \neq a_{l-r}''$). Thus

$$s_1 := (....(\zeta((\zeta((\zeta s) \star h_{a_1,a_1})) \star h_{a_1,a_{l-1}} \star h_{a_1,a_{l-2}})) \star ...) \star h_{a_1,a_{l-r}}) \sim$$
$$t_1 := (....(\zeta((\zeta((\zeta t) \star h_{a_1,a_1})) \star h_{a_1,a_{l-1}} \star h_{a_1,a_{l-2}})) \star ...) \star h_{a_1,a_{l-r}}) \ (\kappa)$$

with

$$s_1(x_1, ..., x_l) = s(x_{l-r+1}, ..., x_l, h_{a_1,a_{l-r}}(x_1), ..., h_{a_1,a_l}(x_{l-r}))$$

and

$$t_1(x_1, ..., x_r) = t(h_{a_1,a_{l-r+1}}(x_2), ..., h_{a_1,a_{l-1}}(x_r), h_{a_1,a_l}(h_{a_1,a_{l-r}}(x_l))) = c^r_\infty(\mathbf{x}),$$

since $h_{a_1,a_l} \star h_{a_1,a_{l-r}} = c^1_\infty$ and because of $a_1 \neq a_{1-r}$.
From this and from $\kappa_\infty \subseteq \kappa$, it follows that $\kappa \cap \kappa_a \neq \kappa_0$, in contradiction to our assumption. Therefore (b) holds.

(c): Let $\kappa_a(\varrho) \subset \kappa \subseteq \kappa_1(\varrho)$. Then $\kappa \nsubseteq \kappa_a$. By Lemma 20.9.2, (b) we have $\kappa_\infty \subseteq \kappa$. From $\kappa_\infty \subseteq \kappa \subseteq \kappa_1(\varrho)$ it follows that $\kappa = \kappa_\infty$ or $\kappa = \kappa_1(\varrho)$ (by (b)). Since the first case is not possible, (c) holds.

(d) and (e) follow from (c). ■

Lemma 20.9.6 *Let $\emptyset \subset \varrho \subset E_k$ or let ϱ be an h-ary relation on E_k with the properties $h \geq 2$ and*

$$\forall (a_1, ..., a_h) \in \varrho \ \exists t \in P_k \cap pPOL_k\varrho : \ Im(t) \subseteq \{a_1, ..., a_h\}. \qquad (20.12)$$

Then for an arbitrary congruence κ of $pPOL_k\varrho$,
(a) $(\kappa \nsubseteq \kappa_a(\varrho) \cup \mu(\varrho) \ \wedge \ \kappa \subseteq \kappa_a) \implies \kappa = \kappa_a$,
(b) $(\kappa \nsubseteq \kappa_1(\varrho) \cup \mu(\varrho) \ \wedge \ \kappa \nsubseteq \kappa_a) \implies \kappa = \kappa_1$.
(c) $(\kappa \nsubseteq \kappa_a(\varrho) \ \wedge \ \kappa \subseteq \kappa_a) \implies \kappa_a(\varrho) \cup \mu_0(\varrho) \subseteq \kappa$.

Proof. (a): $\kappa \nsubseteq \kappa_a \cup \mu(\varrho)$ and $\kappa \subseteq \kappa_a$ imply that there are functions $(f^n, g^n) \in \kappa$ and a set $R := \{(r_{j1}, r_{j2}, ..., r_{jn}) \mid i = 1, ..., h\}$ with $\{(r_{1i}, r_{2i}, ..., r_{hi}) \mid i = 1, ..., n\} \subseteq \varrho$, and
1) $R \subseteq D(f)$, $R \subseteq D(g)$ and $f(r_{j1}, r_{j2}, ..., r_{jn}) \neq g(r_{j1}, r_{j2}, ..., r_{jn})$ for certain $j \in \{1, ..., h\}$
or
2) $R \subseteq D(f)$ and $R \nsubseteq D(g)$.
The first case can be reduced to the second case as follows: Let $\widehat{e}_3^3$ be a ternary function of $pPOL_k\varrho$ defined by

$$\widehat{e}_3^3(x,y,z) = \begin{cases} z & \text{if} \quad x = y, \\ \infty & \text{otherwise.} \end{cases}$$

Then

$$\widehat{e}_3^3(f(\mathbf{x}), f(\mathbf{x}), f(\mathbf{x})) = f(\mathbf{x}) \sim \widehat{e}_3^3(f(\mathbf{x}), g(\mathbf{x}), f(\mathbf{x})) =: g'(\mathbf{x})\ (\kappa)$$

with $R \not\subseteq D(g')$.
Consequently, we can assume that $R \subseteq D(f)$ and $R \not\subseteq D(g)$.
Let $(a_1, ..., a_h) \in \varrho$. Then the functions $t_1, ..., t_h$ with

$$t_i \begin{pmatrix} a_1 \\ a_2 \\ . \\ . \\ a_h \end{pmatrix} = \begin{pmatrix} r_{1i} \\ r_{2i} \\ . \\ . \\ r_{hi} \end{pmatrix}$$

and $t_i(x) = \infty$ for $x \notin \{a_1, ..., a_h\}$ $(i = 1, 2, ..., n)$ belong to $pPOL_k\varrho$. Then, from $(f, g) \in \kappa$ it follows

$$f(t_1(x), ..., t_n(x)) =: f'(x) \sim g(t_1(x), ..., t_n(x)) =: g'(x)\ (\kappa),$$

where $\{r_1, ..., r_h\} \not\subseteq D(g')$ and $(f'(a_1), ..., f'(a_h)) \in \varrho$ for all $(r_1, ..., r_h) \in \varrho$. By Lemma 20.9.2 we have $\kappa_a(\varrho) \subseteq \kappa$ and thus $(g', c_\infty^1) \in \kappa$ and $(f', c_\infty^1) \in \kappa$. (20.12), implies the existence of a function $t^1 \in P_k \cap pPOL_k\varrho$ with $Im(t^1) \subseteq \{a_1, ..., a_h\}$. Using this function and the fact that $f'' = f' \star t$, we obtain another function of $P_k \cap pPOL_k\varrho$ with $(f'', c_\infty^1) \in \kappa$. Thus $(\Delta(e_2^2 \star f''), \Delta(e_2^2 \star c_\infty^1)) = (e_1^1, c_\infty^1) \in \kappa$ and $\kappa = \kappa_a$.
(b): It is easy to check that $\kappa \not\subseteq \kappa_1(\varrho) \cup \mu(\varrho)$ and $\kappa \not\subseteq \kappa_a$ imply $\kappa \cap \kappa_a \not\subseteq \kappa_a(\varrho) \cup \mu(\varrho)$. Hence (b) follows from (a) and $\kappa_\infty \subseteq \kappa$.
(c): Obviously, (c) holds, if $U(\varrho) = \emptyset$. Thus we can assume $U(\varrho) \neq \emptyset$.
Because of Lemma 20.9.5, (a) we have to show that $\mu_0(\varrho) \subseteq \kappa$. Let κ be a congruence of $pPOL_k\varrho$ with $\kappa \subseteq \kappa_a$ and $\kappa \not\subseteq \kappa_a(\varrho)$. Then there are two different n-ary functions f, g with $(f, g) \in \kappa$ and $\rho_j := (r_{1j}, r_{2j}, ..., r_{hj}) \in \varrho$ $(j = 1, ..., n)$ with $(r_{i1}, r_{i2}, ..., r_{in}) \in D(f)$ for all $i \in \{1, ..., h\}$ or $(r_{i1}, r_{i2}, ..., r_{in}) \in D(g)$ for all $i \in \{1, ..., h\}$.
We can assume that $\kappa \neq \kappa_a$. Then by (a) we have $\kappa \subseteq \kappa_a(\varrho) \cup \mu(\varrho)$, i.e., $f(r_1, ..., r_n) = g(r_1, ..., r_n)$ holds. Because of $f \neq g$ there exists an n-tuple $\mathbf{a} := (a_1, ..., a_n) \in E_k^n$ with $f(\mathbf{a}) \neq g(\mathbf{a})$. It was already shown in proof of (a) that we can assume $\mathbf{a} \notin D(g)$ and $\mathbf{a} \in D(f)$.
Let $(\alpha_1, ..., \alpha_h) \in \varrho$ and $a \in U(\varrho)$. Then the functions $h_1, ..., h_n$ defined by $(h_i(\alpha_1), ...h_i(\alpha_h)) = r_i$, $h_i(a) = a_i$ and $h_i(x) := \infty$ otherwise belong to $pPOL_k\varrho$. By this we have

$$f'(x) := f(h_1(x), ..., h_n(x)) \sim g(h_1(x), ..., h_n(x)) =: g'(x)\ (\kappa),$$

where $f'(\alpha_i) = g'(\alpha_i)$, $f'(a) = f(a_1, ..., a_n)$ and $a \notin D(g')$. Obviously, if ϱ is a unary relation or because of (20.12) (if ϱ is an areflexive relation and

$h \geq 2$) $pPOL_k\varrho$ contains a function $t \in P_k$ with $Im(t) \subseteq \{\alpha_1, ..., \alpha_n\}$. Thus the function t' defined by

$$t'(x) := \begin{cases} t(x) & \text{if} \quad x \notin U(\varrho), \\ a & \text{otherwise}, \end{cases}$$

belongs to $pPOL_k\varrho$ and we get $f' \star t' =: f'' \sim g'' := g' \star t'$ (κ) and $e_2^2(f''(x), x) = e_1^1(x) \sim \widehat{e}_1^1(x) := e_2^2(g''(x), x)$ (κ) with

$$\widehat{e}(x) := \begin{cases} x & \text{if } x \notin U(\varrho), \\ \infty & \text{if } x \in U(\varrho). \end{cases}$$

Let p^m be an arbitrary function of $pPOL_k\varrho$ and let

$$p_a(\mathbf{x}) := \begin{cases} p(\mathbf{x}) & \text{if} \quad \mathbf{x} \in (E_k \backslash U(\varrho))^m, \\ a & \text{otherwise}, \end{cases}$$

and

$$p_\infty(\mathbf{x}) := \begin{cases} p(\mathbf{x}) & \text{if} \quad \mathbf{x} \in (E_k \backslash U(\varrho))^m, \\ \infty & \text{otherwise}. \end{cases}$$

From $(e_1^1, \widehat{e}_1^1) \in \kappa$ it follows $e_1^1 \star p_a = p_a \sim \widehat{e}_1^1 \star p_a = p_\infty$ (κ) and $e_2^2(p_a(\mathbf{x}), p(\mathbf{x})) = p(\mathbf{x}) \sim e_2^2(p_\infty(\mathbf{x}), p(\mathbf{x})) = p_\infty(\mathbf{x})$ (κ). Consequently, two m-ary functions p, q with $p(\mathbf{x}) = q(\mathbf{x})$ for $\mathbf{x} \in (E_k \backslash U(\varrho))^m$ are κ-kongruent, i.e., $\mu_0(\varrho) \subseteq \kappa$. ∎

Lemma 20.9.7 *Let ϱ be an areflexive h-ary relation with $h \geq 2$ and let $(0, 1, ..., h-1) \in \varrho$. Further, for every $(r_{1j}, r_{2j}, ..., r_{hj}) \in \varrho$, $j = 1, ..., n$, there are unary functions $h_1, ..., h_n \in P_k \cap pPOL_k\varrho$ with $(h_1(i), h_2(i), ..., h_n(i)) = (r_{i1}, r_{i2}, ..., r_{in})$ for all $i \in \{0, ..., h-1\}$. Then for every congruence κ of $pPOL_k\varrho$:*

(a) $\mu_0(\varrho) \subset \kappa \backslash \kappa_1(\varrho) \subseteq \mu(\varrho) \implies \mu(\varrho) \subseteq \kappa$;

(b) $\kappa_a(\varrho) \cup \mu_0(\varrho) \subset \kappa \subseteq \kappa_a(\varrho) \cup \mu(\varrho) \implies \kappa = \kappa_a(\varrho) \cup \mu(\varrho)$;

(c) $\kappa_1(\varrho) \cup \mu_0(\varrho) \subset \kappa \subseteq \kappa_1(\varrho) \cup \mu(\varrho) \implies \kappa = \kappa_1(\varrho) \cup \mu(\varrho)$.

Proof. (a): Let κ be a congruence of $pPOL_k\varrho$ with $\mu_0(\varrho) \subset \kappa \backslash \kappa_1(\varrho) \subseteq \mu(\varrho)$. Then there are two κ-kongruent functions f^n, g^n, such that for certain h-tuple $(r_{11}..., r_{h1})$, ..., $(r_{1n}, ..., r_{hn}) \in \varrho$ there are an n-tuple $(\alpha_1, ..., \alpha_n) \in E_k^n$ and an element $\alpha \in E_k$ with $f(\alpha_1, ..., \alpha_n) = \alpha$, $(\alpha_1, ..., \alpha_n) \notin D(g)$ and

$$f \begin{pmatrix} r_{11} & ... & r_{1n} \\ . & ... & . \\ . & ... & . \\ r_{h1} & ... & r_{hn} \end{pmatrix} = \begin{pmatrix} a_1 \\ . \\ . \\ a_h \end{pmatrix} = g \begin{pmatrix} r_{11} & ... & r_{1n} \\ . & ... & . \\ . & ... & . \\ r_{h1} & ... & r_{hn} \end{pmatrix}.$$

By assumption, there are functions $h_1, ..., h_n \in P_k \cap pPOL_k\varrho$ with $\{(h_1(i), ..., h_n(i)) \mid i \in E_h\} = \{(r_{j1}, r_{j2}, ..., r_{jn}) \mid j \in \{1, 2, ..., h\}\}$. Let t^m be an arbitrary function of $pPOL_k\varrho$. Then, with the help of

$D_t :=$

$$\left\{\mathbf{x} \in E_k^m \mid \exists \mathbf{a}_1, ..., \mathbf{a}_h : \mathbf{x} \in \{\mathbf{a}_1, ..., \mathbf{a}_h\} \wedge \begin{pmatrix} \mathbf{a}_1 \\ \cdot \\ \cdot \\ \mathbf{a}_h \end{pmatrix} \subseteq \varrho \wedge t \begin{pmatrix} \mathbf{a}_1 \\ \cdot \\ \cdot \\ \mathbf{a}_h \end{pmatrix} \in \varrho \right\}$$

one can define the functions t_i by

$$t_i(\mathbf{x}) := \begin{cases} h_i(t(\mathbf{x})) & \text{if} \quad \mathbf{x} \in D_t, \\ \alpha_i & \text{if} \quad \mathbf{x} \notin D_t \wedge \mathbf{x} \in D(t), \\ \infty & \text{otherwise} \end{cases}$$

$(i = 1, 2, ..., n)$. These functions belong to $pPOL_k\varrho$ and we get

$$u(f(t_1(\mathbf{x}), ..., t_n(\mathbf{x})) = t(\mathbf{x}) \sim t'(\mathbf{x}) = u(g(t_1(\mathbf{x}), ..., t_n(\mathbf{x})) \ (\kappa)$$

with

$$u(x, y) := \begin{cases} y & \text{if} \quad x \in \{a_1, ..., a_h, \alpha\}, \\ \infty & \text{otherwise,} \end{cases}$$

and

$$t'(x) := \begin{cases} t(\mathbf{x}) & \text{if} \quad \mathbf{x} \in D_t, \\ \infty & \text{otherwise.} \end{cases}$$

Thus $\mu(\varrho) \subseteq \kappa$.

(b) and (c) follow from (a). ■

Lemma 20.9.8 *Let $\varrho \subseteq E_k^h$ be an areflexive coherent relation with $h \geq 2$. Then*

(a) $\forall (a_1, ..., a_h) \in \varrho \ \exists t \in P_k \cap pPOL_k\varrho : \ Im(t) \subseteq \{a_1, ..., a_h\}$

and w.l.o.g.

(b) one can assume that ϱ fulfills the assumptions of Lemma 20.9.7.

Proof. By definition of ϱ (see Section 20.5) we can assume that there is a subgroup G_ϱ of S_h and a surjective function $\varphi : E_k \longrightarrow \{0, 1, ..., h-1\}$ with the following properties:

$$\varrho = \{(\pi(0), \pi(1), ..., \pi(h-1)) \mid \pi \in G_\varrho\},$$

ϱ is symmetric in respect to every $\pi \in G_\varrho$ and

$$\forall (a_0, ..., a_{h-1}) \in \varrho : \ (\varphi(a_0), ..., \varphi(a_{h-1})) \in \varrho.$$

For an arbitrary $\mathbf{a} := (a_0, ..., a_{h-1}) \in \varrho$ we consider the function $\varphi_\mathbf{a} : \{0, 1, ..., h-1\} \longrightarrow \{a_0, ..., a_{h-1}\}$ with $\varphi_\mathbf{a}(i) = a_i$ for all $i \in \{0, ..., h-1\}$. Further, let $h_\mathbf{a} := \varphi_\mathbf{a} \star \varphi$. Then $Im(h_\mathbf{a}) = \{a_0, ..., a_{h-1}\}$ and $h_\mathbf{a} \in pPOL_k\varrho$, since for every $\mathbf{b} := (b_0, .., b_{h-1}) \in \varrho$ there exists a permutation $\pi_\mathbf{b} \in G_\varrho$ with $\varphi(\mathbf{b}) = (\varphi(b_0), ..., \varphi(b_{h-1}))$. Thus we obtain $h_\mathbf{a}(\mathbf{b}) = \varphi_\mathbf{a}(\varphi(\mathbf{b})) = \varphi_\mathbf{a}((\pi_\mathbf{b}(0), ..., \pi_\mathbf{b}(h-1))) = (a_{\pi_\mathbf{b}(0)}, ..., a_{\pi_\mathbf{b}(h-1)}) \in \varrho$ because of symmetry of ϱ with respect to every permutation $\pi \in G_\varrho$. Consequently, (a) and (b) hold. ■

Theorem 20.9.9 *([Lau-D 91]) Let $\emptyset \subset \varrho \subset E_k$ or let $\varrho \subseteq E_k^h$ an areflexive coherent relation with $h \geq 2$ and $U(\varrho) = \emptyset$. Then $pPOL_k\varrho$ has exactly 8 congruences with the following congruence lattice:*

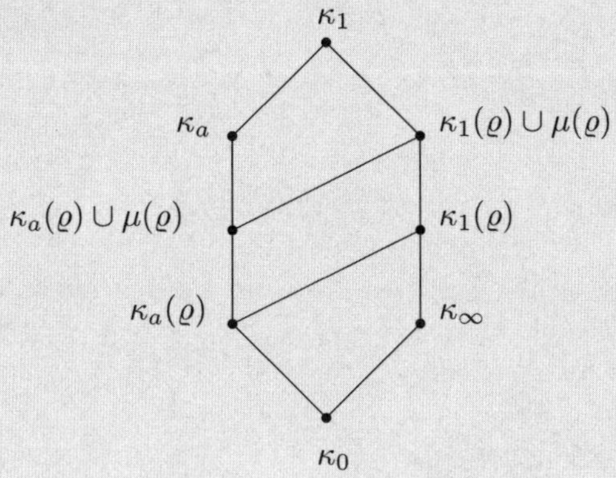

Proof. Since $\mu(\varrho) = \mu_0(\varrho)$ for all ϱ with $\emptyset \subset \varrho \subset E_k$ and for all areflexive relations with $U(\varrho) = \emptyset$, our theorem follows from Lemmas 20.9.2 and 20.9.5–20.9.8. ■

The following theorem also follows from Lemmas 20.9.2, 20.9.5–20.9.8:

Theorem 20.9.10 *([Lau-D 91]) Let $\varrho \subseteq E_k^h$ be an areflexive relation with $h \geq 2$ and $U(\varrho) \neq \emptyset$. Then $pPOL_k\varrho$ has exactly 10 congruences, and the congruence lattice of $pPOL_k\varrho$ is given by*

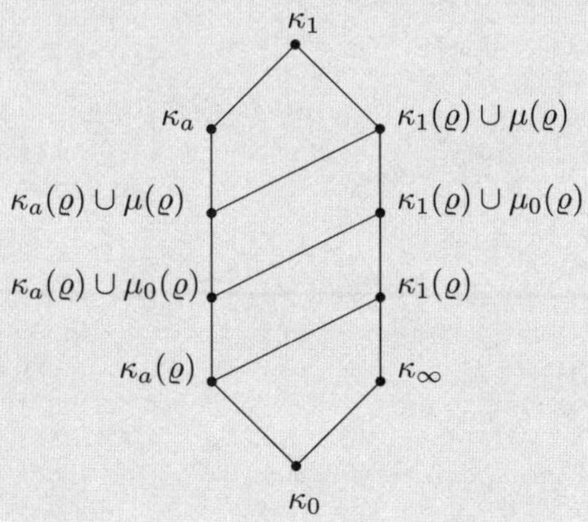

■

References

[Ago-D-H 83] Agoston, I.; Demetrovics; J., Hannak, L.: On the number of clones containing all constants. Math. Soc. J. Bolyai, Szeged (Hungary), 43. Lec. in universal algebra, 21–25 (1983)

[Ale-V 94] Alekseev, V. B.; Voronenko, A. A.: On some closed classes in partial two-valued logic. (Russian, English) Discrete Math. Appl. **4**, No. 5, 401–419 (1994); translation from Diskretn. Mat. **6**, No. 4, 58–79 (1994)
Алексеев, В. Б.; Вороненко, Л. Л.: О некоторых замкнутых классах в частичной двузначной логике. Дискретная Математика **6**, 4, 58–79 (1994)

[Ass 59,72,81] Asser, G.: Einführung in die mathematische Logik. Teil I–III. Teubner Leipzig 1959 (überarbeitete Fassung 1982), 1972, 1981

[Bag-D 80] Bagyinszki, J.; Demetrovics, J.: The structure of the maximal classes in prime-valued logics. C. R. Math. Rep. Acad. Sci. Canada **2**, 209–213 (1980)

[Bag-D 82] Bagyinszki, J.; Demetrovics, J.: The lattice of linear classes in prime-valued logics. Banach Center Publications (Warszawa) 7/1982, 105–123

[Bai 67a] Bairamov, R. A.: On Sheffer functions in many-valued logic. (Russian) Problems in the Theory of Electronic Digital Computers, No. 3, pp. 62–71. Akad. Nauk Ukrain. SSR, Kiev 1967
Байрамов, Р. А.: О функциях Шеффера в многозначной логике. Сб. "Вопросы теории Электронных цифровых матем. машин". Семинар АН УССР, вып. 3, Киев 1967, стр. 62–71

[Bai 67b] Bairamov, R. A.: On one serie of precomplete classes in k-valued logic. (Russian) Kibernetika **1**, 7–9 (1967). English Translation: Cybernetics **3**, 6–7 (1967)
Байрамов, Р. А.: Об одной серии предпольных классов в k-значной логике. Кибернетика **1** (1967), 7–9.

[Bak-P 75] Baker, K. A.; Pixley, A. F.: Polynomial interpolation and the chinese remainder theorem for algebraic systems. Math. Z. **143**, 165–174 (1975)

[Bak 77] Baker, K. A.: Finite equational bases for finite algebras in congruence-distributive equational classes. Adv. in Math. **24**, 207–243 (1977)

[Ban-G 90] Bandemer, H.; Gottwald, S.: Einführung in Fuzzy-Methoden. Akademie-Verlag, Berlin 1990

[Ber 80] Berman, J.: A proof of Lyndon's finite basis theorem. Discrete Math. **29**, 229–233 (1980)

[Bij-T 91] Bijev, G.; Todorov, K.: On the representation of abstract semigroups by transformation semigroups: computer investigations. Semigroup Forum **43**, 253–256 (1991)

[Bir 35] Birkhoff, G.: On the structure of abstract algebras. Proc. Camb. Phil. Soc. **50**, 433–454 (1935)

[Bir 35] Birkhoff, G.: Subdirect unions in universal algebra. Bull. Amer. Math. Soc. **50**, 764–768 (1944)

[Bir 48] Birkhoff, G.: Lattice theory. Rev. ed. American Mathematical Society Colloquium Publications. 25. New York, 285 p., 1948 (see also: Birkhoff, G.: Lattice theory. Third (new) ed. AMS Colloquium Publications, Vol. 25, Providence, R.I.: American Mathematical Society, 418 p., 1967)

[Blo 72] Blochina, G. N.: On predicate description of Post classes (Russian). Diskret. Analiz **16**, 16–29 (1970)
Блохина, Г. Н.: О предикатном опицании классов Поста. Дискретный анализ **16**, 16–29 (1970)

[Bod-K-K-R 69] Bodnarchuk, V. G.; Kaluzhnin, L. A.; Kotov, V. N.; Romov, B. A.: Galois theory for Post algebras I–II. (Russian) Kibernetika, No. 3, 1–10 (1969), No. 5, 1–9 (1969). English translation: Cybernetics (1969), 243–252 and 531–539
Боднарчук, В. Г.; Калужнин, Л. А.; Котов, В. Н.; Ромов, Б. А.: Теория Галуа для алгебр Поста I–II. Кибернетика 3, 1–10 (1969), 5, 1–9 (1969)

[Bör-P 97] Börner, F., Pöschel, R.: Partial clones. Beiträge der Jahrestagung "Algebra und Grenzgebiete". Preprint-Reihe der PH Güstrow 1990

[Bör-H-P 91] Börner, F.; Haddad, L.; Pöschel, R.: Minmal partial clones. Bull. Aust. Math. Soc. **44**, No. 3, 405–415 (1991)

[Bör 97] Börner, F.: Clones of partial functions. In: Denecke, K. et al., General algebra and applications in discrete mathematics. Proceedings of the conference on general algebra and discrete mathematics, Potsdam 1996. Aachen: Shaker Verlag. Berichte aus der Mathematik. 35–52 (1997)

[Bör-H 98] Börner, F.; Haddad, L.: Maximal partial clones with no finite basis. Algebra Univers. **40**, No. 4, 453–476 (1998)

[Bul-K-S-S 95] Bulatov, A.; Krokhin, A.; Safin, K.; Sukhanov, E.: On the structure of clone lattices. General Algebra and Discrete Mathematics. Heldermann-Verlag, Berlin 1995, 27–34

[Bul-L-S 96] Bulatov, A.; Lau, D.; Strauch, B.: The cardinalities of sublattices of depth 2 in the lattices of clones on a 3-elementary set. Preprint Universität Rostock

[Bul 98a] Bulatov, A. A.: Polynomial reducts of modules I. Rough classification. Mult.-Valued Log. **3**, No. 2, 135–154 (1998)

[Bul 98b] Bulatov, A. A.: Polynomial reducts of modules II. Algebras of primitive and nilpotent functions. Mult.-Valued Log. **3**, No. 3, 173–193 (1998)

[Bul 99a] Bulatov, A. A.: Sublattices of a lattice of clones of functions on a 3-element set. I. (Russian, English) Algebra Logika **38**, No. 1, 3–23 (1999); translation in Algebra Logic **38**, No. 1, 1–11 (1999)

[Bul 99b] Bulatov, A. A.: Sublattices of the lattice of clones of functions on a 3-element set. II. (Russian, English) Algebra Logika **38**, No. 3, 269–295 (1999); translation in Algebra Logic **38**, No. 3, 144–158 (1999)

[Bul-K-J 2000] Bulatov, A. A.; Krokhin, A. A.; Jeavons, P.: Constraint satisfaction problems and finite algebras. Montanari, Ugo (ed.) et al., Automata, languages and programming. 27th international colloquium, ICALP 2000, Geneva, Switzerland, July 9–15, 2000. Proceedings. Berlin: Springer. Lect. Notes Comput. Sci 1853, 272–282 (2000)

[Bul-K-S-S-S 2001] Bulatov, A.; Krokhin, A.; Safin, K.; Semigrodskikh, A.; Sukhanov, E.: On the structure of clone lattices II. Multi.-Valued Log. **7**, No. 5–6, 379–389 (2001)

[Bul 2001] Bulatov, A. A.: Conditions satisfied by clone lattices. Algebra Univers. **45**, No. 1–2, 237–241 (2001)

[Bul 2002] Bulatov, A. A.: Polynomial clones containing the Mal'tzev operation of the groups $\mathbb{Z}_{p^2}$ and $\mathbb{Z}_p \times \mathbb{Z}_p$. Multi. Val. Logic, Vol. 8(2), p. 193–221 (2002)

[Bul-I 2002] Bulatov, A. A.; Idziak, P. M.: Counting Mal'tzev clones on small sets. Discrete Mathematics (submitted)

[Bur 67] Burle, G. A.: The classes of k-valued logics containing all one variable functions. (Russian) Diskret. Analiz **10**, 3–7 (1967)
Бурле, Г. А.: Классы k-значной логики, содержащие все функции одной переменной. Дискретный анализ **10**, 3–7 (1967)

[Bur-D-H-L 85] Burosch, G.; Dassow, J.; Harnau, W.; Lau, D.: On subalgebras of an algebra of predicates. J. Inf. Process Cybern. EIK **21**, 1/2, 9–22 (1985)

[Bur-S 81] Burris, S.; Sankappanavar, H. P. : A course in universal algebra. Graduate Texts in Mathematics, Vol. 78. New York, Heidelberg, Berlin: Springer-Verlag (1981)

[But 60] Butler, J. W.: On complete and independent sets of operations in finite algebras. Pacific. J. Math. **10**, 1169–1179 (1960)

[Coh-J-G 2003] Cohen, D.; Jeavons, P.; Gault, R.: New tractable classes from old. Constraints **8**, No. 3, 263–282 (2003)

[Coh 65] Cohn, P. M.: Universal Algebra. Harper & Row, New York, 1965

[Com 66] Comtet, L.: Recouvrement, bases de filtre et topologies d'un ensemble fini. C. R. Acad. Sci. Paris, Sér. A **262**, 1091–1094 (1966)

[Csá 83a] Csákany, B.: Three-element groupoids with minimal clones. Acta Sci. Math. **45**, 111–117 (1983)

[Csá 83b] Csákany, B.: All minimal clones on the three-element set. Acta Cybernet. **6**, 227–238 (1983)

[Csá 86] Csákany, B.: On conservative minimal operations. Universal algebra, Colloq., Szeged/Hung. 1983, Colloq. Math. Soc. János Bolyai **43**, 49–60 (1986)

[Csá 2002] Csákany, B.: Minimal clones. Lectures held in Tále, Slovakia. Summer School on General Algebra and Ordered Sets. 15 p. (2002)

[Cze 99] Czédli, G.: Two minimal clones whose join is gigantic. Publ. Math. **55** (1999), No. 1-2, 155–159

[Cze-H-K-P-S 2001] Czédli, G.; Halas, R.; Kearnes, K. A.; Pálfy, P.P.; Szendrei, Á.: The join of two minimals clones and the meet of two maximal clones. Algebra Univers. **45**, No. 2/3, 161–178 (2001)

[Das 81] Dassow, J.: Completeness problems in the structural theory of automata. Math. Forschung Bd. 7, Akademie-Verlag, Berlin 1981

[Dav 68] Davies, R. O.: On n-valued Sheffer functions. Preprint: Leicester 1968, Université de Montréal 1974

[Dav-P 90] Davey, B. A.; Priestley, H.A.: Introduction to lattices and order. Cambridge University Press, Cambridge 1990

[Dem-H 79a] Demetrovics, J.; Hannak, L.: The cardinality of closed sets in precomplete classes in k-valued logics. Acta Cybernetica **4**, 3, 273–277 (1979)

[Dem-H 79b] Demetrovics, J., Hannak, L.: The cardinality of selfdual closed classes in k-valued logics. MTA SzTAKI Közlemenyek **23**, 8–17 (1979)

[Dem-H-M 80a] Demetrovics, J.; Hannak, L.; Marcenkov, S. S.: Some remarks on the structure of P_3. C. R. Math. Rep. Acad. Sci. Canada 2, 215–219 (1980)

[Dem-H 83] Demetrovics, J.; Hannak, L.: The number of reducts of a preprimal algebra. Algebra Universalis, **16**, 178–185 (1983) **127**, 44–46 (1959)

[Dem-M 83] Demetrovics, J.; Mal'cev, I. A.: On the depth of infinitely generated subalgebras of Post's iterative algebra P_3. Colloquium on Universal Algebra, Szeged 1983. Colloq. Math. Soc. János Bolyai 43, 85–96 (1986)

[Dem-M 84] Demetrovics, J.; Mal'tsev, I. A.: Essential minimal TC-clones in three-valued logics. (Russian, English summary) Közlemények, Magy. Tud. Akad., Számitástechnikai Autom. Kut. Intéz. 31, 115–151 (1984)

[Dem-M 86] Demetrovics, J.; Mal'tsev, I. A.: Essentially minimal TC-clones on the three-element base set. C.R. Math. Acad. Sci., Soc. R. Can. 8, 191–196 (1986)

[Dem-M 89] Demetrovics, J.; Mal'tsev, I. A.: On construction of a Burle clone on a three-element set. (Russian: Деметрович, Jу.; Мальцев, И. А.: О сроении клона бурле на трехЭлементном множестве.) Acta Cybern. 9, No. 1, 1–25 (1989)

[Den 82] Denecke, K.-D.: Preprimal algebras. Akademie-Verlag, Berlin 1982

[Den-L 86] Denecke, K.; Lau, D.: Kongruenzen auf Klons und vollinvariante Kongruenzen relativ freier Algebren II. Rostock. Math. Kolloq. 29, 4–20 (1986)

[Den-P 88] Denecke, K.; Pöschel, R.: The characterization of primal algebras by hyperidentities. General algebra, Dedicated Mem. of Wilfried Nöbauer, Contrib. Gen. Algebra **6**, 67–87 (1988)

[Den-T 96] Denecke, K.; Todorov, K.: Allgemeine Algebra und Anwendungen. Shaker Verlag, Aachen 1996

[Den-W 2000] Denecke, K.; Wismath, S. L.: Hyperidentities and clones. Algebra, Logic and Applications 14. Amsterdam: Gordon and Breach Publishers. xi, 314 p. (2000)

[Den-W 2002] Denecke, K.; Wismath, S. L.: Universal algebra and applications in theoretical computer science. Boca Raton, FL: Chapman & Hall/CRC (2002)

[Den-E-W 2004] Denecke, K. (ed.); Erné, M. (ed.), Wismath, S. L. (ed.): Galois connections and applications. Mathematics and its applications (Dortrecht) 565. Dortrecht: Kluwer Acvademic Publishers. xvi, 501 p. (2004)

[Eps-F-R 74] Epstein, G.; Frieder, G.; Rine, D. C.: The development of multiple-valued logic as related to computer science: A historical summary. Computer, September 1974 (And, a brief in Proc. 1074 Intern. Symp. on Multiple-Valued Logic. West Virginia University, Morgantown, W. Va.)

[Ern 82] Erné, M.: Einführung in die Ordnungstheorie. B.I.-Wissenschaftsverlag, Mannheim 1982

[Fos 59] Foster, A. L.: An existence theorem for functionally complete universal algebras. Math. Z. **71**, 69– 82 (1959)

[Fos 70] Foster, A. L.: Congruence relations and functional completeness in universal algebras, structure theory of hemi-primals I. Math. Z. **113**, 293–308 (1970)

[Fos-P 64] Foster, A. L.; Pixley, A. F.: Semi-categorial algebras I, II. Math. Z. **83**, 147–169 (1964); **85**, 169–184 (1964)

[Gor 73] Gorlov, V. V.: On congruences in closed Post classes. (Russian) Mat. Zametki **13**, 725–734 (1973)
Горлов, В. В.: О конгруЭнциях на замкнутых классах Поста. Мат. заметки **13**, 725–734 (1973)

[Gor 77a] Gorlov, V. V.: On closed classes of k-valued logic whose congruences are trivial all of them. (Russian) Mat. Zametki **22**, 499–509 (1977)
Горлов, В. В.: О замкнутых классах k-значной логики, все конгруЭнции которых тривиальны. Мат. Заметки **22**, 499–509 (1977)

[Gor 77b] Gorlov, V. V.: On closed classes of k-valued logic, all superclasses of which have only trivial congruences. Sov. Math. Dokl. **18**, 625–628 (1977); translation from Dokl. Akad. Nauk SSSR 234, 273–276
Горлов, В. В.: О замкнутых классах k-значной логики, все надклассы которых имеют только тривиальные конгруЭнции. Докл. АН СССР 234, 273–276 (1977)

[Gor 79] Gorlov, V. V.: A sufficient condition for closed classes of k-valued logic to have only trivial congruences. Mat. Sb., N. Ser. 110 (152), 551–578 (1979)
Горлов, В. В.: Достаточное условие, при котором замкнутые классы k-значной логики имеют только тривиальных конгруЭнции. Мат. Сборник 110 (152), 551–578 (1979)

[Gor-L 82] Gorlov, V. V.; Lau, D.: On a family of M-classes of k-valued logic and congruences on submaximal classes of P_3 (Russian: Горлов, В. В., Лау, Д.: Об одном семействе M-классов k-значной логики и конгруЭнциях на субмакцимальных классах P_3.) J. Infor. Process. Cybern. EIK **18**, 12, 669–686 (1982)

[Gor-L 83a] Gorlov; V. V.; Lau, D.: Über Automorphismen auf Funktionenalgebren. Rostock. Math. Kolloq. 23, 35–42 (1983)

[Gor-L 83b] Gorlov, V. V.; Lau, D.: Congruences on closed sets of self-dual functions in many-valued logics and on closed sets of linear functions in prime-valued logics. Közl. MTA Szamitastech. Automat. Kutato Int. Budapest **29**, 31–39 (1983)

[Got 89] Gottwald, S.: Mehrwertige Logik. Eine Einführung in Theorie und Praxis. Akademie-Verlag Berlin 1989

[Gni 65] Gnidenko, W. M.: Ermittlung der Ordnung der prävollständigen Klassen von Funktionen der dreiwertigen Logik. In: Probleme der Kybernetik 8, Akademie-Verlag, Berlin 1965, 379– 384

[Grä 68] Grätzer, G.: Universal algebra. D. van Nostrand & Co., Princeton N.Y., 1968

[Grü 83a] Grünwald, N.: Bestimmung sämtlicher abgeschlossenen Mengen aus $P_{3,2}$, deren Projektion F_8^n ist. Rostock. Math. Kolloq. **23**, 5–26 (1983)

[Grü 83b] Grünwald, N.: Beschreibung aller abgeschlossenen Mengen aus $P_{3,2}$, deren Projektion F_8^n ist, mit Hilfe von Relationen. Rostock. Math. Kolloq. **23**, 27–34 (1983)

[Grü 84] Grünwald, N.: Strukturaussagen über den Verband der abgeschlossenen Mengen von $P_{k,2}$, insbesondere von $P_{3,2}$. Dissertation A, Universität Rostock, 1984

[Had-R 86] Haddad, L.; Rosenberg; I. G.: An interval of finite clones isomorphic to $(P(N), \subseteq)$. C. R. Math. Rep. Acad. Sci. Canada (6) **8**, 375–379 (1986)

[Had 88] Haddad, L.: Maximal partial clones determined by quasi-diagonal relations. J. Inf. Process. Cybern. **24**, No. 7/8, 355–366 (1988)

[Had-R 89] Haddad, L.; Rosenberg, I. G.: Maximal partial clones determined by the areflexive relations. Discrete Appl. Math. **24**, No. 1–3, 133–143 (1989)

[Had-R 91] Haddad, L.; Rosenberg, I. G.: Partial Sheffer Operations. Europ. J. Combinatorics **12**, 9–21 (1991)

[Had-R 94] Haddad, L.; Rosenberg, I. G.: Finite clones containing all permutations. Can. J. Math. Vol. **46** (5), 951–970 (1994)

[Had-R 95] Haddad, L.; Rosenberg, I. G.: Partial clones containing all permutations. Bull. Aust. Math. Soc. **52**, No. 2, 263– 278 (1995)

[Had 98] Haddad, L.: On the depth of the intersection of two maximal partial clones. Mult.-Valued Log. **3**, No. 4, 259–270 (1998)

[Had-F 98a] Haddad, L., Fugère, J.: Intersections of maximal partial clones, I. Mult.-Valued Log. **3**, No. 2, 97- 109 (1998)

[Had-F 98a] Haddad, L.; Fugère, J.: Intersections of maximal partial clones, II. Mult.-Valued Log. **3**, No. 1, 55–75 (1998)

[Had-L 2000a] Haddad, L.; Lau, D.: Pairwise intersections of Slupecki type maximal partial clones. Beitr. Algebra Geom. **41**, No. 2, 537–555 (2000)

[Had-L 2000b] Haddad, L.; Lau, D.: Families of finitely generated maximal partial clones. Mult.-Valued Log. **5**, No. 3, 201–228 (2000)

[Had-L 2003] Haddad, L.; Lau, D.: Uncontable families of partial clones containing maximal clones. Beiträge zur Algebra und Geometrie/Contributions to Algebra and Geometry (submitted), 28 p. (2003)

[Had-L 2006] Haddad, L.; Lau, D.: A criterion for partial Sheffer functions. In preparation

[Had-L-R 2002] Haddad, L.; Lau, D.; Rosenberg, I. G.: Intervals of partial clones containing maximal clones. J. Autom. Lang. Comb. (submitted)

[Had-M-R 2002] Haddad, L.; Machida, H.; Rosenberg, I. G.: Maximal and minimal partial clones. J. Autom. Lang. Comb. **7**, No. 1, 83–93 (2002)

[Had-S 2002] Haddad, L.; Simons, G. E.: Maximal partial clones of 4-valued logic. Mult.-Valued Log. **8**, No. 2, 193–221 (2002)

[Had-S 2003] Haddad, L.; Simons, G. E.: Intervals of Boolean partial clones. Italian J. Pure & Appl. Math. (submitted), 23. p. (2003)

[Har 74a] Harnau, W.: Die Definition der Vertauschbarkeitsmengen in der k-wertigen Logik und das Maximalitätsproblem. Z. Math. Logik u. Grundl. Math. **20**, 339–352 (1974)

[Har 74b] Harnau, W.: Die vertauschbaren Funktionen der k-wertigen Logik und ein Basisproblem. Z. Math. Logik u. Grundl. Math. **20**, 453–463 (1974)

[Har 75] Harnau, W.: Über Kettenlängen der teilweise geordneten Menge Φ_k der Vertauschbarkeitsmengen der k-wertigen Logik. Math. Nachr. **68**, 289–297 (1975)

[Har 76] Harnau, W.: Eine Verallgemeinerung der Vertauschbarkeit in der k-wertigen Logik. EIK **12**, 33–43 (1976)

[Har 83] Harnau, W.: Ein verallgemeinerter Relationen- und ein modifizierter Superpositionsbegriff für die Algebra der mehrwertigen Logik. Dissertation B, Universität Rostock 1983

[Her 55] Hermes, H.: Einführung in die Verbandstheorie. Springer, Berlin 1955

[Ihr 93] Ihringer, Th.: Allgemeine Algebra. B.G. Teubner Stuttgart 1993

[Ihr 94] Ihringer, Th.: Diskrete Mathematik. B.G. Teubner Stuttgart 1994

[Ihr 2003] Ihringer, Th.: Allgemeine Algebra. Mit einem Anhang über Universelle Coalgebra von H. P. Gumm) Neuauflage. Berliner Studienreihe zur Mathematik 10. Lemgo: Heldermann Verlag xii, 218 S. (2003)

[Jab 54] Jablonskij, S. V.: On funktional completeness in the three-valued calculus (Russian). Dokl. Akad. Nauk SSSR **95**, 1153–1155 (1954)
Яблонский, С.В.: О функционалъной полноте в трехзначном исчислении. Докл. АН СССР **95**, 1153–1155 (1954)

[Jab 58] Jablonskij, S. V.: Functional constructions in many-valued logics (Russian). Tr. Mat. Inst. Steklova **51** 5–142 (1958)
Яблонский, С.В.: Функционалъные построения в k-значные логике. Труды мат. инст. им. В. А. Стеклова **51**, 5–142 (1958)

[Jab-G-K 70] Jablonski, S. W.; Gawrilow, G. P.; Kudrjawzew, W. B.: Boolesche Funktionen und Postsche Klassen. Akademie-Verlag, Berlin 1970

[Jab 74] Jablonskij, S. V.: The structure of the upper neighborhood for predicate-describable classes in P_k (Russian). Dokl. Akad. Nauk SSSR 218, 304–307 (1974). English translation: Sov. Math., Dokl. 15, 1353–1356 (1974).
Яблонский, С. В.: Строение верхней окрестности для предикатно-описуемых классов в P_k. Докл. АН СССР **218**, 304–307 (1974)

[Jab-L 80] Jablonski, S. W.; Lupanow, O. B. (Hsg.): Diskrete Mathematik und mathematische Fragen der Kybernetik. Akademie-Verlag 1980

[Jan-M 59] Janov, Ju. I.; Mučnik, A. A.: Existence of k-valued closed classes without a finite basis (Russian). Dokl. Akad. Nauk. SSSR **127**, 44–46 (1959).
Янов, Ю. И; Мучник, А. А.: О существовании k-значных замкнутых классов не имеюших конечного базиса. Докл. АН СССР **127**, 44–46 (1959)

[Jea 98] Jeavons, P.: On the algebraic structure of combinatorial problems. Theoretical Computer Science 200 (1998), 185–204

[Jea-C-P 98] Jeavons, P.; Cohen, D.; Pearson, J.: Constraints and universal algebra. Ann. of Math. Artif. Intell. **24**, No. 1–2, 51–67 (1998)

[Jez-Q 95] Jezek, J.; Quackenbush, R.: Minimal clones of conservative functions. Int. J. Algebra Comput. **5**, No. 6, 615–630 (1995)

[Jun 2005] Jungnickel, D.: Graphs, networks and algorithms. Second edition. Springer-Verlag Berlin Heidelberg 2005

[Kea 95] Kearnes, K. A.: Minimal clones with Abelian representations. Acta Sci. Math. **61**, No. 1-4, 59–76 (1995)

[Kea-S 99] Kearnes, K. A.; Szendrei, Á.: The classification of commutative minimal clones. Discuss. Math., Algebra Stoch. Methods **19**, No. 1, 147–178 (1999)

[Kno 85] Knoebel, A.: The equational classes generated by single functionally precomplete algebras. Memoirs of the Amer. Math. Soc. **57**, 332, Providence, Rhode Island, 1985

[Kol 74] Kolpakov, V. I.: Subalgebras of monotone functions in Post algebras (Russian). Diskret. Analiz, Novosibirsk **24**, 30–45 (1974)
Колпаков, В. И.: Подалгебры монотонных функций в алгебрах Поста. Дискретный анализ **24**, 30–45 (1974)

[Kra 45] Krasner, M.: Generalisation et analogues de la theorie de Galois. Congres de la Victoire de l'Ass. France Avancem. Sci., 54–58 (1945)

[Kra 87/68] Krasner, M.: Endotheorie de Galois abstraite. Sem. P. Dubreil (Algebre et Theorie des nombres) 22e annee, Fasc. 1, 1968/69, no. 6

[Kre-G-S 88] Kreiser, L.; Gottwald, S.; Stelzner, W.: Nichtklassische Logik. Eine Einführung. Akademie-Verlag Berlin 1988

[Kro 98] Krokhin, A.: Intervals in clone lattices. (Russian) Dissertation, Ekaterinburg 1998

[Kro-R 2003] Krokhin, A.; Rosenberg, I. G.: A monoidal interval of clones of selfdual functions. Beitr. Algebra Geom. (submitted), 18 p. (2003)

[Kru 73] Kruse, R. L.: Identities satisfied by a finite ring. J. Algebra **26**, 298–318 (1973)

[Kud 70] Kudrjavcev, V. B.: On the coverings of precomplete classes of k-valued logic. (Russian) Diskret. Analiz, Novosibirsk **19**, 32–44 (1970)
Кудрявцев, В. Б.; О покрытиях предпольных классов k-значной логики. Дискретный анализ **19**, 32–44 (1970)

[Kuz 59] Kuznecov, A. V.: Mathematics in USSR during forty years (Russian). Survey articles, 102–115, Moscov 1959
Кузнецов, А.В.: Математика в СССР за 40 лет. т. 1, §13. Алгебра логика и ее обобщения. Москва 1959, 102–115

[Lau 75] Lau, D.: Prävollständige Klassen von $P_{(k,l)}$. Elektron. Informationsverarb. Kybernet. EIK **11**, 10–12, 624 -626 (1975)

[Lau 77a] Lau, D.: Eigenschaften gewisser abgeschlossener Klassen in Postschen Algebren. Dissertation A, Universität Rostock 1977

[Lau 77b] Lau, D.: Kongruenzen auf gewissen Teilklassen von $P_{k,l}$. Rostock. Math. Kolloq. 3, 37- 43 (1977)

[Lau 78a] Lau, D.: Bestimmung der Ordnung maximaler Klassen von Funktionen der k-wertigen Logik. Z. Math. Logik u. Grundl. Math. **24**, 79–96 (1978)

[Lau 78b] Lau, D.: Über die Anzahl von abgeschlossenen Mengen linearer Funktionen der n-wertigen Logik. Elektron. Informationsverarb. Kybernet. EIK **14**, 11, 561–563 (1978)

[Lau 79a] Lau, D.: Congruences on closed sets of k-valued logic. In: Colloq. Math. Soc. J. Bolyai, Vol. **28** Finite algebras and multiple-valued logic, Szeged 1979, 417–440

[Lau 79b] Lau, D.: Automorphismen auf den maximalen Klassen der k-wertigen Logik. Rostock. Math. Kolloq. 12, 13–16 (1979)

[Lau 81] Lau, D.: Ergänzungen zu Kongruenzen auf abgeschlossenen Mengen der mehrwertigen Logik, Preprint 1981, 2 S.

[Lau 82a] Lau, D.: Submaximale Klassen von P_3. J. Inf. Process. Cybern. EIK **18**, 4/5, 227–243 (1982)

[Lau 82b] Lau, D.: Die maximalen Klassen von $Pol_k(0)$. Rostock. Math. Kolloq. 19, 29–47 (1982)

[Lau 84a] Lau, D.: Unterhalbgruppen von $(P_3^1, *)$. Rostock. Math. Kolloq. 26, 55–62 (1984)

[Lau 84b] Lau, D.: Die maximalen Klassen von $Pol_k\{(x, x+1 \; mod \; k) | x \in E_k\}$. Rostock. Math. Kolloq. 25, 23–30 (1984)

[Lau 84c] Lau, D.: Funktionenalgebren über endlichen Mengen. Dissertation B, Universität Rostock, 1984

[Lau 85] Lau, D.: Abgeschlossene Mengen quasilinearer Funktionen in P_3. Rostock. Math. Kolloq. 28, 33–45 (1985)

[Lau 86] Lau, D.: Ein Kriterium für den Nachweis der Abzählbarkeit gewisser Teilverbände des Verbandes der abgeschlossenen Mengen von Funktionen der k-wertigen Logik. Rostock. Math. Kolloq. 30, 11–18 (1986)

[Lau 88a] Lau, D.: Über abgeschlossene Mengen linearer Funktionen in mehrwertigen Logiken. J. Inf. Process. Cybern. EIK **24**, 7/8, 367–381 (1988)

[Lau 88b] Lau, D.: Über abgeschlossene Teilmengen von $P_{k,2}$. J. Inf. Process. Cybern. EIK **24**, 10, 495–513 (1988)

[Lau 88c] Lau, D.: Über abgeschlossene Teilmengen von $P_{3,2}$. J. Inf. Process. Cybern. EIK**24**, 11/12, 561–572 (1988)

[Lau 88d] Lau, D.: Maximale Klassen von $P_k(l)$. Rostock. Math. Kolloq. 34, 71–77 (1988)

[Lau 88e] Lau, D.: Über partielle Funktionenalgebren. Rostock. Math. Kolloq. 33, 23–48 (1988)

[Lau 88f] Lau, D.: Über die Dimension der abgeschlossenen Teilmengen von P_2 (Ergänzungen und Berichtigungen zu Г.Н. Блохина: О предикатном опицании классов Поста. Дискретный анализ **16**, 16–29 (1970)

[Lau-S 90] Lau, D.; Schröder, B.: On the number of closed subsets of linear functions in the 6-valued logic. Beiträge zur Algebra und Geometrie **31** 19–32 (1990)

[Lau 90] Lau, D.: Kongruenzen auf abgeschlossenen Mengen linearer Funktionen in mehrwertigen Logiken. Rostock. Math. Kolloq. **43** 3–16 (1990)

[Lau-D 90] Lau, D.; Denecke, K.: Congruences on maximal partial clones and strong regular varieties generated by preprimal partial algebras, I. Wiss. Z. Pädagog. Hochsch. Karl Liebknecht, Potsdam **34**, No. 1, 117–122 (1990)

[Lau-D 91] Lau, D.; Denecke, K.: Congruences on maximal partial clones and strong regular varieties generated by preprimal partial algebras, II. Demonstr. Math. **24**, No. 1/2, 105–119 (1991)

[Lau 91a] Lau, D.: Ein Vollständigkeitskriterium für durch h-adisch elementare Relationen beschriebene maximale Klassen von P_k. Rostock. Math. Kolloq. **45**, 3–8 (1991)

[Lau 91b] Lau, D.: On closed subsets of Boolean functions (A new proof for Post's theorem). J. Inform. Process. Cybernet. EIK **27**, 3, 167–178 (1991)

[Lau 92a] Lau, D.: A completeness criterion for $P_k(l)$. J. Inform. Process. Cybernet. EIK **28**, 3, 87–112 (1992)

[Lau 92b] Lau, D.: Ein neuer Beweis für Rosenberg's Vollständigkeitskriterium. J. Inform. Process. Cybernet. EIK **28**, 4, 151–197 (1992)

[Lau 92c] Lau, D.: Die maximalen Klassen von $Pol_k E_l$ für $2 \leq l \leq k-1$. Preprint September 1992, 65 S.

[Lau 93] Lau, D.: Ein maximaler, abzählbarer Teilverband von Klassen monotoner Funktionen der 3-wertigen Logik. Preprint, 1993, 59 S.

[Lau 95a] Lau, D.: Die maximalen Klassen von $\bigcap_{a \in Q} Pol_k\{a\}$ für $Q \subseteq E_k$ (Ein Kriterium für endliche semi-primale Algebren mit nur trivialen Unteralgebren). Rostock. Math. Kolloq. **48**, 27–46 (1995)

[Lau 95b] Lau, D.: Die maximalen Klassen von $\bigcap_{\varrho \in Q} Pol_3 \varrho$ für $Q \subseteq \mathfrak{P}(\{0,1,2\})$, Teil I–III, Preprints 1995, Rostock. Math. Kolloq. **51**, 111–126 (1997), **52**, 85–105 (1999), **53**, 3–22 (1999)

[Lau 2006] Lau, D.: On Boolean partial clones. In preparation

[Law 68] Lawvere, F. W.: Some algebraic problems in the context of functorial semantics of algebraic theories. In: MacLane, S. (Ed.): Report of the Midwest Category Seminar II. Lecture Notes in Math. **61**, 41–61, Berlin 1968

[Len 86] Lengvárszky, Zs.: A note on minimal clones. Acta Sci. Math. **50**, 335–336 (1986)

[Lév-P 96] Lévai; L., Pálfy, P. P.: On binary minimal clones. Acta Cybern. **12**, No. 3, 279–294 (1996)

[Lo 63a] Lo CzuKai: On the precompleteness of the classes of functions preserving a partition. (Russian)
Ло Чжу-Кай: О предполноте класов функций сохраняющих пазбиение. Acta Sc. Natur. Univ. Jiliensis **2** (1963)

[Lo 63b] Lo CzuKai: Precomplete sets and rings of linear functions. (Russian) Ло Чжу-Кай: Предполнота множества и кольца линейных функций. ibid. **2** (1963)

[Lo 63c] Lo CzuKai: Precomplete classes determined by binary relations in many-valued logics. (Russian) Ло Чжу-Кай: Предполные классы, определяемые бинарными отношениями в многозначной логике. ibid **4** (1963)

[Lo 64] Lo CzuKai: Precomplete classes defined by normal k-ary relations in k-valued logics (Russian). Ло Чжу-Кай: Предполные классы, опредеяемые нормальными k-арными отношениями, в k-знатчной логике. ibid **3** (1964)

[Lug-W 67] Lugowski, H.; Weinert, H. J.: Grundzüge der Algebra, Teil I–III, Leipzig 1967/68

[Lyn 54a] Lyndon, R. C.: Identities in two-valued calculi. Trans. Amer. Math. Soc. **71**, 457–465 (1954)

[Lyn 54b] Lyndon, R. C.: Identities in finite algebras. Proc. Amer. Math. Soc. **5**, 8–9 (1954)

[Mac 79] Machida, H.: On closed sets of three-valued monotone logical functions. In: Colloquia Mathematica Societatis Janos Bolyai 28, Finite Algebra and multiple-valued logic; Szeged (Hungary) 1979; 441–467

[Mac-R 99] Machida, H.; Rosenberg, I. G.: On gigantic pairs of minimal clones. RIMS Kokyuroku 1093, 87–92 (1999)

[Mac-R 2004] Machida, H.; Rosenberg, I. G.: Centralizers of monoids containing the symmetric group (submitted)

[Mal 52] Mal'tsev, A. I.: Symmetric groupoids. Mat. Sb. **31**, 1, 136–151 (1952) Мальцев, А. И.: Симметрические группоиды. Мат. Сб. **31**, 1, 136–151 (1952)

[Mal 66] Mal'tsev, A. I.: Iterative algebras and Post's varieties (Russian). Algebra i Logika (Sem.) **5**, 5–24 (1966) Мальцев, А. И.: Итеративные алгебры и многообразия Поста. Алгебра и логика **5** (1966), 5–24

[Mal 67] Mal'tsev, A. I.: A strengthening of the theorems of Słupecki and Jablonskij (Russian, English summary). Algebra i Logica (Sem.) **6**, 3, 61–75 (1967) Мальцев, А. И.: Об одном усилении теорем Слупецкого и Яблонского. Алгебра и логика **6**, 3, 61–75 (1967)

[Mal 73] Mal'tsev, I. A.: Some properties of cellular subalgebras of a Post algebra and their basic cells. (Russian, English). Algebra Logic **11**, 315–325 (1972); translation from Algebra Logika **11**, 5, 571–587 (1972) Мальцев, И. А.: Некоторые сбойства клеточных подалгебр Поста и их основных клеток. Алгебра и логика **11**, 5, 571–587 (1972)

[Mal 73a] Mal'tsev, I. A.: Congruences and automorphisms in cells of Post algebras (Russian, English). Algebra Logic **11**, 369–373 (1972); translation from Algebra Logic **11**, 6, 666–672 (1972). Мальцев, И. А.: Конгруенции и автоморфизмы на клетках алгебр Поста. Алгебра и логика **11**, 6, 666–672 (1972)

[Mal 73b] Mal'tsev, I. A.: Some properties of cells of Post algebras. (Russian) Diskret. Analiz, Novosibirsk **23**, 24–31 (1973) Мальцев, И. А.: Некотопые свойства клеток алгебр Поста. Дискретный анализ **23**, 24–31 (1973)

[Mal 76] Mal'tsev, I. A.: On congruences in subalgebras of iterative Post algebras. (Russian) Diskret. Analiz, Novosibirsk **29**, 40–52 (1976)
Мальцев, И. А.: О конгруЭнциях на подалгебрах итеративных алгебр Поста. Дискретный анализ **29**, 40–52 (1976)

[Mal 73] Mal'tsev, I. A.: Homomorphisms of completely restricted expensions of Post algebras. (Russian) Rostock. Math. Kolloq. **11**, 85–92 (1979)
Мальцев, И. А.: Гомоморфизмы бполне ограниценных расширений итератибных алгебр Поста.

[Mar 54] Martin, N. M.: The Sheffer functions of 3-valued logic. J. Symbolic Logic **19**, 45–51 (1954)

[Mar-D-H 80] Marchenkv, S. S.; Demetrovics, J.; Hannak, L.: On closed classes of self-dual functions in P_3. (Russian) Metody Diskretn. Anal. **34**, 38–73 (1980)
Марченков, С. С.; Деметрович, Я.; Ханнак, Л.: О замкнутых классах самодвойственных функций в P_3. In: Методы дискретного анализа и решени комбинаторных задач 34. Москва 1980, 38–73

[Mar 83] Marchenkov, S. S.: On closed classes of self-dual functions of many-valued logic. II. (Russian) Probl. Kibern. **40**, 261–266 (1983)
Марченков, С. С.: О замкнутых классах самодвойственных функций многозначной логики II. Проблемы кибернетики **40**, 261–266 (1983)

[Mar 60] Martynyuk, V. V.: Investigation of some classes in many-valued logics. Probl. Kibernetiki **3**, 49–60 (1960)
Мартынюк, В. В.: Исследование некоторых классов функций в многозначных логиках. Пробл. кибернет. **3**, 49–60 (1960)

[McK-M-T 87] McKenzie, R.; McNulty, G. F.; Taylor, W.: Algebras, Lattices, Varieties, vol. 1. Wadsworth, Belmont (Cal.), 1987

[McK 70] McKenzie, R.: Equational bases for lattice theories. Math. Scand. 27, 24–38 (1970)

[McK 76] McKenzie, R.: On minimal, locally finite varieties with permuting congruence relations. Preprint 1976

[McK 78] McKenzie, R.: Para-primal varieties: A study of finite axiomatizability and definable principal congruences in locally finite varieties. Algebra Universalis **8**, 336–348 (1978)

[McK 96] McKenzie, R.: The residual bounds of finite algebras. J. of Algebra and computation **6**, No. 1, 1–28 (1996)

[Men 85] Menne, A.: Einführung in die formale Logik. Wissenschaftliche Buchgesellschaft, Darmstadt 1985

[Miy 71] Miyakawa, M.: Functional completeness and structure of three-valued logics I – Classification of P_3 –, Res. of Electrotech. Lab., no. 717, 1–85 (1971)

[Miy-S-L-R 87] Miyakawa, M; Stojmenovic, I.; Lau, D.; Rosenberg, I.G.: Classifications and basis enumerations in many-valued logics– a survey –. Proc. 17th. Intern. Symposium on Multiple-valued Logic. Boston, May 1987, 152–160

[Miy 88] Miyakawa, M.: Classifications and basis enumerations in many-valued logic algebras. Reseaeches of the electrotechnical laboratory, No. 889, 1- 201 (1988)

[Miy-S-M-L 90] Miyakawa, M.; Stojmenovic, I.; Mishima, T.; Lau, D.: On the structure of maximal closed sets of $P_{k,2}$. Proc. 20th International Symposium on Multiple-Valued Logic, Charlotte, North Carolina 1990, 254–261

[Mur 65] Murskij, V. L.: The existence in three-valued logic of a closed class with finite basis, not having a finite complete system of identities (Russian, English) Sov. Math. Dokl. **6**, 1020–1024 (1965); translation from Dokl. Akad. Nauk SSR, **163**, 815–818
Мурский, В. Л.: Существование в трехзначной логике замкнутого класса с конечным базисом, не имеющего конечной полной системы тождеств. Докл. АН СССР **163**, 815–818 (1965)

[Nov 73] Novikov, P. A.: Grundzüge der mathematischen Logik. Deutscher Verlag Wiss. Berlin 1973

[Oat-P 65] Oates-Williams, S.; Powell, M. B.: Identical relations in finite groups. J. Algebra **1**, 11–39 (1965)

[Oat 80] Oates-Williams, S.: Murskii's algebra does not satisfy Min. Bull. Austral. Math. Soc. **22**. 199–203 (1980)

[Pal 86] Palfy, P. P.: The arity of minimal clones. Acta Sci. Math. **50**, 331–333 (1986)

[Pan-V 2000] Pantovic, J.; Vojvodic, D.: The cardinality of the set of clones containing unary minimal clones on the three element set. Mult.-Valued Log. **5**, No. 5, 367–371 (2000)

[Pap 94] Papadimitriou, C. H.: Computational complexity. Addison-Wesley, Reading, Mass. (1994)

[Per 69] Perkins, P.: Bases for equational theories of semigroups. J. Algebra **11**, 298–314 (1969)

[Pix 71] Pixley, A.: The ternary discriminator function in universal algebras. Math. Ann. **191**, 167–180 (1971)

[Pös-K 79] Pöschel, R.; Kalužnin, L.A.: Funktionen- und Relationenalgebren. Berlin 1979

[Pos 20] Post, E. L.: Determination of all closed systems of truth tables. Bull. Amer. Math. Soc. **26**, 427 (1920)

[Pos 21] Post, E. L.: Introductions to a general theory of elementary propositions. Amer. J. Math. **43**, 163–185 (1921)

[Pos 41] Post, E. L.: The two-valued iterative systems of mathematical logic. Ann. Math. Studies 5, Princeton Univ. Press 1941

[Qua 71] Quackenbush, R. W.: Demi-semi-primal algebras and Mal'cev-type conditions. Math. Z. **122**, 166–176 (1971)

[Qua 82] Quackenbush, R. W.: A new proof of Rosenberg's primal algebra characterization theorem. In: Finite algebra and multiple-valued logic, Szeged, 1979, Colloq. Math. Soc. Janos Bolyai 28, 603–634 (1981)

[Qua 95] Quackenbush, R. W.: A survey of minimal clones. Aequationes Math. **50**, No. 1-2, 3–16 (1995)

[Rad 83] Radtke, S.: Die Anzahl aller möglichen Halbordnungsrelationen auf einer maximal sechselementigen Menge. Rostock. Math. Kolloq. **23**, 55–61 (1983)

[Rau 96] Rautenberg, W.: Einführung in die Mathematische Logik. Ein Lehrbuch mit Berücksichtigung der Logikprogrammierung. Vieweg, Braunschweig/Wiesbaden 1996

[Res-D 89] Reschke, M.; Denecke, K.: Ein neuer Beweis für die Ergebnisse von E. L. Post über abgeschlossene Klassen Boolescher Funktionen. J. Inform. Process. Cybernet. EIK **25**, 7, 361–380 (1989)

[Ric 78] Richter, M. M.: Logikkalküle. B. G. Teubner Stuttgart 1978

[Rin 84] Rine, D. C. (ed.): Computer science and multiple-valued logic, theory and applications. Rev. ed. Amsterdam–New York–Oxford: North-Holland Publishing Company XIV, 548p. (1984)

[Rom 80] Romov, B. A.: Maximal subalgebras of algebras of partial multivalued logic functions. (Russian, English). Cybernetics **16**, 31–41 (1980); translation from Kibernetica 1980, No. 1, 28–36 (1980)

[Rom 81] Romov, B. A.: The algebras of partial functions and their invariants. (Russian, English). Cybernetics **17**, 157–167 (1981); translation from Kibernetica 1981, No. 2, 1–11 (1981)

[Rom 90] Romov, B. A.: The completeness problem in the algebra of partial functions of finite-valued logic. (Russian, English). Cybernetics **26**, No. 1, 133–138 (1990); translation from Kibernetica 1990, No. 1, 102–106 (1990)

[Ros 65] Rosenberg, I. G.: La structure des fonctions de plusieeurs variables sur un ensemble fini. C. R. Acad. Sci. Paris, Ser. A–B, **260**, 3817–3819 (1965)

[Ros 66] Rosenberg, I. G.: Zu einigen Fragen der Superpositionen von Funktionen mehrerer Veränderlicher. Bul. Inst. Politehn. Iasi, **12** (**16**), 7–15 (1966)

[Ros 69] Rosenberg, I. G.: Über die Verschiedenheit maximaler Klassen in P_k. Rev. Roumaine Math. Pures Appl. **14**, 431–438 (1969)

[Ros 70a] Rosenberg, I. G.: Über die funktionale Vollständigkeit in den mehrwertigen Logiken. Rozpravy Československe Akad. Ved. Řada Mat. Přirod. Věd **80**, 3–93 (1970)

[Ros 70b] Rosenberg, I. G.: Algebren und Relationen. Elektron. Informationsverarbeit. Kybernetik. EIK **6** (1970), 115–124

[Ros 70c] Rosenberg, I. G.: Complete sets for finite algebras. Math. Nachr. **44**, 253–258 (1970)

[Ros 73] Rosenberg, I. G.: The number of maximal closed classes in the set of functions over a finite domain, J. combinat. Theory, Ser. A **14**, 1–7 (1973)

[Ros 74] Rosenberg, I. G.: Completeness, closed classes and relations in multiple-valued logics. Proceedings 1974 Intern. Sympos. on multiple-valued logics, Morgantown, May 29-31, 1974, 1–26

[Ros 75] Rosenberg, I. G.: Composition of functions on finite sets, completeness and relations: A short survey. Univ. Montréal, Preprint CRM-529 (1975) (see also [Ros 77] and [Ros 84])

[Ros 77] Rosenberg, I. G.: Completeness properties of multiple-valued logic algebras. In: Rine, D. C. (ed.): Computer science and multiple-valued logic, theory and applications. North-Holland Publ. Comp., Amsterdam 1977, 144–186

[Ros 78] Rosenberg, I. G.: On generating large classes of Sheffer functions. Aequationes Math. **17**, 164–181 (1978)

[Ros 84] Rosenberg, I.G.: Completeness properties of multiple-valued logic algebras. In: Rine, D. C. (ed.): Computer science and multiple-valued logic, theory and applications. Rev. ed. Amsterdam–New York–Oxford: North-Holland Publishing Company XIV, 548 p., 150–192 (1984)

[Ros-S 85] Rosenberg, I. G.; Szendrei, Á.: Submaximal clones with a prime order automorphism. Acta (Szeged), 49 Fasc. 1–4, 29–48 (1985)

[Ros 86] Rosenberg, I. G.: Minimal clones I: The five types. Lectures in Universal Algebra (L. Szabo, Á. Szendrei eds.), Colloq. Math. Soc. J. Bolyai 43, North Holland, 405–427 (1986)

[Ros-H 87] Rosenberg, I. G.; Haddad, L.: Critere general de completude pour les algebres partielles finies. C.R. Acad Sci Paris t. 304, ser. I No. 17, 507–509 (1987)

[Ros 88] Rosenberg, I. G.: Clones of boolean functions: a survey. S.A.J. Philosophy, 7, No. 2, 90–99 (1988)

[Ros 89] Rosenberg, I. G.: Partial algebras and clones via one-point extension. General algebra, Dedicated Mem. of Wilfried Nöbauer, Contrib. Gen. Algebra 6, 227–242 (1988)

[Ros-H 91] Rosenberg, I. G.; Haddad, L.: Completeness theory of finite partial algebras. Algebra Universalis **29**, 378–401 (1991)

[Ros-M 2001] Rosenberg, I. G.; Machida, H.: Gigantic pairs of minimal clones – characterization and existence. Mult.-Valued Log. **7**, No. 1-2, 129–148 (2001)

[Rou 67] Rousseau, G.: Completeness in finite algebras with a single operation. Proc. Amer. Math. Soc. **18**, 1009–1013 (1967)

[Roz 85] Rozenfel'd, R. A.: On the number of partial orderings on a 6-set. (Russian: Розенфельд, Р. А.: О числе частичных упорядочений на 6-множестве.) Rostocker Math. Kolloq. **28**, 46–48 (1985)

[Sal 60] Salomaa, A. A.: On the composition of functions of several variables ranging over a finite set. Ann. Univ. Turku. Ser. A I **41**, 48 p. (1960)

[Sal 64] Salomaa, A. A.: Some completeness criteria for sets of functions over a finite domain I. Ann. Univ. Turku. Ser. A I **53**, 9 p. (1962)

[Sal 64] Salomaa, A. A.: On infinitely generated sets of operations infinite algebras. Ann. Univ. Turku. Ser. A I **74**, 1–12 (1964)

[Sai 70] Шайн, Б. М.: Рестриктивно-мультипликативные алгебры преобразований. Изв. вузов Мат. 4, 91–102 (1970)

[Sch 74] Schmidt, J.: Mengenlehre (Einführung in die axiomatische Mengenlehre), 1: Grundbegriffe. Mannheim: Bibliographisches Institut 1966, 241 S. (1966); 2. verbesserte und erweiterte Auflage 1974

[Sch 69] Schofield, P.: Independent conditions for completeness of finite algebras with a single generator. J. London Math. Soc. **44**, 413–423 (1969)

[Sch 87] Schöning, U.: Logik für Informatiker. B.I. Hochschultaschenbücher, Mannheim-Wien-Zürich 1987

[Sch 84] Schröder, B.: Über abgeschlossene Mengen von linearen Funktionen in P_{2p} ($p > 2$, prim). Rostock. Math. Koloq. 31, 21–41 (1987)

[Schw 83] Schweigert, D.: On varieties of clones. Semigroup Forum **26**, 275–285 (1983)

[Schw 84] Schweigert, D.: Clone equations and hyperidentities. Univ. Kaiserslautern, Preprint Nr. 86 (1984)

[Sco 64] Scott, W.R.: Group theory. New Yersey 1964

[Sto 87] Stojmenovic, I.: Some combinatorial and algorithmic problems in many-valued logics. University of Novi Sad, 1–150 (1987)

[She 13] Sheffer, H. M.: A set of five independent postulates for Boolean algebras with applications to logical constants. Trans. Amer. Math. Soc. **14**, 481–488 (1913)

[Sko 73] Skornjakow, L. A.: Elemente der Verbandstheorie. Akademie-Verlag, Berlin 1973

[Str 96a] Strauch, B.: Die Menge $\mathcal{M}(M \cap T_0 \cap T_1)$. Preprint Universität Rostock, Juni 1995, 16 S.

[Str 96b] Strauch, B.: Die Menge $\mathcal{M}(S \cap T_0 \cap T_1)$. Preprint Universität Rostock, Dezember 1996, 12 S.

[Str 97a] Strauch, B.: On partial classes containing all monotone and zero-preserving total Boolean functions. Math. Log. Quart. **43**, 510–524 (1997)

[Str 97b] Strauch, B.: Noncountable many classes containing a fixed class of total Boolean functions. In: General Algebra and applications in Discrete Mathematics, Proceedings of "Conference on General Algebra and Discrete Mathematics" (edited by K. Denecke and O. Lüders), Shaker Verlag, Aachen 1997, 177–188

[Sza-S 81] Szabó, L., Szendrei, Á.: Slupecki-type criteria for quasilinear functions over a finite dimensional vector space. Elektron. Informationsverarb. Kybernet. EIK **17**, 601–611 (1981)

[Sza 92] Szabó, L.: On minimal and maximal clones. Acta Cybern. **10**, No. 4, 323–327 (1992)

[Sza 98] Szabó, L.: On minimal and maximal clones. II. Acta Cybern. **13**, No. 4, 405–411 (1998)

[Sze 76] Szendrei, Á.: Idempotent reducts of abelian groups. Acta Sci. Math. (Szeged) **38**, 171–182 (1976)

[Sze 82] Szendrei, Á.: On the idempotent reducts of modules I–II. Universal algebra, Proc. Colloq., Esztergom/Hung. 1977, Colloq. Math. Soc. János Bolyai 29, 753–780 (1982)

[Sze 78] Szendrei, Á.: On closed sets of linear operations over a finite set of square-free cardinality. Elektron. Informationsverarb. Kybernet. EIK **14**, 11, 547–559 (1978)

[Sze 80] Szendrei, Á.: On closed classes of quasilinear functions. Czechoslovak Math. J. **80**, 498–509 (1980)

[Sze 82] Szendrei, Á.: Algebras of prime cardinality with a cyclic automorphism. Arch. Math. (Basel) **39**, 417– 427 (1982)

[Sze 86] Szendrei, Á.: Clones in Universal Algebra. Seminaire de Mathematiques Superieures, vol. 99, Les Presses de l'Universite de Montreal, Montreal 1986

[Tar 86] Tardos, G.: A maximal clone of monotone operations which is not finitely generated. Order **3**, 211–218 (1986)

[Til 92] Tilli, T.: Fuzzy-Logik. Franzis-Verlag, München 1992

[Ugo 88] Ugol'nikov, A. B.: On closed Post classes (Russian). Izv. Vyssh. Uchebn. Zaved., Mat. 1988, No. 7, 79–88 (1988)
Угольников, А. Б.: О замкнутых классах Поста. Известия бузов. математика **7** (1988), 79–88 (1988)

[Wal 2000] Waldhauser, T.: Minimal clones generated by majority operations. Algebra Univers. **44**, No. 1–2, 15–26 (2000)

[Web 35] Webb, D. L.: Generation of any n-valued logic by one binary operator. Proc. Nat. Acad. Sci. **21**, 252–254 (1935)

[Web 36] Webb, D. L.: Definition of Post's generalized negative and maximum in terms of one binary operation. Amer. J. Math. **58**, 193–194 (1936)

[Wec 92] Wechler, W.: Universal algebra for computer scientists. EATCS Monographs on Theoretical Computer Science. 25. Berlin etc.: Springer-Verlag. 339 p. (1992)

[Wer 78a] Werner, H.: Discriminator-algebras. (Studien zur Algebra und ihre Anwendungen, 6), Akademie-Verlag, Berlin 1978

[Wer 78b] Werner, H.: Einführung in die Allgemeine Algebra. B.I. Wissenschaftsverlag, Mannheim 1978

[Whe 61] Wheeler, R. F.: Complete propositional connectives. Z. Math. Logik u. Grundl. Math. 7 (1961), 185–198

[Wil-R 72] Wilde, G.; Raney, Sh.: Computation of the transformation semigroups of three letters. J. Austr. Math. Soc. **14**, 335 (1972)

[Wil 2001] Willard, R.: Extending Baker's theorem. Conference on Lattices and Universal Algebra (Szeged, 1998), Algebra Universalis **45**, No. 2–3, 335–344 (2001)

[Zak 67] Zakharova, E. Yu.: Criterions of completeness of function systems from P_k (Russian). Problemy Kibernet. **18**, 5–10 (1967)
Захарова, Е. Ю.: Критерии полноты систем функций из P_k. Пробл. Кибернет. **18**, 5–10 (1967)

[Zac-K-J 69] Zakharova, E. Yu.; Kudryavtsev, V. B.; Yablonskij, S. V.: On precomplete classes in k-valued logics (Russian). Dokl. Akad. Nauk SSSR **186**, 509–512 (1969). English translation: Soviet Math. Doklady **10**, 618–622 (1969)
Захарова, Е. Ю.; Кудрявцев, В. Б.; Яблонский , С. В.: О предполных классах в k-значных логиках. Докл. АН СССР **186**, 509–512 (1969)

[Zyl 25] Zylinski, M. E.: Some remarks concerning the theory of deduction. Fund. Math. 7, 203–209 (1925)

Glossary

Index

Springer Monographs in Mathematics

This series publishes advanced monographs giving well-written presentations of the "state-of-the-art" in fields of mathematical research that have acquired the maturity needed for such a treatment. They are sufficiently self-contained to be accessible to more than just the intimate specialists of the subject, and sufficiently comprehensive to remain valuable references for many years. Besides the current state of knowledge in its field, an SMM volume should also describe its relevance to and interaction with neighbouring fields of mathematics, and give pointers to future directions of research.

Abhyankar, S.S. **Resolution of Singularities of Embedded Algebraic Surfaces** 2nd enlarged ed. 1998
Alexandrov, A.D. **Convex Polyhedra** 2005
Andrievskii, V.V.; Blatt, H.-P. **Discrepancy of Signed Measures and Polynomial Approximation** 2002
Angell, T. S.; Kirsch, A. **Optimization Methods in Electromagnetic Radiation** 2004
Ara, P.; Mathieu, M. **Local Multipliers of C*-Algebras** 2003
Armitage, D.H.; Gardiner, S.J. **Classical Potential Theory** 2001
Arnold, L. **Random Dynamical Systems** corr. 2nd printing 2003 (1st ed. 1998)
Arveson, W. **Noncommutative Dynamics and E-Semigroups** 2003
Aubin, T. **Some Nonlinear Problems in Riemannian Geometry** 1998
Auslender, A.; Teboulle M. **Asymptotic Cones and Functions in Optimization and Variational Inequalities** 2003
Banasiak, J.; Arlotti, L. **Perturbations of Positive Semigroups with Applications** 2006
Bang-Jensen, J.; Gutin, G. **Digraphs** 2001
Baues, H.-J. **Combinatorial Foundation of Homology and Homotopy** 1999
Böttcher, A.; Silbermann B. **Analysis of Toeplitz Operators** 2nd ed. 2006
Brown, K.S. **Buildings** 3rd printing 2000 (1st ed. 1998)
Chang, K. **Methods in Nonlinear Analysis** 2005
Cherry, W.; Ye, Z. **Nevanlinna's Theory of Value Distribution** 2001
Ching, W.K. **Iterative Methods for Queuing and Manufacturing Systems** 2001
Crabb, M.C.; James, I.M. **Fibrewise Homotopy Theory** 1998
Chudinovich, I. **Variational and Potential Methods for a Class of Linear Hyperbolic Evolutionary Processes** 2005
Dineen, S. **Complex Analysis on Infinite Dimensional Spaces** 1999
Dugundji, J.; Granas, A. **Fixed Point Theory** 2003
Ebbinghaus, H.-D.; Flum J. **Finite Model Theory** 2006
Elstrodt, J.; Grunewald, F. Mennicke, J. **Groups Acting on Hyperbolic Space** 1998
Edmunds, D.E.; Evans, W.D. **Hardy Operators, Function Spaces and Embeddings** 2004
Engler, A.; Prestel, A. **Valued Fields** 2005
Fadell, E.R.; Husseini, S.Y. **Geometry and Topology of Configuration Spaces** 2001
Fedorov, Y.N.; Kozlov, V.V. **A Memoir on Integrable Systems** 2001
Flenner, H.; O'Carroll, L. Vogel, W. **Joins and Intersections** 1999
Gelfand, S.I.; Manin, Y.I. **Methods of Homological Algebra** 2nd ed. 2003
Griess, R.L. Jr. **Twelve Sporadic Groups** 1998
Gras, G. **Class Field Theory** corr. 2nd printing 2005
Hida, H. ***p*-Adic Automorphic Forms on Shimura Varieties** 2004
Ischebeck, F.; Rao, R.A. **Ideals and Reality** 2005
Ivrii, V. **Microlocal Analysis and Precise Spectral Asymptotics** 1998
Jech, T. **Set Theory** (3rd revised edition 2002) corr. 4th printing 2006
Jorgenson, J.; Lang, S. **Spherical Inversion on SLn (R)** 2001
Kanamori, A. **The Higher Infinite** corr. 2nd printing 2005 (2nd ed. 2003)
Kanovei, V. **Nonstandard Analysis, Axiomatically** 2005
Khoshnevisan, D. **Multiparameter Processes** 2002
Koch, H. **Galois Theory of *p*-Extensions** 2002
Komornik, V. **Fourier Series in Control Theory** 2005
Kozlov, V.; Maz'ya, V. **Differential Equations with Operator Coefficients** 1999

Lau, D. **Function Algebras on Finite Sets** 2006
Landsman, N.P. **Mathematical Topics between Classical & Quantum Mechanics** 1998
Leach, J.A.; Needham, D.J. **Matched Asymptotic Expansions in Reaction-Diffusion Theory** 2004
Lebedev, L.P.; Vorovich, I.I. **Functional Analysis in Mechanics** 2002
Lemmermeyer, F. **Reciprocity Laws: From Euler to Eisenstein** 2000
Malle, G.; Matzat, B.H. **Inverse Galois Theory** 1999
Mardesic, S. **Strong Shape and Homology** 2000
Margulis, G.A. **On Some Aspects of the Theory of Anosov Systems** 2004
Miyake, T. **Modular Forms** 2006
Murdock, J. **Normal Forms and Unfoldings for Local Dynamical Systems** 2002
Narkiewicz, W. **Elementary and Analytic Theory of Algebraic Numbers** 3rd ed. 2004
Narkiewicz, W. **The Development of Prime Number Theory** 2000
Onishchik, A.L.; Sulanke, R. **Projective and Cayley-Klein Geometries** 2006
Parker, C.; Rowley, P. **Symplectic Amalgams** 2002
Peller, V. **Hankel Operators and Their Applications** 2003
Prestel, A.; Delzell, C.N. **Positive Polynomials** 2001
Puig, L. **Blocks of Finite Groups** 2002
Ranicki, A. **High-dimensional Knot Theory** 1998
Ribenboim, P. **The Theory of Classical Valuations** 1999
Rowe, E.G.P. **Geometrical Physics in Minkowski Spacetime** 2001
Rudyak, Y.B. **On Thom Spectra, Orientability and Cobordism** 1998
Ryan, R.A. **Introduction to Tensor Products of Banach Spaces** 2002
Saranen, J.; Vainikko, G. **Periodic Integral and Pseudodifferential Equations with Numerical Approximation** 2002
Schneider, P. **Nonarchimedean Functional Analysis** 2002
Serre, J-P. **Complex Semisimple Lie Algebras** 2001 (reprint of first ed. 1987)
Serre, J-P. **Galois Cohomology** corr. 2nd printing 2002 (1st ed. 1997)
Serre, J-P. **Local Algebra** 2000
Serre, J-P. **Trees** corr. 2nd printing 2003 (1st ed. 1980)
Smirnov, E. **Hausdorff Spectra in Functional Analysis** 2002
Springer, T.A.; Veldkamp, F.D. **Octonions, Jordan Algebras, and Exceptional Groups** 2000
Sznitman, A.-S. **Brownian Motion, Obstacles and Random Media** 1998
Taira, K. **Semigroups, Boundary Value Problems and Markov Processes** 2003
Talagrand, M. **The Generic Chaining** 2005
Tauvel, P.; Yu, R.W.T. **Lie Algebras and Algebraic Groups** 2005
Tits, J.; Weiss, R.M. **Moufang Polygons** 2002
Uchiyama, A. **Hardy Spaces on the Euclidean Space** 2001
Üstünel, A.-S.; Zakai, M. **Transformation of Measure on Wiener Space** 2000
Vasconcelos, W. **Integral Closure. Rees Algebras, Multiplicities, Algorithms** 2005
Yang, Y. **Solitons in Field Theory and Nonlinear Analysis** 2001
Zieschang, P.-H. **Theory of Association Schemes** 2005

Printing: Krips bv, Meppel
Binding: Stürtz, Würzburg